Micromechanics and MEMS

IEEE Press
445 Hoes Lane, P.O. Box 1331
Piscataway, NJ 08855-1331

IEEE Press Editorial Board
John B. Anderson, *Editor in Chief*

P. M. Anderson	A. H. Haddad	P. Laplante
M. Eden	R. Herrick	R. S. Muller
M. E. El-Hawary	G. F. Hoffnagle	W. D. Reeve
S. Furui	R. F. Hoyt	D. J. Wells
	S. Kartalopoulos	

Dudley R. Kay, *Director of Book Publishing*
John Griffin, *Senior Editor*
Lisa Dayne, *Assistant Editor*
Linda Matarazzo, *Editorial Assistant*
Denise Phillip, *Associate Production Editor*

IEEE Electron Devices Society, *Sponsor*
Kwok Ng, EDS Liaison to IEEE Press

IEEE Industrial Electronics Society, *Sponsor*
Okyay Kaynak, IE-S Liaison to IEEE Press

Technical Reviewers

Geoffrey Shiflett
University of Southern California

Robert H. Stroud
The Aerospace Corporation

Books of Related Interest from IEEE Press...

The Engineering Handbook
Edited by Richard Dorf (A CRC Press Handbook published in cooperation with IEEE Press)
1996   Cloth   2358 pp   IEEE Order No. PC5652   ISBN 0-8493-8344-7

Electro-Mechanical Engineering: An Integrated Approach
Charles Fraser and John Milne, both of *University of Abertay Dundee*
1994   Cloth   544 pp   IEEE Order No. PC5610   ISBN 0-7803-1142-6

Microsensors
Edited by Richard S. Muller, Roger T. Howe, Stephen D, Senturia, Rosemary L. Smith, and Richard M. White
1991   Cloth   480 pp   IEEE Order No. PC2576   ISBN 0-87942-245-9

Micromechanics and MEMS

Classic and Seminal Papers to 1990

Edited by

William S. Trimmer
Belle Mead Research, Inc.

A Selected Reprint Volume

IEEE Electron Devices Society, *Sponsor*

IEEE Industrial Electronics Society, *Sponsor*

The Institute of Electrical and Electronics Engineers, Inc., New York

This book may be purchased at a discount from the publisher when ordered in bulk quantities. Contact:

IEEE Press Marketing
Attn: Special Sales
445 Hoes Lane, P.O. Box 1331
Piscataway, NJ 08855-1331
Fax: (732) 981-9334

For more information on the IEEE Press, visit the IEEE Home Page:
http://www.ieee.org/

©1997 by the Institute of Electrical and Electronics Engineers, Inc.,
345 East 47th Street, New York, NY 10017-2394

All rights reserved. No part of this book may be reproduced in any form, nor may it be stored in a retrieval system or transmitted in any form, without written permission from the publisher.

Printed in the United States of America

10   9   8   7   6   5   4   3   2

ISBN 0-7803-1085-3
IEEE Order Number: PC4390

Library of Congress Cataloging-in-Publication Data
Micromechanics and MEMS : classic and seminal papers to 1990 / edited
 by William Trimmer.
      p.   cm.
   "A selected reprint volume."
   Includes bibliographical references and index.
   ISBN 0-7803-1085-3 (cloth)
   1. Microelectronics.   2. Electromechanical devices.
3. Micromechanics.   4. Actuators.   I. Trimmer, W.   II. Institute of
Electrical and Electronics Engineers.
TK7874.M4876   1996
621.381—DC20
                                                96-34081
                                                    CIP

Contents

Acknowledgments and Dedication   xi

Introduction   xiii

Comments on Writing an Article   xv

SECTION 1   Early Papers in Micromechanics   1
   1.1 There's Plenty of Room at the Bottom   **3**
      R. Feynman (*Journal of Microelectromechanical Systems,* March 1992).
   1.2 Infinitesimal Machinery   **10**
      R. Feynman (*Journal of Microelectromechanical Systems,* March 1993).
   1.3 The Resonant Gate Transistor   **21**
      H. C. Nathanson, W. E. Newell, R. A. Wickstrom, and J. R. Davis, Jr. (*IEEE Transactions on Electron Devices,* March 1967).
   1.4 Silicon Micromechanical Devices   **38**
      J. B. Angell, S. C. Terry, and P. W. Barth (*Scientific American Journal,* April 1983).
   1.5 Anisotropic Etching of Silicon   **50**
      K. E. Bean (*IEEE Transactions on Electron Devices,* October 1978).
   1.6 Silicon as a Mechanical Material   **58**
      K. E. Petersen, (*Proceedings of the IEEE,* May 1982).
   1.7 Microrobots and Micromechanical Systems   **96**
      W. S. N. Trimmer (*Sensors and Actuators,* September 1989).
   1.8 Small Machines, Large Opportunities   **117**
      K. Gabriel, J. Jarvis, and W. Trimmer (*Report of the NSF Workshop on Microelectromechanical Systems Research,* July 1987).

SECTION 2   Side Drive Actuators   145
   2.1 IC-Processed Electrostatic Micro-Motors   **147**
      L-S. Fan, Y-C. Tai, and R. S. Muller (*IEEE International Electronic Devices Meeting,* December 1988).
   2.2 *IC*-Processed Micro-Motors: Design, Technology, and Testing   **151**
      Y-C. Tai, L-S. Fan, and R. S. Muller (*Proceedings IEEE Micro Electro Mechanical Systems,* February 1989).
   2.3 Surface-Micromachining Processes for Electrostatic Microactuator Fabrication   **157**
      T. A. Lober and R. T. Howe (*Technical Digest IEEE Solid-State Sensor and Actuator Workshop,* June 1988).
   2.4 A Study of Three Microfabricated Variable-Capacitance Motors   **161**
      M. Mehregany, S. F. Bart, L. S. Tavrow, J. H. Lang, S. D. Senturia, and M. F. Schlecht (*Transducers '89, Proceedings of the 5th International Conference on Solid-State Sensors and Actuators and Eurosensors III,* June 1990).
   2.5 Friction and Wear in Microfabricated Harmonic Side-Drive Motors   **168**
      M. Mehregany, S. D. Senturia, and J. H. Lang (*Technical Digest IEEE Solid-State Sensor and Actuator Workshop,* June 1990).
   2.6 Measurements of Electric Micromotor Dynamics   **174**
      S. F. Bart, M. Mehregany, L. S. Tavrow, J. H. Lang, and S. D. Senturia (*Microstructures, Sensors, and Actuators,* November 1990).

SECTION 3   Comb Drive Actuators   185
   3.1 Laterally Driven Polysilicon Resonant Microstructures   **187**
      W. C. Tang, T-C. H. Nguyen, and R. T. Howe (*Proceedings IEEE Micro Electro Mechanical Systems,* February 1989).
   3.2 Electrostatic-Comb Drive of Lateral Polysilicon Resonators   **194**
      W. C. Tang, T-C.H. Nguyen, M. W. Judy, and R. T. Howe (*Transducers '89, Proceedings of the 5th International Conference on Solid-State Sensors and Actuators and Eurosensors III,* June 1990).

3.3 Electrostatically Balanced Comb Drive for Controlled Levitation **198**
W. C. Tang, M. G. Lim, and R. T. Howe (*Technical Digest IEEE Solid-State Sensor and Actuator Workshop*, June 1990).

3.4 Polysilicon Microgripper **203**
C-J. Kim, A. P. Pisano, R. S. Muller, and M. G. Lim (*Technical Digest IEEE Solid-State Sensor and Actuator Workshop*, June 1990).

SECTION 4 Electrostatic Actuators 207

4.1 The Principle of an Electrostatic Linear Actuator Manufactured by Silicon Micromachining **209**
H. Fujita and A. Omodaka (*Transducers '87, The 4th International Conference on Solid-State Sensors and Actuators*, June 1987).

4.2 Design Considerations for a Practical Electrostatic Micro-Motor **213**
W. S. N. Trimmer and K. J. Gabriel (*Sensors and Actuators*, 1987).

4.3 SCOFSS: A Small Cantilevered Optical Fiber Servo System **231**
J. E. Wood, S. C. Jacobsen, and K. W. Grace (Proceedings *IEEE Micro Robots and Teleoperators Workshop*, November 1987).

4.4 Microactuators for Aligning Optical Fibers **237**
R. Jebens, W. Trimmer, and J. Walker (*Sensors and Actuators*, November 1989).

4.5 Large Displacement Linear Actuator **246**
R. A. Brennen, M. G. Lim, A. P. Pisano, and A. T. Chou (*Technical Digest IEEE Solid-State Sensor and Actuator Workshop*, June 1990).

4.6 Multi-Layered Electrostatic Film Actuator **251**
S. Egawa and T. Higuchi (*Proceedings IEEE Micro Electro Mechanical Systems*, February 1990).

4.7 Movable Micromachined Silicon Plates With Integrated Position Sensing **257**
M. G. Allen, M. Scheidl, R. L. Smith, and A. D. Nikolich (*Transducers '89, Proceedings of the 5th International Conference on Solid-State Sensors and Actuators and Eurosensors III*, June 1990).

4.8 Micro Electro Static Actuator With Three Degrees of Freedom **261**
T. Fukuda and T. Tanaka (*Proceedings IEEE Micro Electro Mechanical Systems*, February 1990).

4.9 The Modelling of Electrostatic Forces in Small Electrostatic Actuators **267**
R. H. Price, J. E. Wood, and S. C. Jacobsen (*Technical Digest IEEE Solid-State Sensor and Actuator Workshop*, June 1988).

4.10 Silicon Electrostatic Motors **272**
W. S. N. Trimmer, K. J. Gabriel, and R. Mahadevan (*Transducers '87, The 4th International Conference on Solid-State Sensors and Actuators*, June 1987).

4.11 Electrostatic Actuators for Micromechatronics **276**
H. Fujita and A. Omodaka (*Proceedings IEEE Micro Robots and Teleoperators Workshop*, November 1987).

4.12 Electric Micromotors: Electromechanical Characteristics **286**
J. H. Lang, M. F. Schlecht, and R. T. Howe (*Proceedings IEEE Micro Robots and Teleoperators Workshop*, November 1987).

4.13 Electroquasistatic Induction Micromotors **294**
S. F. Bart and J. H. Lang (*Proceedings IEEE Micro Electro Mechanical Systems*, February 1989).

4.14 A Perturbation Method for Calculating the Capacitance of Electrostatic Motors **300**
S. Kumar and D. Cho (*Proceedings IEEE Micro Electro Mechanical Systems*, February 1990).

SECTION 5 Magnetic Actuators 307

5.1 Magnetically Levitated Micro-Machines **309**
R. Pelrine and I. Busch-Vishniac (*Proceedings IEEE Micro Robots and Teleoperators Workshop*, November 1987).

5.2 Fabrication and Testing of a Micro Superconducting Actuator Using the Meissner Effect **314**
Y-K. Kim, M. Katsurai, and H. Fujita (*Proceedings IEEE Micro Electro Mechanical Systems*, February 1990).

5.3 Room Temperature, Open-Loop Levitation of Microdevices Using Diamagnetic Materials **320**
R. E. Pelrine (*Proceedings IEEE Micro Electro Mechanical Systems*, February 1990).

SECTION 6 Harmonic Motors 325

6.1 An Operational Harmonic Electrostatic Motor **327**
W. Trimmer and R. Jebens (*Proceedings IEEE Micro Electro Mechanical Systems*, February 1989).

6.2 The Wobble Motor: An Electrostatic Planetary-Armature, Microactuator **331**
S. C. Jacobsen, R. H. Price, J. E. Wood, T. H. Rytting, and M. Rafaelof (*Proceedings IEEE Micro Electro Mechanical Systems,* February 1989).

6.3 An Electrostatic Top Motor and Its Characteristics **339**
M. Sakata, Y. Hatazawa, A. Omodaka, T. Kudoh, and H. Fujita (*Transducers '89, Proceedings of the 5th International Conference on Solid-State Sensors and Actuators and Eurosensors III,* June 1990).

6.4 Operation of Microfabricated Harmonic and Ordinary Side-Drive Motors **344**
M. Mehregany, P. Nagarkar, S. D. Senturia, and J. H. Lang (*Proceedings IEEE Micro Electro Mechanical Systems,* February 1990).

SECTION 7 Other Actuators   353

Thermal

7.1 Micromechanical Silicon Actuators Based on Thermal Expansion Effects **355**
W. Riethmüller, W. Benecke, U. Schnakenberg, and A. Heuberger (*Transducers '87, The 4th International Conference on Solid-State Sensors and Actuators,* June 1987).

7.2 CMOS Electrothermal Microactuators **359**
M. Parameswaran, Lj. Ristic, K. Chau, A. M. Robinson, and W. Allegretto (*Proceedings IEEE Micro Electro Mechanical Systems,* February 1990).

7.3 Electrically-Activated, Micromachined Diaphragm Valves **363**
H. Jerman (*Technical Digest IEEE Solid-State Sensor and Actuator Workshop,* June 1990).

7.4 Study on Micro Engines—Miniaturizing Stirling Engines for Actuators and Heatpumps **368**
N. Nakajima, K. Ogawa and I. Fujimasa (*Proceedings IEEE Micro Electro Mechanical Systems,* February 1989).

Shape Memory Alloy

7.5 A Micro Rotary Actuator Using Shape Memory Alloys **372**
K. J. Gabriel, W. S. N. Trimmer, and J. A. Walker (*Sensors and Actuators,* 1988).

7.6 Millimeter Size Joint Actuator Using Shape Memory Alloy **379**
K. Kuribayashi (*Proceedings IEEE Micro Electro Mechanical Systems,* February 1989).

7.7 Reversible SMA Actuator for Micron Sized Robot **385**
K. Kuribayashi and M. Yoshitake (*Proceedings IEEE Micro Electro Mechanical Systems,* February 1990).

7.8 Characteristics of Thin-Wire Shape Memory Actuators **390**
P. A. Neukomm, H. P. Bornhauser, T. Hochuli, R. Paravicini, and G. Schwarz (*Transducers '89, Proceedings of the 5th International Conference on Solid-State Sensors and Actuators and Eurosensors III,* June 1990).

7.9 Shape Memory Alloy Microactuators **396**
M. Bergamasco, P. Dario, and F. Salsedo (*Transducers '89, Proceedings of the 5th International Conference on Solid State Sensors and Actuators and Eurosensors III,* June 1990).

Impact

7.10 Micro Actuators Using Recoil of an Ejected Mass **401**
T. Higuchi, Y. Hojjat, and M. Watanabe (*Proceedings IEEE Micro Robots and Teleperators Workshop,* November 1987).

7.11 Precise Positioning Mechanism Utilizing Rapid Deformations of Piezoelectric Elements **406**
T. Higuchi, Y. Yamagata, K. Furutani, and K. Kudoh (*Proceedings IEEE Micro Electro Mechanical Systems,* February 1990).

7.12 Tiny Silent Linear Cybernetic Actuator Driven by Piezoelectric Device With Electromagnetic Clamp **411**
K. Ikuta, S. Aritomi, and T. Kabashima (*Proceedings IEEE Micro Electro Mechanical Systems '92* February 1992).

7.13 Experimental Model and IC-Process Design of a Nanometer Linear Piezoelectric Stepper Motor **417**
J. W. Judy, D. L. Polla, and W. P. Robbins (*Microstructures, Sensors, and Actuators,* November 1990).

Piezoelectric

7.14 Zinc-Oxide Thin Films for Integrated-Sensor Applications **424**
D. L. Polla and R. S. Muller (*Technical Digest IEEE Solid-State Sensors Workshop,* June 1986).

7.15 A Micromachined Manipulator for Submicron Positioning of Optical Fibers **427**
A. M. Feury, T. L. Poteat, and W. S. Trimmer (*Technical Digest IEEE Solid-State Sensors Workshop,* June 1986).

7.16 Ultrasonic Micromotors: Physics and Applications **428**
R. M. Moroney, R. M. White, and R. T. Howe (*Proceedings IEEE Micro Electro Mechanical Systems,* February 1990).

Section 8 Valves and Pumps 435

- 8.1 A Microminiature Electric-to-Fluidic Valve **437**
  M. J. Zdeblick and J. B. Angell (*Transducers '87, The 4th International Conference on Solid-State Sensors and Actuators,* June 1987).
- 8.2 The Fabrication of Integrated Mass Flow Controllers **440**
  M. Esashi, S. Eoh, T. Matsuo, and S. Choi (*Transducers '87, The 4th International Conference on Solid-State Sensors and Actuators,* June 1987).
- 8.3 Normally Close Microvalve and Micropump Fabricated on a Silicon Wafer **444**
  M. Esashi, S. Shoji, and A. Nakano (*Proceedings IEEE Micro Electro Mechanical Systems,* February 1989).
- 8.4 A Thermopneumatic Micropump Based on Micro-Engineering Techniques **450**
  F. C. M. Van de Pol, H. T. G. Van Lintel, M. Elwenspoek, and J. H. J. Fluitman (*Transducers '89, Proceedings of the 5th International Conference on Solid-State Sensors and Actuators and Eurosensors III,* June 1990).
- 8.5 Variable-Flow Micro-Valve Structure Fabricated with Silicon Fusion Bonding **455**
  F. Pourahmadi, L. Christel, K. Petersen, J. Mallon, and J. Bryzek (*Technical Digest IEEE Solid-State Sensor and Actuator Workshop,* June 1990).
- 8.6 A Pressure-Balanced Electrostatically-Actuated Microvalve **459**
  M. A. Huff, M. S. Mettner, T. A. Lober, and M. A. Schmidt (*Technical Digest IEEE Solid-State Sensor and Actuator Workshop,* June 1990).
- 8.7 Micromachined Silicon Microvalve **464**
  T. Ohnstein, T. Fukiura, J. Ridley, and U. Bonne (*Proceedings IEEE Micro Electro Mechanical Systems,* February 1990).

Section 9 Fluidics 469

- 9.1 Microminiature Fluidic Amplifier **471**
  M. J. Zdeblick, P. W. Barth, and J. B. Angell (*Technical Digest IEEE Solid-State Sensor and Actuator Workshop,* June 1990).
- 9.2 A Planar Air Levitated Electrostatic Actuator System **473**
  K. S. J. Pister, R. S. Fearing, and R. T. Howe (*Proceedings IEEE Micro Electro Mechanical Systems,* February 1990).
- 9.3 Liquid and Gas Transport in Small Channels **478**
  J. Pfahler, J. Harley, and H. H. Bau (*Microstructures, Sensors, and Actuators,* November 1990).
- 9.4 Squeeze-Film Damping in Solid-State Accelerometers **487**
  J. B. Starr (*Technical Digest IEEE Solid-State Sensor and Actuator Workshop,* June 1990).
- 9.5 A Micromachined Floating-Element Shear Sensor **491**
  M. A. Schmidt, R. T. Howe, S. D. Senturia, and J. H. Haritonidis (*Transducers '87, The 4th International Conference on Solid-State Sensors and Actuators,* June 1987).
- 9.6 A Multi-Element Monolithic Mass Flowmeter With On-Chip CMOS Readout Electronics **495**
  E. Yoon and K. D. Wise (*Technical Digest IEEE Solid-State Sensor and Actuator Workshop,* June 1990).
- 9.7 Environmentally Rugged, Wide Dynamic Range Microstructure Airflow Sensor **499**
  T. R. Ohnstein, R. G. Johnson, R. E. Higashi, D. W. Burns, J. O. Holmen, E. A. Satren, G. M. Johnson, R. E. Bicking, and S. D. Johnson (*Technical Digest IEEE Solid-State Sensor and Actuator Workshop,* June 1990).

Section 10 Surface Micromachining 503

- 10.1 Polycrystalline Silicon Micromechanical Beams **505**
  R. T. Howe, R. S. Muller (*Journal of the Electrochemical Society: Solid State Science and Technology,* June 1983).
- 10.2 Integrated Fabrication of Polysilicon Mechanisms **509**
  M. Mehregany, K. J. Gabriel, and W. S. N. Trimer (*IEEE Transactions on Electron Devices,* June 1988).
- 10.3 Integrated Movable Micromechanical Structures for Sensors and Actuators **514**
  L-S. Fan, Y-C. Tai, and R. S. Muller (*IEEE Transactions on Electron Devices,* June 1988).
- 10.4 Polysilicon Microbridge Fabrication Using Standard CMOS Technology **521**
  M. Parameswaran, H. P. Baltes, and A. M. Robinson (*Technical Digest IEEE Solid-State Sensor and Actuator Workshop,* June 1988).
- 10.5 Process Integration for Active Polysilicon Resonant Microstructures **524**
  M. W. Putty, S-C. Chang, R. T. Howe, A. L. Robinson, and K. D. Wise (*Sensors and Actuators,* November 1989).
- 10.6 Fabrication of Micromechanical Devices From Polysilicon Films With Smooth Surfaces **532**
  H. Guckel, J. J. Sniegowski, T. R. Christenson, S. Mohney, and T. F. Kelly (*Sensors and Actuators,* November 1989).

10.7 Selective Chemical Vapor Deposition of Tungsten for Microelectromechanical Structures **538**
   N. C. MacDonald, L. Y. Chen, J. J. Yao, Z. L. Zhang, J. A. McMillan, D. C. Thomas, and K. R. Haselton (*Sensors and Actuators*, November 1989).

SECTION 11 Bulk Micromachining **549**

11.1 Fabrication of Hemispherical Structures Using Semiconductor Technology for Use in Thermonuclear Fusion Research **551**
   K. D. Wise, T. N. Jackson, N. A. Masnari, M. G. Robinson, D. E. Solomon, G. H. Wuttke, and W. B. Rensel (*Journal of Vacuum Science Technology*, May/June 1979).

11.2 Micromachining of Silicon Mechanical Structures **555**
   G. Kaminsky (*Journal of Vacuum Science Technology*, July/August 1985).

11.3 Strings, Loops, and Pyramids—Building Blocks for Microstructures **565**
   H. H. Busta, A. D. Feinerman, J. B. Ketterson, and R. D. Cueller (*Proceedings IEEE Micro Robots and Teleoperators Workshop*, November 1987).

11.4 Corner Compensation Structures for (110) Oriented Silicon **570**
   D. R. Ciarlo (*Proceedings IEEE Micro Robots and Teleoperators Workshop*, November 1987).

11.5 A Study on Compensating Corner Undercutting in Anisotropic Etching of (100) Silicon **574**
   X-P. Wu and W. H. Ko (*Transducers '87, The 4th International Conference on Solid-State Sensors and Actuators*, June 1987).

11.6 A New Silicon-on-Glass Process for Integrated Sensors **578**
   L. J. Spangler and K. D. Wise (*Technical Digest IEEE Solid-State Sensor and Actuator Workshop*, June 1988).

11.7 Mechanisms of Anodic Bonding of Silicon to Pyrex® Glass **582**
   K. B. Albaugh, P. E. Cade, and D. H. Rasmussen (*Technical Digest IEEE Solid-State Sensor and Actuator Workshop*, June 1988).

11.8 Silicon Fusion Bonding for Pressure Sensors **584**
   K. Petersen, P. Barth, J. Poydock, J. Brown, J. Mallon Jr., and J. Bryzek (*Technical Digest IEEE Solid-State Sensor and Actuator Workshop*, June 1988).

11.9 Low-Temperature Silicon-to-silicon Anodic Bonding With Intermediate Low Melting Point Glass **588**
   M. Esashi, A. Nakano, S. Shoji, and H. Hebigushi (*Transducers '89, Proceedings of the 5th International Conference on Solid-State Sensors and Actuators and Eurosensors III*, June 1990).

11.10 Fusing Silicon Wafers With Low Melting Temperature Glass **592**
   L. A. Field and R. S. Muller (*Transducers '89, Proceedings of the 5th International Conference on Solid-State Sensors and Actuators and Eurosensors III*, June 1990).

11.11 Silicon Fusion Bonding for Fabrication of Sensors, Actuators and Microstructures **596**
   P. W. Barth (*Transducers '89, Proceedings of the 5th International Conference on Solid-State Sensors and Actuators and Eurosensors III*, June 1990).

11.12 Scaling and Dielectric Stress Compensation of Ultrasensitive Boron-Doped Silicon Microstructures **604**
   S. T. Cho, K. Najafi, and K. D. Wise (*Proceedings IEEE Micro Electro Mechanical Systems*, February 1990).

11.13 Field Oxide Microbridges, Cantilever Beams, Coils and Suspended Membranes in SACMOS Technology **610**
   D. Moser, M. Parameswaran, and H. Baltes (*Transducers '89, Proceedings of the 5th International Conference on Solid-State Sensors and Actuators and Eurosensors III*, June 1990).

11.14 Micromachining of Quartz and its Application to an Acceleration Sensor **614**
   J. S. Daniel, F. Michel, and G. Delapierre (*Transducers '89, Proceedings of the 5th International Conference on Solid State Sensors and Actuators and Eurosensors III*, June 1990).

SECTION 12 LIGA **621**

12.1 Fabrication of Microstructures using the LIGA Process **623**
   W. Ehrfeld, P. Bley, F. Götz, P. Hagmann, A. Maner, J. Mohr, H. O. Moser, D. Münchmeyer, W. Schelb, D. Schmidt, and E. W. Becker (*Proceedings IEEE Micro Robots and Teleoperators Workshop*, November 1987).

12.2 Deep X-Ray and UV Lithographies for Micromechanics **634**
   H. Guckel, T. R. Christenson, K. J. Skrobis, D. D. Denton, B. Choi, E. G. Lovell, J. W. Lee, S. S. Bajikar, and T. W. Chapman (*Technical Digest IEEE Solid-State Sensor and Actuator Workshop*, June 1990).

SECTION 13 Computer Aided Design 639

- 13.1 OYSTER, a 3D Structural Simulator for Micro Electromechanical Design 641
  G. M. Koppelman (*Proceedings IEEE Micro Electro Mechanical Systems,* February 1989).
- 13.2 A CAD Architecture for Microelectromechanical Systems 647
  F. Maseeh, R. M. Harris, and S. D. Senturia (*Proceedings IEEE Micro Electro Mechanical Systems,* February 1990).
- 13.3 CAEMEMS: An Integrated Computer-Aided Engineering Workbench for Micro-Electro-Mechanical Systems 653
  S. Crary and Y. Zhang (*Proceedings IEEE Micro Electro Mechanical Systems,* February 1990).
- 13.4 CAD for Silicon Anisotropic Etching 655
  R. A. Buser and N. F. de Rooij (*Proceedings IEEE Micro Electro Mechanical Systems,* February 1990).

SECTION 14 Metrology 657

- 14.1 Can We Design Microbotic Devices Without Knowing the Mechanical Properties of Materials? 659
  S. D. Senturia (*Proceedings IEEE Micro Robots and Teleoperators Workshop,* November 1987).
- 14.2 The Use of Micromachined Structures for the Measurement of Mechanical Properties and Adhesion of Thin Films 664
  M. Mehregany, M. G. Allen, and S. D. Senturia (*Technical Digest IEEE Solid-State Sensors Workshop,* June 1986).
- 14.3 Mechanical Property Measurements of Thin Films Using Load-Deflection of Composite Rectangular Membrane 667
  O. Tabata, K. Kawahata, S. Sugiyama, and I. Igarashi (*Proceedings IEEE Micro Electro Mechanical Systems,* February 1989).
- 14.4 Fracture Toughness Characterization of Brittle Thin Films 672
  L. S. Fan, R. T. Howe, and R. S. Muller (*Transducers '89, Proceedings of the 5th International Conference on Solid State Sensors and Actuators and Eurosensors III,* June 1990).
- 14.5 Spiral Microstructures for the Measurement of Average Strain Gradients in Thin Films 675
  L-S. Fan, R. S. Muller, W. Yun, R. T. Howe, and J. Huang (*Proceedings IEEE Micro Electro Mechanical Systems,* February 1990).
- 14.6 Polysilicon Microstructures to Characterize Static Friction 679
  M. G. Lim, J. C. Chang, D. P. Schultz, R. T. Howe, and R. M. White (*Proceedings IEEE Micro Electro Mechanical Systems,* February 1990).
- 14.7 Study on the Dynamic Force/Acceleration Measurements 686
  A. Umeda and K. Ueda (*Transducers '89, Proceedings of the 5th International Conference on Solid State Sensors and Actuators and Eurosensors III,* June 1990).
- 14.8 Anomalous Emissivity from Periodic Micro Machined Silicon Surfaces 690
  P. J. Hesketh, B. Gebhart, and J. N. Zemel (*Technical Digest IEEE Solid-State Sensors Workshop,* June 1986).

AUTHOR INDEX 693

SUBJECT INDEX 697

ABOUT THE AUTHOR 701

EDITOR'S NOTES ON THE SECOND PRINTING 702

Acknowledgments and Dedication

I want to thank the people who have supported and believed in me: my wife Ann, sons Scott and Mark, parents Amy and Lynn Wilson, the Trimmers, the La Mottes, the Schulzes, and many friends. Their understanding while I followed my own dreams is much appreciated.

I dedicate this book to my father, Harry Allison Trimmer. I wish he had lived to share my joys.

Introduction

IN any field, the early papers in a field provide a valuable perspective. The original researchers in a field have no guidelines, no sense of the correct problems and approaches. Instead, they are trying to define the field. These people tend to be innovators and synthesizers, and draw from a wide range of fields and insights to define fruitful areas of study. Their papers convey a sense of excitement and exploration.

Reading the original papers in a field is important for several reasons. First, the reader gains an appreciation of why the field has developed in certain directions. Often, as the field develops, the original approaches should be modified, and the scientist who understands the original assumptions is more apt to make the breakthroughs and extensions. Buried in these original papers are plums for exploration that the earlier researchers did not have the time and tools to develop. Second, if one wants to do scholarly work, knowing the literature is a prerequisite. This is a matter of intellectual honesty. I see many papers claiming as new and novel results that have been in the literature for years. Publishers of this type of paper risk their reputations with people who do know and understand the literature. Third, a careful reading of the literature helps one's own work to be well considered and within the context of one's discipline.

The field of micromechanics, or microelectromechanical systems (MEMS), is developing rapidly. We are in the process of writing the papers that future scientists and engineers will consider as the original papers of micromechanics. Unfortunately, many of the original papers are already difficult to find. The papers, especially before the 1990s, tended to be published in hard-to-obtain conference proceedings, or scattered throughout a wide array of journals in related fields. Now, publications such as the Journal of Microelectromechanical Systems (published jointly by the IEEE and ASME) are archiving these papers, so they are available to the scholar.

This book makes the early papers in micromechanics conveniently available to scholars, scientists, engineers, students, users, and people interested in this new technology. Fortunately, this literature makes fewer assumptions about what the reader knows, and a wide range of people can profit from reading these articles. It is especially hoped that students studying micromechanics will read these early papers.

The papers in this book are from 1990 and earlier. They combine to make a reasonably sized book one can conveniently use. After 1990, the literature expands rapidly, and including papers from this later material would necessitate leaving out earlier interesting articles. The literature after 1990 is also more readily accessible.

The field of small mechanical devices has many names: micromechanics, microdynamics, micromechatronics, microsystem technology (mst), micro engineering, micro electro mechanical systems (MEMS), etc. I prefer the term micromechanics, because it broadly connotes the science and engineering of small mechanical devices. The first workshop in this area was called the Micro Robots and Teleoperators Workshop. ["Micro Robots and Teleoperators Workshop, An Investigation of Micromechanical Structures, Actuators and Sensors," IEEE Catalog No. 87TH0204-8, Hyannis, MA, 9–11 November 1987] The subsequent workshops in this series have been called the MEMS Workshops. The acronym MEMS is now widely used to denote the field.

The "micro" used in the above names is sometimes capitalized, sometimes not, sometimes hyphenated, sometimes stands as a separate word, and sometimes combined with other words without hyphens. I find it easier to read the profusion of micro words when "micro" stands as a separate word: micro metrology, micro switch, micro capsule, micro vocabulary. It seems easier for the reader to distinguish between micro pump and micro plug than between micropump and microplug. However, some words, like "micromechanics," are used so frequently that the reader easily recognizes them. With due respect to those who prefer hyphens, or those who feel "micro" is a prefix, this book on micromechanics will discuss micro devices. (The "micro" will usually stand as a separate word, such as micro devices. With frequent usage, phrases will be combined into one word, such as micromechanics.)

Comments on Writing an Article

IN editing the *Journal of Microelectromechanical Systems* and this book, I have examined many hundreds of articles, some well written, some poorly written. Below are some thoughts on how to effectively write an article.

Communication is the prime consideration. Good grammar, sentence structure, vocabulary, and organization have evolved to aid effective communication.

Who is your audience? I visualize three people: a novice, someone knowledgeable in science who knows little about the field, and an expert. Then, as I write, I talk to these three individuals. Each paragraph and each section ideally has something of interest for each reader. When an extended section is needed, for perhaps the expert, the novice can be warned, "The rest of this section contains a detailed examination of the boundary conditions at infinity."

Many of your most important readers will spend only a minute or two reading your article. You want them to understand the salient points, so they will remember and look up your article when it becomes important. The first page of your article is critical. (Having a good figure on the first page of your article that clearly conveys the major thrust of your work is most helpful.)

Before writing, try explaining your article to an intelligent friend who knows little about the field. This will sharpen your sensitivity to the assumptions and jargon you and your colleagues use. If you cannot explain your work to this friend, spend some time thinking about what you are really doing.

Now, before engaging the word processor, take the journal chosen for publication. Glance through several issues, and choose the articles you think are especially well written. Then, use these articles to help plan a strategy to most effectively present your ideas. Finally, sit down with the Information for Authors, and read it carefully.

The abstract, introduction, and conclusion seem to be three places requiring identical information. This is not true.

The abstract will be read by people deciding *if* they want to read your article. What information about your work will help the correct people select your work from the many abstracts they are scanning? What are the key aspects? How does your work differ from the literature?

The introduction is a road map. After reading the introduction, the reader should be able to scan the article, and pick out the important facts.

After reading your whole article, readers may still not grasp the significance. The conclusion is your chance to put the article in perspective. At this point, you can assume the reader knows the details; she or he needs the broad view.

Following the article's introduction is usually a section on the previous work in the field. Many authors use this section to show that their work is clever, and the rest of the literature is stupid. There are several things these authors have forgotten. First, who do they think is going to review their paper? Second, their portrayal of the literature makes them, not the literature, seem simple and sophomoric. Third, the author is missing a chance to communicate important information. People learning the field appreciate a well-written review of previous work. Help them. If your references allow them to discover the literature, they will forever consider you one of the experts in the field.

While writing the body of the article, I stop every few paragraphs and visualize the three people I have chosen for the audience. If someone's head is nodding, I speak to them, tell them what they want to know, and then rewrite to include material of interest to all readers. If you are not writing for your audience . . .

Changing the length and complexity of your sentences and paragraphs keeps your writing lively and interesting. Long sentences add a richness and complexity.

Short paragraphs emphasize.

The scientific literature is immutable. You cannot go back; you cannot change. Researchers a hundred years hence will still be reading the great works. Make yours one.

Section 1

Early Papers in Micromechanics

IN college, I had the pleasure of hearing a talk by Richard Feynman. His descriptions of the micro world were intriguing. Only Feynman could have given a talk like this. He described new vistas for play, and science, and engineering, and novel new applications. The first paper, "There's Plenty of Room at the Bottom," is a transcript of a 1959 talk where Feynman describes this micro domain. I hope it inspires you, as it did me.

While doing the research to reprint "There's Plenty of Room at the Bottom" in The Journal of Microelectromechanical Systems, I discovered an unpublished talk given by Richard Feynman at JPL in 1983. In this talk, Feynman updates his ideas about the micro world. Again, he was insightful. Many of the things he predicts in this 1983 speech have happened, and many other predictions are plums for future researchers. Steve Senturia took the rough transcription of the talk, and turned it into the excellent article, "Infinitesimal Machinery," reprinted here.

The following papers give an excellent overview of the capabilities of silicon fabrication techniques to make micromechanical devices. "The Resonant Gate Transistor" describes the use of sacrificial material to free micromechanical devices from the silicon substrate. (In a private communication, Harvey Nathanson said gold was used as the structural material, and photoresist was used as the sacrificial material in one experiment, and nickel used as the sacrificial material in another experiment.) The papers "Silicon Micromechanical Devices" and "Anisotropic Etching of Silicon" discuss the structure of silicon and etching properties. The paper by Kurt Petersen was instrumental in enticing many researchers to use "Silicon as a Mechanical Material."

The paper "Micro Robots and Micromechanical Systems" discusses how different physical phenomena change as they are scaled into the micro domain. This is a useful guide for selecting the optimum approach. The final paper in this section, "Small Machines, Large Opportunities: A Report on the Emerging Field of Microdynamics," is the report of a three-part NSF workshop held in 1987 and 1988. This report contains the ideas of the experts in the micromechanics field on the important directions for the field and the important research areas and applications.

There's Plenty of Room at the Bottom

Richard P. Feynman

I imagine experimental physicists must often look with envy at men like Kamerlingh Onnes, who discovered a field like low temperature, which seems to be bottomless and in which one can go down and down. Such a man is then a leader and has some temporary monopoly in a scientific adventure. Percy Bridgman, in designing a way to obtain higher pressures, opened up another new field and was able to move into it and to lead us all along. The development of ever higher vacuum was a continuing development of the same kind.

I would like to describe a field, in which little has been done, but in which an enormous amount can be done in principle. This field is not quite the same as the others in that it will not tell us much of fundamental physics (in the sense of, "What are the strange particles?") but it is more like solid-state physics in the sense that it might tell us much of great interest about the strange phenomena that occur in complex situations. Furthermore, a point that is most important is that it would have an enormous number of technical applications.

What I want to talk about is the problem of manipulating and controlling things on a small scale.

As soon as I mention this, people tell me about miniaturization, and how far it has progressed today. They tell me about electric motors that are the size of the nail on your small finger. And there is a device on the market, they tell me, by which you can write the Lord's Prayer on the head of a pin. But that's nothing; that's the most primitive, halting step in the direction I intend to discuss. It is a staggeringly small world that is below. In the year 2000, when they look back at this age, they will wonder why it was not until the year 1960 that anybody began seriously to move in this direction.

Why cannot we write the entire 24 volumes of the Encyclopaedia Britannica on the head of a pin?

Let's see what would be involved. The head of a pin is a sixteenth of an inch across. If you magnify it by 25 000 diameters, the area of the head of the pin is then equal to the area of all the pages of the Encyclopaedia Britannica. Therefore, all it is necessary to do is to reduce in size all the writing in the Encyclopaedia by 25 000 times. Is that possible? The resolving power of the eye is about 1/120 of an inch—that is roughly the diameter of one of the little dots on the fine half-tone reproductions in the Encyclopaedia. This, when you demagnify it by 25 000 times, is still 80 angstroms in diameter—32 atoms across, in an ordinary metal. In other words, one of those dots still would contain in its area 1000 atoms. So, each dot can easily be adjusted in size as required by the photoengraving, and there is no question that there is enough room on the head of a pin to put all of the Encyclopaedia Britannica.

Furthermore, it can be read if it is so written. Let's imagine that it is written in raised letters of metal; that is, where the black is in the Encyclopaedia, we have raised letters of metal that are actually 1/25 000 of their ordinary size. How would we read it?

If we had something written in such a way, we could read it using techniques in common use today. (They will undoubtedly find a better way when we do actually have it written, but to make my point conservatively I shall just take techniques we know today.) We would press the metal into a plastic material and make a mold of it, then peel the plastic off very carefully, evaporate silica into the plastic to get a very thin film, then shadow it by evaporating gold at an angle against the silica so that all the little letters will appear clearly, dissolve the plastic away from the silica film, and then look through it with an electron microscope!

There is no question that if the thing were reduced by 25 000 times in the form of raised letters on the pin, it would be easy for us to read it today. Furthermore, there is no question that we would find it easy to make copies of the master; we would just need to press the same metal plate again into plastic and we would have another copy.

HOW DO WE WRITE SMALL?

The next question is: How do we *write* it? We have no standard technique to do this now. But let me argue that it is not as difficult as it first appears to be. We can reverse the lenses of the electron microscope in order to demagnify as well as magnify. A source of ions, sent through the microscope lenses in reverse, could be focused to a very small spot. We could write with that spot like we write in a TV cathode ray oscilloscope, by going across in lines, and having an adjustment which determines the amount of material which is going to be deposited as we scan in lines.

This method might be very slow because of space charge limitations. There will be more rapid methods. We could first make, perhaps by some photo process, a screen which has holes in it in the form of the letters. Then we would strike an arc behind the holes and draw metallic ions through the holes; then we could again use our sys-

MEMS Editor's Note: This manuscript addresses many current research issues. It is the transcript of a talk given by Richard P. Feynman on December 26, 1959, at the annual meeting of the American Physical Society at the California Institute of Technology, and was published as a chapter in the Reinhold Publishing Corporation book, *Miniaturization*, Horace D. Gilbert, Ed. It is reprinted with the consent of Van Nostrand Reinhold, New York, NY 10003.

The author, deceased, was with the California Institute of Technology, Pasadena, CA.

IEEE Log Number 9105621.

tem of lenses and make a small image in the form of ions, which would deposit the metal on the pin.

A simpler way might be this (though I am not sure it would work): We take light and, through an optical microscope running backwards, we focus it onto a very small photoelectric screen. Then electrons come away from the screen where the light is shining. These electrons are focused down in size by the electron microscope lenses to impinge directly upon the surface of the metal. Will such a beam etch away the metal if it is run long enough? I don't know. If it doesn't work for a metal surface, it must be possible to find some surface with which to coat the original pin so that, where the electrons bombard, a change is made which we could recognize later.

There is no intensity problem in these devices—not what you are used to in magnification, where you have to take a few electrons and spread them over a bigger and bigger screen; it is just the opposite. The light which we get from a page is concentrated onto a very small area so it is very intense. The few electrons which come from the photoelectric screen are demagnified down to a very tiny area so that, again, they are very intense. I don't know why this hasn't been done yet!

That's the Encyclopaedia Britannica on the head of a pin, but let's consider all the books in the world. The Library of Congress has approximately 9 million volumes; the British Museum Library has 5 million volumes; there are also 5 million volumes in the National Library in France. Undoubtedly there are duplications, so let us say that there are some 24 million volumes of interest in the world.

What would happen if I print all this down at the scale we have been discussing? How much space would it take? It would take, of course, the area of about a million pinheads because, instead of there being just the 24 volumes of the Encyclopaedia, there are 24 million volumes. The million pinheads can be put in a square of a thousand pins on a side, or an area of about 3 square yards. That is to say, the silica replica with the paper-thin backing of plastic, with which we have made the copies, with all this information, is on an area of approximately the size of 35 pages of the Encyclopaedia. This is only one-fourth as many pages as a copy of the *Saturday Evening Post*. All of the information which all of mankind has ever recorded in books can be carried around in a pamphlet in your hand—and not written in code, but as a simple reproduction of the original pictures, engravings, and everything else on a small scale without loss of resolution.

What would our librarian at Caltech say, as she runs all over from one building to another, if I tell her that, ten years from now, all of the information that she is struggling to keep track of—120 000 volumes, stacked from the floor to the ceiling, drawers full of cards, storage rooms full of the older books—can be kept on just one library card! When the University of Brazil, for example, finds that their library is burned, we can send them a copy of every book in our library by striking off a copy from the master plate in a few hours and mailing it in an envelope no bigger or heavier than any other ordinary air mail letter.

Now, the name of this talk is "There is *Plenty* of Room at the Bottom"—not just "There is Room at the Bottom." What I have demonstrated is that there is room—that you can decrease the size of things in a practical way. I now want to show that there is *plenty* of room. I will not now discuss how we are going to do it, but only what is possible in principle—in other words, what is possible according to the laws of physics. I am not inventing antigravity, which is possible someday only if the laws are not what we think. I am telling you what could be done if the laws are what we think; we are not doing it simply because we haven't yet gotten around to it.

INFORMATION ON A SMALL SCALE

Suppose that, instead of trying to reproduce the pictures and all the information directly in its present form, we write only the information content in a code of dots and dashes, or something like that, to represent the various letters. Each letter represents six or seven "bits" of information; that is, you need only about six or seven dots or dashes for each letter. Now, instead of writing everything, as I did before, on the *surface* of the head of a pin, I am going to use the interior of the material as well.

Let us represent a dot by a small spot of one metal, the next dash by an adjacent spot of another metal, and so on. Suppose, to be conservative, that a bit of information is going to require a little cube of atoms $5 \times 5 \times 5$—that is 125 atoms. Perhaps we need a hundred and some odd atoms to make sure that the information is not lost through diffusion, or through some other process.

I have estimated how many letters there are in the Encyclopaedia, and I have assumed that each of my 24 million books is as big as an Encyclopaedia volume, and have calculated, then, how many bits of information there are (10^{15}). For each bit I allow 100 atoms. And it turns out that all of the information that man has carefully accumulated in all the books in the world can be written in this form in a cube of material one two-hundredth of an inch wide—which is the barest piece of dust that can be made out by the human eye. So there is *plenty* of room at the bottom! Don't tell me about microfilm!

This fact—that enormous amounts of information can be carried in an exceedingly small space—is, of course, well known to the biologists, and resolves the mystery which existed before we understood all this clearly, of how it could be that, in the tiniest cell, all of the information for the organization of a complex creature such as ourselves can be stored. All this information—whether we have brown eyes, or whether we think at all, or that in the embryo the jawbone should first develop with a little hole in the side so that later a nerve can grow through it—all this information is contained in a very tiny fraction of the cell in the form of long-chain DNA molecules in which approximately 50 atoms are used for one bit of information about the cell.

Better Electron Microscopes

If I have written in a code, with $5 \times 5 \times 5$ atoms to a bit, the question is: How could I read it today? The electron microscope is not quite good enough; with the greatest care and effort, it can only resolve about 10 angstroms. I would like to try and impress upon you, while I am talking about all of these things on a small scale, the importance of improving the electron microscope by a hundred times. It is not impossible; it is not against the laws of diffraction of the electron. The wave length of the electron in such a microscope is only $1/20$ of an angstrom. So it should be possible to see the individual atoms. What good would it be to see individual atoms distinctly?

We have friends in other fields—in biology, for instance. We physicists often look at them and say, "You know the reason you fellows are making so little progress?" (Actually I don't know any field where they are making more rapid progress than they are in biology today.) "You should use more mathematics, like we do." They could answer us—but they're polite, so I'll answer for them: "What you should do in order for us to make more rapid progress is to make the electron microscope 100 times better."

What are the most central and fundamental problems of biology today? They are questions like: What is the sequence of bases in the DNA? What happens when you have a mutation? How is the base order in the DNA connected to the order of amino acids in the protein? What is the structure of the RNA; is it single-chain or double-chain, and how is it related in its order of bases to the DNA? What is the organization of the microsomes? How are proteins synthesized? Where does the RNA go? How does it sit? Where do the proteins sit? Where do the amino acids go in? In photosynthesis, where is the chlorophyll; how is it arranged; where are the carotenoids involved in this thing? What is the system of the conversion of light into chemical energy?

It is very easy to answer many of these fundamental biological questions; you just *look at the thing!* You will see the order of bases in the chain; you will see the structure of the microsome. Unfortunately, the present microscope sees at a scale which is just a bit too crude. Make the microscope one hundred times more powerful, and many problems of biology would be made very much easier. I exaggerate, of course, but the biologists would surely be very thankful to you—and they would prefer that to the criticism that they should use more mathematics.

The theory of chemical processes today is based on theoretical physics. In this sense, physics supplies the foundation of chemistry. But chemistry also has analysis. If you have a strange substance and you want to know what it is, you go through a long and complicated process of chemical analysis. You can analyze almost anything today, so I am a little late with my idea. But if the physicists wanted to, they could also dig under the chemists in the problem of chemical analysis. It would be very easy to make an analysis of any complicated chemical substance; all one would have to do would be to look at it and see where the atoms are. The only trouble is that the electron microscope is one hundred times too poor. (Later, I would like to ask the question: Can the physicists do something about the third problem of chemistry—namely, synthesis? Is there a *physical* way to synthesize any chemical substance?)

The reason the electron microscope is so poor is that the *f*-value of the lenses is only 1 part to 1000; you don't have a big enough numerical aperture. And I know that there are theorems which prove that it is impossible, with axially symmetrical stationary field lenses, to produce an *f*-value any bigger than so and so; and therefore the resolving power at the present time is at its theoretical maximum. But in every theorem there are assumptions. Why must the field be axially symmetrical? Why must the field be stationary? Can't we have pulsed electron beams in fields moving up along with the electrons? Must the field be symmetrical? I put this out as a challenge: Is there no way to make the electron microscope more powerful?

The Marvelous Biological System

The biological example of writing information on a small scale has inspired me to think of something that should be possible. Biology is not simply writing information; it is *doing something* about it. A biological system can be exceedingly small. Many of the cells are very tiny, but they are very active; they manufacture various substances; they walk around; they wiggle; and they do all kinds of marvelous things—all on a very small scale. Also, they store information. Consider the possibility that we too can make a thing very small, which does what we want—that we can manufacture an object that maneuvers at that level!

There may even be an economic point to this business of making things very small. Let me remind you of some of the problems of computing machines. In computers we have to store an enormous amount of information. The kind of writing that I was mentioning before, in which I had everything down as a distribution of metal, is permanent. Much more interesting to a computer is a way of writing, erasing, and writing something else. (This is usually because we don't want to waste the material on which we have just written. Yet if we could write it in a very small space, it wouldn't make any difference; it could just be thrown away after it was read. It doesn't cost very much for the material.)

Miniaturizing the Computer

I don't know how to do this on a small scale in a practical way, but I do know that computing machines are very large; they fill rooms. Why can't we make them very small, make them of little wires, little elements—and by little, I mean *little*. For instance, the wires should be 10 or 100 atoms in diameter, and the circuits should be a few thousand angstroms across. Everybody who has analyzed the logical theory of computers has come to the conclusion that the possibilities of computers are very interesting—if they could be made to be more complicated by several orders of magnitude. If they had millions of times as many elements, they could make judgments. They would have time to calculate what is the best way to make the calculation that they are about to make. They could select the method of analysis which, from their experience, is better than the one that we would give to them. And, in many other ways, they would have new qualitative features.

If I look at your face I immediately recognize that I have seen it before. (Actually, my friends will say I have chosen an unfortunate example here for the subject of this illustration. At least I recognize that it is a *man* and not an *apple*.) Yet there is no machine which, with that speed, can take a picture of a face and say even that it is a man; and much less that it is the same man that you showed it before—unless it is exactly the same picture. If the face is changed; if I am closer to the face; if I am further from the face; if the light changes—I recognize it anyway. Now, this little computer I carry in my head is easily able to do that. The computers that we build are not able to do that. The number of elements in this bone box of mine are enormously greater than the number of elements in our "wonderful" computers. But our mechanical computers are too big; the elements in this box are microscopic. I want to make some that are *sub*-microscopic.

If we wanted to make a computer that had all these marvelous extra qualitative abilities, we would have to make it, perhaps, the size of the Pentagon. This has several disadvantages. First, it requires too much material; there may not be enough germanium in the world for all the transistors which would have to be put into this enormous thing. There is also the problem of heat generation and power consumption; TVA would be needed to run the computer. But an even more practical difficulty is that the computer would be limited to a certain speed. Because of its large size, there is finite time required to get the information from one place to another. The information cannot go any faster than the speed of light—so, ultimately, when our computers get faster and faster and more and more elaborate, we will have to make them smaller and smaller.

But there is plenty of room to make them smaller. There is nothing that I can see in the physical laws that says the computer elements cannot be made enormously smaller than they are now. In fact, there may be certain advantages.

Miniaturization by Evaporation

How can we make such a device? What kind of manufacturing processes would we use? One possibility we might consider, since we have talked about writing by putting atoms down in a certain arrangement, would be to evaporate the material, then evaporate the insulator next to it. Then, for the next layer, evaporate another position of a wire, another insulator, and so on. So, you simply evaporate until you have a block of stuff which has the elements—coils and condensers, transistors and so on—of exceedingly fine dimensions.

But I would like to discuss, just for amusement, that there are other possibilities. Why can't we manufacture these small computers somewhat like we manufacture the big ones? Why can't we drill holes, cut things, solder things, stamp things out, mold different shapes all at an infinitesimal level? What are the limitations as to how small a thing has to be before you can no longer mold it? How many times when you are working on something frustratingly tiny, like your wife's wrist watch, have you said to yourself, "If I could only train an ant to do this!" What I would like to suggest is the possibility of training an ant to train a mite to do this. What are the possibilities of small but movable machines? They may or may not be useful, but they surely would be fun to make.

Consider any machine—for example, an automobile—and ask about the problems of making an infinitesimal machine like it. Suppose, in the particular design of the automobile, we need a certain precision of the parts; we need an accuracy, let's suppose, of 4/10 000 of an inch. If things are more inaccurate than that in the shape of the cylinder and so on, it isn't going to work very well. If I make the thing too small, I have to worry about the size of the atoms; I can't make a circle out of "balls" so to speak, if the circle is too small. So, if I make the error, corresponding to 4/10 000 of an inch, correspond to an error of 10 atoms, it turns out that I can reduce the dimensions of an automobile 4000 times, approximately—so that it is 1 mm across. Obviously, if you redesign the car so that it would work with a much larger tolerance, which is not at all impossible, then you could make a much smaller device.

It is interesting to consider what the problems are in such small machines. Firstly, with parts stressed to the same degree, the forces go as the area you are reducing, so that things like weight and inertia are of relatively no importance. The strength of material, in other words, is very much greater in proportion. The stresses and expansion of the flywheel from centrifugal force, for example, would be the same proportion only if the rotational speed is increased in the same proportion as we decreased the size. On the other hand, the metals that we use have a grain structure, and this would be very annoying at small scale because the material is not homogeneous. Plastics and glass and things of this amorphous nature are very much more homogeneous, and so we would have to make our machines out of such materials.

There are problems associated with the electrical part of the system—with the copper wires and the magnetic parts. The magnetic properties on a very small scale are not the same as on a large scale; there is the "domain" problem involved. A big magnet made of millions of domains can only be made on a small scale with one domain. The electrical equipment won't simply be scaled down; it has to be redesigned. But I can see no reason why it can't be redesigned to work again.

Problems of Lubrication

Lubrication involves some interesting points. The effective viscosity of oil would be higher and higher in proportion as we went down (and if we increase the speed as much as we can). If we don't increase the speed so much, and change from oil to kerosene or some other fluid, the problem is not so bad. But actually we may not have to lubricate at all! We have a lot of extra force. Let the bearings run dry; they won't run hot because the heat escapes away from such a small device very, very rapidly.

This rapid heat loss would prevent the gasoline from exploding, so an internal combustion engine is impossible. Other chemical reactions, liberating energy when cold, can be used. Probably an external supply of electrical power would be most convenient for such small machines.

What would be the utility of such machines? Who knows? Of course, a small automobile would only be useful for the mites to drive around in, and I suppose our Christian interests don't go that far. However, we did note the possibility of the manufacture of small elements for computers in completely automatic factories, containing lathes and other machine tools at the very small level. The small lathe would not have to be exactly like our big lathe. I leave to your imagination the improvement of the design to take full advantage of the properties of things on a small scale, and in such a way that the fully automatic aspect would be easiest to manage.

A friend of mine (Albert R. Hibbs) suggests a very interesting possibility for relatively small machines. He says that, although it is a very wild idea, it would be interesting in surgery if you could swallow the surgeon. You put the mechanical surgeon inside the blood vessel and it goes into the heart and "looks" around. (Of course the information has to be fed out.) It finds out which valve is the faulty one and takes a little knife and slices it out. Other small machines might be permanently incorporated in the body to assist some inadequately-functioning organ.

Now comes the interesting question: How do we make such a tiny mechanism? I leave that to you. However, let me suggest one weird possibility. You know, in the atomic energy plants they have materials and machines that they can't handle directly because they have become radioactive. To unscrew nuts and put on bolts and so on, they have a set of master and slave hands, so that by operating a set of levers here, you control the "hands" there, and can turn them this way and that so you can handle things quite nicely.

Most of these devices are actually made rather simply, in that there is a particular cable, like a marionette string, that goes directly from the controls to the "hands." But, of course, things also have been made using servo motors, so that the connection between the one thing and the other is electrical rather than mechanical. When you turn the levers, they turn a servo motor, and it changes the electrical currents in the wires, which repositions a motor at the other end.

Now, I want to build much the same device—a master-slave system which operates electrically. But I want the slaves to be made especially carefully by modern large-scale machinists so that they are one-fourth the scale of the "hands" that you ordinarily maneuver. So you have a scheme by which you can do things at one-quarter scale anyway—the little servo motors with little hands play with little nuts and bolts; they drill little holes; they are four times smaller. Aha! So I manufacture a quarter-size lathe; I manufacture quarter-size tools; and I make, at the one-quarter scale, still another set of hands again relatively one-quarter size! This is one-sixteenth size, from my point of view. And after I finish doing this I wire directly from my large-scale system, through transformers perhaps, to the one-sixteenth-size servo motors. Thus I can now manipulate the one-sixteenth-size hands.

Well, you get the principle from there on. It is rather a difficult program, but it is a possibility. You might say that one can go much farther in one step than from one to four. Of course, this has all to be designed very carefully and it is not necessary simply to make it like hands. If you thought of it very carefully, you could probably arrive at a much better system for doing such things.

If you work through a pantograph, even today, you can get much more than a factor of four in even one step. But you can't work directly through a pantograph which makes a smaller pantograph which then makes a smaller pantograph—because of the looseness of the holes and the irregularities of construction. The end of the pantograph wiggles with a relatively greater irregularity than the irregularity with which you move your hands. In going down this scale, I would find the end of the pantograph on the end of the pantograph on the end of the pantograph shaking so badly that it wasn't doing anything sensible at all.

At each stage, it is necessary to improve the precision of the apparatus. If, for instance, having made a small lathe with a pantograph, we find its lead screw irregular—more irregular than the large-scale one—we could lap the lead screw against breakable nuts that you can reverse in the usual way back and forth until this lead screw is, at its scale, as accurate as our original lead screws, at our scale.

We can make flats by rubbing unflat surfaces in triplicate together—in three pairs—and the flats then become flatter than the thing you started with. Thus, it is not impossible to improve precision on a small scale by the correct operations. So, when we build this stuff, it is necessary at each step to improve the accuracy of the equipment by working for awhile down there, making accurate lead screws, Johansen blocks, and all the other materials which we use in accurate machine work at the higher level. We have to stop at each level and manufacture all the stuff to go to the next level—a very long and very difficult program. Perhaps you can figure a better way than that to get down to small scale more rapidly.

Yet, after all this, you have just got one little baby lathe four thousand times smaller than usual. But we were thinking of making an enormous computer, which we were going to build by drilling holes on this lathe to make little washers for the computer. How many washers can you manufacture on this one lathe?

A Hundred Tiny Hands

When I make my first set of slave "hands" at one-fourth scale, I am going to make ten sets. I make ten sets of "hands," and I wire them to my original levers so they each do exactly the same thing at the same time in parallel. Now, when I am making my new devices one-quarter again as small, I let each one manufacture ten copies, so that I would have a hundred "hands" at the 1/16th size.

Where am I going to put the million lathes that I am going to have? Why, there is nothing to it; the volume is much less than that of even one full-scale lathe. For instance, if I made a billion little lathes, each 1/4000 of the scale of a regular lathe, there are plenty of materials and space available because in the billion little ones there is less than 2 per cent of the materials in one big lathe.

It doesn't cost anything for materials, you see. So I want to build a billion tiny factories, models of each other, which are manufacturing simultaneously, drilling holes, stamping parts, and so on.

As we go down in size, there are a number of interesting problems that arise. All things do not simply scale down in proportion. There is the problem that materials stick together by the molecular (Van der Waals) attractions. It would be like this: After you have made a part and you unscrew the nut from a bolt, it isn't going to fall down because the gravity isn't appreciable; it would even be hard to get it off the bolt. It would be like those old movies of a man with his hands full of molasses, trying to get rid of a glass of water. There will be several problems of this nature that we will have to be ready to design for.

Rearranging the Atoms

But I am not afraid to consider the final question as to whether, ultimately—in the great future—we can arrange the atoms the way we want; the very *atoms*, all the way down! What would happen if we could arrange the atoms one by one the way we want them (within reason, of course; you can't put them so that they are chemically unstable, for example).

Up to now, we have been content to dig in the ground to find minerals. We heat them and we do things on a large scale with them, and we hope to get a pure substance with just so much impurity, and so on. But we must always accept some atomic arrangement that nature gives us. We haven't got anything, say, with a "checkerboard" arrangement, with the impurity atoms exactly arranged 1000 angstroms apart, or in some other particular pattern.

What could we do with layered structures with just the right layers? What would the properties of materials be if we could really arrange the atoms the way we want them? They would be very interesting to investigate theoretically. I can't see exactly what would happen, but I can hardly doubt that when we have some *control* of the arrangement of things on a small scale we will get an enormously greater range of possible properties that substances can have, and of different things that we can do.

Consider, for example, a piece of material in which we make little coils and condensers (or their solid state analogs) 1000 or 10 000 angstroms in a circuit, one right next to the other, over a large area, with little antennas sticking out at the other end—a whole series of circuits.

Is it possible, for example, to emit light from a whole set of antennas, like we emit radio waves from an organized set of antennas to beam the radio programs to Europe? The same thing would be to *beam* the light out in a definite direction with very high intensity. (Perhaps such a beam is not very useful technically or economically.)

I have thought about some of the problems of building electric circuits on a small scale, and the problem of resistance is serious. If you build a corresponding circuit on a small scale, its natural frequency goes up, since the wave length goes down as the scale; but the skin depth only decreases with the square root of the scale ratio, and so resistive problems are of increasing difficulty. Possibly we can beat resistance through the use of superconductivity if the frequency is not too high, or by other tricks.

Atoms in a Small World

When we get to the very, very small world—say circuits of seven atoms—we have a lot of new things that would happen that represent completely new opportunities for design. Atoms on a small scale behave like *nothing* on a large scale, for they satisfy the laws of quantum mechanics. So, as we go down and fiddle around with the atoms down there, we are working with different laws, and we can expect to do different things. We can manufacture in different ways. We can use, not just circuits, but some system involving the quantized energy levels, or the interactions of quantized spins, etc.

Another thing we will notice is that, if we go down far enough, all of our devices can be mass produced so that they are absolutely perfect copies of one another. We cannot build two large machines so that the dimensions are exactly the same. But if your machine is only 100 atoms high, you only have to have it correct to one-half of one per cent to make sure the other machine is exactly the same size—namely, 100 atoms high!

At the atomic level, we have new kinds of forces and new kinds of possibilities, new kinds of effects. The problems of manufacture and reproduction of materials will be quite different. I am, as I said, inspired by the biological phenomena in which chemical forces are used in a repetitious fashion to produce all kinds of weird effects (one of which is the author).

The principles of physics, as far as I can see, do not speak against the possibility of maneuvering things atom by atom. It is not an attempt to violate any laws; it is something, in principle, that can be done; but, in practice, it has not been done because we are too big.

Ultimately, we can do chemical synthesis. A chemist comes to us and says, "Look, I want a molecule that has the atoms arranged thus and so; make me that molecule." The chemist does a mysterious thing when he wants to make a molecule. He sees that it has got that ring, so he mixes this and that, and he shakes it, and he fiddles around. And, at the end of a difficult process, he usually does succeed in synthesizing what he wants. By the time I get my devices working, so that we can do it by physics, he will have figured out how to synthesize absolutely anything, so that this will really be useless.

But it is interesting that it would be, in principle, possible (I think) for a physicist to synthesize any chemical substance that the chemist writes down. Give the orders and the physicist synthesizes it. How? Put the atoms down where the chemist says, and so you make the substance. The problems of chemistry and biology can be greatly helped if our ability to see what we are doing, and to do things on an atomic level, is ultimately developed—a development which I think cannot be avoided.

Now, you might say, "Who should do this and why should they do it?" Well, I pointed out a few of the economic applications, but I know that the reason that you would do it might be just for fun. But have some fun! Let's have a competition between laboratories. Let one laboratory make a tiny motor which it sends to another lab which sends it back with a thing that fits inside the shaft of the first motor.

HIGH SCHOOL COMPETITION

Just for the fun of it, and in order to get kids interested in this field, I would propose that someone who has some contact with the high schools think of making some kind of high school competition. After all, we haven't even started in this field, and even the kids can write smaller than has ever been written before. They could have competition in high schools. The Los Angeles high school could send a pin to the Venice high school on which it says, "How's this?" They get the pin back, and in the dot of the "i" it says, "Not so hot."

Perhaps this doesn't excite you to do it, and only economics will do so. Then I want to do something; but I can't do it at the present moment, because I haven't prepared the ground. I hereby offer a prize of $1000 to the first guy who can take the information on the page of a book and put it on an area 1/25 000 smaller in linear scale in such manner that it can be read by an electron microscope.

And I want to offer another prize—if I can figure out how to phrase it so that I don't get into a mess of arguments about definitions—of another $1000 to the first guy who makes an operating electric motor—a rotating electric motor which can be controlled from the outside and, not counting the lead-in wires, is only 1/64 inch cube.

I do not expect that such prizes will have to wait very long for claimants.[1]

---

[1]Note: This article, "There's Plenty of Room at the Bottom," was originally published in the February 1960 issue of Caltech's *Engineering and Science Magazine,* and subsequently published as a chapter in the Reinhold Publishing Corporation book *Miniaturization,* Horace D. Bilbert, Ed. The first Feynman challenge was won by Tom Newman, a student of R. Fabian Pease. Tom Newman used an electron beam to write the opening page of "A Tale of Two Cities" in an area of 5.9 x 5.9 μm. ("Tiny Tale Gets Grand," *Engineering and Science Magazine,* January 1986; "High Resolution Patterning System with a Single Bore Objective Lens," by T.H. Newman, K.E. Williams, and R.F.W. Pease, *Journal of Vacuum Science Technology* Vol. B5, No. 1, pp. 88-91, Jan./Feb. 1987) The second Prize was presented by Dr. Feynman on November 28, 1960 to William McLellan who built an electric motor a 1/64th of an inch on a side. It weighs 250 micrograms and generates one millionth of a horsepower.

Infinitesimal Machinery

Richard Feynman

Editor's Preface—This speech, delivered in 1983, is a sequel to the famous 1960 speech "There's Plenty of Room at the Bottom" (reprinted in J. Microelectromechanical Systems, vol. 1, no. 1, pp. 60-66, 1992). It is remarkable in many ways. Feynman anticipates the sacrificial-layer method of making silicon micromotors, the use of electrostatic actuation, and the importance of friction and contact sticking in such devices. He explores the persistent problem of finding meaningful applications for these tiny machines, touching a range of topics along the way. And he looks at the future of computation using a register made of atoms, and quantum-mechanical transitions for computation operations.

The speech was presented to a large audience of scientists and engineers at the Jet Propulsion Laboratory (JPL). Following a brief introduction by Al Hibbs, Feynman used slides, hand gestures, and sketches on a blackboard to supplement his pithy language. The style was quite informal. Many of the remarks were intended to evoke laughter, and did so; some were directed specifically at the JPL audience (such as the reference to enhancement of pictures). Working from a video recording of the speech, I had the benefit of seeing the slides, the gestures, and the blackboard sketches. The editor's task, in such a case, is to find a way to capture the essence of Feynman's message and spirit with minimal perturbation of his actual words, making changes only where the lack of visual aids, or the normal fits and starts of oral presentation require some adjustment. I was greatly aided in this task by a transcript of the spoken text prepared and orally smoothed by Nora Odendahl. Where necessary, I have added italicized comments to clarify terms or images being described, but have tried to keep these to a minimum. I have also added topic headings to identify major sections, and have moved some text around to improve readability. As you go through this speech, imagine yourself a part of the audience— if you are moved to laughter in spots, that is appropriate; and if you are moved to think anew about the problems discussed, that indeed was Feynman's goal.

Stephen D. Senturia

Introduction of Richard Feynman by Al Hibbs—Welcome to the Feynman lecture on "Infinitesimal Machinery." I have the pleasure of introducing Richard, an old friend and past associate. He was educated at MIT and at Princeton, where he received a Ph.D. in 1942. In the War he was at Los Alamos, where he learned how to pick combination locks—an activity at which he is still quite skillful. He next went to Cornell, where he experimented with swinging hoops. Then, both before and during his time at Caltech, he became an expert in drumming, specializing in complex rhythms,

This manuscript is based on a talk given by Richard Feynman on February 23, 1983, at the Jet Propulsion Laboratory, Pasadena CA. It is reprinted with the permission of his estate, Carl Feynman executor.
The author, deceased, was with the California Institute of Technology, Pasadena, CA.
IEEE Log Number 9210135.

particularly those of South America and recently those of the South Pacific. At Caltech, he learned to decode Mayan hieroglyphs and took up art, becoming quite an accomplished draftsman—specializing in nude women. And he also does jogging.

Richard received the Nobel prize, but I believe it was for physics and not for any of these other accomplishments. He thinks that happened in 1965, although he doesn't remember the exact year. I have never known him to suffer from false modesty, so I believe he really has forgotten which year he got the Nobel prize.

WHEN Dick Davies asked me to talk, he didn't tell me the occasion was going to be so elaborate, with TV cameras and everything—he told me I'd be among friends. I didn't realize I had so many friends. I would feel much less uncomfortable if I had more to say. I don't have very much to say—but of course, I'll take a long time to say it.

Revisiting "There's Plenty of Room at the Bottom"

In 1960, about 23 years ago, I gave a talk called "There's Plenty of Room at the Bottom," in which I described the coming technology for making small things. I pointed out what everybody knew: that numbers, information, and computing didn't require any particular size. You could write numbers very small, down to atomic size. (Of course you can't write something much smaller than the size of a single atom.) Therefore, we could store a lot of information in small spaces, and in a little while we'd be able to do so easily. And of course, that's what happened.

I've been asked a number of times to reconsider all the things that I talked about 23 years ago, and to see how the situation has changed. So my talk today could be called "There's Plenty of Room at the Bottom, Revisited."

As I mentioned in the 1960 talk, you could represent a digit by saying it is made of a few atoms. Actually, you'd only have to have to use one atom for each digit, but let's say you make a bit from a bunch of gold atoms, and another bit from a bunch of silver atoms. The gold atoms represent a one, and the silver atoms a zero. Suppose you make the bits into little cubes with a hundred atoms on a side. When you stack the cubes all together, you can write a lot of stuff in a small space. It turns out that all the books in all the world's libraries could have all their information—including pictures using dots down to the resolution of the human eye—stored in a cube $1/120$ inch on a side.

Reprinted from *Journal of Microelectromechanical Systems*, R. Feynman, "Infinitesimal Machinery," Vol. 2, No. 1, pp. 4-14, March 1993. © Carl Feynman, Executor of the Richard Feynman estate.

That cube would be just about the size you can make out with your eye—about the size of a speck of dirt.

If, however, you used only surfaces rather than the volume of the cubes to store information, and if you simply reduce normal scale by twenty-five thousand times, which was just about possible in those days, then the *Encyclopedia Britannica* could be written on the head of a pin, the Caltech library on one library card, and all the books in the world on thirty-five pages of the *Saturday Evening Post*. I suggested a reduction of twenty-five thousand times just to make the task harder, because due to the limitations of light wavelength, that reduction was about ten times smaller than you could read by means of light. You could, of course, read the information with electron microscopes and electron beams.

Because I had mentioned the possibility of using electron beams and making things still smaller, six or eight years ago someone sent me a picture of a book that he reduced by thirty thousand times. In the picture, there are letters measuring about a tenth of a micron across [*passes the picture around the audience*].

I also talked in the 1960 lecture about small machinery, and was able to suggest no particular use for the small machines. You will see there has been no progress in that respect. And I left as a challenge the goal of making a motor that would measure 1/64 of an inch on a side. At that time, the idea that I proposed was to make a set of hands—like those used in radioactive systems—that followed another set of hands. Only we make these "slave" hands smaller—a quarter of the original hands' size—and then let the slave hands make smaller hands and those make still smaller hands. You're right to laugh—I doubt that that's a sensible technique. At any rate, I wanted to get a motor that couldn't be made directly by hand, so I proposed 1/64 of an inch.

At the end of my talk, Don Glaser, who won the Nobel prize in physics—that's something that's supposed to be good, right?—said, "You should have asked for a motor 1/200 inch on a side, because 1/64 inch on a side is just about possible by hand." And I said, "Yeah, but if I offered a thousand-dollar prize for a motor 1/200 inch on a side, everybody would say 'Boy, that guy's a cheapskate! Nobody's ever going to do that.'" I didn't believe Glaser, but somebody actually did make the motor by hand!

As a matter of fact, the motor's very interesting, and just for fun, here it is. First look at it directly with your eye, to see how big it is. It's right in the middle of that little circle—it's only the size of a decimal point or a period at the end of a sentence. Mr. McLellan, who made this device for me, arranged it very beautifully, so that it has a magnifier you can attach—but don't look at it through the magnifier until you look at it directly. You'll find you can't see it without the magnifier. Then you can look through the magnifier and turn this knob, which is a little hand generator which makes the juice to turn the motor so you can watch the motor go around [*gives the McLellan motor to the audience to be passed around*].

What We Can Do Today

Now I'd like to talk about what we can do today, as compared to what we were doing in those days. Back then, I was speaking about machinery as well as writing, computers, and information, and although this talk is billed as being about machinery, I'll also discuss computers and information at the end.

My first slide illustrates what can be done today in making small things commercially. This is of course one of the chips that we use in computers, and it represents an area of about three millimeters by four millimeters. Human beings can actually make something on that small a scale, with wires about six microns across (a micron is a millionth of a meter, or a thousandth of a millimeter). The tolerances, dimensions, and separations of some of the wires are controlled to about three microns. This computer chip was manufactured five years ago, and now things have improved so that we can get down to about one-half micron resolution.

These chips are made, as you know, by evaporating successive layers of materials through masks. [*Feynman uses "evaporating" as a generic term for all semiconductor process steps.*] You can create the pattern in a material in several ways. One is to shine light through a mask that has the design that you want, then focus the light very accurately onto a light-sensitive material and use the light to change the material, so that it gets easier to etch or gets less easy to etch. Then you etch the various materials away in stages. You can also deposit one material after another—there's oxide, and silicon, and silicon with materials diffused into it—all arranged in a pattern at that scale. This technology was incredible twenty-three years ago, but that's where we are today.

The real question is, how far can we go? I'll explain to you later why, when it comes to computers, it's always better to get smaller, and everybody's still trying to get smaller. But if light has a finite wavelength, then we're not going to be able to make masks with patterns measuring less than a wavelength. That fact limits us to about a half a micron, which is about possible nowadays, with light, in laboratories. The commercial scale is about twice that big.

So what could we do today, if we were to work as hard as we could in a laboratory—not commercially, but with the greatest effort in the lab? Michael Isaacson from the Laboratory of Submicroscopic Studies (appropriate for us) has made something under the direction of an artist friend of mine named Tom Van Sant. Van Sant is, I believe, the only truly modern artist I know. By truly modern, I mean a man who understands our culture and appreciates our technology and science as well as the character of nature, and incorporates them into the things that he makes.

I would like to show you, in the next slide, a picture by Van Sant. That's art, right? It represents an eye. That's the eyelid and the eyebrow, perhaps, and of course you can recognize the pupil. The interesting thing about this eye is that it's the smallest drawing a human being has ever made. It's a quarter of a micron across—250 millimicrons— and the central spot of the pupil is something like fifteen or twenty millimicrons, which corresponds to about one hundred atoms in diameter. That's the bottom. You're not going to be able to see things being drawn more than one hundred times smaller, because by that time you're at the size of atoms. This picture is as far down as we can make it.

Because I admire Tom Van Sant, I would like to show you some other artwork that he has created. He likes to draw eyes, and the next slide shows another eye by him. This is real art, right? Look at all the colors, the beauty, the light, and so forth—qualities that of course are much more appreciated as art. (Maybe some of you clever JPL guys know what you're looking at, but just keep it to yourselves, eh?)

To get some idea of what you're looking at, we're going to look at that eye from a little bit further back, so you can see some more of the picture's background. The next slide shows it at a different scale. The eye is now smaller, and perhaps you see how the artist has drawn the furrows of the brow, or whatever it is around the eye. The artist now wants to show the eye to us on a still smaller scale, so we can see a little more of the background. So in this next slide, you see the city of Los Angeles covering most of the picture, and the eye is this little speck up in the corner!

Actually, all these pictures of the second eye are LANDSAT pictures of an eye that was made in the desert. You might wonder how someone can make an eye that big—it's two and one-half kilometers across. The way Van Sant made it was to set out twenty-four mirrors, each two feet square, in special locations in the desert. He knew that when the LANDSAT passes back and forth overhead, its eye looks at the land and records information for the picture's pixels. Van Sant used calculations so that the moment the LANDSAT looked at a particular mirror, the sun would be reflecting from the mirror right into the eye of the LANDSAT. The reflection overexposed the pixel, and what would have been a two-foot square mirror instead made a white spot corresponding to an area of several acres. So what you saw in the first picture was a sequence of overexposed pixels on the LANDSAT picture. Now that's the way to make art! As far as I know, this is the largest drawing ever made by man.

If you look again at the original picture, you can see one pixel that didn't come out. When they went back to the desert, they found that the mirror had been knocked off its pedestal, and that there were footprints from a jack rabbit over the surface. So Van Sant lost one pixel.

The point about the two different eyes is this: that Van Sant wanted to make an eye much bigger than a normal eye, and the eye in the desert was 100 000 times bigger than a normal eye. The first eye, the tiny one, was 100 000 times smaller than a normal eye. So you get an idea of what the scale is. We're talking about going down to that small level, which is like the difference in scale between the two-and-one-half-kilometer desert object and our own eye. Also amusing to think about, even though it has nothing to do with going small, but rather with going big— what happens if you go to the next eye, 100 000 times bigger? Then the eye's scale is very close to the rings of Saturn, with the pupil in the middle.

I wanted to use these pictures to tell us about scale and also to show us what, at the present time, is the ultimate limit of our actual ability to construct small things. And that summarizes how we stand today, as compared to how the situation looked when I finished my talk in 1960. We see that computers are well on their way to small scale, even though there are limitations. But I would like to discuss something else—small machines.

Small Machines—How to Make Them

By a machine, I mean things that have movable parts you can control, that have wheels and stuff inside. You can turn the movable parts; they are actual objects. As far as I can tell, this interest of mine in small machines is a misguided one, or more correctly, the suggestion in the lecture "Plenty of Room at the Bottom" that soon we would have small machines was certainly a misguided prediction. The only small machine we have is the one that I've passed around to you, the one that Mr. McLellan made by hand.

There is no use for these machines, so I still don't understand why I'm fascinated by the question of making small machines with movable and controllable parts. Therefore I just want to tell you some ideas and considerations about the machines. Any attempt to make out that this is anything but a game—well, let's leave it the way it is: I'm fascinated and I don't know why.

Every once in a while I try to find a use. I know there's already been a lot of laughter in the audience— just save it for the uses that I'm going to suggest for some of these devices, okay?

But the first question is, how can we make small machines? Let's say I'm talking about very small machines, with something like ten microns (that's a hundredth of a millimeter) for the size of a rotor. That's forty times smaller than the motor I passed around—it's invisible, it's so small.

I would like to shock you by stating that I believe that with today's technology we can easily—I say *easily*—construct motors one fortieth of this size on each dimension. That's sixty-four thousand times smaller than the size of McLellan's motor. And in fact, with our present technology, we can make thousands of these motors at a time, all separately controllable. Why do you want to make them? I told you there's going to be lots of laughter, but just for fun, I'll suggest how to do it—it's very easy.

It's just like the way we put those evaporated layers down, and made all kinds of structures. We keep making the structures a little thicker by adding a few more layers. We arrange the layers so that you can dissolve away a layer supporting some mechanical piece, and loosen the piece. The stuff that you evaporate would be such that it could be dissolved, or boiled away, or evaporated out. And it could be that you build this stuff up in a matrix, and build other things on it, and then other stuff over it. Let's call the material "soft wax," although it's not going to be wax. You put the wax down, and with a mask you put some silicon lumps that are not connected to anything, some more wax, some more wax, and then silicon dioxide or something. You melt out or evaporate the wax, and then you're left with loose pieces of silicon. The way I described it, that piece would fall somewhere, but you have other structures that hold it down. It does seem to me perfectly obvious that with today's technology, if you wanted to, you could make something one-fortieth the size of McLellan's motor.

When I gave the talk called "Plenty of Room at the Bottom," I offered a thousand-dollar prize for the motor—I was single at the time. In fact, there was some consternation at home, because I got married after that, and had forgotten all about the prize. When I was getting married, I explained my financial position to my future wife, and she thought that it was bad, but not so bad. About three or four days after we came back from the honeymoon, with a lot of clearing of my throat I explained to her that I had to pay a thousand dollars that I had forgotten about—that I had promised if somebody made a small motor. So she didn't trust me too much for a while.

Because I am now married, and have a daughter who likes horses, and a son in college, I cannot offer a thousand dollars to motivate you to make movable engines even forty times smaller. But Mr. McLellan himself said that the thousand dollars didn't make any difference—he got interested in the challenge.

Of course, if we had these movable parts, we could move them and turn them with electrostatic forces. The wires would run in from the edges. We've seen how to make controllable wires—we can make computers, a perfect example of accurate control. So there would be no reason why, at the present time, we couldn't make little rotors and other little things turn.

Small Machines—How to Use Them

What use would such things be? Now it gets embarrassing. I tried very hard to think of a use that sounded sensible—or semisensible—you'll have to judge. If you had a closed area and a half wheel that you turned underneath, you could open and shut a hole to let light through or shut it out. And so you have light valves. But because these tiny valves could be placed all over an area, you could make a gate that would let through patterns of light. You could quickly change these patterns by means of electrical voltages, so that you could make a series of pictures. Or, you could use the valves to control an intense source of light and project pictures that vary rapidly—television pictures. I don't think projecting television pictures has any use, though, except to sell more television pictures or something like that. I don't consider that a use—advertising toilet paper.

At first I couldn't think of much more than that, but there are a number of possibilities. For example, if you had little rollers on a surface, you could clean off dirt whenever it fell, and could keep the surface clean all the time.

Then you might think of using these devices—if they had needles sticking out—as a drill, for grinding a surface. That's a very bad idea, as far as I can tell, for several reasons. First, it turns out that materials are too hard when they are dimensioned at this small scale. You find that everything is very stiff, and the grinder has a heck of a job trying to grind anything. There's an awful lot of force, and the grinder would probably grind down its own face before it ground anything else. Also, this particular idea doesn't use the individualization that is possible with small machines—you can individually localize which one is turning which way. If I make all the small machines do grinding, I've done nothing I can't do with a big grinding wheel. What's nice about these machines—if they're worth anything—is that you can wire them to move different parts differently at different times.

One application, although I don't know how to use it, would be to test the circuits in a computer that is being manufactured. It would be nice if we could go in and make contacts at different places inside the circuit. The right way to do that is to design ahead of time places where you could make contacts and bring them out. But if you forgot to design ahead, it would be convenient to have a face with prongs that you could bring up. The small machines would move their little prongs out to touch and make contact in different places.

What about using these things for tools? After all, you could drill holes. But drilling holes has the same problem—the materials are hard, so you'll have to drill holes in soft material.

Well, maybe we can use these tools for constructing those silicon devices. We have a nifty way of doing it now, by evaporating layers, and you might say, "Don't bother me." You're probably right, but I'd like to suggest something that may or may not be a good idea.

Suppose we use the small machines as adjustable masks for controlling the evaporation process. If I could open and close these masks mechanically, and if I had a source of some sort of atoms behind, then I could evaporate those atoms through the holes. Then I could change the hole—by changing the voltages—in order to change the mask and put a new one on for the next layer.

At the present time, it is a painstaking job to draw all the masks for all the different layers—very, very carefully—and then to line the masks up to be projected. When you're finished with one layer you take that layer off and put it in a bath with etch in it; then you put the next layer on, adjust it, go crazy, evaporate, and so on. And that way, we can make four to five layers. If we try to make four hundred layers, too many errors accumulate; it's very, very difficult, and it takes entirely too long.

Is it possible that we could make the surfaces quickly? The key is to put the mask next to the device, not to project it by light. Then we don't have the limitations of light. So you put this machine right up against the silicon, open and close holes, and let stuff come through. Right away you see the problem. The back end of this machine is going to accumulate goop that's evaporating against it, and everything is going to get stuck.

Well then, you haven't thought it through. You should have a thicker machine with tubes and pipes that brings in chemicals. Tubes with controllable valves—all very tiny. What I want is to build in three dimensions by squirting the various substances from different holes that are electrically controlled, and by rapidly working my way back and doing layer after layer, I make a three-dimensional pattern.

Notice that the silicon devices are all two-dimensional. We've gone very far in the development of computing devices, in building these two-dimensional things. They're essentially flat; they have at most three or four layers. Everyone who works with computing machinery has learned to appreciate Rent's law, which says how many wires you need to make how many connections to how many devices. The number of wires goes up as the 2.5 power of the number of devices. If you think a while, you'll find that's a little bit too big for a surface—you can put so many devices on a surface, but you can't get the wires out. In other words, after a while this two-dimensional circuit becomes all wires and no devices, practically.

If you've ever tried to trace lines in two dimensions to make a circuit, you can see that if you're only allowed one or two levels of crossover, the circuit's going to be a mess to design. But if you have three-dimensional space available, so that you can have connections up and down to the transistors, in depth as well as horizontally, then the entire design problem of the wires and everything else becomes very easy. In fact, there's more than enough space. There's no doubt in my mind that the ultimate development of computing machines will end up with the development of a technology—I don't mean my technology, with my crazy machines—but *some* technology for building up three-dimensional circuits, instead of just two-dimensional circuits. That is to say, thick layers, with many, many layers—hundreds and hundreds of them.

So we have to go to three dimensions somehow, maybe with tubes and valves controlled at small scale by machines. Of course, if this did turn out to be useful, then we'd have to make the machines, and they would have to be three-dimensional, too. So we'd have to use the machines to make more machines.

The particular machines I have described so far were just loose pieces that were moving in place—drills, valves, and so forth that only operate in place. Another interesting idea might be to move something over a surface or from one place to another. For example, you could build the same idea that we talked about before, but the things—the little bars or something—are in slots, and they can slide or move all over the surface. Maybe there's some kind of T-shaped slot they come to, and then they can go up and down. Instead of trying to leave the parts in one place, maybe we can move them around on rollers, or simply have them slide.

Electrostatic Actuation

Now how do you pull them along? That's not very hard—I'll give you a design for pulling. [*At the blackboard, Feynman draws a rectangular block with a set of alternating electrodes creating a path for the block.*] If you had, for example, any object like a dielectric that could only move in a slot, and you wanted to move the object, then if you had electrodes arranged along the slot, and if you made one of them plus, and another one minus, the field that's generated pulls the dielectric along. When this piece gets to a new location, you change the voltages so that you're always pulling, and these dielectrics go like those wonderful things that they have in the department store. You stick something in the tube, and it goes whshhhht! to where it has to go.

There is another way, perhaps, of building the silicon circuits using these sliding devices. I have decided this new way is no good, but I'll describe it anyway. You have a supply of parts, and a sliding device goes over, picks up a part, carries it to the right place, and puts it in—the sliding devices assemble everything. These devices are all moving, of course, under the electrical control of computer stuff below them, under their surfaces. But this method is not very good compared to the present evaporation technique, because there's one very serious problem. That is, after you put a piece in, you want to make electrical contacts with the other pieces, but it's very difficult to make good contacts. You can't just put them next to each other—there's no contact. You've got to electrodeposit something or use some such method, but once you start talking about electrochemically depositing something to seal the contact, you might as well make the whole thing the other way by evaporation.

Another question is whether you should use AC or DC to do the pulling: you could work it either way. You could also do the same thing to generate rotations of parts by arranging electrostatic systems for pulling things around a central point. The forces that will move these parts are not big enough to bend anything very much; things are very stiff at this dimensional scale.

If you talk about rotating something, the problem of viscosity becomes fairly important. You'll be somewhat disappointed to discover that if you left the air at normal air pressure in a small hole ten microns big, and then tried to turn something, you'd be able to do it in milliseconds, but not faster. That would be okay for a lot of applications, but it's only milliseconds. The time would be in microseconds, if it weren't for viscous losses.

I enjoy thinking about these things, and you can't stop, no matter how ridiculous things get, so you keep on going. At first, the devices weren't moving—they were in place. Now they can slide back and forth on the surface. Next come the tiny, free-swimming machines.

Mobile Microrobots

What about the free-swimming machine? The purpose is no doubt for entertainment. It's entertaining because you have control—it's like a new game. Nobody figured when they first designed computers that there would be video games. So I have the imagination to realize what the game here is: You get this little machine you can control from the outside, and it has a sword. The machine gets in the water with a paramecium, and you try to stab it.

How are we going to make this game? The first problem is energy supply. Another one is controlling the device. And if you wanted to find out how the paramecium looks to the device, you might want to get some information out.

The energy supply is, I think, fairly easy. At first it looks very difficult because the device is free-swimming, but there are many ways to put energy into the device through electrical induction. You could use either electrical or magnetic fields that vary slowly, generating EMFs inside.

Another way, of course, is to use chemicals from the environment. This method would use a kind of battery, but not as small as the device. The whole environment would be used—the liquid surrounding the device would be the source of a chemical reaction by which you could generate power. Or you could use electromagnetic radiation. With this method you would shine the light on the device to send the signal, or use lower frequencies that go through water—well, not much goes through water but light.

The same methods can be used for control. Once you have a way to get energy in—by electrical induction, for example—it's very easy to put digits or bits on the energy signal to control what the machine is going to do. And the same idea could be used to send signals out. I shouldn't be telling people at JPL how to communicate with things that are difficult to get at or are far away—this is far away because it's so small. You'll figure out a way to send the signals out and get them back again—and enhance the pictures at the end.

It's very curious that what looks obvious is impossible. That is, how are you going to propel yourself through the liquid? Well, you all know how to do that—you have a tail that swishes. But it turns out that if this is a tiny machine a few microns long, the size of a paramecium, then the liquid, in proportion, is enormously viscous. It's like living in a thick honey. And you can try swimming in thick honey, but you have to learn a new technique. It turns out that the only way you can swim in thick honey is to have a kind of an "S" shaped fin. Twisting the shape pushes it forward. It has to be like a piece of a screw, so that as you turn it, it unscrews out of the thick liquid, so to speak. Now, how do we drive the screw?

You always think that there aren't any wheels in biology, and you say, "Why not?" Then you realize that a wheel is a separate part that moves. It's hard to lubricate, it's hard to get new blood in there, and so forth. So we have our parts all connected together—no loose pieces. Bacteria, however, have flagella with corkscrew twists and have cilia that also go around in a type of corkscrew turn. As a matter of fact, the flagellum is the one place in biology where we really do have a movable, separable part. At the end of the flagellum on the back is a kind of a disc, a surface with proteins and enzymes. What happens is a complicated enzyme reaction in which ATP, the energy source, comes up and combines, producing a rotational distortion [*here, Feynman is using his hands to simulate a molecule changing shape and experiencing a net rotation*]; when the ATP releases, the rotation stays, and then another ATP comes, and so forth. It just goes around like a ratchet. And it's connected through a tube to the spiral flagellum that's on the outside.

Twenty years ago when I gave my talk, my friend Al Hibbs, who introduced me today, suggested a use of small devices in medicine. Suppose we could make free-swimming little gadgets like this. You might say, "Oh, that's the size of cells—great. If you've got trouble with your liver, you just put new liver cells in." But twenty years ago, I was talking about somewhat bigger machines. And he said, "Well, swallow the surgeon." The machine is a surgeon—it has tools and controls in it. It goes over to the place where you've got plaque in your blood vessel and it hacks away the plaque.

So we have the idea of making small devices that would go into the biological system in order to control what to cut and to get into places that we can't ordinarily reach. Actually, this idea isn't so bad, and if we back off from the craziness of making such tiny things, and ask about a device that is more practical today, I think it is worth considering having autonomous machines—that is, machines that are sort of robots. I would tether the machines with thin wires—swallowing wires isn't much. It's a little bit discouraging to think of swallowing those long tubes with the optics fibers and everyting else that would have to go down so the guy can watch the inside of your duodenum. But with just the little wires, you could make the device go everywhere, and you could still control it.

Even the wires are really unnecessary, because you

could control the machine from the outside by changing magnetic fields or electric induction. And then we don't have to make the motors, engines, or devices so very tiny as I'm talking about, but a reasonable size. Now it's not as crazily small as I would like—a centimeter or one half of a centimeter—depending on what you want to do the first few times, the scale will get smaller as we go along, but it'll start that way. It doesn't seem impossible to me that you could watch the machine with X-rays or NMR and steer it until it gets where you want. Then you send a signal to start cutting. You watch it and control it from the outside, but you don't have to have all these pipes, and you aren't so limited as to where you can get this machine to go. It goes around corners and backs up.

I think that Hibbs's "swallowable surgeon" is not such a bad idea, but it isn't quite appropriate to the tiny machines, the "infinitesimal machines." It's something that should be appropriate for small machines on the way to the infinitesimal machines.

Making Precise Things from Imprecise Tools

These machines have a general problem, and that's the refinement of precision. If you built a machine of a certain size, and you said, "Well, next year I want to build one of a smaller size," then you would have a problem: you've only got a certain accuracy in dimensions. The next question is, "How do you make the smaller one when you've only got that much accuracy?" It gets worse. You might say, "I'll use this machine to make the smaller one," but if this machine has wobbly bearings and sloppy pins, how does it make an accurate, beautiful, smaller machine?

As soon as you ask that question, you realize it's a very interesting question. Human beings came onto the earth, and at the beginning of our history, we found sticks and stone—bent sticks and roundish funny stones, nothing very accurate. And here we are today, with beautifully accurate machines—you can cut and measure some very accurate distances.

How do you get started? How do you get something accurate from nothing? Well, all machinists know what you do. In the case of large machinery, you take the stones, or whatever, and rub them against each other in every which way, until one grinds against the other. If you did that with one pair of stones, they'd get to a position at which, no matter where you put them, they would fit. They would have perfectly matched concave and convex spherical surfaces.

But I don't want spherical surfaces—I want flat surfaces. So then you take three stones and grind them in pairs, so that everybody fits with everybody else. It's painstaking and it takes time, but after a while, sure enough, you've got nice flat surfaces. Someday, when you're on a camping trip, and everything gets boring, pick up some stones. Not too hard—something that can grind away a little bit, such as consolidated or weak sandstones. I used to do this all the time when I was a kid in Boston. I'd go to work at MIT and on the way pick up two lumps of snow, hard snow that was pushed up by the snowplow and refrozen. I'd grind the snow all the way till I got to MIT, then I could see my beautiful spherical surfaces.

Or, for example, let's say you were making screws to make a lathe. If the screw has irregularities, you could use a nut that's breakable; you would take the nut apart and turn it backwards. If you ran the screw back and forth through the nut, both reversed and straight, soon you would have a perfect screw and a perfect nut, more accurate than the pieces you started with. So it's possible.

I don't think any of these things would work very well with the small machines. Turning things over and reversing and grinding them is so much work, and is so difficult with the hard materials, that I'm not really quite sure how to get increased precision at the very small level.

One way, which isn't very satisfactory, would be to use the electrostatic dielectric push-pull mechanism. If this device were fairly crude in shape, and contained some kind of a point or tooth that was used for a grinder or a marker, you could control the position of the tooth by changing the voltage rather smoothly. You could move it a small fraction of its own irregularity, although you wouldn't really know exactly what that fraction was. I don't know that we're getting much precision this way, but I do think it's possible to make things finer out of things that are cruder.

If you go down far enough in scale, the problem is gone. If I can make something one-half of a percent correct, and the size of the thing is only one hundred atoms wide, then I've got one hundred and not one hundred and one atoms in it, and every part becomes identical. With the finite number of atoms in a small object, at a certain stage, objects can only differ by one atom. That's a finite percentage, and so if you can get reasonably close to the right dimensions, the small objects will be exactly the same.

I thought about casting, which is a good process. You ought to be able to manufacture things at this scale by casting. We don't know of any limitation—except atomic limitations—to casting accurate figures by making molds for figures that match the originals. We know that already, because we can make replicas of all kinds of biological things by using silicone or acetate castings. The electron microscope pictures that you see are often not of the actual object, but of the casting that you've made. The casting can be done down to any reasonable dimension.

One always looks at biology as a kind of a guide, even though it never invents the wheel, and even though we don't make flapping wings for airplanes because we thought of a better way. That is, biology is a guide, but not a perfect guide. If you are having trouble making smooth-looking movable things out of rather hard materials, you might make sacs of liquid that have electric fields in them and can change their shapes. Of course, you would then be imitating cells we already know about. There are probably some materials that can change their shape under electric fields. Let's say that the viscosity depends on the electric field, and so by applying pressure, and then weakening the material in different places with electric fields, the material would move and bend in various ways. I think it's possible to get motion that way.

Friction and Sticking

Now we ask, "What does happen differently with small things?" First of all, we can make them in very great numbers. The amount of material you need for the machines is very tiny, so that you can make billions of them for any normal weight of any material. No cost for materials—all the cost is in manufacturing and arranging the materials. But special problems occur when things get small—or what look like problems, and might turn out to be advantages if you knew how to design for them.

One problem is that things stick together by molecular attraction. Now friction becomes a difficulty. If you were to have two tungsten parts, perfectly clean, next to each other, they would bind and jam. The atoms simply pull together as if the two parts were one piece. The friction is enormous, and you will never be able to move the parts. Therefore you've got to have oxide layers or other layers in between the materials as a type of lubricant—you have to be very careful about that or everything will stick.

On the other hand, if you get still smaller, nothing is going to stick unless it's built out of one piece. Because of the Brownian motion, the parts are always shaking; if you put them together and a part were to get stuck, it would shake until it found a way to move around. So now you have an advantage.

At the end of it all, I keep getting frustrated in thinking about these small machines. I want somebody to think of a good use, so that the future will really have these machines in it. Of course, if the machines turn out to be any good, we'll also have to make the machines, and that will be very interesting to try to do.

Computing with Atoms

Now we're going to talk about small, small computing. I'm taking the point of view of 1983 rather than of 1960, and will talk about what is going to happen, or which way we should go.

Let's ask, what do we need to do to have a computer? We need numbers, and we need to manipulate the numbers and calculate an answer. So we have to be able to write the numbers.

How small can a number be? If you have N digits, you know the special way of writing them with base two numbers, that is, with ones and zeros. Now we're going to go way down to the bottom—atoms! Remember that we have to obey quantum-mechanical laws, if we are talking about atoms. And each of these atoms is going to be in one of two states—actually, atoms can be in a lot of states, but let's take a simple counting scheme that has either ones or zeros. Let's say that an atom can be in a state of spin up or of spin down, or say that an ammonia molecule is either in the lowest or the next lowest state, or suppose various other kinds of two-state systems. When an atom is in the excited state—a spin up—let's call it a "one"; a "zero" will correspond to spin down. Hereafter when I say a one, I mean an atom in an excited state. So to write a number takes no more atoms than there are digits, and that's really nothing!

Reversible Gates

Now what about operations—computing something with the numbers? It is known that if you can only do a few operations of the right kind, then by compounding the operations again and again in various combinations, you can do anything you want with numbers.

The usual way of discussing this fact is to have these numbers as voltages on a wire instead of states in an atom, so we'll start with the usual way. [*Feynman draws a two-input AND gate at the blackboard.*] We would have a device with two input wires A and B, and one output wire. If a wire has a voltage on it, I call it a "one"; if it has zero voltage, it's a "zero." For this particular device, if both wires are ones, then the output turns to one. If either wire is one, but not both, or if neither is one, the output stays at zero—that's called an AND gate. It's easy to make an electric transistor circuit that will do the AND gate function.

There are devices that do other things, such as a little device that does NOT—if the input wire is a one, the output is a zero; if the input wire is a zero, the output is one. Some people have fun trying to pick one combination with which they can do everything, for example, a NAND gate that is a combination of NOT and AND—it is zero when both input wires are ones, and one when either or both inputs are not ones. By arranging and wiring NAND gates together in the correct manner, you can do any operation. There are a lot of questions about branchings and so forth, but that's all been worked out. I want to discuss what happens if we try to do this process with atoms.

First, we can't use classical mechanics or classical ideas about wires and circuits. We have atoms, and we have to use quantum mechanics. Well, I love quantum mechanics. So, the question is, can you design a machine that computes and that works by quantum-mechanical laws of physics—directly on the atoms—instead of by classical laws.

We find that we can't make an AND gate, we can't make a NAND gate, and we can't make any of the gates that people used to say you could make everything out of. You see immediately why I can't make an AND gate. I've only got one wire out and two in, so I can't go backwards. If I know that the answer is zero, I can't tell what the two inputs were. It's an irreversible process. I have to emphasize this fact because atomic physics is reversible, as you all know, microscopically reversible. When I write the laws of how things behave at the atomic scale, I have to use reversible laws. Therefore, I have to have reversible gates.

Bennett from IBM, Fredkin, and later Toffoli investigated whether, with gates that are reversible, you can do everything. And it turns out, wonderfully true, that the irreversibility is not essential for computation. It just happens to be the way we designed the circuits.

It's possible to make a gate reversible in the following cheesy way, which works perfectly. [*Feynman now draws a block with two inputs, A and B, and three outputs.*] Let's suppose that two wires came in here, but we also keep the problem at the output. So we have three outputs: the A that we put in, the B that we put in, and the answer. Well, of course, if you know the A and the B along with the answer, it isn't hard to figure out where the answer came from.

The trouble is that the process still isn't quite reversible, because you have two pieces of information at the input, that is, two atoms, and three pieces of information at the output. It's like a new atom came from somewhere. So I'll have to have a third atom at the input [*he adds a third input line, labeled C*]. We can characterize what happens as follows:

Unless A and B are both one, do nothing. Just pass A, B, and C through to the output. If A and B are both one, they still pass through as A and B, but C, whatever it is, changes to NOT C. I call this a "controlled, controlled, NOT" gate.

Now this gate is completely reversible, because if A and B are not both ones, everything passes through either way, while if A and B are both ones on the input side, they are both ones on the output side too. So if you go through the gate forward with A and B as ones, you get NOT C from C, and when you go backward with NOT C at the output, you get C back again at the input. That is, you do a NOT twice, and the circuit, or atom is back to itself, so it's reversible. And it turns out, as Toffoli has pointed out, that this circuit would enable me to do any logical operation.

So how do we represent a calculation? Let's say that we have invented a method whereby choosing any three atoms from a set of N would enable us to make an interaction converting them from a state of ones and zeros to a new state of ones and zeros. It turns out, from the mathematical standpoint, that we would have a sort of matrix, called M. Matrix M converts one of the eight possible combination states of three atoms to another combination state of the three atoms, and it's a matrix whose square is equal to one, a so-called unitary matrix. The thing you want to calculate can be written as a product of a whole string of matrices like M—millions of them, maybe, but each one involves only three atoms at a time.

I must emphasize, that, in my previous example with AND gates and wires, the wires that carried the answer after the operation were new ones. But the situation is simpler here. After my matrix operates, it's the same register—the same atoms—that contain the answer. I have the input represented by N atoms, and then I'm going to change them, change them, change them, three atoms at a time, until I finally get the output.

The Electron as Calculating Engine

It's not hard to write down the matrix in terms of interactions between the atoms. In other words, in principle, you can invent a kind of coupling among the atoms that you turn on to make the calculation. But the question is, how do you make the succession of three-atom transformations go bup-bup-bup-bup-bup in a row? It turns out to be rather easy—the idea is very simple. [*Feynman draws a row of small circles, and points often to various circles in the row through the following discussion.*]

You can have a whole lot of spots, such as atoms on which an electron can sit, in a long chain. If you put an electron on one spot, then in a classical world it would have a certain chance of jumping to another spot. In quantum mechanics, you would say it has a certain amplitude to get there. Of course, it's all complex numbers and fancy business, but what happens is that the Schrödinger function diffuses: the amplitude defined in different places wanders around. Maybe the electron comes down to the end, and maybe it comes back and just wanders around. In other words, there's some amplitude that the electron jumped to here and jumped to there. When you square the answer, it represents a probability that the elelctron has jumped all the way along.

As you all know, this row of sites is a wire. That's the way electrons go through a wire—they jump from site to site. Assume it's a long wire. I want to arrange the Hamiltonian of the world—the connections between sites—so that an electron will have zero amplitude to get from one site to the next because of a barrier, and it can only cross the barrier if it interacts with the atoms [of the registers] that are keeping track of the answer. [*In response to a question following the lecture, Feynman did write out a typical term in such a Hamiltonian using an atom-transforming matrix M positioned between electron creation and annihilation operators on adjacent sites.*]

That is, the idea is to make the coupling so that the electron has no amplitude to go from site to site, unless it disturbs the N atoms by multiplying by the matrix $M2$, in this case, or by $M1$ or $M3$ in these other cases. If the electron started at one end, and went right along and came out at the other end, we would know that it had made the succession of operations $M1$, $M2$, $M3$, $M4$, $M5$—the whole set, just what you wanted.

But wait a minute—electrons don't go like that! They have a certain amplitude to go forward, then they come back, and then they go forward. If the electron goes forward, say, from here to there, and does the operation $M2$ along the way, then if the electron goes backwards, it has to do the operation $M2$ again.

Bad luck? No! $M2$ is designed to be a reversible operation. If you do it twice, you don't do anything; it undoes what it did before. It's like a zipper that somebody's trying to pull up, but the person doesn't zip very well, and zips it up and down. Nevertheless, wherever the zipper is at, it's zipped up correctly to that particular point. Even though the person unzips it partly and zips it up again, it's always right, so that when it's finished at the end, and the Talon fastener is at the top, the zipper has completed the correct operations.

So if we find the electron at the far end, the calculation is finished and correct. You just wait, and when you see it, quickly take it away and put it in your pocket so it doesn't back up. With an electric field, that's easy.

It turns out that this idea is quite sound. The idea is very interesting to analyze, to see what a computer's limitations are. Although this computer is not one we can build easily, it has got everything defined in it. Everything is written: the Hamiltonian, the details. You can study the limitations of this machine, with regard to speed, with regard to heat, with regard to how many elements you need to do a calculation, and so on. And the results are rather interesting.

Heat in a Quantum Computer

With regard to heat: everybody knows that computers generate a lot of heat. When you make computers smaller, all the heat that's generated is packed into a small space, and you have all kinds of cooling problems. That is due to bad design. Bennett first demonstrated that you can do reversible computing—that is, if you use reversible gates, the amount of energy needed to operate the gates is essentially indefinitely small if you wait long enough, and allow the electrons to go slowly through the computer. If you weren't in such a hurry, and if you used ideal reversible gates—like Carnot's reversible cycle (I know everything has a little friction, but this is idealized)—then the amount of heat is zero! That is, essentially zero, in the limit—it only depends on the losses due to imperfections.

Furthermore, if you have ordinary reversible gates, and you try to drag the thing through as quickly as you can, then the amount of energy lost at each fundamental operation is one kT of energy per gate, or per decision, at most! If you went slower, and gave yourself more time, the loss would be proportionately lower.

And how much kT do we use per decision now? 10^{10} kT! So we can gain a factor of 10^{10} without a tremendous loss of speed, I think. The problem is, of course, that it depends on the size that you're going to make the computer.

If computers were made smaller, we could make them very much more efficient. It hadn't been realized previous to Bennett's work that there was, essentially, no heat requirement to operate a computer if you weren't in such a hurry. I have also analyzed this model, and get the same results as Bennett with a slight modification, or improvement.

If this device is made perfectly, then the computer could work ballistically. That is, you could have this chain of electron sites and start the electrons off with a momentum, and they simply coast through and come out the other end. The thing is done—whshshsht! You're finished, just like shooting an electron through a perfect wire.

If you have a certain energy available to the electron, it has a certain speed—there's a relation between the energy and the speed. If I call this energy that the electron has kT, although it isn't necessarily a thermal energy, then there's a velocity that goes with it, v_T, which is the maximum speed at which the electron goes through the machine. And when you do it that way, there are no losses. This is the ideal case; the electron just coasts through. At the other end, you take the electron that had a lot of energy, you take that energy out, you store it, and get it ready for shooting in the next electron. No losses! There are no kT losses in an idealized computer—none at all.

In practice, of course, you would not have a perfect machine, just as a Carnot cycle doesn't work exactly. You have to have some friction. So let's put in some friction.

Suppose that I have irregularities in the coupling here and there—that the machine isn't perfect. We know what happens, because we study that in the theory of metals. Due to the irregularities in the positions or couplings, the electrons do what we call "scattering." They head to the right, if I started them to the right, but they bounce and come back. And they may hit another irregularity and bounce the other way. They don't go straight through. They rattle around due to scattering, and you might guess that they'll never get through. But if you put a little electric field pulling the electrons, then although they bounce, they try again, try again, and make their way through. And all you have is, effectively, a resistance. It's as if my wire had a resistance, instead of being a perfect conductor.

One way to characterize this situation is to say that there's a certain chance of scattering—a certain chance to be sent back at each irregularity. Maybe one chance in a hundred, say. That means if I did a computation at each site, I'd have to pass a hundred sites before I got one average scattering. So you're sending electrons through with a velocity v_T that corresponds to this energy kT. You can write the loss per scattering in terms of free energy if you want, but the entropy loss per scattering is really the irreversible loss, and note that it's the loss **per scattering, not per calculation step** [*heavily emphasized, by writing the words on the blackboard*]. The better you make the computer, the more steps you're going to get per scattering, and, in effect, the less loss per calculation step.

The entropy loss per scattering is one of those famous $\log_2$ numbers—let me guess it is Boltzmann's constant, k, or some such unit, for each scattering if you drive the electron as quickly as you can for the energy that you've got.

If you take your time, though, and drive the electron through with an average speed, which I call the drift speed, v_D (compared to the thermal speed at which it would ordinarily be jostling back and forth), then you get a decrease in the amount of entropy you need. If you go slow enough, when there's scattering, the electron has a certain energy and it goes forward-backward-forward-bounce-bounce and comes to some energy based on the temperature. The electron then has a certain velocity—thermal velocity—for going back and forth. It's not the velocity at which the electron is getting through the machine, because it's wasting its time going back and forth. But it turns out that the amount of entropy you lose every time you have 100% scattering is simply a fraction of k— the ratio of the velocity that you actually make the electron drift compared to how fast you could make it drift. [*Feynman writes on the board the formula: $k(v_D/v_T)$.*]

If you drag the electron, the moment you start dragging it you get losses from the resistance—you make a current. In energy terms, you lose only a kT of energy for each scattering, not for each calculation, and you can make the loss smaller proportionally as you're willing to wait longer than the ideal maximum speed. Therefore, with good design in future computers, heat is not going to be a real problem. The key is that those computers ultimately have to be designed—or should be designed—with reversible gates.

We have a long way to go in that direction—a factor of 10^{10}. And so, I'm just suggesting to you that you start chipping away at the exponent.

Thank you very much.

Editor's Note

IT was with great delight that I found an unpublished talk by Richard Feynman on Infinitesimal Machinery.

I thank the many people who have made possible the publication of this talk. Helen Tuck, of the California Institute of Technology, kindly helped me in some earlier research by suggesting several people to contact. One of these, Ralph Leighton, of Friends of Tuva, mentioned a talk by Dr. Feynman during the early eighties, and suggested I contact Jurrie J. van der Woude at the Jet Propulsion Laboratory. Mr. van der Woude not only remembered the talk, but was able to find the original video recording.

The process of transforming the video recording to a publishable transcript started with Mike Wall, a graduate student in Princeton University's Physics Department. Mike spent many hours transcribing the talk. A professional editor, Nora Odendahl, reworked this text, and changed the manuscript from a verbal transcript to smoothly reading text.

Steve Senturia took the text next, and technically edited the work. As he mentions in his introduction, he strove to sharpen and clarify the technical content, without losing Feynman's lively style. As you will see, he succeeded well.

Several other people have made comments and helped with the text. Carl Feynman, Richard Feynman's son, was kind enough to read the text and make comments. William Athas, of the USC Information Sciences Institute, read the text and made suggestions on the low-power computing section. Robert Lyons, of Bell Laboratories, made the photograph of Feynman by the black board from the video recording.

Here follows the introduction by Steve Senturia, and the talk by Richard Feynman. Enjoy.

WILLIAM TRIMMER
Editor

The Resonant Gate Transistor

HARVEY C. NATHANSON, MEMBER, IEEE, WILLIAM E. NEWELL, SENIOR MEMBER, IEEE,
ROBERT A. WICKSTROM, AND JOHN RANSFORD DAVIS, JR., MEMBER, IEEE

Abstract—A device is described which permits high-Q frequency selection to be incorporated into silicon integrated circuits. It is essentially an electrostatically excited *tuning fork* employing field-effect transistor "readout." The device, which is called the resonant gate transistor (RGT), can be batch-fabricated in a manner consistent with silicon technology. Experimental RGT's with gold vibrating beams operating in the frequency range 1 kHz $<f_0<$ 100 kHz are described. As an example of size, a 5-kHz device is about 0.1 mm long (0.040 inch). Experimental units possessing Q's as high as 500 and overall input-output voltage gain approaching +10 dB have been constructed.

The mechanical and electrical operation of the RGT is analyzed. Expressions are derived for both the beam and the detector characteristic voltage, the device center frequency, as well as the device gain and gain-stability product. A batch-fabrication procedure for the RGT is demonstrated and theory and experiment corroborated. Both single- and multiple-pole pair band pass filters are fabricated and discussed. Temperature coefficients of frequency as low as 90–150 ppm/°C for the finished batch-fabricated device were demonstrated.

Glossary

A = area
C_{fb} = capacitance between input force plate and resonator = $\epsilon_0 A/\delta_e$
C_{fs} = capacitance between force plate and substrate
D = damping constant
F = force
GS = gain-frequency stability product
I_D = drain current
I_{PO} = channel pinch-off current
K = dynamic spring "constant"
K_0 = mechanical spring constant
L = beam length
L_c = channel length
M, D = equivalent mass and damping constant of resonator
N_A = substrate doping
Q = quality factor of resonator = $K/\omega D$
Q_B = beam-induced channel charge
Q_D = surface depletion region charge
Q_S = built-in oxide charge
R_L = load resistance
V_a = channel voltage
V_P = polarization voltage
V_{PI} = characteristic pull-in voltage
V_{PO} = channel pinch-off voltage
V_T = threshold voltage
W = channel width
W_{ox} = oxide thickness
Y = Young's modulus
d = beam thickness
e_d = drain signal voltage
e_{in} = sinusoidal input voltage of frequency ω
e_{out} = sinusoidal output voltage of frequency ω
f_0 = resonant frequency
g_{m_0} = clamped-gate FET transconductance
q = electronic charge = 1.6×10^{-19} coulombs
r_d = drain resistance
ψ = frequency stability factor
δ = sinusoidal resonator deflection
δ_e = equilibrium spacing between resonator and substrate
δ_0 = equilibrium spacing when $V_P = 0$
$\epsilon_{si}, \epsilon_{ox}, \epsilon_0$ = permittivity of silicon, silicon dioxide, and air, respectively
ρ = material density
ρ_s = substrate resistivity
μ = amplification factor of equivalent clamped-gate FET
μ_n = channel mobility
ϕ_p = bulk Fermi level in substrate = $(E_f - E_i)$
ω_0 = $2\pi f_0$ = angular mechanical resonant frequency
ω_r = $\sqrt{K/M}$ = angular resonant frequency

I. Introduction

EXPLOITATION of the unique capabilities of silicon integrated circuits in one major area—namely, digital circuits—is at a stage where further progress will be self-sustaining. There are other areas, however, which could similarly benefit from these capabilities, but where the application of integrated circuit technology has not progressed as rapidly. The classic problem which has hindered the entry of integrated circuits into many nondigital systems is the lack of a compatible tuning element. The difficulties encountered in trying to obtain high-Q tuned integrated circuits are by now well known [1].

A practical integrated tuning device must satisfy various constraints such as small size, capability of high-Q, and the possibility of batch-fabrication. Reasonable manufacturing tolerances are also an inherent and

---

Manuscript received October 3, 1966; revised December 19, 1966. The work reported here was supported in part by the U. S. Air Force Avionics Laboratory, Electronics Technology Division, Wright-Patterson AFB, under Contract AF-33(615)-3442.
The authors are with the Westinghouse Research Laboratories, Pittsburgh, Pa.

unavoidable problem which must be given thorough consideration in *all* high-Q tuned circuits, as may readily be seen from a simple example. If a Q of 100 is required in a typical application, the corresponding passband is only 1 percent of the resonant frequency. Therefore, a variation of the order of 1 percent in any parameter which significantly influences the resonant frequency will shift the desired signal completely out of the passband. Conventional tuned circuits overcome the tolerance problem through the choice of only very stable components for the critical elements and the incorporation of adjustable components (e.g., the trimmer capacitors on an IF transformer) to permit periodic compensation for any residual drift. On the other hand, diffused components in silicon integrated circuits are noted for relatively poor stability and the incorporation of adjustable components (to say nothing of the actual process of adjustment) defeats many of the desired advantages of integrated circuits. Therefore, it is mandatory that the most stable and most controllable parameters of an integrated circuit be used as the critical tuning elements.

The tolerance problem is further aggravated if the resonant feedback loop includes an active element because Q is then also extremely sensitive to component tolerances. In fact, a given percentage variation in a component often causes a variation in Q which is of the order of Q times as great [2].

The use of a passive mechanical resonator seems to offer the most promising solution which can satisfy all of these constraints [3]. The Q of these resonators is determined by losses and should be relatively stable. The resonant frequency is determined primarily by geometrical dimensions, which are among the easiest parameters to control accurately in integrated circuits. Therefore, although the tolerance problem is not avoided, it is at least minimized.

The resonant gate transistor (RGT) is a new, novel, mechanically resonant tuning device which is compatible with silicon integrated circuits and which promises to solve the tuning problem over the frequency range from about 0.5 kHz to 1 MHz [4], [5]. The three essential elements in the RGT are as follows:

1) An input transducer to convert the input electrical signal into a mechanical force. The present RGT utilizes electrostatic attraction, although devices using the piezoelectric effect, magnetostatic attraction, or other effects instead can be imagined.

2) A mechanical resonator which is sufficiently isolated from its surroundings so that the desired Q can be obtained. A variety of singly or multiply resonant suspended beam resonators are possible.

3) An output transducer to sense the motion of the mechanical resonator and generate a corresponding electrical signal. The present RGT uses a field-effect transistor (FET). The possible use of a bipolar transistor has been analyzed (see Appendix I), and other types of transducers can be conceived.

Figure 1 shows one practical geometry which is presently being investigated. A description of detailed device operation has been published elsewhere [4]. In Fig. 1 a metal beam electrode, clamped on one end to an insulating oxide, is fabricated parallel to and suspended *over* the surface of a silicon slice. Underneath the beam at its end is an insulated input force plate. Voltages applied to this plate exert electrostatic forces on the beam electrode causing it to vibrate. Only at the *mechanical resonance frequency* of the beam is the vibration appreciable. Vibrations of the beam are detected as variations of field-effect induced charge in the channel region of an MOS-type detector underneath the middle of the beam. Output is extracted at the "drain" of the device. A polarization voltage V_P controls both the magnitude of the signal force and the magnitude of the induced detector charge, making the gain of the device proportional to V_P^2. Figure 2 shows a photomicrograph of an actual RGT having a clamped-clamped beam resonator 0.5 mm in length, along with the tuning curve of this structure. The fundamental resonant frequency of this device is about 60 kHz, and Q's of 100 to 500 have been obtained.

In this paper, a simplified and generalized model of the RGT will be analyzed, and the work to date on practical devices will be summarized. The presentation is directed to those who seek to understand the operation of the RGT as well as its capabilities and limitations in integrated circuits.

II. Theoretical

A. Electromechanical Analysis of a Simplified Device Model

Once the basic technology of batch-fabrication of suspended, mechanically resonant beams is mastered, a wide variety of geometries will be feasible for the RGT. For maximum usefulness, the initial analysis should therefore be as general as possible, even if some accuracy must be sacrificed. The first simplification depends on the fact that many resonant structures can be adequately represented by a lumped mass and spring having the same resonant frequency, the same effective spring constant, and the same Q.

A second important consideration is that the mathematics of this initial analysis should be kept as simple as possible to avoid obscuring the basic simplicity of the RGT. The worst mathematical complication in an accurate analysis arises because, during beam deflection, the spacing between the resonator and the substrate is a function of position along the beam. Therefore, an effective value of spacing which is independent of position has been used in the analysis.

The resulting model of the RGT which has been analyzed is shown in Fig. 3 together with the required circuit connections. The analysis begins with the equation for the electrostatic force on the vibrating mass:

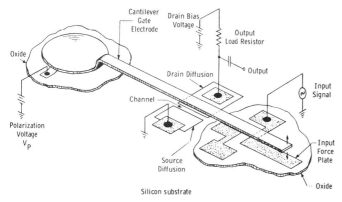

Fig. 1. Geometry and circuit connections of an RGT with a C-F resonant beam.

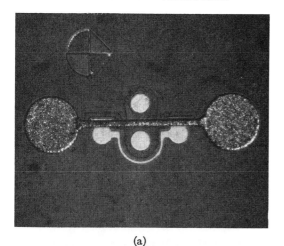

(a)

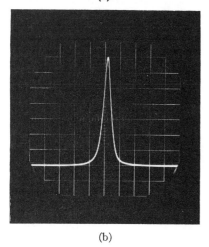

(b)

Fig. 2. Photograph and typical frequency response of an RGT with a C-C beam resonator.

$$\text{electrostatic force} = -\frac{1}{2}(V_P + e_{\text{in}}\sin\omega t)^2 \frac{\partial C_{fb}}{\partial \delta_e}$$

$$= \frac{1}{2}\left[V_P^2 + 2V_P e_{\text{in}}\sin\omega t + \frac{e_{\text{in}}^2}{2}(1-\cos 2\omega t)\right]\frac{\epsilon_0 A}{\delta_e^2}. \quad (1)$$

The first term in the bracket in (1) represents the constant component of force caused by the polarization

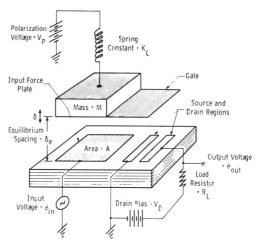

Fig. 3. Simplified model of the RGT.

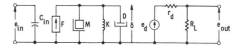

Fig. 4. Equivalent circuit for the RGT.

voltage. This component leads to a dependence of δ_e on V_P which will be discussed later. The third term is very small in normal linear operation and will be neglected. The second term is the signal component of force:

$$F\sin\omega t = \frac{V_P \epsilon_0 A}{\delta_e^2} e_{\text{in}}\sin\omega t. \quad (2)$$

This force causes a corresponding sinusoidal deflection of the beam resonator, the amplitude and phase of which are given by the standard resonance equation:

$$\delta = \frac{\frac{F}{K}}{1 - \frac{\omega^2}{\omega_r^2} + j\frac{1}{Q}\frac{\omega}{\omega_r}}. \quad (3)$$

The deflection of the resonator is sensed by the surface FET, which will be discussed in detail in the next section. Suffice it to say that the variation in the polarization field caused by the vibration of a gate held at constant voltage is equivalent to that of a conventional FET in which the gate is fixed in position but the voltage changes. Therefore,

$$e_d = -\mu\delta_e\delta\frac{\partial}{\partial \delta_e}\left(\frac{V_P}{\delta_e}\right) = \mu\frac{\delta}{\delta_e}V_P. \quad (4)$$

The equivalent circuit which represents (2) to (4) is shown in Fig. 4. The overall voltage transfer function at resonance with matched load is then

$$\left(\frac{e_{\text{out}}}{e_{\text{in}}}\right)_r = -jQV_P^2\frac{\mu R_L}{r_d + R_L}\frac{\epsilon_0 A}{K\delta_e^3}$$

$$= -j\frac{Q\mu V_P^2}{2}\frac{\epsilon_0 A}{K\delta_e^3}. \quad (5)$$

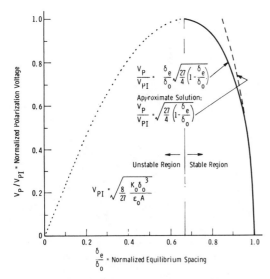

Fig. 5. Variation in beam displacement with polarization voltage for the simplified model of the RGT.

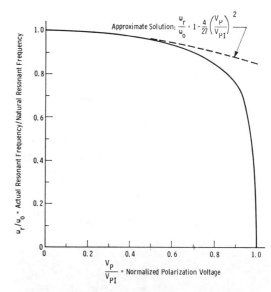

Fig. 6. Variation in resonant frequency with polarization voltage for the simplified model of the RGT.

Away from resonance, the transfer function decreases in the same way as that of a conventional LC tank circuit.

Equation (5) can be simplified, but first we must return to consider the constant component of force in (1). The equilibrium position of the resonator occurs when this force is counterbalanced by the restoring force of the spring:

$$\text{net force} = \frac{1}{2}\frac{\epsilon_0 A V_P^2}{\delta_e^2} - K_0(\delta_0 - \delta_e) = 0. \quad (6)$$

Equation (6) gives δ_e as a function of V_P. However, there is a maximum value of V_P above which the spring force is unable to maintain equilibrium, and the spacing immediately goes to zero. The maximum value of V_P we define as the pull-in voltage V_{PI}. Setting the derivative of (6) equal to zero shows that $V_P = V_{PI}$ when $\delta_e = 2\delta_0/3$. Therefore, from (6),

$$V_{PI} = \sqrt{\frac{8}{27}\frac{K_0 \delta_0^3}{\epsilon_0 A}}. \quad (7)$$

Using (7), (6) may be put into the normalized form

$$\frac{V_P}{V_{PI}} = \frac{\delta_e}{\delta_0}\sqrt{\frac{27}{4}\left(1 - \frac{\delta_e}{\delta_0}\right)}. \quad (8)$$

A plot of (8) is shown in Fig. 5. In normal operation, V_P/V_{PI} has a value in the vicinity of 0.5, and $\delta_e/\delta_0 \approx 1$. Equation (8) can then be approximated by the simpler equation:

$$\frac{V_P}{V_{PI}} \approx \sqrt{\frac{27}{4}\left(1 - \frac{\delta_e}{\delta_0}\right)} \quad \text{when } \frac{\delta_e}{\delta_0} \approx 1. \quad (9)$$

Another result of the interaction between the electrostatic force and the spring force is that the dynamic spring "constant" is not constant, but varies with V_P.

Differentiating (6) gives

$$K = \frac{\partial}{\partial \delta_e}\left[\frac{1}{2}\frac{\epsilon_0 A V_P^2}{\delta_e^2} - K_0(\delta_0 - \delta_e)\right]$$

$$= K_0 - \frac{\epsilon_0 A V_P^2}{\delta_0^3} = K_0\left(3 - 2\frac{\delta_0}{\delta_e}\right). \quad (10)$$

The variation in K also causes a shift in the actual resonant frequency ω_r away from the mechanical resonant frequency ω_0. According to (9) and (10),

$$\frac{\omega_r}{\omega_0} = \sqrt{\frac{K/M}{K_0/M}} = \sqrt{3 - 2\frac{\delta_0}{\delta_e}}$$

$$\approx 1 - \frac{4}{27}\left(\frac{V_P}{V_{PI}}\right)^2. \quad (11)$$

The approximate form of (11) results from using (9). Both the exact and the approximate forms of (11) are plotted in Fig. 6. Although the resonant frequency approaches zero as the polarization voltage approaches V_{PI}, the practical tuning range is at most about 15 percent below ω_0. This variation in resonant frequency with voltage may be either an advantage or a disadvantage—it provides a means for fine tuning, but necessitates voltage regulation if ultimate frequency stability is required.

The quantitative frequency stability with respect to the polarization voltage may be defined as

$$\psi \equiv \frac{1}{Q\dfrac{V_P}{\omega_r}\dfrac{\partial \omega_r}{\partial V_P}} \approx -\frac{27}{8Q}\frac{\omega_r}{\omega_0}\left(\frac{V_{PI}}{V_P}\right)^2$$

$$\approx -\frac{27}{8Q}\left(\frac{V_{PI}}{V_P}\right)^2. \quad (12)$$

The factor Q is included to relate the stability to the bandwidth rather than to the center frequency. For

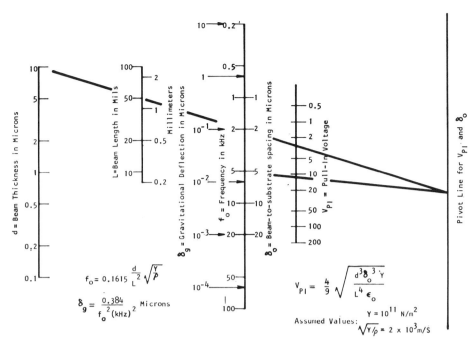

Fig. 7. Nomograph for C-F (cantilever) beam resonator.

a typical case where $V_P/V_{PI} = 0.5$ and $Q = 100$, $\psi = -0.135$. Thus a 1-percent variation in V_P would cause the resonant frequency to change by 7.4 percent of a bandwidth ($= 1/0.135$).

Now that the pull-in voltage has been derived, the equation for the transfer function at matched load can be further simplified. Substituting (7) into (5) gives

$$\left(\frac{e_{\text{out}}}{e_{\text{in}}}\right)_r = -j \frac{4}{27}\left(\frac{V_P}{V_{PI}}\right)^2 Q\mu \frac{K_0}{K}\left(\frac{\delta_0}{\delta_c}\right)^3$$

$$\approx -j\frac{4}{27}\left(\frac{V_P}{V_{PI}}\right)^2 Q\mu \quad \text{for } \frac{V_P}{V_{PI}} \ll 1. \quad (13)$$

Equation (13) shows the fundamental factors which determine the gain capabilities of the RGT. The j factor indicates that at resonance there is a 90° phase shift between the input and the output voltages. Because of this phase shift, conventional 0° or 180° feedback changes the resonant frequency without changing the Q or tending toward instability. If it is desirable to change Q by means of feedback, the loop must include an additional $\pm 90°$ of phase shift.

The constant factor 4/27 in (13) results from the particular simplified geometry which has been assumed, and would be different for each practical geometry. However, it is useful as it stands for estimating the order of magnitude of the gain capability of the device.

The third factor in (13) indicates that the gain is strongly dependent on the polarization voltage. Typically $V_P/V_{PI} \approx 0.5$, so this factor is of the order of 0.25. Note that there is a tradeoff between gain and frequency stability which enters into the choice of the proper polarization voltage. Increasing V_P increases the gain but decreases the frequency stability. From (12) and (13) it is seen that the gain-frequency stability product for a matched load is dependent only on the amplification factor of the FET:

$$GS = \left|\frac{e_{\text{out}}}{e_{\text{in}}}\right|_r |\psi| \approx \frac{R_L \mu}{r_d + R_L}$$

$$= \frac{\mu}{2} \quad \text{for } R_L = r_d. \quad (14)$$

Because the gate-to-substrate spacing in the RGT is considerably greater than in a conventional FET, the amplification factor is considerably smaller, as will be discussed later.

The factor in (13) which makes a near-unity-gain seem feasible for the RGT is Q. Depending on the application, Q may be determined by other requirements such as bandwidth. However, for most applications in which the RGT is attractive, Q's of at least 50 are required. If unity gain is not feasible, the attenuation can be compensated by one or more extra stages of amplification. More important than gain is the noise figure of the device. Since the overall noise figure of the RGT with amplification can never be better than the RGT alone, it is most important that surface noise and signal feedthrough in the device be kept as low as possible.

In Fig. 7 we have generated a nomograph from the equations derived in this section. This nomograph permits the determination of the center frequency of a cantilever f_0, the characteristic pull-in voltage V_{PI}, and the gravitational beam deflection δ_g of the RGT versus the three RGT parameters L, d, and δ_0.

As an example of the use of these nomographs, a line drawn through the points $d = 10$ microns and $L = 0.050$ inch in Fig. 7 intersects the f_0 axis at the point $f_0 \sim 2$ kHz. Extending the line to the right and pivoting it at the indicated pivot line, we see that a beam-to-substrate spacing of 5 microns results in a predicted V_{PI} of about 12 volts.

Closer inspection of the implicit relationships between the variables in Fig. 7 reveals that for fixed values of f_0 and δ_0, a thinner, shorter beam will result both in a smaller device and in a lower value of V_{PI} and therefore in a lower required value of V_P needed to achieve a given overall device gain. For instance, using the above example where $f_0 = 2$ kHz, $\delta_0 = 5$ microns, and $L = 0.020$ inch, we can reduce V_{PI} to about 4.5 volts while still keeping $f_0 \sim 2$ kHz. Therefore, depending on the geometries which prove most practical, RGT's are potentially useful on relatively low voltages. Furthermore, if the spacing can be sufficiently reduced to yield a pull-in voltage of one volt or less, it is conceivable that the RGT can be made to operate solely on contact potential differences between the various electrodes of the structure. For instance, if the input plate is aluminum and the beam is gold, the difference in vacuum work functions for these two materials (~ 0.5 volt) provides an effective bias of $V_P \sim 0.5$ volt in (2) for the electrostatic force. A similar work function difference exists for the gold beam–silicon couple encountered at the detector.

In conclusion, in this section we have presented the simple design theory of the RGT. Throughout the discussion we have assumed that the detector is a surface FET of voltage gain μ. We have derived expressions for the device gain and the gain-stability product based on this value of μ.

For circuit applications it is often sufficient to consider the surface detector as nothing more than a "black box." However, for device design purposes, we treat the "substrate-controlled" detector on the RGT with greater detail in Section II-B. We derive general expressions for pinch-off voltage, transconductance, and amplification factor for this detector, and illustrate these expressions with a simple overall device design example. The surface-controlled detector represents probably the simplest, yet most effective way (see Appendix I) of detecting beam vibration consistent with silicon-based technology.

B. The Field-Effect Detector

The mechanical vibration δ of the resonant beam electrode as a function of input frequency is given by (3). There are a number of ways in which this vibration can be used to effect a voltage output. Since the beam is already polarized to a voltage V_P to provide an input force, vibrations of the beam electrode produce a varying field perpendicular to the semiconductor surface, suggesting some form of electrostatic "readout." For this reason, detection of beam movement in the RGT is accomplished by field-effect modulation of the channel of a conventional, "normally on" type of MOS transistor [6]. Movement of the beam at constant voltage V_P exerts a variable surface field perpendicular to the channel of the MOS transistor, modulating the channel conductance and therefore the output of the device.

Note that by choosing an MOS region containing a *normally on* channel region, it is not necessary to have a large inverting field on the beam electrode, which might place an unnecessary restriction on the quiescent *mechanical* operating point of the beam electrode.

In the order for the operating "μ" of the RGT to be high, it is desirable that the VI characteristic of the detector be pentode-like, i.e., that it "pinches off." Since the effective surface gate of the detector (the beam electrode) is several microns away from the silicon surface, gate-controlled pinch-off occurs at a high voltage. In order to obtain low-voltage pinch-off in these comparatively "gateless" structures, substantial substrate pinch-off must be employed. It will be shown that for normal oxide surface charge obtained during steam oxidation of p-type silicon, values of substrate resistivities in the range 5–15 ohm·cm P can result in quite usable pentode-like detector characteristics.

We have already discussed the equivalent amplification factor μ of the surface detector in dealing with the gain [see (13)] and gain-stability product [see (14)] of the RGT. In this section, we derive the fixed-gate transconductance g_{m_0} of the detector. The quantity g_{m_0} has been found experimentally to be somewhat easier to predict and control than the product $g_{m_0} r_d = \mu$ because of the marked dependence of r_d on nonuniformities and lateral fringing effects at the semiconductor surface.

Since we assume the detector region to be "normally on" and reasonably far away from the beam electrode, a derivation of g_{m_0} involves the combined effect of both fixed oxide charge and substrate pinch-off.

We assume a detection region channel W units wide by L_c units long. Following a recent derivation of MOS characteristics where substrate pinch-off was predominant [7], we write the mobile charge in the channel as the algebraic sum of three components: 1) a charge component induced by the beam field, Q_B (coulombs/cm^2); 2) a charge component due to the built-in oxide charge, Q_S; and 3) a subtractive charge due to ionized depletion region acceptors, Q_D:

$$Q_{\text{mobile}} = Q_B - Q_D + Q_S \quad (\text{coulombs/cm}^2). \quad (15)$$

Relating charge densities to appropriate mechanisms [7], we find

$$Q_{\text{mobile}} = \epsilon_0 \left[\frac{V_P - V_a(x)}{\delta_e} \right] - \epsilon_{si} \sqrt{\frac{2qN_A}{\epsilon_{si}} [V_a(x) + 2\phi_p]} + Q_S \quad (16)$$

where $V_a(x)$ is the channel voltage, a function of position in the region between source and drain. Other terms are defined in the Glossary.

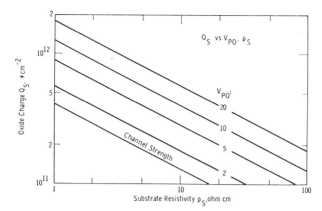

Fig. 8. Detector pinch-off voltage as a function of oxide charge and substrate resistivity (assuming $\mu_p \sim 500$ cm$^2 \cdot$volt$^{-1} \cdot$sec^{-1}).

Following Shockley's gradual channel approximation [8] we set the drain current I_D equal to

$$(\mu_n W/L_c) \int_0^{V_D} Q_{\text{mobile}} dV.$$

Thus,

$$I_D = \frac{\mu_n W \epsilon_0}{L_c \delta_e} \left[\left[V_P V_D - \frac{V_D^2}{2} \right] \right.$$
$$- \frac{4}{3} V_T \phi_p \left\{ \left[1 + \frac{V_D}{2\phi_p} \right]^{3/2} - 1 \right\}$$
$$\left. + \frac{\delta_e Q_S V_D}{\epsilon_0} \right] \quad (17)$$

where, for simplicity, we define the threshold voltage

$$V_T \equiv 2 \frac{\epsilon_{si}}{\epsilon_0} \sqrt{\frac{qN_A \phi_p}{\epsilon_{si}}} \delta_e.$$

We allow the beam-to-substrate distance δ_e to approach infinity in (17), and calculate the value of drain voltage at which $\partial I_D/\partial V_D = 0$. This is the substrate controlled pinch-off voltage V_{PO} since the beam, being so far away, cannot itself cause pinch-off. This simplified situation is encountered when δ_e is greater than a few bulk extrinsic Debye lengths.

Setting $\partial I_D/\partial V_D = 0$ in (17) and assuming at pinch-off $V_{Po} \gg 2\phi_P$, we find

$$\lim_{\delta_0 \to \infty} [V_D|_{\partial I_D/\partial V_D = 0}] = V_{PO} = \frac{Q_S^2}{2q\epsilon_{si}N_A} \text{ volts.} \quad (18)$$

Equation (18) relates the pinch-off voltage V_{PO} to the surface oxide charge Q_S (coulombs/cm^2) and bulk doping N_A (cm^{-3}). In Fig. 8 we plot Q_S versus ρ_s, the silicon substrate resistivity for parameters of pinch-off voltage. We assume a p-type substrate where $\mu_p = 500$ cm$^2 \cdot$volt$^{-1} \cdot$s^{-1}. As can be seen in this figure, common values of inversion layer content ($2-5 \times 10^{11}$ charges·cm^{-2}) require p-type substrate resistivities in the range of 5-10 ohm·cm to keep V_{PO} in the range of 2-10 volts.

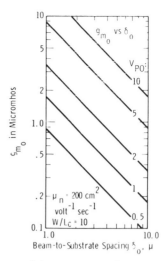

Fig. 9. Fixed-beam detector transconductance as a function of pinch-off voltage and beam-to-substrate spacing (assuming $W/L_c = 10$).

Low values of V_{PO} permit pentode-like operation and therefore relatively high detector gain without excessively large supply voltages.

We have also included in Fig. 8 a curve marked "channel strength." We have obtained this curve from (18) by setting $V_{PO} = 2\phi_p \sim 1$ volt (a good estimate for lightly doped substrate material). It can be shown that this bottom curve represents in a simplified form the oxide channel strength in ohm·cm as originally defined by Atalla et al. [9]. The equation of this curve is given, from (18), as

$$\text{channel strength} = \frac{4\epsilon_{si}\phi_p}{\mu_p Q_S^2} \text{ ohm·cm.} \quad (19)$$

It will be remembered that the channel strength of an MOS system containing a positive oxide charge $+Q_S$ is defined as the *minimum* p-type substrate resistivity which will just be inverted by the presence of Q_S. For instance, the channel strength of a system containing 2×10^{11} charges/cm^2 is about 5 ohm·cm p-type. Equation (19) is important in the fabrication of the RGT in that for a given process-dependent Q_S value, it gives the minimum value of substrate resistivity consistent with a *normally on* RGT detector.

Returning to our calculation of g_{m_0}, we assume that V_{PO} is independent of V_P within operating limits, and, in addition, that $V_P \gg V_{PO}$. We find from (17) that g_{m_0}, the device "fixed-beam" transconductance, equals

$$g_{m_0} \equiv \frac{\partial I_{PO}}{\partial V_P} = \mu_n \frac{W}{L_c} \epsilon_0 \frac{V_{PO}}{\delta_e} \text{ mho.} \quad (20)$$

In Fig. 9 we plot the transconductance of the detector versus the beam-to-substrate spacing for various practical values of pinch-off voltage. As can be seen for V_{PO} near 5 volts, g_{m_0} is on the order of 1–2 μmho, a relatively low value due mainly to the large value of δ_0 required by device fabrication.

The impedance level of the device is inversely proportional to the pinch-off current of the detector. Using the value V_{PO} in (18), we obtain from (17), again assuming $\delta_0 \gg$ bulk extrinsic Debye length,

$$I_{PO} = \frac{1}{3} \mu_n \frac{W}{L_c} Q_S V_{PO} = \frac{1}{6} \frac{W}{L_c} \frac{Q_S^3}{q N_A \epsilon_{si}}. \quad (21)$$

The cubic dependence of I_{PO} on Q_S indicates the necessity of close control on the effective oxide charge if one hopes to obtain predictable detector impedance levels in the RGT.

Assuming an expression for output impedance similar to Shockley's [8] it can be shown that

$$\mu = g_{m_0} r_d \approx \left(\frac{\epsilon_0 L_c}{\epsilon_{si} \delta_0} \right). \quad (22)$$

Since in the ordinary MOS transistor

$$\mu \approx \frac{\epsilon_{ox} L_c}{\epsilon_{si} W_{ox}} \text{ (reference [6])}$$

it can be seen that the amplification factor of the RGT. is on the order $[(1/3) \cdot (W_{ox}/\delta_0)]$ times as *small* as the μ of an MOS. For a typical $W_{ox} \approx 2000$ Å and $\delta_0 \approx 50\,000$ Å, a factor of 75 reduction in μ over an MOS of equivalent channel length is predicted. However, (13) indicates that the mechanically resonant beam system essentially amplifies the input voltage by a factor of Q at resonance, making overall gains on the order 0.1–1.0 or greater still possible.

We illustrate an example of the design of an RGT. Suppose we wish to fabricate a 5-kHz RGT which will exhibit a voltage gain of -20 dB (loss) at a polarization voltage of $V_P = +15$ volts with an output load impedance of 20 kΩ. In order to achieve reasonable frequency stability [see (14)], we will, in addition, require that $(V_P/V_{PI})^2 \approx 0.1$ under operating conditions.

From Fig. 7 we find that, among others, a beam 0.020 inch long by 4 microns thick will resonate at about 5 kHz. Note that this choice of L and d is somewhat arbitrary. Any attempt at optimum design involves a complicated relationship between the beam aspect ratio and its loss mechanism. Optimization will not be treated in this paper. If we choose $V_{PI} \sim 47$ volts, Fig. 7 indicates that δ_0 must be ≈ 8.5 microns, giving $(V_P/V_{PI})^2 = 0.1$. Now for an insertion loss of 20 dB with $R_L = 20K$, we find, from (13),

$$\text{resonant gain} = \frac{4}{27} \left(\frac{V_P}{V_{PI}} \right)^2 \frac{Q\mu}{1 + \frac{r_d}{R_L}} = \frac{1}{10}. \quad (23)$$

Assuming $R_L \ll r_d$, we find that for a 20-dB insertion loss, the relation between g_{m_0} and Q is given by

$$\frac{\mu}{r_d} = g_{m_0} = \frac{340 \, \mu\text{mho}}{Q}. \quad (24)$$

At the present stage of RGT technology it is hard to predict what value of Q will obtain in a given device. Experience has shown that in the 1–10 kHz frequency range, values of Q in the 50–200 range are typical for plated gold. Assuming $Q \sim 100$, we find that $g_{m_0} \sim 3.4$ μmho. If we make the detector width W about $\frac{1}{4}$ of the beam length L, then $W \approx 0.005$ inch. If the channel length L_c equals about 6 microns, then $W/L_c = 20$. From Fig. 9 for g_{m_0} (remembering to adjust for the fact that $W/L_c = 20$ by halving the required g_{m_0}), we see that V_{PO} must be about 8 volts. From Fig. 8 a substrate resistivity of 15 ohm·cm p-type will result in a pinch-off of 8 volts when $Q_S \sim 3 \times 10^{11}$ positive oxide charges/cm², a reasonably typical value obtained during oxidation. From (21), $I_{PO} = 25 \, (W/L_c) \, \mu$A or $I_{PO} \sim 0.5$ mA. By operating at the knee of the detector VI characteristic, it can be shown that the required supply voltage for the 20-kΩ load is then about 18 volts, making a common V_P and V_d supply practical.

In the conclusion, Figs. 7, 8, and 9 permit the design of an RGT of a specific frequency, gain, output impedance level, and operating voltage. Prime independent variables are beam dimensions d and L, δ_0, N_A, Q_S, and W/L_c (we assume a gold beam). In the next section we show these design equations to be in good agreement with experiment.

III. Experimental

In this section, we describe a method for batch-fabricating RGT's consistent with silicon integrated technology. We describe the behavior of a typical device and discuss the range of parameters we have achieved at present. We confirm some theoretical predictions of Section II and, finally, discuss recent experimental results on multiple-beam coupled-mode devices.

A. Fabrication of Resonant Beam Structure

Fabrication of the detection region of the RGT is conventional. After the $N+$ source-drain regions are diffused and an insulating oxide regrown, contact windows are opened in the oxide [Fig. 10(I)]. In Fig. 10(II) a metal bilayer is deposited over the slice. The important properties of this layer are that it 1) must adhere well to the oxide, and 2) should be easily electroplated by the metal from which the cantilever beam will be made. In Fig. 10(III) the input and tuning gates, as well as the source-drain contact land areas, are formed via the deposition of a thin metal flash. In Fig. 10(IV), via a photoresist process, we deposit the metal spacer layer onto which we will plate the beam electrode. The thickness of this layer will correspond closely to the nominal beam-to-substrate distance δ_0 in the final device. At this stage a photoresist layer is spun over the slice and exposed such that the resist is removed down to bare metal everywhere the beam and its pad are to be plated. There is a marked flexibility in the process at this stage. Either one or a number of beam areas of varying shapes and lengths can be photo-delineated. Suspended me-

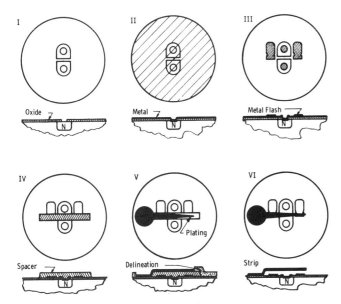

Fig. 10. Resonant beam fabrication process.

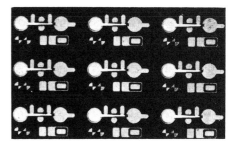

Fig. 11. Batch-fabricated C-F RGT's ($L \approx 0.017$ inch).

chanical webs for coupling purposes can also be fabricated at this time, as will be discussed further on in this section. In Fig. 10(V) an appropriate metal is plated in the delineated areas to form the beam electrode. Although we have used gold because of its easy plateability and relative absence of residual stress, a number of different metals or alloys are available, some having lower internal loss and lower temperature coefficients of Young's modulus. After plating, the top resist layer is sprayed-stripped, followed by a quick metal spacer etch, then a pad-contact etch. A final rinse in a nonpolar organic solvent completes the process, resulting in a suspended beam electrode [Fig. 10(VI)].

B. Experimental Measurements

B-1) General Device Characteristics: We have thus far fabricated two general categories of devices, those with clamped-free beams and those with clamped-clamped beams. The latter exhibit a center frequency 6.27 times higher than the former for equal beam length and thickness.

The general characteristics of the clamped-free devices were discussed elsewhere [5]. In summary, the C-F devices previously reported had a gold beam electrode 0.038 inch long and about 10 microns thick. Spacing from the substrate was about 8–12 microns. In a typical run of these devices, the following average parameters were observed: $f_0 = 2.8$ kHz, $Q = 90$, and gain $= -20$ to -40 dB at $V_P = 85$ volts. It is interesting to note that of a run of 23 devices fabricated at the same time on the same chip, a mean center frequency deviation of ± 8 percent was observed. This reproducibility of center frequency is presently encouraging.

In Fig. 11 we show a section of a slice of some more recently fabricated C-F RGT's using a tapered cantilever design. An auxiliary MOS transistor is fabricated as a buffer output amplifier beside each unit. Fundamental frequencies of these devices, where $L \cong 0.017$ inch, depend, of course, on beam thickness. By varying d, f_0's in the range from 7 through 24 kHz have been obtained. As an example, a group of 12-kHz devices using this geometry operated with a $V_P = +30$ volts and showed a gain of -8 dB at a 30-kΩ load impedance. Q's of about 60–100 were in evidence.

In an attempt to make higher-frequency RGT's which approach the 1/10th-MHz range of operation, clamped-clamped devices as shown in Fig. 2 were fabricated. These devices have gold beams of length 0.019 inch with thicknesses in the range of 3 to 8 microns. The detector in these devices is located at the center of the beam electrode, which is an antinode for the fundamental vibration. On either side of the MOS detector are input force plates, which are connected together by a shorting band as shown in Fig. 2. A grounded N^+ guard ring, an extension of the source contact, prevents direct feedthrough of input gate signals to the output drain contact. Feedthrough can occur because of capacitive coupling of the input gate to the relatively high sheet resistant channel inversion layer.

In a typical run, a 10 000 Å thermal oxide is grown upon 10 ohm·cm P chemically polished $\langle 111 \rangle$ silicon. The oxide is opened for source-drain diffusion, and a $POCl_3$, 2.5-micron, n-type diffusion is carried out at 1100°C. Oxide is regrown over the n-type regions and, at this stage, channel content may be adjusted by baking the slices in an appropriate ambient at 350°C for times on the order of one-half hour. Channel pinch-off currents in the range 0.05 through 0.5 mA are typically employed. At this stage, the device is ready for cantilever formation, as described in Section III-A.

In Fig. 12 we plot the output voltage of a clamped-clamped 30-kHz RGT versus input frequency at a constant input voltage of one volt. A voltage insertion loss of about -20 dB is observed at resonance. A bandwidth of 450 Hz implies a Q of 67. From the nomograph in Fig. 7, an f_0 of 30 kHz indicates, for an $L = 0.019$-inch beam, a gold thickness (assuming the plated gold possesses bulk mechanical properties) of about 3.4 microns. This implies a weight gain of 2.8 mg during the gold-plating operation on a one-inch-diam slice. (The plated-to-nonplated area ratio of the cantilever mask used in making these devices was 0.085.) An observed weight gain of 3.3 mg after plating the slice from which this 30-kHz device came indicates either a lack of plating uniformity over the whole slice, or that the acoustical velocity $\sqrt{Y/\rho}$ of the plated gold is only about 85 percent of the accepted bulk gold value.

Operation of C-C RGT's at higher-order modes has also been demonstrated. For instance, a 21-kHz RGT has been shown to exhibit tuning peaks at 21, 72.4, 135.2, 215, and 410 kHz, in fair agreement with the modal ratios $1:2.7:5.4:8.9\cdots$ predicted by Rayleigh [10] for C-C beams. Deviations from the Rayleigh ratios are due in part to changes in the effective value of V_{PI} associated with the different modes. This leads to a different percentage "detuning" at constant V_P for each mode [see (11)], modifying the experimentally observed ratios. Because of the symmetric geometry of the C-C RGT series shown in Fig. 2, gains in the third mode ($5.4\ f_0$) usually exceed the gain observed in second-mode operation ($2.7\ f_0$). A nonoptimized "gain" of -17 dB has been seen in the third mode at a frequency of 250 kHz in a $L=0.019$-inch device.

B-2) Amplitude of Beam Vibration: It is possible to obtain a figure for the sinusoidal beam vibration necessary to obtain a given output voltage from these C-C devices. This is important since nonlinearities may result if, for reasonable output voltages (say, 100 mV), the amplitude of beam vibration at resonance is an appreciable percentage of δ_0.

Consider a 40-kHz C-C RGT which possesses a drain resistance of 10^5 ohms, a beam-to-substrate spacing of 5 microns, and an insertion loss of about 6 dB when operated in a normal manner with $V_P=60$ volts, $V_D=60$ volts, and $R_L=36$ K. Now, by applying a nonresonant ac input signal *directly* to the cantilever beam (while holding the dc beam bias at $+60$ volts to maintain quiescent operation), it is possible to obtain the value of the amplification factor for the "fixed-gate" MOS transistor [see (4)]. The measurement is made at $f\ll f_0$.

For this particular device, a "fixed-gate" μ of 0.7 was obtained at 0.1 through 5 kHz. From (4) it can be seen that the beam movement δ necessary to obtain an output of 100 mV is given simply by

$$\delta = \frac{r_d + R_L}{R_L}\frac{\delta_e}{\mu V_P} e_{\text{out}} = \frac{100+36}{36}\frac{5\times 10^{-6}}{0.7\times 60}\frac{1}{10}$$
$$= 450\ \text{Å}. \quad (25)$$

The inherently small displacements of the beam electrode in this 40-kHz device indicate that creep and fatigue processes which occur in the flexure of ordinary metals should not be severe in the RGT. Only long-term life tests (which have not yet been made) can confirm this hypothesis.

As a check on the experimentally determined value of μ, we note that the MOS detector on this RGT had a pinch-off voltage of $V_{PO}=7$ volts. Correcting for the fact that W/L_c in this device is 21, we find from Fig. 9 that the theoretical "fixed-beam" transconductance

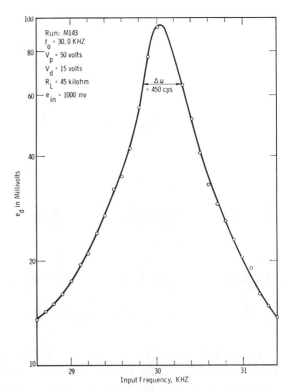

Fig. 12. Response characteristic of a 30 kHz C-C RGT.

g_{m_0} should be about 5.2 μmho. The calculated amplification factor is then

$$\mu = g_{m_0} r_d = (5.2\times 10^{-6})(10^5) = 0.52 \quad (26)$$

which compares favorably with the experimentally observed value of $\mu=0.7$ seen above.

B-3) Variation of Gain and Resonant Frequency with Polarization Voltage: Equations (11) and (13) show that the resonant frequency and gain of an RGT can be expected to vary with polarization voltage. These relationships were substantiated by measurements on an RGT which had a resonant frequency near 32 kHz. The geometry of this device is as shown in Fig. 2, where the beam dimensions are length$=19$ mils, width$=0.5$ mil, and thickness≈ 3 microns. The beam-to-substrate spacing is approximately 4 microns. Based on these values, the nomograph in Fig. 7 predicts a pull-in voltage of about 75 volts.

Experimental measurements were made using a constant input signal of 0.965 volt and a 40 000-ohm load resistor. The results are plotted in Fig. 13. It can be seen that both resonant frequency and gain vary approximately parabolically with polarization voltage, as predicted by (11) and (13). Quantitative correlation of the theory and the measurements depends on the pull-in voltage, which is somewhat risky to obtain experimentally because of possible damage to the device. However, if the pull-in voltage is 75 volts, (11) predicts a 10-percent shift in frequency at a polarization voltage of

$$V_P = V_{PI}\sqrt{\frac{2.7}{4}} = 75\times 0.822 = 61\ \text{volts}. \quad (27)$$

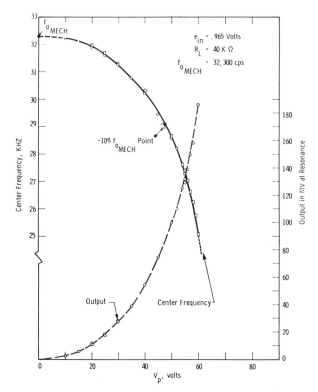

Fig. 13. Experimental variation of output voltage and resonant frequency with polarization voltage. (Output and center frequency versus V_p for #148 C-C RGT.)

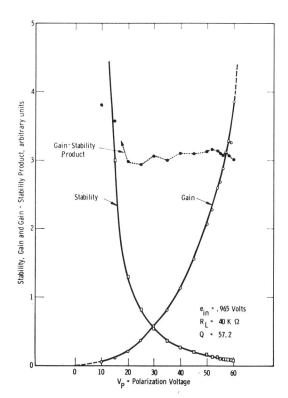

Fig. 14. Experimental variation of gain, frequency stability, and gain-stability product with polarization voltage. [Gain-stability product for 32.3-kHz RGT versus V_p (#148).]

Figure 13 shows that the 10-percent shift actually occurred at 47.5 volts. In view of the simplifications involved in the theoretical model, and the inaccuracies involved in measuring the beam thickness and the beam-to-substrate spacing, this agreement is quite satisfactory.

In Fig. 14 the gain, stability, and gain-stability product have been plotted, in arbitrary units, for the above device. In calculating $|\psi|$ [in (12)] we have used the approximate form $|\psi| \sim (1/2Q)(1 - f_r/f_0)$. This form assumes parabolic detuning. It is exact only when $V_P \ll V_{PI}$ and is simpler to use inasmuch as it does not involve a possibly inaccurate graphical determination of the term $\partial f_r/\partial V_P$.

For V_P greater than 20 volts, the GS product is seen to be constant to about ± 3 percent over 1.5 decades in gain as predicted by (14). Quantitatively, the average value of the GS product obtained experimentally for this device is

$$GS = |e_{\text{out}}/e_{\text{in}}|_r |\psi| = 0.006.$$

When we calculate μ from GS taking into account the presence of a parallel load reactance of $-j\,35\,k\Omega$ associated with the measuring equipment, we find that the experimental value of μ for the RGT detector is on the order of 0.03. This value of μ is about a factor of 4–5 less than the theoretical value of μ calculated from either $g_{m_0} r_p$ [in (20)] or $\mu = \epsilon_0 L_c/\epsilon_{si} \delta_e$ [in (22)].

It is presently thought that the reason for the discrepancy between the theoretical voltage gain for the RGT [for instance, (13)] and the experimental result is the fact that, in the simplified model presented in Section II, no account is taken of *where* the input gate and detector are situated under the cantilever beam. Obviously, if the input gate is made quite short and is crowded very close to the clamped end of the beam, the gain of the RGT will be severely reduced although this reduction will not explicitly appear in the gain expression [see (13)]. In the simplified analysis it has been assumed that both the input plate and the detector are located at the antinode of the cantilever. This assumption avoids the difficulties inherent in the distributed nature of the beam deflection, but can lead to inaccuracies of factors of 2 or more when precise comparisons between experiment and theory are made. We can, of course, define "gain-reduction factor" which depends on gate and detector geometry in a complicated manner. The calculation of this factor will not be discussed in this paper. Suffice it to say, the factor is less than 1 and is constant for a given device geometry. It is about 0.5 for the C-F device illustrated in Fig. 11. Measurements of the ratio of the beam versus input gate device gain in the C-C RGT in Figs. 2 and 14 indicate a geometrical factor of 0.25–0.15 for these devices. In other words, the fact that the input plates in this device are off to the side represents a loss of about 12 dB over the ideal gain expression, which accounts for the factor of 4 discrepancy of μ in Fig. 14.

The "geometrical factor" can best be calculated with new geometries using beam deflection theory. This "factor" represents an important detail describing device utility. However, it is the basic agreement of the qualitative behavior of the RGT with the simplified theory presented in this section which indicates that the role of V_P on device behavior is satisfactorily understood.

B-4) Temperature Sensitivity of Resonant Frequency:
One primary reason for going to a mechanical resonator for achieving high-Q integrated tuning is the fact that since a mechanical resonance is passive, it cannot oscillate with small changes in beam parameters. We contrast the stability of a mechanical system with the stability of an active feedback RC filter where high selectivity is obtained only by working close to the verge of oscillation [2]. Here, small changes in loop parameters brought on by temperature fluctuations can easily quench Q or, if the loop gain approaches one, initiate self-oscillation.

Experiments on RGT's have shown gain variations as small as 1 percent over a 50°C range, indicating that gross variations of Q with temperature are probably not of primary importance at this stage of device development.

More important is the variation of Young's modulus of gold with temperature. Values of $(1/Y)\,\partial Y/\partial T$ of of about 240 ppm/°C have been reported for bulk gold [11]. Since $f_0 \propto \sqrt{Y}$, the fundamental lower limit on frequency stability in the RGT should be about 120 ppm/°C when gold cantilevers are used. There is no guarantee of course that plated gold will be as good as bulk gold from a stability standpoint. Better cantilever materials and/or temperature compensation of V_P can, however, substantially improve this figure. The possibility of electroplating beams of low-temperature coefficient alloys such as Ni Span-C[1] is still an open question because of the control in percent Ni content needed to hold $(1/Y)\partial Y/\partial T$ at below 10 ppm/°C($\sim$27 percent $\pm \frac{1}{4}$ percent).

In Fig. 15 we illustrate a representative series of f_0 versus temperature curves made on a nominally 8-kHz C-F RGT with a gold-plated cantilever about $L\sim 0.017$ inch long and $d\sim 4$ microns thick. Gold plating was carried out at 50°C using a slightly alkaline plating solution and dc plating techniques. We first note the gross decrease in f_0 with V_P, *independent* of temperature. This is just the electrostatic "tuning" effect expressed by (11). In addition, for a given value of V_P, the variation of f_0 with T is illustrated over the range $30° \leq T \leq 80°C$.

Some generalizations can be made from Fig. 15:

1) C-F devices investigated to date decrease in center frequency with increasing temperature.

2) The percentage decrease in f_0 depends on V_P— higher values of V_P result in higher temperature coefficients.

[1] International Nickel Co., Inc.

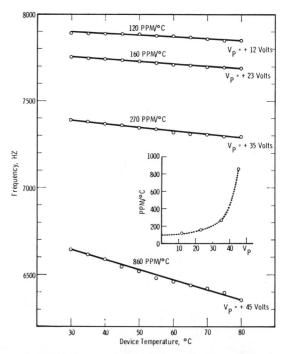

Fig. 15. Experimental variation in resonant frequency with temperature, at various polarization voltages.

3) A plot of ppm/°C versus V_P does not seem to extrapolate through the origin at $V_P=0$ (see insert), but approaches about 100 ppm/°C.

We divide the observed variation of ppm/°C with V_P into 1) a constant variation, $\sim$100 ppm/°C, associated with the intrinsic properties of the gold beam itself (probably due to Young's modulus) and independent of V_P; and 2) an "electrostatic" contribution to instability varying as $V_P{}^n$ where $n \approx 2$ at low V_P. The exact reason for a temperature instability varying as $V_P{}^2$ is not known at present. It is thought that either reversible stress relief resulting in changes in δ_e, or the stronger variation of δ_e (and therefore f_0 with Y as V_P approaches V_{PI}) may be causing this effect. Further experimentation is necessary to ascertain the dominant mechanism.

In any case, it is important to note that when we calculate V_{PI} for the above structure [48+ volts using (27)], we find that when $(V_P/V_{PI})\sim 0.5$, the temperature coefficient of frequency is about 160 ppm/°C. This value is already within range of the stability of evaporated thin-film resistors used in the production of thin-film oscillators on glass [12]. Preliminary results are therefore encouraging, and selection of a better beam material is in process.

In Table I we have tabulated some possible metals which could be employed as the beam electrode. The metals are listed in order of increasing sensitivity of Young's modulus to temperature in ppm/°C. Note that of the metals which promise relative ease of plating, gold is about a factor of 2 more sensitive to temperature than, say, palladium. Current work is therefore being directed toward perfecting a process for obtaining palladium beams for RGT's in the hope of reducing the

TABLE I
CHANGES OF YOUNG'S MODULUS WITH TEMPERATURE
IN ppm/°C FOR VARIOUS MATERIALS

	Y kg/mm²	$-\dfrac{dY}{dT}$ kg/mm²/C°	$-\dfrac{1}{Y}\dfrac{dY}{dT}$ ppm/C°
tungsten	41 500	4.2	101
palladium	12 400	1.3	105
molybdenum	33 600	4.3	128
tantalum	19 000	2.5	132
platinum	17 300	2.3	133
beryllium	29 000	4.1	141
iridium	54 000	10.6	196
gold	7 900	1.9	240
iron	21 500	5.6	260
rhodium	38 600	10.5	272
copper	13 100	3.9	298
manganese	20 000	6.3	315
silver	8 200	3.4	415
zinc	9 400	4.0	425
aluminum	7 200	3.1	430
zirconium	7 000	3.4	485
titanium	10 500	6.4	610
tin	5 500	7.7	1400

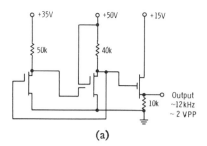

(a)

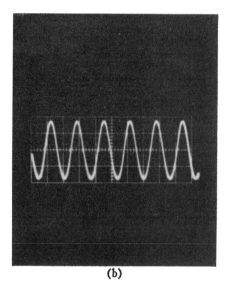

(b)

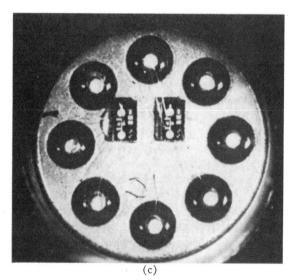

(c)

Fig. 16. Twin RGT oscillator requiring no external 90° phase shift network. (a) Circuit. (b) Ouput waveform. (c) Twin RGT oscillator mounted on TO-5 header.

inherent frequency shift of the device to about 50 ppm/°C.

B-5) Integrated Oscillator: Thus far only the simple open-loop bandpass properties of the RGT have been mentioned. In this brief section, one particularly important area of application is described—namely, the construction of monolithic audio oscillators.

It will be recalled that the integration of audio-frequency RC oscillators on a silicon slice using conventional components has been impractical because of the high RC products needed to obtain the necessary phase shift to permit oscillation. Using the RGT, a number of low-frequency monolithic oscillators are feasible. The only supplementary components needed are the load resistors of the FET. The relatively low value of the load resistors and the elimination of any capacitors needed in the device results in a low-frequency integrated oscillator that is compatible with current semiconductor fabrication techniques.

Because the output of the RGT is in quadrature (90°) with the device input at resonance [see (3)], it is necessary to introduce an additional 270° phase shift around the device to get it to oscillate at its open-loop frequency. A phase shift of 180° is easy to obtain. However, the additional 90° phase shift is just as hard to obtain, circuitwise, for the RGT as it is to obtain in a small volume for more conventional RC oscillators.

One simple solution compatible with monolithic technology is to use *two* RGT's tuned to the same frequency. Since both devices exhibit a 90° phase shift at resonance, two devices in series plus 180° additional phase shift can result in closed-loop oscillatory behavior. For the 180° phase shift, we take advantage of the phase reversal at the output when the signal is applied directly on the cantilever rather than on the input force plate.

In Fig. 16(a) is illustrated a simple circuit employing two RGT's selected to be about 12 kHz at the quiescent voltages indicated. Note that only external resistors of values quite compatible with integrated circuits are needed. We illustrate the output waveform of the oscillator in Fig. 16(b) and present a photograph of the final oscillator encapsulated in a TO-5 header in Fig. 16(c). This device oscillated at about 12 kHz with a voltage output ~2 volts peak-to-peak. Higher-frequency oscillators of this type (47 kHz) have exhibited output voltages of up to 15 volts peak-to-peak at reasonably low distortion level.

An obvious problem associated with the above approach is the necessity of fabricating the two cantilever frequencies to within a bandwidth of each other. Since the two cantilevers of each oscillator are side by side throughout the entire fabrication process, it seems reasonable to assume that they both will be exposed to the same treatment. Once techniques improve to the point where the beams are well matched, there should be no reason why complex oscillators cannot be implemented by double-RGT techniques.

B-6) Coupled Multipole Filters: The bandpass characteristics of the RGT are determined almost entirely by the resonant modes of the mechanical structure of the device. Simple cantilever beams possessing a single fundamental resonance have been studied thus far in this paper. In many situations, a more complex bandpass characteristic is desirable. A telemetry filter requiring both wide bandwidth and steep rejection slopes is one example. By designing the shape of the resonant member in the RGT to resonate at two or more closely spaced frequencies [13] we can implement sophisticated bandpass behavior in a batch-fabricated manner.

One simple system exhibiting doubly tuned behavior consists of two identical, loosely coupled C-F cantilever beams. Such a system will "split" into two frequencies, and can exhibit useful pass response.

We have fabricated C-F twin-beam RGT's containing a coupling web near the clamped end, as diagramed in Fig. 17(a). Under Beam 1 are two input electrodes used to excite the beam electrostatically. Vibrations of Beam 1 are mechanically transmitted to Beam 2 via the coupling web, causing Beam 2 to vibrate. Under Beam 2 is the channel of the FET detector, the source and drain contacts of which appear in the figure. Since the input electrodes are under Beam 1 while the channel is under Beam 2, only collective vibrational modes result in signal transmission from input to output. When the beams are tuned to exactly the same frequency, the two modes of collective vibration occur 1) when both beams go up and down together and 2) when one goes up while the other goes down. The modal frequency is higher under condition 2) than under 1) because the coupling web is stretched and acts as an additional spring tending to "stiffen" the system. If the web length is adjusted properly, the two frequencies associated with modes 1) and 2) can be made about f_0/Q apart. Smooth-topped bandpass properties are then possible.

In Fig. 17(b) we present a finished coupled RGT. This device has two identical beams of length = 9.5 mils, thickness ≈ 4 microns, and beamwidth ≈ 0.5 mil. The coupling web in these devices is 0.5 mil wide and 0.3 mil long measured from the clamped end. Beam-to-substrate spacings δ_0 of about 10 microns for these devices have resulted in values of V_{PI} of about 300 volts.

In Fig. 17(c) we illustrate the response of one of these devices when it has been properly tuned. A double-peak characteristic is in evidence. Peak frequencies 275 Hz apart at 22 kHz correspond to a coupling coefficient of

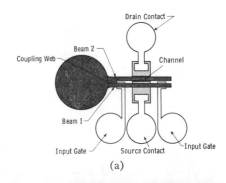
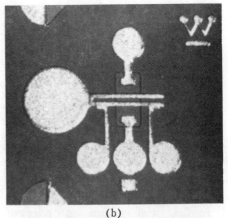
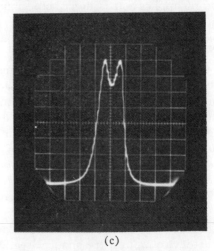

Fig. 17. Doubly resonant C-F bandpass RGT. (a) Device geometry. (b) Photograph of an actual device ($L = 9.5$ mils). (c) Frequency response (experimental), where $f_{center} \approx 22$ kHz, bandwidth ≈ 300 Hz.

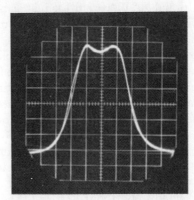

Fig. 18. Doubly resonant bandpass RGT, where $f_{center} \sim 132$ kHz, bandwidth ≈ 1000 Hz.

$K = 1.25$ percent. Measurement of the Q of the individual peaks in these devices, obtained by detuning the structure, gives a $Q \sim 130$. Thus $kQ \sim 1.62$. It can be shown classically that a $(KQ)^2 \approx (1.6)^2 = 2.6$ should result in a theoretical midband ripple of about 1 dB, which is close to the midband ripple (1.9 dB) observed in Fig. 17(c).

These devices were operated at $V_P \approx -210$ volts with $R_L = 70$ $k\Omega$ and $V_D = +30$ volts. An output of 87 mV in the passband with an input of 580 mV indicates a loss of -16 dB at resonance, which is the lowest we have observed for a coupled RGT to date. The properly tuned device was also free from spurious resonances.

In Fig. 18 we show the response of another type of coupled RGT. This device is similar in construction to the device illustrated in Fig. 17 except that both ends of the cantilevers are affixed to the oxide surface. The center frequency of this essentially C-C device is about 132 kHz with a bandpass of about 1000 Hz.

It should be pointed out that the shape of the bandpass characteristic in these devices was quite dependent on V_P. By changing V_P a few tens of volts, it was possible to tune the response of the device from 1) two well-separated low-gain peaks, through 2) a region where, as the peaks came together, double-peak high-gain bandpass behavior occurred, through a region where 3) the peaks began to diverge again with another accompanying decrease in gain. It is felt that the variation of the bandpass characteristics of the device with V_P is due to a differential electrostatic tuning caused by the lack of electrostatic ground-plane symmetry for the two beams in the structure illustrated in Fig. 17(a). The very high values of Q in the structure make tuning difficult since when K must be small, small differentials in frequency between the two beams have a large effect on bandpass shape. In any case, the results of the twin-beam structure are encouraging, and work will begin on triple-beam electrostatically symmetric RGT's in the near future. The prospect of being able to batch-fabricate complex bandpass filters is an exciting one.

IV. Discussion

The RGT holds much promise for solving the problem of tuning integrated circuits. The device is quite small, the typical volume of its mechanical resonator being about 10^{-9} cubic inches. Even including the volume of the necessary silicon chip to contain the FET detector, the device represents quite a considerable decrease in volume at a given frequency and Q over existing state-of-the-art means of achieving high-Q tuning. Since the fabrication process is basically compatible with silicon-based technology, it is possible that eventually not only size but substantial cost savings may be realized once optimum fabrication methods are developed.

Before long-range predictions about the usefulness of the RGT are made, however, certain problems associated with the device must be resolved.

We have demonstrated that it should be possible to make RGT's with temperature coefficients of center frequency about half that associated with Young's modulus for the beam material. However, if we restrict ourselves to the plating of pure materials, 50 ppm/°C seems to be about the best practical temperature stability obtainable, from Table I. By more sophisticated fabrication techniques, such as bimetal plating, alloy plating, or simply the assembly of prepunched low-coefficient beams onto the silicon slice, it may be possible to increase the temperature stability of the device. It is also possible that diode-type voltage compensation of V_P with temperature may also increase device stability. The latter method is currently being investigated. In any case, the temperature stability of the mechanical Q of the RGT already represents a large qualitative improvement in inherent device stability over, say, the RC feedback approach to tuning. Ultimate limits of both Q and f_0 stability in the RGT are still not understood.

As far as reproducibility and predictability of center frequency is concerned, much work still has to be done. We are encouraged by the approximately 20–30 percent spread in f_0 over any given slice, considering the laboratory methods of fabrication we are using. On the other hand, no attempt has been made to *predetermine* the mean center frequency of the resonators on a given slice to say 5 percent, by current-time integration of plating current. The problem of fabricating predictable and precise beam frequencies is compounded when the semiconductor die contains a number of different beams, especially when the beams have different aspect ratios. More work on attempting to hold the center frequency as constant as possible over the entire slice must be done, not so much for single RGT's where simple frequency range "bin-type" classification is effective, but for multiple beam/frequency RGT's where the yield drops to zero rapidly as the variance increases.

Other problems of the RGT which must be solved are as follows:

1) The role of residual beam stress after plating which can result in beam curl, and therefore unpredictable values of δ_0.

2) The roles of annealing and fatigue on long-term device behavior.

3) Control of Q_S, and therefore the detector impedance level.

4) The role of the free oxide surface as far as excess device noise and ultimate V–I quiescent stability is concerned.

5) The role of feedthrough in degrading the ultimate dynamic range of the RGT.

V. Conclusions

We have demonstrated batch-fabricated transistors exhibiting high-Q tuning characteristics in the range 1–100 kHz. Both single and multiresonant structures have been constructed. Theory and experiment corre-

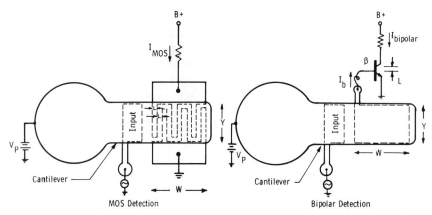

Fig. 19. Geometries of two possible basic detection schemes for the RGT.

late well, indicating a good understanding of the basic device properties. Temperature sensitivity of the device is shown to be under control. Although more work is necessary to permit quantity production of the RGT, progress has been relatively rapid on the development of such individual device characteristics as gain, frequency range, size, and multipole pass behavior.

Appendix I
Comparison of Bipolar and MOS Transistors for Detecting Cantilever Movement in Resonant Gate Devices

The vibrations of a polarized, mechanically resonant cantilever can be detected either by field-effect modulation of an MOS transistor or, alternately, by using the cantilever as a vibrating capacitor to modulate the base current in a bipolar transistor of current gain β. For equivalent geometries, it is shown in this Appendix that the ratio of output current in the two detection schemes is given by

$$\frac{I_{\text{MOS}}}{I_{\text{bipolar}}} = \frac{1}{8\pi} \frac{qE_{\text{gap silicon}}}{kT} \frac{\text{resonant period (seconds)}}{\text{transistor base lifetime}},$$

indicating that for frequencies below about 10 MHz, field-effect detection results in a greater output signal.

We define comparative geometries in Fig. 19. Underneath a cantilever Y units in breadth is an active detection region W units wide. In the case of the MOS transistor, the region W by Y contains a convoluted channel L units long, the periphery P of which is about $\sim Y(W/2L)$. In the case of the bipolar transistor, the active region is simply a vibrating capacitor of area WY.

The following assumptions are made in this comparison:

a) The cantilever ac motion δ, its equilibrium position δ_e, and polarization voltage V_P are the same for both the MOS and the bipolar transistor.

b) The output resistance of both types of transistors is sufficiently high that the ac output current times the load resistance (assumed equal for the two systems) solely determines system voltage gain.

c) Noise and stability differences in the two detection schemes are neglected.

We define the change in capacitive stored charge Q_0 associated with the cantilever moving a distance δ as

$$Q_0 = \epsilon_0 \frac{V_P}{\delta_e} (WY) \left[\frac{\delta}{\delta_e}\right]. \qquad (28)$$

Referring first to the left side of Fig. 19, we note that this additional charge appears directly within the MOS active region. The output current is then Q_0 divided by the transit time in the MOS transistor $T_{\text{MOS}} = L_c^2/\mu_n V_{PO}$:

$$I_{\text{MOS}} = \frac{Q_0}{T_{\text{MOS}}}. \qquad (29)$$

In the case of the bipolar transistor, the same charge Q_0 enters the base region at a certain *rate* Q_0/τ_0, where τ_0 is the period of the resonance frequency (seconds). Because of the finite base lifetime τ_D (seconds), however, base recombination occurs at a rate $\sim Q_s/\tau_D$, where Q_s is the steady-state base charge. As a result, the steady-state base charge is given by

$$Q_s = \frac{\tau_D}{\tau_0} Q_0,$$

which for $\tau_D/\tau_0 \ll 1$ (low resonant frequency) is considerably less than the value Q_0 *all* of which appears in the channel of the MOS device. The charge Q_s moves across the base of the transistor with a transit time $T_{\text{bipolar}} = L_c^2/2D$:

$$I_{\text{bipolar}} = \frac{(\tau_D/\tau_0)Q_0}{T_{\text{bipolar}}}. \qquad (30)$$

Dividing (29) by (30), including the effects of area difference implied in Fig. 1, and taking the Einstein relation $D/\mu = kT/q$, gives

$$\frac{I_{\text{MOS}}}{I_{\text{bipolar}}} \cong \frac{1}{8\pi} \frac{qV_{PO}}{kT} \frac{\tau_0}{\tau_D}. \qquad (31)$$

Since the relative transit times for the MOS and bipolar transistors differ by a factor of qV_{PO}/kT for equivalent base widths, a value of V_{PO} on order E_G, the silicon bandgap, results in an advantage of about 40 to 1 in favor of field-effect demodulation. This comparison is not entirely fair, however, since conventional transistor base lengths (formed by parallel double-diffusion) are typically two to three times thinner than typical MOS channel lengths (formed by lateral diffusion). However, the MOS still has in its favor a factor of about 5 in transit time even when fabrication advantages of the bipolar are considered.

The effect of the term τ_0/τ_D in (31) is to give the MOS transistor additional advantage at low frequencies. For instance, for a 100-kHz resonant device, the term τ_0/τ_D is $\sim$100 for $1 = \mu$s base material.

Since it would be necessary to resort to ultrahigh-β transistors in order to compete with the inherent efficiency of the field-effect detection process, noise and stability problems in both MOS and bipolar detection should be about equivalent. This is because low-current, high-β transistors are essentially surface-recombination-velocity-controlled and therefore susceptible to many of the surface problems of field-effect surface devices. Output impedances of both the MOS and the bipolar transistor are controlled essentially by the same voltage-dependent Early effect, leading to little advantage of one over the other at small active lengths. Problems associated with the precise control of channel conductance in the MOS transistor are admittedly quite real, but with advancing surface technology this aspect may become less important.

In summary, below about 10 MHz (for typical bipolar base lifetimes), significantly greater output signal is achieved using field-effect-type detection in RGT's than would be obtained utilizing bipolar-type detection systems of equivalent geometry. This result has important implications for future design of resonant gate devices.

Acknowledgment

The authors would like to acknowledge the assistance of E. A. Halgas, G. J. Machiko, E. M. Black, and H. B. Shaffer in the fabrication and evaluation of experimental devices. Helpful discussions on the gold electroplating process have been held with M. P. Lepselter of the Bell Telephone Laboratories.

Bibliography

References

[1] W. E. Newell, "Tuned integrated circuits—a state-of-the-art survey," *Proc. IEEE*, vol. 52, pp. 1603–1608, December 1964.
[2] W. E. Newell, "Selectivity and sensitivity in functional blocks," *Proc. IRE (Correspondence)*, vol. 50, p. 2517, December 1962.
[3] W. E. Newell, "Ultrasonics in integrated electronics," *Proc. IEEE (Special Issue on Ultrasonics)*, vol. 53, pp. 1305–1309, October 1965.
[4] H. C. Nathanson and R. A. Wickstrom, "A resonant-gate silicon surface transistor with high-Q bandpass properties," *Appl. Phys. Lett.*, vol. 7, pp. 84–86, August 15, 1965.
[5] H. C. Nathanson, W. E. Newell, and R. A. Wickstrom, "Tuning forks sound a hopeful note," *Electronics*, vol. 38, pp. 84–87, September 20, 1965.
[6] S. R. Hofstein and F. P. Heiman, "The insulated-gate field-effect transistor," *Proc. IEEE*, vol. 51, pp. 1190–1202, September 1963.
[7] H. C. Nathanson, "A high field triode," *Solid-State Electron.*, vol. 8, pp. 349–363, 1965.
[8] W. Shockley, "A unipolar field-effect transistor," *Proc. IRE*, vol. 40, pp. 1365–1376, November 1952.
[9] M. M. Atalla, E. Tannenbaum, and E. J. Schiebner, "Stabilization of silicon surfaces by thermally grown oxides," *Bell Sys. Tech. J.*, vol. 38, p. 761, May 1959.
[10] Lord Rayleigh, *Theory of Sound*, vol. 1. London: Macmillan, 1894.
[11] Von Werner Köster, "Die Temperaturabhandigkeit des Elastizitätsmoduls reiner Metalle," *Zeit. für Metallkunde*, vol. 39, pp. 1–8, 1948.
[12] N. Schwartz and R. W. Berry, *Physics of Thin Films*, vol. 2, G. Hass and R. E. Thun, Eds. New York: Academic Press, 1964, pp. 393–398.
[13] See, for instance, W. P. Mason and R. N. Thurston, "A compact electromechanical band-pass filter for frequencies below 20 kilocycles," *IRE Trans. on Ultrasonics Engineering.*, vol. UE-7, pp. 59–70, June 1960.

General References

[14] J. J. O'Connor, "A 400 cps tuning fork filter," *Proc. IRE*, vol. 48, pp. 1857–1865, November 1960.
[15] F. Dostal, "The increasing applications of tuning forks and other vibrating metal resonators in frequency control systems," *Proc. 19th Ann. Symp. on Frequency Control*, pp. 59–77, 1965.

Silicon Micromechanical Devices

Tiny valves, nozzles, pressure sensors and other mechanical systems can be chemically etched in a wafer of single-crystal silicon. Such devices can be mass-produced much as microelectronic circuits are

JAMES B. ANGELL, STEPHEN C. TERRY, PHILLIP W. BARTH

In the past 30 years the element silicon has become familiar as the material from which electronic components and systems are made. In the form of microelectronic chips it can be found in everything from dishwashers to the Space Shuttle control systems. The exploitation of silicon's electrical properties has been accompanied by a less publicized exploration of its other properties and their potential uses. This research has led to the development of a technology called micromachining that allows silicon to be made into mechanical devices almost as small as microelectronic ones.

Micromachining starts with the same batch-fabrication techniques that have made silicon integrated-circuit chips inexpensive. The techniques make it possible to fabricate many chips at once, so that the cost of production is spread over many chips and the cost per chip is low. To this repertory of techniques micromachining adds chemical etching techniques for forming three-dimensional shapes such as pits, holes, pyramids, trenches, hemispheres, cantilevers, diaphragms, needles and walls. A wide variety of mechanical devices can be constructed from combinations of these structural elements.

Among the silicon micromechanisms that have been built are valves, springs, mirrors, nozzles, connectors, printer heads, circuit boards, heat sinks and sensors for properties such as force, pressure, acceleration and chemical concentration. Even a device as complex as a gas chromatograph, an instrument for identifying and measuring gases in an unknown mixture, can be built on a disk of silicon a few centimeters in diameter.

Much of the interest in micromachining derives from the need for cheaper and more versatile sensors. Until recently the electronic components of automatic-control systems and measuring instruments were more expensive than the sensors. With the advent of the microprocessor the cost of electronics began to fall dramatically. Batch-fabricated silicon sensors could replace expensive hand-assembled sensors much as batch-fabricated silicon circuits have replaced vacuum-tube assemblies.

The latest development in this area is the integrated sensor, a silicon chip that includes both a sensor and associated signal-conditioning electronics. Integrated sensors are less expensive than sensors and the necessary electronic components fabricated separately. In addition, because the signal from an integrated sensor is less vulnerable to noise and leakage, integrated sensors perform better than discrete sensors and have a wider variety of applications.

Silicon is a semiconductor, one of the elements that lie between the metals and the nonmetals in the periodic table of the elements. It is distinguished from other semiconductors, however, in that it can be readily oxidized. Silicon forms a surface layer of silicon dioxide (SiO_2) when it is exposed to steam. The layer is chemically inert and electrically insulating; essentially it is a glass. Oxide layers are used to protect areas of the silicon during the batch fabrication of microelectronic devices, which is why most such devices are made out of silicon rather than out of other semiconductors such as germanium. Micromachining also makes extensive use of silicon dioxide surface layers.

The silicon employed in the electronics industry takes the form of single crystals whose production is in itself a sophisticated technology. Large "boules," or single crystals of silicon, 10 centimeters in diameter and a meter long, are grown and sliced into disk-shaped wafers from .2 to .5 millimeter thick. The wafers are polished to a mirror finish to remove flaws created by sawing. The homogeneous crystal structure of the material gives it the electrical properties needed in microelectronic circuits. As it turns out, silicon in this form also has desirable mechanical properties.

Silicon, which is directly below carbon in the periodic table, forms the same type of crystal as diamond, although the interatomic bonds are somewhat weaker. Single-crystal silicon is brittle and can be cleaved like a diamond, but it is harder than most metals. In addition single-crystal silicon is surprisingly resistant to mechanical stress. In both tension and compression it has a higher elastic limit than steel; on the other hand, when the limit is reached, silicon fractures whereas steel deforms inelastically. Finally, as a single crystal silicon remains strong under repeated cycles of tension and compression, whereas polycrystalline metals tend to weaken and break because stresses accumulate at the intercrystal boundaries.

Both micromachining and microelectronic fabrication begin with photolithography, the photographic technique used to transfer copies of a master pattern onto the surface of a silicon wafer. The first step in photolithography is to grow a thin layer of oxide on the wafer surface by heating it to between 800 and 1,200 degrees Celsius in an atmosphere of steam. (Dry oxygen also works, but steam is much faster.) Next, a thin layer of an organic polymer that is sensitive to ultraviolet radiation, called a photoresist, is deposited on the oxide surface. A photomask, generally a glass plate on which there is a metal pattern, is placed in contact with the photoresist-coated surface, and the wafer is exposed to ultraviolet radiation. The metal on the mask is opaque to ultraviolet wavelengths, whereas the glass is transparent. The radiation causes a chemical reaction in the exposed areas of the photoresist. If the photoresist is of the type called positive, the chemical reaction weakens the polymer. If the photoresist is of the type called negative, the reaction strengthens the polymer. The wafer is rinsed in a developing solution that removes either the exposed areas or the unexposed areas of photoresist, leaving a pattern of bare and photoresist-coated oxide on the wafer surface. The photoresist pattern is either the positive or the negative image of the pattern on the photomask.

The wafer is then placed in a solution

Reprinted with permission from *Scientific American Journal*, J. Angell, S. Terry, and P. Barth, "Silicon Micromechanical Devices," Vol. 248, pp. 44-55, April, 1983. © Scientific American, Inc./W. H. Freeman and Co.

SILICON ACCELEROMETER developed by Lynn M. Roylance of Stanford University is a miniature instrument that exploits both the mechanical and the electrical properties of silicon. It is made by selective chemical etching of a "wafer" cut from a large single crystal of silicon. The accelerometer, shown here in a scanning electron micrograph, is essentially a mass of silicon with a thickness equal to that of the wafer, suspended at the end of a thin silicon beam. An electrical resistor is fabricated in the underside of the beam; its resistance changes when the beam flexes, yielding a measure of acceleration. The mass is freed from the surrounding material by etching completely through the silicon at the periphery of the mass. The beam is defined by etching away all but a thin layer of silicon in that part of the wafer. The packaged device covers an area of two millimeters by three millimeters and is .6 millimeter thick, small enough to be sutured to the heart to measure its acceleration. Hundreds of accelerometers can be fabricated simultaneously on one silicon wafer.

of hydrolfluoric acid, which attacks the oxide but not the photoresist of the underlying silicon. The photoresist protects the oxide areas it covers. Once the exposed oxide has been etched away the remaining photoresist can be stripped off with hot sulfuric acid, which attacks the photoresist but not the oxide or the silicon. The final result is a pattern of oxide on the wafer surface that duplicates the photoresist pattern and is therefore either a positive or a negative copy of the pattern on the photomask. The oxide pattern itself serves as a mask in subsequent processing steps.

At this point micromachining diverges from microelectronic fabrication. In processing an electronic device the oxide pattern serves as a mask during the "doping" of the wafer with impurities such as boron or phosphorus. When the wafer is heated, the impurity atoms deposited on its surface diffuse a short distance into the silicon underlying the openings in the oxide, creating the shallow conductive or resistive regions of which microelectronic devices are composed. In micromachining the oxide is used as a mask during chemical etching. The etchants attack the silicon underlying openings in the oxide layer, excavating deep three-dimensional pits in the wafer.

Two types of etchants are employed in the micromachining of silicon. Isotropic etchants etch the silicon crystal at the same rate in all directions and create gently rounded shapes. Anisotropic etchants, which are also known as orientation-dependent or crystallographic etchants, etch at different rates in different directions in the crystal lattice; they can form well-defined shapes with sharp edges and corners. The most useful isotropic etchants are mixtures of hydrofluoric, nitric and acetic acids (HNA etchants). The anisotropic etchants are all hot alkaline solutions such as aqueous potassium hydroxide (KOH), aqueous sodium hydroxide (NaOH) and a mixture of ethylenediamine, pyrocatechol and water known as EDP.

Micromachining makes more extensive use of the anisotropic etchants than of the isotropic ones. A typical micromachining operation, the etching of an array of holes completely through the wafer, illustrates why. An isotropic etchant moves both downward and outward from an opening in the oxide, undercutting the oxide mask and enlarging the etched pit while deepening it. The pit created is at least twice as wide as it is deep, so that the openings in the oxide must be placed at least as far apart (edge to edge) as the wafer is thick if separate holes are to be etched through the wafer. On the other hand, an anisotropic etchant, if it is properly employed, does not undercut an oxide mask and creates pits with well-defined side walls. The pits deepen without widening. Thus an anisotropic etchant can create closely spaced arrays of holes; the edges of the openings in the oxide mask can be placed as close together as the limits of photolithography allow.

The shape of a hole formed by an anisotropic etchant depends on the orientation of the atomic planes in the silicon wafer. Just as various one-dimensional rows of corn stalks can be identified at different angles in a two-dimensional cornfield, so various two-dimensional planes of atoms lie at different angles in a three-dimensional crystal. The planes of atoms in the crystal are identified by means of a coordinate system assigned to the crystal.

Each silicon atom bonds with four neighboring atoms. The bonding structure is tetrahedral, that is, each atom lies at the center of a tetrahedron defined by the four atoms with which it is bonded. The bonding structure is an inconvenient basis for a coordinate system because the atomic bonds are not at right angles to one another. It is more convenient to think of the silicon crystal as being composed of stacked layers of repeating cubes. Each cube has an atom at each corner and at the center of each face and interlocks with four neighboring cubes. This interlocking face-centered cubic structure can serve as the basis of a coordinate system. Any atom can be designated as the point of origin of a coordinate system whose axes lie along the edges of the cube to which the atom belongs. The basic unit of measure is the length of the edge of one cube.

A vector, or direction in the crystal, is described by three coordinates known as the Miller indexes. For example, a [110] vector points diagonally across the face of a unit cube. The coordinates of a given vector also designate the entire set of atomic planes perpendicular to that direction. Thus the notation (100) de-

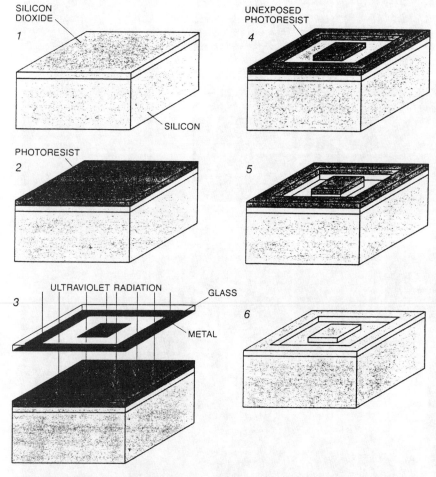

OXIDE PATTERNS on the wafer surface protect selected areas during the wet chemical etching that shapes a micromechanical device. First, a layer of silicon dioxide (SiO_2) is grown on the wafer surface; silicon reacts with steam at temperatures between 800 and 1,200 degrees Celsius to form the oxide. The oxidized wafer is coated with a polymer sensitive to ultraviolet radiation, called a photoresist. Next the wafer is placed under a master mask (a metal pattern on a glass plate) through which it is exposed to radiation. Photoresist in exposed areas weakens and is rinsed away in a developing solution. Then the wafer is etched in an acid that attacks the uncovered oxide but not the remaining photoresist or the silicon. The result is a pattern of openings in the oxide that duplicates the metal pattern on the glass plate. In subsequent processing steps the oxide pattern serves as a mask for the etching of the underlying silicon.

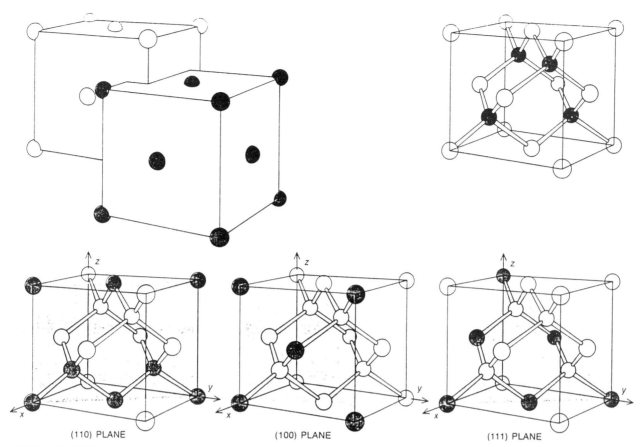

(110) PLANE   (100) PLANE   (111) PLANE

CRYSTALLINE SILICON shares with diamond the crystal structure called interlocking face-centered cubic. An atom lies at each corner and in the center of each face of the unit cube, and the cubes interlock in such a way that several atoms from neighboring cubes lie within each unit cube. The axes of the unit cube provide a rectilinear coordinate system that makes it possible to specify directions and planes within the crystal. A crystalline direction is designated by three coordinates called the Miller indexes, which are integer multiples of the length of one edge of a unit cube. The same set of indexes designates the planes perpendicular to the direction. Crystalline orientation is important in the fabrication of micromechanical devices because some of the etchants employed attack different directions in the crystal at different rates. Most such anisotropic etchants progress rapidly in the crystal direction perpendicular to the (110) plane and less rapidly in the direction perpendicular to the (100) plane. The direction perpendicular to the (111) plane etches very slowly, if at all.

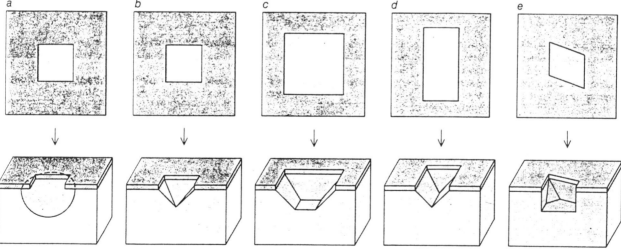

ANISOTROPIC ETCHANTS create faceted holes composed of the crystal planes that are etched slowest, unlike isotropic etchants, which invariably produce a gently rounded hole (a). The shape of an anisotropically etched hole is determined by the crystalline orientation of the wafer surface, the shape and orientation of the openings in the mask on that surface and the orientation dependence of the etchant itself. A square opening oriented along the <110> directions of a <100> wafer yields a pyramidal pit with {111} side walls (b). If the opening is larger, the point of intersection of the {111} planes is deeper, and a flat-bottomed pit can be created by stopping the etch before that depth is reached (c). A rectangular opening on the same wafer gives a V-shaped groove (d). Holes with parallel side walls can be created by etching a wafer that has a different surface orientation. In a <110> wafer two sets of {111} planes are perpendicular to the surface although not to each other. If an oxide opening on a <110> wafer is properly oriented, etching creates a hole with vertical side walls. The side walls that intersect at acute angles are bridged by still other planes. These are just a few of the many shapes that can be etched.

scribes all the planes perpendicular to the *x* axis. Because the crystal structure is symmetrical, the *x*, *y* and *z* directions are interchangeable, and a generalized notation can be introduced to describe all equivalent directions and sets of planes. The notation <110> designates the diagonals across any face of a unit cube, and the notation {110} designates the sets of planes perpendicular to all <110> vectors. The different brackets and parentheses distinguish planes from directions and distinguish generalized planes and directions from particular planes and directions.

The mechanisms that underlie the dependence of the etch rate of an anisotropic etchant on crystal direction are not well understood. The differences in etch rate depend on temperature and on the composition of the etchant, and they also seem to be related both to the density of atomic bonds on an exposed silicon plane and to the radius of the hydrated hydroxyl ions in the etching solution.

Experience has shown, however, that exposed {111} planes are etched slowly, if at all, by the anisotropic etchants. In contrast, {110} planes are usually etched rapidly, and {100} planes are etched at an intermediate rate. Simple etched holes may have only {111} and {100} facets. Other higher-order planes, such as {221} planes, are also etched at an intermediate rate and may appear as facets of complex etched shapes.

The type of hole an anisotropic etchant forms in a wafer is determined by the crystal orientation of the wafer surface as well as by the orientation dependence of the etchant. The two types of silicon wafers most useful for anisotropic etching are <100> wafers (wafers in which {100} planes are parallel to the surface and a <100> direction is perpendicular to it) and <110> wafers (in which {110} planes are parallel to the surface and a <110> direction is perpendicular to it).

The orientation, size and shape of the oxide opening on the wafer surface also play a part in determining the type of hole formed, as an example illustrates. If a square hole is opened in an oxide layer on a <100> wafer, treatment with anisotropic etchant can create an upside-down pyramidal pit. Four sets of {111} planes intersect the surface of a <100> wafer along the two perpendicular <110> directions that lie in the surface plane. During anisotropic etching the exposed {100} surface plane is etched downward at a constant rate, giving the pit a flat bottom at the start of the etch process. At the edges of the hole four inward-slanting {111} planes (at angles of 55 degrees to the surface) form the walls of the pit. As etching continues, more of the {111} planes is exposed and the area of the flat bottom shrinks. Eventually the {111} planes meet at a point and the flat bottom is gone. The etching then stops because no readily attacked planes are exposed.

The square oxide opening must be precisely aligned (within one or two degrees) with the <110> directions on the wafer surface to obtain pyramidal pits that conform exactly to the oxide mask rather than undercutting it. Most silicon wafers have a flat edge parallel to a <110> direction in the wafer. The square mask pattern is lined up with this flat edge during the photolithography step that precedes the etching.

The size of the oxide pattern determines not only the area of the pit on the wafer surface but also the depth of the pit. The larger the square oxide opening, the deeper the point at which the {111} side walls of the pit intersect. If the oxide opening is large enough, the {111} planes do not intersect within the wafer. The etched pit therefore extends all the way through the wafer, creating a small square opening on the bottom surface.

The influence of the shape of the oxide opening on the geometry of the etched pit is straightforward. If the opening in the oxide on a <100> wafer

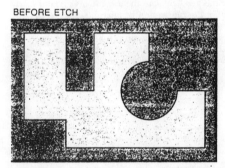  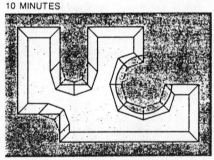 

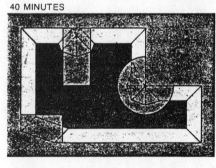  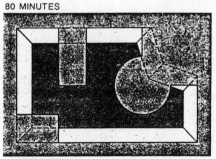 

OXIDE CANTILEVER can be formed by undercutting an oxide etching mask. Convex oxide shapes are undercut by an anisotropic etchant, although properly oriented concave shapes are not. The rectangular oxide opening shown here is a concave shape; the oxide tabs protruding into the opening are convex shapes. If the downward progress of the etchant is limited by an etch-stop layer, a prolonged anisotropic etch ultimately yields an oxide tab overhanging a shallow rectangular pit with {111} side walls and a flat bottom, namely the etch-stop layer. Cantilevers like the one formed when the oxide shape at the upper left is completely undercut have been used to make accelerometers, arrays of electrostatically controlled switches and arrays of mirrors. Undercutting makes it difficult to etch a hole into which a right-angle silicon corner projects. The silicon under the right-angle oxide shape at the lower left is rounded off during etching. A roughly right-angle silicon corner is formed under the circular oxide shape at the upper right at one point during the etch. The corner is eroded as the etch continues, however. Timed etches of patterns with circular tabs can be employed to make holes with projecting right-angle corners.

is rectangular rather than square, etching creates a long trench rather than an inverted pyramid. The walls and the ends of the trench are {111} planes, and if the etch goes to completion, the trench has a *V*-shaped cross section.

Pyramidal pits, *V*-shaped grooves and other faceted pits and holes are one kind of structural element of which micromechanical devices are composed. A second structural unit is a thin membrane of silicon. A membrane can be formed by etching a wafer for a period just short of what it would take to etch through it. It is difficult to make diaphragms of uniform thickness by this method, however, because the thickness of the diaphragm is determined by that of the wafer, and wafer thickness typically varies by 10 micrometers or more both across a single wafer and from one wafer to another.

Membranes of more precisely defined thickness and shape can be created by a technique that exploits another property of anisotropic etchants. The rate at which the solutions etch a wafer depends not only on crystal orientation but also on the extent to which the silicon has been doped with impurity atoms such as boron, phosphorus, arsenic or antimony.

The etch rate of a doping-dependent etchant (a category that overlaps the categories of isotropic and anisotropic etchants) depends on the type of dopant atoms and their concentration. The isotropic etchant HNA is doping-dependent in some mixture ratios: it etches heavily doped silicon much faster than it etches lightly doped silicon. On the other hand, the anisotropic etchants potassium hydroxide and EDP etch silicon that has been heavily doped with boron much more slowly than they etch silicon lightly doped with boron.

The first step in creating a membrane by anisotropic etching is to create an etch-stop layer of silicon heavily doped with boron on the surface of the wafer. A layer of some substance rich in boron is deposited on the front surface of the wafer, which is then heated in a furnace to between 1,000 and 1,200 degrees C. At these temperatures the boron atoms diffuse into the silicon. The depth of the doped layer is controlled by the temperature and the duration of the diffusion step. If the wafer is then etched from the back with an anisotropic etchant, the etchant eats through to the doped layer and stops, creating a membrane as thick as the doped layer on the surface of the wafer.

If the designer wants to put a membrane some distance below the wafer surface, he dopes the surface in much the same way and then grows a layer of silicon over the doped layer by heating the wafer in an atmosphere of silane gas (SiH_4). The gas decomposes at high tem-

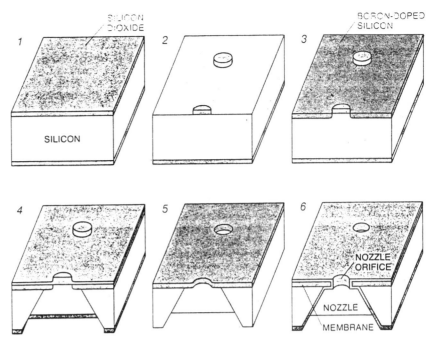

SILICON INK NOZZLE in one design for a high-speed ink-jet printer is a pyramidal pit whose bottom is a thin membrane through which an orifice of precise diameter has been etched. The diameter of the orifice is determined by the first three processing steps. The wafer is oxidized and dots of photoresist are defined on the top surface. The oxide not protected by photoresist is etched away, leaving a pattern of oxide dots. When boron is diffused into the silicon, it creates a layer of boron-doped silicon everywhere except under the dots. Then a square opening is defined in the oxide on the bottom of the wafer and is etched with an anisotropic etchant. The etchant creates a pyramidal pit with {111} side walls but stops when it comes to the doped layer. It eats through the undoped dots in the layer, however, so that when all the oxide is removed, the nozzle orifices penetrate the wafer. Finally, the wafer is reoxidized to give all surfaces a protective coating. Several arrays of nozzles are fabricated simultaneously on one wafer.

perature and deposits silicon on the wafer surface. The new layer, which is called an epitaxial layer, assumes the crystal structure and orientation of the silicon on which it is deposited. The epitaxial layer is typically from five to 20 micrometers thick.

If the wafer is then etched from both sides in an anisotropic etchant, a membrane is created at a depth equal to the thickness of the epitaxial layer. The thickness of the membrane itself is again determined by the thickness of the boron-doped layer.

A third example of a micromechanical structure is the cantilever beam, a thin beam of silicon dioxide supported at only one end. In many applications a cantilever is suspended over a shallow pit in the silicon. The techniques for making a cantilever and a pit were developed by a group working with Kurt E. Petersen at the San Jose Research Laboratory of the International Business Machines Corporation. The techniques exploit two properties of anisotropic etchants: their doping dependence and their tendency to undercut convex shapes in an oxide mask.

The simple geometric oxide openings discussed so far are concave oxide shapes, that is, the line between the surrounding oxide and the opening is always concave with respect to the opening. A concave oxide shape is not undercut by an anisotropic etchant if the opening is properly oriented on the wafer surface. If a finger or a tab of oxide projects from one side of such an opening, however, its boundary is convex with respect to the exposed silicon. Such convex oxide shapes are undercut by an anisotropic etchant.

The first step in making a cantilever over a shallow pit is the formation of a boron etch-stop layer. A layer of epitaxial silicon is laid down over the boron-doped stratum, oxide is grown over the epitaxial silicon and an opening is made in the oxide. The wafer is then etched in an anisotropic etchant. If an oxide tab protrudes into the oxide opening, the etchant begins to undercut it. At the attached end of the oxide tab the etchant eventually runs into a {111} plane and stops etching. Since the etch-stop layer limits the downward progress of the etch, the final result is a cantilever of oxide extending over a shallow depression in the substrate.

In the Integrated Circuits Laboratory at Stanford University work in micromachining began in 1965 when Kensall D. Wise (now at the University of Michigan) and one of us (Angell) developed a closely spaced array of electrodes for recording electric potentials in the brain. The project led to the development of a variety of brain probes. Since then projects in micromachining at Stanford have included pressure sensors, accelerometers, gas chromatographs, miniature thermometer arrays and structures for the dielectric isolation of microelectronic circuits. Micromechanical techniques and devices are also being developed at several other universities and industrial laboratories. The devices we shall describe below include some of our own and some developed by others.

Arrays of nozzles fabricated in silicon have found application in ink-jet printers for digital computers. In one type of ink-jet printer ink is forced through a linear array of nozzles, creating several parallel streams that are broken into droplets as they leave the nozzles. As droplets in each stream pass single file through a charging electrode, each droplet is either given an electric charge or allowed to remain electrically neutral. An uncharged droplet passes undeflected through an electrostatic field and leaves a dot on the paper, but a charged droplet is deflected by the field and instead of hitting the paper is intercepted by a collection gutter. A printed character is formed by selecting droplets from the vertical array of nozzles as the entire mechanism moves horizontally across the paper.

Ernest Bassous, Larry Kuhn and their colleagues at the IBM Thomas J. Watson Research Center did the pioneering work in micromechanical ink-jet nozzle arrays, developing many different nozzle systems. The simplest way to make a nozzle is to etch a pyramidal pit in the silicon. If the intersection point of the {111} planes that form the walls of the

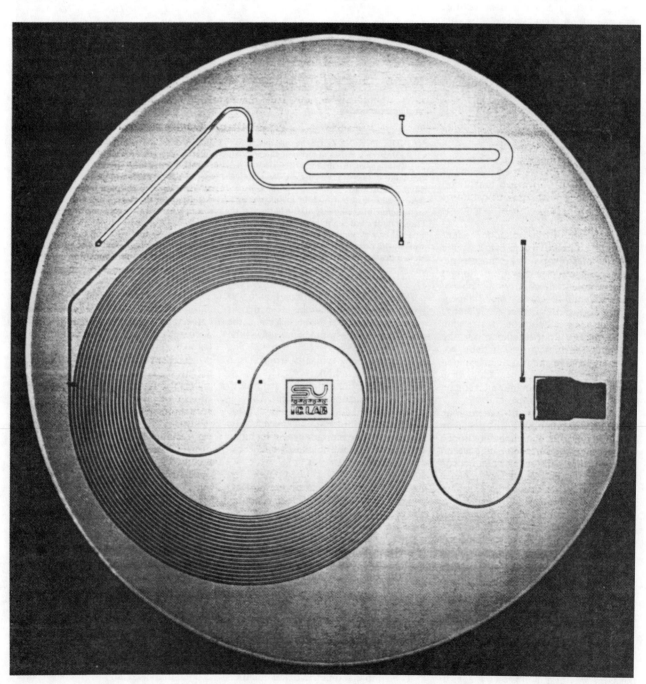

SILICON GAS CHROMATOGRAPH is a complete analytic instrument built on a wafer five centimeters across. Capillary column of the chromatograph is a spiral 1.5 meters long. The column is formed by etching a groove in a wafer and bonding the wafer to glass. This prototype was developed at Stanford University with funds provided by the National Institute for Occupational Safety and Health.

pit is below the bottom surface of the wafer, a small square opening is created. It is difficult to make nozzle orifices of uniform size with this technique, however, in part because the dimensions of the orifice depend on the thickness of the wafer; as noted above, the thickness is hard to control.

One nozzle design solves the problem by putting a membrane three micrometers thick at the bottom of each nozzle cavity. A pyramidal pit, which serves as the cavity, is etched down to the membrane, but the nozzle orifice is defined by a boron-diffusion procedure, not by the geometry of the pit. By this method Bassous and his co-workers have built arrays of eight nozzles with circular orifices exactly 20 micrometers in diameter spaced exactly .3 millimeter apart.

The first step in making the array is to define the nozzle orifices as dots of photoresist on the front surface of a wafer coated with oxide on both sides. All the oxide on the front surface is etched away except for the dots protected by photoresist. Then the wafer surface is doped with boron, which is driven into the silicon to a depth of three micrometers by heating the wafers in a diffusion furnace. Only the silicon under the oxide dots remains undoped.

Next, photolithography and an etching step are used to define a pattern of square openings in the oxide on the back surface of the wafer. The openings are centered under the undoped dots on the front surface. The wafer is etched in a doping-dependent anisotropic etchant, which eats through 99 percent of the wafer's thickness but stops when it comes to the heavily doped boron layer. It etches all the way through the undoped dots, however, creating the nozzle orifices. The diameter of each orifice, the thickness of the surrounding membrane and the placement of the orifices with respect to one another can be controlled with micrometer accuracy because they are defined not by the wafer thickness or the length of the etching step but by photolithography and the boron-diffusion step.

The most elaborate of the micromechanical devices made so far is a gas chromatograph the size of a box of kitchen matches. Two of us (Terry and Angell) built a prototype of the device in 1975. It is now being developed as a commercial product by Microsensor Technology, Inc.

Gas chromatography is an analytic technique for separating, identifying and measuring the quantity of each gas in a mixture. A sample of the mixture is injected by a valve into a long capillary column, through which it is flushed by an inert carrier gas, in most cases helium. The walls of the capillary column are lined with a thin layer of a material such as a silicone oil or a polymer in which different gases have different degrees of solubility. As the component gases pass through the column they are repeatedly adsorbed and desorbed in the lining. Because the time a component gas remains adsorbed depends on its solubility, each gas travels through the column at a different rate and emerges at a specific time. The output stream is passed over a detector that measures a property of the gas such as its thermal conductivity. The output signal generated by the detector is a series of bell-shaped peaks corresponding to the sample gases, separated by flat regions corresponding to the inert carrier gas. Many modern gas chromatographs have an associated microprocessor that identifies each gas by matching its retention time with known retention times and measures the quantity of each gas by calculating the area under the output peak.

Like a conventional gas chromatograph, the miniature one consists of a capillary column, a sample-injection valve and a thermal-conductivity detector. All the parts of the miniature gas chromatograph, however, fit on a single wafer five centimeters in diameter. The volume of the capillary column is much smaller than that of a conventional capillary column. Because a chromatograph operates properly only if the volume of the injected sample gas is much smaller than the volume of the column, it was necessary to design a miniature sample-injection valve to accompany the miniature capillary column. Because the internal volume between the valve and the column and between the column and the detector must also be minimized, the valve and column are fabricated on the same wafer and the detector is a silicon chip mounted on the wafer.

The capillary column is a groove 1.5 meters long, wound into a spiral so that it fits on the wafer. A glass plate bonded to the wafer forms the top surface of the column. At the input end of the column a hole leads to the bottom surface of the

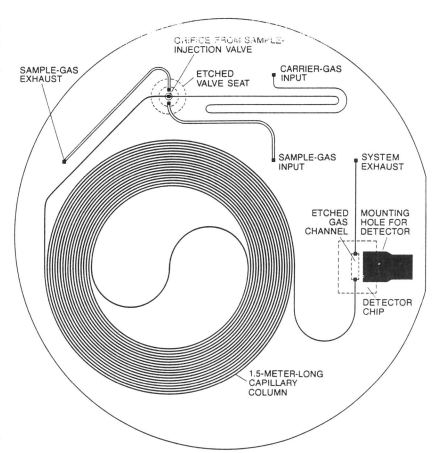

SEPARATION OF GASES in the silicon gas chromatograph is based on differences in the solubility of various gases in a liquid that lines the capillary column. An inert carrier gas flows continuously through the capillary-column channel. When a valve in the channel is opened, a pulse of the gas to be analyzed is fed into the column and flushed through it by the carrier gas. As the gases in the sample pass through the column they are repeatedly adsorbed and desorbed on a thin liquid lining. Each gas is identified by its retention time in the column. As each gas arrives in turn at the end of the column, it passes through a hole to a channel on the back of the wafer over which a thermal-conductivity detector is mounted. Sample gases have a lower thermal conductivity than the helium carrier gas has and cause voltage peaks in the detector output. The volume of each gas is determined from the area under the voltage peak it creates.

wafer. The helium carrier gas enters the column through this hole. A short distance away another hole in the column channel leads to a valve on the back surface of the wafer. The sample gas enters a separate channel through yet another hole, flows through the channel to the valve and is injected by the valve into the capillary column.

The valve seat on the back surface of the wafer consists of a silicon sealing ring, which surrounds both the input and the output orifices, and a silicon seating ring, which surrounds only the output orifice leading to the capillary column. A diaphragm of nickel and Teflon is clamped against the sealing ring. Normally a spring holds the plunger of a solenoid against the diaphragm, pushing it against the seating ring. When the solenoid is actuated, the plunger withdraws and the diaphragm relaxes, allowing gas to flow from the input orifice over the seating ring and into the output orifice. The effective dead volume of the valve is the volume of the capillary-column orifice, which is four nanoliters.

At the output end of the capillary column another etched hole leads to a gas channel etched into the bottom surface of the wafer. The chip on which the thermal-conductivity detector is built is inverted over this channel and clamped to the wafer. The detector is a thin-film metal resistor on a thermally isolating membrane of Pyrex glass in the middle of the chip. A constant electric current is passed through the resistor, and its resistance is monitored. Sample gases have a thermal conductivity lower than that of the carrier gas and remove less heat from the resistor, increasing its resistance and creating voltage peaks. The amplitude of each voltage peak is proportional to the quantity of one gas in the mixture.

The micromachining of the chromatograph wafer begins with the etching of the valve seat. The valve well, the sealing ring and the seating ring are defined by isotropic etches through concentric circular openings in the oxide layer on the back of the wafer. Then the feedthrough holes in the valve and at the ends of the capillary column and the sample-gas channels are formed by an anisotropic etch through square openings defined in the oxide on the front surface of the wafer. Finally the capillary column and the carrier-gas channels are defined by an isotropic etch through an oxide pattern on the front surface.

At this point the capillary column is a shallow spiral groove about 200 micrometers across at the surface of the wafer and 40 micrometers deep. The open spiral is made into an enclosed channel by stripping the oxide off the surface and bonding the wafer to a plate of Pyrex glass. The bonding technique, called anodic bonding, requires no intermediate liquid layer such as a glue or solder. A method of this kind is essential since

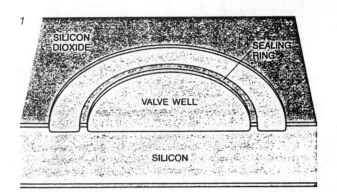

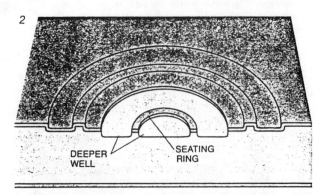

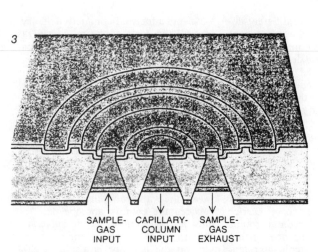

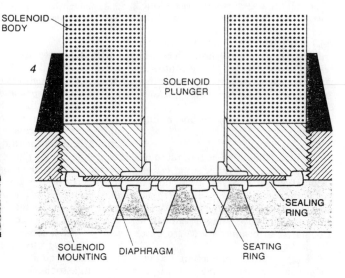

SAMPLE INJECTION VALVE for the miniature chromatograph is fabricated on the same wafer as the capillary column. Two isotropic etching steps define the valve sealing ring and the valve seat, and an anisotropic etch defines the feedthrough holes within the valve well. The first isotropic etch creates a knife-edge sealing ring of silicon raised 10 micrometers above the floor of a well. In the second step a smaller concentric well with another ring—the valve seat—is etched 10 micrometers into the floor of the first well. In the third step the feedthrough holes are formed by an anisotropic etch through square openings in the oxide on the bottom surface of the wafer. The hole within the valve seat opens to the capillary column; the other two holes open to the sample-gas channels. A metal-and-Teflon diaphragm and a solenoid plunger complete the valve. Normally the diaphragm presses against the valve seat, preventing the sample gas from flowing into the column. When the solenoid is actuated, the diaphragm relaxes, allowing pressurized sample gas to flow over the valve seat.

the shallow capillary column might be blocked by extruded glue or solder.

Anodic bonding, which is sometimes called electrostatic bonding or the Mallory process, was developed by George Wallis and Daniel I. Pomerantz of P. R. Mallory & Co. The glass plate and the silicon wafer are placed in contact and heated to about 400 degrees C. A large negative voltage is applied to the top surface of the glass and the silicon is electrically grounded. The glass contains a small amount of sodium; at this temperature the sodium ionizes and becomes mobile. The positive sodium ions move toward the negative electrode, leaving behind bound negative charges in the glass near the gap between the glass and the silicon. As a result the voltage across the gap is nearly equal to the applied voltage. Electrostatic attraction pulls the glass and the silicon into intimate contact and they fuse together, creating a tight seal.

The thermal-conductivity detector developed by John H. Jerman of Stanford for the gas chromatograph is a chip fabricated on a separate wafer. The wafer is oxidized and covered with a layer of Pyrex glass deposited by the technique called sputtering. Thin-film metal resistors are deposited on top of the glass. Then an anisotropic etch through square openings in the oxide on the back of the wafer removes the entire thickness of silicon, leaving a membrane of thermally isolating glass under each set of resistors. The wafer is sawed into chips, to which wires are attached, and the front surface of a chip is clamped over the gas channel at the end of the capillary column.

Because the capillary column has such a small volume, the miniature gas chromatograph requires much smaller amounts of carrier gas than a conventional chromatograph. This and the small size of the unit itself make it possible to build portable chromatography instruments. Indeed, a portable instrument consisting of five miniature chromatographs and a microcomputer is being developed by Microsensor Technology. If the five capillary columns are lined with different materials, the instrument can be made sensitive to about 100 gases. Such an instrument can monitor air quality or measure the energy content of the gas in a natural-gas pipeline.

Micromachining has recently been used to fabricate a high-performance heat sink in a silicon wafer. The heat sink is designed to improve the performance of very-high-speed integrated circuits. In general the power dissipation of an electronic device increases along with its speed, and the heat evolved by the fastest circuits would disrupt their own operation if they were not adequately cooled. Even with special packaging schemes for high-speed circuits,

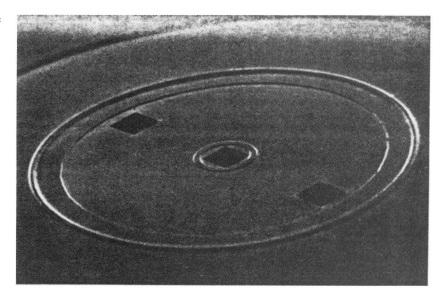

VALVE WELL shown in this scanning electron micrograph is four millimeters in diameter and surrounds two concentric silicon rings: the valve seal (outer complete ring) and the valve seat (inner complete ring). The ring walls are 50 micrometers thick. The orifice in the center leads to the capillary column and the other two openings lead to the input and exhaust channels.

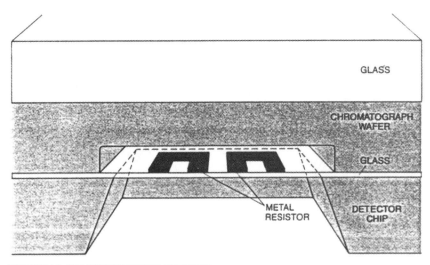

THERMAL-CONDUCTIVITY DETECTOR for the chromatograph is a resistor defined by etching a thin film of metal deposited on a wafer that has been coated with glass. After the resistor is formed the wafer is etched anisotropically from the back, leaving only a thin membrane of glass under the resistor. Because glass is a poor thermal conductor (whereas silicon is a good one), the membrane thermally isolates the resistor. A change in the thermal conductivity of the gas stream flowing over the resistor alters its temperature and hence its electrical resistance.

the upper limit on power dissipation is about 20 watts per square centimeter of chip.

David B. Tuckerman and Fabian Pease of the Stanford Integrated Circuits Laboratory have developed a heat sink that can handle much higher power densities. The idea is to fabricate the high-speed electronic devices on the front surface of a wafer and to cut closely spaced rectangular grooves in the back of the wafer. A liquid coolant is forced through these channels. Silicon itself has very high thermal conductivity, and Tuckerman and Pease have fabricated devices in which the wall thickness, channel width and channel depth give the highest cooling efficiency.

The rectangular grooves are made in much the same way as the V-shaped grooves described above. An anisotropic etch through an oxide pattern consisting of narrow parallel lines creates V grooves if the wafer, like those discussed so far, is a $<100>$ wafer. If the wafer is a $<110>$ wafer, on the other hand, an anisotropic etch through the same oxide pattern can create grooves with vertical side walls and a roughly rectangular cross section. The rectangular grooves are preferable for a heat sink because they can be spaced more closely than V grooves and they expose more surface area to the circulating coolant. After the grooves are etched the wafer is anodically bonded to glass to create enclosed microchannels. A heat sink with etched channels 300 micrometers deep and 50 micrometers wide, separated by silicon fins 100 micrometers wide, cooled by the forced circulation of water, can handle power dissipations of more than 1,000 watts per square centimeter.

The etched heat sink can be constructed only on silicon and only on wafers of a particular orientation. Tuckerman and Pease have also investigated the possibility of creating heat sinks by sawing grooves rather than etching them. A rotary saw blade with a thin metal rim coated with diamond grit can cut channels from 50 to 200 micrometers wide on the back of a wafer. Although the dimensions of the channels cannot be controlled as precisely with sawing as they can be with etching, sawing does make it possible to create heat sinks on silicon wafers of any orientation and on wafers of other materials, such as the compound semiconductor gallium arsenide and sapphire, a substrate for some types of integrated circuit. In addition the wafers can be sawed in a crosshatch pattern, so that the circuits are supported by micropillars rather than microfins and the heat sink is less likely to be clogged by particles in the coolant.

Accelerometers and pressure sensors are two examples of integrated silicon sensors. The smallest accelerometer fabricated up to now is an oxide-cantilever device with an on-chip amplifier. The accelerometer, designed by Petersen, Anne C. Shartel and Norman F. Raley of the IBM San Jose Research Laboratory, consists of an oxide beam suspended over a shallow well formed by the boron etch-stop technique. A metal layer is deposited on the top surface of the oxide cantilever. The metal layer and the flat silicon on the bottom of the well act as two plates of a variable air-gap capacitor. A lump of gold is formed on the free end of the beam by plating. If the silicon chip is moved suddenly, the inertia of the gold weight causes the beam to flex, changing the air gap and hence the capacitance.

The output of the sensor is a voltage proportional to acceleration. The instrument has a sensitivity of two millivolts per g (the unit of acceleration equal to the acceleration of gravity at the earth's surface). It should be noted that the amplifier is an indispensable part of the device. The change in capacitance being detected is very small and would be swamped by larger capacitance changes in any cables attached to the sensor if the signal were not amplified on the chip. This is true in general of small capacitive sensors; signal conditioning of some kind must precede transmission.

The most successful silicon sensors at present are pressure sensors, some of which include integrated electronic devices. There are two basic types of silicon pressure sensors: piezoresistive and capacitive sensors. The electrical resistors on the thin flexible diaphragm of a piezoresistive sensor change resistance when the diaphragm flexes. In a capacitive sensor the flexible diaphragm is one plate of a variable air-gap capacitor.

In most cases the diaphragm of a pressure sensor is anisotropically etched in a $<100>$ silicon wafer. A simple timed etch can be used to make diaphragms from 10 to 50 micrometers thick. Thin-

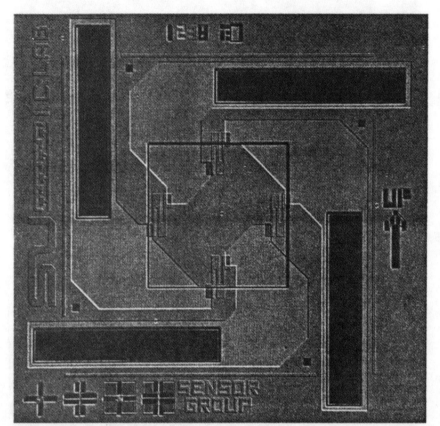

SILICON PRESSURE SENSOR is shown in the middle of its fabrication sequence. The recessed square in the center of the photomicrograph is a thin pressure-sensing diaphragm .5 millimeter on a side. Purple serpentine elements on the diaphragm are resistors made by diffusing boron into the wafer. Green regions, which act as conductors between the resistors, have a higher boron concentration. Black rectangles near the edge of the sensor are incomplete bonding pads. They are deep V-shaped grooves that will be coated with gold, to which lead wires can be attached. When the diaphragm flexes, the resistance of the purple areas changes as a result of the mechanical stress, an effect called piezoresistance. The resistors are positioned so that the resistance of two of them increases and that of the other two decreases; the two-legged resistor configuration is twice as sensitive to pressure changes as a single resistor would be.

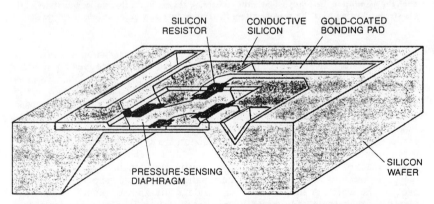

CROSS SECTION of the pressure sensor exaggerates the thickness of the diaphragm in order to show detail. The diaphragm is defined by a shallow anisotropic etch from the top of the wafer and a deep etch from the bottom. When the sensor is complete, the top surface is bonded to a glass plate, creating a sealed cavity. The bottom of the diaphragm is exposed to outside pressure. Since the pressure in the cavity is known, the output is a measure of absolute pressure.

ner diaphragms must be made from etch-stop layers.

The resistors in a piezoresistive sensor are formed by doping some areas of the diaphragm. Two types of dopants are available. Elements to the right of silicon in the periodic table, such as phosphorus, have five valence electrons, whereas silicon has four. The extra electron can become a carrier of electric current. Elements to the left of silicon in the periodic table, such as boron, have three valence electrons. The absence of an electron (a "hole") can also act as a current carrier, although holes are less mobile than electrons. Current will not flow from a phosphorus-doped region to a boron-doped region because of the difference in charge carriers. Thus if the silicon from which the wafer was made was lightly doped with phosphorus, a heavier concentration of boron can be diffused into selected areas of the wafer to make resistors that are electrically isolated from the rest of the diaphragm.

The piezoresistive effect is a change in electrical resistance resulting from mechanical stress; the magnitude of the effect depends on the orientation of the resistors. By fortunate coincidence boron-doped resistors in a <100> wafer are most sensitive to stress when they are aligned with the <110> edges of an anisotropically etched cavity.

A piezoresistive pressure sensor we developed has four resistors on a square diaphragm. The resistors are near the diaphragm edges (the areas of maximum stress) and are aligned with the <110> directions (the directions along which response to stress is greatest). The resistors are joined in the two-legged configuration known as a Wheatstone bridge. When the diaphragm flexes, the resistance of two resistors increases and that of the other two decreases by an equal amount. As a result the bridge is twice as sensitive to pressure changes as a single resistor would be.

The first pressure-sensor chip to include active circuitry (that is, circuitry including transistors that amplify the signal) was designed by John M. Borky and Kensall Wise of the University of Michigan. The circuitry converts the output voltage of a Wheatstone bridge into a frequency-modulated (FM) signal; in other words, changes in pressure are translated into changes in frequency. If a different range of frequencies is assigned to each of several sensors, they can share a transmission line in the same way different FM radio stations share the radio-frequency band.

Capacitive pressure sensors are inherently more sensitive to pressure changes than piezoresistive sensors, but as we pointed out above, the signal from a capacitive sensor can easily be lost in noise. Signal-conditioning circuitry improves the performance of a piezoresistive sensor, but it is crucial to that

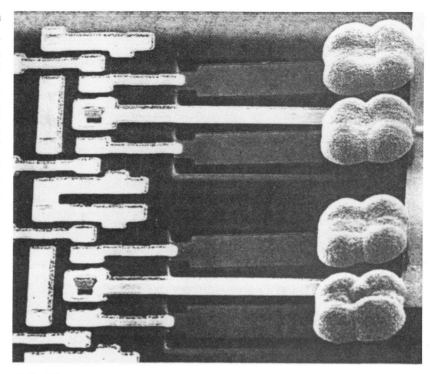

CAPACITIVE ACCELEROMETERS are cantilever beams of silicon dioxide overhanging shallow wells in the silicon. Four accelerometers are shown in this scanning electron micrograph. Bumps of gold at the right end of each cantilever increase the inertia of the beam and therefore the sensitivity of the device. A metal layer on top of the beam, an electrically conductive boron etch-stop layer at the bottom of the well and the air gap between them form a variable capacitor. Under acceleration the beam flexes, changing the capacitance. The circuitry at the left amplifies the signal; the sensor output is a voltage proportional to acceleration.

of a capacitive sensor. Craig S. Sander, James W. Knutti and James D. Meindl of the Stanford Integrated Circuits Laboratory designed the first capacitive pressure sensor with on-chip active circuitry. The diaphragm of this sensor, which is recessed about five micrometers into the silicon wafer, serves as one plate of a capacitor. The other plate is a metal film on a sheet of glass that is anodically bonded to the wafer.

A fixed quantity of gas is sealed in the cavity between the diaphragm and the metal-coated glass plate. The other side of the diaphragm is exposed to the pressure to be measured. When the pressure changes and the diaphragm flexes, the thickness of the insulating air gap between the metal and the silicon changes. The active circuitry on the chip monitors the resulting changes in capacitance, putting out a series of voltage pulses. The time between two pulses is proportional to the measured pressure. This pulse-time-coded signal, like the FM signal from the integrated piezoresistive pressure sensor, is less easily obscured by noise than a simple uncoded voltage would be.

Integrated sensors can meet the requirements of a wider variety of applications than discrete sensors because the signals are relatively immune to noise and leakage. Thus integrated sensors can be put in harsh environments, such as under the hood of an automobile, or in situations where the sensing device must be physically separated from the signal-processing electronics, as in biomedical monitoring. Many of the sensors now under development are designed for such difficult applications. For example, two of us (Barth and Angell) are working on sensors to be implanted in living tissue.

Also under development are chips with several types of sensors, such as a temperature and a pressure sensor, several copies of the same sensor (for reliability) and chips that include some means of sensor recalibration. In the future many integrated sensors will be able to convert analogue signals into digital signals so that they can be connected directly to digital electronics. In general the biggest challenge at present is to determine how much circuitry it is feasible and cost-effective to put on a sensor chip.

At the same time that a great deal of attention is being given to sensors, strictly mechanical silicon devices continue to be developed. Silicon etching techniques are still being refined, and new micromechanical structures will emerge from the laboratories over the next several years. Silicon mechanical devices may one day be nearly as commonplace as silicon electronic devices are now.

Anisotropic Etching of Silicon

KENNETH E. BEAN

Abstract-Anisotropic etching of silicon has become an important technology in silicon semiconductor processing during the past ten years. It will continue to gain stature and acceptance as standard processing technology in the next few years. Anisotropic etching of (100) orientation silicon is being widely used today and (110) orientation technology is emerging. This paper discusses both orientation-dependent and concentration-dependent etching of (100) and (110) silicon. Very exact process control steps may be designed into a process by use of (100) anisotropic and concentration-dependent etching. Also, methods of oxide or nitride pin hole detection in (100) silicon are presented. Mask alignments to obtain different etch front termination in both (100) and (110) silicon are shown. Very high packing density structures, less than 1 μm, are obtained in the (110) technology, and extremely high etching ratios of greater than 650 to 1 are obtained in (110) orientation-dependent etching. Some of the many applications for anisotropic and concentration-dependent etching are described.

ORIENTATION-DEPENDENT ETCHING

WET CHEMICAL ETCHING has been used in silicon semiconductor processing since its beginning in the early 1950's. Isotropic etches, i.e., etches that etch in all crystallographic direction at the same rate, consisting of hydrofluoric, nitric, and acetic acid (HF, HNO_3, and CH_3OOH) were used for etching and chemically polishing in the early work [1]-[4]. In the more recent past, people have become interested in anisotropic or orientation-dependent etches [5]-[7] (ODE). This paper will discuss primarily wet chemical, orientation-dependent, and concentration-dependent etching; however, vapor phase and plasma orientation-dependent etching are also of current interest and many of the same masking and alignment technologies apply. We will describe etch solutions, masking materials, mask alignments, and impurity concentration etch dependence for both (100) and (110) silicon orientations. Some applications for each orientation will be described.

Etches and Masks

Table I gives some of the etches used in both planar and orientation-dependent etching. Planar etch is an etch consisting of hydrofluoric, nitric, and acetic acid (HF $\sim$ 8 percent, $HNO_3 \sim$ 75 percent, $CH_3OOH \sim$ 17 percent). This solution etches silicon at about 5 μm/min at 25°C. The etch rate is approximately equal in all crystallographic directions as the name planar indicates. The ratios of the constituents of planar etch may be altered to change the etch rate drastically. For example "B" etch is a mixture consisting of 1 part HF, 40 parts HNO_3, and 15 parts CH_3OOH which etches (111) silicon very slowly, $\sim$0.15 μm/min, and (100) at $\sim$0.20 μm/min at 25°C (approximately $\frac{1}{25}$ the rate of planar).

Another etch made up of these same chemicals is Dash etch [3], [6], [8]. This etch is somewhat orientation-dependent and is used for crystal defect delineation or definition in (111) silicon. It etches p or n silicon in the ⟨100⟩ direction at $\sim$1300 Å/min, at 25°C, and at 46 Å/min in the ⟨111⟩ direction at this same temperature. However, this etch also etches heavily doped silicon (p^+- or n^+-type greater than 5×10^{18}) much faster ($\sim$2.5 μm) than lightly doped p- or n-type silicon. (See Table I.) Therefore, it becomes concentration-dependent. An even more orientation-dependent etch used for etching (100) silicon is 250-g potassium hydroxide, 200-g normal propanol, and 800-g deionized water [9]. This etch can be masked by photoresist for shallow 25°C etching or by use of SiO_2 or Si_3N_4 for deep (>20 μm) etching at 80°C. The etch rate in the ⟨100⟩ direction is at least one hundred times faster than in the ⟨111⟩ direction if the mask is properly aligned with the trace of the ⟨111⟩ plane at the (100) surface, see Figs. 15 and 16. This alignment is also parallel and perpendicular with the ⟨110⟩ direction or the flat on the (100) slice. This etch attacks oxide at $\sim$28 Å/min, however, Si_3N_4 is not measurably attacked even after 5-hours etching at 80°C.

Another orientation-dependent etch that may be used for ⟨100⟩ direction etching to advantage is ethylenediamine 255 cm^3, pyrocatechol 45 g, and water 120 cm^3, hereafter referred to as EDA [7], [9]-[11], [21]. This solution etches ⟨100⟩ silicon at 1.1 μm/min at 100°C and SiO_2 at $\sim$ 8 Å/min. The slow etch rate of SiO_2 and the non-sodium or potassium nature of this etch makes it desirable for polysilicon etching in MOS-type device processing. This etch is also concentration and/or type sensitive and stops or greatly reduces in etch rate at a p+ interface or junction. This etch may also be used for ⟨110⟩ direction etching, hydrazine may also be added to this

TABLE I
CHEMICAL ETCHING OF SILICON

ETCHES	CHARACTERISTICS	COMPOSITION	RATE AND REMARKS
PLANAR ETCH	ETCH UNIFORMITY	HF - HNO_3 - HAc $\sim$ 8% $\sim$75% $\sim$17%	$\sim$5 μ/MIN ⟨111⟩ RT
1-3-10	ETCHES P^+ OR N^+ SILICON "STOPS" AT P^- OR N^-	HF HNO_3 HAc 1 3 10	$\sim$3 μ/MIN ⟨100⟩ RT
(100) ODE	ETCHES [100] $\sim$100 X [111] DIRECTION	KOH - NORMAL PROPANOL H_2O	$\sim$1 μ/MIN AT 80°C IN ⟨100⟩ SILICON STOPS AT P^{++} INTERFACE
(110) ODE	ETCHES [110] 600 X [111] DIRECTION	KOH - H_2O	$\sim$.8 μ/MIN AT 80°C IN ⟨110⟩ SILICON
ETHYLENEDI-AMINE	ORIENTATION DEPENDENT AND CONCENTRATION DEPENDENT	ETHYLENEDI-AMINE - CATECHOL - H_2O (HYDRAZINE)	$\sim$1.1 μ/MIN AT 100°C IN [100]. STOPS ETCHING AT P^{++} INTERFACE. VERY SLOW ETCHING OF SiO_2

solution for faster etching. Another etch solution which is very orientation-dependent in the ⟨110⟩ direction etching is potassium hydroxide and water (KOH ~35 percent by weight, and H_2O) at 80°C. This solution etches silicon six hundred times faster in the ⟨110⟩ direction than in the ⟨111⟩ direction when the mask is properly aligned with the trace of the ⟨111⟩ plane at the (110) surface or 35.26° off from the (110) flat. (Not parallel and perpendicular to the (110) flat or direction in this case [20].) The ⟨110⟩ silicon etch rate is ~0.8 μm/min and the SiO_2 etch rate is ~30 Å/min.

Silicon belongs to the diamond cubic crystal structure. In the cubic structure the crystalographic directions are perpendicular to the crystal planes. There are several low-index planes that we may choose to work on in this diamond cubic structure. Fig. 1 shows examples of some of these planes and their position in the cubic structure. Numbers one, three, and seven are the three index planes most commonly used for silicon device processing. Fig. 2 shows one of each of the low-index planes in the cube, that is, the (100), the (110), and the (111). One of the important parameters in orientation-dependent etching of silicon is the atomic lattice packing density and available bonds in the crystallographic plane. In Fig. 3 we show photographs of the silicon diamond cubic structure model taken from three different crystal directions. The ⟨111⟩ direction shows a very high atomic packing density in the (111) plane. The ⟨100⟩ direction photo is taken of the same model but moved around 54.75° from the ⟨111⟩ direction, so we are now looking at a (100) plane. It is readily evident that the atomic packing density is considerably less dense. The ⟨110⟩ direction photo is taken by moving around 90° from the original ⟨111⟩ direction. The atomic packing density is very low in this plane when compared to the {111} plane. From this information one would expect that the etch rate, or epitaxial deposition growth rate, would be considerably faster in the ⟨110⟩ direction than in the ⟨100⟩ direction and even more so than in the ⟨111⟩ direction.

When we examine the crystal projection of the silicon (100) plane, Fig. 4, we see that the {100} surface has four-fold symmetry. That is, the high atomic packing density {111} planes are equal distance out from the {100} surface plane at equal angles and they are 90° to each other. Also we can determine that the {111} planes are intersecting the {100} plane at a rather steep angle, which is 54.74°, and that they are in ⟨110⟩ directions. (See the four {110} planes at the periphery of the projection. These are 90° to the {100} surface plane.)

When we grow an oxide, or deposit a silicon nitride masking film on the {100} silicon slices and then align a mask parallel and/or perpendicular to these ⟨110⟩ directions (as indicated in Fig. 5), as in standard photolithographic processing, we can open up narrow lines, through the mask material, which are aligned parallel with the trace of the high atomic packed {111} planes. When we place these slices in an orientation-dependent etch such as KOH, normal propanol, and H_2O, the etching will proceed in the ⟨100⟩ direction (into the slice) until the etch front hits the {111} planes, intersecting the (100)

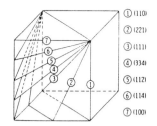

Fig. 1. Crystalography of silicon substrate orientations.

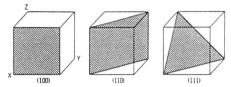

Fig. 2. Low crystallographic index planes of silicon.

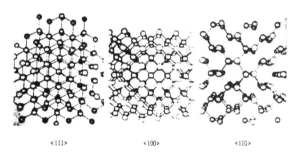

Fig. 3. Model showing three low-index directions of silicon.

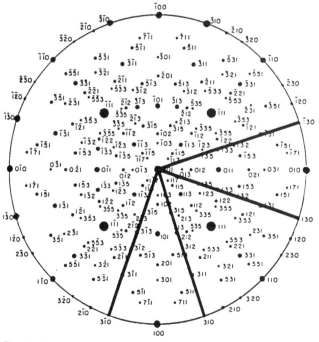

Fig. 4. Standard (001) projection for a face centered cubic crystal.

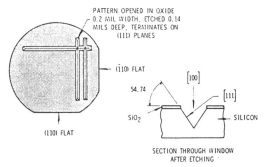

Fig. 5. Mask alignment on (100) silicon.

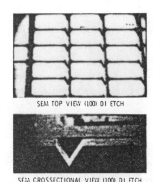

Fig. 6. Top view and cross-sectional view of (100) ODE silicon.

Fig. 7. ODE etch pit in (100) silicon.

Fig. 8. (a) (100), ODE, 2040× 65° SEM of 10-μm squares on 20-μm centers. (b) (100) ODE 1240× 65° SEM of 10-μm squares containing mask oxide.

Fig. 9. Corner undercutting in (100) silicon versus ODE depth. (a) 23 μm deep. (b) 54 μm deep.

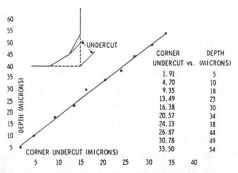

Fig. 10. Corner undercutting versus ODE depth in (100) silicon.

plane at the edge of the mask opening, and then the etching will stop as in Fig. 6. Also see data in Figs. 13 and 14. The etch depth to oxide opening width ratio is 0.707. That is if we open an oxide pattern line 10 μm wide it will etch 7.07 μm deep and then for practical purposes stop etching. Due to this effect one can use ODE on (100) silicon as a method for determining pin hole density in oxide films. When one uses ethylene diamine pyrocatechol and water (EDA) (255 cm^3, 54 g, 120 cm^3, respectively) at 100°C the etch front will propagate into the (100) silicon to form a perfect square topped inverted pyramid, through the pin hole (see Fig. 7). If one uses a square window frame pattern opened in the oxide or nitride mask and the same etch conditions, a square based pyramid structure can be etched as in Fig. 8(a) and (b). Fig. 8(a) was etched using a thin 1000-Å oxide mask and 8-min etch time in EDA as above. In this case, the thin oxide mask is left intact on the pyramids. In Fig. 8(b), an additional 30-s etch in planar etch was used to point the pyramids and lift off the oxide. The pyramids are 8 μm high, and have 10 μm/square bases on 20-μm centers. In the case where one wishes to etch V-grooves for isolation purposes or for VMOS structures, edge undercutting of the mask is negligible, however, corner undercutting may become a problem in deep, >20 μm, etch depths for isolation structures [12]-[14]. Fig. 9 shows the effect of corner undercutting on the relatively fast etch {331} planes, which are exposed on outside (convex) corners in masked arrays. Corner undercutting does not occur at inside (concave) corners. In Fig. 9 it can be seen that corner undercutting is related to etch depth. Fig. 10 plots this relationship of corner undercutting to etch depths for {100} silicon using the KOH, propanol, H$_2$O etch at 80°C. From this data one can design corner compensation into the mask to prevent the faceted or rounded corners. Fig. 11 shows an example wherein mask corner compensation was used to produce square corners in arrays that require 50-μm-deep etching. Note the corner compensation and oxide mask pattern pro-

Fig. 11. Corner compensation of oxide mask on ODE (100) structure.

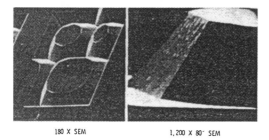

Fig. 12. Corner faceting in (100) ODE, 50 μm deep.

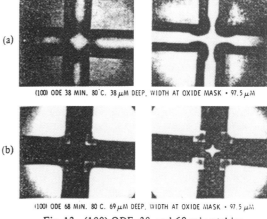

Fig. 13. (100) ODE, 38- and 68-min etching.

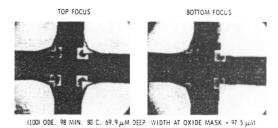

Fig. 14. (100) ODE, 98-min etching.

jecting out into space above the ODE V-groove. With corner compensation, underetching will leave a tip of silicon sticking out at each corner, see Fig. 13(a), and overetching of the designed compensation-to-depth ratio will, of course, undercut the corners. Fig. 12 is a 180X SEM of a {100} ODE silicon slice etched 50 μm deep without corner compensation of the mask. Note the {331} faceting or rounding of the outside corners. The double four-fold symmetry of the {331} planes in ⟨310⟩ directions, see Fig. 4, is evident in the corner faceting of this pattern. Note the octagon shape at the top of the smaller mesas. These {331} planes intersect the {100} silicon surface at an angle of 46.5°, which may be advantageously used in some cases, and will be further discussed later. Also note the exact crystallographically sharp 90° corners at the inside corners of the array. Fig. 12 also shows a higher magnification view of these {331} facets. Again note the exact crystalographic structure of the ODE front.

As was previously mentioned, the ODE stops when the etch depth reaches the depth at which the two {111} planes at the edge of the mask opening intersect. If the mask is properly aligned with the {110} direction no undercutting occurs along the mask edges (only at convex corners). Fig. 13 shows evidence of this statement. Fig. 13(a) shows a (100) silicon slice with a corner compensated oxide mask. This mask has oxide open lines 97.5 μm wide at the silicon-oxide interface. After etching 38 min at 80°C the etch depth measured 38 μm deep and the width opening at the silicon-oxide interface measured 97.5 μm. Note the tip of silicon projecting out under the compensated corner oxide. (The corner compensation was designed for 50-μm etch depth.) The slices were placed back in the ODE for another 30 min at 80°C, thus giving a total ODE time of 68 min. At this time the etch depth measured 69 μm deep and the linewidth opening at the silicon-oxide interface measured 97.5 μm, see Fig. 13(b), indicating no undercutting of the oxide mask. The ODE should "V" or bot-

tom out at this depth and time. That is 97.5 × 0.707 = 68.9. The slices were again placed back into the ODE for an additional 30 min. After this additional 30-min etch, the depth measured 69.9 μm and the linewidth opening at the silicon-oxide interface again measured 97.5 μm, indicating no undercutting of the oxide mask even after 48-min overetching. The total etch time was 98 min. The complete lack of undercutting indicates the etch stopping ability of the {111} planes when the (100) slice is properly aligned. Fig. 14 shows the ODE slice after a total of 98-min etching. Note in Fig. 13(b) the faceting of the corners under the corner compensation due to the overetching for 18 min. The compensation was designed for 50-μm etch depth. Also note the "V" bottom in the bottom focus photo of Fig. 13(b) indicating the completion of the ODE. A replica of the trace of the {111} planes can be seen in the corner compensation oxide in both Fig. 13(b) and Fig. 14. This replica is due to a change in mask oxide thickness, and shows that the {111} planes blocked the corner undercutting for some period of time before breaking down to allow corner undercutting, or faceting, to the faster etching {331} planes. The importance of proper alignment cannot be overemphasized in the use of orientation-dependent etching. If a mask is aligned off the desired crystalographic direction undercutting will occur until the etch front reaches the slow etch {111} planes. This will occur in a series of plate-like steps, see Kendall [15]. Fig. 15 shows a top view of two (100) silicon mesas with corner compensated oxide mask, ODE 50 μm deep. Undercutting can be detected as a function of mask alignment. Measurements were made from slices mis-

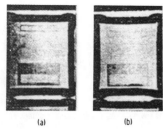

Fig. 15. Undercutting versus mask alignment accuracy in (100) ODE. (a) 1.5° off alignment. (b) 3.9° off alignment.

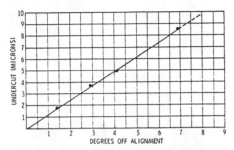

Fig. 16. Undercutting versus alignment accuracy.

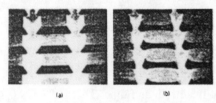

Fig. 17. ODE of (100) silicon. (a) ⟨310⟩ alignment. (b) ⟨110⟩ alignment.

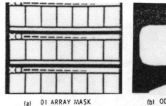

Fig. 18. Effect of mask corner compensation on ODE and ODD.

Fig. 19. Effect of mask corner compensation of ODE and ODD.

oriented from 1° to 7° and ODE 50 μm deep using (100) ODE at 80°C. The data are plotted in Fig. 16. As in the case of Kendall, for (110) silicon, the (100) data also give a linear plot of undercutting versus degrees off alignment. Note also the plot projects through zero indicating no undercutting for a properly aligned mask on (100) silicon. The U/D (undercut to depth) in this case is 0.10 at 4° off orientation whereas the U/D for {110} silicon is 0.10 at 3.5° off orientation. This indicates a slightly less stringent dependence on alignment accuracy for {100} silicon. If we reexamine the silicon crystal structure projected in Fig. 4 we see the double four-fold symmetry of the {331} planes, mentioned and identified in the discussion of Fig. 12. These planes have been identified as the prevalent, fast-etching, corner-undercutting planes when using either KOH, propenal and water, or EDA in our work. Others have identified the corner faceting planes as {221} planes when etching with EDA [7], [16]. These relatively low (46°) angles, with respect to the (100) planes, may be advantageously used in cases wherein one wished to run metallization up and down these planes [17]. Fig. 17 shows examples of two (100) silicon slices ODE at the same time in the same solution and having the same mask pattern. Slice in Fig. 17(a) had the mask pattern aligned parallel with the {310} direction, which gives etch termination on the 46° {331} planes. Slice in Fig. 17(b) had the mask pattern aligned parallel with the normally used ⟨110⟩ direction. The mask did not have corner compensation; however, note the perfect 90° corners in the ⟨310⟩ aligned slice and the corner undercutting on the ⟨110⟩ aligned slice. It should be pointed out that edge undercutting is much greater in the ⟨310⟩ direction alignment. The ratio is 18-μm undercutting per 25-μm etch depth. There is no corner undercutting in the ⟨310⟩ aligned slice, due to the fact that the faster etching {331} planes are now the etch terminating sides, or masking edge.

The effects of orientation-dependent etching, etch mask corner compensation, and orientation-dependent deposition in graphically displayed in Figs. 18 and 19. Fig. 18(a) is a mask array for dielectric isolation processing [12], [14] of a seven-element array wherein each element is dielectrically isolated from its neighboring element. The eleven narrow rectangles above the seven-element array are thickness indicators designed into each circuit bar. These indicators reveal the thickness of the single-crystal layer in each finished circuit or array [13], [14]. At the left end of the thickness indicators there are two alignment squares, one with and one without corner compensation. These squares are initially the same size. After ODE, 53 μm deep, the uncompensated corner square has etched to an octagonal shape, consisting of eight {331} planes as sides. (See Fig. 18(b).) Note the large difference in size or surface area at this point in processing. The undercutting of the masking oxide is evident in this photograph. Fig. 19 shows the left side of this same array after the slice has completed dielectric isolation processing. The moats have been filled with polycrystalline silicon, the slice inverted and ground and polished back to the isolated silicon islands. Note the near perfect square from the compensated mask area and the near perfect octagon from the uncompensated mask area. Also note the first two thickness indicators showing through. (Fig. 19(a) is looking at the back side of the slice with respect to Fig. 18(a).) Fig. 19(b) shows the effect of orientation-dependent deposition of silicon after the slice, as in Fig. 19(a), has had an epitaxial deposition which is actually simultaneous deposition of single- and polycrystal silicon. The

Fig. 20. Orientation-dependent deposition on (100) silicon through circular oxide mask openings. 1000× and 2500× magnification.

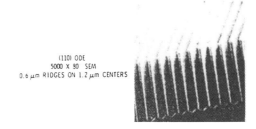

Fig. 22. 5000× 80° SEM of (110) ODE silicon 0.6-μm ridges on 1.2-μm centers.

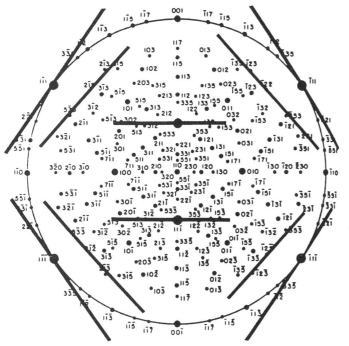

Fig. 21. Standard (110) crystal projection for a face centered cubic crystal with {111} and {221} planes designated.

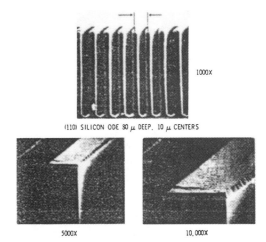

Fig. 23. 1000×, 5000×, and 10 000× SEM's of (110) ODE silicon. 80-μm etch depth, 5-μm ridges on 10-μm centers.

Fig. 24. 240× 80° SEM of 10° off (110) ODE silicon. 5-μm ridges on 10-μm centers, 160 μm deep.

two squares are the compensated and uncompensated square areas. The octagon, {331} plane bounded, figure has grown back to a near perfect square bounded by {111} planes. Note the (100) stacking fault in the mask corner compensated square. Another example of orientation-dependent deposition [13], [18], [19] is shown in Fig. 20. These near perfect squares are grown through an oxide mask with 5-μm-diameter circles, open for silicon nucleation. The 5-μm-diameter circles are on 25-μm centers; therefore, one can see the lateral spreading effect of the (100) oriented epitaxial deposition. The final width of the "Epi Top" in this case is ~18 μm. The epitaxial linear growth is ~6.5 μm on each side of the original 5-μm-diameter mask opening.

(110) Orientation-Dependent Etching

When we examine the crystal projection, Fig. 21, of the very open lattice ⟨110⟩ direction of silicon, shown in Fig. 3, we find that the high packing density {111} planes are now 90° to the (110) surface plane, however, they are no longer 90° to each other. This indicates that we can etch vertical moats with straight wall {111} side termination. We have found that we have a >600 to 1 etching ratio in the ⟨110⟩ to ⟨111⟩ direction [14]-[19]. This means that we can etch very narrow, for example, 0.6-μm wide, moats on very tight packing densities, for example, 1 μm or less center to center, straight down into or even completely through a (110) orientation silicon slice. In this (110) mirror image symmetry we also have two relative lowangle {111} planes intersecting the (110) surface at 35.26° The effect of these will be seen and discussed later. (See Fig. 22.) Fig. 22 is a 500×, 80° SEM of a cleaved (cross-sectional) (110) silicon slice that was etched using 50 percent KOH, 50 percent D.I. water etch at 80°C. The oxide mask was opened by e-beam lithography and had 0.6-μm-wide openings on 1.2-μm centers. Some undercutting, due to a slight misalignment, is evident at the oxide mask surface. We routinely

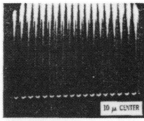

Fig. 25. 570× 85° SEM pointed ridges on 10-μm centers, (110) ODE.

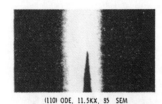

Fig. 26. 11 500× 85° SEM pointed ridge.

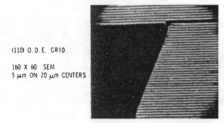

Fig. 27. 160× 60° SEM cross-sectional and top view of (110) silicon slice, ODE from top and bottom surface simultaneously.

Fig. 28. 500× SEM, (110) ODE aperture grid, apertures 5 μm × 5 μm on 20-μm centers etched through 270-μm-thick slice.

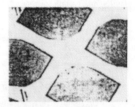

Fig. 29. 2000× SEM (110) ODE aperture grid. 5 μm on 20-μm centers etched through 270-μm-thick slice.

obtain etch ratios greater than 600 to 1 using this etch system, that is etching 600 μm deep in the ⟨110⟩ direction while undercutting less than 1 μm in the ⟨111⟩ direction. See Fig. 23. Fig. 23 is a series of three SEM's of a (110) ODE cross section. The etch depth is 80 μm, the mask pattern is 5-μm-wide moats open on 10-μm centers. The top SEM is at 1000× magnification. Note the low-angle {111} planes at the bottom of the moats. The lower left SEM is of the same structure at 5000× magnification and the lower right SEM is at 10 000× magnification. The original 6000-Å-thick oxide mask is readily seen at the 5 and 10 thousand magnifications. These structures may be tilted at any angle by simply cutting the (110) silicon slice at the desired angle off the ⟨110⟩ direction. Fig. 24 is a 240×, 80° SEM cross-sectional view of an ODE (110) slice that was sawed 10° off the (110) plane. These very narrow ridges can be diffused and oxidized by normal silicon semiconductor processing. They may also be pointed to extremely sharp points as is shown in Figs. 25 and 26. This is accomplished by etching in Dash-type etch for approximately 10 min after moat etch is completed. Fig. 25 is a SEM, cross-sectional view, at 570× magnification. When an array such as this is pointed, the top (110) surface becomes a black body, with no surface reflection. Fig. 26 is an eleven and one half thousand × (11.5 thousand ×) SEM of one of these pointed ridges. Even at this magnification it is difficult to estimate a radius of curvature. Very fine, high density X-Y structures, or sieves, can be made by etching the (110) slice with masks aligned with the 90° (111) traces on both sides (top and bottom) of the slice. This pattern forms a rhombus with 109.47° obtuse angles. Fig. 27 is a 60° SEM showing both top and cross-sectional views of a (110) silicon slice, etched from both sides simultaneously. A quadrant was cleaved out of the slice for photographic purposes. Fig. 28 is a 500× magnification, SEM, of the top surface of the (110) slice etched from both sides simultaneously. The slice was 269.2 μm (10.6 mils) thick. The mask opens 5-μm-wide moats on 20-μm centers on both sides of the slice, resulting in 5 μm × 5 μm rhombic openings when etched to completion. Fig. 29 is a 2000× SEM of the structure. This structure offers new possibilities in electrical X-Y addressing, and optical mechanical designs. In an experiment, a rhombic window frame mask was purposely aligned 109° off the proper alignment of the sides with the {111} traces at the (110) surface. The resulting etched structure had undercutting at all four corners. (See Fig. 30.) However, it was observed that the acute angle of the masked structure undercut to a pair of planes that are 90° to each other and 76° from the (110) surface plane. These planes were identified as {221} planes that are 45° off a normal (110) flat, see Fig. 21, and may be used for vertical etching of 90° corner structure (squares, rect-

Fig. 30. (110) silicon ODE 24 μm deep. Rhombus pattern aligned 35.26° off (110) flat. Acute angle undercut to 90° crystal facet.

{110} SILICON O.D.E. ETCHED 24 μm DEEP
ALIGNED 45° OFF OF THE (1̄10) FLAT
90° CORNERS - NO MASK COMPENSATION
UNDERCUTTING WAS 17.8 um.
ETCH TERMINATION ON {221} PLANES
{221} INTERSECTS THE {110} SURFACE AT 76°

Fig. 31. (110) silicon ODE 24 μm deep. Square pattern aligned 45° off (110) flat. 90° corners with no mask corner compensation, etch termination on {221} planes.

TABLE II
SOME APPLICATIONS FOR (100) AND (110) ANISOTROPIC ETCHING IN SILICON DEVICE PROCESSING

(100)	(110)
RADIATION HARDENED CIRCUITS	HIGH VOLTAGE DIODE ARRAYS
ELECTRONIC PRINTER	VERTICAL MULTIPLE JCT SOLAR CELL
CROSSPOINTS	WAVE GUIDES
ISO PLANAR	SENSISTOR
POLY PLANAR	IR DETECTORS
V MOS	METALLIZATION TEMPLATES
J FET ARRAYS	HIGH VALUE CAPACITORS
DI PROCESS THICKNESS INDICATOR	OPTICAL COLLIMATORS
SOLID STATE PRESSURE TRANSDUCER	BLACK BODIES
SOLAR CELL ANTI-REFLECTING SURFACE	

angles, L shapes, etc.) in (110) silicon. A square, 90° corner pattern was aligned in this manner and etched 24 μm deep. Fig. 31 shows the resulting 90° corners after etching. The undercutting to depth ratio is 18 to 24 μm, approximately the same as that of the ⟨310⟩ direction alignment on the (100) silicon surface. Again perfect 90° corners are obtained without mask corner compensation.

Some of the many applications for anisotropic etching in semiconductor device processing are listed in Table II. This is only a partial listing and a small beginning in the use of this versatile technology.

ACKNOWLEDGMENT

The author wishes to thank his fellow workers, J. Powell, L. Webb, R. Staring, D. Yeakley, B. Housewright, T. Powell, Dorothy Splawn, and W. Runyan for their many contributions to this work. Also special thanks to W. Runyan and Hettie Smith, Emily Apperson, M. White, and R. Stein for critique and preparation of the manuscript.

REFERENCES

[1] D. R. Turner, "Electropolishing of silicon in hydrofluoric acid solutions," *J. Electrochem. Soc.*, vol. 105, pp. 402-408, 1958.
[2] R. R. Stead (Texas Instruments Incorporated), "Etching of semiconductor materials," U.S. Patent 2 871 110, Jan. 1959; U.S. Patent 2 927 011, Mar. 1960; U.S. Patent 2 973 253, Feb. 1961.
[3] B. Schwartz and H. Robbin, "Chemical etching of silicon, Part I, The system HF, HNO_3, H_2O and $HC_2H_3O_2$," *J. Electrochem. Soc.*, vol. 106, no. 6, pp. 505-508, 1959; Part II, *J. Electrochem. Soc.*, vol. 107, no. 2, pp. 108-111, 1960; Part III, *J. Electrochem. Soc.*, vol. 108, no. 4, pp. 365-372, 1961.
[4] W. C. Dash, "Copper precipitation on dislocation in silicon," *J. Appl. Phys.*, vol. 27, pp. 1193-1195, 1956.
[5] J. M. Crishal and A. L. Harrington, "A selective etch for elemental silicon," *Electrochem Soc. Extended Abstract* (Spring Meeting, 1962, Los Angeles, CA), Abstr. no. 89.
[6] R. M. Finne and D. L. Klein, "A water amine complexing agent system for etching silicon," *E.C.S. J.*, vol. 114, no. 9, Sept 1967.
[7] D. B. Lee, "Anisotropec etching of silicon," *J. Appl. Phys.*, vol. 40, no. 1, Oct. 1969.
[8] H. Muraoka and T. Y. Sumitomo, *Controlled Preferential Etching Technology*, in *Semiconductor Silicon 1973* (Electrochemical Society, H. R. Huff and R. R. Burgess, Eds.).
[9] J. B. Price, "Anisotropic etching of silicon with $KOH-H_2O$ isopropyl alcohol," in *ECS Semiconductor Silicon 1973*, pp. 339-353.
[10] J. C. Greenwood, *Ibid.*, vol. 116, no. 9, Sept. 1969.
[11] A. Bogh, *Ibid*, vol, 118, no. 2, Feb. 1971.
[12] U. S. Davidson and F. Lee, "Dielectric isolated integrated circuit substrate processes," *Proc. IEEE*, vol. 57, no. 9, pp. 1532-1537, Sept. 1969.
[13] K. E. Bean and P. S. Gleim, "The influence of crystal orientation on silicon semiconductor processing," *Proc. IEEE*, vol. 57, no. 9, pp. 1469-1476, Sept. 1969.
[14] K. E. Bean and W. R. Runyan, "Dielectric isolation: Comprehensive current future," *Electrochem. Soc. J.*, vol. 124, no. 1, pp. 5C-12C, Jan. 1977.
[15] D. L. Kendall, *Appl. Phys. Lett.*, vol. 26, no. 4, p. 195, Feb. 15, 1975.
[16] M. J. Declercq, "A new CMOS technology using anisotropic etching of silicon," *IEEE J. Solid-State Circuits*, vol. SC-10, no. 4, pp. 191-197, Aug. 1975.
[17] K. E. Bean, J. S. Crabbe, J. G. Hoffman, and F. D. Malone, "An improved radiation hardened JFET array," in *Semiconductor Silicon 1973* (Electrochem. Soc.), H. R. Huff and R. R. Burgess, Eds.
[18] R. K. Smeltzer, D. L. Kendall, and G. L. Varnell, "Vertical multiple junction solar cell fabrication," in *Conf. Rec. Tenth IEEE Photovoltaic Specialists Conf.*, p. 194, 1973.
[19] K. E. Bean, R. L. Yeakley, and T. K. Powell, "Orientation dependent etching and deposition of silicon," in *ECS Extended Abstracts Spring Meet.* (May 1974), vol. 71-1.
[20] D. F. Weirauch, *J. Appl. Phys.*, vol. 46, p. 1478, 1975.
[21] E. Bassous and E. F. Baran, *J. Electroch. Soc.*, vol. 125, no. 8, Aug. 1978.

Silicon as a Mechanical Material

KURT E. PETERSEN, MEMBER, IEEE

Abstract—Single-crystal silicon is being increasingly employed in a variety of new commercial products not because of its well-established electronic properties, but rather because of its excellent mechanical properties. In addition, recent trends in the engineering literature indicate a growing interest in the use of silicon as a mechanical material with the ultimate goal of developing a broad range of inexpensive, batch-fabricated, high-performance sensors and transducers which are easily interfaced with the rapidly proliferating microprocessor. This review describes the advantages of employing silicon as a mechanical material, the relevant mechanical characteristics of silicon, and the processing techniques which are specific to micromechanical structures. Finally, the potentials of this new technology are illustrated by numerous detailed examples from the literature. It is clear that silicon will continue to be aggressively exploited in a wide variety of mechanical applications complementary to its traditional use as an electronic material. Furthermore, these multidisciplinary uses of silicon will significantly alter the way we think about all types of miniature mechanical devices and components.

I. INTRODUCTION

IN THE SAME WAY that silicon has already revolutionized the way we think about electronics, this versatile material is now in the process of altering conventional perceptions of miniature mechanical devices and components [1]. At least eight firms now manufacture and/or market silicon-based pressure transducers [2] (first manufactured commercially over 10 years ago), some with active devices or entire circuits integrated on the same silicon chip and some rated up to 10 000 psi. Texas Instruments has been marketing a thermal point head [3] in several computer terminal and plotter products in which the active printing element abrasively contacting the paper is a silicon integrated circuit chip. The crucial detector component of a high-bandwidth frequency synthesizer sold by Hewlett-Packard is a silicon chip [4] from which cantilever beams have been etched to provide thermally isolated regions for the diode detectors. High-precision alignment and coupling assemblies for fiber-optic communications systems are produced by Western Electric from anisotropically etched silicon chips simply because this is the only technique capable of the high accuracies required. Within IBM, ink jet nozzle arrays and charge plate assemblies etched into silicon wafers [5] have been demonstrated, again because of the high precision capabilities of silicon IC technology. These examples of silicon micromechanics are not laboratory curiosities. Most are well-established, commercial developments conceived within about the last 10 years.

The basis of micromechanics is that silicon, in conjunction with its conventional role as an electronic material, and taking advantage of an already advanced microfabrication technology, can also be exploited as a high-precision high-strength high-reliability mechanical material, especially applicable wherever miniaturized mechanical devices and components must be integrated or interfaced with electronics such as the examples given above.

The continuing development of silicon micromechanical applications is only one aspect of the current technical drive toward miniaturization which is being pursued over a wide front in many diverse engineering disciplines. Certainly silicon microelectronics continues to be the most obvious success in the ongoing pursuit of miniaturization. Four factors have played crucial roles in this phenomenal success story: 1) the active material, silicon, is abundant, inexpensive, and can now be produced and processed controllably to unparalleled standards of purity and perfection; 2) silicon processing itself is based on very thin deposited films which are highly amenable to miniaturization; 3) definition and reproduction of the device shapes and patterns are performed using photographic techniques which have also, historically, been capable of high precision and amenable to miniaturization; finally, and most important of all from a commercial and practical point of view, 4) silicon microelectronic circuits are batch-fabricated. The unit of production for integrated circuits—the wafer—is not *one* individual saleable item, but contains hundreds of identical chips. If this were not the case, we could certainly never afford to install microprocessors in watches or microwave ovens.

It is becoming clear that these same four factors which have been responsible for the rise of the silicon microelectronics industry can be exploited in the design and manufacture of a wide spectrum of miniature mechanical devices and components. The high purity and crystalline perfection of available silicon is expected to optimize the *mechanical* properties of devices made from silicon in the same way that *electronic* properties have been optimized to increase the performance, reliability, and reproducibility of device characteristics. Thin-film and photolithographic fabrication procedures make it possible to realize a great variety of extremely small, high-precision mechanical structures using the same processes that have been developed for electronic circuits. High-volume batch-fabrication techniques can be utilized in the manufacture of complex, miniaturized mechanical components which may not be possible by any other methods. And, finally, new concepts in hybrid device design and broad new areas of application, such as integrated sensors [6], [7] and silicon heads (for printing and data storage), are now feasible as a result of the unique and intimate integration of mechanical and electronic devices which is readily accomplished with the fabrication methods we will be discussing here.

While the applications are diverse, with significant potential impact in several areas, the broad multidisciplinary aspects of silicon micromechanics also cause problems. On the one hand, the materials, processes, and fabrication technologies are all taken from the semiconductor industry. On the other hand, the applications are primarily in the areas of mechanical en-

---

Manuscript received December 2, 1981; revised March 11, 1982. The submission of this paper was encouraged after the review of an advance proposal.

The author was with IBM Research Laboratory, San Jose, CA 95193. He is now with Transensory Devices, Fremont, CA 94539.

TABLE I

	Yield Strength (10^{10} dyne/cm^2)	Knoop Hardness (kg/mm^2)	Young's Modulus (10^{12} dyne/cm^2)	Density (gr/cm^3)	Thermal Conductivity (W/cm°C)	Thermal Expansion (10^{-6}/°C)
*Diamond	53	7000	10.35	3.5	20	1.0
*SiC	21	2480	7.0	3.2	3.5	3.3
*TiC	20	2470	4.97	4.9	3.3	6.4
*Al$_2$O$_3$	15.4	2100	5.3	4.0	0.5	5.4
*Si$_3$N$_4$	14	3486	3.85	3.1	0.19	0.8
*Iron	12.6	400	1.96	7.8	0.803	12
SiO$_2$ (fibers)	8.4	820	0.73	2.5	0.014	0.55
*Si	7.0	850	1.9	2.3	1.57	2.33
Steel (max. strength)	4.2	1500	2.1	7.9	0.97	12
W	4.0	485	4.1	19.3	1.78	4.5
Stainless Steel	2.1	660	2.0	7.9	0.329	17.3
Mo	2.1	275	3.43	10.3	1.38	5.0
Al	0.17	130	0.70	2.7	2.36	25

*Single crystal. See Refs. 8, 9, 10, 11, 141, 163, 166.

gineering and design. Although these two technical fields are now widely divergent with limited opportunities for communication and technical interaction, widespread, practical exploitation of the new micromechanics technology in the coming years will necessitate an intimate collaboration between workers in *both* mechanical *and* integrated circuit engineering disciplines. The purpose of this paper, then, is to expand the lines of communication by reviewing the area of silicon micromechanics and exposing a large spectrum of the electrical engineering community to its capabilities.

In the following section, some of the relevant mechanical aspects of silicon will be discussed and compared to other more typical mechanical engineering materials. Section III describes the major "micromachining" techniques which have been developed to form the silicon "chips" into a wide variety of mechanical structures with IC-compatible processes amenable to conventional batch-fabrication. The next four sections comprise an extensive list of both commercial and experimental devices which rely crucially on the ability to construct miniature, high-precision, high-reliability, *mechanical* structures on silicon. This list was compiled with the primary purpose of illustrating the wide range of demonstrated applications. Finally, a discussion of present and future trends will wrap things up in Section VIII. The underlying message is that silicon micromechanics is not a diverging, unrelated, or independent extension of silicon microelectronics, but rather a natural, inevitable continuation of the trend toward more complex, varied, and useful integration of devices on silicon.

II. Mechanical Characteristics of Silicon

Any consideration of mechanical devices made from silicon must certainly take into account the mechanical behavior and properties of single-crystal silicon (SCS). Table I presents a comparative list of its mechanical characteristics. Although SCS is a brittle material, yielding catastrophically (not unlike most oxide-based glasses) rather than deforming plastically (like most metals), it certainly is not as fragile as is often believed. The Young's modulus of silicon (1.9×10^{12} dyne/cm^2 or 27×10^6 psi) [8], for example, has a value approaching that of stainless steel, nickel, and well above that of quartz and most other borosilicate, soda-lime, and lead-alkali silicate glasses [9]. The Knoop hardness of silicon (850) is close to quartz, just below chromium (935), and almost twice as high as nickel (557), iron, and most common glasses (530) [10]. Silicon single crystals have a tensile yield strength (6.9×10^{10} dyne/cm^2 or 10^6 psi) which is at least 3 times higher than stainless-steel wire [8], [11]. In practice, tensile stresses *routinely* encountered in seed crystals during the growth of large SCS boules, for example, can be over 18 000 psi (40-kg boule hanging from a 2-mm-diameter seed crystal, as illustrated in Fig. 1). The primary difference is that silicon will yield by fracturing (at room temperature) while metals usually yield by deforming inelastically.

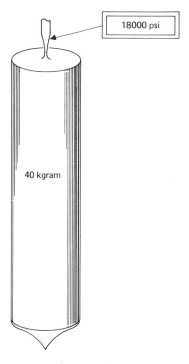

Fig. 1. Stresses encountered commonly in silicon single crystals are very high during the growth of large boules. Seed crystals, typically 0.20 cm in diameter and supporting 40-kg boules, experience stresses over 1.25×10^8 Pa or about 18 000 psi in tension.

Despite this quantitative evidence, we might have trouble intuitively justifying the conclusion that silicon is a strong mechanical material when compared with everyday laboratory and manufacturing experience. Wafers do break—sometimes without apparent provocation; silicon wafers and parts of wafers may also easily chip. These occurrences are due to several factors which have contributed to the misconception that silicon is mechanically fragile. First, single-crystal silicon is normally obtained in large (5–13-cm-diameter) wafers, typically only 10–20 mils (250 to 500 μm) thick. Even stainless steel of these dimensions is very easy to deform inelastically. Silicon chips with dimensions on the order of 0.6 cm × 0.6 cm, on the other hand, are relatively rugged under normal handling conditions unless scribed. Second, as a single-crystal material, silicon has a tendency to cleave along crystallographic planes, especially if edge, surface, or bulk imperfections cause stresses to concentrate and orient along cleavage planes. Slip lines and other flaws at the edges of wafers, in fact, are usually responsible for wafer breakage. In recent years, however, the semiconductor industry has attacked this yield problem by contouring the edges of wafers and by regularly using wafer edge inspection instruments, specifically designed to detect mechanical damage on wafer edges and also to assure that edges are properly contoured to avoid the effects of stress concentration. As a result of these quality control improvements, wafer breakage has been greatly reduced and the intrinsic strength of silicon is closer to being realized in practice during wafer handling. Third, chipping is also a potential problem with brittle materials such as SCS. On whole wafers, chipping occurs for the same qualitative reasons as breaking and the solutions are identical. Individual die, however, are subject to chipping as a result of saw- or scribe-induced edge damage and defects. In extreme cases, or during rough handling, such damage can also cause breakage of or cracks in individual die. Finally, the high-temperature processing and multiple thin-film depositions commonly encountered in the fabrication of IC devices unavoidably result in internal stresses which, when coupled with edge, surface, or bulk imperfections, can cause concentrated stresses and eventual fracture along cleavage planes.

These factors make it clear that although high-quality SCS is intrinsically strong, the apparent strength of a particular mechanical component or device will depend on ... crystallographic orientation and geometry, the number and size of surface, edge, and bulk imperfections, and the stresses induced and accumulated during growth, polishing, and subsequent processing. When these considerations have been properly accounted for, we can hope to obtain mechanical components with strengths exceeding that of the highest strength alloy steels.

General rules to be observed in this regard, which will be restated and emphasized in the following sections, can be formulated as follows:

1) The silicon material should have the lowest possible bulk, surface, and edge crystallographic defect density to minimize potential regions of stress concentration.

2) Components which might be subjected to severe friction, abrasion, or stress should be as small as possible to minimize the total number of crystallographic defects in the mechanical structure. Those devices which are never significantly stressed or worn could be quite large; even then, however, thin silicon wafers should be mechanically supported by some technique—such as anodic bonding to glass—to suppress the shock effects encountered in normal handling and transport.

3) All mechanical processing such as sawing, grinding, scribing, and polishing should be minimized or eliminated. These operations cause edge and surface imperfections which could result in the chipping of edges, and/or internal strains subsequently leading to breakage. Many micromechanical components should preferably be separated from the wafer, for example, by etching rather than by cutting.

4) If conventional sawing, grinding, or other mechanical operations are necessary, the affected surfaces and edges should be etched afterwards to remove the highly damaged regions.

5) Since many of the structures presented below employ anisotropic etching, it often happens that sharp edges and corners are formed. These features can also cause accumulation and concentration of stress damage in certain geometries. The structure may require a subsequent isotropic etch or other smoothing methods to round such corners.

6) Tough, hard, corrosion-resistant, thin-film coatings such as CVD SiC [12] or Si_3N_4 should be applied to prevent direct mechanical contact to the silicon itself, especially in applications involving high stress and/or abrasion.

7) Low-temperature processing techniques such as high-pressure and plasma-assisted oxide growth and CVD depositions, while developed primarily for VLSI fabrication, will be just as important in applications of silicon micromechanics. High-temperature cycling invariably results in high stresses within the wafer due to the differing thermal coefficients of expansion of the various doped and deposited layers. Low-temperature processing will alleviate these thermal mismatch stresses which otherwise might lead to breakage or chipping under severe mechanical conditions.

As suggested by 6) above, many of the structural or mechanical disadvantages of SCS can be alleviated by the deposition of passivating thin films. This aspect of micromechanics imparts a great versatility to the technology. Sputtered quartz, for example, is utilized routinely by industry to passivate IC chips against airborne impurities and mild atmospheric corrosion effects. Recent advances in the CVD deposition (high-temperature pyrolytic and low-temperature RF-enhanced) of SiC [12] have produced thin films of extreme hardness, essentially zero porosity, very high chemical corrosion resistance, and superior wear resistance. Similar films are already used, for example, to protect pump and valve parts for handling corrosive liquids. As seen in Table I, Si_3N_4, an insulator which is routinely employed in IC structures, has a hardness second only to diamond and is sometimes even employed as a high-speed, rolling-contact bearing material [13], [14]. Thin films of silicon nitride will also find important uses in silicon micromechanical applications.

On the other end of the thin-film passivation spectrum, the gas-condensation technique marketed by Union Carbide for depositing the polymer parylene has been shown to produce virtually pinhole-free, low-porosity, passivating films in a high polymer form which has exceptional point, edge, and hole coverage capability [15]. Parylene has been used, for example, to coat and passivate implantable biomedical sensors and electronic instrumentation. Other techniques have been developed for the deposition of polyimide films which are already used routinely within the semiconductor industry [16] and which also exhibit superior passivating characteristics.

One excellent example of the unique qualities of silicon in the realization of high-reliability mechanical components can be found in the analysis of mechanical fatigue in SCS structures. Since the initiation of fatigue cracks occurs almost exclusively at the *surfaces* of stressed members, the rate of fatigue depends strongly on surface preparation, morphology, and defect density. In particular, structural components with highly polished surfaces have higher fatigue strengths than those with rough surface finishes as shown in Fig. 2 [17]. Passivated surfaces of polycrystalline metal alloys (to prevent intergrain diffusion of H_2O) exhibit higher fatigue strengths than unpassivated surfaces, and, for the same reasons, high water vapor content in the atmosphere during fatigue testing will significantly decrease fatigue strength. The mechanism of fatigue, as these effects illustrate, are ultimately dependent on a surface-defect-initiation process. In polycrystalline ma-

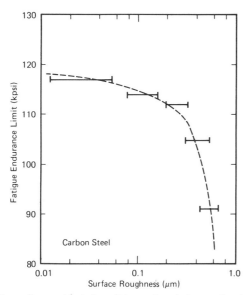

Fig. 2. Generally, mechanical qualities such as fatigue and yield strength improve dramatically with surface roughness and defect density. In the case of silicon, it is well known that the electronic and mechanical perfection of SCS surfaces has been an indispensable part of integrated circuit technology. Adapted from Van Vlack [17].

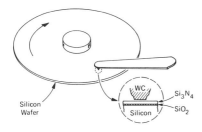

Fig. 3. A rotating MNOS disk storage device demonstrated by Iwamura et al. [21]. The tungsten-carbide probe is in direct contact with the nitride-coated silicon wafer as the wafer rotates at 3600 r/min. Signals have been recorded and played back on such a system at video rates. Wear of the WC probe was a more serious problem than wear of the silicon disk.

terials, these surface defects can be inclusions, grain boundaries, or surface irregularities which concentrate local stresses. It is clear that the high crystalline perfection of SCS together with the extreme smoothness and surface perfection attainable by chemical etching of silicon should yield mechanical structures with intrinsically high fatigue strengths [18]. Even greater strengths of brittle materials can be expected with additional surface treatments [9]. Since hydrostatic pressure has been shown to increase fatigue strengths [19], any film which places the silicon surface under compression should decrease the initiation probability of fatigue cracks. Si_3N_4 films, for example, tend to be under tension [20] and therefore impart a compressive stress on the underlying silicon surface. Such films may be employed to increase the fatigue strength of SCS mechanical components. In addition, the smoothness, uniformity, and high yield strength of these thin-film amorphous materials should enhance overall component reliability.

A new rotating disk storage technology which has recently been demonstrated by Iwamura et al. [21] not only illustrates some of the unique advantages derived from the use of silicon as a mechanical material but also indicates how well silicon, combined with wear-resistant Si_3N_4 films, can perform in demanding mechanical applications. As indicated in Fig. 3, data storage was accomplished by an MNOS charge-storage process in which a tungsten carbide probe is placed in direct contact with a 3-in-diameter silicon wafer, rotating at 3600 r/min. The wafer is coated with 2-nm SiO_2 and 49-nm Si_3N_4, while the carbide probe serves as the top metal electrode. Positive voltage pulses applied to the metal probe as the silicon passes beneath will cause electrons to tunnel through the thin SiO_2 and become trapped in the Si_3N_4 layer. The trapped charge can be detected as a change in capacitance through the same metal probe, thereby allowing the signal to be read. Iwamura et al. wrote and read back video signals with this device over 10^6 times with little signal degradation, at data densities as high as 2×10^6 bits/cm^2. The key problems encountered during this experiment were associated with wear of the tungsten carbide probe, not of the silicon substrate or the thin nitride layer itself. Sharply pointed probes, after scraping over the Si_3N_4 surface for a short time, were worn down to a 10-μm by 10-μm area, thereby increasing the active recording surface per bit and decreasing the achievable bit density. After extended operation, the probe continued to wear while a barely resolvable 1-nm roughness was generated in the hard silicon nitride film. Potential storage densities of 10^9 bits/cm^2 were projected if appropriate recording probes were available. Contrary to initial impressions, the rapidly rotating, harshly abraded silicon disk is not a major source of problems even in such a severely demanding mechanical application.

III. MICROMECHANICAL PROCESSING TECHNIQUES

Etching

Even though new techniques—and novel applications of old techniques—are continually being developed for use in micromechanical structures, the most powerful and versatile processing tool continues to be etching. Chemical etchants for silicon are numerous. They can be isotropic or anisotropic, dopant dependent or not, and have varying degrees of selectivity to silicon, which determines the appropriate masking material(s). Table II gives a brief summary of the characteristics of a number of common wet silicon etches. We will not discuss plasma, reactive-ion, or sputter etching here, although these techniques may also have a substantial impact on future silicon micromechanical devices.

Three etchant systems are of particular interest due to their versatility: ethylene diamine, pyrocatechol, and water (EDP) [22]; KOH and water [23]; and HF, HNO_3, and acetic acid CH_3OOH (HNA) [24], [25]. EDP has three properties which make it indispensable for micromachining: 1) it is anisotropic, making it possible to realize unique geometries not otherwise feasible; 2) it is highly selective and can be masked by a variety of materials, e.g., SiO_2, Si_3N_4, Cr, and Au; 3) it is dopant dependent, exhibiting near zero etch rates on silicon which has been highly doped with boron [26], [27].

KOH and water is also orientation dependent and, in fact, exhibits much higher (110)-to-(111) etch rate ratios than EDP. For this reason, it is especially useful for groove etching on (110) wafers since the large differential etch ratio permits deep, high aspect ratio grooves with minimal undercutting of the masks. A disadvantage of KOH is that SiO_2 is etched at a rate which precludes its use as a mask in many applications. In structures requiring long etching times, Si_3N_4 is the preferred masking material for KOH.

HNA is a very complex etch system with highly variable etch rates and etching characteristics dependent on the silicon dopant concentration [28], the mix ratios of the three etch

TABLE II

Etchant (Diluent)	Typical Compositions	Temp °C	Etch Rate (μm/min)	Anisotropic (100)/(111) Etch Rate Ratio	Dopant Dependence	Masking Films (etch rate of mask)	References
HF HNO_3 (water, CH_3COOH)	10 ml 30 ml 80 ml	22	0.7–3.0	1:1	$\leq 10^{17} cm^{-3}$ n or p reduces etch rate by about 150	SiO_2 (300Å/min)	24,25,28,30
	25 ml 50 ml 25 ml	22	40	1:1	no dependence	Si_3N_4	
	9 ml 75 ml 30 ml	22	7.0	1:1	-----	SiO_2 (700Å/min)	
Ethylene diamine Pyrocatechol (water)	750 ml 120 gr 100 ml	115	0.75	35:1	$\geq 7 \times 10^{19} cm^{-3}$ boron reduces etch rate by about 50	SiO_2 (2Å/min) Si_3N_4 (1Å/min) Au,Cr,Ag,Cu,Ta	20,26,27,35, 43,44
	750 ml 120 gr 240 ml	115	1.25	35:1			
KOH (water, isopropyl)	44 gr 100 ml	85	1.4	400:1	$\geq 10^{20} cm^{-3}$ boron reduces etch rate by about 20	Si_3N_4 SiO_2 (14Å/min)	23,32,33,36, 37,38,42
	50 gr 100 ml	50	1.0	400:1			
H_2N_4 (water, isopropyl)	100 ml 100 ml	100	2.0	----	no dependence	SiO_2 Al	40,41
NaOH (water)	10 gr 100 ml	65	0.25–1.0	----	$\geq 3 \times 10^{20} cm^{-3}$ boron reduces etch rate by about 10	Si_3N_4 SiO_2 (7Å/min)	34

components, and even the degree of etchant agitation, as shown in Fig. 4 and Table II. Unfortunately, these mixtures can be difficult to mask, since SiO_2 is etched somewhat for all mix ratios. Although SiO_2 can be used for relatively short etching times and Si_3N_4 or Au can be used for longer times, the masking characteristics are not as desirable as EDP in micromechanical structures where very deep patterns (and therefore highly resistant masks) are required.

As described in detail by several authors, SCS etching takes place in four basic steps [30], [31]: 1) injection of holes into the semiconductor to raise the silicon to a higher oxidation state Si^+, 2) the attachment of hydroxyl groups OH^- to the positively charged Si, 3) the reaction of the hydrated silicon with the complexing agent in the solution, and 4) the dissolution of the reacted products into the etchant solution. This process implies that any etching solution must provide a source of holes as well as hydroxyl groups, and must also contain a complexing agent whose reacted species is soluble in the etchant solution. In the HNA system, both the holes and the hydroxyl groups are effectively supplied by the strong oxidizing agent HNO_3, while the flourine from the HF forms the soluble species H_2SiF_6. The overall reaction is autocatalytic since the HNO_3 plus trace impurities of HNO_2 combine to form additional HNO_2 molecules.

$$HNO_2 + HNO_3 + H_2O \rightarrow 2HNO_2 + 2OH^- + 2h^+.$$

This reaction also generates holes needed to raise the oxidation state of the silicon as well as the additional OH^- groups necessary to oxidize the silicon. In the EDP system, ethylene diamine and H_2O combine to generate the holes and the hydroxyl groups, while pyrocatechol forms the soluble species $Si(C_6H_4O_2)_3$. Mixtures of ethylene diamine and pyrocatechol

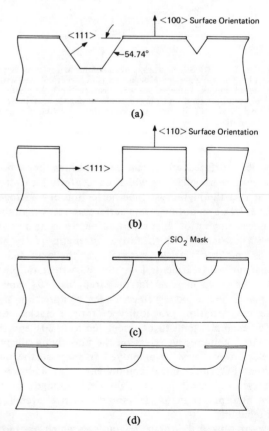

Fig. 4. A summary of wet chemically etched hole geometries which are commonly used in micromechanical devices. (a) Anisotropic etching on (100) surfaces. (b) Anisotropic etching on (110) surfaces. (c) Isotropic etching with agitation. (d) Isotropic etching without agitation. Adapted from S. Terry [29].

without water will not etch silcon. Other common silicon etchants can be analyzed in the same manner.

Since the etching process is fundamentally a charge-transfer mechanism, it is not surprising that etch rates might be dependent on dopant type and concentration. In particular, highly doped material in general might be expected to exhibit higher etch rates than lightly doped silicon simply because of the greater availability of mobile carriers. Indeed, this has been shown to occur in the HNA system (1:3:8) [28], where typical etch rates are 1-3 μm/min at p or n concentrations $>10^{18}$ cm^{-3} and essentially zero at concentrations $<10^{17}$ cm^{-3}.

Anisotropic etchants, such as EDP [26], [27] and KOH [32], on the other hand, exhibit a different preferential etching behavior which has not yet been adequately explained. Etching decreases effectively to zero in samples heavily doped with boron ($\sim 10^{20}$ cm^{-3}). The atomic concentrations at these dopant levels correspond to an average separation between boron atoms of 20-25 Å, which is also near the solid solubility limit (5×10^{19} cm^{-3}) for boron *substitutionally* introduced into the silicon lattice. Silicon doped with boron is placed under tension as the smaller boron atom enters the lattice substitutionally, thereby creating a local tensile stress field. At high boron concentrations, the tensile forces became so large that it is more energetically favorable for the excess boron (above 5×10^{19} cm^{-3}) to enter interstitial sites. Presumably, the strong B-Si bond tends to bind the lattice more rigidly, increasing the energy required to remove a silicon atom high enough to stop etching altogether. Alternatively, since this etch-stop mechanism is *not* observed in the HNA system (in which the HF component can readily dissolve B_2O_3), perhaps the boron oxides and hydroxides initially generated on the silicon surface are not soluble in the KOH and EDP etchants. In this case, high enough surface concentrations of boron, converted to boron oxides and hydroxides in an intermediate chemical reaction, would passivate the surface and prevent further dissolution of the silicon. The fact that KOH is not stopped as effectively as EDP by p$^+$ regions is a further indication that this may be the case since EDP etches oxides at a much slower rate than KOH. Additional experimental work along these lines will be required to fully understand the etch-stopping behavior of boron-doped silicon.

The precise mechanisms underlying the nature of chemical anisotropic (or orientation-dependent) etches are not well understood either. The principal feature of such etching behavior in silicon is that (111) surfaces are attacked at a much slower rate than all other crystallographic planes (etch-rate ratios as high as 1000 have been reported). Since (111) silicon surfaces exhibit the highest density of atoms per square centimeter, it has been inferred that this density variation is responsible for anisotropic etching behavior. In particular, the screening action of attached H_2O molecules (which is more effective at high densities, i.e., on (111) surfaces) decreases the interaction of the surface with the active molecules. This screening effect has also been used to explain the slower oxidation rate of (111) silicon wafers over (100). Another factor involved in the etch-rate differential of anisotropic etches is the energy needed to remove an atom from the surface. Since (100) surface atoms each have two dangling bonds, while (111) surfaces have only one dangling bond, (111) surfaces are again expected to etch more slowly. On the other hand, the differences in bond densities and the energies required to remove surface atoms do not differ by much more than a factor of two among the various planes, so it is difficult to use

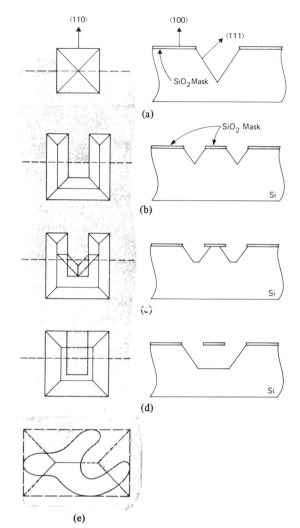

Fig. 5. (a) Typical pyramidal pit, bounded by the (111) planes, etched into (100) silicon with an anisotropic etch through a square hole in an oxide mask. (b) Type of pit which is expected from an anisotropic etch with a slow convex undercut rate. (c) The same mask pattern can result in a substantial degree of undercutting using an etchant with a fast convex undercut rate such as EDP. (d) Further etching of (c) produces a cantilever beam suspended over the pit. (e) Illustration of the general rule for anisotropic etch undercutting assuming a "sufficiently long" etching time.

these factors alone to explain etch rate differentials in the range of several hundred or more [33] which is maintained over a relatively large temperature range. This implies that some screening effects must also play a role. It seems likely that the full explanation of anisotropic etching behavior is a combination of all these factors.

Since anisotropic etching will be a particularly useful tool in the micromachining of structures described below, some detailed descriptions of the practical engineering aspects of this complex subject are deserved.

Consider a (100) oriented silicon wafer covered with SiO_2. A simple rectangular hole etched in the SiO_2 (and oriented on the surface in the (110) directions) will result in the familiar pyramidal-shaped pit shown in Fig. 5(a) when the silicon is etched with an anisotropic etchant. The pit is bounded by (111) crystallographic surfaces, which are invariably the slowest etching planes in silicon. Note that this mask pattern consists only of "concave" corners and very little undercutting of the mask will occur if it is oriented properly. Undercutting due to mask misalignment has been discussed by several workers including Kendall [33], Pugacz-Muraszkiewicz [34], and Bassous [35]. The more complicated mask geometry shown in Fig. 5(b) includes two convex corners. Convex corners, in general, will be undercut by anisotropic etches at a rate determined by the magnitude of the maximum etch rate, by the etch rate ratios for various crystallographic planes, and by the amount of local surface area being actively attacked. Since the openings in the mask can only support a certain flux of reactants, the net undercut etch rate can be reduced, for example, by using a mask with very narrow openings. On the other hand, the undercut etch rate can be increased by incorporating a vertical etch stop layer (such as a heavily boron-doped buried layer which will limit further downward etching); in this case, the reactant flux from the bottom of the etched pit is eventually reduced to near zero when the etch-stopping layer is exposed, so the total flux through the mask opening is maintained by an increased etch rate in the horizontal direction, i.e., an increased undercut rate.

In Fig. 5(b), the convex undercut etch rate is assumed to be slow, while in Fig. 5(c) it is assumed to be fast. Total etching time is also a factor, of course. Convex corners will continue to be undercut until, if the silicon is etched long enough, the pit eventually becomes pyramidal, bounded again by the slow etching (111) surfaces, with the undercut portions of the mask (a cantilever beam in this case) suspended over it, as shown in Fig. 5(d). As an obvious extension of these considerations [34], a general rule can be formulated which is shown graphically in Fig. 5(e). If the silicon is etched long enough, any arbitrarily shaped closed pattern in a suitable mask will result in a rectangular pit in the silicon, bounded by the (111) surfaces, oriented in the (110) directions, with dimensions such that the pattern is perfectly inscribed in the resulting rectangle.

As expected, different geometries are possible on other crystallographic orientations of silicon [35]-[38]. Fig. 4 illustrates several contours of etched holes observed with isotropic etchants as well as anisotropic etchants acting on various orientations of silicon. In particular, (110) oriented wafers will produce vertical etched surfaces with essentially no undercut when lines are properly aligned on the surface. Again, the (111) planes are the exposed vertical surfaces which resist the attack of the etchant. Long, deep, closely spaced grooves have been etched in (110) wafers as shown in Fig. 6(a). Even wafers not exactly oriented in the (110) direction will exhibit this effect. Fig. 6(b) shows grooves etched into a surface which is 10° off the (110) direction—the grooves are simply oriented 10° off normal [36]. Note also that the four vertical (111) planes on a (110) wafer are not oriented 90° with respect to each other, as shown in the plan view of Fig. 6(c).

Crystallographic facet definition can also be observed after etching (111) wafers, even though long times are required due to the slow etch rate of (111) surfaces. The periphery of a hole etched through a round mask, for example, is hexagonal, bounded on the bottom, obviously, by the (111) surface [39]. The six sidewall facets are defined by the other (111) surfaces; three slope inward toward the center of the hole and the other three slope outward. The six inward and outward sloping surfaces alternate as shown in Fig. 7.

Electrochemical Etching

While electrochemical etching (ECE) of silicon has been studied and basically understood for a number of years [45]-[47], practical applications of the technique have not yet been fully realized. At least part of the reason ECE is not now a popular etching procedure is due to the fact that previous

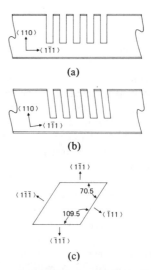

Fig. 6. Anisotropic etching of (110) wafers. (a) Closely spaced grooves on normally oriented (110) surface. (b) Closely spaced grooves on misoriented wafer. (c) These are the orientations of the (111) planes looking down on a (110) wafer.

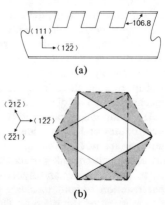

Fig. 7. Anisotropic etching of (111) silicon surfaces. (a) Wafer cross section with the steep sidewalls which would be found with grooves aligned along the (122) direction. (b) Top view of a hole etched in the (111) surface with three inward sloping and three undercut sidewalls, all (111) crystallographic planes.

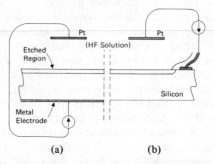

Fig. 8. Uniform electrochemical etching of wafer surfaces has been practiced in the past by making electrical contact either to the back (a) or to the front (b) of the wafer (with suitable protection for the current carrying leads). A positive voltage applied to the silicon causes an accumulation of holes at the silicon/solution interface and etching occurs. A negatively biased platinum electrode in the HF-based solution completes the circuit.

implementations of ECE offered no real advantage over the conventional, isotropic, dopant-dependent formulations discussed in the preceding section. As shown by Fig. 8(a) and (b), in typical ECE experiments electrical contact is made to the front or back of the wafer (the contacted region suitably protected from the etching solution, e.g., with wax or a special holding fixture) and the wafer is either totally immersed or is slowly lowered into the solution while a constant current flows between the positively biased silicon electrode and the negative platinum electrode. Since etching is still, principally, a matter of charge transfer, the fundamental steps are the same as discussed above. The etchants employed, however, are typically HF/H_2O solutions. Since H_2O is not as strong an oxidizing agent as HNO_3, very little silicon etching occurs (<1 Å/min) when the current flow is zero. Oxidation, then, is promoted by applying a positive voltage to the silicon which causes an accumulation of holes in the silicon at the Si/solution interface resulting in an accumulation of OH^- in the solution at the interface. Under these conditions, oxidation of the silicon surface proceeds very rapidly while the oxide is readily dissolved by the HF. Holes, which are transported to the negative platinum electrode (cathode) as H^+ ions, are released there in the form of hydrogen gas bubbles. In addition, excess hole-electron pairs can be created at the silicon surface by optical excitation, thereby increasing the etch rate.

Since the oxidation rate is controlled by current flow and optical effects, it is again clear that the etching characteristics will depend not only on dopant type and resistivity but also on the arrangement of p and n layers in the wafer interior. In particular, ECE has been employed successfully to remove heavily doped substrates (through which large currents are easily conducted) leaving behind more lightly doped epi-layer membranes (which conduct smaller currents, thereby etching more slowly) in all possible dopant configurations (p on p^+, p on n^+, n on p^+, n on n^+) [48], [49].

Localized electrochemical jet etching has been used to generate small holes or thinned regions in silicon wafers. A narrow stream of etchant is incident on one side of a wafer while a potential is applied between the wafer and the liquid stream. Extremely rapid etching occurs at the point of contact due to the thorough agitation of the solution, the continual arrival of fresh solution at the interface, and the rapid removal of reacted products.

A more useful electrochemical procedure using an anisotropic etchant has been developed by Waggener [50] for KOH and more recently by Jackson et al. [51] for EDP. Instead of relying on the electric current flowing through the solution to actively etch the silicon, a voltage bias on an n-type epitaxial layer is employed to stop the dissolution of the p-type silicon substrate at the n-type epitaxial layer. This technique has the advantage of retaining all the anisotropic etching characteristics of KOH and EDP without the need for a buried p^+ layer. Such p^+ films, while serving as simple and effective etch-stop layers, can also introduce undesirable mechanical strains in the remaining membrane which would not be present in the electrochemically stopped, uniformly doped membrane.

When ECE is performed at very low current densities, or in etchant solutions highly deficient in OH^- (such as concentrated 48-percent HF), the silicon is not fully oxidized during etching and a brownish film is formed. In early ECE work, the brownish film was etched off later in a conventional HNA slow silicon etch, or the ECE solution was modified with H_2SO_4 to minimize its formation [47]. This film has since been identified as single-crystal silicon permeated with a dense network of very fine holes or channels, from much less than 1 μm to several micrometers in diameter, preferentially oriented in the direction of current flow [52], [53]. The thickness of the layer can be anywhere from micrometers up to many mils. Porous silicon, as it is called, has a number of interesting properties. Its average density decreases with increasing applied current

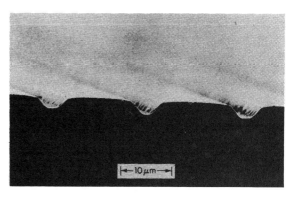

Fig. 9. SEM profile of laser-etched grooves [56]. The horizontal bar indicates 10 μm. Conditions were 100 torr Cl_2, 5.5-W multiline argon-ion laser, $f/10$ focusing, single scan at 90 μm/s. Photo courtesy of D. Ehrlich.

density to as low as 10 percent of normal silicon. Since it is so porous, gases readily diffuse into the structure so that the high-temperature oxidation, for example, of a relatively thick (~4-μm) porous silicon layer can be completed in a very short time (30 min at 1100°C) [52]. Several studies have been undertaken to determine the feasibility of using such deeply oxidized porous silicon regions as a planarizing, deep IC isolation technique [54]. The porous regions are defined by using Si_3N_4 masking films which are attacked relatively slowly by the concentrated HF ECE solution. Problems, however, encountered in the control and elimination of impurities trapped in the porous silicon "sponge-like" material, stress-related effects, and enhanced leakage currents in devices isolated by this technique have been difficult to overcome. Mechanical devices, on the other hand, may not be restricted by these disadvantages.

Besides magnifying the effective thermal oxidation rates, porous silicon can also be chemically attacked at enormously high rates. As expected, the interiors of the pores provide a very large surface area for exposure to the etchant solution. Wafers covered with 100-μm-thick porous silicon layer, for example, will actually shatter and explode when immersed in fast-etching HNA solutions.

Gradations in the porosity of the layer can be simply realized by changing the current with time. In particular, a low current density followed by a high current density will result in a high-porosity region covered with a low-porosity film. Since the porous region is still a single crystal covered with small holes (reported to be near 100 Å on the surface), it is not surprising that single-crystal epitaxial layers have been g own over porous silicon regions, as demonstrated by Unagami and Seki [55]. Once the thickness of the epi-layer corresponds to several times the diameter of the surface pores, it has been verified that the layer will be a uniform single crystal since the crystallinity of the substrate was maintained throughout, despite its permeation with fine holes.

A relatively new tool added to the growing list of micromechanical processing techniques is laser etching. Very high instantaneous etch rates have been observed when high-intensity lasers are focussed on a silicon surface in the presence of some gases. In particular, 20-30 MW/cm^2 of visible argon-ion laser radiation, scanned at rates of 90 μm/s in atmospheres of HCl and Cl_2 produced 3-μm-deep grooves [56], as shown in Fig. 9. At least part of the etching reaction occurs solely as a of local thermal effects. It has been known for some time that silicon will be vigorously attacked by both these gases at temperatures above about 1000°C. Recent experiments in laser annealing have verified that silicon can easily be raised above the melting point at these power densities. There is still some controversy concerning the magnitude of photochemical effects, which might aid in the dissociation of the chlorine-based molecules and enhance the etch rate. In a typical reaction, for example,

$$4HCl + Si_{solid} \rightarrow 2H_2 + SiCl_4.$$

Although many applications in the area of IC fabrication have been suggested for laser etching, the fact that the laser must be scanned over the entire wafer and the etching therefore takes place "serially," net processing time per wafer will necessarily by very high in these applications. For example, a 20-W laser at a power density of 10^7 W/cm^2 etching a 1-μm layer will require over 100 h to completely scan a 4-in-diameter wafer even if etch rates of 100 μm/s are realized. Laser etching is clearly applicable only in special micromachining processing requirements such as the various contours which may be required in print-heads, recording-heads, or other miniature mechanical structures integrated with electronics on the same silicon ship. Versatile as they are, conventional, isotropic, anisotropic, electrochemical, and ion-etching processes exhibit a limited selection of etched shapes. On the other hand, the significant key advantage of laser etching is that nearly any shape or contour can be generated with laser etching in a gaseous atmosphere simply by adjusting the local exposure dose continuously over the etched region. Such a capability will be extremely useful in the realization of complex mechanical structures in silicon.

Epitaxial Processes

While the discussion up to this point has concentrated on material removal as a micromachining technique, material addition, in the form of thin film deposition or growth, metal plating, and epitaxial growth are also important structural tools. Deposited thin films have obvious applications in passivation, wear resistance, corrosion protection, fatigue strength enhancement (elaborated on in Section II), and as very thin, high-precision spacers such as those employed in hybrid surface acoustic wave amplifiers and in other thin-film devices. On the other hand, epitaxy has the important property of maintaining the highly perfect single-crystal orientation of the substrate. This means that complex vertical and/or horizontal dopant distributions (i.e., fast and slow etching regions for subsequent micromachining by etching) can be generated over many tens of micrometers without compromising the crystal structure or obviating subsequent anisotropic processes. Etch-stop layered structures are important examples and will be considered in more detail in Section VI. Fig. 10(a), however, briefly illustrates two simple configurations: hole A is a simple etch-stop hole using anisotropic etching and a p$^+$ boron-doped buried layer while hole B is a multilevel hole in which the epi-layer and a portion of the lightly doped substrate have been anisotropically etched from the edge of the p$^+$ buried region. One obvious advantage of these methods is that the depth of the hole is determined solely by the thickness of the epi-layer. This thickness can be controlled very accurately and measured even before etching begins. Such depth control is crucial in many micromechanical applications we will discuss later, particularly in fiber and integrated optics.

Where the goal of IC manufacturing is to fabricate devices as small as possible (indeed, diffusions deeper than a few micrometers are very difficult and/or time-consuming), a necessary feature of most micromechanical processing techniques is the

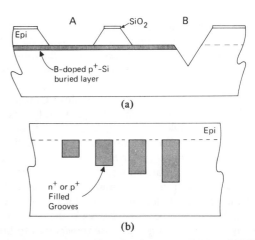

Fig. 10. (a) Since anisotropic etchants such as KOH and EDP exhibit reduced etch rates on silicon heavily doped with boron, many useful structures have been realized by growing epi over a diffused region to form a buried etch-stop layer. (b) Diagram showing how epitaxial silicon could be grown preferentially [57] in vertical-walled grooves. Doped grooves with large cross sections (>25 × 25 μm) can then be buried beneath an ordinary epi-layer.

ability to generate structures on the order of tens or even hundreds of micrometers. Both etching and epitaxial deposition possess this property. Epitaxial silicon can be grown at rates of 1 μm/min, so that layers even greater than 100 μm are readily attainable. In addition, the process parameters can be accurately controlled to allow the growth of complex three-dimensional patterns. For example, since the growth rate depends critically on temperature and gas-mixing dynamics, increased deposition rates can be observed at the *bottom* of deep, narrow, anisotropically etched grooves. In this way, Runyan *et al.* [57] (and later Smeltzer) were able to completely fill 10-μm-wide grooves (up to 100 μm deep) epitaxially with negligible silicon growth over the rest of the wafer surface. The simultaneous addition of HCl gas during the growth process is required to obtain these unusual results. Since HCl gas is an isotropic silicon etchant at these temperatures, the silicon which is epitaxially grown on the outer surface is immediately etched away in the flowing gas stream. Silicon grown in the poorly mixed atmosphere of the grooves, however, etches at a much slower rate and a net growth occurs in the groove. Heavily doped, buried regions extending over tens of micrometers are easily imagined under these circumstances as indicated in Fig. 10(b). After refilling the grooves with heavily doped silicon, the surface has been lightly etched in HCl and a lightly doped layer grown over the entire wafer. These results could not be obtained by conventional diffusion techniques. One implementation of such structures which has already been demonstrated is in the area of high-power electronic devices [58], to be discussed below in more detail. Such a process could also be used in mechanical applications to bury highly doped regions which would be selectively etched away at a later stage to form buried channels within the silicon structure.

Finally, a limited amount of work has been done on epitaxial growth through SiO$_2$ masks. Normally under these conditions, SCS will grow epitaxially on the bare, exposed crystal while polycrystalline silicon is deposited on the oxide. This mixed deposit has been used in audio-frequency distributed-filter, electronic circuits by Gerzberg and Meindl at Stanford [59]. At reduced temperatures, however, with HCl added to the H$_2$ and SiCl$_4$ in the gas stream no net deposits will occur on the SiO$_2$ while faceted, single-crystal, epitaxial pedestals will grow on the exposed regions since polysilicon is etched by the HCl at a faster rate than the SCS [60]. Such epitaxial projections may find use in future three-dimensional micromechanical structures.

Thermomigration

During 1976 and 1977, Anthony and Cline of GE laboratories performed a series of experiments on the migration of liquid eutectic Al/Si alloy droplets through SCS [61]-[67]. At sufficiently high temperatures, Al, for example, will form a molten alloy with the silicon. If the silicon slice is subjected to a temperature gradient (approximately $50°C/cm$, or $2.0°C$ across a typical wafer) the molten alloy zone will migrate toward the hotter side of the wafer. The migration process is due to the dissolution of silicon atoms on the hot side of the molten zone, transport of the atoms across the zone, and their deposition on the cold side of the zone. As the Al/Si liquid region traverses the bulk, solid silicon in this way, some aluminum also deposits along with the silicon at the colder interface. Thermomigration hereby results in a p-doped trail extending through, for example, an n-type wafer. The thermomigration rate is typically $3\ \mu m/min$ at $1100°C$. At that temperature, the normal diffusion rate of Al in silicon will cause a lateral spread of the p-doped region of only $3-5\ \mu m$ for a migration distance of $400\ \mu m$ (the full thickness of standard silicon wafers).

Exhaustive studies by Anthony and Cline have elucidated much of the physics involved in the thermomigration process including migration rate [62], p-n junction formation [64], stability of the melt [65], effect of dislocations and defects in the silicon bulk, droplet morphology, crystallographic orientation effects, stresses induced in the wafer as a result of thermomigration [67], as well as the practical aspects of accurately generating, maintaining, and characterizing the required thermal gradient across the wafer. In addition, they demonstrated lamellar devices fabricated with this concept from arrays of vertical junction solar cells, to high-voltage diodes, to negative-resistance structures. Long migrated columns were found to have smaller diameters in (100) oriented wafers, since the droplet attains a pyramidally tapered point whose sides are parallel to the (111) planes. Migrated lines with widths from 30 to $160\ \mu m$ were found to be most stable and uniform in traversing 280-μm-thick (100) wafers when the lines were aligned along the (110) directions. Larger regions tended to break up into smaller independent migrating droplets, while lines narrower than about $30\ \mu m$ were not uniform due to random-walk effects from the finite bulk dislocation density in the wafer. Straight-line deviations of the migrated path, as a result of random walk, could be minimized either by extremely low ($\ll 100/cm^2$) or extremely high ($>10^7/cm^2$) dislocation densities. On the other hand, the dislocation density in the recrystallized droplet trail is found to be essentially zero, not unexpected from the slow, even, liquid-phase epitaxy which occurs during droplet migration. Dopant density in the droplet trail corresponds approximately to the aluminum solid solubility in silicon at the migration temperature $\sim 2 \times 10^{19}\ cm^{-3}$, which corresponds to $\rho = 0.005$ $\Omega \cdot cm$. The p-type trail from a 50-μm-diameter aluminum droplet migrated through a 300-μm-thick n-type wafer would, therefore, exhibit less than 8-Ω resistance from front to back and would be electrically well-isolated from other nearby trails due to the formation of alternating p-n junctions, as shown in Fig. 11.

Nine potential sources of stress (generated in the wafer from the migrated regions) have been calculated by Anthony and

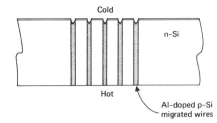

Fig. 11. In some applications of silicon micromechanics, it is important to connect the circuitry on one side of a wafer to mechanical structures on the other side. Thermomigration of Al wires, discussed extensively by Anthony and Cline [61]-[67], allows low-resistance (<8-Ω), close-spaced (<100-μm) wires to be migrated through thick (375-μm) wafers at reasonable temperatures ($\sim 1100°C$) with minimal diffusion ($<2\ \mu m$).

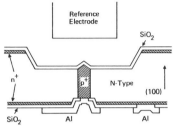

Fig. 12. Structure of the gate-controlled diode of Wen and Zemel [69]. Circuitry is on the bottom (protected) side of the wafer, while the sensor electrode is on the top. The p^+ feedthrough was accomplished by thermomigration of Al from the circuit side to the sensor side of the wafer. For ionic concentration measurements, an appropriate ion-sensitive membrane must be deposited over the oxide on the sensor side. Figure courtesy of C. C. Wen.

Cline. Maximum stresses intrinsic to the process (i.e., those which are present even when processing is performed properly) are estimated to be as high as $1.39 \times 10^9\ dyne/cm^2$, which can be substantially reduced by a post-migration thermal anneal. Although the annealed stress will be about two orders of magnitude below the yield point of silicon at room temperature, it may increase the susceptibility of the wafer to fracture and should be minimized, especially if a large number of migrated regions are closely spaced.

One obvious utilization of thermomigration is the connection of circuitry on one side of a wafer to a mechanical function on the other side. Another application may be the dopant-dependent etching of long narrow holes through silicon. Since the work of Anthony and Cline, the thermomigration process has been used to join silicon wafers [68] and to serve as feedthroughs for solid-state ionic concentration sensors (see Fig. 12) [69]. Use of thermomigrated regions in power devices is another potential application. Even more significantly, laser-driven thermomigration has been demonstrated by Kimerling *et al.* [70]. Such a process may be extremely important in practical implementations of these migration techniques, especially since the standard infrared or electron-beam heating methods used to induce migration are difficult to control uniformly over an entire wafer.

Field-Assisted Thermal Bonding

The use of silicon chips in exposed, hostile, and potentially abrasive environments will often require mounting techniques substantially different from the various IC packaging methods now being utilized. First reported by Wallis and Pomerantz in 1969, field-assisted glass–metal thermal sealing [71] (sometimes called Mallory bonding after P. R. Mallory and Co., Inc., where Wallis and Pomerantz were then employed) seems to

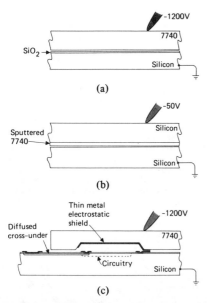

Fig. 13. Field-assisted thermal bonding can be used to hermetically bond (a) 7740 glass to silicon (bare or oxidized) or (b) silicon to silicon simply by heating the assembly to about 300°C and applying a voltage. Glass can be bonded to IC chips (c) if the circuitry is first protected by etching a shallow (~10-μm) well in the glass and depositing a grounded metal shield inside the well [76].

fulfill many of the requirements for bonding and mounting micromechanical structures. The technique is simple, low temperature, high strength, reliable, and forms hermetic seals between metals and conventional alkali-silicate glasses [72]. It is also very similar to well-known high-temperature thermal bonds where the cohesive metal–oxides, which are generated during the heating process, readily mix with the viscous glass. In the case of silicon, a glass slide is placed over a polished wafer (bare *or* thermally oxidized), the assembly is heated to about 400°C, and a high voltage (~1200 V) is applied between the silicon and the metal contact to the other side of the glass. If the sample is not too large, the metal contact may be a simple point probe located near one corner as shown in Fig. 13(a). Since the negative electrode is applied to the glass, ionic conduction causes a drift of positive ions away from the glass/Si interface into the bulk of the glass. The depletion of positive ions at the interface results in a high electric field across the air gap between the two plates. Electrostatic forces here, estimated to be higher than 350 psi, effectively clamp the pieces locally, conforming the two surfaces to obtain the strong, uniform, hermetic seal characteristic of field-assisted thermal bonding. The bonding mechanism itself has been the subject of some controversy, as discussed recently by Brownlow [73]. His convincing series of deductions, however, suggest that the commonly observed initial current peak at the onset of bonding is actually dissipated in the newly formed, narrow space-charge region in the glass at the interface. This high energy-density pulse, in the early stages of bonding, was shown to be capable of increasing the interfacial temperature by as much as 560°C, more than enough to induce the familiar, purely thermal glass/metal seal. Brownlow shows how this model correlates well with several other features observed during the bonding process.

From a device viewpoint, it is important to recognize that the relative expansion coefficients of the silicon and glass should match as closely as possible to alleviate thermal stresses after the structure has cooled. This aspect of field-assisted bonding also has the obvious advantage of yielding integrated mechanical assemblies with very small mechanical drifts due to ambient temperature variations. Corning borosilicate glasses 7740 and 7070 have both been used successfully in this regard. In addition, Brooks *et al.* [74] have even bonded two silicon wafers by sputtering approximately 4 μm of 7740 glass over one of the wafers and sealing the two as already described, with the negative electrode contacting the coated wafer as shown in Fig. 13(b). Since the glass is so thin, however, the sealing voltage was not required to be above 50 V.

A high degree of versatility makes this bonding technique useful in a wide variety of circumstances. It is not necessary to bond to bare wafers, for example; silicon passivated with thermal oxide as thick as 0.5 μm is readily and reliably bonded at somewhat higher voltage levels. The bonding surface may even be partially interrupted with aluminized lines, as shown by Roylance and Angell [75], without sacrificing the integrity or hermeticity of the seal since the aluminum also bonds thermally to the glass. In addition, glass can be bonded to silicon wafers containing electronic circuitry using the configuration shown in Fig. 13(c) [76]. The circuitry is not affected if a well is etched in the glass and positioned over the circuit prior to bonding. A metal film deposited in the well is grounded to the silicon substrate during actual bonding and serves as an electrostatic shield protecting the circuit. Applications of all these aspects will be presented and expanded upon in the following sections.

IV. GROOVES AND HOLES

Even simple holes and grooves etched in a silicon wafer can be designed and utilized to provide solutions in unique and varied applications. One usage of etched patterns in silicon with far-reaching implications, for example, is the generation of very high precision molds for microminiature structures. Familiar, pyramidal-shaped holes anisotropically etched in (100) silicon and more complex holes anisotropically etched in (110) silicon were used by Kiewit [77] to fabricate microtools such as scribes and chisels for ruling optical gratings. After etching the holes in silicon through an SiO_2 mask, the excess SiO_2 was removed and very thick layers of nickel–phosphorus or nickel–boron alloys were deposited by electroless plating. When the silicon was completely etched away from the thick plated metal, miniature tools or arrays of tools were accurately reproduced in the metal with geometrically well-defined points having diameters as small as 50 nm. The resulting metal tools had a hardness comparable to that of file steel.

Similar principles were employed by Wise *et al.* [78] to fabricate miniature hemispherical structures for use as thermonuclear fusion targets. In these experiments, a large two-dimensional array of hemispherical holes was etched into a silicon wafer using an HNA isotropic solution, approximately as shown in Fig. 4(c). After removing the SiO_2/Cr/Au etch mask, polymer, glass, metal, or other thin films are deposited over the wafer, thereby conforming to the etched hemispherical shapes. When two such wafers are aligned and bonded, the silicon mold can be removed (either destructively by etching or nondestructively by using a low adhesion coating between the silicon and the deposited film). The resulting molded shape is a thin-walled spherical shell made from the deposited material. Fig. 14 is the process schedule for a simple metal hemishell demonstrated by Wise *et al*.

The potential of making arrays of sharp points in silicon itself by etching was employed in a novel context by Thomas and Nathanson [79], [80]. They defined a very fine grid

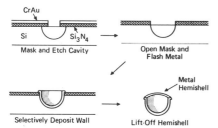

Fig. 14. Fabrication sequence for free-standing metal hemishells using an isotropic silicon-etching technique [78]. Typical dimensions of the hemishell are 350-μm diameter with a 4-μm-thick wall. Courtesy of K. D. Wise.

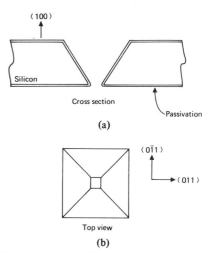

Fig. 15. (a) Cross section and (b) top view of anisotropically etched silicon ink jet nozzle in a (100) wafer developed by E. Bassous et al. [5], [43], [8].

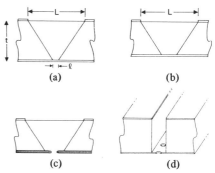

Fig. 16. A number of different methods have been developed for fabricating silicon ink jet nozzles. (a) and (b) show the errors in final nozzle size which occurs when the wafer thickness varies. (c) shows a p^+ membrane structure. This design yields round nozzles and also minimizes the effects of wafer thickness variations. Nozzles can be more closely spaced by using the p^+ membrane technique on a (110) wafer, as shown in (d) [35].

(typically 25 μm center to center) in an SiO_2 mask, then isotropically etched the silicon exposed in the grid lines with an HNA mixture. The isotropic etch undercuts each square segment of the oxide grid uniformly around its periphery. If the etching is quenched just after the oxide segments are completely undercut and fall from the surface, a large array of very sharply tipped silicon points is obtained. Point diameters were estimated to be about 20 nm. These silicon points, at densities up to 1.5×10^5 cm^2, were used by Thomas and Nathanson as efficient, uniform, photosensitive field emitter arrays which were imaged onto a phosphor screen closely spaced to the wafer. A more complex extension of this fabrication technique will be described below in the section on Thin Cantilever Beams.

Ink Jet Nozzles

Since anisotropic etching offers a powerful method for controlling undercutting of masks during silicon etching, these techniques are important candidates for etching high-resolution holes clear through wafers as Bassous et al. [5], [43], [81], [82] first realized and pursued extensively; see Fig. 15. Patterns etched clear through wafers have many potential applications, as will be seen below, but one of the simplest and most commercially attractive is in the area of ink jet printing technology [83], [86]. As shown in Fig. 16(a), the geometry of the pyramidal hole in (100) silicon can be adjusted to completely penetrate the wafer, the square hole on the bottom of the wafer forming the orifice for an ink jet stream. The size of the orifice (typically about 20 μm) depends on the wafer thickness t and mask dimension L according to $l = L - (2t/\tan\theta)$, where $\theta = 54.74°$ is the angle between the (100) and (111) planes. In practice, the dimension l is very difficult to control accurately because 1) wafer thickness t is not easy to control accurately and 2) small angular misalignments of a square mask will result in an effective L which is larger than the mask dimension [43], thereby enlarging l as shown in Fig. 16(b). The angular misalignment error can be eased by using a round mask (*diameter L*) which will give a square hole $L \times L$ independent of orientation, as described in Section III (and Fig. 5(e)) by the general rule of anisotropic undercutting.

Membrane structures have also been used in ink jet nozzle designs not only to eliminate the effects of wafer thickness variations, but also to permit more densely packed orifices as well as orifice shapes other than square. In one technique described by Bassous et al. [35], the wafer surface is highly doped with boron everywhere but the desired orifice locations. Next, the wafer is anisotropically etched clear through with EDP as described above, using a mask which produces an l which is 3 to 5 times larger than the actual orifice. Since EDP does not attack silicon which is highly doped with boron, a p^+ silicon membrane will be produced, suspended across the bottom of the pit with an orifice in the center corresponding to the location previously left undoped; see Fig. 16(c). The use of a membrane can also be extended to decrease the minimum allowed orifice spacing. Center-to-center orifice spacing is limited to about 1.5 times the wafer thickness when the simple square geometries of Figs. 15, 16(a)-(c) are employed, but can be much closer using membranes. Orifice spacings in *two* dimensions can be made very small by using (110) oriented wafers and etching vertical-walled *grooves* (as described in Section III) clear through the wafer, aligned to rows of orifices on the other side fabricated by this membrane technique. The result, shown in Fig. 16(d), is a number of closely spaced rows containing arbitrarily spaced holes in a long, narrow rectangular p^+ membrane [35].

Deep grooves or slots etched clear through (110) silicon have been used by Kuhn et al. [87] in another important ink jet application. At a characteristic distance from the ink jet orifice, the ink stream, which is ejected under high pressure, begins to break up into well-defined droplets at rates of about 10^5 drops per second as a result of a small superimposed sinusoidal pressure disturbance. A charge can be induced on individual droplets as they separate from the stream at this point by passing the jet through a charging electrode. Once charged, the drops can be electrostatically deflected (like an electron beam) to strike the paper at the desired locations. Kuhn et al. etched

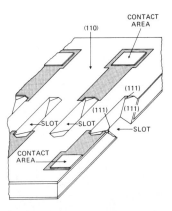

Fig. 17. Grooves anisotropically etched clear through a (110) wafer were employed as charge electrode arrays by Kuhn et al. [87] in an ink jet printing demonstration. A charge can be induced on individual ink droplets as they pass through the grooves by applying a voltage to the walls of the groove. Subsequently, drops are "steered" to the paper after traveling through a high electric field. Figure courtesy of L. Kuhn.

several grooves clear through (110) silicon, doped the walls of the grooves so they would be conductive, and defined contact pads connected to the doped sidewalls of the grooves, as shown in Fig. 17. By arranging for the streams to pass through these grooves right at the breakoff points, the grooves can be operated as an array of independent charge electrons. In the design of large, linear arrays of closely spaced ink jet orifices (typical spacing is less than 250 μm), where high precision miniaturized structures are required, silicon micromechanics can provide useful and viable structural alternatives, as long as the usual materials considerations (such as materials compatibility, fatigue, and corrosion) are properly taken into account.

In an effort to integrate ink jet nozzle assemblies more efficiently and completely, another experimental structure was demonstrated in which nozzle, ink cavity, and piezoelectric pressure oscillator were combined using planar processing methods [88]. Orifice channels were first etched into the surface of a (110) oriented wafer as shown in Fig. 18, using an isotropic HNA mixture. After growing another SiO₂ masking layer, anisotropic (EDP) etching was employed to etch the cavity region as well as a deep, vertical-walled groove (which will eventually become the nozzle exit face) clear through the wafer. The wafer must be accurately aligned to properly etch the vertical grooves according to the pattern in Fig. 19. After etching, the silicon appears as seen in Fig. 20(a). The individual chips are separated from the wafer and thick 7740 glass (also containing the supply channel) is anodically bonded to the bottom of the chips. Next, a thin 7740 glass plate (125 μm thick), serving as the pump membrane, is aligned to the edge of the nozzle exit face and anodically bonded to the other side of the silicon chip. The exit orifice, after anodic bonding, is shown in Fig. 20(b). Once the piezo-plate is epoxied to the thin glass plate, a droplet stream can be generated, exiting the orifice at the edge of the chip and parallel to the surface, as shown in Fig. 21.

This planar integrated structure was deliberately specified to conform to the prime requirement of silicon micromechanical applications—no mechanical machining or polishing and minimum handling of individual chips to keep processing and fabrication costs as low as possible. Even though the drops are ejected from the edge of the wafer in this design, the exit face is defined by crystallographic planes through anisotropic etch-

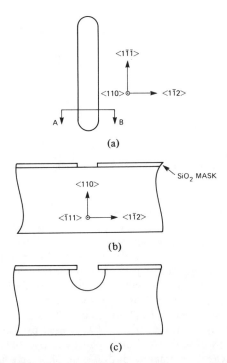

Fig. 18. Orientation and cross section of the isotropically etched nozzle for the planar ink jet assembly after etching. (a) Top view of nozzle channel. (b) Cross section AB before silicon etch. (c) After silicon etch. Typical channel depth is 50 μm.

ing. Any other nozzle design in which drops are to be ejected parallel to the surface would require an expensive polishing step on the edge of the chip to obtain the necessary smoothness which occurs automatically in this design as a result of inexpensive, planar, batch-processed, anisotropic etching.

Miniature Circuit Boards and Optical Benches

The packing density of silicon memory and/or circuitry chips can be greatly increased by using silicon essentially as miniature pluggable circuit boards. Two-dimensional patterns of holes have been anisotropically etched clear through two wafers, which are then bonded together such that the holes are aligned as illustrated in Fig. 22. When the resulting cavities are filled with mercury, chips with beam-lead, plated, or electromachined metal probes can be inserted into both sides of the minicircuit board. Such a packaging scheme has been under development for low-temperature Josephson-junction circuits [89]. Dense circuit packaging and nonpermanent die attachment are the primary advantages of this technique. In the case of Josephson-junction circuits, there is an additional advantage in that the entire computer—substrates for the thin-film circuits, circuit boards, and structural supports—are all made from silicon, thereby eliminating thermal mismatch problems during temperature cycling.

Perhaps the most prolific application of silicon anisotropic etching principles is miniature optical benches and integrated optics [90]-[102]. Long silicon V-grooves in (100) wafers are ideal for precise alignment of delicate, small-diameter optical fibers and permanently attaching them to silicon chips. Two cleaved fibers can be butted together this way, for example to accuracies of 1 μm or better. In addition, a fiber can be accurately aligned to some surface feature

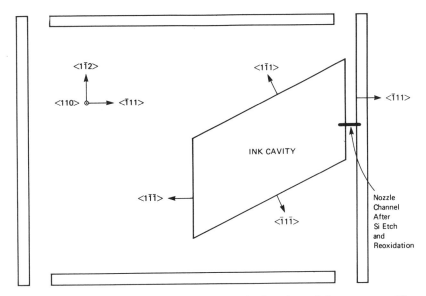

Fig. 19. Orientation of the anisotropically etched ink cavity and deep grooves. After EDP etching, all the (111) surfaces will have flat, vertical walls. Typical cavity size is about 0.5 cm.

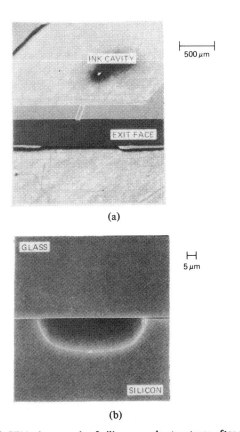

Fig. 20. (a) SEM photograph of silicon nozzle structures after the EDP etch, ready for anodic bonding. Note the nozzle channel which connects the ink cavity to the flat, vertical walls of the exit face. (b) SEM photograph of the ink jet orifice after anodic bonding; glass membrane on top, silicon on bottom.

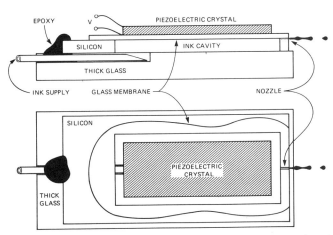

Fig. 21. Schematic of completed nozzle structure showing thick and thin glass plates anodically bonded to either side of the silicon, ink supply line, and piezoelectric ceramic epoxied to the thin glass plate. From [88].

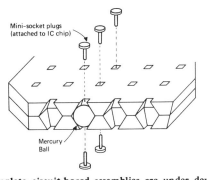

Fig. 22. Complete circuit-board assemblies are under development to optimize the packaging and interconnection of cryogenic Josephson-junction circuits and computers [89]. Miniature socket arrays are created by bonding together two silicon wafers with anisotropically etched holes and filling the cavity with mercury. Miniature plugs attached to the circuit chips themselves are inserted into both sides of the "circuit board." Silicon is used because it can be micromachined accurately, wiring can be defined lithographically, and thermal mismatch problems are alleviated.

[96], [97], [99], [101], [102]. In Fig. 23(a), a fiber output end is butted up against a photodiode, which can then be integrated with other on-chip circuitry; fiber arrays, of course, are also easily integrated with diode arrays. In Fig. 23(b), a fiber core is accurately aligned to a surface waveguiding layer,

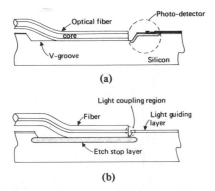

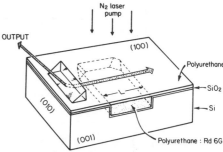

Fig. 23. Silicon is rapidly becoming the material of choice for manipulating fiber-optic components. Two examples are shown here. (a) Coupling a fiber output to a diode detector using an etched V-groove for simple and accurate fiber alignment. (b) Coupling a fiber output to a deposited thin-film optical waveguide using a buried etch-stop layer to obtain precise vertical alignment.

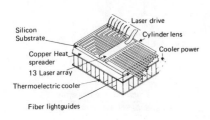

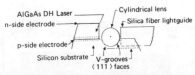

Fig. 24. The most advanced fiber-optic coupling scheme was designed and demonstrated by Crow et al. [100]. The output from an array of solid-state lasers was focussed into a corresponding array of optical fibers using another fiber, aligned between the laser array and the output fibers, as a cylindrical condenser lens. All the fibers are aligned by pressing them into accurately aligned V-grooves anisotropically etched into the silicon. Figure courtesy of J. Crow.

by resting the fiber on a buried etch-stop diffusion over which an epitaxial layer has been grown to an accurate thickness.

The most ambitious use of silicon as a mini-optical bench is the GaAs laser-fiber array developed by Crow et al. [100]. In this assembly, the light outputs from a perpendicular array of GaAs lasers, mounted on the silicon surface in Fig. 24, are coupled into an optical fiber aligned parallel to the array by one V-groove. This first fiber serves as a cylindrical lens to focus the highly divergent laser light into a perpendicular array of fibers corresponding to the laser array. The linear fiber bundle can now be maneuvered, swept, or positioned independently of the laser package. In addition, this scheme couples the laser light into the fibers very efficiently, while the silicon substrate has the important advantages of serving as an efficient heat sink for the laser array, can be processed to provide isolated electrical contacts and, potentially, on-chip driving electronics to each individual laser in the array.

In addition to fiber alignment aids, such V-grooves, when passivated with SiO_2 and filled with a spun-on polymer, have also been employed as the light-guiding structures themselves [91], [92]. A similar, highly innovative device demonstrated by Hu and Kim also made use of anisotropically etched and

Fig. 25. The high-precision structures of which SCS is inherently capable have included the laser resonator shown here which was demonstrated by Hu and Kim [98]. In this case, sidewalls defined by (100) crystallographic planes have become the perfectly flat and parallel surfaces necessary for the aligned mirrors of a thin-film laser cavity. Figure courtesy of C. Hu.

filled waveguides [98]. When a shallow rectangular well, oriented parallel to the (010) and (001) directions, is etched into a (100) silicon wafer using KOH, the sidewalls of the etched well are defined by these planes and are vertical to the surface. Since the two facing walls of the cavity are ideal, identical crystallographic planes, they are perfectly parallel to each other and normal to the wafer surface. After the wafer is oxidized and spun with a polymer containing a laser dye, the two reflecting, parallel walls of the etched hole (with the dye in between) form a laser cavity. This waveguide laser was optically pumped with a pulsed nitrogen laser by Hu and Kim. Some of the radiation in the cavity itself is coupled out through leakage modes to the thin, excess layer of polymer covering the wafer surface around the laser cavity, as shown in Fig. 25. The output radiation is, of course, in the form of surface guided waves and can be coupled out by conventional integrated optics prism or grating methods.

Gas Chromatograph on a Wafer

One of the more ambitious, practical, and far-reaching applications of silicon micromechanical techniques has been the fully integrated gas chromotography system developed at Stanford by S. Terry, J. H. Jerman, and J. B. Angell [29], [103]. The general layout of the device is illustrated in Fig. 26(a). It consists of a 1.5-m-long capillary column, a gas control valve, and a detector element all fabricated on a 2-in silicon wafer using photolithography and silicon etching procedures. Isotropic etching is employed to generate a spiral groove on the wafer surface 200 μm wide, 40 μm deep, and 1.5 m long. After the wafer is anodically bonded to a glass plate, hermetically sealing the grooves from each other, the resulting 1.5-m-long capillary will be used as the gas separation column. Gas input to the column is controlled by one valve fabricated integrably on the wafer along with the column itself. The valve body is etched into the silicon wafer in three basic steps. First a circular hole is isotropically etched to form the valve cylinder. A second isotropic etch enlarges the valve cylinder while leaving a circular ridge in the bottom of the hole which will serve as the valve seating ring. Finally, holes are anisotropically etched clear through the wafer in a manner similar to ink jet nozzles such that the small orifice exists in the center of the seating ring (see Fig. 26(b)). The flexible valve sealing diaphragm, initially made from a silicon membrane, is now a thin (5-15-μm) nickel button flexed on or off by a small electrical solenoid. Both the valve body and sealing diaphragm are coated with parylene to provide conformal leak-tight sealing surfaces. The sensor, located in the output line of the column,

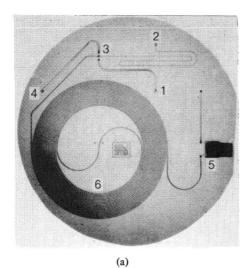

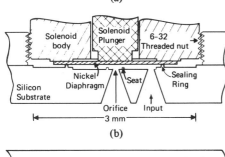

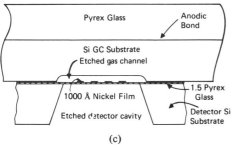

Fig. 26. The most ambitious project utilizing the mechanical properties of silicon is the Stanford gas chromatograph [29], [103]. (a) Overall view of the full silicon wafer showing 1) sample input, 2) purge input, 3) valve region, 4) exhaust of unused sample, 5) sensor region, 6) separation column. The various etched grooves are sealed by anodically bonding a glass plate over the entire wafer. A cross section of the valve assembly is drawn in (b) including the valve cavity, seating ring, and input orifice etched into the silicon as well as the thin nickel diaphragm. The thin-film thermal detector in (c) is also silicon based, consisting of a metal resistor evaporated on SiO_2, thermally isolated by etching the silicon from beneath. Figures courtesy of J. Jerman and S. Terry.

is also based on silicon processing techniques. A thin metal resistor is deposited and etched in a typical meandering configuration over a second oxidized silicon chip. Next, the silicon is anisotropically etched from the back surface of the wafer leaving an SiO_2 membrane supported over the etched hole. This hole is aligned so that the metal resistor is positioned in the center of the membrane and thus thermally isolated from the silicon substrate as shown in Fig. 26(c). The gases separated in the column are allowed to flow over the sensor before being exhausted.

Operation of the column proceeds as follows. After completely purging the system with the inert carrier gas, which flows continuously through port 2 at a pressure of about 30 psi, the valve 3 is opened and the unknown gas sample (held at a pressure higher than the purge gas) is bled into the

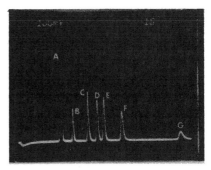

Fig. 27. Example of an output from the miniature gas chromatograph shown in Fig. 26. A) nitrogen; B) pentane; C) dichloromethane; D) chloroform; E) 111-trichloroethane; F) trichloroethylene; G) toluene. Photo courtesy of J. Jerman and S. Terry.

column through port 1 while the narrow purge supply line appears as a high impedance path to the direction of the sample flow. After introducing a sample with a volume as low as 5 nl, the valve is closed again and purge gas flushes the sample through the column 6. Since the etched capillary is filled with a gas chromatography liner, the various molecular constituents of the sample gas traverse the column at different rates and therefore exit the system sequentially. The sensor element 5 detects the variations in thermal conductivity of the gas stream by biasing the thin, deposited metal resistor at a fixed current level and monitoring its resistance. A burst of high thermal conductivity gas will remove heat from the resistor more efficiently than the low conductivity carrier gas and a small voltage pulse will be detected. A typical signal is shown in Fig. 27. Such a small chromatograph can only operate properly if the sample volume is much smaller than the volume of the column. For this reason, it is essential to fabricate the ultra-miniature valve and detector directly on the wafer with the column to minimize interfering "dead space."

A complete, portable gas chromatograph system prototype is being developed by the Stanford group which will continuously monitor the atmosphere, for example, in a manufacturing environment and identify and record 10 different gases with 10 ppm accuracy—all within the size of a pocket calculator.

Miniature Coolers

Besides the Stanford gas chromatograph, the advantageous characteristics of anodic bonding are being employed in even more demanding applications. Recognizing the proliferation of cryogenic sensing devices and circuits based on superconducting Josephson junctions, W. A. Little at Stanford has been developing a Joule-Thomson minirefrigeration system initially based on silicon anisotropic etching and anodic bonding [104]. As shown in Fig. 28, channels etched in silicon comprise the gas manifold, particulate filter, heat exchanger, Joule-Thomson expansion nozzle, and liquid collector. The channels are sealed with an anodically bonded glass plate and a hypodermic gas supply tubing is epoxied to the input and output holes. Such a refrigerator cools down the region near the liquid collector as the high-pressure gas (after passing through the narrow heat exchange lines) suddenly expands into the liquid collector cavity. Little has derived scaling laws for such Joule-Thomson minirefrigeration systems, which show that cooling capacities in the 1-100-mW range at 77 K, cool down rates on the order of seconds, and operating times of 100's of hours (with a single gas cylinder) are attainable using a total channel length of about 25 cm, 100 μm in diameter—dimensions simi-

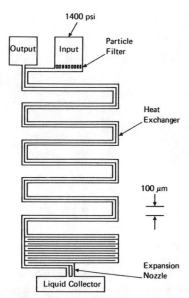

Fig. 28. Grooves etched in silicon have been proposed for the construction of miniature cryogenic refrigerators. In the Joule–Thomson system here, high pressure N_2 gas applied at the inlet expands rapidly in the collection chamber, thereby cooling the expansion region. An anodically bonded glass plate seals the etched, capillary grooves. Adapted from W. A. Little [104].

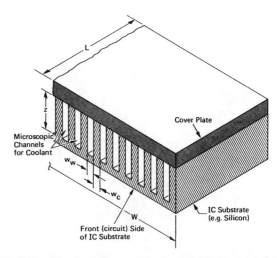

Fig. 29. Schematic view of a compact heat sink incorporated into an integrated circuit chip [105]. For a 1-cm² silicon IC chip, using water as the coolant, the optimum dimensions are approximately $w_w = w_c = 57$ μm and $z = 365$ μm. The cover plate is 7740 glass anodically bonded to the silicon, and the channels are anisotropically etched into the (110) wafer with a KOH-based etchant. Thermal resistances less than 0.1 C/W were measured. Figure courtesy of D. Tuckerman.

lar to the gas chromatograph design discussed previously. These lines, however, must not only withstand the thermal shocks of repeated heating and cooling, but also survive the high internal gas pressures (as high as 1000 psi) which occur simultaneously. SCS can be designed to work well in this application because of its high strength. In addition, the glass/silicon bond is ideal not only because of its strength, but also because the nature of the bonding process presupposes an excellent match in thermal coefficients of expansion of the two materials. One disadvantage of silicon in this application is its very high thermal conductivity, even at low temperatures, which limits the attainable temperature gradient from the (ambient) inlet to the liquid collection chamber. Similar all-glass devices have already found use in compact, low-temperature IR sensors and will likely be employed in other scientific instruments from high-sensitivity magnetometers and bolometers to high-accuracy Josephson-junction voltage standards.

As the cycle times of conventional room-temperature computer mainframes and the level of integration of high-speed semiconductor bipolar logic chips continue to increase, the difficulty of extracting heat from the chips in the CPU is rapidly creating a serious packaging problem. Faster cycle times require closer packing densities for the circuit chips in order to minimize signal propagation times which are already significant in today's high-speed processors. This increased packing density is the crux of the heat dissipation problem. Maximum power dissipation capabilities for conventional multichip packaging assemblies have been estimated at 20 W/cm². In response to these concerns, a new microcooling technology has been developed at Stanford by Tuckerman and Pease which makes use of silicon micromachining methods [105]. As shown in Fig. 29, a (110) oriented wafer is anisotropically etched to form closely spaced, high aspect ratio grooves about $\frac{3}{4}$ of the way through the wafer. A glass plate with fluid supply holes is anodically bonded over the grooves to provide sealed fluid channels through which the coolant is pumped. Input and output manifolds are also etched into the silicon at the same time as the grooves. The circuitry to be cooled is located on the opposite side of the wafer. Over a 1-cm² area, a thermal resistance of about 0.1°C/W was measured for a water flow rate of 10 cm³/s, for a power dissipation capability of 600 W/cm² (at a typical temperature rise above ambient of 60°C). This figure is 30 times higher than some previously estimated upper limits.

The use of silicon in this application is not simply an extravagant exercise. Tuckerman and Pease followed a novel optimization procedure to derive all the dimensions of the structure shown in Fig. 29. For optimal cooling efficiency, the fins should be 50 μm wide with equal 50-μm spaces and the height of the fins should be about 300 μm. Fortuitously, these dimensions correspond closely to typical silicon wafer thicknesses and to typical anisotropically etched (110) structures easily realized in practice. Besides the fact that the fabrication of such miniature structures would be extremely difficult in materials other than silicon, severe thermal mismatch problems are likely to be encountered during temperature cycling if a heat-sink material other than silicon were employed here.

The microcooling technique of Tuckerman and Pease is a compact and elegant solution to the problem of heat dissipation in very dense, very-high-speed IC chips. Advantages of optimized cooling efficiency, thermal and mechanical compatibility, simplicity, and ease of fabrication make this an attractive and promising advance in IC packaging. Bipolar chips with 25 000 circuits, each operating at 10 mW per gate (250 W total) are not unreasonable projections for future CPU's, now that a practical cooling method, involving silicon micromechanics, has been demonstrated.

Applications to Electronic Devices

Various isotropic and anisotropic etching procedures have been employed many times in the fabrication of IC's and other silicon electronic devices [106], [107]. In particular, silicon etching for planarization [108], for isolation of high-voltage devices [109], [110], or for removing extraneous regions of a chip to reduce parasitics [111], [112], and in VMOS [113] (more recently UMOS [114]) transistor structures are well-

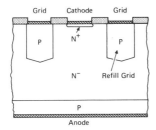

Fig. 30. The deep grid structure of a vertical-channel field-controlled thyristor [58] was accomplished by anisotropically etching deep grooves in the (110) wafer and growing p-doped silicon in the grooves by the epitaxial refill process of Runyan et al. and Smeltzer [57] which is shown in Fig. 10(b). Figure courtesy of B. Wessels.

known and some are used extensively in commercial products. Two areas of application in this category deserve special comment in this section, however. The first is a novel technique for producing very deep, doped regions for high-power electronic devices and is based on the epitaxial groove-filling process first demonstrated by Runyon et al. and Smeltzer [57] and shown schematically in Fig. 10(b). High-voltage high-power devices require deep diffusions not only to accommodate larger space-charge regions in the silicon (for increased breakdown voltages) but also to carry the larger currents for which such devices are designed. It is not unusual, for example, to schedule high-temperature diffusion cycles lasting over 100 h during some stages in the fabrication of high-power electronic devices. Furthermore, the geometries of such structures are limited because lateral diffusion rates are approximately equal to the vertical rates, i.e., diffusion in silicon is an isotropic process. By anisotropically etching grooves in (110) n-type silicon and refilling them epitaxially with p-type SCS, a process is obtained which appears effectively as an anisotropic diffusion. In this way, very deep, high aspect ratio, closely spaced diffused regions have been realized for high-speed vertical-channel power thyristors such as those demonstrated by Wessels and Baliga [58] (illustrated in Fig. 30), as well as for more complex buried-grid, field-controlled power structures [115]. Similar types of "extended" device geometries have been demonstrated by Anthony and Cline [64] using aluminum thermomigration (see Fig. 11). These micromachining techniques offer another important degree of freedom to the power device designer, which will be increasingly exploited in future generations of advanced high-power devices and IC's.

A second electronic device configuration employing the micromechanical principles discussed here is the V-groove multijunction solar cell [116]. The basic device configuration and a schematic processing schedule are shown in Fig. 31. Fabrication is accomplished by anodically bonding an SiO_2-coated silicon wafer to 7070 glass, anisotropically etching long V-grooves the full length of the wafer completely through the wafer to the glass substrate, ion-implanting p and n dopants into the alternating (111) faces by directing the ion beam at alternate angles to the surface, and finally evaporating aluminum over the entire surface at normal incidence such that the overhanging oxide mask prevents metal continuity at the top of the structure, while adjacent p and n regions at the bottom are connected in series. Solar conversion efficiencies of over 20 percent are expected from this device in concentrated sunlight conditions when the light is incident through the glass substrate. Advantages of these cells are ease of fabrication (one masking step), high voltage (~70 V/cm of cells), long effective light-absorption length (and therefore high

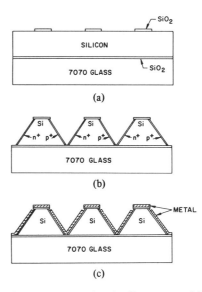

Fig. 31. Major fabrication steps for the V-groove, multijunction solar cell [116]. (a) Grow silicon dioxide layer, field assist bond oxidized wafer to glass, etch pattern windows in silicon dioxide. (b) Anisotropically etch silicon down to 7070 glass substrate, implant n^+ and p^+ regions at an angle, anneal implants. (c) Deposit metallization and alloy. Figure courtesy of T. Chappell.

efficiency) because of multiple internal reflections, no light-blocking metal current collection grid on the illuminated surface, and excellent environmental protection and mounting support provided by the glass substrate. Silicon solar cells based on this technique offer dramatic improvements over present single-crystal designs and may eventually be of commercial value.

V. SILICON MEMBRANES

While the micromechanical devices and components discussed in the preceding section were fabricated exclusively by rather straightforward groove and hole etching procedures, the following applications require some additional processing technologies; in particular, dopant-dependent etching for the realization of thin silicon membranes, which have been discussed in Section III.

X-Ray and Electron-Beam Lithography Masks

An early application of very thin silicon membrane technology which is still very much in the process of development is in the area of high-precision lithography masks. Such masks were first demonstrated by Spears and Smith [117] in their early X-ray lithography work and later extended by Smith et al. [118]. Basically, the procedure consists of heavily doping the surface of the silicon with boron, evaporating gold over the front surface, etching the gold with standard photolithographic or electron-beam techniques to define the X-ray mask pattern, and finally etching away most of the silicon substrate from the back side of the wafer (except for some support grids) with EDP [119]. Since heavily boron-doped silicon is not as rapidly attacked by EDP (or KOH), a self-supporting membrane is obtained whose thickness is controlled by the boron diffusion depth, typically 1–5 μm. Since the boron enters the silicon lattice substitutionally and the boron atoms have a smaller radius than the silicon, this highly doped region tends to be under tension as discussed in Section III. When the substrate is etched away, then the member becomes stretched taut and appears smooth and flat with no wrinkles, cracks, or bowing. X-rays are highly attenuated by the gold layers but not by the thin silicon "substrate" [120], [121]. Several variations on this scheme have been reported. Bohlen et al. [122], for example, have taken the X-ray design one step further by plasma etching completely through the remaining thin p^+ silicon regions not covered by gold and using the mask structure for electron-beam proximity printing.

These same basic principles were employed as early as 1966 by Jaccodine and Schlegel [123] to fabricate thin membranes (or windows) of SiO_2 to measure Young's modulus of thermally grown SiO_2. They simply etched a hole from one side of an oxidized Si wafer to the other (using hot Cl_2 gas as the selective etchant), leaving a thin SiO_2 window suspended across the opposite side. By applying a pressure differential across this window, they succeeded in measuring its deflection and determining Young's modulus of the thermally grown SiO_2 layer. Such measurements were later expanded upon by Wilmsen et al. [124]. Finally, Sedgwick et al. [119] and then Bassous et al. [125] fabricated these membrane windows from silicon and Si_3N_4 for use as ultra-thin electron-beam lithography "substrates" (to eliminate photoresist line broadening due to electron backscattering exposures from the substrate) for the purpose of writing very high resolution lines and for use in generating high-transparency X-ray masks. Thin, unsupported silicon nitride windows also have the advantage, in these applications, of being in tension as deposited on the silicon wafer, in the same way that boron-doped silicon membranes are in tension. SiO_2 membranes, such as those studied by Jaccodine and Schlegel [123] and by Wilmsen et al. [124], on the other hand, are in compression as deposited, tend to wrinkle, bow, and distort when the silicon is etched away, and are much more likely to break.

Circuits on Membranes

The potential significance of thin SCS membranes for electronic devices has been considered many times. Anisotropic etching, together with wafer thinning, were used by Rosvold et al. [111] in 1968 to fabricate beam-lead mounted IC's exhibiting greatly reduced parasitic capacitances. The frequency response of these circuits was increased by a factor of three over conventional diffused isolation methods. Renewed interest in circuits on thinned SCS membranes was generated during the development of dopant-dependent electrochemical etching methods. Theunissen et al. [45] showed how to use ECE both for beam-lead, air-gap isolated circuits as well as for dielectrically isolated circuits. Dielectric isolation was provided by depositing a very thick poly-Si layer over the oxidized epi, etching off the SCS substrate electrochemically, then fabricating devices on the remaining epi using the poly-Si as an isolating dielectric substrate. Meek [49], in addition to extending this dielectric isolation technique, realized other unique advantages of such thin SCS membranes, both for use in crystallographic ion channeling studies, as well as large-area diode detector arrays for use in low parasitic video camera tubes.

A backside-illuminated CCD imaging device [126] developed at Texas Instruments depends fundamentally on the ability to generate high-quality, high-strength, thin membranes over large areas. Since their double level aluminum CCD technology effectively blocked out all the light incident on the top surface of the wafer, it was necessary to illuminate the detector array from the backside. In addition, backside illumination improves

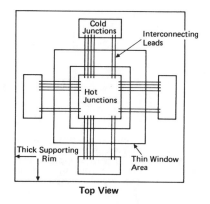

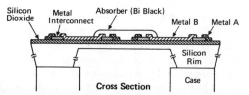

Fig. 32. Thermopile detector fabricated on a silicon membrane [127]. The hot junctions of the Au–poly-Si thermocouples are located in the central region of the membrane, while the cold junctions are located on the thick silicon rim. Efficient thermal isolation, small size, and a large number of integrated junctions result in high sensitivity and high-speed detection of infrared radiation. Figure courtesy of K. D. Wise.

spatial sensing uniformity and eliminates inference problems associated with front illumination through transparent layers. The high absorption coefficient of silicon in the visible, however, required the imager to be subsequently thinned from the backside (after circuit fabrication) to about 10 μm for efficient collection of photogenerated carriers. It was found that thin, highly uniform membranes could be realized over areas greater than 1 cm^2 with no deleterious effect on the sensitive CCD array and that these membranes exhibited exceptional strength, durability, and resistance to vibration and thermal cycling. Several such large-area CCD imaging arrays (800 × 800 pixels) will be installed in the space telescope scheduled to be launched by the Space Shuttle in 1985.

An important aspect of thin insulating membranes is that they provide excellent thermal isolation for thin-film devices deposited on the membrane. Lahiji and Wise [127] have demonstrated a high-sensitivity thermopile detector based on this principle. They fabricated up to 60 thin-film thermocouples (Bi-Sb and Au–polycrystalline Si), wired in series on a 2 mm × 2 mm × 1 μm SiO_2/p^+-Si membrane. Plan and cross-sectional views of this device are shown in Fig. 32. Hot junctions are arranged in the central membrane region while cold junctions are spaced over the thick periphery of the chip. When the membrane is coated with a thin thermal absorbing layer, sensitivities up to 30 V/W and time constants below 10 ms were observed for chopped 500 C black-body radiation incident from the etched (or bottom) surface of the wafer. Such low-mass, thermally isolated structures are likely to be commercially developed for these and related applications.

One thermally isolated silicon structure, in fact, is already commercially available. The voltage level detector of a high-bandwidth ac frequency synthesizer (Models 3336A/B/C) manufactured by Hewlett-Packard [4] is shown in Fig. 33. Two thin silicon cantilever beams with larger masses suspended in the center have been defined by anisotropic etching. The

Fig. 33. A high-bandwidth, thermal rms voltage detector [4] fabricated on silicon employs two cantilever beams with matching temperature-sensitive diodes and heat dissipation thin-film resistors on each. This device is used in the output-voltage regulation circuitry of the HP Model 3330 series of frequency synthesizers. Photo courtesy of P. O'Neil.

central masses of each beam are thermally isolated from each other and from the rest of the substrate. Fabricated on each isolated silicon island are a temperature-sensing diode and a thin-film heat-dissipation resistor. When a dc control current is applied to one resistor, the silicon island experiences a temperature rise which is detected by the corresponding diode. Meanwhile, a part of the ac output signal is applied to the resistor on the second island resulting in a similar temperature rise. By comparing the voltages of the two temperature-sensitive diodes and adjusting the ac voltage level until the temperatures of the two diodes match, accurate control of the output ac rms voltage level is obtained over a very large frequency range. This monolithic, silicon thermal converter offers the advantages of batch-fabrication, good resistor and diode parameter matching, while minimizing the effects of ambient thermal gradients. In addition, the masses of the islands are small, the resulting thermal time constants are therefore easy to control, and the single chip is simple to package.

In some applications, great advantages can be derived from electronic conduction normal to SCS membranes. In particular, Huang and van Duzer [128], [129] fabricated Schottky diodes and Josephson junctions by evaporating contacts on either side of ultrathin SCS membranes produced by p^+ doping and anisotropic etching. As thin as 400 Å, the resulting devices were characterized by exceptionally low series resistances, one-half to one-third of that normally expected from epitaxial structures. For Josephson junctions, the additional advantage of highly controllable barrier characteristics, which comes for free with silicon, could be of particular value in microwave detectors and mixers.

Large-area Schottky diodes on SCS membranes with contacts on either side have also found use as dE/dx nuclear particle detectors by Maggiore et al. [130]. Since the diodes (membranes) are extremely thin, 1-4 μm, the energy loss of particles traversing the sample is relatively small. This means that heavier ions, which typically have short stopping distances, can be more readily detected without becoming implanted in the silicon detector itself. Consequently, higher sensitivities, less damage, and longer lifetimes are observed in these membrane detectors compared to the more conventional epitaxial detectors.

Thin, large-area, high-strength SCS membranes have a number of other applications related to their flexibility. Guckel et al. [131] used KOH and the p^+ etch-stop method to generate up to 5-cm^2 membranes as thin as 2-4 μm. They mounted the structure adjacent to an electroded glass plate and caused it to vibrate electrostatically at the mechanical resonant frequency. Since the membrane is so large (typically 0.8 $\times$ 0.8 cm), the resonant frequency is in the audio range 10-12 kHz, yet the Q is maintained at a relatively high value, 23 000 in vacuum, 200 in air.

Pressure Transducers

Certainly the earliest and most commercially successful application of silicon micromechanics is in the area of pressure transducers [132]. In the practical piezoresistive approach, thin-film resistors are diffused into a silicon wafer and the silicon is etched from the backside to form a diaphragm by the methods outlined in Section III. Although the silicon can be etched isotropically or anisotropically from the backside (stopping the etching process after a fixed time), the dimensional control and design flexibility are dramatically improved by diffusing a p^+ etch-stop layer, growing an epitaxial film, and anisotropically etching through the wafer to the p^+ layer. As Clark and Wise showed [133], the membrane thickness is accurately controlled by the epi thickness and its uniformity is much improved. The resistors are located on the diaphragm, near the edges where the strains are largest. A pressure differential across the diaphragm cause deflections which induce strains in the diaphragm thereby modulating the resistor values. Chips containing such membranes can be packaged with a reference pressure (e.g., vacuum) on one side. The first complete silicon pressure transducer catalog, distributed in August 1974 by National Semiconductor, described a broad line of transducers in which the sensor chip itself was bonded to another silicon wafer in a controlled atmosphere, as shown in Fig. 34(a), so that the reference pressure was maintained within the resulting hermetically sealed cavity. This configuration was also described in 1972 by Brooks et al. [74] who employed a modified, thin-film anodic bond (as shown in Fig. 13(b)) to seal the two silicon pieces. Silicon eutectic bonding techniques (Au, Au–Sn) and glass-frit sealing are also used frequently in these applications. The National Semiconductor transducer unit is mounted in a hybrid package containing a separate bridge detector, amplifier, and thick-film trimmable resistors. The configuration of Fig. 34(a) suffers from the fact that the pressure to be sensed is incident on the top surface of the silicon chips where the sensitive circuitry is located. Although relatively thick parylene coatings [15] cover the membrane and chip surfaces of this silicon transducer line, it is clear that a different mounting technique is required for many applications in which the unknown pressure can be applied to the less-sensitive backside.

Presently, Foxboro, National Semiconductor, and other companies frequently mount chips in a manner similar to that shown in Fig. 34(b) such that the active chip surface is now the reference side. Chips are bonded both to ceramic and to stainless-steel assemblies. Many commercial sensor units are not yet even hybrid package assemblies and signal conditioning is accomplished by external circuitry. Recently, however, the

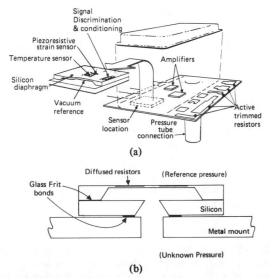

Fig. 34. Piezoresistive pressure transducers have been the earliest and most successful mechanical applications of silicon. At least eight firms now manufacture such sensors, rated for pressures as high as 10 000 psi. (a) Hybrid sensor package marketed by National Semiconductor. The resistor bridge on the silicon diaphragm is monitored by an adjacent detector/amplifier/temperature-compensation chip and trimmable thick-film resistors. Figure courtesy of National Semiconductor Corporation. A cross section of a typical mounted sensor chip is shown in (b). Chip bonding methods include eutectic bonding, anodic bonding, and glass-frit sealing.

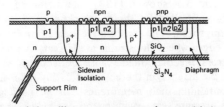

Fig. 35. Piezoresistive silicon pressure transducers with integrated detection and signal conditioning circuitry are now available commercially. Borky and Wise [134] have fabricated a pressure sensor (shown here in cross section) in which the bipolar circuitry is located on the deflectable diaphragm itself. Figure courtesy of K. D. Wise

Microswitch division of Honeywell has been marketing an integrated pressure transducer chip which incorporates some of the required signal-conditioning circuitry as well as the piezoresistive sensing diaphragm itself. A further indication of future commercial developments along these lines can be seen in the fully integrated and temperature-compensated sensors demonstrated by Borky and Wise [134], and by Ko et al. [135]. A cross-sectional view of the membrane transducer fabricated by Borky and Wise, Fig. 35, shows how the signal-conditioning circuitry was incorporated on the membrane itself, thereby minimizing the chip area and providing improved electrical isolation between the bipolar transistors.

Several companies supply transducers covering a wide range of applications; vacuum, differential, absolute, and gauge as high as 10 000 psi. Specific areas of application include fluid flow, flow velocity, barometers, and acoustic sensors (up to about 5 kHz) to be used in medical applications, pneumatic process controllers, as well as automotive, marine, and aviation diagnostics. In addition, substantial experience in reliability has been obtained. One of Foxboro's models has been cycled from 0 to 10 000 psi at 40 Hz for over 5×10^9 cycles (4 years) without degradation.

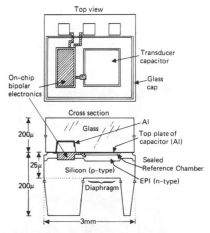

Fig. 36. One silicon diaphragm pressure transducer fully integrated with on-chip electronics is the capacitive sensor assembly demonstrated by Sander et al. [136] at Stanford. The design of this device has been directed toward implantable, biomedical applications. An etched glass plate, bonded to the silicon according to Fig. 13(c), hermetically seals the circuitry and also contains the top capacitor electrode. Figure courtesy of J. Knutti.

Few engineering references are available in the open literature concerning the design of silicon pressure transducers. In a recent paper, however, Clark and Wise [133] developed a comprehensive stress-strain analysis of these diaphragm sensors from a finite-element approach. Dimensional tolerances, piezoresistive temperature coefficients, optimum size and placement of resistors, the effects of potential process-induced asymmetries in the structure of the membranes, and, of course, pressure sensitivities have been considered in their treatment.

The sensitivities and temperature coefficients of membrane-based, capacitively coupled (CC) sensors were also calculated by Clark and Wise and found to be substantially superior to the piezoresistive coupled (PC) sensors. For the geometry and mounting scheme, they proposed, however, (with the very thin—2-μm—capacitive electrode gap exposed to the unknown gas), it was concluded that overriding problems would be encountered in packaging and in maintaining the electrode gap free of contaminants and condensates.

Recently, however, a highly sophisticated, fully integrated capacitive pressure sensor has been designed and fabricated at Stanford by C. Sander et al. [136]. As shown in Fig. 36, the device employs many of the micromechanical techniques already discussed. A silicon membrane serves as the deflectable element; wells etched into the top 7740 glass plate are used both as the spacer region between the two electrodes of the variable capacitor and as the discharge protection region above the circuitry, the principle of which was discussed in Section III (Fig. 13(c)). Field-assisted thermal bonding seals the silicon chip to the glass plate and assures the hermeticity of the reference chamber (which is normally kept at a vacuum level). The frequency-modulated bipolar detection circuitry is designed to charge the capacitive element with a constant current source, firing a Schmitt trigger when the capacitor reaches a given voltage. Clearly the firing rate of the Schmitt trigger will be determined by the value of the capacitor—or the separation of the capacitor plates. Perhaps one of the more significant aspects of this pressure transducer design is that the fabrication procedure was carefully planned to satisfy the primary objectives and advantages of silicon micromechanics. In particular, the silicon wafer and the large glass plate are

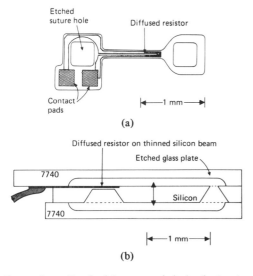

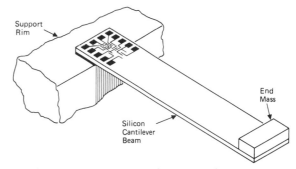

Fig. 37. Research at Stanford has extended the basic piezoresistive pressure sensor concept to complex strain sensor and accelerometer geometries for biomedical implantation applications. The strain sensor (a) contains a diffused piezoresistive element as well as etched suture loops on either end. Figure adapted from [137]. The accelerometer (b) is a hermetically sealed silicon cantilever beam accelerometer [75] sandwiched between two anodically bonded glass plates for passivation and for protection from corrosive body fluids. Figure adapted, courtesy of L. Roylance.

Fig. 38. The mechanical resonant frequency of a silicon cantilever beam was excited in the "Resonistor" by applying a sinusoidal current signal (at 1/2 of the resonant frequency) to a resistor on the silicon surface. These thermal fluctuations cause periodic vibrations of the beam which are detected by on-chip piezoresistive sensors. The signal from the sensor was employed in a feedback loop to detect and stabilize resonant oscillations. The function proposed for the "Resonistor" was a tuned, crystal oscillator. Adapted from Wilfinger et al. [138].

both processed using conventional IC techniques, both plates are anodically bonded, and only then is the entire assembly diced up into completed, fully functional transducer chips. Inexpensive batch fabrication methods, as required for practical, commercial silicon IC applications, are followed throughout.

Other Piezoresistive Devices

The principle of piezoresistance has been employed in other devices analogous to pressure transducers. J. B. Angell and co-workers at the Stanford Integrated Circuits Laboratory have advanced this technique to a high level of creativity. His group has been particularly concerned with *in vivo* biomedical applications. Fig. 37(a), for example, shows a silicon strain transducer etched from a wafer which has been successfully implanted and operated in the oviduct of a rabbit for periods exceeding a month [137]. Its dimensions are 1.7 X 0.7 mm by 35 μm thick. Two bonding pads on the left portion of the element make contact to a u-shaped resistor diffused along the narrow central bar. Two suture loops at both ends are also etched in the single-crystal transducer to facilitate attachment to internal tissue. Similar miniature strain transducers, etched from silicon, are now available commercially.

A cantilever beam, microminiature accelerometer, also intended for *in vivo* biomedial studies, is shown in Fig. 37(b). It was developed by Roylance and Angell [75] at Stanford and represents more than an order of magnitude reduction in volume and mass compared to commercially available accelerometers with equivalent sensitivity. Sutured to the heart muscle, it is light enough (<0.02 g) to allow high-accuracy high-sensitivity measurements of heart muscle accelerations with negligible transducer loading effects. It is also small enough (2 X 3 X 0.06 mm) for several to fit inside a pill which, when swallowed, would monitor the magnitude and direction of the pill's movement through the intestinal tract, while telemetry circuitry inside the pill transmits the signals to an external receiver.

Fabrication of the silicon sensor element follows typical micromechanical processing techniques—a resistor is diffused into the surface and the cantilever beam is separated from the surrounding silicon by etching from both sides of the wafer using an anisotropic etchant. The thickness of the thinned region of the beam, in which the resistor is diffused, is controlled by first etching a narrow V-groove on the top surface of the wafer (whose depth is well defined by the width of the pattern as in the case of ink jet nozzles) and, next, a wider V-groove on the bottom of the wafer. When the etched holes on either side meet (determined by continual optical monitoring of the wafer), etching is stopped. The remaining thinned region corresponds approximately to the depth of the V-groove on the top surface. The final form is that of a very thin (15-μm) cantilever beam active sensing element with a silicon (or gold) mass attached to the free end, surrounded by a thick silicon support structure. A second diffused resistor is located on the support structure, but adjacent to the active piezoresistor for use as a static reference value and for temperature compensation. The chip is anodically bonded on both sides to two glass plates with wells etched into them. This sealed cavity protects the active element by hermetically sealing it from the external environment, provides mechanical motion limits to prevent overdeflection, yet allows the beam to deflect freely within those limits. Resonant frequencies of 500 to 2000 Hz have been observed and accelerations of less than $10^{-3}g$ have been detected. Such devices would be extremely interesting in fatigue and yield stress studies.

An early micromechanical device with a unique mode of operation was demonstrated by Wilfinger et al. [138], and also made use of the piezoresistive effect in silicon. As shown in Fig. 38, a rectangular silicon chip (typically 0.9 X 0.076 X 0.02 cm) was bonded by one end to a fixed holder, forming a silicon cantilever beam. Near the attached edge of the bond, a circuit was defined which contained a heat-dissipating resistor positioned such that the thermal gradients it generated caused a deflection of the beam due to thermal expansion near the (hotter) resistor, relative to the (cooler) backside of the chip. These deflections were detected by an on-chip piezoresistive bridge circuit, amplified, and fed back to the heating resistors to oscillate the beam at resonance. Since the beam

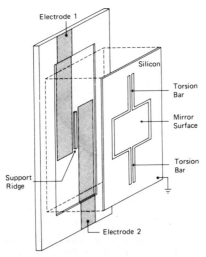

Fig. 39. Exploded view of silicon torsion mirror structures showing the etched well, support ridge, and evaporated electrodes on the glass substrate. From [139].

has a very-well-defined resonant frequency and a high Q (>2000), the output from the bridge exhibits a sharp peak when the heated resistor is excited at that mechanical resonant frequency. This oscillator function has been demonstrated in the range of 1.4 to 200 kHz, and stable, high-Q oscillations were maintained in these beams continuously for over a year with no signs of fatigue.

Silicon Torsional Mirror

This section closes with the description of a device which is not actually a membrane structure, but is related to the strain-measurement mechanisms discussed above and has important implications concerning the future capabilities and potential applications of SCS micromechanical technology. The device is a high-frequency torsional scanning mirror [139] made from SCS using conventional silicon processing methods. An exploded view, shown in Fig. 39, indicates the silicon chip with the anisotropically etched mirror and torsion bar pattern, as well as the glass substrate with etched well, central support ridge, and electrodes deposited in the well. After the two pieces are clamped together, the silicon chip is electrically grounded and a high voltage is applied alternately to the two electrodes which are very closely spaced to the mirror, thereby electrostatically deflecting the mirror from one side to the other resulting in twisting motions about the silicon torsion bars. If the electrode excitation frequency corresponds to the natural mechanical torsional frequency of the mirror/torsion bar assembly, the mirror will resonate back and forth in a torsional mode. The central ridge in the etched well was found to be necessary to eliminate transverse oscillations of the mirror assembly. A cross-sectional view of the torsional bar and of the mirror deflections is shown in Fig. 40. The well-defined angular shapes in the silicon, which are also seen in the SEM (scanning-electron microscope) photograph in Fig. 41(a) (taken from the backside of the silicon chip), result, of course, from the anisotropic etchant. Fig. 41(b) gives typical device dimensions used in the results and the calculations to follow.

Reasonably accurate predictions of the torsional resonant frequency can be obtained from the equation [140]

$$f_R = \frac{1}{2\pi} \sqrt{\frac{12KEt^3}{\rho l b^4 (1+\nu)}} \quad (1)$$

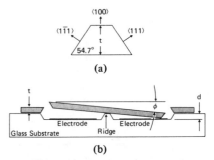

Fig. 40. (a) Cross section of the anisotropically etched torsion bar where $t = 134$ μm. (b) Cross section of the mirror element defining the deflection angle ϕ, where $d = 12.5$ μm and a voltage is applied to the electrode on the right.

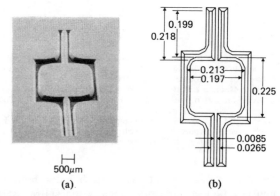

Fig. 41. (a) SEM of typical torsion mirror (tilted 60°) and (b) measured dimensions of 15-kHz mirror element (in cm). The SEM photo is a view of the mirror from the back surface where the electrostatic fields are applied.

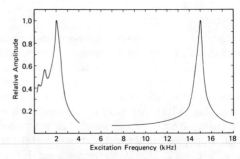

Fig. 42. Deflection amplitude versus drive frequency for two mirrors with differing resonant frequencies.

where E is Young's modulus of silicon ($E = 1.9 \times 10^{12}$ dyne/cm^2), t is the thickness of the wafer ($t \sim 132$ μm), ρ is the density of silicon ($\rho = 2.32$ g/cm^3), ν is Poisson's ratio ($\nu = 0.09$) [141], l is the length of the torsion bar, b is the dimension of the square mirror, and K is a constant depending on the cross-sectional shape of the torsion bar ($K \sim 0.24$). For these parameters, we calculate $f_R = 16.3$ kHz, compared to the experimental value of 15 kHz for the device shown in Fig. 41. The resonant behavior of two experimental torsional mirrors is plotted in Fig. 42.

While complex damping mechanisms, including viscous air-damping and proximity effects due to closely spaced electrodes [142], dominate the deflection amplitudes near resonance, close agreement between theory and experiment can be obtained at frequencies far enough below resonance and at deflection angles small compared to the maximum deflection angle $\phi_{\max} = 2d/b$, illustrated in Fig. 43. Under these restric-

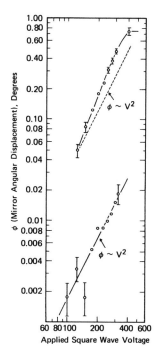

Fig. 43. Experimental deflections of torsion mirror. Resonant displacements are shown at the top, off-resonance at the bottom. Note departure from square-law dependence at resonance.

tions, it can be shown that [143]

$$\phi = \frac{\epsilon_0 V^2 l b^3 (1+\nu)}{16 K E d^2 t^4} A \qquad (2)$$

where ϵ_0 is the free-space dielectric permittivity, V is the applied voltage, d is the steady-state electrode/silicon separation, and A is an areal correction factor ($A \sim 0.8$) due to the fact that the active electrode area is somewhat less than half the area of the mirror. We can see from the lower curve in Fig. 43, that the square-law dependence on voltage is confirmed by the data and that the observed deflection amplitudes are only about 20 percent below those predicted by (2). As expected, nonlinearities in the deflection forces are also evident in Fig. 43 during operation at resonance, since the square-law dependence is not maintained.

Optically, silicon possesses an intrinsic advantage over common glass or quartz mirrors in high-frequency scanners because of its high E/ρ ratio, typically 3 times larger than quartz. Using the mirror distortion formulation of Brosens [144], $\frac{1}{3}$ smaller distortions are expected in rapidly vibrated silicon mirrors, compared to quartz mirrors of the same dimensions.

Of prime importance in the study of mechanical reliability is the calculation of maximum stress levels encountered. The maximum stress of a shaft with the trapezoidal cross section of Fig. 38(a) occurs at the midpoint of each side and is given by [145]

$$\tau_{\max} \simeq \frac{16 K E}{(1+\nu)} \left(\frac{dt}{bl} \right) \qquad (3)$$

when the torsion bars are under maximum torque ($\phi = \phi_{\max}$). For our geometry, this corresponds to about 2.5×10^9 dyne/cm^2 (36 000 psi), or more than an order of magnitude below the fracture stresses found in the early work of Pearson et al. [11]. Reliability, then, is predicted to be high.

This initial prediction of reliability was verified in a series of life tests in which mirrors were continuously vibrated at resonance, for periods of several months. Despite being subjected to peak accelerations of over 3.5×10^6 cm/s^2 (3600 g's), dynamic stresses in the shaft of over 2.5×10^9 dyne/cm^2 (36 000 psi), 30 000 times a second for 70 days ($\sim 10^{11}$ cycles) no stress cracking or deterioration in performance was detected in the SEM for devices which had been properly etched and mounted. After a dislocation revealing etch on this same sample, an enhanced dislocation density appeared near the fixed end of only one of the torsion bars. Since this effect was observed only on one bar, it was presumed to be due to an asymmetry in the manual mounting and gluing procedure, resulting in some unwanted traverse oscillations.

These calculations and observations strongly indicate that silicon mechanical devices, such as the torsion mirror described here, can have very high fatigue strengths and exhibit high reliability. Such results are not unexpected, however, from an analysis of the mechanisms of fatigue. It is well known that, whatever the process, fatigue-induced microcracks initiate primarily at free surfaces where stresses are highest and surface imperfections might cause additional stress concentration points [19]. Since etched silicon surfaces can be extremely flat with low defect and dislocation damage to begin with, SCS structures with etched surfaces are expected, fundamentally, to possess enhanced fatigue strengths. In addition, the few microcracks which do develop at surface dislocations and defects typically grow during those portions of the stress cycle which put the surface of the material in tension. By placing the surface of the structure under constant, uniform compression, then, enhanced fatigue strengths have been observed in many materials. In the case of silicon, we have seen how thin Si$_3$N$_4$ films, while themselves being in tension, actually compress the silicon directly underneath. Such layers may be expected to enhance even further the already fundamentally high fatigue strength of SCS in this and other micromechanical applications.

A comparison of the silicon scanner to conventional, commercial electromagnetic and piezoelectric scanners is presented in Table III. The most significant advantages are ease of fabrication, low distortion, and high performance at high frequencies.

VI. THIN CANTILEVER BEAMS

Resonant Gate Transistor

Micromechanics as a silicon-based device technology was actually initiated by H. C. Nathanson et al. [147], [148] at Westinghouse Research Laboratories in 1965 when he and R. A. Wickstrom introduced the resonant gate transistor (RGT). As shown in Fig. 44, this device consists of a plated-metal cantilever beam, suspended over the channel region of an MOS transistor. Fabrication of the beam is simply accomplished by first depositing and delineating a spacer layer. Next, photoresist is applied and removed in those regions where the beam is to be plated. After plating, the photoresist is stripped and the spacer layer is etched away, leaving the plated beam suspended above the surface by a distance corresponding to the thickness of the spacer film. Typical dimensions employed by Nathanson et al. were, for example, beam length 240 μm, beam thickness 4.0 μm, beam-to-substrate separation 10 μm.

Operating as a high-Q electromechanical filter, the cantilever beam of the RGT serves as the gate electrode of a surface MOSFET. A dc voltage applied to the beam biases the transistor at a convenient operating point while the input signal electrostatically attracts the beam through the input force plate,

TABLE III

	Silicon Mirror		Electromagnetic[a]		Piezoelectric[a]	
Fabrication procedure	Batch fabrication of two lithographically processed plates		Complex mechanical assembly of many parts		Two bonded ceramic plates with separate mirror attached	
Frequency scan angle	15 kHz ±1°	50 kHz [b] ±2°	1 kHz ±30°	15 kHz ±2°	1 kHz ±5°	40 kHz ±0.1°
Power	<0.1 W dissipated in drive circuitry		≈0.5 W dissipated in assembly		<0.1 W dissipated in drive circuitry	
Relative distortion	1/3 (silicon mirror)		1 (quartz mirror)		1 (quartz mirror)	
Reliability	≈10^{12} cycles demonstrated		Very high		Very high	
Other	High voltage		High power Heavy assembly		High voltage Off-axis mirror Creep and hysteresis	

[a] See Ref. 146.
[b] Projected Performance.

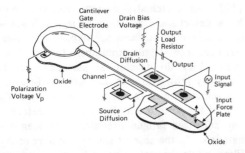

Fig. 44. The earliest micromechanical cantilever beam experiments were conceived at Westinghouse and based on the plated-metal configuration shown here. Operated as an analog filter, the input signal causes the plated beam to vibrate. Only when the signal contains a frequency component corresponding to half the beam mechanical resonant frequency are the beam motions large enough to induce an output from the underlying MOS structure [147], [148]. Figure courtesy of H. Nathanson.

thereby effectively increasing the capacitance between the beam and the channel region of the MOS transistor. This change in capacitance results in a variation of the channel potential and a consequent modulation of the current through the transistor. Devices with resonant frequencies (f_R) from 1 to 132 kHz, Q's as high as 500, and temperature coefficients of f_R as low as 90 ppm°C were described and extensively analyzed by Nathanson *et al.* They constructed high-Q filters, coupled multipole filters, and integrated oscillators based on this fabrication concept. Since the electrostatically induced motions of the beam are only appreciable at the beam resonant frequency, the net Q of the filter assembly is equivalent to the mechanical Q of the cantilever beam. Typical ac deflection amplitudes of the beams at resonance for input signals of about 1 V were ~50 nm.

Practical, commercial utilization of RGT's have never been realized for a number of reasons, some of which relate to technology problems, and some having to do with overall trends in electronics. The most serious technical difficulties discussed by Nathanson *et al.* are 1) reproducibility and predictability of resonant frequencies, 2) temperature stability, and 3) potential limitations on lifetime due to fatigue. The inherent inaccuracies suffered in this type of selective patterned plating limited reproducibility to 20-30 percent over a given wafer in the studies described here. It is not clear if this spread can be improved to much better than 10 percent even with more stringent controls. Temperature stability was related to the temperature coefficient of Young's modulus of the plated beam material, about 240 ppm for gold (the temperature coefficient of f_R is about half this value). Although this problem could be solved, in principle, by plating low-temperature coefficient alloys, such experiments have not yet been demonstrated. Lifetime limitations due to fatigue is a more fundamental problem. Although the strain experienced by the cantilever beam is small (~10^{-5}), the stability of a polycrystalline metal film vibrated at a high frequency (e.g., 100 kHz) approaching 10^{14} times (10 years) is uncertain. Indeed, it is known, for example, that polycrystalline piezoelectric resonators will experience creep after continued operation in the 10's of kilohertz. (Single-crystal or totally amorphous materials, on the other hand, exhibit much higher strengths and resistance to fatigue.) These technological difficulties, together with trends in electronics toward digital circuits, higher frequencies of operation (>1 MHz for D/A and A/D conversion), higher accuracies, and lower voltages have conspired to limit the usefulness of devices like the RGT. The crux of the problem is that the RGT filter, while simpler and smaller than equivalent all-electronic circuits, was forced to compete on a basis which challenged well-established conventions in circuit fabrication, which did not take real advantage of its unique mechanical principles, and which pitted it against a very powerful, fast-moving, incredibly versatile all-electronic technology. For all these reasons, conceptually similar devices, which will be discussed below, can only hope to be successful if they 1) provide functions which cannot easily be duplicated by *any* conventional analog or digital circuit, 2) satisfactorily solve the inherent problems of mechanical reliability and reproducibility, and 3) are fabricated by techniques totally compatible with standard IC processing since low-cost high-yield device technologies are most likely only if well-established batch fabrication processes can be employed.

Micromechanical Light Modulator Arrays

The first condition was addressed during the early 1970's when several attempts to fabricate two-dimensional light-modulator arrays were undertaken with various degrees of short-lived success [149]–[152]. Conventional silicon circuits

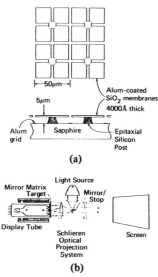

Fig. 45. Taking advantage of the excellent mechanical behavior of thermally grown SiO_2, Westinghouse fabricated large arrays (up to 400 000) of deflectable oxide flaps (0.35 µm thick) and demonstrated an electron-beam addressed image projection display with intrinsic image storage capability [153], [154]. (a) Top view and cross section of oxide flap geometry. The electron-beam addressing and optical projection assembly schematic is shown in (b). Figure courtesy of H. Nathanson.

certainly cannot modulate light. These membrane light modulator arrays were composed of thin (~200-nm) perforated metal sheets, suspended over support structures (~5 µm high) on various substrates. Small segments of the metal sheet behaved as independent mirror elements and could be electrostatically deflected by a scanned electron-beam or by voltages applied to underlying deposited electrodes. The preliminary device demonstrated by van Raalte [149], for example, contained 250 000 individually deflectable thin metal regions. Clearly, the light valve function had not previously been successfully implemented by any other thin-film technology.

Nevertheless, these metal-film designs did not satisfy the second and third requirements. Not only were they potentially susceptible to metal fatigue (gold has a particularly low fatigue strength), but the principles of fabrication were tedious, complex, and intrinsically difficult. In addition, portions of the fabrication process were not even remotely compatible with conventional IC techniques.

During their studies of silicon point field-emitter arrays discussed in Section IV [79], [80], Thomas and Nathanson found that large arrays of deformagraphic elements could also be built using totally conventional and compatible IC processing methods [153], [154]. In addition, while previous workers employed metal films as the deformagraphic material, they used thermally grown amorphous SiO_2 films with fatigue strengths expected to be substantially greater. The structure consisted of a 5-µm silicon epitaxial layer grown over sapphire and covered with a 350-nm-thick thermal oxide. First the oxide is etched into the pattern shown in Fig. 45(a), which defines the final cloverleaf shape of the deflectable oxide membranes. Next, a much thinner oxide is grown and etched away only along the grid lines separating individual light-valve elements. The crucial step involves isotropic etching of the silicon from underneath the oxide elements, undercutting the thin film until only narrow pedestals remain supporting the oxide structures. Finally, the thin oxide remaining in the four slits of each self-supported SiO_2 plate is etched away and the entire array is rinsed and dried. Evaporation of a thin (30-nm) aluminum film makes the oxide beams highly reflective and deposits a metal grid through the openings in the oxide onto the sapphire substrate.

Writing an image on the Mirror-Matrix Display tube is done by raster scanning a modulated electron beam across the target. Each individual mirror element will become charged to a voltage dependent on the integrated incident electron current. Typically, the electron-beam voltage is specified to cause a large secondary electron emission coefficient. The secondaries are collected by a metal grid closely spaced to the target (100 µm) and the mirror metallizations become charged positively. When a negative voltage is then applied to the stationary metal grid on the sapphire surface, electrostatic forces mechanically deflect those elements which have become charged. The pattern of deflected and undeflected beams is illuminated from the backside through the sapphire substrate and the reflected light is projected in a Schlieren optical system such that the deflected beams appear as bright spots on a ground-glass screen. A schematic of the optical and electron-beam illumination systems is shown in Fig. 45(b). Since the metallizations on the mirror surfaces are electrically well insulated from the metal grid on the sapphire surface, charge images can be stored on the deformagraphic array for periods up to many days. At the same time, fast erasure can be accomplished by biasing the external grid negatively and flooding the array with low-energy electrons which equalize the varying potentials across the mirror surfaces.

The particular cloverleaf geometry implemented in the mirror matrix target (MMT) design exhibits very high contrast ratios since the deflected "leaves" scatter light from the optical system at 45° from the primary diffracted radiation. An opaque, cross-shaped stop placed at the focal point of the imaged light beam will only pass that portion of the light reflected from bent "leaves." Contrast ratios over 10:1 have been attained in this way. Over and above the optical advantages, however, the mirror matrix target satisfies all the basic requirements for a potentially practicable micromechanical application. The principles of fabrication are completely compatible with conventional silicon processing; it performs a function (light modulation and image storage) not ordinarily associated with silicon technology; and it greatly alleviates the potential fatigue problems of previous metal-film deformagraphic devices by using amorphous SiO_2 as the active, cantilever beam material. Unfortunately, one of the fabrication problems of the MMT was the isotropic Si etch. Since this step did not have a self-stopping feature, and the dimensions of the post are critical, small etch-rate variations over an array could have a dramatic impact on the yield of a large imaging array.

The MMT has never become a viable, commercial technology primarily because it did not offer a great advantage over conventional video monitors. Image resolution of both video and deformagraphic displays, for example, depend primarily on the number of resolvable spots of the electron-beam system itself (not the form of the target) and, therefore, should be identical for equivalent electron optics and deflection electronics. In addition, since electron-beam scan-to-scan positioning precision is typically not high, it would be very difficult to write on single individual mirrors accurately scan after scan. This means a single written picture element would actually spread, on the average, over at least four mirrors. For an optical image resolution of 10^6 PEL's, then, at least 4×10^6 mirrors would be required—a formidable task for high-yield photolithography. Finally, the potential savings in the deformagraphic systems due to their intrinsic storage capability is more than offset by the expensive mirror matrix target itself (equivalent in area to at least 20 high-density IC chips when yield criteria are taken into account) as well as the imaging lamp and optics, which are required *in addition to* the electron tube and its associated high-voltage circuits. Historically, high-density displays have always been exceptionally difficult and demanding technologies. While silicon micromechanics may eventually find a practical implementation for displays (especially if silicon-driving circuitry can be integrated on the same chip, matrix addressing the two-dimensional array of mirrors—all electronically), we need not be that ambitious to find important, novel, much-needed device applications amenable to silicon micromechanical techniques.

SiO₂ Cantilever Beams

The full impact of micromechanical cantilever beam techniques was not completely evident until the development of controlled anisotropic undercut etching, as described in Section III. Flexible, fatigue-resistant, amorphous, insulating cantilever beams, co-planar with the silicon surface, closely spaced to a stationary deflection electrode, and fabricated on ordinary silicon with conventional integrated circuitry on the same chip can be extremely versatile electromechanical transducers. A small, linear array of voltage-addressable optical modulators was the first demonstration of this new fabrication technique [155]. Since the method forms the basis of other devices discussed in the rest of this section, we will describe it in some detail. The silicon wafer is heavily doped with boron in those regions where cantilever beams are desired. Since this film will serve as an etch-stop layer during fabrication of the beams, as well as a deflection electrode during operation of the device, the dopant concentration at the surface must be high enough initially to effectively inhibit subsequent anisotropic etching even after the many high-temperature cycles required to complete the structure and its associated electronic circuitry. These additional high-temperature steps tend to decrease the original surface concentration. A level of 7×10^{19} cm^{-3} is usually taken to be the minimum peak value necessary to stop the etchant. Next, an epitaxial layer is grown on the wafer to a thickness corresponding approximately to the desired electrode separation. Since the buried p$^+$ region is so heavily doped, the electrical quality of the epi grown over these regions may be poor and electronic devices probably should not be located directly above them. At this point, any necessary electronics can be fabricated on the epi-layer adjacent to the buried regions. After depositing (e.g., Si$_3$N$_4$) or growing (e.g., SiO$_2$) the insulating material to be used for the cantilever beam, a thin metal film (typically 30 to 40 nm of gold on chrome) is evaporated and delineated to form the upper electrode. Before silicon etching, care must be taken to insure that the aluminum circuit metallization is adequately passivated since the hot EDP solution will attack these films. Finally, the insulator is patterned as shown in Fig. 46(b), the exposed silicon is etched in EDP to undercut the insulator and free the cantilever beams, and the wafer is carefully rinsed and dried. Since further lithography steps are virtually impossible after the beam is freed, the anisotropic etch is always the last processing step.

As long as the patterns are correctly oriented in (110) directions, undercutting can be completely avoided in unwanted areas as described in Section III. Fig. 47 shows an optical

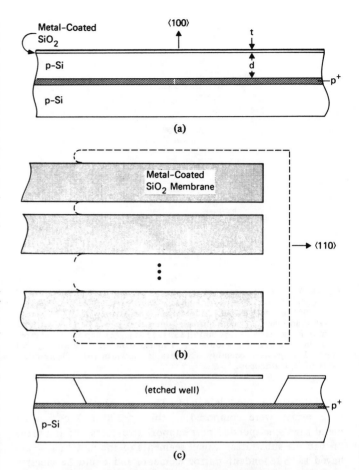

Fig. 46. (a) Cross section of the layers in the linear, cantilever-beam light modulator array. (b) The metal is etched into lines and the oxide is etched from between the lines within the dashed region (top view). (c) The silicon is anisotropically etched from under the oxide to release the beams. From [155]. Each beam can be independently deflected by applying a voltage between the top metallization and the p$^+$ layer.

modulator array fabricated in this way. Since the patterns have been oriented correctly, only the beams are undercut, while the periphery of the etched rectangular hole is bounded by the (111) planes. At the same time, the buried p$^+$ layer stops etching in the vertical direction. The dimensions of the beams in this first demonstration array were 100 μm long made from SiO$_2$ 0.5 μm thick. These are spaced 12 μm from the bottom of the well. A Cr-Au metallization, about 50 nm thick, serves as the top electrode.

Each individual beam can be deflected independently through electrostatic attraction simply by applying a voltage between the top electrode and the buried layer. For low voltages, the deflection amplitude varies approximately as the square of the voltage as indicated in Fig. 48. Both experiment and detailed calculations have shown that once the membrane tip is moved approximately a third of the way down into the etched well, the beam position becomes unstable and it will spontaneously deflect the rest of the way as a result of the rapid buildup of electrostatic forces near the tip. This effect was also analyzed by Nathanson *et al.* [147], [148] for the plated gold RGT beam. In the present geometry, the threshold voltage for spontaneous deflection can be shown to be approximately [156]

$$V_{th} = \sqrt{\frac{3Et^3 d^3}{10\epsilon_0 l^4}} \qquad (4)$$

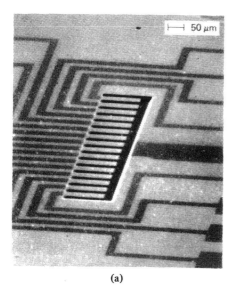

Fig. 47. SEM photographs of a completed 16-element light modulator array, ready for bonding. Note the high degree of flatness and uniformity of the beams. The SiO$_2$ beams are 100 µm long, 25 µm wide, and 0.5 µm thick.

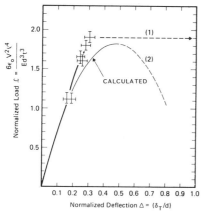

Fig. 48. Calculated beam deflection is plotted here as the lighter line. At the threshold voltage (the peak in the calculated curve), the membrane tip will spontaneously deflect the rest of the way to the bottom of the pit, since its position is unstable along line (2). Experiment data, including the observed spontaneous threshold deflection (along line (1)), are also shown.

where l is the beam length, t is the thickness, ρ is the density, E is Young's modulus, ϵ_0 is the free-space permittivity, and d is the electrode separation. The 16-element array shown in

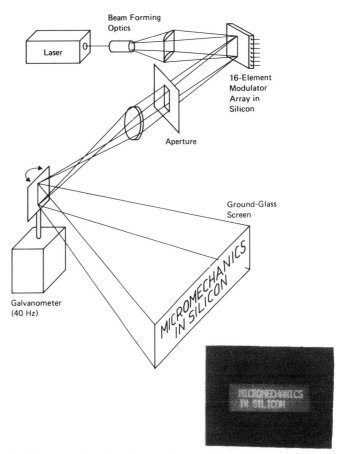

Fig. 49. Schematic of simple optical display employing the 16-beam array in Fig. 47. Each beam is operated as a single modulating element for the corresponding horizontal line in the display. The inset shows a photograph of the display projected onto the ground glass screen.

Fig. 47 was operated below threshold in the simple optical system shown in Fig. 49. A laser was focussed on all 16 mirrors simultaneously while an external aperture was adjusted such that only light reflected from bent membranes was allowed to pass. The resulting vertical line of light, now modulated at 16 points, was then scanned horizontally by a galvanometer to produce a two-dimensional display by rapidly deflecting each membrane independently in the correct sequence. An example of this display is given in the inset in Fig. 49.

Certainly, resonant frequency is an important parameter of such cantilever beams. The first mechanical resonance can be accurately calculated from [157]

$$f_R = 0.162 \frac{t}{l^2} \sqrt{\frac{E}{\rho}} K \qquad (5)$$

where K is a correction factor (close to one) depending on the density, Young's modulus, and thickness ratios of the metal layer to the insulating layer. For the array given above, $f_R = 45$ kHz. The highest resonant frequency yet observed is 1.25 MHz for the 8.3-µm-long, 95-nm-thick SiO$_2$ beams shown in Fig. 50(a).

Once the resonant frequency and the dimensions of a beam are measured, (5) can readily be used to determine Young's modulus of the insulating beam material. Fig. 50(b) shows an array of beams of various lengths fabricated specifically for such measurements. The vibrational amplitude of a reflected optical signal is plotted in Fig. 51 as a function of vibrational frequency for beams of five different lengths. Note that an

TABLE IV

	Insulator Thickness (Å)	Chromium Thickness (Å)	Assumed Insulator Density (g/cm^3)	Measured Young's Modulus (10^{12} dyn/cm^3)	Published Bulk Young's Modulus (10^{12} dyn/cm^2)
SiO$_2$ (thermal-wet)	4250	150	2.2	0.57	
SiO$_2$ (thermal-dry)	3250	150	2.25	0.67	
SiO$_2$ (sputtered)	4000	100	2.2	0.92	0.7 (fused silica)[a]
Si$_3$N$_4$ (CVD)	3500	100	3.1	1.46	1.5[b]
Si$_3$N$_4$ (sputtered)	2900	100	3.1	1.3	
7059 glass (sputtered)	4200	50	2.25	0.52	0.6 (typ.)[3]
Nb$_2$O$_5$ (sputtered)	8400	50	4.47	0.85	1.6 (Nb$_2$O$_3$)[d]
α-SiC (glow discharge)	8800	50	~3.0	0.85	4.8 (single x-tal)[e]
Cr (sputtered)	---	---	7.2	1.8	2.8[f]

[a] See Refs. 164 and 165.
[b] See Ref. 164.
[c] See Ref. 165.
[d] See Ref. 166.
[e] See Ref. 10.
[f] See Ref. 163.

(a)

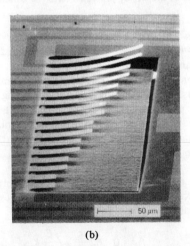

(b)

Fig. 50. Various beam sizes have been fabricated from various materials for the measurements of Young's modulus. The beams in (a) are 8 μm long and less than 0.1 μm thick—they have a 1.2-MHz mechanical resonant frequency. The array in (b) is made from dry thermal SiO$_2$ and the beams range in length from 118 to 30 μm.

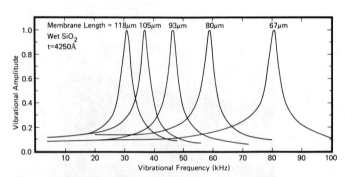

Fig. 51. Resonant behavior measurements of five SiO$_2$ beams with different lengths to determine Young's modulus of the thermally grown oxide layer.

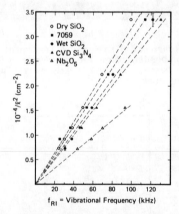

Fig. 52. Determination of Young's modulus according to (5). See Table IV for a complete listing of the films and their parameters. From [158].

electrostatically deflected cantilever beam is a simple frequency doubler because the beam experiences an attractive force for positive as well as negative voltage swings. The resonant frequency is easily detected and can be shown, in Fig. 52, to follow the l^{-2} dependence predicted by (5). A series of measurements on a wide variety of deposited, insulating thin films was undertaken by Petersen and Guarnieri [158], the results of which are tabulated in Table IV. In the past, Young's modulus has been an extremely difficult parameter to measure in thin films due to the fragility of the samples and the concomitant problems in their handling and mounting as well as minutely deforming them and monitoring their motions [159]-[161]. This new, micromechanical technique, on the other hand, is simple, accurate, applicable to a wide range of materials and deposition methods, and useful in many thin-film microstructure studies. Chen and Muller, for example, have fabricated various composite p$^+$-Si/SiO$_2$ beams [162], as indicated in Fig. 53, studying their mechanical stability. Such measurements provide additional insight into the nature

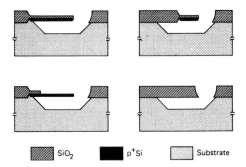

Fig. 53. Miniature cantilever beams of several configurations, including p+-Si layers, were fabricated by Jolly and Muller [162] using anisotropic etching together with the p+ etch-stop technique, as shown here in cross section. The warpage and stability of the beams were studied as a function of beam length (15-160 μm), width (5-40 μm), and thickness (0.2-0.8 μm). Figure courtesy of R. Muller.

of deposited films, how they differ from the corresponding bulk materials, how they depend on deposition conditions, and how multilayer films interact mechanically.

Integrated Accelerometers

While pressure transducers have been the first commercially important solid-state mechanical transducers, it is likely that accelerometers will become the next. Substantial literature already exists in this area, including the relatively large silicon cantilever beam piezoresistive device of Roylance and Angell which was described in Section V. Closely related to this structure in terms of fabrication principles is the folded cantilever beam accelerometer developed jointly by Signetics Corporation and Diax Corporation [167]. Initial work on the folded beam silicon accelerometer concentrated on piezoresistive strain sensors diffused into the silicon beam, while later studies by Chen et al. employed deposited piezoelectric ZnO sensors on identical silicon devices [168]. None of these demonstrations incorporated signal detection or conditioning circuitry on the sensor chip itself, however, as has been accomplished in pressure transducers.

Cantilever beam accelerometers, such as those mentioned above, made by etching clear through a wafer must address serious packaging problems. Special top and bottom motion-limiting plates, for example, must be included in the assembly to prevent beam damage during possible acceleration overshoots. A more problematical issue (which is also a concern in silicon pressure transducers) is the potential for unintentional, residual stresses resulting in hysteresis or drift in the detected signal. Such stresses could be developed, for example, from temperature-coefficient mismatches between the silicon and the various packaging and/or bonding materials.

Small oxide cantilever beams etched directly on the silicon surface alleviate many of the handling, mounting, and packaging problems of solid-state accelerometers since the beam itself is already rigidly attached to a thick silicon substrate. Furthermore, since the active beam element occupies just the top surface of the silicon, potential strains induced during mounting and packaging of the substrate will have little effect on the detector itself. An example of this type of accelerometer is the ZnO/SiO$_2$/Si composite beam structure demonstrated by Chen et al. [169]. As shown in Fig. 54, beam motions induce a voltage across the thin piezoelectric ZnO which is detected and amplified by adjacent, on-chip MOS circuitry. Careful device design and ZnO thin-film deposition techniques have yielded devices with low drift and hysteresis, potential

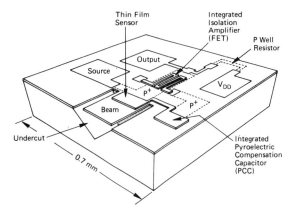

Fig. 54. Layout of the integrated, planar-processed PI-FET accelerometer demonstrated by Chen et al. [169]. As the cantilever beam vibrates, voltages induced in the thin-film, ZnO piezoelectric sensor element are detected by the on-chip MOS isolation amplifier. Figure courtesy of R. Muller.

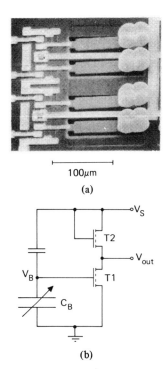

Fig. 55. The integrated accelerometer shown here in (a) consists of an SiO$_2$ cantilever beam sensor (loaded with a gold mass for increased sensitivity) coupled to an MOS detection circuit. The capacitance of the beam (typically 3.5 fF) is employed in a voltage divider network (b) from which small variations in the beam capacitance drive the detection transistor. From [170].

areas of concern when dealing with piezoelectric materials. By completely encapsulating the ZnO layer, for example, charge-storage times on the order of many days were observed. An advantage of the piezoelectric cantilever beam approach is that no buried p+ layer is required. One disadvantage is that ZnO is not yet an established commercial technology and may be difficult to adapt to standard IC fabrication procedures.

Another cantilever beam accelerometer [170] also integrated with on-chip circuitry is shown in Fig. 55. As the circuit schematic indicates, the capacitor formed by the p+-Si buried layer and the beam metallization are used in a capacitive voltage divider network to bias an MOS transistor in an active operating region, similar to the operating mode of the RGT. Motions

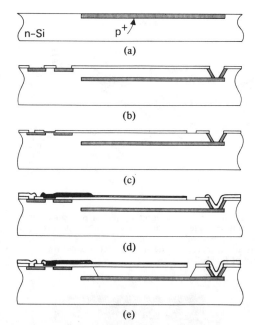

Fig. 56. Abbreviated processing schedule for the integrated accelerometer shown in Fig. 55. (a) Diffusion of p^+ ground electrode and etch stop. (b) Implantation of source, drain, and the sidewalls of the anisotropically etched via. (c) Growth of gate oxide, opening of contact holes, and field oxide etch which defines the geometry of the cantilever beam. (d) Metallization steps including the Al circuit metal, the protective gold overcoat on the Al, and the thin Cr–Au beam metal. (e) Final step in which the silicon is anisotropically etched in a self-stopping procedure to undercut and release the metal-coated SiO_2 cantilever beam.

of the chip normal to the surface can result in movements of the cantilever beam which will modulate the source–drain voltage of the transistor by varying the voltage bias on its gate. Fabrication proceeds according to the schedule shown in Fig. 56. After diffusing the boron-doped buried layer in selected areas and growing n-type epi over the wafer, conventional Al-gate p-channel transistors are defined adjacent to the p^+ regions. Electrical contact to the buried layer is accomplished by doping the sidewalls of a hole anisotropically etched down to the heavily doped region. This diffusion is done at the same time as the source–drain diffusion by a boron-ion-implant step. Next, Al is deposited and etched to form the metal interconnections—even down in the hole used to contact the buried p^+ layer.

Another very thin sputtered Cr–Au layer is deposited and etched for the electrode covering the cantilever beam. This same film is also employed as the plating base to selectively plate a protective Au layer over the already defined aluminum conductors, since the EDP etchant will attack thin Al films. Additional thick gold bumps have also been plated on the ends of the cantilever beams to increase their deflection amplitude with acceleration. Finally, the silicon is anisotropically etched in EDP to free the SiO_2 cantilever beams as described above.

The sensitivity of an SiO_2 beam of thickness t, width b, and length l with a concentrated load at the tip of mass M can be shown to be approximately

$$\Delta V/g = V_s \left(\frac{C_0}{3 C_{eff}} \right) \frac{740 l^4}{E d b^2 t^3} M \text{ (V/g of acceleration)} \quad (6)$$

where d is the epi thickness, ϵ_0 is the free-space permittivity, and ρ and E are the density and Young's modulus of SiO_2, respectively. V_s is the circuit supply voltage and C_{eff} is the effective circuit capacitance in the voltage divider network (~120 fF). Also

$$C_0 = \frac{\epsilon_0 l b}{d} \simeq 3.5 \text{ fF} \quad (7)$$

is the equilibrium capacitance between the beam and the buried layer. In good agreement with these calculations, sensitivities of 2.2 mV/g have been measured with loads of 0.35 μg, V_s = −22 V, and beam dimensions of 105 μm by 25 μm and t = 0.5 μm, d = 7 μm. Typical beam motions are about 60 nm/g of acceleration at the beam tip. Capacitance variations as small as 10 aF (corresponding to an acceleration of 0.25 g) have been detected with this sensor. Clearly, with such minute equilibrium capacitances, these and other similar miniature solid-state accelerometers *require* on-chip integrated detection circuitry simply to maintain parasitics at a tolerable level. This trend toward higher levels of integration appears to be a continuing feature in many areas of sensor development.

Electromechanical Switches

In 1972, Frobenius et al. [171] reported a lithographically fabricated *threshold* accelerometer, conceptually different from the analog devices described above. Based on the plated beam technique developed for the RGT, the tip of the plated beam is allowed to make intermittent contact with another metallization on the surface during large accelerations of the chip normal to its surface. Although this was the first demonstration of a photolithographically generated micromechanical electrical switch, the previously mentioned fatigue problem associated with plated-metal deflectable elements, in addition to its uncertain applicability, halted further development along these lines.

The insulator beam techniques used above for light modulators and accelerators, however, also lend themselves very well to the fabrication of small, fast, integrable, voltage-controlled electrical switching devices which are illustrated schematically in Fig. 57. Such four-terminal switches are not possible with the techniques described by Frobenius because the use of the *insulating* portion of the beam is crucial. Functional micromechanical switching was first realized [156] with a device like that shown schematically in Fig. 57(a). Construction of these switches follows the processing sequence shown in Fig. 58. The first few steps are identical to those employed in the optical modulator array; dope the surface with boron for the vertical etch-stop layer, grow an epitaxial spacer film (about 7 μm), deposit or grow the insulator for the cantilever beam (350-nm SiO_2), deposit a thin Cr–Au metallization (50 nm), delineate the metal lines and the insulator patterns to define the shape of the membrane, which is shown in cross section in Fig. 58(a). Plated metal crossovers and fixed electrode contact points are fabricated in a manner similar to the beams in the RGT and the threshold accelerometer of Frobenius et al. First a photoresist layer (PR1) is applied and patterned to provide contact holes through which gold will be plated and to define mesas over which gold will be plated to form the crossover and fixed electrode structures (Fig. 58(b)). Next, a second Cr–Au layer (0.3 μm) is evaporated over the entire wafer to serve as a plating base, and a second photoresist layer (PR2) is applied to define windows through which gold is selectively, electrochemically plated to a thickness of about 2 μm (Fig. 58(c)). After stripping the photoresist layers and excess plating base, the cantilever beams are released by etching the exposed silicon in EDP for about 20 min, then rinsed and dried (Fig. 58(d)).

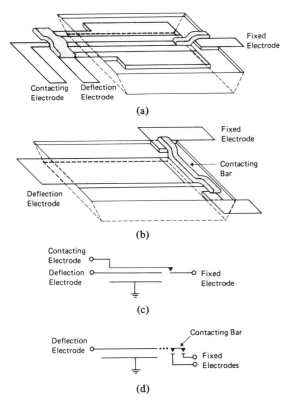

Fig. 57. Two designs of micromechanical switches. (a) The single-contact low-current design. (b) The double-contact configuration. (c) and (d) Suggested circuit representations of these devices. From [156].

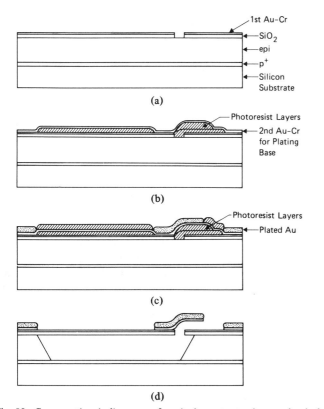

Fig. 58. Cross-sectional diagrams of a single-contact micromechanical switch at various stages during the fabrication procedure. (a) After first metal etch and oxide etch. (b) After evaporation of Au–Cr plating base. (c) After selective Au plating through photoresist holes. (d) Finished structure after photoresist stripping, removal of excess plating base, and EDP etch.

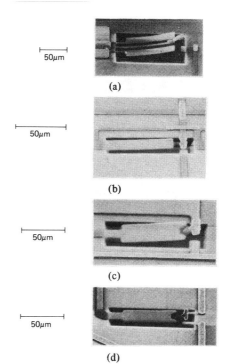

Fig. 59. Four different micromechanical switch designs. (a) Single-contact. (b) Double-contact with a contact bar as shown in Fig. 57(b). (c) and (d) Double-contact designs with two orientations of the fixed electrodes. From [172].

The SEM photographs in Fig. 59 show four different switch configurations, which can be classified as single- or double-contact designs [172]. Electrically, these devices behave as ideal, four-terminal, fully isolated, low-power, voltage-controlled switches. As a voltage is applied between the deflection electrode and the p^+ ground plane, the cantilever beam is deflected and the switch closes, connecting the contact electrode and the fixed electrode. Oscilloscope traces of pulsed switching is shown in Fig. 60. Single-contact switches are simple to operate and were the first to be demonstrated. However, their current-carrying capability is limited to less than 1 mA since all surface metallizations, which extend the length of the cantilever beam (including the current-carrying signal line leading up to the contact bar) must be thin to minimize its influence on the beam's mechanical characteristics. Much higher currents can be switched in the double-contact design since all the signal lines can be plated thick. With one extra masking step, even the fixed electrode regions under the contact bar can be plated prior to the application of the photoresist spacing layers.

While little consideration was given to the details of the contact electrode design in these early switch studies, it is clear that any practical utilization of the devices will rely critically on the ability to maximize current-carrying capability, contact force, reliability, and lifetime by optimizing the metallurgy and the geometry of the contacting electrodes. Gold, for example, is easy to electroplate and is certainly corrosion resistant but may be a poor choice as a contact metal because of its ductility and its self-welding tendency. Equally important is the design and configuration of the contacting surfaces for optimum electrical performance. Clearly, the development of micromechanical switches is in an early stage, yet the versatility of lithography and thin-film processing techniques permit a high degree of engineering design options for such micromechanical structures.

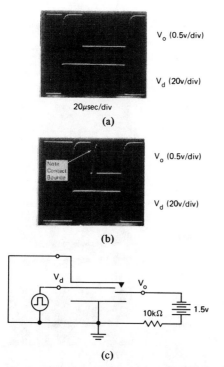

Fig. 60. (a) and (b) Oscilloscope traces of typical pulsed switching behavior for the circuit shown in (c). The deflection voltage in (a) is just above the switching threshold (~60 V), while the voltage in (b) has been increased to about 62 V. Note the delay time (~40 μs) between the application of the deflection voltage and the actual time the switch is closed. For higher deflection voltages (shown in the second oscilloscope trace), delay times are reduced and contact bounce effects are observed.

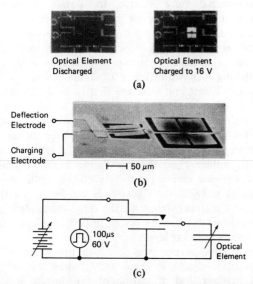

Fig. 61. A charge-storage application of micromechanical switches is illustrated by the circuit of (a) and (b). The capacitor leaves can be charged (and, therefore, deflected) by increasing the charging voltage to 16 V and pulsing the switch on for 100 μs. They will remain stored in this deflected position, shown at the right in (c), until the switch is pulsed again with the charging voltage reduced to zero, thereby releasing the leaves back to their discharged, or undeflected, positions.

Applications of these micromechanical switches range from telephone and analog signal switching arrays, to charge-storage circuits, to temperature [172] and magnetic field sensors. Fig. 61 shows an optical storage cell illustrating the high off-state impedance of which these micromechanical switches are capable. In this device, the charge-storage capacitor is an MMT-type deflectable cloverleaf element. When the switch is activated and the cloverleaf element is charged up to 16 V, the four leaves deflect and appear bright in a dark-field illuminated microscope. Furthermore, cells can be stored for many hours in either the charged (bright) or discharged (dark) condition.

The optimum design of these switches for particular operating parameters involves tradeoffs between the three primary performance criteria of speed, current-carrying capacity, and switching voltage. While the switch resonant frequency (which is related to the switching speed) is given by

$$f_R = \frac{1}{2\pi} \sqrt{\frac{Ebt^3}{4l^3(M + 0.23m)}} \qquad (8)$$

where m is the mass of the oxide beam and M is the mass of the plated contact bar, the switching voltage can be expressed approximately as given in (4), repeated here

$$V_{th} = \sqrt{\frac{3Et^3d^3}{10\epsilon_0 l^4}}. \qquad (9)$$

From these relationships, we can understand the conflicting requirements for high performance on all aspects of device operation. Since both f_R and V_{th} depend in similar ways on the dimensions of the cantilever beam, high-speed devices imply high switching voltages. While this problem can be alleviated somewhat by reducing the electrode separation d (or epi thickness), the hold-off voltage across the contact electrodes will also be reduced. At the same time, high current capacity implies a large contact metallization M, which also limits switching speed through (8). For the geometries described here, it seems unlikely that resonant frequencies above 200–300 kHz can be realized at voltages less than 20 V in low-current (<1-mA) applications. At the other end of the spectrum, current levels above 1 A might be difficult to obtain in a single device. These micromechanical switch geometries, therefore, are not *very* high speed, nor *very* high current devices, but rather seem to fill a niche between transistors and conventional electromagnetic relays. Other micromechanical switch geometries, however, possibly related to the torsion mirror described in Section V, may be possible which have different ranges of performance.

The advantages of these switches are that they can be batch-fabricated in large arrays, they exhibit extremely high off-state to on-state impedance ratios, the off-state coupling capacitance is very small, switching and sustaining power is extremely low, switching speed is at least an order of magnitude faster than relays, and other electronic devices can easily be integrated on the same chip. Their most serious disadvantages seem to be a relatively high switching voltage (near 50 V) and relatively low current-carrying capability (probably less than 1 A). The ideal applications would be in systems requiring large arrays of medium-current switches or drivers with very low internal resistances. It is still difficult, for example, to integrate large arrays of bipolar drivers with internal resistances less than about 10 Ω.

VII. Silicon Head Technology

An area of particular importance which has already begun to feel the impact of silicon micromechanics is silicon head technology. Heads of all types have common features which are ideally solved by silicon micromechanical techniques. They

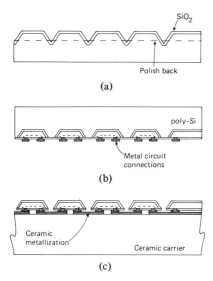

Fig. 62. Fabrication of the thermal print head found in several products manufactured by Texas Instruments is shown here. Based on earlier dielectric isolation schemes, SCS islands are embedded in a polysilicon substrate by a process of etching grooves, depositing SiO$_2$ followed by the thick poly-Si substrate (250 μm or more), and polishing back the single crystal just until the grooves are penetrated. Next, circuitry is fabricated in the single-crystal islands and the chip is bonded to a ceramic carrier. Finally, the polysilicon is selectively removed. Adapted from Bean and Runyan [175].

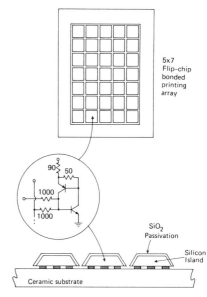

Fig. 63. A 5 × 7 array of silicon islands forms the complete print head in the T.I. Silent 700 computer terminals. Each island is about 25 μm thick and contains the bipolar power latching circuit shown, yet this thin silicon IC chip is also forcefully abraded against the paper while simultaneously thermally cycled to 260°C. Adapted from [174].

typically must be moved rapidly across some surface such as paper, magnetic media, or optical disk and so must be lightweight for high speed, accurate movement, and tracking abilities. In addition, it is always valuable, especially in multichannel heads, to have decoding and driving electronics available as close to the head as possible both to limit flexible wire attachments and to increase signal-to-noise ratios. As printing and recording technologies strive for higher and higher resolution, thin-film methods are being intensely pursued for the design and fabrication of active head elements. Micromechanics offers the potential of using thin-film methods not only for active head elements, but also for the precise micromachining of the passive structural assemblies on which the active head elements are located. Batch-fabrication, materials compatibility, simplified electronic interfaces, increased reliability, and low cost result from this strategy.

Ink jet nozzles have already been extensively discussed in Section IV and it seems probable that some variations on these designs will find their way into products in the 1980's. We will not describe these ink jet applications of silicon head technology further in this section.

Texas Instruments Thermal Print Head

Exploitation of silicon head technology has been successfully implemented in the thermal print head manufactured and sold by Texas Instruments [3]. New generations of high-speed (120 characters per second) print heads are now employed in the TI 780 series computer terminals and the 400 element Tigris plotter/printer [173], [174]. Based on the dielectric isolation process described by Bean and Runyan [175], a very abbreviated schematic of the fabrication procedure is illustrated in Fig. 62. Anisotropic etching is used to etch a grid in a silicon wafer which defines a 5 × 7 array of SCS mesas, for example. After growing an insulating film (SiO$_2$) over the surface, a thick (10–15-mils) layer of polysilicon is deposited, the wafer is turned upside down, and the single crystal is mechanically polished (or etched) back just until the original mesas are reached as shown in Fig. 62(b). At this stage, typical dimensions of the single-crystal mesas are 250 μm square by 25 μm thick. Conventional bipolar circuit fabrication methods are now used to define circuitry on each individual mesa as well as some peripheral circuitry in the single-crystal region surrounding the mesa array. The circuit metallization is completed by depositing solder or beam-lead connections wherever the circuit makes electrical contact to the ceramic substrate. Finally, the wafer is diced up, each individual die is flip-chip bonded to a ceramic substrate containing thick-film printed metallurgy, and the thick polysilicon layer is completely etched away. The resulting structure is shown in Fig. 63. Each mesa contains a latching circuit, a power transistor, and a power-dissipation resistor. The fabrication technique is designed to thermally isolate the mesas from each other (to inhibit thermal crosstalk) and to provide a controlled rate of heat transport to the ceramic substrate. Under normal operation, the latches on the 35 mesas are all set high or low in a pattern corresponding to the alphanumeric character to be printed, then the thermal dissipation supply voltage is pulsed and the activated mesa circuits are heated to about 260°C for 33 ms, causing dark spots to appear on the thermally sensitive paper in contact with the silicon print head.

In complete contrast to presently accepted practices for handling IC chips, in which the active silicon circuit is carefully sealed up, hidden from view, and diligently protected in sophisticated packages, it is important to realize that this thermal print head design places the silicon in direct, unprotected, and abrasive contact with the paper. Clearly, our traditional views of silicon chips as fragile components, always requiring delicate handling and careful protection, need to be modified. In addition, the micromachining techniques discussed in this review are certainly not limited to the fabrication of thermal print heads and it is likely that other electro-printing methods will also make use of the principles of silicon head technology.

VIII. CONCLUSION

While inexpensive microprocessors proliferate into automobiles, appliances, manufacturing equipment, instruments, and office machines, the engineering difficulties and the rapidly increasing expense of interfacing digital electronics with sensors and transducers is demanding more attention and causing lengthy development times. In addition, the general technical trends in instrumentation, communication, and input/output (I/O) devices continue to be in the direction of miniaturization for improved performance [1], [6], [7], [176] and reliability. As we have seen in this review, by employing the low-cost, batch-fabrication techniques available with silicon micromachining methods, many of these sensor and transducer I/O functions can be integrated on silicon, alongside the necessary circuitry, using common processing steps, and resulting in systems with improved performance and more straightforward implementation. Significantly, the interest of the engineering community has risen dramatically in response to these recent trends in I/O technology. Three special issues of the IEEE TRANSACTIONS ON ELECTRON DEVICES (September 1978 "Three-Dimensional Semiconductor Device Structures," December 1979 "Solid-State Sensors, Actuators, and Interface Electronics," and January 1982 "Solid-State Sensors, Actuators, and Interface Electronics") have been devoted to micromachining and integrated transducer techniques. At least three feature articles have appeared in major trade journals within the past year [177]-[179]. The first issue of a new international journal *Sensors and Actuators* devoted to research and development in the area of solid-state transducers was published in November 1980.

Why are silicon-related I/O applications enjoying such a sudden technical popularity? A large part of the answer comes from the nature of the new, microprocessor-controlled electronics market, in which the costs of the sensors, transducers, and interface electronics exceed the cost of the microprocessor itself [1], [6]. This surprising turn of events was undreamed of 7-8 years ago. Since silicon technology has been so successful in the development of sophisticated, yet inexpensive, VLSI electronic circuits, it seems an obvious extension to employ the same materials and the same fabrication principles to lower the costs of the sensors and transducers as well. The purpose of this review has been to illustrate how this extension might be accomplished for a broad range of applications.

From this applications point of view, it is clear that microprocessors and other computing systems will continue to be employed in every conceivable consumer and commercial product. Since the interface and transducer functions will largely determine the costs of such products, those who have learned to fabricate sensors, transducers, and interfaces in a cost-effective manner will be successful in the areas of computer-controlled equipment, from robots, industrial process controllers, and instruments, to toasters, automobiles, and bathroom scales, to displays, printers, and storage devices.

Given the rapidly accelerating demand for silicon-compatible sensors and transducers, and the corresponding success in demonstrating these various components, it is evident that silicon will be increasingly called upon, not only in its traditional electronic role, but also in a wide range of mechanical capacities where miniaturized, high-precision, high-reliability, and low-cost mechanical components and devices are required in critical applications, performing functions not ordinarily associated with silicon. We are beginning to realize that silicon isn't just for circuits anymore.

ACKNOWLEDGMENT

This paper has benefited greatly from the many technical contributions, helpful discussions, and suggestions provided by R. Muller, K. D. Wise, E. Bassous, P. Cade, J. Knutti, A. Shartel, V. Hanchett, J. Hope, and F. Anger.

REFERENCES

[1] Proceedings from the topical meeting on "The Limits to Miniaturization," for optics, electronics, and mechanics at the Swiss Federal Institute of Technology, Lausanne, Switzerland, Oct. 1980.
[2] Commercial devices have been available for some time. See, for example, the National Semiconductor catalog, *Transducers; Pressure and Temperature* for Aug. 1974.
[3] Texas Instruments Thermal Character Print Head, EPN3620 Bulletin DL-S7712505, 1977.
[4] P. O'Neill, "A monolithic thermal converter," *Hewlett-Packard J.*, p. 12, May 1980.
[5] E. Bassous, H. H. Taub, and L. Kuhn, "Ink jet printing nozzle arrays etched in silicon," *Appl. Phys. Lett.*, vol. 31, p. 135, 1977.
[6] W. G. Wolber and K. D. Wise, "Sensor development in the microcomputer age," *IEEE Trans. Electron Devices*, vol. ED-26, p. 1864, 1979.
[7] S. Middelhoek, J. B. Angell, and D.J.W. Noorlag, "Microprocessors get integrated sensors," *IEEE Spectrum*, vol. 17, p. 42, Feb. 1980.
[8] A. Kelly, *Strong Solids*, 2nd ed. (Monographs on the Physics and Chemistry of Materials). Oxford, England: Clarendon, 1973.
[9] E. B. Shand, *Glass Engineering Handbook*. New York: McGraw-Hill, 1958.
[10] *CRC Handbook of Chemistry and Physics*, R. C. Weast, Ed. Cleveland, OH: CRC Publ., 1980.
[11] G. L. Pearson, W. T. Reed, Jr., and W. L. Feldman, "Deformation and fracture of small silicon crystals," *Acta Metallurgica*, vol. 5, p. 181, 1957.
[12] K. Kuroiwa and T. Sugano, "Vapor-phase deposition of beta-silicon carbide on silicon substrates," *J. Electrochem. Soc.*, vol. 120, p. 138, 1973.
S. Nishino, Y. Hazuki, H. Matsunami, and T. Tanaka, "Chemical vapor deposition of single crystalline beta-SiC films on silicon substrate with sputtered SiC intermediate layer," *J. Electrochem. Soc.*, vol. 127, p. 2674, 1980.
[13] C. W. Dee, "Silicon nitride: Tribological applications of a Ceramic material," *Tribology*, vol. 3, p. 89, 1970.
[14] E. Rabinowicz, "Grinding damage of silicon nitride determined by abrasive wear tests," *Wear*, vol. 39, p. 101, 1976.
[15] T. E. Baker, S. L. Bagdasarian, G. L. Kix, and J. S. Judge, "Characterization of vapor-deposited paraxylylene coatings," *J. Electrochem. Soc.*, vol. 124, p. 897, 1977.
[16] L. B. Rothman, "Properties of thin polyimide films," *J. Electrochem. Soc.*, vol. 127, p. 2216, 1980.
[17] Adapted from Van Vlack, *Elements of Materials Science*. Reading, MA: Addison Wesley, 1964, p. 163.
[18] J. H. Hobstetter, "Mechanical properties of semiconductors," in *Properties of Crystalline Solids* (ASTM Special Technical Publication 283). Philadelphia, PA: ASTM, 1960, p. 40.
[19] G. Sinclair and C. Feltner, "Fatigue strength of crystalline solids," presented at the Symposium on Nature and Origin of Strength of Materials at the 63rd Annual Meeting of ASTM (ASTM Publisher, 1960, p. 129).
[20] C. M. Drum and M. J. Rand, "A low-stress insulating film on silicon by chemical vapor deposition," *J. Appl. Phys.*, vol. 39, p. 4458, 1968.
[21] S. Iwamura, Y. Nishida, and K. Hashimoto, "Rotating MNOS disk memory device," *IEEE Trans. Electron Devices*, vol. ED-28, p. 854, 1981.
[22] R. M. Finne and D. L. Klein, "A water-amine-complexing agent system for etching silicon," *J. Electrochem. Soc.*, vol. 114, p. 965, 1967.
[23] H. A. Waggener, R. C. Kragness, and A. L. Taylor, *Electronics*, vol. 40, p. 274, 1967.
[24] H. Robbins and B. Schwartz, "Chemical etching of silicon, II. The system HF, HNO_3, $HC_2H_3O_2$," *J. Electrochem. Soc.*, vol. 106, p. 505, 1959.
[25] B. Schwartz and H. Robbins, "Chemical etching of silicon, IV. Etching technology," *J. Electrochem. Soc.*, vol. 123, p. 1903, 1976.

[26] J. C. Greenwood, "Ethylene diamine-catechol-water mixture shows preferential etching of p-n junctions," *J. Electrochem. Soc.*, vol. 116, p. 1325, 1969.

[27] A. Bohg, "Ethylene diamine-pyrocatechol-water mixture shows etching anomaly in boron-doped silicon," *J. Electrochem. Soc.*, vol. 118, p. 401, 1971.

[28] H. Huraoka, T. Ohhashi, and Y. Sumitomo, "Controlled preferential etching technology," in *Semiconductor Silicon 1973*, H. R. Huff and R. R. Burgess, Eds. (The Electrochemical Society Softbound Symposium Ser., Princeton, NJ, 1973), p. 327.

[29] S. C. Terry, "A gas chromatography system fabricated on a silicon wafer using integrated circuit technology," Ph.D. dissertation, Department of Electrical Engineering, Stanford University, Stanford, CA, 1975.

[30] W. Kern, "Chemical etching of silicon, germanium, gallium arsenide, and gallium phosphide," *RCA Rev.*, vol. 29, p. 278, 1978.

[31] S. K. Ghandhi, *The Theory and Practice of Microelectronics*. New York: Wiley, 1968.

[32] J. B. Price, "Anisotropic etching of silicon with potassium hydroxide-water-isopropyl alcohol," in *Semiconductor Silicon 1973*, H. R. Huff and R. R. Burgess, Eds. (The Electrochemical Society Softbound Symposium Ser., Princeton, NJ, 1973), p. 339.

[33] D. L. Kendall, "On etching very narrow grooves in silicon," *Appl. Phys. Lett.*, vol. 26, p. 195, 1975.

[34] I. J. Pugacz-Muraszkiewicz, "Detection of discontinuities in passivating layers on silicon by NaOH anisotropic etch," *IBM J. Res. Develop.*, vol. 16, p. 523, 1972.

[35] E. Bassous, "Fabrication of novel three-dimensional microstructures by the anisotropic etching of (100) and (110) silicon," *IEEE Trans. Electron Devices*, vol. ED-25, p. 1178, 1978.

[36] K. E. Bean, "Anisotropic etching of silicon," *IEEE Trans. Electron Devices*, vol. ED-25, p. 1185, 1978.

[37] A. I. Stoller, "The etching of deep vertical-walled patterns in silicon," *RCA Rev.*, vol. 31, p. 271, 1970.

[38] D. Kendall, "Vertical etching of silicon at very high aspect ratios," *Annu. Rev. Materials Sci.*, vol. 9, p. 373, 1979.

[39] W. K. Zwicker and K. K. Kurtz, "Anisotropic etching of silicon using electrochemical displacement reactions," in *Semiconductor Silicon 1973*, H. R. Huff and R. R. Burgess, Eds. (The Electrochemical Society Softbound Symposium Ser., Princeton, NJ, 1973), p. 315.

[40] D. B. Lee, "Anisotropic etching of silicon," *J. Appl. Phys.*, vol. 40, p. 4569, 1969.

[41] M. J. Declercq, L. Gerzberg, and J. D. Meindl, "Optimization of the hydrazine-water solution for anisotropic etching of silicon in integrated circuit technology," *J. Electrochem. Soc.*, vol. 122, p. 545, 1975.

[42] D. F. Weirauch, "Correlation of the anisotropic etching of single crystal silicon spheres and wafers," *J. Appl. Phys.*, vol. 46, p. 1478, 1975.

[43] E. Bassous and E. F. Baran, "The fabrication of high precision nozzles by the anisotropic etching of (100) silicon," *J. Electrochem. Soc.*, vol. 125, p. 1321, 1978.

[44] A. Reisman, M. Berkenblit, S. A. Chan, F. B. Kaufman, and D. C. Green, "The controlled etching of silicon in catalyzed ethylene diamine-pyrocatechol-water solutions," *J. Electrochem. Soc.*, vol. 126, p. 1406, 1979.

[45] A. Uhlir, "Electrolytic shaping of germanium and silicon," *Bell Syst. Tech. J.*, vol. 36, p. 333, Mar. (1956).
D. R. Turner, "Electropolishing silicon in hydroflouric acid solutions," *J. Electrochem. Soc.*, vol. 105, p. 406, 1958.

[46] M.J.J. Theunissen, J. A. Appels, and W.H.C.G. Verkuylen, "Application of electrochemical etching of silicon to semiconductor device technology," *J. Electrochem. Soc.*, vol. 117, p. 959, 1970.
R. L. Meek, "Anodic dissolution of n^+ silicon," *J. Electrochem. Soc.*, vol. 118, p. 437, 1971.

[47] C. D. Wen and K. P. Weller, "Preferential electrochemical etching of p^+ silicon in an aqueous $HF-H_2SO_4$ electrolyte," *J. Electrochem. Soc.*, vol. 119, p. 547, 1972.

[48] H.J.A. van Dijk and J. de Jonge, "Preparation of thin silicon crystals by electrochemical thinning of epitaxially grown structures," *J. Electrochem. Soc.*, vol. 117, p. 553, 1970.

[49] R. L. Meek, "Electrochemically thinned n/n^+ epitaxial silicon—Method and applications," *J. Electrochem. Soc.*, vol. 118, p. 1240, 1971.

[50] H. A. Waggener, "Electrochemically controlled thinning of silicon," *Bell Syst. Tech. J.*, vol. 50, p. 473, 1970.

[51] T. N. Jackson, M. A. Tischler, and K. D. Wise, "An electrochemical etch-stop for the formation of silicon microstructures," *IEEE Electron Device Lett.*, vol. EDL-2, p. 44, 1981.

[52] Y. Watanabe, Y. Arita, T. Yokoyama, and Y. Igarashi, "Formation and properties of porous silicon and its applications," *J. Electrochem. Soc.*, vol. 122, p. 1351, 1975.

[53] T. Unagami, "Formation mechanism of porous silicon layer by anodization in HF solution," *J. Electrochem. Soc.*, vol. 127, p. 476, 1980.

[54] T. C. Teng, "An investigation of the application of porous silicon layers to the dielectric isolation of integrated circuits," *J. Electrochem. Soc.*, vol. 126, p. 870, 1979.

[55] T. Unagami and M. Seki, "Structure of porous silicon layer and heat-treatment effect," *J. Electrochem. Soc.*, vol. 125, p. 1339, 1978.

[56] D. J. Ehrlich, R. M. Osgood, and T. F. Deutsch, "Laser chemical technique for rapid direct writing of surface relief in silicon," *Appl. Phys. Lett.*, vol. 38, p. 1018, 1981.

[57] W. R. Runyan, E. G. Alexander, and S. E. Craig, Jr., "Behavior of large-scale surface perturbations during silicon epitaxial growth," *J. Electrochem. Soc.*, vol. 114, p. 1154, 1967.
R. K. Smeltzer, "Epitaxial deposition of silicon in deep grooves," *J. Electrochem. Soc.*, vol. 122, p. 1666, 1975.

[58] B. W. Wessels and B. J. Baliga, "Vertical channel field-controlled thyristors with high gain and fast switching speeds," *IEEE Trans. Electron Devices*, vol. ED-25, p. 1261, 1978.

[59] L. Gerzberg and J. Meindl, "Monolithic polycrystalline silicon distributed RC devices," *IEEE Trans. Electron Devices*, vol. ED-25, p. 1375, 1978.

[60] P. Rai-Choudhury, "Chemical vapor deposited silicon and its device applications," in *Semiconductor Silicon 1973*, H. R. Huff and R. R. Burgess, Eds. (The Electrochemical Society Softbound Symposium Ser., Princeton, NJ, 1973), p. 243.

[61] H. E. Cline and T. R. Anthony, "Random walk of liquid droplets migrating in silicon," *J. Appl. Phys.*, vol. 47, p. 2316, 1976.

[62] ——, "High-speed droplet migration in silicon," *J. Appl. Phys.*, vol. 47, p. 2325, 1976.

[63] ——, "Thermomigration of aluminum-rich liquid wires through silicon," *J. Appl. Phys.*, vol. 47, p. 2332, 1976.

[64] T. R. Anthony and H. E. Cline, "Lamellar devices processed by thermomigration," *J. Appl. Phys.*, vol. 48, p. 3943, 1977.

[65] ——, "Migration of fine molten wires in thin silicon wafers," *J. Appl. Phys.*, vol. 49, p. 2412, 1978.

[66] ——, "On the thermomigration of liquid wires," *J. Appl. Phys.*, vol. 49, p. 2777, 1978.

[67] ——, "Stresses generated by the thermomigration of liquid inclusions in silicon," *J. Appl. Phys.*, vol. 49, p. 5774, 1978.

[68] T. Mizrah, "Joining and recrystallization of Si using the thermomigration process," *J. Appl. Phys.*, vol. 51, p. 1207, 1980.

[69] C. C. Wen, T. C. Chen, and J. M. Zemel, "Gate-controlled diodes for ionic concentration measurement," *IEEE Trans. Electron Devices*, vol. ED-26, p. 1945, 1979.

[70] L. C. Kimerling, H. J. Leamy, and K. A. Jackson, "Photoinduced zone migration (PIZM) in semiconductors," in *Proc. Symp. on Laser and Electron Beam Processing of Electronic Materials* (The Electrochemical Society Publisher, Electronics Division), vol. 80-1, p. 242, 1980.

[71] G. Wallis and D. I. Pomerantz, "Field-assisted glass-metal sealing," *J. Appl. Phys.*, vol. 40, p. 3946, 1969.

[72] P. B. DeNee, "Low energy metal-glass bonding," *J. Appl. Phys.*, vol. 40, p. 5396, 1969.

[73] J. M. Brownlow, "Glass-related effects in field-assisted glass-metal bonding," IBM Rep. RC 7101, May 1978.

[74] A. D. Brooks and R. P. Donovan, "Low-temperature electrostatic silicon-to-silicon seals using sputtered borosilicate glass," *J. Electrochem. Soc.*, vol. 119, p. 545, 1972.

[75] L. M. Roylance and J. B. Angell, "A batch-fabricated silicon accelerometer," *IEEE Trans. Electron Devices*, vol. ED-26, p. 1911, 1979.

[76] J. H. Jerman, J. M. Pendleton, L. N. Rhodes, C. S. Sanders, S. C. Terry, and G. V. Walsh, "Anodic bonding," Stanford University Lab. Rep. for EE412, 1978.

[77] D. A. Kiewit, "Microtool fabrication by etch pit replication," *Rev. Sci. Instrum.*, vol. 44, p. 1741, 1973.

[78] K. D. Wise, M. G. Robinson, and W. J. Hillegas, "Solid-state processes to produce hemispherical components for inertial fusion targets," *J. Vac. Sci. Technol.*, vol. 18, p. 1179, 1981.
K. D. Wise, T. N. Jackson, N. A. Masnari, M. G. Robinson, D. E. Solomon, G. H. Wuttke, and W. B. Rensel, "Fabrication of hemispherical structures using semiconductor technology for use in thermonuclear fusion research," *J. Vac. Sci. Technol.*, vol. 16, p. 936, 1979.

[79] R. N. Thomas and H. C. Nathanson, "Photosensitive field emission from silicon point arrays," *Appl. Phys. Lett.*, vol. 21, p. 384, 1972.

[80] R. N. Thomas, R. A. Wickstrom, D. K. Schroder, and H. C. Nathanson, "Fabrication and some applications of large area silicon field emission arrays," *Solid-State Electron.*, vol. 17, p. 155, 1974.

[81] E. Bassous, "Nozzles formed in mono-crystalline silicon," U.S. Patent 3 921 916, 1975.

[82] E. Bassous, L. Kuhn, A. Reisman, and H. H. Taub, "Ink jet nozzle," U.S. Patent 4 007 464, 1977.

[83] R. G. Sweet, "High frequency recording with electrostatically

deflected ink jets," *Rev. Sci. Instrum.*, vol. 36, p. 131, 1965.
[84] F. J. Kamphoefner, "Ink jet printing," *IEEE Trans. Electron Devices*, vol. ED-19, p. 584, 1972.
[85] R. D. Carnahan and S. L. Hou, "Ink jet technology," *IEEE Trans. Ind. Appl.*, vol. IA-13, p. 95, 1977.
[86] Special Issue on Ink Jet Printing, *IBM J. Res. Develop.*, vol. 21, 1977.
[87] L. Kuhn, E. Bassous, and R. Lane, "Silicon charge electrode array for ink jet printing," *IEEE Trans. Electron Devices*, vol. ED-25, p. 1257, 1978.
[88] K. E. Petersen, "Fabrication of an integrated, planar silicon ink-jet structure," *IEEE Trans. Electron Devices*, vol. ED-26, p. 1918, 1979.
[89] W. Anacker, E. Bassous, F. F. Fang, R. E. Mundie, and H. N. Yu, "Fabrication of multiprobe miniature electrical connector," *IBM Tech. Discl. Bull.*, vol. 19, p. 372, 1976.
S. K. Lahiri, P. Geldermans, G. Kolb, J. Sokolwski, and M. J. Palmer, "Pluggable connectors for Josephson device packaging," *J. Electrochem. Soc. Extended Abstr.*, vol. 80-1, p. 216, 1980.
[90] L. P. Boivin, "Thin-film laser-to-fiber coupler," *Appl. Opt.*, vol. 13, p. 391, 1974.
[91] C. C. Tseng, D. Botez, and S. Wang, "Optical bends and rings fabricated by preferential etching," *Appl. Phys. Lett.*, vol. 26, p. 699, 1975.
[92] W. T. Tsang, C. C. Tseng and S. Wang, "Optical waveguides fabricated by preferential etching," *Appl. Opt.*, vol. 14, p. 1200, 1975.
[93] C. C. Tseng, W. T. Tsang, and S. Wang, "A thin-film prism as a beam separator for multimode guided waves in integrated optics," *Opt. Commun.*, vol. 13, p. 342, 1975.
[94] W. T. Tsang and S. Wang, "Preferentially etched diffraction gratings in silicon," *J. Appl. Phys.*, vol. 46, p. 2163, 1975.
[95] —, "Thin-film beam splitter and reflector for optical guided waves," *Appl. Phys. Lett.*, vol. 27, p. 588, 1975.
[96] J. S. Harper and P. F. Heidrich, "High density multichannel optical waveguides with integrated couplers," *Wave Electron.*, vol. 2, p. 369, 1976.
[97] H. P. Hsu and A. F. Milton, "Single mode optical fiber pick-off coupler," *Appl. Opt.*, vol. 15, p. 2310, 1976.
[98] C. Hu and S. Kim, "Thin-film dye laser with etched cavity," *Appl. Phys. Lett.*, vol. 29, p. 9, 1976.
[99] H. P. Hsu and A. F. Milton, "Flip-chip approach to endfire coupling between single-mode optical fibres and channel waveguides," *Electron. Lett.*, vol. 12, p. 404, 1976.
[100] J. D. Crow, L. D. Comerford, R. A. Laff, M. J. Brady, and J. S. Harper, "GaAs laser array source package," *Opt. Lett.*, vol. 1, p. 40, 1977.
[101] H. P. Hsu and A. F. Milton, "Single-mode coupling between fibers and indiffused waveguides," *IEEE J. Quantum Electron.*, vol. QE-13, p. 224, 1977.
[102] J. T. Boyd and S. Sriram, "Optical coupling from fibers to channel waveguides formed on silicon," *Appl. Opt.*, vol. 17, p. 895, 1978.
[103] S. C. Terry, J. H. Jerman, and J. B. Angell, "A gas chromatograph air analyzer fabricated on a silicon wafer," *IEEE Trans. Electron Devices*, vol. ED-26, p. 1880, 1979.
[104] W. A. Little, "Design and construction of microminiature cyrogenic refrigerators," in *AIP Proc. of Future Trends in Superconductive Electronics* (University of Virginia, Charlottesville, 1978).
[105] D. B. Tuckerman and R.F.W. Pease, "High-performance heat sinking for VLSI," *IEEE Electron Device Lett.*, vol. EDL-2, p. 126, 1981.
[106] K. E. Bean and J. R. Lawson, "Application of silicon orientation and anisotropic effects to the control of charge spreading in devices," *IEEE J. Solid-State Circuits*, vol. SC-9, p. 111, 1974.
[107] M. J. Declerq, "A new C-MOS technology using anisotropic etching of silicon," *IEEE J. Solid-State Circuits*, vol. SC-10, p. 191, 1975.
[108] H. N. Yu, R. H. Dennard, T.H.P. Chang, C. M. Osburn, V. Dilonardo, and H. E. Luhn, "Fabrication of a miniature 8 k-bit memory chip using electron beam exposure," *J. Vac. Sci. Technol.*, vol. 12, p. 1297, 1975.
[109] C.A.T. Salama and J. G. Oakes, "Nonplanar power field-effect transistors," *IEEE Trans. Electron Devices*, vol. ED-25, p. 1222, 1978.
[110] K. P. Lisiak and J. Berger, "Optimization of nonplanar power MOS transistors," *IEEE Trans. Electron Devices*, vol. ED-25, p. 1229, 1978.
[111] W. C. Rosvold, W. H. Legat, and R. L. Holden, "Air gap isolated micro-circuits-beam-lead devices," *IEEE Trans. Electron Devices*, vol. ED-15, p. 640, 1968.
[112] L. A. D'Asaro, J. V. DiLorenzo, and H. Fukui, "Improved performance of GaAs microwave field-effect transistors with low inductance via-connections through the substrate," *IEEE Trans. Electron Devices*, vol. ED-25, p. 1218, 1978.
[113] T. J. Rodgers and J. D. Meindl, "Epitaxial V-groove bipolar integrated circuit process," *IEEE Trans. Electron Devices*, vol. ED-20, p. 226, 1973.
[114] E. S. Ammar and T. J. Rodgers, "UMOS transistors on (110) silicon," *IEEE Trans. Electron Devices*, vol. ED-27, p. 907, 1980.
[115] B. J. Baliga, "A novel buried grid device fabrication technology," *IEEE Electron Device Lett.*, vol. EDL-1, p. 250, 1980.
[116] T. I. Chappell, "The V-groove multijunction solar cell," *IEEE Trans. Electron Devices*, vol. ED-26, p. 1091, 1979.
[117] D. L. Spears and H. I. Smith, "High resolution pattern replication using soft X-rays," *Electron. Lett.*, vol. 8, p. 102, 1972.
[118] H. I. Smith, D. L. Spears, and S. E. Bernacki, "X-ray lithography: A complementary technique to electron beam lithography," *J. Vac. Sci. Technol.*, vol. 10, p. 913, 1973.
[119] T. O. Sedgwick, A. N. Broers, and B. J. Agule, "A novel method for fabrication of ultrafine metal lines by electron beams," *J. Electrochem. Soc.*, vol. 119, p. 1769, 1972.
[120] P. V. Lenzo and E. G. Spencer, "High-speed low-power X-ray lithography," *Appl. Phys. Lett.*, vol. 24, p. 289, 1974.
[121] C. J. Schmidt, P. V. Lenzo, and E. G. Spencer, "Preparation of thin windows in silicon masks for X-ray lithography," *J. Appl. Phys.*, vol. 46, p. 4080, 1975.
[122] H. Bohlen, J. Greschner, W. Kulcke, and P. Nehmiz, "Electron beam step and repeat proximity printing," in *Proc. Electrochem. Soc. Meet.* (Seattle, WA, May 1978).
[123] R. J. Jaccodine and W. A. Schlegel, "Measurement of strains at Si-SiO$_2$ interface," *J. Appl. Phys.*, vol. 37, p. 2429, 1966.
[124] C. W. Wilmsen, E. G. Thompson, and G. H. Meissner, "Buckling of thermally grown SiO$_2$ thin films," *IEEE Trans. Electron Devices*, vol. ED-19, p. 122, 1972.
[125] E. Bassous, R. Feder, E. Spiller, and J. Topalian, "High transmission X-ray masks for lithographic applications," *Solid-State Technol.*, vol. 19, p. 55, 1976.
[126] G. A. Antcliffe, L. J. Hornbeck, W. W. Chan, J. W. Walker, W. C. Rhines, and D. R. Collins, "A backside illuminated 400 × 400 charge-coupled device imager," *IEEE Trans. Electron Devices*, vol. ED-23, p. 1225, 1976.
[127] G. R. Lahiji and K. D. Wise, "A monolithic thermopile detector fabricated using integrated-circuit technology," in *Proc. Int. Electron Devices Meet.* (Washington, DC), p. 676, 1980.
[128] C. L. Huang and T. van Duzer, "Josephson tunnelling through locally thinned silicon," *Appl. Phys. Lett.*, vol. 25, p. 753, 1974.
—, "Single-crystal silicon-barrier Josephson junctions," *IEEE Trans. Magn.*, vol. MAG-11, p. 766, 1975.
[129] —, "Schottky diodes and other devices on thin silicon membranes," *IEEE Trans. Electron Devices*, vol. ED-23, p. 579, 1976.
[130] C. J. Maggiore, P. D. Goldstone, G. R. Gruhn, N. Jarmie, S. C. Stotlar, and H. V. Dehaven, "Thin epitaxial silicon for dE/dx detectors," *IEEE Trans. Nucl. Sci.*, vol. NS-24, p. 104, 1977.
[131] H. Guckel, S. Larsen, M. G. Lagally, G. Moore, J. B. Miller, and J. D. Wiley, "Electromechanical devices utilizing thin Si diaphragms," *Appl. Phys. Lett.*, vol. 31, p. 618, 1977.
[132] O. N. Tufte, P. W. Chapman, and D. Long, "Silicon diffused-element piezoresistive diaphragms," *J. Appl. Phys.*, vol. 33, p. 3322, 1962.
A.C.M. Gieles and G.H.J. Somers, "Miniature pressure transducers with silicon diaphragm," *Philips Tech. Rev.*, vol. 33, p. 14, 1973.
Samaun, K. D. Wise, and J. B. Angell, "An IC piezoresistive pressure sensor for biomedical instrumentation," *IEEE Trans. Biomed. Eng.*, vol. BME-20, p. 101, 1973.
W. D. Frobenius, A. C. Sanderson, and H. C. Nathanson, "A microminiature solid-state capacitive blood pressure transducer with improved sensitivity," *IEEE Trans. Biomed. Eng.*, vol. BME-20, p. 312, 1973.
[133] S. K. Clark and K. D. Wise, "Pressure sensitivity in anisotropically etched thin-diaphragm pressure sensors," *IEEE Trans. Electron Devices*, vol. ED-26, p. 1887, 1979.
[134] J. M. Borky and K. D. Wise, "Integrated signal conditioning for silicon pressure sensors," *IEEE Trans. Electron Devices*, vol. ED-26, p. 1906, 1979.
[135] W. H. Ko, J. Hynecek, and S. F. Boettcher, "Development of a miniature pressure transducer for biomedical applications," *IEEE Trans. Electron Devices*, vol. ED-26, p. 1896, 1979.
[136] C. S. Sander, J. W. Knutti, and J. D. Meindl, "A monolithic capacitive pressure sensor with pulse-period output," *IEEE Trans. Electron Devices*, vol. ED-27, p. 927, 1980.
[137] H. C. Tuan, J. S. Yanacopoulos, and T. A. Nunn, "Piezoresistive force sensors for observing muscle contraction," *Stanford Univ. Electron. Res. Rev.*, p. 102, 1975.
[138] R. J. Wilfinger, P. H. Bardell, and D. S. Chhabra, "The resonistor: A frequency selective device utilizing the mechanical resonance of a silicon substrate," *IBM J. Res. Develop.*, vol. 12, p. 113, 1968.
[139] K. E. Petersen, "Silicon torsional scanning mirror," *IBM J. Res. Develop.*, vol. 24, p. 631, 1980.

[140] A. Higdon and W. D. Stiles, *Engineering Mechanics, Volume II: Dynamics.* Englewood Cliffs, NJ: Prentice-Hall, 1961, p. 555.
[141] H. F. Wolf, *Silicon Semiconductor Data.* New York: Pergamon, 1969.
[142] W. E. Newell, "Miniaturization of tuning forks," *Science*, vol. 161, p. 1320, 1968.
[143] S. P. Timeshenko and J. M. Gere, *Mechanics of Materials.* New York: Van Nostrand Reinhold, 1972, p. 516.
[144] P. J. Brosens, "Dynamic mirror distortions in optical scanning," *Appl. Opt.*, vol. 11, p. 2987, 1972.
[145] R. J. Roark and W. C. Young, *Formulas for Stress and Strain.* New York: McGraw-Hill, 1975.
[146] L. Beiser, "Laser scanning systems," in *Laser Applications Volume 2*, M. Ross. Ed. New York: Academic Press, 1974.
[147] H. C. Nathanson and R. A. Wickstrom, "A resonant-gate silicon surface transistor with high-Q bandpass properties," *Appl. Phys. Lett.*, vol. 7, p. 84, 1965.
[148] H. C. Nathanson, W. E. Newell, R. A. Wickstrom, and J. R. Davis, Jr., "The resonant gate transistor," *IEEE Trans. Electron Devices*, vol. ED-14, p. 117, 1967.
[149] J. A. van Raalte, "A new Schlieren light valve for television projection," *Appl. Opt.*, vol. 9, p. 2225, 1970.
[150] K. Preston, Jr., "A coherent optical computer system using the membrane light modulator," *IEEE Trans. Aerosp. Electron. Syst.*, vol. AES-6, p. 458, 1970.
[151] L. S. Cosentino and W. C. Stewart, "A membrane page composer," *RCA Rev.*, vol. 34, p. 45, 1973.
[152] B. J. Ross and E. T. Kozol, "Performance characteristics of the deformagraphic storage display tube (DSDT)," presented at the IEEE Intercon., Mar. 1973.
[153] J. Guldberg, H. C. Nathanson, D. L. Balthis, and A. S. Jensen, "An aluminum/SiO_2 silicon on sapphire light valve matrix for projection displays," *Appl. Phys. Lett.*, vol. 26, p. 391, 1975.
[154] R. N. Thomas, J. Guldberg, H. C. Nathanson, and P. R. Malmberg, "The mirror matrix tube: A novel light valve for projection displays," *IEEE Trans. Electron Devices*, vol. ED-22, p. 765, 1975.
[155] K. E. Petersen, "Micromechanical light modulator array fabricated on silicon," *Appl. Phys. Lett.*, vol. 31, p. 521, 1977.
[156] —, "Dynamic micromechanics on silicon: Techniques and devices," *IEEE Trans. Electron Devices*, vol. ED-25, p. 1241, 1978.
[157] J. P. Den Hartog, *Mechanical Vibrations.* New York: McGraw-Hill, 1956.
[158] K. E. Petersen and C. R. Guarnieri, "Young's modulus measurements of thin films using micromechanics," *J. Appl. Phys.*, vol. 50, p. 6761, 1979.
[159] J. W. Beams, J. B. Freazeale, and W. L. Bart, "Mechanical strength of thin films of metals," *Phys. Rev.*, vol. 100, p. 1657, 1955.
[160] C. A. Neugebauer, "Tensile properties of thin, evaporated gold films," *J. Appl. Phys.*, vol. 31, p. 1096, 1960.
[161] J. M. Blakely, "Mechanical properties of vacuum-deposited gold," *J. Appl. Phys.*, vol. 35, p. 1756, 1964.
[162] R. D. Jolly and R. S. Muller, "Miniature cantilever beams fabricated by anisotropic etching of silicon," *J. Electrochem. Soc.*, vol. 127, p. 2750, 1980.
[163] *Metals Reference Handbook*, C. J. Smithels, Ed. London, England: Butterworths, 1976.
[164] R. E. McMillan and R. P. Misra, "Insulating materials for semiconductor surfaces," *IEEE Trans. Elec. Insulat.*, vol. EI-5, p. 10, 1970.
[165] *Corning Laboratory Glassware Catalogue*, 1978.
[166] J. F. Lynch, C. G. Ruderer, and W. H. Duckworth, *Engineering Properties of Selected Ceramic Materials.* The American Ceramic Society, OH, 1966.
[167] Air Force Tech. Rep. AFAL-TR-77-152, Air Force Avionics Lab., Wright-Patterson Air Force Base, OH, 45433.
[168] P. Chen, R. S. Muller, T. Shiosaki, and R. M. White, "Silicon cantilever beam accelerometer utilizing a PI-FET capacitive transducer," *IEEE Trans. Electron Devices*, vol. ED-26, p. 1857, 1979.
[169] P. Chen, R. Jolly, G. Halac, R. S. Muller, and R. M. White, "A planar processed PI-FET accelerometer," in *Proc. Int. Electron Devices Meet.* (Washington, DC, Dec. 1980), p. 848.
[170] K. E. Petersen, A. Shartel, and N. Raley, "Micromechanical accelerometer integrated with MOS detection circuitry," *IEEE Trans. Electron Devices*, vol. ED-29, p. 23, Jan. 1982.
[171] W. D. Frobenius, S. A. Zeitman, M. H. White, D. D. O'Sullivan, and R. G. Hamel, "Microminiature ganged threshold accelerometers compatible with integrated circuit technology," *IEEE Trans. Electron Devices*, vol. ED-i9, p. 37, 1972.
[172] K. E. Petersen, "Micromechanical membrane switches on silicon," *IBM J. Res. Develop.*, vol. 23, p. 376, 1979.
[173] B. Boles, "Thermal printing applications and technology," presented at the Invitational Computer Conf., Sept. 1979.
[174] M. L. Morris, *Thermal Line Printers*, Texas Instruments product bulletin.
[175] K. E. Bean and W. R. Runyan, "Dielectric isolation: Comprehensive, current and future," *J. Electrochem. Soc.*, vol. 124, p. 5c, 1977.
K. E. Bean, "Chemical vapor deposition applications in microelectronics processing," *Thin Solid Films*, vol. 83, p. 173, 1981.
[176] H. C. Nathanson and J. Guldberg, "Topologically structured thin films in semiconductor device operation," in *Physics of Thin Films Volume 8.* New York: Academic Press, 1975.
[177] "System designers fish for microcomputer-compatible sensors and transducers," *EDN Mag.*, p. 122, Mar. 20, 1980.
[178] "Sensing the real world with low cost silicon," *Electronics*, p. 113, Nov. 6, 1980.
[179] "Ultraminiature mechanics," *Machine Des.*, p. 112, Jan. 8, 1981.

MICROROBOTS AND MICROMECHANICAL SYSTEMS

W. S. N. TRIMMER

AT&T Bell Laboratories, Holmdel, NJ 07733 (U.S.A.)

(Received January 15, 1988; in revised form July 11, 1988; accepted August 28, 1988)

Abstract

The domain of micromechanical systems is an extensive, useful and yet largely unexplored area. It is likely that, in many applications, small mechanisms will prove to be faster, more accurate, gentler and less expensive than the macro systems presently used. This paper explores the advantages of micromechanical systems and analyzes the scaling of forces in the micro domain.

1. Introduction

Over the past 20 years we have witnessed a revolution in the miniaturization of electronic components from the dimensions of inches to microns. Now we have the potential to accomplish a similar revolution with many mechanical components [1]. Moreover, as we shrink various devices, fundamental scaling properties enhance the performance of these small systems. In this paper we shall examine what happens as we scale the dimensions of motors and other mechanical systems into the micro domain. (An excellent overview of the micro field is given in an NSF report on microdynamics [2].)

There are several reasons for developing micro systems. Systems with small dimensions have advantages in handling small parts, advantages that include speed, accuracy and gentleness. Intellectually, one is also drawn to this extensive and relatively unexplored domain. Benefits will probably accrue in such areas as electronics assembly, medicine and space exploration. Micro systems potentially have numerous advantages in performing micro tasks that are at, or even beyond, the limits of human manipulation and patience.

The actuators we use in the macro world often depend upon magnetic forces. As we shall see, these magnetic forces scale poorly into the micro domain. Fortunately, several other forces scale well. Scaling theory shows that electrostatics, pneumatics, surface tension and biological forces are all strong enough to be useful in the small domain.

This paper explores the field of microrobotic and micromechanical systems. The first portion of the paper examines the advantages, extent and applicability of micro systems. The second portion analyzes what type of forces will be useful to power these micromechanical systems. 'Microrobotic' is used to refer to small, automated systems; 'micro teleoperator' refers to a small system that is remotely controlled by a human; and 'micromechanical systems' refers to *all* small mechanical systems.

2. Advantages of micromechanical systems

This Section discusses what sizes of mechanical systems are appropriate to handle small objects. For example, what sized tooling is appropriate for assembling optical fibers, lasers and light detectors used in fiber optic communication. There are advantages in using micro tooling that is more commensurate in size with these small parts than the macro tooling presently used. These advantages include:

(a) *Higher throughput.* The time for an operation scales with the system's linear dimension raised to roughly the first power, s^1. That is, a system one-tenth the size of the original has the potential of performing the task ten times faster. (Later in the analysis, it will be shown that the exponent ranges from 0.5 to 1.5 for most force laws.)

(b) *Higher accuracy.* Problems due to phenomena such as the temperature coefficient of expansion, deflection and vibration become less troublesome as the size of the system decreases. Despite these advantages, it is curious that people generally consider smaller mechanical systems to be less accurate, indeed often toys. This need not be the case; the precision with which small machines can be made may ultimately be limited by atomic dimensions [3].

(c) *Gentleness.* The small forces and masses associated with small systems make them more gentle.

(d) *Improved performance.* Because the cost of materials scales as s^3 (the volume), a small system can be built with very expensive materials that have desirable properties. This freedom to use exotic materials often allows the designer to improve performance.

(e) *Floor space.* An important consideration in manufacturing plants is the cost of floor space. Small systems reduce this cost and may also enable an operator to tend more machines because of their proximity.

Well-made mechanical systems tend to have an accuracy (deviation from the ideal) and resolution (least motion increment) of a part in 10^5, give or take an order of magnitude or two. To obtain the precision movements usually needed to handle small parts, macro designers resort to devices such as massive granite tables and temperature controls. Not only do these massive mechanical systems need to be made with extreme precision, but they also tend to be slow. In contrast, a centimeter-sized system with a similar part in 10^5 resolution has an incremental motion of a tenth of a micron, and it is potentially faster handling millimeter-sized components.

Electronics exemplifies a field where numerous advantages have accrued from miniaturization. The small size of VLSI electronic circuits, for example, allows designers to use exotic and expensive materials having desirable properties. A case in point: while one can use gold for interconnecting VLSI components, it would be prohibitively expensive for discrete components. Another advantage is convenience: if the word processor I am using were made of vacuum tubes, it would fill my office instead of a corner of my desk. There are also speed advantages in moving the electronic components closer together. Thus, I believe that the advantages in micro structures, which now seem so obvious in electronics, will someday be obvious in micromechanical systems as well.

Personal experiences have led me to an appreciation of using small mechanical systems to handle small parts. A system on which I collaborated used a meter-sized robot to handle millimeter-sized chips with an accuracy of about ten microns. This macro system had several disadvantages. The forces needed to move the robot's mass were much larger than the forces that would destroy the chip, and the robot could easily crush the chip. A lot of time was spent building fixturing that was compliant yet still maintained the high accuracy needed. A one centimeter robot would have about one millionth the mass, and hence require forces roughly a millionth of those needed to move the macro robot. Temperature changes of the macro robot were also troublesome. A one degree change of temperature led to about 10 microns expansion, the total error budget of our task. The larger robot also spent most of the time moving the chip from one work station to another. Because of the size of the robot and associated equipment, these workstations had to be several feet apart. If a centimeter-sized robot and associated equipment could have been used, the distances and transit times could have been dramatically reduced. Finally, finding space for this robot and its four by six foot table in a clean room was difficult. This thousand-to-one difference between the size of the macro robot and the chips is equivalent to using a bulldozer to move sugar cubes. The corresponding accuracy requirement is equivalent to positioning the sugar cube to within a hair's width. On the larger scale of the bulldozer, it is easier to see the advantages of using tooling commensurate with the parts being handled.

3. The large and small

The domain of small mechanical systems is an extensive area ripe for exploration. There are many intriguing things to learn and useful properties to discover [4]. To help display the range of sizes of mechanical systems, those available to mankind are logarithmically plotted in Fig. 1. A logarithmic plot has been chosen because each time the scale of the system under investigation is changed by several orders of magnitude, there are new phenomena to study and exploit. The scale shows the size of objects in Ångstroms (10^{-10} m). Atoms, which are probably the smallest mechanical particles we will use in the near future, are roughly 1 Å in diameter. Man is about 10^{10} Å (meter sized), and the universe about 10^{37} Å in diameter. For

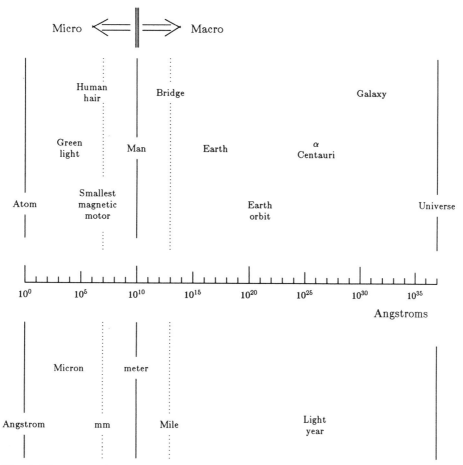

Fig. 1. The range of mechanical objects is shown plotted on a logarithmic scale. Objects range from atoms (roughly Ångstrom sized) to man (about a meter or 10^{10} Ångstroms in size) to the Universe ($\approx 10^{37}$ Ångstroms). Micromechanical systems span about a quarter of the scale. Since mechanical systems larger than a few miles will probably not be built in the near future, micro systems are the majority of the scale available for exploration.

convenience, this scale is divided into the micro domain, which contains those systems 10 cm in size and smaller, and the macro domain, which contains meter-sized and larger systems. The micro domain represents about a quarter of the total scale. However, since mechanical systems larger than a few miles long probably will not be built in the near future, the micro domain represents the major portion of the scale presently available for investigation. The vertical dotted lines delineate the normal range of mechanical systems. The whimsical photograph (Fig. 2) of an ant holding a 900 micron silicon gear helps to challenge our preconceptions graphically. The gear was made by Gabriel, Mehregany and Trimmer [5].

Fig. 2. Photograph of an ant holding a 900 micron gear made by Mehregany, Gabriel and Trimmer. Photograph by C. LaGreca. Ant caught by Scott Trimmer (age 8).

4. Small mechanical systems of interest

In building small robots and systems, one is faced with problems similar to those faced by early macro machine makers. One cannot easily obtain a micro motor, gear or screw. Building these systems will take ingenuity. Given the large number of things that can be done, however, an important early task is deciding what systems have the most utility, and the greatest chance of success. Several areas seem especially promising: electronics assembly, surgery, space exploration and micromachining.

Electronics assembly is becoming increasingly difficult for macro tooling for at least three reasons. First, new generations of electronics generally require more interconnections. Secondly, to reduce the propagation delay between devices, designers are moving components closer together, and they are decreasing the inter-device capacitance by reducing the size of the interconnecting lands and pads. Thirdly, many of the new technologies and materials make the devices more susceptible to damage. Smaller tooling excels in handling these difficulties, because it is inherently faster, more accurate and gentler. Many electronics assembly tasks could be performed nicely on a tea saucer if millirobotic systems were available. The desire to make more complex miniature circuits will surely challenge assembly concepts.

Despite the many advances in medicine, many operations still require opening a large wound in the patient's body. Yet the actual repair task usually does not require an opening this large. The development of minute actuators, sensors and other mechanical parts will help facilitate building micro manipulators and teleoperators that can be inserted into the body through a small incision or needle.

Space exploration is limited by the ability to move mass out of the earth's gravitational field. Reducing the size, and hence the mass, of the mechanical systems will certainly increase the number of missions that can be performed.

New techniques are available to make micro structures, actuators and sensors [6 - 22]. Rapid progress in this area has been made on micromachining silicon using anisotorpic etches developed for the electronics industry. Modifying these techniques makes it possible to fabricate very fine mechanical structures, and this silicon technology can be used to build the components needed for these micromechanical structures. For example, it is possible to build completely assembled mechanical structures on a silicon wafer [23, 24]. During the processing of the silicon wafer, structural and sacrificial layers are deposited on top of each other in selected areas. After processing, the sacrificial layers are etched away, leaving structural material in any desired shape. Using these techniques, complete gear boxes including the housing and supporting shafts can be built on silicon wafers. These nano gear boxes are typically half a millimeter in size. Several groups are also working on silicon-based microactuators to power these tiny devices [25 - 34]. While the technology is not presently available, one can envision making components on the atomic scale.

Should an industrial base capable of building micromachines be developed? Material costs (and possibly floor space costs) may be lower. One also assumes that the developers of new micro systems will take advantage of new automation and robotic techniques, and so micro assembly systems will be more efficient than the older macro systems they replace. What are the uses of micro capabilities? Some of the readers will probably help to answer this question.

5. The scaling of mechanical systems

When designing micromechanical systems, the first thing to decide is what forces should be used for activation. How the different forces scale into the micro domain is described below. The details of this analysis are given in the Appendix.

A matrix formalism is used to describe the scaling results. This nomenclature is designed to show a number of different cases and scale sizes in a simple format. The size of the system is represented by a single scale variable s, which represents the linear scale of the system. The choice of s for a system is a bit arbitrary. s could be the separation between the plates of a capacitor, or it might be the length of one edge of the capacitor. Once chosen, however, it is assumed that all dimensions of the system are scaled down in size as s is decreased. Nominally $s = 1$; if s is then changed to 0.1, all the dimensions of the system are decreased by a factor of ten. A number of different cases are shown in one equation. For example,

$$F = \begin{bmatrix} s^1 \\ s^2 \\ s^3 \\ s^4 \end{bmatrix} \quad (1)$$

shows four cases for the force law. The top scales as s^1, next scales as s^2, etc. The scaling of the time required to perform an operation, as discussed later, is

$$t = \begin{bmatrix} s^{1.5} \\ s^1 \\ s^{0.5} \\ s^0 \end{bmatrix} \quad (2)$$

The top time scaling, $s^{1.5}$, always refers to the case where force scales as s^1, etc. This notation is used consistently throughout this paper. A dash $[-]$ means that this case does not apply. (Equation (2) is a general relationship that holds whenever the forces scale as in eqn. (1). This is shown below.)

5.1. Magnetic forces

This Section describes the scaling of magnetic forces into the micro domain. Three cases are examined: constant current density, constant heat flow per unit surface area and constant temperature rise across the windings. It is assumed that the magnetic forces are generated by the interaction of two wires carrying current. (The scaling results of the interaction between a wire and permanent magnet are given in braces $\{s^n\}$.)

In the first case, the current density is assumed to be constant $[s^0]$, and hence a wire with one-tenth the cross-sectional area carries one-tenth the current. The heat generated per volume of windings is constant for this case. The force generated for this constant current case scales as $[s^4]$ ($\{s^3\}$), *i.e.*, when the system decreases by a factor of ten in size, the force generated magnetically decreases by a factor of ten thousand. Clearly this is not a strong micro force.

Since heat can be more easily conducted out of a small volume, it is possible to run isolated small motors with higher current densities. However, increasing the current density makes the motors much less efficient. If the heat flow per unit surface area of the windings is constant, the current density in the wires scales as $[s^{-0.5}]$. This increase in current density for small systems increases the force generated, and the force scales as $[s^3]$ ($\{s^{2.5}\}$).

A third possible constraint on the magnetic system as it is scaled down in size is the maximum temperature that the wire and insulation can withstand. If the system parameters are scaled so that there is a constant temperature difference between the windings and surrounding environment, then the current scales as $[s^{-1}]$, and the force scales as $[s^2]$ ($\{s^2\}$). As will be discussed later, forces that scale as $[s^2]$ are useful in small systems. However, below there are described several other forces that also scale as $[s^2]$, which do not waste the large amounts of power dissipated by this magnetic case.

Using the matrix notation above, the currents required for the different force scaling laws are given by:

$$J = \begin{bmatrix} - \\ s^{-1} \\ s^{-0.5} \\ s^0 \end{bmatrix} \qquad (3)$$

In the case of thin films, current densities may be limited even more than assumed by the above analysis by electromigration phenomena.

5.2. Electrostatic forces

Electrostatic actuators have a distinguished history [35], but are not in general use for motors. Electrostatic forces, however, become significant in the micro domain and have numerous potential applications. The exact form of the scaling of electrostatic forces depends upon how the E field changes with size. Generally, the breakdown E field of insulators increases as the system becomes smaller [36, 37]. Two cases will be examined here: (1) constant E field ($E = [s^0]$); (2) an E field that increases slightly as the system becomes smaller ($E = [s^{-0.5}]$).

For the constant electric field ($E = [s^0]$), the force scales as $[s^2]$. When E scales as $[s^{-0.5}]$, then the force has the even better scaling of $F = [s^1]$. When the size of the system is decreased, both of these force laws give higher accelerations and smaller transit times.

There are several other interesting forces. Biological forces from muscle are proportional to the cross-section of the muscle, and scale as $[s^2]$. Pneumatic and hydraulic forces are caused by pressures (P) and also scale as $[s^2]$. Surface tension has an absolutely delightful scaling of $[s^1]$ because it depends upon the length of the interface.

5.3. The unit cube

Below is a discussion of how the above force laws affect the acceleration, transit time, power generation and power dissipation as one scales to smaller domains. In going from here to there as quickly as possible with a certain force, one wants to accelerate for half the distance, and then decelerate. The mass of the object scales as $[s^3]$ (density is assumed to be intensive, or to not change with scale). Now the equations of dynamics give

$$a = F/m = [s^F][s^{-3}] \tag{4}$$

$$t = (2x/a)^{1/2} = (2xm/F)^{1/2} \tag{5}$$

$$t = ([s^1][s^3][s^{-F}])^{1/2}$$

where s^F represents the scaling of the force F. Here only the time to accelerate has been calculated, but an equal time is needed to decelerate, and both these times scale in the same way. For the forces given in eqn. (1), the accelerations and transit times can be expressed as

$$a = \begin{bmatrix} s^{-2} \\ s^{-1} \\ s^{0} \\ s^{1} \end{bmatrix} \tag{6}$$

$$t = \begin{bmatrix} s^{1.5} \\ s^{1} \\ s^{0.5} \\ s^{0} \end{bmatrix} \tag{7}$$

Even in the worst case, where $F = [s^4]$, the time required to perform a task remains constant when the system is scaled down. Under more favorable force scaling, for example, the $F = [s^2]$ scaling case, the time required decreases as $[s]$ with the scale. A system ten times smaller can perform an operation ten times faster. This is an observation that we know intuitively: small things tend to be quick.

Inertial forces tend to become insignificant in the small domain, and in many cases kinematics may replace dynamics. This leads to interesting control strategies.

5.4. Power generated and dissipated

As the scale of a system is changed, one wants to know how the power produced depends upon the force laws. For example, consider the unit cube above, which is first accelerated and then decelerated. The power, P, or work done on the object per unit time is

$$P = Fx/t \tag{8}$$

The scaling of each of the terms on the right is known.

$$P = \begin{bmatrix} s^1 \\ s^2 \\ s^3 \\ s^4 \end{bmatrix} \begin{bmatrix} s^1 \\ s^1 \\ s^1 \\ s^1 \end{bmatrix} \begin{bmatrix} s^{-1.5} \\ s^{-1} \\ s^{-0.5} \\ s^0 \end{bmatrix} \qquad (9)$$

$$P = \begin{bmatrix} s^{0.5} \\ s^2 \\ s^{3.5} \\ s^5 \end{bmatrix}$$

The power that can be produced per unit volume ($V = [s^3]$) is

$$\frac{P}{V} = \begin{bmatrix} s^{-2.5} \\ s^{-1} \\ s^{0.5} \\ s^2 \end{bmatrix} \qquad (10)$$

When the force scales as $[s^2]$, then the power per unit volume scales as $[s^{-1}]$. For example, when the scale decreases by a factor of ten, the power that can be generated per unit volume increases by a factor of ten. For force laws with a higher power than s^2, the power generated per volume degrades as the scale decreases. There are several attractive force laws that behave as s^2, and one should try to use these forces when designing small systems.

For the magnetic case, one may be concerned about the power dissipated by the resistive loss of the wires. The power due to this resistive loss, P_R, is

$$P_R = I^2 R = (JA)^2 \frac{\rho l}{A} \qquad (11)$$

where A is the cross section of the wire, ρ is the resistivity of the wire, and l is the length of the wire. This gives

$$P_R = J^2 [Al] \rho \qquad (12)$$

where Al is the volume. The resistivity scales as $[s^0]$ and the volume scales as $[s^3]$, and from eqn. (3) above,

$$J = \begin{bmatrix} - \\ s^{-1} \\ s^{-0.5} \\ s^0 \end{bmatrix} \qquad (13)$$

Hence the power dissipated scales as:

$$P_R = \begin{bmatrix} - \\ s^1 \\ s^2 \\ s^3 \end{bmatrix} \quad (14)$$

$$\frac{P_R}{V} = \begin{bmatrix} - \\ s^{-2} \\ s^{-1} \\ s^0 \end{bmatrix} \quad (15)$$

For the magnetic case where force scales as $[s^2]$, the power that must be dissipated per unit volume scales as $[s^{-2}]$, or, when the scale is decreased by a factor of ten, a hundred times as much power must be dissipated within the volume. This magnetic case is bad if one is concerned about power density or the amount of cooling needed. However, future advances in the field of superconductors may reduce or eliminate this problem.

5.5. Summary of the scaling results

The force has been found to scale in one of four different ways: $[s^1]$, $[s^2]$, $[s^3]$, $[s^4]$. If the scale size is decreased by a factor of ten, the forces for these different laws decrease by ten, one hundred, one thousand, and ten thousand respectively. Obviously, one wants to work with force laws that behave as $[s^1]$ or $[s^2]$. The different cases that lead to these force laws, the accelerations, the transit times and the power generated per unit volume are given below.

(a) $[s^1]$: surface tension; electrostatics where $E = [s^{-0.5}]$.

(b) $[s^2]$: electrostatics where $E = [s^0]$; pressure forces; biological forces; magnetics where $J = [s^{-1}]$.

(c) $[s^3]$: magnetics where $J = [s^{-0.5}]$.

(d) $[s^4]$: magnetics where $J = [s^0]$.

$$F = \begin{bmatrix} s^1 \\ s^2 \\ s^3 \\ s^4 \end{bmatrix} \quad a = \begin{bmatrix} s^{-2} \\ s^{-1} \\ s^0 \\ s^1 \end{bmatrix} \quad t = \begin{bmatrix} s^{1.5} \\ s^1 \\ s^{0.5} \\ s^0 \end{bmatrix} \quad \frac{P}{V} = \begin{bmatrix} s^{-2.5} \\ s^{-1} \\ s^{0.5} \\ s^2 \end{bmatrix} \quad (16)$$

Force laws that behave as $[s^1]$ or $[s^2]$ are the most promising. The acceleration increases for these laws as one scales down the system. The power that can be produced per unit volume also increases for these two laws. The surface tension scales advantageously, $[s^1]$, however, it is not clear how to use this force in most applications. Biological forces also scale well, $[s^2]$, but may be difficult to implement. Electrostatic and pressure-related forces appear to be useful forces in the small domain.

6. Conclusion

Watch makers have been the repository of fine mechanical skills, and their micro art is declining. Why then, should the micromechanical area start to grow now? The answer, in a word, is electronics: the same forces that are decreasing the usage of mechanical watches. Rapid progress has been made by the electronics industry using etching techniques to fabricate small mechanical parts. Once these micro structures are cut free from their silicon substrate and become independent components, the parts can be used to build a wide array of mechanical systems. The electronics industry is also placing increasingly severe demands on assembly tools. The future of assembly is with fast, accurate and gentle tooling. Electronics is already developing the micro sensors and intelligence needed to construct micro systems. This development of micro sensors and intelligence is important because, when a micro system breaks, it is difficult to use macro hands to find and fix the problem. What is now needed is the development of micro structures and actuators.

Other reasons for exploring the micro domain include the needs of medicine, space exploration and intellectual curiosity.

Given the desire to make micro systems, how does one start? Ten-centimeter sized systems can probably be made with modified macromachines: small electric motors, solenoids, encoders, etc. An interesting development in this size domain is robotic hands. Each finger is a small robot. Recently, hands have been developed using both magnetic electric motors and pneumatic actuators [38 - 41]. The centimeter- and millimeter-sized systems, however, will be very difficult to build with scaled-down versions of our macro equipment. They will probably require the development of new technologies. The following approach is suggested. First, build the necessary actuators and sensors, and then integrate these into systems.

As an existence proof and demonstration of the capabilities of small mechanical systems, nature proudly displays a wide array of small biological systems such as bacteria and cells.

Acknowledgements

This article is the result of discussions with many people over a number of years, and I express my gratitude.

R. Feynman's talk at Occidental College twenty years ago was the germination of my interest in micromechanical structures. J. Jarvis and O. G. Lorimor have been supportive, and a source of sound advice. The collaborative effort of A. Feury, K. Gabriel, R. Jebens, R. Mahadevan, M. Mehregany, T. Poteat and J. Walker in making micromechanical systems has been a source of excitement and satisfaction. Numerous useful comments have been given by A. Bobeck, R. Soneira, R. Stroud and R. Trenner, and, of course, Ann Trimmer's patience has made this possible.

References

1 R. P. Feynman, *There's Plenty of Room at the Bottom*, Reinhold, New York, 1960.
2 *Small Machines, Large Opportunities: A Report on the Emerging Field of Microdynamics*, A National Science Foundation report of the Workshop on Microelectromechanical Systems Research, 1980. (Copies of this report are available through G. A. Hazelrigg of the National Science Foundation, Washington, DC.)
3 K. E. Drexler, *Engines of Creation*, Anchor Press/Doubleday, New York, 1986.
4 T. A. McMahon and J. T. Bonner, *On Size and Life*, Scientific American Books, New York, 1983.
5 K. J. Gabriel, W. S. N. Trimmer and M. Mehregany, Micro gears and turbines etched from silicon, *4th Int. Conf. Solid-State Sensors and Actuators (Transducers '87), Tokyo, Japan, June 2 - 5, 1987*, pp. 853 - 856.
6 P. W. Barth, P. J. Shlichta and J. B. Angell, Deep narrow vertical-walled shafts in (110) silicon, *Proc. 3rd Int. Conf. Solid-State Sensors and Actuators (Transducers '85), Philadelphia, PA, June 11 - 14, 1985*, pp. 371 - 373.
7 E. Bassous, H. H. Taub and L. Kuhn, Ink jet printing nozzle arrays etched in silicon, *Appl. Phys. Lett.*, 31 (1977) 135 - 137.
8 E. Bassous, Fabrication of novel three-dimensional microstructures by the anisotropic etching of (100) and (110) silicon, *IEEE Trans. Electron Devices*, ED-25 (1978) 178 - 193.
9 K. E. Bean, Anisotropic etching of silicon, *IEEE Trans. Electron Devices*, ED-25 (1978) 1185 - 1193.
10 W. Ehrfeld, P. Bley, F. Gotz, P. Hagmann, A. Maner, J. Mohr and N. O. Moser, Fabrication of microstructures using the LIGA process, *IEEE Micro Robots and Teleoperators Workshop, Hyannis, MA, Nov. 1987*.
11 L.-S. Fan, Y.-C. Tai and R. S. Muller, Integrated movable micromechanical structures for sensors and actuators, *IEEE Trans. Electron Devices*, ED-35 (1988) 724 - 730.
12 W. E. Feely, Micro-Structures, *IEEE Solid-State Sensor and Actuator Workshop, Hilton Head Island, SC, June 1988*, pp. 13 - 15.
13 A. M. Feury, T. L. Poteat and W. S. Trimmer, A micromachined manipulator for submicron positioning of optical fibers, *1986 IEEE Solid-State Sensor Workshop, Hilton Head, SC, June 1986*.
14 H. Guckel, S. Larsen, M. G. Lagally, B. Moore, J. B. Miller and J. D. Wiley, Electromechanical devices utilizing thin silicon diaphragms, *Appl. Phys. Lett.*, 31 (1977) 618 - 619.
15 H. Guckel, D. W. Burns, C. C. G. Visser, H. A. C. Tilmans and D. Deroo, Fine-grained polysilicon films with built-in tensile strain, *IEEE Trans. Electron Devices*, ED-35 (1988) 800.
16 R. T. Howe and R. S. Muller, Polycrystalline silicon micromechanical beams, *J. Electrochem. Soc.*, 130 (1983) 1420 - 1423.
17 M. Mehregany, M. G. Allen and S. D. Senturia, The use of micromachined structures for the measurement of mechanical properties and adhesion of thin films, *Proc. 1986 Solid-State Sensor Workshop, Hilton Head, SC, June 1986*.
18 M. Mehregany, J. J. Gabriel and W. S. N. Trimmer, Integrated fabrication of polysilicon mechanisms, *IEEE Trans. Electron Devices*, ED-35 (1988) 719 - 723.
19 K. E. Petersen, Silicon as a mechanical material, *IEEE Trans. Electron Devices*, ED-70 (1982) 420 - 457.
20 K. E. Petersen, Fabrication of an integrated, planar silicon ink-jet structure, *IEEE Trans. Electron Devices*, ED-26 (1979) 1918 - 1920.
21 K. E. Petersen, Micromechanical light modulator array fabricated on silicon, *Appl. Phys. Lett.*, 31 (1977) 521 - 523.
22 K. D. Wise, M. G. Robinson and W. J. Hillegas, Solid-state process to produce hemispherical components for inertial fusion targets, *J. Vac. Sci. Technol.*, 18 (1981) 1179 - 1182.

23 Long-Sheng Fan, Yu-Chong Tai and R. S. Muller, Pin joints, gears, springs, cranks, and other novel micromechanical structures, *4th Int. Conf. Solid-State Sensors and Actuators (Transducers '87), Tokyo, Japan, June 2 - 5, 1987*, pp. 849 - 852.

24 K. J. Gabriel, W. S. N. Trimmer and M. Mehregany, Micro gears and turbines etched from silicon, *4th Int. Conf. Solid-State Sensors and Actuators (Transducers '87), Tokyo, Japan, June 2 - 5, 1987*, pp. 853 - 856.

25 H. Fujita and A. Omodaka, The principle of an electrostatic linear actuator manufactured by silicon micromachining, *4th Int. Conf. Solid-State Sensors and Actuators (Transducers '87), Tokyo, Japan, June 2 - 5, 1987*, pp. 861 - 864.

26 H. Fujita and A. Omodaka, The fabrication of an electrostatic linear actuator by silicon micromachining, *IEEE Trans. Electron Devices, ED-35* (1988) 731 - 734.

27 M. Esashi and T. Matsuo, The fabrication of integrated mass flow controllers, *4th Int. Conf. Solid-State Sensors and Actuators (Transducers '87), Tokyo, Japan, June 2 - 5, 1987*, pp. 839 - 843.

28 J. H. Lang and M. F. Schlecht, Electrostatic micromotors: electromechanical characteristics, *IEEE Micro Robots and Teleoperators Workshop, Hyannis, MA, Nov. 1987*.

29 R. H. Price, J. E. Wood and S. C. Jacobsen, The modelling of electrostatic forces in small electrostatic actuators, *IEEE Solid-State Sensor and Actuator Workshop, Hilton Head Island, SC, June 1988*, pp. 131 - 135.

30 W. Riethmuller and W. Benecke, Thermally excited silicon microactuators, *IEEE Trans. Electron Devices, ED-35* (1988) 758 - 763.

31 W. S. N. Trimmer, K. J. Gabriel and R. Mahadevan, Silicon electrostatic motors, *4th Int. Conf. Solid-State Sensors and Actuators (Transducers '87), Tokyo, Japan, June 2 - 5, 1987*, pp. 857 - 860.

32 W. S. N. Trimmer and K. J. Gabriel, Design considerations for a practical electrostatic micro motor, *Sensors and Actuators, 11* (1987) 189 - 206.

33 J. E. Wood, S. C. Jacobsen and K. W. Grace, SCOFFS: a small cantilevered optical fiber servo system, *IEEE Micro Robots and Teleoperators Workshop, Hyannis, MA, Nov. 1987*.

34 M. J. Zdeblick and J. B. Angell, A microminiature electric-to-fluidic valve, *4th Int. Conf. Solid-State Sensors and Actuators (Transducers '87), Tokyo, Japan, June 2 - 5, 1987*, pp. 827 - 829.

35 O. D. Jefimenko, *Electrostatic Motors*, Electret Science Company, Star City, 1973.

36 R. B. Comizzoli, Bulk and surface conduction in CVD SiO_2 and PSG, *J. Electrochem. Soc., 123* (1976) 386 - 391.

37 J. Thewlis, *Encyclopaedic Dictionary of Physics*, Pergamon, New York, 1961, p. 326.

38 H. Hanfusa and H. Asads, A robot hand with elastic fingers and its application to assembly processes, *IFAC Symp. on Information and Control Problems in Manufacturing Technology, Tokyo, Oct. 1978*, pp. 127 - 138.

39 S. C. Jacobsen, J. E. Wood, D. F. Knutti and K. B. Biggers, The UTAH/M.I.T. dextrous hand: work in progress, *Int. J. Robotic Res., 3* (1984) 21 - 50.

40 M. T. Mason and J. K. Salisbury, Jr., *Robot Hands and the Mechanics of Manipulation*, M.I.T. Press, Cambridge, 1985.

41 Y. Nakano, M. Fujie and Y. Hosada, Hatachi's robot hand, *Robotics Age*, (July) (1984) 18 - 20.

Appendix

A. Introduction

In describing mechanical systems, one is interested in the behavior of the force laws. The scaling of current, and hence the magnetic force, will be described in Section B, and the scaling of the electrostatic force will be examined in Section C. The equations used are Maxwell's equations, the electromagnetic force law $F = q(E + v \times B)$, and the equations of dynamics. The MKSA (SI) system of units is used throughout this Appendix.

B. Scaling of magnetic forces

This Section examines how forces that are generated magnetically scale. Three cases will be examined: constant current density, constant heat flow per unit surface area and constant temperature rise across the windings.

The scaling of the magnetic force can be calculated from the force law and one of Maxwell's equations:

$$dF_b = I_b \, dl_b \times B_a \tag{A1}$$

$$\nabla \times B - c^{-2} \frac{\partial E}{\partial t} = \mu_0 J \tag{A2}$$

Figure A1 below shows a typical configuration. The F is the magnetic force, I is the total current through the wire, B is the magnetic induction, E is the electric field, t is the time, l measures the distance along the wires, and d is the separation between the wires. Integrating eqn. (A2) over the area of the 'd' loop, and using Stokes's theorem, one has

$$\int (\nabla \times B) \cdot dA = \int \left(c^{-2} \frac{\partial E}{\partial t} + \mu_0 J \right) \cdot dA = \int B \cdot dl = 2\pi dB = c^{-2} \int \frac{\partial E}{\partial t} \cdot dA + \mu_0 I_a \tag{A3}$$

or

$$B_a = \frac{\mu_0}{2\pi d} I_a + \frac{1}{2\pi d c^2} \int \frac{\partial E}{\partial t} \cdot dA \tag{A4}$$

The B is equal to two terms, the current, I_a, term, and an $\dot{E}$ term. The vector nature of the force equation, $dF_b = I_b \, dl_b \times B_a$, and the values of the dot and cross products do not change as the system is scaled to smaller sizes, so we do not need to be concerned with the vector part of this equation in our scaling analysis. The example under consideration in Fig. A1 has been chosen so that the dot and cross products are equal to one. The force equation (A1) can now be written

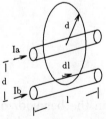

Fig. A1. Two wire segments of length l are shown separated by d and carrying currents I_a and I_b. The forces on the wires are given by eqns. (A5) and (A6).

$$dF_b = \frac{\mu_0}{2\pi d} I_a I_b \, dl_b + \frac{1}{2\pi dc^2} I_b \left[\int \frac{\partial E}{\partial t} \cdot dA \right] dl_b \tag{A5}$$

Integrating along a length l of the wire $\int_0^l dF$,

$$F_b = \frac{\mu_0}{2\pi} I_a I_b \frac{l}{d} + \frac{1}{2\pi dc^2} I_b \left[\int \frac{\partial E}{\partial t} \cdot dA \right] l \tag{A6}$$

Now as all the dimensions of the system are scaled by the scale factor, s, the first term scales as l^2. (When l doubles, so does d, and their effect cancels.) It is more difficult to see how the second term scales. The first term is examined for three cases in the Subsections below: (1) constant current density; (2) constant heat flow through the surface of the wire, (3) constant temperature rise. These three assumptions lead to very different forces and power dissipations. The fourth Subsection discusses the forces resulting from the interaction of a permanent magnet and electromagnet. This force is given by eqn. (A1), $dF_b = I_b \, dl_b \times B_a$. The fifth Subsection relates the second term, $\int (\partial E/\partial t) \cdot dA$, to the three current cases.

B.1. Constant current density

Here the current density is assumed to be an intensive variable: its value does not change as the scale size is changed. In terms of our notation, $J = [s^0]$. Figure A2 shows the end of one of the wires. The total current is

$$I = \int J \cdot dA = JA \tag{A7}$$

The area scales as $A = [s^2]$, and

$$I = [s^0] \cdot [s^2] = [s^2] \tag{A8}$$

$$F = \frac{\mu_0}{2\pi} I_a I_b \frac{l}{d} = [s^4] \tag{A9}$$

The force scales as s^4. If the scale size decreases by a factor of 10, the force decreases by a factor of 10 000. This regime is quite unattractive.

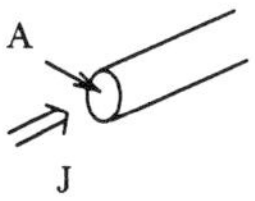

Fig. A2. A wire with cross-section A and current density J carries a current $I = JA$.

B.2. Constant heat flow through the surface of the wire

In this case there are two intensive variables, the heat flow out of the wire per unit wire surface area ($\dot{Q}/A_s$), and the resistivity, ρ, of the wire. The heat flow out of the wire must be equal to the power dissipated in the wire, $\dot{Q} = P = I^2 R$. Figure A3 below shows a piece of the wire. The A_e is the area of the end of the wire, A_s is the surface area, and r is the radius. Now,

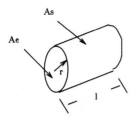

Fig. A3. A piece of wire is resistively heated by the current flowing through the cross-section A_e, and cooled by the heat flowing through the surface area A_s.

$$\frac{\dot{Q}}{A_s} = \rho = [s^0] \quad \text{(intensive variables)} \tag{A10}$$

$$\dot{Q} = P = I^2 R = I^2 \left(\frac{\rho l}{A_e}\right) \tag{A11}$$

where the resistance equals the resistivity, ρ, times (l/A_e). Scaling $\dot{Q}/A_s$ and using $A = [s^2]$,

$$\frac{\dot{Q}}{A_s} = [s^0] = \frac{I^2 \rho l}{A_s A_e} = I^2 [s^{-3}] \tag{A12}$$

$$I = [s^{1.5}], \quad J = \frac{I}{A_e} = [s^{-0.5}] \tag{A13}$$

The force scales as I^2 and

$$F = [s^3] \tag{A14}$$

For this case, when the scale decreases a factor of ten, the force decreases by a factor of 1000 and the current density increases by about a factor of three. As will be shown later, this increase in current density rapidly increases the resistive power dissipation.

B.3. Constant temperature rise of the wire

There is a maximum temperature that the wire and insulation can withstand. This Subsection assumes a maximum temperature rise between the coil and the surrounding ambient, independent of the scale size. The current density and the force scaling are calculated.

A section of a wire is shown in Fig. A4. The radius of the wire is r_s, the length of the section under consideration is l. For the integration an interior radius r_1 and an exterior radius r_2 are used. The equation of heat conduction is

$$\frac{dQ}{dt} = -kA \frac{dT}{dx} \tag{A15}$$

where Q is the heat flow, t is the time, k is the thermal conductivity, A is the area, T is the temperature and x is distance. For a cylinder of radius r_1

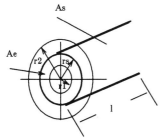

Fig. A4. The temperature rise between the center of a wire and its surface, at radius r_s, can be found by integrating eqn. (A15).

inside the wire, dQ/dt is equal to the resistive heat dissipated within this cylinder:

$$\frac{dQ}{dt} = P = I^2 R = [JA_{e1}]^2 \frac{\rho l}{A_{e1}} = J^2 \rho A_{e1} l = J^2 \rho \pi r_1^2 l \tag{A16}$$

where $R = \rho l / A_{e1}$ (ρ is the resistivity), $A_{e1} = \pi r_1^2$ and $I = JA_{e1}$. In eqn. (A15), A is the area of the cylinder surface, $A_s = 2\pi r_1 l$ and dT/dx becomes dT/dr. Rewriting eqn. (A15):

$$\frac{1}{2} \int_{r=0}^{r_s} J^2 \rho r_1 \, dr_1 = \int_{T_0}^{T_s} -k \, dT \tag{A17}$$

Integrating

$$J^2 \rho \left[\frac{1}{4} r_s^2 \right] = -k[T_s - T_0] = k \, \Delta T \tag{A18}$$

The temperature rise in this case is intensive, $\Delta T = [s^0]$, and scaling the above equation,

$$\begin{aligned} J^2 \rho [s^2] &= [s^0] \\ J &= [s^{-1}] \\ I &= JA = [s^1] \\ F &= [s^2] \end{aligned} \tag{A19}$$

Outside the wire, the total heat flux is constant, and eqn. (A15) becomes

$$\frac{dQ}{dt} = P = J^2 \rho \pi r_s^2 l \tag{A20}$$

and the integral (A17) becomes

$$\frac{1}{2} J^2 \rho r_s^2 \int_{r_s}^{r_f} \frac{1}{r_2} \, dr_2 = \int_{T_s}^{T_f} -k \, dT \tag{A21}$$

Integrating

$$\frac{1}{2} J^2 \rho r_s^2 [\log_e r_f - \log_e r_s] = -k[T_f - T_s] = -k \, \Delta T \tag{A22}$$

$$\Delta T = -\frac{J^2 \rho r_s^2}{2k} \left[\log_e \frac{r_f}{r_s} \right] \tag{A23}$$

Equation (A23) scales the same as eqn. (A18) except for the $\log_e(r_f/r_s)$ factor. When the system is scaled down, the ratio of r_f to r_s must remain constant, and the $\log_e$ term remains constant. (This says that the distance to the heat sink, r_f, must scale.) Hence, (A23) scales exactly the same way as (A18), and the scaling laws for this case are given by (A19).

For this case of constant temperature rise, when the scale is decreased a factor of ten, the current density increases by a factor of ten, and the force decreases by a factor of one hundred.

B.4. A wire and permanent magnet

The force between a current-carrying wire and a permanent magnet is given by eqn. (A1)

$$dF_b = I_b \, dl_b \times B_a \tag{A1}$$

The scaling of I_b for the three cases has been derived in the previous Subsections and is given by s^2 for the constant current density case, $s^{1.5}$ for the constant heat flow case and s^1 for the constant temperature rise case. The dl_b is a length and scales as s, and the maximum B_a that can be conveniently produced depends upon the saturation field of the material used, and is an intensive variable $[s^0]$. Hence the force for these three cases scales as s^3 for the constant current density, $s^{2.5}$ for the constant heat flow and s^2 for constant temperature rise. For comparison, the force between two current-carrying wires scales as: s^4 for the constant current density, s^3 for the constant heat flow, and s^2 for constant temperature rise.

B.5. The $\int (dE/dt) \cdot dA$ term in the force equation

The second term in the force equation raises interesting possibilities. One hopes that this term leads to better scaling properties. Unfortunately, this term is just the displacement current, and as will be seen in the capacitor example below, this term is equivalent to the current term discussed above.

Equations (A1), (A3) and (A4) are reproduced below for reference.

$$dF_b = I_b \, dl_b \times B_a \tag{A1}$$

$$\int (\nabla \times B) \cdot dA = \int \left(c^{-2} \frac{\partial E}{\partial t} + \mu_0 J \right) \cdot dA = \int B \cdot dl = 2\pi dB = \mu_0 I_a + c^{-2} \int \frac{\partial E}{\partial t} \cdot dA \tag{A3}$$

or

$$B_a = \frac{\mu_0}{2\pi d} I_a + \frac{1}{2\pi d c^2} \int \frac{\partial E}{\partial t} \cdot dA \tag{A4}$$

The equality of these two terms can be shown for a capacitor by considering the integral of $B \cdot dl$ around the plates of the capacitor. In Fig. A5, the large circle is the loop, and the smaller circles are the plates of the capacitor. This is equivalent to the surface integral of $\nabla \times B$ over the area of the loop. If this surface cuts through one of the wires shown, then the first term incorporating I_a is appropriate, and since the E field is zero away from the capacitor plates, the second term is zero. If the surface passes through the capacitor plates, which are shown as the two disks, then there is zero current passing through the surface, and only the dE/dt term is non-zero. The plates of the capacitor are, of course, infinitely close, etc. The two terms are equivalent, and either produces the same B field.

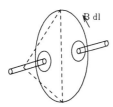

Fig. A5. The integral of $B \cdot dl$ around the larger loop is equal to the integral of the time derivative of the E field, and the current density J, as shown in eqn. (A3). When integrating around the capacitor plates, shown as small circles, either of the two terms can be made to vanish, and hence they must scale in the same way.

C. Scaling of electric forces

Electrostatic forces scale well into the small domain. Two cases are examined: first, where the electric field is constant, $[s^0]$; secondly, where the electric field increases slightly as the scale size is decreased, $[s^{-0.5}]$.

The electrostatic forces are calculated from the Lagrangian and the potential energy. The Lagrangian is

$$\frac{\partial L}{\partial x} - \frac{d}{dt} \frac{\partial L}{\partial \dot{x}} = 0$$

$$L = T - U \tag{A24}$$

If the potential energy is only a function of x, $U = U(x)$, and the kinetic energy is $T = \frac{1}{2} m \dot{x}^2$, then

$$-\frac{\partial U}{\partial x} - \frac{d}{dt} m\dot{x} = 0 = -\frac{\partial U}{\partial x} - F$$

$$F = -\frac{\partial U}{\partial x} \tag{A25}$$

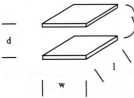

Fig. A6. The forces between the parallel plates of a capacitor of width w, length l and separated by a distance d are given by eqn. (A25).

For a parallel plate capacitor, such as that shown in Fig. A6, the potential energy can be calculated as follows:

$$U = \frac{1}{2} CV^2$$

$$C = \epsilon_0 \frac{wl}{d} \qquad V = Ed \tag{A26}$$

$$U = \frac{1}{2} \epsilon_0 wldE^2$$

and explicitly,

$$F = -\frac{1}{2} \epsilon_0 \frac{\partial}{\partial x}[wldE^2] \tag{A27}$$

$$F = [s^2]E^2 \tag{A28}$$

Now,

$$E = s^0 \quad \text{or} \quad s^{-0.5} \tag{A29}$$

and from eqn. (A28) above,

$$F = s^2 \quad \text{or} \quad s^1 \tag{A30}$$

The electrostatic force scales as $[s^2]$ or as $[s^1]$, assuming the E field scales as $[s^0]$ or $[s^{-0.5}]$. When the scale size is decreased by ten, the electrostatic force decreases by a factor of one hundred or ten.

Small Machines, Large Opportunities:
A Report on the Emerging Field of Microdynamics

Report of the Workshop on Microelectromechanical Systems Research

KAIGHAM GABRIEL, AT&T BELL LABORATORIES
JOHN JARVIS, AT&T BELL LABORATORIES
WILLIAM TRIMMER, AT&T BELL LABORATORIES

SUMMARY

The miniaturization of electronics has produced a far-reaching technological revolution. Now mechanics is poised on the brink of a similar miniaturization, and its own revolution. Researchers are working toward creating microdynamical systems, the microscale derivatives of conventional large-scale electromechanical systems.

The technology of microdynamics is based on that of microelectronics but calls for important advances over it. The goal is to make fully assembled devices and systems that can do what large-scale electromechanical systems cannot do as well, as cheaply, or at all. Potential applications for microdynamics include medical and surgical equipment, scientific instruments, manufacturing equipment, and consumer products. Some of the most exciting possibilities are: tools for microsurgery, improved prosthetic devices, cell sorters, instruments for use in spacecraft, systems for fiberoptics communications, equipment for electronics manufacture, and larger, more quickly accessible computer memories.

Already, researchers have been able to use the techniques of silicon micromachining to fabricate microsensors, now widely used in automobiles. Researchers have also made a variety of movable mechanical microcomponents. The largest task ahead is the invention and refinement of microactuators to power the microdynamical systems formed of these components.

The intertwined goals of research in microdynamics include developing materials for fabricating system components; modifying electronics mask design and etching techniques for the purposes of microdynamics; creating simulation tools for predicting the results of microdynamic fabrication processes; using various power sources and designs for microactuators; designing complete and functional microsystems; measuring and understanding physical phenomena that govern the micro realm in which these systems will operate; and achieving useful interaction between microdynamical systems and the macro world.

Reprinted from *Report of the NSF Workshop on Microelectromechanical Systems Research*, pp. 1-31, July 1987.

Bringing the concept of microdynamics to fruition will require not only research on specific processes and products but also interchange among researchers in different disciplines. Just as urgently, it will require the participation of both academe and industry in studying the science of the micro world, and in training scientists and engineers to work in the field of microdynamics.

Moreover, even though work in microdynamics to date has taken advantage of the technology of silicon integrated circuit fabrication, the new field will need materials, equipment, and processes that existing electronics facilities cannot provide. This need will increase as microdynamical systems leave the research stage and become part of accepted technology.

The Workshop recommends, therefore, a two-part approach to the microdynamics program. The first part, logically and chronologically, is for NSF to provide funding for individuals and groups engaged in researching the field at universities, and to solicit other funding agencies to do the same. The second part is for NSF to assist research groups in obtaining and sharing large equipment and, upon the development of an adequate technology base, to participate in establishing a national research laboratory for the microdynamics program.

Researchers now working in microdynamics predict that the field will grow quickly as a large-scale effort is directed toward exploring the micro realm. By encouraging this growth, the United States will have the chance to stay ahead of overseas competition in a new and potentially vital branch of science and industry.

1. INTRODUCTION

Three decades ago, electronics required the piece-by-piece assembly of bulky vacuum tubes and wires; today, it involves the simultaneous manufacture of hundreds of circuits, each containing millions of components, on wafers six inches in diameter. At present, mechanical assembly is at the same stage that electronics was thirty years ago: It is equally ready for a revolution in technology and size.

This committee's research area includes development of an engineering science base and technology for the design, analysis, fabrication, and operation of devices and systems that have active mechanical components and that typically measure less than 1 mm^3. The capabilities of conventional manufacturing techniques set the boundaries for the largest of these devices; on the small side biological systems are probably the asymptote. The gear of a mechanical watch is massive by comparison to the smallest gear now fabricated: a micro gear the diameter of a human hair, with a shaft 1/5 that width, and notched teeth the size of red blood cells.

Like their full-size counterparts, the components of these devices and systems, the gears, cranks, levers, springs, and so on, can rotate and slide in continuous motion with respect to one another. Some systems can emit or detect acoustic or electromagnetic fields, and some can carry out optical, fluid, thermal, or chemical processes. Thus, we call these systems **microdynamical systems.**

This new technology, originating in microelectronics, offers the possibility of fabricating fully assembled, low-cost mechanical devices and systems that can perform greatly improved or entirely new functions. Microdynamical systems hold promise for fields as diverse as medicine (e.g., arthoscopic, eye, and catheter surgery); information storage and retrieval; fiberoptic communications; and unmanned interplanetary space exploration. Like microelectronics, microdynamics could lead to products as advanced beyond present ones as a compact disc is beyond a long-playing record, and as fundamentally different.

2. BACKGROUND

Forty years have passed since the microelectronics revolution began with the invention of the transistor at Bell Laboratories, and progress in the way we conceive of and use electronics is likely to continue at a rapid pace. One area of progress is in silicon micromachining techniques; in recent years these techniques have provided the basis for a viable and growing sensor industry. This industry's greatest commercial success is pressure sensors for automobiles.

Solid-state sensors, whose function is to measure and describe the environment, consist of geometric forms connected to a silicon base, or substrate. They change physical energy to electrical signals through the flexing of cantilevers or diaphragms. To go beyond sensors and develop components that can change the environment, researchers have again relied on silicon micromachining, with pieces now released from the substrate and able to go through large motions relative to their size. Creating integrated **systems** that combine microscale electronic and mechanical components, at once independent and interconnected, is the next step. Crucial to building these systems is the integrated microactuator, the microscale analog of electric (and other power source) motors so common in our macroscopic world.

The methods and materials of microelectronics used in early microdynamical systems include lithography, silicon and the materials deposited on the silicon surface to make microcircuits, and the vast body of knowledge that enables technicians to craft these materials with submicron precision to do specific functions. In addition, microdynamics researchers hope to follow the example of microelectronics by reaping the potential benefits of batch fabrication processes, especially the benefit of cost reduction.

The challenges and rewards of applying and extending concepts developed for microelectronics design and fabrication to the new area of microdynamical systems are the subject of this report. Yet even though silicon technology predominates in current use, microdynamics may well develop other materials and methods for system manufacture and assembly, such as high-precision molding techniques and advanced micro versions of conventional machine shop techniques. Microdynamics will require a broad range of materials and technologies to fulfill its rich potential.

3. OPPORTUNITIES

What can we do with these new and, we hope, inexpensive microdynamical systems? At least four types of applications may well benefit from the continued development of these devices and systems: medical apparatus and services, instruments for scientific research, industrial equipment, and consumer products.

The potential uses of microdynamics technology in medicine and surgery are manifold, as present medical science already suggests. Catheter based diagnostic and surgical procedures, such as arthoscopic surgery and angioplasty, are now in use and have proven better than traditional approaches to the same problems. Yet the size of present-day instruments limits the possibilities of such surgical procedures. For example, catheters in use today often measure 5 mm wide, half the diameter of a pencil; smaller surgical tools could permit less traumatic invasive procedures and enable the surgeon to use multiple instruments simultaneously on a given area.

The many new therapeutic approaches based on microsurgery push not only conventional instrument fabrication but also the dexterity of surgeons to their limits. With microfabricated tools, some of them teleoperated (remotely manipulated and sensed), more surgeons could perform highly delicate operations; surgeons could work longer on a given operation; and individual surgeons could continue to work over a longer period in their careers.

Scientific instrumentation is a second area that will benefit from development of microdynamics technology and may in turn contribute to that same development. Just as scientific instruments that use electronics improved when transistors replaced vacuum tubes, so any scientific instrument that uses mechanics could improve with the incorporation of microdynamical systems. For example, analytical instrumentation could work with smaller samples if implemented with microdynamical systems, which would be particularly advantageous if the instrument were on board a spacecraft. Microdynamical systems would make possible a variety of instruments to measure, and perhaps to manipulate, fundamental physical properties.

The third area, manufacturing, consists of two broad but complementary endeavors, discrete and process, which describe the difference between making a car and making steel. The discrete branch includes the traditional assembly line and all other lines in which individual components are assembled into products. Process industries involve the continuous (or nearly continuous) conversion of raw materials into new materials or energy: for example, fermentation, distillation, oil refinement, and electric power generation. As mechanical components and electronics continue to shrink, the benefits of the microdynamics approach to discrete manufacturing are obvious; to name just one example, the aligning of fibers, detectors, and lasers in assembling fiberoptics systems. In terms of process industries, one can envision arrays of microdynamical systems making new composite materials with specified properties, just as machinery on a larger scale creates fiberglass.

The greatest impact of microdynamics will most likely be in consumer products. A strong market exists for small, portable, high-performance audio, video, and computing equipment; like surgical tools, some of these items have already thoroughly exploited conventional technology, as, for example, the memory capacity of computer hard disks is now doing. Hobbies and games, such as trains and animated toys, also call for highly functional, inexpensive miniature mechanical devices and may themselves serve as springboards for more serious applications, such as robots.

These consumer products require an interface between the mechanically passive electronics world - the display of information generated - and the physically perceived world - the realm of effect. Increasing the richness of these interactions will provide demand for and application of microdynamical systems. In this area, unlike in medical applications, cost is especially important: the microsystems must be made inexpensively and in large quantities. A mature and well-developed technology will be necessary to provide the robust and functionally attractive microsystems whose costs are low enough for inclusion in mass-produced consumer goods.

In all these categories or indeed in any specific application, work on microdynamics will prove self reinforcing: each advance in microdynamics technology will increase its range of applications. For example, building very small systems for use in scientific instruments will require better understanding of the underlying science, which will in turn contribute to microdynamics. Microdynamical components and systems will use their own technologies in their own manufacture and thereby lead to improvements in the process or in the equipment. Microdynamical systems manufacture might also lead to advances in the microelectronics used to make the systems themselves, and could be used as a supplement to the batch fabrication of microelectronics parts. The snowballing effect of increased knowledge and more sophisticated microdynamics can only prove beneficial to the whole of microtechnology.

Below is a list of applications of microdynamical devices that appear feasible. These applications will result from systems and components that are likely to be demonstrated in laboratories during early research programs or that may appear as the initial product offerings in the next few years.

- Prosthetic devices, made lighter, cheaper, better articulated, with a greater range of function; or even prosthetic/artificial organs.

- Ingestible or implantable "smart pills" that have sensors and are combined with dose regulating drug dispensers. By actively dispensing the drugs these would differ from current passive devices; the receptacle would also be the dispenser.

- Connectors for repairing blood vessels. Silicon has the advantage of being inert in biological systems.

- Micro-manipulation of biological materials: for example, sorting of individual cells in order to make diagnostic tests, such as counting the number of white blood cells, or in order to perform artificial insemination.

- Catheter-based medical diagnosis. The flexible fiberoptics-based imaging systems already developed are limited by the inability of visible light to penetrate solid tissues. Researchers are now working on mechanical sector ultrasonic imaging systems that will form high-resolution, two-dimensional images of the tissues surrounding the catheter or endoscope. Scaling down these systems, which currently measure about 4 mm in diameter, will require microactuators. The imaging systems will extend the usefulness of current catheter-based instruments, and will provide clinical information not just on coronary arteries, but also on such conditions as gastrointestinal carcinomas or pancreatic tumors.

- Catheter-based medical therapies, primarily angioplasty, an alternative to cardiac bypass surgery. Auger-like cutters, with motors on the order of 1 mm or less, would be particularly useful for angioplasty.

- Clean room instrumentation. The cost of keeping a laboratory or production facility free of particulates increases with the amount of "clean" space required, so reducing the size of the equipment, and hence the amount of clean space needed, would reduce costs.

- Micro-optical benches, small versions of the stable surface used as a research tool in optics.

- Micro-optical systems capable of providing high-speed beam positioning, deformable structures, scanning, shutters, all used in fiberoptics communications.

- Force-balanced transducers: use of microactuators in force-balance or nulling configurations would make sensors more precise and durable.

- Valves for fluid control: micro-sized servo valves, which are sensor-controlled valves, are probably a prerequisite for using hydraulic or pneumatic energy sources in microactuators.

- Microdynamical materials transport and manipulation: micro analogs of existing macro systems, for use in such applications as electronic components assembly.

- Photonic components and systems assembly, for aligning optical fibers.

- Low-inertia and high-speed mechanical structures, which, on a small scale, would operate faster, require less force, and be more efficient.

- Filters and tooling for process industries that use extrusion. These industries could include the biochemical and drug industries, or the making of biological products. With this very small equipment, achieving a high degree of separation could be done economically.

- Instruments for spacecraft and other applications with severe size and weight constraints. For example, equipment used to test solids and gasses in space would be smaller.

- Compliant or active probe microarrays for microelectronic manufacture or biological investigations. These probes would make electrical contact and test microelectronic chips, or work as electrodes to pick up neural activity.

- Electric switches and relays, which operate faster when they are smaller.

- Deformable structures and distributed actuators providing manufacturable analogs of muscle, peristaltic pumps, active skin, valves, controlled friction, micro positioning. Distributed actuators would enable various parts of a surface to act; for example, one might create a filter with many actuators that would work cooperatively to change the size of the mesh.

- Electronic fuel injectors for automobile engines.

- Movable elements implementing mass storage system data access, for speeding access time in non-electronic memories.

- Printer mechanisms made smaller and less expensive.

- Electromechanical sensors arranged in large numbers and multiplexed on simple analog or digital conduits for the reliable detection of parameters such as strain, position, force, pressure, acceleration, and spatial orientation.

- Micro telepresence systems, which relay sensory and motor commands down to the microdomain and back again.

- Arrays of any of the discrete microdynamical devices described above, that is, multiple and parallel small systems.

In addition, there is the prospect of a sufficiently low cost per unit that microdynamical systems can be employed in single-use applications. "Throw-away" items with these systems could be used for example, in surgical applications or for inexpensive consumer goods.

To introduce the concept of microdynamics to a larger audience, researchers could focus on designing tools and mechanisms for the earliest possible use: tools for tissue cutting, tissue manipulation, and ultraprecise positioning; structures for optical positioning and electro-optic interaction.

4. GENERAL RESEARCH NEEDS

Much of the early work in microdynamics has been an empirical and simultaneous exploration of both design and process. Researchers are not yet able to reproduce and transfer fabrication processes from one laboratory to another; moreover, researchers take diverse and sometimes incompatible approaches to the field. The large numbers of system concepts to be explored, the great variety in processes, and the blending of materials at the micro scale, especially when fabrication involves chemical modification, all contribute to the difficulty of formulating theoretical or simulation approaches for predicting the results from fabricating a particular system.

Researchers need, therefore, to pursue both theoretical studies and further experimentation and implementation. These two endeavors have a synergistic relationship and together will provide the **science base** for microdynamic processing and design. Detailed analysis of microstructures and computer modeling programs will become increasingly useful as this science base grows.

In addition to building a science base, a second general theme will be devising the **system concepts** for micromachines that can use components with much more parameter variation than is encountered in the macro world. In large mechanical systems, components are typically expensive and relatively precise, whereas in microsystems they are typically inexpensive and relatively imprecise. For microdynamics, researchers turn to the analogy of electronics; here, pairs or sets of transistors fabricated adjacent to one another are relatively identical, even though small variations in the transistors occur from wafer to wafer and batch to batch. The close matching of parameters in these adjacent pairs or sets makes production of reliable and predictable circuits easier. As in electronics, a goal for microdynamics will be to make systems with components that match in a given system, though they may vary from system to system.

A third theme is the **interface** between the macro world and the microdynamical world. Coupling power and motion between the two realms is problematic. For example, the human hand can rarely (and only with the greatest difficulty) manipulate, repair, or assemble components that are tens of microns in size, so small that they are nearly invisible to the unaided eye and likely to be blown away with a breath. We need research on ways to bridge the gap between mechanical devices scaled to human manipulation and mechanical devices scaled in microns.

Not in the short term, but eventually, we will need to understand theoretical limits to the minimum size of microdynamical systems. Biology suggests that as system size decreases, microdynamics will reach a demarcation point at which a fundamentally different science and technology become necessary. Knowing whether the only limits are the sizes of atoms, or whether the limits are higher, will be helpful.

5. PROPOSED RESEARCH AGENDA

In keeping with these stated needs, researchers in microdynamics have identified a number of specific topics for study, all of which contribute to one another. The starting point is the area that has received the most attention and that has generated the most concrete results: microfabrication of components. Work with microfabrication immediately raises questions about the procedure of silicon micromachining, derived from electronic fabrication, and the requirements for materials used. Closely related are design tools for microfabrication: computer modeling programs that will predict the results of a given procedure and the results of changes in that procedure.

Combining microcomponents into functioning micromachines will require study of the *sine qua non* of microdynamics: a microactuator. At the same time, the field will need research on the architecture and performance of complete microdynamical systems, as well as on the systems' ultimate goal, interaction with the physical world. Finally, underlying all of the above, the field will require investigating and measuring the physical phenomena that either contribute to or impede the realization of microdynamical systems. In all these activities, the science, systems work, and engineering are interdependent and will need to remain closely coupled.

5.1 Fabrication: Silicon Micromachining, Materials, Mask Design, and Etching

The silicon planar technology of electronics is the foundation of techniques for fabricating microdynamical structures. **Silicon micromachining** involves depositing flat layers of material, using a mask to pattern each layer lithographically, and using each patterned layer either as part of the circuit or part of the formation of the next layer. The advantages of the silicon planar process include the availability of a variety of metal, semiconductor, and insulating films whose properties have been extensively studied; highly refined deposition and patterning equipment; proven electronic device fabrication sequences; and the economies of batch fabrication. However, these advantages come at the cost of very expensive fabrication equipment and facilities, which also require high operating budgets.

Adding "bulk" and "surface" micromachining processes has extended conventional microelectronics technology. The former, which are the basis for present commercial applications such as sensors, involve carving up the thin wafer of single-crystal silicon. Bulk micromachining employs orientation-dependent chemical etches such as EDP, KOH, and hydrazine in conjunction with several etch-stop techniques to sculpt microstructures from the silicon substrate.

Surface micromachining, on the other hand, is the carving of layers put down sequentially. It uses the selective etching of sacrificial thin films to form free-standing or even completely released thin-film microstructures, which are the basis of microdynamics. The most studied process has used polycrystalline silicon (polysilicon) structural layers and silicon oxide sacrificial layers. Recently, researchers have modified this basic process to fabricate rotating and sliding or translating polysilicon microstructures, including rotary springs, linear rotary cranks, three-gear gear trains, and a prototype electrostatic motor. Figure 1 shows an example of a crank mechanism that incorporates both rotating and sliding hubs. Figure 2 shows a gear train and a pair of microtongs.

Microdynamical systems containing elements with one or more degrees of freedom (that is, numbers of independent motions) challenge the capacities of the **materials** used. In particular, thin-film microstructures unattached to the substrate are (due to their planar form) subject to deformation through residual stress in the film or through thermal stress caused by differential thermal expansion. In the crank structure shown in Figure 1, for example, warpages of less than 1% of the length of the arms will cause the mechanism to stick. If microdynamic structures are to handle significant loads, researchers need to know the Hooke's Law limits (the point to which the material can stretch while showing proportionality of stretch to applied force) and the ultimate strength of the materials. In addition, fatigue limits (the point at which repeated bending causes the material to fail) are a necessary design parameter.

The motion of microdynamical elements with respect to one another introduces a new set of materials specifications, involving friction, wear, abrasion, and lubrication. Researchers need thorough studies to understand the effect of surface morphology (or shape) and surface chemistry on the amount of friction between thin-film surfaces. Early investigations have shown that microstructures can adhere to some surfaces with strong surface bonds. However, the meaning of friction at micro levels is still unclear; it is less predictable, since the cause of adherence is unknown.

Most likely, conventional silicon microelectronic materials cannot meet the requirements for control of friction, warpage, wear, fatigue limits, and so on. Microdynamical systems will require a variety of new materials to be incorporated into microfabrication processes, as well as new methods of synthesizing familiar materials. To gain the advantages of three-dimensional microdynamic systems, with significant (>10 μm) thickness perpendicular to the plane of the substrate, researchers will need techniques for depositing thick films. The surface of a wafer in process, after the thick films have been etched, will have much rougher topography than is typical in microelectronic processing. Therefore, researchers also need to find techniques for depositing coatings that conform to and protect the shape of structures on substrates roughened by etching.

Among the materials that could be of interest for bearing surfaces (surfaces that rub against one another) are silicon carbide, diamond, boron nitride, amorphous carbon, tungsten, and non-stoichiometric silicon nitride. The design of composite microdynamical structures, incorporating insulating and conducting thin films, will require much better understanding and control of residual stress in order to prevent warpage of the structures. Polymer films may be useful as structural layers or as outermost layers that play the role of bearings. Fluids for lubrication, mechanical coupling, or enhancement of electric energy density may need to be incorporated into the microdynamical structure. Photonic materials (for example, semiconductor lasers, optical fibers, photodetectors) are the object of substantial research, and of efforts to incorporate them into the silicon planar process. These materials may have applications for position-sensing in microdynamical systems. Finally, magnetic thin films and the emerging high-temperature superconductors may also find a place in microdynamical systems.

As in material synthesis, lithography and pattern transfer are critically dependent on silicon processing technology. To an even greater extent than material synthesis, though, lithography and pattern transfer (the removal or deposition of material as defined by planar mask patterns) require extensive modifications and new developments. These changes will be needed for both steps: exposure of the mask design and etching.

Current **mask design and creation** programs were written in response to the fabrication requirements of silicon-based *electronic* devices and are now highly optimized for the technology of microelectronics. Many of the program features are, at best, useless and, at worst, contrary to the needs of silicon mechanical device fabrication. Current programs typically include the ability to define, align, and replicate rectangular features arranged in Manhattan grids. While alignment and replication are also necessary in the creation of microdynamical device masks, pattern features other than rectangular, for example, curvilinear and freeform, will be necessary for making items such as springs, gears, and bearings.

Initially, programmers can modify the software that controls the mask-making printers to approximate non-rectangular features with rectangular features. Such modification can satisfy current needs and may prove adequate and cost-effective for many years. Yet using this software to make the calculations for non-rectangular patterns is costly in design time and process time, and like pictures created with a typewriter, the results are less than satisfactory. Ultimately, the mask-making printers themselves must be modified to print non-rectangular features. On simpler, less accurate machines, doing this may merely entail defining a non-rectangular mask-feature "print wheel," which, like a daisy wheel, attaches to a standard printer but provides a new "font" (or fonts) for the microdynamical device's mask patterns.

Related to the material synthesis and thick-film deposition requirements are corresponding requirements in the **etching** of such materials. Thicker material depositions will necessitate the development of deeper etches. Even though existing silicon etch technologies are capable of making deep etches relative to current material thicknesses, the deeper etches typically result in rough surfaces. Smoother surfaces can be obtained, but usually by using standard shallow etches for prohibitively long etch times, thereby tying up expensive equipment and shortening the equipment's lifespan.

Etching processes specifically designed for microdynamical systems will need to address the inherently three-dimensional nature of the components. Exposure systems based on E-beams, X-ray lithography (similar to the German LIGA process), laser-assisted etching, and holographic exposure systems are all potential candidates for three-dimensional pattern transfer methods.

Finally, the current precision of pattern definition and transfer are inadequate for microdynamical systems because the walls of the components tend to become imprecise at greater etched depths. The separation between components, relative to the size of the components, presents another difficulty, as this separation is large compared to that in macro systems. The amount of clearance between parts will determine the types of systems possible in the micro domain.

5.2 Design Tools

Microfabrication technologies, based on batch fabrication, lithography, and selective etching, impose new constraints on the design process. The conventional iterative fabrication of system components, which involves sequential refinement and modification, is widely used for testing concepts in mechanical systems design. However, the fact that a microfabricated design consists of a mask set and a process sequence makes this approach inap-

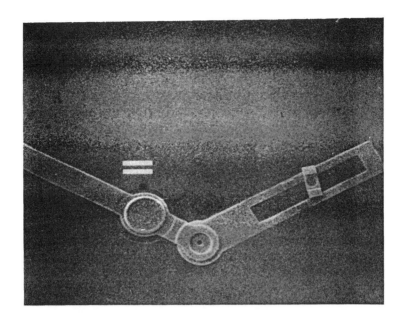

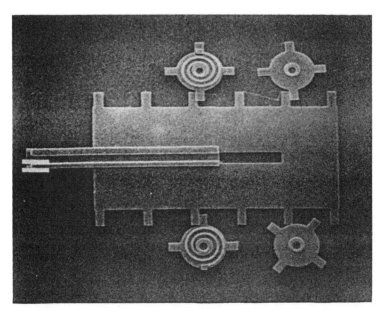

Figure 1

(a) A crank-slot combination. The slot is 103 μm long and 20 μm wide. The diameter of the two pin points is 50 μm. (b) A gear-slide combination. The movable slide measures 210 by 100 μm and has a guide at the center so only translational movement is allowed. (Berkeley Sensor & Actuator Center, University of California at Berkeley)

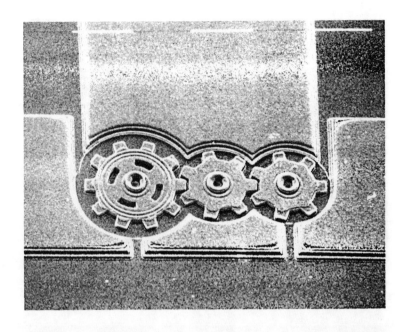

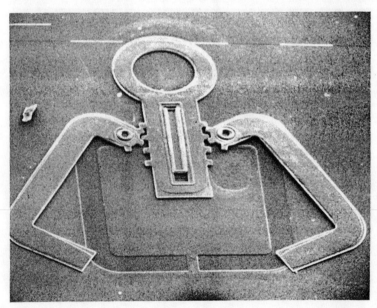

Figure 2

(a) A three-gear gear train with gear ratios of 1.4:1.0:1.0. The smallest gear has a 125 μm diameter and 20 μm diameter shaft. (b) A pair of micro-tongs demonstrating linear-to-rotary motion conversion, with a maximum jaw opening of 400 μm. (AT&T Bell Laboratories, Holmdel, NJ)

propriate. The entire design must be completed before fabrication is begun, and the cost of mid-course changes in process can be high because functions are interlocked so that changing one aspect of the design may affect other aspects. Making a change is a lengthy procedure; the "recipes" for process sequences of VLSI circuits can be up to 200 steps long.

Thus, simulating designs before they are fabricated, as is done in electronics and in large-scale mechanics, is highly beneficial. A set of computer-aided design tools can reduce overall cost and/or time between conception and prototype and improve designs for better performance. However, for several reasons, these simulation tools are not now readily used in microdynamical systems design.

A design, mask set, and process sequence for a microfabricated part cannot be completed until many system issues - such as system architecture and partitioning, signal processing methods - and packaging methods are worked out. Although we have good tools for mask layout, mechanical drafting, process simulation, and finite element simulation, we do not have a merged, or compatible, set of simulation tools that permit both electrical and mechanical evaluation of device and package design, at varying levels of simulated detail.

A second, and more critical, need is to develop a database of mechanical, physical, and dynamical properties of the constituent materials. Ideally, these various properties can be tabulated and incorporated into the design frame so that a designer can anticipate the effects of process or geometry variation on the resulting structure and its performance.

Among the essential issues that microdynamical devices raise are three-dimensional modeling; large electric fields between structural components; large thermal effects; gradients in material properties on a small dimensional scale; fluid dynamics in and around small structures; and tribology (friction and lubrication). Addressing these issues may well exceed the capabilities of existing CAD tools.

5.3 Microactuators

Microdynamics researchers exploring microactuation are finding, as is generally the case in engineering, that achievable technology dictates the devices they will use and the phenomena they will exploit. Microdynamical elements produced by IC-derived processes are measured in micrometers, and in these dimensions the forces of electrostatics predominate, even though these same forces may be insignificant in large mechanical systems.

Very large electric fields, greater than 100 MV/m, can be sustained in dielectric thin films (insulating films). The breakdown strength of air in micron-sized gaps can also be greater than 100

MV/m for smooth surfaces. The chief implication of these high fields is that electrostatic motors can have energy densities in the gap comparable to those in conventional magnetostatic motors, without the need for a return path, and with the added advantage of compatibility with conventional electronic materials and processes. Moreover, electrostatic actuation can be implemented with voltage-switching devices and circuits compatible with microdynamical systems.

Researchers thoroughly studied electrostatic motors several decades ago, but the low macroscopic breakdown field strength in air and the technological hurdles involved in evacuating the gaps caused them to set aside the concept. Yet several of these motor designs may well prove feasible for microactuators, including variable capacitance, induction, and permanent electret motors.

Thus, one of the primary research tasks in microdynamics is the development of electrostatic motors, including research on the possible scaling down of earlier macro designs. (An example of a current project is SCOFSS, a small, cantilevered optical fiber electrostatic servo system.) Researchers need to determine mechanical impedances, speed/torque curves, efficiencies, and other essential motor parameters. They also need to study frictional effects; wear properties of pairs of materials (such as silicon nitride on silicon, and others) that can be reproduced for designs; lubricants and lubricating procedures; and bearing design and fabrication.

Contributing to the utility of this type of motor is the development of feedback sensors measuring force, position, and velocity. Such sensors enable closed-loop or servoed operation of the motors, the sources of controllable mechanical energy, that power the parts that interact with the physical world. The goal is production of sensors, actuators, and associated electronics in a compatible, simplified processing sequence.

Even though electrostatics is the focus of much present research, other energy sources offer some possibilities: optical, magnetic, thermal, pneumatic, ultrasonic agitation, hydraulic, and aerodynamic. Researchers will need to consider force, power, thermodynamics, and coupling efficiencies for each of the power sources.

Researchers will also study material properties as prime movers for microdynamic systems. These are the properties of shaped memory alloys, magnetostriction, surface tension, piezoelectrics, and thermal expansion. Study of microfluidics will cover flow in channels, and design and operation of valves and orifices. Use of vibrational resonances for operating both sensors and actuators is yet another option under investigation.

5.4 Microcharacterization and Metrology

In addition to working directly toward the *creation* of useful micromachines, researchers will also need to work on techniques for the *study* of microdynamical systems. For example, researchers expect to use surface probes and electron microscopes, the tools of semiconductor technology, to learn more about surface characterization of microstructures. High spatial-resolution surface microscopy and surface analytical techniques, possibly scanning tunneling microscopy, will help determine the roles that surface chemistry, surface topography, and surface charge play in such phenomena as friction and wear of microstructures.

Another issue is the motion, under mechanical and electrostatic forces, of microstructures over surfaces. Microdynamical characterization must include the development of microscopic techniques to measure the speed and acceleration of microstructures and to characterize the deflection and bending of these structures under the action of internal and external forces. Since electrostatic forces are currently the major focus of microactuator investigations, the development of techniques to measure electric field, charge, and potential distributions that vary over time and change over submicron dimensions will be important for observing microstructures in motion.

Moving microstructures create heat, wear surfaces, generate particles, and change shape. Early research projects will apply current analytical techniques to characterizing and monitoring these phenomena, but soon researchers will need to modify and tailor these analytical methods to characterize microdynamical devices more accurately.

In electronics, technicians use electrical test structures to observe integrated circuit processes and to predict the operational characteristics of completed devices. To date, microdynamics researchers have developed only a few such test structures, but they expect to make innovative on-chip test structures that will perform the important task of defining both the mechanical and electrical characteristics of microdynamical systems.

5.5 Science Base

Microstructure science, which deals with objects from the atomic scale (1 angstrom) up to that of macroscopic systems (1 cm), has become a major research area. It includes solid state physics, chemistry, and the biological sciences. During the past ten years, the revolutionary progress in this area has included the fabrication of submicron structures and of quantum well devices with materials arranged in layers one atom thick.

While much of the work in microstructure sciences is readily transferable to microdynamical sciences, researchers in microdynamics anticipate new science as they begin to investigate microstructures in motion. They will need models that accurately predict the mechanical properties of microdynamical structures' surfaces. Studies of fatigue, stress, distortion, and stability of microstructures fabricated of various materials, and to which high electric fields are applied, should prove challenging and useful.

Because a large surface-to-volume ratio is a distinctive attribute of a microdynamical structure, surface physics and surface chemistry will play major roles in the fabrication, characterization, and operation of microdynamical devices. Surface charge and surface roughness may control frictional wear and will help determine the maximum electric fields that can be applied to a structure. Researchers will need to explore the relationship between surface structure and the shape and size of microdynamical elements.

Since the conceptual boundaries between materials, components, and systems are blurring (if not completely disappearing) as the scale of systems is reduced, interaction among practitioners of different disciplines - electrical engineering, mechanical engineering, materials sciences, solid state physics, chemistry, medicine, and biomedical sciences - is increasingly important. Interdisciplinary research programs are well suited to provide the range of expertise needed for academic exploration in this new field.

5.6. System Architecture and Performance

The next step, of course, is creating complete microdynamical systems. Microdynamics, with its merged electronics and mechanical devices, creates the possibility of local sensing and control of function, supplemented perhaps by overall global control from a central location. The **partitioning** of microdynamic systems, both from a component point of view (hybrid versus fully integrated), and from a control point of view (local sensing and local control versus global sensing and/or global control,) will depend on how, in light of technological capabilities, desired system performance can be achieved at an acceptable cost.

Whereas some degree of local sensing may be done compactly and at relatively low cost, the distribution of large amounts of digital computing power may not be cost-effective because of the silicon area required; using computing support larger than the mechanical system itself defeats the purpose of miniaturization. We will need to develop efficient methods of local sensing and control of actuators.

One example of system architecture would be active and passive components connected in parallel/series combinations in order to prescribe the flow, effort, power, and impedance relationships within a machine. Using appropriate arrangements, machine behavior can be made to the system, rather than being achieved only via feedback control techniques.

Distributed systems would use arrays of microactuators. Numerous sensor/effector groupings, positioned in close proximity, may more effectively manage energy flows in order to permit distributed control of mechanical, electromagnetic, thermal, and chemical systems. Merging a number of discrete microactuators could make possible distributed macro-actuators, which would in turn need control architecture, power distribution, sensing and command configuration, and overall control algorithms.

In addition to systems architecture, researchers need to explore the ways in which microdynamical systems will perform. **Size** is an immediate consideration; the micron-to-millimeter-size scale of microfabricated structures creates new possibilities for mechanical system design. The components can have low mass and, therefore, low inertia, small thermal mass, and high aspect ratios either for heat removal or thermal isolation. Flow at the scale of such structures is at low Reynolds numbers (the viscous forces dominate the inertial forces). Microstructures with large area/volume ratios would be useful both for distributed actuators, in which effective actuation can be achieved at modest forces distributed over large areas, and for materials processing, such as catalyzed chemical synthesis, in which large surface areas may be required.

Microactuator systems include prime movers, transmissions, and end effectors. At present, microfabricated prime movers are limited to small displacements, even though the forces may be large. Microactuator systems will need both design control of the mechanical impedance of a component, so that the tradeoff between force and displacement can be optimized for a given application, and increased total actuator mechanical power output. In addition, since both discrete indexed motion and continuous motion have applications, researchers must develop each type of actuator.

Sensors have applications in their own right, but they are critical in implementing overall control of an actuator system. Component precision, which is limited by microfabrication capabilities, and device noise define the ultimate achievable performance of a feedback loop controlling an actuator. Systems development will, therefore, require continued research on sensors, sensor precision, and sensor noise.

Distributed detectors can provide distributed sensing for precise and reliable determination of a system's state. **Measurement** of the performance of devices and microsystems will require the development of new tools and calibration methods. One consideration will be perturbation of the system caused by the measurement itself.

System integrity, that is, topics such as reliability, fault detection, maintainability, and manufacturability, will depend on new data about and insights into material properties, failure mechanisms, and predictability and repeatability of process sequences. The multiplicity of working parts may mean that systems can tolerate the failure of some parts.

Inevitably, though, the ultimate question is the **interface to the macro world.** Interaction between microsystems and the macro world require development of technologies in several areas: manipulation and assembly; inspection and testing; measurement and calibration; power sources and their connections; input and output for data and command signals; telepresence, the human-machine interface; packaging; coupling to actuator output, either for end use, or as an input to a macro-actuator; and fault detection and correction.

6. FACILITIES AND FUNDING NEEDS

Clearly, further progress in microdynamics will require a publicly accessible science base, as well as focused research on specific processes or products. Whereas industry can serve the latter need, academe will best respond to the need for microdynamic sciences and thereby also help to supply trained practitioners of this new art.

Just as clearly, an emerging technology such as microdynamics requires sufficient funds to ensure success. The NSF role in funding the microdynamics program has two parts. First, as the agency initiating the program, NSF should provide long-term funding to individual principal investigators or small research groups in universities. NSF should make early funding or seed money available in 1988 to stimulate innovative proposals, to attract talented graduate students to the program, and to generate the results required to write more focused, long-range research proposals.

Second, NSF should work to interest other government agencies and the private sector in the microdynamics program and should solicit their support of it. Since microdynamics technology has potential applications in space, defense, health, and the industrial and consumer sectors of the economy, NSF can turn to a wide range of other funding agencies.

Setting an immediate budget for the program is crucial. NSF should announce the available budget soon and give first priority to funding individual research groups for a minimum of three years. In establishing minimum funding and making initial grants, NSF should recognize the recipients' need for extensive travel, sometimes with graduate student assistants, to consult other researchers; need for interdisciplinary meetings such as the November 1987, IEEE Workshop on Micro Robots and Teleoperators; and need for using the laboratory facilities of other researchers' universities. Some restructuring of funding may be necessary to meet these needs.

For the first three years of an established microdynamics program, researchers should make maximum use of available facilities and outside services. Many standard processes are available now in university research groups, and can serve small, focused projects. Access to more sophisticated instruments, such as electron-beam lithography and mask-making equipment, can be purchased from industrial service companies or from national research laboratories.

The NSF should also support purchases by small research groups to supplement available equipment, processes, and services. A moderate NSF equipment budget during the first few years of the program would help address the immediate equipment needs of microdynamics researchers. To make best use of this budget, NSF should encourage research groups to write shared equipment proposals. Such joint equipment purchases would promote shared use of the installed equipment base, motivate investigators to form interdisciplinary research groups, and foster development of processing standards.

We also recommend the creation of a new independent microdynamics research program. As the microdynamics program begins to generate new devices and integrated systems, demand for facilities and equipment will increase. Though some of the microdynamics processes and equipment are remnants or variations of silicon integrated circuit technology, microdynamics requires new materials, thicker material layers, and different processes to make released structures. These processes and materials are not available from silicon integrated circuit facilities. Furthermore, because changing just one step in the long VLSI circuit process can affect that whole process, the use of standard electronics facilities for microdynamics manufacture can disrupt the electronics production line.

Another consideration is the economics of microdynamics. Researchers can explore new devices in many types of facilities, from relatively primitive laboratories to fully equipped clean rooms comparable to integrated circuit manufacturing plants. However, understanding the performance of manufactured parts and, in turn, understanding the economic benefit to society from microdevices, require the context of, and therefore access to, the type of facilities actually used for manufacturing.

To meet the many needs of this emerging technology, and to meet the demand for more complicated integrated microdynamic systems, the committee proposes the creation of microdynamics laboratories for research and teaching. Models for such a facility include the National Magnet Laboratory, or the NSF-sponsored Engineering Research Centers.

The proposed center would provide design tools and fabrication for prototype device concepts. A resident staff would maintain an established set of documented process sequences; maintain a CAD system adapted for use in microdynamical system design; assist potential users of the facility in defining overall design parameters consistent with available process sequences; train potential users to operate the CAD system; review designs prior to commitment to fabricate; carry out fabrication and/or assist facility users in fabrication; and maintain a series of diagnostic test procedures for characterizing process sequences.

Since the microsystems fabricator must use different materials and either non-standard versions of standard processes or process steps not available in typical electronics factories, the microdynamics laboratory must duplicate some capabilities of a standard integrated circuit prototype line, in addition to having highly flexible capabilities for handling a variety of deposited dielectrics, metals, and polymers. The laboratory must also develop and maintain micromachining capabilities, diagnostic methods for micromachining process control, and wafer handling and lithography methods for the non-planar structures that are intrinsic to micromachining. Furthermore, the center must have equipment and processes for packaging, since microdynamical systems cannot be tested effectively until after first-level packaging. Packaging operations will include substrate design and preparation and bonding and sealing techniques, and they may greatly expand the scope of materials and materials-processing steps with which the laboratory must contend.

Since merging designs in prototype fabrication runs may prove difficult and prototype runs may prove expensive, each investigator seeking to use the center may have to face high costs for fabrication done by the resident staff, or may have to commit personnel from the investigator's own staff to the fabrication sequence in the microdynamics center. This latter model somewhat resembles the National Magnet Laboratory, in which the resident staff maintains the magnets, but individual investigators build and operate their own experiments.

One way of establishing the proposed center would be to locate operations at a single facility with a resident staff. Such a laboratory would more easily provide: a link between the CAD tools and the actual process, compatibility among all process steps, and a single source for documentation. To attract a skilled staff, though, the center would have to maintain its own research program as well as provide services for outside researchers.

Alternatively, the center could be distributed among several different locations, possibly a mix of university and industrial facilities. Advantages of the distributed scheme would include: using the infrastructure already in place, providing multiple capabilities for critical processing steps, having a larger pool of experts involved in the center's activities, and making a given institution's unique capabilities available to users from other institutions. Disadvantages would include the difficulty of standardizing and documenting processes, and of managing design and fabrication sequences for investigators who are not involved at one of the center's sites. At present, no well established example of a successful distributed laboratory for innovative processing exists, although individual industrial firms and universities have successfully collaborated in prototype fabrication by carrying out process steps at more than one facility.

Researchers in microdynamics are still formulating the specifications for the proposed national laboratory, whether centralized or distributed. With grants to small groups for projects and equipment purchases, progress in the field will continue, even before the establishment of the center. However, the creation at an appropriate future date of a dedicated facility for the microdynamics program will enable this new technology to grow as effectively and as quickly as possible.

Cost estimates for the national microdynamics laboratory assume that investigators from other institutions will not have to pay an access fee but will have to support their own expenses, including visits to the center. Therefore, the overall plan must include funding for research programs as well as for the facilities.

The microdynamics laboratory's budget should provide at least five years of funding and include funds for equipment, laboratory space, and a five year operating budget. We estimate the five-year cost for a centralized facility at $50 million, and roughly half that amount for a distributed facility. A distributed facility could offer much of the same processing as a centralized facility, but a centralized facility would better promote the standardization and integration of microdynamics, and the gathering of a core of researchers who will shape the future of microdynamics.

7. THE OUTLOOK FOR MICRODYNAMICS

Microdynamics may well create profound changes in medicine, science, communications, and industry. With the emergence of microdynamics in the United States, this nation could lead the way among its foreign competitors in a new and important technology. In the past, the United States has relinquished mechanical design and manufacture to other countries; now, we have an opportunity to reenter the industry and give it new life.

To take advantage of this opportunity, our country must provide broad support for the field of microdynamics. Early support will bring rewards: since the study and manufacture of microdynamical systems will depend upon the development of those very systems, researchers forsee that even small gains in microdynamics technology will lead to its exponential growth. Moreover, the technology will spread quickly because it is currently based on widely available silicon processing.

Therefore, to gain benefits already envisioned and benefits yet to be imagined, our best course is to launch the microdynamics program now. Academic research programs are important to the development of this field; they will provide valuable research results, a flow of skilled personnel from academia to industry, and a source of expertise for consulation with industry. Cooperation between academic research programs and the emerging industry should be encouraged as a means of enhancing technology transfer.

8. BIBLIOGRAPHY

For more information, the following sources can be consulted:

Four special issues of the IEEE Transactions on Electron Devices detail work primarily on integrated sensors but also include some actuator technology: December 1979, January 1982, July 1985 and June 1988.

The journal *Sensors and Actuators,* published by Elsevier Sequoia, is a source of articles on micro sensor and actuator technologies.

The proceedings of the 4th International Conference on Solid-State Sensors and Actuators, "Transducers '87," contain several papers on microactuator and micro structure/ technology in addition to extensive coverage of solid-state sensors.

The Proceedings of the IEEE Micro Robots and Teleoperators Workshop, held in Hyannis, Massachusetts during November 1987 contain many articles pertinent to microdynamical systems. Reprints of this volume are available from the same sources that supplied this report.

The following articles treat many issues important to the development of microdynamical systems:

Introduction

(1) R. P. Feynman,"There's Plenty of Room at the Bottom" in *Miniaturization,* ed. H. Gilbert, Reinhold, New York, 1960.

(2) O. D. Jefimenko, *Electrostatic Motors,* Electret Science Company, Star City, 1973.

(3) K. E. Petersen, "Silicon Sensor Technologies," *Technical Digest,* IEEE International Electron Devices Meeting, Washington, D.C., December 1985, pp. 2-7.

(4). S. Middelhoek and A. C. Hoogerwerf, "Smart Sensors, When and Where?," *Sensors and Actuators,* 8, 1985, pp. 39-48.

(5) H. Yamasaki, Approaches to Intelligent Sensors, *Proceedings,* Fourth Sensor Symposium, IEE of Japan, pp. 69-76, Tokyo, 1984.

(6) R. S. Muller, "Strategies for Sensor Research," *Technical Digest,* Transducers '87, 4th International Conference on Solid-State Sensors and Actuators, 107-111, Tokyo, June 1987.

(7) W.H. Ko, "Frontiers in Solid-State Biomedical Transducers," *Technical Digest,* IEEE International Electron Devices Meeting, Washington, D.C., December 1985, pp. 112-115.

(8) R. T. Howe, "Resonant Microsensors," *4th Int. Conference on Solid-State Sensors and Actuators,* pp. 843-848, Tokyo, June 1987.

(9) K. E. Petersen, "Silicon as a Mechanical Material," *Proceedings of the IEEE,* 70, May 1982, pp. 420-457.

Fabrication Processes

(10) E. Bassous, "Fabrication of Novel Three-Dimensional Structures by the Anisotropic Etching of (100) and (110) Silicon," *IEEE Transactions on Electron Devices*, ED-25, September 1978, pp. 1178-1185.

(11) A. Reisman, M. Berkenblit, S. A. Chan, F. B. Kaufman, and D. C. Green, "The Controlled Etching of Silicon in Catalyzed Ethylene Diamine-Pyrocatechol-Water Solutions," *J. of the Electr. Soc.*, 6, August 1979, pp. 1406-1414.

(12) H. Seidel, "The Mechanism of Anisotropic Silicon Etching and its Relevance for Micromechanics," *Technical Digest*, 4th International Conference on Solid-State Sensors and Actuators, 120-125, Tokyo, June 1987.

(13) C. D. Fung and J. R. Linkowski, "Deep Etching of Silicon Using Plasma," *Micromachining and Micropackaging of Transducers*, edited by C. D. Fung, P. W. Cheung, W. H. Ko, and D. G. Fleming. Amsterdam: Elsevier Science Publishers, 1985, pp. 159-164.

(14) R. S. Muller, "Technologies for Integrated Sensors," *Technical Digest*, 1984 Government Microcircuit Applications Conference, Las Vegas, October 1984, pp. 276-279.

(15) W. H. Ko, J. T. Suminto, and G. J. Yen, "Bonding Techniques for Microsensors," *Micromachining and Micropackaging of Transducers*, edited by C. D. Fung, P. W. Cheung, W. H. Ko, and D. G. Fleming. Amsterdam: Elsevier Science Publishers, 1985, pp. 41-61.

(16) B. E. Burns, P. W. Barth, and J. B. Angell, "Fabrication Technology for a Chronic In Vivo Pressure Sensor," *Technical Digest*, IEEE International Electron Devices Meeting, San Francisco, December 1984, pp. 210-212.

(17) K. E. Petersen, "Dynamic Micromechanics on Silicon: techniques and Devices," *IEEE Transactions on Electron Devices*, ED-25, October 1978, pp. 1241-1250.

(18) R. T. Howe, "Polycrystalline Silicon Microstructures," *Micromachining and Micropackaging of Transducers*, edited by C. D. Fung, P. W. Cheung, W. H. Ko, and D. G. Fleming. Amsterdam: Elsevier Science Publishers, 1985, pp. 169-187.

(19) H. Guckel and D. W. Burns, "A Technology for Integrated Transducers," *Technical Digest*, IEEE International Conference on Solid-State Sensors and Actuators, Philadelphia, June 1985, pp. 90-92.

(20) E. W. Becker, W. Ehrfeld, P. Hagmann, A. Maner and D. Munchmeyer, "Fabrication of Microstructures with High Aspect Ratios and Great Structural Heights by Synchrotron Radiation Lithography, Galvanoforming, and Plastic Moulding (LIGA process)," *Microelectronic Engineering* 4, 1986, pp. 35-56

Materials

(21) E. Obermeier, P. Kopystynski, and R. Niessl, "Characteristics of Polysilicon Layers and Their Application in Sensors," *Technical Digest*, 1986 Solid-State Sensors Workshop, Hilton Head Island, June 2-5, 1986.

(22) R. S. Muller, "Heat and Strain-Sensitive Thin-Film Transducers," *Sensors and Actuators*, 4, December 1983, pp. 173-182.

(23) M. Sekimoto, H. Yoshihara, and T. Ohkuso, "Silicon Nitride Single-Layer X-Ray Mask," *J. Vacuum Science and Tech.*, 21, 1982, pp. 1017-1021

(24) H. Guckel, D. W. Burns, C. R. Rutigliano, D. K. Showers, and J. Uglow, "Fine Grained Polysilicon and its Application to Planar Pressure Transducers," *Technical Digest-Transducers '87*, 4th International Conference on Solid-State Sensors and Actuators, Tokyo, June 1987, pp. 277-282

Applications

(25) L. Roylance and J. B. Angell, "A Batch-Fabricated Silicon Accelerometer," *IEEE Transactions on Electron Devices*, ED-26, December 1979, pp. 1911-1917.

(26) S .D. Senturia, "Microfabricated Structures for the Measurement of Mechanical Properties and Adhesion of Thin Films," *Technical Digest- Transducers '87*, 4th International Conference on Solid-State Sensors and Actuators, Tokyo, June 1987, pp. 11-16.

(27) P. L. Chen, R. S. Muller, and A. P. Andrews, "Integrated Silicon PI-FET Accelerometer with Proof Mass," *Sensors and Actuators*, 4, pp. 119-126.

(28) K. Najafi, K. D. Wise, and T. Mochizuki, "A High-Yield IC-Compatible Multichannel Recording Array," *IEEE Transactions on Electron Devices*, ED-32, 1985, pp. 1206-1211.

(29) K. E. Petersen, J. Brown, and W. Renken, "High-Precision, High-Performance Mass-Flow Sensor with Integrated Laminar Flow Micro-Channels," *Technical Digest*, IEEE International Conference on Solid-State Sensors and Actuators, Philadelphia, June 1985, pp. 361-363.

(30) S.C. Chang and D. B. Hicks, "Tin Oxide Microsensors," *Technical Digest*, IEEE International Conference on Solid-State Sensors and Actuators, Philadelphia, June 1985, pp. 381-384.

(31) R. T. Howe and R. S. Muller, "Resonant-Microbridge Vapor Sensor," *IEEE Transactions on Electron Devices*, ED-33, 1986, pp. 499-507.

(32) H. Fujita and A. Omodaka, "The Principle of an Electrostatic Linear Actuator Manufactured by Silicon Micromachining," *Technical Digest-Transducers '87*, 4th International Conference on Solid-State Sensors and Actuators, Tokyo, June 1987, pp. 861-864.

(33) K. J. Gabriel and W. S. N. Trimmer, "Micro Gears and Turbines Etched from Silicon," *Technical Digest- Transducers '87*, 4th Int. Conference on Solid-State Sensors and Actuators, Tokyo, June 1987, pp. 853-856.

(34) A. M. Feury, T. L. Poteat and W. S. N. Trimmer, "A Micromachined Manipulator for Submicron Positioning of Optical Fibers," *IEEE Solid State Sensors Workshop*, Hilton Head, June 1986

(35) W. S. N. Trimmer and K. J. Gabriel, "Design Considerations for a Practical Electrostatic Micro Motor," *Sensors and Actuators*, 11, 1987, pp. 189-206.

(36) S. Bart, T. Lober, R. Howe, J. Lang and M. Schlect, "Design Considerations for Micromachined Electric Actuators," *Sensors and Actuators*, 14, 1988, pp. 269-292,

(37) J. E. Wood, S. C. Jacobsen and K. W. Grace, "SCOFFS: A Small Cantilevered Optical Fiber Servo System," *Proceedings IEEE Micro Robots and Teleoperators Workshop*, Hyannis, November 9-11, 1987

(38) L.S. Fan, Y.C. Tai, and R.S. Muller, "Integrated Movable Micromechanical Structures for Sensors and Actuators," *IEEE Transactions on Electron Devices*, ED-35, June 1988, pp. 724-73

(39) M. Mehregany, K. J. Gabriel and W. S. N. Trimmer, "Integrated Fabrication of Polysilicon Mechanisms," *IEEE Transactions on Electron Devices*, ED-35, June 1988, pp. 719-723.

ACKNOWLEDGEMENTS

This report could not have been produced without the encouragement, support and participation of many people. The panel members and NSF attendees, Roger Brockett, Steve Charles, George Hazelrigg, Roger Howe, Ralph Hollis, Frank Huband, Steve Jacobsen, Noel MacDonald, Howard Moraff, Richard Muller, Theodore Pilkington, Kurt Petersen, Stephen Senturia, Robert Stengel and John Wood, met three times over the course of eight months, expending significant energy to formulate and integrate their ideas and opinions. In addition to the regular panel members, the meetings were enhanced by discussions with Dan Cho, Joseph Mathias, Richard Price, Barry Royce, Robert Stroud, Allen Stubberud and Bob Thomas.

The support of the NSF is gratefully acknowledged. The editors also wish to thank AT&T Bell Laboratories for its support.

Section 2

Side Drive Actuators

FRICTION is a major problem in designing micro motors. The coefficient of friction between two smooth, small surfaces tends to be large, and even worse, quite variable. The development of good micro bearings will be a major advance in the field.

These problems make friction a major hurdle in designing electrostatic micro motors. One's first inclination is to suspend a disk containing electrodes, over a substrate containing appropriate drive electrodes as shown in Fig. 2.1. By applying voltages between the substrate electrodes and the disk electrodes, forces are generated which tend to pull the electrodes together. The tangential component of this force causes the disk to rotate until the substrate and disk electrodes align. Unfortunately, along with the tangential force tending to align the electrodes, there is a clamping force pushing the disk firmly onto the substrate. Now our variable friend friction has his way, and the disk may rotate a few steps, or may not budge. Until good micro bearings are developed, this clamping force will remain a dominant problem.

The side drive motor has a clever solution to this problem. As shown in Fig. 2.2, the electrodes are placed around the rim of

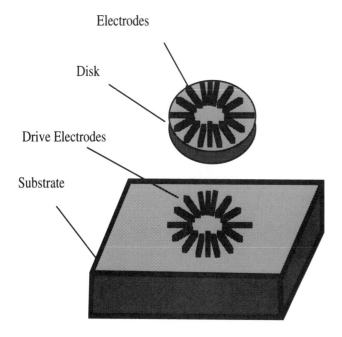

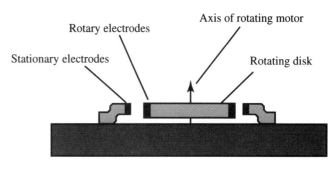

Fig. 2.2

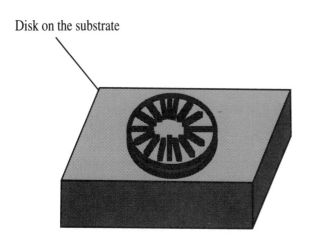

Fig. 2.1

the disk. By placing electrodes symmetrically on opposite sides of the disk, there are equal and opposite forces that tend to counterbalance each other. The major components of these forces are not clamping the disk to the substrate. The coplanar disk and electrode structure of the side drive actuators are conveniently fabricated using surface micromachining. Please note that irregularities in the electrodes and asymmetries in the electric field caused by the substrate do produce small, unbalanced forces that must be supported by the bearing.

An alternate strategy to overcome the clamping force is to place the disk between two structures, the roof above and the substrate below, as shown in Fig. 2.3. Electrodes on the roof and substrate are used to drive the electrodes on the rotor. Here, the attractive forces between the roof and substrate tend to counterbalance each other, and there is no net clamping force. Unfortunately, in some early designs, the roof was a thin layer built over the disk. This roof is deflected down by the electrostatic forces until it touches the disk, stopping the motor. Designs with a thick roof would mitigate this problem. An advantage of this motor geometry is the higher capacitance between the electrodes, and hence a higher torque than the side drive actuator can provide. Note that if the rotor is grounded and voltages are applied to the

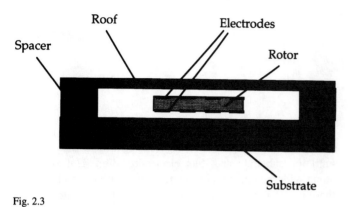

Fig. 2.3

top and bottom electrodes, the position of the rotor is unstable. (A small perturbation moving the rotor up will cause the forces to change in such a way that they tend to force the rotor up.) If the rotor is not attached to some voltage source, but left floating, the forces of instability are less. The problems of static-charge buildup also cause unbalanced forces.

In the first paper, "IC-Processed Electrostatic Micro-motors," the development of the first side drive motor is described. The second paper, "IC-Processed Electrostatic Micro-Motors: Design, Technology, and Testing," describes the design and fabrication of the side drive motor, and discusses the dynamics of the motion of the micromotor. Next, "Surface-Micromachining Processes for Electrostatic Micro-actuator Fabrication" investigates the processing constraints on the structural-sacrificial surface micromachining used to make polysilicon side drive motors. "A Study of Three Microfabricated Variable-capacitance Motors" compares the side drive motor to two similar motor geometries, the top drive and the harmonic motor. The next paper, "Friction and Wear in Microfabricated Harmonic Side-Drive Motors," discusses the twin problems of friction and wear. The final paper, "Measurements of Electric Micromotor Dynamics," uses a stroboscopic method to measure the dynamics of an electrostatic motor.

IC-Processed Electrostatic Micro-motors

Long-Sheng Fan, Yu-Chong Tai, and Richard S. Muller

Berkeley Sensor and Actuator Center
An NSF/Industry/University Cooperative Research Center
Department of EECS and the Electronics Research Laboratory
University of California, Berkeley CA 94720

ABSTRACT

We describe the design, fabrication, and operation of several micro-motors that have been produced using integrated-circuit processing [3]. Both rotors and stators for these motors, which are driven by electrostatic forces, are formed from 1.0-1.5 µm-thick polycrystalline silicon. The diameters of the rotors in the motors we have tested are between 60 and 120 µm. Motors with several friction-reducing designs have been fabricated using phosphosilicate glass (PSG) as a sacrificial material [4,5] and either one, or three polysilicon depositions.

INTRODUCTION

Recent publications have discussed possible designs for micro-motors [1,2] based on electrostatic-drive principles. Using technology derived from IC manufacturing processes, we have built and tested several electrostatically driven rotating motors and driven them both in stepwise fashion and through continuous revolutions. Included among the motors are structures with 4 and 8 rotor poles and 6, 12, and 24 stator poles. Typical gaps between the rotors and stators in this first realization of operating micro-motors are 2 µm or greater. The technology for the motors, which have rotors that turn on stationary axles fixed to the silicon substrate, is based upon the processes described in [4, 5].

Design has been based on many assumptions because fundamental parameters and mechanical behavior at the micrometer scale are largely unknown at this time. Extensive research is now underway in our laboratory to determine mechanical and electrical behavior for these very small structures. We describe here the results of initial tests of the motors as well as some of the design ideas that we have incorporated in them. Separate sections of this paper are devoted first to discussion of the stepping motors, and second to discussion of the synchronous motors.

STEPPING MOTORS

To design stepping motors for micropositioning, large starting torque, small frictional force, and fine angle resolution are desirable. Means to optimize these characteristics in micro-motors have been explored in our design. Figure 1 shows the top view and cross sectional view of a stepping motor that implements these features.

Starting Torque: The design of a rotating motor begins with an estimation of the torque exerted by the electric field. This can be expressed in terms of the derivative of the stored energy which, for a given bias V between rotor and stator, is conveniently represented as $1/2 C V^2$, where C represents the capacitance across the driving electrodes that have voltage V across them. To find the rotor torque T, we take the derivative with respect to the rotor angle θ.

$$T(\theta) = \frac{1}{2} V^2 \frac{\partial C(\theta)}{\partial \theta} \quad (1)$$

Torque values are of the order of pNm for voltages of order 100V and typical micro-motor dimensions. The angular dependence of this torque, which must overcome frictional restraint in order to cause the motor to rotate, is being studied in terms of electrostatic field plots such as the one shown in Fig. 2. The field-mapping program for Fig. 2 is a two-dimensional approximation. The regions of high equipotential-line density in Fig. 2 indicate locations of high electric fields. The effects of electrostatic drive are to maximize the field energy by aligning the rotor poles to the "active" stator poles. The rotor and stator in Fig. 2 are designed so that the next-to-be activated pole pair approximately aligns to the maximum torque position after each step, assuring a large starting torque for the next step.

Figure 4 shows bias and phasing scheme of these stepping motors. The size of the step for the motor is a function of the number of stators n_s and rotors n_r. Its size Θ in radians is given by:

$$\Theta = 2\pi \left(\frac{1}{n_s} - \frac{1}{n_r} \right) \quad (2)$$

We only need six stator poles for a twenty-four-step angular resolution.

Friction Reduction: By designing the motors so that the plane of the rotor is slightly lower (typically 0.5 µm) than the plane of the stator, the electrical field will have a component perpendicular to the substrate which will give rise to a force that tends to lift the rotor from this surface and thus to reduce friction. The purpose of this levitation force is not to overcome gravitational force (the rotor has a mass measured in tens of nanograms), but rather to overcome electrical attraction of the rotor to the substrate.

Another technique to reduce friction is to provide bushings to support the rotor. Friction-reducing bushings have been incorporated in several of the motors. In one design, hemispherical bushings extend from the under-surface of the rotor to provide a small-area contact with the substrate as shown in figure 3. To make the bushings, circular holes are patterned in the resist covering the sacrificial-oxide layer [4,5] and this layer is then isotropically etched before depositing polysilicon to form the rotor. The resultant extended hemispheres of polysilicon reduce friction between the two surfaces.

Yet another approach to reducing friction between the surfaces has been to use silicon nitride as the hub material in several of the stepping-motor designs. Our measurements have shown a smaller coefficient of friction to characterize silicon nitride in contact with polycrystalline silicon than is the case with two layers of polycrystalline silicon. We have therefore fabricated the stepping micro-motors using silicon nitride as the hub material.

Fabrication and Test

The stepping motors have been made using a process similar to that described for making pin joints in references 4 and 5. For the micro-motors, however, silicon nitride is deposited in place of the second layer of polycrystalline silicon, and a silicon nitride/silicon dioxide composite layer is used for electrical isolation. Figures 5 and 6 show a stepping-motor with six stator poles and eight rotor poles; from Eq. (2), it will rotate with 15° steps. The polysilicon rotor is 60 µm in diameter and 1 µm-thick. The silicon-nitride hub is 1 µm-thick, and a 2-µm-wide lateral air gap separates rotor and stator. A silicon-nitride axle and four polysilicon-hemispherical bushings (2.0 µm in diameter) are used to reduce friction between the rotor and its supports. The polysilicon bushings, which reduce the contact area for the rotor, are built over a thin layer of silicon dioxide (500 nm), covered by silicon nitride (1 µm). This composite layer and bushing have also been designed to increase electrical isolation between the rotor and the substrate.

At this time, these motors have been moved with manually switched voltages. They have been tested at manual-switching rates of 12 rpm, and require starting voltages of roughly 120 V from stator to stator. Using Eq. (1), we estimate, therefore, that the frictional forces are in the tens of nN range.

SYNCHRONOUS MOTOR

Figure 7 shows the top view of a synchronous motor having 12 stators and a 4-pole rotor so that it will function basically as a three-phase motor. Figure 8 shows the cross section along line AA' in Fig. 7 and makes apparent several friction-reducing schemes of the design. First, the rotor is supported by a flanged hub [4] that shortens the lever arm to the force point on the shaft. Second, both the stator and rotor are designed as coplanar structures to minimize the electrostatic force toward the substrate. Third, a ground plane is designed underneath both the rotor and the stator which is meant to have the same voltage as the rotor, thereby eliminating vertical electrostatic force. The ability to change both the ground-plane voltage and that of the stator provides an extra level of control for this motor design. Also apparent in Fig. 8 are silicon-nitride spacers at the vertical ends of the rotor and stator. The spacers interfacing with the hub act as solid lubricants to reduce friction and to provide wear protection for the rotor.

Fabrication and Test

In the motor shown in Fig. 9, the rotor diameter is 120 µm, and the gap between the stator and the rotor is 4 µm. Figure 10 shows a similar but smaller motor with a 24-pole stator and a 8-pole rotor. The diameter of the rotor is 60 µm, and the stator-rotor gap is 2 µm. The motors in Figs. 9 and 10 both have cross sections as shown in Fig. 8, and are fabricated with a three-layer-polysilicon process.

To make the motors, the starting wafers are first covered with two insulating layers: a film of thermally grown silicon dioxide (300 nm) and a 1 µm overlay layer of silicon nitride. The first layer of polysilicon(300 nm) is deposited and patterned to provides a grounding plate for the rotor. A 2.2 µm phosphosilicate glass(PSG) is then deposited to be the sacrificial layer[4,5]. After the anchor opening in the PSG, a second layer of polysilicon(1.5 µm) is deposited, thermally oxidized(100 nm polysilicon dioxide), and patterned to form both the stator and rotor. A 340-nm film of silicon nitride is then deposited as the spacer material. Anisotropic reactive-ion-etch (RIE) of the nitride then forms the spacers. Flange opening is done by PSG wet etching, followed by deposition of a second layer of sacrificial PSG(700 nm) to refill the rotor undercut. The hub anchor is then opened, followed by the deposition and patterning of the third layer of polysilicon(1.5 µm). Sacrificial PSG etch in buffered HF finishes the process.

The synchronous motors have been tested in air under two drive conditions. First, the motors were tested with only one pair of the stators and the rotor biased as shown in Fig. 11. The rotor is grounded and voltage is applied to two stators across a diameter of the motor. The remaining stators are unconnected. Under these bias conditions, stepwise rotations have taken place at voltages ranging from 100 to 400 V. However, corona, or even electrical breakdown of air, has sometimes occurred at biases above 300V. Since the minimum distance between stator and rotor, or stator and ground plane, is about 2 µm, electric fields as high as $1.5 \times 10^8 \, Vm^{-1}$ can, therefore, be imposed without causing breakdown.

A second bias arrangement under which the motors have been tested makes use of three-phase signals as shown in Fig. 12. Under the 3-phase bias condition, motors of the type shown in Fig. 9 were set into continuous rotation and observed using a video camera. Using a 200 V drive voltage, the motor rotation is about 50 rpm, at 350 V, the maximum rotation is about 500 rpm.

CONCLUSIONS

Electrical drive of rotating *IC*-fabricated micromechanisms has been demonstrated with examples of stepping and three-phase synchronous drive micro-motors described. Design features to minimize friction and to maximize torque in micro-motors are discussed. Typical drive voltages for present designs exceed 100 V. Manually switched motors have tested at speeds up to 12 rpm. Synchronous motors have been driven at speeds to 500 rpm. Experimental evaluation is continuing aimed mainly at understanding frictional effects.

Acknowledgements We thank Profs. D.J. Angelakos, R.M. White, and Paul T. Yang for valuable discussion, and K. Voros, R. Hamilton, D. Giandomenico and the staff of the Berkeley Microfabrication Laboratory for assistance in processing.

REFERENCES

[1] W.S. Trimmer and K.J. Gabriel, "Design Considerations for a Practical Electrostatic Micromotor," *Sensors and Actuators*, 11, pp. 189-206, (March, 1987).

[2] S.F. Bart, T.A. Lober, R.T. Howe, J. H. Lang, and M.F. Schlecht, "Design Considerations for Microfabricated Electric Actuators," *Sensors and Actuators*, 14, pp. 269-292 (July, 1988).

[3] *Patent Pending*.

[4] R.S. Muller, L-S. Fan, and Y-C. Tai, "Micromechanical Elements and Methods for Their Fabrication," U.S. Patent 4,740,410, issued to the Regents of the University of California, April 26, 1988.

[5] L-S. Fan, Y-C. Tai, and R.S. Muller, "Integrated Movable Micromechanical Structures for Sensors and Actuators," *IEEE Trans. Electr. Devices*, ED-35, pp. 724-730, (June, 1988).

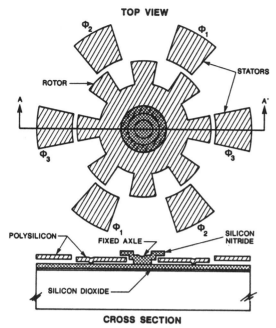

Figure 1 Top view and cross section of a stepping micromotor.

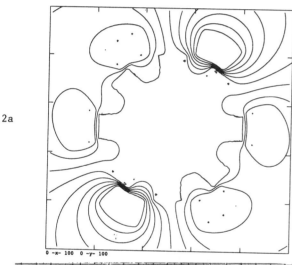

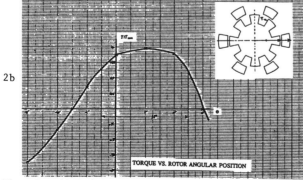

Figure 2 (a) Equipotential lines corresponding to a symmetrical biasing scheme from a two-dimensional simulation. (b) Torque versus rotor angular position from two-dimensional simulations.

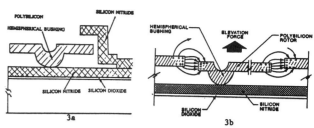

Figure 3 (a) Silicon-nitride axle and hemispherical bushing. The bushing provides a small-area contact with silicon nitride layer on substrate. (b) Levitation force to overcome substrate attraction.

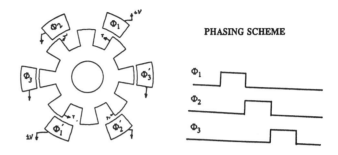

Figure 4 Biasing and phasing scheme of a stepping motor.

Figure 5 SEM photograph of a twenty-four-step stepping micromotor. The gap between the stator and rotor is 2 μm.

Figure 6 Photograph of the same motor in figure 5. Six contact pads are connected to the six stator poles.

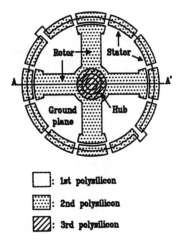

☐ : 1st polysilicon
▦ : 2nd polysilicon
▨ : 3rd polysilicon

Figure 7   Top view of a synchronous motor.

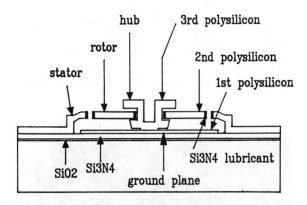

Figure 8   Cross section of line AA' in Fig. 1.

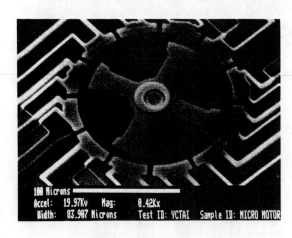

Figure 9   SEM photograph of a 12-stator, 4-rotor-pole micro motor. The gap between the stator and rotor is 6 µm.

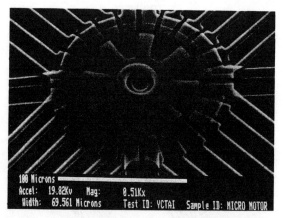

Figure 10   SEM photograph of a 24-stator, 8-rotor-pole micro motor. The gap between the stator and rotor is 4 µm.

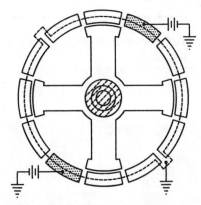

Figure 11   Stator-pair drive configuration. Only a pair of the stators are biased and the ground plane is grounded.

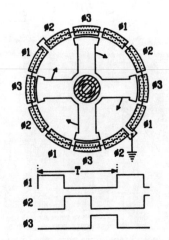

Figure 12   Three-phase drive configuration. All the stators are connected into three-phase scheme and the ground plane is grounded.

IC-Processed Micro-Motors: Design, Technology, and Testing

Yu-Chong Tai, Long-Sheng Fan, and Richard S. Muller

Berkeley Sensor and Actuator Center
An NSF/Industry/University Cooperative Research Center
Dep't of EECS and the Electronics Research Laboratory
University of California, Berkeley CA 94720

ABSTRACT

Micro-motors having rotors with diameters between 60 and 120 µm have been fabricated and driven electrostatically to continuous rotation. These motors were built using processes derived from *IC* micro-circuit fabrication techniques. Initial tests on the motors show that friction plays a dominant role in their dynamic behavior. Observed rotational speeds have thus far been limited to several hundred rpm which is a small fraction of what should be achievable if only natural frequency were to limit the response. Experimental starting voltages are at least an order-of-magnitude larger than had been expected (60 volts at minimum and above 100 V for some structures). Observations of asynchronous as well as synchronous rotation between the driving fields and the rotors can be explained in terms of the torque/rotor-angle characteristics for the motors.

I. INTRODUCTION

Design considerations for micro-motors have been discussed recently [1,2]. Successful implementation of both stepping and synchronous motors and initial testing results were presented by our group at the 1988 Int. Electron Devices Meeting [3]. In this paper, we describe new aspects of these motors, including theoretical design, advanced polysilicon technology, and phenomenological mechanisms in several micro-motors that have been produced using *IC* processing [4].

II. DESIGN

Shown in Fig. 1 is an SEM photograph of a three-phase micro-motor with 12 stators and a 4-pole rotor that was fabricated using a three-layer-polysilicon process. The radius of the rotor is 60 µm and the air gap between the rotor and stator is 2 µm. Fig. 2 shows a schematic cross section of the motor.

Several design features can be seen from Fig. 2. First, the micro-motor is built on a composite layer composed of 300 nm of silicon dioxide and 1 µm of silicon nitride. This layer helps to prevent electrical breakdown between the motor and the silicon substrate. The 1-µm nitride layer is also needed to protect the oxide from HF etching while the sacrificial layer is being dissolved. Experimentally, we have found that this layer can withstand more than 500 V before breakdown occurs. A second feature is a ground plane, formed of first-layer polysilicon, situated beneath both the rotor and the stator. The ground plane, which acts as an electrostatic shield between the rotor and the substrate, improves the performance of the motor by eliminating vertical Coulombic forces on the rotor which result in increased frictional drag. Another advantage of the ground plane is that it provides the possibility of biasing the rotor since it is in electrical contact with the rotor through its hub. This provides an extra freedom for control of the motor compared to a floating rotor design-- a freedom that might be useful to increase the electrical drive on start-up, for example. Another effective friction-reducing feature in the motor shown in Fig. 2 is the use of silicon-nitride spacers at the vertical walls of the rotor and the stators. Figure 3 magnifies the hub and stator parts to illustrate this feature. The nitride spacer between the rotor and the hub forms a lubricating as well as a wear-protecting material with a smaller coefficient of friction and better wear-withstanding characteristics than are found for polysilicon.

III. FABRICATION

The fabrication process is based on the pin-joint technology, reported in [5,6]. In this discussion, we emphasize techniques to fabricate the design features discussed in the last section. Figure 4 shows the major steps in the motor fabrication. In Fig. 4a formation of the 300-nm thick ground plane on top of the oxide-nitride composite layer is indicated. After this layer is deposited, the wafers are covered with 2.2 µm of phosphosilicate glass(PSG) which acts as the sacrificial layer. After opening anchor spots for the stator, a 1.5 µm second-layer polysilicon is deposited and oxidized to have a 100 nm thermal

Reprinted from *Proceedings IEEE Micro Electro Mechanical Systems*, pp. 1-6, February 1989.

oxide on top as shown in Fig. 4b. The polysilicon-oxide composite layer then is plasma-etched to form the rotor and stators. A 340-nm LPCVD nitride is then deposited (Fig. 4c) and reactive-ion-etched(RIE) to form the nitride spacers(Fig. 4 d). The thin oxide layer on the polysilicon is used to protect the polysilicon during the nitride RIE etch. Once the nitride spacers are formed, the process follows the outline already reported [5,6], to form the axle-bushing hub. The completed motor (prior to being freed from the sacrificial support layer) is shown in Fig. 4e.

IV. EXPERIMENTAL PROCEDURE

The setup used to drive the motors is shown in Fig. 5. Its function is as follows: a square-wave, extracted from a pulse generator, is sent to a TTL counter which is linked to a TTL decoder. The decoder sends the three-phase signals to the a high-voltage driver and also feeds back a signal to restart the counter after each three counts.

Three voltage phases are therefore repeatedly generated at the driver circuit and sent through current-limiting resistors to the micro-motors. The frequency of the phase voltage is controlled by the square-wave generator and the magnitude of the phase voltage is controlled by the voltage supplied to the driver.

The voltage phases supplied to the micro-motors are shown in Fig. 6. The twelve stator poles are energized in a three-phase pattern. During tests, the phase voltage is first set to a constant value and the phase frequency then is varied from high to low until rotational motion of the motor is observed (through a video attachment on the microscope). Motor movements are video-recorded to determine the speed of rotation and other dynamical properties.

V. MOTOR DYNAMICS

Two-dimensional numerical simulators are used to determine the static-torque/rotor-position characteristics. The simulation results of the torque/position characteristics associated with 4 equi-positioned stators energized at 100 V (and the remaining 8 grounded) is shown in Fig. 7. The torque on the rotor is obtained by multiplying the simulated torque density times the thickness of the rotor which neglects any three-dimensional effects. Figure 7 has a span of only 90 degrees because the torque/position characteristic is periodic for this 4-pole rotor.

A natural frequency f_N, which arises from energy exchange between the electric field and the moving rotor, is an important motor parameter. The frequency f_N can be calculated by considering the rotor mechanics near an equilibrium ($T = 0$) point and assuming zero friction. Then

$$f_N = \frac{1}{2\pi}\sqrt{\frac{T'}{I}} \quad (1)$$

where I is the rotational inertia of the rotor and $-T'$ is the slope of the T vs. θ curve at $T = 0$.

For the motor of Fig. 1, the rotational inertia of the rotor is 2.01×10^{-20} kg-m^2 and T' is approximated as 1.83×10^{-11} Nt-m/rad. Hence f_N = 1.5 kHz (excitation voltage = 100 V). Note that the T vs. θ curve is roughly linear only between $\pm 15°$.

When the starting angle θ_s is outside the $\pm 15°$ range, the oscillating frequency will differ from f_N because $T(\theta)$ depends on θ. In general, the oscillating frequency can be calculated from

$$f(\theta_s) = \frac{1}{4\int_0^{\theta_s} \frac{d\theta}{\sqrt{2I^{-1}\int_{\theta_s}^{\theta} T(\eta)d\eta}}} \quad (2)$$

here η is a dummy variable and $T(\eta)$ is the torque in Fig. 7. It is straightforward to show that $f(\theta_s)$ in Eq. (2) reduces to f_N in Eq. (1) inside the $\pm 15°$ range where $T(\theta) \approx -T' \times \theta$. Equation (2) shows that the oscillating frequency is a function of the starting angle θ_s. As seen in Fig. 7, $|T' \times \theta| \geq |T(\theta)|$ for all θ in [0, θ_s]. Hence,

$$f_N = \frac{1}{4\int_0^{\theta_s} \frac{d\theta}{\sqrt{2I^{-1}\int_{\theta}^{\theta_s} T'\eta d\eta}}} \geq \frac{1}{4\int_0^{\theta_s} \frac{d\theta}{\sqrt{2I^{-1}\int_{\theta_s}^{\theta} T(\eta)d\eta}}} = f(\theta_s) \quad (3)$$

Equation (3) shows that f_N (in our case 1.5 kHz for 100 V applied to a single phase of 4 stators) is the maximum oscillating frequency.

Since the electrostatic torque acting on the rotor is proportional to the square of the voltage, the natural frequency at a different phase voltage V_p is

$$f_N(V_p) = 1.5\left(\frac{V_p}{100}\right)^2 \text{ kHz} \quad (4)$$

As an example, if V_p equals 200 V, f_N is 6.1 kHz.

The natural frequency of the motor sets an upper bound for the rate at which the the rotor can respond to a switched field on the stators. We can argue, for example, that $\frac{1}{4f_N}$ is the minimum time for the rotor to travel from a starting angle θ_s to $\theta = 0$ if the initial angular velocity is zero as is the case for start/stop-mode stepping motors. In terms of a maximum rotational speed ω_{max}, we have

$$\omega_{max} = \frac{1}{\frac{1}{4f_N} \times n} \times 60 = \frac{240 f_N}{n} \text{ rpm} \quad (5)$$

where n is the number of steps per revolution. As an example, with $V_p = 200$ V and $n = 12$, the maximum speed of our stepping motor is predicted to be 120 krpm using Eq. (5).

Neglected in the analysis presented thus far are loading effects (other than inertia) on the rotor. These effects, principally friction, reduce the effective maximum frequency below the calculations in Eqs. (4) and (5). From our measurements, the frictional effects are completely dominant because the observed maximum rotational speed at 200 V is 150 rpm. With friction, the term $T(\eta)$ in Eq. (2) should be replaced by $T_{net}(\eta)$, the net torque on the rotor which is smaller than $T(\eta)$. At this point insufficient information is available about the frictional mechanisms in our motors to go further with this analysis.

The foregoing discussion refers to open-loop drive of the motors, which is the mode used for our experiments. The natural frequency limitation on rpm could be exceeded with closed-loop control, which would be possible with the addition of a rotor-position sensor.

VI. SYNCHRONOUS AND ASYNCHRONOUS MOTION

Another surprising result from dynamical studies of the motors in addition to the frictional effects discussed above is the observation of asynchronous as well as synchronous motion in the motors. Specifically, we observe that it is possible for the rotor to rotate in opposition to the rotational sense of the phasing field. This behavior, we believe, can be understood in terms of the dynamic balance between electrostatic-drive torque and that due to friction. It is easiest to introduce this concept by considering the motor driven by only one phase of the three-phase voltage drive.

Figure 8 (a) shows the voltage connections for single-phase drive of the motor of Fig. 1 as well as a defining condition for the initial position of the rotor ($\theta = 30°$). The angle θ in Fig. 8 (a) is measured from a rest ($T = 0$) position. Shown in Fig. 8 (b) (which is similar to Fig. 7) is the corresponding normalized torque/position characteristic. Indicated by straight lines on Fig. 8 (b) are assumed constant frictional-torque values which, of course, oppose the applied torque T.

Consider that the horizontal stator (in Fig. 8 (a)) is energized. Since T exceeds the frictional value, the rotor will begin to rotate in the clockwise direction and will accelerate until it reaches angle P at which the frictional torque equals the the driving torque. As θ decreases toward zero, the rotor will decelerate due to friction. Whether the rotor passes through $\theta = 0$ depends upon specific values for frictional and driving torque as well as on the rotor inertia. If it passes $\theta = 0$, it will continue to decelerate between $\theta = 0$ and $\theta = Q$. The rotor may stop anywhere between angles P and Q, but if it can pass point Q, it will decelerate more strongly because the driving torque is reversed in this region. If the rotor stops between $\theta = Q$ and $\theta = -30°$, it will accelerate again, but this time in a counter-clockwise direction. Once again, the rotor will repeat a similar acceleration-deceleration-stop cycle except that the direction of rotation is different. Several cycles may occur, but eventually the rotor will stop between $\theta = P$ and $\theta = Q$.

The final resting place of the rotor after the cyclic motion discussed above (when the motor is energized by a single phase) indicates whether synchronous or asynchronous rotation will occur in the motor when it is driven by three phases.

a) Three-phase synchronous mode

Normalized three-phase torque/position characteristics of the motor are shown in Fig. 9. In Fig. 9, the drive voltages are phased $\pm 30°$ apart because the motor has twelve stator poles. The starting position is again assumed at $\theta = 0$. In Fig. 9, $\theta = P$ and $\theta = Q$ are the angles at which frictional and driving torque are in balance for phase A and $\theta = R$ is one of the corresponding angles for phase B. We conclude that if a static three-phase excitation of the stator field results in a final resting position for the rotor between $\theta = P$ and $\theta = R$ or any of the corresponding $\pm \frac{m\pi}{6}$-shifted regions, the motor will operate synchronously (where m is an integer).

The explanation is in the following. Based on the discussion for single-phase excitation above, the rotor would tend to come to rest inside the frictional band, either between $\theta = P$ and $\theta = R$, or between $\theta = R$ and $\theta = Q$. If the former, then the rotor will rotate counter-clockwise because of the positive, larger-than-friction, driving torque from phase B. A similar behavior would be repeated with respect to phase C so that the rotor would rotate synchronously with the field following phases ...ABCABCABC....

b) Three-phase asynchronous mode

The motor and field rotation will be in opposite directions, however, if the rotor rest position under the conditions described (in section *(a)* above) occurs between $\theta = R$ and $\theta = Q$. or its $\pm \frac{m\pi}{6}$-shifted regions. The reason is as follows. If, after phase A, the rotor rests between $\theta = R$ and $\theta = Q$, the next phase (B) will not be able to rotate the rotor because inside that region the driving torque produced by this phase is smaller than the maximum frictional torque. However, phase C will apply a negative torque, which can overcome the friction on the rotor and make it rotate in a clockwise direction, opposite to the rotation of the stator fields. For continuous motion phase C must leave the rotor at rest between $\theta = R"$ and $\theta = Q"$. Then, next phase A will not be able to turn the rotor because of the friction. However, the next cycle of phase B will again cause clockwise rotation. A feature of this motion is that the rotor rotates at only one-half the frequency of the driving fields because it is driven only during a half cycle of one phase; for an applied signal ...ABCABC..., the rotor rotates ...ACBACB....

Experimentally, both modes have been observed. In some cases, it appears that both synchronous and asynchronous behavior alternates spontaneously. It is believed this behavior can be explained by irregular friction in real motors, since friction determines the critical angle values for the rotational variations described above.

VII. CONCLUSIONS

Initial tests of electrostatically-driven *IC*-processed micromotors show that frictional effects play a dominant role in their dynamic behavior. Rotational speeds in these motors are only a small fraction of what should be achievable if only natural frequency were to limit the response of the rotor. The experimental starting voltages are at least an order-of-magnitude larger than had been expected (60 volts at minium and above 100 V for some structures).

Observations of asynchronous as well as synchronous rotation between the driving fields and the rotors are explained in terms of the torque/rotor-angle characteristics.

Acknowledgements We thank Profs. D.J. Angelakos, R.M. White, D. Giandomenico, and Paul T. Yang for valuable discussion, and K. Voros, R. Hamilton, and the staff of the Berkeley Microfabrication Laboratory for assistance in processing.

REFERENCES

[1] W.S. Trimmer and K.J. Gabriel, "Design Considerations for a Practical Electrostatic Micromotor," *Sensors and Actuators,* 11, 189-206, (March, 1987).

[2] S.F. Bart, T.A. Lober, R.T. Howe, J. H. Lang, and M.F. Schlecht, "Design Considerations for Microfabricated Electric Actuators," *Sensors and Actuators,* 14, 269-292 (July, 1988).

[3] L.S. Fan, Y.C. Tai, and R.S. Muller,"IC-processed Electrostatic Micro-motors," *1988 IEEE Int. Electr. Devices Meeting,* 666-669, San Francisco, CA, December 11-14, 1988.

[4] *Patent Pending*.

[5] R.S. Muller, L.S. Fan, and Y.C. Tai, "Micromechanical Elements and Methods for Their Fabrication," U.S. Patent 4,740,410, issued to the Regents of the University of California, April 26, 1988.

[6] L.S. Fan, Y.C. Tai, and R.S. Muller,"Integrated movable micromechanical structures for sensors and actuators," *IEEE Trans. Electro. Devices,* vol. **ED-35**, pp.724-730, June 1988.

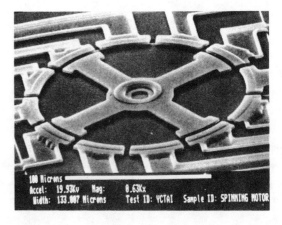

Figure 1. SEM photograph of a fabricated micromotor; rotor diameter = 120 μm.

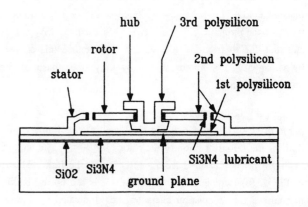

Figure 2. Schematic cross section of the micro-motor in Fig. 1.

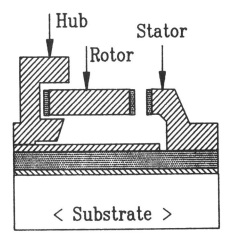

: polysilicon

: silicon nitride

: silicon dioxide

Figure 3. Cross section showing construction of the hub, rotor, and stator for the micro-motor of Fig. 1.

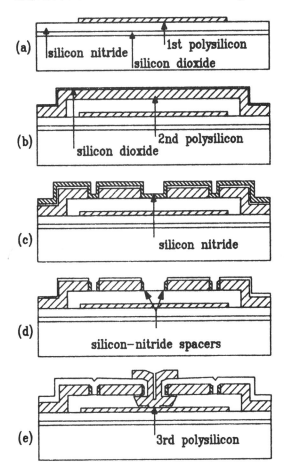

Figure 4. Major fabrication steps to construct the micro-motors.

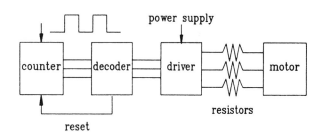

Figure 5. Setup for 3-phase drive of the micro-motors.

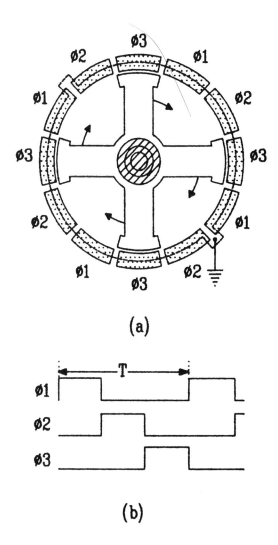

Figure 6. Electrical connections for 3-phase operation of the motor.

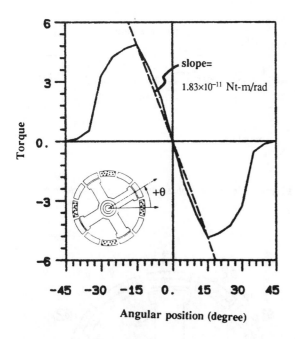

Figure 7. Torque/position characteristics of the micro-motor as obtained from two-dimensional simulation.

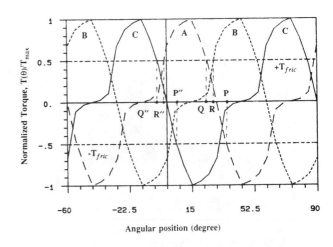

Figure 9. Three-phase torque/position characteristics.

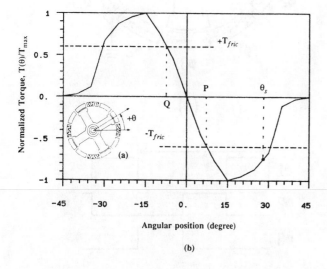

Figure 8. Single phase torque/position characteristics.

156

Surface-Micromachining Processes for Electrostatic Microactuator Fabrication

Theresa A. Lober and Roger T. Howe[†]

Microsystems Technology Laboratories
Dept. of Electrical Engineering and Computer Science
Massachusetts Institute of Technology
Cambridge, Massachusetts 02139

ABSTRACT

Surface-micromachining, the selective etching of multiple layers of deposited thin films, is essential for silicon microactuator fabrication. This paper presents a study of the etch characteristics of three forms of hydrofluoric acid, used as the micromachining etchant for fabricating a rotory variable-capacitance micromotor structure from polycrystalline silicon (poly-Si) thin films. Low pressure chemically vapor deposited SiO_2 (LPCVD LTO) is used to form the sacrificial layers and LPCVD Si_3N_4 is used as an electrical isolation layer for the micromachined structure. 7:1, NH_4F:HF buffered HF (BHF), 48 wt% concentrated hydrofluoric acid (LHF), and HF vapor in air (VHF) are evaluated at 20° C for etch rate and selectivity, and attack of four types of LPCVD poly-Si films: $POCl_3$-doped or undoped, with or without 1150° C annealing. Both BHF and VHF are found to damage $POCl_3$-doped poly-Si films, possibly by attack of grain boundaries. VHF is found to be effective for micromachining suspended structures since it avoids deformation of microstructures due to surface tension effects during drying after wet etching with LHF or BHF.

INTRODUCTION

Movable micromechanical parts, such as gears and turbines, have been fabricated recently from poly-Si thin films [1,2]. A means of applying forces to these movable structures is necessary if they are to perform useful mechanical work. Force-scaling calculations and materials compatibility with silicon technology argue for electrostatic actuation of micromechanical structures [3,4]. Of the various electrostatic motor configurations, the variable-capacitance type is attractive for an initial microfabrication study because it requires only conducting and insulating films [3].

Figure 1 illustrates the basic structure of the preliminary rotary, two-pole, three phase variable capacitance micromotor structure [3]. The sandwich of structural poly-Si and sacrificial LTO layers is shown schematically in Fig. 1(a) prior to the micromachining etch, while Fig 1(b) depicts the released structure. The segmented stator consists of overhanging poly-Si cantilevers which are dielectrically isolated from the silicon surface by a sandwich of LPCVD Si_3N_4 and thermal SiO_2. The poly-Si stator film is also used to fabricate a pin bearing to further confine the poly-Si rotor. In this preliminary structure, a torque can be applied to the rotor by sequential phasing of the voltage on the stator electrodes. The vertical electrostatic force attracting the rotor to the stator [3] may cause friction and electrical shorting of the stator pieces by conduction through the rotor.

This paper investigates surface-micromachining technology for fabricating the simple micromotor structure shown in Fig. 1. Characterization of this technology is critical, since the micromotor structural design is constrained by the etch rate and selectivity of the etchant used to release the microstructure by removing the sacrificial layers. The etch system must completely remove the oxide spacer films while leaving sufficient insulating Si_3N_4 and preserving the surface condition of the conducting, poly-Si structural layers. Additionally, the etching process should not produce residues or contribute to structural deformation through surface tension effects.

[†]Berkeley Sensor & Actuator Center, Dept. of Electrical Engineering and Computer Sciences. University of California, Berkeley, California 94720

Two types of wafers are fabricated for this study. The first contains complete micromotor structures and singly and doubly-supported poly-Si beams. The second type contains only patterned micromotor rotors from a single poly-Si layer deposited on various films on the substrate. Four poly-Si process conditions are evaluated: undoped, unannealed and annealed; and doped, unannealed and annealed. Three micromachining etchants are compared in this study to release the poly-Si structures: 48 wt% hydrofluoric acid (LHF), 7:1, NH_4F: HF buffered HF (BHF), and HF vapor in air (VHF). Relative etch rates of unpatterned LPCVD LTO layers, LTO spacer layers, and the LPCVD Si_3N_4 insulating layer are measured by releasing the microstructures fabricated on the first set of sample wafers with the three HF etchants. Structural attack of LPCVD poly-Si films is evaluated as a function of process conditions and underlying films by exposing the second group of sample structures to the three HF etchants. Together, these experiments are utilized to determine the potential of poly-Si to act as both an electronic and structural material for micromachined structures.

FABRICATION PROCESS

The process sequence for fabrication of the preliminary micromotor structure and the first set of wafer samples, with the exception of the final micromachining etch, is a subset of the MIT Baseline CMOS process [5]. Compatibility with this IC process constrains the film deposition conditions, thicknesses, and doping levels of the microstructures. Silicon wafers (n-type, 2 ohm-cm resistivity) of <100> orientation are first oxidized to grow a 400 Å-thick thermal stress-relief SiO_2 layer. This is followed by a 1500 Å-thick LPCVD silicon nitride film, deposited by reacting dichlorosilane and ammonia at 800° C and a pressure of 400 mT. The first LTO spacer layer, 0.5 μm-thick, is then deposited at 400° C and 250 mT with a reaction of silane and oxygen. Silane is reacted at 250 mT and 625° C to deposit the first LPCVD polysilicon structural layer, 5000 Å-thick, on the LTO spacer. This polysilicon film is phosphorus-doped during a 60 minute, 925° C diffusion cycle using a $POCl_3$ liquid diffusion source in a nitrogen ambient. Rotor structures are then patterned and anisotropically etched in the polysilicon film using a CCl_4 plasma. The second LPCVD LTO spacer, 1 μm-thick, is deposited on these patterned structures, and trenches are patterned and plasma etched through both LTO layers to expose the nitride-coated substrate. A final poly-Si film, either 0.5 or 1 μm-thick, is then deposited and doped

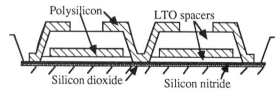

Figure 1(a). Sandwich of films creating the micromotor

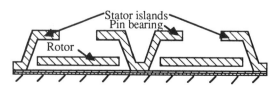

Figure 1(b). Released micromotor structure

during a second POCl₃ diffusion cycle. This second poly-Si layer is patterned and plasma etched to define the 1 μm-thick micromotor pin bearing and stator islands, as well as 0.5 μm-thick singly and doubly-supported micro-beams. Finally, the sandwich of patterned LTO and poly-Si films is annealed in a nitrogen ambient at 1150° C for 20 minutes to reduce residual stress in the polysilicon films. The completed set of wafers is split into three groups; each group to be released by one of the HF etchants.

The second set of test structures, utilized to characterize the attack of thin poly-Si films by HF the micromachining etchants, is fabricated using a subset of the first process sequence. Micromotor rotor patterns are plasma etched in 5000Å-thick poly-Si films for four film process conditions: undoped, unannealed and annealed; and doped, unannealed and annealed. Each type of film is deposited on the following materials:

1) 1 μm-thick, 1.8 wt% $POCL_3$-doped thermal oxide;
2) 1 μm-thick LTO;
3) 1500 Å-thick silicon nitride;
4) 1000 Å-thick undoped thermal oxide;
5) lightly doped single crystal silicon.

As with the micromotor wafers, the second set of test wafers is split into three groups for exposure to the three micromachining etchants. Wafers to be etched in BHF or LHF are placed into a 1000 ml teflon beaker containing the etch bath. After etching, wafers are rinsed first in deionized (DI) water for 5 minutes, then in methanol for 2 minutes, and then dried in air under a chemical hood. Wafers to be etched with VHF are suspended 1 cm above the surface of 1000 ml of LHF in an air flow of 28 l/min. No rinsing or drying of VHF-released samples is required; however, removal of nonvolatile etch products may require a DI water rinse.

EXPERIMENTAL RESULTS

Timed etches are completed on the micromotor sample wafers and the singly and doubly-supported beams to determine HF, BHF, and VHF etch rates for unpatterned LTO and Si_3N_4, and LTO spacer layers under 5000Å-thick polysilicon films. The etch front, and thus the etch rate, of spacer layers under 5000 Å-thick poly-Si structures can be visually monitored, since polysilicon is transparent at that thickness. Table I tabulates the measured etch rates, which indicate that VHF provides the highest etch selectivity, ≈ 90:1, between undercutting LTO spacer layers and unpatterned silicon nitride.

Material / Etchant	Silicon nitride	Unpatterned LTO	LTO spacer
BHF	7 - 12 Å/min	700 Å/min	570 Å/min
LHF	140-175 Å/min	10500 Å/min	7800 Å/min
VHF	85-125 Å/min	15000 Å/min	8400 Å/min

Table I. Etch rates of the three micromachining etchants

Timed etches using the three HF etchants are also completed with the second set of wafers, containing the poly-Si rotor structures, after which the samples are inspected for attack of the poly-Si films by each etch system. Three distinct phenomena are observed: *i)* LHF does not noticeably attack poly-Si films, regardless of doping and annealing history or the underlying film; *ii)* BHF and VHF attack $POCl_3$-doped, 5000 Å-thick poly-Si films to a degree dependent on annealing condition and underlying materials; and *iii)* undoped poly-Si films, unannealed or annealed, are not attacked by any of the etchants. By "attack," is meant nonuniform, isolated etching completely through the poly-Si film, appearing to occur at grain boundaries. The SEM in Fig. 2(a) illustrates this condition on a doped and annealed poly-Si rotor, deposited on LTO, which was exposed to VHF for 15 minutes. Figure 2(b) depicts the same film after VHF exposure for 30 minutes.

Table II compares the degree of poly-Si attack by BHF and VHF for each of the five layers used as the underlying material for the poly-Si films, as listed above. The greatest degree of attack occurs to annealed films deposited on $POCl_3$-doped oxide, while no noticeable attack occurs to films

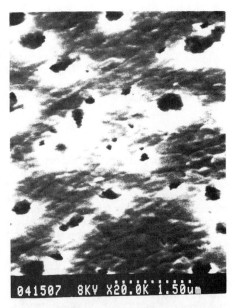

Figure 2(a). Poly-Si film attack after 15 min VHF exposure

Figure 2(b). Poly-Si film attack after 30 min VHF exposure

deposited on single crystal silicon or undoped thermal oxide. Attack of poly-Si films deposited on LTO is worse than for films deposited on Si_3N_4, and annealed films are more susceptible to the nonuniform etch than unannealed films. These observations suggest that the poly-Si attack mechanism is dependent on grain configuration and size, which is determined by the material underlying the poly-Si layer and the temperature of the poly-Si deposition.

Underlying material	Doped unannealed	Doped annealed
Silicon	—	—
1000 Å thermal oxide	—	—
1500 Å silicon nitride	+	+
1 μm LTO	+++	++++
1 μm doped thermal oxide	++++	+++++

Table II. Attack of poly-Si for different underlying layers

The following reaction models describe the etching of silicon dioxide and silicon nitride, respectively, when water is present as a catalyst:

$$SiO_2 + 4HF + 2H_2O \rightarrow SiF_4 + 4H_2O \qquad (a)$$

$$3H_2O + 2Si_3N_4 + 18HF \rightarrow 3SiF_4 + 8NH_3 + 3SiOF_2. \qquad (b)$$

These models describe both liquid and vapor etching of the films. Products of the silicon dioxide etch are volatile; they are removed by either the liquid solution of BHF and LHF or the air flow of VHF. The silicates ($SiOF_2$) produced by the silicon nitride etch are removed by liquid etch solutions, but not by the VHF air flow; thus, vapor etching of silicon nitride results in the deposition of a uniform layer of silicate "dust" on the nitride surface. Fig. 3 illustrates this dust in the field regions of the SEM, showing a poly-Si film on a nitride surface. This dust appears to mask the nitride from oxidation, thus slowing its etch rate below that for LHF, as shown in Table I [6]. The dust particles are removed by DI rinsing of the etched samples, but not by air flow.

Figure 4(a). 100-250 μm-long LHF-released cantilevers

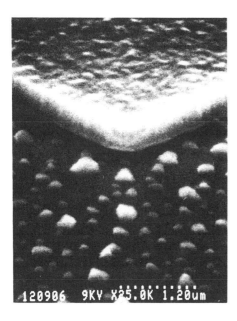

Figure 3. Silicate dust from VHF-etched nitride

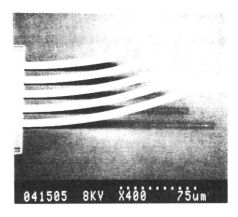

Figure 4(b). 100-250 μm-long VHF-released cantilevers

Use of VHF rather than either of the liquid etchants to release suspended structures appears to eliminate possible structural deformations due to surface tension effects during drying of the microstructures. Comparison of the LHF-etched cantilever beams shown in Fig. 4(a), ranging in length from 100-250 μm, and the same VHF-etched geometries in Fig. 4(b) suggests that the removal of liquid etch and rinse solutions from under suspended structures may cause deflections unattributable to built-in bending moments, and may produce surface conditions causing the suspended structures to stick to the substrate. During the vapor etch of unsupported structures, however, insufficient air flow may permit condensation under and around the releasing structure. Figure 5 illustrates this condition with a VHF-released poly-Si rotor which is stuck to the silicon substrate by the condensed liquid at its periphery.

Figure 5. VHF-released poly-Si rotor

CONCLUSIONS

The selectivity of all three etchants between LTO and Si_3N_4 is found to be insufficient to release the preliminary micromotor structure while preserving the Si_3N_4 dielectric isolation under the stator anchors. In addition, BHF and VHF attack the $POCl_3$ doped poly-Si motor components. The SEM shown in Fig. 6 of a micromotor structure released in LHF indicates that the rotor and stator structures are completely released and that the nitride and oxide films have been removed from the field areas. Although VHF release avoids surface tension-related structural deflections, it produces an accumulation of silicate dust within the micromotor periphery which is difficult to remove without a liquid rinse.

Figure 6. Poly-Si micromotor structure released with LHF

Given its attack by BHF and VHF, $POCl_3$ doped and 1150° C annealed poly-Si films are not attractive for micromachining applications. Heavy doping of the film is done in the preliminary fabrication sequence to achieve high conductivity for charge induction in the rotor structure, while high temperature annealing is completed to reduce built-in stresses of the micromachined structures. Thus, additional research is needed to develop poly-Si process conditions and micromachining etch systems for microactuator fabrication. For undoped, unannealed poly-Si films, VHF provides an attractive technique for releasing suspended micromachined structures while avoiding structural deflection due to liquid-induced surface tension effects.

ACKNOWLEDGEMENTS

The authors would like to acknowledge Prof. Rosemary L. Smith for initially suggesting the use of vapor HF as a micromachining etchant, and Prof. Stephen D. Senturia for many helpful technical discussions and a critical reading of the manuscript. This work was supported in part by the Power Electronic Collegium of the MIT Laboratory for Electromagnetic and Electronic Systems, and by the National Science Foundation, under contract ECS-8614328. Fabrication of all test structures was completed in the MIT Integrated Circuits Laboratory of the Microsystems Technology Laboratories. The authors thank Paul Maciel, Joe Walsh, and Mike Schroth for process development assistance.

REFERENCES

[1] L.S. Fan, Y.C. Tai, and R.S. Muller, "Pin joints, gears, springs, cranks, and other novel micromechanical structures," in <u>Technical digest, International Conference on Solid-State Sensors and Actuators,</u> Tokyo, Japan, pp. 849 - 852, 1987.

[2] K.J. Gabriel, W.S.N. Trimmer, and M. Mehregany, "Micro gears and turbines etched from silicon," in <u>Technical digest, International Conference on Solid-State Sensors and Actuators,</u> Tokyo, Japan, pp. 853 - 856, 1987.

[3] S.F. Bart, T.A. Lober, R.T. Howe, J.H. Lang, and M.F. Schlect, "Design considerations for microfabricated electric actuators," <u>Sensors and Actuators</u>, at press.

[4] W.S.N. Trimmer and K.J. Gabriel, "Design considerations for a practical electrostatic micro-motor," <u>Sensors and Actuators,</u> 11, vol. 11, pp. 189 - 206, 1987.

[5] P.K. Tedrow and C.G. Sodini, "The MIT twin-well CMOS process," Microsystems Technology Laboratories Report, 1988.

[6] T.A. Lober, "A microfabricated electrostatic motor design and process," MIT EECS Dept., S.M. Thesis, 1988.

A Study of Three Microfabricated Variable-capacitance Motors

M. MEHREGANY, S. F. BART, L. S. TAVROW, J. H. LANG, S. D. SENTURIA and M. F. SCHLECHT

Microsystems Technology Laboratories, Laboratory for Electromagnetic and Electronic Systems, Department of Electrical Engineering and Computer Science, Massachusetts Institute of Technology, Cambridge, MA 02139 (U.S.A.)

Abstract

This paper discusses the design, microfabrication, operating principles and experimental testing of three types of rotary variable-capacitance micromotors. The advantages and disadvantages of these motors are discussed. The three motor types are top-drive, side-drive and harmonic side-drive. In this work, the micromotors are surface micromachined using heavily-phosphorus-doped polysilicon for the structural material, deposited oxide for the sacrificial layers and LPCVD nitride for electrical isolation. Frictional forces associated with electric pull-down forces on the rotor are dominant in the side-drive and harmonic side-drive motors fabricated and tested to date. Air drive and electric excitation have been used in studying these effects. Side-drive micromotors have been successfully operated by a three-phase electrical signal with the rotors air-levitated. With air levitation, successful operation is achieved at bipolar excitations greater than 80 V across 4 µm air-gap motors having eight rotor and twelve stator poles, with only half of the stator poles excited. Motor operation is sustained indefinitely.

1. Introduction

The principles of design, fabrication and operation of electric micromotors have been investigated in recent years [1–4]. This paper discusses related issues in three types of microfabricated variable-capacitance micromotors including top-drive, side-drive and harmonic side-drive designs. These micromotors are fabricated using polysilicon surface micromachining technology in which heavily phosphorus-doped LPCVD polysilicon is used as the structural material, deposited oxide is used as the sacrificial spacer and silicon-rich nitride is used as the electrical insulator. Air drive [5] and electric excitation are used in conjunction with the side-drive and harmonic side-drive micromotors to study the electric interactions of the rotor with the stator and the substrate. Three-phase electric signals are used successfully to spin the side-drive motors while the rotor is air-levitated.

2. Top-drive, Side-drive and Harmonic Side-drive Micromotors

This section presents a general description of the three microfabricated electric motor designs considered in this work. In all cases, the operation of the motors relies on the storage of electrical energy in a variable rotor-stator capacitance. The change in this capacitance as a function of rotor position is proportional to the output torque of the motor.

Figure 1 is an SEM photograph of a three-phase, variable-capacitance, top-drive micromotor. The motor is 300 µm in diameter and is fabricated using a two-level polysilicon process [1, 6, 7]. In this motor design, electrical energy is stored in the air gap formed by the overlap of the stator and rotor poles. The air gap-dimension

Fig. 1. A typical three-phase, variable-capacitance, top-drive micromotor.

Fig. 2. A typical three-phase, variable-capacitance, side-drive micromotor.

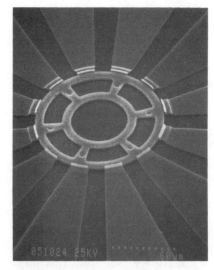

Fig. 3. A typical variable-capacitance, harmonic side-drive micromotor with no center bearing.

(typically 1–2 µm) is controlled by adjusting the thicknesses of the oxide spacer layers above and below the rotor, and the depth of the rotor bushings. The motor operation relies on tangential electrostatic forces tending to align the rotor poles under the excited stator poles.

Figure 2 is an SEM photograph of a three-phase, variable-capacitance, side-drive micromotor. This motor is 100 µm in diameter and is also fabricated using a two-level polysilicon process. Electrical energy is stored in the air gap formed between the rotor and stator pole sidewalls. In contrast to the top-drive design, in which the air-gap dimension is controlled by deposited film thicknesses, the air-gap dimension in the side-drive design is controlled by photolithography and pattern delineation. The motor operation relies on tangential electrostatic forces tending to align the rotor poles with the excited stator poles.

Figure 3 is an SEM photograph of a variable-capacitance harmonic side-drive micromotor, sometimes referred to as a wobble motor. This motor is fabricated using a two-level polysilicon process identical to that of the top-drive micromotor and is 100 µm in diameter. There is no center bearing and the rotor is held in place by a slight overlap (typically 1 µm) of the stator poles. The outer radius of the rotor is 1 µm smaller than the inner radius of the stator poles. This dimension corresponds to the thickness of the oxide spacer covering the sidewalls of the rotor. Electrical insulation of the underside of the stator overhangs is required for this motor design since the rotor contacts the excited stator electrodes during motor operation. Motor operation relies on normal electric forces attracting the rotor to a pair of excited stator poles [8–10]. As the electric excitation travels around the stator poles, the rotor can be made to roll on the inside diameter of the stator poles. Figure 4 is an SEM photograph of a harmonic side-drive motor fabricated in a process similar to that of the side-drive motor. In this case, the center bearing must be fabricated in a way to provide the rotor with the clearance necessary for reaching the stator poles. This is possible if the oxide spacer thicknesses are selected appropriately. This design requires electrical isolation of the rotor-stator sidewalls since the rotor is in contact with the excited stator electrodes during motor operation.

Fig. 4. A typical variable-capacitance, harmonic side-drive micromotor with a center bearing.

3. Fabrication Issues

Polysilicon surface micromachining has been used in the fabrication of passive micromechanisms [6, 7] and micromotors [3, 4]. In general, heavily-phosphorus-doped LPCVD polysilicon is used as the structural material, while deposited oxide (LTO in this work) is used for the sacrificial spacer and silicon nitride is used as the electrical insulator. The fabrication process for the side-drive motors is discussed here with emphasis on the key issues. For the fabrication of top-drive and the harmonic side-drive motors, similarities and differences are pointed out.

Figure 5 shows cross-sectional sketches of a side-drive micromotor at different fabrication process steps. Starting with a silicon substrate, the electrical isolation is first established (Fig. 5(a)). This isolation layer is required to survive the final HF release step of the micromotor and withstand several hundred volt biases in the final device. In this work, we have experimented with two isolation schemes which use stochiometric LPCVD silicon nitride (Si_3N_4) or LPCVD silicon-rich silicon nitride (Si_xN_y) as the HF-resistant isolation layer. Si_3N_4 films have high residual tensile stress which can lead to tensile cracking in thicker films. This property in conjunction with the high HF etch rate of these films limits their potential as isolation layers in micromotor technology [11]. Si_xN_y films have been shown to be a superior alternative in this application [3, 4]. These films can be deposited with very low residual stress [12] and their HF etch rate is significantly lower than the Si_3N_4 films. We have consistently used 2.0-μm-thick Si_xN_y and 1.0-μm-thick thermal oxide/1.0-μm-thick Si_xN_y sandwich layers for successful electrical isolation. This isolation scheme has eliminated electrical breakdown through the substrate in our micromotors.

It has been shown [3, 4] that a conductive polysilicon plate under the rotor and in electrical contact with it enhances motor operation by eliminating electrical pull-down forces on the rotor. This electrostatic shield plate, which is located under the rotor, may be incorporated by deposition and patterning of a thin conductive polysilicon layer (3500 Å in this work) after electrical isolation layer deposition (not shown in Fig. 5). At this point, a 2-μm-thick LTO layer is deposited as the first sacrificial layer. The bushings [6] are patterned and time-etched (nearly 1.5 μm deep) into the oxide prior to patterning and opening the stator anchors (Fig. 5(a)). Note that the bushing is shown as a continuous ring in Fig. 5 but may not be restricted to this design (e.g., the micromotor of Fig. 2 uses six small circular bushings). Next, a 2.5-μm-thick LPCVD polysilicon layer is deposited followed by heavy phosphorus doping. The rotor and stator are patterned in this polysilicon layer (Fig. 5(b)) by reactive-ion-etching (RIE). This step defines the airgap which is the most important performance parameter of the micromotor. In our initial side-drive micromotor designs (Fig. 2), the first LTO layer was patterned into a circular pedestal under the rotor, requiring the stator poles to be patterned over this step (Figs. 5(a, b)). Figure 6 is an SEM photograph showing a close-up of the stator poles patterned over the LTO step in a micromotor similar to that in Fig. 2 (before release). Note that the polysilicon at the foot of the step is cleared long after the polysilicon layer is etched through on the flat surfaces. The required over-etch in this case leads to an enlargement of the air gaps, degrading the performance of the final motor. This problem may be solved by redesigning the stator anchors to eliminate underlying oxide steps where the poly is to be patterned as shown in Fig. 7. Figure 8 shows a close-up of the 1-μm air gap in this motor which is patterned into a 2.2-μm-thick polysilicon film.

At this point a 1-μm-thick LTO layer is deposited and patterned to open the bearing anchor (Fig. 5(c)). The bearing is then formed in a 1-μm-thick polysilicon film which is doped heavily with phosphorus (Fig. 5(d)). The bearing is patterned in an anisotropic RIE etch followed by a short isotropic etch clearing the remaining polysilicon at the foot of the steps. At this point, the devices are released in a 1:1 HF:DI solution for 15 min, rinsed in DI water, rinsed in methanol and dried by a nitrogen gun. We have found that

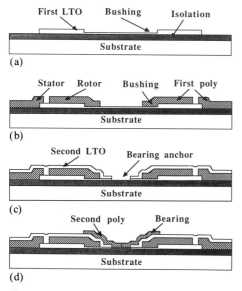

Fig. 5. Basic side-drive motor fabrication process: (a) after first spacer layer is patterned; (b) after rotor and stator are patterned; (c) after second spacer layer is patterned; (d) completed device.

Fig. 6. A close-up view of the stator electrodes patterned over the LTO step in a micromotor (not released) similar to that of Fig. 2.

Fig. 8. SEM photograph showing a close-up of the air gap (1 μm wide) of the motor in Fig. 7.

Fig. 7. SEM photograph showing the rotor and stator of an optimized side-drive motor design (2.2-μm-thick polysilicon).

incorporation of the bushings leading to reduced rotor/substrate contact eliminates long-term rotor stiction problems. Devices tested up to 5 months after release are readily spun by a jet of air.

The top-drive motor fabrication process differs from that of the side-drive one in three ways. First, the rotor is patterned in the first polysilicon deposition while the stator poles are fabricated in the second polysilicon layer (Fig. 1). Second, the air gap is defined by the LTO thicknesses and the rotor bushing depth, all of which are easy parameters to control. Third, proper planarization is required after the second LTO deposition (which in turn requires an excessively thick oxide deposition). Otherwise, the stator poles fabricated over rotor poles are at a higher elevation than the remaining stator poles (Fig. 1). The planarization requirement is a severe fabrication disadvantage of this motor as compared to the other two designs.

The fabrication requirements for the two harmonic side-drive motors in Figs. 3 and 4 are similar to those of the top-drive and side-drive motors, respectively. Note that in the case of the harmonic side-drive motor of Fig. 3, no planarization is required since all of the stator poles overlap the rotor. Furthermore, in the harmonic side-drive motor of Fig. 4, the second LTO deposition thickness is selected so as to ensure that the rotor can reach the excited stator electrodes.

4. Electric Issues

A general capacitive equivalent circuit shown in Fig. 9 is best suited for discussing the electrical characteristics of the three motor designs under study. The capacitances of most importance in motor operation are C_{SR} and C_{BR}. $C_{SR}(\theta, z)$ denotes the variable rotor-stator capacitance associated with the air gap, where θ specifies rotor

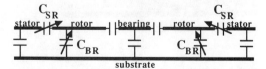

Fig. 9. A capacitive equivalent circuit for the variable-capacitance motors of Figs. 1–4.

angular position and z specifies rotor vertical position. $C_{BR}(z)$ denotes the rotor-substrate capacitance and consists of two capacitances in parallel associated with the rotor bushings and the remaining rotor area (i.e., the bushing and the rotor surfaces are at different heights). The bearing-rotor capacitance may be neglected since the rotor is free to contact the bearing and does so during operation. The stator-substrate capacitances do not affect motor operation (in our work, electrical breakdown has been observed to take place in the air gaps before isolation breakdown can take place).

Note that the top-drive motor can provide higher output torque as compared to the other two designs due to larger changes in its rotor–stator capacitance (Fig. 1). However, associated with C_{SR} and C_{BR} there are large vertical forces tending to clamp the rotor up to the stator poles or down to the substrate. This rotor instability in the top-drive motor is a severe limitation. Furthermore, the stator pole cantilevers are flexible and may be pulled down by electrostatic forces during operation.

In the side-drive design, the rotor instability is in the lateral direction. However, this rotor instability is structurally compensated by the bearing which keeps the rotor nearly centered between the excited electrodes. The output torque is drastically reduced due to the small rotor-stator overlap capacitances. There can be vertical forces on the rotor associated with C_{SR} and C_{BR}. The vertical force component associated with C_{SR} tends to align the rotor with the stator electrodes in the vertical direction. Since upon release the rotor in our devices is 0.5 μm lower than the stator electrodes, this force tends to pull the rotor up. However, we have found that the vertical force associated with C_{BR} pulling the rotor down in our devices is dominant (we will not consider the vertical pull-up force associated with C_{SR} on the rotor in the current devices). To clarify the origin of the pull-down force associated with C_{BR} on the rotor, consider a bipolar voltage excitation on opposing stator poles while the rest of the stator poles and the substrate are at electrical ground. Since, due to the bearing clearance, the rotor may not be centered between the excited stator poles, its potential is not necessarily zero. This vertical force on the rotor may be eliminated by insuring ohmic contact of the rotor to the substrate (or equivalently an electrostatic shield plate under the rotor [3, 4]).

In the harmonic side-drive motor, lateral rotor instability is not an issue. The output torque of the motor is higher since motor operation relies on normal electric forces rather than tangential ones. A gear-reduction between the electric signal frequency and the rotor speed is inherent in this design [8]. The two C_{SR} capacitances are associated with adjacent electrodes rather than opposing ones. Vertical force issues on the rotor are similar to those of the side-drive motor.

5. Results

In this work, the top-drive micromotors are designed with four rotor and twelve stator or six rotor and eighteen stator (Fig. 1) poles. Three rotor radii of 50, 100 and 150 μm are implemented with 10, 15 and 25 μm radial overlap with the stator poles, respectively. Initial devices did not include electrical isolation on the underside of the stator poles, which was incorporated in later devices. Electrical testing of the initial devices typically resulted in motor destruction due to the rotor shorting out the opposing stator electrodes. The later devices which did include electrical isolation on the underside of the stator poles were not thoroughly tested due to the major fabrication and design shortcomings of this motor, namely, required planarization and rotor instability.

The side-drive motors are fabricated with eight rotor and twelve stator or twelve rotor and eighteen stator poles (i.e., 3:2 designs; see Figs. 2 and 7). The rotor radius is 50 μm and its thickness is 2.5 μm. The smallest lateral air gap achieved in initial devices was 4 μm because of the over-etch problem discussed above. As indicated in Fig. 8, this problem has been solved in devices currently being processed. Air gaps as small as 1 μm have been achieved. The rotors include structural bushings that reduce contact area with the substrate. The bushing heights are selected (typically 1.5 μm) such that, upon release, the rotor is at a lower level than the stator electrodes by 0.5 μm. Two bearing designs are investigated which include that discussed above (hereafter referred to as type I) and another reported in refs. 3 and 4 (hereafter referred to as type II). For the type II bearing designs, the nitride solid lubrication [3, 4] is not used here.

The 3:2 design of the side-drive motors was selected to enhance motor performance. Initial motor designs (Fig. 2) are clearly not optimized. Current 3:2 side-drive motor designs (Fig. 7) are being optimized following guidelines used in the design of variable-reluctance motors [13]. In the design of Fig. 7, the rotor pole width to pole pitch ratio is 0.4 while the gap spacing is as small as possible. Furthermore, for a linear unsaturated system (which applies to an air-gap variable-capacitance motor even more closely than to a variable-reluctance motor), it has been shown [14] that the above rotor pole width to rotor pole pitch ratio and a rotor pole pitch to gap-size ratio of 8 represents a nearly optimum geometry. These results are based on the assumption that the motor is infinitely long in the axial direction, which is not the case for the micromotors described here. However, they remain a reasonable guide for choosing rotor and stator tooth geometries.

The harmonic side-drive motors are designed with twelve (Fig. 3) and eighteen stator (Fig. 4) poles. The rotor radius is 40, 50 and 60 μm while the stator pole pattern radius is 1 to 6 μm larger than that of the rotor in each case. The rotor is 2.5 μm thick and incorporates bushings similar to that of those of the side-drive micromotors.

Electrical testing of the side-drive and harmonic side-drive motors has concentrated on diagnostics identifying the key issues affecting motor operation. Frictional forces originating from the vertical electric forces on the rotor have found to be dominant in our devices. Device testing has been carried out on motors similar to ones shown in Figs. 2 and 4 fabricated in the same process on the same wafers. For these experiments the motor bearing in Fig. 4 is identical to that of the side-drive motor in Fig. 2 and this motor is not used in the wobble mode (i.e., operated as a side-drive motor with a non-salient rotor). In each case, we have fabricated the motors using the two bearing designs identified above (i.e., types I and II). The motors with type I bearing design are studied both with and without electrostatic shielding of the rotor. All motors with type II bearing design include electrostatic shielding of the rotor. Air drive using a micropositioned biological recording micropipette [5] and electrostatic excitation are used to study motor operation. All experiments described below were carried out on 4-μm air-gap motors similar to those in Figs. 2 and 4.

For devices with type I bearing and no electrostatic shield plate, vertical forces on the rotor are sufficient to prevent motor operation with electric excitation. Occasional rotor movement is observed at bipolar voltages above 100 V if the bearing and the substrate are electrically grounded. The rotor may be discharged if it touches the bearing and the vertical force is momentarily eliminated until the rotor loses its contact with the bearing again. In experiments conducted in which the motor in Fig. 4 is first spun with air (with speeds recorded as high as 10 000 rps using the set-up reported in ref. 5), the rotor can be slowed down to a total stop with constant bipolar excitation of 120 V on two opposing stator poles. Similar devices which include rotor electrostatic shielding and incorporate type I or type II bearing designs are slowed down but not stopped under similar test conditions. These experiments demonstrate that electrostatic shielding of the rotor in our devices has reduced the pull-down force on the rotor but has not eliminated it. Lack of proper electrical contact between the rotor and the shield plate may be responsible for the existing vertical forces on the rotor.

It is possible to adjust the air drive to effectively levitate the rotor without spinning it. This approach may be used to compensate for the electrostatic pull-down of the rotor when electric excitation is applied. With air levitation applied in this manner, three-phase bipolar electrical excitation has led to successful motor operation. These experiments have concentrated on the 8:12 rotor:stator-pole side-drive motors with only half of the stator poles excited. In each experiment, air levitation was adjusted to just overcome the rotor pull-down forces. For devices with type I bearing design and no shield plate, motor operation is achieved with a bipolar voltage excitation larger than 140 V. For devices with type I bearing design, which include a rotor electrostatic shield plate, motor operation is achieved with a bipolar voltage excitation larger than 100 V. For devices with type II bearing design and rotor electrostatic shield plate, motor operation is achieved with a bipolar voltage excitation greater than 80 V.

The top speed of the motors has not been measured currently but is greater than 300 rpm. Motor operation is sustained indefinitely. In our devices, electric breakdown across the air gaps has been the main cause of failure. This breakdown takes place randomly at bipolar excitations greater than 160 V.

6. Conclusions

Three microfabricated variable-capacitance motor designs have been considered in this work. The top-drive motor architecture suffers from design and fabrication difficulties associated with rotor instability and required planarization steps, respectively. However, it can provide higher output torque than the other two designs. The side-drive motor design overcomes these design and fabrication difficulties but its output torque is drastically reduced. The harmonic side-drive motor design recovers some of the output torque for an increase in fabrication complexity (i.e., required rotor-stator isolation).

Frictional effects originating from electrostatic pull-down forces on the rotor have been dominant in the devices fabricated and tested to date. The presence of pull-down forces on the rotor in the devices with a shield plate indicates the lack of effective electrical contact between the rotor, bearing and the shield plate. Air drive and levitation have been used to study these undesired forces and successfully operate the motors.

Acknowledgements

The authors wish to acknowledge the initial work of Theresa Lober and Professor Roger T. Howe on the top-drive micromotor, technical discussions with Professor Martin A. Schmidt and assistance from Daniel Sobek. The authors wish to thank Dr Kaigham J. Gabriel for his suggestions on the air-drive instrumentation and Dr Ramaswamy Mahadevan for his input on numerical simulation of the field problems involved. This work was supported in part by the National Science Foundation, Charles Stark Draper Laboratory, Inc. and an IBM Fellowship (Mehregany). Fabrication was carried out in the MIT Integrated Circuit Laboratory (ICL). The authors wish to thank the ICL staff, in particular Michael Schroth and Joe Walsh, for their assistance. The authors wish to express their appreciation for starting wafers with silicon-rich nitride from the Berkeley Sensor and Actuator Center.

References

1 S. F. Bart, T. A. Lober, R. T. Howe, J. H. Lang and M. F. Schlecht, Design considerations for micromachined electric actuators, *Sensors and Actuators, 14* (1988) 269-292.
2 S. F. Bart and J. H. Lang, An analysis of electroquasistatic induction micromotors, *Sensors and Actuators, 20* (1989) 97-106.
3 L. S. Fan, Y. C. Tai and R. S. Muller, IC-processed electrostatic micro-motors, *Proc. 1988 IEEE Int. Electron Devices Meet., IEEE Electron Devices Soc., San Francisco, CA, U.S.A., Dec. 11-14, 1988*, pp. 666-669.
4 Y. C. Tai, L. S. Fan and R. S. Muller, IC-processed micro-motors: design, technology, and testing, *Proc. Micro Electro Mechanical Systems, IEEE Robotics and Automation Council, Salt Lake City, UT, U.S.A., Feb. 20-22, 1989*, pp. 1-6.
5 K. J. Gabriel, F. Behi, R. Mahadevan and M. Mehregany, In situ measurement of friction and wear in integrated polysilicon micromechanisms, *Sensors and Actuators, A21-A23* (1990) 184-188.
6 M. Mehregany, K. J. Gabriel and W. S. N. Trimmer, Integrated fabrication of polysilicon mechanisms, *IEEE Trans. Electron Devices, ED-35* (1988) 719-723.
7 L. S. Fan, Y. C. Tai and R. S. Muller, Integrated movable micromechanical structures for sensors and actuators, *IEEE Trans. Electron Devices, ED-35* (1988) 724-730.
8 W. S. N. Trimmer and R. Jebens, Harmonic electrostatic motors, *Sensors and Actuators, 20* (1989) 17-24.
9 S. C. Jacobsen, R. H. Price, J. E. Wood, T. H. Rytting and M. Rafaelof, A design overview of an eccentric-motion electrostatic microactuator (the Wobble Motor), *Sensors and Actuators, 20* (1989) 1-15.
10 H. Fujita and A. Omodaka, The fabrication of an electrostatic linear actuator by silicon micromachining, *IEEE Trans. Electron Devices, ED-35* (1988) 731-734.
11 T. A. Lober and R. T. Howe, Surface-micromachining processes for electrostatic microactuator fabrication, *Tech. Digest, IEEE Solid-State Sensor and Actuator Workshop, Hilton Head Island, SC, U.S.A., June 6-9, 1988*, pp. 59-62.
12 M. Sekimoto, H. Yoshihara and T. Ohkubo, Silicon nitride single-layer X-ray mask, *J. Vac. Sci. Technol., 21* (1982) 1017-1021.
13 M. R. Harris, A. Hughes and P. J. Lawrenson, Static torque production in saturated doubly-salient machines, *Proc. IEEE, 122* (1975) 1121-1127.
14 K. C. Mukherji and S. Neville, Magnetic permeance of identical double slotting, *Proc. IEEE, 118* (1971) 1257-1268.

FRICTION AND WEAR IN MICROFABRICATED HARMONIC SIDE-DRIVE MOTORS

Mehran Mehregany, Stephen D. Senturia, and Jeffrey H. Lang

Microsystems Technology Laboratories
Laboratory for Electromagnetic and Electronic Systems
Department of Electrical Engineering and Computer Science
Massachusetts Institute of Technology
Cambridge, Ma 02139

ABSTRACT

This paper describes polysilicon variable-capacitance rotary harmonic side-drive micromotors, presents results from operational and frictional studies of these motors, and for the first time, reports *in situ* quantitative studies of wear under electric excitation. Voltages as low as 26V across 1.5 μm gaps are sufficient for operating these motors. Frictional force estimates at the bushings, 0.15 μN, and in the bearing, 0.04 μN, are obtained from measurements of stopping voltages. Extended operation of these motors to near 100 million wobble cycles at excitation frequencies of 10,000 rpm and 25,000 rpm, for operational durations of 150 hours and 71 hours, respectively, are studied. The results indicate that bearing wear is significant and results in changes in the gear ratio of the motors by as much as 20%. Typical gear ratios are near 90 at the start of motor operation and decrease to near 70 as the bearings wear out.

1. INTRODUCTION

Operation of variable-capacitance ordinary side-drive micromotors was first reported by Fan, Tai, and Muller [1,2]. Operation of similar micromotors using an air-levitation assist was reported by us shortly thereafter in [3]. However, in [3], failure in motor operation persisted without the air-levitation assist. At MEMS 1990 [4], we identified the native oxide of polysilicon as the cause of this motor operation failure, asserted that micromotor release and testing techniques are the determining factors for successful electrical operation of the motors, reported the operation of ordinary side-drive motors without air levitation, and presented results from frictional studies of the ordinary side-drive motors.

Additionally, at MEMS 1990 [4], we reported the development and demonstration of novel rotary variable-capacitance harmonic side-drive (hereafter referred to as wobble) micromotors. This paper further describes the wobble micromotors reported in [4], presents results from operational and frictional studies of these motors, and for the first time, reports *in situ* quantitative studies of wear under electric excitation.

2. MICROFABRICATED WOBBLE MOTORS

Figure 1 is a SEM photograph of a typical released wobble micromotor studied in this paper. Figure 2 is a schematic drawing of the cross-sectional view of the motor prior to release, which is when the low-temperature oxide (LTO) layers are dissolved in HF. Motor structural components are fabricated from heavily-phosphorus-doped polysilicon. After release, the rotor is supported on the bushings and is free to rotate about a center-pin bearing contacting an electric shield under the rotor. During motor operation, the rotor is intended to be in electrical contact with the shield positioned beneath it through mechanical contact at the bearing or at the bushing supports. As reported in [4], native oxide formation on the polysilicon surfaces can disrupt the intended electrical contacts and lead to the clamping of the rotor to the shield beneath it via electrostatic attraction, preventing motor operation. Proper release and testing techniques aimed at minimizing the native oxide formation are required for successful motor operation, as discussed in detail in [4], and are briefly review in Section 3.

The central feature of the design presented here is that the rotor wobbles around the center bearing post rather than the outer stator for conventional wobble motors [5-8]. A schematic

Fig. 1. A 12 stator-pole, 2.5 μm-gap, 100 μm-diameter, wobble micromotor with bushing 2 (i.e., the three small rotor indentations near the bearing).

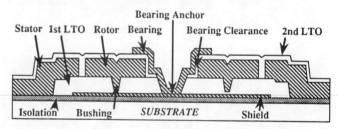

Fig. 2. Motor cross-section after fabrication completion and prior to release.

drawing showing the plan view of the design used here is shown in Fig. 3. In this design, the rotor lateral movement (i.e., wobble distance) is equal to the clearance in the bearing which is the difference between the bearing and inner rotor radii. This clearance, which we estimate to be near 0.3 μm, is specified by the thickness of the second sacrificial oxide. Furthermore, by ensuring a bearing clearance smaller than the nominal air-gap size, which is greater than 1.5 μm in our motors, the need for insulation between the rotor and the stator is eliminated. In this design, the rotor rotates in the same direction as the excitation signal and the point of contact.

A significant advantage of the wobble motor is that its drive torque is proportional to the motor gear ratio, n. The gear ratio relates the angle of rotation of the stator excitation to the angle of rotation of the rotor. For the wobble motor design presented here, n is given by

$$n = \frac{r_b}{\delta}, \qquad (1)$$

where r_b is the radius of the bearing and δ is the bearing clearance. Let θ designate the angle between the diameter containing the contact point and the energized electrode, then the wobble motor drive torque, $T_w(\theta)$, due to a single excited stator pole at angular position θ, estimated from in-plane two-dimensional simulations when the rotor is electrically grounded, is given by

$$T_w(\theta) = \frac{\epsilon_o t V^2}{2} \tau_w(\theta) \, n, \qquad (2)$$

where ϵ_o is the permitivity of air, t is the rotor thickness, and V is the applied excitation [8]. In (2), $\tau_w(\theta)$ is a normalized torque calculated by the simulations and specified by the motor design which includes effective air-gap size, pole width, and pole pitch [8]. The effective air-gap size is the actual distance of the rotor to the excited stator pole and is a function of the nominal air-gap size and the bearing clearance. Note that the wobble motor drive torque is multiplied by the gear ratio, n. The torque obtained from multiply-excited stators can be derived by superposition.

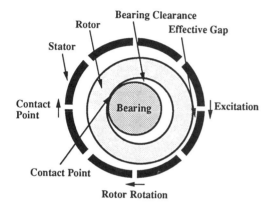

Fig. 3. The wobble micromotor using a center-pin bearing to eliminate the need for stator insulation.

A small bearing clearance in the wobble motor design used here is desirable because it reduces slip during motor operation, reduces rotor wobble, and reduces the risk of breakdown in the air gap. Recognizing that smaller bearing clearances decrease drive torque by increasing the effective air-gap size (see Fig. 3), we have increased the bearing radius to enhance the drive torque through further increase of the gear ratio.

3. FABRICATION AND TESTING

The wobble motor presented here, using a small bearing clearance and eliminating the need for rotor-stator insulation, simplifies fabrication and allows for a single fabrication process suitable for both wobble and ordinary side-drive motors. Details of this fabrication process, the polysilicon native oxide issue, as well as proper release and testing techniques are documented in detail elsewhere [4,9] and summarized here.

The motor fabrication process is discussed here in reference to the cross-sectional schematic of Fig. 2. Initially, substrate isolation is established using a 1 μm-thick LPCVD silicon-rich nitride layer over a 1 μm-thick thermally grown SiO_2 film [1-3]. A thin (3500Å) LPCVD polysilicon film is deposited, heavily doped with phosphorus, and patterned to form the shield. Stator anchors and bushing molds are patterned in the first sacrificial LTO layer which is 2.3 μm thick. The rotor, stator, and air gaps are patterned into a 2.5 μm-thick, heavily-phosphorus-doped, polysilicon layer using RIE. The final rotor-stator polysilicon thickness is 2.2 μm since a patterned thermally grown oxide mask (not shown in Fig. 2) is used for the RIE etch of the polysilicon. Note that the inside radius of the rotor patterned at this point would correspond to the bearing radius in the final device. A bearing radius of 18 μm is used for the wobble motors. The second sacrificial LTO layer is deposited, providing an estimated 0.3 μm coverage on the rotor inside radius side-walls, and patterned to open the bearing anchor. A 1 μm-thick LPCVD polysilicon film is deposited, heavily doped with phosphorus, and patterned to form the bearing. The motor is released by dissolving the sacrificial LTO in HF. During motor operation, the bushings are intended to provide electrical contact between the rotor and the shield as they slide over the shield.

The shield which is positioned under the rotor was suggested in [2] to be effective in the elimination of the rotor clamping forces by electrostatic shielding of the rotor from the substrate. However, even with electric shields incorporated in our devices, air levitation was found to be necessary in our previous work [3] to overcome frictional forces associated with the electrostatic clamping of the rotor to the shield beneath it. The rotor clamping was attributed to a lack of proper electrical contact between the rotor, the shield, and the bearing. This electrical contact is necessary to ensure that the rotor and the shield are at the same electric potential, thereby eliminating the clamping forces otherwise caused by the electric field between them. However, we believe that native oxide formation on the bushings, the bearing, and the shield surfaces results in a loss of contact, leading to an electric field between the rotor and the shield. This results in a clamping of the rotor to the shield (which was compensated for by air levitation in our previous work [3]). The reader is referred to [4] for a detailed discussion of these issues, including supporting studies, and to [10] for a possible model for calculating the electrostatic rotor clamping forces. With proper release and testing techniques which are summarized below and are directed at minimizing the formation of this oxide, we can readily operate the motors without air levitation for extended periods of time.

Our release method consists of 15 minutes in a 49% (by weight) commercially-available HF solution, 2 minutes in a DI rinse, 5 minutes in a 1:1 H_2SO_4:H_2O_2 (pirana) clean, 10 minutes in a DI rinse, 30 seconds in the HF solution, 90 seconds DI rinse, and a nitrogen dry. The initial 15 minute oxide etch in HF is required to release the various motors on the die. The 5 minute pirana clean is effective in cleaning the exposed polysilicon surfaces from organic as well as ionic contaminants. The second HF etch removes the thin oxide formed on the polysilicon surfaces during the pirana clean. The final DI rinse time is critical. Rinse times under 60 seconds do not produce reliable results. Long rinse times may be detrimental since the native oxide growth rate is greater in DI water than in air. A 90 second rinse time has been typically used in this work with good results.

For all experimental studies reported here, the above release process is used. Immediately after release, the motors are stored and tested in nitrogen. In nitrogen, the rotor clamping failure mechanism is eliminated. In this case, the operation of wobble motors is repeatable (see measurement results in the next section) and has been studied for extended times up to six days (see Section 5).

4. OPERATIONAL AND FRICTIONAL STUDIES

Design parameter permutations including bushing style and air-gap size are incorporated into the fabrication of wobble motors. Electrical actuation for all of these design permutations has been confirmed. The wobble motor measurements reported in this section were carried out on 12 stator-pole, 100 μm-diameter motors (similar to that in Fig. 1) with 1.5 μm and 2.5 μm gaps. For these motors, the stator pole width and pitch are 27 and 30 degrees, respectively.

The wobble motors are studied using a six-phase, unipolar, square-wave excitation. Therefore, six stator poles are excited (independent of the total number of stator poles available) with a center-to-center angular separation of 60 degrees (e.g., every other stator pole for the motor in Fig. 1). This stator pole excitation arrangement results from a limitation in our power supply which can excite only six independent phases. Two signal excitations were studied which differ in the number of the stator poles excited at one time. In the first case, excitation A, the excitation followed the pattern '000001', '000011', '000010', '000110', and so on, where a '0' indicates a grounded stator pole and a '1' indicates an excited one. In the second case, excitation B, the excitation followed the pattern '000001', '000010', '000100', and so on. Unless otherwise specified, the rotor is electrically grounded through the shield. Note that excitation A approximately simulates the condition in which all twelve stator poles are excited with a signal excitation similar to that of excitation B. Excitation voltages as high as 150V across 1.5 μm-wide air gaps (i.e., electric field intensities of 1×10^8 V/m) are routinely used without electric field breakdown in the gaps. Using excitation B, the wobble motors have been operated at speeds estimated up to 700 rpm.

Measurement of starting and stopping voltages is performed on the wobble motors when operating in nitrogen. These measurements are performed by operating the motor at near 6 rpm independent of the excitation pattern. This requires that excitation A be switched twice as fast as excitation B, or approximately 10 ms and 20 ms, respectively. We measure the minimum voltage required for sustaining motor operation (stopping voltage) as well as the minimum voltage require to restart the motor (starting voltage). We have found that the starting and stopping voltages are nearly the same when the excitation is applied within a few seconds of previous rotor movement. Therefore, we insert a 30 second delay for the starting voltage measurements. At the slow operational speeds used for the stopping and starting voltage measurements, the rotor motion is step-wise since the settling time associated with the rotor transient response is much shorter than the switching times above. Therefore, the stopping and starting voltage measurements are only different in the length of time during which the rotor is at rest.

Freshly released motors which have not been previously operated are used in the measurements. The motors are typically operated for two minutes before taking measurements. The measured stopping voltages are reproducible to within 3% in the same testing session. The scatter becomes 10% when the same motor is tested in different sessions with a re-release prior to each session, or for devices on different dies. The starting voltage measurements are far more scattered. This scatter can be as high as 30% in the same session and from session to session, which indicates the complexity of static friction characteristics in the micromotors even after only 30 seconds at rest.

We have used two bushing designs in the wobble motors which are at an equal radius (lever arm), 29 μm, but have different apparent areas of contact. The apparent area of contact is an estimated 180 μm^2 for bushing 1 (which is of a continuous ring shape) and 6 μm^2 for bushing 2 (Fig. 1).

Table I summarizes the experimentally measured stopping and starting voltages as well as the corresponding estimated torques. All values listed in Table I are averages of the two bushing designs since no dependence was found on the bushing apparent area of contact, in agreement with the results reported in [4,11]. Each entry in Table I is an average computed from data on three motors (per bushing type) from different dies of the same wafer tested at different times. The wobble motors operate even when the rotor is not electrically grounded. Therefore, it is not necessary to ground the shield (and therefore the rotor) as long as the rotor is in electrical contact with the shield. This is to be expected since as long as the shield and the rotor are in electrical contact, regardless of their electric potential, there is no electric field and hence no clamping force between them. However, as seen from lines 5 and 6 in Table I, higher voltages are required for initiating and sustaining motor operation. The reason is that without grounding, the rotor floats to a potential in the direction of the excited stator voltage so that a reduced voltage appears across the gap. Torque is thereby reduced for a given voltage. Since the rotor potential is not known in this case, corresponding torque values cannot be estimated from this type of measurement.

The measured voltages are used in conjunction with two-dimensional field simulations in the plane of the motors to estimate torque. Specifically, single-phase drive-torque curves for

ID	Gap (μm)	Excit. Pattern	Rotor Bias	Stopping		Starting	
				Voltage (V)	Torque (pN-m)	Voltage (V)	Torque (pN-m)
1	1.5	A	GND	26	5	35	9
2	1.5	B	GND	33	6	52	14
3	2.5	A	GND	37	4	50	7
4	2.5	B	GND	49	5	70	10
5	2.5	A	FLT	61	-	81	-
6	2.5	B	FLT	80	-	105	-

Table I. Measured stopping and starting voltages as well as estimated torques for the wobble micromotors.

the motors studied here are calculated from the wobble motor analysis in [8]. A bearing clearance of 0.25 μm is assumed, accounting for the roughness of the rotor inside radius (see Section 5 for details). Since the air gap of the micromotors is comparable to the rotor thickness, the contribution of the fringing fields in the axial direction must be accounted for in interpreting the experimental results. A correction factor calculated from field simulations in the axial plane of the motor (see [9] for details) is used to approximately account for the contribution of the fringing fields in the axial direction. These correction factors are applied to the single-phase drive-torque curves since the system may be considered approximately invariant in the angular direction over a stator pole span. For the 1.5 μm air-gap motors here, the corresponding correction factor is 1.95 while for the 2.5 μm air-gap motors this value is 2.27 [9].

Furthermore, in estimating torque values from the voltage measurements, an experimentally measured value for the gear ratio, n, is used rather than the value predicted by Eqn. (1). In the low excitation voltage range, corresponding to the stopping and starting voltage measurements, the rotor slips frequently, resulting in an apparent gear ratio higher than that predicted by Eqn. (1). The experimentally measured gear ratio in this voltage range is 85 for excitation A and 100 for Excitation B. Note that excitation B produces twice as much rotor angular rotation per excitation step than excitation A and therefore more slip. As a side point, it is worth mentioning that for motor speeds in the range of 5 rpm to 360 rpm, the measured gear ratio is constant.

In our torque analysis, the stopping rotor position has been assumed to be at the zero torque position with respect to the excited stator pole (or poles for excitation A). Therefore, the rotor starting position with respect to the next excited stator pole (or poles for excitation A) is known. In practice, it is difficult to experimentally determine the rotor position since the wobble distance (or rotor lateral motion) is very small. The assumed rotor position and the measured excitation voltage are used in conjunction with the field analysis described above to estimate the torques in Table I. Note that the wobble motor model used here predicts consistent torques from the stopping voltage measurements with variations in air-gap size and excitation pattern. In all recorded cases, the starting torques are higher than the stopping torques. Furthermore, the required starting torques are not as consistent as those of stopping torques, in agreement with results in [4]. The estimated averages for the stopping torque, 5±1 pN-m, and starting torque, 10±3 pN-m, are independent of the apparent bushing contact area. The average starting torque is twice that of the average stopping torque, in agreement with similar studies from the ordinary side-drive motors reported in [4].

In [4], we used stopping voltage measurements on ordinary side-drive motors to study frictional behavior of varying bushing styles and sizes. The estimated bushing frictional force in that study was 0.15 μN. For both ordinary and harmonic side-drive motors, the bushing operations produce sliding friction. Therefore, we have used a value of 0.15 μN for the bushing frictional force to estimate the bearing frictional force for the wobble motors. The lever arm of the bushings in the wobble motors is 29 μm while the bearing radius (or the lever arm of the bearing frictional force) is 18 μm. Since the frictional torques at the bearing and the bushings should add up to 5 pN-m, which is the average stopping torque, the bearing frictional force can be estimated at 0.04 μN. Due to its larger lever arm, the bushing frictional force accounts for 87% of the total frictional torque of 5 pN-m. Note that the bearing frictional force estimate reported here is consistent with that obtained for the side-drive motors in [4]. In other words, as would be expected, the rolling action in wobble motor bearing produces less friction than that of the sliding action in the side-drive motor bearing.

5. WEAR STUDIES

Because the gear ratio of the wobble micromotor is equal to the bearing radius divided by the bearing clearance (see Eqn. (1)), under extended operation, changes in the gear ratio (reflected in changes in motor speed) can be a direct measure of wear in the bearing. We have used this concept to measure wear in the bearing of the wobble motors. Our experimental procedure is as follows. Freshly released motors are used and motor speed is measured periodically from frame by frame examination of a recorded video. A standard 8mm video system is used which exposes 30 frames/sec. The motor speed is measured by counting the number of frames required for the motor to make five turns. Note that the speed measurement accuracy increases by increasing the number of motor turns for which the frames are counted.

The following definitions are needed before proceeding. Excitation frequency designates the number of cycles per unit time that the electrical signal travels on the stator and is related to the motor speed by the gear ratio. A wobble cycle is when the point of contact in the bearing has made one complete turn. Since the stator excitation and the point of contact rotate at the same rate, the number of wobble cycles can be calculated by multiplying the excitation frequency by the duration of operation. For a given excitation frequency, Ω, for any two points in time during motor operation (designated by subscripts 1 and 2),

$$\Omega = n_1 \omega_1 = n_2 \omega_2, \quad (3)$$

where ω is the motor speed. Therefore, the gear ratio as a function of wobble cycles can be calculated from the speed measured as a function of time. Note that as the bearing wears out, the motor speeds up which corresponds to a reduction in the gear ratio. In theory, the gear ratio is related to the bearing clearance by Eqn. (1). However, in practice, we have found that the gear ratio, n, is given by:

$$n = G(V) + \frac{r_b}{\delta}, \quad (4)$$

where $G(V)$ is an excitation dependent offset in the gear ratio, accounting for rotor slip. For any two points in time during motor operation, the following expression can be derived:

$$\frac{\delta_2 - \delta_1}{\delta_1} = \frac{\delta_2}{r_b}(n_1 - n_2). \quad (5)$$

An implicit assumption in deriving Eqn. (5) is that the slip term remains the same.

We have studied the extended operation of four identical wobble motors (in nitrogen) under three variations of excitation B involving differing voltage amplitude and frequency. A 90V excitation at 10,000 rpm frequency, corresponding to 10,000 wobble cycles per minutes, was used to test the first two motors. Figure 4 shows the data for the gear ratio as a function of wobble cycles for these two motors, Motor 1 and Motor 2. Motor 1 was continuously operated for nearly 30 million wobble cycles over 50 hours at which time it was dismantled for SEM inspection. Motor 2 was continuously operated for nearly 90 million wobble cycles over 150 hours at which time it was dismantled for SEM inspection. Figure 5 shows similar data for the gear ratio as a function of wobble cycles for Motor 3 and Motor 4. Both motors were continuously operated for nearly 100 million wobble cycles using an excitation frequency of 25,000 rpm, which corresponds to a 2.5 times increase in the operational speed as compared to the tests of Fig. 4. A 90V excitation was used for Motor 3 and a 120V excitation for Motor 4.

Our measurements indicate three regions in the bearing wear characteristics of the wobble motors. The first region corresponds to the initial 0.3-0.5 million wobble cycles and is the burn-in period. At the start of the burn-in period, the gear ratio is typically near 90. As Figs. 4 and 5 indicate, except for Motor 2, the gear ratio increased (or the motor slowed down) in this region for the remaining three motors. It is likely that we missed detecting an increase in the gear ratio for Motor 2 by not taking enough data points in the initial stage. Furthermore, three of the four motors stopped once and the fourth (Motor 3) stopped several times in this region. After a stop, a momentary jet of nitrogen directed at the rotor was used to restart the motors. It is possible that the relatively larger number or size of the particles, produced during the initial (as compared to later) wear of the asperities on the rotor inner radius, is responsible for the observed behavior during the burn-in period. By filling in the bearing clearance or increasing slip in the bearing, these wear particles can lead to an apparent increase in the gear ratio and cease motor operation.

After the burn-in period, the motors continuously operate until the experiment is ended at a desired point. In the second region, which we call the break-in period, the general behavior for all four motors is similar, even though the detailed shape of the response varies. During the break-in period, the gear ratio decreases (i.e., the motor speeds up) significantly as the rotor inner radius asperities wear out, smoothing that surface and increasing the bearing clearance. By the end of the break-in period, the gear ratio is typically 70 to 76, reaching a plateau. This corresponds to nearly a 20% change in the gear ratio from the start of motor operation. In terms of motor speed, for example, Motor 1 operates at near 109 rpm at the start of operation and at near 137 by the end of the break-in period. Motor 3 operates at near 286 rpm at the start of operation and at near 357 rpm by the end of the break-in period. Note that the break-in period is under 30 million wobble cycles for three of the motors and up to 70 million wobble cycles for Motor 3.

After the break-in period, the gear ratio reaches a plateau at which point we enter the third region in the wear characteristics. Note that in this region, for Motor 2 and Motor 4, the gear ratio eventually increases (the motor speed decreases) with further operation. This gear ratio increase is attributed to an increase in rotor slip. This is corroborated by the fact that increasing the excitation voltage in this region of operation returns the gear ratio to the plateau value for the corresponding motor. We have not been able to identify the cause of rotor slip in this region of operation. It may be attributed to increased friction at the bushings or increased slip in the bearing. SEM inspection of the bushing surfaces after the wear tests have not given additional information. Note that Eqn. (5) is no longer accurate under these conditions since the slip term is now increasing significantly with motor operation.

As seen from Fig. 4, for two identical motors under identical excitations, the detailed wear characteristics during the break-in period may be significantly different. However, note that Motor 4 (Fig. 5) exhibits the most rapid wear during the break-in period as compared to the remaining three motors. This is to be expected, since the higher excitation voltage used in that test as compared to the remaining three, results in higher contact forces in the bearing. Furthermore, comparing Figs. 4 and 5 suggests that the higher operational speed results in increased wear.

As seen above, by the end of the break-in period, the overall gear ratio decreases (indicating a speed increase), corresponding to an increase of several hundred angstroms in the bearing clearance. The bearing clearance which is specified by the second sacrificial oxide thickness is estimated at 0.3 μm. However, the effective bearing clearance may be different due to the roughness of the RIE patterned rotor inside radius sidewalls (which in turn results in roughness of the bearing post walls). SEM inspection of unoperated motors reveal that the rotor inner radius roughness is as high as 600 Å. SEM inspection of the inner

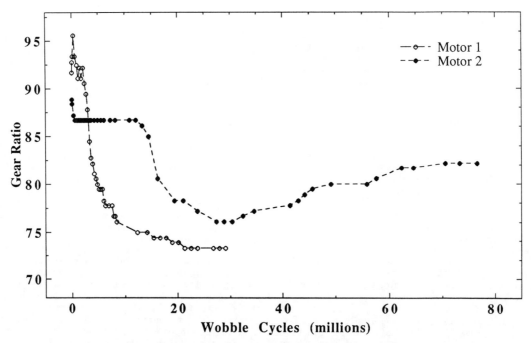

Fig. 4. Wear data for Motor 1 and Motor 2 operated at an excitation voltage of 90V and frequency of 10,000 rpm. Note that a decrease in gear ratio indicates an increase in speed.

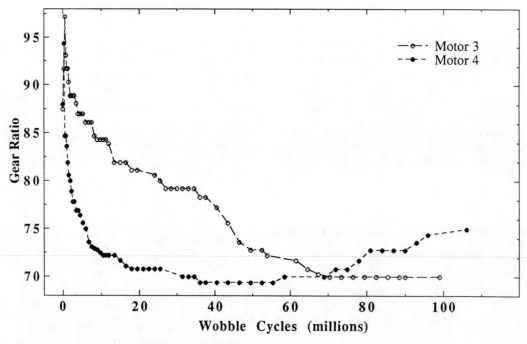

Fig. 5. Wear data for Motor 3 and Motor 4 operated at excitation voltages of 90V and 120V, respectively, and an excitation frequency of 25,000 rpm.

rotor radius of Motor 2 after 30 million wobble cycles shows the inner rotor radius to have been smoothed at least by a factor of three. The results are similar for the remaining three motors.

The wear data in Figs. 4 and 5 corroborate the SEM inspection results. The bearing clearance is very nearly 0.3 μm at the end of the break-in period, when the rotor inner radius asperities are worn out. Therefore, using Eqn. (5) in conjunction with the wear data for each motor, the bearing clearance at the start of motor operation can be calculated. For the motors tested here, the average bearing clearance at the start of motor operation is 2400±100 Å. The overall increase in the bearing clearance from the start of motor operation can be calculated (Eqn. (5)) at 500-700 Å for the different motors. On the average, these numbers correspond to a 25% increase in the bearing clearance by the end of the break-in period.

These results are further corroborated by wear studies of the ordinary side-drive motors. We have operated these motors at two rotor speeds of 160 and 2500 rpm for a total of 40,000 rotor revolutions. SEM inspection of the bushing surfaces and the rotor inner radius reveal scattered wear particles several hundred angstroms in size. Figure 6 is a SEM photo showing a magnified view of the bushing surface (which slides on the shield). Note the wear particles which are adhered to the bushing surface. We have not detected wear on the polysilicon shield which the bushings slide over during motor operation. Figure 7 is a SEM photo showing a portion of the rotor inner radius from the backside. Again, wear particles are observed adhering to the area adjacent to the bearing surface. Unoperated motors which have been dismantled for SEM inspection exhibit clean surfaces with no particles.

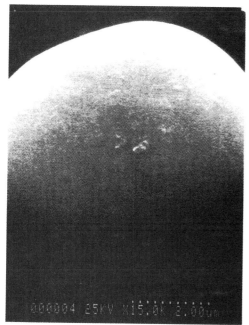

Fig. 6. SEM photo showing magnified view of the bushing surface of an ordinary side-drive motor after 40,000 cycles of operation at 2,500 rpm.

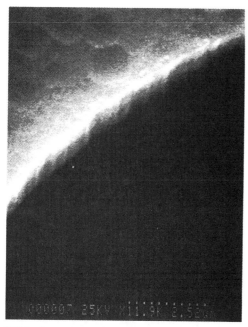

Fig. 7. SEM photo showing a portion of the rotor inner radius from the backside for an ordinary side-drive micromotor after 40,000 cycles of operation at 2,500 rpm.

6. CONCLUSION

A novel, polysilicon, rotary, variable-capacitance, wobble micromotor was described. Voltages as low as 26V across 1.5 μm gaps are sufficient for operating the wobble motors. Frictional force estimates at the bushings, 0.15 μN, and in the bearing, 0.04 μN, were obtained from measurements of stopping voltages. Extended operation of wobble motors up to and over 100 million wobble cycles was studied. Since the gear ratio of the wobble micromotor depends on the bearing clearance, under extended operation, changes in the gear ratio can be a direct measure of wear in the bearing. The results indicate that bearing wear is significant and can result in changes in the gear ratio by as much as 20%. Typical gear ratios are near 90 at the start of motor operation and decrease to near 70 as the bearing wears out.

ACKNOWLEDGMENTS

The authors wish to thank Dr. K.J. Gabriel of AT&T Bell Laboratories for designing the six-phase power supply used in this work. The authors wish to acknowledge technical discussions with S. Bart and Prof. M.A. Schmidt of MIT. This work was supported in part by the National Science Foundation under grant #ECS-8614328 and an IBM Fellowship (Mehregany). Fabrication was carried out in the MIT Integrated Circuit Laboratory (ICL). The authors wish to express their appreciation for silicon-rich nitride wafers from Berkeley Sensor and Actuator Center.

REFERENCES

[1] L.S. Fan, Y.C. Tai and R.S. Muller, Proc. of 1988 IEEE Int. Electr. Devices Meeting, San Francisco, CA, Dec. 1988, p. 666.

[2] Y.C. Tai, Ph.D. Thesis, Department of Electrical Engineering and Computer Science, University of California-Berkeley, June 1989.

[3] M. Mehregany, S.F. Bart, L.S. Tavrow, J.H. Lang, S.D. Senturia and M.F. Schlecht, Proc. of The 4th Int. Conf. on Solid State Sensors and Actuators (Transducers'89), Montreux, Switzerland, June 1989, p. 106. Complete paper to appear in Sensors & Actuators A21, 173 (1989).

[4] M. Mehregany, P. Nagarkar, S.D. Senturia and J.H. Lang, Proc. of The 3rd Micro Electro Mechanical Systems, Napa Valley, CA, Feb. 1990, p. 1.

[5] W.S.N. Trimmer and R. Jebens, Proc. of Micro Electro Mechanical Systems, Salt Lake City, Utah, Feb. 1989, p. 13.

[6] S.C. Jacobsen, R.H. Price, J.E. Wood, T.H. Rytting and M. Rafaelof, Proc. of Micro Electro Mechanical Systems, Salt Lake City, Utah, Feb. 1989, p. 17.

[7] H. Fujita and A. Omodaka, IEEE Trans. Electron Devices 35, 731 (1988).

[8] R. Mahadevan, Proc. of The 3rd IEEE Workshop on Micro Electro Mechanical Systems, Napa Valley, CA, Feb. 1990, p. 120.

[9] M. Mehregany, S.F. Bart, L.S. Tavrow, J.H. Lang and S.D. Senturia, 36th National Symposium of the American Vacuum Society, Boston, MA, Oct. 1989, p. 108. Complete paper to appear in J. Vac. Sci. Tech. A, July-August 1990.

[10] J.H. Lang, Proc. of The 3rd Toyota Conference, Aichi-ken, Japan, Oct. 1989, p. 9.1.

[11] M.G. Lim, J.C. Chang, D.P. Schultz, R.T. Howe and R.M. White, Proc. of The 3rd IEEE Workshop on Micro Electro Mechanical Systems, Napa Valley, CA, Feb. 1990, p. 82.

MEASUREMENTS OF ELECTRIC MICROMOTOR DYNAMICS

Stephen F. Bart, Mehran Mehregany[1], Lee S. Tavrow, Jeffrey H. Lang, and
Stephen D. Senturia

Laboratory for Electromagnetic and Electronic Systems
Microsystems Technology Laboratories
Department of Electrical Engineering and Computer Science
Massachusetts Institute of Technology
Cambridge, Massachusetts

ABSTRACT

This paper describes dynamic measurements in rotary, variable-capacitance, side-drive micromotors using a stroboscopic dynamometry technique. This technique has allowed, for the first time, the capture of detailed micromotor transients. The combination of the data from these experiments with parameter estimation techniques has resulted in the first determination of micromotor friction parameters and drive torque amplitude, independent of simulated drive torque values.

The dynamometry technique uses a strobe flash which is triggered from a phase excitation signal after a known time delay. This acts essentially as a video shutter allowing the position of the rotor as a function of the time delay to be recorded and measured. A dynamic model is developed that includes an electrostatic drive term, a velocity dependent viscous drag term, and a coulomb friction term that is dependent on the square of the drive voltage and the sign of the velocity. From the position-versus-time data, coefficients for this model are estimated using non-linear least-square-error estimation. It is shown that both viscous drag and coulomb friction terms are required if the model is to accurately fit all the experimental data. The motor dynamics are shown to have little dependence on rotor bushing apparent area of contact.

1 INTRODUCTION

The surface micromachining of polysilicon has recently been applied to the fabrication of many types of electroquasistatic microactuators, including several working variable-capacitance (VC) rotary micromotors [5, 18, 12, 20]. Initial work on micromotor design began with feasibility studies [3, 2, 14, 8, 21] and evolved into the development of design rules [11, 1] and the determination of critical fabrication procedures [13]. Detailed modeling of the electric drive torque of micromotors has already been addressed [1, 11, 18, 5, 2].

Early micromotor designs that could not be electrically driven without the assistance of air-levitation [12] raised the issue of friction; its mechanism as well as magnitude were unknown for micromotors. The first type of friction to be studied was static friction [9, 13, 19]. The first experimental predictions of dynamic friction came from the work on micromotor dynamics by Tai and Muller [19] and Gabriel et al. [6]. Gabriel et al. examined rotors that where driven by air jets and not by electrostatic forces [6]. It is unclear whether these air jets affected the dynamic behavior that was observed. In the experiments by Tai and Muller, only the final position and the number of velocity

[1]Current address: Dept. of Electrical Eng. and Applied Physics, Case Western Reserve Univ., Cleveland, OH 44106.

Reprinted with permission from *Microstructures, Sensors, and Actuators*, S. Bart, M. Mehregany, L. Tavrow, J. Lang, and S. Senturia, "Measurements of Electric Micromotor Dynamics," DSC-Vol. 19, pp. 19-29, November 1990. © American Society of Mechanical Engineers.

Figure 1: A scanning electron micrograph of a 50 μm rotor radius, 1.5 μm gap micromotor with 12 stator and 8 rotor poles. The annular indentation in the rotor near the center bearing is the bushing.

reversals during a rotor step transient were captured [19]. This data led to the conclusion that the rotor dynamics were underdamped, but was insufficient to develop a detailed model for the full dynamic behavior.

Here, through the use of stroboscopic dynamometry, a sampled time-versus-position history has been captured for rotor step transients as a function of drive voltage [1]. The purpose is to examine this transient behavior and use it as a tool to explore the drive and friction torques in the VC ordinary side-drive (radial-gap) micromotors (VCMs) previously reported in [11, 13]. By examining the effect of different friction terms in a dynamic model, we have been able to gain a further understanding of dynamic micromotor behavior.

Figure 1 shows a micromotor that is typical of the motors studied in this work. Figure 2 shows a schematic drawing of a cross-sectional view of the motor. A polysilicon surface micromachining technique using heavily-phosphorus-doped LPCVD poly-crystalline silicon for the structural components, undoped low-temperature LPCVD silicon-dioxide for the sacrificial layers, and LPCVD silicon nitride for electrical isolation has been the basis for micromotor fabrication. The reader is referred to previous work [11, 13] for a complete description of the fabrication sequence used to produce the micromotor shown in Fig. 1.

The rotor is supported on the bushings and is free to rotate about a center-pin bearing. During motor operation, the rotor is intended to be in electrical contact with the shield beneath it via the bushing or the bearing. This electrical contact insures that the rotor and the shield are at the same electric potential, thereby eliminating the clamping forces otherwise caused by the electric field between them. As reported previously [13], native oxide formation on the polysilicon surfaces can prevent motor operation by disrupting the intended electrical contact. Proper release and testing techniques aimed at minimizing the native oxide formation were reported previously [13] and used in the experiments described here.

The rotor shown in Figure 1 has a radius of 50 μm. The gap between an aligned rotor and stator pole is 1.5 μm. The concentric ring bushing supports the rotor 1.8 μm above the substrate and is located at a radius of 21 μm. The rotor thickness is 2.2 μm, leading to a rotor moment of inertia of

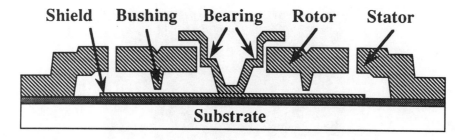

Figure 2: Diagram of the micromotor cross-section after fabrication and release.

2.2×10^{-20} Kg-m^2. The clearance between the rotor inner diameter and the bearing pin is 0.3 μm. The motor design has 12 stator and 8 rotor poles, each 18 degrees in width. The space between stator poles is 12 degrees, and between rotor poles is 27 degrees. Every fourth stator pole is electrically connected to form three stator pole sets or phases. Although all the rotor poles are a single piece of conducting material, they can be thought of as being two pole sets. This pole geometry allows one rotor pole set to be partially aligned with the next stator phase in the excitation sequence when the previous rotor pole set becomes fully aligned, thus improving torque coverage [11]. Note that the angular distance that the rotor must travel to move from alignment with one stator phase to the next is 15 degrees.

In the following section, the dynamometry apparatus and technique will be described. We then discuss a mathematical model which has been used to simulate the dynamic behavior of the rotor transient. This model has several free parameters which are estimated using non-linear least-square-error estimation in the next section. Finally, these parameters imply physical interpretations which are discussed.

2 EXPERIMENTAL METHOD AND DATA

The following section describes the stroboscopic dynamometry technique used to obtain dynamic measurements of the micromotor rotor. Stroboscopic techniques have been used because they require no special integrated sensors or circuitry, do not produce any added load torques on the rotor, and are relatively immune to noise. The apparatus and experimental technique described below were used to obtain angular position-versus-time data for rotor step transients. In a second experiment, a minor modification of the apparatus allowed the capture of undriven or "spin-down" rotor motion. Typical data from both experiments are presented below.

The micromotors are driven open-loop with a three-phase, bipolar, square-wave of voltage. The length of time that a given phase pulse is excited (20 ms) is long compared to the settling time of the rotor. This drive excitation is supplied by a programmable oscillator which drives six bipolar output stages grouped in three phases. Each output stage is capable of driving 10 pF to ± 200 V with a rise time of 2 μs and a fall time to zero of 26 μs. Connection to the micromotor stators is made via microprobes.

To make the stroboscopic measurements, the microscope light source on a normal microprobe station is replaced by a General Radio model 1531-AB Strobotac. An adjustable delay is triggered by one phase of the drive circuit. After the prescribed delay, the strobe flashes. Thus by sweeping the delay time, the transient of an operating motor can be visualized and recorded. The strobe flash, which has a rated persistence of 3 μs on the brightest setting, allows the rotor to be effectively stopped in time even at it's highest velocity. To record the position of the rotor, a CCD camera is mounted on the microscope and connected to a video-tape recorder for later frame-by-frame analysis. In order to account for propagation or other delays, the full system delay is calibrated using a commercial photodiode.

The experiment is run by adjusting the delay time to successive values and recording the rotor image onto a segment of video tape. The actual position verses strobe delay time must be extracted from the video recording by measuring the rotor's position in a number of video frames for each delay time. Examples of the data obtained in this way are shown in Figs. 3–5 for a single micromotor. Each data point shown in these figures represents the mean value of 10 rotor position measurements made from 10 video frames. Note that each video frame captures a different, but equivalent rotor transient. Thus, nonuniform rotor motions, as well as the precision of reading the position values from the video display (approximately 0.5 degree), cause errors. The error bars in Figs. 3–5 represent plus and minus one standard deviation for the 10 measurements. The position data is normalized to the final resting position in order to reduce the effect of misalignment of the measuring scale.

In order to examine undriven or "spin-down" dynamics, only a single modification of the previous setup is required. During the step-response measurement, each phase is turned off only when the next phase is turned on. To create a spin-down test, the phase is quenched after a known excitation time. Note that this time must be significantly shorter than the rotor settling time. In the apparatus used here, the signal that normally drives the output stage is interrupted by a controllable monostable, which allows the high-voltage to be quenched after 40 μs (this includes the drive circuit fall-time). Video recording and data extraction were performed as described above. An example of a spin-down test data set is shown in Fig. 6. The data points for $t < 40$ μs show the rotor position during its initial acceleration when the applied voltage is still on (± 95V).

3 ANALYTICAL MODEL

The behavior of the rotor step transients shown in Figs. 3–5 looks, at first glance, quite similar to that produced by a second order system. However, a simple, second-order model was found to be insufficient to correctly predict the data. In fact, another damping mechanism is required. One common damping mechanism, characteristic of solids in contact, is provided by coulomb (sliding) friction, which is dependent on normal forces [15]. However, an alternative possible mechanism, which is a function of the area of contact, can act between very smooth surfaces [15].

To examine whether contact area dependent damping occurs in the micromotors described here, an additional step-transient data set was taken in which only the area of the bushing contact was altered. The micromotor used in Figs. 3–6 has a concentric ring bushing which has an area of approximately 130 μm^2 [10]. A reduced area bushing, which consists of 3 small dimples with a total area of approximately 6 μm^2 [10], was also tested. The total rotor area is approximately 3800 μm^2. The two corresponding data sets are shown in Fig. 7. The similarity of these curves seems to indicate that the dynamic friction does not depend on the bushing contact area to a large degree. This conclusion agrees with previous work on static friction [9, 6].

Returning then to normal forces, gravity is one mechanism. Also, an electrostatic attraction could provide a normal force between the rotor and either the substrate or the bearing pin. Such an electrostatic attraction could be caused by trapped charges in the rotor material [19] or by a bias voltage that the rotor might acquire due to imperfect grounding [1, 7], which might be caused by the growth of native oxide on the polysilicon surfaces [13]. Finally, an axially off-center rotor would experience a radial force (side-pull) tending to pull the rotor into contact with the bearing [1]. The normal force caused by a rotor bias voltage would depend on the square of the bias voltage, which would, in turn, be proportional to the square of the drive voltage. The radial side-pull force would also depend on the square of the drive voltage. Normal forces caused by trapped charges and by gravity would be constant.

Based on the above arguments, we propose the following model:

$$J\ddot{\theta} = -B\dot{\theta} - (C_1 + C_2 V^2) \cdot \text{sgn}(\dot{\theta}) + AV^2 \Upsilon(\theta) \qquad (1)$$

where J is the rotor moment of inertia, B is a coefficient of viscous drag, C_1 and C_2 represent a constant and a voltage dependent kinetic friction term, respectively, and $\dot{\theta}$ and $\ddot{\theta}$ are the angular velocity and the angular acceleration, respectively. The final term on the right represents the drive torque where A is the amplitude normalized to the drive voltage, V, squared and $\Upsilon(\theta)$ represents the shape of the drive torque as a function of the rotor position. Note that this relation does not model static friction, which is often considered to be important [19, 13, 9]. However, since the minimum delay that can be achieved in the experiment described here is about 15 μs, the effects of such a friction would be very difficult to resolve. Therefore, this omission is appropriate.

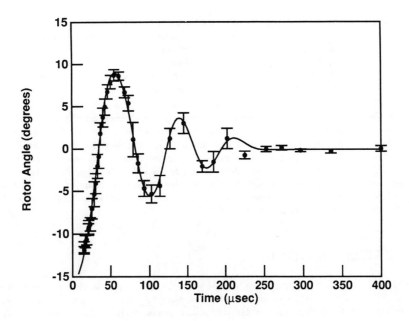

Figure 3: The dots indicate the average of 10 rotor step-transient measurements with $V = \pm 95$ volts. The error bars indicate plus and minus one standard deviation. The solid line is a computed fit based on the parameters given in Table 1.

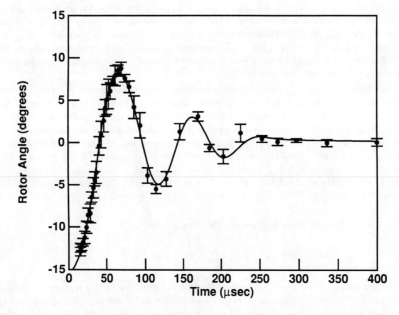

Figure 4: The dots indicate the average of 10 rotor step-transient measurements with $V = \pm 80$ volts. The error bars indicate plus and minus one standard deviation. The solid line is a computed fit based on the parameters given in Table 1.

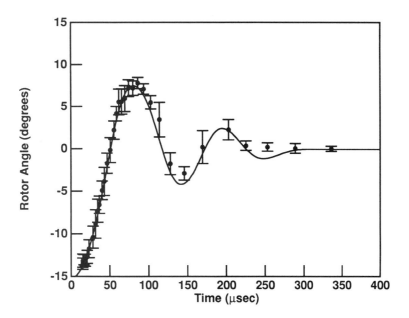

Figure 5: The dots indicate the average of 10 rotor step-transient measurements with $V = \pm 65$ volts. The error bars indicate plus and minus one standard deviation. The solid line is a computed fit based on the parameters given in Table 1.

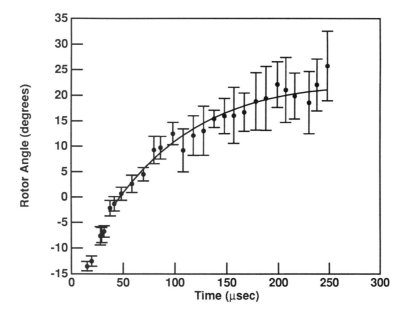

Figure 6: The dots indicate the average of 10 rotor spin-down measurements. Prior to 40 μs, a drive voltage of ± 95 volts was on (or being switched off). The error bars indicate plus and minus one standard deviation. The solid line is a computed fit based on the parameters given in Table 1.

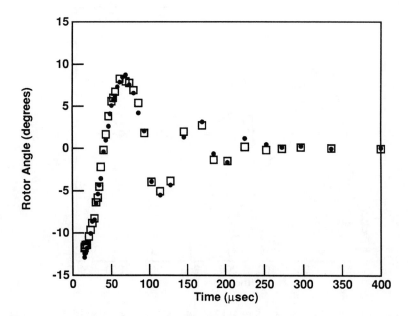

Figure 7: The dots indicate the same step-transient data as is shown in Fig. 4, in which the rotor has a concentric ring bushing. The squares indicate the same experiment run on a rotor which is identical except that its bushing consists of 3 small dimples with about 4% of the area of the concentric ring bushing.

4 PARAMETER ESTIMATION

The dynamic model described by Eqn. (1) remains incomplete without values for its parameters. The aim of this section is to perform a non-linear least-square-error fit of the model to the data in order to estimate these parameters. Not only will this yield values for physical parameters, but it will also show the relative importance of the terms in the model. This will help to determine the relative importance of the physical mechanisms that underlie the respective terms, and hence improve understanding of the system dynamics.

Since the data consists only of position-versus-time data, $\theta[t]$, we wish to cast the least-square-error analysis from which the parameters are derived into a form which requires only this form of data. Thus, we wish to minimize the error function χ^2, where

$$\chi^2 = \sum_t \left\{ \mathcal{W}[t] \left(\tilde{\theta}([t], \bar{P}) - \theta[t] \right)^2 \right\} \tag{2}$$

and where $\tilde{\theta}$ is a set of positions, one for each value of t, which are obtained by numerically integrating Eqn. (1), given a parameter set, $\bar{P} = \{A, B, C_1, C_2\}$. The coefficients $\mathcal{W}[t]$ are a set of weights,

$$\mathcal{W}[t] = \frac{1}{\sigma^2[t]} \tag{3}$$

where $\sigma[t]$ is the standard deviation of the θ value at a given t. Such a curve fitting criterion is standard practice [17].

The values of $\theta[t]$ and $\mathcal{W}[t]$ comprise the experimental data. To calculate $\tilde{\theta}([t], \bar{P})$, the values of V, J and $\Upsilon(\theta)$ are required in addition to $\bar{P}$. V is a given experimental parameter. From the geometry of the rotor, J is easily computed to be 2.2×10^{-20} Kg-m^2. For the computations described below the drive torque shape is computed from

$$\Upsilon(\theta) = \frac{-\text{sgn}(\theta)}{1 - \frac{\pi}{90\theta + \pi}} \tag{4}$$

Table 1: Parameters estimated from the data shown in Figures 3–6.

Parameter (units)	Parameter Value	Parameter Standard Deviation
A (N-m/V^2)	1.4×10^{-15}	2×10^{-17}
B (N-m-s)	2.2×10^{-16}	2×10^{-17}
C_1 (N-m)	9.4×10^{-14}	1×10^{-13}
C_2 (N-m/V^2)	1.3×10^{-16}	1×10^{-17}
$\theta_{V=95}[t=0]$ (deg)	-15.5	0.2
$\theta_{V=80}[t=0]$ (deg)	-15.2	0.2
$\theta_{V=65}[t=0]$ (deg)	-15.2	0.2
$\theta_{V=0}[t=41\mu s]$ (deg)	-1.6	0.9
$\dot{\theta}_{V=0}[t=41\mu s]$ (rad/sec)	5070	490

where θ is the angular position in radians and defined to be zero when the rotor and stator are aligned. This expression is an analytic curve fit to a drive torque shape that was computed using a two-dimensional finite-element field solving package [11]. The accuracy of the parameter estimation is, to some degree, dependent on the accuracy of this estimated torque shape. However, it was determined empirically that the dependence of the estimated parameters on this shape was very weak. Furthermore, since the spin-down test is undriven (i.e., $V = 0$ during times of interest), it is independent of the drive torque shape and C_2. Thus, combining this data further reduces the effect of torque shape errors on the estimation of B and C_1.

Also needed to compute $\tilde{\theta}([t], \bar{P})$ are the initial position and velocity of the rotor. In the case of the driven step-response data, the rotor is performing a stepping motion, coming to rest in near alignment with each phase. Thus, the initial velocity is expected to be zero. The initial position of each step corresponds to the rotor being aligned with the previous phase, which is 15 degrees behind the final position of the presently excited phase. However, since the alignment of our angle measuring device was subject to error, the initial rotor position is treated as a parameter. As long as the value that is obtained is not farther than the alignment tolerance from -15 degrees, the difference can be taken as alignment error. In the case of the spin-down test, both the initial position and the initial velocity are only approximately known, so they are both treated as parameters to be estimated.

The technique for estimating the parameters is to minimize the error function, χ^2, over the parameter space, $\bar{P}$. The minimization is carried out numerically using a Marquardt gradient-expansion algorithm [4]. For ease of numerical computation an inverse tangent function with an appropriately sharp transition is used in place of sgn($\dot{\theta}$) in Eqn. (1). The estimated parameters, based on the combined data from Figs. 3–6, are shown in Table 1. These parameters are then used to numerically integrate Eqn. (1) to obtain the estimated curves shown in Figs. 3–6.

The value of χ^2 obtained for the values in Table 1, where we have estimated 9 parameters from 135 data points, is 0.5, indicating a good fit. Since this fit is the result of a search along the χ^2 hypersurface rather than an exact solution, there is no analytic form for the uncertainties in the estimated parameters. For the case of a model which is linear in its parameters, and for measurement errors which are normally distributed, the standard deviations of the estimated parameters are given by the diagonal terms in the error covariance matrix, which is the inverse of the χ^2 curvature matrix [4]. These parameter standard deviations are the root-sum-squares of the products of the standard deviation of each data point multiplied by the effect that data point has on the determination of the parameter. This relation can be considered a reasonable definition of the parameter uncertainty for non-linear solutions as well [4]. These standard deviation values are given in Table 1.

5 DISCUSSION

The fit of the model to the data shown in Figs. 3–6 is the result of considerable experimentation with the choice of terms in the model of Section 3. In this section, we will examine the estimated parameter values with the intent of exposing those physical mechanisms that are likely to be important in micromotor systems.

For a ±95-volt drive excitation, use of the parameter A indicates a maximum drive torque amplitude of 12.9 pN-m. Two-dimensional finite-element electrostatic simulations in the plane of the rotor yield a corresponding value of 8 pN-m [11, 13]. When these two-dimensional simulations are corrected to account for out-of-plane fringing fields, they yield a maximum drive torque value of 20.6 pN-m [1, 11]. Since the three-dimensional nature of our micromotors is difficult to model exactly, our experiments and simulations can be considered to be quite close.

The coefficient of viscous drag, B, can be compared to the value given by an analytic solution for a disk in a housing under laminar flow [16]. Such a solution yields

$$B = \frac{\pi \mu R^4}{s} = 2 \times 10^{-16} \text{ N-m-s} \tag{5}$$

where μ is the absolute viscosity of air (1.83×10^{-5} Kg/m-s), R is the rotor radius, and s is the gap between the rotor and the housing, taken here to be the bushing height, 1.8 μm. This value of B is essentially identical to the estimated value shown in Table 1. Note that the Reynolds number, $\rho \dot{\theta} R^2 / \mu$, where ρ is the air density, is on the order of unity for the maximum rotor velocity encountered in these experiments of approximately 5,000 rad/s. Hence the laminar flow criterion is easily met. This low Reynolds number also indicates that the viscous drag is not significantly affected by rotor acceleration.

The friction torques are assumed to correspond to a physical mechanism which supplies a normal force between the rotor and contacting surfaces at the bushing or the bearing. In order to compare the estimated values of friction torque to analytic values for these normal forces, a friction coefficient must be assumed. Although no definitive value has yet to appear, we will assume a friction coefficient of 0.3 here. This value is in the range of values that have been proposed recently [19, 6].

The constant friction term corresponds to a physical mechanism which supplies a constant normal force. The gravitational force on the rotor is approximately 2×10^{-10} N. Multiplying this force by a friction coefficient of 0.3 and the bushing radius (21 μm), results in a frictional torque due to gravity which is about 75 times smaller than the value of C_1. Trapped charge on the rotor has been proposed as a mechanism that could account for the increased friction [19]. Alternatively, it is possible that our estimate of C_1 is too large. Since the standard deviation of the estimated C_1 is slightly larger than the parameter itself, it seems clear that the level of error in our experiment precludes resolving the small value of C_1 accurately. In fact, it is possible that the parameter C_1 is not required at all. To see if this is so, Eqn. (1) was integrated using the parameters in Table 1 except that $C_1 = 0$. The curves obtained were virtually identical to those plotted in Figs. 3–5. This indicates that a constant contact friction, as represented by C_1, is not a significant component of the damping forces on the rotor.

If the voltage dependent friction term is neglected, the simultaneous fit of the model to all four data sets in Figs. 3–6 is poor. The need for a friction term which is dependent on V^2 implies a normal force between two contacting surfaces which is electrical in origin. If the rotor is not axially centered between the stator poles, there is a net radial force (often termed "side-pull" in macroscopic motors) tending to close the smaller gap. Frictional forces in the bearing caused by this mechanism would be proportional to V^2. In addition, if the rotor is not in proper electrical contact with the shield due to the polysilicon native oxide, it can develop a bias voltage which would yield a rotor-to-shield attractive force [1, 7, 13]. Frictional forces associated with this mechanism would also be proportional to V^2. The following paragraphs compute order of magnitude values for the torques that could be expected from these two mechanisms.

For the side-pull mechanism, the order of magnitude of the rotor-to-stator radial force can be estimated assuming that the rotor-stator gap behaves as a simple parallel plate capacitor. Assuming that the rotor is off-center by the full 0.3 μm allowed by the bearing clearance, the radial force is approximately 5×10^{-11} N/V^2. If we treat the bearing as generating simple sliding friction at a lever arm of 13 μm (i.e., the bearing-pin radius) with a friction coefficient of 0.3, we find a value of approximately 2×10^{-16} N-m/V^2 which is within a factor of 2 of the estimated value of C_2. Since we have not accounted for fringing fields, which are known to be important [1, 11], or for the effect of rolling at the rotor bearing contact, the closeness of these values should not be construed as proof that this mechanism is the cause of the voltage-dependent friction term. Nonetheless, it seems clear that this mechanism could be important as has also been reported previously [19].

For the clamping mechanism, the rotor-to-shield attractive force can also be estimated using simple parallel plate capacitor approximations [1]. If the native oxide layer that electrically isolates the rotor is thin, then the rotor-to-shield capacitance will be dominated by the bushing-to-shield capacitance.

Assuming that the area of the bushing contact is 130 μm^2 and that the force acts at the bushing lever arm of 21 μm with a friction coefficient of 0.3, a value of $C_2 = 5 \times 10^{-18}$ N-m/V^2 is found. This is much smaller than the estimated value of C_2 in Table 1. Note, however, that this value depends strongly on the thickness of the insulating layer which was assumed to be 10 Å, but is not known with accuracy. Also, this frictional torque should be dependent on the bushing area [1].

It should be noted that both of these torque mechanisms would have a dependence on the rotor angular position. However, in the motor design discussed here, which has two sets of rotor poles, this type of angular dependence is expected to be small, and is not required by the dynamic model presented here to fit the data well.

6 CONCLUSION

The stroboscopic dynamometry technique described here has allowed, for the first time, the capture of a full micromotor step transient. Undriven or "spin-down" motor dynamics have also been examined. The analysis of the data from these experiments using parameter estimation techniques has resulted in the first determination of micromotor friction parameters and drive torque amplitude, independent of simulated drive-torque amplitude values. It has further allowed the development of a minimal dynamic model that gives insight into the important mechanisms that underlie micromotor dynamic behavior.

Although the dynamic model used here is not necessarily the only appropriate form, the micromotor step response and undriven dynamics are well modeled by an equation of motion which contains electrostatic drive, viscous drag and coulomb friction terms. The estimated viscous friction coefficient is very close to the value given by a simple analysis for a disk spinning in a housing under laminar air flow. The coulomb friction requires a term which is dependent on the drive voltage squared. Simple calculations indicate that bearing friction caused by radial forces may offer the best explanation for this friction, but that bushing friction due to rotor-to-substrate attraction may also play a role. A constant coulomb friction term can be omitted without significantly altering the predicted dynamics. This implies that mechanisms which result in a constant friction torque, such as gravity, do not have a large effect on micromotor dynamics. It was further demonstrated, by the comparison shown in Fig. 7 that the frictional drag has a weak, if any, dependence on the bushing contact area.

ACKNOWLEDGMENTS

The authors wish to thank Charles E. Miller of the Stroboscopic Light Lab at MIT, for his help in developing the experimental apparatus. The work reported in this paper was supported in part by the United States National Science Foundation under grant ECS-8614328 and an IBM fellowship (Mehregany).

REFERENCES

[1] BART, S. F., *Modeling and Design of Electroquasistatic Microactuators*, PhD thesis, Massachusetts Institute of Technology, Cambridge, MA, 1990.

[2] BART, S. F., AND LANG, J. H., "An analysis of electroquasistatic induction micromotors," *Sensors and Actuators 20*, 1/2 (1989), pp. 97–106.

[3] BART, S. F., LOBER, T. A., HOWE, R. T., LANG, J. H., AND SCHLECHT, M. F., "Design considerations for microfabricated electric actuators," *Sensors and Actuators 14*, 3 (1988), pp. 269–292.

[4] BEVINGTON, P. R., *Data Reduction and Error Analysis for the Physical Sciences*, McGraw-Hill Book Company, New York, 1969.

[5] FAN, L. S., TAI, Y. C., AND MULLER, R. S., "IC-processed electrostatic micromotors," *Sensors and Actuators 20*, 1/2 (1989), pp. 41–47.

[6] GABRIEL, K. J., BEHI, F., MAHADEVAN, R., AND MEHREGANY, M., "*In Situ* friction and wear measurements in integrated polysilicon mechanisms," *Sensors and Actuators A21* (1990), pp. 184–188.

[7] LANG, J. H., "Initial thoughts on the dynamics and control of electric micromotors," In *Proceedings of the 3rd Toyota Conference* (Aichi-ken, Japan, Oct., 1989), pp. 9.1–9.14.

[8] LANG, J. H., AND BART, S. F., "Toward the design of successful electric micromotors," In *Technical Digest of the IEEE Solid-State Sensor and Actuator Workshop* (Hilton Head Island, SC, June 6–9, 1988), pp. 127–130.

[9] LIM, M. G., CHANG, J. C., SCHULTZ, D. P., HOWE, R. T., AND WHITE, R. M., "Polysilicon microstructures to characterize static friction," In *Proceedings of the IEEE Micro Electro Mechanical Systems Workshop* (Napa Valley, CA, February 11–14, 1990), pp. 82–88.

[10] MEHREGANY, M., *Microfabricated Silicon Electric Mechanisms*, PhD thesis, Massachusetts Institute of Technology, Cambridge, MA, 1990.

[11] MEHREGANY, M., BART, S. F., TAVROW, L. S., LANG, J. H., AND SENTURIA, S. D., "Principles in design and microfabrication of variable-capacitance side-drive motors," *Journal of Vacuum Science and Technology A 8*, 4 (Jul/Aug, 1990), pp. 3614–3624.

[12] MEHREGANY, M., BART, S. F., TAVROW, L. S., LANG, J. H., SENTURIA, S. D., AND SCHLECHT, M. F., "A study of three microfabricated variable-capacitance motors," *Sensors and Actuators A21* (1990), pp. 173–179.

[13] MEHREGANY, M., NAGARKAR, P., SENTURIA, S. D., AND LANG, J. H., "Operation of microfabricated harmonic and ordinary side-drive motors," In *Proceedings of the 3rd IEEE Workshop on Micro Electro Mechanical Systems* (Napa Valley, CA, 1990), pp. 1–8.

[14] PRICE, R. H., WOOD, J. E., AND JACOBSEN, S. C., "Modeling considerations for electrostatic forces in electrostatic microactuators," *Sensors and Actuators 20*, 1/2 (1989), pp. 107–114.

[15] RABINOWICZ, E., *Friction and Wear of Materials*, John Wiley and Sons, Inc., New York, 1965.

[16] SCHLICHTING, H., *Boundary-Layer Theory*, seventh ed, McGraw-Hill Book Company, New York, 1979.

[17] SCHWEPPE, F. C., *Uncertain Dynamic Systems*, Prentice-Hall, Englewood Cliffs, NJ, 1973.

[18] TAI, Y. C., AND MULLER, R. S., "IC-processed electrostatic synchronous micromotors," *Sensors and Actuators 20*, 1/2 (1989), pp. 49–55.

[19] TAI, Y. C., AND MULLER, R. S., "Frictional study of IC-processed micromotors," *Sensors and Actuators A21* (1990), pp. 180–183.

[20] TAVROW, L. S., BART, S. F., SCHLECHT, M. F., AND LANG, J. H., "A LOCOS process for an electrostatic microfabricated motor," *Sensors and Actuators A23* (1990), pp. 893–898.

[21] TRIMMER, W. S. N., AND GABRIEL, K. J., "Design considerations for a practical electrostatic micro motor," *Sensors and Actuators 11*, 2 (1987), pp. 189–206.

Section 3

Comb Drive Actuators

THE comb drive actuator is a simple and ingenious device. Imagine that alternate tines are cut off two combs and interleaved, as shown in Fig. 3.1. One comb is free to move, the other is stationary. When a voltage is applied, the moving comb moves into the stationary comb. By changing the voltage, the position of the moving comb can be controlled.

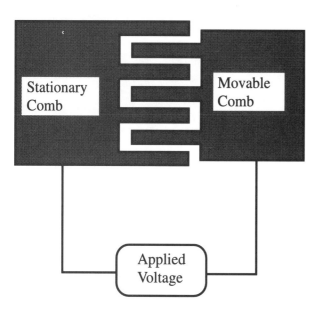

Fig. 3.1

The movable comb can be supported by a flexible spring, and large motions can be obtained. The energy is the electrostatic force times the distance the comb travels, and the long travel of this actuator increases the energy that can be produced. When this actuator is used at resonance, the distance traveled and the energy of the comb increases to even larger values.

In early comb drive actuators, a cantilevered beam was used to provide the suspension for the moving comb. However, as shown in Fig. 3.2, this causes the tip of the cantilever beam, and hence the comb, to move in an arc. This curving motion will cause the opposing comb teeth to touch and short out. An improvement is the crab leg suspension. Two cantilever beams are connected together such that the deviations from each arc cancel each other out, as shown to the right in Fig. 3.2.

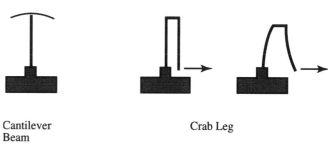

Cantilever Beam   Crab Leg

Fig. 3.2

In the first paper, "Laterally Driven Polysilicon Resonant Microstructures," the fabrication and operation of the comb drive is described. Silicon surface micromachining with a single sacrificial and a single structural layer is used. Both linear and rotary actuators are described. The second paper, "Electrostatic-comb Drive of Lateral Polysilicon Resonators," describes the resonant properties of comb drive actuators. In "Electrostatically Balanced Comb Drive for Controlled Levitation," problems and solutions are discussed for the unwanted attraction of the moving comb toward the substrate. The fourth paper, "Polysilicon Microgripper," makes a clever pair of tweezers using surface micromachining, and uses the comb drive actuators to drive the tweezers.

Laterally Driven Polysilicon Resonant Microstructures

William C. Tang, Tu-Cuong H. Nguyen, and Roger T. Howe

University of California at Berkeley
Department of Electrical Engineering and Computer Sciences
and the Electronics Research Laboratory
Berkeley Sensor & Actuator Center
Berkeley, California 94720

Abstract

Interdigitated finger (comb) structures are demonstrated to be effective for electrostatically exciting the resonance of polysilicon microstructures parallel to the plane of the substrate. Linear plates suspended by a folded-cantilever truss and torsional plates suspended by spiral and serpentine springs are fabricated from a 2 μm-thick phosphorus-doped LPCVD polysilicon film. Resonance is observed visually, with frequencies ranging from 18 kHz to 80 kHz and quality factors from 20 to 130. Simple beam theory is adequate for calculating the resonant frequencies, using a Young's modulus of 140 GPa and neglecting residual strain in the released structures.

I. INTRODUCTION

Resonant micromechanical structures are used as transducing elements in a wide variety of sensors. These devices utilize the high sensitivity of the frequency of a mechanical resonator to physical or chemical parameters which affect its potential or kinetic vibrational energy [1-4]. Microfabricated resonant structures for sensing pressure [5-7], acceleration [8], and vapor concentration [9] have been demonstrated.

Mechanical vibration of microstructures can be excited in several ways, including piezoelectric films [5], thermal expansion [8,10], electrostatic forces [1,6], and magnetostatic forces [7]. Vibration can be detected by means of piezoelectric films [5], piezoresistive strain gauges [10], optical techniques [10,11], and capacitively [1,6,9]. Electrostatic excitation combined with capacitive (electrostatic) detection is an attractive approach for silicon microstructures because of simplicity and compatibility with micromachining technology [1,2].

Both crystalline silicon [6] and polycrystalline silicon (polysilicon) [9] microstructures have been driven and sensed electrostatically by means of fixed electrodes forming parallel-plate capacitors with the structure. However, there are several drawbacks to the parallel-plate-capacitor drive and sense of micromechanical structures. The electrostatic force is nonlinear unless the amplitude of vibration is limited to a small fraction of the capacitor gap. In addition, the quality factor Q of the resonance is very low at atmospheric pressure because of squeeze-film damping in the micron-sized capacitor gap [9]. If the microstructure is resonated in vacuum, the efficient parallel-plate excitation and the very high intrinsic Q lead to steady-state excitation voltages with mV-level amplitudes. Such low voltage levels complicate the design of the sustaining amplifier [12].

In this paper, we describe the design, fabrication, and initial testing of an electrostatic comb structure for exciting and sensing the vibration of polysilicon microstructures *parallel* to the plane of the substrate. Interdigitated fingers and the mechanical structure are etched in a single LPCVD polysilicon film, which has been deposited on a patterned sacrificial oxide layer. The drive capacitance is linear with displacement of the structure, resulting in a force which is independent of the vibration amplitude. Moreover, the vibration amplitude can be of the order of 10 μm for certain comb and structure designs. The use of weaker fringing fields to excite resonance is advantageous for high-Q structures (resonating in vacuum), since this results in larger steady-state excitation voltages. Furthermore, the quality factor for lateral vibration at atmospheric pressure is substantially higher than for vibration normal to the substrate [13,14]. Couette flow in the gap between the structure and the substrate occurs for lateral motion of the structure, which is much less dissipative than squeeze-film damping [13]. Another significant advantage of the lateral-drive concept is that a variety of elaborate geometric structures, such as differential capacitive excitation and detection, can be incorporated without an increase in process complexity. This flexibility is being exploited in current efforts to develop a microactuator technology based on lateral resonant structures [15].

The theory of the electrostatic comb drive is derived first and used to obtain electromechanical transfer functions for generalized lateral structures. We next present the design and modeling of a linear resonant structure with a truss suspension and a torsional resonant structure using a spiral-spring suspension or a serpentine-spring suspension. The fabrication process, a straightforward application of surface micromachining technology, is then described. Initial measurements of resonant frequencies and quality factors are compared with analytical and finite-element calculations.

II. ELECTROSTATIC COMB DRIVE

Figure 1 shows the layout of a linear resonant structure which can be driven electrostatically from one side and sensed capacitively at the other side with interdigitated finger (comb) structures. Alternatively, the structure can be driven differentially (push-pull) using the two combs, with the motion sensed by the impedance shift at resonance [12]. In analyzing the electromechanical transfer function, we consider the former, two-port configuration. The motion is sensed by detecting the short-circuit current through the time-varying interdigitated capacitor with a *dc* bias [1].

The driving force and the output sensitivity are both proportional to the variation of the comb capacitance C with the lateral displacement x of the structure. A key feature of the electrostatic-comb drive is that $(\partial C/\partial x)$ is a constant, independent of the displacement x, so long as x is less than the finger overlap (Fig. 1). Therefore, electrostatic-comb drives can have linear electromechanical transfer functions for large displacements, in contrast to parallel-plate capacitive drives.

Reprinted from *Proceedings IEEE Micro Electro Mechanical Systems*, pp. 53-59, February 1989.

At the sense port, harmonic motion of the structure in Fig. 1 results in a sense current i_s which is given by

$$i_s = V_S (\partial C / \partial x)(\partial x / \partial t), \quad (1)$$

where V_S is the bias voltage between the structure and the stationary sense electrode. At the drive port, the static displacement x as a function of drive voltage is given by

$$x = \frac{F_x}{k_{sys}} = \frac{\frac{1}{2} v_D^2 (\partial C / \partial x)}{k_{sys}}, \quad (2)$$

where F_x is the electrostatic force in the x-direction, k_{sys} is the system spring constant, and v_D is the drive voltage.

For a drive voltage $v_D(t) = V_P + v_d \sin(\omega t)$, the time derivative of x is

$$\frac{\partial x}{\partial t} = \frac{(\partial C / \partial x)}{2 k_{sys}} \frac{\partial (v_D^2)}{\partial t} \quad (3)$$

$$= \frac{(\partial C / \partial x)}{2 k_{sys}} \left[2\omega V_P v_d \cos(\omega t) + \omega v_d^2 \sin(2\omega t) \right],$$

where we have used the fact that $(\partial C / \partial x)$ is a constant for the interdigitated-finger capacitor. The second-harmonic term on the right-hand-side of Eqn. (3) is neglible if $v_d \ll V_P$. Furthermore, if a push-pull drive is used, this term results in a common-mode force and is cancelled to first order. At mechanical resonance, the magnitude of the linear term in Eqn. (3) is multiplied by the quality factor Q [1,9], from which it follows that the magnitude of the transfer function $T(j\omega_r) = X / V_d$ relating the phasor displacement X to phasor drive voltage V_d at the resonant frequency ω_r is:

$$\left| \frac{X}{V_d} \right| = V_P \frac{Q}{k_{sys}} (\partial C / \partial x). \quad (4)$$

The transconductance of the resonant structure is defined by $G(j\omega) = I_s / V_d$. Its magnitude at resonance can be found by substitution of Eqn. (4) into the phasor form of Eqn. (1):

$$\left| \frac{I_s}{V_d} \right| = \omega V_P V_S \frac{Q}{k_{sys}} (\partial C / \partial x)^2. \quad (5)$$

A planar electrode extends under the the comb and plate in Fig. 1, which can be grounded or set to a dc potential in order to minimize parasitic capacitive coupling between the drive and sense ports. An additional function of this electrode is to suppress the excitation of undesired modes of the structure.

III. RESONANT STRUCTURE DESIGN

We now consider the mechanical design of two classes of lateral resonant structures which demonstrate the power and flexibility of the lateral drive approach: linear resonant plates with folded supporting beams and torsional resonant plates with differential drive and sense ports.

A. Linear Resonant Plates

Figure 1 shows the layout of a linear resonant plate with a 50 μm-long folded-beam suspension. Motivations for this truss suspension are its large compliance and its capability for relief of built-in residual strain in the structural film. The folded cantilever beams are anchored near the center, thus allowing expansion or contraction of the four beams along the y-axis (Fig. 1). Both the average residual stress in the polysilicon film and stress induced by large-amplitude plate motion should be largely relieved by this design. In addition, the long effective support lengths result in a highly compliant suspension. Plates with 200 μm-long trusses are resonated with amplitudes as large as 10 μm.

An accurate analytical expression for the fundamental lateral resonant frequency, f_r, can be found using Rayleigh's Method:

$$f_r = \frac{1}{2\pi} \left[\frac{k_{sys}}{(M_p + 0.3714 M)} \right]^{\frac{1}{2}}, \quad (6)$$

where M_p and M are the masses of the plate and of the supporting beams, respectively. For the folded-beam structure, an analytical expression k_{sys} can be found by assuming that the trusses joining the folded beam segments are rigid:

$$k_{sys} = 24 E I / L^3 = 2 E h (W/L)^3, \quad (7)$$

where $I = (1/12) h W^3$ is the moment of inertia of the beams. Residual strain in the structure is neglected in finding this expression. Combining Eqs. (6) and (7), it follows that

$$f_r = \frac{1}{2\pi} \left[\frac{2 E h (W/L)^3}{(M_p + 0.3714 M)} \right]^{\frac{1}{2}}. \quad (8)$$

The quality factor Q is estimated by assuming that Couette flow underneath the plate is the dominant dissipative process [14]:

$$Q = \frac{d}{\mu A_p} (M k_{sys})^{\frac{1}{2}}, \quad (9)$$

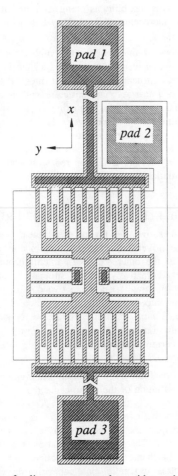

Fig. 1. Layout of a linear resonant plate with comb structures on both ends and a 50 μm-long folded-beam on each side.

where μ is the absolute viscosity of air (1.8×10^{-5} $N \cdot s \cdot m^{-2}$), and d is the offset between the plate and the substrate. Quality factors for lateral motion are much higher than for motion normal to the substrate [9,16].

B. Torsional Resonant Plates

Another class of structures is driven into torsional resonance by a set of concentric interdigitated electrodes. Figure 2 shows one of the designs with two Archimedean spirals as supporting beams. Figure 3 is a scanning-electron micrograph (SEM) of another design using four serpentine springs. The structures are supported only at the center, enabling some relaxation of the built-in residual stress in the polysilicon film. An advantage of the torsional approach is that four or more pairs of balanced concentric comb structures can be placed at the perimeter of the ring, allowing a high degree of flexibility for differential drive and sense. Since both the drive and the sense ports are differentially balanced, excitation of undesired oscillation modes is avoided and signal corruption by feedthrough is minimized. As with the lateral structure, extensive ground planes are utilized.

The torsional spring constant of the Archimedean spiral is given by [17]:

$$k_\theta = \frac{EhW^3}{12L} \quad (\mu N \cdot \mu m \cdot rad^{-1}), \qquad (10)$$

where L is the length of the spiral. As was done for the the lateral resonant structures, residual strain in the spiral spring is neglected in the analysis. This assumption will be reexamined in the discussion of the measured resonant frequencies.

The torsional resonant frequency, f_θ, is evaluated by replacing k_{sys} in Eq. (6) with the torsional spring constant, k_θ, and the masses, M_p and M, with the mass moments of inertia, J_p and J:

$$f_\theta = \frac{1}{2\pi} \left[\frac{k_\theta}{(J_p + 0.3714J)} \right]^{\frac{1}{2}}. \qquad (11)$$

The value of J can be found by evaluating the following integral over an appropriate limit:

$$J = \int r^2 \, dM = \rho h \iint r^3 \, d\theta \, dr, \qquad (12)$$

where ρ is the density of polysilicon ($2.3 \times 10^3 kg \cdot m^{-3}$).

The quality factor is estimated similarly to Eq. (9) by assuming Couette flow underneath the plate, and is given by

$$Q = \frac{d(Jk_\theta)^{\frac{1}{2}}}{\mu \int r^2 dA_p} \qquad (13)$$

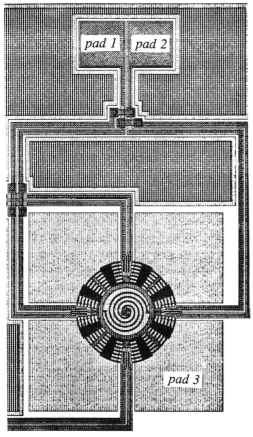

Fig. 2. Layout of a torsional resonant plate with 4 pairs of balanced concentric comb structures supported by two spirals.

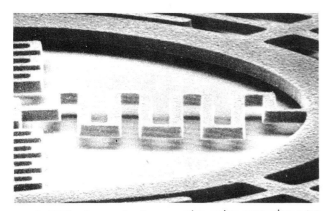

Fig. 3. SEM micrograph of a serpentine spring supporting a torsional resonant plate.

IV. FABRICATION PROCESS

The structures are fabricated with the four-mask process illustrated in Fig. 4. A significant advantage of this technology is that all the critical features are defined with one mask, eliminating errors due to mask-to-mask misalignment. The process begins with a standard POCl$_3$ blanket n+ diffusion, which defines the substrate ground plane, after which the wafer is passivated with a layer of 1500 Å-thick low-pressure chemical-vapor-deposited (LPCVD) nitride deposited on top of a layer of 5000 Å-thick thermal SiO$_2$ (Fig. 4(a)). Contact windows to the substrate ground plane are then opened (Fig. 4(b)) using a combination of plasma and wet etching.

The next steps involve deposition and definition of the first polysilicon layer. A layer of 3000 Å-thick, *in situ* phosphorus-doped polysilicon is deposited by LPCVD at 650°C then patterned with the second mask (Fig. 4(c) and 4(d)). This layer serves as a second electrode plane, the interconnection to the n+ diffusion, and for standoff bumps to prevent sticking of the second polysilicon layer to the substrate after the final micromachining step. (Standoff bumps are not included in the initial process run.) A 2 μm-thick LPCVD sacrificial phosphosilicate glass (PSG) layer is deposited and patterned with the third mask, as shown in Figs. 4(e) and 4(f), which defines the anchors of the microstructures.

The 2 μm-thick polysilicon structural layer is then deposited by LPCVD (undoped) at 605°C (Fig. 4(g)). The structural layer is doped by depositing another layer of 3000 Å-thick PSG (Fig. 4(h)) and then annealing at 950°C for one hour. This doping process is designed to dope the polysilicon symmetrically by diffusion from the top and the bottom layers of PSG. A stress-annealing step is then optionally performed at 1050°C for 30 minutes in N_2. The annealing temperature is lower than 1100°C in order to avoid loss of adhesion between the PSG and the Si_3N_4 [12,16].

After stripping the top PSG layer by a timed etch in 10:1 HF, the plates, beams, and electrostatic comb drive and sense structures are defined in the final masking step (Fig. 4(i)). The structures are anisotropically patterned in a CCl_4 plasma by reactive-ion etching, in order to achieve nearly vertical sidewalls. Figure 4(j) illustrates the final cross section after the wafer is immersed in 10:1 diluted HF to etch the sacrificial PSG. The wafer is rinsed repeatedly with DI water for at least 30 minutes after the micromachining step is completed and then dried in a standard spin dryer.

Surface-micromachined polysilicon structures can become stuck to the substrate after the final drying process [18]. The yield of free-standing structures is zero on wafers for which the 30-minute stress anneal at 1050°C is omitted. When the stress anneal is included in the process, 70% of the structures are free-standing. The 30% which are initially attached to the substrate could be freed easily with a probe tip; the high flexibility of the structures allows manipulation without breakage. No amount of probing, however, could free any of the unannealed structures.

A series of clamped-clamped microbridges is used to estimate the average residual strain in the structural polysilicon film from the minimum buckling length [19]. The moment of the residual strain is qualitatively studied by a series of clamped-free cantilever beams. Since the microbridges have "step-up" anchors, it is expected that end effects will have to be modeled carefully to obtain an accurate value of the residual strain [20]. Moreover, the "stiction" of the diagnostic microbridges and cantilevers to the substrate during drying is also a source of error in calculating the strain and its moment [21].

For the unannealed samples, the microcantilevers have a tendency to deflect and attach to the substrate for lengths greater than 150 μm. We interpret the buckling length of about 120 μm for microbridges using the simple clamped-clamped Euler's criterion [19] and estimate the strain as about 10^{-3}. Annealed samples have apparently undeflected cantilevers under optical and SEM observation and have a buckling length of about 220 μm, indicating a residual strain of about 3×10^{-4}. These estimated values are typical of residual strain for phosphorus-doped polysilicon.

Figures 5–8 are scanning-electron micrographs of the completed structures.

V. EXPERIMENTAL MEASUREMENTS

The resonant frequencies, quality factors, and transfer function of the structures with beam lengths of 80 μm or longer can be found by visual observation under 25X magnification. Sinusoidal and dc bias voltages are applied to the structures via probes contacting the numbered polysilicon pads in Figs. 1 and 2. For the linear structures, the sinusoidal drive voltage is applied to one set of fixed electrode fingers via *pad 3*, while a *dc* bias is supplied to *pad 1* (connected to the dormant sense fingers) and *pad 2* (connected to the first-level polysilicon ground plane and to the suspended structure). The diffused ground plane is left floating in the initial measurements. The dormant fingers are biased to eliminate electrostatic attraction between them and the resonant structure. For the torsional structures (Fig. 2), the sinusoidal voltage is applied to the drive fingers via *pad 2*, and the dormant sense fingers, ground plane, and resonant structure are biased via *pad 1* and *pad 3*.

In order to provide large-amplitude lateral motion in air for visual observation, dc biases of up to 40 V and driving voltage amplitudes (zero to peak) of up to 10 V are used. Resonant frequencies are determined by maximizing the amplitude of vibration, which can be as large as 10 μm for the linear structures with the longest support beams. The measured resonant frequencies for the linear structures are listed in Table I and those for the torsional structures are listed in Table II. The results include measurements from two different electrostatic comb designs (type A and type B), which are described in the figure with Table I.

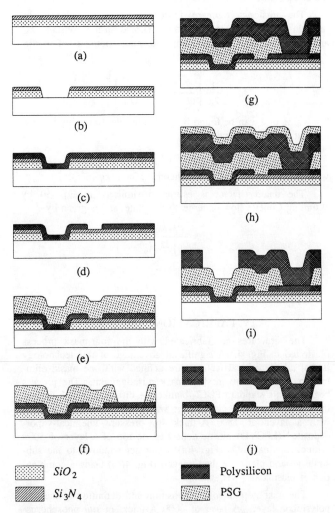

Fig. 4. Process sequence of a lateral resonant structure. (a) Deposition of LPCVD nitride on top of a layer of thermal SiO_2. (b) Contact windows to substrate n+ diffusion. (c) Deposition of *in situ* P-doped LPCVD polysilicon. (d) Patterning of first polysilicon layer. (e) Deposition of sacrificial PSG. (f) Anchor windows for second polysilicon layer. (g) Deposition of the undoped LPCVD polysilicon structural layer. (h) Deposition of second layer of PSG for doping symmetry. (i) Patterning of second polysilicon layer. (j) Final cross section after wet etching of sacrificial PSG.

Fig. 5. SEM micrograph of a linear resonant structure with 100 μm-long beams.

Fig. 6. SEM micrograph of the linear comb structure.

Fig. 7. SEM micrograph of two two-turn Archimedean spirals supporting a torsional resonant plate.

Fig. 8. SEM micrograph of the concentric comb structure.

TABLE I
Predicted and Measured Resonant Frequency Values
of the Linear Resonant Structures

Beam length [μm]	Type A		Type B	
	Predicted [kHz]	Measured [kHz]	Predicted [kHz]	Measured [kHz]
80	75.5	75.0 ± 0.05	71.8	72.3 ± 0.05
100	53.7	54.3 ± 0.05	51.1	50.8 ± 0.05
120	40.6	41.1 ± 0.1	38.7	39.4 ± 0.1
140	32.0	32.0 ± 0.2	30.5	30.0 ± 0.2
160	26.0	25.9 ± 0.2	24.8	25.0 ± 0.2
180	21.7	21.5 ± 0.3	20.7	20.3 ± 0.3
200	18.4	18.2 ± 0.3	17.6	17.5 ± 0.3

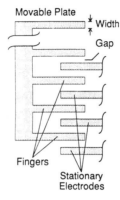

Types A and B Features

Features	A	B
# of fingers	9	11
Width [μm]	4	4
Gap [μm]	3	2
Fitted $\partial C/\partial x$ [aFμm^{-1}]	58	150

TABLE II
Predicted and Measured
Resonant Frequency Values
of the Torsional Resonant Structures

Supporting Beam Type	Predicted [kHz]	Measured [kHz]
Spiral	10.5	9.7 ± 0.3
Serpent	60.7	59.4 ± 0.2

The calculated resonant frequencies in Tables I and II are found from Eqns. (8) and (9) with the Young's modulus adjusted to give the best fit to the experimental data. For the serpentine-spring torsional structures, a finite-element program [22] is used to find the effective spring constant. The best-fit value for Young's modulus is $E = 140$ GPa for both linear and torsional resonant structures. From Table I and II, the calculated and measured resonant frequencies are in close agreement for all the lateral structures.

Initial visual measurements of the quality factor Q are plotted in Fig. 9 for the linear resonant structures. The visual measurement of Q is especially difficult for structures with small vibration amplitudes, which is reflected in the larger error bars for these points. The calculated quality factors from Eqn. (9) are consistently higher than the measured values, indicating that the assumption of pure Couette flow is an oversimplification for these structures. However, the calculated values of Q are of the correct magnitude and may be useful for design. The highest measured Q is about 130 for a structure with 80 μm-long folded-beam suspension.

The magnitude of the electromechanical transfer function is measured by estimating the amplitude of vibration from the envelope of the blurred vibrating structure at a given drive voltage and bias voltage. Figure 10 is a comparison of the experimental results for the linear resonant structures with the calculated values from Eqn. (4), for which the capacitance variation ($\partial C / \partial x$) for the two types of comb drives is a fitting parameter. The type A drive is found to have $\partial C / \partial x = 58$ aFμm^{-1} and the type B has $\partial C / \partial x = 150$ aFμm^{-1}. These experimental results are about a factor of five less than the variation found by a simple two-dimensional simulation using a Green's function approach. However, the proximity of the comb to the substrate was neglected in the simulation.

Electrical characterization of the structures is being pursued. Since there is no integrated buffer circuit, the sense current tends to be swamped by feedthrough [12,16]. The differential drive and sense torsional two-port resonant structure is attractive for minimizing feedthrough.

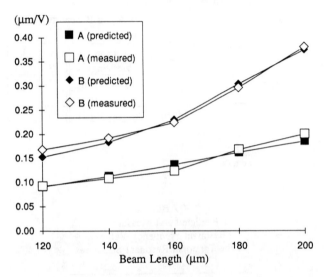

Fig. 10. Predicted and measured magnitude of the electromechanical transfer function for the linear resonant structures. The values are obtained at $V_p = 40$V. The fitting parameter, $\partial C / \partial x$, is found to be 58 aFμm^{-1} for type A comb drive and 150 aFμm^{-1} for type B.

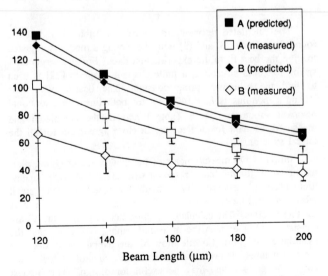

Fig. 9. Predicted and measured Q for types A and B linear resonant structures with different beam length.

VI. CONCLUSIONS

This paper has demonstrated the feasibility and outlined some of the design issues for the electrostatic comb drive and detection of laterally resonant polysilicon microstructures. From the initial visual measurements, it appears that the first-order theory for the transfer function and resonant frequency are adequate for predicting the observed behavior. However, the quality factor calculated from Couette drag due to shear flow beneath the structure consistently overestimates Q. It is likely that dissipation due to the air movement in the interdigitated fingers is a significant cause of the lower measured quality factors.

The measured resonant frequencies are accurately predicted for all the linear and torsional structures by simple beam theory, using a Young's modulus of 140 GPa and neglecting the residual strain in the released structures. The excellent fit with the measured frequencies over a range of truss dimensions and for the spiral and serpentine springs suggests that the residual strain is effectively relieved in these structures. Otherwise, it would be expected that a shift in frequency from the simple unstrained theory would be observed, with a greater shift for the shorter beams [11]. The value of 140 GPa is less than that measured for fine-grained, undoped polysilicon [11] and greater than that measured for phosphorus-doped, rapid-thermal-annealed polysilicon [23].

An important topic for further research is the simulation and verification of the resonant modes for these complex microstructures [24]. At atmospheric pressure, those modes with motion normal to the substrate are heavily damped, which greatly relaxes the design constraints on the electrostatic comb. For operation in vacuum, design of the drive structure to ensure excitation of a single mode will be challenging, due to the high intrinsic Q of polysilicon microstructures.

Acknowledgements

The authors wish to thank Y.-C. Tai for help with process development, L.-S. Fan for advice on polysilicon properties and for the Green's function solver, S. Hoagland, and R. M. Hamilton, and K. Voros of the UC Berkeley Microfabrication Facility for invaluable assistance in processing the devices.

REFERENCES

[1] R. T. Howe, "Resonant microsensors," *Technical Digest*, 4th International Conference on Solid-State Sensors and Actuators, Tokyo, Japan, June 2-5, 1987, 843-848.

[2] M. A. Schmidt and R. T. Howe, "Resonant structures for integrated sensors," *Technical Digest*, IEEE Solid-State Sensor Workshop, Hilton Head Island, South Carolina, June 2-5, 1986.

[3] R. M. Langdon, "Resonator sensors – a review," *J. Phys. E., Sci. Inst.*, **18**, 103-115, (1985).

[4] E. P. EerNisse and J. M. Paros, "Practical considerations for miniature quartz resonator force transducers," *Proceedings*, 28th ISA International Instrumentation Symposium, 1982, 33-44.

[5] J. G. Smits, H. A. C. Tilmans, and T. S. J. Lammerink, "Pressure dependence of resonant diaphragm pressure sensors," *Technical Digest*, 3rd International Conference on Solid-State Sensors and Actuators, Philadelphia, Pennsylvania, June 11-14, 1985, 93-96.

[6] J. C. Greenwood, "Etched silicon vibrating sensor," *J. Phys. E., Sci. Inst.*, **17**, 650-652, (1984).

[7] K. Ikeda, *et al*, "Silicon pressure sensor with resonant strain gauges built into diaphragm," *Technical Digest*, 7th Sensor Symposium, Tokyo, Japan, May 30-31, 1988, 55-58.

[8] D. C. Satchell and J. C. Greenwood, "Silicon microengineering for accelerometers," *Proc. Inst. of Mech. Eng.*, **1987-2**, Mechanical Technology of Inertial Devices, Newcastle, England, April 7-9, 1987, 191-193.

[9] R. T. Howe and R. S. Muller, "Resonant-microbridge vapor sensor," *IEEE Trans. on Electron Devices*, **ED-33**, 499-506, (1986).

[10] W. Benecke, *et al*, "Optically excited mechanical vibrations in micromachined silicon cantilever structures," *Technical Digest*, 4th International Conference on Solid-State Sensors and Actuators, Tokyo, Japan, June 2-5, 1987, 838-842.

[11] D. W. DeRoo, "Determination of Young's modulus of polysilicon using resonant micromechanical beams," M.S. Report, Dept. of Electrical and Computer Engineering, University of Wisconsin – Madison, January 1988.

[12] M. W. Putty, "Polysilicon resonant microstructures," M.S. Thesis, Dept. of Electrical Engineering and Computer Science, The University of Michigan, Ann Arbor, Mich., September 1988.

[13] M. A. Schmidt, "Microsensors for the measurement of shear forces in turbulent boundary layers," Ph.D. Thesis, Massachusetts Institute of Technology, Cambridge, Mass., May 1988.

[14] M. A. Schmidt, R. T. Howe, S. D. Senturia, and J. H. Haritonidis, "Design and calibration of a microfabricated floating-element shear-stress sensor," *IEEE Trans. on Electron Devices*, **35**, 750-757, (1988).

[15] A. P. Pisano, "Resonant-structure micromotors," *Technical Digest*, IEEE Micro Electromechanical Systems Workshop, Salt Lake City, Utah, February 20-22, 1989.

[16] R. T. Howe, "Integrated silicon electromechanical vapor sensor," Ph.D. Thesis, Dept. of Electrical Engineering and Computer Sciences, University of California at Berkeley, December 1984.

[17] A. M. Wahl, *Mechanical Springs*, 1st edition, (Cleveland: Penton), 1945.

[18] T. A. Lober and R. T. Howe, "Surface-micromachining processes for electrostatic microactuator fabrication," *Technical Digest*, IEEE Solid-State Sensor and Actuator Workshop, Hilton Head Island, South Carolina, June 6-9, 1988, 59-62.

[19] H. Guckel, T. Randazzo, and D. W. Burns, "A simple technique for the determination of mechanical strain in thin films with applications to polysilicon," *J. Appl. Phys.*, **57**, 1671-1675, (1985).

[20] T. A. Lober, J. Huang, M. A. Schmidt, and S. D. Senturia, "Characterization of the mechanisms producing the bending moments in polysilicon micro-cantilever beams by interferometric deflection measurements," IEEE Solid-State Sensor and Actuator Workshop, Hilton Head Island, South Carolina, June 6-9, 1988, 92-95.

[21] D. W. Burns, "Micromechanics of integrated sensors and the planar processed pressure transducer," Ph.D. Thesis, Dept. of Electrical and Computer Engineering, University of Wisconsin – Madison, May 1988.

[22] SuperSAP, Algor Interactive Systems, Inc., Essex House, Pittsburgh, Penn. 15206.

[23] M. W. Putty, S.-C. Chang, R. T. Howe, A. L. Robinson, and K. D. Wise, "Modelling and characterization of one-port polysilicon resonant microstructures," *Technical Digest*, IEEE Micro Electromechanical Systems Workshop, Salt Lake City, Utah, February 20-22, 1989.

[24] M. V. Andres, K. W. H. Foulds, and M. J. Tudor, "Sensitivity of a frequency-out silicon pressure sensor," *Technical Digest*, Eurosensors, 3rd Symposium on Sensors & Actuators, Cambridge, England, September 22-24, 1987, 18-19.

Electrostatic-comb Drive of Lateral Polysilicon Resonators

WILLIAM C. TANG, TU-CUONG H. NGUYEN, MICHAEL W. JUDY and ROGER T. HOWE

University of California at Berkeley, Department of Electrical Engineering and Computer Sciences and the Electronics Research Laboratory, Berkeley Sensor & Actuator Centre*, Berkeley, CA 94720 (U.S.A.)

Abstract

This paper investigates the electrostatic drive and sense of polysilicon resonators parallel to the substrate, using an interdigitated capacitor (electrostatic comb). Three experimental methods are used: microscopic observation with continuous or stroboscopic illumination, capacitive sensing using an amplitude-modulation technique and SEM observation. The intrinsic quality factor of the phosphorus-doped low-pressure chemical-vapor-deposited (LPCVD) polysilicon resonators is 49 000 ± 2000, whereas at atmospheric pressure, $Q < 100$. The finger gap is found to have a more pronounced effect on comb characteristics than finger width or length, as expected from simple theory

1. Introduction

Mechanical resonators are highly sensitive probes for physical or chemical parameters which alter their potential or kinetic energy [1]. Silicon resonant microsensors for measurement of pressure [2], acceleration [3], and vapor concentration [4] have been demonstrated. Recently, polysilicon micromechanical structures have been resonated electrostatically parallel to the plane of the substrate by means of one or more interdigitated capacitors (electrostatic combs) [5, 6]. Some advantages of this approach are (i) less air damping on the structure, leading to higher quality factors, (ii) linearity of the electrostatic-comb drive and (iii) flexibility in the design of the suspension for the resonator. For example, folded-beam suspensions can be fabricated without increased process complexity, which is attractive for releasing residual strain and for achieving large-amplitude vibrations [5, 6]. Such structures are of particular interest for resonant microactuators [7].

*An NSF/Industry/University Cooperative Research Center.

This paper reports the initial characterization of the electrostatic-comb drive and additional measurements on polysilicon resonators. Test structures are fabricated from a single layer of LPCVD polysilicon using a simple five-mask process [5, 6]. Variations in the finger lengths, widths and gaps between fingers in the comb are incorporated into a series of test structures. Observations of resonating structures under an optical microscope and a scanning electron microscope are used to measure directly the electromechanical transfer function and the quality factor of the mechanical resonance, Q. Finally, the motional current in the sense capacitor is found without on-chip circuitry by means of a carrier modulation technique.

2. Electrostatic-comb Drive

Figure 1 is an SEM of a linear resonant plate with two electrostatic-comb drives. The circuit configuration for resonating the device is shown in Fig. 2, where V_P is the drive d.c. bias and V_d is the a.c. drive voltage. The derivative of the drive capacitance with respect to lateral displacement, $\partial C/\partial x$, is constant for the comb drive for displacements much less than the finger overlap. Therefore, the electromechanical transfer function relating the phasor vibrational amplitude

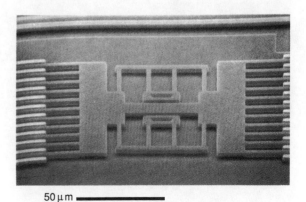

Fig. 1. SEM of a linear resonant plate with two electrostatic-comb drives.

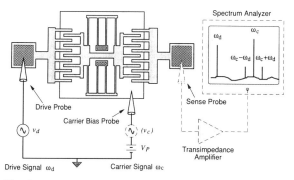

Fig. 2. Circuit schematic indicating the electrical connections necessary for driving a lateral resonant microstructure into resonance. The dotted lines correspond to additional circuitry required for motional current sensing via electromechanical modulation.

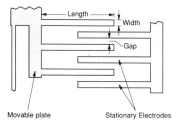

Fig. 3. Comb-structure dimensions.

X to the a.c. drive voltage V_d at resonance is given by [5, 6]

$$\left| \frac{X}{V_d} \right| = V_P \frac{Q}{k_{sys}} (\partial C/\partial x) \quad (1)$$

where Q and k_{sys} (system spring constant) are the mechanical characteristics of the resonator. The transconductance of the resonant structure, defined by $G(j\omega) = I_s/V_d$, where I_s is the phasor current in the sense electrode, is given by [5, 6]

$$\left| \frac{I_s}{V_d} \right| = \omega V_P V_S \frac{Q}{k_{sys}} (\partial C/\partial x)^2 \quad (2)$$

where V_S is the bias voltage between the structure and the stationary sense electrode.

In order to design the comb drive using these results, values are needed for $\partial C/\partial x$ and the quality factor Q of the resonance. The derivative $\partial C/\partial x$ is a function of the finger width and length, the comb gap, the polysilicon thickness, and the offset from the substrate (Fig. 3). The effects of different finger widths, lengths, and gaps are studied for the specific case of a 2 μm thick polysilicon resonator with 200 μm long folded flexures, which is suspended 2 μm above the substrate. The quality factor is determined by viscous drag from Couette flow under the resonant structure [5, 6] and by damping between the comb fingers. The latter contribution is evaluated using measurements on these structures.

3. Technique for Characterizing Resonant Microstructures

Several techniques have been developed to characterize resonant microstructures. They include visual techniques, in which vibrating plates are observed under high magnification (provided by a scanning electron microscope and optical microscopes) under continuous or stroboscopic illumination, and an electrical technique, which promises high accuracy and convenience.

Visual determination of resonant frequency and quality factor requires large driving voltages in air to provide sufficient vibrational amplitudes. Typical d.c. bias voltages V_P are 40 to 50 V, with a.c. drive-voltage amplitudes of about 10 V. Under continuous illumination, amplitudes are estimated by observing the envelope of the vibrating structures. By strobing the light source at a frequency 100 times less than that of the a.c. drive, the mode shape of the resonating structure can be observed.

Measurement of the current induced in the interdigitated sense electrode (Fig. 2) by motion of the structure is difficult without an on-chip buffer circuit [8]. However, this can be accomplished by superimposing a high-frequency a.c. signal on top of the d.c. bias which is applied to the structure. This signal serves as a carrier which is modulated by the time-varying sense capacitance. As a result, electrical feedthrough from fixed parasitic capacitors and the sense current due to the vibrating structure are separated in the frequency domain, as shown in Fig. 2.

4. Experimental Results

The resonant frequencies of the set of resonators with different comb geometries are listed in Table 1 with the comb dimensions defined in Fig. 3. The values obtained from both optical and electrical techniques are in close agreement. The calculated resonant frequencies in Table 1 are found using Rayleigh's method [5, 6]

$$f_r = \frac{1}{2\pi} \left[\frac{2Eh(W/L)^3}{(M_p + 0.3714M)} \right]^{1/2} \quad (3)$$

where h, W and L are the thickness, width and length of the supporting beams, respectively, and M_p and M are the masses of the plate and the beams. This equation assumes that the folded structure allows release of the residual compres-

TABLE 1. Calculated and measured resonant frequencies of a set of comb-drive structures suspended with 200 µm long beams

Comb characteristics				Resonant frequencies (kHz)			
No. of fingers	Finger length (µm)	Finger width (µm)	Gap (µm)	Calculated	Measured		
					Optical ±0.05	Strobe ±0.05	Electrical ±0.05
12	20	2	2	23.4	22.9	23.1	22.8
12	30	2	2	22.6	22.3	22.4	22.9
12	40	2	2	21.9	22.1	22.0	22.0
12	50	2	2	21.3	21.5	21.6	21.6
12	40	3	2	20.4	20.9	20.5	20.3
12	40	4	2	19.1	19.2	19.3	19.3
12	40	5	2	18.1	18.8	18.4	18.0
12	40	2	3	21.3	21.1	21.2	21.4
12	40	2	4	20.8	20.5	20.7	21.0
12	40	2	5	20.2	20.0	19.9	19.8

sive strain in the polysilicon film. A best-fit value for the Young's modulus for these structures is $E = 150$ GPa. An earlier processing run with a similar process has a best-fit Young's modulus of 140 GPa, somewhat lower than that from the data of Table 1.

Optical and electrical measurements of the quality factor Q are plotted in Fig. 4 for a set of 12-finger test structures with different finger gaps. The fingers are 40 µm long, with an overlap of 20 µm and 2 µm × 2 µm cross sections. An important observation from Fig. 3 is that Q is low for structures with either small finger gaps or widely separated fingers.

By measuring the electromechanical transfer function at resonance, eqn. (1) together with the calculated spring constant k_{sys} yields values of the derivative of drive capacitance with displacement. Figure 5 is a plot of $\partial C/\partial x$ as a function of the finger gap, which shows the expected sharp increase with reduced gaps. Figures 4 and 5 provide the empirical basis for designing electrostatic-comb drives.

The resonant behavior of an 11-finger comb structure is observed at a pressure of 1×10^{-7} Torr

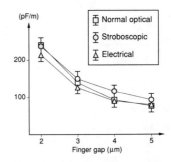

Fig. 5. Plots of $\partial C/\partial x$ vs. finger gap comparing results obtained via optical and electrical measurement techniques.

in a scanning electron microscope. Figure 6 is an SEM of the vibrating structure suspended by a pair of folded beams 140 µm long. The motion of the structure is lateral to the substrate, without any indication of torsional or vertical motion. It is important to note that the resonant frequency of the vertical mode is identical to that of the

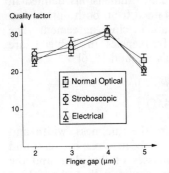

Fig. 4. Plots of quality factor vs. finger gap comparing optical and electrical measurement techniques.

Fig. 6. SEM of a vibrating microstructure showing no indication of any torsional or vertical motion under high vacuum (10^{-7} Torr).

designed lateral mode due to the square cross section of the suspensions. The electrostatic comb, with underlying ground plane, is therefore capable of cleanly driving just the lateral mode of the structure. Finally, the structure is observed to elevate about 200 nm upon application of the d.c. bias, an effect which warrants further study.

In the SEM, this structure resonates at $f_r = 31\,636.91 \pm 0.02$ Hz for a d.c. bias of 5 V. The quality factor is evaluated with both time domain and frequency domain methods

$$Q \simeq 1.43 T f_r \quad \text{and} \quad Q = \frac{f_r}{f_2 - f_1}$$

where T is the time for the oscillation amplitude to drop from 90% to 10% of its full amplitude after stopping the drive and $(f_2 - f_1)$ is the -3 dB bandwidth. The values of Q are $49\,000 \pm 2000$ and $50\,000 \pm 5000$ for the time and frequency domain methods, respectively. The drive efficiency for this design in a vacuum is measured to be 20 ± 2 μm/V under a d.c. bias of 5 V.

5. Conclusions

This paper has demonstrated three experimental methods for characterizing the electrostatic-comb drive of lateral polysilicon resonators. Transfer function and quality-factor measurements obtained with both optical and electrical techniques agree within the estimated experimental errors. Additional simulation and experimental studies are needed to fully characterize the comb drive; however, the initial results presented here provide empirical guidelines. The gap between comb fingers is found to be the most important design parameter for both the quality factor and the drive efficiency for operation at atmospheric pressure.

Observations of the resonator in the SEM demonstrate that the electrostatic comb is capable of selectively exciting only the lateral mode of oscillation. By strobing either the electron beam or the video signal, it is hoped that the motion of the structure can be observed more precisely.

Acknowledgements

The authors wish to thank D. S. Eddy of General Motors Research Laboratories for initiating discussion of electrostatic drive, J. J. Bernstein of the Charles Stark Draper Laboratory for suggesting the modulation technique, Professor R. M. White and T. A. Faltens of the University of California at Berkeley for initiating and helping with the SEM observations, and D. P. Schultz for the optical observations. This project is funded by General Motors Research Laboratories through the Berkeley Sensor and Actuator Center.

References

1 M. A. Schmidt and R. T. Howe, Resonant structures for integrated sensors, *IEEE Solid-State Sensor Workshop, Hilton Head Island, SC, U.S.A., June 2–5, 1986.*
2 K. Ikeda, H. Kuwayama, T. Kobayashi, T. Watanabe, T. Nishikawa and T. Yoshida, Silicon pressure sensor with resonant strain gages built into diaphragm, *Proc. 7th Sensor Symp., Tokyo, Japan, May 30–31, 1988*, pp. 55–58.
3 D. W. Satchell and J. C. Greenwood, A thermally-excited silicon accelerometer, *Sensors and Actuators, 17* (1989) 241–245.
4 R. T. Howe and R. S. Muller, Resonant-microbridge vapor sensor, *IEEE Trans. Electron Devices, ED-33* (1986) 499–506.
5 W. C. Tang, T.-C. H. Nguyen and R. T. Howe, Laterally driven polysilicon resonant microstructures, *IEEE Micro Electro Mechanical Systems Workshop, Salt Lake City, UT, U.S.A., Feb. 20–22, 1989*, pp. 53–59.
6 W. C. Tang, T.-C. H. Nguyen and R. T. Howe, Laterally driven polysilicon resonant microstructures, *Sensors and Actuators, 20* (1989) 25–32.
7 A. P. Pisano, Resonant structure micro-motors: historical perspective and analysis, *Sensors and Actuators, 20* (1989) 83–89.
8 M. W. Putty, S. C. Chang, R. T. Howe, A. L. Robinson and K. D. Wise, Process integration for active polysilicon resonant microstructures, *Sensors and Actuators, 20* (1989) 143–151.

ELECTROSTATICALLY BALANCED COMB DRIVE FOR CONTROLLED LEVITATION

William C. Tang, Martin G. Lim, and Roger T. Howe*

University of California at Berkeley
Department of Electrical Engineering and Computer Sciences
and the Electronics Research Laboratory
Berkeley Sensor & Actuator Center
Berkeley, California 94720

Abstract

This paper is an experimental study of the levitating force (*normal* to the substrate) associated with interdigitated capacitor (electrostatic comb) lateral actuators. For compliant suspensions, normal displacements of over 2 μm for a comb bias of 30 V are observed. This phenomenon is due to electrostatic repulsion by image charges mirrored in the ground plane beneath the suspended structure. By electrically isolating alternating drive-comb fingers and applying voltages of equal magnitude and opposite sign, levitation can be reduced by an order of magnitude, while reducing the lateral drive force by less than a factor of two.

INTRODUCTION

Surface-micromachined polysilicon resonators which are driven by interdigitated capacitors (electrostatic combs) have several attractive properties [1-3]. Vibrational amplitudes of over 10 μm are possible with relatively high quality-factors at atmospheric pressure, in contrast to structures which move normal to the surface of the substrate. The comb-drive capacitance is linear with displacement, resulting in an electrostatic drive force which is independent of vibrational amplitude. Electrostatic combs have recently been used for the static actuation of friction test-structures [4] and microgrippers [5].

Potential applications of lateral resonators include resonant accelerometers and rate gyroscopes, as well as resonant microactuators [6]. For efficient mechanical coupling between a vibrating pawl and a toothed wheel, it is essential that both structures remain co-planar. However, 2 μm-thick polysilicon resonators with compliant folded-beam suspensions have been observed to levitate over 2 μm when driven by an electrostatic comb biased with a DC voltage of 30 V. This effect must be understood in order to design functioning resonant microactuators, with the possibility that levitation by interdigitated combs may offer a convenient means for selective pawl engagement. In this paper, the electrostatic forces responsible for levitation are described, along with a modified comb design with independently biased fingers. For appropriate drive voltages, the levitation effect can be nearly eliminated. Experimental measurements of this effect are reported for a variety of comb structures.

VERTICAL LEVITATION AND CONTROL METHODS

Vertical Levitation

Successful electrostatic actuation of micromechanical structures requires a ground plane under the structure in order to shield it from relatively large vertical fields [7,8]. In previous studies of the electrostatic-comb drive, a heavily doped polysilicon film underlies the resonator and the comb structure. However, this ground plane contributes to an unbalanced electrostatic field distribution, as shown in Fig. 1 [9]. The imbalance in the field distribution results in a net vertical force induced on the movable comb finger. The positively biased drive comb fingers induce negative charge on both the ground plane and the movable comb fingers. These like charges yield a vertical force which repels or levitates the structure away from the substrate. The net vertical force, F_z, can be evaluated using energy methods:

$$E = q\Phi \tag{1}$$

where E is the stored electrostatic energy, q is the charge induced on the movable finger, and Φ is the potential. Differentiating with respect to the normal direction z yields

$$F_z = \frac{\partial E}{\partial z} = q\frac{\partial \Phi}{\partial z} + \Phi\frac{\partial q}{\partial z}. \tag{2}$$

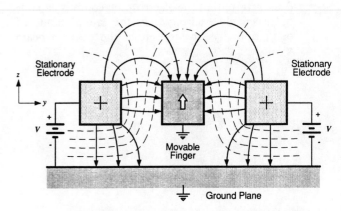

Fig. 1. Simplified cross section of a comb drive with the same voltage applied to all drive comb fingers, including potential contours (dashed lines) and the electric field distribution (solid lines).

Reprinted from *Technical Digest IEEE Solid-State Sensor and Actuator Workshop*, pp. 23-27, June 1990.

* Present Address: Xerox Palo Alto Research Center, 3333 Coyote Hill Road, Palo Alto, CA 94304

However, we have that

$$q \neq 0 \text{ and } \frac{\partial \Phi}{\partial z} \neq 0, \quad (3)$$

and thus,

$$F_z \neq 0. \quad (4)$$

Whether this force causes significant static displacement or excites a vibrational mode of the structure depends on the compliance of the suspension and the quality factor for vertical displacements.

Modified Comb Design

There are several means to reduce the levitation force. By eliminating the ground plane and removing the substrate beneath the structures, the field distribution becomes balanced. Alternatively, a top ground plane suspended above the comb drive will achieve a balanced vertical force on the comb. Both of these approaches require much more complicated fabrication sequences. A simpler solution is to modify the comb drive itself. Reversing the polarity on alternating drive fingers results in an altered field distribution, as shown in Fig. 2. Continuing with the energy analysis from (2) and noting that we now have

$$\frac{\partial \Phi}{\partial z} = 0 \text{ and } \Phi = 0, \quad (5)$$

it follows that

$$F_z = 0. \quad (6)$$

Controlling levitation may be achieved by changing the comb drive or by modifying the ground plane. Various structures have been developed to alternate the polarity at every stationary drive finger (Fig. 3 (a)), every other finger, every fourth finger (Fig. 3 (b)) and even every sixth drive finger. Three different designs addressing the role of the ground plane are implemented. The conventional ground plane (Fig. 3 (a)) is compared to comb drives with a recessed ground plane and a ground plane which extends only beneath the comb fingers (Fig. 3(c)).

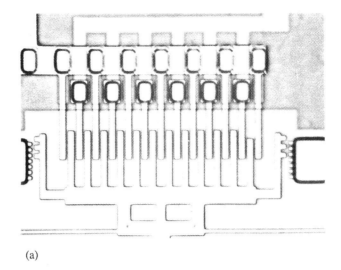

(a)

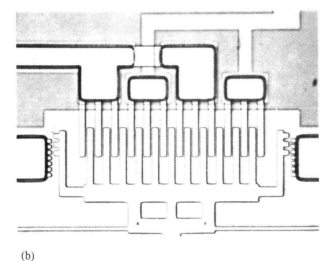

(b)

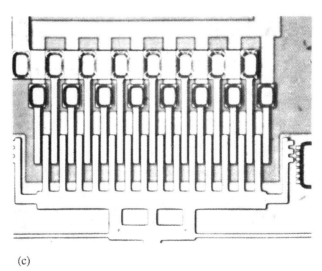

(c)

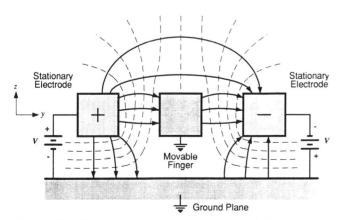

Fig. 2. Simplified cross section of a comb drive with equal and opposite voltages applied to alternating drive comb fingers.

Fig. 3. Optical micrograph of modified comb drives having two drive electrodes consisting of (a) alternating drive fingers and (b) sets of three drive fingers. (c) Modified comb drive with striped ground plane extending only under comb fingers.

EXPERIMENTAL TECHNIQUES

Levitation amplitudes are recorded from low-voltage SEM pictures at various DC biases on the combs. All structures are wired together to make possible the measurement of a number of structures in a single SEM session. The angle of tilt and magnification are fixed for comparison between different structures. Vertical displacements are evaluated by accurately measuring the SEM images with a set of standard linewidths. An SEM of a levitated structure is shown in Fig. 4.

Although measuring the vertical displacement as a function of the comb voltages initially demonstrates levitation control, it is more useful to calculate the vertical force for a particular comb geometry (gap between fingers, film thickness, and offset from the substrate.) In order to do this, the Young's modulus of polysilicon is found by fitting the measured lateral resonant frequency of the structure to the formula from Rayleigh's method [1,2]. Due to difficulties with the polysilicon plasma etching process, the cross section of the suspension is trapezoidal, with the width at the top of the beam $a = 2.2$ µm and the width at the bottom of the beam $b = 2.8$ µm, for a film thickness $h = 1.94$ µm. The expression for lateral resonant frequency of the test resonator in terms of structural dimensions and Young's modulus is [10]:

$$f_r = \frac{1}{2\pi}\left[\frac{2Eh[a^3 + (b^4-4ba^3+3a^4)/4(b-a)]}{L^3(M_p+0.25M_t+0.343M_b)}\right]^{\frac{1}{2}} \quad (7)$$

where L is the length of the folded beam (400 µm), M_p is the plate mass, M_t is the mass of the outer connecting trusses, and M_b is beam mass. Therefore, E can be expressed as

$$E = \frac{2\pi^2 f_r^2 L^3 (M_p+0.25M_t+0.343M_b)}{h[a^3 + (b^4-4ba^3+3a^4)/4(b-a)]} \quad (8)$$

Using a polysilicon density of 2.3×10^3 kg·m^{-3}, the Young's modulus is found to be $E = 150$ GPa for this process run, which is consistent with earlier results [1,2]. The vertical spring constant is found from the Young's modulus and the geometry of the suspension:

$$k_z = \frac{2h^3(a^2+4ab+b^2)E}{3(a+b)L^3} = 86 \text{ nN·µm}^{-1}. \quad (9)$$

Finally, the levitation force is given by $F_z = k_z \Delta z$ where Δz is the vertical displacement.

Similarly, the lateral (drive) force of the comb is found by observing the lateral displacement versus voltage, using a vernier with a 0.5 µm scale. The lateral force, F_l, is then evaluated with $F_l = k_x \Delta x$, where Δx is the lateral displacement, with the lateral spring constant given by

$$k_x = \frac{2h}{L^3}\left[a^3 + \frac{(b^4-4ba^3+3a^4)}{4(b-a)}\right]E = 140 \text{ nN·µm}^{-1}. \quad (10)$$

EXPERIMENTAL RESULTS

Levitation for unbalanced comb drives

Levitation is observed by applying a voltage of 0 to 25 V to all drive fingers and is plotted for the test structure in Fig. 5. The vertical displacement increases with applied voltage and reaches an equilibrium near 20 V where the attractive forces between the displaced interdigitated fingers offset the repulsive electrostatic forces between the ground plane and movable fingers. The suspension restoring force also retards levitation. The initial negative deflection for a grounded comb, shown in Fig. 5, cannot be attributed to gravity. Charging effects in the dielectric films are a possible source of this offset displacement. We define the vertical drive capacity, γ_v, as the levitation force per square of the applied voltage: $\gamma_v = F_v V^{-2}$ [N·V^{-2}]. Using the technique described in the

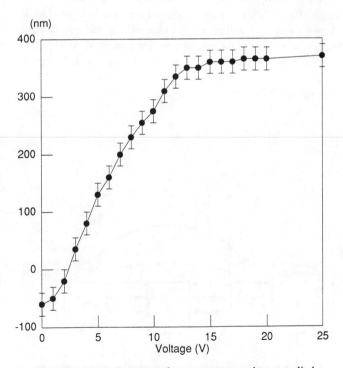

Fig. 4. SEM micrograph of a comb drive with 10V DC bias on the drive. The levitation is 510 ± 20 nm by direct linewidth measurement. Note that the drive fingers, because of the positive bias, appear darkened in the SEM.

Fig. 5. Vertical displacement for a common voltage applied to both electrodes of a modified comb-drive structure with 18 movable comb fingers and 19 fixed drive fingers, each of which is 40 µm long, 4 µm wide, and 2 µm thick. The gap between fingers is about 2 µm and nominal distance from substrate is also 2 µm.

previous section, the levitation force is estimated to be 900 pN·V^{-2} of DC bias at $z = 0$ on this particular 19 drive-finger comb.

Levitation for modified comb drives

Figure 6 is a plot of the levitation resulting from holding one set of an alternating drive fingers at 15V and varying the other set of electrodes from -15V to 15V. The structure is the same as that tested in Fig. 5. As expected, negative voltages in the range of -10 V to -15 V suppress the lifting behavior. As the disparity between the magnitudes of the voltages increases, more lifting occurs, with the limiting case of +15 V applied to all drive fingers yielding the same vertical displacement as found in Fig. 5.

Drive capacity vs levitation control

We define the lateral drive capacity, γ_l, of an electrostatic comb drive as the lateral force per square of the applied voltage: $\gamma_l = F_l V^{-2} [N \cdot V^{-2}]$. Experimental results show that the drive capacity of the balanced comb drive is less than that of the unbalanced one with same geometry and number of fingers. The suppression ranges from 50% to 90% of the unbalanced drive, depending on the interdigitation method and ground plane design.

Both γ_l and γ_v for various balanced comb designs normalized to each drive finger are tabulated in Table I. The normalized value of γ_v at $z = 0$ for the unbalanced combs is 47 ± 1 pN·V^{-2} per drive finger, and the normalized γ_l being 16 ± 1 pN·V^{-2} per drive finger. None of the comb designs completely eliminates the levitation as predicted by idealized theory, which assumes evenly spaced comb fingers with vertical sidewalls and neglects native oxide films, which may serve as a charge-trapping dielectric layer.

Table I

Normalized vertical and lateral drive capacities per drive finger

Type	γ_v at $z = 0$ ± 1 pN·V^{-2}	γ_l ± 1 pN·V^{-2}
A	3	10
B	26	8
C	30	11
D	35	12
E	36	14
F	38	14

A is a structure with striped ground plane and the others are with conventional ground plane. The electrodes of B alternate every drive comb finger; C, every other finger; D, every third finger; E, every fourth finger; and F, every sixth finger. Comb fingers are 40 μm long, 4 μm wide, and 2 μm thick. The gap is nominally 2 μm and the comb is offset from the substrate by 2 μm.

Several qualitative observations were noted during testing of these structures. First, although vertical levitation can be induced with DC bias, none of the structures responds to vertical AC excitation even in vacuum. Therefore, we are unable to obtain the vertical resonant frequencies. Second, those structures with recessed ground planes behave unpredictably, especially when a step change is applied to the DC bias. They have a tendency to be pulled down and become stuck to the substrate, which may be due to charging effects in the underlying dielectric passivation layers.

CONCLUSIONS

We have successfully evaluated and quantified levitation induced by electrostatic comb drives by direct tests in an SEM, which provides insights into designing structures for controlled out-of-plane motions. Among the various designs studied in this experiment, the best levitation suppression is obtained by alternating the two electrodes at every drive finger with a striped ground plane underneath the comb structure. Levitation can be further reduced by designing structures with vertically stiff suspensions. If controlled levitation is desired, soft suspensions can be used together with ratioed differential and common mode voltages applied to the two electrodes. Vertical AC excitation, charging effects, and etching processes for achieving vertical sidewalls are topics for further research.

Acknowledgements

The authors wish to thank Charles Hsu for his assistance in the fabrication process and the staff at the Berkeley Microfabrication Laboratory. This project was funded by the Berkeley Sensor & Actuator Center, an NSF/Industry/University Cooperative Research Center.

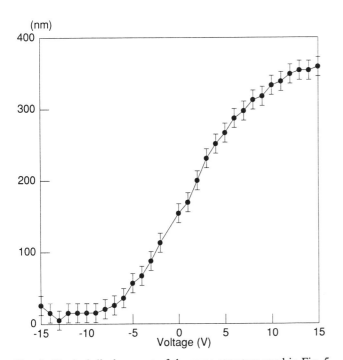

Fig. 6. Vertical displacement of the same structure used in Fig. 5 for varying voltage on one electrode of a modified comb structure, with the other electrode fixed at +15 V.

REFERENCES

[1] W. C. Tang, T.-C. H. Nguyen, and R. T. Howe, "Laterally driven polysilicon resonant microstructures," *Technical Digest,* IEEE Micro Electromechanical Systems Workshop, Salt Lake City, Utah, February 20-22, 1989, 53-59.

[2] W. C. Tang, T.-C. H. Nguyen, M. W. Judy, and R. T. Howe, "Electrostatic-comb drive for lateral polysilicon resonators," *Technical Digest,* 5th International Conference on Solid-State Sensors and Actuators, Montreux, Switzerland, June 25-30, 1989, 138-140.

[3] R. A. Brennen, A. P. Pisano, and W. C. Tang, "Multiple Mode Micromechanical Resonators," *Technical Digest,* IEEE Micro Electromechanical Systems Workshop, Napa Valley, Calif., February 12-14, 1990, 9-14.

[4] M. G. Lim, J. C. Chang, D. P. Schultz, R. T. Howe, and R. M. White, "Polysilicon Microstructures to Characterize Static Friction," *Technical Digest,* IEEE Micro Electromechanical Systems Workshop, Napa Valley, Calif., February 12-14, 1990, 82-88.

[5] C. J. Kim, A. P. Pisano, R. S. Muller, and M. G. Lim, "Polysilicon Microgripper," *Technical Digest,* IEEE Solid-State Sensor and Actuator Workshop, Hilton Head Island, S.C., June 4-7, 1990.

[6] A. P. Pisano, "Resonant-structure micromotors," *Technical Digest,* IEEE Micro Electro Mechanical Systems Workshop, Salt Lake City, Utah, February 20-22, 1989, 44-48.

[7] Y.-C. Tai, L. S. Fan, and R. S. Muller, "IC-processed micromotors: design, technology, and testing," *Technical Digest,* IEEE Micro Electro Mechanical Systems Workshop, Salt Lake City, Utah, February 20-22, 1989, 1-6.

[8] M. Mehregany, P. Nagarkar, S. D. Senturia, and J. H. Lang, "Operation of microfabricated harmonic and ordinary side-drive motors," *Technical Digest,* IEEE Micro Electromechanical Systems Workshop, Napa Valley, Calif., February 12-14, 1990, 1-8.

[9] Maxwell Solver, v. 4.20, Ansoft Corp., 4 Station Square, 660 Commerce Court Building, Pittsburgh, Pa. 15219.

[10] J. Gere and S. Timoshinko, *Mechanics of Materials*. 2nd ed., Belmont: Wadsworth, Inc., 1984.

POLYSILICON MICROGRIPPER

Chang-Jin Kim, Albert P. Pisano, Richard S. Muller, and Martin G. Lim*

Berkeley Sensor & Actuator Center
An NSF/Industry/University Cooperative Research Center
University of California, Berkeley, California 94720

ABSTRACT

A polysilicon, electrostatic, comb-drive microgripper has been designed, fabricated, and tested. Its main features are a flexible, cantilever comb-drive arm with a bidirectional actuation scheme and an over-range protector. Three different electromechanical models are developed and, along with fabrication constraints, are employed to design the microgripper and to simulate its performance. Experiments have demonstrated that a gripping range of 10μm can be accommodated with an applied potential of 20V. The motion dependence on drive voltage has been measured and compared with model prediction. The gripper motion is observed to be smooth, stable, and controllable. Measurements were carried up to the maximum of the voltage source (50V).

1. INTRODUCTION

Recent developments in micromechanics have led to success in building new types of microactuators [1, 2], elements which are essential building blocks for many micromechanical systems. Microgrippers capable of handling tiny objects have direct applications as end-effectors for currently available "macro"-sized micromanipulators. Mehregany, Gabriel, and Trimmer [3] have fabricated silicon tongs to be actuated by external means. Chen and co-workers [4] demonstrated tungsten microtweezers actuated by a voltage applied across two cantilevers about 2μm apart when fully open. These tweezers had a closure voltage of roughly 150V.

We report a prototype surface micromachined polysilicon microgripper. To avoid frictional effects and the relatively large clearance of micro-joints, the present microgripper is based on flexure structures. An electrostatic comb driver [2] has been modified for the gripper to obtain a reasonably large static movement. The microgripper has been modeled in three different ways. A simple, lumped-beam model has been used to predict the gripping range and force. Testing shows the required driving voltage of about 20V to accommodate a gripper range of 10μm, which is adequate to handle objects of the size range of micrometers, including most bacteria and animal cells. The measured gripper motion as a function of drive voltage is reported along with model predictions.

* Present address: Xerox Palo Alto Research Center
Palo Alto, California

2. DESIGN

The microgripper consists of two movable gripper tines driven by three electrostatic comb drivers shown in Fig. 1. The central gripper tines, labeled *drive arm* on Fig. 1, extend to sections labeled *extension arm* and *gripper tip*. The two electrode drivers, labeled *open driver* and *close driver* are attached to the substrate and actuate the gripper by bending the drive arm through the electrostatic fringe-field force of the comb-actuation mechanism. Lengths of drive and extension arms, L_{dr} and L_{ext}, are 400 and 100μm, respectively.

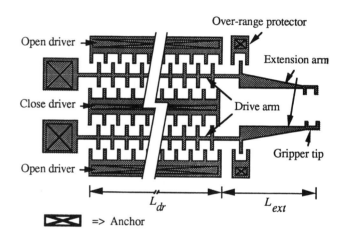

Fig. 1 Schematic of the microgripper

Electrostatic comb-drive actuation has been selected because of its stable drive over relatively long distances. It stands in contrast to actuation by fields applied across a variable-gap capacitor which results in unstable drive available over only very short distances (about 2μm, in practice). The generated fringe-field force by comb actuators, however, is generally smaller than the direct force across parallel-plate capacitors. To increase force and obtain the maximum displacement with this comb-drive mechanism, a cantilever with a large number of comb teeth has been designed.

The provision of separate open- and close-drivers doubles the effective gripping range for a given maximum voltage. The extension arm separates manipulated objects from the electric field in the comb area and also increases the gripping range. To reduce deformation while gripping objects, the extension arm is designed to be rigid compared to the flexible-drive arm. The over-range protector is at the same potential as the gripper tine. If a high voltage is applied during operation, this protector prevents contact between the drive arm and open drivers and thus enables repeatable operation even with electrical overdrive.

Configuration of the interdigitated combs used for the microgripper is shown in Fig. 2. The width of the comb teeth and beam b is 2μm, the gap between the drive-arm comb teeth and driver-comb teeth d is 3μm, and the length of the comb teeth is 10μm. There is no initial tooth overlap.

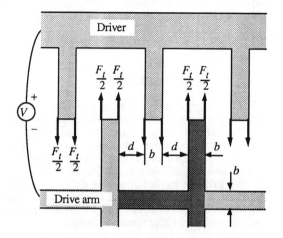

Fig. 2 Comb configuration

3. MODELS

Three electromechanical models of the microgripper have been investigated in which (a) deflections of beams are assumed to be very small compared to the beam lengths so that the rotation effect of the beam can be neglected, and (b) lengths of teeth are assumed to be much larger than the beam width which implies rigid-body behavior of the teeth as parts of the beams.

3.1 Concentrated-force model

Actuation is due to the fringe-field electrostatic force between the driver and drive-arm teeth. The drive arm is assumed to be a combination of m successive comb elements (Fig. 3). The actuation force along each side of an individual comb tooth is [2]

$$\frac{F_t}{2} = \frac{\varepsilon t}{2d} V^2 \quad (1)$$

where ε is the permittivity of air, d is the gap between the drive-arm comb teeth and driver-comb teeth, and t is the thickness of the gripper. The pitch of the comb is p, and p' is the length of the flexible part of each comb element shown darker in Fig. 2. The gripper-tip displacement δ is calculated by equations (1) - (6), which are based on the free-body diagram of Fig. 3.

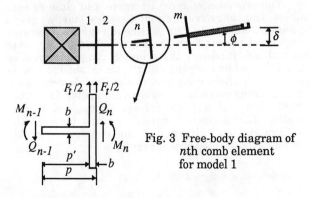

Fig. 3 Free-body diagram of nth comb element for model 1

$$\delta = \sum_{n=1}^{m} \delta_n + \phi_m L_{ext} \quad (2)$$

$$\delta_n = p\phi_{n-1} + b(\phi_n - \phi_{n-1}) + \frac{3bp'^2 + 2p'^3}{6E_Y I} Q_{n-1} + \frac{p'^2}{2E_Y I} M_n - \frac{bp'^2}{4E_Y I} F_t \quad (3)$$

$$\phi_n = \phi_{n-1} + \frac{2bp' + p'^2}{2E_Y I} Q_{n-1} + \frac{p'}{E_Y I} M_n - \frac{bp'}{2E_Y I} F_t \quad (4)$$

$$Q_n = Q_{n-1} - F_t \quad (5)$$

$$M_n = M_{n-1} + \frac{b}{2} F_t - pQ_{n-1} \quad (6)$$

where $n = 1, 2 \ldots m$, and $\delta_0 = \phi_0 = Q_m = M_m = 0$. E_Y is modulus of elasticity, and I is the second moment of inertia of the cross section of the flexible portion of the comb element.

3.2 Uniformly distributed-force model

Another model assumes a uniformly distributed load by averaging the fringe-field electrostatic force over the entire drive arm. This assumption should be admissible as long as the drive arm has a dense population of comb teeth. This model is similar to the concentrated-force model but employs a distributed load $w = F_t/p$ as shown in Fig. 4. The gripper-tip displacement δ is calculated by equations (1), (2), (7)-(10).

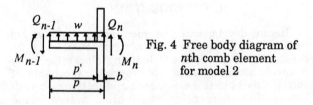

Fig. 4 Free body diagram of nth comb element for model 2

$$\delta_n = p\phi_{n-1} + b(\phi_n - \phi_{n-1}) + \frac{3bp'^2 + 2p'^3}{6E_Y I} Q_{n-1} + \frac{p'^2}{2E_Y I} M_n - \frac{6b^2 p'^2 + 8bp'^3 + 3p'^4}{24 E_Y I} w \quad (7)$$

$$\phi_n = \phi_{n-1} + \frac{2bp' + p'^2}{2E_Y I} Q_{n-1} + \frac{p'}{E_Y I} M_n - \frac{3b^2 p' + 3bp'^2 + p'^3}{6E_Y I} w \quad (8)$$

$$Q_n = Q_{n-1} - pw \quad (9)$$

$$M_n = M_{n-1} + \frac{p^2}{2} w - pQ_{n-1} \quad (10)$$

3.3 Lumped-beam model

A third model averages the compliance of the drive arm. We consider that p is very small so that an infinitesimal analysis can be applied. A comb element under pure bending moment M is shown in Fig. 5.

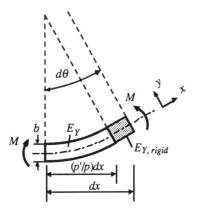

Fig. 5 A comb element under pure bending moment

The average strain along the comb element, dx, is (assuming $E_{Y,rigid} \rightarrow \infty$)

$$\varepsilon_x = \frac{\sigma_x}{E_Y}\left(\frac{p'}{p}\right) \quad (11)$$

From geometry, however, the strain is also

$$\varepsilon_x = \frac{d\theta}{dx} y \quad (12)$$

Moment equilibrium with eqns. (11) and (12) gives

$$M = \int_A y\sigma_x dA = \frac{E_Y}{(p'/p)}\frac{d\theta}{dx}\int_A y^2 dA = \frac{E_Y I}{(p'/p)}\frac{d\theta}{dx} \quad (13)$$

where $I = tb^3/12$. Eqn. (13) implies that the whole drive arm can be treated as a homogeneous flexible beam with flexural rigidity, $E_Y I(p/p')$. Using this lumped-beam model, the deflection along the gripper tine is

$$\delta(x) = \frac{wx^2}{24 E_Y I(p/p')}(x^2 - 4L_{dr}x + 6L_{dr}^2), \text{ for } 0 \leq x \leq L_{dr}$$

$$\delta(x) = \frac{wL_{dr}^4}{8 E_Y I(p/p')} + \frac{wL_{dr}^3}{6 E_Y I(p/p')}(x - L_{dr}), \text{ for } L_{dr} \leq x \leq L_{dr} + L_{ext} \quad (14)$$

The comparison of the predictions of these three models in Fig. 6 shows that the lumped-beam model, as simple as it is, generates numerical results that are very similar to the two more complicated models. The results shown in Fig. 6 are for an applied potential of 20V (F_t = 2.5nN).

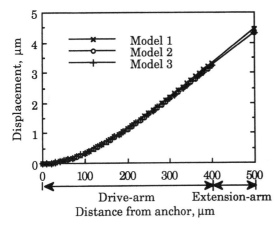

Fig. 6 Comparison of three different models

3.4 Gripping force

The gripping force F_{gr} can be estimated by assuming a small gripper deflection and a rigid object. The initial displacement of the gripper tip before reaching the object is δ_0. Using eqn. (14) of the previous lumped-beam model in Fig. 7, we can obtain gripping force as in eqn. (15). We can calculate, for example, that when δ_0 is 2μm a drive voltage of 30V generates 44nN of gripping force. This force is almost 200 times the weight of a silicon cube 10μm on a side, and would be very effective in restraining it.

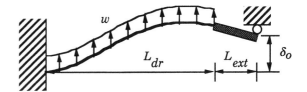

Fig. 7 Model for gripping-force estimation

$$F_{gr} = \frac{\frac{w}{24 E_Y I(p/p')}L_{dr}^3(3L_{dr} + 4L_{ext}) - \delta_0}{\frac{L_{dr}}{E_Y I(p/p')}\left(\frac{L_{dr}^2}{3} + L_{dr}L_{ext} + L_{ext}^2\right)} \quad (15)$$

The predicted deformation of the neutral axis of the gripper arm before and after the gripper tip reaches the object is shown in Fig. 8 as a function of the drive voltage.

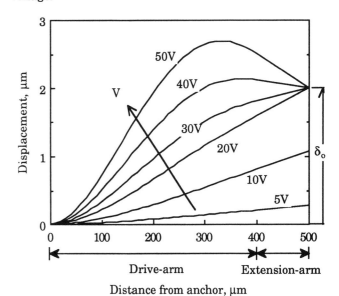

Fig. 8 Simulated neutral-axis deformation of a gripper arm during gripping

4. FABRICATION AND TESTING

Fabrication has been based on a four-mask process as described in Lim et al. [5]. A picture of a completed microgripper, including temporary breakaway support beams attached to the necks of the gripper tips, built-in scale next to the gripper tips, and over-range protectors at the left ends of the extension arms, is shown in Fig. 9. Dimples along each gripper tine prevent the gripper from sticking to the substrate during the final step of freeing from the sacrificial PSG layer [1]. A closeup view of the gripper tips along with the built-in scale is shown in Fig. 10. The fabricated microgrippers are 2.2µm thick and have 2.1µm-wide beams and comb teeth.

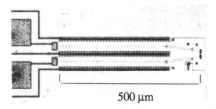

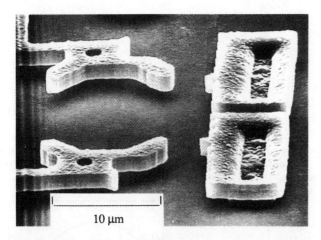

Fig. 10 SEM picture of gripper tip and built-in scale

After freeing the gripper tip from temporary support beams using a probe tip under the microscope, a dc bias was applied between the drive arm and open- or close-driver. The displacements of gripper tips have been measured with the built-in scales to give the results shown in Fig. 11. Also given on Fig. 11 are predictions of the lumped-beam model. The effect of finite beam thickness on the actuation force has been considered through a FEM analysis. The movement of the microgripper is observed to be smooth and stable, and the displacement is repeatable and controllable.

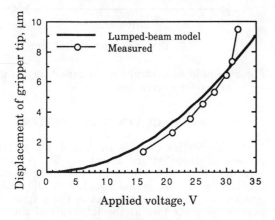

Fig. 11 Displacements of gripper tip

An applied potential of 20V resulted in 2-3µm movements of each gripper tip in both the open and close direction. Defining gripping range as the difference between the maximum-tip separation and minimum-tip separation for a given maximum applied voltage, we obtained about 10µm of gripping range with 20V as the maximum applied potential. In Fig. 11, voltages higher than 20V are to show roughly quadratic dependence of displacement on applied voltage. The assumptions for the models are not valid for large displacements, and the measured displacements deviate from the predicted values at high voltages. For gripper opening, the over-range protector worked as intended, resulting in repeatable gripper motion even after excessive driving voltages. Up to 50V (maximum of power source) were applied, but the over-range protector prevented further displacements at potentials higher than 32V. For gripper closing, higher voltages than that required for complete gripper closure resulted in tine bending comparable to that shown in Fig. 8.

5. CONCLUSIONS

In this research, we have demonstrated a polysilicon, electrostatic comb-drive microgripper. The electrostatic comb driver, developed originally for resonators [2], has been modified and used to provide static drive for a microgripper. The microgripper features a flexible, cantilever comb-drive arm with a bidirectional actuation and an over-range protector. Performance of the microgripper under gripping force has been simulated using three different models. Experiments showed that the microgripper has met its design goal of providing gripper-tip movement of several micrometers at drive voltages less than 20V. The gripper motion was observed to be smooth, stable, and controllable.

REFERENCES

[1] L.-S. Fan, Y.-C. Tai, and R. S. Muller, "IC-processed Electrostatic Micro-motors." *IEEE Int. Electron Devices Meeting,* pp. 666-669, San Francisco, CA, Dec. 1988.

[2] W. C. Tang, T. H. Nguyen, and R. T. Howe, "Laterally Driven Polysilicon Resonant Structures," *Proc. IEEE Micro Electro Mechanical Systems Workshop,* pp. 53-59, Salt Lake City, UT, Feb. 1989.

[3] M. Mehregany, K.J. Gabriel, and W. S. Trimmer, "Integrated Fabrication of Polysilicon Mechanisms," IEEE Trans. on Electron Devices, Vol. 35, No. 6, pp. 719-723, June 1988.

[4] L. Y. Chen, Z. L. Zhang, J. J. Yao, D. C. Thomas, and N. C. MacDonald, "Selective Chemical Vapor Deposition of Tungsten for Microdynamic Structures," *Proc. IEEE Micro Electro Mechanical Systems Workshop,* pp. 82-87, Salt Lake City, UT, Feb. 1989.

[5] M. G. Lim, J. C. Chang, D. P. Schultz, R. T. Howe, and R. M. White, "Polysilicon Microstructures for Characterize Static Friction." *Proc. IEEE Micro Electro Mechanical Systems Workshop,* pp. 82-88, Napa Valley, CA, Feb. 1989.

Section 4

Electrostatic Actuators

ELECTROSTATIC actuators are governed by the equation

$$F = -\frac{\partial U}{\partial x}$$

where F is the electrostatic force generated between the plates of the actuator, U is the energy contained in the electrostatic field, and the derivative is with respect to the motion, x, of one of the actuator plates. Fig. 4.1 shows the configuration for two plates. The simplest assumption is the parallel plate capacitor approximation, which ignores edge effects. The energy under this assumption is

$$U = -\frac{1}{2} C V^2 = -\frac{\varepsilon w L V^2}{2d}$$

The force, F, along the d direction (pulling the plates together) is given by

$$F_d = -\frac{\partial U}{\partial d} = -\frac{\varepsilon w L V^2}{2d^2} = -\frac{1}{2} \varepsilon A E^2$$

The force in the w direction (the motion shown as x in Fig. 4.1) is given by

$$F_w = -\frac{\partial U}{\partial w} = \frac{\varepsilon L V^2}{2d} = \frac{\varepsilon L d V^2}{2d^2} = \frac{1}{2} \varepsilon L d E^2$$

where ε is the permittivity ($\varepsilon = 8.85 \times 10^{-12}$ F/m for empty space), A is the area of the plates, E is the electrostatic field (usually equal to about 100 V across a micron, or 10^8 V/m).

The electrostatic force of the plates is most easily thought of as a force per unit area, or a pressure. For F_d, this area is the area of the plates, and for F_w, the area is the cross section between the plates. For a typical electrostatic actuator with 100 V across a 1-μm air gap, this pressure is 4.42×10^4 N/m². Hence, the pressure generated by an electrostatic actuator, $P = F_d/A$, is about 6 lbf/in², or approximately half an atmosphere. As an example, the force generated by 100 V between two plates 100 μm on a side, and separated by a micron is approximately 4.42×10^{-4} N, or 2×10^{-3} lb.

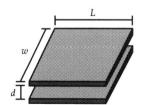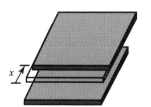

Fig. 4.1

To move a block of material 100 μm on a side, the electrostatic force is appreciable. It can generate accelerations of over 100,000 gs, quite sufficient for most applications. However, to counterbalance another pressure (for example, in a valve), one often wants pressures larger than half an atmosphere.

How can the electrostatic force be increased? A material of higher dielectric permittivity can be used between the plates. This increases the force, but may also increase the viscous drag. A higher voltage can be used across the plates. The increases in force scale as the square of the voltage. Electrical breakdown, reliability, and available power supplies limit the maximum voltage.

Electrostatic actuation is an excellent source of force for many applications, but not a universal solution.

The papers in this section describe a wide variety of ways to make electrostatic actuators. The first paper, "The Principle of an Electrostatic Linear Actuator Manufactured by Silicon Micromachining," describes a way to couple motion off a silicon wafer using a cylindrical roller that is attracted to electrodes on the wafer. The second paper, "Design Considerations for a Practical Electrostatic Micro-Motor," examines electrostatic actuators using parallel plates. "SCOFSS: A Small Cantilevered Optical Fiber Servo System," discusses how to control the motion of a thin, flexible beam using multiple electrodes, and also discusses the use of electrets. The fourth paper, "Microactuators for Aligning Optical Fibers," discusses the use of curved electrodes to generate larger electrostatic forces. This actuation is used to build a fiber switch. In the fifth paper, "Large Displacement Linear Actuator," electrostatic actuation over a small gap is mechanically coupled to produce a large motion. In "Multi-Layered Electrostatic Film Actuator," a thin film actuator induces image charges on alternating layers of stacked films. Movable suspended silicon plates are electrostatically moved in "Movable Micromachined Silicon Plates with Integrated Position Sensing" and the motion detected by changes in capacitance. The next paper, "Micro Electro Static Actuator with Three Degrees of Freedom," describes a three-degree-of-freedom actuator suitable for use as a micro manipulator. The last six papers, "The Modeling of Electrostatic Forces in Small Electrostatic Actuators," "Silicon Electrostatic Motors," "Electrostatic Actuators for Micromechatronics," "Electric Micromotors: Electromechanical Characteristics," "Electroquasistatic Induction Micromotors," and "A Perturbation Method for Calculating the Capacitance of Electrostatic Motors" discuss the design implications of electrostatic motors.

THE PRINCIPLE OF AN ELECTROSTATIC LINEAR ACTUATOR MANUFACTURED BY SILICON MICROMACHINING

HIROYUKI FUJITA, AKITO OMODAKA

Institute of Industrial Science
The University of Tokyo
22-1, Roppongi 7-chome, Minato-ku,
Tokyo 106 JAPAN

ABSTRACT

The actuator consists of a plane wafer with striped electrodes covered with insulation layer and thin cylindrical rollers. The roller is placed parallel to striped electrodes on the insulated surface of the plane. Voltage applied between the roller and the electrode forms electric field which attracts the roller to the electrode. If voltage is applied to some of the striped electrodes at appropriate position, the roller begins rotating. As the roller rotates, the voltage on the electrode is switched forward to keep the roller moving farther. The electric field and the force between striped electrodes and the roller are calculated. The dependence of these values on the dimension of each electrode is discussed. The design and assembly of an enlarged model are described. The performance of the model actuator is measured experimentally.

INTRODUCTION

Motion control in future will need more flexibility, more intelligence, and higher degree of freedom. The higher degree of freedom requires small, light and high precision actuators. It is not easy to fill the need with mere improvement of conventional actuators.

We propose an alternative way of driving, the electrostatic force. The idea of the electrostatic drive is old[1] but the real application has been restricted by its low power density. The advantage of the electrostatic force appears in sub-mm range for the following reasons:
(1) Thin insulation layers withstand higher electric stress than thick layers. Breakdown strength as high as 2 MV/cm is easily obtained in SiO_2 thin film. The power density of the electrical field reaches 7×10^5 J/m^3; this value is equal to that of 1.3 T magnetic field.
(2) It is not the volume but the surface force and needs only two electrodes separated by an insulator. The silicon micromachining technology enables us to manufacture minute electrode patterns and thin insulation layers of high quality.
(3) It is not driven by current but by voltage. Voltage switching is far easier and faster than the current switching. Energy loss associated with Joule heating is also small.

CONFIGURATION OF THE ACTUATOR

The actuator consists of a plane wafer with striped electrodes embedded in an insulation layer and a cylindrical electrode (roller) as shown in Fig.1 and Fig.2. The position of the roller is determined by position detectors. The striped electrodes on the one side of the roller are activated to form an electric field against the roller (Fig.5). The electrostatic force attracts the roller and makes it rotate. As the roller passes over a striped electrode, the electrode is discharged and another in the forward direction is charged to keep the movement. An ESLAC (electrostatic linear actuator) consists of

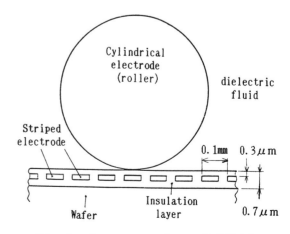

Fig.1 Cross section of ESLAC.

rollers sandwiched by the plane wafers which displace linearly. The approximate dimensions of ESLAC are shown in Fig.1. Breakdown strength of insulator is over 1MV/cm, which gives applied voltage in the order of 10V.

ELECTRIC FIELD ANALYSIS

Analytical solution

Let us derive a 'handy' analytical solution on following assumptions:
(1) The two dimensional analysis can be applied.
(2) Striped electrodes can be replaced by two half infinite planes; the voltage of one plane is U and the other is grounded (Fig.3 (a)). The roller lies along the boundary of two planes.
(3) The electric field in the right half plane in Fig.3 (a) can be replace by the field in a cylinder-to-plane system (the meshed region in Fig.3 (b)). The field in the left half plane is assumed to be zero.
(4) The permittivity of the insulation layer and fluid is uniform (ϵ).

On the assumptions above, the problem becomes the calculation of the field in the cylinder-to-plane system. One can obtain the field with either the conformal transformation or the image method.[2] Electric field strength $E(\theta)$ is given as follows with the angle θ in Fig.3 (b) and $\alpha = r/d$.

$$E(\theta) = \frac{\sqrt{1+2\alpha}}{\alpha(1+2\alpha \cos^2\frac{\theta}{2})\,[\ln\{(\sqrt{1+2\alpha}+1)^2/2\alpha\}]} \cdot \frac{U}{d} \quad (1)$$

The Maxwell stress p is equal to $E^2/2$. Note that the electrical field vector is normal to the electrode surface and that p acts in the direction from the center of the cylinder to outward. Integrating two elements of p vectors in X-direction (parallel to the plane) and Y-direction (normal to the plane) over the surface, one obtains the driving force Fx and the sticking force Fy on the roller of length ι.

$$F_x = \frac{\epsilon \iota U^2}{r[\ln\{(\sqrt{1+2\alpha}+1)^2/2\alpha\}]^2} \,,\quad F_y = \frac{-\pi\alpha}{2\sqrt{1+2\alpha}} F_x \quad (2)$$

The dependence of Fx and Fy on the roller radius r is given in Fig.4 with d=50nm, U=5V, ι=1cm and the relative permittivity equals to 40. Fx is independent of r. Fy is proportional to the half power of r. The line W in Fig.4 is the weight of a solid steel cylinder. The surface force Fx and Fy exceeds the body force W in small dimension. The electrostatic drive seems attractive in this region.

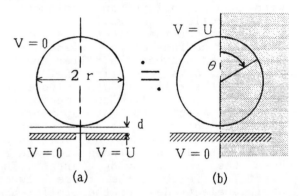

Fig.3 Substitution by a cylinder-to-plane system for analytical solution.

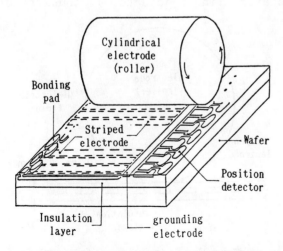

Fig.2 Schematic picture of ESLAC.

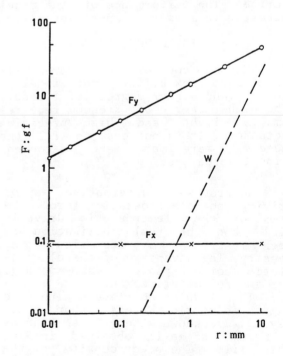

Fig.4 Electrostatic force calculated by the analytical solution for d=50nm, V=5V and ϵ^*=40.

Numerical solution

The analytical solution suggests the possibility of electrostatic driving in sub-mm range. More detailed analysis was conducted by the surface charge method, a numerical computation method for electric field.[3,4] The field distributions in various configurations was determined with the fixed insulation thickness d=50nm and the applied voltage U=5V. Electric field vectors on the roller are shown in Fig.5 with r=0.2mm, w=82μm, s=38μm. The field was the strongest in the vicinity of the srtiped electrode at high potential. The maximum strength was 0.26 MV/m. Table 1 shows the scale effect for a similar shape. Reduction in size increased the driving force.

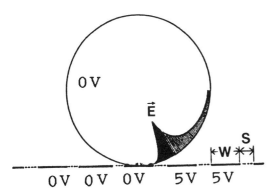

Fig.5 Field distribution on a roller obtained by the surface charge method. r = 0.2mm, w = 0.082mm, s = 0.038mm.

case	V	d	r	w	s	E_{max}	F_x	F_y
unit	V	Å	μm	μm	μm	MV/m	mgf	mgf
1	5	500	10	1.6	0.4	69.2	390	1350
2	5	500	100	16	4	20.0	33	249
3	5	500	1000	160	40	0.55	1.1	6.2

Table 1 Results obtained by surface charge method. See Fig.5 for w and s.

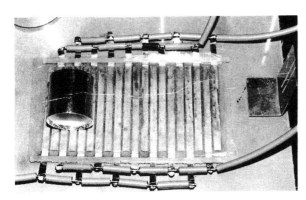

Fig.6 Enlarged model of ESLAC.

ENLARGED MODEL

Design of the model

We built an enlarged model of ESLAC to demonstrate the performance. The picture is shown in Fig.6. Striped electrodes were rectangular pieces of aluminum tape attached on a plate insulator. Multi-electrode and two-electrode models (Fig.7) were built. Insulation sheet of 0.2 mm thickness covered the electrodes. The roller was a thin shelled plastic pipe of 50mm in both length and diameter wrapped with aluminum foil. The roller touched a grounding wire. AC voltage up to 5kV was applied to one electrode of the two-electrode model. For the multi-electrode model, 16 electrodes were divided in 4 phases (Fig.8). Voltage was applied to each phase in turn through relays controlled manually. When the relay opened the charge escaped through the 10MΩ resistor.

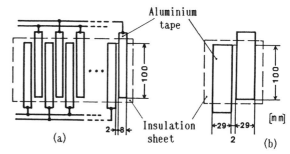

Fig.7 Dimension of striped electrodes.
(a) multi-electrode model.
(b) two-electrode model.

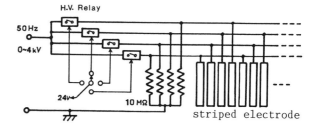

Fig.8 Power supply.

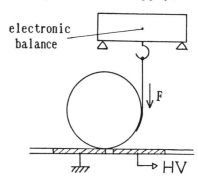

Fig.9 Experimental setup for measuring the driving force.

Performance of the model

Driving force was measured by a electronic balance connected to the roller on the two-electrode model (Fig.9). For the results in Fig.10, different marks represent different runs. Solid marks mean the onset of partial discharge between the roller and the surface of the insulation film. The vertical scale is the square root of force because the theoretical value given by eq.(2) is proportional to V^2. Up to the discharge onset voltage, the measured values lie very close to the theoretical line (the break line) by eq.(2). The force decreased and fluctuated at higher voltage. Surface charge depositing on the insulation film worked unfavorably.

Continuous driving of the roller was achieved by the multi-electrode model. The roller began rotating at V=1.5kV. The movement was stepwise because of the manual switching. The average velocity of the roller vs. voltage is shown in Fig.11; the instantaneous velocity was higher.

CONCLUSIONS

A novel micro actuator, ESLAC, was proposed. Electric field analysis showed the advantage of the actuator appeared in the sub-mm range. The performance of ESLAC was confirmed by the enlarged model. Promising results was obtained with the model.

The authors wish to thank Prof.F. Harashima and Prof. T. Ikoma in University of Tokyo for their stimulative discussions, Dr. S. Matsumoto in Toshiba Corp. for the aids in the field computation, and Mr. H. Tanaka and Mr. J. Ohtani in University of Tokyo for the experimental help.

REFERENCES

1) J. G. Trump : Elect. Engin., 66, 525 (1947).
2) H. Prinz : Hochspannungs felder, R. Oldenbourg, Verlag, Munchen. (1969).
3) T. Kouno : IEEE Trans. on Elec. Insul., EI-21, 869 (1986).
4) H. Singer : Bull BEV/VSE, No.65, 739 (1974).

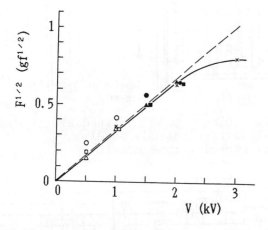

Fig.10 Driving force vs. voltage. The break line is calculated by eq. (2).

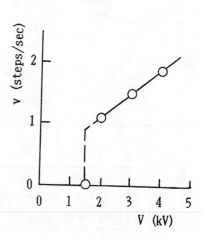

Fig.11 Average velocity of the roller vs. voltage.

Design Considerations for a Practical Electrostatic Micro-Motor

W. S. N. TRIMMER and K. J. GABRIEL

AT&T Bell Laboratories, Holmdel, N.J. 07733 (U.S.A.)

(Received August 12, 1986; accepted in revised form October 10, 1986)

Abstract

Magnetic motors and actuators dominate the large-scale motion domain. For smaller, micro-mechanical systems, electrostatic forces appear more attractive and promising than magnetic forces. Despite their distinguished history, electrostatic motors have found few practical applications because of the high voltages and mechanical accuracies traditionally required. This paper explores the design of electrostatic motors utilizing the advances in silicon technology. Using silicon wafers, and the associated insulators, conductors, anisotropic etching and fine-line photolithographic techniques, it is possible to develop large electrostatic fields with moderately high voltages (≈ 100 V) across insulators of well-controlled thickness. We present two preliminary designs and numerical simulations: one for a linear electrostatic motor and one for a rotary electrostatic motor.

1. Introduction

Electrostatic motors are well suited as an actuating force for small, micro-scale devices. These motors have a long history. Andrew Gordon [1] and Benjamin Franklin [2] built electrostatic motors in the 1750s, 100 years before the advent of magnetic electric motors. The first capacitor electrostatic motor was developed by Karl Zipernowsky in 1889 [3]. A review of early electrostatic motors is given by Oleg Jefimenko [4]. In this review, Jefimenko describes the history of electrostatic motors from Gordon's electric bells, to modern motors that can be powered from atmospheric electricity. Despite their distinguished history, electrostatic motors have found few practical applications because of the high voltages and mechanical accuracies required. The advances in silicon technology micro-machining have changed the prospects for electrostatic motors. Using silicon wafers, and the associated insulators, conductors and anisotropic etching techniques, it is possible to develop large electrostatic fields using only ≈ 100 V across thin insulators of well-controlled thickness. The large number of fine conducting metal pole faces that can be defined photolithophryly on silicon substrates increases the force the electrostatic motor can generate.

In this paper we explore the notion of an electrostatic motor further and present two preliminary designs; one for a linear electrostatic motor and one for a rotary electrostatic motor. These designs are preliminary structures, and are intended to demonstrate and study the feasibility of electrostatic forces for actuation. We are in the process of building motors based on these designs.

The paper is divided into five sections. The first presents the elemental physical structure and theoretical development common to both designs. The second section describes the advantages of silicon for the construction of these motors. The third and fourth sections discuss the linear and rotary motor designs in detail, to which the theoretical development of the first section is applied. The third section also gives a specific linear motor design. The final section presents some specific designs of rotary motors, along with numerical simulations of their performance.

2. Theoretical development

Two conducting parallel plates separated by an insulating layer create a capacitor with a capacitance given by

$$C = \epsilon_r \epsilon_0 w \frac{l}{d} \tag{1}$$

where w is the width of the plates, l is the length of the plates, d is the separation between the two plates and ϵ_0 and ϵ_r are the free space and relative permittivities. If a voltage V is applied across these two plates, the potential energy of this capacitor is

$$U = -\frac{1}{2} CV^2 = -\frac{\epsilon_r \epsilon_0 w l V^2}{2d} \tag{2}$$

(The minus sign is a result of including the energy lost by the voltage source.) The force in any of the three directions (w, l, d) is given by the negative partial derivative of the potential energy in each of the three directions. To calculate the force in the w direction, consider the plates offset so they overlap by x, as shown in Fig. 2. In terms of the overlap x, the potential energy of the capacitor is now

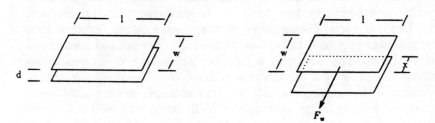

Fig. 1. Two conducting parallel plates forming a capacitor.

Fig. 2. Two parallel plates overlapped by x.

$$U = -\frac{\epsilon_r \epsilon_0 x l V^2}{2d} \quad (3)$$

and taking the derivative with respect to x gives the force

$$F_w = -\frac{\partial U}{\partial x} = \frac{1}{2}\frac{\epsilon_r \epsilon_0 l V^2}{d} \quad (4)$$

The forces in the l and d directions are similarly

$$F_l = -\frac{\partial U}{\partial l} = \frac{1}{2}\frac{\epsilon_r \epsilon_0 w V^2}{d} \quad (5)$$

and

$$F_d = -\frac{\partial U}{\partial d} = -\frac{1}{2}\frac{\epsilon_r \epsilon_0 w l V^2}{d^2} \quad (6)$$

(Note that the above derivation uses a parallel plate capacitor assumption to calculate the potential energy. This is a reasonable assumption when the plates overlap as shown in Fig. 2. For example, when one plate is moved in the x direction, the potential energy stored in the overlapping parallel plate region changes, while the fringing areas remain nearly constant. The fringing effects at the edges of the plates are very difficult to calculate, and it is fortunate that they do not change significantly as the plates move when the plates have a substantial overlap. When the plates do not overlap, or overlap completely, the fringing effects become important, and the above equations are only rough approximations.)

Equations (4) and (5) suggest that if two plates of a parallel capacitor are slightly displaced with respect to each other, a force parallel to the plates is developed, tending to realign the plates. It is this aligning force that we will exploit to make linear and rotary electrostatic motors.

Equation (4) shows that the aligning force F_w is proportional to the length of the edge l. In the calculations later in this paper the force per unit edge, f_w, will be used to calculate the force produced by the motors, where

$$f_w = \frac{F_w}{l} = \frac{1}{2}\frac{\epsilon_r \epsilon_0 V^2}{d} = \frac{1}{2}\epsilon_r \epsilon_0 d E^2 \quad (7)$$

The force depends upon the electric field intensity squared, $E = V/d$, hence one wants to use the maximum E field possible.

3. The advantages of a small electrostatic motor made of silicon

When handling small parts, there are advantages in using small, commensurately sized manipulators. Small systems tend to move small parts more quickly than large systems. Because of reduced problems with thermal expansion, vibration, etc., it is easier to obtain high accuracy with smaller systems, and the reduced floor space used by small systems is also advantageous.

The electric motors normally used in macro systems are electromagnetic. Unfortunately, electromagnetic forces do not scale well into the micro domain. (This can be seen from Ampere's law, $\int B \cdot dl = \mu_0 \oint J \cdot dA$. As an electromagnetic motor is scaled down in size, the area, dA, decreases more rapidly than the line integral for the magnetic field, dl, and it becomes difficult to obtain the needed current, $J \cdot dA$, in very small electromagnetic motors). Electrostatic forces, as will be seen by the examples in this paper, scale well. The performance of electrostatic motors will be explored in the following section.

The purpose of this paper is to explore how to make practical electrostatic motors. These motors require very smooth surfaces, thin high quality insulators and very narrow conducting pole faces. In developing the technology needed to produce VLSI electronics, silicon processing has developed the techniques to satisfy the above needs. The surface of silicon wafers are flat to several microns, common insulating layers can withstand E fields of several million volts per centimeter [5, 6] and conducting pole faces a few microns wide can be deposited on the silicon substrate. The next section describes a linear motor made on silicon.

4. Linear electrostatic motor

A simple linear electrostatic motor has a pair of differentially misaligned parallel plate capacitors as shown in Fig. 3.

When a voltage is applied to the misaligned plates A-A', a force (calculated in the previous section) is exerted, which aligns plates A-A' as seen in Fig. 4.

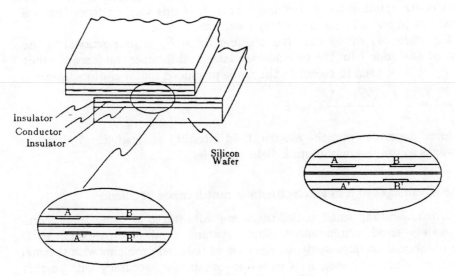

Fig. 3. A linear electrostatic motor made on a silicon wafer.

Fig. 4. Electrostatic plates after electrostatic forces have aligned plates A-A'.

Now, plates B-B' are misaligned and in a position to be activated to cause a motion (and force) back to the original position. Although the forces and motions in such a simple design are relatively small, they are useful in demonstrating the principles of the linear electrostatic motor proposed below.

By differentially 'phasing' the conductive and non-conductive portions of the top and bottom plates of many capacitors, we can sustain motion in either direction and also increase the force generated by a linear electrostatic motor. Figure 5 shows a portion of one possible configuration of plates for such a motor.

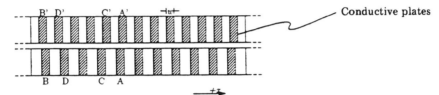

Fig. 5. Phase relationship between the electrostatic plates.

Given the relative positions of the top and bottom conducting plates, applying a voltage across plates A-A' (and all other plates in the same relative position) creates a force tending to align the plates. If the top capacitor plate is held stationary and the bottom is free to move, this force will cause a motion of $w/3$ in the $+x$ direction (w is the width of one capacitor plate). Figure 6 shows the relative positions of the top and bottom motor pieces after plates with the relative alignments of plates A-A' have been energized and aligned.

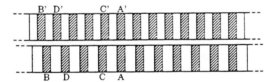

Fig. 6. The new position of the motor after plates A-A' have been energized.

Conversely, applying a voltage across plates B-B' (and all other plates in the same relative position) causes a motion of $w/3$ in the $-x$ direction. Note that in the former case, plates C-C' would then be in a position to continue the $-x$ motion. This differential phasing of the capacitor plates is the same as used for electromagnetic motors.

From the previous section, the aligning force per unit length was shown to be given by eqn. (5). For a linear electrostatic motor of length L and a differential spacing as shown below, we can fit $L/2.3w$ individual capacitors (the $0.3w$ allows for the phase shift between plates), for a total capacitor edge length of $lL/2.3w$.

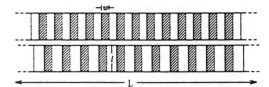

Fig. 7. Dimensions of the electrostatic motor.

Since only a third of the available capacitors are charged at any one time, the total, active capacitor length generating force is given by $lL/6.9w$.

Thus, the force generated by such a linear electrostatic motor is

$$F = \frac{f_w lL}{6.9w} = \frac{1}{2} \frac{\epsilon_r \epsilon_0 lLV^2}{6.9wd} \tag{8}$$

For a linear electrostatic motor formed on silicon, assuming $w = 6\ \mu m$, $l = 5$ mm, $d = 1\ \mu m$, $L = 5$ cm and $V = 100$ V, 1.87 Newtons (0.42 lb) of force is generated by a structure that is approximately 0.375 g in mass.

5. Theoretical model

This section develops a theoretical model for the rotary electrostatic motor. The parameters we are interested in are the force, torque, power, alignment tolerances and the capacitance.

5.1. Torque

Figure 8 shows the rotor (or stator) of an electrostatic motor. The annular ring between r_i and r_o is filled with radial pole faces that are used to produce torque. The torque is given by the familiar relationship

$$\tau = r \mathbf{x} F \tag{9}$$

Because r and F are always perpendicular in this design, this becomes $\tau = rF$. The incremental torque produced by the incremental ring, $r\,dr$ is

$$d\tau = r\,dF = r\left[\frac{F_w}{l}\right]dl = r\left[\frac{1}{2}\epsilon_r\epsilon_0 \frac{V^2}{d}\right]dl \tag{10}$$

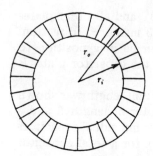

Fig. 8. The pole faces of the rotary motor between radii r_i and r_o.

The total length of the pole faces, dl, in this incremental ring is

$$dl = \frac{dA}{np} = \frac{2\pi r\, dr}{np} \tag{11}$$

where dA is the area within the ring, p is the pitch of the pole faces, n is the number of motor phases (in a three-phase motor, only one third of the poles are actively producing torque), and dr is the width of the incremental ring. The incremental torque becomes

$$d\tau = \frac{\pi \epsilon_r \epsilon_0 V^2}{dnp} r^2\, dr \tag{12}$$

Integrating to obtain the torque,

$$\tau = \int_{r_i}^{r_o} d\tau = \int_{r_i}^{r_o} \frac{\pi \epsilon_r \epsilon_0 V^2}{dnp} r^2\, dr = \frac{\pi \epsilon_r \epsilon_0 V^2}{3dnp}[r_o^3 - r_i^3] \tag{13}$$

The torque the motor produces is given by eqn. (13), where r_o is the outer radius of the motor pole faces, r_i is the inner radius, p is the pitch of the pole faces, and n is the number of phases of the motor. Because the torque depends upon r^3, the outer part of the motor produces the majority of the torque. For example, when $r_i/r_o = 0.7$, two thirds of the maximum torque is produced. Later it will be shown that increasing r_i/r_o decreases the sensitivity of the motor to misalignments.

5.2. Power

The power, P, generated by the motor can be easily calculated from the above torque:

$$P = \tau\omega = \tau 2\pi f \tag{14}$$

where ω is the angular frequency and f is the frequency at which the motor is rotating. The frequency of rotation increases as the motor becomes smaller. To find the general type of scaling for ω, consider Fig. 9. Two rotating masses are connected by a rod. At rod fracture, the centripetal force needed to keep the masses rotating is equal to the yield strength of the rods,

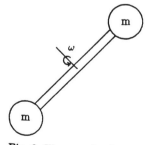

Fig. 9. The angular frequency ω can be increased as the scale is decreased.

$$F = m\omega^2 r = KA \tag{15}$$

where K is the yield strength. Scaling this equation gives

$$\omega^2 = \frac{KA}{mr} = \frac{[s^2]}{[s^3][s^1]} = [s^{-2}] \tag{16}$$

$$\omega = [s^{-1}] \tag{17}$$

Thus the frequency of rotation scales as s^{-1} for this example. In this particular scaling example, the linear velocity along the edge of the rotor remains constant as the motor is scaled down in size. This increase in angular velocity for smaller systems substantially increases the power obtainable from a small motor.

The power a motor can generate is given by eqn. (14), where τ is the torque and f is the frequency of rotation. As motors become smaller, the inherent increase in f helps to offset the decrease in torque.

5.3. Alignment tolerances

If the centres of the stator and rotor exactly coincide, their pole faces will be correctly aligned. However, misalignments of the rotor centre as shown in Fig. 10 in either the δx or the δy directions will cause misalignments of the pole faces on the stator and rotor. Figure 11 shows a three-phase motor with the stator broken into three sectors, P_1, P_2 and P_3. When the stator and rotor centres are correctly aligned, and the pole faces of sector P_1 are aligned with the rotor, then the pole faces of sectors P_2 and P_3 are out of phase by 120°. By applying voltages to P_1, P_2 and P_3 in the correct sequence, the motor can be made to rotate in either direction.

In the analysis below, it is assumed that all the pole faces within a sector must be aligned within 1/12 of the pitch, p (distance between adja-

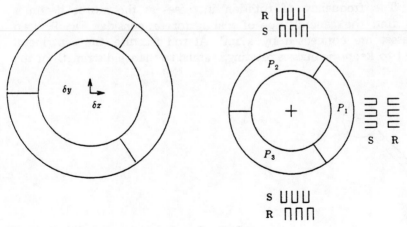

Fig. 10. Possible rotor misalignments δx and δy.

Fig. 11. Phase shift between the rotor (R) and stator (S) pole faces.

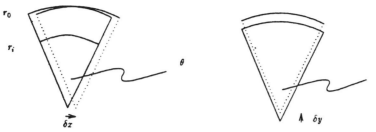

Fig. 12. Misalignment of the rotor by δx.   Fig. 13. Misalignment of the rotor by δy.

cent pole faces). All the pole faces within a section will be pulling in the correct direction, even with an alignment error of 1/6 of the pitch. The allowable displacements in the δx and δy directions have been calculated. The details of this analysis are given in the Appendix.

There are three sources of alignment error. Two of these sources of misalignments are due to displacements in the δx direction, and the third is due to misalignments in the δy direction. The resultant allowable errors are presented below as δx_1, δx_2 and δy.

The relationship between δx_1 and the misalignment of the pole faces, e, is given by,

$$\delta x_1 = e\left(\frac{r_o + r_i}{r_o - r_i}\right) \qquad (18)$$

where $e = p/12$ and r_o and r_i are the outer and inner radii of the annular ring containing the pole faces as shown in Fig. 12 (In Fig. 11, these are the circles containing P_1, P_2 and P_3.)

The relationship between δx_2 and e is given by

$$\delta x_2 = e \frac{1}{[1 - \cos(\theta/2)]} \qquad (19)$$

where θ is the angle of the sector shown in Fig. 12.

The displacement δy results when the entire sector moves up as shown in Fig. 13, and is given by

$$\delta y = \frac{e}{\sin(\theta/2)} \qquad (20)$$

5.4. Capacitance

From eqn. (1) the capacitance between the parallel faces of the poles is given by

$$C = \epsilon_r \epsilon_0 \frac{A}{d} \qquad (21)$$

where A is the area (wl) and d is the separation. If n is the number of phases in which the motor has been divided, r_o and r_i are the outer and inner radii of the annular ring containing the pole faces, and half of the annular ring is covered with pole faces, then the area is

$$A = \pi(r_o^2 - r_i^2)\frac{1}{2}\frac{1}{n} \tag{22}$$

and

$$C = \epsilon_r\epsilon_0\frac{\pi(r_o^2 - r_i^2)}{2nd} \tag{23}$$

5.5. Pole face width versus separation

The coefficient of friction limits the width of the pole faces. As the pole face becomes wider, the force pulling the two plates together increases. However, the force trying to slide one plate across the other depends upon the edge, and it remains constant. When the force pulling the plates together times the coefficient of static friction becomes larger than the force trying to slide the plates, the motor stops moving.

Figure 14 shows the clamping force F_+ and the force F_s just needed to break the static friction. The angle between these forces is θ. The value of F_s is given by

$$F_s = \mu N \quad \text{or} \quad \mu = F_s/N; \quad N = F_+ \tag{24}$$

and the angle θ is given by

$$\tan\theta = F_s/N = \mu \tag{25}$$

The electrostatic forces acting on the plates can be seen in Fig. 15. The F_+ is again the clamping force, and $F_=$ is the force trying to slide the plates. The value of θ is

$$\tan\theta = F_=/F_+ \tag{26}$$

and using eqns. (4) and (6),

$$\tan\theta = d/x \tag{27}$$

For the motor to operate, the force of static friction must be less than the electrostatic force sliding the plates:

$$\mu = F_s/F_+ < F_=/F_+ = d/w \tag{28}$$

For typical silicon surfaces μ is about 1/4, and

$$\mu = 1/4 < d/w \tag{29}$$

If the separation between the plates is greater than one fourth of the pole face width, then the motor can overcome the coefficient of static friction. Obviously if the plates are lubricated, μ decreases, and the plates can be allowed to come closer together.

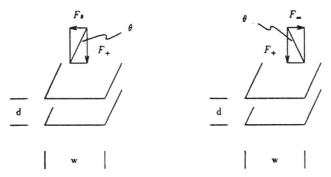

Fig. 14. Clamping force F_+ and frictional force F_s.

Fig. 15. Clamping force F_+ and sliding force F_s.

6. Numerical examples

This section examines the theoretical model of the rotary motor by exploring several numerical examples. The initial guess at designing a rotary electrostatic motor has serious alignment problems. Subsequent examples show the development of a reasonable motor. Table 6 shows how this reasonable motor changes as the size of the motor changes.

Table 1 shows the initial guess. Before discussing its results, the format of the Table will be described. The first row gives the separation between pole faces, the width of the poles and the pitch (separation between adjacent lands). The next row has the inner and outer radii of the annular ring of the

TABLE 1

Initial guess

1 μm	3 μm	6 μm
Separation d	Width w	Pitch p
1 mm	2.5 mm	3
Radius r_i	Radius r_o	Phases n
100	130 V	7
rps	Voltage	ϵ_r
560 mW	8.9×10^{-4} N m	0.48 N
Power	Torque	Force$_=$
1.2 μm	1 μm	0.5 μm
δx_1	δx_2	δy
170 pf	9 cm g	1.4 N
Capacitance	Torque	Force$_+$

pole faces, and the number of independently driven sectors of pole faces (the number of motor phases). Last is the revolutions per second at which the rotor is turning, the voltage applied to the pole faces and the relative permittivity. The last three rows give the results. The first row gives the power output of the motor (neglecting frictional forces, which will of course reduce this value), the torque and the total force generated parallel to the direction of motion. Next are the allowable misalignments in the x and y directions. The last row shows the capacitance, the torque (given this time in the more commensurate cgs units rather than MKS, $1 \text{ N m} \sim 1.02 \times 10^4 \text{ cm g}$) and the force clamping the rotor and stator together.

This electrostatic motor is 0.5 cm in diameter, with a 1 μm separation between the poles and pole faces 3 μm wide. Most of the face of the motor is covered with pole faces (r_i = 1 mm, r_o = 2.5 mm). The rotor is assumed to be spinning at 100 rps. The motor produces 0.5 W, but the alignment tolerances are micron and sub micron.

Table 2 shows the same motor, except that only a thin annular ring of pole faces at the edge is used. The allowable tolerance δx_1 has been increased from 1.2 μm to 4.5 μm. The power has been reduced by a factor of two. In Table 3 the alignment terms that depend upon θ have been improved by using twelve phases. The allowable misalignment in δx_2 is now large, 14 μm. The misalignment δy has been improved, but can be improved more by doubling the land width and pitch. This is shown in Table 4. δy is now almost 4 μm. Note that δx is now 9 μm. One can gain more power by decreasing r_i, which decreases δx_1 but not δy. Table 5 gives the results. The power produced is 58 milliwatts per phase. Since each motor phase will be developing torque about a third of the time, there are an average of four sectors providing torque at any one time. Hence the motor can produce four times the power

TABLE 2

Easier alignment using a narrower band of pole faces

1 μm Separation d	3 μm Width w	6 μm Pitch p
2 mm Radius r_i	2.5 mm Radius r_o	3 Phases n
100 rps	130 V Voltage	7 ϵ_r
290 mW Power	4.64×10^{-4} N m Torque	0.2 N Force$_=$
4.5 μm δx_1	1 μm δx_2	0.5 μm δy
73 pf Capacitance	4.7 cm g Torque	0.6 N Force$_+$

TABLE 3

Number of sectors increased from 3 to 12

1 μm Separation d	3 μm Width w	6 μm Pitch p
2 mm Radius r_i	2.5 mm Radius r_o	12 Phases n
100 rps	130 V Voltage	7 ϵ_r
290 mW Power	4.6×10^{-4} N m Torque	0.2 N Force$_=$
4.5 μm δx_1	14 μm δx_2	2 μm δy
18 pf Capacitance	4.6 cm g Torque	0.6 N Force$_+$

TABLE 4

Easier alignment using wider lands

1 μm Separation d	6 μm Width w	12 μm Pitch p
2 mm Radius r_i	2.5 mm Radius r_o	12 Phases n
100 rps	130 V Voltage	7 ϵ_r
140 mW Power	2.3×10^{-4} N m Torque	0.1 N Force$_=$
9 μm δx_1	30 μm δx_2	3.8 μm δy
18 pf Capacitance	2.4 cm g Torque	0.6 N Force$_+$

of a single phase, or about 230 milliwatts. The initial design, which had more stringent alignment tolerances, produced 500 milliwatts.

The behaviour of the final design in Table 5 as it is scaled to different sizes is given in Table 6. The unlisted parameters in Table 6 are the same as those given in Table 5. The millimeter-sized motor produces about 10 milliwatts, the centimeter-sized motor produces about 1 watt, and the 10 centimeter motor develops about 100 watts.

The examples above are not meant as final designs, but as guides. Indeed, there appears to be a fertile area for further exploration.

TABLE 5

Final electrostatic motor design

1 μm Separation d	6 μm Width w	12 μm Pitch p
1.5 mm Radius r_i	2.5 mm Radius r_o	12 Phases n
100 rps	130 V Voltage	7 ϵ_r
230 mW Power	3.7×10^{-4} N m Torque	0.18 N Force$_=$
4 μm δx_1	30 μm δx_2	3.8 μm δy
32 pf Capacitance	3.7 cm g Torque	1 N Force$_+$

TABLE 6

The final design scaled to different sizes

Diameter	Power	Torque	Force	rps
1 mm	10 mW	3×10^{-6} N m	7×10^{-3} N	500 Hz
5 mm	230 mW	3.7×10^{-4} N m	0.18 N	100 Hz
1 cm	920 mW	3×10^{-3} N m	0.72 N	50 Hz
10 cm	92 W	3 N m	73 N	5 Hz

7. Conclusion

Magnetic motors and actuators dominate the large-scale macro domain. For smaller micro-systems, however, electrostatic forces are more promising. Electrostatic forces become relatively larger as the scale is decreased. It is also easy to bring a few hundred volts into a small volume, and control this power with small electronic components. Silicon appears to be an especially promising technology for electrostatic actuators because of the good insulators available, the precise pole face dimensions that can be fabricated, and the smooth surfaces that can easily be produced.

There are, however, critical issues that must be resolved before electrostatic motors can be produced. First, pin-hole-free dielectrics must be deposited on the silicon substrate. There are several promising candidate materials, which appear to have both the desired electrical properties (able to sustain electric fields significantly larger than the largest fields discussed in this paper) and mechanical properties. We are currently collaborating with electronic device designers on the fabrication of these materials. Secondly, the moving parts of the electrostatic motors proposed here must either have intrinsic lubrication or be lubricated by other agents. Prototype motors can be run 'dry', and depend on the lubrication of the silicon surfaces (silicon and certain of its related derivates are known to have a coefficient of static friction of ≈ 0.25, and provided the separation between the plates is greater than a quarter of the pole face width, the motors will run dry). In addition, we have also explored several 'wet' lubricants, which could serve as a lubricating agent for the electrostatic motors. Earlier work on a silicon and piezoelectric inch worm [7] has solved similar bearing problems.

The electrostatic motors simulated here have attractive properties. For example, the motor 5 mm in diameter and 1 mm thick produces a fraction of a watt output using 130 V, and a linear motor weighing a third of a gram can produce a fraction of a pound of force. These motors will be useful in a number of micro-applications.

References

1 P. Benjamin, *A History of Electricity*, John Wiley, New York, 1898, pp. 506 - 507.
2 J. Sparks (ed.), *The Works of Benjamin Franklin*, Vol. 5, Whittemore, Niles, and Hall, Boston, 1856, p. 301.
3 Zipernowsky electrostatic motor, *Electrical World*, 14 (1889) 260.
4 Oleg D. Jefimenko, *Electrostatic Motors*, Electret Scientific Company, Star City, 1973.
5 L. Marton, *Advances in Electronics and Electron Physics*, Academic Press, New York, 1969, Vol. 26, p. 409.
6 N. J. Chou and J. M. Eldridge, *J. Electrochem. Soc.: Solid State Science*, (Oct.) (1970) 1288.
7 A. M. Feury, T. L. Poteat and W. S. N. Trimmer, A micromachined manipulator for submicron positioning of optical fibers, *IEEE Solid-State Sensors Workshop, IEEE Electron Devices Society, Hilton Head Island, SC, June, 1986*.

Biographies

Kaigham J. Gabriel was born in Baghdad, Iraq in 1955. He received his B.S.E.E from the University of Pittsburgh in 1977 and his S.M. (1979) and Sc.D. (1983) degrees in electrical engineering from M.I.T. in Cambridge, Mass. He is a member of the technical staff in the Robotics Technology Research Department of AT&T Bell Laboratories in Holmdel, New Jersey, U.S.A. He is interested in teleoperators, dexterous manipulation and man–machine interaction.

William Stuart Newberry Trimmer was born in Long Beach, California, in 1943, received his B.A. from Occidental College in Los Angeles and his Ph.D. from Wesleyan University in Connecticut. He is a member of the technical staff of the Robotic Systems Research Department of AT&T Bell Laboratories, Holmdel, New Jersey, U.S.A. He is interested in making small mechanical structures for use in building micro-robots.

Appendix

This Appendix will examine the misalignment of the rotor and stator pole faces for the rotary electrostatic motors. When the centres of the rotor and stator are offset by δx and δy, as shown in Fig. A1, the pole faces in one sector cannot all simultaneously coincide. If the misalignment is too much, some pole faces will be pulling while others are pushing, and the efficiency of the motor will be decreased. In the following calculations it will be assumed that the pole faces must be aligned with a tolerance of 1/12 the pitch, p. Figure A2 shows a three-pole motor and magnified views of the stator, S, and rotor, R, pole faces. The P_1 pole faces are aligned. P_2 and P_3 are misaligned by $\pm 120°$, or a third of the pitch p.

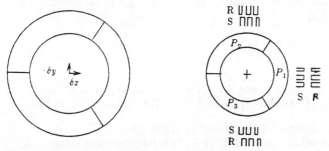

Fig. A1. Possible rotor misalignments δx and δy.

Fig. A2. Phase shift between the rotor (R) and stator (S) pole faces.

The δx misalignments cause two sources of error. The first is the angular misalignments shown in Fig. A3. The dark lines show, say, the stator, and the dashed lines show the rotor. The displacement δx means the pole faces of the rotor and stator make an angle of ψ with respect to each other, and the pole faces are misaligned by an error e. (The pole faces are between radius r_o and r_i.) Then

$$\sin \psi = \frac{e}{(r_o - r_i)/2} = \frac{\delta x}{(r_o + r_i)/2} \tag{A.1}$$

The allowable δx is

$$\delta x_1 = e \left(\frac{r_o + r_i}{r_o - r_i} \right) \tag{A.2}$$

where $e = p/12$.

The other source of misalignment from a δx is shown in Fig. A4. When the pole faces in Region 1 are moved δx, the phase between the stator and rotor is changed by some $Phase_1$. Because the pole faces in Region 2 are at an angle, the effective pitch becomes $\tilde{p}$ instead of p, and the $Phase_2$ is different. The effective pitch in Region 2 is

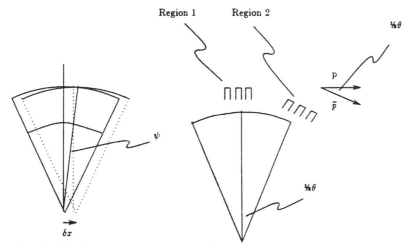

Fig. A3. Misalignment of the rotor by δx causing angle ψ.

Fig. A4. Change in the effective pitch of the pole faces.

$$\tilde{p} = \frac{p}{\cos(\theta/2)} \tag{A.3}$$

The phases and phase differences between these regions are

$$Phase_1 = 2\pi \frac{\delta x}{p} \tag{A.4}$$

$$Phase_2 = 2\pi \frac{\delta x}{\tilde{p}} = 2\pi \frac{\delta x}{p} \cos(\theta/2)$$

$$\delta Phase = 2\pi \frac{e}{p} = 2\pi \frac{\delta x}{p} [1 - \cos(\theta/2)] \tag{A.5}$$

δx_2 is given by

$$\delta x_2 = e \frac{1}{[1 - \cos(\theta/2)]} \tag{A.6}$$

where $e = p/12$. The misalignment due to δy can be seen in Fig. A5. A movement δy causes the pole faces along the edges to be misaligned by e, and

$$\sin(\theta/2) = \frac{e}{\delta y} \tag{A.7}$$

$$\delta y = \frac{e}{\sin(\theta/2)} \tag{A.8}$$

where $e = p/12$.

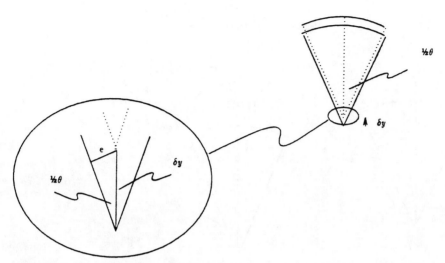

Fig. A5. Misalignments of the rotor by δy.

SCOFSS: A SMALL CANTILEVERED OPTICAL FIBER SERVO SYSTEM

J.E. Wood[1], S.C. Jacobsen[2] and K.W. Grace[2]

Center for Engineering Design
[1] Department of Bioengineering
[2] Department of Mechanical and Industrial Engineering
University of Utah
Salt Lake City UT 84112

Abstract. The Center for Engineering Design has constructed a 100 micron scale electrostatically actuated cantilevered fiber-optic beam for purposes of investigating analysis, fabrication and control of micro electro-mechanical systems. The quartz beam is coated with a polymer, which is then implanted with electrons in an SEM. Laser light, passed through the fiber optic, projects onto an LEPD to give horizontal and vertical displacement readings of the tip of the beam. This coordinate data is fed back to a two-degree-of-freedom model-based position-reference controller (with velocity damping) which applies optimized potentials to conductive driver plates underneath the fiber to move it towards the desired position. The importance of multi-mode flexions of the beam are considered in the dynamic analysis. Fabrication and control of smaller devices, which utilize the interaction of net-charge armatures with voltage-controlled bases having FET sensor elements, are now anticipated.

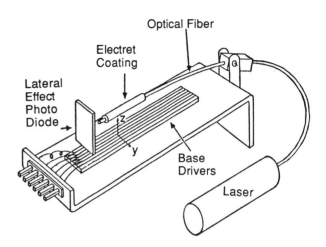

Fig. 1. Schematic of SCOFSS system, with laser, fiber-optic beam (with electret region), LEPD, and driver plates. The MINC computer, and the amplifiers which drive the plates, are not illustrated.

1. Introduction

Small or micro electro-mechanical systems (MEMS) can often deliver numerous performance advantages over their larger-scale counterparts. These typically include higher bandwidth of operation, higher energy densities, and the dominance of controllable interactive forces over passive (e.g., gravity, elastic, and image) forces. In particular, the Center for Engineering Design is interested in controlling the movement of micron-scale objects and in sensing their positions. As an intermediate- (or mini-) scale prototype for testing control and sensing principles, we have constructed a Small Cantilevered Optical Fiber Servo System (SCOFSS) as a precursor to smaller-scaled systems and to fully levitated devices (i.e., no elastic forces). This system has proven useful for experimentally verifying the modelling of different force contributions, such as image forces in conductors, gravitational forces, controller forces, and elastic forces. This system also involves fabrication issues not unlike those expected for other planned smaller systems, including implantation of electrons into polymers (which are typically not charge-compensated by metalization), servo control of a small object, micro-chip design and fabrication, and accurate sensing of a small object. Applications of related devices might be in simple acceleration sensors, spatial light modulators, or actuators [Jacobsen, 1987].

2. Experimental Apparatus

The SCOFSS system is illustrated in Fig. 1. The cantilevered beam is a quartz fiber-optic of length $L = 5$ cm and of radius $a = 70$ microns (including cladding), mounted in a chuck, with a 5 mW laser source applied at the proximal (supported) end. The laser, as it spreads from the beam tip with a cone angle of 30°, projects a spot approximately 2500 microns in diameter onto a Lateral Effect Photo Diode (LEPD), whose active sensor plane is located about $t = 0.45$ cm beyond the beam tip. The LEPD responds to the centroid of the spot, with about one micron resolution.

The cantilever beam has an $l = 1.5$ cm region, which terminates at $b = 4.5$ cm from the supported side, dip-coated with a charge-retentive polymeric material such as Teflon (FEP or TFE), polycarbonate (PC), polystyrene (PS), or polyimide (PI). The coatings, preferably PC or TFE depositions (PC is used in the experiments herein), are layered up to about 3-4 microns in thickness. The fiber and coating are then irradiated in an SEM (at about 30 keV), thus forming an *electret* region over length l. Effective linear charge densities on the order of $\lambda = -0.1$ nanoCoulombs/centimeter (nC/cm) are typically achieved.

The charged region of the beam is then positioned over a set of parallel conductive driver strips, each of width $w = 192$ microns (on $c = 200$ micron centers) and length $s = 2$ cm. Thus, if the shallow angles of arching are neglected (see Sec. 3.4) and only small deflections are allowed, then the system can be considered as a rod with two degrees of freedom (y,z) to be controlled.

The other elements of the system include a MINC (DEC) computer which implements the control law, and specifies the appropriate potentials (up to ±150 volts by high impedance amplifiers) to be applied to the driver strips.

3. System Forces

The forces acting on the fiber-optic beam can be catagorized as "passive" (uncontrolled) and "active" (controlled). In order to put all forces on a common footing, we consider all vector forces to be the forces (or loads) which, if acting uniformily over the electreted region of the filament, would produce a deflection equivalent to the actual force (or load). The individual forces are detailed below.

3.1. Passive System Forces

Four forces affect the beam when the conductive driver strips are grounded. These are:

(1) The *elastic* force (F_e): This is a restoring force pointing towards some *equilibrium* reference position (this need not be the unstrained position). For observed beam deflections away from the reference position (y_0, z_0), there will be some force F_e which can be considered to be acting uniformly over the electreted region to produce that deflection. The force components are given by [Timoshenko and Young, 1968; Price, 1985]

Reprinted from *Proceedings IEEE Micro Robots and Teleoperators Workshop*, November 1987.

$$F_e(y) = \frac{c^*}{\alpha}(b^*)^{-3}(E^*)(I^*)(\psi - \psi_o)$$

$$F_e(z) = \frac{c^*}{\alpha}(b^*)^{-3}(E^*)(I^*)(\zeta - \zeta_o) \quad (1)$$

where $c^* = -13.6$ dynes, $\alpha = 0.25$ is the loading factor (see discussion below), $b^* = b/(1 \text{ cm})$, E^* is the beam modulus normalized by $E = 7.2 \times 10^{11}$ dynes/cm^2, and I^* is the beam area moment of inertia normalized by $I = \pi a^4/4 = 1.9(10^{-9})$ cm^4. Typical values are $b^* = 4.5$, $I^* = 1$, and $E^* = 1$. Also, y and z are the horizontal (y) and vertical (z) displacements of point b on the beam, normalized by 100 microns. Normally, ψ_o and ζ_o would be the normalized reference coordinates of the beam *attachment* point above the substrate. However, as discussed in Sec. 3.4, after tilting the beam, ψ_o and ζ_o become the initial equilibrium location of the point defined by b^*.

A nominal value of $\alpha = 1/4$ was used in the above elastic deflection equation in order to make a meaningful force sum with the image and control forces (and the "corrected" gravity force) which act on the electret which covers the distal third of the beam (α was calculated exactly for each coated fiber, based on the actual length and position of the electret region).

(2) The *image force* (F_i): This is an attractive force between the electreted region (l) of the filament and the conductive base. It has components (using the thin filament approximation)

$$F_i(y) = 0$$

$$F_i(z) = \frac{k^* (\lambda^*)^2 (l^*)}{\zeta} \quad (2)$$

where $k^* = -9.0$ dynes, and where λ^* is the linear charge density normalized by 0.1 nC/cm, and l^* is the length of the electron-charged region along the filament normalized by 1 cm. Typical values are $\lambda^* = -0.5$ and $l^* = 1.5$, which at the usual reference height of $\zeta = 5.2$ (see Sec. 3.4), yield an image force of 0.65 dynes.

Since we tilt the beam at the support end so that it becomes level at the distal end (see Fig. 1 and Sec. 3.4), the initial elastic deflection of the beam, due to both the image force at the reference height and the gravitational force, is nulled out. The *effective image force* is then given by

$$F_i^{eff}(z) = F_i(z) - F_i^o(z_o) \quad (3)$$

where it is noted that the effective image force is zero at the reference height, negative below it and positive above it.

(3) The *gravitational force* (F_g): This force, distributed over beam length L, is normal to the base in actual use, with components

$$F_g(y) = 0$$

$$F_g(z) = d^* (g^*)(\rho^*)(V^*) \quad (4)$$

where $d^* = -2.0$ dynes, g^* is the acceleration due to gravity normalized by $g = 980$ cm/sec^2, ρ^* is the mass density normalized by $\rho = 2.65$ grams/cm^3 (quartz), and V^* is the filament volume normalized by $V = \pi a^2 L = 7.7 (10^{-4})$ cm^3. Typical values are $g^* = \rho^* = V^* = 1$. The weight of the polymer (PC electret) deposition is neglected.

The *effective gravity force* of the beam, i.e., that weight distributed over electret region l which gives the same *static* deflection at b as the beam weight distributed over total length L, is calculated to be

$$F_g^{eff} = (0.6) F_g(z) \quad (5)$$

using standard beam theory. However, it is noted that part of the initial tilt setting used herein (see Sec. 3.4) cancels out the elastic displacement of the beam due to gravity. The height-independent gravity term is thus not of concern in the operation of the controller. In the general controller discussions below (which could presume an untilted beam), the gravity force vector F_g should be taken to be the effective gravity force F_g^{eff}. In the *tilted* beam case herein, at the *reference* position, we can set $F_g = F_i = F_e = 0$.

(4) The *dielectrophoretic* (DEP) force (F_{dep}): The DEP force [see Jones and Kallio, 1979] can be divided into "image-related" and "control-related" components (see Sec. 3.2). Both types were analyzed for this system, and were found to be relatively insignificant at the size-scale dealt with herein (at smaller size-scales, the DEP forces become more important). Thus, they are not included in the modelling.

3.2. Active System Forces

Two force types (DEP and drivers) have components affected by the voltages applied to the substrate strips. However, as indicated above, the control-related components of the DEP force are ignorable, and are thus not discussed.

(1) The *driver force* (F_D): Application of non-zero potentials to the driver strips adds forces (over region l) to the beam in two controllable directions (y,z). The dependence of these driver forces can be expressed as

$$F_D = f(\lambda, D_i, X_{act}) \quad (6)$$

where λ is the linear charge density on the filament, D_i is the set of five plate voltages (i=1,2,...5), and X_{act} is the actual 2D position of the electreted region of the filament with respect to the plates. The plates, modelled as finite in width and infinite in length, are considered to be set in an infinite (grounded) conductive plane. A derivation of the details of the above relationship is not presented [Price, 1985], although it is noted that the forces are a linear homogeneous function of the strip potentials, with terms of the form

$$F_{Di} = \lambda \; D_i \; f'(X_{act}) \quad (7)$$

where f' is a nonlinear function of position.

3.3 Stability Considerations

In the absence of elastic forces, there is no set of fixed plate potentials which can keep the filament in a stable "levitated" state [Earnshaw, 1842]. However, in principle, a time-varying set of plate potentials can be applied to suspend the filament stably.

With the addition of inherently stabilizing elastic forces (i.e., a cantilevered system), *passive static stability* is achieved for certain equilibrium heights above the conductive base. Different equilibrium heights are obtained by different sets of driver potentials. For beam equilibrium locations sufficiently close to the plates, the system becomes *passively unstable* wherein, for small perturbations, the sum of image, driver and gravity forces dominates the elastic force and the beam becomes latched down to the base [Petersen, 1978].

3.4 Beam Alignment

The alignment of the beam over the driver strips was important to proper SCOFSS operation. Lateral alignment, such that the beam was parallel to the strips, was performed in two steps. First, coarse alignment was achieved using an optical microscope looking down on the beam and strips. Second, sinusoidal inputs were applied to the central three drive plates. Fine lateral alignment was achieved when the beam no longer showed an oscillatory lateral component of displacement on the oscilloscope.

Vertical alignment was more involved than the lateral alignment.

First, the droop (at b = 45,000 microns) due to gravity alone is about 200 microns, with a tip angle of about 0.35°. Second, the droop (at b) due to the image force, for parameters $\zeta = 5.2$ (for this choice, see below), $\lambda^* = -0.5$ and $l^* = 1.5$, is about 110 microns. The combined droop is thus about 310 microns. To compensate this, the beam was then tilted up (like a fly rod; see Fig. 1) such that the distal portion of the beam was parallel to the drive strips. This procedure essentially nulls out the static elastic force due to gravity and the image force at the reference position, and removes any parallax error at the LEPD for the equilibrium reference position.

Vertical alignment then proceeds by adjusting the beam as it responds to sinusoidal inputs. Specifically, when the *three* central plates are grounded, and the two groups of *three* plates on either side of the central group are driven at the same potential, there is a *null force point* at $y_0 = 0$ and $z_0 = (3c/2)\sqrt{3} \sim 520$ microns, which serves as a useful calibration point. To use this, the outer two groups of three plates are driven sinusoidally while the height of the cantilevered end of the fiber is adjusted. When the amplitude of the beam responses are minimal or zero, the beam height is taken to be at the theoretical value of 520 microns. This point is also the equilibrium position of the beam for the normal *five*-plate operation described below. Because the field lines generated by the five plate control configuration are different than those produced for the nine plate calibration process, there is then no null condition at (y_0, z_0).

4. Closed Loop Servo Controller

The closed loop control of the beam enables precise (micron order) positioning of the beam. The control loop can be broken into four basic components: the state estimator, the control law, the driver, and the plant (which can be either a computer model or the actual cantilevered filament). The elements are shown schematically in Fig. 2.

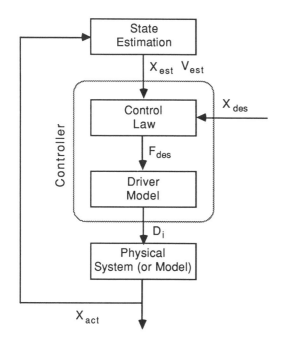

Fig. 2. Schematic of principal elements of SCOFSS controller. The input is X_{des}. In the real (physical) case, the actual system position (X_{act}) is fed back to the controller via the state estimator which produces a measured (estimated) position and velocity. For model simulations, the output is instead acceleration, which must be integrated twice to get the position state. See text for details.

4.1. State Estimator

The *state estimator* in the SCOFSS system is accomplished via an LEPD reading of a laser spot emanating from the beam tip. The output of the LEPD is a (y,z) coordinate reading, approximating the position of the beam tip, which is then transmitted to two 12-bit A/D converters of the computer controller (a DEC MINC). The LEPD coordinate reading is then corrected for beam tip angle by

$$y_{est} = \beta \Delta y + y_0$$
$$z_{est} = \beta \Delta z + z_0 \quad (8)$$

where y_{est} and z_{est} estimate the 2D position of the beam at point b, where $\beta = 0.75$ (nominal value; calculated exactly for each fiber), and where Δy and Δz are the LEPD-observed displacements away from the reference position.

The corrected position (X_{est}) signal is then differentiated (via analog circuitry) to produce an estimated velocity which is then run through a first-order filter before going to the controller where it is used in the damping term of the control law (V_{est}; see Sec. 4.2).

In earlier SCOFSS prototypes, some of the parallel base strips were used as sensors, with their potentials free-floating, thus reflecting the proximity of the charged filament. However, capacitive coupling to the sapphire base and adjacent polysilicon driver strips, and low signal-to-noise ratios plagued this approach. Future MEMS devices will eliminate the LEPD, and instead use the output of sensitive FET gates set beneath the filament (Jacobsen, et al., 1987). For such devices there would be equations for converting FET currents into estimates of electret positions (after the influence of the near-by driver strips had been subtracted out). With the LEPD, no such equations are needed.

4.2. Control Law

The *control law* is a two-degree-of-freedom (2DF) proportional formulation, with velocity damping (PD controller with position reference only). The controller prescribes a desired total force (F_{des}) to be applied to the filament, based on the position of the beam relative to the desired position, and on the actual velocity of the beam (see the vector representation in Fig. 3).

The control law is, in vector form, given by

$$F_{des} = K_p(X_{des} - X_{est}) - K_v V_{est} \quad (9)$$

where X_{des} is the desired set point, X_{est} is the beam position (at point b) as determined by the LEPD (corrected), and V_{est} is the estimated 2D velocity. K_p and K_v are, in general, diagonal matrices [2 by 2] of position and velocity gains, respectively. In the actual implementation, they were simply scalars.

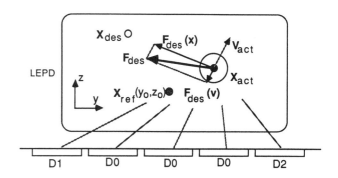

Fig. 3. Schematic of control law model. The fiber diameter is represented by the circle at X_{act} (the 2D location of the electreted region of the fiber). The fiber has a static equilibrium at X_{ref}. F_{des} has two vector components: $F_{des}(x)$ which points towards the desired position (X_{des}), and $F_{des}(v)$ which is opposite in direction to the actual velocity V_{act} at the tip. In practice, X_{act} and V_{act} are replaced by X_{est} and V_{est} respectively.

4.3. Driver

The *driver* takes into account models of the image force (F_i), the gravity force (F_g), and elastic force (F_e), then calculates the potentials to be applied to the driver strips in the base, so as to generate a total net vector force on the filament equal to the desired force (F_{des}) determined by the control law. Since only two-degrees of freedom are being controlled, in principle, only two conductive strips are needed to effect movement in all directions. However, we chose to actuate the filament with five driver strips. This affords more force and versatility, but it also introduces a third-order redundancy. By constraining the central three strips to be of equivalent potential, the level of redundancy is reduced to one. This is resolved (using a Lagrange multiplier approach) by minimizing the sum of squares of the outer two independent potentials (D_1 and D_2) and the potential (D_0) common to the middle three strips (see Fig. 3). Specifically, we minimize the cost functional

$$J = \{D_0^2 + D_1^2 + D_2^2\}. \quad (10)$$

This was intended to reduce, with minimal computational overhead, the occurrence of saturation of the amplifiers (which were limited to ±150 volts per strip), thereby extending the region of control.

The vector force to be applied by the driver strips is then (see the force polygon of Fig. 4)

$$F_D = F_{des} - \{F_g + F_i + F_e\}. \quad (11)$$

This is a model-based controller formulation, albeit quasi-static, since the dynamics (inertial terms) of the beam do not enter into the corrective formulation. The potentials (the three D_i) generating F_D are then determined as linear combinations of the F_D components, subject to the above cost functional. In the case of amplifier saturation, F_D might not be achievable. Also, recall that for the tilted beam case, $F_g = 0$ and F_i is referenced to the equilibrium position. Moreover, F_e and F_i are based on the actual (fed back) position of the beam. No feedforward control algorithms [An, et al., 1987], with inertial terms, were tried, although they should be expected to enhance SCOFSS performance.

The outputs of the driver box (the three potentials) are then transmitted through D/A converters in the MINC to the strip amplifiers.

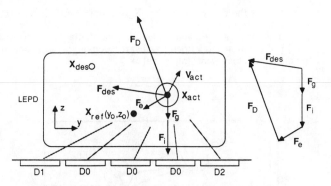

Fig. 4. Schematic of driver model. F_D is the force to be produced by the driver strips so as to yield the net (desired) force F_{des} (see Fig. 3). The corresponding force polygon is also shown.

4.4. Mechanical System

The *mechanical system* consists of the cantilevered quartz fiber optic beam which is effected by the model-based controller described above. The sum of all forces acting on the filament (intrinsic and air damping are neglected; Newell, 1968) gives the dynamics

$$F_{act} = F_D + F_g + F_i + F_e \sim F_{des} \quad (12)$$

where F_{act} is the actual force, which should approximate F_{des} (amplifier saturation may affect the accuracy of the approximation).

In the case of *simulations*, F_{act} yields the "actual" acceleration (A_{act}) via

$$A_{act} = F_{act} / m_{dyn}^{eff} \quad (13)$$

which is then integrated to get the new velocity and position. The m_{dyn}^{eff} of the system is not the static mass (weight) of Sec. 3.1, but rather a mass which yields the same natural frequency as the first mode of the actual beam (using a beam stiffness determined by Eq. 1 and a natural frequency determined by Eq. 16). For the system parameters herein, and the first mode eigenvalue (see Sec. 7.1), we have

$$m_{dyn}^{eff} = (0.4)\, m \quad (14)$$

where $m = \rho A L$ is the total beam mass.

The actual filament (with controller off) is highly underdamped (with a damping ratio of $\xi = 0.0125$, and thus a quality factor of $Q = 1/(2\xi) = 40$), with an observed first natural frequency of roughly $f_{n1} = 40$ Hz. With the controller on, the effective damping parameters of the beam change significantly (see Sec. 6.3).

5. Servo Operation

The cantilevered system can be run in trajectory or regulator modes, as described below.

5.1. Trajectory Operation

The experiments presented below were run in trajectory mode. That is, the desired set point was changed in time (the trajectory), while the beam tried to follow. In some cases, set points representing a desired pattern were transmitted to the controller at differing speeds (trajectory scaling) while the actual track of the beam was monitored. Some of the results, with the set points kept within the *passively stable* region, are discussed below.

Set points were also input via a manually operated 2DF joystick. The display of the actual beam trajectory was monitored on an oscilloscope.

5.2. Regulator Operation

An alternative mode of operation is to specifiy a fixed set-point. External disturbances are then measured by noting the magnitude of the plate voltages required to keep the fiber tip at (or near) the choosen set-point. Such a mode was not investigated extensively, although it might be used as a simple 2D accelerometer.

6. Experiments

The SCOFSS system has been operated in several modes, with a variety of filament charges, over a range of speeds. Some of these protocols are detailed below. The complete control loop cycle rate through the MINC is on the order of 600 Hz.

6.1. Linear Charge Density

The effective linear charge density within the polycarbonate polymer coating on the filament was necessarily calculated for each experiment involving control (since F_i and F_D are λ-dependent). This was done by first applying ±37 volts to the central three driver plates (in open-loop mode) and monitoring the up and down deflections of the beam. Then, by using the open-loop model of forces, a value for λ was determined which caused a theoretical deflection of the beam which best matched the observed deflection(s). This numerical value of λ was then assigned to the controller model.

6.2 Control Space

The maximum 2D region in which the beam tip can be moved is limited by the amplifiers (which saturate) and the optimization algorithm for the driver plates. The region also depends on the filament charge density (λ). A typical region is shown in Fig. 5.

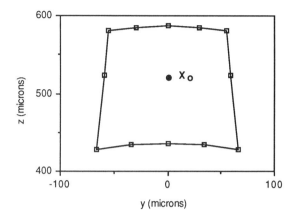

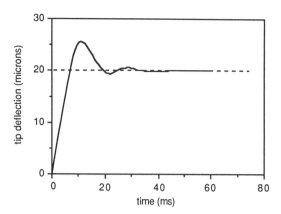

Fig. 6. Step-input response. The initial height of the filament was 520 microns with a 20 micron vertical step input applied. The controller position gain was $K_p = 3.8 \times 10^{-5}$ dynes/micron, while the velocity gain was $K_v = 4.0 \times 10^{-7}$ dynes-sec/micron. Data taken from an oscilloscope photograph.

Fig. 5. Region of maximum controlled beam excursion (5-plate control). X_o denotes the equilibrium reference position of the beam tip with all plates grounded. The linear charge density of the filament was $\lambda = -0.069$ nC/cm. The maximum possible plate voltages were ±150 volts, with the optimization algorithm in effect. The lower bound of the region represents the transition to the unstable region (with feedback on). At a height (measured from the plates to the center of the beam) of about 440 microns, the beam makes the transition to latchdown. The corresponding plate voltage is about +70 volts applied to the three central strips. The controller could have applied increasingly positive potentials to increase the attraction to the strips, but this just puts the filament deeper into the unstable zone.

It is noted that there is no noticeable steady-state error for step inputs. This is because the reasonably accurate models of gravity, image, elastic and voltage-dependent forces have obviated any need for residual position error to compensate load variations.

6.4. Sinusoidal input

In this test, sine-waves, discretized into twelve steps per period, were transmitted at a range of frequencies, from 0 to 25 Hz. (this is well below the observed first natural frequency of the beam; $f_{n1} \sim 40$ Hz.). This data basically demonstrates the tracking ability of the model-based controller stored in the MINC. Typical error traces are shown in Fig. 7.

The tapered shape of the region is probably attributable to the drop in producable lateral field strength at increasing heights. The effects of the controller optimization algorithm on region shape were not analyzed.

Increasing λ (making it more negative) increases the lateral extent of deflection and the maximum height which can be achieved (it produces more repulsive force when the plates are maximally negative) while at the same time increasing the minimum height which can be stably achieved (since the destabilizing image force increases by λ^2, while the control force is proportional to only λ). The opposite applies, within limits, for a decrease in λ. If λ is sufficiently low (of order -0.05 nC/cm or smaller) the beam cannot be forced by the controller into a passively unstable region (because of amplifier limitations). In fact, there may be passively stable regions which are inaccessible.

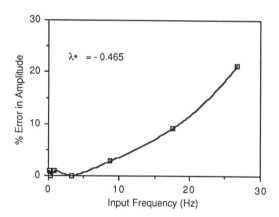

Fig. 7. Sinusodial errors versus frequency. The input amplitude is ±10 microns. The % error is the difference in actual and desired amplitudes, divided by the desired amplitude. The linear charge density was $\lambda = -0.047$ nC/cm. Data taken from an oscilloscope photograph.

7. Higher Modes

In order to assess the importance of higher flexural modes on the dynamics and controllability of the beam, computer simulations of a multi-mode beam were investigated.

7.1. Modal analysis

Beam vibration is a problem handled by classical texts [Timoshenko, et al., 1974]. In short, a mode of free vibration is presented for every root (ε) of the equation

$$\cos(\varepsilon)\cosh(\varepsilon) + 1 = 0. \qquad (15)$$

6.3 Step-input

A typical step-input response of the system, with the controller on, is shown in Fig. 6. The overshoot of the step is about 30%, which, if the system is considered to be second-order-linear, yields an effective damping ratio of $\xi = 0.36$ [Dorf, 1986].

These roots or eigenvalues (ε_1=1.875, ε_2=4.694, etc.) in turn yield the natural modal frequencies

$$\omega_i = \left(\frac{\varepsilon_i}{L}\right)^2 \sqrt{\frac{EI}{\rho A}} \qquad (16)$$

where EI = 1.36 (10^3) dynes-cm^2, and A = 1.54 (10^{-4}) cm^2. Our investigation addressed only the first two modes, while ignoring higher modes. Using the "assumed modes method" presented by Craig [1981], we examined the significance of the modal amplitudes over a range of operating frequencies. (For further discussion on our analysis, see Price et al., 1987.)

7.2. Modal Significance

The above 2-mode model was investigated using a computer simulation. The magnitude of the tip deflections, for modes 1 and 2, over a range of frequencies, are shown in Fig. 8. It is noted that the amplitude of mode 2 is insignificant compared to mode 1, and thus mode 2 can be discounted as a significant source of beam instability and error for the LEPD estimate of position. Thus, a simple first-mode model of beam dynamics suffices.

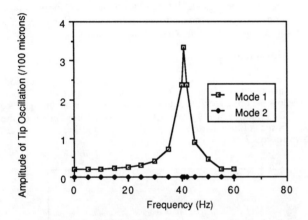

Fig. 8. Simulation of tip deflections, for modes 1 and 2, versus the sinusoidal input frequency applied to the driving plates.

7.3. Controller delays

The role of delays, such as exist between the time the MINC computer receives the state estimate (from the LEPD) and the time the actual potential appears at the plates (about a 1.6 ms delay), was investigated using the above 2-mode model [Grace, 1987]. Only for significant delays (order of one loop period), did the system go unstable. And, when the system did go unstable, because of gains and/or delays, the second-mode vibrations remained small (less than 4%) compared to the first-mode displacements.

8. Discussion and Conclusions

The SCOFSS system has served as a useful test bed for addressing issues of fabrication, analysis and control of micro electro-mechanical systems. It has been shown that sufficient forces can be generated between electreted regions and conductive elements driven at reasonable voltages to control at least a cantilevered system. Clearly, further analysis can be given to this system. However, it is the intention of the CED to start designing and fabricating new MEMS devices.

Acknowledgments

This work was supported by grants from the System Development Foundation, and from the Defense Advanced Research Projects Agency. The authors would like to thank Prof. Richard Price for his analytical contributions and his editorial comments on this paper. We would also like to thank Pratap Khanwilkar who was involved with the experimentation, Nelson Clayton, Eric Simon and Todd Shelby for the electreted fibers, Edward Cetron for his assistance in the implementation of the MINC and VAX computers, Todd Johnson and Karl Wilhelmsen for their assistance in the implementation of the electronics and experimental protocols, Richard Phillips for assistance on the design of the SOS strips, Tom Hansen for the fabrication of the SOS strips, and Prof. Sanford Meek for his help on beam analysis and experimentation issues.

References

An, C.H., Atkeson, C.G., Griffiths, J.D. and Hollerbach, J.M. (1987): "Experimental Evaluation of Feedforward and Computed Torque Control", IEEE Conf. Robotics and Automation, Raleigh, NC, March 31-April 3, 165-168.

Craig, R.R. (1981): *Structural Dynamics*, Wiley.

Dorf, R.C. (1986): *Modern Control Systems*, 4th ed., Addison-Wesley, Reading, Mass.

Earnshaw, S. (1842): "On the Nature of Molecular Forces which Regulate the Constitution of the Luminiferous Ether", Trans. Cambridge Phil. Soc., 7:97-114.

Grace, K.W. (1987): "Automatic Control of a Small Cantilevered Optical Fiber", M.S. Thesis, Mech. Engr. Dept., University of Utah.

Jacobsen, S.C. (1987): "Electric Field Machine", U.S. Patent No. 4,642,502, February 10.

Jacobsen, S.C., McCammon, I.D., Biggers, K.B. and Phillips, R.P. (1987): "Tactile Sensing System Design Issues in Machine Manipulation", IEEE Conf. Robotics and Automation, Raleigh, NC, March 31-April 3, 2087-2096.

Jones, T.B. and Kallio, G.A. (1979): "Dielectrophoretic Levitation of Spheres and Shells", J. of Electrostatics, 6:207-224.

Newell, W.E. (1968): "Miniaturization of Tuning Forks", Science, 161:1320-1326.

Petersen, K.E. (1978): "Dynamic Micromechanics on Silicon: Techniques and Devices", IEEE Trans. Electron Devices, ED-25(10):1241-1250.

Price, R.H. (1985): "Electrostatic and Elastic Forces", personal communication.

Price, R.H., Jacobsen, S.C., and Khanwilkar, P.S. (1987): "Oscillatory Stabilization of Micromechanical Systems", accompanying paper.

Timoshenko, S. and Young, D.H. (1968): *Elements of Strength of Materials*, 5th ed., Van Nostrand, New York.

Timoshenko, S., Young, D.H. and Weaver, W. (1974): *Vibrational Problems in Engineering*, Wiley.

Microactuators for Aligning Optical Fibers

R. JEBENS, W. TRIMMER and J. WALKER

Robotics Systems Research Department, AT&T Bell Laboratories, Holmdel, NJ 07733 (U.S.A.)

Abstract

This paper describes two microactuators used to align fiber optics. One, an actuator using a thin strand of shape memory alloy, is used to align an input fiber with one of two output fibers. This component is useful for switching fiber-optic signals. The second is an electrostatic actuator capable of switching optical fibers, and also of making fine adjustments to correct for misalignments.

1. Introduction

Telecommunications systems are using optical fibers 125 µm in diameter to transmit thousands of telephone calls. These fibers must be aligned with high reliability to other fibers, lasers and light detectors with micron accuracy, and in some instances with tenth of a micron accuracy. As these fibers are being used more extensively, it is imperative to find inexpensive ways to align, maintain alignment and switch fibers. The development of reliable microactuators and microstructures to align and maintain alignment of these fiber-optic components is an important goal with a potentially large economic impact.

Earlier investigations of micromechanical systems have led us to develop techniques for making microactuators and small, accurate structures [1–3]. This paper will discuss our use of these techniques to develop two different ways to align and multiplex light-wave fibers.

Shape memory alloy (SMA) material has the useful property of returning to a trained shape when heated above a critical temperature. In the process, it can exert force, and do work. Below this temperature the material is soft, and easily deformed. Earlier, this material has been utilized to control devices [4–9] such as microturbines and microtweezers. Using SMA, a simple fiber multiplexer has been made. It consists of upper and lower plates, the fibers, a single SMA wire and a bias spring (which can be eliminated).

Electrostatic forces are being investigated by several research groups for use in microactuators [10–21]. Even though electrostatic forces are small in the macro world, they can become a predominant force for micro systems [22]. Electric fields have been used to move a fiber between two V-groove stops in a configuration similar to that of the SMA actuator. Electrostatics also has the potential of making the submicron motions needed to align single-mode fibers.

2. A SMA Fiber Switch

Many applications require fiber switches. Figure 1 shows a typical loop application. When the terminal is powered and operating, one wants the network output connected to the terminal receiver, and the terminal transmitter connected to the network input. However, when the terminal is turned off, or fails for any reason, the fiber switches *must* move to the bypass mode shown in dotted lines in Fig. 1. This requires two single-pole, double-throw fiber switches.

Figure 2 shows an implementation of this switch. The network input and output are the two gray fibers to the left. A bias spring pushes both fibers up against V-grooves in the top plate. The upper gray fibers to the right are also pressed into the same V-grooves, and this aligns the gray fibers to the right with the gray fibers to the left. The gray fibers to the right are the ends of a loop, and in this position of the switch the network input is connected directly back to the network output, bypassing the terminal. When the loop of SMA (only partially shown) is heated by a current, it tries to achieve a maximum radius of curvature, and hence pushes the input and output fibers down against the bottom V-grooves. This aligns the moving fibers with the white fibers to the right. One white fiber channels the light coming from the network output to the terminal receiver, and the other white fiber channels the terminal transmitter back to the network input. As long as the current flows through the SMA wire, the terminal is connected to the fiber-optic network. When the terminal is turned off, or fails for any reason, the SMA cools, and the

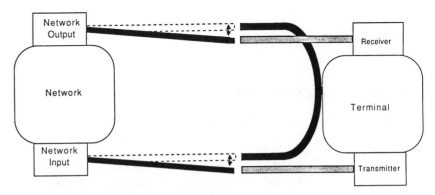

Fig. 1. A fiber switch to connect a terminal to an optical fiber loop.

fibers are returned to their original position against the upper V-grooves.

These V-grooves are made by the use of a 90° single-point diamond tool in a modified Hauser jig borer. Using one-way motion of this machine as a planar, grooves were cut in lucite using 8 μm as a roughing cut, and 0.25 μm as a finish depth of cut. Typically, grooves cut by this method can yield accuracies better than 0.05 μm with surface roughness of below 0.01 μm. These grooved plates may be replicated using electroforming techniques and subsequent injection molding with a suitably filled resin to lower the coefficient of expansion. Using this technique, V-grooves can be inexpensively made with comparable or better accuracy than optically defined anisotropically etched silicon V-grooves. As an example, in the RCA Video Disc system (cut with a diamond tool in copper), the video carrier mechanically cut in the groove has a peak-to-peak depth of 875 Å, with a recovered carrier signal-to-noise ratio of better than 50 dB at a 30 kHz bandwidth. This signal-to-noise indicates an average surface roughness on the order of one Ångstrom rms noise in the pressed record surface [23].

Using V-grooves to align fibers depends on the fiber core being concentric with the outside of the fiber. For multimode fibers with a large core, the alignment required is only a few microns, and can easily be achieved with typical fibers and V-grooves. The problem of using single-mode fibers aligned by means of V-grooves is less clear, as it depends on the insertion loss allowed in the optical system. Single-mode fibers can be made with

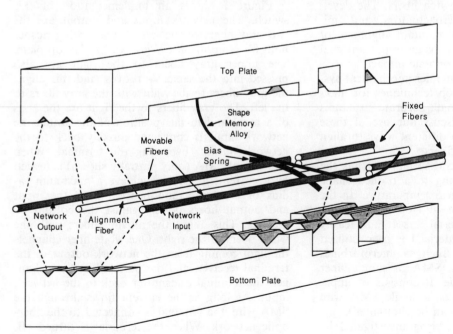

Fig. 2. A fiber-optical switch using a shape memory alloy actuator.

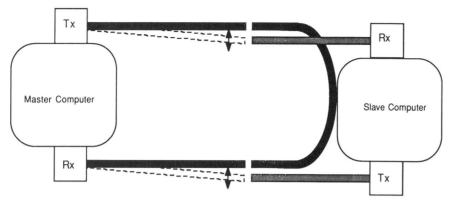

Fig. 3. A fiber switch used to connect a slave computer onto a loop.

the core concentric with the cladding to a third of a micron [24]. If the total misalignment between the single-mode fibers is one micron, the resulting insertion loss is about one dB [25]. There are systems (such as routing signals within a computer where fiber losses are small) where a dB loss in a switch is allowable, and V-groove alignment is an attractive approach for these systems. On systems requiring minimum insertion loss, alignments are typically made to about one tenth of a micron. This is beyond what can be done using V-grooves. A possible exception may be to use the same section of fiber for both ends with the rotational orientation of the fiber maintained.

Another requirement is that the end diameters be in contact with the V-groove walls with sufficient force to ensure contact. If the fixed fiber is clamped in the V-groove with sufficient force to deform the groove walls, some misalignment will result. Thus the area of the V-groove used for alignment should be removed somewhat from areas of high clamping forces, although some smaller force is still needed to keep the fiber in contact with the V-groove.

During assembly, the fibers were placed on the bottom plate. An external jig holds the fibers in place, making assembly easier. When the top plate is attached, the fibers are held securely in place at the ends of the plates. The plates are relieved to give the movable fibers room to move up and down. When the movable fibers are up, they are pressed into the same V-grooves on the top plate as the upper fixed fibers, causing them to align. When the movable fibers are pressed down, they rest in the same V-grooves on the bottom plate as the bottom fixed fibers. After the top and bottom plates are clamped together, a 0.003 inch SMA wire is slid through the same groove as the bias spring, and the ends of the wire are looped up and attached to the top plate. It is this loop that expands when the SMA is heated.

To demonstrate this switch, an AT&T ODL RS 232-1 fiber-optic modem is connected to the master computer. Its output is Tx, and its input is Rx as shown in Fig. 3. When the fiber switch is off, the loop bypasses the slave computer, and the master computer communicates to itself. When the fiber switch is powered, the movable fiber is switched to a second ODL RS 232-1 fiber-optic modem connected to the slave computer. Now the master communicates with the slave, and the slave communicates with the master. An implementation of the switch is shown in Fig. 4. The fiber switch is in the foreground sitting on a larger

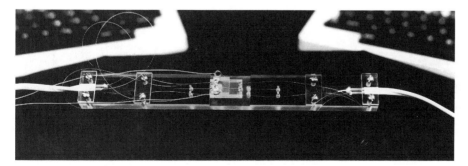

Fig. 4. Photograph of the switch in the foreground, and the master and slave computers in the background. The fiber switch is sitting on a plastic block. Electrical leads and fibers can be seen entering the switch.

plastic block. The fibers and the electrical wires to drive the SMA can be seen entering the switch, which is about 1 by 0.5 by 0.25 inches. In the background are the master and slave computers.

An Electrostatic Aligner and Switch

Moving the fiber electrostatically is perhaps the simplest way to align and multiplex optical fibers actively. However, earlier configurations that have been used to move fibers, such as that shown in Fig. 5, require high voltages (typically 400 V) [26], which are incompatible with many applications. The required voltage can be greatly reduced by using the electrode configuration shown in Fig. 6. Here the fiber is separated from the electrode by a thin insulator, typically a micron thick, and the fiber moves by wrapping around the curved electrode. This coupling of the electrostatic energy to move a fiber is more efficient, because part of the fiber is always in a high-field, high-force, region close to the electrode. In the electrode arrangement shown in Fig. 5, the fiber is far from the electrode, and hence in a region of low force, for most of its travel.

The electrostatic fiber switch we made is sketched in Fig. 7. A photograph of the bottom plate sitting on the alignment jig is shown in Fig. 8. The curved electrode was made in the aluminum bottom plate by diamond-point machining a V-groove of varying depth. The curve generated was

depth = constant + (depth variation)
$$\times (x^4 - 4x^3 + 6x^2)/3$$

where x is the fraction of the distance along the groove. This is the equation of a uniformly loaded beam, fixed at one end. It was chosen so that the fiber position would be unstable, and the fiber would snap between the undeviated and fully deviated positions as the voltage is changed. The V-grooves at the top and bottom plates can then

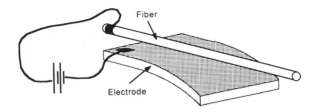

Fig. 6. A fiber touching a curved electrode. Lower voltages are required to move the fiber here than in Fig. 5.

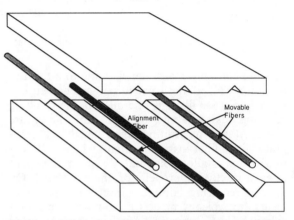

Fig. 7. A double-pole electrostatic switch. The curved electrodes in the bottom plate were made by diamond-point machining.

Fig. 8. The bottom plate of the electrostatic switch. The two outer grooves gradually get deeper, giving the two movable fibers a curved surface to wrap around.

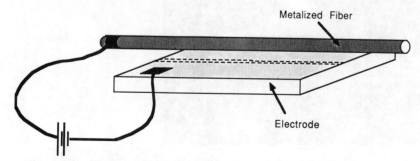

Fig. 5. A fiber separated from the actuating electrode.

be used to align the moving fiber to the fixed output fibers. Active alignment may be possible by utilizing electrode shapes that allow gradual fiber deflections, and these will be discussed below.

The voltage required to move the fiber is (please see the next Section for the derivation and assumptions)

$$V = \left(\frac{k \cosh^{-1}(S/R)}{8\pi\epsilon}\right)^{1/2} \frac{1}{r}$$

where k is the spring constant, S/R is the ratio of fiber–electrode separation to the fiber diameter and r is the radius of curvature of the electrode. The switch built had a 1 μm thick insulation layer, and a curved groove 1.8 cm long. From the above equation, at least 60 V will be needed to move the switch initially. In operation, 120 to 140 V were required to move the switch initially. (Because the curved electrode was actually fabricated using 2 μm steps instead of a continuous curve, this higher voltage is expected.) The fiber then stayed aligned in the second position until the voltage dropped below 40 V. These voltages can be lowered by using a thinner insulator or a longer length of movable fiber.

Electrostatic forces can also be used to provide the 5 to 10 μm motions needed to align single-mode fibers. Using an electrode with a different characteristic curve, such as $y = x^n$ where $n < 2$, the amount of fiber wrapped around the curved electrode changes with voltage, and by changing the applied voltage, the fiber can be moved into alignment with a stationary fiber. One possible arrangement consists of having separate electrodes on the two sides of a properly curved V-groove, and applying voltages on these orthogonal surfaces to move the fiber in two dimensions. This arrangement could also be used to make fine adjustments on the fiber position after it has been moved into position by another actuator.

Electrostatic Analysis

This Section analyzes the use of electrostatics to align and multiplex optical fibers. A fiber, coated with a conductor, that wraps around a curved electrode will be assumed. Applying voltages between the fiber and electrodes will cause the fiber to move. Rather than attempt to design specific systems, this Section will examine the theoretical limitations placed on the design by energy considerations. This analysis shows that fibers can be electrostatically aligned with tens of volts, and can be multiplexed with less than a hundred volts.

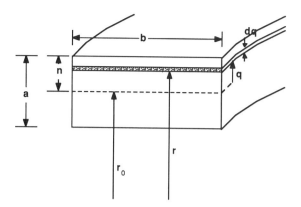

Fig. 9. The cross-section of a deflected beam.

To bend a section of fiber into an arc of radius r_0 requires some energy E_k. This energy must be equal to or less than the electrostatic energy, E_C, available to bend the fiber. Equating E_k and E_C gives the limits of what one can design. Note that even if the electrostatic energy is larger than the energy required to bend the fiber, this does not guarantee that the fiber will bend, but only guarantees that the energy needed to bend the fiber is available. For example, if there is a small piece of dust or irregularity on the curved electrode, this can stop the fiber from wrapping around the electrode.

Calculating the energy of a fiber bent into an arc is easily done with the help of Fig. 9. The force on the small cross-section shown at radius r is

$$\Delta F = EA \frac{\Delta l}{l}$$

Here E is Young's modulus, $\Delta l/l$ is the fractional change in the length of this section and A is the cross-section ($b\,dq$). The energy required to stretch this section from rest to maximum extension, $\Delta l_{\max}$, is

$$dE_s = \int_0^{\Delta l_{\max}} \Delta F_s \, d\Delta l$$

$$= \int_0^{\Delta l_{\max}} \frac{Eb\,dq}{l} \Delta l \, d\Delta l$$

$$= \frac{Eb\,dq}{l(\approx l_0)} \frac{\Delta l_{\max}^2}{2}$$

Since

$$\Delta l_{\max} = l_0 q/r_0$$

where l_0 is the length of the section, we have

$$dE_s = \frac{Ebl_0 q^2}{2r_0^2} dq$$

Integrating the sections in the top half (assuming the neutral line is in the center of the fiber) gives the energy stored in half the fiber, $E_k/2$:

$$E_k/2 = \int dE_s$$

$$= \int_{q=0}^{n} \frac{Ebl_0}{2r_0^2} q^2 \, dq$$

$$= \frac{Ebl_0}{6r_0^2} n^3$$

and

$$E_k = \frac{Ebl_0 a^3}{24 r_0^2}$$

since $n = a/2$.

The deflection of the fiber can be calculated from Fig. 10:

$$x^2 + y^2 = r_0^2$$

and at the end of the fiber,

$$x_e = l_0$$

$$y_e = r_0 - \Delta y$$

$$\frac{1}{r_0} = \frac{2\Delta y}{l_0^2}$$

giving

$$E_k = \frac{1}{2}\left(\frac{Eba^3}{3l_0^3}\right) \Delta y^2 = k \, \Delta y^2 / 2$$

The spring constant k is

$$k = Eba^3 / 3l_0^3$$

A square cross-section was analyzed to simplify the presentation. The equation of k for the circular cross-section of the fiber differs from the above equation by a constant. An experimentally determined value of k will be used later in this paper. Note that the value of k depends upon l_0^3.

The electrostatic energy can be calculated from the capacitance, C, and the voltage, V. For a fiber of outer radius R and length l_0 with a conductive coating on the surface, and an electrode a distance S away, as shown in Fig. 11, the capacitance in MKS is [27]

$$C = \frac{2\pi \epsilon l_0}{\cosh^{-1}(S/R)}$$

and the energy available is

$$E_C = CV^2/2$$

$$= \frac{\pi \epsilon l_0}{\cosh^{-1}(S/R)} V^2$$

(The capacitance of the fiber far from the electrode is small, and will be ignored to simplify the calculation. For example, in the fiber switch built, the electrostatic energy of the fiber in the far position is down by a factor of ten.)

To move the fiber, the electrostatic energy available has to be equal to or greater than the energy needed to bend the fiber to the corresponding radius of curvature. We will set

$$E_C = E_k$$

$$\frac{\pi \epsilon l_0}{\cosh^{-1}(S/R)} V^2 = k \, \Delta y^2 / 2$$

and

$$V = \left(\frac{k \cosh^{-1}(S/R)}{2\pi \epsilon l_0}\right)^{1/2} \Delta y$$

or rewriting in terms of the radius of curvature r

$$V = \left(\frac{k l_0^3 \cosh^{-1}(S/R)}{8\pi \epsilon}\right)^{1/2} \frac{1}{r}$$

Two examples will be calculated. First, the case of fiber multiplexing where one fiber is moved

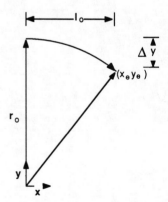

Fig. 10. A fiber of length l_0 being deflected a distance Δy.

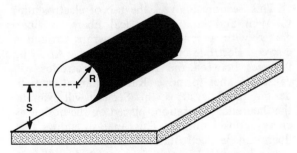

Fig. 11. A fiber of radius R, separated from an electrode by a distance S.

between two output fibers; and secondly, the case of fiber alignment, where one fiber is moved by precise amounts to align it with another fiber. Here it is assumed that the fiber is wrapping around a mandrel of constant radius, r. For fiber multiplexing, one must move the diameter of a fiber. Because there can be an electrode above and below the fiber, each electrode has to move the fiber half its diameter, or about 70 μm. For the alignment case one has to move the fiber approximately 10 μm. Assuming a 1 cm length of fiber in air, and a 2 μm insulator on the electrodes, the values for the multiplexing case (and the aligning case) are:

$l = 10^{-2}$ m

$\epsilon = 8.85 \times 10^{-12} \dfrac{F}{m}$

$\Delta y = 7 \times 10^{-5} (10^{-5})$ m

$S = 6.2 \times 10^{-5}$ m

$R = 6 \times 10^{-5}$ m

$\cosh^{-1}(S/R) = 0.257$

$k = 1.9 \dfrac{N}{m}$

giving a voltage of

$V = 65$ V (9.3 V).

Hence the minimum voltage needed to move a 1 cm fiber between two output fibers is 65 V. The smaller motions needed to align fibers require about 10 V. If the length of the fiber doubles to 2 cm, the required voltages decrease by a factor of four. If instead of air, a fluid of relative dielectric permittivity equal to four is used, the voltages can be decreased by an additional factor of two.

Practical Considerations

There are some issues that become apparent after performing the experiments. While the fiber switches described above worked without failure during our laboratory test, we feel these issues must be addressed before these switches are developed into products. The first is the friction of the fiber against the V-groove as the fiber is being guided into alignment. In the case of the SMA-actuated device, if the fiber is off center, there is an additional tangential friction against the bias spring and the SMA. This tangential motion requires that the SMA actuator produce a larger force than that required simply to deflect the fiber. In addition, when the fiber has seated in the V-groove, the bending couples formed by the bias spring, the SMA wire and the V-groove ends should not be sufficient to bend the fiber end away from the V-groove walls.

In the case of the electrostatically deflected fiber there is a similar type of situation. The tangential friction of the fiber on the V-groove wall resists the fiber sliding. Since the fiber will undoubtedly touch one V-groove wall first, the perpendicular wall must have sufficient attractive force to overcome the sliding friction on the first wall. This clamping force can be limited by using a sufficiently thick insulator. Other possibilities are to put an electrode at the bottom of the V-groove, or to split the V-groove electrode and switch the potential from side to side. A second issue is the compliance of the V-groove. While the fibers must be in solid contact with the V-groove, the forces on the fiber must not distort the groove. The last issue is the problem of longitudinal differential expansion between the fiber and supporting grooves, changing the optical gap between the fiber ends. If these devices are to be molded from plastic (say, from an electroformed replica of the master) a suitably filled resin with a low coefficient of thermal expansion to match the fiber's coefficient should be used. The shortest length of V-grooves possible should also be used to reduce the differential expansion.

Conclusions

Structures were made that show the feasibility of using SMA wire or electrostatic fields to deflect optical fibers. This motion can be used to align a fiber to one of two fibers. The electrostatic deflection can also be used to align actively single-mode fibers. Single-point diamond machining (and associated molding technology) has been shown to be a viable alternative to etching V-grooves in silicon for holding and aligning fibers.

Acknowledgements

We thank K. J. Gabriel and M. Mehregany for many stimulating discussions on electrostatics and shape memory alloys, G. L. Miller for initially drawing our attention to the shape memory alloys, R. Pimpinella for his insights into fiber optic systems and J. Jarvis for his encouragement.

References

1 K. J. Gabriel, W. S. N. Trimmer and J. A. Walker, A micro rotary actuator using shape memory alloys, *Sensors and Actuators*, 15 (1988) 95–102.
2 A. M. Feury, T. L. Poteat and W. S. N. Trimmer, A micromachined manipulator for submicron positioning of optical fibers, *IEEE Solid-State Sensors Workshop, Hilton Head Island, SC, U.S.A., June 1986.*
3 W. S. N. Trimmer and K. J. Gabriel, Design considerations for a practical electrostatic micro motor, *Sensors and Actuators*, 11 (1987) 189–206.
4 A. Golestaneh, Shape-memory phenomena, *Phys. Today*, (Apr.) (1984) 62–70.
5 K. Kuribayashi, A new actuator of a joint mechanism using TiNi alloy wire, *Int. J. Robotics Res.*, 4 (1988) 47–58.
6 L. McD. Schetky, Shape memory effect alloys for robotic devices, *Robotic Age*, (July) (1984) 13–17.
7 L. McD. Schetky, Shape memory alloys, *Sci. Am.*, (Nov.) (1979) 74–82.
8 J. Yaeger, A practical shape-memory electromechanical actuator, *Mech. Eng.* 106 (1984) 51–55.
9 K. Kuribayashi, Millimeter size joint actuator using shape memory alloy, *Proc. IEEE Micro Electro Mechanical Systems, Salt Lake City, UT, U.S.A., Feb. 1989*, pp. 139–144.
10 H. Fujita and A. Omodaka, The principle of an electrostatic linear actuator manufactured by silicon micromachining, *4th Int. Conf. Solid-State Sensors and Actuators (Transducers '87), Tokyo, Japan, June 2–5, 1987*, pp. 861–864.
11 H. Fujita and A. Omodaka, The fabrication of an electrostatic linear actuator by silicon micromachining, *IEEE Trans. Electron Devices*, ED-35 (1988) 731–734.
12 O. D. Jefimenko, *Electrostatic Motors*, Electret Science Company, Star City, 1973.
13 J. Lang, M. Schlect and R. Howe, Electric micromotors: electromechanical characteristics, *Proc. IEEE Micro Robots and Teleoperators Workshop, Hyannis, MA, U.S.A., Nov. 1987.*
14 R. Mahadevan, Capacitance calculations for a single-stator, single-rotor electrostatic motor, *IEEE Micro Robots and Teleoperators Workshop, Hyannis, MA, U.S.A., Nov. 1987.*
15 R. H. Price, J. E. Wood and S. C. Jacobsen, The modelling of electrostatic forces in small electrostatic actuators, *IEEE Solid-State Sensor and Actuator Workshop, Hilton Head Island, SC, U.S.A., June 1988.*
16 J. Wood, S. Jacobsen and K. Grace, SCOFFS: a small cantilevered optical fiber servo system, *Proc. IEEE Micro Robots and Teleoperators Workshop, Hyannis, MA, U.S.A. Nov. 1987.*
17 S. F. Bart and J. H. Lang, Electroquasistatic induction micromotors, *Proc. IEEE Micro Electro Mechanical Systems Workshop, Salt Lake City, UT, U.S.A., Feb. 1989*, pp. 7–12.
18 S. C. Jacobsen, R. H. Price, J. E. Wood, T. H. Rytting and M. Rafaelof, The wobble motor: an electrostatic, planetary-armature, microactuator, *Proc. IEEE Micro Electro Mechanical Systems Workshop, Salt Lake City, UT, U.S.A., Feb. 1989*, pp. 17–24.
19 W. C. Tang, T-C. H. Nguygen and R. T. Howe, Laterally driver polysilicon resonant microstructures, *Proc. IEEE Micro Electro Mechanical Systems Workshop, Salt Lake City, UT, U.S.A., Feb. 1989*, pp. 53–59.
20 Y. Kim, M. Katsurai and H. Fujita, A proposal for a superconducting actuator using Meissner effect, *Proc. IEEE Micro Electro Mechanical Systems Workshop, Salt Lake City, UT, U.S.A., Feb. 1989*, pp. 107–112.
21 M. Sakata, An electrostatic microactuator for electromechanical relay, *Proc. IEEE Micro Electro Mechanical Systems Workshop, Salt Lake City, UT, U.S.A., Feb. 1989*, pp. 149–151.
22 W. S. N. Trimmer, Microrobots and micromechanical systems, *Sensors and Actuators*, 19 (1989) 267–287.
23 J. Guarracini, J. H. Reisher, J. L. Walentine and C. A. Whybark, Micromachining video disc groves and signals, *RCA Rev.*, 43 (1982) 66.
24 D. P. Jablonowski, Fiber manufacture at AT & T with the MCVD process, *J. Lightwave Technol.*, LT-4 (1986) 1016–1019.
25 L. G. Cohen, Lightguide Materials, Diagnostics and Applications group, AT&T Bell Laboratories, Murray Hill, NJ, U.S.A., personal communication.
26 J. Corbin and S. Levy, Electrostatic optical switch with electrical connection to coated optical fiber, *U.S. Patent 4 152 043* (Bell Laboratories Assignee)
27 L. V. Bewley, *Two Dimensional Field in Electrical Engineering*, Macmillan, New York, 1948, pp. 45–46.

Biographies

Robert W. Jebens received a BS with distinction in physics from Worcester Polytechnic Institute in 1960. He began his graduate work in physical metallurgy at Cornell University and finished at Rutgers University, receiving a MS degree in physics in 1963. He was a member of the technical staff at the RCA David Sarnoff Research Center for 23 years, much of that time performing research in high density optical, electron beam and mechanical recording processes and equipment. He is presently a member of the technical staff in the Robotics Systems Research Department of A&T Bell Laboratories in Holmdel, NJ, U.S.A. His research interests include high-precision miniature sensors, actuators and mechanisms fabricated through extensions of conventional mechanical techniques.

William Stuart Newberry Trimmer was born in Long Beach, California, in 1943, received his B.A. from Occidental College in Los Angeles and his Ph.D. from Wesleyan University in Connecticut. Before working in industry, he was department chairman of the physics department at the College of Wooster. He has worked for Johnson and Johnson designing acoustical imaging systems for early detection of breast cancer, and has worked for the Singer Corporate Laboratory developing electro-optical systems. He is presently a member of the technical staff at AT&T Bell Laboratories, Holmdel, NJ, U.S.A. He has co-organized the first workshop on micro robots and teleoperators (Hyannis, 1987), and co-organized a NSF workshop and report on microdynamics. He is interested in making small mechanical structures for use in builiding micromechanical systems.

James A. Walker received the B.S.E.E. degree from Rutgers University, New Brunswick, NJ in 1984, and is presently finishing work toward his M.S. degree, also at Rutgers University.

He is a member of technical staff in the Robotics Systems Research Department of AT&T Bell Laboratories in Holmdel, NJ. His research interests include microelectromechanical systems, with primary emphasis on related silicon-based fabrication technology.

LARGE DISPLACEMENT LINEAR ACTUATOR

Reid A. Brennen, Martin G. Lim, Albert P. Pisano, and Alan T. Chou

Berkeley Sensor & Actuator Center
Electronics Research Laboratory
Department of Electrical Engineering and Computer Science
University of California, Berkeley CA 94720

ABSTRACT

In this paper, a tangential drive (T-drive) polysilicon, linear actuator is presented which produces large magnitude tangential motion by flexing a microstructure in an essentially straight line with moderate input voltage. These devices have been designed, fabricated, and successfully tested to have substantial displacements even for static positioning. The working principle of the T-drive is that the strong electrostatic forces of attraction between a fixed bar and a free bar are converted to large amplitude tangential motion by the parallelogram flexural suspension of the free bar. Operating devices are capable of tangential displacements as great as 32 μm. This displacement is stable and easily varied, since the tangential displacement can be controlled by adjusting the potential between the fixed electrode and the traversing bar. Static displacements are detectable for voltages as low as 15 VDC. Typical T-drives have free bars 200 μm long, 12 μm wide, with flexural suspensions 450 μm long and 2 μm wide. All component thicknesses are 2 μm. Theoretical models for both the flexural suspension and electrostatic forces have been derived and they predict the relation between tangential displacement and excitation voltage.

INTRODUCTION

Within the last two years, several designs of micro-actuators have emerged for specific applications. Among these, one may find shape-memory allow diaphragms for micropumps [1], cantilevered, electrostatically actuated structures to guide probe tips for scanning tunneling microscopes [2], bi-stable mechanical actuators for non-volitile logic elements [3], electrostatic linear actuators for micro friction evaluation [4], a linear micromotor for magnetic disk drives [5], electrostatically-actuated tweezers [6] as well as grippers [7]. A common design theme among these and other micro-actuators are the use of electrostatic, piezoelectric, or shape-memory properties of materials to actuate the micro structures through relatively small displacements (0.01 to 10 μm) with respect to the microactuator size. Although actuation amplitudes and forces for these devices seem to be adequate, the design of higher performance microactuators will require the development of larger amplitude motion with no compromise of actuation force.

Among the electrostatic actuators, one may find two basic designs. The first basic design utilizes parallel plate capacitors with one moving plate that is allowed to displace in the direction of the major field lines, yielding a large-force, small-amplitude actuator. The second basic design utilizes the fringe field of capacitors to drive the moving plate parallel to the fixed plate and perpendicular to the major field lines [8]. This results in a low-force, large amplitude actuator.

In this paper is described the design and performance of a large-force, large-amplitude electrostatic actuator that exhibits the best features of previous electrostatic actuator designs. Thus, the moving capacitor plate is allowed to move parallel to the electric field lines (generating large force) as well as displace parallel to the fixed capacitor plate (generating large amplitude motion). This new electrostatic actuator design utilizes tangential motion of the moving capacitor plate, and thus, is called the T-drive.

Unoptimized designs of T-drive actuators show that low voltage, large force, and large displacement linear actuation is possible.

DESIGN

The force-generating components of the T-drive are made up of two parallel bars which are separated by a small gap. When a voltage is applied across the gap, an electrostatic force is created that acts perpendicular to, and between, the bars. For convenience, this direction is called the normal direction. By fixing one of the bars and attaching the other bar (the free bar) to one side of a parallelogram flexure suspension, Fig. 1, this force can be utilized for motion both parallel and perpendicular to the fixed bar by adjusting the geometry of the suspension. The parallelogram suspension is used in order to constrain the two bars to remain parallel. If the suspension beams of the flexure are not parallel to the normal direction, the total perpendicular force generated, F_{total}, can be decomposed into two force components: one component parallel to and the other component perpendicular to the suspension beams. The tangential force component acts at the end of the flexures and deflects them, resulting in lateral motion of the free bar. This lateral movement can only occur when the initial orientation of the beams is not parallel to F_{total}, i.e. the normal direction. Gravity acting on a simple pendulum serves as an analogy to the concept of the T-drive.

The two critical design features of the T-drive are the flexure geometry, which controls the gap distance as the structure displaces, and the free bar length, which affects the magnitude of the electrostatic force (Fig. 1). The flexure geometry parameters include the offset angle of the beams, the beam dimensions (length, width, and thickness), and the initial gap distance between the free and fixed bars. Larger initial angles of the beams result in a more rapid decrease in gap width and a more rapid increase in applied force as the free bar moves laterally. However, if the initial angle is too large, the free bar will contact the fixed bar as it displaces, and thus short the circuit preventing further movement. The rate of approach of the free bar toward the fixed bar is a function of beam length. The initial gap between the bars determines whether the bars will contact and, if so, what maximum actuator displacement is possible before they meet. The width, thickness, and length of the beams determine the stiffness of the flexure suspension and the magnitude of F_{total} is proportional to the free bar length.

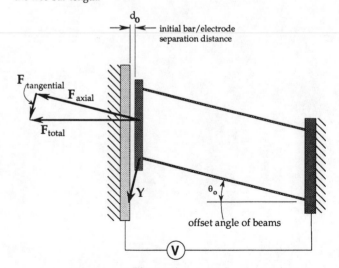

Figure 1

Schematic of Tangential Drive Linear Actuator. Note that it is $F_{tangential}$ that deflects the structure in the Y-direction.

Reprinted from *Technical Digest IEEE Solid-State Sensor and Actuators*, pp. 135-139, June 1990.

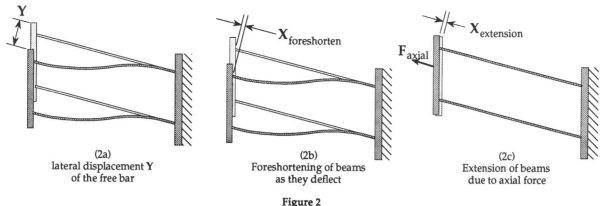

Figure 2
Types of beam displacements
(Shaded structures are in displaced position)

(2a) lateral displacement Y of the free bar

(2b) Foreshortening of beams as they deflect

(2c) Extension of beams due to axial force

THEORY

Using a modified parallel-plate capacitance model, the total electrostatic force F_{total}, generated between the facing sides of the bars is estimated by:

$$F_{total} = C_1 \frac{\varepsilon_o A}{2} \left(\frac{V}{d}\right)^2 \quad (1)$$

where A is the capacitor plate area, V is the applied potential, d is the gap between the bars, ε_o is the permittivity of air, and C_1 is a constant factor. The factor C_1 is based on results from a electrostatic finite element package, Maxwell [9] and improves the estimate for F_{total} by including fringe fields in the capacitance analysis. For our model, C_1 is set to 1.15. Clearly the factor, C_1 must be recalculated if the design of the device is modified.

As the free bar moves laterally, it approaches the fixed bar. The rate of approach is governed by three separate phenomena (Fig. 2). The first, due to the initial angle of the support beams, is the approach of the free bar as it moves laterally a distance Y (Fig. 2a). The second is the foreshortening, X_f, in the *axial* direction, of the support beams as they deflect (Fig. 2b). The third is the axial extension, X_e, of the support beams as the axial component of the total force is applied (Fig. 2c). If θ_o is the angle of the beams with respect to the normal direction, then the gap, d, can be expressed as:

$$d = d_o - Y\sin\theta_o + (X_f - X_e)\cos\theta_o \quad (2)$$

where d_o is the initial gap. Using small displacement beam theory, expressions for Y, X_f, and X_e can be derived [10], and, expanding each of the terms in Eq. 2:

$$d = d_o - \frac{F_{total} L \cos^2\theta_o}{b t E} \quad (3)$$

$$+ \frac{\tan\theta_o \sin\theta_o}{4\lambda}\left\{-\lambda L\left(1+\gamma^2\right) + \sinh(\lambda L)\left[\cosh(\lambda L)\left(1+\gamma^2\right) - 2\gamma\sinh(\lambda L)\right]\right\}$$

$$\text{where } \gamma = \frac{\cosh(\lambda L) - 1}{\sinh(\lambda L)} \quad (4)$$

$$\lambda = \left[\frac{F_{total} \cos\theta_o}{E I}\right]^{\frac{1}{2}} \quad (5)$$

and L is the length of the flexure beams, b is the width of the beams, t is the thickness of the beams, E is Young's modulus of elasticity for polysilicon, and I is the area moment of inertia of the beams.

Figure 3
SEM of a single bar T-drive structure. The L-shaped breakaway support is removed before testing.

Figure 4
SEM of the free bar of a single bar T-drive. The scale and pointer can be seen between the two suspension beams on the left. The dimples can be seen in the center of the free bar. Note the vertical side walls of the polysilicon in the gap.

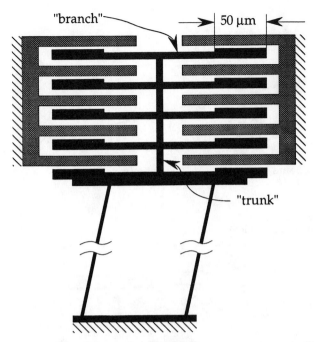

Figure 5
Schematic of Multiple bar T-drive Structure
Note the 50 mm long sections on each branch which act as the parallel plate attractors.

Figure 6
SEM of the free bar component of a multiple bar T-drive.

By solving equations (1) and (3) simultaneously, values for F_{total} and d can be found for given values of the applied voltage V. Once F_{total} is known, the lateral displacement Y of the free bar can be determined, by:

$$Y(F_{total}) = \frac{\tan\theta_o}{\lambda}\left[\lambda L - \sinh(\lambda L) + \frac{(\cosh(\lambda L) - 1)^2}{\sinh(\lambda L)}\right] \quad (6)$$

Two designs were implemented to demonstrate the T-drive concept. The first design was very similar to the schematic in Fig. 1. This design used a single free bar attached to the parallelogram suspension (Figs. 3 and 4). In this design, the electrostatic attracting force is linearly proportional to the length of the free bar. The second design increased the effective bar length by using multiple free bar/fixed bar pairs which form a "tree" (Figs. 5 and 6). The odd-shaped bars or "branches" attached to the central "trunk" of the tree of the suspension each have a wider 50 μm long section which has a 2 μm distance from the free bar. All other parts of the branch, including the back side of the 50 μm section were separated by a distance of at least 6 μm gap from the fixed bars. In this manner, the 50 μm section acted as a constant attraction surface area; as the structure displaced, the effective area of the parallel "plates" did not change.

Both of the structure types used a scale and pointer method to measure displacements. Due to the long overhang lengths of the structures, sometimes as great as 600 μm long, breakaway supports were used to provide added stiffness to the structure during fabrication (seem in Fig. 3). The breakaway supports were mechanically removed by physically rupturing them with test station probes. Small dimples were introduced into the free bars to prevent large area contact between the free bars and the substrate, thereby keeping the released structure free from surface tension adhesion during rinsing and drying.

FABRICATION and EXPERIMENTAL METHODS

The structures were fabricated using the process developed by Lim [4]. The structural layer defining the device was 2 μm thick phosphorus-doped polycrystalline silicon. The T-drive structures were fabricated with 1.7 μm wide, 400 and 446 μm length beams and with beam angles ranging from 4.6° to 5.2°. The initial gap between the free and fixed bars was 2.1 μm. A ground plane was used beneath the free bar and its suspension to prevent electrostatic interference from the substrate.

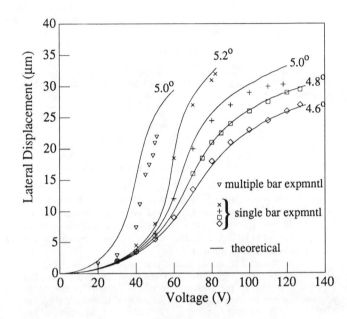

Figure 7
Lateral displacement Y versus applied voltage for various T-drive structures. The corresponding angle of the beams with respect to the primary direction is shown next to each set of data.

Static displacement measurements as a function of applied voltage were performed for several structures by use of the scale and pointer. This method allowed a ± 0.5 μm accuracy. The applied voltages ranged from 0 to 127 V. Out-of-plane deflections of the free bar were made using a microscope focusing method with an accuracy of about ±2 μm for small deflections. All testing was performed in air observed under under 1000X magnification both directly through a microscope and on a video monitor. The beam widths of the structures were measured using SEM images.

RESULTS and DISCUSSION

The theoretical determination of static displacement was dependent on the accuracy of both Young's modulus and the suspension beam widths. After measuring the beam widths, the value of Young's modulus used in the theoretical model was varied until theoretical displacements were found that corresponded to the experimental data. Using this method, the Young's modulus was found to be 105 ±15 GPa.

Static displacement amplitudes as a function of voltage were measured for several single and multiple bar structures and are plotted in Fig. 7. The displacements of the structures were smoothly controlled by varying the voltage. The T-drive designs with larger suspension beam angles deflected more for a given voltage. The larger beam angle resulted in a larger tangential component of the total electrostatic force

Table 1 Maximum Displacements attained for Single and Multiple-Bar T-drive structures each with its beams at different angles with respect to the normal direction.

	Maximum displacement (μm) (voltage at maximum displacement)			
	Angle of beams			
	4.6°	4.8°	5.0°	5.2°
single bar T-drive (446 μm beam)	27.0 (127V)	29.5 (127V)	30.5 (118)*	32.0 (82V)*
Multiple bar T-drive (446 μm beam)	22.0 (57V)†	20.0 (53V)†	22.0 (50)†	21.0 (45)†

* At higher voltages, the free bar contacted the fixed bar.
† At higher voltages, one of the branches deflected and contacted the fixed bar.

since $F_{tangential} = F_{total} \sin\theta$. This larger $F_{tangential}$ reduced the gap more for a given voltage than that of the lesser-angle beam designs. This increased F_{total} and correspondingly, $F_{tangential}$. Consequently, the displacements of the larger angle beam designs increased faster than the lower voltage designs as the voltage was increased.

For a given voltage, the single bar structures deflected less than the multiple bar structures. The experimental results bear out the expected proportionality between the length of the attracting surfaces (i.e. the bar lengths) and F_{total}.

The maximum static displacement amplitudes attained for the structures are shown in Table 1. For an applied voltage of 127V, a static displacement of 32 μm was measured for a single bar T-drive with 446 μm beams at an angle of 5.2°. The 32 μm displacement was the greatest displacement measured for any of the T-drive structures. The maximum displacement increased with an increase in θ_0. Geometrical constraints limited the maximum displacement of the structures with large θ_0. In these cases the bars contacted before the voltage reached 127V.

There was a significant difference between the maximum attainable displacements of the single and multiple bar structures. The multiple bar design of the tree end was inherently more compliant than the single bar structures. It was noted during testing that the branches deflected towards their respective fixed bars due to the large forces generated as the gap became smaller. When the gap decreased to a certain amount, the force balance between the electrostatic force and the resisting force in one of the branches would become unstable and the branch contact the fixed bar. The complexity of the "tree" may have contributed to the less than predicted displacements.

During testing of the single and multiple bar T-drives, it was noted that the free bar section of the devices levitated out of plane. The amplitude of the levitation varied from device to device depending on the physical dimensions of each device. For most of the devices, the out-of-plane levitation increased as the applied voltage increased until about 30 V was reached, after which the levitation decreased and was not perceptible above about 50 V. This levitation is due to the interaction of the ground plane and the free and fixed bars [11].

Theoretical calculations reveal that the forces that the T-drive would be able to generate are dependent on how much the suspension has been displaced before the force is exerted. This effect is due to the elastic forces required to deflect the suspension. For example, for one multiple bar design, an initial displacement of 10 μm would require approximately 36V (Fig. 8). Assuming the structure at this point contacts the the structure upon which it acts and the T-drive does not displace further, an increase of 8V would provide a force of 2 μN (1 μN is the force exerted by a 350 μm cube of silicon in gravity). Lower initial displacements reduce the rate at which the applied force would increase with voltage.

The differences in displacements between the theoretical model and the experimental data can be attributed to several causes. First, the structures not only have compliant suspension beams, but the free bars could also deform. As noted above, this was of especial concern for the multiple bar T-drive. Further, in the multiple bar free bar/fixed bar design, only part of each branch was near to the fixed bar, the nominal gap being 2 μm. However, the branch was surrounded by other features that were about 6 μm distant that were at the opposite potential. Since an electrostatic force was generated by this gap, the total force was actually less than the theoretical force.

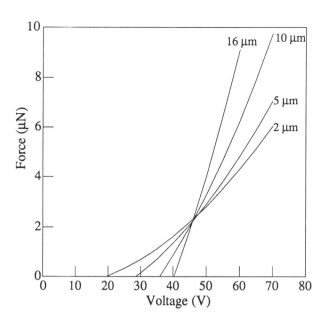

Figure 8
Theoretical force exerted by the T-drive versus applied voltage at fixed lateral displacements. The fixed displacement is noted by each curve.

The model had several limitations. As noted above, the beam theory used to describe the T-drive suspension was valid only for small displacements, or more accurately, for small changes of slope of the beams. These T-drive suspension beams did have significant changes of slope when they deflected. The effect of this increased slope was to effectively increase the stiffness of the system [12]. A second limitation of the model was that the model assumed perfect geometric dimensions whereas the actual structures had dimensions that varied slightly with the fabrication process. Since large displacements only occurred when the gap had become very small, the uniformity of the gap was very important.

The motion generated by the T-drive is not perfectly linear, but since the free bar travels in a very large radius arc as the beams deflect, but this non-linearity is very small.

CONCLUSION

Electrostatic large displacement structures have been developed, fabricated, and tested. The T-drive concept demonstrates that direct electrostatic attraction can be used to generate large displacements. These structures attained static displacements of up to 32 μm with an applied voltage of 82V and 22 μm with an applied voltage of 50V. Structures with longer beams show promise for even greater displacements. A theoretical model has been developed which closely describes the experimental data. Using this model, a value has been calculated for Young's modulus of 105 GPa. Further, it has been shown that relatively large forces can be applied with small increases in voltage once the free bar undergoes a large displacement and the gap between the free and fixed bars is reduced.

ACKNOWLEDGEMENT

The authors would like to thank the industrial members of the Berkeley Sensor & Actuator Center for the support and funding provided for this project.

REFERENCES

[1] J.D. Busch and A.D. Anderson, "Prototype Micro-Valve Actuator", *Proc. IEEE Micro Electro Mechanical Systems Workshop*, Napa Valley, CA, Feb. 1989.

[2] S. Akamine, T.R. Albrecht, M.J. Zdeblick, and C.F. Quate, "Microfabrication of Integrated Scanning Tunneling Microscopes", *Transducers '89, 5th International Conference on Solid-State Sensors and Actuators*, Montreux, Switzerland, 25-30 June 1989.

[3] P.L. Bergstrom, T. Tamagawa, and D.L. Polla, "Design and Fabrication of Micromechanical Logic Elements",*Proc. IEEE Micro Electro Mechanical Systems Workshop*, pp. 15-20, Napa Valley, CA, Feb. 1989.

[4] M. G. Lim, J. C. Chang, D. P. Schultz, R. T. Howe, and R. M. White, "Polysilicon Microstructures for Characterize Static Friction." *Proc. IEEE Micro Electro Mechanical Systems Workshop*, pp. 82-88, Napa Valley, CA, Feb. 1989.

[5] Martin G. Lim, Roger T. Howe, and Roberto Horowitz, "Design and Fabrication of a Linear Micromotor for Magnetic Disk File Systems", ASME Winter Annual Meeting, San Francisco, December 10-15, 1989

[6] L. Y. Chen, Z. L. Zhang, J. J. Yao, D. C. Thomas, and N. C. MacDonald, "Selective Chemical Vapor Deposition of Tungsten for Microdynamic Structures," *Proc. IEEE Micro Electro Mechanical Systems Workshop*, pp. 82-87, Salt Lake City, UT, Feb. 1989.

[7] Chang-Jin Kim, Albert P. Pisano, Richard S. Muller, and Martin G. Lim, "Polysilicon Microgripper", *to be presented at IEEE Solid-State Sensor and Actuator Workshop*, Hilton Head Island, South Carolina, June 4-7, 1990.

[8] William C. Tang, Tu-Cuong H. Nguyen, Michael W. Judy, and Roger T. Howe, "Electrostatic-Comb Drive of Lateral Polysilicon Resonators", *Transducers '89, 5th International Conference on Solid-State Sensors and Actuators*, Montreux, Switzerland, 25-30 June 1989.

[9] Ansoft Corporation, Pittsburgh, PA 15123 *Maxwell Users Guide*, January, 1989

[10] Raymond J. Roark and Warren C. Young, Formulas for Stress and Strain, 5th Ed., McGraw-Hill, New York, 1975

[11] William C Tang, Martin G. Lim, and Roger T. Howe, "Electrostatically Balanced Comb Drive",*to be presented at IEEE Solid-State Sensor and Actuator Workshop*, Hilton Head Island, South Carolina, June 4-7, 1990

[12] Egor P. Popov, Introduction to Mechanics of Solids, Prentice-Hall, Inc., Englewood Cliff, New Jersey, 1968

MULTI-LAYERED ELECTROSTATIC FILM ACTUATOR

Saku Egawa and Toshiro Higuchi

Institute of Industrial Science
University of Tokyo
7-22-1 Roppongi, Minato-ku, Tokyo 106, Japan

ABSTRACT

There is a strong need for forceful and small size actuators. The authors propose the multi-layered electrostatic film actuator, aiming at realizing high force/weight ratio actuators applicable for large scale machines. It has numerous films with plenty of small electrodes stacked and connected. In order to build this structure, an unique electrostatic actuator named the image charge stepping actuator have been developed. It utilizes image charges induced on high resistor to produce force. It have been fabricated on polymer films and a prototype of the multi-layered actuator with four film actuators have been constructed. The fabricated multi-layered actuator can produce force of 0.4N.

INTRODUCTION

At the present time, mechanical systems whose sizes are about one centimeter to one meter are most often actuated by electromagnetic motors. However, conventional motors are much heavy and large compared with human or other animals' muscles. This brings about a serious problem especially in designing robots because in the robots, their actuators must often carry weight of actuators by themselves. In order to overcome this problem, various new actuators have been developed, such as actuators using shape memory alloy [1], pneumatic muscle [2, 3], and mechanochemical actuators using gels [4].

The electrostatic actuator is a kind of electric motors which has been long studied [5]. Several electrostatic motors whose dimensions and structures are analogous to conventional electromagnetic motors were developed [6, 7], but they could not produce enough force for actual use. This is because they did not take full advantage of electrostatic force. Advantages of electrostatic actuators will be summarized that they suit for miniaturization. They need only thin plane electrodes to produce electric field and do not require large and heavy three dimensional structures such as iron cores and coils in electromagnetic motors. In addition, maximum force per unit volume increases as size of an actuator is reduced, since breakdown field strength of air increases in very small gaps, and interfacial area per volume is enlarged while force is in proportion to the area [8, 9].

Recently, several micro electrostatic motors for micro mechanisms have been designed and developed using silicon micromachining technique [8, 9, 10, 11, 12, 13]. In the field of micro mechanisms, electrostatic force is considered to be advantageous. However, not only in micro mechanisms, but also in ordinary scale machines, electrostatic force can show its talent. The actuator which comprises a large number of small actuator elements and which has large interfacial area [9, 14] can take advantage of miniaturization and at the same time can obtain large force, adding the forces of the element actuators. Theoretically, the total force of such an actuator can be increased to any extent by reducing size of the elements and increasing the number of them, so that high force/volume ratio can be realized. This actuator will be applicable for various machines.

Aiming at realizing forceful and small size actuators, we have been developing the multi-layered electrostatic film actuator, in which numerous films with plenty of small electrodes are stacked and connected [14]. In order to build this structure we have developed an unique actuator named the image charge stepping actuator [14]. In this paper, we describe the concept of the multi-layered actuator, the principle of image charge stepping actuator, its fabrication on polymer films, and stacking of the films. Experimental results are also provided.

MULTI-LAYERED ELECTROSTATIC FILM ACTUATOR

As described above, large force can be produced by an electrostatic actuator with plenty of small elements (electrodes) and large interfacial area. For the material of this actuator, thin fibers or films seem desirable because they allow large surface per volume. Surface area per volume of fibers is larger than that of films but fabrication technique of electrodes on fibers has not been established and its development will be very hard. On the other hand, conventional photo lithography process is applicable to films. Thus we have adopted films, especially polymer films other than silicon substrates which are widely used in recent micro mechanisms, because polymer films are more flexible, less brittle, less expensive, and easier to get large thin films.

When using films, attention should be paid to the fact that films bend and wrinkle easily. However, films will not much be deformed when they are simply pulled straight. Therefore, for film actuators, linear motion is preferable. Considering this, we have adopted a structure shown in Fig. 1, where films are bound into two arrays and they are interleaved. This actuator resembles muscle in the sense that it has interleaved structure and that it produces tensile force. It may be called "Electrostatic Artificial Muscle." Although this actuator looks alike MIPA of Jacobsen et al. [9], there are some differences. They imply the use of silicon or some other hard materials while we use polymer films. Our actuator operates on a principle different from that of MIPA, which is described later.

Effect of reducing size of elements is explained as follows. Electrostatic force acting on an object can be calculated by integrating the Maxwell's stress over a closed surface surrounding the object. Since the Maxwell's stress is in proportion to energy density of electric field E, force of one unit of actuator is estimated as

$$f_u \propto \tfrac{1}{2}\varepsilon E^2 s \propto E^2 l_0^2$$

where ε is permittivity, s is effective interfacial area of the unit actuator, and l_0 is the characteristic dimension. Thus force density per area of the film becomes

$$f_s = f_u/s_u \propto E^2$$

where s_u is area of one unit. This indicates that force produced by a film is definite regardless of size of its elements if strength of field keeps constant. However, thickness of the actuator h can be reduced in proportion to size of elements. Thus the number of actuators which can be stacked in a unit height increases, that is, force density per volume is in inverse proportion to l_0.

$$f_v = f_s/h \propto E^2/l_0$$

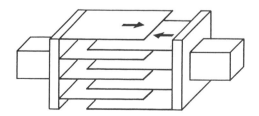

Fig. 1 Schematic Figure of Multi-Layered Film Actuator

(Electrostatic Artificial Muscle)

Thus, as size of elements is reduced, force of an actuator of a certain volume increases. In addition, since applied voltage V is in proportion to El_0, it can be decreased and the voltage control circuit become easier. Moreover, field E can be increased as already mentioned, obtaining larger force multiplied by square of increase of E. These indicate the advantage of reducing size of elements.

BUILDING MULTI-LAYERED FILM STRUCTURE WITH CONVENTIONAL ACTUATORS

There are several types of electrostatic actuators. Which of them fits best for the multi-layered actuator?

The variable capacitance motor [6, 8, 10, 12] is the most conventional electrostatic actuator. In its linear form, it has parallel striped electrodes on both of two parts which shift relatively to each other. (For simplification, here we call the parts *stator* and *slider*, although it is not absolutely defined which one moves.) Many problems should be solved to fabricate this actuator on polymer films and to stack these film actuators.

(a) It produces normal attractive force besides tangential motive force. To maintain its air gap between the stator and the slider, it requires bearings, whose fabrication on films is difficult. Another way to cancel the normal force is to put the slider between two stators. However, this approach is ineffective for film actuators because it is hard to keep a narrow air gap over large area since the normal force will bend films easily. If the stator and the slider touch each other without lubrication, the normal force will cause so large friction that the slider can no longer move. If the films are hardened by thickening in order to be able to keep the air gap, force density decreases and the advantage of using films disappears. Technique for fabricating bearings on films or some other way to keep the gap and reduce friction should be developed.

(b) Electrodes on the slider and the stator have to be aligned so that they run parallel with each other. Otherwise, the actuator cannot function because its force will be canceled in total since direction of force may vary on places. Alignment of films with long and narrow-pitch electrodes requires very high precision. Moreover, when this actuator is stacked, difficulty increases since films of every actuator have to be aligned respectively. Technique for high precision fabrication and assembly should be developed.

(c) Since films are extensible, they are tend to be imprecise in their shape. Thus displacement between electrodes of the slider and the stator may vary on different places. Therefore, local feedback is required to attain good performance. Fabrication technique of sensors and voltage switching circuits on films should be developed.

The induction motor [6, 7, 13] is another type of the electrostatic actuator. It has a slider made of or coated with slightly conductive material. When traversing electric field produced by the stator is applied on the slider, electric charge is induced on the slider with some lag behind the field. The displacement between the field and the charge causes motive force. Since the induction motor has no electrodes on the slider, it does not require alignment between the stator and the slider nor feedback control. However it has normal attractive force as same as the variable capacitance motor. This causes same problem as (a) above.

There are also a family of actuators classified as *the variable gap electrostatic actuator* [15]. In these actuators, their movable parts are in rolling contact with their stators [9, 11]. Therefore friction in these actuators is small. However, this configuration is probably not applicable to film actuators.

IMAGE CHARGE STEPPING ACTUATOR

Since actuators devised up to the present have problems in constructing the multi-layered film structure, we have developed an unique actuator using high resistor, named *the image charge stepping actuator*, which is free from many of those problems. Fig. 2 shows the basic structure and operating principle of the image charge actuator. Its stator has striped electrodes covered with isolative material. The electrodes are wired so that every three electrodes have same potential. The slider has no electrodes but consists of isolator and high resistor layers. It is driven through the charge induction stage (1 and 2 in Fig. 2) and the actuation stage (3 and 4) as below:

(1) The slider is just put on the stator. The slider has no electric charge at first. Positive and negative voltages are applied on two

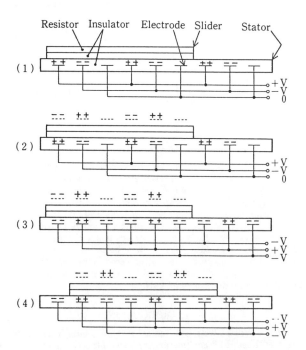

Fig. 2 Structure and Principle of Image Charge Stepping Actuator

groups (phases) of electrodes respectively as shown in Fig. 2 (1).

(2) Electric field produced by the electrodes causes movement of charges in the resistor which continues until the induced charges neutralize the field in the resistor. In the equivalent state, the induced charges can be indicated by a mirror image of charges on the electrodes, displayed as broken lines in Fig. 2 (2).

(3) After charges are sufficiently induced, the signs of voltages applied on the two groups at the first step are inverted and negative voltage is applied on the third group as shown in Fig. 2 (3). This causes an instant change of charges on the electrodes. However, charges in the resistor cannot move in a short time, obstructed by the high resistance. Thus, charge configuration shown in Fig. 2 (3) appears. Although it requires complex calculation to exactly predict produced force, its direction can be estimated. In this configuration, the negative charge on the third group repels the upper right negative charge in the resistor and attracts the upper left positive charge. Tangential motive force which thrusts the stator to the right is produced as the sum of these two forces. In addition, normal repulsive force is produced because the induced charges on the resistor have same signs as the charges on the electrodes just below. This force reduces friction between the stator and the slider. It helps the slider to move.

(4) The stator halts after it moves about one pitch of electrode. This procedure can be repeated, shifting the voltage pattern applied to the electrodes.

The image charge actuator fits for the multi-layered actuator as described below.

(a) It does not need alignment between films, feedback control, nor high precision in dimensions of electrodes. This is because, instead of having electrodes on the slider, the image charge actuator utilizes induced image charges, which act as a precise copy of the electrodes on the stator. The image charges can also be regarded as short life electrets temporarily made on the slider.

(b) It can function even though it has no bearing and the slider is in contact with the stator because the normal repulsive force reduces friction at the moment when the motive force thrusts the slider. When the slider is stationary, large friction is produced contrarily, which holds the slider tightly. Moreover, since there is no air gap, small separation between the electrodes and the resistor can be easily realized over large area.

(c) The multi-layered actuator with small number of layers can be constructed simply by stacking and connecting image charge actuators fabricated on films. It requires no complex additional structures since its layers can bear weight of upper layers with the normal repulsive force.

The image charge actuator resembles the induction motor. However, the induction motor uses continuously traversing field and it always has normal attractive force. The image charge actuator changes field discontinuously. It enables the repulsive force to be produced. Moreover, slide of charges in the resistor is smaller than that of the induction motor. Therefore, energy loss in the resistor is smaller.

RESISTIVITY CONTROL

The most difficult problem in making actuators using high resistor is to control its resistivity [13]. In the image charge actuator, it should be selected so that the time for charging and discharging are in the range of about 10 times to 100 times as much as the time for traversing one pitch. If resistivity is too high, it takes too much time for charge induction. If resistivity is too low, the induced charges disappear before the slider moves in the actuation stage.

To calculate the first estimate of the required resistivity, the simplest model of one actuator unit shown in Fig. 3 can be used. The lower plates of the two capacitors represent the two electrodes on which voltage is applied in the induction stage. The third electrode is omitted because it has smaller effect here. Considering dimensions of this structure gives

$$C = \varepsilon \frac{L w_C}{d_C}, \quad R = \rho \frac{d_R}{L w_R}$$

where L is length of electrodes and ρ is volume resistivity. Thus, time constant of charging and discharging becomes

$$\tau = \tfrac{1}{2} CR = \varepsilon \rho \frac{w_C d_R}{2 d_C w_R}$$

To obtain estimation of exponent of resistivity, we consider dimensions w_C, d_C, w_R, and d_R to be all approximately equal to the characteristic dimension l_0 and ignore the coefficient 1/2. Thus,

$$\tau \approx \varepsilon \rho$$

The right side is known as relaxation time of free charges in resistor. When very thin coating is used for the resistor layer instead of bulk resistor, surface resistivity σ characterizes it. Then,

$$R = \sigma d_R / L$$

Thus,

$$\tau = \varepsilon \sigma \frac{w_C d_R}{2 d_C} \approx \varepsilon \sigma l_0$$

If we assume, considering the actually fabricated actuator, $\tau \approx 1$ [sec], $l_0 \approx 1$ [mm], and $\varepsilon \approx \varepsilon_0 \approx 10^{-11}$ [F/m], the required resistivity is obtained.

$$\rho \approx 10^{11} \, [\Omega/\text{m}], \quad \sigma \approx 10^{14} \, [\Omega/\Box]$$

This volume resistivity is higher than that of semiconductor and lower than that of common isolators. Some polymers such as nylon are in this domain but resistivity of nylon is dependent on temperature and humidity. Technique for its stabilization or new material should be developed.

Modifying surface resistivity by coating is a convenient way to control resistivity. It is easier to apply a coating than to produce a bulk film. However, the required surface resistivity for 1mm scale actuator is very high. It is higher than that of typical antistatic coating, which gives slight conductivity on isolative material to avoid collection of static charges. It is difficult to precisely make and maintain such high surface resistance because it is easily affected by environment such as humidity and contamination, and it easily gets mechanical damages. Soaking whole the actuator in dielectric liquid may help to control the environment to stabilize resistivity. It may also facilitate lubrication and heat transfer.

Control technique of resistivity is the key to development of the image charge actuator. Further study is required for this problem.

FABRICATION

We have fabricated the image charge actuator on PET (Poly Ethylene Terephthalate) films. Fig. 4 illustrates its cross section. To produce the slider, striped pattern of 0.42mm pitch is drawn at first using laser beam on a copper plated PET film. Then it is

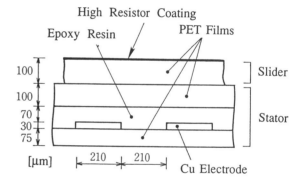

Fig. 4 Cross Section of Image Charge Actuator Fabricated On PET Film

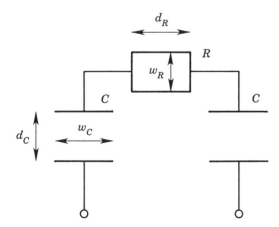

Fig. 3 Simple Model of One Unit of Image Charge Actuator

Fig. 5 Photo of Fabricated Film Actuator with 0.42mm Pitch Electrode

etched to form the electrodes. The electrodes are covered with epoxy resin and a PET film. For the isolator layer of the slider, a PET film is used. A soft (high resistive) antistatic agent is coated on the film as the resistor layer. Fig. 5 shows the fabricated actuator. White dots seen on the slider are holes picked on the slider to prevent, by leading air, the stator and slider from sticking together.

When constructing the multi-layered actuator, film actuators are stacked in such a order as illustrated in Fig. 6. This configuration is preferable since, as well as the motive force, the normal repulsive force which reduces friction is produced on all the sliding surfaces. In this configuration, separation between layers is required to prevent the adjacent layers from affecting each other.

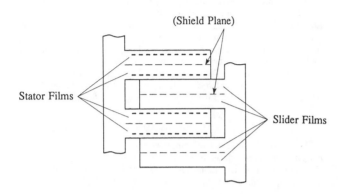

Fig. 6 Construction of Multi-Layered Actuator

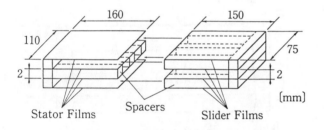

Fig. 7 Multi-Layered Actuator with Four Layers

The required minimum separation is several times as much as pitch of electrodes. In addition to spatial separation, shield planes can be added, which completely reject electrical interference.

We have constructed a prototype of the multi-layered actuator by stacking four of the film actuators as shown in Fig. 7 and 8. Its effective interfacial area is 8cm×12.2cm per one layer and 390 cm^2 in total. Since about 2mm separation of layers is required this time, spacers of 2mm high are inserted. When the pitch is sufficiently reduced, spacers can be eliminated because the required separation becomes so small that thickness of the film itself can include it.

EXPERIMENTS

We applied ±800V on the film actuator through mechanical relays to drive it. It typically takes about 10 seconds to sufficiently induce charges on the slider which has no charges at first. The time constant varies with temperature, humidity, and aging of the resistor coating. It ranges from about 3 seconds to 30 seconds. Considering the relation between time constant and resistivity, surface resistivity of the coating is estimated as 10^{15} Ω/□. After the charges are once induced, the time required for induction decreases to less than one second since the charge induced in the preceding induction process can be utilized repeatedly. The slider can move at minimum cycle time of 0.2 seconds when no load is applied, with ratio of voltage application time for charge induction and actuation stage selected to 9:1.

Fig. 9 shows a step movement of the slider observed by an optical instrument. The slider traverses distance of one pitch in 10 milliseconds. Since the charging time is too long compared with the traversing time, it can be noted that use of more appropriate material for the resistor layer will reduce initial charging time.

The multi-layered actuator has been operated with the same voltage applied. The spacers function effectively to avoid interference between the layers. Without the spacers, it occasionally reveals some improper operations such as rotation of films on their surfaces. We have measured the motive force produced by the multi-layered actuator. Some weights were used to exert a load on the actuator as shown in Fig. 10. Produced force is calculated as a sum of weight of the slider and the maximum load against which the slider can move. Force of this actuator stalls if its cycle time is decreased. In this experiment, it is fixed at 2 seconds. Measurement was done for the actuators with one, two, and four layers respectively. The force is approximately in proportion to the number of layers. This implies that the layers do not affect each other and their forces are effectively added. The actuator with four layers produces about 0.4N. Considering dimensions of the actuator, force density per unit interfacial area is 0.001N/cm^2 and that per unit volume is 0.004N/cm^3.

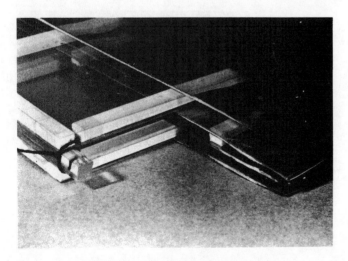

Fig. 8 Photo of Fabricated Multi-Layered Actuator

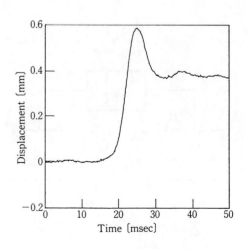

Fig. 9 Step Movement of Slider

Fig. 10 Measurement of Force by Applying Weight

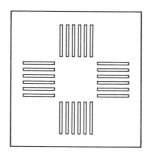

Fig. 11 Example of Electrode Pattern for 3 DOF Actuator

VARIATIONS

In this chapter we provide some variations of material, scale, and configuration of the image charge actuator.

A. Nylon Coating for Resistor Layer

As already mentioned, nylon has appropriate resistivity for the image charge actuator. We have tested performance of nylon coating. At first, nylon 6 pellets are dissolved in m-cresol. Then it is applied on a PET film to form a coating of about 5μm thick. The slider with nylon coating occasionally works properly. However, its resistivity easily varies with humidity and it gives no reliable result. The resistivity ranges from one hundredth to one hundred times of desirable value. Some stabilization technique should be developed or nylon coating can be used only in precisely controlled environment.

B. 0.1mm Pitch Film Actuator

A finer pitch actuator has been fabricated. Its stator has 0.1mm pitch electrodes on a 25μm thick polyimide film covered with polyimide. Reports on this actuator will appear in succeeding papers.

C. 3 DOF Actuation and 20μm Pitch Actuator

Since the slider of the image charge actuator is homogeneous in its extent, it can be placed at any arbitrary angle to the stator. This enables an actuator with three degrees of freedom to be realized. Fig. 11 shows an example of electrodes on the stator of a 3 DOF actuator with four groups of electrodes. Its slider can move to an arbitrary position and orientation in the plane, selecting the groups to be activated and determining their direction of thrust. Although this actuator does not fit for stacked structure, it will be useful for such applications as thin positioning tables and handling machines for thin materials. A 3 DOF actuator with 20μm pitch electrodes on alumina substrate have been designed. It will be fabricated before long.

D. Use of Photo Conductor

The resistor layer may be replaced by a photo conductor layer. Resistivity of the photo conductor is reduced by applying light while inducing charges. During the actuation stage, the light is cut off to prohibit the charges from running off. Advantages of the use of photo conductor include: (a) Performance of the actuator will be less affected by environment than when high resistor is used. When using photo conductor, the lower its resistivity under light is, in the shorter time charge induction can be done, and the higher its resistivity in the dark is, the more tightly the charges are fixed. Therefore, photo conductor with wider range of resistivity is preferable. If its resistivity under light is sufficiently low and that in the dark is sufficiently high, performance of the actuator will not change even if its resistivity somewhat varies. (b) Application of light reduces charge induction time. This improves time ratio of induction and actuation and the cycle time can be decreased.

The way to applying light is a problem in the use of photo conductor. When only one layer of the image charge actuator is used for applications such as positioning tables, light can be illuminated from above. In the multi-layered actuator, its films themselves may lead the light like optical fibers.

CONCLUSIONS

We have proposed the multi-layered electrostatic film actuator, which may be called electrostatic artificial muscle. To build that actuator, the image charge stepping actuator have been developed. As a prototype, we have fabricated an image charge actuator with 0.42mm pitch electrodes on PET films. Using four of this actuator, a multi-layered actuator have been constructed. Its performance have been examined by making it pull up some weights. It can operate under 0.4N load. The film actuators do not affect each other and the their force are effectively added. Though the obtained force density of $0.004N/cm^3$ is not yet satisfactory for ordinary scale machines such as robots, it is expected to be improved, reducing size of its elemental structure.

REFERENCES

[1] K. Ikuta, M. Tsukamoto, and S. Hirose, "Shape Memory Alloy Servo Actuator System with Electric Resistance Feedback and Application for Active Endoscope," *Proc. 1988 IEEE International Conference on Robotics and Automation*, pp. 427-430, Philadelphia, PA, Apr. 24-29, 1988

[2] K. Inoue, "Rubbertuators and Applications for Robots," *The 4th International Symposium on Robotics Research*, pp. 57-63, Aug. 14, 1987, MIT Press

[3] G. B. Immega, "ROMAC Actuators for Micro Robots," *Proc. IEEE Micro Robots and Teleoperators Workshop*, Hyannis, MA, Nov. 1987

[4] T. Tanaka, "Gels," *Scientific American,* Vol. 244, No. 1, pp. 110-123, Jan. 1981

[5] O. D. Jefimenko, "Electrostatic Motors," *Electrostatics and its Applications* (editor, A. D. Moore), Chap. 7, pp. 131-147, John Wiley & Sons, Inc., New York, 1973

[6] B. Bollée, "Electrostatic Motors", *Philips Technical Review,* Vol. 30, No. 6/7, pp. 178-194, 1969

[7] S. D. Choi and D. A. Dunn, "A Surface-Charge Induction Motor," *Proc. of the IEEE,* Vol. 59, No. 5, pp. 737-748, May 1971

[8] J. H. Lang, M. F. Schlecht, and R. T. Howe, "Electric Micromotors: Electromechanical Characteristics," *Proc. IEEE Micro Robots and Teleoperators Workshop,* Hyannis, MA, Nov. 9-11, 1987

[9] S. C. Jacobsen, R. H. Price, J. E. Wood, T. H. Rytting, and M. Rafaelof, "The Wobble Motor: An Electrostatic, Planetary-Armature, Microactuator," *Proc. IEEE Micro Electro Mechanical Systems Workshop,* pp. 17-24, Salt Lake City, UT, Feb. 20-22, 1989

[10] W. S. N. Trimmer and K. J. Gabriel, "Design Considerations for a Practical Electrostatic Micro-Motor," *Sensors and Actuators,* Vol. 11, No. 2, pp. 189-206, Mar. 1987

[11] H. Fujita and A. Omodaka, "The Fabrication of an Electrostatic Linear Actuator by Silicon Micromachining," *IEEE Trans. Electron Devices,* Vol. ED-35, No. 6, pp. 731-734, Jun. 1988

[12] Y. C. Tai, L. S. Fan, and R. S. Muller, "IC-Processed Micro-Motors: Design, Technology, and Testing," *Proc. IEEE Micro Electro Mechanical Systems Workshop,* pp. 1-6, Salt Lake City, UT, Feb. 20-22, 1989

[13] S. F. Bart and J. H. Lang, "Electroquasistatic Induction Micromotors," *Proc. IEEE Micro Electro Mechanical Systems Workshop,* pp. 7-12, Salt Lake City, UT, Feb. 20-22, 1989

[14] T. Higuchi and S. Egawa, "Electrostatic Actuator with Resistive Slider," *1989 National Convention Record I.E.E. Japan,* Section 6, pp. 191-192, Apr. 4-6, 1989 (in Japanese)

[15] H. Fujita, A. Omodaka, and M. Sakata, "Variable Gap Electrostatic Actuators," *Technical Digest of the 8th Sensor Symposium,* 1989, pp. 145-148, 1989

Movable Micromachined Silicon Plates with Integrated Position Sensing

MARK G. ALLEN*, MARTIN SCHEIDL and ROSEMARY L. SMITH**

Microsystems Technology Laboratories, Massachusetts Institute of Technology, Cambridge, MA 02139, (U.S.A.)

ALEKS D. NIKOLICH

The Charles Stark Draper Laboratory, Inc., Cambridge, MA 02139 (U.S.A.)

Abstract

A process for fabricating large silicon plates of varying thicknesses suspended by thin, flexible polyimide beams has been developed. The plates have been integrated with an on-chip ring oscillator which provides a read-out of the plate position. The silicon plates can be moved electrostatically and the position of the plate determined by the frequency shift of the ring oscillator. Initial testing of the device yields a full-scale frequency modulation of approximately 2%, corresponding to a full-scale frequency shift of 13 kHz from a zero-deflection frequency of 680 kHz.

Introduction

The technique of micromachining has been used to make a host of movable and flexible structures for sensing and actuation purposes (see, e.g. refs. 1–3). Recently, miniature deflectable reflecting plates for use as mirrors in fiber optic switching and display applications have been fabricated [4–7]. As many of these structures are micromachined from silicon, and as silicon is a stiff material, relatively large driving voltages may be required for large motions. In this work we have used silicon plates supported by flexible polyimide beams in the hope of reducing the driving forces required for large-scale deflection.

Structures using polyimide-supported plates as thermally isolated structures for gas sensors have been described by Stemme [8]. In that structure, polyimide was chosen for its thermal, not mechanical, properties. Recently, the mechanical properties of polyimide have been studied with the ultimate goal of using this material for structural members in microsensors and microactuators

[9, 10]. Quantitative knowledge of the Young's modulus, residual stress and debond energy (adhesion) has now permitted the design and fabrication of polyimide-based microsensors and microactuators. Other desirable properties of polyimide are its mechanical flexibility (relative to silicon), its planarization properties, and its compatibility with integrated circuit processing. This paper describes the design, fabrication and testing of square silicon plates suspended by polyimide beams. A read-out scheme which allows on-chip sensing of the plate position is also reported. Possible applications for this device currently under investigation include feedback-stabilized electrostatically positionable mirrors (as described above) as well as a dynamically rebalanced accelerometer using the silicon plate as a proof mass.

Implementation

The polyimide-supported silicon plates described below are square, 2 mm on a side, 10 μm in thickness, and supported at one end by a polyimide beam loaded in a torsional configuration (see Fig. 1). The mechanical response of these structures has been described previously [11]. A schematic of the structure with integrated read-out is shown in Fig. 2. Two metal pads can be seen on the movable plate: a drive electrode for electrostatic positioning and a sense electrode for position sensing. The silicon plate is attracted electrostatically by application of a voltage between the drive electrode on the movable plate and a second plate underneath the device. Variation of this voltage causes a change in the electrostatic attraction force and therefore a change in the angle of deflection of the movable plate. The plate position is determined using a capacitive read-out scheme. A variable capacitor is formed by the series connection of two capacitors, one variable and one fixed. The first capacitor (variable) is formed by the sense electrode on the movable plate and an overhead conducting plate at floating potential. The second

---

*Present address: Georgia Institute of Technology, School of Electrical Engineering, Atlanta, GA 30332, U.S.A.

**Present address: University of California at Davis, Department of EECS, Davis, CA 95616, U.S.A.

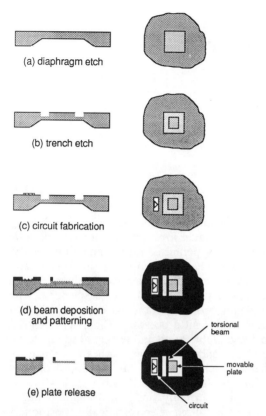

Fig. 1. Process sequence for device fabrication.

capacitor is formed between this same overhead plate and a fixed U-shaped electrode on the chip surface. The total coupling capacitance between the sense electrode on the movable plate and the fixed electrode on the chip surface is therefore given by the series connection of these two capacitors. It should be noted that it is not necessary to make electrical contact to the overhead conducting plate. As the drive voltage is increased, the movable plate is deflected downward and the total coupling capacitance between the sense electrode and the fixed U-shaped electrode decreases. This capacitance change is sensed by an on-chip ring oscillator (Fig. 3), which contains this variable capacitor in its feedback circuit. Changes in the plate position can therefore be detected by changes in the ring oscillator frequency.

Fabrication

The devices described above were fabricated using standard bulk micromachining techniques. The starting material was a single-side polished $\langle 100 \rangle$ n-type silicon wafer of 2 in diameter and 11 ± 1 mil thick. A 1.3-μm-thick masking oxide was grown at 1100 °C in steam for 3 h. The front of the wafer was protected with photoresist and the oxide on the backside of the wafer was patterned into a series of square holes 3 mm on a side using a buffered oxide etch (BOE) (see Fig. 1). The photoresist was stripped in 3:1 sulfuric acid: hydrogen peroxide solution and the wafers were etched in 20 wt.% potassium hydroxide (KOH) solution at 56 °C for 11 h. The etch rate of this solution was 20–21 μm/h, forming 3-mm square diaphragms approximately 40 μm in thickness (Fig. 1(a)). The front-side oxide was then patterned using an infrared aligner to align the front-side pattern to the diaphragms and the wafers were etched in the above KOH solution for 30 min to form trenches 10 μm deep. During this etch the diaphragms were also etched 10 μm from the back, resulting in a 30-μm-thick square 2 mm on a side surrounded by a 1-mm-wide trench 20 μm thick (Fig. 1(b)). All masking oxide was then

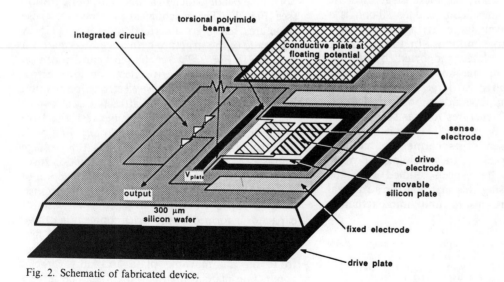

Fig. 2. Schematic of fabricated device.

stripped in BOE and the wafers were cleaned using a standard RCA cleaning procedure.

The three-stage ring oscillator circuit for readout of the plate position was implemented at this point using a self-aligned polysilicon-gate PMOS process with 10 μm design rules (Fig. 1(c)). After the circuit process was completed (including sintering of patterned metal), the mechanical structure process was continued as follows. An adhesion promoter layer consisting of a 0.5% solution of γ-triethoxyaminopropylsilane in 95% methanol/5% water was spin-coated on the front of the wafer at 5000 rpm for 30 s. The polyimide (DuPont PI-2555) was then spun on the front side of the wafer as its polyamic acid precursor from solution in N-methylpyrrolidone at 2000 rpm for 120 s. The polyamic acid was baked at 120 °C for 10 min to drive off the solvent and induce partial imidization and reaction with the adhesion promoter. Photoresist (KTI 820-20) was then spin-coated at 2500 rpm for 30 s on top of the polyimide and baked at 90 °C for 25 min. The photoresist and partially imidized polyamic acid were simultaneously patterned by the resist developer (a solution of tetramethylammonium hydroxide in water) to form the supporting beams (Fig. 1(d)). The resist was then stripped using acetone and methanol, and the patterned polyamic acid beams were fully imidized by baking at 400 °C in nitrogen for 45 min. The thickness of the post-cured polyimide was approximately 4 μm. A second level of metal consisting of 100 Å of chromium and 1500 Å of gold was sputter-deposited and patterned using standard etchants to form the sense and drive electrodes and to connect these electrodes to the integrated circuit. Finally, the wafers were blanket-etched from the back using a sulfur hexafluoride plasma etch until the silicon in the trenches surrounding the plates was etched away, releasing the beams and center plates (Fig. 1(e)). Figure 4 shows a photomicrograph of a fabricated device.

Testing and Results

The chip was packaged in a standard flat-pack carrier. A 3-mil-thick piece of polyimide (Kapton) adhesive tape was placed on the bottom of the flat-pack to provide insulation for the electrostatic drive, and the chip was mounted on the tape using a commercially available cyanoacrylate adhesive (Permabond 200). A front-side substrate contact was made by etching a small area of field oxide using concentrated hydrofluoric acid and bonding directly to the exposed silicon. A glass microscope cover slip approximately 5 × 5 mm on a side coated with sputtered indium–tin oxide was used as the external cover plate for capacitively coupling the ring oscillator signal from the movable plate to the fixed U-shaped electrode. This cover plate was positioned with the indium–tin oxide side facing the movable plate so as to cover the movable plate and U-shaped electrode except for the edge of the plate with the torsional polyimide beam. The outer edges of the cover plate were supported on the polyimide in the field regions of the chip; thus, the cover plate was separated from the movable plate by the thickness of the polyimide film (approximately 4 μm).

The feedback loop of the on-chip ring oscillator (Fig. 3) was completed using a 47K external resistor (R_1) and a 2 pf external capacitor (C_0) in parallel with the variable (movable plate) capacitor C. For the testing described below, the sense and drive pads on the movable plate were tied together electrically. The chip was bonded and packaged with cover plate as described above, and terminal voltages of $V_{DD} = -6$ V, $V_{GG} = -10$ V and V_{SB} (substrate bias) $= +5$ V were applied. An off-chip comparator was used to buffer the oscillator signal and the frequency of oscillation was measured using a digital frequency counter. Under these conditions, a free-running (zero-deflection) ring oscillator frequency of approximately 682 kHz was observed. A positive voltage relative to the circuit ground was applied to the metal case of the flat-pack carrier which caused the movable plate (held near circuit ground by the on-chip oscillator) to deflect downward towards the case. This downward deflection decreased the coupling capacitance and increased the frequency of the ring oscillator, as expected. Figure 5 shows a plot of the ring oscillator frequency as a function of the applied deflecting voltage. A nonlinear relationship is observed due to the nonlinear force–voltage electrostatic relationship. A full-scale

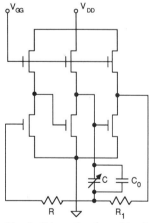

Fig. 3. Schematic of on-chip ring oscillator. R is an on-chip resistor; C is the variable capacitor formed between the movable plate and the fixed electrode as described in the text; R_1 and C_0 are external components used to set the oscillation period, as described in the text.

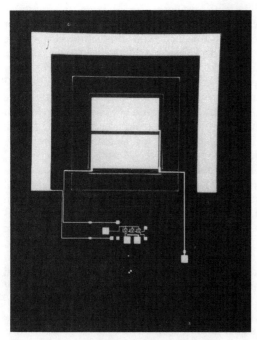

Fig. 4. Photomicrograph of fabricated device.

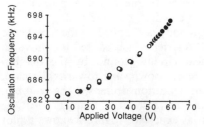

Fig. 5. Ring oscillator frequency as a function of attracting voltage on plate: (●) sweeping voltage from low to high; (○) sweeping voltage from high to low.

(approximately 70 μm) plate deflection corresponding to the maximum applied voltage of 60 V yields a full-scale frequency shift of 13 kHz, corresponding to a full-scale frequency modulation of approximately 2%. Application of drive voltage to the movable plate without the indium–tin oxide cover plate in place resulted in deflection of the plate with no corresponding change in oscillator frequency.

Conclusions

A polyimide-supported electrostatically deflectable silicon plate with integrated sensing of the plate position has been demonstrated. The device has been implemented using standard bulk micromachining techniques and a self-aligned polysilicon-gate PMOS process. A full-scale deflection of the plate results in a frequency shift of approximately 2%, which is easily observable with an off-chip comparator/buffer and a digital frequency counter. Under current investigation is the use of the plate position read-out for feedback stabilization of the plate, and the closed-loop response of the feedback-stabilized device to applied acceleration.

Acknowledgements

This work was supported by a graduate fellowship from the International Society for Hybrid Microelectronics (MGA) and by The Charles Stark Draper Laboratory. Microfabrication was carried out in the Microsystems Technology Laboratories, and in the Microelectronics Laboratory, a central facility of the Department of Materials Science and Engineering, supported in part by the National Science Foundation under contract number DMR-84-18718. Technical discussions with Professors Martin Schmidt and Stephen Senturia of MIT and sample fabrication assistance from Douglas Young of MIT are also gratefully acknowledged.

References

1 K. E. Petersen, Silicon sensor technologies, *IEDM Tech. Digest*, (1985) 2–7.
2 W. Riethmueller and W. Benecke, Thermally excited silicon microactuators, *IEEE Trans. Electron Devices, ED-35* (1988) 758–762.
3 K. E. Petersen, Micromechanical membrane switches on silicon, *IBM J. Res. Dev., 23* (1979) 376.
4 L. J. Hornbeck, 128 × 128 Deformable mirror device, *IEEE Trans. Electron Devices, ED-30* (1988) 539–545.
5 K. Gustafsson and B. Hoek, Fiberoptic switching and multiplexing with a micromechanical scanning mirror, *Tech. Digest, 4th Int. Conf. Solid-State Sensors and Actuators (Transducers '87), Tokyo, Japan, June 2–5, 1987*, pp. 212–215.
6 K. E. Petersen, Silicon torsional scanning mirror, *IBM J. Res. Dev., 24* (1980) 671.
7 K. E. Petersen and C. R. Guarnieri, Micromechanical light modulator array fabricated on silicon, *Appl. Phys. Lett., 31* (1977) 521.
8 G. Stemme, An integrated gas flow sensor with pulse-modulated output, *Tech. Digest, 4th Int. Conf. Solid-State Sensors and Actuators (Transducers '87), Tokyo, Japan, June 2–5, 1987*, pp. 364–367.
9 M. G. Allen, M. Mehregany, R. T. Howe and S. D. Senturia, Microfabricated structures for the *in situ* measurement of residual stress, Young's modulus, and ultimate strain of thin films, *Appl. Phys. Lett., 51* (1987) 241.
10 M. Mehregany, R. T. Howe and S. D. Senturia, Novel microstructures for the *in situ* measurement of mechanical properties of thin films, *J. Appl. Phys., 62* (1987) 3579.
11 M. G. Allen, M. Scheidl and R. L. Smith, Design and fabrication of movable silicon plates suspended by flexible support, *Proc. Micro Electro Mechanical Systems Workshop, Salt Lake City, UT, U.S.A., Feb. 20–22, 1989*, pp. 76–81.

MICRO ELECTRO STATIC ACTUATOR WITH THREE DEGREES OF FREEDOM

Toshio Fukuda*, Takayasu Tanaka**

* Nagoya Univ., Dept. of Mech. Eng.
Furo-cho, Chikusa-ku, Nagoya, 464-01, JAPAN
PHONE: <052> 781-5111 EX.4478,3301
FAX: <052> 781-9243

** Science University of Tokyo, Dept. of Mech. Eng.
1-3 Kagura-zaka, Shinjuku-ku, Tokyo 162, JAPAN

Abstract

This paper deals with a newly developed micro electro static actuator with three degrees of freedom, which can be applied to a micro-manipulator. This actuator has a flat surface with quad electrodes, all of which can be regulated by the voltages. In this paper, two types of microactuators, version 1 and version 2, made from different processes, are reported consisting of similar structure and function.

1. Introduction

Recently, many types of robotic manipulators have been studied, some of which are aiming at handling small objects ranging from millimeter to micrometer order of size. Grasping and clamping are required to handle much smaller objects. We have already reported a method of the bilateral control method of such micro-manipulators, requiring the contractive position and force controls, and also a visual recognition method of objects for the micro-manipulator/1,2/. It is advantageous to develop a small manipulator itself in size for handling small objects rather than to use the conventional relatively larger size of manipulator. Thus, the actuator employed at such a micro manipulator is required to be much smaller than ever and to have more degrees of freedom. In this sense, it turns out to be difficult to use a ordinary electro magnetic motors due to lack of the technology of making smaller size of actuators. Therefore, the development of a new actuator which replaces the electro-magnetic motor has been long desired.

The micro-manipulator which have already described by the authors used a PMN type of the piezoelectric actuator with one degree of freedom on the part of gripper. In order to increase the versatility of the micro manipulator, it is necessary to develop smaller size of actuators, while having much more degrees of freedom. Hence, to solve the problem in this work, we have used an electrostatic actuator which is considered to be more advantageous, when being miniaturized, and to have much more degrees of freedom.

In the study to have more degrees of freedom of manipulators, many of them have developed actuators with one degree of freedom so far and been placed individually by assembling them to make more degrees of freedom, which is one of crucial points of developing miniaturized actuators with more degrees of freedom.

In this paper, two types of microactuators are shown as version(1) and version(2). Both have similar structures and functions, but the latter has simpler structure than the former: The micro actuators developed in this work have a quad-type of the electrostatic micro actuator structure with three degrees of freedom. At present, this actuator has a millimeter order of size, but will be able to be miniaturized to much smaller micrometer of size without any changes of the fabrication methodology. A micro probe made of the silicon is attached at the tip of the actuator. This paper describes the motion of the actuator tested in the static mode.

2. Electro Static Force Actuator

When some voltages are applied between a couple of conductive plates shown in Fig.1, it is well known that electron is charged between the plates. This potential attracts these plates each other. This force F is called the electro static force, and is calculated as follows:

$$F = \frac{\varepsilon S V^2}{2d^2} \quad (1)$$

where F: electrostatic force,
V: applied voltage
ε: dielectric constant
d: distance between electrodes.
s: the area of electrode

The electro static actuator uses the force and/or torque after changed into displacement. Some studies of the electro static force actuator were described/3,4,5/, but there have been few actuators developed so far which have more degrees of freedom as shown in this paper.

3. Specification, Requirement and Design Concept

It is aimed to develop a micro actuator, so that the micromanipulator can handle/grip one unit of micro order of biological cells one by one, considering the application of micro manipulator in the field of bio technology. Thus, the amount of displacement at the tip of the probe is required to be from a few ten micro meters to hundreds micrometers to grasp small biological objects.

When miniaturizing a manipulator in an ordinary method, its conventional mechanism and structure will be complicated, so that it is less advantageous for producing and making maintenance of micro manipulator. Therefore, a manipulator with simple mechanism and structure as well as several degrees of freedom is required.

4. Basic Structure of the Actuator
4.1 Microactuator: version 1

The basic structure of the micro actuator, version1, is shown in Fig. 2. One microactuator consist of a probe, a probe support, movable electrode, and stator electrode. The lower electrode support has a concave structure with edges around and four stator electrodes are attached on the concave flat bottom with the electrical insulation. Thus, voltage can be applied to these stator electrodes independently.

Figures 3 and 4 show the overview of the

actuator. The size of the upper electrode is 21x21 millimeter, and has a quad type of the structure with 0.1 millimeter wide slit, made by the wire cut machining method. The upper electrode consisting of flexible four parts can make relative motion against the lower electrode. Then making the center of the electrode as a suspending point of four electrodes, these movable electrodes can move individually.

The four legs of the probe support are fixed on the movable electrodes, respectively. Then, the tip of the probe can be moved in three axial directions arbitrarily in accordance with the behavior of movable electrodes. Table 1 shows the physical properties of the actuator.

The displacement of the flexible movable electrode due to the electro-static force is calculated for each electrode as follows (see Fig. 5):

$$v = \frac{5Fl^3}{48EI} \quad (2)$$

where, F: electro-static force,
v: displacement
E: Young's modules,
I: moment of inertia of area

It is assumed here that the movable electrode is a cantilever, and that electro-static force effects only at the center of the cantilever. This formula calculates the displacement at the center of the cantilever.

At the tip of the probe made of silicon a force sensor is built by doping impurities as shown in Fig. 6. The resistance of the diffusion resistor formed at the tip of the probe will change due to the displacement caused by applied forces at the tip. Figures 7 and 8 show the relationship between the changing ratio of the resistance and the applied force at the tip and also the relationship between the changing ratio of resistance and the displacement of the tip of the probe, respectively/5/.
At the root of the probe where dose not deform, a diffusion resistor is fabricated for the comparison with the reference diffusion resistance and also the bridge circuit for measurement can be fabricated.

4.2 Microactuator: version 2

Figure 9 shows the Micro-Actuator. This Micro-Actuator has three degrees of freedom, at each side, and is used as the end effector for a robot. Fig. 10 shows the number of respective electrodes, and mode.

The movable electrode is driven by the electrostatic force by the applying a voltage between stator electrode and movable electrode suspended by springs.

Hence, consider the relation of electrostatic force and elasticity of the spring.

The configuration of electrodes is also assumed to be the quad type and the area of all section is 10 X 7 millimeter, and maximum applied voltage is 300 voltage. Fig.11, 12, and 13 show configuration of electrodes, spring, and model of the spring.

The stator electrode (Fig.11) which consist of print circuit board(PCB) 0.6 millimeter thick was made by photo fabrication process. First, photo resist was painted on PCB. Secondly, contact the mask firm on resist coated metal, and print . And lastly, etch the pattern resisted PCB, after the development. At the same time, circuit was made from electrode to the supply pad, respectively. There is the danger of short circuit between stator and movable electrode, stator electrode was coated with insulator.

Movable electrode (Fig.12) which consist of thin brass plate 50 micrometer thick was made by the same process with stator electrode. The elasticity of the spring K is changed widely by the primary distance between the electrodes d_0.

In this paper, d_0, h, and l are assumed to be 100 micrometer, 50 micrometer, and 7 millimeter, which are determined in consideration of capacity of machining. Thus, k and b are calculated.

5. Modes of Motion

Figure 10 shows the mode of motion of the micro actuator. The tip of the probe can move differently in accordance with the different applied voltage at each electrode, no. 1, 2, 3 and 4. It will be easily understood that three different types of the motion mode can be generated: the voltages to the electrodes No. 1 and No. 2 are applied from the state without any voltage applied to them. Then the voltages are switched to apply to the electrodes No. 3 and No. 4 from the electrodes, No. 1 and No. 2. Then again by swithching the applied voltages to No. 1 and No. 2, the tip of the probe can make a rotational motion around the Y axis, which is called "the Mode 1" in this paper. Likewise, apply the voltages to the electrodes, No. 2 and No. 3, then switch to the electrodes of the applied voltages to the electrodes No. 1 and No. 4. It will cause the rotational motion around the x axis, which is called "the Mode 2" here. Then by applying voltages to all electrodes and switching to zero, the translational motion can be generated in the z direction, which is called "the Mode 3" here.

6. Experimental Method and Results
6.1 Experimental Method

The motion of the probe tip of the versions1 and 2 are measured for each mode of the motion by using the noncontact eddy current type of displacement sensors. The applied voltages to electrodes are also regulated to 100 V, 150 V and 200 V. For the case of the mode 1 and the mode 3, the tip of probe covered by an Aluminum film is fixed under the noncontact displacement sensor and then the displacement along the Z direction is measured.

For the case of the mode 2, a pair of the noncontact sensors are located shown in Fig. 14. The tip of the probe is covered with the T-type of Aluminum film and the displacements at both ends are measured, by which the rotational angle is obtained. The angle can be computed as follows:

$$\theta = 2\tan^{-1}\left(\frac{\Delta t_1 + \Delta t_2}{2l}\right) \quad (3)$$

where
$\Delta t_1, \Delta t_2$: displacements of the z direction measured by the both noncontact sensors
l: distance between sensors.

The sensor signals are passed through the low pass filter and then are recorded in the pen recorder shown in Fig. 15.

6.2 Experimental Results

Measured data by the recorder for the version1 are shown in Figs. 16, 17, and 18. Figure 16 shows the experimental result of the mode 1 for the case of the input voltage of 150 V. Similarly, Fig. 17 does the result of the mode 3 for the case of the input voltage of 150 V.
Figure 18 shows the measured data of the mode 2.

From these experimental results, for the case of the Mode 1 and Mode 3, 8.2 micrometer and 3.9 micrometer displacement in maximum were obtained in the z direction, respectively. For the case of the Mode 2, the rotational angle around the x obtained to be $8.5*10^{-3}$ in maximum. Figure 19 shows the relationship between the applied voltages(100V, 150V, 200V) and the displacements in the z direction. Figure 20 shows the relationship between the applied voltages and the rotational angle around the x axis for the experiment of the Mode 2. The curved lines in these figures were drawn as a quadratic approximation with respect to applied voltages. From the formula for the calculation of the displacement of a cantilever, it is well known that the displacement of the cantilever is in proportion to an applied load. In this work, the load can be replaced with the electro static force between the electrodes. It can be considered that the displacement of the cantilever will be proportional to the square of the applied voltage since the electro static force is also proportional to the square of the applied voltage.

In this paper, the data are measured at three points, but it will appear more evidently with more experiments that the displacement and the rotational angle are the quadratic function with respect to the applied voltage. In this actuator, the thickness of the movable electrode was 0.1 millimeter, but it will be able to produce larger displacement and rotational angle by making the upper electrode thinner than this.

Figure 21 shows the block diagram of test system for the version2. Electrodes could be driven by the modulated electrostatic drive voltage amplified high voltage amplifier with a square wave drive from a function generator. Displacement of the tip of the micro actuator was detected optical fiber sensor.

Experimental results of the displacement are shown in Table 2.

7.Conclusions

In this paper, we have considered the actuator with multiple degrees of freedom for the micro manipulator. We have proposed and fabricated the electro static force micro actuator with three degrees of freedom for micro manipulator. By the experiments, it could be recognized to make some displacement micrometer order of amount and verified to move. In addition, improvement and problems including the dynamics properties of the actuator must be solved in future.

The authors would like to thank Mr. Y. Haishi and Mr. Y. Nagahama, Mitsubishi Electronics Corp. for his superb help in machining the movable electrode for Micro-Actuator.

This work was supported by the Memorial Science and Technology Foundation of Tokyo Electric Corp.

References

1) T. Fukuda, K. Tanie, T. Mitsuoka,"A Study on Control of a Micromanipulator(1st Report,The basic feature of Micro Gripper and one method of Bilateral control)", Proc. of Micro Robots and Teleoperators Workshop, Hyannis, MA, Nov. 1987.

2) T. Fukuda, O. Hasegawa, "Creature Recognition and Identification by Image Processing Based on Expert System",1989 IEEE Int'l Conf. on SMC. , pp.837-842 (1989).

3) H. Fujita, A. Omodaka,"Electrostatic Actuators for Micromechatronics", Proc.of Micro Robots and Teleoperators Workshop, Hyannis, MA, Nov. 1987, pp.83-92.

4) Jacobsen,S.C., et al, "The Wobble Motor: An electrostatic, Planetary-armature, Microactuator", Proc. IEEE workshop of MEMS, pp.17-24, (1989).

5) S. Syoji, M. Esashi, _. Matsuo,: Micro ISFET, 22th Japan ME Society Meeting M-44p,(1983).

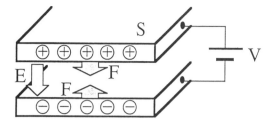

Fig.1 Principle of the electrostatic force actuator

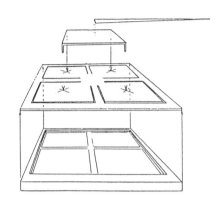

Fig.2 Structure of the Manipulator

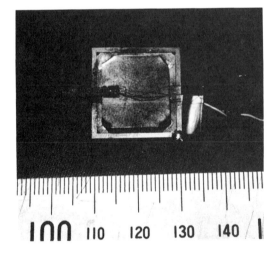

Fig. 3 Overview of the micro-actuator version 1 No.1

Fig. 4 Overview of the micro-actuator version 1 No.2

Table 1 Physical properties

Item	Unit	Measure
Length	mm	30
Width	mm	21
Thickness	mm	3.5
Weight	g	4.5
Driving Voltage	DC V	200

Fig. 9 View of micro-actuator version 2

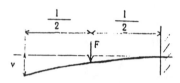

Fig. 5 Deflection of the cantilever

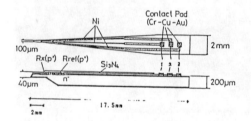

Fig. 6 Structure of the probe

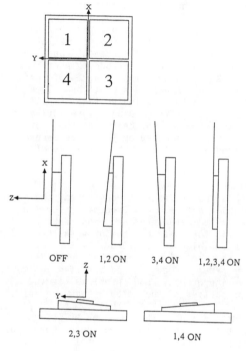

Fig. 10 Moving mode

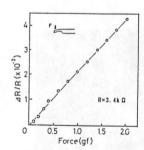 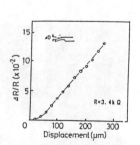

Fig. 7 Relationship between the changing ratio of resistance and the force applied at the tip of the probe

Fig. 8 Relationship between the changing ratio of resistance and the displacement of the tip of the probe

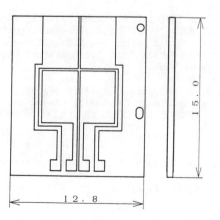

Fig. 11 Stator electrode

264

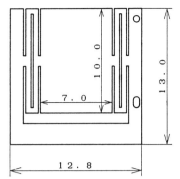

Fig. 12 Movable electrode

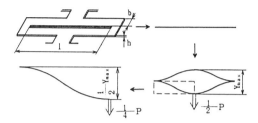

Fig. 13 Model of the spring

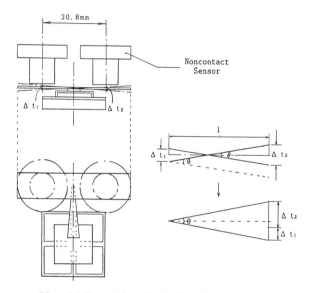

Fig. 14 Measuring Method (Mode 2)

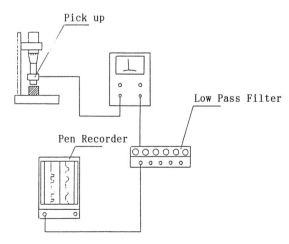

Fig. 15 Equipment of the experiments

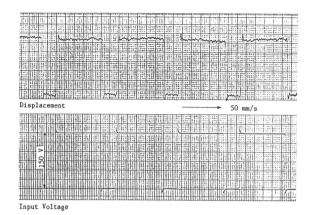

Fig. 16 Mode1 (150V)

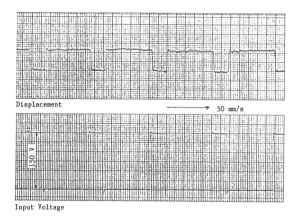

Fig. 17 Mode3 (150V)

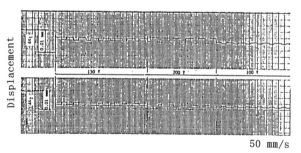

Fig. 18 Mode2

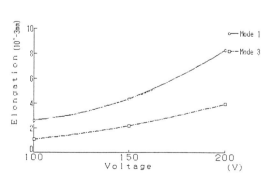

Fig. 19 Relationship between applied voltage and displacement of tip of the probe

265

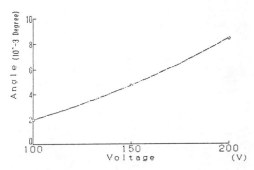

Fig. 20 Relationship between applied voltage and rotational angle

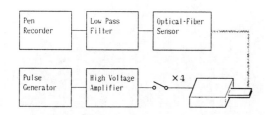

Fig. 21 Test system

Table 1 Experimental results

(Unit μm)

Driving Voltage	Electrode Number	
	1,2	1,2,3,4
150V	4.5	7.5
200V	26.1	18.8
230V	50.0	26.3

THE MODELLING OF ELECTROSTATIC FORCES IN SMALL ELECTROSTATIC ACTUATORS

Richard H. Price,[1] John E. Wood,[2] and Stephen C. Jacobsen[3]

Center for Engineering Design
[1]Department of Physics
[2]Department of Bioengineering
[3]Department of Mechanical and Industrial Engineering
University of Utah
Salt Lake City UT 84112

Abstract. Electrostatic force generation may offer distinct advantages over more familiar magnetostatics at size scales approaching microns. The fabrication of very small electrostatic actuators is becoming technologically feasible, but is extremely difficult, so that mathematical modelling of actuator designs is likely to be very important in the advancement of this technology. Modelling involves difficulties not only in finding solutions (typically numerical) to a mathematical problem, but more important, it requires that the mathematical problem be well formulated. This in turn requires an understanding of, and an intuition for, what electrostatic effects are likely to be relevant, as well as an appreciation for the behavior of materials in electrostatic interactions and for the impact on other machine components (bearings, beam loads, etc.). The well established lore of magnetostatics is not of much use as a guide in this task for several reasons: Magnetic materials tend to be either highly permeable (i.e., ferromagnetic) or to have no magnetic effect. By contrast there are no electrostatically inert materials; the relative dielectric constant ε of any solid (of normal density) is of order 2 or greater, and thus any solid element of an electrostatic configuration has a significant influence on the field. Also, the sources of magnetic fields, currents or magnetization, can be specified with some confidence, while the sources of the electrostatic field, electric charge and polarization, are much more elusive and subject to change. It is the purpose of this paper to point out some of the effects that must be taken into account if a mathematical model is to give an adequate representation of the behavior of an actual system. To do this we sketch a brief list of the types of electrostatic elements and interactions (conductors, dielectrics, compensated and uncompensated electrets, ferroelectrics, image forces, dielectrophoretic forces, etc.) and we use this list as background for discussing some electrostatic effects that may be important in the design or modelling of actuators. For some of these effects, applicable results are reported from experimental investigations carried out with a small (several hundred micron scale) electrostatically actuated device ("SCOFSS") built to study aspects of microelectromechanical design, and of control via electrostatic actuation.

1. INTRODUCTION

On macroscopic size scales, conveniently produced electric fields are much weaker than the magnetic fields in motors. On a size scale of microns, however, this is no longer true. Strong electric fields can be produced by low voltages, and are not limited by the 3×10^6 volt/m coronal discharge point of air. There are strong motivations for using such electrostatic fields as the basis for actuation in small devices. [For recent reviews of these motivations see e.g., Jacobsen, 1987, 1988; Lang, et al., 1987; Trimmer and Gabriel, 1987; and Fujita and Omodaka, 1987.] Not the least of these motivations is the example set by nature: muscular motion arises from electrostatic force, albeit via covalent bonding.

Work on microelectrostatic actuators, wherein a force or torque is transmitted out of the device, is well underway in a number of research groups, but we know of no device to date which can be considered useful. We believe that success will arrive in the next few years, but will require the development of new intuition and insights. Although a formal analogy can be drawn [Lang, et al., 1987; Layland, 1969] between magnetic and electric actuators, in practice most of the intuition of magnetic motors is inapplicable to electrostatic actuators.

Effective actuator design will require considerable effort in modelling and will lead to new intuition. The relationship of understanding to modelling is a two-way street; modelling leads to improved understanding, but the formulation of a model requires understanding for two quite distinct reasons: For one thing, considerable insight is needed to choose configurations that are plausible actuator candidates and that merit study. The second reason is more subtle: Any mathematical model is a simplified approximation of what exists and happens in the real world. If the model is to represent the real world with adequate accuracy, it must be formulated with some understanding of what is and what is not crucial in the interactions. As we shall explain presently, this question is a more difficult one in electrostatics than in magnetostatics.

The differences between electrostatics and magnetostatics can be divided, though not perfectly, into two categories: (i) materials, and (ii) field behavior. Clearly the brief nature of this article allows neither a broad nor a deep review of these subjects. We will be able to discuss only very superficially a number of issues. To reinforce the basic message of this paper, we shall also present several examples of material and conceptual issues for which the obvious answer, or the common knowledge, is wrong.

The discussion will be organized into categories based on the type of electrostatic element involved. In several instances we shall illustrate material and field phenomena with examples of our own laboratory work in electrostatics, especially with an electrostatically driven Small Cantilevered Optical Fiber Servo System (SCOFSS) described elsewhere [Wood, et al., 1987; Price, et al., 1987].

2. ELECTROSTATIC ELEMENTS

2.1 CONDUCTORS

2.1.1 Conductors and Potentials

A mathematically ideal conductor has enough mobile charge to shield its interior from electric fields. Electrodes, conducting elements set (by a voltage source) to a specified potential, behave in practice more-or-less in accord with simple idealized theory. A very basic and crucial difference between magnetic and electrostatic modelling is that the production of electric fields by electrodes is very different from the generation of magnetic fields, in which the sources of the field, either the currents or the magnetization of magnetic material, are specified. The electrostatic analogy of fixing the field in this manner would be the specification of electric charge distribution. (See the discussion below of electrets.) When electrodes are used, however, potentials, not charge distributions, are fixed. (Charge distributions on an electrode change, for instance, when a nearby charged armature moves; see the discussion below of image forces.)

This difference is fundamental to a basic design difficulty of electrostatic actuators. With magnetic sources, strong magnetic fields can be created over large regions of space. Consider, for example, a planar solenoid formed by two parallel sheets of oppositely directed current; as long as the sheets are large (compared to the separation between them), the field strength depends only on the surface current density (Amps/cm) in the sheets and not on their separation. Similarly, if electrostatic fields were produced with parallel sheets of fixed charge (parallel planar electrets, see below), the situation would be like that of magnetostatics; provided that the plates were large the field strength would not depend on separation, and the field, in principle, could be produced over a large region. However, for fields generated with electrodes, this is not true. Fixed voltages produce fields highly dependent on geometry. Strong fields require small gaps, and the general design principles of magnetostatics are inapplicable.

The specification of fields via voltages can also give rise to somewhat surprising effects and to errors in models and intuition. Figure 2 shows a simplified description of the SCOFSS apparatus (see Fig. 1) used to study electrostatic actuation. A relatively long (5 cm) and narrow (140 micron diameter) quartz optical fiber is "permanently" negatively charged with implanted electrons. This fiber moves under the influence of the electric fields produced by the conducting strips, of width $2d$, on a silicon substrate, to which the fiber is parallel.

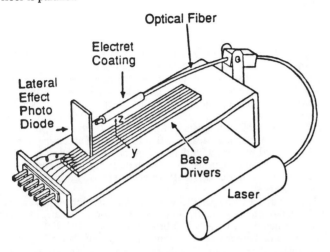

Fig. 1. Schematic of SCOFSS system, with laser source, cantilevered fiber-optic beam (with electret region), lateral-effect photodiode for measuring beam position, and voltage-controlled driver plates.

If the strips were nonconducting, and if positive charge were implanted in strips B and D, with the other strips left uncharged, it is clear that at any fiber height h the effect of the strips would be to attract the fiber. Suppose now that the strips are electrodes, that A, C, E are grounded and that B and D are set to a positive potential. This would induce positive charge on B and D, and would seem to create a field, at least qualitatively the same as that in the example of nonconductive strips. In fact the field produced is quite different; at large heights the field is attractive, as expected, but at small heights, over the center of strip C, the field is repulsive. It turns out that for B and D positive, negative charge is induced in strip C so that the electric field vectors point upward near the surface of C. At a height of $d\sqrt{3}$ above the center of C there is no force due to the imposed voltage. (We have in fact used this no-force condition as a technique for calibrating fiber heights.) This example underscores the fact that by setting the potentials rather than the charges, we are removed by an intuitive step from the sources of the field. Consequently, the charge distributions (which typically get concentrated at corners and edges) do not usually turn out to be what we want them to be for actuation.

Conductors with no connection to external reference voltages are said to be floating. If, for example, strip C in Fig. 2 had no wire connection, it would be floating and its potential would, in principle, be determined by initial conditions, presumably the condition that the net charge is originally and always zero. The laboratory reality is very different. The floating conductor is capacitively coupled to its environment, and in the real, imperfect, world it is also resistively coupled. The resulting RC time constant is rarely more than one second (see the discussion, below, of insulators) and the conductor therefore will float only for this time before coming to an equilibrium determined by its resistive coupling to the elements of its environment (the other strips, the substrate, etc.). Strip C in Fig. 2, for example, will within several seconds come to a voltage close to that imposed on B and D, but somewhat lower since C is also coupled to low voltage elements in the environment.

When sources (currents and magnetization for magnetostatics, charge density for electrostatics) are specified, there is no ambiguity about the meaning of the source. On the other hand, when a field is created by 10 volts being applied to a conductor, there can be considerable confusion lurking in the question "10 volts with respect to what?" This is especially true because in electrostatics, unlike magnetostatics, we can have monopolar sources. In three-dimensions this does not usually cause a problem since, in practice, it is understood that voltages are specified with respect to spatial infinity. This is no longer the case when (as in SCOFSS) the configuration is very long compared to its transverse dimensions, and can be approximated as two dimensional. For a two-dimensional source with a nonzero monopole, there is an infinite potential difference between spatial infinity and the source of the field.

An example of the confusion that this can cause is the configuration shown in Fig. 3: two long circular cross section wires, of radius a, length L, and separation s. For $L >> s$ the electrostatic configuration can be idealized as two-dimensional. If one wire is charged to voltage V_0 and the other to $-V_0$, it is straightforward to compute the (opposite) charge per unit length induced on the two wires and the force of attraction, per unit length, between them. The well-known result [Jackson, 1975; Prob. 1.7] is

$$\text{force/length} = \frac{1}{2s}\left(\frac{V_0}{\ln(s/a)}\right)^2. \tag{1}$$

(This assumes $s >> a$. The more general answer, without this constraint, can also be given in closed form in terms of elementary functions, but is less instructive here [Jackson, 1975; Prob. 2.4; Price and Phillips, 1988].) This problem which has no source monopole, has no ambiguities. Contrast this with the case that both wires are at $+V_0$. A valid *mathematical* solution of this problem is simply that the potential is V_0 everywhere, but this solution is clearly inappropriate since it predicts no electric fields, and therefore no induced charge and no forces between the wires. Physical intuition dictates that if both wires are charged to $+V_0$ they will repel each other. Yet there seems no mathematical alternative, since the only elements in this electrostatic configuration are the two wires (both at potential V_0) and infinity which must either be at V_0 (the inappropriate solution) or at infinite potential (physically acceptable, but of no help in resolving the question of the charges and forces).

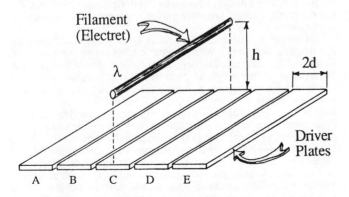

Fig. 2. Fixed-charge filament (electret) interacting with a set of five driver plates.

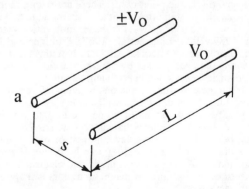

Fig. 3. Interacting conductive rods at potentials $\pm V_0$.

It is amusing how such a simple configuration is so paradoxical, and it is instructive how unexpected is the resolution: Except for the equal potential case, the configuration in Fig. 3 is incompletely defined by the geometry and the wire voltages. The force between the wires is determined by the distance d to the nearest large external conductor and the potential difference between V_0 and that conductor. If V_0 is the difference between the external potential and the common potential of the two wires, the force per unit length is [Price and Phillips, 1988]

$$\text{force/length} = \frac{1}{2s}\left(\frac{V_0}{\ln(s/a) + 2\ln(d/s)}\right)^2. \quad (2)$$

The fact that the force depends on passive elements in the environment underscores how unintuitive electrostatics can be. It should be noted that no similar surprise can arise in magnetostatics. When we have isolated the important elements of a magnetic configuration, distant passive elements cannot enter in this subtle way.

Our work in measuring field strengths generated by the SCOFSS electrodes required these insights. For the 200 micron wide by 2 cm long electrodes, our measurements of the potential field in the neighborhood of the electrodes were in good agreement with simple two-dimensional theoretical predictions, only out to about 500 microns. At larger distances the influence of environment (the location and potentials of the mounting hardware, etc.) became significant.

2.1.2 Conductors and Image Forces

For the understanding of and design of electromechanical actuators, sensors, etc., it is crucial to understand a phenomenon that is present whenever charges and conductors are present, the phenomenon of image forces.

If, say, a positive charged particle is moved to a location at a distance h above a grounded conductor, negative charge will be induced on the conductor (in order for it to remain at ground) and the negative charge will be concentrated in that position on the conductor closest to the positive charge. If the conductor is an infinite plane (in practice, if its edges are at distances $\gg h$) the fields generated by the charge arrangement induced on the plane are exactly the same as the field due to a mirror *image* negative charge located a distance h below the plane. This shows, in particular, that the charged particle will be attracted to the plane with a force that varies as $1/h^2$. If the grounded conductor is spherical, the effects of induced charge can again be represented with an image charge, although with slightly more complexity. For more generally shaped grounded conductors we use the expression "image force" somewhat loosely to describe the attraction of a charged body to the conductor, though the force of attraction cannot always be represented with a simple finite set of fictitious charges.

Image forces are crucial to electrostatic design for a number of reasons, and they have no real analog in magnetostatics. One reason for their importance is that they are unavoidable; any source of field (fixed charge, electrode, etc.) external to a conductor is attracted to the conductor by image forces. In the SCOFSS configuration pictured in Fig. 1, for example, the charged fiber is attracted to the plane of the conducting strips with a force (proportional to the square of the fiber charge) that is independent of the applied voltages. The reason that image forces are particularly of concern in microelectromechanical design is that as distances decrease image forces increase more quickly than controllable forces (e.g., the force of an electrode on an electreted armature) and can overwhelm the controllable forces. Furthermore, relative surface roughness is likely to be a problem in microelectromechanical design. A gap between electrically active elements of 1 or 2 microns may have large fractional variability. If the gap is modelled as constant, it is unlikely that the model will give results in good agreement with reality. Applied fields may perhaps be treated adequately by an approximation of constant average gap, but not image forces. Not only are they strongly distance-dependent, and formally infinite at zero gap, but they are also destabilizing. At small distances the strong image forces act to make the distances smaller yet, and thereby to create a positive feedback loop which ends with surfaces sticking, conductors shorting out, etc.

2.2 DIELECTRICS

2.2.1 Dielectrics and Polarization

There is an important difference in material properties for magnetostatics and electrostatics, that can easily be overlooked in modelling. Some materials (due to collective phenomena) have enormously strong magnetic effects; ferromagnetic materials have a small-field incremental, (relative) permeability μ typically in the range 10-10^4. Most materials, however, are paramagnetic or diamagnetic, have values of $|\mu-1|$ smaller than 10^{-2} [Jackson, 1975, Sect. I.4] and, as a good approximation, can be considered to be impermeable, i.e., to be magnetically inert. In modelling magnetic configurations, therefore, it is usually straightforward to recognize which elements must be included as part of the active model, and which elements serve other (structural, etc.) functions. By contrast there is no solid material (of ordinary density) which is electrostatically inert. Molecules have fairly high electrical (as opposed to magnetic) polarizability, and no solid material has a low-frequency relative dielectric constant ε less than 2 (the value for Teflon®). Materials used in microfabrication typically have values larger than 2 ($\varepsilon = 3.8$ for SiO_2) or much larger ($\varepsilon = 27$ for Ta_2O_5). Dielectric materials, even with a dielectric constant as small as 2, have profound effects on external electrical fields. A simple example of this is the interaction of a charged particle with a planar face of a large (compared to the distance to the particle) dielectric. The presence of the particle induces a distribution of polarization in the dielectric which in turn generates a field external to the dielectric. It turns out that the external field thus produced can be represented with an image charge at the same location as if the dielectric were a conductor, but with charge reduced in magnitude by the reduction factor $(\varepsilon-1)/(\varepsilon+1)$.

The form of this reduction factor is somewhat specific to the planar geometry, but the nature of the result is general: A dielectric acts qualitatively in the same manner as an uncharged, floating, conductor, but with an effect smaller by a reduction factor. This reduction factor goes to unity in the limit $\varepsilon \to \infty$ (the dielectric effectively becomes a conductor); the reduction factor vanishes when $\varepsilon \to 1$ (the vacuum limit). The implication of this general behavior is that even for a material with $\varepsilon = 2$, the effect on the field due to a dielectric structure is (assuming the planar reduction factor) one third of the effect that a geometrically identical floating conductor would have, and thus this effect cannot be ignored in modelling.

2.2.2 Dielectrophoretic Forces

Because dielectrics, at least as idealized here, cannot be electrically charged, the forces on dielectrics differ somewhat in their nature from those on charged materials, or charged conductors. There is no force directly caused by an external electric field. Rather, the electric field induces a polarization in the dielectric (the generalization of image charge) and makes the dielectric element, in effect, an electrical dipole $\mathbf{p}$. Because the energy of interaction is $-\mathbf{p}\cdot\mathbf{E}$, the dipole (i.e., the dielectric material) is drawn to regions of stronger $\mathbf{E}$ field strength. (For exceptions to this see, e.g., Epstein, 1965.) The dielectric material therefore couples not to the $\mathbf{E}$ field, but to the gradient of the $\mathbf{E}$ field with what is called the "dielectrophoretic" (DEP) force. Dielectrophoretic forces can be nonnegligibly strong, and their existence has consequences for modelling. We would like to be able to think of simple charged particles and (as in the case of SCOFSS) fibers, but chargeable materials (in fact all nonconductive materials) are dielectric, and experience DEP forces. These forces may overwhelm the direct E-field-to-charge coupling and produce motions totally different from those predicted by a model which does not include DEP effects. As a rule of thumb the condition for DEP forces to be negligible compared to E-field-to-charge coupling is

$$\left(\frac{\varepsilon-1}{\varepsilon+2}\right)\frac{E\ \text{vol}}{Q\ d} \ll 1, \quad (3)$$

where "vol" is the volume of the charged particle, Q is its charge, E is the electric field due to external sources, and d is the characteristic length scale on which E changes. If the particle charge is proportional to its volume, this relation suggests that DEP forces increase in importance as actuator size decreases.

2.3 Insulators

In simple models of electrostatic configurations, elements appear which are electrically insulated from each other. But in reality perfect insulation is impossible. As pointed out above in connection with conductors, this means that a floating conductive element will have constant charge only for a time small compared to its RC time constant, where C is the typical capacitance between the element and other elements and R is the characteristic resistance through which the element is connected to other elements. Capacitance scales as the first power of length, so that for microchip sizes capacitive coupling will be quite small. The SCOFSS electrodes, for example, are macroscopic (2 cm) in one dimension, but even for these the capacitance is no more than 1-10 picofarads. For a configuration which is micro sized (say 10 microns) in three dimensions, the capacitance will be a small fraction of a picofarad.

These small capacitances mean that even with the highest resistivity materials, fields will not be "static" for indefinite periods. In practice, careful design is needed to achieve RC constants of more than a few seconds. A consequence of this is that truly "static" fields can be produced only by electrodes, not by isolated sources.

Models of actuators which use other elements (floating conductors, dielectrics, etc.) will be valid only if they are meant to operate at frequencies (e.g., commutation rates) well above the reciprocal of the relevant RC constant. For the high resistance considerations of microelectromechanical devices the question of resistivity is a difficult one, involving properties almost impossible to model. The resistivity of a high-resistivity material such as Teflon® depends so strongly on ambient humidity, surface contamination, how the material has been handled, etc. that characterizing resistance with a simple single number, such as $10^{19}\Omega$-cm, is a convenient, but misleading fiction.

The very concept of resistivity -- a linear relationship between electric field and current density, based on a diffusive picture of current flow -- is often inappropriate. At sufficiently high field strength the relationship will be decidedly nonlinear. This is most familiar when it is catastrophic and results in breakdown, arc discharge, etc. In microelectromechanical devices coronal discharge in air is an obvious problem. It has been pointed out by several authors [Lang, et al, 1987; Fujita and Omodaka 1987] that the breakdown strength (e.g., 3×10^6 volt/m for air) characterizing macroscopic lengthscales does not apply to the gaps in microelectromechanical devices, since the avalanche phenomenon responsible for coronal discharge is suppressed by the small (< 10 micron) lengthscales [see von Hippel, 1954]. Air breakdown can be avoided, of course, simply by evacuating all strong field regions in a device. In this case, however, breakdown will occur in a different form: as field emission from the surface of the material, typically at electric field strengths of 10^8-10^9 volts/m. At very small lengthscales still other physical effects enter; in particular, electron wavefunction overlap and tunneling, as in an STM.)

One of the applications of the modelling of actuators is a study of the onset of dielectric breakdown, coronal discharge, field emission, etc. From computations of electric field strengths, predictions can be made whether or where breakdown will occur. Designs can be refined to control breakdown, e.g., by eliminating sharp edges of conductors. The common wisdom is that the electric field strength on a conductor is greatest at the point of greatest curvature. This is a useful rule of thumb, but is in no sense strictly valid [Price and Crowley, 1985]. Surface curvature is an easily computed local quantity whereas electric field strength is nonlocal; it requires a global solution of the electrostatic problem, and is much more difficult to determine, requiring mathematical modelling and, typically, finite element solutions.

It should be noted that the design and modelling difficulties posed by finite resistivity have no analog in standard magnetic devices. While the source (charge) of the electrostatic field can drift through materials of finite resistance, the sources of magnetic fields (currents confined to wires; magnetization in permanent magnets) do not.

2.4 Electrets and Ferroelectrics

Charge can be fixed, to some extent, on any nonconducting solid, but there are certain materials in which the charge density that can be achieved is high, and for which the process of charge implantation is relatively controllable. Such materials (typically polymers, e.g., Teflon®, polycarbonate, polyimide, etc.), when charged with electrons, through the use of a scanning electron microscope, a coronal discharge arc, the wet electrode apparatus, etc., are called electrets [Sessler, 1982], and are of considerable interest for microelectromechanical design both because they need no wire connections, and because they hold the promise of high field strengths without small electrode gaps. Especially interesting is the possibility of implanting high spatial resolution charge patterns on electret materials, thereby creating an easily fabricated structure with high field gradients.

A distinction must be made between two types of electrets: compensated and uncompensated. A compensated electret is a sheet of electretable polymer with a metalized coating on one side. During the charging process this metal layer is grounded so that it acquires a positive surface charge density equal in magnitude to that of the negative charge implanted in the polymer. The resulting two-layer sandwich is equivalent to a dipole sheet. Compensated electrets are relatively well studied, have commercial applications (electret microphones, etc.) and are reasonably stable. Because of their dipole structure, however, they are less interesting as armatures or field generators in actuators than are uncompensated electrets.

Relatively little work has been done with uncompensated electrets. In our own work we have been able to achieve surface charge densities of 10 nanocoulombs/cm^2 in Teflon®. This is equivalent to a field of order 10^6 volts/m and it seems clear that air breakdown, along with other means of charge motion and compensation, is involved in limiting the surface field strength [Roos, 1969].

Our own experimental studies with electrets have focussed on high spatial resolution charge density patterns. We found effects suggestive of transient surface conductivity, and of the influence of the dielectric properties of the electreted polymer. At this point we can only be sure that there are physical processes going on which are not part of any simple model of charge trapping in polymers. Certainly an improved understanding of the nature of electrets (especially high spatial resolution, uncompensated electrets) is needed before their fabrication becomes predictable via design tools.

Unlike an electret, a ferroelectric material cannot hold a significant amount of implanted charge. But due to its crystal structure (typically that of perovskite) it can take on a quasipermanent electrical polarization. A ferroelectric is the electrical analog of a permanent magnet. Like a permanent magnet it can have no monopole, but the strength of its dipole field can be imposing. For lithium niobate at saturation polarization, the fields are equivalent to those produced by a surface charge density of 3×10^5 nanocoulombs/cm^2. This is equivalent to a Maxwell pressure of 10^8 psi. The forces that would be produced by such fields in a micromechanical setting are amusing. If a 70 micron radius fiber of lithium niobate were polarized perpendicular to the fiber axis, and placed with the axis 100 microns from a metal plane, the image force of attraction would be 10 tons per centimeter of fiber.

No such forces, or *any* forces, are observed in working with ferroelectrics, but this can be ascribed to the fact that metal electrodes are usually deposited on opposite faces of ferroelectrics during the poling process, and these electrodes, like the metallic layer in a compensated electret, acquire charge so that the external field of the ferroelectric-electrode system vanishes. But ferroelectrics can be poled without attached electrodes. In such a case will enormous forces be produced? Again, too simple a model does not take into account the reality of materials. The extraordinary fields predicted for a saturated ferroelectric would breakdown air and arc from one poled face to the opposite. In vacuum, surface currents or field emission would quickly reduce the external fields. So fields corresponding to surface charge densities of 3×10^5 are impossible, but some uncanceled external field can be maintained. The possibility cannot be ruled out that such a field is usefully strong.

3. CONCLUSIONS

We have reviewed the basic elements of electrostatic devices (conductors, dielectrics, insulators, electrets and ferroelectrics) and have shown that the modelling of electrostatic forces is a very different, and in many ways a more difficult, undertaking than the modelling of magnetic interactions; the difference between the "real-world" behavior of electrostatic configurations and "textbook" models is much greater than for magnetic configurations. Models must take into account the realities of materials, in particular the fact that resistivity is never infinite, and that any structural part of an actuator will affect fields through its dielectric properties. We have seen that in this sense (the behavior of materials) as well as in the most basic conceptual viewpoint (fields determined by potentials rather than by direct sources) electrostatics and magnetostatics are at opposite poles. It would seem that much new understanding is needed for, and will emerge from, the modelling of electrostatic devices that will be carried out in the next few years.

ACKNOWLEDGMENTS

This work was supported by grants from the System Development Foundation and from the Defense Advanced Research Projects Agency.

REFERENCES

Epstein, L. (1965): "Electrostatic Suspension," Am. J. Phys. 33(5):406-407.

Fujita, H. and Omodaka, A. (1987): "Electrostatic Actuators for Micromechatronics," IEEE Micro Robots and Teleoperators Workshop, Hyannis, MA, Nov. 9-11.

Jackson, J.D. (1975): *Classical Electrodynamics*. Wiley, New York.

Jacobsen, S.C. (1987): "Electric Field Machine," U.S. Patent No. 4,642,504, February 10.

Jacobsen, S.C. (1988): "Electric Field Machine," U.S. Patent No. 4,736,127, April 5.

Lang, J.H., Schlecht, M.F. and Howe, R.T. (1987): "Electric Micromotors: Electromechanical Characteristics," IEEE Micro Robots and Teleoperators Workshop, Hyannis, MA, November 9-11.

Layland, M.W. (1969): "Generalized Electrostatic Machine Theory," IEE Proceedings, 116:403-405.

Price, R.H., Jacobsen, S.C. and Khanwilkar, P.S. (1987): "Oscillatory Stabilization of Micromechanical Systems," IEEE Micro Robots and Teleoperators Workshop, Hyannis, MA, November 9-11.

Price, R.H. and Phillips, R.P. (1988): "The Force between Two Charged Wires." Paper in preparation. Note that eq.(2) assumes $d < L$.

Price, R.H. and Crowley, R.J. (1985): "The Lightning-Rod Fallacy," Am. J. Phys. 53(9):843-848.

Roos, J. (1969): "Electrets, Semipermanently Charged Capacitors," J. Appl. Phys. 40(8):3135-3139.

Sessler, G.M. (1982): "Polymeric Electrets," Chap. 6 in *Electrical Properties of Polymers*, Ed. D.A. Seanor, Academic Press, New York.

Trimmer, W.S.N. and Gabriel, K.J. (1987): "Design Considerations for a Practical Electrostatic Micro-Motor," Sensors and Actuators, 11(2):189-206.

von Hippel, A.R. (1954): *Dielectrics and Waves*. J. Wiley, New York.

Wood, J.E., Jacobsen, S.C. and Grace, K.W. (1987): "SCOFSS: a Small Cantilevered Fiber Optic Servo System," IEEE Micro Robots and Teleoperators Workshop, Hyannis, MA, November 9-11.

SILICON ELECTROSTATIC MOTORS

W.S.N. Trimmer, K.J. Gabriel and R. Mahadevan

AT&T Bell Laboratories
Holmdel, NJ 07733

ABSTRACT

Magnetic motors and actuators dominate the large-scale motion domain. For smaller, micro-mechanical systems, electrostatic forces appear more attractive and promising than magnetic forces. In the large-scale domain, despite their distinguished history, electrostatic motors have found few practical applications because of the high voltages and mechanical accuracies traditionally required. This paper explores the design of electrostatic motors utilizing the advances in silicon technology. Using silicon wafers, and the associated insulators, conductors, anisotropic etching and fine-line photolithographic techniques, it is possible to develop large electrostatic fields with moderately high voltages (~ 100V) across insulators of well-controlled thickness (~ 1 μm). Preliminary designs and theoretical predictions for a linear electrostatic motor and a rotary electrostatic motor are presented.

INTRODUCTION

The motor most commonly used in the macro world is the ubiquitous electric motor which is, in fact, an electromagnetic motor. While this motor is successful in producing large scale motions, it is poorly suited to small mechanisms that produce small motions. Electrostatic forces are well suited as an actuating force for small, micro-scale devices. These motors have a long history. Andrew Gordon[1] and Benjamin Franklin[2] built electrostatic motors in the 1750's, 100 years before the advent of electromagnetic motoers. The first capacitor electrostatic motor was developed by Karl Zipernowsky in 1889.[3] A review of early electrostatic motors is given by Oleg Jefimenko.[4] In this review, Jefimenko traces the history of electrostatic motors from Gordon's electric bells, to modern motors which can be powered from atmospheric electricity.

If centimeter, millimeter and smaller motors were available, there would be many useful applications. For example, we are interested in building micro robotic (computer controlled) and micro teleoperator (human controlled) systems and are pursuing ways of making small motors, gears, shafts, and other components. These micro mechanical systems have advantages which include speed, accuracy, gentleness, and reduced floor space. As an example, consider a robotic system that was built to handle small semiconductor chips with ten micron accuracy. The robot weighs a hundred pounds and the system occupies over thirty square feet of floor space. This large massive system makes it difficult to handle the fragile millimeter sized chips and necessitates moving the chips large distances between work spaces. The thermal expansion of the large robot also makes it difficult to obtain the desired ten micron accuracy. One of our goals is to build small, accurate, and quick micro robotic and micro teleoperator systems, that can be used in a wide range of applications.

A key development necessary in the construction of micromechanical systems is a micro actuator. Electrostatic motors that operate at reasonable voltages (10 - 100 volts) have not been fabricated because they require thin, high-quality insulators, narrow pole faces, and flat surfaces. Silicon processing and the electronics industry provide mature technologies which excel in precisely these areas.

We propose, and are in the process of building, silicon electrostatic motors comprised of two silicon chips; a stator and a rotor. To make the stator chip, a silicon wafer is coated with an insulating material, the electrostatic pole faces are deposited, and a second insulating layer is deposited on top of the pole faces. The rotor chip is made in precisely the same manner and placed on top of the stator so that the rotor and stator pole faces are registered with respect to one another. By applying a voltage between the conductive pole faces on the stator and rotor, attractive forces can be generated which cause the rotor to move.

THEORY

If a voltage V is applied across two conducting parallel plates separated by an insulating layer,

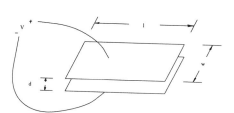

Figure 1. Parallel plate capacitor

the potential energy of this capacitor is given by

$$U = \frac{1}{2}CV^2 = \frac{\epsilon_r \epsilon_o w l V^2}{2d}. \quad (1)$$

The magnitude of the force tending to realign the plates due to a displacement in the w direction is given by

$$F_w = \frac{1}{2}\frac{\epsilon_r \epsilon_o l V^2}{2d}. \quad (2)$$

The magnitude of the force clamping the rotor and stator together is

$$F_d = -\frac{1}{2}\frac{\epsilon_r \epsilon_o w l V^2}{d^2}. \quad (3)$$

It is this aligning force that is exploited in linear and rotary electrostatic motors.

Silicon processing technology enables one to: (1) define many fine lands and hence increase the edge length, l and (2) deposit thin, high-quality insulators which allows very small plate separations d and relatively large dielectric constants, ϵ_r. Hence, as can be seen from Equation 2, large forces can be generated using large values of l and small values of d for excitation voltages, V, in the range of 10 - 100 volts.

LINEAR ELECTROSTATIC MOTOR

A simple linear electrostatic motor has pairs of differentially misaligned parallel plate capacitors. The stator of such a linear motor is shown in Figure 2.

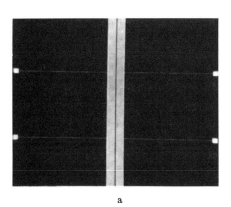

a

b

Figure 2. Linear electrostatic motor. Light bands are conducting lands, dark bands are insulating separators. Several sections of the motor are shown in Figure 2 a, and a detail showing the conducting lands is shown in Figure 2 b.

Differential phasing of the conductive and non-conductive portions of the top and bottom plates of many capacitors is used to sustain motion in either direction and increase the number of edges and hence, the force generated by the motor.

A scaled model of the linear electrostatic motor was used to study the variation of capacitance with position and separation. If a fixed voltage, V, is used to drive the electrostatic motor, the forces exerted on the rotor are proportional to the rates of change of capacitance. Hence, the capacitance measurements can be used to estimate the tangential and clamping forces experienced by the rotor. The width and pitch of lands used in the scaled model were 0.635cm and 1.27cm, respectively, while the separation d was varied between 0.015cm and 0.075cm.

The experimentally measured capacitance $C(x,d)$, is plotted as a function of position x, for different separations d, in Figure 3.

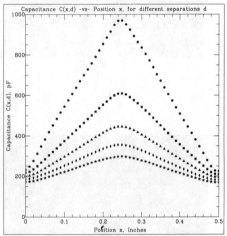

Figure 3. Measured capacitance as a function of position x, for different separations d. ● d = 0.006 in; ■ d = 0.012 in; ▲ d = 0.018 in; ♦ d = 0.025 in; ★ d = 0.032 in.

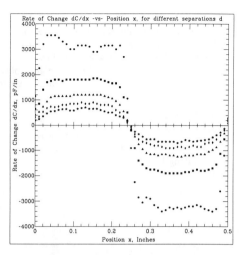

Figure 4. Rate of change of capacitance with position, plotted as a function of position x for different separations d. This corresponds to the desired tangential force exerted on the rotor.

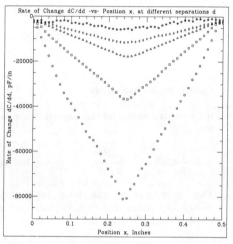

Figure 5. Rate of change of capacitance with separation, plotted as a function of position x for different separations d. This corresponds to the undesirable vertical clamping force between the rotor and stator.

The capacitance measurements are then utilized to compute the rate of change of capacitance with position, dC/dx, and the rate of change of capacitance with separation dC/dd. These are shown plotted in Figures 4 and 5, respectively. From to clamping force increases with d. Hence, there is a design compromise involved in the choice of the separation distance d. The ratio of tangential force to clamping force found from measurements is seen to be lower than that predicted by the simple parallel plate approximation of Equations 2 and 3.

As a numerical example, the predicted force from such a motor with $w = 6$ microns, $p = 12$ microns, $l = 5$ mm, $d = 1$ micron, $L = 5$ cm and $V = 100$ volts is 1.87 newtons (0.42 lbs). This force can be generated by a structure which is approximately 0.375 grams in mass.

ROTARY ELECTROSTATIC MOTOR

Using the above principles, one can also build a rotary electrostatic motor. For illustration, an 5 mm diameter electrostatic motor is discussed. This motor has the same land width, pitch, and separation (w = 6 microns, p = 12 microns, d = 1 micron) as the linear motor example. The stator chip is shown in Figure 6. There are two annular rings, the outer for driving the motor, and the inner for sensing the position. Each ring is divided into six sectors. When the rotor chip is placed on top of the stator, the lands of two opposing sectors can be made to align. The alignment of the other four sectors are either ahead or behind by 120 degrees. By applying a positive voltage to a sector that is 120 degrees ahead, and a negative voltage to the opposite sector with the same phase, charge is inductively coupled in the rotor chip, causing the motor to turn. Voltages are brought on the chip by probing the square pads surrounding the motor.

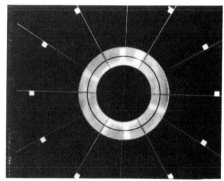

Figure 6. A photograph of of a rotary electrostatic motor stator.

The torque produced by the motor is given by

$$\tau = \frac{\pi \epsilon V^2}{3 d\, \mathbf{n}\, p} \left[r_o^3 - r_i^3 \right] \quad (4)$$

Here r_i and r_o are the inner and outer radius of the annular ring of lands, and $\mathbf{n}$ is the number of phases. The motor's torque goes as the third power of the radius, and the outer region of the motor produces most of the torque. For example, the outer third of the motor produces more than two thirds of the maximum possible torque. By using only the outer ring of the motor, aligning the stator and rotor chips becomes easier. This also leaves the inner region free for encoders.

For proper operation of the motor, the rotor and stator need to be correctly aligned to each other. With the above design of 6 micron lands, and a 5 mm diameter motor, the two chips need to be aligned to within 4 microns. This can be achieved with patience and care.

The 5 mm diameter motor is capable of producing a substantial amount of power in this small domain. If the motor is rotating at 100 revolutions per second, and is being driven with 130 volts, the motor can produce a quarter of a watt of power. Of course, some of this power will be consumed by frictional losses. This example and the rotary and linear examples are not meant as final designs, but as guide posts. Indeed, there appears to be a fertile area for further exploration.

CONCLUSIONS

Magnetic motors and actuators dominate the large-scale macro domain. For micro mechanical systems, however, electrostatic forces appear more promising. Silicon processing technology appears to be an especially promising avenue for the development of these motors because of the high-quality insulators and the fine-line photolithography techniques available, and the flat surfaces which can be produced easily.

REFERENCES

[1] P. Benjamine, *A History of Electricity*, John Wiley and Sons, New York, 1898, pp 506, 507.

[2] J. Sparks, Ed., *The Works of Benjamine Franklin* Whittemore, Niles, and Hall, Boston, 1856, Vol. 5, p. 301.

[3] Zipernowsky, *Electrostatic Motor*, Electrical World, Vol. 14, p. 260, 1889.

[4] Oleg D. Jefimenko, *Electrostatic Motors*, Electret Scientific Company, Star City, 1973.

Electrostatic Actuators for Micromechatronics

Hiroyuki Fujita, Akito Omodaka

Institute of Industrial Science
The University of Tokyo
22-1, Roppongi 7-chome, Minatoku,
Tokyo 106 JAPAN

The electrostatic drive is more advantageous for small, micro scale devices than traditional electromagnetic devices. Silicon micromachining techniques for VLSI manufacturing have enabled us to fabricate micro electrostatic actuators. This paper deals with the topics of electrostatic actuators such as basic concepts, some examples, analysis methods, process considerations, and the measurement of the performance. As a particular example an electrostatic linear actuator is discussed in some details. The microactuator consists of a plane wafer with striped electrodes covered with an insulation layer and thin cylindrical rollers. When voltage was applied between some of the striped electrodes and the roller the electric field between them attracted the roller and made it rotate. The silicon micromachining techniques such as lithography, PCVD, and etching were utilized to fabricate the plane wafer.

Introduction

The word "mechatronics" is the blend of mechanics and electronics; this field deals with a inseparably combined system of mechanical and electronic components. Because the higher level integration of mechanism and electronic circuitry is a today's trend especially for small size systems, miniaturization of motors and actuators are required in the same way as the sensors and the control devices. The expected application of micromechatronics is to handle minute parts and devices especially on microelectronic chips.[1]

The electrostatic drive is more advantageous for micro scale devices than the traditional electromagnetic drive[2]. The idea of electrostatic drive has a long history[3,4] but the real application was limited by its low power density. As the micromachining techniques developed rapidly, the electrostatic drive attracted renewed attention. Silicon micromachining techniques for VLSI manufacturing have enabled us to fabricate precise electrode patterns and good insulation films. Electrostatic actuators on silicon wafers are not only light and compact but also suitable for integrated "smart" actuators and the element of micro manipulators. Therefore this paper deals with the topics of electrostatic actuators such as basic concepts, some examples, analysis methods, process considerations, and the measurement of the performance. As a particular example an electrostatic linear actuator is discussed.

Electrostatic energy conversion

Why micromechanists favor electrostatic force?

The advantage of the electrostatic drive in sub-mm range is based on the followings reasons:

(1) Thin insulation layers withstand higher electric field strength than thick layers. Breakdown strength as high as 2 MV/cm is easily obtained in SiO_2 thin film[5]. The power density of the electric field reaches 7×10^5 J/m^3; this value is equal to the power density of 1.3 T magnetic field. The contracting pressure by the field reaches 1.3 MPa. Thin insulation layers of high quality are readily obtained by the silicon process. Voltage at most 100 V across the insulator is high enough to generate strong field mensioned above.

(2) The electrostatic force is the surface force and obeys a favorable scaling law in small dimension[1,2]. Furthermore it needs only two thin electrode separated by an insulator. The manufacturing process for the silicon device is well suited for making minute and precise electrode patterns. The electric field can also be confined between the electrodes. On the other hand, electromagnetic force is the body force and needs considerable "ampere turns" to produce enough magnetic field.

(3) The electrostatic actuator is driven essentially by voltage. Voltage switching is far easier and faster than the current switching. Energy loss associated with Joule heating is also small.

Table 1 summarizes the advantages together with the compatibility of the electrostatic drive with the silicon micromachining.

A brief historical review

Many kinds of electrostatic motors were proposed so far[3]. An electric bell by Gordon had a pendulum oscillating between the positively and negatively charged bells. Gordon also invented an electric fly that was driven by corona wind from corona discharge on needle tips. Franklin made a spark motor named the electric wheel. The rotating electrodes of the motor received homo charge from one of the stationary electrodes at high potential and was attracted by the other stationary electrode of opposite polarity. A corona motor had an insulator disk on which charge was supplied by corona discharge set on stationary needles. The disk was rotated by the attractive force between the charge and the stationary electrode of opposite polarity. An electret motor utilized the force between an electret disk and electrodes at high potential[6]. A dielectric motor used the charge conduction in a dielectric fluid[7,8,9].

Of particular interest is a variable capacitance motor. When a variable capacitor with a moving element is energized, the element is attracted to the direction along which the capacitance increases. An example of this type was the Trump's vacuum insulated electrostatic motor[4]. Most of the micro actuators described below are based on the principle.

The variable capacitance actuator

Force and energy recovery: Suppose a simple circuit consisting of a constant voltage supply and a variable capacitance actuator. The force, F_x, in x direction is given as follows:

$$F_x = -\frac{\partial U}{\partial x} = \frac{V^2}{2} \cdot \frac{\partial C}{\partial x} \quad (1)$$

Where U is the energy both in the capacitor and from the power source, V is the applied voltage (fixed), and C is the capacitance. When the actuator moves from x_1 to x_2 with the corresponding capacitance value of C_1 and C_2, the supplied energy from the source equals to $V^2(C_2-C_1)$. One half of the energy is converted to the mechanical work, W, and the other half is stored as the potential energy, U_c, in the capacitor. Namely,

$$W = U_c = \frac{1}{2} \cdot V^2(C_2 - C_1) \quad (2)$$

Tabel 1 Advantages of elctrostatic drive and its compatibility micromachining

Features of electrostatic drive	advantages as a micro-actuator	compatibility with micro-machining
surface force	dominant in small dimension (increase in surface to volume ratio)	fabrication of minute electrode
	only thin film is required	precise patterning of thin film
	light weight	use of light materials
voltage/charge driving	easy control with high switching speed	switching of voltage by gate devices
	low power consumption and high efficiency	easy to integrate
need for high electric field	thin films withstand high field strength	thin insulation film of high quality
	no breakdown in gas occurs below Paschen's minimum	recision manufacturing of small gap separation

The charge in the capacitor should be discharged as soon as possible when the capacitance reaches the maximum. The potential energy is recovered to increase the efficiency.

If the actuator has multiple phase, that means it consists of successively energized capacitors, the energy may be transferred to the next capacitor. The circuit shown in Fig.1 (a) enables the operation. The capacitor C_2 is charged up to V at t=0. At first, the switch S_2 is closed and the charge flows through an inductance L as shown in Fig.1 (b). When the voltage across C_2 becomes zero, the switch S_2 is opened and S_1 is closed. Another capacitance C_1 is charged up to $\sqrt{C_2/C_1}$ V. The voltage decreases to V as the capacitance C_1 increases. The value of the inductance L is determined by the recovering time and the energy loss in the circuit. The loss is given as follows:

$$P_{DL} = \frac{\pi R_L C_2 V^2 (\sqrt{C_1} + \sqrt{C_2})}{8\sqrt{L}} \quad (3)$$

More generally, the energy in C_2 is recovered to the power supply which has enough large inductance.

Energy loss by charging current: Suppose an equivalent circuit in Fig.2 (a) to calculate the energy loss in the resistance. The resistance includes that of the switch and of the wire. The capacitance is assumed to change from C_1 at t=0 to C_2 at $t=T_s$ linearly (Fig.2 (b)):

$$C = C_1 \cdot (1+at) \quad (4)$$

$$a = \frac{-C_1 + C_2}{C_1 T_s} \quad (5)$$

The current, i, flows during the movement of the actuator ($0<t<T_s$) is given by the following formula:

$$i = \frac{aC_2 V}{1+aC_1 R} \cdot \{1 - (1+at)^{-\frac{1}{aC_1 R} - 1}\} \quad (6)$$

If the electrical time constants $C_1 R$ and $C_2 R$ are much shorter than T_s the current is approximated as follows:

$$i \approx I \cdot \{1 - (1+at)^{-\frac{T}{R(C_2 - C_1)}}\} \quad (7)$$

$$I = \frac{V(C_2 - C_1)}{T_s} \quad (8)$$

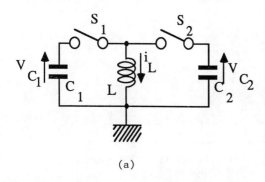

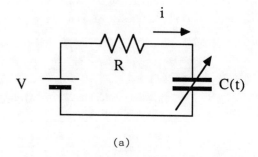

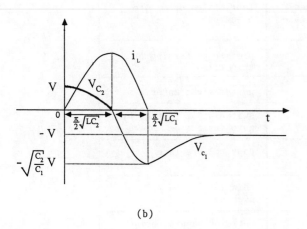

Fig.1 An energy recovery circuit (a) and the time dependence of voltage and current (b).

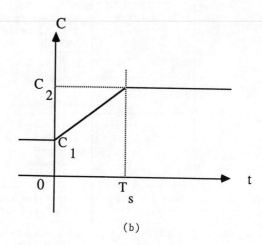

Fig.2 An equivalent circuit for variable capacitance motors (a) and assumed time dependence of capacitance (b).

The average current, I, is the key parameter to discuss the loss.

When the actuator stops at $t=T_s$, the current exhibits a well known exponential decay:

$$i \approx I\{1 - (\frac{C_2}{C_1})^{-\frac{1}{aC_1R}-1}\} \exp(-\frac{t-T_s}{C_2R}) \quad (9)$$

Integrating the Joule loss, Ri^2, over the whole period, we have the lost energy per step, E_{DR}. The average power loss, P_{DR}, is given by E_{DR}/T_s. Therefore,

$$P_{DR} \approx RI^2(1 - \frac{3C_1R}{2T_s} + \frac{C_2R}{T_s}) \quad (10)$$

The first term at the right hand side is a steady state term. The rests are due to the electrical transient in the R-C circuit. As discussed later, the period T_s decreases as the output power per volume of the actuator increases; in this case the transient term may not be neglected.

<u>Numerical example</u>: The simplest linear actuator[1] with three phases looks like Fig.3. Note that the number of the pairs of the electrodes in a phase may be very large. Suppose we have an actuator of 10mm square, with the gap separation of 1μm, and the applied voltage of 100V. The relative permittivity of the insulator is assumed to be 4. As shown in Fig.3 the width of the electrode is equal to the separation between them. The voltage is applied to a phase when the overlapping of the upper and lower electrodes exceeds one third of the width, until they fully overlap. The maximum and the minimum values of the capacitance (see Fig.2 (b)) are 600pF (C_2) and 200pF (C_1).

Two output specifications, the output power and the force, are needed to determine the further detail. The force, F, corresponds to the width of the each electrode. The displacement, x_2-x_1, of each phase is calculated by the force and the increment of the capacitor, C_2-C_1, using eq. (1). For F=1N (~100gf), x_2-x_1=2μm and for F=0.1N (~10gf), x_2-x_1=20μm. That gives the electrode pitches of 6μm and 20μm, respectively[1].

Fig.3 A simple model of the electrostatic linear motor

The output power, P, determines the switching interval of the phase, T_s, average current, I, and the loss. P is given by the formula[4]:

$$P = \frac{V^2}{2} \cdot \frac{\partial C}{\partial t} \quad (11)$$

Therefore T_s equals to $V^2(C_2-C_1)/2P$. For P=100mW, T_s=20μs and for P=1W, T_s=2μs. The average current, I, equals to 2P/V; I=2mA for P=100mW and I=20mA for P=1W. Substituting the values and R=100 ohm into eq. (10), we have P_{DR} equals to 0.4mW for P=100mW and 40mW for P=1W. The relative loss, P_{DR}/P, is proportional to the output power if the dimension of the actuator is fixed.

Electrostatic microactuators

Driving of cantilevers

First utilization of electrostatic force in silicon micro mechanism was to bend or tilte micro beams. The topic was described in Petersen's excellent review[10]. A thin cantilever suspended over a shallow pit was fabricated by anisotropic etching of silicon[11,12] and deflected by the electrostatic force between the cantilever and an electrode at the bottom of the pit. The resonance frequency of the vibration exceeded 1 MHz[10]. The configuration was used, for example, for a mechanical filter[13], a light modulator array[14], and electromechanical switch[15]. A mirror supported by two torsion bars etched from silicon wafer was attached to a glass substrate. The mirror was tilted electrostatically, that made a light scanning mirror[16]. Applications as sensors are quite a few, e. g. resonance micro sensors[17] and a vacuum sensor[18].

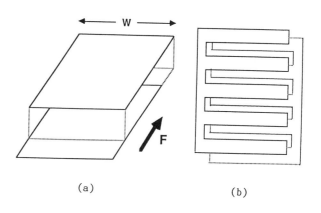

Fig.4 An electrostatic motor with misaligned parallel plates (a) and modified electrode pattern to increase driving force (b).

Electrostatic micro-motor

Suppose a two-parallel-plate capacitor of width w with small misalignment (Fig.4). The force aligning the plates is given by eq. (1). In order to maximize the force the width w should be as large as possible. It can be achieved by dividing the plate into many parallel stripes (Fig.4 (b)). Trimmer, et al. realized such a structure on silicon[1,19]. They divided a stator disk into 6 sectors. Each sector has the calm like electrode in Fig.4 (b). The position of "teeth" in each sector, however, is slightly different as shown in Fig.5. When a rotor disk having equally separated "teeth" was placed on the stator, the "teeth" on the stator and the rotor align in two opposed sectors and misalign in other sectors. Voltage up to 100 V was applied to drive the rotor between opposed sectors with misalignment. The voltage is switched to next sectors as the rotor rotates. SiO_2 insulation layer covers the surface to achieve good insulation property and smooth interface.

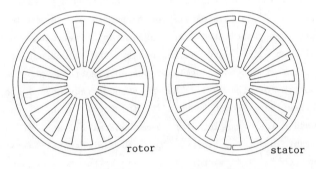

Fig.5 Rotor and stator of the electrostatic motor.

Electrostatic linear actuator (ESLAC)

The actuator[20] consists of a plane wafer with striped electrodes embedded in an insulation layer and a cylindrical electrode (roller) as shown in Fig.6 and Fig.7. The position of the roller is determined by position detectors. The striped electrodes on the one side of the roller are activated to form an electric field against the roller. The electrostatic force attracts the roller and makes it rotate. As the roller passes over a striped electrode, the electrode is discharged and another in the forward direction is charged to keep the movement. An ESLAC (electrostatic linear actuator) consists of rollers inserted in between a C-shaped wafer and a slider (Fig.8).

The dimensions of striped electrodes are shown in Fig.9. Forty lines and spaces of approximately 30μm in width were formed. The length of the line was 7.7 mm.

The electrodes was covered with SiO_2 layer. Metal electrodes on the insulation layer are for voltage supply to striped electrodes, for grounding of the roller, and for detecting roller position (Fig.10). Rollers of 7 mm in length and 0.2 - 0.8 mm in diameter were placed parallel to the striped electrode on the wafer. Figure 11 shows the model actuator with two rollers and a metal plate on them.

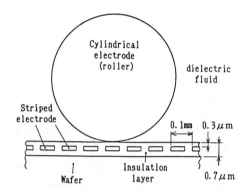

Fig.6 Cross section of ESLAC.

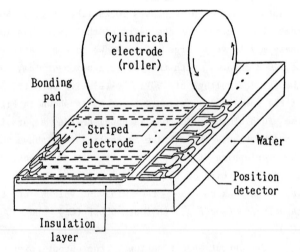

Fig.7 Driving mechanism of the roller.

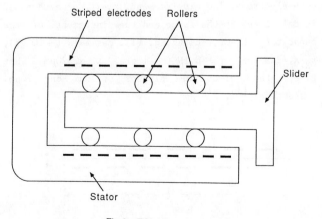

Fig.8 ESLAC

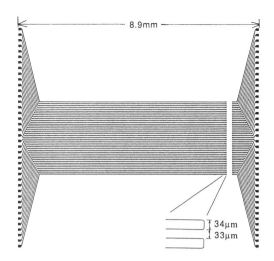

Fig.9 Dimensions of striped electrodes.

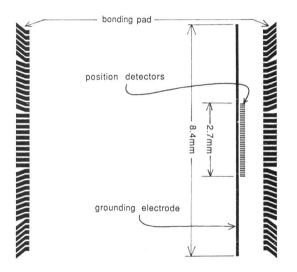

Fig.10 Dimensions of surface electrodes.

Fig.11 Picture of the model of ESLAC.

Electrostatic linear motor (artificial muscle model)

Yano, et al.[21] proposed a electrostatic linear motor based on the contraction mechanism of muscle. Muscle contraction occurs as thick and thin filaments in the muscle slide past each other. The thick filaments composed of myosin has protrusions with a particular separation. The protrusion called a cross bridge is believed to be a micro capacitor of 10 nm electrode separation. Thin filament composed of actin has the helical arrangement of the globular actin molecules. The twist pitch of the helix is 37 nm, while the separation of the cross bridge is 43 nm. When the helix slides between the "electrodes" of the micro capacitor, the capacitance changes with the movement. It reaches the maximum where the actin molecules pile up vertically. The minimum occurs where the molecules lie flat. The schematic pictures in Fig.12 shows how driving force is produced by switching on and off the appropriate capacitors.

An engineering implementation of the principle is shown in Fig.13. A stator has parallel plate capacitors and brushes attached to them. A slider consists of alternately placed high and low permittivity regions and has source electrode at high and ground potentials. Only appropriate capacitors which draws in the high permittivity parts are charged up through brushes.

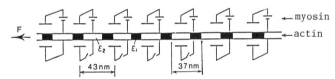

Fig.12 Electrostatic linear motor model of muscle.

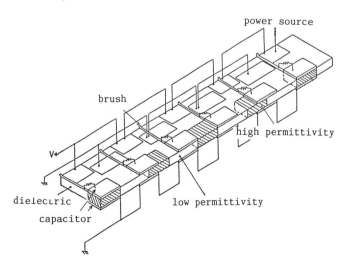

Fig.13 Electrostatic linear motor based on muscle contraction mechanism.

281

Fabrication and evaluation of an electrostatic motor

The detailed procedure of making and evaluating on electrostatic actuator is described in this chapter. The ESLAC is chosen as an example. Available techniques and common difficulties are discussed.

Calculation of the performance

Once electric field on the driving electrode is determined, it is rather easy to obtain the force **F** by the formula:

$$F = \int_s dF \tag{12}$$

$$dF = (\varepsilon/2)E^2 \mathbf{n}\, dS \tag{13}$$

Where ε is the permittivity of the media, S is the electrode surface, and **n** is the normal vector of the surface. The integration (2) should be performed for each vector component.

The electric field can be determined by either analytical[22] or numerical method[23]. The analytical method is applicable to rather simple geometries. The solution, if any, is very useful to have rough estimation.

In the case of ESLAC, the roller-to-plane electrode system is simplified and replaced by a cylinder-to-plane system (Fig.14)[20]. The field in the cylinder-to-plane system can be calculated by either the conformal transformation or the image method[22]. The field strength E on the cylinder, the driving force F_x, and the sticking force F_y are given as follows:

$$E(\theta) = \frac{\sqrt{1+2\alpha}}{\alpha(1+2\alpha\cos^2\frac{\theta}{2})\,[\ln\{(\sqrt{1+2\alpha}+1)^2/2\alpha\}]} \cdot \frac{U}{d} \tag{14}$$

$$F_x = \frac{\varepsilon l U^2}{r[\ln\{(\sqrt{1+2\alpha}+1)^2/2\alpha\}]^2} \tag{15}$$

$$F_y = \frac{\pi\alpha}{2\sqrt{1+2\alpha}} \cdot F_x \tag{16}$$

Where θ is the angle shown in Fig.14 (b), α equals to r/d, r and l is the radius and the length of the roller, and d is the insulation thickness. Figure 15 shows a result.

The numerical method is more versatile. We can determine almost any field with three dimensional combination of electrodes and dielectrics. The finite element method, the charge simulation method, and the surface charge method are most commonly used. For some methods computer codes are available. The field distribution on a roller obtained by the surface charge method is shown in Fig.16[20].

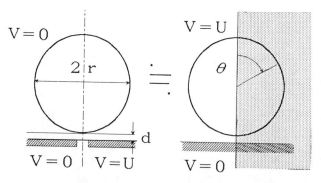

Fig.14 Substitution by a cylinder-to-plane system for analytical solution.

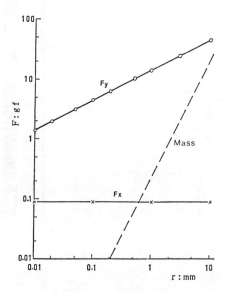

Fig.15 Electrostatic force calculated by the analytical solution for d=50nm, V=5V and ε^*=40.

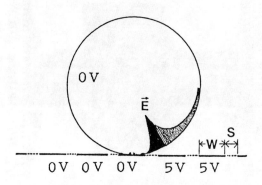

Fig.16 Filed distribution on a roller obtained by surface charge method.
The roller diameter, W, and S, are equal to 0.2mm, 0.082mm, and 0.038mm.

The kinetics of ESLAC without any load is calculated. Figure 17 (a) shows an ESLAC consisting of a plate and rollers on a substrate. The weight, the radius, and the moment of inertia of the roller are M, r, and J. The weight of the plate per roller is m. Only the movement in the horizontal direction (parallel to the substrate) is analyzed. The horizontal forces on the roller are the electrostatic driving force, F_x, the frictional force, F_r, between the roller and the substrate, and the reactive force, F_r', the counter part of which accelerate the plate. Note that the roller rotates without slip and that the plate moves twice as fast as the roller. The equation of motion for the roller is given as follows:

$$\frac{d^2x}{dt^2} = \frac{F_x}{M+4m+J/r^2} \qquad (17)$$

If the roller is a thin pipe, the moment of inertia J is equals to Mr^2. Substituting J in eq. (17), we have:

$$\frac{d^2x}{dt^2} = \frac{F_x}{2(M+2m)} \qquad (18)$$

It means the effective mass of the actuator is the sum of twice the mass of the roller and four times that of the slider. With the external force (load), F, the numerator of Eq. (18) should read F_x-2F.

The fabrication process

There are couple of requirements for the process and its product; (1) compatibility with VLSI process, (2) batch fabrication for low cost and reliability, (3) high insulation strength, (4) tough material property, (5) smooth and wear free surface, and (6) fine patterning and alignment. Fan, et al.[24] and Gabriel, et al.[25] introduced the manufacturing technique of micro structures that satisfies the second and the sixth requirements. The electrostatic drive could be easily built into the structure. Friction, wear, and mechanical reliability (the fourth and fifth requirements) seem the major difficulties.

The flow diagram of the fabrication process of ESLAC is shown in Fig.18. Thermal oxide layer was grown on a well rinsed silicon wafer. A conductive layer made of WSi was sputtered on the oxide layer. The electrode pattern shown in Fig.9 was formed on the conductive layer by photolithography and wet etching. The PCVD process was used to cover the electrodes with SiO_2. Through holes were opened at appropriate positions on the insulation layer. Aluminum electrodes shown in Fig.10 were formed by vacuum coating and lithography. After the wafer was bonded on the package, bonding pads of the striped electrode and the package pins were connected with gold wires.

Problems we encountered so far are the surface roughness and occasional poor insulation of the SiO_2 layer, high contact resistance between the roller and the aluminum grounding electrode, and the fragile bonding of wires.

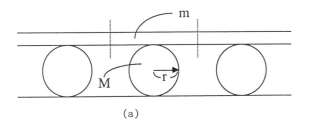

(a)

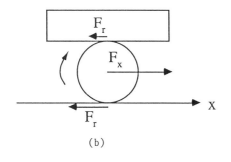

(b)

Fig.17 Model of ESLAC to calculate its kinetics (a) and the force acting to the roller (b).

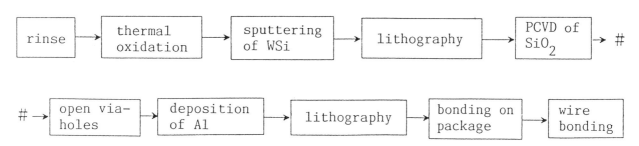

Fig.18 Fabrication process.

Performance measurement

Micro actuators are so small that the measurement itself is a problem. Scale models are useful for demonstration and for confirming the calculation[1,2]. Nonecontact measurements such as optical, image, and indirect measurements are preferred. Figures 19 and 20 show the measurement of driving performance of a scale model of ESLAC. The upper pictures are the streak images of the movement. The velocity was determined by digital image processing. At 3.1kV the movement was stepwise. The instantaneous velocity is 4 cm/s. The maximum and the average acceleration determined from the slopes of Fig.19(b) were 0.43 and 0.25 m/s^2. Substituting M= 5g, m=0 and F_x= 0.42 gf (obtained by eq. (15)) into eq. (18), we have a theoretical value of 0.42 m/s^2. The agreement is excellent.

The preliminary results obtained by the micro models are as follows. The applied voltage on each striped electrode was controlled by the parallel interface of a microcomputer. Continuous rotation of the roller was observed with the roller of 0.4 mm in diameter and the applied voltage above 80 V. In the configuration shown in Fig.11 the plate of 10 mg was linearly driven with applied voltage of 100 V. The insulation thickness between the roller and the electrode was 250 nm. The average electrical field strength was 4 MV/cm.

The driving force on the solid steel roller of 0.8 mm in diameter was roughly estimated by declining wafer and measuring the critical angle at which the driving force was balanced with the component of gravity parallel to the wafer. The force per unit length of the roller was ~20 mgf/cm at 100 V.

Conclusion

The electrostatic microactuators were reviewed. The compatibility of the actuator with the VLSI technology suggests us to have a fully integrated system of the sensors, logics, control circuits and the actuators on a silicon chip. In such a system, which is called a smart actuator, the actuation of the microstructures by the electrostatic force will play a major role. Micromanipulators for bioengineering, medical application, optics, and the fabrication of electric devices will utilize smart actuators. Micro robots or the autonomous moving sensors are the next step. In order to satisfy such a requirement, further intensive works and inventions are needed now.

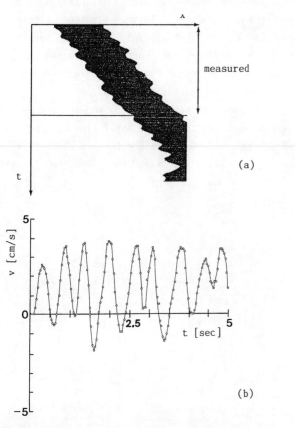

Fig.19 Velocity measurement of the enlarged model at 3.1kV by image processing. (a) streak image. (b) roller velocity.

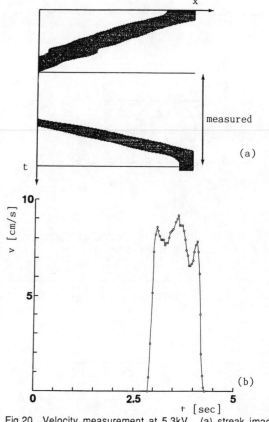

Fig.20 Velocity measurement at 5.3kV. (a) streak image. (b) roller velocity.

REFERENCES

[1] W. S. N. Trimmer and K. J. Gabriel, "Design consideration for a practical electrostatic motor", Sensors & Actu., Vol. 11, pp. 189-206 (1987).

[2] B. Bollee, "Electrostatic Motors", Philips Tech. Review, Vol. 30, pp.178, 1969.

[3] Handbook of Electrostatics, Ohm-Sha Publ. Co., 1981, ch. 19, pp. 651-675, in Japanese.

[4] J. G. Trump, "Electrostatic Sources of Electric Power", Elect. Engin., Vol. 66, pp. 525-534, June 1947.

[5] N. J. Chou and J. M. Eldridge, "Effects of material and processing parameters on the dielectric strength of thermally grown SiO2 films", J. Electrochem. Soc., Vol. 117, pp. 1287-1293, 1970.

[6] K. Kudo, "Construction of Electret Motors", Proc. of national convention of Inst. of Electrostatics Japan, 1980, pp. 137-139, in Japanese.

[7] P. E. Secker and I. N. Scialom, "A Simple Liquid Dielectric Motor", Journal of Applied Physics, Vol. 39, pp. 2957-2961, May 1968.

[8] P. E. Secker and M. R. Belmont, "A miniature multipole liquid-immersed dielectric motor", J. Phys. D: Appl. Phys., Vol. 3, pp.216-220, 1970.

[9] S. D. Choi and D. A. Dunn, "A Surface-Charge Induction Motor", Proc. of the IEEE, Vol. 59, pp.737-748, May 1971.

[10] K. E. Petersen, "Silicon as a Mechanical Material", Proc. of the IEEE, Vol. 70, pp. 420-457, May 1982.

[11] E. Bassous, "Fabrication of Novel Three-Dimensional Micro-structurs by the Anisotropic Etching of (100) and (110) Silicon", IEEE Transactions on electron devices, Vol. ED-25, pp.1178-1185, Oct. 1978.

[12] K. Bean, "Anisotropic Etching of Silicon", IEEE Transactions on electron devices, Vol. ED-25, pp.1185-1193, Oct. 1978.

[13] R. J. Wilfinger, P. H. Bardell, and D. S. Chhabra, "The resonistor: A frequency selective device utilizing the mechanical resonance of a silicon substrate", IBM J. Res. Develop., Vol. 12, pp.113, 1968.

[14] J. Guldberg, H. C. Nathanson, D. L. Balthis, and A. s. Jensen, "An aluminum/SiO_2 silicon on sapphire light valve for projection displays", App. Phys. Lett., Vol. 26, pp.391, 1975.

[15] K. E. Peterson, "Dynamic Micromechanics on Silicon: Techniques and Devices", IEEE Transactions on Electron Devices, Vol. ED-25, pp.1241-1250, Oct. 1978.

[16] K. E. Petersen, "Silicon Torsional Scanning Mirror", IBM J. Res. Develop., Vol. 24, pp.631, 1980.

[17] R. T. Howe, "Resonant Microsensors", Proc. of The 4th Internatinal Conf. on Solid-State Sensors and Actuators, 1987, pp.843-848.

[18] Y. Kawamura, K. Sato, T. Terasawa, and S. tanaka, "Si Cantilever- Oscillator as a Vacuum Sensor", ibid, pp.283-286.

[19] A. W. van Herwaarden and P. M. Sarro, "Floating-Membrane Thermal Vacuume Sensor", ibid, pp. 287-290.

[20] H. Fujita and A. Omodaka, "The principle of an electrostatic linear actuator manufactured by silicon micromachining", ibid, pp. 861-864.

[21] M. Yano and S. Tukida, "The mechanism of muscle contraction—Microscopic electrostatic linear motor", Chemistry Today, Vol. 188, pp. 18-22, 1980, in Japanese.

[22] H. Printz, Hochspannungsfelder. München:R. Oldenbourg Verlag, 1969.

[23] T. Kouno, "Computer Calculation of Electrical Field", IEEE Trans. on Elect. Insul., Vol. EI-21, pp. 869-872, Dec. 1986.

[24] L. S. Fan, Y. C. Tai, and R. S. Muller, "Pin joints, Gears, Springs, Cranks, and Other Novel Micromechanical Structures", Proc. of The 4th Internatinal Conf. on Solid-State Sensors and Actuators, 1987, pp.849-852.

[25] K. J. Gabriel, W. S. N. Trimmer and M. Mehregany, "Micro Gears and Turbines Etched From Silicon", ibid, pp.853-856.

ELECTRIC MICROMOTORS: ELECTROMECHANICAL CHARACTERISTICS

Jeffrey H. Lang & Martin F. Schlecht
Laboratory For Electromagnetic And Electronic Systems
Department Of Electrical Engineering And Computer Science
Massachusetts Institute Of Technology
Cambridge, MA 02139

Roger T. Howe
Electronics Research Laboratory
Department Of Electrical Engineering And Computer Sciences
University Of California
Berkeley, CA 94720

Abstract

Planar rotary and linear motors can now be fabricated using silicon micromachining. These motors, refered to here as micromotors, have a gap separation on the order of 1 μm, and lateral dimensions on the order of 100 μm or more. This paper examines the electromechanical characteristics of these micromotors. To begin, an analysis of electromagnetic to mechanical energy conversion is presented to demonstrate that electric drive is preferable to magnetic drive in micromotors. This is followed by a discussion of the geometry, design and operation of electric micromotors which emphasizes the duality of electric and magnetic motors. Next, a prototype design for the rotary variable-capacitance micromotor is considered as an example. Rotor speeds over 10^5 $rad \cdot s^{-1}$ and accelerations over 10^9 $rad \cdot s^{-2}$ appear possible for this micromotor. Finally, some of the research problems posed by the development of micromotors are identified.

(1) Introduction

Silicon micromachining was developed over the last decade as a means of accurately fabricating very small structures without the assembly of discrete components. It generally involves the selective etching of a silicon substrate [3,25] or deposited thin-film layers of related materials [15,13], or perhaps both. Recently, silicon micromachining was applied to the fabrication of structures which include a rotary or linear bearing, thereby allowing unrestricted motion of the moving component in one degree of freedom [11,2]. These bearings enabled the development of electrically-driven motors, refered to here as micromotors [2]. Micromotors have a planar geometry with a gap separation on the order of 1 μm, and lateral dimensions on the order of 100 μm or more. Their very small size, particularly their micron-scale gap separation, and the characteristics of silicon micromachining combine to provide micromotors with electromechanical characteristics which are significantly different from those of conventional motors. These electromechanical characteristics, and their influence on the design and operation of micromotors, are examined in this paper.

The rotary structure in Figure 1 illustrates the essence of the micromotors considered here. This structure comprises a disk, or rotor, which is pinned to a subtrate, or stator, by a central bearing which restricts the lateral and axial motion of the disk. The disk is shown suspended above the substrate to illustrate its freedom to rotate about the bearing. This entire structure can be micromachined from silicon and related materials using the three-mask sequence of deposition and etching steps summarized in Figure 2. Initially, a sacrificial layer of silicon dioxide is deposited on the substrate. After depositing a layer of silicon nitride and plasma etching the disk with a central circular window using the first mask, the structure appears as shown in Figure 2(A). A second sacrificial layer of silicon dioxide is deposited next, and a second circular window, concentric with the first, is etched through both silicon dioxide layers using the second mask. The deposition and etching of a second silicon nitride layer using the third mask then forms the bearing. The structure now appears as shown in Figure 2(B). Finally, the structure is immersed in buffered hydrofluoric acid to remove the silicon dioxide layers and free the disk. To avoid subsequent warping of the rotor and bearing due to residual stress, the silicon nitride should be non-stochiometric silicon rich [33]. Alternatively, polysilicon could be used. Extensions of these steps allow the inclusion of planar electrodes on the substrate and disk with which the structure can be electrified to produced a torque on the disk.

The design and fabrication of a practical micromotor is considerably more complex than shown in Figures 1 and 2 [2], and many alternatives surely exist. Nonetheless, these figures illustrate several important characteristics of the micromotors which can be fabricated using silicon micromachining. Silicon micromachining is ideally suited to the fabrication of a very smooth micron-scale gap between the rotor and stator of a micromotor. This is essential if substantial torques or forces are to be produced by an electric, as opposed to magnetic, micromotor. This is in turn important because the materials and steps of silicon micromachining are most compatible with electric micromotors. Silicon micromachining can also accurately fabricate and align all micromotor components through its planar lithographic and etching steps, which is necessary to maintain close tolerance on the bearing and provide suitable rotor balance and stability. Finally, silicon micromachining can be compatible with the fabrication of the integrated electronics which are necessary to stabilize and control the micromotors [11].

The preceding characteristics, and their influence on the design and operation of micromotors is examined in the remainder of this paper. In Section 2, an analysis of electromagnetic to mechanical energy conversion is presented to demonstrate that electric drive is preferable to magnetic drive in micromotors. Section 3 then follows with a discussion of the geometry, design and operation of electric micromotors which emphasizes the general duality of electric and magnetic motors. In Section 4, the design and performance of a prototype rotary variable-capacitance micromotor is considered as an example. Finally, Section 5 identifies some of the research problems posed by the development of micromotors.

(2) Micromotor Drive

An electromagnetic motor may usually be treated as either magnetic or electric in character whenever its largest dimension is much less than the wavelength of light at its frequency of operation [22]. For a rotary micromotor such as that shown in Figure 1, having a radius of 300 μm and operating at an electrical frequency of 1 MHz, this condition is satisfied by six orders of magnitude. Thus, the micromotors considered here may be treated as either magnetic or electric in character, depending on their dominant form of energy storage.

Most conventional electromagnetic motors are magnetic in character. However, micromotors with a very small gap are sufficiently different from conventional motors that they are subject to a different set of design and operating constraints. A reappraisal of their material selection, fabrication, torque production and efficiency, is therefore warranted. This reappraisal, provided below, demonstrates that micromotors should be electric in character, in contrast to conventional electromagnetic motors. Here, the focus is on motors which store electromagnetic energy in one or more passive gaps. Electromagnetic motors employing the expansion of piezoelectrically or thermally active materials are not considered.

The torque or force produced by an electromagnetic motor is the negative spatial rate at which it converts stored electromagnetic energy to mechanical energy with its electromagnetic states held constant [43,22]. Since the structures of magnetic and electric motors can provide equivalent spatial rates, a comparison of the stored energy, or energy density, available for such conversions provides a reasonable metric for comparing the torque or force produced by these motors. In both types of motors, most of this energy is stored in their gaps.

The energy density in the gap of a magnetic motor is $\frac{1}{2}\mu_o^{-1}B^2$, where μ_o is the permeability of free space and B is the magnitude of the magnetic flux density in the gap. In conventional magnetic motors, B is limited to approxi-

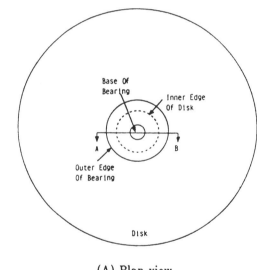

(A) Plan view.

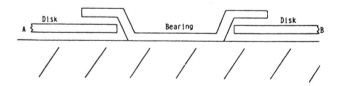

(B) Cross section view through A-B.

Figure 1: Basic micromotor structure.

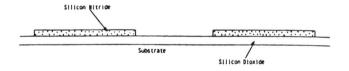

(A) After deposition and etching of disk.

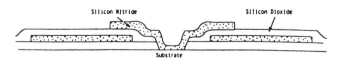

(B) After deposition and etching of bearing.

Figure 2: Fabrication steps of basic micromotor structure.

mately 1.5 T by the saturation of the magnetizable material which defines the gap [12]. Thus, the stored energy density is limited to approximately $9.0 \times 10^5 \ J \cdot m^{-3}$. Although this energy density is independent of the size of the motor, it may well be unachievable in a micromotor. The current density in the windings which produce the magnetic flux is limited by the sheet resistance resulting from their thin-film fabrication, electromigration phenomena, and perhaps by thermal constraints. The properties of any thin-film magnetic materials which are compatible with silicon micromachining, and flux leakage due to the planar micromotor geometry will likely further reduce the achievable energy density.

The energy density in the gap of an electric motor is $\frac{1}{2}\epsilon_o E^2$, where ϵ_o is the permittivity of free space and E is the magnitude of the electric field in the gap. In a conventional electric motor with an air gap, E is limited to approximately $3.0 \times 10^6 \ V \cdot m^{-1}$ by the electrical breakdown of air [21,9,32]. Thus, the stored energy density is limited to approximately $40 \ J \cdot m^{-3}$. This limit is more than four orders of magnitude below that for a magnetic motor, which explains why conventional electric motors are not competitive with magnetic motors.

The very small gap of the micromotors considered here dramatically alters the previous comparison. As either the gap separation or the gas density in the gap decreases, fewer ionizing gas collisions are available to sustain electrical breakdown across the gap [7,41]. The pioneering work reported in [24] describes the important result of this phenomenon, namely a dramatic increase in the breakdown electric field as the product of gap separation and pressure decreases [24,21]. Hence, the breakdown field increases dramatically as gap separation is reduced at a given gas pressure.

The breakdown strength of air at standard density begins to increase as the gap separation approaches 4 μm [9,32], a phenomenon that has already been observed at work in several micromachined silicon structures. One study reports a sustained electric field of $3.2 \times 10^7 \ V \cdot m^{-1}$ across a gap of 12.5 μm [25]. An earlier study using carefully machined metal surfaces reports a breakdown electric field of $1.7 \times 10^8 \ V \cdot m^{-1}$ [5]. Since field emission from a planar surface does not occur until the electric field reaches $10^9 \ V \cdot m^{-1}$ [41], higher electric fields may be possible. However, surface roughness will locally concentrate the electric field, resulting in a lower average breakdown electric field. The smooth surfaces made possible by silicon micromachining are therefore very important to the successful operation of an electric micromotor.

For the purposes of comparison, a maximum electric field of $3.0 \times 10^8 \ V \cdot m^{-1}$ is assumed. This results in an energy density of $4.0 \times 10^5 \ J \cdot m^{-3}$, which is comparable to the maximum obtainable in conventional magnetic motors, and probably exceeds that which could realistically be achieved by a magnetic micromotor. Thus, the torque or force density of an electric micromotor is competitive with that of a magnetic micromotor, in marked contrast to the case of conventional motors.

A reappraisal of efficiency also appears to favor electric micromotors over magnetic micromotors. Static excitation of a magnetic motor requires static currents through its windings, leading to persistent conduction losses. However, static excitation of an electric motor requires static voltages across its gap, which can be sustained with little loss. Therefore, an electric motor is inherently less lossy than a magnetic motor at standstill. This argument continues as the speed of the motors increases until reactive power dominates the excitation of the motor, which should occur at comparable speeds in both types of micromotors. Thus, from standstill to some speed, electric motors are inherently more efficient. Further, most magnetic motors contain magnetizable material which exhibit increasing eddy-current and hysteresis losses as speed increases. Since electric motors need not contain polarizable material, the corresponding loss is avoided, and it follows that electric micromotors appear inherently more efficient than their magnetic counterparts at all speeds.

Finally, and perhaps most importantly, electric micromotors are compatible with the materials and steps of silicon micromachining [2,11], which in turn appears compatible with the fabrication of electronic circuits [11]. Only conducting and insulating films are required in an electric micromotor, in contrast to magnetic micromotors which likely require magnetizable films. Thick magnetizable films may actually be necessary to confine the magnetic fields in a magnetic micromotor, a requirement that is yet incompatible with the planar lithography used in silicon micromachining. In addition, micromotors require solenoidal windings which could be difficult to fabricate. The compatibility of micromotor fabrication with the fabrication of electronics is also important because active electronic circuits are likely essential on the same substrate as the micromotor. Electric micromotors appear more compatible in this regard.

(3) Micromotor Design And Operation

The structure in Figure 1 provides the basis for a rotary electric micromotor in which the substrate is the stator and the disk is the rotor. Comparable structures exist for linear micromotors [2]. What remain to be developed are refinements of this structure including the electrification of the stator and rotor in such a way that a net torque is produced to rotate the rotor about the bearing. This development is presented with a review of the basic operation of an electric motor. The review is presented in terms of linear motors, yet, it should be clear that it applies to rotary motors as well.

Figure 3(A) shows an incremental section of two linear surfaces with matched but offset distributions of alternating electric charge. Let the lower charge distribution reside on the stator and the upper charge distribution reside on the rotor. In general, a rotor can move across a stator with linear motion, the rotary motion of a disk as in Figure 1, or with the rotary motion of a drum. In addition, the vectors in Figure 3(A) indicate the predominant electric forces acting between the rotor and the stator. These attractive forces pull the rotor vertically down and horizontally to the right. It is the horizontal component of this force that propels the rotor in the desired manner. Note that an equally plausible means of propulsion could involve charge distributions which produce repulsive forces.

As the rotor in Figure 3(A) translates to the right, its charge distribution begins to align with that of the stator, and the horizontal force acting on the rotor diminishes. Therefore, in order to sustain the horizontal force, and hence desired rotor motion, it is necessary to advance the stator charge distribution in synchronism with that of the rotor so as to maintain the horizontal offset between the two as shown in Figure 3(A). To do so, the stator is constructed with conducting electrodes on which the charge can be independently excited. A properly phased alternating excitation of these electrodes then produces the desired traveling stator charge distribution. Implementation of the proper phasing may require measurements of rotor position and an actively controlled excitation, depending on the type of micromotor. Alternatively, the rotor charge distribution could be retarded. However, this requires brushes or slip rings, which may be difficult to fabricate in a micromotor.

In addition to the desired horizontal force, the rotor in Figure 3(A) experiences a vertical force which acts to pull it against the stator. This force is likely several orders of magnitude greater than the propulsion force, and poses a problem for micromotors. In a conventional electric motor, the strength of the bearings together with the stiffness of the rotor would prevent such a collapse. However, a thin-film micromotor rotor may be relatively flexible and the bearing in Figure 1 can not sustain axial load. To address this problem, a second stator is constructed above the rotor, as shown in Figure 3(B), and is excited in synchronism with the lower stator. The vertical forces between the upper stator and the rotor balance those between the lower stator and the rotor, while all horizontal forces reinforce to provide a greater force for propulsion. Note that if the reinforcement of horizontal forces is to occur properly, the electrodes on the two stators must be accurately aligned. This is easily accomplished with silicon micromachining since lithographic steps are easily aligned and no assembly of discrete micromotor components is required.

A mobile charged conducting body in an electric field created by stationary external conductors each with fixed total charge can never be in stable equilibrium [10,6,34]. While it is not yet clear how this theorem extends to systems with an uncharged conducting body or external conductors with fixed voltage, it is apparent that one must be prepared for unstable rotor equilibria. The rotor in Figure 3(B), for example, is in vertical equilibrium when it is centered between the two stators, but this equilibrium is unstable if the stators are excited by voltage sources. In general, it appears that some means of rotor stabilization is required in micromotors to avoid striking between the rotor and stator. Whether this stabilization should come from active differential excitation of the two stators, from an aerodynamic shaping of the rotor, or from some other means is an important topic for future research, as is the more fundamental topic of the instability itself. It is interesting to note that different micromotors exhibit different rotor instabilities, offering hope that there exist micromotors which are easily stabilized.

What distinguishes the various types of electric motors is the manner in which their charge distributions are produced. Many possible motors exist, and several which do not require brushes or slip rings are described below. Fortunately, all electric motors have more familiar magnetic counterparts through a duality [20], simplifying their invention. The permanent electret motor is the dual of the permanent magnet motor. It has permanently-polarized charge embedded in its rotor, and its stator is excited to maintain the desired offset of its charges from those on the rotor. The electric hysteresis motor is the dual of the magnetic hysteresis motor, and employs hysteresis in the polarization of its rotor to maintain the proper offset between its excited stator charges and induced rotor charges. The electric induction motor is the dual of the magnetic induction motor. It too employs induced rotor charges, but in a resistive rotor so that charge relaxation maintains the desired charge offset. Lastly, the variable-capacitance motor is the dual of the variable-reluctance motor. Its rotor charge is also induced, but this charge is defined by a saliency in conducting electrodes embedded in the rotor. As in the permanent electret motor, its stator charges must be excited to maintain the proper offset from the rotor charges.

There is a surprisingly long history to electric motor analysis and design, as evidenced by the references in [18]. Early studies of various hysteresis motors were reported in 1893 [1] and 1931 [35]. Early studies of the induction motor were reported in 1893 [42] and 1896 [27]. More recently, this motor was studied and optimized in different forms [5,19,40,8]. Early studies of the variable-capacitance motor were reported in 1907 [26] and in 1933 [39]. More recently, this motor was studied in [5,37,38]. The latter two studies described a silicon motor in which the stator and rotor electrodes were lithographically defined by thin films

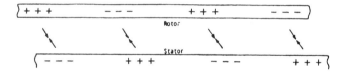

(A) With one stator.

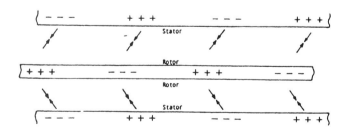

(B) With two stators.

Figure 3: Micromotor charge distributions and forces.

deposited on separate substrates, which were then diced and assembled. Finally, the electromechanical dynamics of many electric motors, and generalized structures such as those in Figure 2, were studied in depth in [22].

(4) Micromotor Performance

The performance of an electromagnetic motor is often limited by the temperature rise caused by its internal losses, assuming that the electronics which power it can deliver the required excitation. A comprehensive analysis of thermal and electronic limitations is beyond the scope of this paper. It is useful, however, to evaluate the performance of a micromotor without considering the aforementioned limitations. The resulting performance yields insight into the ultimate performance that can be achieved and therefore the scope of possible applications. Additionally, this performance generates design goals for other system components, such as heat sinks, and control and power electronics.

By way of example, this section analyzes the performance of the rotary variable-capacitance micromotor. The induction micromotor is treated elsewhere with similar results [2]. Here, the micromotor is assumed to be excited with properly synchronized phase voltages, and its output torque is assumed to be balanced by its rotor inertia when computing maximum acceleration, and by the viscous drag created by the air surrounding its rotor when computing maximum speed. Operating the micromotor in an vacuum eliminates viscous drag as a major retarding torque. In the absence of such a torque, motor speed is limited by electronic and thermal constraints, or by rotor fracture due to mechanical stress.

The electrode design for the example variable-capacitance micromotor is illlustrated in edge-view in Figure 4. The micromotor employs a three-phase configuration of planar wedge-shaped electrodes on the rotor and two symmetric stators. For simplicity, electrical connections are shown to the stator electrodes of only one phase, and these connections are shown to be independent between the stators. This independence could be used to stabilize the rotor, however, for the present purposes, the two independent stator phases are assumed to be externally connected in parallel. As parameters for this micromotor, an outer rotor and electrode radius R_o of 300 μm, an inner electrode radius R_i of 200 μm, a number of rotor electrodes M of 60, a rotor thickness T of 1 μm, a gap separation D of 1 μm, and a maximum gap voltage V_o of 100 V are assumed. The maximum electrode voltage is based on a maximum electric field in the gap of 10^8 $V \cdot m^{-1}$, which is somewhat less than the breakdown field predicted for gap separations on the order of 1 μm, and an order of magnitude below the field emission limit [21,9,41,32,25,5].

The viscous drag torque τ_v for a planar rotor in air can be approximated by considering the continuum analysis of a rotating disk in a housing [31]. At atmospheric pressure, the molecular mean free path in air is about 70 nm, a factor of 15 less than the gap width of 1 μm, implying that a continuum model for air flow is appropriate. For the case of laminar flow, the viscous drag torque is

$$\tau_v = \frac{\pi \omega \eta R_o^4}{D} \quad (1)$$

where $\eta = 1.83 \times 10^{-5}$ $kg \cdot m^{-1} \cdot s^{-1}$ is the viscosity of air and ω is the rotational speed of the rotor. Rotational speeds computed using this relation must be checked to ensure that they are consistent with laminar flow. This requires a Reynold's number, R_e, below 3×10^5, where

$$R_e = \frac{\rho \omega R_o^2}{\eta} \quad (2)$$

and $\rho = 1.2$ $kg \cdot m^{-3}$ is the mass density of air[31]. Thus, the torque predicted by (1) will be valid for rotational speeds up to 5×10^7 $rad \cdot s^{-1}$ for the example micromotor.

The maximum rotational speeds achievable could be very large if the micromotor is run in a vacuum, and its rotor levitated to eliminate friction torques. Mechanical stress in the rotor might then reach the yield stress of the rotor, and limit the maximum rotational speed. The average radial bursting stress for a rotating disk is described in [29]. Setting this stress equal to the rotor yield stress σ_y, a limiting rotational velocity is computed to be

$$\omega^2 = \frac{\sigma_y}{0.075 \rho_r R_o^2} \quad (3)$$

where ρ_r is the mass density of the rotor. For a silicon nitride rotor, the yield stress is 1.4×10^{10} $kg \cdot m^{-1} \cdot s^{-2}$ and the mass density is 3.1×10^3 $kg \cdot m^{-3}$ [25], yielding a maximum rotational speed of 2.6×10^7 $rad \cdot s^{-1}$.

Rotor inertia is also important because it limits micromotor acceleration. For the planar micromotor rotor, the rotor inertia J_r is given by

$$J_r = \frac{1}{2} \pi \rho_r T R_o^4 \quad (4)$$

Again assuming a nitride rotor, $J_r = 3.9 \times 10^{-17}$ $kg \cdot m^2$, for the example micromotor.

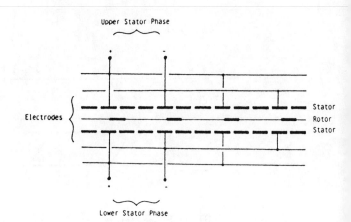

Figure 4: Electrode layout and connection.

The electric torque τ_e produced by a variable-capacitance motor is determined from the spatial rate of change of its stored electric energy or coenergy [43,39,22,5]. The stored electric coenergy $\mathcal{W}'_e$ is used here, thus

$$\tau_e = \frac{\partial \mathcal{W}'_e}{\partial \theta} \qquad (5)$$

where the partial derivative is taken with the stator phase voltages held constant. For a variable-capacitance motor with independent phases, such as the micromotor shown in Figure 4, the stored electric coenergy for each phase $\mathcal{W}'_{e_p}$ is

$$\mathcal{W}'_{e_p} = \frac{1}{2} C_p(\theta) V_p^2 \qquad (6)$$

where $C_p(\theta)$ is the phase capacitance, θ is the rotational rotor position and V_p is the phase voltage. The total coenergy is the sum of the phase coenergies. Substituting (6) independently into (5) for each phase gives the electric torque for each phase τ_{e_p} as

$$\tau_{e_p} = \frac{1}{2} V_p^2 \frac{dC_p}{d\theta} \qquad (7)$$

Since the total stored electric coenergy is the sum of those from the individual phases, the total electric torque is the sum of the individual phase torques.

For the present purposes, only an average electric torque for the variable-capacitance micromotor is required. Thus, the torques of the independent phases are time averaged and summed to yield a total average electric torque. For each phase, the excitation voltage is assumed to be constant at $2V_o$ when the phase capacitance is increasing and zero otherwise. This is the synchronous excitation required for maximum torque production. Under the assumption that the minimum phase capacitance is much smaller than the maximum phase capacitance as a function of rotational position, the time average electric torque $<\tau_{e_p}>$, found by averaging (7), is given by

$$<\tau_{e_p}> = \frac{M \; Maximum(C_p(\theta)) \; V_o^2}{\pi} \qquad (8)$$

Since the example micromotor has three phases, the maximum time average electric torque for it is 3 times that given by (8). Finally, for the example micromotor,

$$Maximum(C_p(\theta)) = \frac{\epsilon_o \pi (R_o^2 - R_i^2)}{-6D} \qquad (9)$$

where a parallel connection of the two independent stators is assumed. This capacitance is 0.2 pF for the example micromotor. Combining (8) and (9), and multiplying by 3 for the three phases, yields

$$<\tau_e> = \frac{M\epsilon_o(R_o^2 - R_i^2)V_o^2}{2D} \qquad (10)$$

for the total average electric torque. This torque evaluates to $1.3 \times 10^{-7} N \cdot m$.

The above results can now be combined to assess the performance of the example micromotor. To begin, balancing (4) against (10) yields a maximum rotational speed of $2.9 \times 10^5 rad \cdot s^{-1}$. This speed satifies both the laminar flow and yield stress conditions given by (2) and (3), respectively, and corresponds to a rotor tip velocity of 86 $m \cdot s^{-1}$ and a power output of 0.038 W. Next, dividing (10) by (4) yields a maximum micromotor acceleration of $3.3 \times 10^9 \; rad \cdot s^{-2}$. This is at least four orders of magnitude greater than the acceleration of conventional motors. Finally, combining (1) and (4) yields the inertial-viscous settling time $\mathcal{J}_r \omega / \tau_v$ of 85 μs. In summary, the performance of a micromotor is be considerably different than the performance of a conventional motor.

The variable-capacitance motor is a synchronous motor so that its power electronics must be capable of providing uniform voltage pulses at the synchronous rate. For the example micromotor, the power electronics must excite the capacitive load represented by the phase capacitance and parasitics with a 200 V pulse train having a 500 ns pulse width. A reasonable estimate for the total load capacitance is 1 pF. In order to obtain good square pulse characteristics, the resistive-capacitive time constant of the motor phase electrodes must be small compared to 500 ns. Assuming a sheet resistance of 100 $\Omega/\square$ for degenerately doped 250 nm-thick polysilicon electrodes[36], an average electrode width of 7.5 μm after accounting for an inter-electrode spacing of 1 μm, and an electrode length of 100 μm, the electrode resistance is 1350 Ω and the time constant is 1.2 ns, which is small compared to the required pulse width.

The ohmic losses associated with charging and discharging the rotor electrodes can be approximated by calculating $I^2 R$, where I is the rms current flowing into the phase capacitance. The average power dissipation is 40 μW or about 0.1% of the available mechanical power. It seems reasonable to assume that this power dissipation will not cause heat dissipation problems, especially in light of air flow in the motor gap.

(5) Conclusion

This paper explored the design, fabrication and operation of micromotors. A process was described for fabricating these micromotors using silicon micromachining. The particular characteristics of the micromotors and silicon micromachining then guided the design and performance analysis of the micromotors. In summary, the design and performance of practical micromotors appear to be considerably different from conventional motors.

For several reasons, electric drive is preferable to magnetic drive for planar micromotors. Electric micromotors require only conducting and insulating films, are more efficient, and have more useful energy densities due to the higher electric fields sustainable in micron-sized gaps. Electric torques or forces can be coupled to the rotor by sequentially exciting planar electrodes located on the substrate or an overhanging stator. Fortunately, the design of electric micromotors can benefit from their duality relationship with the well-understood conventional magnetic motors. A simplified performance analysis for an example rotary variable-capacitance micromotor indicates that rotor accelerations over $10^9 \; rad \cdot s^{-2}$ and rotor velocities over $10^5 rad \cdot s^{-1}$ are possible.

If a practical micromotor technology is to be developed, several areas need further study. First, the physics of electric breakdown in planar micron and submicron gaps must be understood. In particular, the effect of surface coatings and electrode surface morphology on the breakdown field are of practical importance in developing a micromotor fabrication technology. The control of residual stress in thin-film composites is a second important research area, since eccentric stress distributions will warp the rotor and upper stator. Much further research is necessary to understand friction, wear [28], and static charging, or triboelectricity, in micromachined bearings. Finally, the stability of the micromotor rotor must be understood. Some micromotors will be inherently more stable than others. Further, it may be possible to design air bearings that stabilize the rotor. To do so, the flow regime that characterizes micromotors, where the gap separation is only an order of magnitude larger than the mean free path of the gas molecules, must be understood.

An implicit assumption in Figure 1 is that relaxation of residual stress in the released silicon nitride or polysilicon disk does not cause it to warp and bind against the substrate and bearing. Polysilicon films, for example, have a compressive residual stress that varies through the thickness of the film [17,14]. As a result, polysilicon cantilever beams have a built-in bending moment and deflect. High-temperature annealing reduces this average compressive stress as well as its variation through the film [14,16]. With silicon nitride, more flexibility is available by varying its composition [33]. A major challenge in fabricating thin-film micromotors will be the control of residual stress to achieve planar rotors with micron-sized electrode gaps.

Given advances in the above areas, intensive research in micromotor design and fabrication is still necessary. Reliable starting and high-performance operation of synchronous variable-capacitance motors requires sensing the relative position of the rotor electrodes with respect to those on the stator [5]. Also, active stabilization of the rotor requires sensing the vertical position of the rotor between the upper and lower stators. Capacitive readout of the rotor position is an attractive approach for starting and levitation. However, the sub-picofarad sensing capacitors are best measured with on-chip electronics. Another incentive for integration of the micromotor structure with active electronics is the minimization of parasitic capacitance in the motor power electronics. In order to achieve the goal of a fully integrated micromotor, the electronics fabrication and micromotor fabrication must be merged. Here, MOS electronics are a plausible candidate for this merge. Integrated high-voltage switches are required to achieve maximum micromotor speed. The fabrication technology needed to integrate conventional MOS circuitry with these high-voltage switches has been developed recently for smart power electronics [4,23,30]. These, or similar switches must also be merged with the micromotor fabrication. It is likely that the design of both the electronic and electromechanical components of the system will be compromised by integration. However, the enhanced capabilities of effective position sensing, rotor levitation, and miniaturization may compensate and yield a superior system performance.

Finally, if micromotors are to prove useful beyond self-contained applications such as controllable mirror mounts, controllable shutters, or gyroscopes, a means of harnessing their mechanical output torque must be developed. Reference [11] provides possible solutions to this problem.

Acknowledgments

The authors acknowledge the contributions of Martin A. Schmidt and Stephen D. Senturia of MIT, who developed the initial fabrication process for the present micromotor bearing, and Stephen F. Bart and Theresa A. Lober of MIT, who helped develop and analyze the present micromotors [2]. This paper was written with support from the National Science Foundation in the form of Grant 8614328-ECS, and a Presidential Young Investigator Award (RTH).

References

[1] R. Arno. Über ein rotirendes elektrisches Feld und dursh elektrostatische Hysteresis bewirkte Rotationen. *Electrotechnische Zeitschrift*, 14:17–18, 1893.

[2] S. F. Bart, T. A. Lober, R. T. Howe, J. H. Lang, and M. F. Schlecht. Microfabricated electric actuators. Submitted to *Sensors and Actuators*, 1987.

[3] K. E. Bean. Anisotropic etching of silicon. *IEEE Transactions on Electron Devices*, ED-25:1185–1193, 1978.

[4] H. Becke. Approaches to isolation in high voltage integrated circuits. In *Technical Digest of the IEEE International Electron Devices Meeting*, pages 724–727, Washington, DC, 1985.

[5] B. Bollée. Electrostatic motors. *Philips Technical Review*, 30(6/7):178–194, 1969.

[6] W. Braunbek. Freely suspended bodies in electric and magnetic fields. *Zeitschrift für Physik*, 112, 1939.

[7] S. C. Brown. *Basic Data of Plasma Science*. John Wiley and Sons, Inc., New York, NY, 1959.

[8] S. D. Choi and D. A. Dunn. A surface-charge induction motor. *IEEE Proceedings*, 59(5):737–748, 1971.

[9] T. W. Dakin, G. Luxa, G. Oppermann, J. Vigreux, G. Wind, and H. Winkelnkemper. Breakdown of gases in uniform fields. *Electra*, 32:61–82, 1974.

[10] S. Earnshaw. On the nature of the molecular forces which regulate the constitution of the luminiferous ether. *Transactions of the Cambridge Philosophical Society*, 7, 1842.

[11] L. Fan, Y. Tai, and R. S. Muller. Pin joints, gears, springs, cranks and other novel micromechanical structures. In *Technical Digest of the 4th International Conference on Solid-State Sensors and Actuators*, pages 849–852, Tokyo, Japan, June 1987.

[12] A. E. Fitzgerald, Jr. C. Kingsley, and S. D. Umans. *Electric Machinery*, pages 14–16. McGraw-Hill Book Company, New York, NY, fourth edition, 1983.

[13] H. Guckel and D. W. Burns. Fabrication techniques for integrated sensor microstructures. In *Technical Digest of the IEEE International Electron Devices Meeting*, pages 176–179, Los Angeles, CA, 1986.

[14] H. Guckel, T. Randazzo, and D. W. Burns. A simple technique for the determination of mechanical strain in thin films with applications to polysilicon. *Journal of Applied Physics*, 57:1671–1675, 1985.

[15] R. T. Howe. Polycrystalline silicon micromachining. In C. D. Fung, P. W. Cheung, W. H. Ko, and D. G. Fleming, editors, *Micromachining and Micropackaging of Transducers*, pages 169–187, Elsevier, 1985.

[16] R. T. Howe and R. S. Muller. Polycrystalline and amorphous silicon micromechanical beams: annealing and mechanical properties. *Sensors and Actuators*, 4:447–454, 1983.

[17] R. T. Howe and R. S. Muller. Polycrystalline silicon microstructures. *Journal of the Electrochemical Society*, 130:1420–1423, 1983.

[18] O. D. Jefimenko. Electrostatic motors. In A. D. Moore, editor, *Electrostatics and its Applications*, chapter 7, pages 131–147, John Wiley and Sons, Inc., New York, NY, 1973.

[19] C. Kooy. Torque on a resistive rotor in a quasi electrostatic rotating field. *Appl. Sci. Res.*, 20:161–172, February 1969.

[20] M. W. Layland. Generalized electrostatic machine theory. *IEE Proceedings*, 116:403–405, 1969.

[21] J. M. Meek and J. D. Craggs, editors. *Electrical Breakdown of Gases*. John Wiley and Sons, Inc., Chichester, 1978.

[22] J. R. Melcher. *Continuum Electromechanics*, chapter 2–5. MIT Press, Cambridge, MA, 1981.

[23] W. G. Meyer, G. W. Dick, K. E. Olson, K. H. Lee, and J. A. Shimer. Integrable high-voltage CMOS: devices, process application. In *Technical Digest of the IEEE International Electron Devices Meeting*, pages 732–735, Washington, DC, 1985.

[24] F. Paschen. Ueber die zum Funkenübergang in Luft, Wasserstoff und Kohlensäure bei verschiedenen Drucken erforderliche Potentialdifferenz. *Annalen der Physik*, 37:69–96, 1889.

[25] K. E. Petersen. Silicon as a mechanical material. *IEEE Proceedings*, 70(5):420–457, May 1982.

[26] W. Petersen. *Elektrostatische Maschinen*. PhD thesis, Technical University of Darmstadt, 1907.

[27] G. Quinke. Ueber Rotationen im constanten electrischen. *Annalen der Physik*, 59:417–486, 1896.

[28] E. Rabinowicz. Grinding damage of silicon nitride determined by abrasive wear tests. *Wear*, 39:101–107, 1976.

[29] R. J. Roark and W. C. Young. *Formulas for Stress and Strain*, page 572. McGraw-Hill Book Company, New York, NY, fifth edition, 1975.

[30] K. Sakamoto, T. Okabe, M. Kimura, K. Satonaka, and T. Nishimura. Technology for a 250V monolithic complementary MOSFET LSI with n+ buried-layer protected CMOS logic. In *Technical Digest of the IEEE-Japan Society of Applied Physics VLSI Technology Symposium*, pages 21–22, San Diego, CA, May 1986.

[31] H. Schlichting. *Boundary-Layer Theory*, pages 647–652. McGraw-Hill Book Company, New York, NY, 1979.

[32] W. O. Schumann. *Elektrische Durchbruchfeldstärke von Gasen*, page 25. Springer, Berlin, 1923.

[33] M. Sekimoto, H. Yoshihara, and T. Ohkubo. Silicon nitride single layer X-ray mesh. *Journal of Vacuum Science and Technology*, 21(4), November-December 1982.

[34] J. A. Stratton. *Electromagnetic Theory*. McGraw-Hill Book Company, New York, NY, 1941.

[35] K. Strobl. *Der electrostatische Asynchronmotor*. PhD thesis, Technical University of Vienna, 1931.

[36] S. M. Sze, editor. *VLSI Technology*, page 104. McGraw-Hill Book Company, New York, NY, 1983.

[37] W. S. N. Trimmer and K. J. Gabriel. Design considerations for a practical electrostatic micro motor. *Sensors and Actuators*, 11(2):189–206, 1987.

[38] W. S. N. Trimmer, K. J. Gabriel, and R. Mahadevan. Silicon electrostatic motors. In *Technical Digest of the 4th International Conference on Solid-State Sensors and Actuators*, pages 857–860, Tokyo, Japan, June 1987.

[39] J. G. Trump. *Vacuum Electrostatic Engineering*. PhD thesis, Massachusetts Institute of Technology, 1933.

[40] J. Ubbink. Optimization of the rotor surface resistance of the asynchronous electrostatic motor. *Appl. Sci. Res.*, 22:442–448, September 1970.

[41] A. R. von Hipple. *Molecular Science and Molecular Engineering*, chapter 3 and 5. John Wiley and Sons, Inc., New York, NY, 1959.

[42] W. Weiler. *Z. Phys. Chem. Unterricht*, 6:194–195, 1893.

[43] H. H. Woodson and J. R. Melcher. *Electromechanical Dynamics, Part 1*, chapter 3 and 4. John Wiley and Sons, Inc., New York, NY, 1968.

ELECTROQUASISTATIC INDUCTION MICROMOTORS

Stephen F. Bart & Jeffrey H. Lang

Laboratory for Electromagnetic and Electronic Systems
Department of Electrical Engineering and Computer Science
Massachusetts Institute of Technology
Cambridge, MA, 02139, USA

ABSTRACT

This paper studies the steady-state operation of the electroquasistatic induction micromotor (IM). A rotary pancake IM compatible with surface micromachining serves as an example. A model is developed to predict the electric potential, field and free charge within the IM. The model also predicts the motive torque and transverse force of electric origin acting on its rotor. The torque is balanced against bushing friction and windage to determine rotor velocity. Here, the bushing friction is modeled as a function of the transverse force acting on the rotor.

The model is used to study IM performance and its dependence on IM dimensions and material properties. For example, IM performance is predicted to be a complex function of axial IM dimensions and a strong function of rotor conductivity. The study also reveals that IM performance can differ significantly from that of the variable-capacitance micromotor. For example, the dependence of motive torque and transverse force on velocity, and the excitation and control requirements can all be significantly different. Their dependence on micromotor geometry, dimensions and material properties can also be significantly different.

(1) INTRODUCTION

The surface micromachining of silicon-related materials has recently been applied to the fabrication of very small mechanical structures which exhibit unrestrained motion over at least one degree of freedom, and which do not require assembly of discrete parts [3]. The use of these structures as the basis for electroquasistatic micromotors has been considered from a variety of viewpoints [12,13,6,4,5,7,2], and functional micromotors have begun to appear [11]. In general, this work has focused on variable-capacitance micromotors. However, the possibility of alternative micromotors has been discussed [6,5,2], and the purpose of this paper is to study one such alternative, namely the induction micromotor.

Due to its simplicity, the variable-capacitance micromotor (VCM) has received most attention to date. Its construction uses only good insulators and conductors, and simple open-loop voltage pulses are sufficient for its excitation. However, the VCM is not without inherent disadvantages. First, in order to obtain the variable stator-rotor capacitance necessary for its operation, its rotor must be physically salient. Planarization over this saliency may require undesirable fabrication steps. Second, in order to obtain large variations in capacitance, and hence improved motive torque, its stator-rotor gap must be relatively small. This too may pose fabrication problems. Third, it is a synchronous motor [14], consequently, even good performance may ultimately require rotor position feedback which may become a significant disadvantage. Finally, its rotor suffers a transverse force of electric origin which may lead to undesirable bushing friction and wear [6,5,2].

With the preceding discussion in mind, the induction micromotor (IM) becomes an interesting alternative. First, the IM uses charge relaxation rather than physical saliency to establish its rotor charge distribution. Consequently, its rotor can be a smooth uniform conductor; the rotor can even be a fluid so fluid pumps are possible. However, the conductivity of IM materials strongly affects performance so fabrication difficulties associated with VCM rotor saliency are traded for those associated with conductivity control in the IM. Second, IM and VCM performance exhibit different dependencies on micromotor geometry and dimensions. Thus, fabrication constraints may favor one micromotor over the other for a given geometry or application. Third, the IM is an asynchronous motor [14] so good performance can be achieved without rotor position feedback. Finally, the transverse forces of electric origin acting on the IM and VCM rotor are quite different. This too may favor one micromotor over the other. For example, this difference can significantly affect bushing friction and wear [5]. In summary, the IM is an appealing alternative to the VCM, and should be studied.

With the above comments serving as motivation, this paper studies a reasonably general rotary pancake electroquasistatic IM; extensions to other geometries are discussed. The electric potential, field and free charge within the IM are predicted. These results are used to predict the corresponding steady-state rotor motive torque and transverse force. Extension of this modeling to consider dynamics is also discussed. An example with realistic IM parameters is provided to illustrate performance and study performance dependencies.

(2) IM OPERATION & MODELING

This section outlines a two-dimensional electroquasistatic electromechanical model of the IM shown in Figure 1. The model predicts the steady-state motive torque and transverse force of electric origin acting on the IM rotor. The details of its development are given in the Appendix.

As shown in Figure 1, the IM rotor is an annular disk having uniform permittivity and conductivity. It rotates with constant velocity between two coplanar concentric arrays of stator electrodes which support azimuthally-traveling potential waves of equal temporal and spatial frequency. However, these potential waves may have different complex amplitudes. The two stator-rotor gaps also have uniform permittivity and conductivity. All material properties and dimensions are independent. Thus, the IM studied here is reasonably general.

As the potential waves travel around the stators, they induce free electric image charges of opposite sign on the two neighboring rotor surfaces. These image charges travel in synchronism

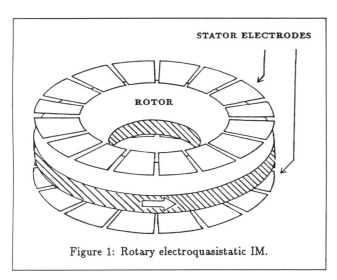

Figure 1: Rotary electroquasistatic IM.

surface micromachining; its geometry is as in Figure 1, and is similar to VCM geometries studied in [12,13,6,4,5,7,2]. The active outer and inner radii of the IM are 50 μm and 35 μm, respectively. The two sets of stator electrodes are excited with single-spatial-harmonic azimuthally-traveling potential waves which are in phase. Both potential waves have an amplitude of 100 V, a temporal frequency of 50 kHz and an azimuthal periodicity of 4. The rotor has a thickness of 1 μm and a conductivity of 10^{-4} S; its resistivity is 10^6 Ω-cm. The rotor is axially offset between the two sets of stator electrodes so that the upper and lower stator-rotor gaps have axial lengths of 0.7 μm and 1.3 μm, respectively. A circular rotor bushing with a radius of 22 μm maintains the axial rotor offset, and is held in contact with the upper stator by the transverse force acting on the rotor. The contact between the bushing and the upper stator results in a friction torque equal to the normal force scaled by a kinetic friction coefficient of 0.5, and multiplied the bushing radius [5].

The model described above is used to predict the performance of the example IM. The results are shown in Figure 2. This figure shows three torque curves, one for motive torque, bushing-friction load torque and windage load torque. The windage load torque is modeled under the assumption that air in laminar flow fills the two stator-rotor gaps [2]. The two torques of electric origin are characteristic of induction interactions [14,9]. The motive torque vanishes at the synchronous velocity of 12.5 krps because the image charges on the rotor are not displaced from the potential waves traveling on the stator electrodes. For lesser velocities, the image charges lag behind the potential waves, and the resulting motive torque is positive. For greater velocities, the image charges are carried ahead of the potential waves by the rotor motion, and the motive torque is negative. The bushing-friction load torque, however, peaks at the synchronous velocity because it is at this velocity that the image charges are aligned with the potential waves. This results in the greatest transverse force on the rotor. Note that for negative velocities, the IM actually operates as a brake, while for velocities which exceed the synchronous velocity, the IM operates as a generator.

Steady-state operation occurs when the motive torque is balanced by the two load torques. As seen from Figure 2, this occurs approximately at a velocity of 10 krps. Note that the the

with the potential waves, but, following charge relaxation in the rotor and stator-rotor gaps, they lag behind. The resulting azimuthal displacement between the potential waves and the image charges gives rise to a motive torque acting on the rotor. If the dominant charge relaxation time is too long, then very little image charge is induced. Yet, if this time is too short, then the image charges are only slightly displaced from the potential waves. Both conditions reduce motive torque. Thus, there exists an optimal set of IM material properties for a given IM geometry and excitation. IM design is therefore inherently more complex than VCM design because VCM performance is optimized by using the best available insulators and conductors.

The details of the IM model are given in the Appendix, and are only outlined here. Laplace's equation is first solved in the rotor and rotor-stator gaps to yield the potential in these regions; the corresponding electric field and image charges follow directly. This solution involves self-consistently splicing the resulting potential and electric field of the three regions together at their boundaries, and matching the outermost potentials to the excitation imposed with the stator electrodes. Next, the electric field is used to evaluate the electroquasistatic stress tensor on the rotor surface. Finally, this stress tensor is integrated over the rotor surface to yield the motive torque and transverse force of electric origin acting on the rotor. To simplify this modeling, the rotary IM is treated as an equivalent two-dimensional linear IM. The justification for this simplification is given in the Appendix.

The IM shown in Figure 1 has one of many possible geometries. Alternative micromotors might be linear, might have their stators at the edge of the rotor, or might have a single-sided stator, and so on. To the extent that these alternatives can be modeled as an equivalent two-dimensional linear IM, they can be modeled as is the Appendix. Further, the rotor can be a fluid in laminar flow, and the stator-rotor gaps can be filled with similar fluids. Thus, by appropriately interchanging directions, and redefining dimensions and material properties, the model developed in the Appendix becomes considerably more general.

(3) AN IM EXAMPLE

To illustrate and study IM performance, an example is presented here. The example IM appears to be compatible with

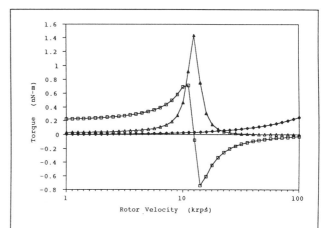

Figure 2: IM performance; □ denotes motive torque, △ denotes bushing-friction load torque, and ◇ denotes windage load torque.

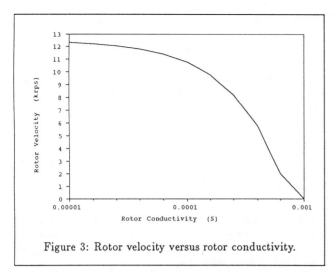

Figure 3: Rotor velocity versus rotor conductivity.

windage load torque has little influence on this balance, which emphasizes the necessity of minimizing the bushing-friction load torque. One possible method of minimization is to fabricate bushings which fill the majority of the stator-rotor gaps [5].

(4) PERFORMANCE DEPENDENCIES

This section studies the dependence of IM performance on several material properties and dimensions. The model outlined in Section 2, and detailed in the Appendix guides this study. The results of this study begin to illuminate how IM performance and design are related.

(4.1) Rotor Conductivity

The motive torque and transverse force of electric origin acting on the rotor are both functions of rotor conductivity. For example, the shapes of the two corresponding curves in Figure 2 spread away from the synchronous velocity as rotor conductivity increases; they shrink toward the synchronous velocity as rotor conductivity decreases. The maximum values of these curves, however, are not functions of rotor conductivity. This is characteristic of induction interactions [14, Chapter 4], and is discussed further in the Appendix. As the motive torque and transverse force change with changing conductivity, so too does the torque balance which establishes the steady-state rotor speed. The result of these changes is shown in Figure 3 which displays steady-state velocity as a function of rotor conductivity for the IM described in Section 3.

The IM example of Section 3 assumes a rotor conductivity of 10^{-4} S which is the approximate conductivity of undoped poly-silicon [10]. However, the conductivity of poly-silicon is strongly dependent on the poly-silicon deposition process, and is therefore hard to control, particularly when low conductivities are desired [1,8]. Consequently, the results of Figure 3 indicate that it may be difficult to reliably fabricate an IM with pre-specified performance. Further, the conductivity of undoped to lightly-doped poly-silicon is a strong function of temperature. For example, it rises to approximately 10^{-2} S at 100°C [1]. Again, tight control over rotor conductivity and performance may prove to be difficult.

One way to diminish the problem of conductivity control, aside from using new materials, is to design or operate the IM in such a way that higher conductivities, which are presumably easier to control, are in fact desirable. Several possibilities for this exist. First, as discussed in the Appendix, an increase in conductivity can be compensated by an increase in the temporal frequency of the stator excitation. In other words, by operating the IM at higher electrical frequencies, higher conductivities can be tolerated. Second, a higher rotor conductivity can be effectively reduced with conduction barriers fabricated into the rotor. This, however, may introduce fabrication problems associated with planarization or fine-scale lithography. Third, a thin highly-conducting layer could be deposited on an insulating rotor to yield a lower apparent conductivity; note that this possibility is not well modeled by the Appendix. Finally, a combination of these possibilities may be most appropriate.

(4.2) IM Axial Dimensions

Both the motive torque and transverse force are functions of IM axial dimensions. However, they are different functions of these dimensions so they can, perhaps, be independently manipulated during design. In general, their dependence over a wide range of operating conditions is quite complex. However, assuming that the rotor velocity is near the synchronous velocity of the IM, as in the balanced operation of Figure 2, the dependence is quite simple.

If the rotor velocity is near the synchronous velocity, then the slip between the rotor and the stator potential waves is small. In this case, the rotor appears highly conducting, and the two stator-rotor gaps can be viewed as approximately independent capacitors. Consequently, the contribution to motive torque from each rotor image charge distribution is approximately inversely proportional to the axial length of the corresponding stator-rotor gap, much like for a VCM. Similarly, the transverse force contribution is inversely proportional to the square of the axial length of the corresponding stator-rotor gap. This dependence is illustrated in Figure 4. This figure is identical to Figure 2 with the exception that the axial length of both stator-rotor gaps has been doubled. Note that the motive torque at low speed has changed little from the example in Section 3, so the starting of both motors is identical. Further, the rotor velocity at which the three torques in Figure 2 balance has changed little in Figure 4. However, the transverse rotor force has reduced by a factor of 4, so that bushing wear, for example, may be significantly reduced. Such interactions between design and performance are typical of the IM and other micromotors.

(4.3) Relative Charge Relaxation Times

In the IM of Section 3, the two stator-rotor gaps are insulators. Consequently, the rotor has the faster charge relaxation time, and image charges of opposing sign are induced by the potential waves excited with the stator electrodes. If this situation is reversed, that is, if the rotor has a slower charge relaxation time than the two stator-rotor gaps, then image charges of like sign are induced. This would be the case, for example, if the rotor were insulating and conducting fluids filled the two stator-rotor gaps. One significant consequence of this is that the two torque curves of electric origin in Figure 2 reverse sign. There are then three velocities at which the three torques balance in steady state, one near the synchronous velocity, one **far above the synchronous velocity and one negative velocity** [9, Section 5.14]. Interestingly, the balance which occurs near

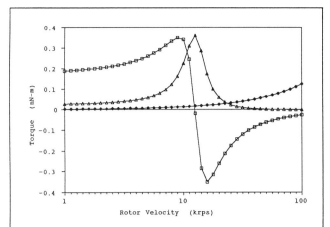

Figure 4: IM performance; □ denotes motive torque, △ denotes bushing-friction load torque, and ◇ denotes windage load torque. The upper and lower stator gaps are 1.4μm and 2.6μm, respectively, here.

the synchronous velocity is unstable, while the other two are stable. Thus, even if the stator excitation is temporally static, that is, if the potential waves do not travel, rotor motion can still be driven. A second significant consequence is that the rotor is repelled from each stator in the transverse direction. This tends to center the rotor between the two stators and may eliminate bushing friction and wear.

(4.4) Excitation & Control Issues

The excitation and control of an IM promises to have advantages over that of a VCM. As outlined above, the IM is an asynchronous micromotor which produces steady-state motive torque at all velocities except the synchronous velocity. In contrast, the VCM is a synchronous motor which produces steady-state motive torque only at the synchronous velocity. Thus, the VCM probably requires position feedback to control its excitation if good performance is desired. The IM does not necessarily require such feedback. Additionally, as discussed above, it is possible to design an IM that produces motive torque with a static excitation. The excitation required by this IM is particulary simple.

(5) SUMMARY & CONCLUSIONS

This paper studied the steady-state operation of the electroquasistatic induction micromotor (IM). A rotary pancake IM compatible with surface micromachining served as an example. A model was developed to predict the electric potential, field and free charge within the IM. The model also predicted the motive torque and transverse force of electric origin acting on its rotor. This torque was balanced against bushing friction and windage to determine rotor velocity, where the bushing friction was modeled as a function of the transverse force acting on the rotor. Extensions to other IM geometries and the inclusion of excitation dynamics were also discussed.

The model was used to study IM performance and its dependence on IM dimensions and material properties. For example, IM performance was predicted to be a complex function of axial IM dimensions and a strong function of rotor conductivity, and supporting examples were provided. The study also revealed that IM performance can differ significantly from that of the variable-capacitance micromotor. For example, the dependence of motive torque and transverse force on velocity, and the excitation and control requirements can all be significantly different. Their dependence on micromotor geometry, dimensions and material properties can also be significantly different.

One possible disadvantage of the IM is its performance dependence on rotor conductivity. Several ways to diminish this disadvantage were discussed. On the other hand, an appealing advantage is that it can, in principle, be used to pump fluids. This may become one of its most interesting applications. In any event, the IM is sufficiently different from the VCM that it deserves further study and development.

ACKNOWLEDGEMENTS

The authors wish to thank T. A. Lober, M. Mehregany, J. R. Melcher, M. F. Schlecht, S. D. Senturia, and L. S. Tavrow, all of MIT, for their many contributions to this paper. The work reported in this paper was supported by the United States National Science Foundation under grant ECS-8614328.

REFERENCES

[1] J. M. Andrews. Electrical conduction in implanted polycrystalline silicon. *Journal of Electronic Materials*, 8(3):227–247, 1979.

[2] S. F. Bart, T. L. Lober, R. T. Howe, J. H. Lang, and M. F. Schlecht. Design considerations for microfabricated electric actuators. *Sensors and Actuators*, 14(3):269–292, July 1988.

[3] L. S. Fan, Y. C. Tai, and R. S. Muller. Pin joints, gears, springs, cranks, and other novel micromechanical structures. In *Technical Digest, International Conference on Solid-State Sensors and Actuators*, pages 849–852, Tokyo, Japan, June 1987.

[4] H. Fujita and A. Omodaka. Electrostatic actuators for micromechatronics. In *IEEE Micro Robots and Teleoperators Workshop*, Hyannis, MA, 1987.

[5] J. H. Lang and S. F. Bart. Toward the design of successful electric micromotors. In *Technical Digest of the IEEE Solid-State Sensor and Actuator Workshop*, pages 127–130, Hilton Head Island, SC, June 6–9 1988.

[6] J. H. Lang, M. F. Schlecht, and R. T. Howe. Electric micromotors: electromechanical characteristics. In *Proceedings of the IEEE Workshop on Micro Robots and Teleoperators*, Hyannis, MA, November 9–11 1987. Reprinted in the *Proceedings of the ASME Annual Winter Meeting, DSC-6*, 403–410, Boston, MA, December 13–18, 1987.

[7] T. A. Lober and R. T. Howe. Surface-micromachining processes for electrostatic microactuator fabrication. In *Technical Digest of the IEEE Solid-State Sensor and Actuator Workshop*, pages 59–62, Hilton Head Island, SC, June 6–9 1988.

[8] N. C. C. Lu, L. Gerzberg, C. Y. Lu, and J. D. Meindl. Modeling and optimization of monolithic polycrystalline silicon resistors. *IEEE Trans. on Electron Devices*, ED-28(7):818–830, July 1981.

[9] J. R. Melcher. *Continuum Electromechanics*, chapter 2–5. MIT Press, Cambridge, Massachusetts, 1981.

[10] J. Y. W. Seto. The electrical properties of polycrystalline silicon films. *Journal of Applied Physics*, 46(12):5247–5254, 1975.

[11] Y. C. Tai, L. S. Fan, and R. S. Muller. IC processed micro-motors: design, technology, and testing. In *Technical Digest, IEEE–MEMS Workshop*, Salt Lake City, Utah, Feb. 1989.

[12] W. S. N. Trimmer and K. J. Gabriel. Design considerations for a practical electrostatic micro motor. *Sensors and Actuators*, 11(2):189–206, 1987.

[13] W. S. N. Trimmer, K. J. Gabriel, and R. Mahadevan. Silicon electrostatic motors. In *Technical Digest of the 4th International Conference on Solid-State Sensors and Actuators*, pages 857–860, Tokyo, Japan, June 1987.

[14] H. H. Woodson and J. R. Melcher. *Electromechanical Dynamics, Part 1*, chapter 3 and 4. John Wiley and Sons, Inc., New York, 1968.

(A) APPENDIX: IM MODELING

Following [9, Section 5.14], this appendix develops a model of the IM shown in Figure 1. Expressions are developed for the electric potential, field and free charge everywhere within this IM. Expressions are also developed for the motive torque and transverse force of electric origin acting on its rotor.

Rather than develop a three-dimensional model of the IM in Figure 1, a simpler model is developed here. To do so, the rotary IM is straightened into an equivalent linear IM. This is justified if its inner radius is much larger than its axial length multiplied by its azimuthal electrical periodicity. In this case, the radial, azimuthal and axial cylindrical coordinates of the original rotary IM, namely r, θ and z, become the Cartesian coordinates of the equivalent linear IM. In truth, θ becomes a normalized Cartesian coordinate; the actual coordinate is $r\theta$. Next, the potential, electric field and free charge are assumed to be uniform in the r direction. Thus, fringing electric fields are ignored, and the model becomes two-dimensional. This is justified if the difference between the inner and outer active IM radii is much larger than the IM axial length.

Modeling now procedes with reference to Figure 5. This figure shows the Cartesian coordinate system used for modeling, gives the axial lengths and electrical properties of the rotor and the two stator-rotor gaps, and labels the surfaces at the boundaries of the rotor and two stator-rotor gaps. Note that each ϵ is a permittivity and each σ is a conductivity.

(A.1) Potential, Electric Field And Free Charge

The driving sources of potential, electric field and free charge within the IM are the imposed stator potentials. These potentials are assumed to take the form

$$\Phi^u(\theta, t) = \Re \left\{ \hat{V}^u e^{j(\omega t - m\theta)} \right\} \quad (1)$$

$$\Phi^l(\theta, t) = \Re \left\{ \hat{V}^l e^{j(\omega t - m\theta)} \right\} \quad (2)$$

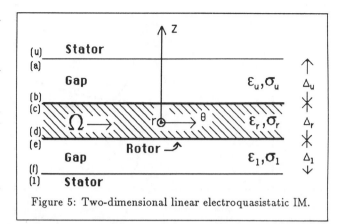

Figure 5: Two-dimensional linear electroquasistatic IM.

where Φ is potential, t is time, ω is temporal frequency, m is azimuthal periodicity, a superscript "$\wedge$" denotes a complex amplitude, and $\Re\{\cdot\}$ takes the real part of its argument. Note the use of superscripts to denote the surfaces labeled in Figure 5. Since all IM material is assumed to be electrically linear, the IM potential, electric field and free charge takes the same form as the potentials in (2). Thus, for constant z,

$$\Box(\theta, t) = \Re \left\{ \hat{\Box} e^{j(\omega t - m\theta)} \right\} \quad (3)$$

everywhere within the IM where $\Box$ denotes potential, an electric field component, or free surface charge.

Since the permittivity and conductivity of the rotor and two stator-rotor gaps are all uniform, free charge exists only at the surfaces which bound these regions. Thus, Laplace's equation governs the potential within the IM. For each of the three regions in Figure 5, Laplace's equation can be solved and the result can be used to express the electric field components at the region boundaries as functions of the potential at those boundaries [9, Section 2.16]. In the upper stator-rotor gap, rotor, and lower stator-rotor gap,

$$\begin{bmatrix} \hat{E}_z^a \\ \hat{E}_z^b \end{bmatrix} = \begin{bmatrix} -\gamma \coth(\gamma \Delta_u) & \frac{\gamma}{\sinh(\gamma \Delta_u)} \\ \frac{-\gamma}{\sinh(\gamma \Delta_u)} & \gamma \coth(\gamma \Delta_u) \end{bmatrix} \begin{bmatrix} \hat{\Phi}^a \\ \hat{\Phi}^b \end{bmatrix} \quad (4)$$

$$\begin{bmatrix} \hat{E}_z^c \\ \hat{E}_z^d \end{bmatrix} = \begin{bmatrix} -\gamma \coth(\gamma \Delta_r) & \frac{\gamma}{\sinh(\gamma \Delta_r)} \\ \frac{-\gamma}{\sinh(\gamma \Delta_r)} & \gamma \coth(\gamma \Delta_r) \end{bmatrix} \begin{bmatrix} \hat{\Phi}^c \\ \hat{\Phi}^d \end{bmatrix} \quad (5)$$

$$\begin{bmatrix} \hat{E}_z^e \\ \hat{E}_z^f \end{bmatrix} = \begin{bmatrix} -\gamma \coth(\gamma \Delta_l) & \frac{\gamma}{\sinh(\gamma \Delta_l)} \\ \frac{-\gamma}{\sinh(\gamma \Delta_l)} & \gamma \coth(\gamma \Delta_l) \end{bmatrix} \begin{bmatrix} \hat{\Phi}^e \\ \hat{\Phi}_f \end{bmatrix} \quad (6)$$

respectively, where

$$\gamma = \frac{m}{r} \quad (7)$$

and E is electric field. Similarly,

$$\hat{E}_\theta = j\gamma \hat{\Phi} \quad (8)$$

on all surfaces labeled in Figure 5.

The potential and electric field solutions above are spliced together with boundary conditions. First, potential is continuous across each boundary. Thus,

$$\hat{\Phi}^u = \hat{\Phi}^a \quad ; \quad \hat{\Phi}^b = \hat{\Phi}^c \quad ; \quad \hat{\Phi}^d = \hat{\Phi}^e \quad ; \quad \hat{\Phi}^f = \hat{\Phi}^l \quad . \quad (9)$$

From (8), this also states that tangential electric field is continuous across each boundary. Second, the discontinuity of normal displacement flux at a boundary must equal the free surface charge at that boundary. Thus,

$$\hat{\sigma}_f^{bc} = \epsilon_u \hat{E}_z^b - \epsilon_r \hat{E}_z^c \tag{10}$$

$$\hat{\sigma}_f^{de} = \epsilon_r \hat{E}_z^d - \epsilon_l \hat{E}_z^e \tag{11}$$

where σ_f is free electric surface charge. Third, charge conservation is imposed at the boundaries. Thus,

$$j(\omega - m\Omega)\hat{\sigma}_f^{bc} + \sigma_u \hat{E}_z^b - \sigma_r \hat{E}_z^c = 0 \tag{12}$$

$$j(\omega - m\Omega)\hat{\sigma}_f^{de} + \sigma_r \hat{E}_z^d - \sigma_l \hat{E}_z^e = 0 \tag{13}$$

where Ω is the rotational velocity of the rotor.

The above equations constitute a complete set of linear equations. Their solution completely describes the potential, electric field and free charge at all surfaces of the IM, and hence within the entire IM [9, Section 2.16]. These equations can be solved explicitly with reasonable ease, but this is not done here for the sake of brevity.

Equations (12) and (13) are particularly important because they are the only equations in which dynamics appear. All other equations are static. Correspondingly, two comments are in order. First, it is in these equations that the effects of more complex temporal dependencies of the stator excitation appear. In particular, the $j\omega$ terms arise from $\partial \sigma_f/\partial t$ terms. Thus, if V^u, V^l or ω become functions of time through control, then (12) and (13) become correspondingly more complex. Second it is in (12) and (13), and (10) and (11), that the electrical properties of the IM materials appear. The other equations involve only geometry, which is independent from dynamics. In certain situations (10) through (13) can offer considerable insight to IM operation. Consider, for example, the IM in Section 3 in which $\sigma_u = \sigma_l = 0$. In this case, substitution of (10) and (11) into (12) and (13) reveals that the charge relaxation dynamics are characterized by the term $j(\omega - m\Omega)\epsilon_r/\sigma_r$. This term is a normalized slip frequency, and it plays the same role as the slip in traditional induction interactions [14]. If this normalized slip is kept constant, for example, during manipulation of electrical frequency, rotor velocity and rotor conductivity, then IM performance is unchanged.

The above results assume a single-spatial-harmonic traveling potential wave excitation. A realistic micromotor, however, could not implement such an excitation because infinitely-fine individually-excited stator electrodes are impractical. Thus, the stator potential waves will include spatial harmonics of the fundamental, which have multiples of m as their azimuthal periodicity. In this case, since the IM is electrically linear, the resulting potential, electric field and free charge in the IM can be constructed as a superposition of their harmonic components, each derived individually as above, but with appropriately different V^u, V^l and m.

(A.2) Motive Rotor Torque

The motive torque of electric origin acting on the rotor, τ, is found by integrating the appropriate components of the electroquasistatic stress tensor [14], multiplied by the rotor radius, over the surface of the rotor. Here, the contributions to this integral from the r-directed inner and outer surfaces may be omitted because the stress tensor vanishes on them; the stress tensor vanishes there because E_r vanishes everywhere within the two-dimensional IM. Let $T_{\theta z}$ be the θ-directed component of the stress tensor on a z-directed surface. Then,

$$\tau = \int_{R_i}^{R_o} \int_0^{2\pi} (T_{\theta z}^b - T_{\theta z}^e) r^2 \, d\theta \, dr \tag{14}$$

where R_o and R_i are the outer and inner active IM radii,

$$T_{\theta z}^b = \epsilon_u E_\theta^b E_z^b \tag{15}$$

$$T_{\theta z}^e = \epsilon_l E_\theta^e E_z^e \tag{16}$$

and a superscript "$*$" denotes complex conjugation. Because of the averaging performed by the integration over θ in (14),

$$\tau = \pi \int_{R_i}^{R_o} \Re\{\epsilon_u \hat{E}_\theta^b \hat{E}_z^{b*} - \epsilon_l \hat{E}_\theta^e \hat{E}_z^{e*}\} r^2 \, dr \tag{17}$$

Finally, it is important to note that the integral in (17) is complicated because all electric quantities depend on γ which in turn depends on r. Numerical integration is usually required.

(A.3) Transverse Rotor Force

The transverse force of electric origin acting on the rotor, f, is also found by integrating the appropriate components of the electroquasistatic stress tensor over surface of the rotor. Here, the inner and outer surfaces may again be omitted because the stress tensor again vanishes on them; again the stress tensor vanishes there because E_r vanishes everywhere within the two-dimensional IM. Let T_{zz} be the z-directed component of the stress tensor on a z-directed surface. Then,

$$f = \int_{R_i}^{R_o} \int_0^{2\pi} (T_{zz}^b - T_{zz}^e) r \, d\theta \, dr \tag{18}$$

where

$$T_{zz}^b = \frac{\epsilon_u}{2}(E_z^b E_z^b - E_\theta^b E_\theta^b) \tag{19}$$

$$T_{zz}^e = \frac{\epsilon_l}{2}(E_z^e E_z^e - E_\theta^e E_\theta^e) \tag{20}$$

Again, because of the averaging performed by the integration over θ in (18),

$$f = \frac{\pi}{2} \int_{R_i}^{R_o} \Re\{\,\epsilon_u(\hat{E}_z^b \hat{E}_z^{b*} - \hat{E}_\theta^b \hat{E}_\theta^{b*}) \\ - \epsilon_l(\hat{E}_z^e \hat{E}_z^{e*} - \hat{E}_\theta^e \hat{E}_\theta^{e*})\} r \, dr \tag{21}$$

Again, it is important to note that the integral in (21) is complicated because all electric quantities depend on γ which in turn depends on r, and usually requires numerical integration.

A PERTURBATION METHOD FOR CALCULATING THE CAPACITANCE OF ELECTROSTATIC MOTORS

Suresh Kumar and Dan Cho
Department of Mechanical and Aerospace Engineering
Princeton University
Princeton, NJ 08544

ABSTRACT

Recent advances in micromachining technology have made possible the fabrication of a number of electrostatic devices, including variable capacitance micromotors. These variable capacitance micromotors employ one or more sets of conducting plates that act as capacitors. An accurate estimate of the capacitance between these sets of plates for various configurations is necessary to design and simulate such devices. Capacitance calculations can be performed using a variety of methods, including finite element techniques. An efficient method of calculating the capacitance is to reduce the problem to a set of integral equations that are then approximated by a set of algebraic equations. This method was previously used to calculate the capacitance of a linear micromotor with an infinite set of conducting lands on both the rotor and the stator, which introduces spatial periodicity that greatly simplifies the problem. Recent micromotor designs, however, employ only a few sets of conducting lands per phase, which makes it unclear if the previous results can be applied. In this paper, a new perturbation method for solving the integral equations and thereby estimating the capacitance is presented. This method does not make the assumption of infinite sets of conducting lands to introduce spatial periodicity. The perturbation method is computationally efficient and can be used for any design, including those that use very few repeated sets of conducting lands. The results obtained by this method are compared with the previous results which include experimental data. Finally, the method is extended to rotary micromotors and to three dimensional problems.

1. INTRODUCTION

Integrated-circuit processing and other micromachining techniques have been used to fabricate a number of tiny mechanical devices including microsensors and micromotors [1-3]. Variable capacitance micromotors are of the simplest and most efficient type of micromotors, and design considerations for these devices have been studied [4,5]. A variable capacitance micromotor consists of a rotor and a stator made of an electrically insulating material, on which many conducting lands are fabricated. The force needed to drive the motor is generated when an excitation voltage is applied to these lands. This force is proportional to the gradient of the capacitance between the plates [4]. Hence, it is necessary to obtain an accurate estimate of the capacitance and its gradient in order to design and study these devices. An iterative method for estimating the capacitance of a micromotor is discussed in this paper. This method is used to compute the capacitance for various configurations, and its accuracy and efficiency are shown.

2. PROBLEM STATEMENT

Figure 1a represents the configuration of the lands of a rotary micromotor, and Figure 1b represents that of a linear micromotor. Each of the devices has M lands per phase on both the rotor and the stator, and an electric potential of +V is applied to each of the M rotor lands, while an electric potential of -V is applied to each of the M stator lands. Further, it is assumed that the corresponding lands are offset by a distance X in the case of the linear micromotor, and by an angle Θ in the case of the rotary micromotor. The problem is to determine the total capacitance between the lands of a phase as a function of the offset of the lands.

This problem can be solved in the most general case using finite element methods. However, such a solution would involve determining the electric field at all points in the gap, as well as far away from the lands themselves, which makes the method quite inefficient.

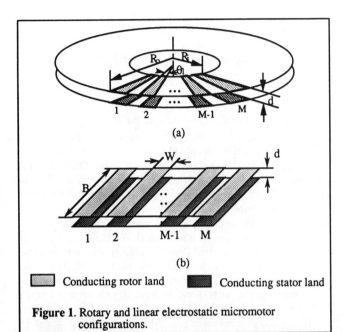

Figure 1. Rotary and linear electrostatic micromotor configurations.

The total capacitance between the lands is, by definition

$$C = \frac{\sum_{i=1}^{M} \int_{A_i} \rho_i \, dA_i}{2V} \qquad (1)$$

where ρ_i is the charge density distribution on the ith land, which has an area A_i. Thus one could find the capacitance if the charge distribution is known. The charge density at any point P on the lands is, in general, a function of both r and θ (or x and y). However, the charge density distribution along the radial (or the lateral) direction can be assumed to be constant if the conducting lands are long and slender. Refering to Figure 1, this condition can be expressed as

$$\theta_l (R_o - R_i) \gg \theta_l R_i \gg d \quad \text{(Rotary Micromotor)} \qquad (2\text{-a})$$

$$B \gg W \gg d \quad \text{(Linear Micromotor)} \qquad (2\text{-b})$$

Then, the problem becomes two dimensional, since the charge distribution needs to be determined only along one direction. The equivalent two dimensional problem is illustrated in Figure 2. This paper deals with the solution of the two dimensional problem first. The methods developed are then extended to the three dimensional case.

The potential at any point $P_1(x_1, y_1)$ due to an elemental line of charge is given by

$$\phi(x_1, y_1) = -\frac{\rho}{4\pi\varepsilon} \ln\left[(x_1 - x_2)^2 + (y_1 - y_2)^2\right] \qquad (3)$$

where ε is the dielectric constant of the medium surrounding the

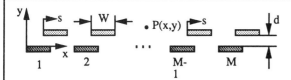

Figure 2. Two dimensional approximation of the electrostatic micromotor. x and y are the global coordinates. s is the local coordinate for each land.

lands, and ρ is the charge per unit length of the line of charge which passes through the x-y plane at $P_2(x_2,y_2)$ and is perpendicular to it. Using this, we can express the potential at any point $P(x,y)$ due to the conducting lands shown in Figure 2 as follows

$$\phi(x,y) = \sum_{k=1}^{M} \left\{ \int_0^W \frac{-\rho_{R_k}(s)}{4\pi\varepsilon} \ln\left([x-x_{R_k}(s)]^2 + [y-d]^2\right) ds \right.$$

$$\left. + \int_0^W \frac{-\rho_{S_k}(s)}{4\pi\varepsilon} \ln\left([x-x_{S_k}(s)]^2 + y^2\right) ds \right\} \quad (4)$$

where $\rho_{R_k}(s)$ is the charge distribution on the kth rotor land, and $\rho_{S_k}(s)$ is the charge distribution on the kth stator land, each expressed in the local coordinate, s. The coordinate values $x_{R_k}(s)$ and $x_{S_k}(s)$ are expressed with respect to the global coordinate system. The global coordinate values can be expressed in terms of the local coordinates using the following equation

$$x_{R_k}(s) = X - 2W(k-1) - s \quad \text{and} \quad x_{S_k}(s) = 2W(k-1) - s$$

$$\forall k \in [1,M] \quad \text{and} \quad \forall s \in [0,W] \quad (5)$$

The charge distribution is determined by approximating it as a piecewise constant function. Each land is divided into n segments, and the charge density over each segment is assumed constant. This is expressed in the following condition

$$\rho_{R_k}(s) = \rho_{R_{ki}} \quad \text{and} \quad \rho_{S_k}(s) = \rho_{S_{ki}},$$

$$\text{for } s \in \left(s_i - \frac{\Delta w_i}{2}, s_i + \frac{\Delta w_i}{2}\right), \quad i=1,2,\ldots,n \quad (6)$$

The center of each segment is located at s_i, and each segment is Δw_i in width. Using equation (6), equation (4) can be written as

$$\phi(x,y) = \sum_{k=1}^{M} \left\{ \sum_{i=1}^{n} \rho_{R_{ki}} \int_{-\Delta w_i}^{\Delta w_i} \frac{-1}{4\pi\varepsilon} \ln\left([x-x_{R_k}(s_i)-s]^2 + [y-d]^2\right) ds \right.$$

$$\left. + \sum_{i=1}^{n} \rho_{S_{ki}} \int_{-\Delta w_i}^{\Delta w_i} \frac{-1}{4\pi\varepsilon} \ln\left([x-x_{S_k}(s_i)-s]^2 + y^2\right) ds \right\} \quad (7)$$

Define "influence factors", $F_{R_{ki}}(x,y)$ and $F_{S_{ki}}(x,y)$, to denote the potential at $P(x,y)$ due to unit charge density on the ith segment of kth rotor and lands

$$F_{R_{ki}}(x,y) = \int_{-\Delta w_i}^{\Delta w_i} \frac{-1}{4\pi\varepsilon} \ln\left([x-x_{R_k}(s_i)-s]^2 + [y-d]^2\right) ds$$

$$F_{S_{ki}}(x,y) = \int_{-\Delta w_i}^{\Delta w_i} \frac{-1}{4\pi\varepsilon} \ln\left([x-x_{S_k}(s_i)-s]^2 + y^2\right) ds \quad (8)$$

Using equation (8), equation (7) can be written compactly as

$$\phi(x,y) = \sum_{k=1}^{M} \left\{ \sum_{i=1}^{n} \rho_{R_{ki}} F_{R_{ki}}(x,y) + \sum_{i=1}^{n} \rho_{S_{ki}} F_{S_{ki}}(x,y) \right\} \quad (9)$$

This simplification converts the problem of finding 2M unknown functions (cf: equation (4)) to that of finding 2nM unknown values (cf: equation (9)). Since the potential, $\phi(x,y)$, is known at each point on the lands, we can evaluate equation (9) at the centers of the n segments in each rotor-stator pair. This yields the following 2nM equations from which the 2nM unknown values, $\rho_{R_{ki}}, \rho_{S_{ki}}$, (i = 1..n, k = 1..M) can be solved

$$\phi(x_{R_{lj}},d) = \sum_{k=1}^{M} \left\{ \sum_{i=1}^{n} F_{R_{lj}R_{ki}} \rho_{R_{ki}} + \sum_{i=1}^{n} \rho_{R_{lj}S_{ki}} F_{S_{ki}} \right\} = +V$$

$$\phi(x_{S_{lj}},d) = \sum_{k=1}^{M} \left\{ \sum_{i=1}^{n} \rho_{R_{ki}} F_{S_{lj}R_{ki}} + \sum_{i=1}^{n} F_{S_{lj}S_{ki}} \rho_{S_{ki}} \right\} = -V,$$

$$l=1,2,\ldots,M, \quad j=1,2,\ldots,n \quad (10)$$

For notational simplicity, the above equations are written as

$$[F]_{2nM \times 2nM} \{\rho\}_{2nM} = \{V\}_{2nM} \quad (11)$$

Here, the vector of charge densities, $\{\rho\}_{2nM}$, is constructed from the n charge densities over the segments of each rotor-stator pair as shown below

$$\rho_{R_k} = \begin{Bmatrix} \rho_{R_{k1}} \\ \vdots \\ \rho_{R_{kn}} \end{Bmatrix}, \quad \rho_{S_k} = \begin{Bmatrix} \rho_{S_{k1}} \\ \vdots \\ \rho_{S_{kn}} \end{Bmatrix},$$

$$\rho_k = \begin{Bmatrix} \rho_{R_k} \\ \rho_{S_k} \end{Bmatrix}, \quad \{\rho\}_{2nM} = \begin{Bmatrix} \rho_1 \\ \vdots \\ \rho_M \end{Bmatrix} \quad (12)$$

Similarly, the vector of electric potentials is constructed form the potentials at each point. Defining the vector $\mathbf{v}_0$ to be

$$\{v_0\}_{2n} = \{+V \ldots +V, -V \ldots -V\}^T \quad (13\text{-a})$$

we can construct $\{V\}_{2nM}$ in the following way

$$\{V\}_{2nM} = \begin{Bmatrix} v_0 \\ v_0 \\ \vdots \\ v_0 \end{Bmatrix} \quad (13\text{-b})$$

Finally, the matrix $[F]_{2nM \times 2nM}$ is constructed from a number of sub-matricies, each of dimension 2n x 2n, as follows

$$[F]_{2nM \times 2nM} = \begin{bmatrix} F_{11} & F_{12} & F_{13} & \cdots & F_{1M} \\ F_{21} & F_{22} & F_{23} & \cdots & F_{2M} \\ \vdots & \vdots & \vdots & \cdots & \vdots \\ F_{M1} & F_{M2} & F_{M3} & \cdots & F_{MM} \end{bmatrix} \quad (14)$$

The submatrix $[F_{kl}]_{2n \times 2n}$ represents the effect of the charge distribution due to the lth pair of lands on the potential of the kth pair. Thus, each submatrix can be partitioned into rotor-stator effects as follows

$$[F_{kl}]_{2n \times 2n} = \begin{bmatrix} F_{R_k R_l} & F_{R_k S_l} \\ F_{S_k R_l} & F_{S_k S_l} \end{bmatrix} \quad (15)$$

The coefficients of each submatrix is determined from equation (8), using the coordinates of the centers of the segments of the ith rotor-stator pair in the place of (x,y)

$$F_{R_k R_l}(i,j) = \int_{-\Delta w_j}^{\Delta w_j} \frac{1}{4\pi\varepsilon} \ln\left[(x_{R_{ki}} - x_{R_{lj}} - s)^2\right] ds$$

$$F_{R_k S_l}(i,j) = \int_{-\Delta w_j}^{\Delta w_j} \frac{1}{4\pi\varepsilon} \ln\left[(x_{R_{ki}} - x_{S_{lj}} - s)^2 + d^2\right] ds$$

$$F_{S_k R_l}(i,j) = \int_{-\Delta w_j}^{\Delta w_j} \frac{1}{4\pi\varepsilon} \ln\left[(x_{S_{ki}} - x_{R_{lj}} - s)^2 + d^2\right] ds$$

$$F_{S_k S_l}(i,j) = \int_{-\Delta w_j}^{\Delta w_j} \frac{1}{4\pi\varepsilon} \ln\left[(x_{S_{ki}} - x_{S_{lj}} - s)^2\right] ds$$

$$\forall i \in [1,n], \forall j \in [1,n], \forall k \in [1,M], \forall l \in [1,M] \quad (16)$$

where $x_{R_{ki}}, x_{R_{lj}}, x_{S_{ki}}$ and $x_{S_{lj}}$ are the coordinates of the centers of the segments expressed in the global coordinate system.

The above set of equations contain $2nM$ unknowns, which can become very large even for a moderate number of lands. Direct solution of these equations, using Gaussian elimination, would require nearly $(2nM)^3$ floating point operations [6], which may not be practical. For example, calculating the capacitance of a design having 7 sets of lands per phase by dividing each land into 15 segments will involve over 3×10^6 floating point multiplications. The number of unknowns can be reduced to half by considering the anti symmetry of the problem.

$$\rho_{R_{ki}} = -\rho_{S_{(M-k+1)(n-i+1)}} \quad (17)$$

However, even this does not reduce the complexity of the problem significantly. Hence, an approximate method which requires less computing is sought.

The problem can be greatly simplified if the number of lands are very large. As the number of lands tends to infinity, spatial periodicity is introduced in the problem, making the charge distribution on each land identical. Thus, the number of unknowns to be solved reduces dramatically to $2n$ (or n, taking equation (17) into account). This approach has been used by Mahadevan [7] to compute the capacitance of an electrostatic micromotor with a very large number of lands per phase. However, the number of pairs of lands required to validate the assumption of identical charge distributions for each land is not clear from such a study. Many recent micromotor designs[2,3] use only a few lands, and it is not clear if the results could be extended to such designs.

In this paper, a new technique for solving the complete set of equations for an arbitrary number of lands is presented. The method is used to compute the capacitance for a number of cases, and the accuracy as well as the efficiency of the method are shown.

3. PERTURBATION METHOD

To solve the charge density for the general case of M pairs of lands, consider equation (11). As we have already remarked, the number of unknowns is quite large, and hence a direct solution through Gaussian elimination is not desirable. However, a straightforward application of popular iterative methods such as the Gauss-Seidel method to this problem is not possible, since the matrix F is neither sparse nor diagonally dominant [8]. Moreover, F is not symmetric positive definite, when the segments of each land are not chosen to be of equal width. If the land segments are chosen proportionally smaller at the edges, so as to decrease n required to obtain the same level of accuracy, we see that

$$F(i,j) \neq F(j,i) \text{ for } i \neq j, \forall i \in [1,2nM], \forall j \in [1,2nM] \quad (18)$$

Thus, a direct application of many iterative schemes designed to handle non-diagonally dominant matrices are not guaranteed to converge [8]. Hence, it becomes necessary to use some physical insight of the problem to approximate it for an easier solution.

First, it can be seen that of the M^2 submatricies of F, there are only $2M-1$ distinct ones. Specifically,

$$F_{kl} = F_{k-l} \quad \forall k \in [1,M], \forall l \in [1,M] \quad (19)$$

Hence, the set of equations can be written as

$$\begin{bmatrix} F_0 & F_1 & F_2 & \cdots & F_{M-1} \\ F_{-1} & F_0 & F_1 & \cdots & F_{M-2} \\ \vdots & \vdots & \vdots & \cdots & \vdots \\ F_{1-M} & F_{2-M} & F_{3-M} & \cdots & F_0 \end{bmatrix} \begin{Bmatrix} \rho_1 \\ \rho_2 \\ \vdots \\ \rho_M \end{Bmatrix} = \begin{Bmatrix} v_0 \\ v_0 \\ \vdots \\ v_0 \end{Bmatrix} \quad (20)$$

where ρ_k and v_0 are defined in equations (12) and (13).

Equation (20) is solved by first assuming a charge distribution based on some knowledge of the problem, and then using an iterative scheme to improve on it. We assume that the charge on each pair is determined primarily by the presence (or absence) of its immediate neighbors. Thus, we assume that all interior pairs have the same charge distribution, which is assumed to be equal to the charge distribution on the middle pair for the case when M=3. The end pairs, in turn, are assumed to have an initial charge distribution based on the end pairs for the case when M=3. Thus, if $^M\rho_k$ denotes the charge distribution on the kth pair when the total number of pairs is M, the initial charge distribution is assumed to be

$$^M\rho_k = \begin{cases} ^3\rho_1 & k = 1 \\ ^3\rho_2 & 2 \leq k \leq M-1 \\ ^3\rho_3 & k = M \end{cases} \quad (21)$$

Given a charge distribution, $\rho_1^p, \ldots, \rho_M^p$, obtained after p iterations, we can obtain a better estimate of the distribution, $\rho_1^{p+1}, \ldots, \rho_M^{p+1}$, based on the previous iteration. This is achieved by first calculating a residual vector (defined below) for each of the M pairs

$$r_k^p = v_0 - \sum_{i=1}^{M} F_{i-k} \rho_i^p \quad k = 1,2,\ldots,M \quad (22)$$

Defining an error vector for each pair as

$$e_k^p = \rho_k - \rho_k^p \quad k = 1,2,\ldots,M \quad (23)$$

where ρ_k is the true charge distribution of the kth pair, equation (20) can be written in terms of the residual and error vectors

$$\begin{bmatrix} F_0 & F_1 & F_2 & \cdots & F_{M-1} \\ F_{-1} & F_0 & F_1 & \cdots & F_{M-2} \\ \vdots & \vdots & \vdots & \vdots & \vdots \\ F_{1-M} & F_{2-M} & F_{3-M} & \cdots & F_0 \end{bmatrix} \begin{Bmatrix} e_1^p \\ e_2^p \\ \vdots \\ e_M^p \end{Bmatrix} = \begin{Bmatrix} r_1^p \\ r_2^p \\ \vdots \\ r_M^p \end{Bmatrix} \quad (24)$$

Consider the kth pair

$$\sum_{l=1}^{M} F_{l-k} e_l^p = r_k^p \quad (25)$$

The residual on the kth pair is built up by the contributions of the errors from each of the M pairs. Thus, the contribution of the error in the kth pair to its own residual can be written as a fraction, κ, of the total value

$$F_0 e_k^p = \kappa r_k^p \quad 0 < \kappa < 1, \quad k \in [1,M] \quad (26)$$

As a conservative estimate, the contribution of each pair to the residual is assumed to be the same. This gives

$$F_0 e_l^p \approx F_{l-k} e_l^p \quad l = 1,2,..M \quad k \in [1,M] \quad (27)$$

Thus, we can get an estimate of κ

$$\kappa = \frac{1}{M} \quad (28)$$

This is then used to get the next iteration

$$\rho_k^{p+1} = \rho_k^p + \frac{1}{M} F_0^{-1} r_k^p, \quad k = 1,2,\ldots,M \quad (29)$$

The process can be stopped after P iterations, when the charge density is sufficiently accurate. A suitable test can be based on an estimate of the residual.

$$\max_{k=1,\ldots,M} \| r_k^P \|_\infty < \varepsilon \quad (30)$$

where ε is a desired accuracy.

Equations (21-30) describe the perturbation method used to calculate the charge distribution on the conducting lands of an electrostatic motor without inverting equation (20) directly. This method converges linearly when the initial approximation is sufficiently accurate. The rate of convergence, or the convergence factor η, can be defined as

$$\eta = \lim_{p \to \infty} \eta_p = \lim_{p \to \infty} \frac{C - C_p}{C - C_{p-1}} \quad (31)$$

where C is actual value of the capacitance, and C_p is the estimated value after p iterations. The right hand side of equation (29) will approach a constant value when the estimate is sufficiently accurate, and this can be used to accelerate the convergence by providing an improved estimate C_e

$$C_e = \frac{C_{p+1} - \eta_p C_p}{1 - \eta_p} \quad (32)$$

The perturbation technique involves the inversion of only one n by n matrix and this is far more efficient than the direct solution of equation (20). The initial charge distribution can be found either by solving equation (20) for the case when M=3 or for a single pair of lands. The former is a better approximation, requiring fewer iterations, but requires more operations for determining the initial approximation. The number of floating point multiplications needed to determine the charge distribution in the former case is $n^3+PM(M+10)n^2$, while the number of floating point operations needed when Gaussian elimination is used is $M^3n^3/3$. Thus, the perturbation method is more efficient when the total number of iterations, p, satisfies the following condition

$$p < \frac{(M^3-10)\,n}{3M(M+1)} \quad (33)$$

It will be shown later that the above condition is easily satisfied even when the desired accuracy is better then 1%.

The perturbation method can be easily extended to solve three dimensional problems. Consider a rotary micromotor that does not satisfy condition(1). This will be the case, for example, when the lands are etched so as to cover most of the surface area of the rotor. In such cases, it may not be correct to assume that all the lands have the same charge distribution. The perturbation technique, however, does not make that assumption, and it is easily applied to the problem at hand.

Figure 3 shows the conductor configuration for a rotary micromotor. Each land is divided into segments, both along the radial and circumferential directions, and it is assumed that the charge density is constant over the area of each segment. The segments are numbered as shown, and the center of the ith segment of the kth rotor and stator lands are assigned the coordinate values $(x_{R_{ki}}, y_{R_{ki}}, z_{R_{ki}})$ and $(x_{R_{ki}}, y_{R_{ki}}, z_{R_{ki}})$ respectively. The set of equations to be solved can then be constructed similar to the two dimensional case. The individual coefficients of the F matrix change for this problem, and equation (15) can be extended for the three dimensional case as

$$F_{R_k R_l}(i,j) = \frac{1}{4\pi\varepsilon} \int_{\Delta A_i} \frac{1}{\sqrt{(x_{R_{ki}} - x_{R_{lj}} - s_x)^2 + (y_{R_{ki}} - y_{R_{lj}} - s_y)^2}} dA$$

$$F_{R_k S_l}(i,j) = \frac{1}{4\pi\varepsilon} \int_{\Delta A_i} \frac{1}{\sqrt{(x_{R_{ki}} - x_{S_{lj}} - s_x)^2 + (y_{R_{ki}} - y_{S_{lj}} - s_y)^2 + d^2}} dA$$

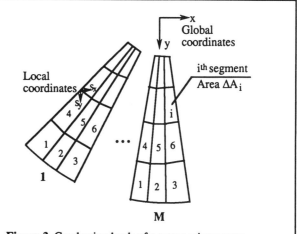

Figure 3. Conducting lands of a rotary micromotor. Each land is divided into n segments.

$$F_{S_k R_l}(i,j) = \frac{1}{4\pi\varepsilon} \int_{\Delta A_i} \frac{1}{\sqrt{(x_{S_{ki}} - x_{R_{lj}} - s_x)^2 + (y_{S_{ki}} - y_{R_{lj}} - s_y)^2 + d^2}} dA$$

$$F_{S_k S_l}(i,j) = \frac{1}{4\pi\varepsilon} \int_{\Delta A_i} \frac{1}{\sqrt{(x_{S_{ki}} - x_{S_{lj}} - s_x)^2 + (y_{S_{ki}} - y_{S_{lj}} - s_y)^2}} dA$$

$$\forall\, i \in [1,n],\ \forall\, j \in [1,n],\ \forall\, k \in [1,M]\ \text{and}\ \forall\, l \in [1,M] \quad (34)$$

where s_x and s_y are the local coordinates for each land. Equation (34) is used to generate the coefficients, and the perturbation technique described in equations (21-30) can be then applied.

4. RESULTS

The perturbation method was used to determine the capacitance for a number of cases, involving various ratios of rotor-stator gap to land width and for various number of lands per phase. This section describes the results.

An electrostatic motor with a single pair of lands were studied first. Equation (20) was solved for various conductor geometries, and by dividing each land into segments of both equal and unequal width. Figure 4 shows the charge density distribution for a typical case. The conductor geometry considered was $d/W=0.1$. Each lands was divided into 15 equal segments, and equation (20) was solved for various values of rotor offset. It can be seen that the charge density displays a singularity at the edges of the lands, in accordance with the implicit assumption of zero land thickness. The charge density distribution gets skewed as the rotor offset increases, with a higher charge density over the overlapping regions. This shape of the distribution can be exploited to get better accuracy for the same number of segments.

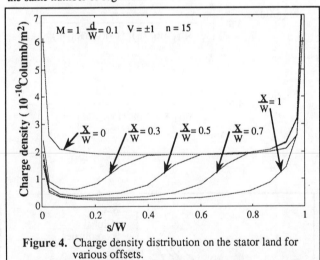

Figure 4. Charge density distribution on the stator land for various offsets.

Figure 5 shows the variation of the relative error in the capacitance as the number of segments is increased. The dotted lines show the errors for the case when the each land is divided into equal segments. The solid lines show the errors when the segments are chosen to be wider near the center and smaller near the edges (where the charge density changes rapidly), in accordance with equation (33).

$$\Delta w_i = \frac{W}{2}\left\{\sin\left[\frac{i\pi}{n}\right] - \sin\left[\frac{(i+1)\pi}{n}\right]\right\} \quad (35)$$

It can be seen that dividing the land segments unequally results not only in a faster rate of convergence, but the error in estimating the capacitance is smaller for the same number of segments compared to the case when segments of equal sizes are used. Figure 6 shows the

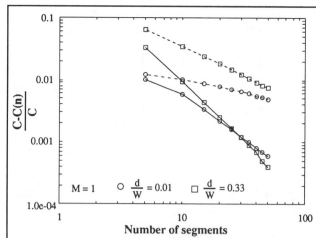

Figure 5. Errors in estimating the capacitance for one pair of lands. The offset between the rotor and the stator land is zero.

convergence of the capacitance estimates as a function of the number of segments for various gap separations. The land width in each case is unity, and the rotor offset is zero. Each land is divided into segments of unequal size, using equation (35). It can be seen that dividing each land into 15 segments is sufficient to get an accuracy of 99%. Based on these results, the number of segments, n was chosen to be 15, with the width of each segment determined by equation (35).

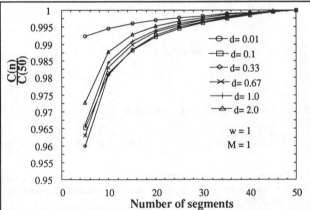

Figure 6. Convergence of the capacitance estimates for one pair of lands. The offset between the rotor and stator land is zero.

The perturbation method was studied in detail for a conductor geometry of $d/W=0.1$, for various rotor offsets, and for $M=1$ to $M=10$. Figure 7 shows the accuracy of the initial approximation of the charge distribution, as given by equation (21). The case of full rotor offset ($X=W$) was considered, since this is the most critical case. The solid line denotes the exact charge distribution for 5 pairs of lands, obtained by solving equation (20) using Gaussian elimination. Circles indicate the initial approximation, determined the solution of the charge distribution of the case when $M=3$ which was obtained using equation (20). It can be seen that the initial approximation is itself an excellent one, and the error in the charge distribution is very small.

Figure 8 shows the convergence of the capacitance values for various cases. The convergence factor η_p calculated using equation (31) is plotted for each iteration. The actual values of the capacitance for each case were determined by the direct solution of equation (20). The convergence factor reaches a constant value very quickly, indicating the accuracy of the initial estimate. Figure 9 shows the relative error in the capacitance for each iteration, expressed as a percentage. The error is less than 1% in most cases after 5 iterations. Note as a comparison, that for a case with $M=7$, the number of iterations could be as large as 30 before the direct

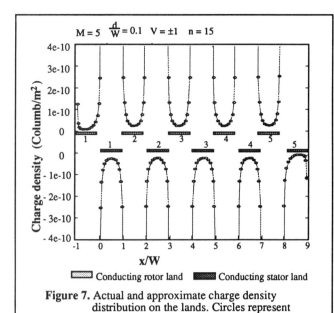

Figure 7. Actual and approximate charge density distribution on the lands. Circles represent the values obtained by the initial approximation.

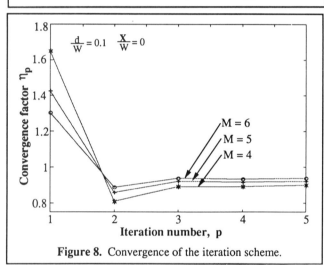

Figure 8. Convergence of the iteration scheme.

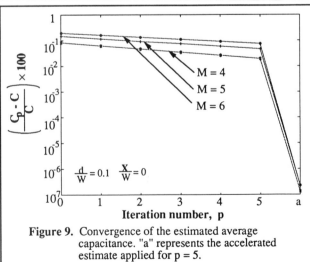

Figure 9. Convergence of the estimated average capacitance. "a" represents the accelerated estimate applied for p = 5.

solution becomes as efficient as the perturbation method. The Figure also shows the dramatic reduction in the error when the acceleration scheme given by equation (32) was applied after 5 iterations.

Figure 10 shows the variation in the capacitance with the rotor offset for M=1 to M=10. Each of the estimates were obtained by using the perturbation method for 5 iterations, and using equation (32) after that. The greatest variation in capacitance, as the number of pairs are increased, results when the rotor is completely offset ($X = \pm W$). It can also be seen that the capacitance converges to a constant value (depending on the rotor offset) when the number of pairs M=10. Figure 11 shows this trend for the case when the rotor is fully offset. The capacitance value for $M \geq 10$ is nearly identical to the case when the number of lands is infinite. The capacitance value when the number of lands is infinite was calculated using the method described by Mahadevan [7].

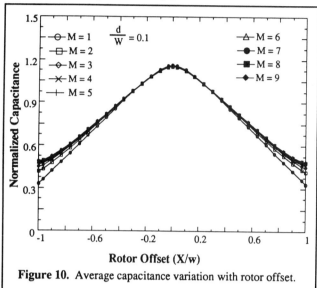

Figure 10. Average capacitance variation with rotor offset.

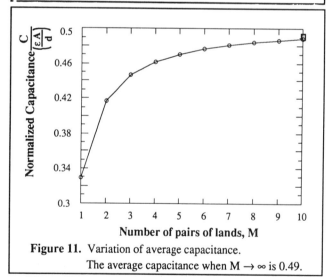

Figure 11. Variation of average capacitance.
The average capacitance when $M \to \infty$ is 0.49.

Figure 12 compares the results obtained by the perturbation method with experimental values given in Mahadevan [7]. Ten pairs of lands, each 6.0 inches (152.4 mm) long, 0.25 inches (6.35 mm) wide with 0.5 inches (12.7 mm) pitch, and a mylar sheet with relative dielectric constant of 1.8 was used in Mahadevan's experiments. The experiment considered many separation gap distances, and results for 0.007inches (0.18mm), 0.018inches (0.46mm), 0.025inches (0.64mm) are shown. The plots compare the experimental results with those calculated by Mahadevan and those obtained using the perturbation method. The plots show a good agreement between the values predicted by the two methods. The plots also show good agreement between experimental and predicted values, if the constant offset between the curves (attributed to stray capacitances in the measurement setup [7]) is neglected. Note that the method used by Mahadevan [7], which assumes an infinite number of lands, is quite accurate if there are at least 10

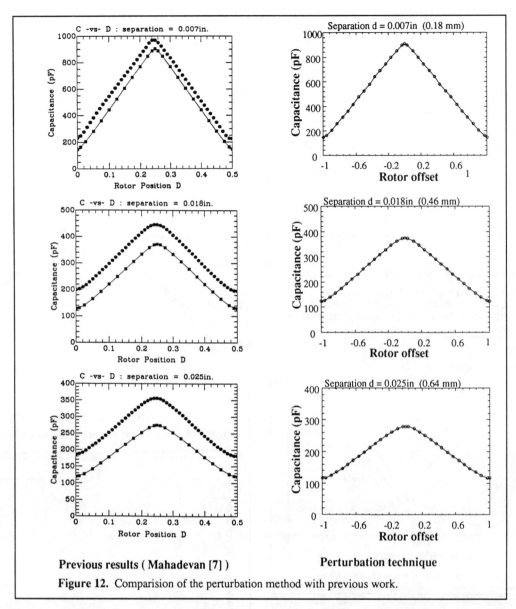

Figure 12. Comparision of the perturbation method with previous work.

lands per phase, and when the computational problem can be reduced to that of a two dimensional case.

5. CONCLUSIONS

An efficient method for calculating the capacitance of electrostatic micromotors is developed in this paper. The method is computationally very efficient compared to the direct solution, employing Gaussian elimination, and the estimated values converge after only a few iterations. The results of the method was compared to the available experimental data, which showed an excellent agreement. The perturbation method can be applied to micromotor designs with an arbitrary number of lands and to three dimensional problems.

ACKNOWLEDGEMENT

This work was supported by an IBM Manufacturing Research Fellowship and a grant from the NJ Advanced Technology Center in Surface Processed Materials.

REFERENCES

[1] T. A. Lober and R. T. Howe, "Surface-Micromachining Processes for Electrostatic Microactuator Fabrication", *Technical Digest of the IEEE Solid State Sensor and Actuator Workshop*, Hilton Head Island, July 1988.

[2] M. Mehregany, et al., "A Study of Three Microfabricated Variable Capacitance Motors", *Proc. of the 5th International Conference on Solid-State Sensors and Actuators and Eurosensors III*, Montreau, Switzerland, June 1989.

[3] Y. Tai, L. Fang, and R. Muller, "IC-Processed Micro-Motors: Design, Technology and Testing", *Proc. IEEE Micro Electro Mechanical Systems Workshop*, Salt Lake City, Feb. 1989.

[4] W. S. N. Trimmer, and K. J. Gabriel, "Design Considerations for a Practical Electrostatic Micro-Motor", *Sensors and Actuators*, 11, 1987.

[5] J. H. Lang, M. F. Scheldt, and R. H. Howe, "Electric Micromotors: Electromechanical Characteristics", *Proc. IEEE Micro Robots and Tele Operators Workshop*, Hyannis, Nov. 1987.

[6] K. E. Atkinson, *An Introduction to Numerical Analysis*, John Wiley and Sons, New York, 1978.

[7] R. Mahadevan, "Capacitance Calculations for a Single Stator, Single Rotor Electrostatic Motor", *Proc. of the IEEE Micro Robots and Tele Operators Workshop*, Hyannis, Nov. 1987.

[8] L. A. Hageman, and D. M. Young, *Applied Iterative Methods, Academic Press*, New York, 1981.

Section 5

Magnetic Actuators

THERE is a controversy over whether magnetic actuators or electrostatic actuators are more suitable for the micro domain. Actually, the suitability of an actuator depends greatly on the application. Applications differ on the amount of power available, suitable voltages and currents, temperature requirements, size constraints, and so on. Can one type of actuation satisfy all these requirements?

Magnetic actuators, in contrast to electrostatic actuators, tend to use lower voltages, more power, and are sensitive to the magnetic properties of the materials used. Small electromagnets have difficulty generating large magnetic fields. Often, the inclusion of a permanent magnet in the system greatly increases the overall magnetic field and the force generated by the actuator. Properly designed, magnetic actuators can make strong actuators with a large holding torque. In developing micromechanical systems, the designer is well advised to consider magnetic actuators.

The article "Micro Robots and Micromechanical Systems," in Section 1, describes how forces generated by electromagnets scale as systems become smaller. If the current density (current per unit area) remains the same in the wire windings, the magnetic forces decrease rapidly for small electromagnetic systems. However, because large amounts of heat can be removed from a small volume without causing large temperature differences, one can aggressively increase the current density through the windings, and still keep the motor cool. Increasing the current density makes electromagnetic forces competitive with other micro actuators.

Increasing the current density raises other considerations. Increasing the current density greatly increases the resistive losses in the coils, and causes low efficiencies. This additional power loss matters little in many applications, because the actual power used by the micro actuator is insignificant compared to the power needed for the logic and signal conditioning electronics. If, however, many small magnetic actuators are close to each other, each actuator heats its neighbor, and it is harder to keep the system cool. Electromigration also limits the maximum current density.

In the first paper, "Magnetically Levitated Micro-Machines," the energetics of magnetic actuators are considered, and a magnetic manipulator proposed. The second paper, "Fabrication and Testing of a Micro Superconducting Actuator Using the Meissner Effect," uses a superconductor to support a permanent magnet. Driving sections of the superconductor normally causes the permanent magnet to move. In the third paper, "Room Temperature, Open-Loop Levitation of Microdevices Using Diamagnetic Materials," a diamagnetic material is suspended and moved by a magnetic field.

MAGNETICALLY LEVITATED MICRO-MACHINES

Ron Pelrine and Ilene Busch-Vishniac
Dept. of Mechanical Engineering
University of Texas at Austin
Austin, Texas 78712
(512) 471-3038

Abstract

The use of magnetic levitation (maglev) as a microdrive has been studied. A wide variety of micromechanical systems are possible with maglev technology. Advantages for factory automation include very light and small manipulators, integration of transport with precision processes, and high tolerance for adverse environments. A prototype system using air core electromagnets driving a permanent magnet manipulator has been built. Preliminary results include 2 degrees of freedom at ±1.5 μm relative accuracy and 15 moves per second.

Introduction

Magnetic levitation (maglev) can be used to drive micro-manipulators with negligible friction. A schematic of a magnetically driven micro-robotic system is shown in Fig. 1. These micro-robotic systems may offer performance advantages for research tools, such as microinjectors for biology and scanning tunneling microscopes. However, the greatest potential of maglev micro-machines lies in the area of manufacturing systems. The emphasis of our current work is on maglev micromechanical systems suitable for factory automation.

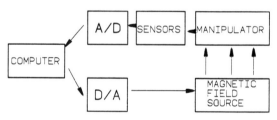

FIG. 1: SCHEMATIC OF MAGLEV SYSTEM

For the purpose of this paper, factory automation can be divided into 3 main types of operations: material and tool transport, mechanical processing (such as machining or alignment processes), and non-mechanical processing. For transport, maglev allows large motion ranges. This can be accomplished, for example, using a track of drive coils, so that the transport parts of the system are similar to DC linear motors.[1] For mechanical processing, maglev can reduce or eliminate friction. Without friction, precision manipulators can be made very small and light, reducing vibration, wear, and particulate generation while increasing delicacy. Note that there are no a priori tradeoffs between the manipulator size and the size of the precision workspace, as is the case for piezoelectric drive.[2] In non-mechanical processing, maglev manipulators offer robustness in adverse chemical, nuclear, and temperature environments; part of the reason for this is that the manipulators are not articulated, i.e. they are single rigid bodies, so a wide variety of protective coatings can be applied.

Near-term applications of maglev micro-robotics are primarily in the semiconductor industry. Examples include probing individual components in microchips, assembling hybrid chips, and micro-machining. There are also potentially a number of long-term applications including chip repair, integration of micro-scale components with incompatible processing sequences, assembly of 3-D microcircuits, and design of high speed mineral and cell sorters.

System Fundamentals

The energy, E, of a small magnetic dipole, u, in a magnetic field, B, is given by:

$$E = -(u \cdot B) \qquad (1)$$

There are a variety of ways to generate both u and B, leading to a corresponding variety of magnetically levitated micro-mechanical systems. In general there will be a complex relationship between u and B, possibly involving hysteretic, geometric, and nonlinear material properties. Fortunately there are a number of cases where u is essentially independent of, or simply related to, B. For example, with systems using strong permanent magnets or air core electromagnets to generate the dipole moment(s), u is independent of B. In this case it is straightforward using Lagrangian mechanics and Eq. (1) to derive the force F and the torque T exerted on the dipole by the field as:

$$F = (\nabla B) \cdot u \qquad (2)$$
$$T = u \times B \qquad (3)$$

When the dipole is induced by the magnetic field, as, for example, with bulk superconductors or soft ferromagnets, the force and torque laws will generally be more complex.

The question of the degree of open vs. closed loop control appears to be an important issue for maglev micromechanics. Levitation which requires a great deal of feedback is usually much more expensive and difficult technically than one which does not, because most of the cost and complexity of a many degree of freedom micromanipulator is in the controlling electronics, not the mechanics. It is therefore desirable to have a significant amount of open loop control, even if some feedback is required for precision. However, Earnshaw's Theorem states that there are no stable points for a dipole acted upon only by open loop magnetic forces unless the material is diamagnetic (i.e. the dipole moment u is induced by the magnetic field B in the opposite direction).[3] This implies that, with the exception of diamagnetics, one needs to use non-magnetic forces if the system is to have open loop stability. Using diamagnetic materials, such as superconductors or normal conductors with induced eddy currents, stability and possibly good precision can be achieved open loop in free space without the use of sensors.[4]

These observations motivate the introduction of non-magnetic forces into systems which do not have open loop stability using only magnetic forces. Consider, for example, our prototype system in which the manipulator uses a set of rigidly connected permanent magnets, while the B field is generated by air core electromagnets. By

Earnshaw's Theorem, this system is unstable using only open loop magnetic forces. Open loop stability can be achieved, however, by letting the manipulator interact with solid surfaces, liquid surfaces using buoyancy or surface tension, or possibly other environments. Note that in such a system friction may still be kept negligible. This is true even using a solid surface, where the actual motion of the manipulator off of the surface might resemble the hopping of an insect.

The various operating environments mentioned above will lead to systems with different properties. When one considers the various ways to generate **u** and **B**, and the possible operating environments, one can see that magnetically levitated micro-manipulators form a very broad class of machines. This is particularly true in light of the various system geometries and many orders of magnitude in size that are possible. Hence, micro-levitation has fewer restrictions because of material or power costs than alternative macro-scale machines.

Design Theory

Because of the large number of possible maglev micro-mechanical systems, it is useful to examine how the design goals influence the choice of system and system parameters. Our experimental work has concentrated on systems with precision workspace. For this paper, we define a precision micro-robot as a computer controlled, multi-degree of freedom micro-manipulator whose accuracy is much smaller than any of the system component dimensions. Note that the definition makes no statement about absolute accuracy, only its relation to system components. The size of the workspace is not a good indication of precision for maglev micro-mechanics since a system with smaller component dimensions may actually have a larger workspace by using more electromagnets. Thus, for a given absolute accuracy and range, the designer theoretically has the choice between precision systems made with large components and "coarse" systems made with small components. Other factors, such as payload or delicacy requirements, may limit this choice. Our present experimental work has concentrated on using precision micro-robots to get high accuracy because it was felt to be a more easily implemented design. However, where appropriate, the coarse system may offer distinct advantages, such as more delicacy with a greater degree of open loop control. Recently some possible designs have emerged that could use photolithography to lay down macroscopic areas of micro-scale electromagnets. (See the discussion of single wire drive below.) Hence highly accurate coarse systems may be an interesting direction for future research.

The scale of the system components also plays a major role in the power requirements for levitation. The downward scaling properties of magnetic levitation are favorable enough to permit technical solutions which are totally impractical on large scales. For example, the use of air core electromagnets is quite reasonable in micro-levitation, but not on larger scales because they require considerably more power than iron core electromagnets. However, since air core electromagnets are linear, have no hysteresis, and have no iron core to pick up unwanted magnetization from other external fields, the control advantages can outweigh the power disadvantages on a micro scale.

To examine the scaling laws, consider Maxwell's Equations:

$$\nabla \cdot \mathbf{E} = 4\pi q \quad \text{(c.g.s. units)} \quad (4)$$

$$\nabla \cdot \mathbf{B} = 0 \quad (5)$$

$$\nabla \times \mathbf{E} = -(1/c)\partial \mathbf{B}/\partial t \quad (6)$$

$$\nabla \times \mathbf{B} = (4\pi/c)\mathbf{J} + (1/c)\partial \mathbf{E}/\partial t \quad (7)$$

where **E** is the electric field, **J** is the current density, q is the charge density, and c is the velocity of light. Equation (7) implies that as the system component size, r, is uniformly reduced, the gradients of **B** scale as the current density **J** in the driving electromagnets. For the case where **u** is independent of **B** and proportional to the manipulator mass, as with permanent magnet manipulators, Eq. (2) can be used to conclude that the force per unit mass, **f**, scales as **J** also. At this point one is free to prescribe the form of either **J** or **f**. The simplest requirement is that the magnitude of **f** be scale invariant, so that the magnetic forces are always some fixed multiple of the manipulator's weight. With this assumption, **J** must also be scale invariant. Since the steady state power loss, P, in the electromagnet is equal to the volume integral of $|\mathbf{J}|^2$ times the resistivity of the windings of the electromagnet, it follows that P scales as the volume, or $P \propto r^3$.

The cubic power law makes it clear why levitation is easier on small scales: the power necessary for levitation falls rapidly with decreasing r. There are thermal benefits implied by the cubic law as well. The heat that can be dissipated from an electromagnet typically scales as $r^n(T_C-T_R)$, where T_C is the core temperature of the electromagnet and T_R is room temperature. The exponent n is equal to 1 for a heat transfer mode limited by conduction, and equal to 2 for a mode limited by convection with a scale invariant heat transfer coefficient. (Generally the heat transfer coefficient will increase with decreasing r, so n=2 is a conservative estimate.) It follows from this model that with a cubic scaling law for power, the temperature difference, (T_C-T_R), must decrease with decreasing r, scaling as r^{3-n}. A low core temperature in the electromagnet is advantageous both for greater reliability and reduced accuracy problems associated with thermal expansion.

The requirement that **f** be scale invariant, which led to the cubic scaling law for power, is the most straightforward restriction that can be applied, but other restrictions are possible. One natural choice is that P scale as r^n, where n is defined as above. This is equivalent to saying that the power driving the electromagnets is limited only by the requirement that the core temperature be held at some safe operating value. In this case one gets a scaling law for **f** as follows: $\mathbf{f} \propto r^{(n-3)/2}$ and $\mathbf{J} \propto r^{(n-3)/2}$. Since **f** is the acceleration due to magnetic forces, the manipulator has higher accelerations the smaller the system's components are made. Note that these scaling laws have assumed that **u** is independent of **B**. Similar scaling laws can be derived for the case where **u** is induced by **B**, though they are generally not quite as favorable.

The transient power of the system should be considered as well as the steady state power. The transient power necessary to overcome the inductance of the electromagnets scales as $|\mathbf{J}|^2 r^5/\tau$, where τ is the desired risetime of the electromagnet. For the 1 cm or smaller components considered in this paper, the transient power typically is on the order of the steady state power or much less. As can be seen from the scaling law, however, the transient power requirements in larger systems can be a significant concern.

As mentioned previously, the scaling laws allow iron core electromagnets to be replaced by air core electromagnets on scales of the order of 1 cm. This is roughly the size at which air core electromagnets will levitate permanent magnets a distance of 1 cm without special cooling. At still smaller scales, on the order of 1 mm, it becomes practical to replace the air core electromagnets with single wires or sheets of wires. That is, the magnetic field gradients surrounding a single wire running on the order of an amp of current will lift a small permanent magnet about a millimeter. "Single wire drive" is attractive for very small micro-robots because large numbers of drivers could be fabricated cheaply using photolithography.

The geometry of the system components is important in the design of maglev micro-drives because the geometry of the manipulator and magnetic field source strongly influence power consumption and controllability. The optimum geometry is also influenced by operational requirements, particularly the need to interface transport paths with precision, multidegree of freedom workspaces. Maglev micro-manipulators can be made as stand alone devices, without the need for transport paths. These devices are probably the easiest to design, and operationally the fastest. However, the ability to integrate transport with precision is a powerful advantage in complex systems, especially if the end effectors wear. Current research has concentrated primarily on geometries which allow precision-to-transport interfacing.

Computer modeling programs have been written to calculate the power requirements of a system composed of a set of permanent magnet dipoles and air core electromagnets. Similar programs model single wire drivers. These programs can compute the forces and torques resulting from given electromagnet currents at different manipulator locations, and inversely the currents necessary to give a desired force or torque. Control problems arise when it is necessary to apply large forces or torques from individual electromagnets in order to get a small desired net force or torque. This happens, for example, when two electromagnets produce nearly counter-opposing forces. In these situations a small error in the driving current can lead to a large percentage error in the resulting force/torque. Our modeling effort has produced some geometries that are controllable and use a reasonable amount of power. They are discussed in detail in the next section, and should be regarded as proof of feasibility rather than optimal.

The control strategy for the maglev systems depends on whether the system has open loop stability. Without open loop stability, one needs to guarantee a sufficient amount of feedback for stability. For simple situations, such as transport, one could use analog feedback circuitry, but for multi-degree of freedom micro-robots, a computer with digital control is generally necessary. With open loop stability the manipulator can be put into a hold mode. For example, in our prototype micro-robot system, the manipulator can be pulled into a solid surface and held. A hold mode is desirable for two reasons: it reduces the need for real time control, thus making control of many manipulators by the same computer practical, and many tasks, such as etching or masking in photolithography, require the manipulator to maintain one position for a relatively long period of time. Without a hold mode the manipulator would be constantly using the resources of the most expensive part of the system, the computer.

Besides open loop stability, a desirable attribute in any robotic system is relative motion open loop accuracy. A 1 part in 10 relative motion open loop accuracy means that the manipulator can move open loop from a known position to within 10% of an intended position. Maglev micro-robots have the potential to give extremely good relative motion open loop accuracy because for some systems there is no hysteresis, the drive has a highly linear dependence on the currents, and the fairly compact, rigid body manipulator does not have the complex dynamics associated with extended, articulated structures.

There are two additional aspects of control we are currently studying: control of a macro-robot/micro-robot pair, and sensing systems. The macro/micro combination is a useful match, combining a slow, imprecise robot with a large motion range, with a fast, precise micro-robot with a relatively small motion range in order to extend the workspace.[5] The control problem is complex because of the number of degrees of freedom involved (up to 12).

All of the feedback control strategies require sensing. One immediate problem is that the small size of the manipulator precludes most off-the-shelf sensors, particularly for compact multi-degree of freedom manipulators. Our main focus has been on optical sensors. The present prototype uses thermally-matched, infrared emitting diode/phototransistor pairs. Parts of the micro-robot interrupt the infrared beam, and the phototransistors measure the degree of shadowing. This set-up has the disadvantages of being nonlinear because of the geometry, and having residual thermal drift. Current plans call for replacing the phototransistor with a lateral effect cell, and either putting the infrared emitting diode on the robot or projecting an infrared spot on the lateral effect cell through a pinhole in the robot. Lateral effect cells can be very linear and insensitive to temperature changes.[6]

Prototype Systems

The prototype systems built have demonstrated the basic feasibility of precision maglev micro-machines. However, the experiments to date are preliminary, as will be clear from the discussion. Hence the measured performance represents a *lower* bound on the prototype's capabilities.

The scale used in the prototype micro-robots was 0.5 cm; this was both the coil diameter in the electromagnets, and the center-to-center spacing of the permanent magnets on the manipulator. The manipulator geometry, shown in Fig. 2, was chosen in conjunction with the electromagnet geometry, shown in Fig. 3. We used a triangular manipulator design, with a rare-earth permanent magnet located on each vertex of an equilateral triangle. Two manipulators have been tried, one having a 0.05 g mass and the other a 0.4 g mass. These are fairly heavy manipulators for this scale, in which it should be possible to drive manipulators of a few milligrams. Manipulators lighter than a few milligrams may suffer from significant bending due to lack of stiffness. Both manipulators used copper legs of 0.01 cm diameter to support the manipulator when using a hold mode against the solid surface. Ideally the leg diameter would be scaled to keep the contact area proportional to the mass. The fact that the same diameter legs were used on both manipulators may account for the greater amount of feedback required for the smaller manipulator, as well as its faster settling time. The manipulators and electromagnets were constructed by hand, with tolerances probably no better than ±10% in any dimension. Although maglev micro-robots are capable of high precision positioning, the robots themselves require only low precision fabrication. This observation raises some interesting possibilities for on-line replication in advanced systems.

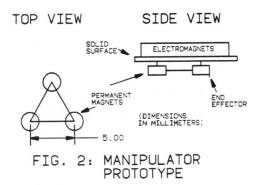

FIG. 2: MANIPULATOR PROTOTYPE

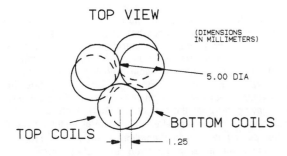

FIG. 3: ELECTROMAGNET GEOMETRY

The electromagnets were wound using 40 micron diameter magnet wire. The maximum electromagnet current used was 20 ma, easily supplied by small signal op amps. The time-averaged power was 0.27 W, with a peak power of 0.54 W. The time-averaged power can probably be pushed up to a few watts with only a modest effort in cooling, so the present speed of the system is fairly low compared to what is practical. The geometry was chosen on the basis of computer modeling and is shown in Fig. 3. It consists of two sets of coils, 3 large top coils with an axial length of 0.2 cm and 3 small bottom coils with an axial length of 0.1 cm.

The system has been tested primarily with the manipulator hanging below the assembly of electromagnets. Other orientations are possible, but with this orientation one can use a dither-type, open loop control. To understand dither control, notice that for a given set of currents through the coils, there is generally a unique equilibrium $(x, y, œ)$ position and orientation for the manipulator on the solid surface, providing the currents are overall attractive and form a bound state. If the solid surface were frictionless, the manipulator would settle to its equilibrium state. The surface actually does have friction, but it can be reduced significantly by dithering the coil currents, that is, by switching the currents between the given set of values and completely off. During the off state the manipulator falls away from the surface due to gravity, typically 10 microns for a 1kHz dither. In this way the manipulator can work its way close to equilibrium, with the closeness determined by stiction at high frequencies, and by scattering at low frequencies. With our system the dither motion range is greater than 2 millimeters. The infrared emitting diode/phototransistor sensors have a range of 1.5 millimeters. Within the sensed range, dither control had an accuracy of about ±50 μm.

Considerable improvement might be made in the dither control using more power, other surfaces (glass and brass have been tried), and better choices for the dither currents and frequencies. However, the dither control is inherently slow, averaging 3-5 moves per second at the above stated accuracy, because as the manipulator approaches its equilibrium, the driving force towards equilibrium decreases. In order to increase speed and accuracy, we have tried a more powerful control strategy using time sequences of coil currents to move the manipulator fixed distances, or steps. Step control works by first finding the time sequences of coil currents which will yield a given size step in each of the sensor coordinate directions. This is done at each point in a grid of points in the workspace, and the time sequences are stored in memory. The step calibration is done using an automated search routine. Step control is implemented by estimating the number of steps along each coordinate direction which are necessary to get to the desired position, based on the latest sensor readings. Each time sequence of currents ends with coil currents that correspond to a hold mode. These holding currents at different locations are just the currents found from the dither control. It is important to realize that with either a dither or a step control strategy, all the motion takes place open loop, and sensor readings are only made with the manipulator in a hold mode.

In practice it is advantageous to use more than one step size, typically a coarse step and a fine step. Using a 10 μm coarse step, and a 2 μm fine step, we have measured relative accuracies in x,y of ±1.5 microns, the limit of our sensors. Accuracies in this case are quoted as relative because slow (5 μm per hour) thermal drifts are known to occur in the sensors, and absolute calibration has yet to be obtained. Motion range with step control has been restricted to a 0.6mm by 0.6mm workspace by a software problem which limits the amount of memory for the time sequences to 64 kilobytes. With this motion range and relative accuracy, an average speed of 15 moves per second was measured, moving between randomly chosen points in the workspace. The system required an average of 11 feedback cycles, i.e. sensor readings, per move.

Only recently a third sensor has been added, resulting in a 3 DOF closed loop system. One experiment used a 40 μm/4 mrad coarse step size with a 2 μm/0.2 mrad fine step and gave a performance of ±2 μm and ±0.2 mrad relative accuracy over a workspace measuring 0.6mm x 0.6mm x 20 mrads. The average speed was 10 moves per second, using about 10 feedback cycles per move.

It is usually desirable to have electrical feedback from a robot end effector. There are two ways to get end effector feedback on a maglev micro-robot. One way is to use a thin wire attached to the manipulator as a tether. A twisted pair of 25 micron wires have been attached to the 0.4 gram manipulator with little effect on performance. A better method of supplying end effector feedback would be to use the solid surface as a contact pad. In this scheme the material to be processed is positioned below the robot and the robot makes two contacts at once: one with the end effector on the work surface, and another on a contact pad. With our prototype, two of the legs of the robot remain in contact with the solid surface to provide electrical continuity, while the third leg is pivoted down to bring the end effector into contact with the material. The pivot method, unlike a tether, does not restrict the motion range, but additional work is needed.

Most of our focus has been on precision micro-robots. Transport has been demonstrated with air core electromagnets using manipulators with 1, 2, and 3 permanent magnets. Single wire drive using a 1 permanent magnet manipulator, and multiple wire drive using 1, 2, and 3 permanent magnet manipulators have also been demonstrated for transport. More work is needed on transport paths and their interfacing to precision workspaces, particularly with single and multiple wire drives. Transport paths are often similar to DC linear motors, and may benefit from a more detailed study of the design of these motors.

Conclusion

Maglev drive allows a rich variety of micromechanical devices. A prototype of one maglev system, using air core electromagnets driving permanent magnet manipulators, has been studied and built. Preliminary experiments have verified that this design can achieve micron level precision at attractive speeds. Maglev micromachines have a number of advantages over competing technologies, particularly the ability to integrate transport with precision. This ability could lead to integrated microfactories, in some ways analogous to integrated microcircuits.

Acknowledgements

The authors would like to acknowledge Wanjun Wang for work with lateral effect cells, Myung C. Jeong for work on the macro/micro robot control problem, and Walter Keene for work on single wire drive. The authors would also like to thank the Bosque and National Science Foundations for support. One author, Ron Pelrine, extends personal thanks to his parents for their encouragement and support, and to Carl Deckard for many invaluable discussions.

References

1. I. Boldea, S. A. Nasar, Linear Motion Electromagnetic Systems, John Wiley & Sons, Inc., New York, 1985, p. 1-21.
2. Dieter W. Pohl, "Some Design Criteria in Scanning Tunneling Microscopy," IBM J. Res. Develop., Vol. 30, No. 4, July 1986, p.417-427.
3. Malcolm McCaig, Permanent Magnets in Theory and Practice, John Wiley & Sons, Inc., New York, 1977, p. 249-250.
4. E. R. Laithwaite, Propulsion Without Wheels, Hart Publishing Company, Inc., New York, 1966, p.179-199.
5. R. L. Hollis, et. al., "Robotic Circuit Board Testing using Fine Positioners with Fiber-Optic Sensing," Proc. Int'l Conf. on Industrial Robots, Tokyo, Japan, Sept. 11-13, 1985.
6. D. J. W. Noorlag, S. Middelhoek, "Two-Dimensional Position-Sensitive Photodetector with High Linearity made with Standard I.C.-Technology," Solid State and Electron Devices, May 1979, Vol. 3, No. 3, p. 75-82.

FABRICATION AND TESTING OF A MICRO SUPERCONDUCTING ACTUATOR

USING THE MEISSNER EFFECT

Yong-Kweon Kim*, Makoto Katsurai* and Hiroyuki Fujita

Institute of Industrial Science
University of Tokyo
7-22-1, Roppongi, Minato-ku
Tokyo 106, Japan

*Department of Electrical Engineering
Faculty of Engineering, University of Tokyo
7-3-1, Hongo, Bunkyo-ku
Tokyo 113, Japan

Abstract

A micro superconducting actuator was fabricated and tested. The levitation and driving was achieved by the Meissner effect and the Lorentz force. For the stator, a YBaCuO superconducting film and etched copper film on polyimide was used. For the slider, an Nd-Fe-B permanent magnet was used. Non-contact and remote measurements were made on the levitation heights by using a laser displacement meter, and on the slider's movement by using an image processing techniques. As the results, the 8mg slider stably levitated 0.99mm above the stator at 280G. The levitation height was decreased by the increase of the weight of the slider and the attractive driving force. The levitation height exhibits a hysteresis characteristic as a function of the applied magnetic field. The slider was driven by switching a current loop manually. The slider was moved 3.3mm in 0.46 seconds while the current was 0.4A. During that time, the average velocity was 7.1mm/sec and the maximum acceleration was $0.14 m/sec^2$. The slider was synchronized to the driving current if the current was over a certain value. The measured forces were compared with the values calculated by the image method and the measurement of magnetic susceptibility of the superconductor. The measured levitating forces were 60 to 70% of the calculated forces. The drag force, including that produced by eddy currents, was estimated to be about $10^{-5}N$, approximately half of the driving force. The resistive loss was estimated to be 176mW at 0.4A. The double layered structure of this actuator is discussed.

1. Introduction

In micro machines, frictional forces are a barrier to many applications. It was reported that frictional forces hindered the rotor from synchronizing to the driving frequency and caused asynchronous movements in early electrostatic micro motors[1]. The coefficients of dynamic friction were estimated to range from 0.2 to 0.5[2,3], which are the same as the coefficient of static friction of brake materials on cast iron (approximately 0.4). Levitating micro machines were proposed to avoid friction[4,5]. Levitation using the Meissner effect of high-Tc superconductors has some advantages over other forms of levitation.

In this paper, a micro superconducting actuator is fabricated and tested. The actuator is levitated and driven by the Lorentz force. In Section 2, levitation and driving methods are briefly reviewed and compared. In Section 3, the design and fabrication of the actuator are discussed. In Section 4, the measurement of the levitation and driving characteristics are described. In Section 5 and 6, the results of experiments are described and discussed.

2. Levitation and driving

In this paper, we choose the levitation using the Meissner effect of superconductors and the Lorentz force as the levitating force and the driving force, respectively, to achieve non-contact micro machines. In this section, some levitating and driving forces are reviewed and compared with the levitation using the Meissner effect and the Lorentz force.

2.1 Levitation using the Meissner effect

Non-contact levitation systems have many advantages; principally because they are frictionless and compatible with harsh environments[4,6]. Levitation can be obtained by various methods; magnetic forces, electrostatic forces and hydrodynamic forces. Magnetic levitation can be achieved by using a permanent magnet, an electromagnet (including superconducting magnets), and a diamagnetic body[6]. This classification of magnetic levitation is based on what object is used to produce a force in a magnetic field. In this paper, our discussion about levitation is confined to levitation using a diamagnetic body, i.e. a superconductor.

The levitation using a diamagnetic body has previously been studied[7,8], and shown that the levitating force is proportional to a relative magnetic permeability, μ_r, of a diamagnetic body and the gradient of the square of the magnetic field as shown below.

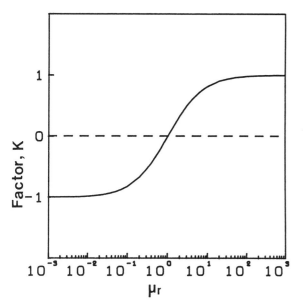

Fig.1 The influence on the force by the relative permeability. The factor K is $(\mu_r-1)/(\mu_r+1)$.

$$F = \frac{1}{2}\mu_0(1-\mu_r)\frac{\partial H^2}{\partial Z} \quad (1)$$

Levitation using graphite or bismuth (whose magnetic susceptibility, χ, is about -10^{-6}, where $\mu_r = 1+\chi$) was experimentally examined[7,8]. It was also reported that it should be possible to levitate a permanent magnet in micron-size above graphite or bismuth[8]. But, the levitation using a diamagnetic body consumes a large amount of energy. For instance, levitation of a small piece of bismuth, weighing 8mg, needed a 2.3T magnetic field.

As shown in Fig.1, the repulsive force acting on a permanent magnet above a diamagnetic body is reduced by a factor of $(\mu_r-1)/(\mu_r+1)$ as μ_r increases from zero to unity. Therefore, if a diamagnetic body has a μ_r near zero, a permanent magnet will be levitated at low magnetic fields. It means that, for the same magnetic field, a heavier magnet can be levitated above a superconductor than above a normal diamagnetic body (such as graphite). Using a superconductor as a diamagnetic body, a 4x4x10mm³ permanent magnet was levitated above a concave disc of lead cooled by liquid helium[9]. Theoretically, a magnetic pressure, 400mgf/mm² is produced on the superconductor when the magnetic field is 1kG at the surface of the superconductor.

The levitation using the Meissner effect of superconductors is accomplished by simple structures including cooling systems, and the magnet is levitated stably without any sensors or controllers. In contrast, the magnetic and electrostatic levitation systems need several sensors and complicated control systems. Although the levitation system using the Meissner effect should be cooled and insulated thermally, it has a potential feasibility for micro machines because the levitating force is proportional to a surface area and the simplicity of the levitating method would eliminate associated complex control circuits.

2.2 Driving force

The Lorentz force, electrostatic force and the force produced by the Meissner effect[5] can all drive micro machines. In Table 1, these forces are compared as driving force. Lorentz force driving has advantages over other forces in that the force is able to work repulsively or attractively in a vertical direction. In addition, it is easy to fabricate a slider and a stator to use the Lorentz force as the driving force. Its disadvantages are that the force is proportional to the volume of the conductor, and the superconducting state may be transited to the normal state by Joule heating and the magnetic field produced by the current. Electrostatic force driving is suitable for micro machines[10] because the force is a surface force, and a stator and a slider can be fabricated by IC process techniques. However, electrodes on the slider should ideally be connected

Table 1 Comparison among driving methods applicable to actuators with levitation using Meissner effect.

	electromagnetic force	electrostatic force	force produced by Meissner effect
scaling	S^3	S^2	S^2
effect on levitation height by dirving force	upwards and downwards	downwards	downwards
structure for the stator	wires	electrodes	patterned superconductors
structure for the slider	patterned permanent magnets	electrodes	patterned permanent magnets
remarks	·Resistive loss. ·The magnetic field of the driving current may exceed the critical field H_{c_1}.	·Electrodes on the slider should be grounded. ·If not grounded, the force decreases.	·The driving force is limited under about 1/10 of the levitating force. ·The magnetic permeability affects both the driving and the levitating forces.

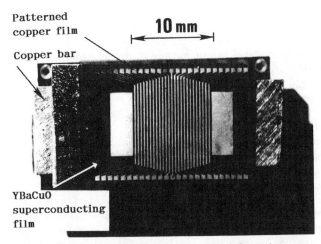

Fig.2 Schematic drawing of the stator of a micro superconducting actuator using the Meissner effect.

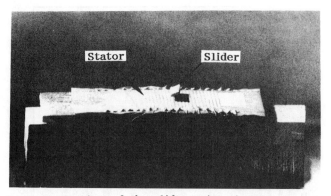

Fig.3 Levitation of the slider using the Meissner effect above the stator.

to ground because, if not, the driving force is reduced. It is difficult to ground the electrodes of a slider which is levitated. The driving method using the force produced by the Meissner effect, obtained by the control of the state of superconductors, was proposed and presented earlier[5]. This drive is based on the energy difference between the superconducting and the normal state as shown as below.

$$E = F_s(0) - F_n = -\frac{1}{2}\mu_0 H_c^2, \quad (2)$$

where F is Gibbs free energy per unit volume, F_n is Gibbs free energy of the normal state, $F_s(0)$ is Gibbs free energy of the superconducting state when the applied to a superconductor is zero, and Hc is the critical magnetic field[11]. For instance, the energy difference is $40J/m^3$ when Hc is 100G.

This method has advantages in that the force is a surface force, and the superconductors are used for both levitation and driving. However, the superconductors have to be patterned, and the characteristics of the superconductor may be degraded during the patterning processes. The strength of this force is restricted by the strength of Hc, and is about a tenth of the levitating force. If the superconductor is not a perfect diamagnetic body (as shown in high-Tc superconductors), it is also influenced by the magnetic susceptibility of the superconductor.

3. Design and fabrication

In this paper, we choose the force produced by the Meissner effect of high-Tc superconductors as a levitating force and the Lorentz force as the driving force. A YBaCuO superconducting film, deposited on a silver plate by a low pressure plasma spraying method[12], was used. The critical values of temperature and current density are 90K and $513A/cm^2$ at 77K, respectively. The thickness of the film is 0.1mm. The levitation height, corresponding to the levitating force, is proportional to the root of the superconductor's thickness. Since a permanent magnet, weighing approximately 200mg, was levitated above the 1.3mm thick YBaCuO superconductor[5], a weighing 20-30mg magnet can be levitated above a 0.1mm thick superconducting film.

As a driving force, the Lorentz force is chosen because it is easy to fabricate a slider and a stator. Since the influence of magnetic susceptibility has little effect on the driving force, the driving force increases as the external current increases. The Lorentz force is comparable with than the electrostatic force in the submillimeter range.

The wires to drive a current were fabricated using copper film rolled on a polyimide substrate (Espanex SFC-250-05R) and wet etching. The thicknesses of the copper and polyimide are 20 and 23μm, respectively. The copper film was etched away by isotropic wet etching. The etchant of copper was a solution of $FeCl_3$(30g) and H_2O(100ml). Since the etching is isotropic and the copper is undercut, the etched line-width of the mask has to be narrower to compensate for the undercutting to obtain the line width desired. As shown in Fig.2, the obtained line width and separation of the copper conductors on the stator are 186 and 210μm, respectively. The rolled film was fixed by a bonding agent on the superconducting film. The polyimide film electrically insulates between the copper and the superconducting film, and prevents the superconducting film from contacting water and degrading the superconducting characteristics. The copper bar was located below the superconducting film to cool the superconducting film by heat conduction.

An Nd-Fe-B magnet was used as a slider, measuring $0.3 \times 1 \times 4 mm^3$ and weighing 8mg. Its coercivity and permeance coefficient are 1000kA/m at 300K and 0.53, respectively. The magnet was fabricated by micro cutting and mechanical rubbing.

4. Experimental details

It is necessary to measure the characteristics of superconducting actuators remotely and without contact, because it is difficult to measure the characteristics of micro machines without interfering with them. Especially, in the case of levitating superconducting actuators, the surroundings of the superconductors should be kept at low temperatures, and it is easy to disturb the slider in levitation

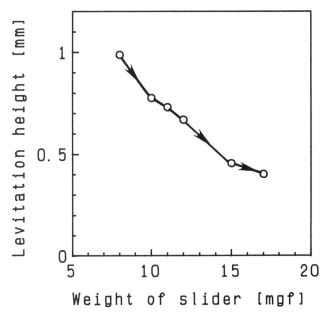

Fig.4 Levitation height versus weight of slider.

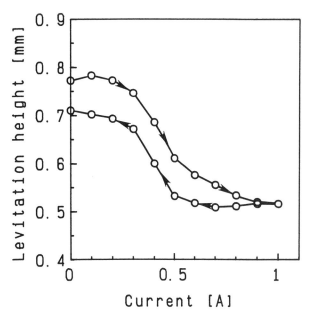

Fig.5 Levitation height versus driving current.

when the characteristics are being measured.

The slider is levitated above the stator which is cooled by liquid nitrogen as shown in Fig.3. The levitation height of the slider is measured using a laser displacement meter (Keyence LC-2100,2320). The instrument, whose resolution is 0.5μm, detects variation of the reflective angle of a laser beam and displays displacements. The instrument is separated by 5cm from the superconducting film, enough not to disturb each other thermally. The wavelength of the laser beam is 670nm (red).

The slider is driven by the current flowing through the copper lines on the stator. Since the current flows around a loop formed by two neighboring lines, the slider is attracted by the current loop and driven by the attractive force. The current is switched manually. The movement of the slider is monitored by a video camera and recorded on video tapes. The replayed pictures are sent to a computer equipped with an image processing board (Photoron FDM 98-4). Using the image processing board, the same scan line of each picture is collected every 1/30 seconds, and stored as one picture composing 256 lines. One line is composed of 256 pixels. Since the actuator in this paper is driven linearly, the stored pictures mean the position of the slider versus time for 8.5 seconds(256/30). The slider and stator are painted white and black, respectively, to strengthen the contrast of the pictures. The stored pictures are transformed into binary-toned pictures. By analyzing the pictures, we obtain the position of slider and also obtain the velocity and acceleration by differentiating the position respect to time.

5. Results

5.1 Levitation height

The levitation height, corresponding to the levitating force, was measured using an Nd-Fe-B magnet (weight: 8mg, thickness: 270μm). The levitation height from the top of the stator to the bottom of the slider was measured when the weight of the slider was increased from 8 to 17mg. As shown in Fig. 4, as the weight of the slider increased, the levitation height decreased. When the weights of slider were 8 and 17mg, the heights were 0.99 and 0.40mm, respectively. The magnetic fields, not including superconductors, are 280 and 650G, respectively, when the measuring points are separated 0.99 and 0.40mm from the center of magnet. These values are apparently more than the lower critical magnetic field, H_{c1}, of the superconducting film (approximately 50G). The levitation forces per unit area of the slider produced by the Meissner effect were 2 and 4.25mgf/mm², respectively, when the height were 0.99 and 0.40mm.

Besides the weight of the slider, the levitation height is influenced by the attractive driving force and the hysteresis of the superconductor's magnetic susceptibility. The slider is attracted by the current of the stator. As the current increases and the attractive force increases, the slider moves closer to the stator. Since the magnetic susceptibility of high-Tc superconductors has a hysteresis over the H_{c1}, the levitation height also exhibits a hysteresis characteristic as a function of the applied magnetic field to the superconductors. The influence of the attractive driving force and the superconductor's magnetic hysteresis on the levitation height was examined experimentally for this actuator.

As the current was increased from 0 to 1A, and then decreased from 1 to 0A, the levitation height was measured at every 0.1A. As the applied current increased, the levitation height decreased. When the current is 1A, the levitation height is 0.52mm, compared with 0.78mm when the current is zero. It appears that the difference of the height tends to saturate although the current becomes larger. The reason is that the repulsive force increases proportionally by the Meissner effect when the gap becomes narrower and the applied magnetic field to supercon-

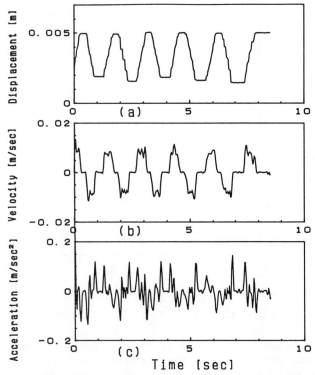

Fig.6 The slider's position, velocity and acceleration versus time at 0.4A when the driving current is decreasing. (a)position (b)velocity (c)acceleration.

ductors becomes large. Also shown in Fig.5 is the hysteresis of levitation height. The initial height when the current increases from zero to 0.1A is larger than the final height when the current returns to zero. The difference between the two values, when the current is zero, is 0.07mm. The superconductor is in the intermediate state and some magnetic flux penetrate the superconductor when the magnetic flux is applied to the superconductor above Hc_1. The magnetic flux penetration causes the hysteresis and may cause a drag force hindering the slider's movement.

5.2 Driving force

The movement of the slider was recorded on video tape and analyzed by an image processing board. Figure 6 shows the slider's position, velocity and acceleration versus time when driven by a current 0.4A. In Fig.6, the slider was moved 3.3mm peak-to-peak displacement corresponding to 9 steps of the slider. The average velocity developed during the peak-to-peak displacement is 7.1mm/sec and the maximum acceleration is 0.14m/sec². The weight of the slider is 9mg. The displacements correspond to 0.4A, when the current is decreasing (Fig.7). If the current increases, the driving force becomes large due to the increase of the current and the magnetic field. Since the increase of the current causes the gap to decrease, the magnetic field applied to the superconductor is increased. If the current decreases and the driving force becomes small, the slider may not be synchronized to the driving current because the magnetic flux penetration hinders the slider's movement.

The peak-to-peak displacements versus driving current were measured. As shown in Fig.7, when the

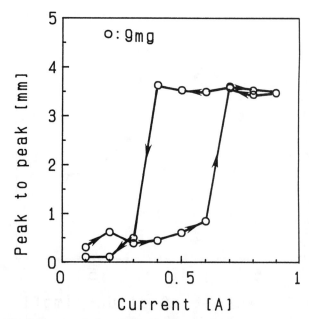

Fig.7 The peak-to-peak displacement versus driving current.

increasing current reaches 0.7A, the peak-to-peak displacement abruptly increases to 3.6mm. The displacements were almost constant for currents above 0.7A. On the other hand, the displacements remained constant when the current was decreased until the current reached 0.3A. The abrupt change in the peak-to-peak displacement means that a drag force exists in this actuator. It can be seen that, for currents between 0.3 and 0.7A, the peak-to-peak displacements when the current is decreasing are larger than when the current is increasing. The reason is that the driving force during the current increase is larger than during the current decrease. This is due to the hysteresis of the levitation height(Fig.5).

6. Discussion

The magnetic flux penetration of high-Tc superconductors affects the levitating force. We tried to estimate the influence of magnetic flux penetration on the levitating force by the relative permeability, μ_r. The levitating force can be calculated using the image method, taking the superconductor as a diamagnetic body which has a uniform magnetic susceptibility over the whole volume. The factor K affecting the force is $(\mu_r-1)/(\mu_r+1)$. The magnetic susceptibility of the superconductor versus applied magnetic field were measured by a VSM (Vibrating sample magneto flux meter). The measured relative permeabilities of the superconductor were 0.92 at 280G and 0.95 at 390G. The magnetic fields correspond to that of the slider's weight, 8 and 10mg in Fig.4. The forces were calculated, based on the experimental data of the levitation height and the relative permeability. For these cases, the percentage of the measured force to the calculated force were 61 and 71%. However, if the measured forces are compared with the calculated force when $\mu_r=0$ (the case of a perfect diamagnetic material), the percentages are 2.5 and 1.8%.

In driving, the magnetic flux penetration also causes a drag force. As shown in Fig.7, the peak-to-peak displacement abruptly changed by the increase or decrease of the current. The drag force in this actuator, including that produced by eddy currents, was about 10^{-5}N. The estimated value was half of the driving force when the current was 0.4A (descending curve of Fig.7). It was observed that the drag force had a spatial non-uniformity of magnetic properties on the superconducting film. Further characterization of the drag force in high-Tc superconductors is needed before further application of such superconductors in micro machines.

The resistive loss was estimated to be 176mW when the current was 0.4A. The resistance from the main power was 1.1Ω at 77K. In this actuator, the loss and associated heat were not large enough to transit the superconductor into the normal state.

In this actuator, the driving wires are installed on the superconducting film. But, it is not good for both the levitating and the driving force. In the levitating force, the driving force works in the attractive force. Also in the driving force, the driving force produced by the Lorentz force is small because the vertical field component to the surface of the superconductor is almost zero (if the superconductor is a perfect diamagnetic body). If the driving wires layer (non-superconducting film) is separated and placed above the superconducting film, the magnet is levitated between the superconducting film and the driving wires layer. The attractive driving force produced from the upper layer will add to the levitating force and increase itself by the increase of the vertical field component.

7. Conclusions

A micro superconducting actuator was fabricated and tested. The levitation and driving was achieved by the Meissner effect and the Lorentz force. For the stator, a YBaCuO superconducting film and etched copper film on polyimide was used. For the slider, an Nd-Fe-B permanent magnet was used. By using a laser displacement meter and image processing techniques, non-contact and remote measurements were made on the levitation height and the slider's movement.

Results indicated that the 8mg slider stably levitated 0.99mm above the stator at 280G. The levitation height was decreased by the increase of the weight of the slider and the attractive driving force. The levitation height exhibits a hysteresis characteristic as a function of the applied magnetic field.

The slider was driven by switching a current loop manually. The slider was actuated 3.3mm in 0.46 seconds while the current was 0.4A. During that time, the average velocity was 7.1mm/sec and the maximum acceleration was 0.14m/sec². The slider was synchronized to the driving current if the current was over the certain value. The values were different during current increase and current decrease. It is believed that magnetic flux penetration caused a drag force.

The measured forces were compared with the values calculated by the image method and the measurement of magnetic susceptibility of the superconductor. The measured levitating forces were 60 to 70% of the calculated forces. The drag force, including that produced by eddy currents, was estimated to be about 10^{-5}N --- half of the driving force. The characteristics of the drag force will be examined in future studies. The problem of a drag force may be solved by the improvement of the high-Tc superconductor's properties.

In conclusion, the micro superconducting actuator using the Meissner effect appears feasible for micro machines. The slider levitated stably without any sensors or controllers. It moved fast and was synchronized to the driving currents if the currents were sufficiently large to overcome the drag force.

Acknowledgments

Authors wish to thank Steel Research Center of NKK for preparations of the superconducting films.

References

[1] Y.-C.Tai, et al.,IC-processed micro motors: design, technology, and testing, Proc.IEEE Micro Electro Mechanical Systems Workshop, Salt Lake City, Utah, U.S.A., February, 1989, pp.1-6.

[2] K.J.Gabriel, et al.,In Situ measurement of friction and wear in integrated polysilicon micromechanisms, Abstracts of The 5th International Conference on Solid-State Sensors and Actuators & Eurosensors III, Montreux, Switzerland, June, 1989, pp.109-111.

[3] Y.-C.Tai and R.S.Muller, Frictional study of IC-processed micromotors, ibid, pp.108-109.

[4] R.Perline and I.B.Vishniac, Magnetically levitated micro-machines, Proc.IEEE Micro Robots and Tele-operators Workshop, Hyannis, Massachusetts, U.S.A., November, 1987.

[5] Y.-K.Kim, et al.,A proposal for a superconducting actuator using Meissner effect, Proc.IEEE Micro Electro Mechanical Systems Workshop, Salt Lake City, Utah, U.S.A., February, 1989, pp.107-112.

[6] B.V.Jayawant,Electromagnetic levitation and suspension techniques, Edward Arnold, London, 1981, p.1.

[7] W.Braunbeck,Free suspension of diamagnetic bodies in magnetic fields, Zeitschrift fur Physik, Vol.112, No.11 (1939) pp.764-769.

[8] A.H.Boerdijk,Levitation by static magnetic fields, Phillips Technical Review, Vol.18, No.4 (1956) pp.125-127.

[9] V.Arkadiev,A floating magnet, Nature, Vol.160, No.2 (1967) p.330.

[10] Fujita and A.Omodaka,Electrostatic actuators for micromechanics, Proc.IEEE Micro Robots and Tele-operators Workshop, Hyannis, Massachusetts, U.S.A., November, 1987

[11] M.Tinkham,Introduction to superconductivity, Mc Graw-Hill, New York, 1975,pp.13-14.

[12] K.Tachikawa, et al.,Preparation of high-Tc superconducting thick films and power conducting tubes by a low-pressure pressure thick spraying, IEEE Trans. on Magnetics, Vol.MAG-25, No.2 (1989) 2029-2032.

ROOM TEMPERATURE, OPEN-LOOP LEVITATION OF MICRODEVICES USING DIAMAGNETIC MATERIALS

Ronald E. Pelrine

Mechanical Research Laboratory
SRI International
333 Ravenswood Avenue, Menlo Park, CA 94025

ABSTRACT

A well known property of superconductors is their ability to levitate magnets because of their diamagnetism (the tendency of a material to oppose applied magnetic fields). It is less well known that diamagnetic levitation, albeit at a weaker level, is possible at room temperatures using ordinary materials such as bismuth and graphite. This paper explores the use of diamagnetic levitation for micromechanical bearings. The simple system of a magnetic sphere supported by a diamagnetic material has an analytic solution for the vertical force; this solution is used to illustrate the fundamental properties of diamagnetic levitation. Work by previous researchers to diamagnetically levitate magnets with masses as large as tens of grams using external fields, as well as micrometer-scale levitation of diamagnetic materials is discussed. Calculations presented in this paper suggest modern magnets on the order of 100 to 300 µm will self-levitate over diamagnetic materials without supporting external fields. This suggests a number of noncontact micromechanical bearing applications, such as for microrobotics and highly sensitive sensors. Diamagnetic bearings are typically limited to light-load applications. To alleviate this limitation, a noncontact, hybrid diamagnetic air bearing is proposed for light-load, low-speed to high-load, high-speed use without the start-up problems normally associated with self-pumping air bearings. A preliminary analysis of the hybrid bearing concept suggests that squeeze film effects may allow virtually the full drive load to be applied at start-up, provided lateral acceleration is high.

INTRODUCTION

Friction and wear are important issues in the field of micromechanics. Both issues primarily relate to surface contact between solids, and in particular to the surface contact that occurs in bearing surfaces supporting the load of the mechanism or machine. One approach towards eliminating surface contact that works in some situations is to use a bending beam or flexure bearing to support the load, as is used, for example, with cantilever-type accelerometers. However, flexure bearings are not possible for many types of micromachines such as rotary motors and large motion linear translators. An attractive approach for these type of machines would be to design a bearing which has no surface contact and consequently no friction or wear in its bearings.

A variety of bearings are known to not have solid-to-solid surface contact: Liquid and air bearings are often used on large-scale machines to support the load on a film of liquid or air. While these type of bearings can probably be used in many micromechanical situations, they do have drawbacks. Liquid bearings can have a tremendous viscous drag at small scales, so their use may be limited to slow-speed applications. Air bearings have considerably less viscous drag, but the air must be pumped into the bearing, either by a separate air pump or by using a self-pumping design. A separate air pump is unattractive because of the greater complexity involved and the difficulty in fabricating, attaching, and maintaining a clean, microscale, air hose. Self-pumping air bearings are possible, but they are ineffective at slow speeds because of insufficient pumping action.

The other type of noncontact bearings that can be used are magnetic or electric suspension bearings. These bearings support a load that is floating freely; i.e., the load is levitated. Two important features of magnetic and electric suspension bearings are that they can work in vacuum, and they can maintain relatively large gaps between solid surfaces of the machine. The latter feature may prove particularly attractive for micromachines operating in unfiltered air, because small-gap bearings may fail because of dust. The drawbacks of the better known magnetic and electric suspension bearings are that they often require feedback, low temperatures (e.g., superconductor levitation), or relatively high power or speed (e.g., induced current levitation) or both. This paper discusses diamagnetic levitation, an old, but not well known type of magnetic suspension that requires no feedback for stability (open-loop stability), uses no power to levitate, operates at room temperature, and can be used with a variety of materials on small scales.

DIAMAGNETIC LEVITATION

As a magnet approaches a diamagnetic material, magnetic dipole moments are induced in the diamagnetic material that oppose the applied field. These induced dipole moments lead to short-range magnetic forces that tend to push away the magnet. The force, **F**, on a diamagnetic material can be written as

$$\mathbf{F} = \mathbf{grad}\left(\int \chi H^2/2 \, dV\right) , \qquad (1)$$

where H is the magnetic field, χ is the susceptibility of the material, and the integral is taken over the material's volume. In the cases of interest here, the diamagnetism is fairly weak, i.e., $|\chi| \ll 1$. In these cases, a good approximation is to calculate the magnetic field, H, as if the diamagnetic material is not there, and then use this approximate H in Eq. (1).

Numerical solutions are generally required to calculate the magnetic field and the force. Figure 1 illustrates a simple case that has an analytic solution and provides insight into diamagnetic levitation. In Figure 1, a uniformly magnetized sphere vertically magnetized is positioned over a diamagnetic disk of thickness, t, and radius, r. This situation would describe a spherical rare earth magnet over polycrystalline bismuth, for example. As is well known from magnetostatics [1], the magnetic field outside the sphere is the same as the field from a point dipole centered inside the sphere and with the same dipole moment as the total dipole moment of the sphere. It follows that the vector magnetic field **H** at a position **x** measured from the sphere's center can be written as

$$\mathbf{H} = [3(\mathbf{x} \cdot \mathbf{m})\mathbf{x} - (x^2)\mathbf{m}]/x^5 \quad (2)$$

$$|\mathbf{m}| = (4\pi R^3/3)pM \quad , \quad (3)$$

where **m** is the total dipole moment, p is the density of the sphere, M is the magnetization per unit mass, and R is the sphere's radius. Substituting Eqs. (2) and (3) into Eq. (1) results in an integral that has an exact, albeit algebraically complicated, solution for the vertical force, F_z. For simplicity, only the case of r, t = ∞ is considered here. In this case,

$$F_z = -\chi(4\pi^3 R^6/3)(pM)^2/z^4 \quad . \quad (4)$$

A more useful quantity than the vertical force is the effective vertical acceleration, a_z; i.e., the vertical force divided by the mass

$$a_z = -\chi \pi^2 R^3 p M^2 / z^4 \quad . \quad (5)$$

Eq. (5) is very useful for understanding the important parameters of diamagnetic levitation. For self levitation to be possible, a_z must be greater than or equal to the acceleration due to gravity, or 980 cm/s². The susceptibility of the diamagnetic material bismuth, for example, is -1.2×10^{-5} cgs units (ergs/(gauss-oersted-cm³). For a state-of-the-art neodymium-iron permanent magnet, p = 7.4 g/cm³, and M is about 130 ergs/gauss-gram. It follows that a neodymium-iron permanent magnet of about 150 µm in radius will just barely (z = R) self-levitate over bismuth.

Permanent magnets much larger than 150 µm in size can also be levitated using diamagnetic materials. This is done by using a second fixed magnet to support part or all of the weight of the levitated magnet, with the diamagnetic material mainly acting to stabilize the levitation. The photo in Figure 2 shows a structure consisting of a frame attached to which are three, 1-mm, cubical neodymium-iron magnets levitated above bismuth. Not shown in the photo is a larger magnet above the levitated structure to supply the main lifting force. The levitated magnetic structure shown in Figure 2 has the same geometry as multidegree-of-freedom maglev microrobots that used solid surface bearings in previous work [2,3]. Magnets as large as 20 g have been levitated by researchers in the field [4], and calculations suggest much larger masses can be levitated (alternatively, a series of such levitated magnets can be linked together). It should also be possible to use an electrostatic force to support the main weight of the magnet, with the diamagnetic material again stabilizing the system.

FIGURE 2 DIAMAGNETIC LEVITATION OF MAGNETIC STRUCTURE USING AN EXTERNALLY APPLIED SUPPORT FIELD

The author has also used semiconductor silicon as the diamagnetic material to levitate a 1-mm cubical magnet with gaps of approximately 50 to 100 µm. In this experiment, a large (10 cm × 20 cm × 3 cm) ceramic magnet provided an external supporting field, with a 1.5-mm-thick silicon layer between the ceramic magnet and the levitated magnet to provide open-loop stability. The silicon layer consisted of four, 380-µm-thick silicon wafers, although single wafers or chips of silicon should be adequate using thicker wafers or smaller levitated magnets. Based on the available literature, the author believes this work may be the first reported example of diamagnetic levitation using silicon, and this work provides a proof-of-principle demonstration that diamagnetic levitation can be used directly on silicon chips.

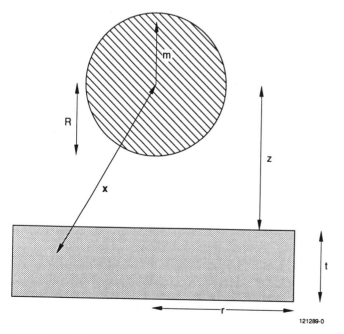

FIGURE 1 SPHERICAL MAGNET OVER DIAMAGNETIC MATERIAL

A number of interesting experiments have also been performed using the reverse system, i.e. levitating a diamagnetic material over a fixed magnet. Kendall, et. al., have reported levitating micrometer-size particles of graphite in magnetic fields [5]. Particles that are this small exhibit Brownian motion, and Kendall, et. al., have suggested using the phenomena for vacuum gauges [5].

DIAMAGNETIC BEARINGS

It will be assumed in the following discussion that the magnets are being self-levitated, although much of what follows can be applied when an external field is used to assist the levitation. Much of the discussion is also relevant to the diamagnetic material being levitated.

Typically, one wants a bearing to resist motion in some degrees of freedom while allowing free motion in others. The levitated permanent magnet shown in Figure 1 can be prevented from rotating about the axes in the plane of the diamagnetic surface either by applying a vertical, homogeneous magnetic field, or more simply, by assembling three or more such magnets in a structure such as the one shown in Figure 2. The stiffness with respect to such rotations and vertical translations can be easily calculated using Eq. (4) or (5). For example, if the bearing uses neodymium-iron magnets with a radius of 10 μm and a gap of 3 μm (z = 13 μm) with a bismuth material, Eq. (5) yields an effective acceleration of about 5 g; i.e., the bearing can support 5 times the weight of the magnets. Surface contact would be made (z = 10 μm) when the bearing is loaded at 15 times the weight of the magnets. Under uniform scaling of both the gap and the magnets, Eq. (5) implies the effective acceleration is inversely proportional to the scale length.

Translations in the plane of the diamagnetic surface as well as rotations about the vertical axis can be controlled with additional electrostatic or magnetic elements. These can be passive, as in the case of permanent magnets fixed on the surface, or active, as in the case of current-carrying conductors or electrodes.

Bearings that can support effective accelerations of 5 g or more would be useful in a number of high-precision micromechanical applications. Microrobot applications may be particularly attractive because diamagnetic bearings do not require fabrication of multiple high-precision bearing surfaces to achieve multiple degrees of freedom. Also, they would not generate particulates, and the absence of friction suggests that high open-loop precision may be possible for lightly loaded robots. The other area of interest is in the field of micromechanical sensors. For example, magnetic objects self-levitated over flat diamagnetic materials should be very sensitive to magnetic fields and accelerations of the substrate.

HYBRID BEARINGS

It is likely that different magnet geometries and diamagnetic materials can significantly increase the bearing pressure (for example, oriented graphite has almost twice the levitation force of bismuth). Nonetheless, unless radically better room-temperature diamagnetic materials are found [6], diamagnetic bearings will tend to be fairly weak compared to both conventional air bearings and to the forces that can be applied using either a magnetic or electrostatic drive on the scales being considered (either drive appears capable of producing effective accelerations at least on the order of 10,000 g on a 10-μm scale using current technology; for electrostatic calculations see Lang, et al. [7]; for magnetic calculations see Pelrine [8]).

To take advantage of the drive technology that is possible on small scales, a hybrid approach is suggested. A significant problem with air and induced current bearings is the surface contact that is required until the velocity is high enough to generate adequate bearing forces. Therefore, they require high start-up force or torque, and the initial surface contact generates particulates and causes wear. A possible solution for some applications is to design the bearing to operate using diamagnetic levitation at low speeds, and to operate as an air or induced-current bearing at high speeds. Such a bearing would support light loading even at rest, heavy loading at high speeds, and would be totally noncontact.

To evaluate the feasibility of a hybrid approach, consider the wedge-film air bearing in Figure 3. For simplicity, the wedge-film bearing is assumed be flat and move in one direction only, although bidirectional, convex wedge-film bearings are also possible [9]. The bearing in Figure 3 consists of a thin, flat plate with permanent magnets attached to provide a diamagnetic bearing at rest. The drive system (electrostatic or magnetic) is assumed to exert both a horizontal force to accelerate the plate and a downward force. Starting from rest, the bearing would operate by gradually increasing the velocity of the plate until sufficient air-bearing forces are established to support higher drive forces and higher loads.

Bearing stability will not be considered here, as this issue depends on the details of the drive system. Stability aside, the load-bearing characteristics of the air bearing can be determined from [9]

$$p(x) = 6uV\theta x(b-x)/((2h+\theta b)(h+\theta(b-x))^2 \quad , \qquad (6)$$

where p(x) is the pressure as a function of distance x from the leading edge, h is the minimum film thickness, b is the width of the plate, u is the viscosity of air (0.0179 centipoise), V is the velocity, and θ is the angle of the plate.

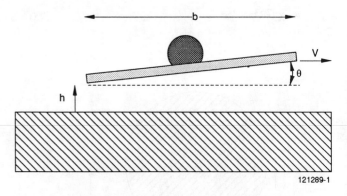

FIGURE 3 HYBRID DIAMAGNETIC AIR BEARING

The model in Eq. 6 assumes that side leakage can be neglected; i.e., the plate is long. As an example, suppose that the air bearing must support a load equal to 10,000 times its weight, as might be the case for a drive system capable of 10,000 g. The bearing load can be determined by integrating the pressure described by Eq. (6). It will be assumed that h = 1 μm is an acceptable film thickness, b = 0.02 cm (200 μm), and the plate is made of 4-μm thick aluminum. The angle θ can be allowed to vary by allowing the plate to rotate about a point (e.g., about the magnet or on a flexure shaft), or set in hardware by rigid shafts. The former method is more likely to be used in practice because it does not require precision fabrication. However, as a simplification, it will be assumed here that θ is set in hardware to 5×10^{-3} radians. With this choice, the maximum film thickness at the leading edge is 2 μm.

The thickness of the drive layer in Figure 3 is assumed to be negligible. Alternatively, the drive layer may be able to operate below the diamagnetic layer, or the drive could be physically separated from the bearing using connecting shafts. Permanent magnets with a 15-μm radius and spaced 60 μm along the length of the plate are sufficient to levitate both their mass and the mass of the plate with a gap of 7 μm, assuming a diamagnetic susceptibility of -1.2×10^{-5} cgs units. The mass density of the bearing can be calculated to be approximately 4×10^{-5} grams per unit length, so the design goal is for the bearing to support a load of about 400 dynes/cm. Numerical integration of Eq. (6) shows that V = 400 cm/s is required to support the design load.

An interesting question is how fast can such a bearing be accelerated from rest without making surface contact. Because the bearing starts at rest with a substantial gap, at start-up the air gap will act as a squeeze-film air bearing. That is, as the plate is pulled downward to the substrate by the drive, the air will resist being squeezed out of the gap. This upward squeeze-film force per unit length of bearing, f_{sq}, can be approximated as [9]

$$f_{sq} = uV_z b^3/g^3 , \qquad (7)$$

where V_z is the vertical velocity and g is the gap.

This upward force can be substantial, and will delay surface contact. The time during this delay can be used to accelerate the bearing up to speed horizontally. For example, assuming only the bearing mass is present and with a downward drive force of 400 dyne/cm, the bearing would contact the surface in approximately 15 μs starting with a 7-μm gap. However, taking into account the upward squeeze-film force, numerical analysis using f_{sq} and the assumed downward drive force shows that after 40 μs the bearing will still have a 2-μm gap. This calculation does not include any additional upward force from the wedge film. If during this time the bearing is being accelerated at 10,000 g horizontally, after 40 μs the bearing will already be at the 400-cm/s velocity necessary to support a full downward force of 400 dyne/cm.

The conclusion is that in cases where high horizontal accelerations are possible, squeeze-film effects may allow full drive loads to be applied to hybrid diamagnetic air bearings at start-up without surface contact.

DIAMAGNETIC DATA

Diamagnetism has been known for many years, and there is a great deal of scientific data on diamagnetic materials. However, some of the more common handbooks appear to contain inconsistencies in susceptibilities, perhaps because inaccuracies have not been corrected. Diamagnetic measurements require high-purity material, and parts-per-million ferromagnetic impurities can sometimes falsify the results. To further complicate matters, diamagnetism can be highly anisotropic. Therefore, caution is recommended when using diamagnetic data, especially for weakly diamagnetic materials.

CONCLUSION

Diamagnetic levitation offers some distinct advantages for micromechanics. It can provide frictionless bearings for micromechanical sensors to measure weak forces resulting from various physical effects. For micromachines, diamagnetic bearings can eliminate wear and stiction, two important factors affecting reliability and control. The chief drawback of diamagnetic levitation is fairly weak force per unit mass, which limits its use as a sole bearing support to situations at relatively low effective accelerations. One way to get around this limitation for bearings that are lightly loaded at low speeds and heavily loaded at high speeds is to use a hybrid bearing, i.e., either a diamagnetic air bearing or a diamagnetic induced current bearing. A preliminary analysis of a diamagnetic, wedge-film air bearing suggests that squeeze-film effects will allow fast start-ups. Such a bearing would eliminate the start-up problems normally associated with self-pumping air bearings.

REFERENCES

[1] J.D. Jackson, *Classical Electrodynamics,* 2nd ed., (John Wiley & Sons, Inc., New York, NY), pp. 194-198:1975.

[2] R. Pelrine, and I. Busch-Vishniac, "Magnetically Levitated Micromachines," Proc. IEEE Micro Robots and Teleoperators Workshop (IEEE, New York, NY):1987.

[3] R. Pelrine, "Maglev Microrobotics: An Approach Toward Highly Integrated Small-Scale Manufacturing Systems," International Electronics Manufacturing Technology Symposium, San Francisco, CA:Sept. 1989.

[4] R.V. Lin'kov, "Diamagnetic Suspension of a Permanent Magnet," *Sov. Phys. Tech. Phys.* 26(6), pp. 635 - 639:June 1981.

[5] B.R. Kendall, M.F. Vollero, and L.D. Hinkle, "Passive Levitation of Small Particles in Vacuum: Possible Applications to Vacuum Gauging," *J. Vac. Sci. & Tech. A*, Vol. 5, No. 4, pt. 4, pp. 2458-2462:July-Aug. 1987.

[6] V.L. Ginzburg, A.A. Gorbatsevich, Yv.V. Kopayev, and B.A. Volkov, "On the Problem of Superdiamagnetism," Solid State Communications, Vol. 50, No. 4, pp. 339-343:April 1984.

[7] J.H. Lang, M.F. Schlecht, and R.T. Howe, "Electric Micromotors: Electromechanical Characteristics," Proc., IEEE Micro Robots and Teleoperators Workshop (IEEE, New York, NY):1987.

[8] R. Pelrine, "Magnetically Levitated Micro-Robotics," Ph.D. dissertation, University of Texas at Austin, pp. 54-58:Dec. 1988.

[9] A.H. Burr, *Mechanical Analysis and Design*, (Elsevier Science Publishing Co., Inc., New York, NY), pp. 28-32:1982.

Section 6

Harmonic Motors

HARMONIC motors are a good example of how things that work well in the micro world may not work well in the macro world, and vice versa.

A harmonic motor has two paths of different lengths that roll upon each other. A cylindrical hole containing a slightly smaller cylindrical rotor is an example of this harmonic motion. As the rotor rolls on the inside of the hole without slipping, it also rotates slightly. This slight rotation of the rotor produces the output motion of the motor.

Figure 6.1 shows the rotation of the cylindrical rotor in the hole. In Fig. 6.1a, the rotor is touching the edge of the hole to the left. The white dot marks a point on the rotor. Figures 6.1b–f show the rotor as it progressively rolls around the hole in a clockwise manner. Note the position of the white dot in Fig. 6.1f after one rotation. The rotor has twisted about its axis slightly in a counterclockwise direction. As the rotor rolls repeatedly around the hole, the twisting motion produces an output torque. (One really has to try this to believe it. Try cutting two rings with different diameters out of paper and rolling one inside the other. The top and bottom of a paper coffee cup work well.)

By applying voltages between the rotor and electrodes in the cylindrical hole, the rotor can be driven electrostatically to rotate, as shown in Fig. 6.1. Each of these rotations about the hole produces another small twist of the rotor axis. This produces an apparent "gear reduction" between the electrical drive to the electrodes and the mechanical shaft output.

Harmonic motors have several advantages. One, the surfaces always roll upon each other, and there is no sliding friction. (Sliding friction in the micro world is usually large, unpredictable, and wastes energy.) Two, because the surfaces can come very close, more electrostatic energy can be generated with reasonable voltages. And three, because of the gear reduction, the motor produces large torque.

Harmonic motors are less suited to our macro world. A large harmonic motor with a large rotor has much in common with an unbalanced washing machine. However, because inertial effects decrease in the micro domain, these unbalanced forces are insignificant for micro harmonic motors.

The idea of the harmonic motor was independently developed by two groups. Their original presentations at the second Microelectromechanical Systems Workshop ("An Operational Harmonic Electrostatic Motor" and "The Wobble Motor: An Electrostatic Planetary-Armature, Microactuator") are the first papers in this section. The next two papers, "An Electrostatic Top Motor and Its Characteristics" and "Operation of Microfabricated Harmonic and Ordinary Side-Drive Motors," describe other ways to fabricate these motors.

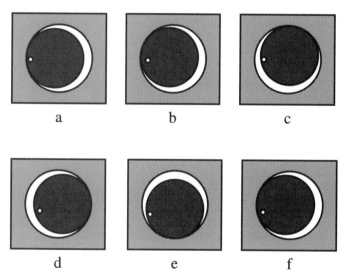

Fig. 6.1

AN OPERATIONAL HARMONIC ELECTROSTATIC MOTOR

W. Trimmer and R. Jebens

Robotics Systems Research Department
AT&T Bell Laboratories
Holmdel, N. J. USA

Abstract

This paper describes an operational harmonic electrostatic motor. A cylindrical rotor is placed inside a hollow cylindrical hole of slightly larger diameter. Electrodes on the circumference of the hole electrostatically attract the rotor and cause it to roll inside the stator. The harmonic motion of the rotor produces a "gear reduction" between the electrical drive frequency and the shaft rotation rate. This motor design has the advantage of increasing the torque of the motor. This motor has several other advantages: First, it uses the clamping force, normally larger than the tangential force used by most electrostatic motor designs, to generate the motion. Second, the sliding friction between the rotor and stator, a source of hindrance for most micro electrostatic motors, helps by keeping the rotor and stator from slipping. Third, this motor uses rolling surfaces that dissipate less energy in friction than sliding surfaces.

1 INTRODUCTION

The field of micro mechanics has many uses in electronic assembly, medical applications, micro space craft, and military applications.[1] [2] [3] Most of these applications require micro actuators. Building these micro actuators has proved to be a challenge for the micro mechanic. This paper describes a working harmonic electrostatic motor. The term harmonic refers to the simple harmonic motion of the center of the rotor. These harmonic motors will be useful for powering micro devices.

Despite the large number of macro electrostatic motor designs that have been proposed,[4] few large scale electrostatic actuators are in use. However in the micro domain, electrostatics appears to be a more advantageous force to use.[5] A number of efforts are under way to build electrostatic micro actuators.[6] [7] [8] [9] [10] [11] [12] Many of these motors suffer from several interrelated problems: friction, clamping forces, lack of good bearings, and the need for gears to match the impedance of the motor to the load. In the motor described below, these features are used to advantage below.

2 HARMONIC ELECTROSTATIC MOTORS

The structure of the harmonic electrostatic motor described in this paper is diagramed in Figure 2.1. A hole is bored through the brass stator, which consists of four segments that are electrically isolated from each other (as shown in Figure 2.2). An insulated aluminum rotor of slightly smaller diameter is placed inside this bored hole.

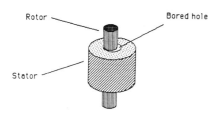

Figure 2.1 The rotor and stator of a harmonic motor.

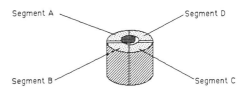

Figure 2.2 The electrically isolated segments of the stator.

The operation of the motor is described in Figure 2.3. Initially a voltage is applied between the rotor and segment A. This potential attracts the rotor to segment A as shown in Figure 2.3A. Next a voltage is applied between the rotor and segment B, causing the rotor to roll towards segment B as shown in Figure 2.3B. Voltages are then applied sequentially to segments C, D, and back to A. (A slightly more complicated sequence is to apply voltage to segments A and B, then B and C, then C and D, then D and A, etc. This sequence has the advantage of always maintaining a force clamping the rotor and stator together and generating larger forces. This sequence was used to drive the experimental motor described in Section 3.) The key to the operation of harmonic motors is the difference in circumference between the bored hole and rotor. When the rotor has rolled around the stator once with out slipping, the rotor has transversed a distance greater than its own circumference. This difference in path length causes a differential rotation between the rotor and stator. Note that the white dot in Figure 2.3 has rotated slightly during one complete electrical cycle (between Figure 2.3A and 2.3E). This "gear ratio" between the electrical drive and shaft rotation rate is $\dfrac{d_i}{d_i - d_o}$ where d_i and d_o are the inner diameter of the rotor and outer diameter of the stator, respectively. Hence, the electrical energy of many electrical cycles is delivered to one rotation of the shaft.

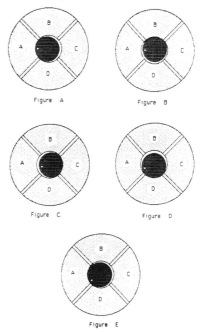

Figure 2.3 The segments of the stator electrostatically driving the rotor.

This motor has several other advantages: First, it uses the clamping force to generate the motion of the rotor. This clamping force is normally larger than the tangential force used in many micro electrostatic motor designs.[13] Second, the sliding friction between the rotor and stator, which is a source of hindrance in many micro electrostatic motors, helps in preventing the rotor and stator from slipping. The surfaces of the stator and rotor could also have fine teeth, to prevent slippage. Third, this motor (in a manner similar to the motor by H. Fujita and A. Omodaka, Hyannis, 1987)[14] uses rolling surfaces that do not dissipate energy in sliding friction.

3 EXPERIMENTAL DETAILS

The harmonic electrostatic motor made used a bored hole an eighth of an inch bored in diameter (eighth inch reamers and dowel pins were readily available). Smaller, and larger, motors of this design are feasible. Larger motors may find applications in commercial products such as auto focus cameras, and smaller motors will be useful in powering micro mechanical devices. The final bored hole diameter after lapping was 0.12520 to 0.12540 inches, and the rotor, with insulating coating, was 0.12440 to 0.12460 inches. The insulating coating was 0.0001 to 0.0002 inches (2 to 5 microns) thick. The length of the stator was 0.5 inches. A pivot bearing, as shown in Figure 3.1 was used to support the rotor. The flexible shaft of the bearing was 0.015 inches in diameter, and about 2.5 inches long. The point of this shaft sat in the bronze cone shown at the bottom of the Figure. The motor was initially operated at 85 volts, and has been excited with up to 250 volts. Before assembly, the measured difference between the diameter of the bored hole and rotor was about 0.00080 inches (0.12520 - 0.12440 inches). After assembly the rotor was observed moving about 0.00050 inches. It is not surprising that the clearance between the parts was larger than the observed motion between them, as any a small protuberance or bend in the rotor or stator will restrict their motion. The measured ratio between the electrical drive frequency and the shaft rotation rate, the "gear ratio", was 315 to 1. The calculated ratio for a 0.00050 clearance, using the equation, $d_i / (d_i - d_o)$, is 250 to 1.

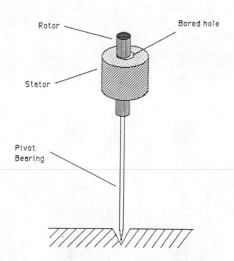

Figure 3.1 The harmonic motor resting on a pivot bearing.

4 ANALYSIS

This section develops an analytic model for the experimental motor described above. First, the electrostatic energy stored between the rotor and an electrode is calculated; then, the change in this energy as the rotor moves is determined. The change in energy is used to calculate the torque and output energy of the motor. Finally two plots of the torque as a function of the shaft angle are presented.

A sketch of the rotor and stator is shown in Figure 4.1. A voltage is applied between the electrode and the rotor (which is assumed to be conductive). The width of the electrode is w and the length l and subtends an angle of θ_E at the center of the stator. As shown in Figure 4.2, the radius of the rotor is r_r, the radius of the stator is r_s, and the angular position of the electrode center is θ. In this Figure, the rotor has been moved to the right until it touches an insulating layer of thickness i on the stator. The distance between the center of the rotor and the center of the stator is 2δ.

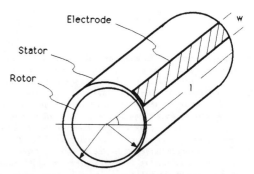

Figure 4.1 An electrode on the stator bore.

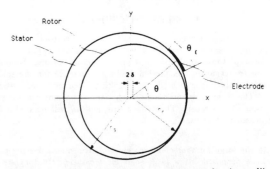

Figure 4.2 A diagram of the rotor and stator, showing radii, angles, and the displacement of the rotor to the right by 2δ.

To calculate the energy stored between the electrode and rotor, the separation, d, between the electrode and rotor as a function of θ is needed. First the intercept of the line at angle θ and the rotor (and similarly for the stator) is calculated, and then the separation, d, between these points on the rotor and stator is calculated. The equations of the rotor, stator, and line at angle θ are:

$$(x_r - \delta)^2 + y_r^2 = r_r^2 \qquad (4.1)$$

$$(x_s - \delta)^2 + y_s^2 = r_s^2 \qquad (4.2)$$

$$\tan\theta = \frac{y}{x} \qquad (4.3)$$

These equations can be solved for x_r, y_r and x_s, y_s, the intersections of the line with the rotor and stator.

$$x_r = \frac{\delta \pm \sqrt{\delta^2 + (r_r^2 - \delta^2) \sec^2(\theta)}}{\sec^2(\theta)} \qquad (4.4)$$

$$y_r = x_r \tan(\theta) \qquad (4.5)$$

$$x_s = \frac{-\delta \pm \sqrt{\delta^2 + (r_s^2 - \delta^2) \sec^2(\theta)}}{\sec^2(\theta)} \qquad (4.6)$$

$$y_s = x_s \tan(\theta) \qquad (4.7)$$

The distance between the rotor and stator, d, is

$$d = \sqrt{(x_s - x_r)^2 + (y_s - y_r)^2} \quad (4.8)$$

Now part of this gap, d, between the rotor and stator is filled with an insulator of thickness i, and the rest is filled with air. If the relative permittivity of the insulator is ϵ_r, then the effective distance between the rotor and stator, d_E, is

$$d_E = d - i\left[\frac{\epsilon_r - 1}{\epsilon_r}\right] \quad (4.9)$$

This effective d_E is used to generate the plots below.

The potential energy stored in the capacitor formed by the rotor and electrode is now calculated.

$$U = \tfrac{1}{2}CV^2 = \frac{\epsilon_o w l V^2}{2 d_E}, \quad (4.10)$$

and the torque produced by the motor is

$$\tau = \frac{dU}{d\phi}. \quad (4.11)$$

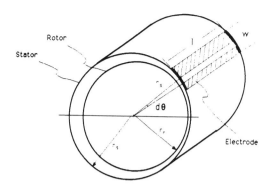

Figure 4.3 The electrode of width, w, and length, l, is broken into elements $R_s\, d\theta$ to integrate for the total electrostatic energy.

$$U = \int_{\theta_s}^{\theta_e} dU = \int_{\theta_s}^{\theta_e} \frac{\epsilon_o l V^2 r_s}{2 d_E}\, d\theta, \quad (4.18)$$

where d is the function of θ given by equations 4.4 to 4.8. Fortunately, as is shown below, this does not need to be integrated to find the torque.

Consider the functions

$$f(x) = \frac{dF(x)}{dx}. \quad (4.19)$$

Now

$$\int_{x_1}^{x_2} f(t)\, dt = F(t)\big|_{x_1}^{x_2} = F(x_2) - F(x_1), \quad (4.20)$$

and

$$\frac{d}{dx}\int_{x_1}^{x_2} f(t)\, dt = \frac{d}{dx}\left[F(x_2) - F(x_1)\right] \quad (4.21)$$

$$= \frac{dF(x_2)}{dx} - \frac{dF(x_1)}{dx} = f(x_2) - f(x_1).$$

From above,

$$\tau = \frac{r_i}{r_i - r_o}\frac{dU}{d\theta} \quad (4.22)$$

$$= \frac{r_i}{r_i - r_o}\frac{d}{d\theta}\left[\frac{\epsilon_o l V^2 r}{2}\right]\int_{\theta_s}^{\theta_e} \frac{1}{d_E}\, d\theta.$$

$$= \left[\frac{r_i}{r_i - r_o}\right]\left[\frac{\epsilon_o l V^2 r}{2}\right]\left(\frac{1}{d_E}\bigg|_{\theta_e} - \frac{1}{d_E}\bigg|_{\theta_s}\right)$$

To find the torque, only the values of d_E at the ends of the electrode need to be known. Numerically integrating the torque over the angle the shaft rotates, ϕ, gives the energy produced by the motor.

$$U_{generated} = \int \tau\, d\phi \quad (4.23)$$

There are two different angles associated with this motor. The angle ϕ is the angle of rotation of the shaft, and the angle θ is the angle to the energized electrode as shown in Figure 4.2. One of the advantages of this motor is the large ratio that can be obtained between θ and ϕ. If r_i and r_o are the radius of the inner and outer circumferences that roll against one another, and n_i and n_o are the number of times the contact point travels around the inner and outer circumference respectively, then the distance traveled by the contact point is

$$2\pi r_o n_o = 2\pi r_i n_i = distance\ contact\ point\ travels. \quad (4.12)$$

The angles through which the contact point and shaft turn are

$$\theta = 2\pi n_o \quad (4.13)$$

$$\phi = 2\pi(n_o - n_i), \quad (4.14)$$

and the ratio of θ to ϕ is

$$\frac{\theta}{\phi} = \frac{n_o}{n_o - n_i} = \frac{r_i}{r_i - r_o}. \quad (4.15)$$

The torque can now be expressed as

$$\tau = \frac{dU}{d\theta}\frac{d\theta}{d\phi} = \frac{dU}{d\theta}\frac{r_i}{r_i - r_o} \quad (4.16)$$

Note that the torque is increased by the "gear ratio" of the motor.

To use equation 4.16 to calculate the torque, the electrostatic potential energy between the electrode and rotor must be calculated. In Figure 4.3 the electrode of length l and width w is shown on the stator. In this Figure, one can also see the varying gap between the rotor and stator. The incremental energy, dU, of the white stripe on the electrode is

$$dU = \frac{\epsilon_o l V^2}{2 d_E}\, dw = \frac{\epsilon_o l V^2}{2 d_E}\, r_s\, d\theta, \quad (4.17)$$

where $dw = r_s\, d\theta$ is the incremental length along the electrode. Because the electrodes are measured in millimeters, and their separations measured in microns, the fringing fields at the edges contain a small fraction of the energy and will be ignored. The integral of the increments, $r_s\, d\theta$, from the start of the electrode, θ_s, to the end of the electrode, θ_e, gives the total energy, U.

The top curve in Figure 4.4 shows the torque for motor described in section 3. Here it is assumed the rotor can move a total distance of 12 microns, the insulator thickness is 4 microns, the relative permittivity, ϵ_r, is 3, and 250 volts is being applied to the two electrodes ahead of the rotor. The energy produced by this motor per shaft rotation is 4.4 millijoules, and the average torque is 7.0×10^{-4} newton-meters (equals 7.14 grams force at a one centimeter radius). In Figure 4.4, the bottom curve is for this motor when $\epsilon_r = 1$.

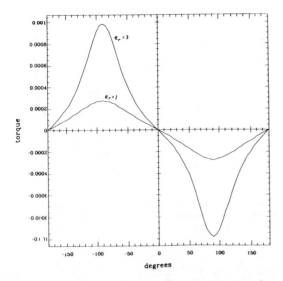

Figure 4.4 A plot of calculated torque vs electrode angle. The upper line assumes an insulator with $\epsilon_r = 3$, and the lower line assumes $\epsilon_r = 1$.

The plot in Figure 4.5 shows the torque from a narrow electrode that is a hundredth of a radian wide (0.573 degrees). The minimum torque from this nearly infinitesimal electrode is 1×10^{-5} newton meters, at about 22 degrees. From this plot it is seen that the electrodes between about 5 and 60 degrees are the most effective in producing torque.

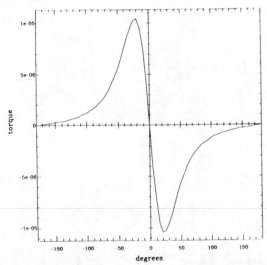

Figure 4.5 Plot of torque vs angle for a very narrow (0.573 degrees) electrode.

The performance of the above motor can be improved by decreasing the thickness of the insulator (electrical breakdown of the insulator limits how thin the insulator can be), increasing the dielectric constant of the insulator, or increasing the voltage. The energy and torque scale as the voltage squared. If the insulator thickness of the experimental motor above is decreased from 4 to 2 microns, the energy and torque are increased to 7.8 millijoules and 1.2×10^{-3} newton - meters. If the dielectric constant of the insulator is increased to 20, the average torque is 29×10^{-4}, and if the dielectric constant is increased to 2000, the peak torque produced by the motor is 0.7 newtons - meters. These dielectric constants are typical for capacitors. Electrolytic capacitors have dielectric constants of between 8 and 27, and ceramic capacitors have dielectric constants of between 2,000 and 6,000.[15] There are many interesting research directions for harmonic motors, however, fabrication techniques on this scale are a challenge. These numbers above are based on approximate theoretical models, and should be considered goals or limits of what can be done.

5 ACKNOWLEDGEMENTS

The authors would like to thank G. Kraft for his superb help in machining the motor components, A Jungreis, R. Mahadevan, M. Mehregany, T. Poteat, and P. Trickey for their stimulating discussions on motors, and J. Jarvis for his encouragement.

REFERENCES

1. Richard Phillips Feynman, *There's Plenty of Room at the Bottom, in Miniaturization,* ed. H. Gilbert, Reinhold Publishing Corp., New York, 1960.

2. *Small Machines, Large Opportunities: A Report on the Emerging Field of Microdynamics,* An National Science Foundation report of the Workshop on Microelectromechanical Systems Research, 1988.

3. The Micro Spacecraft for Space Science Workshop, Sponsored by NASA and the Strategic Defense Initiative Organization, Jet Propulsion Laboratory, Pasadena, California, July 6 and 7, 1988.

4. O. D. Jefimenko, *Electrostatic Motors,* Electret Science Company, Star City, 1973.

5. W. S. N. Trimmer, *Micro Robots and Micro Mechanical Systems,* to be published in *Sensors and Actuators.*

6. H. Fujita, A Omodaka, *The Fabrication of an Electrostatic Linear Actuator by Silicon Micromachining,* IEEE Transactions on Electron Devices, Vol. 35, No. 6, June 1988, p 731 to 734.

7. Hiroyuki Fujita, and Akito Omodaka *The Principle of an Electrostatic Linear Actuator Manufactured by Silicon Micromachining,* Transducers' 87, The 4th International Conference on Solid-State Sensors and Actuators, pp 861 to 864, June 1987.

8. J. H. Lang, M. F. Schlecht, and R. T. Howe, *Electrostatic Micromotors: Electromechanical Characteristics,* IEEE Micro Robots and Teleoperators Workshop, Hyannis, Massachusetts, November 1987.

9. R. H. Price, J. E. Wood, S. C. Jacobsen, *The Modelling of Electrostatic Forces in Small Electrostatic Actuators,* IEEE Solid-State Sensor and Actuator Workshop, Hilton Head Island, SC, June 1988.

10. W. S. N. Trimmer, and K. J. Gabriel, *Design Considerations for a Practical Electrostatic Micro Motor, Sensors and Actuators,* Vol 11, No. 2, p. 189 to 206, 1987.

11. W. S. N. Trimmer, K. J. Gabriel, and R. Mahadevan *Silicon Electrostatic Motors,* Transducers' 87 The 4th International Conference on Solid-State Sensors and Actuators, p 857 to 860, June 1987

12. J. E. Wood, S. C. Jacobsen, K. W. Grace, *SCOFFS: A Small Cantilevered Optical Fiber Servo System,* IEEE Micro Robots and Teleoperators Workshop, Hyannis, Massachusetts, November 1987.

13. R. Mahadevan, *Capacitance Calculations for a Single-Stator, Single-Rotor Electrostatic Motor,* IEEE Micro Robots and Teleoperators Workshop, Hyannis, Massachusetts, November, 1987.

14. H. Fujita, A. Omodaka, *Electrostatic Actuators for Micromechatronics,* IEEE Micro Robots and Teleoperators Workshop, Hyannis, Massachusetts, November, 1987.

15. D. M. Trotter, *Capacitors,* Scientific American, pp 86 to 90, July, 1988.

THE WOBBLE MOTOR: AN ELECTROSTATIC, PLANETARY-ARMATURE, MICROACTUATOR

S. C. JACOBSEN, R. H. PRICE, J. E. WOOD,
T. H. RYTTING and M. RAFAELOF

Center for Engineering Design
University of Utah
Salt Lake City, Utah 84112

1.0 INTRODUCTION

1.1 MICRO ELECTRO MECHANICAL SYSTEMS (MEMS)

Micro Electro Mechanical Systems (MEMS) can be described as machines constructed of small moving subelements which have characteristic dimensions in the range of about 0.5 to 500 microns. The motivation for the development of such devices is the possibility of extending the advantages of small scale, which are currently available in electronic devices, to include mechanical systems with moving parts.

1.2 ADVANTAGES POSSIBLE IN MEMS

Machines composed of *large numbers of small subsystems* can be designed to take advantage of three principal design freedoms:

(1) If a given system is isomorphically reduced in size, the shift in length, area and volume ratios alters the *relative influence* of *various physical effects* which determine the overall operation of the machine. For example: a) By reducing size, the structural stiffness generally increases relative to inertially imposed loads. (b) Electrostatic forces gain against mechanical, inertial and magnetic forces thereby making electrostatic motors more competitive against magnetic alternatives. (c) Various electrical properties are changed; for example, the Paschen effect [16] allows a substantial increase of breakdown voltage in very small gaps. (d) Fluid mechanical processes, and their relationship to surrounding structures, are radically altered. (e) As the size range of the smallest oriented subelement in a machine is decreased, the advantages of close packing of interacting elements (the "structural crystallinity") is enhanced. Such basic alterations in machine structure can potentiate improvements in the speed, precision and functional homogeneity of heat and mass transport processes. (f) Using micron-sized field-generating structures, such as conductors, electrets and magnets, the strength and "graininess" of electromagnetic-field-features can be increased. Strong, textured fields can be used to produce physical interactions not available in larger systems.

(2) If a system consists of large numbers of small subsystems, the internal structure can be *arranged* so that passive and active sub-elements (structures, sensors and actuators) are in appropriate *parallel and series* combinations. Thus the relationships between *flow* (current, force, etc), *effort* (potential, velocity), *power* (flow*effort) and *impedance* (effort / flow) can be prescribed. In this way the behavior of the machine can be *designed-in* as an *intrinsic* system property. This approach can yield better systems which avoid the complex sensing and computation actions typically used to externally impose operational constraints on machine function (as is typically accomplished by feedback control methods).

(3) The use of large numbers of small elements permits a more uniform arrangement of elements throughout a system. With distributed sensing and control, a machine can exhibit: (a) improved robustness through the incorporation of substantial redundancy, and (b) improved precision via better knowledge of system state (i.e., machines can be "locally smart").

1.3 PROJECTIONS FOR MEMS

As the tools necessary to analyze and fabricate micro systems evolve, their application will improve the capabilities of existing systems into which they are integrated. Progress in MEMS can lead to the development of a new class of mechanical, optical, fluid, thermal, and chemical machines which exhibit otherwise impossible improvements in *performance, reliability and economy.*

For example: micro sensors will include active elements which will permit more accurate, null-based sensor operation; advanced optical systems will use high-bandwidth micro servo systems to control light; micro laboratories will be developed in which sensors, valves, pumps, chemical reactors and controllers reside together in a small disposable package; micro production factories, which include micro manipulation subsystems, will be designed in which materials enter, are mechanically processed and then ejected in a finished form.

Efforts should now focus on the capability to efficiently generate and use subelements such as sensors, micro actuators and controllers. This includes their design, fabrication and *integration* into systems with desired internal structure. Important subsystems include: motors; position, velocity, force and pressure sensors; membranes, valves, reactors and pumps; optical switches, reflectors and modulators; mechanical actuators, levers, flexures, bearings and couplings; packages, seals, interfaces and others. All of these elements must by integrable into complex packages using reliable fabrication methods. Finally, a major thrust should be improving the knowledge of potential commercial applications for MEMS so that resources can be focused on accelerating the development process.

The generation of concepts, design tools and fabrication approaches is already moving ahead. Sensors primarily using silicon-based integrated circuit technology are emerging from multiple groups [2,5,6,10,12,17]. Efforts to produce micro actuators are underway using a variety of approaches, including piezoelectric materials, electromagnetically and thermally activated materials, shape memory alloys, electromagnetic-field based machines, and others [3,4,5,6,7,8,9,11,14,15,17,18,20,22].

2.0 MEMS AND ACTUATION SYSTEMS

2.1 USING LARGE NUMBERS OF SMALL ELEMENTS

Actuation systems composed of numerous, small elements could benefit, as biological muscle does, from several important features: (a) achieving high local field strength in the regions between very small *force-producing-elements* (FPEs) (as in the "s" units of natural muscle), (b) using geometrically complex structures to arrange, in parallel, the FPEs such that they closely interact across large interfacial areas (For example, interfacial areas between the actin and myosin filaments in natural human striated muscle reach magnitudes of approximately 10^6 square cm per cubic cm. Thus, muscle can be thought of as an area machine.) and, (c) combining the output of these structures in appropriate parallel-series combinations in order to develop desired relationships between force, displacement and velocity (as in the chains of sarcomeres in natural muscle) [21]. Note that in muscle, numerous sensors (such as spindle organs) are integrated together with contractile structures to monitor lengths and rates of contraction to provide both intrinsic and feedback control of local system functions.

Reprinted from *Proceedings IEEE Micro Electro Mechanical Systems*, pp. 17-24, February 1989.

In mechanical terms, an ideal "macro" actuator would be composed of many micro actuators each of which is internally configured to deploy many FPEs in close proximity across large interfacial areas. The close proximity would allow electromagnetic FPEs to operate in regions of high field strength (due to inverse distance relationships defined by field theory). And, by combining the elements in parallel, the forces generated would be added within the micro actuator. The macro actuator would appropriately connect (combine mechanically) the micro actuators in parallel and series combinations so that the actuator develops desired force, movement and impedance qualities [13].

Such a macro actuator system could achieve major advantages over existing actuators. For example: (a) Large field strengths could be generated and used for force production using smaller voltages. (b) Tradeoffs between output (displacement and force or torque) and impedance could be achieved structurally. Force can be increased by adding FPEs in parallel and displacement can be increased by adding FPEs in series. Gear-type reductions could be avoided, which is desirable since, as they increase force or torque, mechanical output impedance is also increased by the square of the reduction ratio. (c) In the electrostatic case, static loads can be supported with minimal power consumption.

Of course, the development of such complex systems will be a substantial undertaking. Progress will necessarily evolve in stages since many unknown factors, related to concepts, design and fabrication remain to be addressed.

2.2 NEW CHARACTERISTICS FOR MEMS ACTUATORS

MEMS-related actuators may be said to exist already, in the form of driven membranes (e.g., in micro pumps), small-scale piezoelectric positioners, etc. To realize many of the most exciting opportunities in MEMS, however, actuators with different properties are needed. Actuators for MEMS must; (a) be capable of producing large angular or linear displacements, (b) more effectively utilize three dimensional geometries so that the energetic volume and thus power capability is increased, and (c) effectively interface with, and transfer power density to the outside environment. That is, actuators are needed which are the small-scale equivalent of macroscopic motors. As single units such devices will likely find immediate application, in areas such as microsurgery, optics and others. As already mentioned, in arrays, such motors could provide macroscopic motors which offer substantial advantages.

Although magnetic actuation is the basis for essentially all successful macroscopic electric actuators, widespread agreement has emerged that, on sufficiently small size scales, electrostatic actuation has appealing advantages. The arguments for this have been reviewed elsewhere [4,7,11,14,20]. Electrostatic actuation, however, is relatively underdeveloped. Moreover, since designs and viewpoints based on magnetic motor design are not directly applicable, it is far from clear what general configuration might ultimately prove the most suitable micro actuator for each application [19].

2.3 MICROACTUATOR CONFIGURATION ALTERNATIVES

Many alternative configurations, which deploy FPEs across interfacial areas, can be considered as candidates for the production of linear or rotary motion. Four general approaches are briefly introduced below.

Figure 1 shows a simple system of flat, parallel planes which can produce linear actuation. We call this configuration the "multiple interleaved-plane-actuator" or MIPA. Field emitters, consisting of transverse lines of conductive material, are placed on insulating planes and are raised to alternating potentials as shown. Classes of field emitters in the ground structure are switched (thereby performing a commutation function) so that they shift location as a group in order to maintain force as the output plates undergo relative motion. All FPE lines are mechanically in parallel so that the forces add, while the displacement between the ground and output planes is the same as that of the FPEs. The MIPA is an appealing concept on which much work remains to be done (such as achieving practical fabrication in the required size range, maintenance of plate spacing and alignment during operation, friction, force instability of plates, contamination, etc.).

Figure 2 shows another basic linear motion configuration for the deployment of FPEs, consisting of an arrangement of filaments which are interdigitated in a hexagonal manner similar to natural muscle. This "interdigitated filament actuator", or IFA, appears to be a promising scheme with FPEs arranged longitudinally along the fibers so that the FPEs are in close proximity and the ensemble of FPEs is interacting across large interfacial areas [8]. Obviously the IFA presents substantial problems in analysis, design and fabrication and must therefore be evolved carefully and unfortunately slowly.

Figure 3 shows a classical disk-on-bearing system (DOBS) for producing rotary motion. Many such motors have been proposed in the past, and a number are currently under development [3,14,15]. Using polysilicon-etching techniques, operational motors have been constructed in the size range of hundreds of microns with operational speeds apparently less than 1,000 RPM [3]. The DOBS approach has some appeal and, of course, is familiar. But, in systems currently being developed several possible problems are apparent. For example: (a) The fabrication methods utilized limit the geometry and the range of materials available for realization of elements. (b) The fabrication methods which incorporate depth-limited etching restrict motor geometry and length and hence the active interfacial area for FPEs. (c) The DOBS approach requires the use of a relatively large stationary bearing with the associated problems of friction and alignment. (d) The motors, in order to exploit the advantages of small scale, necessarily produce power at very high speeds (but at low torque), thus requiring gear reduction in all but free running applications. Furthermore, and unfortunately, million RPM micro gears, of classical design, may or may not be feasible. (e) Fabrication methods produce surfaces of limited quality (low relative dimensional precision), thus problems can be expected with alignment and unwanted contact between moving and stationary components.

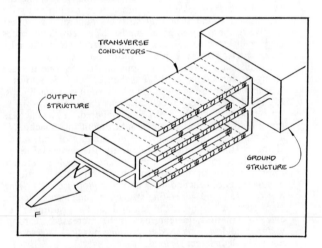

Fig 1. Multiple-Interleaved-Plane Actuator (MIPA).

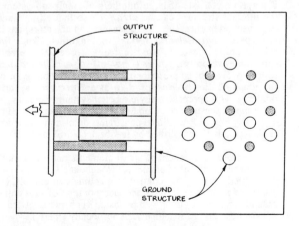

Fig. 2. Interdigitated-Filament-Actuator (IFA).

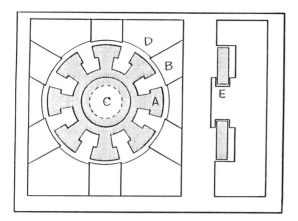

Fig. 3. Disc-on-Bearing System (DOBS).

2.4 PROBLEMS WITH MICRO ACTUATORS

Any design for an electrostatic micromotor must face some very general and very serious difficulties. The generation of strong electrostatic forces, from reasonable voltages, requires that gaps across which fields act, be extremely small (submicron if possible). Friction becomes a crucial consideration for relative motion of surfaces across such gaps (e.g., the motion of the rotor pole past the stator pole in a DOBS as in Fig. 3). Relative surface smoothness is much worse for micro than macroscopic surfaces, and friction, though ill understood, is expected to be a substantial influence. This difficulty is enhanced by the destabilizing tendency, present in any electrostatic device, for conductive or dielectric elements to clamp to any electrically active element ("image force" attraction).

Numerous alternatives have been explored in order to identify specific motor designs which can meet the above challenges. Within our group, work has begun on the development of a simple micro actuator, the wobble motor (WM). More important than just developing another micro motor, the WM project is serving as a construct on which to expand our understanding in areas of analysis, micro-experimentation and micro-fabrication.

3.0 THE WOBBLE MOTOR

3.1 GENERAL CONCEPT

As shown in Fig. 4, the rotor R contacts the stator at the rolling point, RP. The stator walls include electrically conductive, longitudinal segments B, to which voltages can be sequentially applied in order to systematically induce image forces on the electrically conductive but insulated rotor (experimental motors have also functioned with dielectric rotors). The resulting image forces induced in the rotor produce a torque about the rolling point. Reactions at RP are supported either by contact friction or by geometric features such as waves, notches or gear-like teeth. The rotor can be short like a disk, or long like a rod.

The rotor axis, RA, moves in a circular path around and parallel to the stator axis SA, at the wobble frequency ω_w. At steady-state, ω_w matches the frequency ω_s at which the stator segments B are cycled. Note that stator-segment recruitment logic can range from simple, open-loop stepping to full commutation where measurement of rotor position provides the information necessary to effectively coordinate stator-segment excitation. In more complicated systems, where rotor and stator include additional structural detail, improved torque generation schemes can be implemented.

The rotor output frequency, ω_r, is related to the wobble frequency by,

$$\omega_r = \omega_w [(R_s/R_r) - 1] \quad (1)$$

where, depending on the relative radii of the rotor R_r, and the stator R_s, the frequency ω_r can be faster than, equal to, or slower than the frequency ω_w. As R_r and R_s become close, the mechanical reduction of the rotor (ω_w/ω_r) approaches infinity. Since the axis RA of the rotor wobbles in a parallel fashion around the stator axis SA, power take-off from the RA to the SA axis can be easily accomplished by using a flexural member extended out from the rotor or by flexible or pinned linkages, such as Oldham couplings, which function as torque transmission mechanisms which are insensitive to parallel-axis displacement. It should be noted that the design motion of the WM is similar to the classical high frequency mechanical wobbling instability that, in larger systems, produces a sound known as bearing screech.

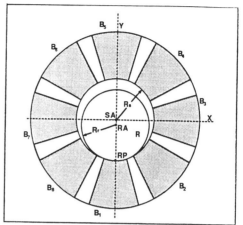

Fig. 4. Geometry of the Wobble Motor.

3.2 ADVANTAGES OF THE WOBBLE APPROACH

Advantages of the wobble configuration include:

(a) The wobble motor kinematics allows the maintenance of a precise small gap determined only by the thickness of the insulator coating on the rotor or stator.

(b) Any conducting or dielectric element of the motor is pulled towards regions of higher electrical field strength. This attraction (the "image force") is the origin of the instability in most microactuator designs, e. g., in the DOBS; in the WM on the other hand this is the driving force.

(c) Bearing friction is substantially reduced (e.g., as compared to DOBS). The nature of rolling friction for small elements is not at all clear, but the macroscopic picture of rolling friction would suggest that the close matching of the rotor and stator diameters avoids significant deformations, and friction should be minimal.

(d) The use of image forces permits, in principle, the use of an electrically passive (unconnected) conductive or dielectric rotor. In practice, charge buildup on insulating coatings appears to be a problem in certain designs, and questions about the charging of insulators must be answered before passive rotor operation is fully reliable.

(e) By lengthening the wobble rotor to form a rod, the interfacial area across which force is generated can be substantially enlarged over disk configured systems. Note that in disk type geometries, where bearing support of the rotor is at least one of the limiting features of the design and where motor thickness is determined by etch depth, force producing interactions occur only at the thin perimeter of the rotor. For these configurations, unlike the wobble motor, the advantages of small scale are not fully exploited since only a negligible fraction of the total surface area is active.

3.3 RELATED CONFIGURATIONS

Of course, the concept of actively rolling members can be extended to a number of other geometries such as; (a) **inverted** wobble motors where the rotor is a hoop-like structure which rolls externally on a cylindrical stator, b) rotors which roll on or in stators which are noncylindrical (planar, conical or spherical) and noncircular (polygonal or geared). Furthermore, many alternatives exist for the combination of wobble motors into parallel and/or series arrangements.

4.0 WOBBLE MOTOR FABRICATION

4.1 BASIC FABRICATION PROCESSES

A number of fabrication approaches are candidates for the construction of micro-actuators and sensors. These approaches can be classified into three broad categories which are:

(1) *Assembly* of mechanical micro parts which are generated in separate processes;

(2) *Liberation* of completely assembled systems from solid blocks, by etching, machining or vaporizing;

(3) *Solidification* of partially or completely assembled components out of liquids, powders or vapors using regionally localized processes such as laser induced polymerization or thermal fusion.

Of course, a specific design could be realized using a combination of these fabrication methods.

4.2 FIVE WOBBLE MOTOR ALTERNATIVES

Five WM designs are being investigated using different fabrication schemes defined as: (1) Direct micro assembly (DMA); (2) Electro discharge machining (EDM); (3) Cylindrical photolithographic etching methods (CPE); (4) Coextrusion of metal and plastic (CMP) and (5) Silicon chip stack (of from one to many chips) with etched holes aligned to form the stator cavity (SCS).

4.2.1 Direct Micro Assembly

Figure 5 shows a DMA system which was fabricated and operated. The stator consists of 32 copper rods (135 microns diameter) insulated and potted in epoxy. The stator is 1,350 microns in diameter and 15-20 mm in length. Rotors have been fabricated of both conductors and insulators with diameters of 1,250 microns and lengths of 15 mm. The motor has been operated using a multi-channel system of stepping amplifiers. Wobble speeds of 250 wobbles per minute (WPM) have been achieved.

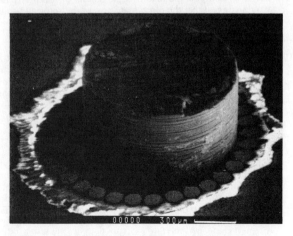

Fig. 5. Direct micro assembly (DMA) Wobble motor.

4.2.2 Electro Discharge Machining

Figure 6 shows the EDM system which, for now, has been selected as the test bed system for the Wobble project. The EDM system can be reliably fabricated out of acceptable materials with satisfactory tolerances and repeatability. As shown in section 5.2, the EDM system is being extensively modelled with a finite-element package [1]. Simulations show good correlation with preliminary experimental measurements.

Fig. 6. Electro discharge machining (EDM) Wobble motor.

4.2.3 Cylindrical Photolithographic Etching

Figure 7 shows the CPE system which includes a quartz rotor approximately 180 microns in diameter and 2,000 microns in length. The rotor operates in a stator cavity 200 microns in diameter and 1,000 microns in length. The stator was formed by first depositing a metallic surface and then etching longitudinal conductors onto a 200 micron glass rod with flared base. The rod was then potted in polyimide resin and later removed by hydrofluoric acid, leaving the conductors and resin to form the inner surface of the stator cavity. The CPL system has operated, however, problems remain with respect to achieving reliable control of rotary motion.

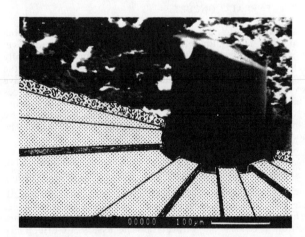

Fig. 7. Cylindrical photolithographic etching (CPE) Wobble motor. A sketch of the conductors under the insulating layer is superimposed on the photo.

4.2.4 Coextrusion of Metal and Plastic

Figure 8 shows the CMP system in which ten 127 micron diameter stainless steel wires were coextruded with a polyethylene thermoplastic to form the stator of a wobble motor. The motivation for this approach is the potentially economic micro fabrication of lengths of motor (Wobble motors by the meter). Since electrical interconnects are not yet developed, the CMP has not yet been experimentally operated.

Fig. 8. Coextrusion of metal and plastic (CMP) Wobble motor.

4.3 THE EDM WOBBLE MOTOR - A SPECIFIC DESIGN

The EDM system includes eight, pie-shaped, steel stator segments interlocked by bonded epoxy films which are approximately 70 microns thick. The inner diameter of the stator is machined to 560 microns. The stator outer diameter is typically 9,525 microns and its length is approximately 5,000 microns. The large base, which includes electrical interconnections, is included for experimental convenience. The wall thickness of the stator could be substantially reduced to make the small size of the motor more apparent (this has been done with one motor, as in Fig. 6 which has a 1,600 micron outer diameter, with no loss whatever in performance).

Aluminum rotors of varying diameters and varying lengths have been used. Rotor insulator coating thickness is a critical parameter (thin is better) and thus a number of coating techniques have been investigated including polymeric films, anodization and other deposited insulators.

The EDM motor is driven by circuits which include a power supply capable of zero to 1,000 volts, connected to eight controlled, high speed, electronic switches. The switches, which directly drive the stator elements of the motor, can be recruited at an adjustable stepping frequency (equal to the wobble frequency in the absence of slip) with a desired duty cycle, which determines driver segment voltage overlap. Since the driver operates as a stepper at a set frequency, regardless of wobble speed, effective torque is not developed unless the rotor operates at synchronous speed. The electronic driver is also configured to accept rotor position information so that when a future WM includes an appropriate sensor, the motor can be operated in a truly commutated mode.

5.0 SIMULATION RESULTS

5.1 MODELLING OF THE EDM MOTOR

Modelling of the EDM WM has been carried out with MAXWELL, an electrostatic finite-element package from the ANSOFT Corporation [1]. (Another approach to the study of an idealized WM is described below.) Figure 9 shows the geometrical input to the problem and defines regions. (For clarity in Figs. 9-11, the rotor is pictured as significantly smaller than the stator cavity; in actual WMs approximate matching of rotor and stator diameters was important for good performance.) The MAXWELL package allows the inclusion of the dielectric effects of the anodization coating A around the conducting rotor core B, and of the air gaps C separating stator segments 1-8. Figure 10 shows the default mesh (the least refined mesh) generated as part of the finite element solution, and Fig. 11 shows electrostatic equipotentials for the solution in the case that only stator segment 1 is driven, with the remaining seven segments, and the rotor, at ground.

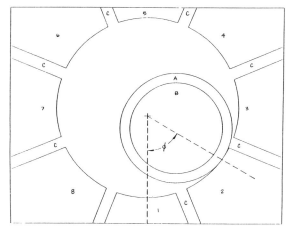

Fig. 9. EDM wobble motor geometry. From one to eight segments can be combined to form "one" electrically active segment.

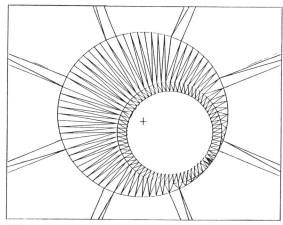

Fig. 10. "Default mesh" (least refined mesh) for the finite-element simulation.

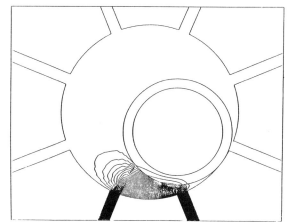

Fig. 11. Electrostatic equipotential lines for the simulation of the WM of Fig. 10.

Figure 12 shows four plots of electrostatic torque computed as a function of the rotor position angle ϕ for a number of cases. In all examples the stator cavity diameter is 571.5 microns, the air gaps 41.1 microns wide the anodization coating has $\varepsilon = 6$, the driving voltage is 300 volts, and the rotor is grounded. Two electrical configurations are represented. For curves A, C, and D (as in Fig. 11) only a single stator segment, segment 1, is driven; for curve B four adjacent segments are driven. (For curve B the rotor position angle ϕ is measured from the midpoint of the four driven segments so that, as in the other curves the torque vanishes at $\phi = 0$ by symmetry.) Curves A and B represent identical constructions, with rotor O.D. 495.3 microns and anodization thickness 30.5 microns; they differ only in the number of driven segments. Comparison of the curves shows the greater peak torque at greater rotor position angle for four driven segments. Curve C shows the considerable increase in torque achieved by increasing the rotor O.D. to 560 microns (11.5 mm less than the stator I.D.). Curve D shows the effect both of the large (560 microns) rotor and of decreasing the anodization thickness to 1 micron. (In this case the maximum torque is so large, 745 dyne-cm/cm, that the results have been reduced by a factor of 10, to fit the graph.)

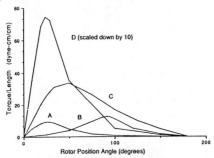

Fig. 12. Simulation results for torque as a function of rotor position angle for different numbers of driven stator segments, different rotor diameters, and different insulator thicknesses.

5.2 MOTOR TORQUE AND RIPPLE

The eight segment EDM motor of Fig. 9 was used to explore recruitment schemes (activation) for stator segments. Two simple schemes were initially used to study performance characteristics such as: torque, ripple, power and efficiency, in terms of various design parameters.

Curve a in Fig. 13 was generated by combining angularly-shifted torque plots from curve A of Fig. 12. In this case, single elements were sequentially activated and deactivated with respect to rotor position at the torque curve crossovers of each element. The simulation thus represents a commutated motor operating against a load at constant speed and at a fixed stator voltage. Curve b in Fig. 13 was similarly generated by combining angularly-shifted torque plots corresponding to curve B of Fig. 12. In this case, single elements were again sequentially activated with respect to rotor position. However, elements were deactivated in a delayed fashion such that, on average, four elements were active at a given time. This case represented a commutated motor, with a wider driver segment, operating against a load at a constant speed and at a fixed stator voltage. The curves show peak torque and ripple for two simple recruitment strategies. Obviously other recruitment strategies, optimized to enhance performance, can be implemented, however their discussion is beyond the intent of this paper.

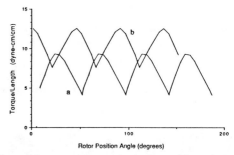

Fig. 13. Motor torque and ripple, with active segments commutated.

5.3 CLOSED FORM SOLUTION FOR THE WM

A closed-form (in contrast to finite-element) solution for the Wobble Motor fields can be found if the following idealizations are made: (a) The dielectric constant of the insulator coating is unity so that the insulator serves the purpose of maintaining a nonzero gap between the conducting rotor and conducting stator, but is itself electrically inert; (b) The stator segments have negligible gaps separating them.

The electrostatic problem then consists of a conducting cylinder (the rotor), interior to a conducting cylindrical surface (the stator) on which the potential is everywhere fixed. This problem can be solved, for example, with bipolar coordinates. Though this idealization omits several important realistic effects (most important, the electrical interaction with the insulator coating), it is useful for exploring the geometric parameter space, especially for extreme geometries (very narrow insulator thickness) in which very strong E-fields with large spatial gradients would pose a challenge for a finite-element approach.

Studies with this solution reveal a complicated interplay of various geometric parameters. Figure 14 shows the "maximized average torque" (MAT): As the rotor rolls an angular distance equal to the angular width of a single stator segment, the average torque is computed. For example, for 45° electrodes, a torque curve like that of curve A in Fig. 12 is computed and the average of the torque is calculated over an angular range of $\Delta\phi = 45°$ near the torque peak. The precise position of the $\Delta\phi = 45°$ range is adjusted to maximize the result. This MAT is the meaningful measure of the average torque a motor could deliver when used as a stepper motor with optimized recruitment of a single electrode.

The results are shown in Fig. 14 for two distinct basic structures: Model 1, with a stator diameter of 400 microns and a rotor diameter of 380 microns, represents what might be built with minor improvement of present approaches. To illustrate the dramatic effect of closely matched diameters, another configuration, Model 2, has stator and rotor diameters of 400 microns and 398 microns. The figure shows the MAT for different insulator thicknesses and for different segment widths (corresponding to a stator of 8, 16, 32, and 64 segments).

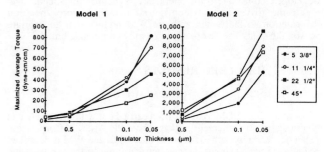

Fig. 14. Maximized average torque (MAT) vs insulator-inverse-thickness for single stator segment motors with four different segment widths.

In both cases the torque corresponds to 300 volts on a single driver segment with the rotor and all other segments grounded. The approximate inverse relationship of torque and insulator thickness is demonstrated by plotting MAT versus 1/(insulator thickness). In all cases the torque increases with a decrease in insulator thickness. (This increase will be somewhat reduced by the electrostatic interaction, omitted in the idealization, with the dielectric insulator.) Even more striking is the enormous torque produced by the model with the closely matched diameters. (For a given insulator thickness, the MAT increases approximately as the inverse of the fractional difference in stator-rotor diameters.)

The complex interplay of geometric parameters appears in the fact that the narrowest (5.375 degrees) electrode produces the greatest MAT for Model 1, but the least torque for Model 2. The reason for this is that for Model 1 the rotor-stator distance is small only near the contact point; the torque peak therefore has a very narrow peak at very small ϕ and is well matched to a narrow electrode. For Model 2, the rotor-stator distance is small everywhere and the torque peak is broad; in this case the peak torque and MAT are optimized by a wider electrode. This example serves to underscore the fact that for the wobble motor geometry there are no simple rules. The optimum value of any parameter depends on the value of the other parameters, and there is much to be gained by insightful design.

6.0 EXPERIMENTAL RESULTS

6.1 COMMENTS

Experimental evaluation of the wobble motor concept is proceeding in three preliminary stages aimed at generating initial, order-of-magnitude estimates of performance, with more precise measurements, and more refined techniques to follow. In stage I, stall torque of an EDM WM was measured. In stage II, the maximum free speed was determined for different driver voltages and recruitment strategies. Stage III is the development of measurement apparatuses which will give a complete dynamic characterization of torque-speed relationships, power and efficiency.

We note that three substantial obstacles were encountered in these preliminary experimental efforts:

(a) The fabrication of repeatable test motors requires careful control of the selection and forming of materials, the application of coatings, microassembly and machining techniques, packaging, etc.

(b) There are no commercially available instruments directly applicable to these measurements, so considerable effort was needed even for simple measurements of stall torque and free speed.

(c) Measurements are hindered by effects not yet fully understood, including charge retention by insulators, electrical arcing and insulator breakdown, electro-acoustical coupling, etc.

6.2 STALL TORQUE MEASUREMENTS

Figure 15 shows two experimental set-ups for the determination of static torque from the EDM WM. In one technique, two glass fibers of 140 micron diameter and 26 mm length were symmetrically attached to the rotor orthogonal to its axis, as shown in Fig. 15b. The four electrodes on one, e.g., the right, half of the WM were driven at voltage V with the remaining four electrodes and the rotor grounded. The rotor was rotated by hand, increasing θ to 90^0. Support was gradually removed, decreasing θ until a position was reached at which internal electrical forces supported the rotor against the known torque caused by the weight of the glass fibers. At each voltage V this procedure was repeated 15 times. Figure 15a shows a second technique still under development; in this approach torque is inferred from the maximum twisting of a taut wire from which the rotor is suspended.

In Fig. 16 the measured torque is plotted against the square of the driver voltage V. Also shown is the predicted maximum torque for the WM configuration of the experiments (stator ID = 571.5 microns; rotor OD = 495.3 microns; anodization thickness = 30.5 microns; four driven stator segments). Note that this is the configuration of curve B of Fig. 12. The fact that the experimental torque, unlike the theoretical prediction, is not proportional to V^2 could be an artifact of the large experimental errors, or it could suggest that charge is being retained in the fairly thick rotor anodization layer.

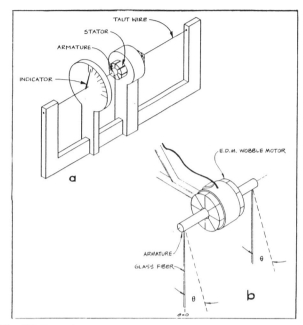

Fig. 15. Experimental apparatus for stall torque measurement using: (a) a taut wire spring, (b) a loaded pendulum.

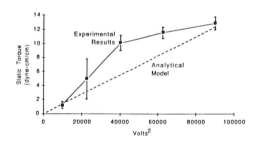

Fig. 16. Plot Stall Torque vs. Voltage (squared)

6.3 FREE-SPEED MEASUREMENT OF THE EDM WOBBLE

The EDM motor shown in Fig. 6 was driven in a stepper mode (i.e. non-commutated) in order to determine the maximum achievable free speed as a function of applied voltage. The motor was mounted in a vertical position so that the rotor rested down against a horizontal friction plate to prevent its movement in axial position. Operating the driver elements of the stator at 300 volts, free speeds of between 300 WPM and 100,000 WPM were routinely observed by stroboscopic measurements. Planned experiments will use new electronics and improved fabrication techniques in order to achieve higher speeds.

The system was also operated within a scanning electron microscope in order to closely examine the dynamic behavior of contact between the rotor and stator.

6.4 TORQUE-SPEED MEASUREMENTS

Of course the dynamic measurement of motor torque and speed is a prime experimental objective. With this information, important characteristics can be quantified such as the motor constant, efficiency, torque-ripple, thermal characteristics, and others. Efforts are now focused on performing these measurements through inferential dynamometry (estimating torque and other measures from motor behavior) and via the development of a micro dynamometer capable of applying specified drag torques to a WM while measuring speed, voltage, current, power and temperature.

7.0 CONCLUSIONS

A variety of micro motor concepts have been evaluated by the CED, with the wobble motor approach being one of those selected for extensive study. Various WM configurations have been analytically evaluated using finite-element methods and closed-form solutions. Important performance characteristics have been estimated such as stall torque, free speed, and alternate strategies for motor control have been examined.

Five motor configurations and a variety of silicon and non-silicon-based fabrication techniques have been examined in detail. Exploratory exercises have actually fabricated motors using: Direct Mechanical Assembly, Electro Discharge Machining, Cylindrical Photolithographic Etching, and Coextrusion of Metal and Plastic. Additionally, motor configurations which could be fabricated using silicon wafer etching techniques are being pursued.

The EDM approach was selected as the motor alternative which could function as an experimental test-bed. Fifteen EDM WMs have been constructed and utilized for experimental purposes. Interestingly, WM explorations have demonstrated that micro systems can be quite easily constructed by means other than etching silicon.

Experiments aimed at generating simple preliminary data have been conducted and results compare reasonably to analytical studies. Most importantly, these exploratory efforts have functioned to educate the research group with respect to the feasibility of various concepts and the relative usefulness of existing fabrication methodologies. Of equal importance have been the preliminary analytical and experimental failures which provided lessons necessary for the avoidance of future traps.

Upcoming efforts will focus on generating analytical tools for the evaluation of new concepts and, through such studies, solidifying the intuitive base necessary for success in designing future systems. Parallel efforts will expand our understanding of fabrication technologies appropriate for the construction of MEMS, and in developing micro mechanical instrumentation necessary to characterize micro actuation systems.

8.0 ACKNOWLEDGMENTS

This work was supported by grants from the Defense Advanced Research Projects Agency (DARPA Contract No. F33615-87-C-5267), the System Development Foundation (SDF Grant No. 348), and the National Science Foundation (ECS-8806449). The authors would also like to thank Prof. Sanford Meek, Dr. William Lee, Russell Wilson, Dale Emery, Shawn Cunningham, Barry Hanover and David Knutti for their technical support.

9.0 REFERENCES

[1] ANSOFT Corp. (1988): "MAXWELL 2D Electrostatic Analysis Package", Pittsburgh, PA.

[2] Clark, J.J. (1988): "CMOS Magnetic Sensor Arrays", *IEEE Solid-State Sensor and Actuator Workshop*, Hilton Head Is., SC, June 6-9.

[3] Fan, L-S, Tai, Y-C and Muller, R.S. (1988): "IC-Processed Electrostatic Micro-motors", *IEEE Intl. Elect. Dev. Mtg.*, San Francisco, CA, December 11-14.

[4] Fujita, H. and Omodaka, A. (1987): "Electrostatic Actuators for Micromechatronics", *Proc. IEEE Micro Robots and Teleoperators Workshop*, Hyannis, MA, November 9-11.

[5] IEEE Micro Robots and Teleoperators Workshop (1987): "Proceedings", Hyannis, MA, November 9-11, IEEE Publication 87T0204-8.

[6] IEEE Solid-State Sensors and Actuator Workshop (1988): "Technical Digest", Hilton Head Is., SC, June 6-9, IEEE Publication 88TH0215-4.

[7] Jacobsen, S.C. (1987): "Electric Field Machine", U.S. Patent No. 4,642,504, February 10.

[8] Jacobsen, S.C. (1988): "Electric Field Machine", U.S. Patent No. 4,736,127, April 5.

[9] Jacobsen, S.C. (1988): "Electric Field Machine", U.S. Patent No. 4,760,302, July 26.

[10] Jacobsen, S.C., Phillips, R.P. and Wood, J.E. (1988): "Systems and Methods for Sensing Position and Movement", U.S. Patent No. 4,767,973, August 30.

[11] Jacobsen, S.C., Wood, J.E. and Price, R.H. (1988): "Micro Positioner Systems and Methods", U.S. Patent No. 4,789,803, December 6.

[12] Jacobsen, S.C., Clayton, N.W., Price, R.H., Wood, J.E. and Lee, W.B. (1988): "Field-based Microsystems for Strain Measurement", *ASME Winter Annual Meeting*, Boston, November 27.

[13] Jacobsen, S.C., Smith, C.C., Biggers, K.B. and Iversen, E.K. (1987): "Behavior Based Design of Robot Effectors", *4th Intl. Sympos. on Robotics Res.*, Santa Cruz, CA, August 10-15.

[14] Lang, J.H., Schlecht, M.F. and Howe, R.T. (1987): "Electric Micromotors: Electromechanical Characteristics", *IEEE Micro Robots and Teleoperators Workshop*, Hyannis, MA, November 9-11.

[15] Muller, R.S. (1988): "Micromotor", *Popular Science*, December, p. 84.

[16] Paschen, F. (1889): "Ueber die zum Funkenubergang in Luft, Wasserstoff und Kohlensaure bei verschiedenen Drucken erforderliche Potentialdifferenz, *Annalen der Physik*, 37:69-96.

[17] Petersen, K.E. (1982): "Silicon as a Mechanical Material", *Proc. IEEE*, 70(5):420-457.

[18] Price, R.H., Jacobsen, S.C. and Khanwilkar, P.S. (1987): "Oscillatory Stabilization of Micromechanical Systems", *IEEE Micro Robots and Teleoperators Workshop*, Hyannis, MA, November 9-11.

[19] Price, R.H., Wood, J.E. and Jacobsen, S.C. (1988): "The Modelling of Electrostatic Forces in Small Electrostatic Actuators", *IEEE Solid-State Sensor and Actuator Workshop*, Hilton Head Is., SC, June 6-9.

[20] Trimmer, W.S.N. and Gabriel, K.J. (1987): "Design Considerations for a Practical Electrostatic Micro-Motor", *Sensors and Actuators*, 11(2):189-206.

[21] Wood, J.E. (1988): "On Statistical-Mechanical Models for the Molecular Dynamics of Striated Muscle", *Proc. Ann. Intl. Conf. IEEE EMBS*, New Orleans, LA, November 4-7.

[22] Wood, J.E., Jacobsen, S.C. and Grace, K.W. (1987): "SCOFSS: A Small Cantilevered Optical Fiber Servo System", *Proc. IEEE Micro Robots and Teleoperators Workshop*, Hyannis, MA, November 9-11.

An Electrostatic Top Motor and its Characteristics

MINORU SAKATA*, YASUYOSHI HATAZAWA**, AKITO OMODAKA, TOMOHIRO KUDOH†
and HIROYUKI FUJITA

Institute of Industrial Science, The University of Tokyo, Tokyo (Japan)

Abstract

Researchers have proposed variable-gas electrostatic actuators (VGEA) such as the electrostatic linear actuator (ESLAC) [1, 2]. As part of this study, we report in this paper another type of VGEA named the electrostatic top motor (ESTOM). Basically, the ESTOM is composed of two parts: one is a concave conical stator with striped electrodes (driving electrodes) and a ground electrode, with an insulation layer covering the driving electrodes; the other is a convex conical rotor with a conductive material covering its surface. A rotor rotates on a stator by applying voltage to the driving electrodes on the stator.

Two types of ESTOM were fabricated: a scale model by a conventional method and a micromodel by semiconductor processes. The stator of the micromodel is 2 mm in cone diameter and has a depth of 0.2 mm. The scale model is a hundred times as large as the micromodel. The scale model was tested to obtain characteristics, such as the relationship between driving voltage and the torque, and so on. The results were compared to the electrostatic field analysis and we discuss the differences. The micromodel is operated by an open-loop control. The rotation of the rotor was observed at an applied voltage of 80–120 V.

1. Introduction

As a driving force of microactuators with sizes less than 1 mm, many types of force can be considered, such as electromagnetic force, the force associated with the phase change of shape memory alloy, the force produced by the piezoelectric effect, and electrostatic force. Within these candidates, electrostatic force is more promising than the other forces because of the following advantages:

(i) electrostatic force dominates at dimensions less than 1 mm because electrostatic force is a surface force;

(ii) electrostatic force is induced only by applied voltage, which makes it easier to control the actuators;

(iii) only a pair of thin films is needed as electrodes to induce an electrostatic force; this makes fabrication processes compatible with semiconductor processes.

Many electrostatic microactuators have been proposed [3–5], however, most of them suffer from friction. This is one of the most serious problems in microactuators because of the following reasons:

(i) bearings and lubricants cannot be applied;

(ii) inertia force is much smaller than the friction force because of mass reduction;

(iii) relative surface roughness is large.

To solve these problems, we propose variable-gap electrostatic actuators (VGEA). Rolling motion instead of sliding motion is adopted in VGEA to reduce the frictional force. A harmonic electrostatic motor [6] and a wobble motor [7] are the cylindrical type of VGEA. The present paper describes the development of a conical VGEA named ESTOM (ElectroStatic TOp Motor).

2. Construction

Figure 1 shows a general idea of the ESTOM. The stator is a concave cone with driving electrodes and a ground electrode. An insulation layer covers only the driving electrodes. The rotor is a convex cone with smaller vertex angle than the stator. Its surface is covered with a conductive material. The rotor is kept in contact with the ground electrode at the cone point, which makes the rotor potential grounded. Also the rotor is kept in contact with the stator along a certain internal generator. Electrostatic force is induced between the rotor and one of the driving elec-

---

*Visiting researcher from OMRON Tateisi Electronics Co.
**Undergraduate research student from Tokyo Denki University.
†Graduate student from the Department of Electrical Engineering, Faculty of Engineering, The University of Tokyo.

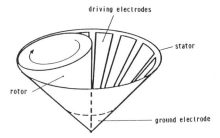

Fig. 1. A general idea of the electrostatic top motor (ESTOM).

trodes to which a voltage is applied. Then the rotor is attracted and rotates in a precessional movement about a cone axis of the stator, by switching the voltage to the other electrodes sequentially.

3. Scale Model

3.1. Numerical Analysis

The electrostatic field between the rotor and the driving electrodes is calculated by the charge superposition method [8] to obtain the torque. Ground and high potentials are applied to the rotor and the electrodes, respectively. The dimensions of the analysis model are shown in Fig. 2. Figure 3 shows the relative positions of the rotor and the driving electrode to which voltage is applied. The electrostatic field is calculated in the configuration.

The torque (T_a) at 2–5 kV was calculated as a function of the aspect ratio (R_a in Fig. 2) of the rotors (Fig. 4). It is clear that the torque is proportional to the square of the voltage. The

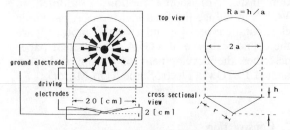

Fig. 2. The dimensions of ESTOM: (a) stator; (b) rotor.

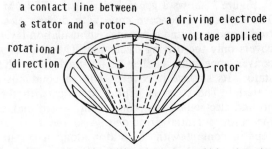

Fig. 3. The position of a driving electrode to which voltage is applied and the rotor.

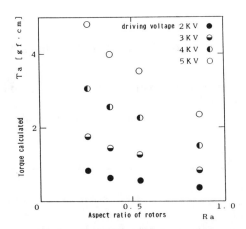

Fig. 4. The relationship between the aspect ratio of the rotor (R_a) and the calculated driving torque (T_a).

torque divided by the coefficient of the rotor's moment of inertia about a generator (A) is adopted as a value which shows the ease of rotation of the rotor. The moment of inertia about a generator (I_a) is calculated by eqn. (1), assuming a uniform surface mass density. The ease of rotation of the rotor is calculated by eqn. (2).

$$I_a = Cp \qquad (1)$$
$$C = 1/4(1 + 3R_a^2)\pi r^4/(1 + R_a^2)^{2.5}$$
$$A = T_a/C \qquad (2)$$

where C = coefficient in cm^4; R_a = aspect ratio; r = length of a generator in cm; p = surface mass density in g/cm^2; T_a = calculated torque in gf cm; A = ease of rotation of the rotor in gf/cm^3.

Figure 5 shows the relationships between C and R_a. Coefficient C has the maximum value at $R_a = 0.33$. Also the relationships between A and R_a are shown in Fig. 6. It can be seen that A, the ease, increases as R_a, the aspect ratio, decreases.

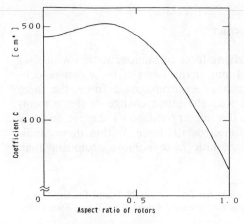

Fig. 5. The relationship between the aspect ratio of the rotor (R_a) and a coefficient C.

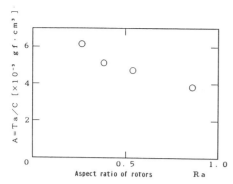

Fig. 6. The relationship between the aspect ratio of the rotor (R_a) and A.

3.2. Driving Experiment

The dimensions of the stator used in the experiments are the same as those of the analysis model in Fig. 2. Twelve dividing electrodes were attached on the surface radially. They were connected into six phases and covered by a PET insulation film. A ground electrode at the center of the cone was left uncovered. Rotors were made of paper covered with aluminum foil. The length of the generator, r, was kept constant. The height, h, and the radius, a, were changed. Parameters of rotors are given in Table 1.

Stall torque (T_e) was measured first. A contact line between the rotor and the stator, which is called 'the generator' in the following, was adjusted at the middle of a particular driving electrode. Voltages from 2 up to 5 kV were applied at the adjacent electrode. The rotor was constrained from moving by a wire attached between a point on the rotor rim and a electric scale. The force multiplied by the distance between the point and the generator gave the torque. The results are shown in Fig. 7.

The rotation speed of the rotor was measured by switching the voltage manually from one electrode to the next. The maximum speeds obtained from ten trials are shown in Fig. 8 at 3, 4 and 5 kV. It is noted that the speed was the apparent rotation speed. The rotor actually rotated at a much slower speed. The ratio of the actual speed to the apparent one depends on the radius of the rotor, a, and that of the stator at the same height

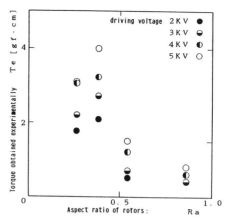

Fig. 7. The relationship between the aspect ratio of the rotor (R_a) and the measured driving torque.

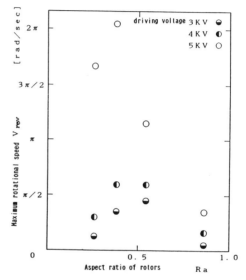

Fig. 8. The relationship between the aspect ratio of the rotor (R_a) and maximum angular speed (V_{rev}).

as the rotor, a_s. The reduction rate is equal to $(a_s - a)/a_s$ [6, 7].

3.3. Discussion of Scale Model

As is shown in Fig. 4, the calculated torque (T_a) increases as the aspect ratio decreases. In the experiment, this tendency was observed for the

TABLE 1. Parameters of rotors

	Length of generator r (cm)	Radius a (cm)	Height h (cm)	Aspect ratio R_a	Moment of inertia (g/cm)	Mass (g)	Reduction ratio
Rotor 1	5.00	3.90	3.32	0.86	12.7	1.81	11/39
Rotor 2	5.00	4.55	2.46	0.54	11.9	1.58	9/91
Rotor 3	5.00	4.68	1.76	0.38	14.1	2.07	3/47
Rotor 4	5.00	4.91	1.25	0.24	15.4	2.28	1/49

larger values of the aspect ratio. For the smallest one, however, the torque decreased a little (Fig. 7). One of the reasons was that the surface charge produced by the discharge was larger at higher voltages and for the rotor with the smallest aspect ratio. This also explains why the torque was not proportional to the square of the applied voltage (Fig. 7).

The rotation speed increased at higher voltages. The ease of rotation of a rotor can be related to the torque divided by the moment of inertia of the rotor. The value, A, which means the ease, increases with a lower aspect ratio of rotors, assuming the same surface mass density for all rotors. But the assumption was not realized in the experiments. Calculated from Table 1, the surface mass densities of the rotors were 0.030 g/cm^2 (rotor 1), 0.022 g/cm^2 (rotor 2), 0.028 g/cm^2 (rotor 3) and 0.030 g/cm^2 (rotor 4) respectively. However, the values calculated from Fig. 7 and Table 1 are consistent with the result in Fig. 8.

In the case of a micromodel, the above disadvantage of an electrostatic force caused by the discharge need not be considered because of the existence of Paschen's minimum.

4. Micromodel

A micromodel was fabricated by semiconductor processes. Figure 9 shows its cross-sectional view and mask pattern. Sixty driving electrodes were patterned by photolithography. The ground electrode was not patterned because of the following reasons:

(i) the gap between the photomask and the stator cone point was 0.2 mm, which made it very difficult to focus a sharp mask image on the photoresist on the stator precisely;

(ii) frictional contact damaged the sputtered or deposited ground electrode easily.

After all process were completed, an aluminum foil connected to the ground by a metal wire was

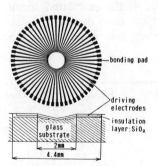

Fig. 9. Mask pattern and cross-sectional view of the micromodel.

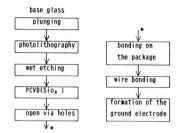

Fig. 10. The flow diagram of the fabrication processes of the micromodel.

placed and aligned on the stator. It was used as the ground electrode.

The flow diagram of the fabrication process is shown in Fig. 10. The base glass was heated. An iron stick with a conical shape at one edge was plunged into the glass to form a concave cone. A conductive later of WSi was sputtered. The driving electrode pattern shown in Fig. 9 was formed on the conductive layer by photolithography and wet etching. The PCVD process was used to cover the driving electrodes with SiO$_2$. The thickness of the SiO$_2$ insulation layer was about 4000 Å. Through holes for bonding, pads were opened. After the base glass was bonded on the package, the bonding pads of the driving electrodes and the package pins were connected with gold wires. The rotor was made of aluminum foil. The aspect ratio of the rotor was about 0.2.

Open control was used to drive a rotor. The rotation of the rotor was observed at applied voltages of 80–120 V.

5. Conclusions

Another type of VGEA, the ESTOM, has been proposed. Two types of ESTOM, a scale model and a micromodel have been fabricated and tested. The relationships between the aspect ratio of the rotors and the torque, and the ease of rotation of a rotor were obtained from the analysis and the experiments on the scale model. With a rotor which has a lower aspect ratio, the ease of rotation of a rotor increases because of its higher reduction rate. Furthermore, the rotation of the rotors was observed in the micromodel of 2 mm diameter at 80–120 V.

Acknowledgement

This work was partly supported by the TORAY Research Foundation.

References

1 H. Fujita, Studies of micro actuators in Japan, *Proc. 1989 IEEE Conf. Robotics and Automation, Scottsdale, AZ, U.S.A., May 1989*, Vol. 3, pp. 1559–1564.
2 H. Fujita, and A. Omodaka, Fabrication of an electrostatic linear actuator by silicon micromachining, *IEEE Trans. Electron Devices, ED-35* (1988) 730–734.
3 L. S. Fan, Y. C. Tai and R. S. Muller, IC processed electrostatic micro-motors. *Sensors and Actuators, 20* (1989) 41–47.
4 J. H. Lang, M. F. Schlecht and R. T. Howe, Electrostatic micromotors: electromechanical characteristics, *IEEE Micro and Teleoperators Workshop, Hyannis, MA, U.S.A., Nov. 1987*.
5 W. S. N. Trimmer and K. J. Gabriel, Design considerations for a practical electrostatic micro-motor. *Sensors and Actuators, 11* (1987) 189–206.
6 W. S. N. Trimmer and R. Jebens, Harmonic electrostatic micromotors, *Sensors and Actuators, 20* (1989) 17–24.
7 S. C. Jacobsen, R. H. Price, J. E. Wood, T. H. Rytting and M. Rafaelof, A design overview of an eccentric-motion electrostatic microactuator (the Wobble Motor), *Sensors and Actuators, 20* (1989) 1–15.
8 T. Kouno, Computer calculation of electrical field, *IEEE Trans. Electr. Insul., EI-21* (1986) 869–872.

OPERATION OF MICROFABRICATED HARMONIC AND ORDINARY SIDE-DRIVE MOTORS

Mehran Mehregany[1], Pradnya Nagarkar[1], Stephen D. Senturia[1], and Jeffrey H. Lang[2]

[1] *Microsystems Technology Laboratories*
[2] *Laboratory for Electromagnetic and Electronic Systems*
Department of Electrical Engineering and Computer Science
Massachusetts Institute of Technology
Cambridge, Ma 02139

ABSTRACT

This paper presents a novel variable-capacitance harmonic side-drive motor and reports the operation of this motor and ordinary variable-capacitance side-drive motors without the need for air-levitation assist. Native oxide formation on motor polysilicon surfaces is identified as the cause of motor operational failure, resulting from the clamping of the rotor to the shield beneath it. With proper release and testing directed at minimizing this oxide formation, we can readily operate the motors. Operational characteristics of the micromotors including the role of rotor electric shielding, speed, and frictional effects are studied. For the side-drive motors, measurements of stopping and starting voltages indicate that the drive torque required to sustain motor operation is 5-7 pN-m while that required to initiate motor operation after a 30 seconds rest is nearly twice as high.

1. INTRODUCTION

Electrical operation of variable-capacitance side-drive micromotors was first reported by Fan, Tai, and Muller [1,2]. Operation of similar micromotors using an air-levitation assist was reported by us shortly thereafter in [3]. In references [2,4], motor operation did not require air levitation; however, this operation was limited to under one minute, at which time the motors ceased to operate. In our previous work, even when electrostatic shielding of the rotor was used as suggested in [2,4], motor operation was only possible when air levitation was used to overcome frictional forces associated with the clamping of the rotor to the electric shield beneath it. This rotor clamping was attributed to electric fields between the rotor and the shield caused by a lack of proper electrical contact between these respective parts. The rotor-shield electrical contact during motor operation is intended to come from mechanical contact of the rotor to the shield or the bearing, all of which are heavily doped polysilicon components; note that the bearing is fabricated in electrical contact with the shield. To date, clear insight into the failure of the electrical contact necessary for electrical motor operation in our work [3] and the short duration of electrical motor operation in [1,2,4] has been lacking.

This paper identifies the native oxide on the motor polysilicon surfaces as the mechanism which prevents the appropriate electrical contact resulting in operational failure, asserts that micromotor release and testing techniques are the determining factors for successful electrical operation of the motors, reports the operation of a novel variable-capacitance harmonic (i.e., wobble) side-drive motor as well as the operation of ordinary side-drive motors without air levitation, and presents experimental estimates of frictional torques in the motors.

2. MICROFABRICATED MOTORS

This section describes the microfabricated harmonic side-drive (hereafter referred to as wobble) and ordinary side-drive (hereafter referred to as side-drive) motors. Center-pin bearings (CPBs) contacting an electric shield under the rotor and extending under the stator poles are used. During motor operation, the rotor is intended to be in electrical contact with the shield positioned beneath it through mechanical contact at the bearing or at the bushing supports (see Section 4 for details).

2.1. Wobble Motors

Figure 1 is a SEM photograph of a typical microfabricated CPB wobble micromotor reported in this work. Operating principles of the wobble motor have been described in [3,5-8]. The new feature of the design presented here is that the rotor wobbles around the center bearing post rather than the outer stator. Schematic drawings showing plan views of the conventional wobble motors in [3,5-8] and the design used here are shown in Fig. 2. In the new design, the wobble distance is equal to the clearance in the bearing which is the difference between the bearing and inner rotor radii. This clearance is specified by the thickness of the second sacrificial oxide which is 0.3 μm in this work. Furthermore, by ensuring a bearing clearance smaller than the nominal air-gap size, which is greater than 1.5 μm in this work, the need for insulation between the rotor and the stator is eliminated.

A significant advantage of the wobble motor is that its drive torque is proportional to the motor gear ratio, n. For the wobble motor design presented here, n is given by

$$n = \frac{\theta}{\phi} = \frac{r_b}{\delta}, \quad (1)$$

where θ is the angle between the diameter containing the contact point and the energized electrode, ϕ is the angle of rotation of the rotor, r_b is the radius of the bearing, and δ is the bearing clearance. The wobble motor drive torque, $T_w(\theta)$, due to a single excited stator pole at angular position θ, estimated from in-plane two-dimensional simulations when the rotor is electrically grounded, is given by

$$T_w(\theta) = \frac{\epsilon_o t V^2}{2} \tau_w(\theta)\, n, \quad (2)$$

where ϵ_o is the permitivity of air, t is the rotor thickness, and V is the applied excitation [8]. In (2), $\tau_w(\theta)$ is a normalized torque calculated by the simulations and specified by the motor design which includes effective air-gap size, pole width, and pole pitch [8]. The effective air-gap size is the actual distance of the rotor to the excited stator pole and is a function of the nominal air-gap size and the bearing clearance. Note that the wobble motor drive torque is multiplied by the gear ratio, n. The drive torque from more general excitation schemes involving more stators can be derived by superposition.

There are two significant differences between the wobble motor design presented here and the conventional wobble motor of [3,5-8] for which the rotor rolls inside the stator. First, in the wobble motor of Fig. 1, the rotor rotates in the same direction as the excitation signal and the point of contact as indicated in Fig. 2(b). In the conventional wobble motor, the point of contact (Fig. 2(a)) and the excitation waveform rotate in the opposite direction of the rotor. Second, holding all other parameters (including stator radius) constant, in the conventional wobble motor, increasing the gear ratio increases τ_w as well, which always leads to a greater output torque. To explain this point, consider the schematic representation of the conventional wobble motor in Fig. 2(a). The gear ratio depends on the rotor outer radius and increases as the rotor outer radius approaches that of the stator. This is accompanied by a decrease in the effective gap size (air-gap separation plus the stator insulation thickness) which in turn increases the electric forces on the rotor, leading to an increase in τ_w. However, the situation is more complicated for the wobble motor in Fig. 2(b). Note that in this motor design, the gear ratio (Eqn. (1)) is determined by the bearing clearance (a process parameter) and the bearing radius (a design parameter). The gear ratio is independent of the rotor outer radius but is inversely proportional to the the bearing clearance. At the same time, the motor effective air-gap size (i.e., nominal air-gap minus the bearing clearance) is linearly proportional to the bearing clearance. However, τ_w is proportional to the reciprocal of the effective air-gap size squared. Therefore, the net effect of increasing the gear ratio by decreasing the bearing clearance is a decrease in the motor drive torque. Note that increasing the bearing radius does not affect τ_w while it increases the gear ratio and therefore the motor drive torque. For the wobble motors in this work, we have selected to enhance motor gear ratio. This is achieved by using small bearing clearances. A small bearing clearance in the wobble motor design used here is desirable because: (1) it reduces slip during motor operation; (2) it reduces rotor wobble; (3) it reduces breakdown possibility in the air gap. Recognizing that smaller bearing clearances reduce drive torque, we have increased the bearing radius to enhance the drive torque through further increase of the gear ratio.

2.2. Side-Drive Motors

We have considered three side-drive motor architectures including stator to rotor pole number ratios of 3:1, 3:2, and 2:1. The former two require three-phase excitation while the latter operates with two-phase excitation. Figures 3, 4, and 5 are SEM photographs of typical 3:1, 3:2, and 2:1 motors fabricated in this work. A detailed comparison of these three motor architectures is documented in [9]. In summary, in comparison with the 3:1 (Fig. 3), the 3:2 architecture (Fig. 4) takes advantage of a second set of rotor poles which are at a high torque position when the first rotor pole set is in its aligned or zero torque position. In this way, the 3:2 architecture pro-

Fig. 1. A CPB, 12 stator pole, 2.5 μm-wide air-gap, 100 μm-diameter, wobble micromotor.

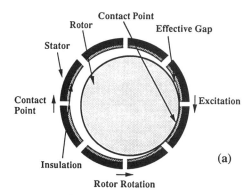

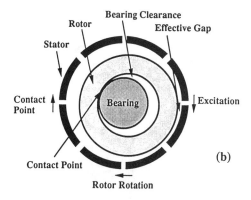

Fig. 2. Two wobble motor designs: (a) Conventional design requiring stator insulation; and (b) the new design using a center bearing to eliminate the need for stator insulation.

vides superior torque coverage with higher minimum torque values. The 2:1 architecture offers a reduction in the complexity of the drive electronics with a sacrifice in performance as measured by torque and torque coverage. The air-gap spacing varies as a function of the reciprocal of angular position,

θ, in our 2:1 motors. This air-gap spacing variation provides for two-phase operation, for which bidirectional motor operation is lost since torque coverage may be provided in one direction only; in Fig. 5, this direction is that of negative θ (i.e., clockwise). We have successfully operated various motors of all three categories and have experimentally observed the superior performance of the 3:2 motors over the other two side-drive architectures.

The side-drive motor drive torque, $T_s(\theta)$, due to a single excited stator phase at orientation θ with respect to the rotor, estimated from in-plane two-dimensional simulations when the rotor is electrically grounded, is given by

$$T_s(\theta) = \frac{\epsilon_o t V^2}{2} \tau_s(\theta), \qquad (3)$$

where ϵ_o is the permitivity of air, t is the rotor thickness, and V is the applied excitation. In (3), $\tau_s(\theta)$ is a normalized torque calculated from the simulations and specified by the motor design which includes air-gap size, pole width, pole pitch, and pole number per phase [9]. The drive torque from more general excitation schemes involving more stator poles can be derived by superposition.

In the side-drive motors, the bearing clearance allows radial displacement of the rotor. When the rotor is electrically grounded and is symmetrically centered in between a set of excited stator poles, the resultant of the radial forces on the rotor due to each excited stator pole is zero. However, when the rotor displaces radially, a net radial force develops which pushes the rotor into the bearing post. Previous work [4,10] indicates that the bearing frictional forces at this contact may be significant. The rotor radial movement may also lead to rotor vertical clamping forces by disturbing excitation symmetry as described in Section 4. Smaller bearing clearances reduce rotor radial movement and the associated frictional forces in the bearing. For the side-drive motors, the bearing radius is minimized to reduce the lever arm of the bearing frictional forces.

Fig. 4. A CPB, 12 stator and 8 rotor pole, 1.5 μm-wide air-gap, 130 μm diameter, 3:2, side-drive micromotor with bushing design 3 (i.e., the three small rotor indentations at the edge of the base for the rotor poles).

Fig. 3. A CPB, 12 stator and 4 rotor pole, 1.5 μm-wide air-gap, 130 μm diameter, 3:1, side-drive micromotor with bushing design 2 (i.e., the three small rotor indentations near the bearing).

Fig. 5. A CPB, 12 stator and 6 rotor pole, 1.5-3.5 μm varying air-gap, 130 μm diameter, 2:1, side-drive micromotor with bushing design 1 (i.e., the circular rotor indentation near the bearing).

3. MICROMOTOR FABRICATION PROCESS

The wobble motor design presented here, using a small bearing clearance and eliminating the need for rotor-stator isolation, simplifies the fabrication process and allows for a single fabrication process suitable for both wobble and side-drive motors. The bearing radius is selected for each motor architecture according to the discussion above. The motor fabrication process is similar to the previously documented work [1-3,9]. Center-pin bearings contacting an electrostatic shield under the rotor and extending under the stator poles are used. Three improved features of the process described here are: (1) The rotor-stator polysilicon is 2.2 μm-thick with air-gaps as small as 1.5 μm anisotropically etched with RIE. The combination of thick rotor-stator polysilicon and small air-gaps enhance motor output torque. (2) The rotor bearing clearance is reduced to 0.3 μm from previously reported values of 0.6-1.0 μm [1-3,10,11]. For the wobble motor, reducing the bearing clearance increases the resulting gear ratio. For the side-drive motors, the frictional model proposed in [4,9] predicts that tighter bearing clearances reduce rotor radial movements which in turn reduce frictional forces in the bearing. Furthermore, tighter bearing clearances reduce rotor radial positioning asymmetry which may lead to substrate clamping of the rotor. (3) No nitride solid lubrication on the inner radius of the rotor has been needed here, in contrast to previous work [1,2,4], thereby simplifying the process.

Figure 6 shows typical micromotor cross-sections during the fabrication process. Initially, substrate isolation is established using a 1 μm-thick LPCVD silicon-rich nitride layer over a 1 μm-thick thermally grown SiO_2 film [1-3]. A thin (3500Å) LPCVD polysilicon film is deposited, heavily doped with phosphorus, and patterned to form the electric shield (Fig. 6(a)). A 2.2 μm-thick low-temperature oxide layer is deposited and patterned in two steps to form the stator anchors and the bushing molds (Fig 6(a)). A 2.5 μm-thick LPCVD polysilicon layer is deposited and heavily doped with phosphorus. The rotor, stator, and air gaps are patterned into this polysilicon layer using RIE (Fig. 6(b)). The thickness of the polysilicon rotor-stator layer is 2.2 μm at this point since a patterned thermally grown oxide mask is used for the RIE etch of the polysilicon. Note that the inside radius of the rotor patterned at this point would correspond to the bearing radius in the final device. A bearing radius of 13 μm is the minimum conservative value in our process and is used for the side-drive motors. This value is determined by the pattern definition and delineation requirements (in the next step) for the bearing anchor which is located inside this trench (Fig. 6(c)). A bearing radius of 18 μm is used for the wobble motors according to the discussion in the previous section. The second sacrificial LTO layer is deposited, providing an estimated 0.3 μm coverage on the rotor inside radius side-walls, and patterned to open the bearing anchor (Fig. 6(c)). A 1 μm-thick LPCVD polysilicon film is deposited, heavily doped with phosphorus, and patterned to form the bearing (Fig. 6(d)). Figure 6(e) shows the motor cross-section after it has been released in HF. During motor operation, the bushings are intended to provide the electrical contact between the rotor and the shield as they slide over the shield.

4. RELEASE AND TESTING

This section reports on the requirements for electrical operation of our micromotors. Electrical operation of CPB side-

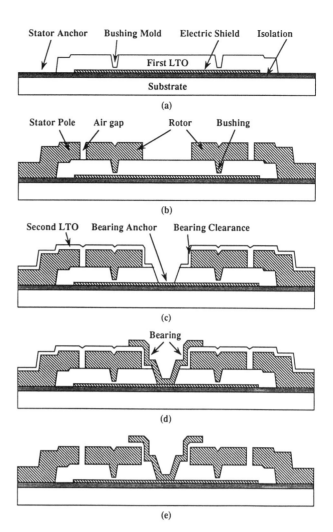

Fig. 6. Micromotor fabrication process: (a) After the first LTO; (b) after the rotor-stator polysilicon; (c) after the second LTO; (d) after the bearing polysilicon; and (e) released motor.

drive micromotors with an air-levitation assist was previously reported by us [3]. In the absence of the electric shield, our studies indicated that substrate clamping of the rotor inhibited motor operation; see [3] for details. In this case, substrate isolation prevents electrical contact between the rotor and the substrate which in turn results in electric fields that lead to the clamping of the rotor to the substrate. We have not been able to achieve successful motor operation in these devices, in contrast to the partial success reported in [1].

An electric shield which is positioned under the rotor was suggested in [2,4] to be effective in elimination of the rotor clamping forces by electrostatic shielding of the rotor from the substrate. However, even with electric shields incorporated in our devices, air levitation was found to be necessary in our previous work [3] to overcome frictional forces associated with the clamping of the rotor to the electric shield beneath it. The rotor clamping was attributed to a lack of proper electrical contact between the rotor, the shield, and the bearing. This work identifies the native oxide on the motor polysilicon surfaces as the cause of this electrical contact failure and asserts that micromotor release and testing techniques are the determining factors for successful electrical operation of the micromotors.

In reference to Fig. 6(e), during motor operation, the rotor is intended to be in electrical contact with the shield through mechanical contacts in the bearing or at the bushing supports. Note that the bearing is fabricated in electrical contact with the shield. This electrical contact ensures that the rotor and the shield are at the same electric potential, thereby eliminating the clamping forces otherwise caused by the electric field between them. However, we believe that native oxide formation on the bushings, the bearing, and the shield surfaces results in a loss of electrical contact, leading to an electric field between the rotor and the shield. This results in a clamping of the rotor to the shield (which was compensated for by air levitation in our previous work [3]). With proper release and testing techniques directed at minimizing the formation of this oxide, we can readily operate the motors without air levitation for extended periods of time.

Our previous release method [3] consisted of 15 minutes in a 49% (by weight) commercially-available HF solution, 5 minutes in a DI rinse, a short methanol rinse, and a nitrogen dry. Typically, 30 minutes elapsed before testing during which the devices were in room atmosphere. This combination of release and testing never resulted in working motors without air levitation.

Our current release method consists of 15 minutes in a 49% (by weight) commercially-available HF solution, 2 minutes in a DI rinse, 5 minutes in a 1:1 $H_2SO_4:H_2O_2$ (pirana) clean, 10 minutes in a DI rinse, 30 seconds in the HF solution, 90 seconds DI rinse, and a nitrogen dry. The initial 15 minute oxide etch in HF is required to release the various motors on the die. The 5 minute pirana clean is effective in cleaning the exposed polysilicon surfaces from organic as well as ionic contaminants. The second HF etch removes the thin oxide formed on the polysilicon surfaces during the pirana clean. The final DI rinse time is critical. Rinse times under 60 seconds do not produce reliable results. Long rinse times may be detrimental since the native oxide growth rate is greater in DI water than in air [12]. A 90 second rinse time has been typically used in this work with good results.

Successful electrical operation of our motors in room atmosphere without air levitation is achieved when freshly released motors are tested immediately. Using our current release process and testing procedure, motor operation durations have been as high as 10 minutes for the side-drive motors and up to 120 minutes for the wobble motors. The initial failure is gradual such that motor operation can be resumed temporarily if the excitation voltage is increased to provide higher drive torque. After total failure, motor operation can be restored if the motor is re-released and tested immediately. Alternatively, air levitation can be used to resume motor operation after total failure. These results suggest that failure of motor operation is due to the clamping of the rotor caused by a lack of electrical contact between the rotor and the shield rather than structural failures such as wear in the bearing. Native oxide formation on the polysilicon surfaces during motor operation is believed to be responsible. When the motors have not been exposed to air after release and are tested in nitrogen, the rotor clamping failure mechanism is eliminated. In this case, both wobble and side-drive motor operations are consistent (see Section 5) at least for several hours. However, when devices are exposed to room atmosphere for 15 minutes before electrical testing, the results are sporadic while, for a time lapse greater than 30 minutes, motor operation can rarely be achieved without air levitation, again suggesting oxide formation in room atmosphere as the cause of failure.

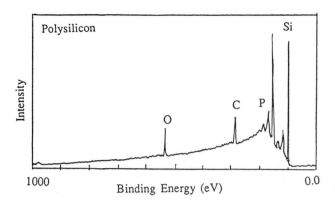

Fig. 7. XPS survey spectrum immediately after surface treatment identical to our current release method.

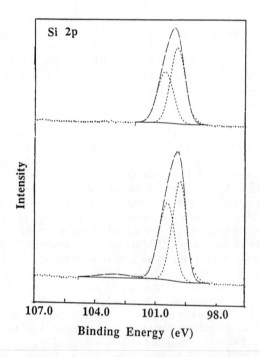

Fig. 8. XPS spectra for the sample in Fig. 7 before (upper spectrum) and after (lower spectrum) a 20 minute exposure to air.

XPS studies of the oxide layer that forms on polysilicon surfaces as a function of surface treatments identical to micromotor release methods are in qualitative agreement with the above results. Figure 7 shows an XPS survey spectrum for a sample, measured immediately after surface treatment, from which O/Si ratios can be estimated. The surface is free of fluorine contamination or other residuals of the release process, except for carbon. Figure 8 shows the silicon 2p envelope for the same sample immediately after surface treatment (upper spectrum) and after a 20 minutes exposure to air (lower spectrum). The broad peak at 103 eV in the lower spectrum is indicative of native oxide formation. Similar XPS studies show that a final methanol rinse (to assist in drying the sample [3,13]) may be detrimental by increasing the O/Si ratio. Furthermore, the O/Si ratio is increased with increasing the final DI rinse time.

These findings are in qualitative agreement with the results in [12], namely that the native oxide growth in air at room temperature requires the coexistence of water and oxygen. Note that in our case, testing in nitrogen eliminated the rotor clamping failure mechanism. In [12], it is further shown that the native oxide growth on bare silicon surfaces in ultra-pure water can be inhibited by lowering the dissolved oxygen concentration in the rinse water. Our XPS studies show an increase in the O/Si ratio with increasing the final DI rinse time in our release process since no attempt was made to control the dissolved oxygen concentration in the DI rinse water.

5. OPERATIONAL CHARACTERISTICS

Design parameter permutations including varying rotor radii, bushing size, bushing style, bushing location, air-gap size, and stator pole number are incorporated into the fabrication of both wobble and side-drive motors. Electrical actuation for many of these design permutations has been confirmed.

5.1. Side-Drive Motors

The side-drive motor measurements reported in this section were carried out on 3:2 motor architectures [9] with 12 stator and 8 rotor poles (hereafter referred to as 12:8 motors) similar to that in Fig. 4. For these motors, the stator pole width and pitch are 18 and 30 degrees while the rotor pole width and pitch are 18 and 45 degrees.

Measurement of starting and stopping voltages for the 12:8 micromotors is performed when the motors are operated in nitrogen. These motors are tested using a three-phase bipolar square-wave excitation, with the rotor electrically grounded through the shield bias. Measurements are done by operating the motor at 3 rpm (i.e., the excitation is switched to a new phase within 800 ms) and measuring the minimum voltage required for sustaining motor operation as well as the minimum voltage require to restart the motor. We have found that the starting and stopping voltages are nearly the same when the excitation is applied within a few seconds of previous rotor movement. Therefore, we insert a 30 second delay for the starting voltage measurements.

When tested in nitrogen, for the 12:8 motors, the measured stopping voltages are reproducible to within 3% in the same testing session. The scatter becomes 15% when the same motor is tested in different sessions with a re-release prior to each session, or for devices on different dies. The starting voltage measurements are far more scattered. This scatter can be as high as 40% in the same session and from session to session. This indicates the complexity of static friction characteristics in the micromotors even after only 30 seconds at rest. Table I shows the experimentally measured stopping and starting voltages for the 12:8 motors. While the stopping voltage measurements differentiate between the three different bushing designs (see below), the starting voltage measurements are averages of the three bushing designs. Except for one, each entry in Table I is an average computed from data on four motors from four different dies of the same wafer tested at different times. The stopping torque for the bushing design 2 of the 130 μm-diameter, 1.5 μm motors is obtained from data on two devices rather than four and may not be reliable. The 100 μm-diameter, 12:8 motors do not include bushing design 3, resulting in the missing entries of the Table I.

The experimentally measured voltages are used in conjunction with two-dimensional finite-element field simulations in the plane of the motors [9] to estimate the torques. Figure 9 shows a typical single-phase drive-torque curve for the 100 μm-diameter, 1.5 μm-gap, 12:8 motors obtained from these simulations. Since the air gap of the micromotors is comparable to the rotor thickness, the contribution of the fringing fields in the axial direction have to be accounted for in interpreting the experimental results. A correction factor calculated from field simulations in the axial plane of the motor (see [9] for details) is used to approximately account for the contribution of the fringing fields in the axial direction. These correction factors are applied to the single-phase drive-torque curves in the regions were there is partial overlap of the rotor/stator poles in the θ direction (i.e., within the ± 15 degrees region). For the 1.5 μm air-gap motors here, the corresponding correction factor is 1.95 while for the 2.5 μm air-gap motors this value is 2.27 [9].

In our measurements, the stopping position of the rotor has always been that of perfect alignment with the final excited stator phase (e.g., phase A in Fig. 9). Therefore, the starting position which is 15 degrees with respect to the next stator phase (e.g., phase B or C in Fig. 9) has also been known. The knowledge of rotor position and excitation voltage in conjunction with the analysis described above is used to estimate the torques in Table I. Note that the stopping voltages do scale correctly with air-gap and motor size for the side-drive motor model used here to estimate the torques.

Note that in all recorded cases, the starting torques are higher than the stopping torques. Furthermore, the required starting torques are not as consistent as those of stopping torques. Therefore, we have not differentiated between the different bushing designs for this measurement. The estimated starting torque is 11$\pm$5 pN-m, calculated by pooling the starting torque data for all of the tested motors.

As shown in Table I, we have used the stopping voltage measurements in the 12:8 motors to study frictional behavior of varying bushing styles, sizes, and locations. Tested motors have incorporated three bushing designs which can be seen in the motors of Figs. 3-5 and are designated by bushing designs 2, 3, and 1, respectively. For designs 2 and 1, the estimated bushing areas vary from 6 to 130 μm^2 while the lever-arm is constant. For designs 2 and 3, the bushings are identical except that the lever-arms vary from 22 to 35 μm. The stopping torques are estimated as 6, 5, and 7 pN-m for bushing designs 1, 2, and 3, respectively. The uncertainty in the stopping torques calculated by pooling the data for each different bushing design is 1 pN-m for all three bushing designs.

For the side-drive motors, both the bearing and the bushing operations produce sliding friction. The data in Table I for bushing designs 2 and 3 can be used to separate the components of frictional torques in the bearing and at the bushings in these motors. This is possible since, in this case, the bushing frictional forces act at different lever arms while the bearing frictional torque is constant. Accordingly, the frictional torque in the bearing is 1.6 pN-m. The corresponding frictional force is 1.2×10^{-7} N acting at a lever-arm of 13 μm (the bearing radius for the side-drive motors). Note that this value corresponds to a rotor position of 15 degrees. The frictional force of the bushings is 1.5×10^{-7} N which is independent of the rotor position. This frictional force acts at bushing lever-arms of 22 and 35 μm for bushings 2 and 3, respectively. The cor-

Rotor Diameter (μm)	Gap (μm)	Stopping Voltage (V)			Stopping Torque (pN-m)			Starting Measurements	
		Bushing 1	Bushing 2	Bushing 3	Bushing 1	Bushing 2	Bushing 3	Voltage (V)	Torque (pN-m)
100	1.5	50	46	-	6	5	-	71	11
130	1.5	46	37	50	6	4	7	66	13
100	2.5	61	58	-	6	5	-	75	9
130	2.5	55	51	57	6	5	7	74	11

Table I. Measured stopping and starting voltages and estimated torques for the 12:8 motors.

responding bushing frictional torques are then 3.4 pN-m for the bushing design 2, and 5.4 pN-m for the bushing design 3.

We have found that, for electrical motor operation, it is not necessary to ground the shield and therefore the rotor as long as the rotor is in electrical contact with the shield. This is to be expected since as long as the shield and the rotor are in electrical contact, regardless of their electric potential, there is no electric field and hence no clamping force between them. Both the wobble and the side-drive motors operate even when the electric shield and therefore the rotor are left at a floating potential. Furthermore, we have found that it is not necessary for the shield to extend under the entire rotor and parts of the stator poles. The motors operate even when a partial shield is used as long as this shield extends under the bushing supports. This behavior suggests the dominance of the bushing/substrate (or shield if the electric contact is lost) capacitance over that of the remaining area of the rotor.

5.2. Wobble Motors

The wobble motor measurements reported here were carried out on 12 stator pole motors similar to that in Fig. 1. For these motors, the stator pole width and pole pitch are 27 degrees and 30 degrees, respectively.

The wobble motors are being studied using a six-phase, unipolar, square-wave excitation. Therefore, six stator poles are excited (independent of the total number of stator poles available) with a center-to-center angular separation of 60 degrees (e.g., every other stator pole for the motor in Fig. 1). The expected gear ratio of our wobble motors is 60, corresponding to a bearing radius of 18 μm and an estimated bearing clearance of 0.3 μm. The experimentally measured gear ratio for typical excitations of 100V is 69, corresponding to a bearing clearance of 0.26 μm which is well within the accuracy of our estimated bearing clearance. Excitation voltages as low as 25V and 36V are sufficient in operating the wobble motors with 1.5 μm and 2.5 μm gaps, respectively.

5.3. Operational Speeds

Excitation voltages as high as 150V across 1.5 μm-wide air gaps (i.e., electric field intensities of 1×10^8 V/m) are routinely used without electric field breakdown in the air gaps. The 12:8, side-drive motors are spun periodically with signal frequencies up 167 Hz corresponding to a rotor speed of 2500 rpm. These motors have occasionally been spun with excitation signal frequencies of 1 KHz corresponding to synchronous rotor speeds of 15,000 rpm. However, the actual rotor speed at these higher excitation frequencies has not been verified due to the limitations in our video recording instrumentation. The wobble motors have been typically operated up to 140 rpm, which because of the gear reduction of 69 reaches the frequency upper limit of our power supply.

6. CONCLUSION

Native oxide formation on the motor polysilicon surfaces has been identified as the cause of operational failure in the electric micromotors. This oxide prevents proper electrical contact between the polysilicon rotor and shield, resulting in electric fields which can lead to the clamping of the rotor to the shield. Micromotor release and testing techniques directed at

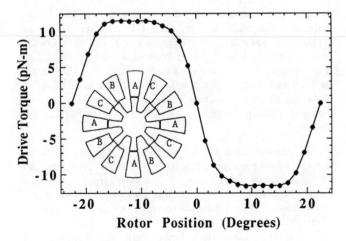

Fig. 9. Single-phase drive-torque characteristics of the 100 μm-diameter, 1.5 μm-gap, 12:8 motors obtained from in-plane two-dimensional field simulations. A 100V stator excitation is assumed for phase A, while the remaining stator phases and the rotor are electrically grounded. The rotor zero degree position with respect to phase A is indicated on the figure.

minimizing this oxide formation have been found to be essential for the successful electrical operation of the micromotors.

Operation of a novel wobble motor design eliminating the need for rotor-stator insulation and operation of side-drive micromotors have been demonstrated. Typical 100 μm-diameter, 1.5 μm-gap wobble motors have been operated at excitation voltages as low as 25V. Side-drive motors with comparable geometry have been operated at excitation voltages as low as 46V. Stopping and starting voltage measurements on the side-drive motors has been used to estimate frictional torques in the bearing and frictional forces at the bushing contacts.

ACKNOWLEDGMENTS

The authors wish to thank Dr. Kaigham J. Gabriel of AT&T-Bell Laboratories for designing the programmable six-phase power supply which was used in testing the micromotors reported in this work. The authors wish to acknowledge technical discussions with S. Bart, L. Tavrow, and Prof. Martin Schmidt of MIT. This work was supported in part by the National Science Foundation under grant #8614328-ECS and through an IBM Fellowship (Mehregany). Fabrication was carried out in the MIT Integrated Circuit Laboratory (ICL). The authors wish to thank the ICL staff, in particular Joe Walsh, for assistance in device fabrication. The authors wish to express their appreciation for silicon-rich nitride wafers from Berkeley Sensor and Actuator Center.

REFERENCES

[1] L.S. Fan, Y.C. Tai and R.S. Muller, Proc. of 1988 IEEE Int. Electr. Devices Meeting, San Francisco, CA, Dec. 1988, p. 666.

[2] Y.C. Tai, L.S. Fan and R.S. Muller, Proc. of Micro Electro Mechanical Systems, Salt Lake City, Utah, Feb. 1989, p. 1.

[3] M. Mehregany, S.F. Bart, L.S. Tavrow, J.H. Lang, S.D. Senturia and M.F. Schlecht, Proc. 4th Int. Conf. on Solid State Sensors and Actuators (Transducers'89), Montreux, Switzerland, June 1989, p. 106. Complete paper to appear in Sensors & Actuators A 173 (1989).

[4] Y.C. Tai, Ph.D. Thesis, Department of Electrical Engineering and Computer Science, University of California-Berkeley, June 1989.

[5] W.S.N. Trimmer and R. Jebens, Proc. of Micro Electro Mechanical Systems, Salt Lake City, Utah, Feb. 1989, p. 13.

[6] S.C. Jacobsen, R.H. Price, J.E. Wood, T.H. Rytting and M. Rafaelof, Proc. of Micro Electro Mechanical Systems, Salt Lake City, Utah, Feb. 1989, p. 17.

[7] H. Fujita and A. Omodaka, IEEE Trans. Electron Devices 35, 731 (1988).

[8] R. Mahadevan, The 3rd IEEE Workshop on Micro Electro Mechanical Systems, Napa Valley, CA, Feb. 1990.

[9] M. Mehregany, S.F. Bart, L.S. Tavrow, J.H. Lang and S.D. Senturia, 36th National Symposium of the American Vacuum Society, Boston, MA, Oct. 1989, p. 108. Complete paper submitted to J. Vac. Sci. Tech. A.

[10] Y.C. Tai and R.S. Muller, Proc. 4th Int. Conf. on Solid State Sensors and Actuators (Transducers'89), Montreux, Switzerland, June 1989, p. 108.

[11] L.S. Tavrow, S.F. Bart, J.H. Lang and M.F. Schlecht, Proc. 4th Int. Conf. on Solid State Sensors and Actuators (Transducers'89), Montreux, Switzerland, June 1989, p. 256.

[12] M. Morita, T. Ohmi, E. Hasegawa, M. Kawakami and K. Suma, Appl. Phys. Lett. 55, 562 (1989).

[13] M. Mehregany, K.J. Gabriel and W.S.N. Trimmer, Trans. Electron Devices 35, 701 (1988).

Section 7

Other Actuators

A WIDE array of physical effects can be used for micro actuation. A number of these micro actuators are discussed in this section. I challenge you to think of other ingenious ways to produce useful forces.

Thermal

The heat transport from a region depends on the temperature gradient, dT/dx. In the micromechanical domain, the distances are generally quite small. Hence, heat transport out of micro regions is usually rapid. With proper design, a small region can be heated and cooled in times measured in microseconds. As a result, actuators depending on temperature can have fast responses. An objection to thermal actuators is the power wasted in the dissipated heat. For some applications, this is a concern. However, often the microcomputer and/or drive electronics to control the micro thermal actuator uses many times the power of the micro actuator.

Shape Memory Alloy

Shape memory alloy (SMA) actuators are controlled by temperature changes. Changing the temperature of an SMA material causes a reversible crystal phase transformation. Below the transformation temperature, the material is in the martensite phase, and it is weak and easily deformed. Above the transformation temperature, the material changes to the austenite phase and becomes strong, exerting large forces in an attempt to return to its memory state. For example, a common form of SMA is a straight wire. Below the martensite temperature, the wire is flexible like solder, and the wire is easily deformed into some shape, like a circle. When the wire is heated, the crystals transform into the austenite phase. Now, the wire behaves just like a piece of spring steel, and the bent wire exerts large forces trying to return to its original straight memory position.

A simple work cycle is as follows. First, the wire is cooled. Now, small forces can be used to distort the wire into the desired shape. Then, the wire is heated, and the large force it exerts trying to return to its memory state is used to do useful work. At the end of this work cycle, the wire is cooled back into the martensite phase, and the next cycle begins.

The SMA effect was observed in 1938 by Alden B. Greninger and V. G. Mooradian in a brass alloy [L. McDonald Schetky, "Shape-Memory Alloys," *Scientific American*, November 1979, p. 74]. G. V. Kurdyumov investigated the martensite phase of this alloy. Later, Thomas A. Read et al. studied SMA effects in gold–cadmium alloys, and demonstrated that forces could be developed by the phase transition. Alloys of iron–platinum, indium–cadmium, iron–nickel, nickel–aluminum, and stainless steel have demonstrated the SMA effect.

In the early 1960s, W. J. Buehler and R. C. Wiley, working at the U.S. Naval Ordnance Laboratory, developed SMA alloys of nickel and titanium. This material is generally referred to as nitinol (Nickel Titanium Naval Ordnance Laboratory), or as TiNi. Because of the superior properties of this alloy, almost all of the recent work has been done with nitinol. By changing the composition of Ti and Ni, the phase transition temperature can be varied from about $-50°C$ to $+166°C$ ["55-Nitinol—The alloy with a memory: Its Physical Metallurgy, Properties, and Applications," C. M. Jackson, H. J. Wagner, and R. J. Wasilewski, NASA-SP 5110, N72-30468, 1972].

A nitinol wire can be stretched 10 to 15% and can develop tensile stresses as high as 1500 MPa (about 200,000 lbf/in^2) ["Shape-memory phenomena," Ahmad A. Golestaneh, *Physics Today,* April 1984, p. 62.]. However, the lifetime and reliability of nitinol at these large strains is poor. In an application requiring millions of cycles and high reliability, the nitinol should be stretched from its memory state only a few percent, and can develop stresses in the range of 200 MPa (about 30,000 lbf/in^2). As the reader can easily see, nitinol is an excellent actuator when large forces are needed.

Using SMA requires some force to distort the material in the soft martensite state, and a way to extract the force in the strong austenite state. In the first paper, "A Micro Rotary Actuator Using Shape Memory Alloys," a thin wire of SMA is used as an actuator. Initially, the wire is twisted about its long axis. When current is passed through one side of the wire, the heat generated causes that side to straighten out, and rotate the center of the wire. By alternately heating the ends, the center undergoes large rotations. This is a simple actuator that generates large forces in the micro domain. A small robotic joint that uses two counteracting SMA actuators is discussed in the next paper, "Millimeter Size Joint Actuator Using Shape Memory Alloy." The following paper by the same author ("Reversible SMA Actuator for Micron Sized Robot") describes the use of reversible SMA, which requires only a single actuator. The paper "Microactuators for Aligning Optical Fibers" (in Section 4, Electrostatic Actuators) discusses the use of SMA actuators for aligning.

Impact

What works well in our macro size domain often is horrid in the micro domain. Good design requires cleverness using the available phenomena. Impact actuators are an excellent example of turning a problem into an advantage.

Friction often sticks micromechanical actuators together so they do not budge. One of the biggest problems of small actuators is overcoming friction. The impact actuator uses friction to good advantage. Figure 2 of "Micro Actuators Using Recoil of an Ejected Mass" shows the action of this actuator. A large mass (microscopically speaking) sits on a surface. Attached to this large mass is an actuator and small mass. In operation, the actuator slowly moves the small mass away, and then rapidly jerks the small mass back. The sharp impact of the return jerk breaks the static friction between the large mass and the surface it sits upon, and the large mass moves over slightly. By doing this repeatedly, the large mass can be moved along the surface in many small steps. This impact mechanism works well under a wide variety of conditions, including dirt. The next three papers ("Precise Positioning Mechanism Utilizing Rapid Deformations of Piezoelectric Elements," "Tiny Silent Linear Cybernetic Actuator Driven by Piezoelectric Device with Electromagnetic Clamp," and "Experimental Model and IC-Process Design of a Nanometer Linear Piezoelectric Stepper Motor") discuss this actuator in more detail.

Piezoelectric

Piezoelectric actuators generally produce very strong forces and very small motions. Larger motions can be obtained by making the piezoelectric material part of a bimorph; however, the forces generated are substantially reduced. Larger motions can also be generated by using the piezoelectric material to make a series of small steps.

The first paper of this subsection ("Zinc-Oxide Thin Films for Integrated-Sensor Applications") describes thin films of zinc oxide. This technology can be integrated with other fabrication processes and generates reasonable forces. The next paper, "A Micromachined Manipulator for Submicron Positioning of Optical Fibers," has two silicon chips as feet at the ends of a piezoelectric bar. Applying a voltage between one foot and the silicon substrate locks that foot. Expanding or contracting the piezoelectric bar then moves the other foot. Repeating this motion causes this "silicon horse" to step across the silicon wafer. The paper "Ultrasonic Micromotors: Physics and Applications" describes an ultrasonic motor. Ultrasonic waves are generated that travel around a ring in different directions, causing the top surface of the ring to move in an elliptical motion. A carefully loaded plate on the top surface only touches the ring at the top of the elliptical motion, driving the load plate in one direction.

MICROMECHANICAL SILICON ACTUATORS BASED ON THERMAL EXPANSION EFFECTS

W. Riethmüller, W. Benecke, U. Schnakenberg, A. Heuberger
Fraunhofer-Institut für Mikrostrukturtechnik
Dillenburger Str. 53
D-1000 Berlin 33/FRG

ABSTRACT

A new type of micromachined silicon actuator has been developed, based on the so-called bimetallic effect. The basic element of the actuator is a cantilever consisting of a silicon-metal sandwich structure, which can be heated via an intermediate electrically isolated polysilicon resistor. For the fabrication only standard microelectronic process technologies and selective anisotropic etching of silicon are used. The result is an absolutely monolithic micromechanical silicon actuator which can work with standard microelectronic voltages. First experimental results are presented and an outline of the fabrication process is given.

INTRODUCTION

Micromechanical sensors based on the deflection of the mechanical element are very thermo-sensitive if sandwich structures of different materials are used. Typical sandwich structures are Si-SiO$_2$ or Si-Si$_3$N$_4$ layers where the dielectric layer is used for the passivation of electrically active devices, e.g. piezoresistors. The temperature cross-sensitivity is due to the difference in the coefficients of thermal expansion of the materials used.

In this work we present a micromechanical silicon actuator based on this bimaterial effect. The basic structure of the actuator element is a fully underetched free standing silicon cantilever with a metal layer on top. An electrically isolated meander-type polysilicon resistor is located between the silicon and the metal layer. If a voltage is applied on the resistor, it leads to a temperature rise in the whole cantilever, corresponding to the dissipated electrical power. Due to the differences in the thermal expansion coefficients of silicon and the metal a deflection of the beam occurs.

Compared to other signal-conversion techniques, e.g. via electrostatic[1] or magnetic forces or to piezoelectric elements this device offers some special features. The fabrication process is very simple and fully compatible with standard microelectronic processes. Due to the possibility of resistance setting and the special way of signal conversion, the device could be operated at standard microelectronic voltage levels. The structure of the mechanical element allows the fabrication of very stable devices with a high production yield and the application under rough conditions. Design consideratons and an outline of the fabrication process of the actuator are described in the following.

DESIGN CONSIDERATIONS

The most important parameters of the actuator are the temperature dependance of the beam-deflection and the corresponding electrical input power. The beam deflection is mainly determined by the differences of the coefficients of thermal expansion and the Young's Modulus of the materials used and the geometry of the mechanical structure.

Using simple theories of beam deflection, one calculates that the thickness of the beam should be as small as possible to increase the temperature dependent deflection at the free end of the cantilever. In our case we decided to use a 4 μm silicon beam. The best result will be obtained for a metal thickness nearly equal to that of the silicon beam. Due to the different values of the Young's Modulus, a small aberration of the metal thickness will be necessary for an optimum result. Because the bending line of the cantilever is a circular arc the absolute deflection increases with the inverse bending radius and with the square of the cantilever length. The width of the cantilever does not affect the behaviour of the element. For the first experimental studies, we realized an actuator structure with a cantilever length of 500 μm and a width of 100 μm for the Si-beam and 80 μm for the metal-layer, respectively. In the case of silicon and gold, the coefficients of thermal expansion differ by a factor of five, in the case of silicon and aluminium they differ by a factor of nine. Due to the fact that Al is etched in anisotropic selective Si-etchants like EDP or KOH, we decided to use gold to reduce the technological expense. For this special design we calculate a maximum beam deflection of 0.12 μm/K.

FABRICATION PROCESS

The etch rate of silicon in anisotropic etchants is very highly dependent on the crystal orientation and the boron (doping) concentration. These properties can be taken as an advantage for the precise shaping of structures in silicon[2,3,4].

An outline of the fabrication process of the silicon actuator is given in Fig. 1. The fabrication process starts with growing a highly boron doped expitaxial layer on standard 4" p-type silicon wafers with (100) orientation and a resistivity of 1-10 Ωcm. The boron concentration was 1.3×10^{20} atoms/cm³, which provides an excellent etch stop in anisotropic solutions. Next a thin LPCVD-Si_3N_4 layer is deposited for electrical isolation of the following polysilicon layer (a). Polysilicon as material for the heating resistor was chosen because of the possibility to define the resistance via a diffusion or implantation process. We have realized two versions with 120 Ω/square and 85 Ω/square via Boron-implantation according to reference[5]. This results to an effective resistance of 20 KΩ and 14.5 KΩ for the meander-type resistor which was defined via lithographical and dry etching techniques (b). A second LPCVD-Si_3N_4 layer was deposited to encapsulate the polysilicon-resistor. The Si_3N_4 is then lithographically structured and etched using dry etching techniques to open the contact holes of the resistor and the outline of the cantilevers (c). After a chromium-gold metallization a lithographical step defines the lateral dimensions of the metal-layer on the cantilever and the bond-pads. The final metal-layer thickness is then formed using electroplating techniques with the resist as a forming mask. With a special heat treatment the gold is made stress free (d). The thin Cr/Au-plating base is then removed and the highly boron-doped layer is isotropically etched in the unprotected areas (e). At least, the cantilevers were underetched with an ethylenediamine-based anisotropic etchant. The result is a fully underetched, free movable cantilever over a pyramidial etch cavity (f). Fig. 2 shows an optical micrograph of the chip bearing four identical cantilevers.

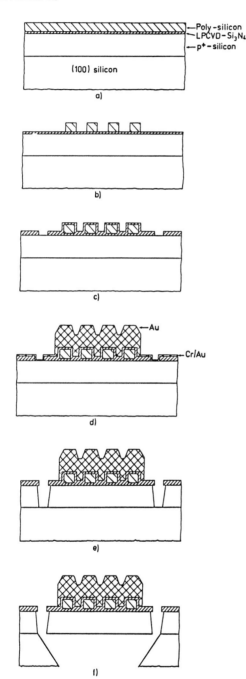

Fig. 1: Sequence of fabrication steps for the silicon actuator

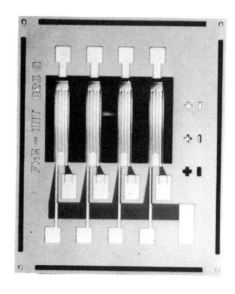

Fig. 2: Optical micrograph of the silicon actuator chip with four identical cantilevers. Chipsize 1.5 x1.25 mm², cantilever-length: 500 µm, -width: 100 µm, Si-thickness: 4 µm, Au-thickness: 1.8 µm

EXPERIMENTAL RESULTS AND DISCUSSION

The completely etched devices show a deflection of 40 - 50 µm upwards at the free end of the cantilever. This is generated by the internal tensile stress of the LPCVD-Si_3N_4 layer. This has been confirmed from other experiments with the same structures only consisting of Si and Si_3N_4. The transfer function of the beam-deflection and the electrically input power for a 4.0 µm thick Si- and a 1.8 µm thick Au-layer is shown in Fig. 3. The deflection increases linear up to an electrical power of 130 mW. For larger input power a saturation is clearly visible. This is due to the fact, that the energy loss increases with increasing temperatures and more electrical energy is needed to hold the cantilever at a fixed temperature. Furthermore the Young's Modu-

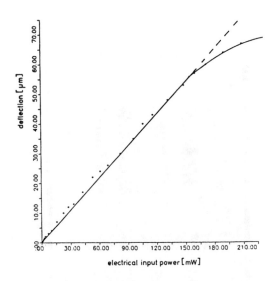

Fig.3. Beam-deflection compared to electrical input power for a cantilever of the device shown in Fig. 2

lus of the metal decreases rapidly with increasing temperature. The melting point of the gold is reached at approximately 300 mW input power. A rough estimation shows that a temperature rise of 100 K corresponds to an input power of approximately 30 mW. With the result of Fig. 3 one can calculate a sensitivity of approximately 0.1 µm/K. This value is in good agreement with the data estimated in the previous chapter. The deviation may be due to the decrease of the mechanical constants with increasing temperatur and the nonuniformity of the heat distribution in the cantilever.

The dynamic behaviour of the cantilever has been tested with a square wave signal. At an electrical input power of 150 mW the static deflection is fully available up to approximately 10 Hz. A motion of the cantilever could be optically observed up to 35 Hz.

CONCLUSION

The cantilever composed of silicon and metal with an intermediate, electrically isolated resistor has been found to be a suitable structure for a micromachined silicon actuator. Further experiments will be done to determine the absolut beam deflection per temperature unit and for more detailed information about the dynamic behaviour of the system.

Devices fabricated in different fabrication processes and other combinations of material will be compared with this first samples. The intension is to increase the beam-deflection per input power unit, to optimise the dynamic behaviour and to adjust the deflection at room temperature.

Application of the basic actuator element discribed are in the field of light modulator, electrical switches, microvalves etc.

ACKNOWLEDGEMENT

We would like to thank C. Dell, H. Sehnert, V. Wallendszus and the staff of our sub-µm-Pilot-CMOS-Technology for their valuable assistance.

REFERENCES

1) K.E. Petersen, "Micromechanical Membrane Switches on Silicon", IBM J. Res. Develop., Vol 23, No. 4, July 1979, 376-385
2) L. Csepregi, R. Hauck, R. Nießl, H. Seidel, "Technologie dünngeätzter Siliziumfolien im Hinblick auf monolithisch integrierbare Sensoren", BMFT-Forschungsbericht T83-089 (Mai 1983)
3) K.E. Petersen, "Silicon as a Mechanical Material", Proc. IEEE, 70 (1982) 420-457
4) H. Seidel and L. Csepregi, "Three Dimensional Structuring of Silicon for Sensor Applications", Sensors and Actuators, 4 (1984) 455-463
5) E. Obermeier, P. Kopystynski, R. Nießl, "Charakteristics of Polysilicon Layers and Their Application in Sensors", IEEE Solid-State Sensor Workshop 1986, Hilton Head Island

CMOS ELECTROTHERMAL MICROACTUATORS

M. Parameswaran, Lj. Ristic, K. Chau, A. M. Robinson and
W. Allegretto*

Department of Electrical Engineering,
*Department of Mathematics and
Alberta Microelectronic Centre
University of Alberta, Edmonton,
Canada T6G 2G7.
Tel: (403) 492 - 4662

ABSTRACT

Fabrication of electrothermal microactuators using a commercial CMOS technology is presented. The effect of differential thermal expansion of adjacent insulating layers is used to produce the mechanical deflection. The field oxide and CVD oxide of the CMOS process form the micromechanical structure. A heating element required to raise the temperature is formed by the polysilicon layer available in the CMOS process. A typical cantilever microactuator exhibited an elastic deflection of 4 µm above the chip surface, and a suspended plate microactuator 1 µm above the chip surface. Maximum frequency response was measured to be 1.5 kHz and 2 kHz for the cantilever and the suspended plate microactuators respectively.

1. INTRODUCTION

Fabrication of miniaturized microelectronic actuators marks one of the recent trends in microelectronics. Microactuators represent one of the key components for potential future applications ranging from micro-robots [1] to miniaturized scanning tunnelling microscope [2]. In general, the mechanical deflection of microelectronic devices is usually associated with the piezoelectric effect. Materials such as ZnO and quartz crystals are commonly used because of this piezoelectric effect and are fabricated into various types of transducers, especially actuators. Present day microelectronics makes extensive use of silicon for the fabrication of circuits as well as sensors. Unfortunately, silicon does not exhibit piezoelectric behaviour which could be used to fabricate microactuators. Silicon and its compounds however, do exhibit a wide range of coefficients of thermal expansion that could be effectively used to fabricate microactuators based on silicon microelectronic processes.

2. ELECTRO-THERMAL ACTUATORS

One of the earliest actuators used for temperature control applications is the bimetal thermostat [3]. In this type of thermostat, two metals of different coefficients of thermal expansion are attached together to form a bimetal strip. This strip, when subjected to a temperature increase, curls towards the metal with the lower coefficient of thermal expansion and produces a mechanical deflection. The differential thermal expansion effect was recently explored in the fabrication of silicon microactuators by Riethmuller et al., [1]. Two thermally conducting layers (P^+-Si and Au) and an electrically isolated heating resistor were used to fabricate a cantilever microactuator. The effect of differential thermal expansion of the P^+-Si and Au layers is used to produce the mechanical deflection. Although the fabrication of this microactuator is compatible with IC technology, many processing steps and materials used to produce it appear to be customized.

On the other hand, fabrication of several micromechanical structures using a commercial CMOS process was successfully demonstrated and is described in [5, 6]. Using this technique, cantilever and suspended plate microactuators can be successfully fabricated and the details are outlined in this paper.

3. ELECTRO-THERMAL ACTUATORS IN CMOS TECHNOLOGY

The first step in fabricating micromechanical structures for CMOS microactuators involves the special layout design for a commercial IC process [5]. A typical layout design for the cantilever structure is illustrated in Fig. 1(a). Although this layout design consists of several layout design rule violations, it does not affect the CMOS process sequence and allows the formation of the basic structure required for the microactuator. Fig. 1(b) shows the cross-sectional view of the cantilever

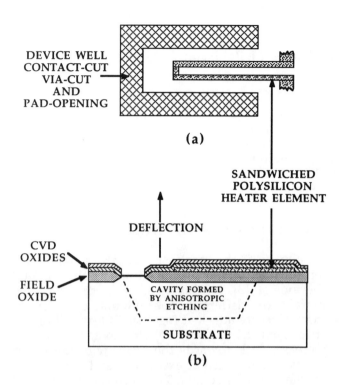

Figure 1. (a) Layout and (b) Cross-sectional view of the cantilever microactuator.

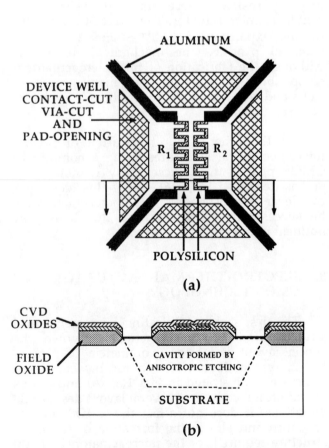

Figure 2. (a) Layout and (b) Cross-sectional view of the suspended plate microactuator.

obtained after the CMOS process. Here, the sandwich structure formed of field oxide (thermal oxide), polysilicon and the phosphorous doped CVD oxide can be seen. The sandwiched polysilicon functions as a heater. Similarly, the layout design for a suspended plate microactuator is illustrated in Fig. 2. Our earlier experiments indicated that CMOS polysilicon heater elements could raise the temperature of a thermally isolated membrane as high as 300°C [7], and this can be effectively used for the functioning of the microactuators.

To form the cantilever and the suspended plate structure, the CMOS die was anisotropically etched using EDP. This micromachining step was performed at a temperature between 95-105°C to retain the aluminum bonding pads without damage [7]. After etching, the die was packaged and bonded for testing. SEM photographs of the cantilever and the suspended plate microactuators are shown in Figures 3 and 4 respectively.

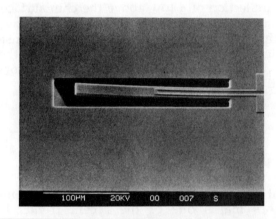

Figure 3. SEM photograph of the cantilever microactuator.

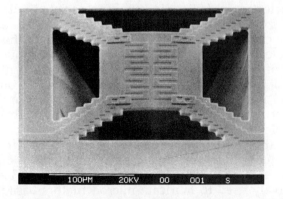

Figure 4. SEM photograph of suspended plate microactuator.

4. EXPERIMENTAL TESTING AND DISCUSSION

4.1 Amplitude response

To evaluate the amplitude response of the microactuators, a dc current was passed through the polysilicon heater and the mechanical deflection was observed and measured using an optical microscope. The movement of the microscope in keeping the deflected microactuator in focus indicated the deflection. The accuracy of measuring deflections using the optical microscope in this manner was estimated at 20%.

<u>Cantilever microactuator:</u>

The field oxide, which forms the bottom layer of the structure, is grown thermally and the following insulation layers (CVD oxides) are grown by LPCVD process. Due to the process difference, the field oxide and the CVD oxides would have different packaging density and hence different coefficients of thermal expansion. This differing coefficients of thermal expansion was observed as a deflection of the free end of the cantilever above the substrate when a current was passed through the heater element. Typical response is shown in Fig. 5. We observed an elastic deflection up to 4 µm above the chip surface as a function of input power. Input powers exceeding 5.6 mW produced irrecoverable plastic deformation. This effect could be possibly due to the excessive amount of heat generated by the heater which anneals the oxides and generates permanent strain.

Following the theory outlined in [3, 4], one of the basic parameters to describe the behaviour of a cantilever microactuator is the conversion factor γ_E, which relates the deflection of the free end of the cantilever to the input electrical power. The conversion factor γ_E for our cantilever microactuator was calculated in the elastic deflection region, between the input powers 2 mW and 5.6 mW, to be 0.97 µm/mW.

<u>Suspended plate microactuator:</u>

The suspended plate structure, which was originally designed for a CMOS gas flow sensor [7], also exhibited an actuator behaviour when current was passed through the polysilicon heater element. Since this suspended plate was supported by the diagonal corners, the structure did not move freely above the surface compared to a cantilever actuator and hence resulted in a smaller deflection. The maximum deflection was estimated to be approximately 1 µm above the chip surface for an input power of 20 mW.

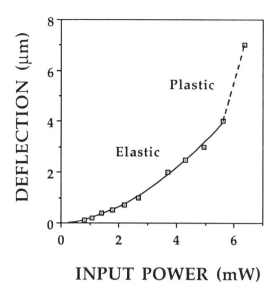

Figure 5. Amplitude response of the CMOS cantilever microactuator

4.2 Frequency response

To determine the frequency response of the microactuators, a square wave signal, with 50% duty cycle, and frequency varying from 1Hz to 10kHz was applied to the heater element. The square wave produced an oscillation of the cantilever. Using the microscope, the extreme positions of the tip of the cantilever were measured as a function of frequency. Typical response of the cantilever microactuator is shown in Fig. 6. At low frequencies the heating and cooling of the oxide sandwich takes place slowly enough and the deflection can follow the input signal, as indicated by the area marked oscillating region of the Fig. 6. When the frequency is increased, the amplitude swing decreases until it reaches a constant value at 1.5 kHz. We believe this effect could be due to the inability of the cantilever structure to cool completely between heating cycles. Hence, the microactuator is unable to follow the input signal and produce a constant deflection beyond 1.5 kHz.

The frequency response of the suspended plate microactuator is also shown in Fig. 6. Since the amplitude of deflection is small, the response is only an estimate, and hence is shown by dashed lines. The oscillation of the actuator can be clearly observed, however, up to a frequency of 2 kHz; beyond this frequency the actuator does not exhibit any noticeable visual movement.

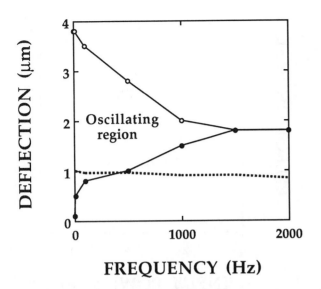

Figure 6. Frequency response of the CMOS microactuators.
_____ Cantilever
------ Suspended plate

The commercial semiconductor fabrication industry does not reveal process parameters such as the thicknesses of various oxides and exact doping levels. Therefore, at this stage, a full theoretical analysis of the microactuators based on a commercial IC process is difficult. The attractive feature, however, is the ease of fabricating such micromechanical devices without the need for a custom fabrication facility. More systematic experiments, employing various design geometries and interaction with the CMOS foundries will allow us to obtain more useful parameters and model these micro-devices for optimum design.

5. CONCLUSION

Fabrication of electrothermal microactuators using commercial CMOS technology has been demonstrated. Two types of microactuators, cantilever and suspended plate, were fabricated and tested. For the geometries of our microactuators, the cantilever underwent a maximum elastic deflection of 4 µm above the chip surface and the suspended plate a maximum deflection of 1µm. The γ_E for the cantilever microactuator was determined to be 0.97 µm/mW. Both types of actuators responded to an input ac signal for frequencies up to 1.5 kHz.

ACKNOWLEDGEMENT

We gratefully acknowledge the services of the Canadian Microelectronics Corporation and the Northern Telecom Electronics Canada Ltd., for fabricating the chips. We wish to thank Mr. G. Fitzpatrick of the Alberta Microelectronic Centre for arranging the experimental set-up and the SEM staff of the Department of Entomology, University of Alberta, for the SEM photographs. This work was supported by the Natural Science and Engineering Research Council of Canada.

REFERENCES

[1] M. Mehregany, K. J. Gabriel and W. S. N. Trimmer, "Integrated Fabrication of Polysilicon Mechanisms", *IEEE Trans. Elec. Dev.*, Vol. 35, June 1988 pp. 719-723.

Long-Sheng Fan, Yu-Chong Tai and R. S. Muller, "Pin Joints, Gears, Springs, Cranks and Other Novel Micromechanical Structures", *Digest of Technical Papers, IEEE. Int. Conf. on Solid State Sensors and Actuators (Transducers' 87)*, Tokyo, Japan, June 1987, pp. 849-852.

[2] S. Akamine, T. R. Albrecht, M. J. Zdeblick and C. F. Quate, "Microfabricated Scanning Tunneling Microscope", *IEEE Elec. Dev. Lett.*, Vol 10, Nov. 1989 pp. 490-492.

[3] R. Griffiths, "Thermostats and Temperature Regulating Instruments", *Charles Griffin and Company, Ltd.*, London, UK, 1934.

[4] W. Riethmuller and W. Benecke, "Thermally Excited Silicon Microactuators", *IEEE Trans. Elec. Dev.*, Vol. 35, June 1988 pp. 758-763.

[5] M. Parameswaran, H. P. Baltes, Lj. Ristic, A. C. Dhaded and A. M. Robinson, "A New Approach to the Fabrication of Micromechanical Structures", *Sensors and Actuators,* (18), 1989 pp. 289-307.

[6] Lj. Ristic, "CMOS Technology: a base for micromachining", *Proc. 17th Yugoslav Conf. on Microelectronics*, Nis, May 9-11, 1989 pp. 741-763.

[7] M. Parameswaran, A. M. Robinson, Lj. Ristic, K. Chau and W. Allegretto, "A Thermally Isolated CMOS Gas Flow Sensor", *Sensors and Materials*, accepted for publication.

Electrically-Activated, Micromachined Diaphragm Valves

Hal Jerman
IC SENSORS
Milpitas, CA 95035

Abstract

Electrically-activated diaphragm valves have been fabricated using heated bimetallic structures to provide the operating force. These valves can be designed to provide fully proportional control of flows in the range of 0-300 cc/min at input pressures from near zero to over 100 PSIG. The valves are batch fabricated using silicon micromachining techniques. By combining these valves with pressure or flow sensing elements, closed loop pressure or flow control is easily accomplished.

Introduction

By using advanced silicon micromachining techniques, high-performance valves have been developed with integral actuators to control gas pressure or flow by the application of an electrical signal. These devices provide the electrical-to-pressure function which complements the pressure-to-electrical function offered by the millions of monolithic silicon pressure sensors fabricated yearly.

Conventional valves for pressure and flow applications have typically used magnetic actuation in the form of solenoids or motors to drive diaphragm or needle-type valves. In a miniature, monolithic silicon-based valve, however, magnetic actuation is not attractive due to the difficulties involved with providing sufficient force. Electrostatic valve actuation has been reported recently,[1] but the forces involved are large only for small electrode gaps. Additionally, only attractive force is feasible, and the $1/x^2$ nature of the force makes fully proportional control difficult. This particular device allowed only low pressure operation, with pressures over about 2 PSIG causing the valve to open even with voltage applied.

Electrostatic actuation is possible using piezoelectric drivers, but the force and deflection properties of materials such as PZT produce either high forces with small deflection for button type actuators, or relatively low forces with high deflection for bimorph type drivers. A button-type driver has been used as a valve actuator by Esashi at Tohoku University, but the complicated assembly and high voltage operation are unattractive for most commercial applications.

For very small structures, thermal actuation becomes more attractive due to the geometrical advantages of scaling. The amount of thermal mass decreases as the volume of material decreases, the thermal loss decreases as the thicknesses decrease, while the forces per unit area remains high. A thermally activated valve using the vapor pressure generated in a heated fluid has been made by Zdeblick at Stanford University.[2] This valve is thermally inefficient due to the large volume of heated fluid, has difficult assembly problems involved with sealing the working fluid in the device, and is restricted to normally-open configurations.

Bimetallic beams[3] have been shown to provide significant force and displacement as integrated actuation elements, however the previously described devices have shown insufficient force to use in most valve applications. The use of bimetallically-driven diaphragms increases considerably the force available from an integrated structure while simultaneously achieving large displacements and symmetric, vertical travel. Although a large variety of combinations of materials can be used, the use of a silicon diaphragm and an aluminum metallic layer is one of the most attractive.

Bimetallic Diaphragm Valve Structure

These valves, such as shown in Fig. 1, consist of a diaphragm actuator with a central boss which mates to an etched silicon valve body. The actuator consists of a circular silicon diaphragm with integral diffused resistors, which acts as one element of the bimetallic structure. An annular aluminum region on that diaphragm forms the other element. By varying the electrical power dissipated in the resistors and thus the temperature of the silicon diaphragm, the displacement of the central boss can be controlled. The actuator chip and the valve body chip make a very compact valve structure without the large, external actuators common in previously developed valves.

In one design a nominally circular silicon diaphragm has been used, and this valve closes with increasing temperature of the diaphragm, making a normally-open valve. A photograph of such a diaphragm structure is shown in Fig. 2.

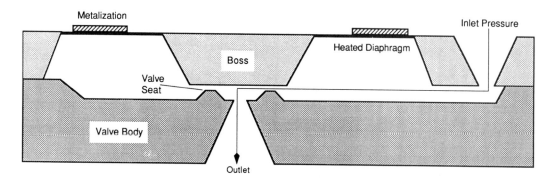

Fig. 1 Cross Section of a Micromachined, Bimetallically-Actuated Diaphragm Valve.

Reprinted from *Technical Digest IEEE Solid-State Sensor and Actuator Workshop*, pp. 65-69, June 1990.

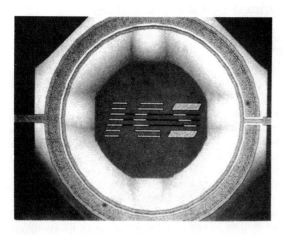

Figure 2. Photograph of typical bimetallic diaphragm with partial backside illumination.

An anisotropic silicon etch has been used to form the diaphragm, the extent of which can be seen from the light transmitted through the diaphragm. The corner compensation has been designed to provide an octagonal approximation to the outer diameter, while the inner diameter achieves a very close approximation to the desired circular shape. The star pattern visible from the center boss is the remnant of the original corner compensation in this non-etch-stopped diaphragm. The thick aluminum layer extends over the outer portion of the diaphragm. A diffused resistor is provided under the aluminum layer, and the aluminum is also used to provide electrical connection to the heater. The valve seat is on a mating silicon wafer, and is very similar to the original valve structure developed by Terry at Stanford, now almost 18 years ago.[4]

The temperature of the bimetallic structure on the diaphragm is used to control the force applied to the central diaphragm boss. Thus, the gap between the valve seat and the bottom surface of the diaphragm boss can be varied, controlling the flow of gas through the valve. There is a gas feedthrough hole in the bimetallic diaphragm chip to allow gas flow through the structure and to equalize the pressure drop across the diaphragm. This minimizes the force necessary to operate the valve.

The flow characteristics of this valve are controlled by the diaphragm and metal thicknesses, the material properties of the metals, the outside diameter, boss ratio, and metal coverage on the diaphragm, and the geometry of the valve seat. The temperature of the bimetallic diaphragm is controlled by varying the power dissipated in heaters either on or in the diaphragm. Gas flows of more than 100 cc/min at pressures over 30 PSIG have been controlled with such a valve, with on/off flow ratios greater than 1000. Prototype valves have been cycled millions of times with no observable change in performance. These valves can be operated as either on-off valves or as proportional control valves by changing the flow geometry and dimensions of the valve elements. Other aspects of the valve operation, such as speed, power dissipation, and magnitude and direction of deflection can be controlled by adjusting the suspension characteristics of the diaphragms.

It is of crucial importance in the design and fabrication of these devices to adequately model the overall flow as a function of applied power to allow the design of this type of valve for a particular application. The analysis of the valves is complicated by the interaction between the gas flow, the diaphragm deflection, and the bimetallic force. These complicated, linked analytical models can now be easily handled using computer programs which numerically solve the set of simultaneous non-linear equations. The individual models used in combination to predict the valve characteristics are described below.

Bimetallic Diaphragm Model

A bimetallic structure consists of a pair of materials, not necessarily metals, with different coefficients of thermal expansion, which are bonded together. As the temperature of the composite is changed, stresses are generated in the structure which can cause useful forces and deflections to be generated by the element. The original analytical model was developed by Timoshenko in a classic paper in 1925.[5] Schematically the structure is shown below:

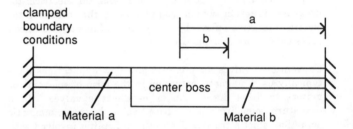

The deflection of this structure at zero force is given by:[6]

$$y(Q=0) = \frac{K_y}{(1+\nu)} \Theta a^2$$

where a is the radius of the diaphragm and K_y is a constant which depends on the bossed ratio, b/a, and Θ is the temperature term:

$$\Theta = \frac{6(\gamma_b - \gamma_a)\Delta T (t_a + t_b)(1 + \nu_e)}{t_b^2 K_1}$$

where γ_a and γ_b are the thermal expansion coefficients of the two materials, t_a and t_b are the respective thicknesses, ν_e is the effective Poisson ratio of the composite beam, and K_1 depends on the relative stiffness of the two components:

$$K_1 = 4 + 6\frac{t_a}{t_b} + 4\left(\frac{t_a}{t_b}\right)^2 + \frac{E_a t_a^3 (1-\nu_b)}{E_b t_b^3 (1-\nu_a)} + \frac{E_b t_b (1-\nu_a)}{E_a t_a (1-\nu_b)}$$

where E_a and E_b are the moduli of elasticity of the two components. The line load at the inner radius, b, at zero displacement is given by:

$$Q_{(y=0)} = \frac{K_q \Theta D_e}{(1+\nu) a}$$

where K_q is a similar geometrical constant and D_e is given by:

$$D_e = \frac{E_a t_a^3}{12(1-\nu_a^2)} K_{2p}$$

with:

$$K_{2p} \approx 6 + \frac{E_b t_b^3 (1-\nu_a^2)}{E_a t_a^3 (1-\nu_b^2)}$$

and the total force is just the line load times the inner circumference:

$$F = Q 2\pi b$$

The factors K_y and K_q are complicated functions of the bossed ratio (b/a) and R_o, the radius beyond which the diaphragm is heated. In this analysis the diaphragm is assumed to be flat at a diaphragm temperature equal to the temperature of the device during metal deposition. Thus by using the appropriate initial metal tension, the initial displacement of the diaphragm can be controlled.

There are a number of different bimetal combinations possible for the structure, but one pair that is attractive is aluminum and silicon. The force and deflection are maximized for pairs which have a large difference in thermal expansion coefficients. It is often desirable to use materials with low thermal conductivities to limit power loss through conduction. Unfortunately both silicon and aluminum have high thermal conductivities, resulting in relatively large power dissipation when these materials are used with thick diaphragms.

For example, for a clamped diaphragm, 2.5 mm in diameter with a 9 μm silicon thickness, a 6 μm aluminum thickness, and a boss ratio (b/a) equal to 0.4, the following deflection characteristics are obtained:

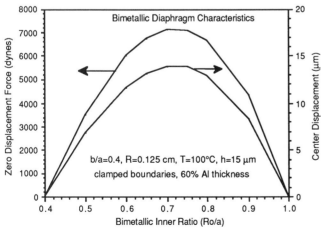

Figure 3. Force and displacement characteristics of a bimetallic diaphragm structure.

Here the deflection and force are maximized for aluminum which covers only about the outer half of the annular diaphragm. Note that for aluminum covering the entire diaphragm there is no force or deflection. The maximum force is about 7100 dynes and the maximum deflection is 13.3 μm. This deflection is downward for the aluminum on the top surface of the device.

Thinner diaphragms provide more deflection at lower force, while thicker diaphragms provide more force at smaller deflections. A silicon thickness between about 6 and 12 μm is optimum for most valve applications.

These first order bimetallic models do not fully take into account the difference in thermal expansion between the heated diaphragm and the supporting frame. In the clamped case the actual boundary condition is described as clamped but moveable. With true clamped boundary conditions, there is a possibility that thin diaphragms will buckle at high temperatures.

Diaphragm Deflection

Flat silicon diaphragms are generally considered to have linear deflection vs. force characteristics. While this is true for the limited deflection of, for example, pressure transducers, in actuator structures the deflection is often a substantial fraction of the thickness of the diaphragm, where the non-linearities become significant. The deflection of any circular diaphragm can be described by an equation with the form:

$$\frac{Q a^2}{\pi E h^4} = A \frac{y}{h} + B \frac{y^3}{h^3}$$

where Q is the applied force, a is the diaphragm radius, E is Young's modulus, h is the diaphragm thickness, and A and B are coefficients which depend on the boss ratio. For non-bossed, flat, clamped diaphragms the value of A is generally taken to be 1.38 and B, 0.41. For bossed structures the diaphragm becomes much stiffer for a given thickness; for a boss ratio of 0.6 each coefficient is increased by a factor of about 20.

Flow Model

The gas flow through the valve is determined by summing the appropriate forces on the valve diaphragm, which determines the deflection of the center of the diaphragm from the relation above. The flow through the valve is then assumed to be due to the restriction formed by the channel between the seating ring and the diaphragm boss. The volumetric, laminar flow through that channel is given by:

$$\text{flow} = \frac{\text{gap}^3 \, w \, (P_{in}^2 - P_{out}^2)}{24 \eta l P_{out}}$$

where w is the effective width of the seating ring, η is the gas viscosity, P_{in} and P_{out} are the absolute input and output pressures, and l the effective length of the seating ring.

This equation is only valid for laminar flow through the channel. The Reynold's number of the flow can be examined to determine if the flow is likely to be laminar. For a nitrogen flow velocity of 30,000 cm/s, nearly at the speed of sound, in a 5 μm channel, at a pressure of one atmosphere, the Reynold's number is only 79, well under the value of 2000 usually taken to be the onset of turbulent flow. Since the channel is 200 μm long, the flow should become well developed not far down the channel.

The average velocity at the output of the channel is found by dividing the volumetric flow given above by the area of the channel. The peak velocity is twice the average velocity for the parabolic flow profile found in laminar flow. For many combinations of high input pressure, low output pressure, and high flow, the average output flow velocity is found to exceed the speed of sound, using the laminar flow equation shown above. It is not possible, however, to obtain supersonic flow in a rectangular channel; that is only possible in a convergent-divergent nozzle.[7] A recent, computational method for the analysis of subsonic, compressible channel flows is available, but it does not extend to the sonic case.[8] Flow through a sonic nozzle becomes only a function of the upstream pressure and the area of the nozzle, at constant temperature. Thus the flow characteristic of the valve changes from a cubic dependence on gap to a linear dependence on gap when sonic flow conditions are achieved.

Experimental Results

A normally open valve has been fabricated using the 2.5 mm diameter actuator shown in Fig. 2. It is mated to a valve body with an annular seat about 400 μm in diameter and 80 μm wide. The gas flow through this structure is typically from the topside of the structure, through a hole connecting the topside of the actuator diaphragm and the backside, across the valve seat, and through a hole in the valve body to the bottom of the valve assembly. By providing the inlet pressure to the topside of the actuator, the pressure tends to help close the valve, reducing the off-flow through the structure at high pressures. When closed, the valve will not suddenly open with inlet pressure pulses, as is the case with the electrostatic valve described by Ohnstein et. al.

The valve flow at a variety of inlet pressures is shown in Fig. 4. This valve has a 10 μm thick diaphragm, a 5 μm thick aluminum layer, and a nominal 4 μm initial gap when the actuator is at room temperature.

Note that with 30 PSIG inputs, the valve actually closes earlier than with lower inlet pressures due to the action of the inlet pressure in helping to close the valve. The valves are fully proportional, with no obvious hysterysis. The leakage flow through this device at 20 PSIG input was about 0.045 cc/min, about a factor of 1,600 less than the full flow.

By using a combination of the models described earlier, the theoretical flow characteristics of such a valve structure can be calculated. The results are shown in Fig 5.

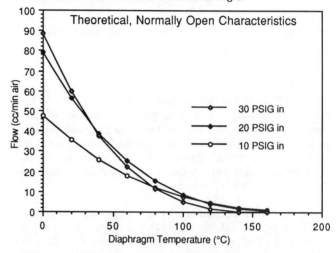

Figure 5. Theoretical flow characteristic of normally open valve structure.

The agreement with the experimental results is very encouraging. Here the initial gap has been selected to roughly match the room temperature 30 PSIG flow. The relatively thick diaphragm and small initial gap of this design results in a low flow, high pressure valve with relatively high power dissipation. By adjusting these parameters and the size of the diaphragms, a family of valves can be made with more optimum low pressure and high flow characteristics.

Pressure Regulator

A similar valve has been operated in a closed loop pressure regulator system to show the ease in using these valves as a computer controlled pressure regulator. The valve was placed between a 30 PSIG gas source and a variable downstream restrictor. The difference between an electrical setpoint and the output of a silicon pressure transducer was used as the error signal to drive the valve open or shut. The restriction was changed to vary the flow through the valve while the electronics system maintained a constant pressure at the head of the restrictor. The results of one such test are shown in Fig. 6.

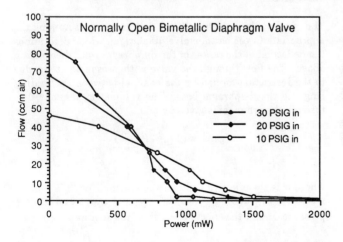

Figure 4. Experimentally determined flow characteristics of a valve with 2.5 mm diameter and 10 μm thick diaphragm.

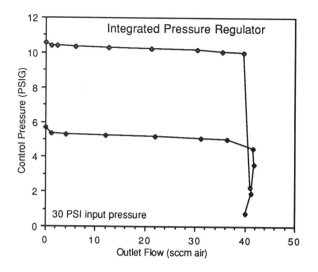

Figure 6. Characteristic of a closed-loop pressure controller using micromachined valve.

Here the electrical setpoint was set to about 10.2 and 5.1 PSI with a 30 PSI inlet pressure. The device provided accurate pressure regulation over a flow range from less than 1 cc/min to about 35 cc/min. Flows greater than 35 cc/min were above the maximum flow through this particular valve at this inlet pressure and thus the output pressure fell to a low value as the outlet restriction was reduced further.

Conclusions

The use of bimetallic driven diaphragms has been shown to provide valve structures with flow and pressure ranges which are useful in a wide variety of applications. The characteristics of the valves can be greatly varied by changing the geometry of the structure, which can be used to optimize a valve for a particular application. The valves are batch fabricated and many hundreds of valves can be made on the same wafer, offering the possibility of a very low cost device. The theory, materials, and fabrication technologies are well understood, which shortens the development time for these devices. It is expected that these valves will greatly influence the types of integrated flow control systems available in the '90's.

References

1 T. Ohnstein, T. Fukiura, J. Ridley and U. Bonne, Micromachined Silicon Microvalve, Proc. IEEE Workshop on Micro-Electro Mechanical Systems (MEMS), pp. 95-98, Feb. 1990.

2 M. J. Zdeblick and J.B. Angell, A Microminiature Electric-to-Fluidic Valve, Proc. 4th Int. Conf. on Solid-State Transducers and Actuators, June 1987, pp 827-829.

3 W. Reithmuller and W. Benecke, Thermally Excited Silicon Microactuators, IEEE Trans. Elec. Dev., v 35 n 6, 1988

4 S.C. Terry, A Gas Chromatography System Fabricated on a Silicon Wafer Using Integrated Circuit Technology, Stanford ICL Technical Report #4603-1, May 1975, p. 41.

5 S.P. Timoshenko, Analysis of Bi-Metal Thermostats, J. Optical Soc. Am., V. 11, 1925, pp. 233-255.

6 R. Roark and W. Young, Formulas for Stress and Strain, McGraw-Hill, 1975, p.165.

7 See, for example, Owczarek, J.A., Fundamentals of Gas Dynamics, Int. Textbook, 1964.

8 Schwartz, L.W., A Pertubation Solution for Compressible Viscous Channel Flows, J. Eng. Math., vol. 21, p. 69-86, 1987.

Study on Micro Engines—Miniaturizing Stirling Engines for Actuators and Heatpumps

Naomasa NAKAJIMA*, Kazuhiro OGAWA** and Iwao FUJIMASA***

* Department of Mechanical Engineering, Faculty of Engineering, The University of Tokyo
** Mitui Engineering and Shipbuilding
*** Advanced Devices Department, Research Center for Advanced Science and Technology, The University of Tokyo

ABSTRACT

This paper presents micro-Stirling engines as a new prospective field of micro-machine study. "Micro-Stirling engine" in this paper means a miniature Stirling engine with a size smaller than a few centimeters cube. The purpose of this study is both to establish the design concept and to realize a micro-actuator and a micro-heatpump. By applying dimensional analysis and computer simulation we investigated how design parameters change when the engine size becomes smaller. Based on the analysis and the simulation we realized a micro-Stirling engine of about 0.05 cc in piston swept volume. The output power is about 10mW at 10Hz vibration. Problems of miniaturizing the engine to a few millimeter cube size are discussed.

INTRODUCTION

The Stirling engine in this paper is not an internal combustion engine like a gasoline engine but an external heating engine. The gas inside the engine is heated from the outside of the cylinder. The advantages of this engine type are:
--thermal efficiency is the best of all the thermal engines,
--any fuel or heat source is available,
--running noise level is low in contrast to gasoline engines.

Another distinctive feature of the engine is reversibility. That is, it acts as a heatpump or a cryocooler if it is driven by a motor.

The typical composition of a Stirling engine [1] is shown in Figure 1. The displacer transfers gas from hot (or cold) space to cold (or hot) space and as a result it raises (or lowers) the pressure of the gas.

The purpose of this study is to realize a micro-Stirling engine with a size smaller than a few centimeters cube. The engine can be applied as a micro-actuator or a micro-heatpump for a micro-machine system.

As for small size Stirling engines, an engine of about 5W output for an artificial heart was developed by Washington University [2]. Hughes Aircraft Corp. in U.S.A. developed a Stirling cycle cryocooler with a size of about 5 cm in diameter and about 14 cm in length [3]. Recently authors designed and realized a Stirling engine of which swept volume (equivalent to exhaust gas volume of a gasoline engine) was 0.11 cc [4].

DIMENSIONAL ANALYSIS

We adopted a dimensional analysis to investigate how design parameters change when the engine size becomes smaller. The results shown in Table 1, were obtained under constant conditions of working gas pressure, temperature distribution, and engine speed. In the table the symbol "ε" means a scale factor(S.F.), and the scale factor per volume(S.F.V.) is equivalent to specific value per power. The table suggests the following considerations for designing a micro-Stirling engine:
a) Heat transfer through walls becomes more effective as surface area per unit volume becomes larger. On the other hand heat insulation between heating wall and cooling wall becomes difficult.
b) A slide mechanism is generally unsuitable, because the friction force becomes larger in comparison with output power.
c) Traditional crank-flywheel or free-piston mechanism is ineffective for a small size engine.
d) The limitation of the engine size is decided mainly by these mechanisms.
e) As the inertia force is negligible for the small size engine, jerky movement of piston and displacer are not harmful.

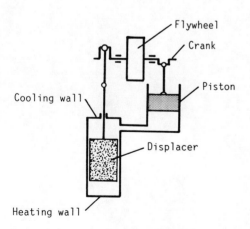

Figure 1. Principle of Stirling Engine concept

DESIGN OF MICRO-STIRLING ENGINE

According to the above suggestions, we designed a micro-Stirling engine with a hysteresis mechanism in place of a crank and flywheel mechanism. Computer simulation [4] was applied to confirm the design and to decide design parameters in detail.

A crank mechanism with flywheel is a method for maintaining phase difference between piston and displacer, but the estimated limit of miniaturization is 9 mm in piston diameter. In order to overcome the limitation of the size of conventional crank-flywheel mechanisms, we investigated a mechanism with hysteresis character.

Figure 2 shows the concept of the micro-Stirling engine. The piston and the displacer are mechanically connected with a spring. A small magnet is situated inside the displacer, and the heating wall and the cooling wall are made of mild steel. The spring and the magnet comprise as a snap action mechanism, which maintains phase difference between piston and displacer, sustaining the reciprocating movement even at relatively low frequency.

Figure 3 shows the engine with about 0.05 cc in piston swept volume. The output power is about 10mW at 10Hz vibration, when the temperature of heating wall and cooling wall are 373 K and 273 K respectively. Figure 4 shows a photograph of the engine.

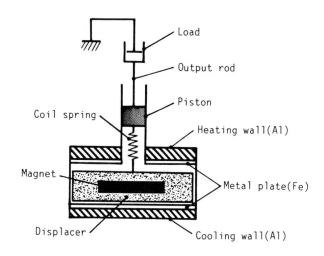

Figure 2. Schematic representation of a magnet-type micro-Stirling Engine

Table 1. Results of the dimensional analysis

		UNIT	S.F.	S.F.V.	NOTES
Scale		m	ε		
Working fluid volume		m^3	ε^3	1	
Mass		kg	ε^3	1	Displacer, flywheel, and so on.
Amount of heat transfer	Thermal conduction	J	ε	ε^{-2}	
	Heat Transfer	J	ε	ε^{-2}	Laminar flow
Stress due to pressure		N/m^2	1	1	Wall, piston rod
Stress due to inertia		N/m^2	ε^2		Link, spring
Power		J	ε^3	1	
Leakeage of working fluid		m^3	ε^2	ε^{-1}	Laminar flow
Fluid loss		J	ε^3	1	Laminar flow
Friction(gravity)		N	ε^3	1	
Friction(pressure)		N	ε^2	ε^{-1}	
Spring constant		N/m	ε	ε^{-2}	
Damping coefficient of fluid		N•sec/m	ε	ε^{-2}	
Inertial mass		kg•m^2	ε^5	ε^2	

S.F. : Scale Factor
S.F.V. : Scale Factor per Volume

POTENTIAL AS MICRO-ACTUATORS

A high output power per weight ratio is desire in actuators for micro robots or micro-submersibles. The ratio of our engine is about one (10mW per 10g), but the ratio may be raised from 10 to 100 times by increasing the gas pressure and the heating wall temperature. We estimated the feasible output power per weight of the micro-Stirling engine as shown in Figure 5, based on the graph of reference [5] and dimensional analysis of micro-Stirling engines. The SMA in the figure indicates actuators made of shape memory alloy. It should be noted that the thermal efficiency of a micro-Stirling engine is about ten times higher than that of shape memory alloy actuators.

Potential application areas for the micro-Stirling engine are micro pump systems to be used inside blood vessels, and cryocoolers to be integrated with IC chips. Both call for further miniaturization down to the size of a few millimeters cube. In order to realize such a engine micro fabrication methods for plastics, ceramics and metals must be established. Also a method of prototype design of such a micro machine must be developed. A process of micro machine

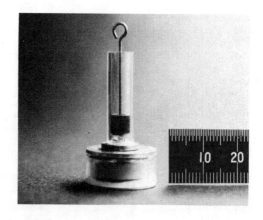

Figure 4. Photograph of the magnet-type micro Stirling Engine

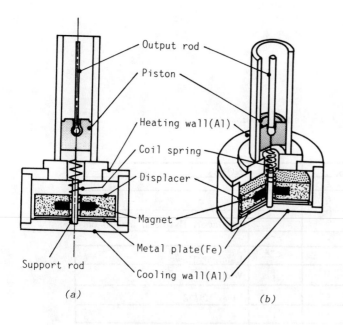

Figure 3. Layout drawing of the magnet-type micro Stirling Engine

(a) Vertical section view

(b) Isometric view

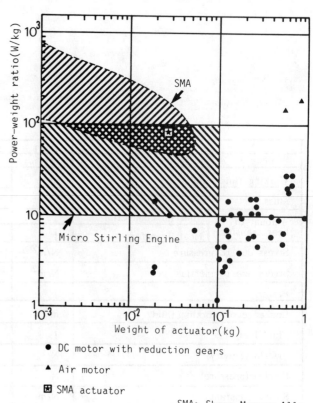

● DC motor with reduction gears
▲ Air motor
✡ SMA actuator

SMA: Shape Memory Alloy

Figure 5. Power-weight ratio of various actuators

(Modified graph of S.Hirose, etc.(5))

design is suggested in Figure 6. Production of a micro-Stirling engine of the required dimensions is only possible with a completely automated assembly process, to be developed in the production design phase. In the preceding prototype design phase, a workstation with manipulators for assembling and tuning single micro machines may prove indispensable.

In the prototype design process performance simulation, design diagnosis, assembling and tuning are important. We think micro mechanical workstation for handling the micro machine is indispensable.

ACKNOWLEDGMENTS

The authors would like to thank Mr. Hiroshi Kanoh, Mr. Kazuhiro Oikawa and Mr. Akira Sakata in the University of Tokyo for their help in design and manufacturing.

REFERENCES

[1] G.Walker: Stirling Engines, Oxford University Press, 1980
[2] M.A.White: Proc. of 18th International Energy Conversion Engineering Conference, pp.694-701,1983
[3] G.Walker: and R.Fauvel: Proc. of 3rd International Stirling Engine Conference(Rome), paper No.51, 1986
[4] K.Ogawa and N.Nakajima: Proc. of 4th International Conference of Stirling Engine(Tokyo), 1988
[5] S.Hirose, K.Ikuta and Y.Umetani: Journal of Japan Robotics Society, 4/2, 15(in Japanese), 1986

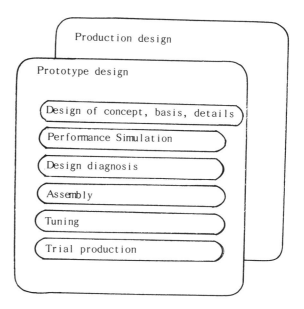

Figure 6. Design flow of the development of micro machines

A Micro Rotary Actuator Using Shape Memory Alloys

K.J. GABRIEL, W. S. N. TRIMMER and J. A. WALKER

AT&T Bell Laboratories, Holmdel, N.J. 07733 (U.S.A.)

(Received May 6, 1987; in revised form April 6, 1988; accepted May 9, 1988)

Abstract

A thin (100 μm diameter) rod of shape memory alloy (Nitinol) wire was clamped at two ends under torsional strain and used as a micro, rotary actuator less than 0.04 cm^3 in volume. Using three electrical connections (the two ends of the wire and a contact in the middle of the wire), the two halves of the wire were differentially heated, which caused repeatable, continuous and directional angular dflections of the wire about its longitudinal axis. In addition, this configuration allows the shape memory alloy to serve as both the actuating mechanism and the mechanical bias (which restores the deformation in the wire after shape recovery). Thus, the biasing takes place with the activation time constant of the wire and not with the thermal cooling time constant as in most previous designs. The present actuator achieves operating bandwidths of approximately 4 Hz, two to three times higher than that of actuators of similar size that rely on bias springs. A micro fluid-valve and a pair of micro-tongs are presented as some applications of such an actuator.

1. Introduction

Shape memory alloys (SMAs) form a group of metals that have interesting shape-recovery characteristics when heated. When these alloys are deformed while below a martensite finish temperature, they recover their original, undeformed shape when heated above an austensite temperature. During the course of the shape recovery, the SMAs produce a combination of both force and displacement. The amount of force and displacement is dependent on the exact geometry of the SMA wire and the amount of heating [1 - 3].

In this paper, we describe a micro-scale actuator made of thin (100 μm) SMA wire (Nitinol) under torsional strain that is capable of creating repeatable, continuous and directional angular deflections. Section 2 describes previous designs of both macro-scale and micro-scale SMA actuators. In Section 3 we present the design for a torsional SMA actuator and report on its dynamic characteristics. A micro fluid-valve and a pair of micro-tongs are presented as applications in Section 4. Finally, Section 5 concludes with discussions of possible improvements to the basic actuator design.

2. Previous work

Actuators made with SMA technology for large-scale applications are well known and most rely on the longitudinal deformation of the SMA. Typically, these actuators employ a SMA member that is deformed in some manner and a bias spring connected to the SMA member. When the SMA member is heated (either directly or as a result of conducting electric current), it performs work. Upon cooling, the bias spring is used to restore the original deformation in the SMA member.

Hashimoto [4] discusses one example of this bias type of SMA actuator. A return spring is mechanically connected in series with the SMA wire and is used to return the SMA wire to its deformed state after cooling. A second type of actuator proposed by Kuribayashi [5] uses two SMA devices connected mechanically in series. Heating of one SMA device shrinks the apparatus in one direction, while heating of the second device shrinks the apparatus in the opposite direction. Actuators of both the bias and differential type have been applied to such macro-scale applications as controlling the movement of robotic end-effector joints and self-regulating greenhouse shutters.

In an application of SMA technology to micro-scale applications, Honma et al. [6] used thin (0.2 mm diameter) TiNi alloy wire to construct a small robotic manipulator. A helical SMA spring approximately 25 mm long was mechanically connected in parallel to a plastic bow that acted as a return spring. The SMA spring was heated by passing a train of electrical pulses through the SMA wire, whose widths controlled the extent of the SMA shape recovery. Utilization of the relation between the recovery strain and electrical resistance change of the TiNi alloy allowed Honma et al. to use the TiNi as a position-sensing element and hence, produce a small actuator with an intrinsic sensor.

In most macro-scale actuators, passive, mechanical bias springs of some sort are used to return the SMA portion of the actuator to its deformed state after cooling. The use of such bias springs in micro-scale applications is undesirable because (1) additional physical size and mass are required for such a spring and (2) the cooling time constant of the SMA member is typically much larger than the heating time constant. Thus, typical operating bandwidths for this type of actuator are less than 1 Hz.

3. Current work

One embodiment of our micro-scale actuator is illustrated in Fig. 1. A thin (100 μm diameter) rod of SMA is clamped at two ends across a 4.5 mm square yoke. The rod is clamped while under torsional strain induced by clamping one end, twisting the rod and clamping the opposite end.

In an ancillary experiment, a 10 cm long, 100 μm diameter piece of SMA wire was twisted and heated by passing a current of 200 mA. The angular deflection resulting from the untwisting during the shape recovery phase was measured and plotted as a function of the number of turns used to establish the deformed, twisted state. As can be seen in Fig. 2, a density of three turns/cm yields the largest rotary excursions without causing irreversible deformation in the SMA.

After clamping the SMA wire with a twist of three turns/cm, the two ends of the rod are each connected to one terminal of a current source with a third terminal (connected to the center of the rod) acting as a common ground (Fig. 1). In addition, a passive mechanical member is attached to the center of the wire to display and utilize the angular deflections of the SMA rod.

The current sources (i_1 and i_2) connected to the rod are pulse-width-modulated, 300 mA, square-wave pulse trains with a repetition period of 200 ms (Fig. 3). Differentially varying the durations of the pulses (or equivalently the duty cycles) causes differential heating of the two SMA halves and hence causes an angular deflection about the longitudinal axis at the center of the SMA rod.

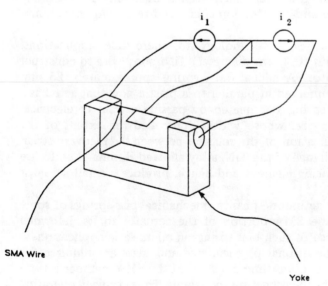

Fig. 1. Diagram of micro rotary actuator showing the clamping yoke, torsionally strained SMA wire and electronics for the differential heating.

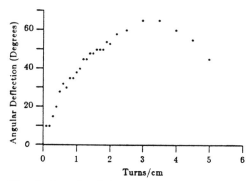

Fig. 2. Angular deflection vs. preclamped twist (turns/cm) for a 10 cm long, 100 μm diameter piece of SMA wire being heated by a 200 mA current.

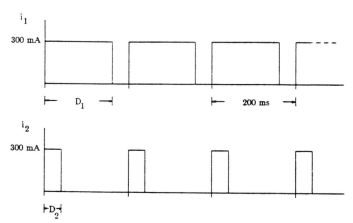

Fig. 3. Current waveforms for controlled angular deflections of the micro actuator.

For the actuator described in this paper, the maximum, clockwise deflection was achieved with a pulse duration of 180 ms for the i_1 pulses and 20 ms for the i_2 pulses. Similarly, maximum counterclockwise deflection was achieved with a pulse duration of 20 ms for i_1 pulses and 180 ms for i_2 pulses. Intermediate angular deflections were achieved by linearly varying the relative pulse durations of i_1 (D_1) and i_2 (D_2). Setting the duration of i_2 pulses to

$$D_2 = 200 - D_1 \text{ ms}$$

appeared to yield a reasonably linear mapping into angular deflections. The rotation of the center of the wire as a function of (D_1, D_2) is shown in Fig. 4. The repeatability of the angular deflections was between 0.5 - 2.0 degrees, with the best repeatability occurring at the two extremes of angular deflection. This repeatability was maintained (on each of several different actuators of the same design) for over three months.

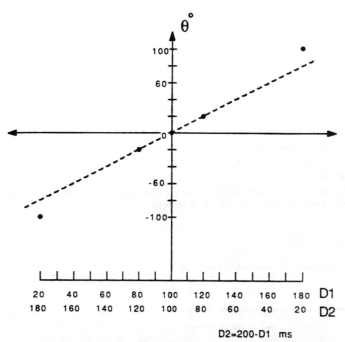

Fig. 4. Graphic representation of the angular deflection of the SMA wire (4.5 mm long, 100 μm in diameter) center as a function of the current waveform durations. Note zero angular deflection at equal duration (100,100) and approximately ±100 degrees deflection at (180,20) and (20,180) respectively.

The actuator can oscillate between clockwise and counterclockwise deflections at a maximum rate of approximately 4 Hz (a function generator was used for these measurements). This is a higher bandwidth of operation than for the actuators described in Section 2.

The higher bandwidth arises because: (1) the activation time constant is shorter due to the thinner diameter wire; (2) the shorter heating time constant is used to rotate the actuator in *both* directions. Most SMA actuators can be moved rapidly in one direction by rapidly heating the wire, but must wait for the wire to cool (which has a longer time constant) before returning to the original position. This use of the SMA against SMA instead of a bias spring reduces the time for one cycle.

4. Applications

A micro fluid-valve and a pair of micro-tongs were constructed using the basic actuator described in the previous section. Figure 5 shows a schematic of the micro fluid-valve. A hollow, glass tube was attached to the center of the rotary actuator and placed in front of a fine air hose. Angular deflections of the wire cause rotation of the attached tube and hence, direct the air flow from the nozzle to various points along the turbine. By directing

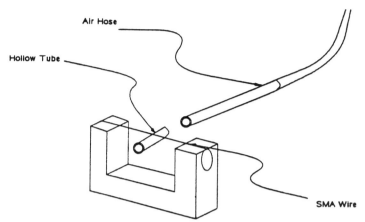

Fig. 5. Micro rotary actuator used as a micro fluid-valve to direct air flow from air hose with a rotating, hollow tube.

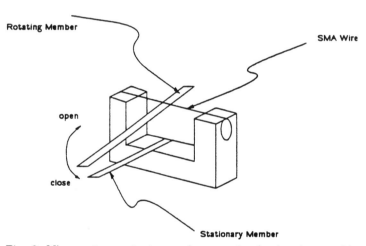

Fig. 6. Micro rotary actuator used as a pair of micro-tongs with a rotating member in opposition to a stationary member.

the air flow over or under the turbine blades, the turbine was made to rotate counterclockwise or clockwise. In addition, by adjusting the point at which the air flow struck the turbine blades, the speed of the turbine could be controlled over a range of ±300 rps with a resolution of ±15 rps.

A pair of micro-tongs was also built as part of a master–slave teleoperator system. A rigid, structural member (approximately 5 mm long, 1 mm wide and 0.5 mm thick) was attached to the center of the SMA wire in an actuator of the type described in Section 3. A second, similar but stationary, rigid member was attached to the actuator yoke to form a pair of micro-tongs (Fig. 6). Angular deflections of the wire cause rotation of the attached member about the wire, resulting in an opening and closing of the tongs.

5. Conclusions

A 0.04 cm^3 micro rotary actuator using SMA under torsional strain has been demonstrated. Torsional deformation of the wire allows: (1) angular motion of the SMA wire to take place during the course of shape recovery; (2) the SMA wire to serve as both the actuating force and the biasing force that restores the deformation in the wire after shape recovery. This enables the present actuator to achieve operating bandwidths of approximately 4 Hz; this is a higher bandwidth than achieved with actuators that rely on bias springs to restore deformation.

A limitation in the current design is the lack of angular position sensing. Both the micro fluid-valve and the micro-tongs are operated open loop using the relatively stationary relationship between pulse duration and angular deflection (Fig. 4). However, any precision application of these micro actuators will require some sort of position sensing. One possibility is to use the method of Honma *et al.* and monitor the change in resistivity of the SMA as a function of the recovery strain.

Another possible modification for precision control of this actuator would be to provide a multiplicity of electrical connections rather than the three connections described here. A multiplicity of connections would allow a pattern of activation (electrical heating) amongst a set of discrete sections spanning the length of the actuating wire, resulting in finer control over the angular deflection.

Acknowledgements

Our thanks to G. L. Miller for originally introducing us to Nitinol. We also thank J. F. Jarvis and S. K. Ganapathy for critical readings of earlier drafts of this paper.

References

1 L. M. Schetky, Shape memory alloys, *Sci. Am.*, (1979) 74 - 82.
2 J. R. Yaeger, A practical shape-memory electromechanical actuator, *Mech. Eng.*, 106 (7) (1984) 51 - 55.
3 C. M. Jackson, 55-Nitinol, the alloy with a memory, its physical metallurgy, properties and applications, *NASA-SP-5110*.
4 I. Hashimoto, Application of shape memory alloy to robotic actuators, *J. Robotic Syst.*, 2 (1) (1985).
5 K. Kuribayashi, A new actuator of a joint mechanism using TiNi alloy wire, *Int. J. Robotics Res.*, 4 (4) (1986) 47 - 58.
6 D. Honma, Y. Miwa and N. Iguchi, Application of shape memory effect to digital control actuator, *Bull. JSME*, 27 (230) (1984) 1737 - 1742.

Millimeter Size Joint Actuator Using Shape Memory Alloy

Katsutoshi KURIBAYASHI
Department of Mechanical Engineering, College of Engineering,
University of Osaka Prefecture
804 Mozu-Umemachi 4-cho, Sakai 591 JAPAN

ABSTRACT

A millimeter size actuator to drive a rotary joint for a small robot was designed and fabricated, using shape memory alloy (=SMA) which has the merits of the big force/weight and the extendability to the micron size. The actuator is designed as the push-pull type which is composed of two SMA sheets of 0.05mmX0.5mm X3mm. First, the theoretical model of the dynamics of the SMA actuator was derived based on the experimental analysis of the dynamics of the larger SMA sheets. Using this model, the design method of the above type of SMA actuator was established. Finally, the theoretical torque vs. the angular displacement of a millimeter size rotary joint driven by the above millimeter size SMA actuator was obtained, which shows maximum 4gf-mm.

INTRODUCTION

A millimeter size actuator is necessary for a small size robot to work under a micro scope and to assemble small mechanical parts. Moreover the development of this actuator will give us a lot of knowledge to design a micron size actuator that is discussed on its necessity in 1st MEMS[1]. So far, some papers[2]-[7] on micro actuators have been reported. However, the rotor speed of the rotary variable capacitance micromotor[2] is too fast to mount it on a robot, although the size 100 μm is enough small. The electrostatic linear actuator of 200μm[3] is difficult to be used for a robot. ROMAC of 10mm which is made of rubber, the rubber tube actuator of 20mm[5] and shape memory alloy actuator of 50mm[6] are not small. The SMA actuator of tortional type of 5mm[7] is rather small, however it is difficult to mount it on a rotary joint.

In this paper, focusing on the following merits of SMA actuator, we will design and fabricate a small SMA actuator for a robot.
1) SMA actuator can generates big force/weight.
2) The control method had been established[8].
3) It can be expected that the smaller the SMA actuator, the faster the response.
4) Micron size of SMA actuator will be made of thin film using micro-machining technology.

Smallest on the market are ϕ70μm wire produced by a drawing method and 50μm thickness sheet produced by a cold rolling mill which guarantee us long life as an actuator during its cyclic use. Using the above materials, we will design and fabricate a millimeter size SMA actuator for a rotary joint of a small manipulator whose structure is like a industrial robot.

SEVERAL TYPES OF MILLIMETER SIZE SMA ACTUATORS

To develop a millimeter size SMA actuator for a rotary joint, SMA actuator will be made from ϕ70μm SMA wire or SMA plate of 50μm thickness, and joints will be made from ceramics plate of thickness 300 μm. SMA plate and ceramics plate will be cut by laser light in order to get SMA actuator and a joint mechanism. To make SMA actuator, it is necessory to memorize the original profile of SMA in a furnace by constrainning its shape, for example, at 500°C for 1 hour. There are many types of SMA actuators related with the original profile of SMA. Typical SMA actuators are displayed in

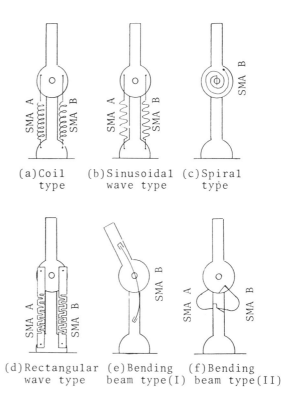

Fig.1 Several SMA actuators for rotary joint mechanisms

Fig.1. To choose one of them as a millimeter size SMA actuator, these actuators are compared with each other on the points of memorizing its original profile, the strength, setting SMA to joint and so forth. The results are shown in Table.1. From the results in Table.1, the SMA actuator of Fig.1(f) was selected to proceed our development.

DYNAMICS OF SMA SHEET AND ITS METHEMATICAL MODEL

Since the dynamics of SMA actuator in Fig.1(f) depends on the dynamics of SMA sheet, its dynamics will be analysed experimentally. However, these millimeter size SMA sheets are too small to analyse its dynamics experimentally with usual measurement equipments without big measurement error. Thus, instead, the larger size of SMA sheets were used as test pieces for the experimental analysis. Based on the experiments, the mathematical model of the dynamics of SMA sheet will be proposed. In later section, the theoretical model will be used to design a millimeter size SMA actuator.

Profile Of SMA Sheet Deformed And Recovered

The experiments of the deformation from the original profile and the recovery to that of SMA sheet were carried out by the method as shown in Fig.2. Their

Table 1 Comparisons among several SMA actuators

Type of SMA actuator	material	Original profile of SMA	Memorizing profile of SMA	Strength	Setting SMA to link	Making joint	Extendability to micron size
Fig.2(a)	wire	coil	X	O	X	O	X
Fig.2(b)	wire, sheet	sinusoidal wave	X	O	X	O	X
Fig.2(c)	wire, sheet	spiral	X	O	X	O	X
Fig.2(d)	sheet	flat	O	X	Δ	Δ	X
Fig.2(e)	sheet	circular arc	Δ	O	O	Δ	Δ
Fig.2(f)	sheet	flat	O	O	O	O	O

O:easy Δ:moderate X:difficult

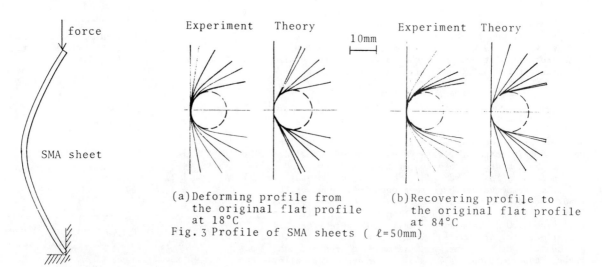

(a) Deforming profile from the original flat profile at 18°C

(b) Recovering profile to the original flat profile at 84°C

Fig. 3 Profile of SMA sheets (ℓ=50mm)

Fig. 2 Schematic diagram of the experimenta method deforming SMA sheet

results are shown in Fig.3 as the profiles during the processes of deforming from the original flat profile of the SMA sheets (a) and recovering to them (b). The SMA sheets used in the experiments are 0.52(mm) *4.98(mm) of length 50(mm), 40(mm), and 30(mm), and their original profiles are linear and flat, with Af=44(°C).

Theoretical model of these profile will be made as follows. Since its serious model is difficult to be derived due to complex properties of elasticity and plasicity of the material, the approximated model will be derived. As easily understood from Fig.3, their profiles can be seperated into two parts. One of them is a circular arc part and the other is linear parts. Moreover, the radius of curvature is constant independent of the deformation, and the deforming or recovering processes of the profile of SMA sheet. Based on the above facts, the relationsip between the radius of curvatures and maximum strain on the surface of SMA sheet can be expressed by next eq.

$$r = \frac{\frac{h}{2}}{\varepsilon_m} \quad (1)$$

The linear parts are assumed to be linear, neglecting the small elastic strain. Therefore, the profile model of SMA sheet is derived as shown in Fig.4. Next, the relationship between the end point (x_E, y_E) and the equations of the circular arc angle θ will be obtained. As the total length of SMA sheet is constant,

$$r\theta + \sqrt{(x_E - r\sin\theta)^2 + (y_E - r(1-\cos\theta))^2} = \frac{\ell}{2} \quad (2)$$

is gotten. As the linear parts is tangent to the circular arc,

$$\frac{x_E - r\sin\theta}{y_E - r(1-\cos\theta)} \cdot (-\tan\theta) = -1 \quad (3)$$

is obtained. From (2) and (3), we get the following eqs.

$$\alpha(\theta\cos\theta - \sin\theta) - \cos\theta + \beta = 0 \quad (4)$$

$$\gamma = \sin\theta + \alpha(1 - \theta\sin\theta - \cos\theta) \quad (5)$$

where

$$\alpha = \frac{r}{\frac{\ell}{2}}, \quad \beta = \frac{x_E}{\frac{\ell}{2}}, \quad \gamma = \frac{y_E}{\frac{\ell}{2}} \quad (6)$$

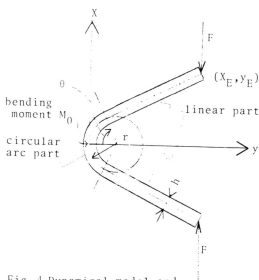

Fig.4 Dynamical model and profile of SMA sheet

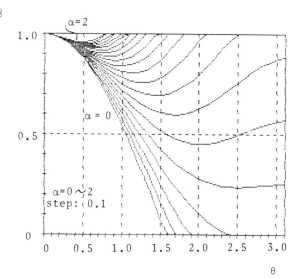

Fig.5 Numerical solutions of eq.(4)

Given X_E, we can get θ by (4),(6) and Y_E from (5),(6) as shown by the folloing flow chart.

$$X_E \xrightarrow{(4),(6)} \theta \xrightarrow{(5),(6)} Y_E$$

However, it is difficult to solve θ from eq.(4) given X_E, because θ is included in the eq.(4) implicitly. Therefore, we solve it graphically by Fig.5 which is the numerical solutions of (4)

Using the above theoretical model, the theoretical profile were calculated, under $\varepsilon_m = 0.0443$ for $\ell = 50, 40$ and 30(mm), which are shown in Fig.3. Their theoretical profiles show rather good agreements with the experimental ones.

Force And Displacement Of SMA Sheet

The force F was loaded to SMA sheet along vertical directions as shown in Fig.2 at the room temperature 18°C, until the circular angle θ becomes $\pi/2$ during the deform ing processes of SMA sheet. Then the SMA sheet was heated up to 84°C holding the vertical desplacement constant at the same state of the end of the before deforming process. After that, the vertical displacement was released to increase slowly, keeping the temperature 84°C. The displacement and the force of SMA sheets of the above processes were recorded on the X-Y recorder, as shown in Fig.6 for $\ell = 50, 40$ and 30(mm). From Fig.6, it can be seen that the larger the ℓ, the larger the displacement, though it makes the force smaller. This reason can be considered as follows. Namely, because the bending moment M_0 at the center of SMA sheet can be expressed as the following eq,

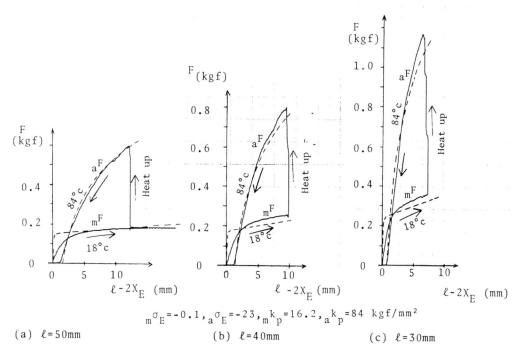

$_m\sigma_E = -0.1, _a\sigma_E = -23, _mk_p = 16.2, _ak_p = 84$ kgf/mm^2

(a) $\ell = 50$mm (b) $\ell = 40$mm (c) $\ell = 30$mm

Fig.6 Force and linear displacement of SMA sheet (the solid line: experiment, the dotted line: theory)

$$M_0 = F \, y_E \tag{7}$$

and because the bending moment M_0 is constant for each θ, independent of ℓ, F becomes smaller due to the large ℓ, that makes Y_E large.

Refering to the experimental results in Fig.6, the theoretical model which describes the relationship between the force and the displacement of SMA sheets, will be obtained. First, we propose the next equations for the bending moments M_0, for the martensite phase,

$$_m M_0 = (_m \sigma_E + _m k_p \theta) Z \tag{8}$$

and for the austenite phase,

$$_a M_0 = (_a \sigma_E + _a k_p \theta) Z \tag{9}$$

where Z is the modulus of section of SMA sheet. In these eqs., σ_E means the elastic stress and $k_p \theta$ means the plastic stress in the SMA sheet. Using the following chart to calculate F, their theoretical curves can be obtained as shown by the dotted lines in Fig.6

```
             (7)
yE  ─────────────────→ F
                       ↑
         (8),(9)       │
θ   ─────────────────→ M0
```

Fig.6 shwons good agreements between the theoretical curves and the experimental ones. Therefore, the theoretical model proposed in this sections was proved valid.

DESIGN METHOD OF A ROTARY JOINT USING SMA SHEET

We will establish the design method of a rotary joint shown in Fig.1(f), by using the equations in 3. The characteristics (torque-angular displacement) of the rotary joint shown in Fig.7(a) will be obtained theoretically. This characteristics depends on both the characteristics (force-linear displacement) of SMA sheet and the geometry of the joint. First, the geometrical relationship between the pushing points P and Q of SMA sheet and the angular displacement γ will be derived. Refering Fig.7(b) which presents the above geometric relationships, the following eqs. are derived.

$$2x_E = R\cos\alpha_2 + r\cos\alpha_1 \tag{10}$$

$$r' = r\sin\alpha_1 \tag{11}$$

$$\gamma + \alpha_1 + \alpha_2 = \pi \tag{12}$$

$$r\sin\alpha_1 = R\sin\alpha_2 \tag{13}$$

From (12) and (13), we can get the next eq.

$$\frac{R}{r}\sin\{\pi - (\gamma + \alpha_1)\} + \sin\alpha_1 = 0 \tag{14}$$

Next, we can easily get the torque τ of the joint as follows.

$$\tau = F \cdot r' \tag{15}$$

If γ and F are given, several valiables can be calculated by the following flow chart.

```
        (14)        (12)        (10)
γ ─────────→ α1 ─────────→ α2 ─────────→ 2xE
              │
              │  (11)        (15)
              └─────────→ r' ─────────→ τ
F
```

The force F can be obtained from X_E by the eqs. in 3. However, as α_1 can not be solved by eq.(14) explicitly, it will be calculated by Fig.8 which shows the numerical solutions of eq.(14).

The above relations are for B side of the joint which can be seen in Fig.7(a). The same relations can be obtained for the reverse side A of the joint in Fig.7(a). The former variables and parameters will be assigned by suffix B, and the latter by A. In order to get the torque of the rotary joint, we define a new variable $\tilde{\gamma}$ whose origin is the central point of the angular displacement of the joint. Thus

$$\tilde{\gamma} = \gamma_B - \overline{\gamma} = -\gamma_A + \overline{\gamma} \tag{16}$$

where $\overline{\gamma}$ indicates the central position of the movable range of the joint. The maximum torque to the counterclockwise derection τ_L can be expressed by

$$\tau_L = _aF_B {}_aR_B^{\,'} - _mF_A {}_mr_A^{\,'} = _a\tau_B - _m\tau_A \tag{17}$$

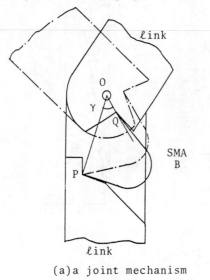
(a) a joint mechanism

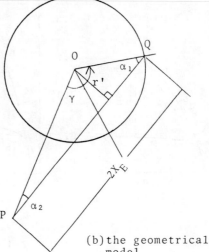
(b) the geometrical model

Fig.7 A joint mechanism shown in Fig.2(f) and its geometrical model

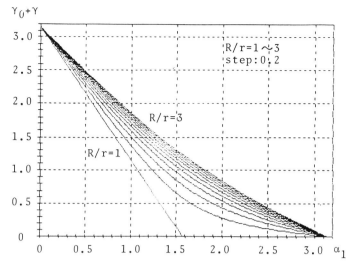

Fig. 8 Numerical solutions of eq.(14)

Given $\tilde{\gamma}$, torque τ_L can be calculated by the following flow chart.

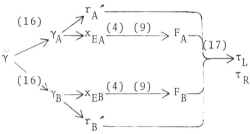

DESIGN OF A JOINT MECHANISM WITH A MILLIMETER SIZE SMA ACTUATOR

The elements of rotary joint mechanism and link of millimeter size were made from ceramics plate of 300 μm thickness by laser light cutting method. They are shown in Photo.1. These elements and shafts of φ 100 μm were assembled to the joint and links by paste as shown in Fig.9. SMA sheets of 50μm* 500μm* 3000μm which are shown in Photo.1 were made from a SMA sheet by the laser light cutting method. The whole joint mechanisms are shown in Fig.9. Dynamics of this SMA actuator were calculated by the eqs. in 3. and 4. The results are shown in Fig.10 and Fig.11. Fig.11 shows the dynamics of the SMA sheet, and Fig.12 shows the torque vs. the angular displacement of the rotary joint. These results indicate that the maximum torque is about 4 gf-mm and the range of the angular displacement is about ±1.4 rad.

CONCLUSIONS

A millimeter size SMA actuator of bending beam type was selected to drive a millimeter size rotary joint made from ceramics plate, considering the easiness to memorize the SMA profile, to set it to a joint, to extend it to micron size, and to make the joint.

The design method of this type of SMA actuator was established by the following steps.
(1) The dynamics of SMA sheet larger than a millimeter size of SMA sheet was analysed experimentally because of the difficulties of experiment for a millimeter size SMA sheet.
(2) To design a millimeter size SMA actuator, theoretical models of the dynamics of SMA sheet were proposed and proved valid.

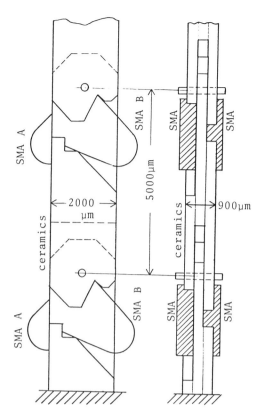

Fig. 9 Joint mechanisms of a robot designed by using millimeter size SMA actuators

(3) The theoretical torque of a rotary joint driven by the SMA sheet was obtained considering the geometry of the joint and the dynamics of SMA sheet.
(4) A millimeter size rotary joint using SMA actuator were designed and its theretical torques vs. the angular displacement were calculated, which indicates maximum 4 gf-mm.

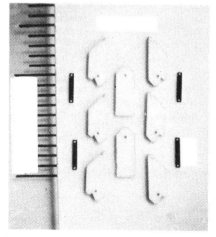

Photo.1 Elements of a millimeter size joints and SMA sheets

ACKNOWLEDGMENT

We would like to thanks to the Corning Japan K.K for a part of the financial support by Corning Research Grant(1988), to the Tokyo Precision MFG, Co, LTD. for the support by the laser light machining, and to Mr. Sigenobu Kishi, an undergraduate student of the University of Osaka Prefecture for the numerical calculation and the experiment.

REFERENCES

[1] W.S.N.Trimmer and K.J.Gabriel,The Proceedings of IEEE Micro Robots and Teleoperator Workshop, IEEE Robotics and Automation Council, Hyannis MA,Nov.9-11 (1987)

[2] J.H.Lang,M.F.Schlecht and R.T.Howe,"Electric Micromotors: Electromechanical Charateristics", The Proceedings of IEEE Micro Robots and Teleoperator Workshop, IEEE Robotics and Automation Council,Hyannis MA,Nov..9-11(1987)

[3] H.Fujita and A Omodaka,"Electrostatic Actuators for Micromechatronics",The Proceedings of IEEE Micro Robots and Teleoperator Workshop, IEEE Robotics and Automation Council, Hyannis MA,Nov.9-11 (1987)

[4] G.B.Immega,"ROMAC Actuators for Micro Robots",The Proceedings of IEEE Micro Robots and Teleoperator Workshop, IEEE Robotics and Automation Council, Hyannis MA,Nov.9-11 (1987)

[5] K.Suzumori,S.Ikuta and H.Tanaka,"Development of Micro-manipulator(1)-FRR-Applied Actuators-,The Preprints of 6th annual Conference of The Robotics Society of Japan,p.275-276(1988)

[6] P.Dario,M.Bergamasco,L.Bernardi and A.Bicchi,"A Shape Memory Alloy Actuating Module for Fine Manipulation",The Proceedings of IEEE Micro Robots and Teleoperator Workshop, IEEE Robotics and Automation Council, Hyannis MA,Nov.9-11 (1987)

[7] J.A.Walker,"A Small Rotary Actuator Based on Tortionally Strained SMA",The Proceedings of IEEE Micro Robots and Teleoperator Workshop, IEEE Robotics and Automation Council, Hyannis MA,Nov.9-11 (1987)

[8] K.Kuribayashi : A New Actuator of Joint Mecanism Using TiNi Alloy Wire, The Int.J. of Robotics Research, Vol.4, PP.47-58 (1986 Winter)

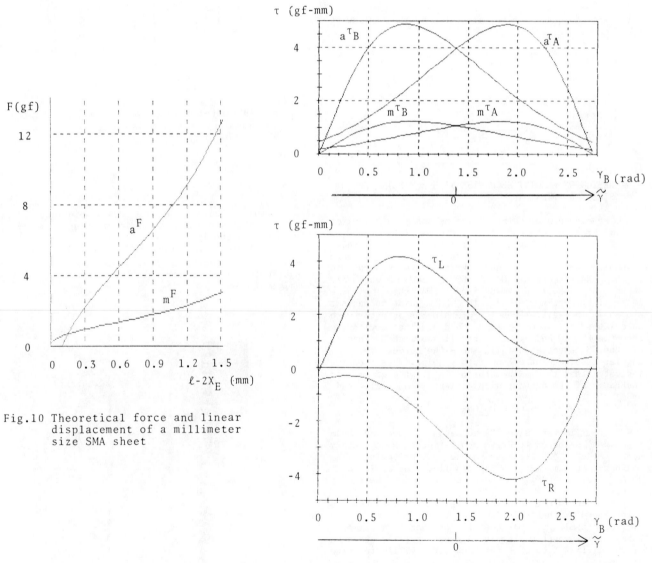

Fig.10 Theoretical force and linear displacement of a millimeter size SMA sheet

Fig.11 Thoretical torque and anguler displacement of a rotary joint driven by millimeter size SMA actuators

Reversible SMA Actuator for Micron Sized Robot

Katsutoshi KURIBAYASHI

Department of Mechanical Engineering,
College of Engineering, University of Osaka
Prefecture 804 Mozu-Umemachi 4-cho, Sakai, 591 JAPAN

Masaaki YOSHITAKE, Soichi OGAWA

Osaka Prefectural Industrial Research Institute,
Econojima, Nishiku, Osaka, 550 Japan

ABSTRACT

Reversible SMA (=RSMA), so called two way and all round SMAs, is very convenient to fabricate a micron sized SMA actuator, because only one SMA material is needed for a SMA actuator. An active bending frame (=ABF) for a robot using a larger sized RSMA sheet whose original shape were memorized round, was made and its shapes were analysed experimentally and theoretically. Based on the theory, a homogeneouse transformation matrix for the ABF was derived. A SCARA Type robot using 2 larger sized ABFs was made and was controlled by electrical current.
Finally it is shown that thin film of reversible TiNi alloy for a micron sized robot could be made by a magnetron sputtering method and by heat treatment methods, and that the frequency response of 10μ m thick thin film of reversible TiNi alloy is about 5 Hz.

INTRODUCTION

Irreversible SMA actuator should be constructed using a couple of SMA materials for a push-pull type or using a SMA material and an ordinary spring for a bias-spring type [1]. Thus it is difficult to fabricate a micron sized SMA actuator by micro machining technologies and to memorize desired shapes.
Recently, reversible SMA (=RSMA), so called two way and all round SMAs, have been developed[2]. For example, RSMA(=reversible TiNi alloy) can be made by the process shown in Fig.1. At first step, TiNi alloy whose component has richer Ni than Ti, is heated 800°C to make β phase. Next the TiNi alloy is constrained at the temperature 400°c for about 4-6 hours to memorize the original shape. During second process, Ti3Ni4 educes due to Ni rich in TiNi. Ti3Ni4 act as a spring component in RSMA actuator. However most structure in RSMA is martensite + austenite phase. Therefore, RSMA actuator is regarded as one kind of bias spring type SMA actuator[1], in which Ti3Ni4 is mixed as a spring component naturally. Thus RSMA actuator does not need any other extra spring to fabricate a bias spring type SMA actuator. This is a big advantage for fabricating micron sized RSMA actuator because we can skip the fabricating process of the bias spring which will bring us a lot of troublesome. Moreover, RSMA sheet memorized roud shape as shown in Fig.1 have another 3 advantages;
1) this sheet has a function of a frame of a robot,
2) this sheet has a function of a joint of a robot,
and
3) this sheet has a function of an actuator.
Therefore RSMA sheet could be used as an excellent element to fabricate a micron sized robot like a mollusc. This kind of element is called an active bending frame(=ABF)[3].
In this paper, an ABF made by using larger RSMA sheets and a robot like a mollusc made by using the above larger ABFs will be explained. The shape of the ABF will be analyzed experimentally and theoretically. Theoretical homogeneous transformation matrix of an ABF which is necessary to control the robot, will be derived. After that, it will be shown that the thin film of RSMA(=reversible TiNi alloy) could be developed, which is a key technology to develop a micron sized robot.

AN ABF USING RSMA

The original shape of RSMA sheet at the temperature over austeneite finish temperature Af which was memorized at the second step in Fig.1, is shown in Fig.2. If the temperature decreases, the profile becomes towards flat shape as shown in Fig.2 and much decreasing temperature cause the shape of concave circular arc. If the temperature of RSMA sheet is under the martensite finish temperature Mf, its shape becomes strongest concave circular arc shown

Fig.1 A making process of reversible SMA

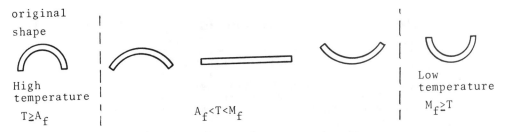

Fig 2 Deforming shape of a reversible SMA sheet for the temperature ranges

in Fig.2. On the contrary, if the temperature of a RSMA sheet is increased from Mf, its shape changes from concave circular arc to convex one. Over the austenite finish temperature, the shape of the RSMA sheet comes back again to the original shape as shown in Fig.2.

The above motion of the RSMA sheet is useful for a mollusc type robot. An ABF using RSMA sheet (=reversible TiNi alloy sheet $0.1*5*40(mm^3)$) is shown in Photo.1. The deformation control of the ABF were carried out by joule's heat flowing electrical current in it. From this photographs, it is understood the shape of the ABF can be controlled by electrical current.

PROFILE OF ABF

To control the position and orientation of the end effector of a robot using many ABFs, the relations between the positions of ends of an ABF and between the tangent angles of ends of an ABF should be known. In order to derive the above relations, at first, the profile of ABF deformed will be analyzed experimentally and theoretically. The results analyzed are shown in Fig.3. Solid lines in Fig.3 show the profiles of ABF obtained by tracing the shapes of RSMA sheet on the photographs. The dotted lines show the theoretical profile under the hypothesis that the profiles of ABF deformed are circular arcs. Comparing the experimental profiles with the theoretical ones shown in Fig.3, they show good agreements. Thus it is concluded the profile of ABF can be approximated by circular arc.

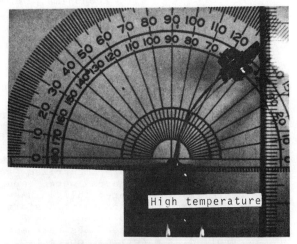

High temperature

Medium temperature

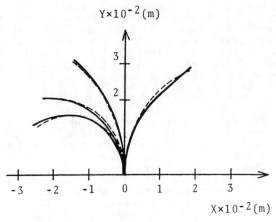

Fig.3 Experimental and theoretical profile of ABF

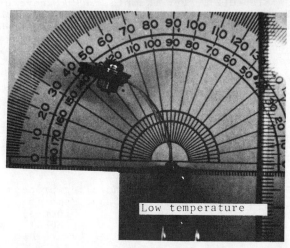

Low temperature

Photo.1 Photographs of the deformed shapes of an ABF using a reversible TiNi alloy sheet($0.1 \times 5 \times 40$ mm^3)

HOMOGENEOUS TRANSFORMATION OF ABF

Based on the above result, the homogeneous transformation matrix which express the position and orientation of an end of ABF observed from the base of the other end of ABF, will be derived. Fig.4 shows the geometrical relations between two ends of ABF taking one of the ends at origin. From this figure, the coordinate of position of an end of ABF is expressed as (X_e, Y_e) and the orientation of the tangent direction of the end is θ ($\theta = 2\phi$). Therefore, the homogeneous transformation matrix of the end can be derived as follows[4].

$$\text{Trans}(X_e, Y_e, 0) \text{Rot}(Z, -2\phi)$$
$$= \text{Trans}\left(\frac{\ell(1-\cos2\phi)}{2\phi}, \frac{\ell\sin2\phi}{2\phi}, 0\right) \text{Rot}(Z, -2\phi)$$
$$= \begin{pmatrix} \cos2\phi & \sin2\phi & 0 & \frac{\ell(1-\cos2\phi)}{2\phi} \\ -\sin2\phi & \cos2\phi & 0 & \frac{\ell\sin2\phi}{2\phi} \\ 0 & 0 & 1 & 0 \\ 0 & 0 & 0 & 1 \end{pmatrix} \quad (1)$$

To prove the fitness of the theory(1), the length l' of the string of ABF and the orientation θ were measured corresponding to the observed angles ϕ of the string of ABF from the photograph. The results are shown in Fig.5. In this figure, solid lines were calculated by the following equations.

$$\ell' = \sqrt{X_e^2 + Y_e^2} = \frac{\ell}{2\phi}\sqrt{2(1-\cos2\phi)} \quad (2)$$

Orientation $= \theta = 2\phi$

Fig.5, shows the good agreement between the experimental results and the theoretical lines, which makes us to conclude the theoretical homogeneous transformation matrix valid.

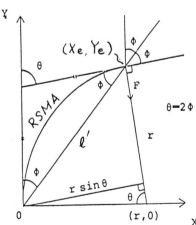

Fig.4 Coordinates system of ABF

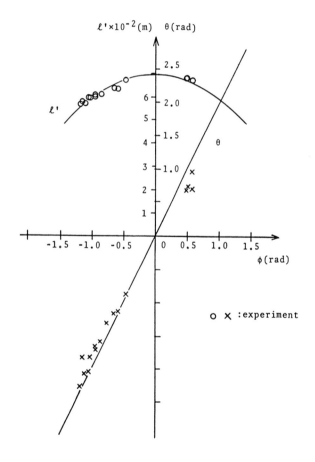

Fig.5 Experimental and theoretical results of the position and orientation of an ABF

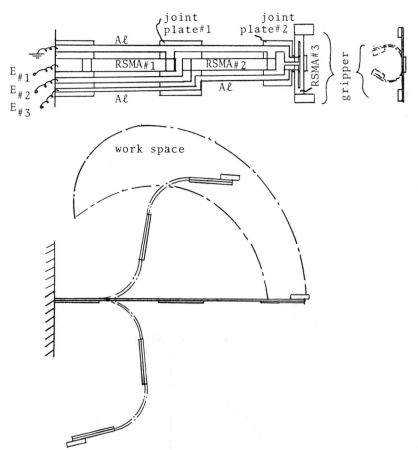

Fig.6 A SCARA robot with three ABFs and its work space

A ROBOT USING ABFS BY RSMA SHEETS

Fig.6 shows a SCARA type robot using 3 ABFs as frames of the robot. Two of them are used to give its robot two degrees of freedom. The other ABF is used as a gripper at the end of the robot. The ends of the RSMA sheets are fixed to the joint plates made from isolation material, and the center part of the latter RSMA sheet for the gripper is fixed to the 2nd joint plate.

In order to control the deformations of ABFs by electrical current, Al sheets are fabricated on its robot as shown in Fig.6. This structure enable us to control the RSMA sheet independently by supplying electrical currents from the base part. To simplify the Al circuit pattern in the robot, the ground line is used as a common line. The robot can be moved vertically as shown in bottom figure in Fig.6. Work space at the top of the robot is shown in same figure.

This structure of the robot shown in Fig.6 are designed so that a micro sized robot similar to the robot in Fig.6 can be fabricated on silicon wafer by micro machining technology.

Photo.2 shows the deformation pattern of a robot made by using two RSMA sheets $0.1*5*40(mm^3)$. ABFs in this robot can be controlled independently by electrical supplied voltages from the base of the robot. These photographs show the high potential of robot using thin film RSMA sheets.

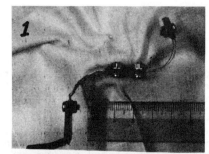
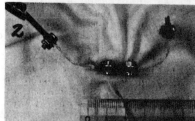
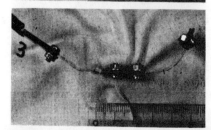

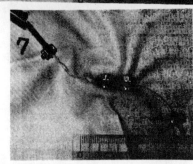

Photo.2 Photographs of the controlled shapes of a SCARA robot with two dgrees of freedom.

THIN FILM OF REVERSIBLE TiNi ALLOY

The possibility to fabricate a micron sized robot similar to that shown in Photo.2 depends on the possibility that a thin film of reversible TiNi alloy could be developed. In this paper, the fact that the thin film of reversible TiNi alloy was developed successfully will be explained as follows.

At first, TiNi thin films 10 μm thick were made on NaCl plates by a magnetron sputtering method. This thin films were separated easily from NaCl plates. At 2nd step, the thin films were heat treated at 800°C for 10 min in the vacuum chamber (5×10^{-7} Torr vacuum). At 3rd step, the thin films were constrained in the crystal glass pipe whose inner diameter 3.7 mm, to memorize circular arc shape as the original shape, at 400°C for 6 hours. Thin film was cut as the shape shown in Fig.7, and was observed at several temperatures. The results are shown in Photo. 3. From this photographs, it is understood that this film of TiNi alloy shows same deformation as shown in Fig. 2. By the rectangular wave electrical current to heat thin film, it was observed that the maximum bending frequency of this thin film is about 5 Hz.

From the above experiment, it is concluded that the thin film of reversible TiNi alloy could be developed. Therefore, next step to develop a micron sized robot should be proceeded by fabricating it on the Si wafer substrate.

CONCLUSIONS

Micron sized robot using reversible SMA is proposed. At first, a larger ABF using larger sized RSMA was analyzed and its homogeneous transformation matrix was derived. A SCARA robot using two larger sized RSMAs was made and, controlled by electrical current. This experiment suggested us it is useful to make this type micron sized robot.

Next, it is established the thin film of reversible TiNi alloy could be developed by a magnetron sputtering method and heat treatment methods, which is a key technology for a micron sized robot.

Finally, it is suggested that the next step for fabricating a robot on the Si wafer substrate should be proceeded.

ACKNOWLEDGMENT

We would like to thanks to the Corning Japan k. k. for a part of the financial support by the Corning Research Grant (1988), and to Mr. Kazuhiko Nishikawa, a graduate student of the University of Osaka Prefecture for the experiment.

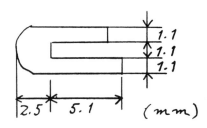

Fig.7 Thin film of R TiNi alloy(10μm thick)

REFERENCES

[1]K.Kuribayashi,"A New Actuator of Joint Mechanism Using Ti-Ni Alloy Wire", The Int. J. of Robotics Research, Vol.4, pp.47-58(1986 Winter)
[2]K.Otsuka and K.Shimizu, "Pseudoelasticity and shape Memory Effects in Alloys", Int. Metals Review, Vol.31, No.3, pp.93-114, 1986.
[3]K.Kuribayashi,"Micro Actuator Using Shape Memory Alloy for Micro Robot", Proc. of Int. conf. on ADVANCED MECHATRONICS by JAPAN Society of Mechanical Engineers, pp.109-114, 1989(Tokyo)
[4]R.P.paul,"Robot Manipultor",MIT Press,1981

High temperature

Low temperature

Photo.3 Photographs of the deformed shapes of a thin film of reversible TiNi alloy

Characteristics of Thin-wire Shape Memory Actuators

P. A. NEUKOMM, H. P. BORNHAUSER, T. HOCHULI, R. PARAVICINI and G. SCHWARZ

Department of Electrical Engineering Design, Swiss Federal Institute of Technology, Zurich (Switzerland)

Abstract

Since 1987 a NiTi wire of only 0.156 mm diameter has been commercially available, which shows promising properties for medium force/medium displacement actuators. The shape memory effects arise from a change from the martensitic to the austenitic phase at a conversion temperature of about 80 °C, causing forces of several newtons and a contraction of several percent. The current to reach this conversion temperature is about 300 mA over 2 s and the driving voltage is about 15 V per meter of wire. Our investigations on mechanical, electrical and thermal characteristics have revealed an interesting feedback feature: the momentary resistance is a measure of the momentary state of contraction. Thus, easily controlled actuators of minimum weight might be realized. The reproducible dynamic operational conditions are: isostatic contraction, 1 N/3.8% to 2.5 N/4%; isometric force, 1 to 4 N. The results of the tensile tests are: tensile strength, 20 N; maximum permissible load, 7.5 N; Young's modulus 34 000 N/mm^2 (standard NiTi) and 17 000 N/mm^2 (trained NiTi). The resistivity is about 44 Ω/m, with a positive temperature coefficient below 47 °C of $+1.5 \times 10^{-3}$/°C and above 100 °C of $+0.34 \times 10^{-3}$/°C and a negative temperature coefficient of -11×10^{-3}/°C during contraction of a standard NiTi at a load of 1.5 N. Endurance tests proved a lifetime of 260 000 cycles at 0.1 Hz, 2 N and 3.5% contraction with a pulse amplitude of 360 mA or 16 V/m.

1. Introduction

The shape memory actuator investigated was a thin NiTi wire of only 0.156 mm diameter, produced by Toki Corporation under the tradename BioMetal. When supplied with a current of about 300 mA, the temperature increased within 2 s above 80 °C, changing the crystalline structure from the martensitic phase to the austenitic phase. The wire contracted by about 4% and was able to pull a load of about 2.5 N. When the current was switched off so that the temperature dropped within 2 s below 40 °C, a pull-back force of about 0.25 N enabled the re-extension to its initial length. Besides all these features, the high power/weight ratio of about 0.5 W/g and the long life of more than 1 million cycles [1] make this NiTi wire very attractive for small and light-weight actuators.

The purpose of this work was to investigate the limiting factors for long-term applications, particularly for implanted controlled actuators. The data presented here are some results of diploma work [2] on mechanical and electrical characteristics, as well as study work [3] on the characteristics of controlled NiTi actuators. Two different types of NiTi wire have been investigated: standard NiTi which reaches maximum length variation after about 50 cycles; and a trained NiTi recently available from the same manufacturer, which has a reproducible length variation already after the second cycle.

2. Mechanical Specifications

The elasticity of extension and the ductile yield of a standard NiTi wire have been measured with a tensile testing machine INSTRON 1122 at 23° and 110 °C (Fig. 1). A similar test with the trained NiTi wire is shown in Fig. 2. The wire behaves as

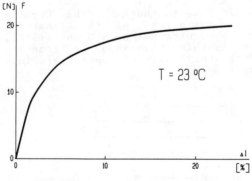

Fig. 1. Tensile test with standard NiTi wire.

TABLE 1. Mechanical specifications of standard and trained NiTi wires

	Standard NiTi wire		Trained NiTi wire
	23 °C	110 °C	25 °C
Young's modulus (at 10 N)	33 400 N/mm^2	34 900 N/mm^2	c. 17 000 N/mm^2
Linear strain (at 10 N)	1.60%	1.56%	c. 3.2%
Ductile yield	20–30%	19–30%	c. 40%
Tensile strength	20–20.4 N	18.4–32.4 N	c. 17 N

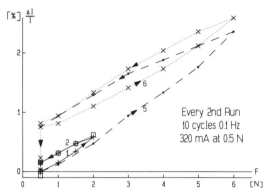

Fig. 2. Elastic and plastic deformation of the trained NiTi wire: excessive stress, but no residual elongation after cycling.

a ductile material, with a more or less linear strain up to 7.5 N and a tensile strength of about 20 N. The new trained NiTi wire exhibits double strain at the same stress, as can be seen in Table 1.

3. Elasticity and Resistivity

The strain and the resistivity versus the applied load of the improved trained NiTi wire have been measured within the limits of elasticity (Figs. 2 and 3). In order to check the reproducibility, ten active cycles at 320 mA and 0.1 Hz were performed at 0.5 N before and after every second run. Test No. 1 was at an increasing loading up to 2 N, then deloading to 0.5 N. A residual elongation of about 0.1% but no residual resistivity change was noticed. After the repeat test, No. 2, no further residual elongation but a drop in resistivity was observed. After cycling, the initial length was completely restored, however, the resistivity remained low. Tests No. 3 and 4 at loading up to 4 N are not shown here; after cycling the initial length was again completely restored, and there was a further drop in resistivity by about 0.2 Ω/m. Test No. 5 shows a plastic deformation of about 0.7% and an additional resistivity drop of 0.1 Ω/m after a load of 6 N. After the final test, No. 6, the plastic deformation was about 0.8% but there was no further change in resistivity. During cycling the initial length was reached again, however, the resistivity dropped another 0.1 Ω/m, so that the total resistivity change was about 0.5 Ω/m or 1.2%.

These experiments demonstrate a relation between stress, strain and resistivity. However, the NiTi wire cannot be used as a very ductile strain gauge meter of satisfactory accuracy.

4. Isostatic Contraction

The contraction versus the applied voltage has been measured at fixed loads between 0.25 and 5 N. The initial length and the absolute contraction depend on many factors, so that we took care to keep the following parameters constant:
- ambient temperature 23 °C, no current of air;
- 3rd cycle of a new NiTi wire measured;
- slow voltage increase and decrease.

Figures 4 to 6 demonstrate the isostatic contraction at loads of 1.8, 2.5 and 5 N of a standard NiTi wire. The minimum pull-back force was 0.25 N, and a load of 5 N led to an irreversible dilatation and degrading. As a comparable measure for the performance, we consider the edge points [4] of martensitic/austenitic start/finish as summarized in Table 2.

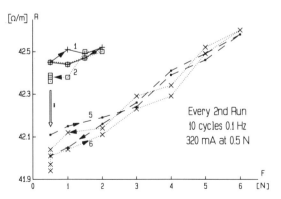

Fig. 3. Resistivity vs. load of the trained NiTi wire: decreased resistivity after excessive stress, also after cycling.

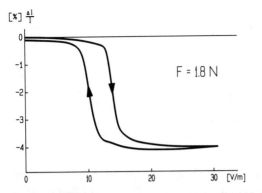

Fig. 4. Contraction vs. voltage at a load of 1.8 N for a standard NiTi wire.

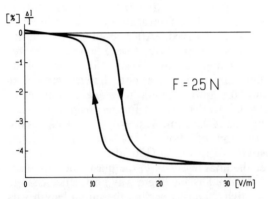

Fig. 5. Contraction vs. voltage at a load of 2.5 N for a standard NiTi wire.

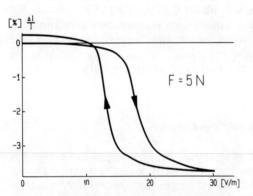

Fig. 6. Contraction vs. voltage at a load of 5 N for a standard NiTi wire: excessive load causes a permanent plastic deformation.

5. Isometric Force

The produced force versus the applied voltage has been measured for a constant length of wire. Again the same interfering parameters as explained in Section 4 have been kept constant. The length of the trained NiTi wire was 100 mm and the stiffness of the test equipment was better than 0.1 mm at 16 N. Figure 7 shows the force response on triangular pulses and Fig. 8 on ordinary rectangular pulses, both at 21 V/m. It is interesting to see that the force response on rectangular pulses is 15–27% stronger compared with

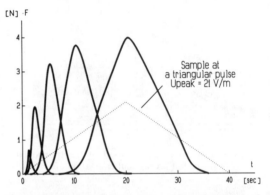

Fig. 7. Isometric force response on triangular pulses of constant amplitude of 21 V/m and variable period duration for a trained NiTi wire.

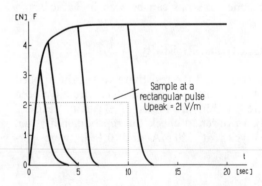

Fig. 8. Isometric force response on rectangular pulses of constant amplitude of 21 V/m and variable period duration for a trained NiTi wire.

TABLE 2. Contraction and supply voltage for a standard NiTi wire

Load (N)	Contraction		Maximum contraction (%)	Extension		Residual contraction (%)
	Start (V/m)	End (V/m)		Start (V/m)	End (V/m)	
0.25	10.8	13.2	2.24	10.1	6.2	+0.12
1.8	13.0	14.9	4	11.1	8.4	+0.08
2.5	13.3	15.3	4.4	11.6	9.6	−0.05
5.0	15.8	19.6	3.75	14.3	11.3	−0.25

TABLE 3. Force versus voltage (pulse period 40 s) for a trained NiTi wire

Voltage (V/m)	Triangular pulse force (N)	Ramp pulse force (N)	Rectangular pulse force (N)
15	1.48	1.52	1.88
18	3.4	3.65	4.2
21	4.0	4.35	4.9

triangular or ramp pulses, particular for low peak voltage. In contrast to the isostatic contraction, no hysteresis could be observed if the pulse period was larger than 50 s. The maximum force versus the applied voltage is given in Table 3.

6. Thermal Contraction and Resistivity Variation

The contraction and the resistance versus the temperature have been investigated from 20 °C up to 170 °C. The contraction of the NiTi wire has been monitored by a pointer free from parallax at 0.1% accuracy and the resistance has been measured using the four-lead technique with an HP 3478 A multimeter of 0.001 ohm accuracy. The climatic test cabinet was a 300 SB/+10JU/40DU from Weiss Klimatechnik. The temperature was changed very slowly and the temperature near the NiTi wire could be measured with 0.5 °C accuracy. Figure 9 shows the contraction versus temperature of a trained NiTi wire at a load of 1 N. The main contraction area (0.5%–4%) is within a narrow temperature gap (66°–90 °C), but the maximum contraction of 5.4% is reached very late at 170 °C. The extension during cooling starts slowly at about 70 °C, and the hysteresis is about 20 °C. Figure 10 displays the resistivity change versus temperature under the same conditions. The resistivity increases from 44.1 Ω/m (20 °C)

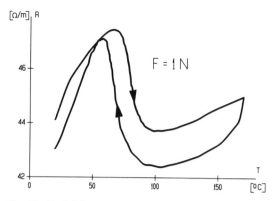

Fig. 10. Resistivity vs. temperature at a load of 1 N for a trained NiTi wire.

during the martensitic phase up to 47.5 Ω/m (68 °C), drops drastically during the transient region to 43.7 Ω/m (99°) and increases slowly to 45.0 Ω/m (170 °C) in the austenitic phase. The temperature coefficient within the transient region is about -6×10^{-3} per °C during contraction. At decreasing temperature, the hysteresis is about 15 °C, but all resistivity values are generally about 1 Ω/m lower, and the temperature coefficient is -6.6×10^{-3} per °C within the transient region. The temperature coefficients of a standard NiTi wire at 1.5 N are -11×10^{-3} per °C and -12×10^{-3} per °C during contraction and extension.

The obvious idea is to use the NiTi wire resistance as a position sensor signal. If there is a reliable relationship between resistance and momentary contraction, the current at constant voltage or, better, the voltage at constant current gives us a feedback signal for a closed-loop controlled actuator system. Actuator and sensor would be combined in the NiTi wire, driven by pulses on a two-lead remote control line.

However, a first glance at Fig. 11 (0.5 N), Fig. 12 (1 N) and Fig. 13 (2 N) reveals a more complex resistance versus momentary length relationship. During the main contraction of about 0.5% to 4% the resistivity drops quite linearly by about 3–4 Ω/m. However, during extension the resistance increases only within the first half of the extension. Thus we may distinguish four different regions:
– 'linear region LC' during contraction with a certain linearity;
– 'defined region DC' during contraction with a one-to-one correspondence;
– 'linear region LE' during extension with a certain linearity;
– 'defined region DE' during extension with a one-to-one correspondence;

The results are summarized in Table 4.

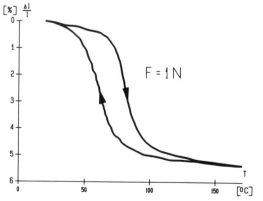

Fig. 9. Contraction vs. temperature at a load of 1 N for a trained NiTi wire.

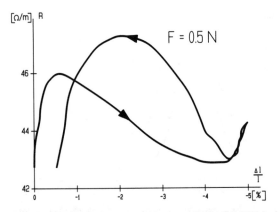

Fig. 11. Resistivity vs. contraction at a load of 0.5 N during slowly changing ambient temperature between 25° and 170 °C for a trained NiTi wire.

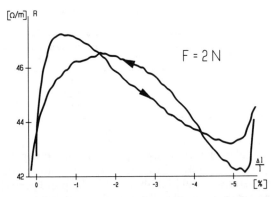

Fig. 13. Resistivity vs. contraction at a load of 2 N, during slowly changing ambient temperature between 25° and 170 °C, for a trained NiTi wire.

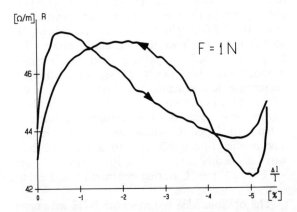

Fig. 12. Resistivity vs. contraction at a load of 1 N during slowly changing ambient temperature between 25° and 170 °C for a trained NiTi wire.

A position-controlled actuator, using the internal NiTi wire resistance as the only position feedback signal, might be developed but only for a continuously increasing contraction at defined loads. As an example, a loaded lever could be lifted step by step quite accurately. Because of the many factors which influence the resistivity, a separate position sensor is recommended for most practical applications.

7. Endurance

Endurance data such as progressive dilatation and fracture depending on the number of cycles are very important for practical applications. Over 20 million cycles of a standard NiTi wire have been reported by the manufacturer [1] at moderate loads and currents. Our goal was to explore the critical limiting factors. At repetitive contractions and extensions at a frequency of 0.1 Hz, a load of 3 N and a pulse peak current of 410 mA, a degrading dilatation has been found after 4000 cycles, and the fracture appeared at 73 000 cycles (see Fig. 14). Table 5 reveals that the peak current over 5 s should not exceed 360 mA, particularly for the new, trained NiTi wire.

TABLE 4. Contraction/extension versus resistivity for a trained NiTi wire

Load	Contraction		Extension	
	LC	DC	LE	DE
0.5 N	0.75 to 3.50% 45.9 to 43.1 Ω/m 0.98 Ω/m (1%)	0.60 to 4.00% 4.60 to 42.8 Ω/m	4.50 to 2.70% 43.0 to 46.9 Ω/m 2.16 Ω/m (1%)	4.60 to 2.00% 43.0 to 47.3 Ω/m
1.0 N	0.75 to 3.50% 47.4 to 44.3 Ω/m 1.13 Ω/m (1%)	0.60 to 4.51% 47.5 to 43.7 Ω/m	4.70 to 3.20% 42.7 to 46.1 Ω/m 2.26 Ω/m (1%)	5.00 to 2.20% 42.4 to 47.1 Ω/m
2.0 N	1.00 to 4.00% 47.2 to 43.7 Ω/m 1.17 Ω/m (1%)	0.70 to 5.00% 47.2 to 43.2 Ω/m	5.00 to 2.70% 42.2 to 46.0 Ω/m 1.65 Ω/m (1%)	5.30 to 1.70% 42.2 to 46.5 Ω/m

TABLE 5. Endurance tests at 0.1 Hz for standard and trained NiTi wires

NiTi type	Current (mA)	Load (N)	Contraction (%)	Number of cycles for 0.5% dilatation	Number of cycles at fracture
Standard	410	3	3.5	4 000	73 000
Trained	410	2	3.8	1 800	23 000
Trained	360	2	3.5	50 000	260 000

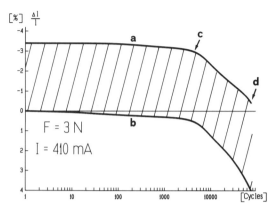

Fig. 14. Endurance test at a load of 3 N and 0.1 Hz for a standard NiTi wire: (a) contracted during a current of 410 mA; (b) extended without current; (c) degrading dilatation of 0.5% at 4000 cycles; (d) fracture after 73 000 cycles.

8. Conclusions

The investigated NiTi wire is a reliable actuator if the load is between 1 to 2 N and if the current × time product of the driving pulses is carefully selected. Monitoring of the wire resistance during contraction is a rough measure of the momentary state of contraction. Disadvantages are the narrow load range and the relatively long reaction time if there is no additional cooling, as well as the limited operational temperature range. Advantages are the high power/weight ratio, the small driving voltages and currents, and the high contraction at considerable loads. Accurate closed-loop controlled actuators with an additional position sensor are possible for many remote-controlled, particularly implantable systems.

References

1 *BioMetal Guidebook*, Toki Corpration, 1988.
2 T. Hochuli and R. Paravicini, Regelbare Aktuatoren mit NiTi-Drähten, *Diploma Work*, Swiss Federal Institute of Technology, Zurich, 1989.
3 H. P. Bornhauser and G. Schwarz, Geregelte NiTi Aktuatoren, *Semester Work*, Swiss Federal Institute of Technology, Zurich, 1989.
4 D. Stöckel, *Legierungen mit Formgedächtnis*, Expert-Verlag, Ehningen, 1989.

Shape Memory Alloy Microactuators

M. BERGAMASCO[a,b], P. DARIO[a,b] and F. SALSEDO[b]

[a]*Scuola Superiore S. Anna, Pisa, and* [b]*Centro 'E. Piaggio', University of Pisa, Pisa (Italy)*

Abstract

Shape memory alloys (SMA) have been investigated as transducing materials for the development of actuators for robotic end-effectors. In this paper, basic concepts on the design and development of SMA actuating systems for robotic applications are discussed.

The thermomechanical characterization of the SMA material is the basis on which the successive steps for the design of the actuating system have been developed. Isometric tests conducted on SMA wire specimens led to the definition of the limit curve concept, i.e. the locus of equilibrium points in the space of the status variables (stress, strain, temperature) which limit the SME behavior of the material. The second step in the design of SMA actuating systems involved the definition of appropriate geometries of SMA active elements in order to obtain the best performances in terms of the required forces and displacements. The definition of the whole actuating system started from the identification of the agonistic–antagonistic configuration as the most appropriate for SMA technology.

As an example of the realization of practical SMA actuating systems, the performances obtained with two different SMA actuators incorporating an active element shaped by coiling thin SMA wires and ribbons, respectively, and arranged according to an agonistic–antagonistic configuration, are presented.

1. Introduction

Shape memory effect (SME) and pseudo-elasticity (PE) are unique properties of shape memory alloys (SMA). SME originates from the thermoelastic martensitic transformation, the crystallographic and thermodynamic aspects of which are reviewed in ref. 1. Thermomechanical characteristics of SMA are described in detail in refs. 1 and 2.

The intrinsic ability of the SMA material to (i) recover the total strain associated with the martensitic transformation and (ii) simultaneously produce a reversion force, can be utilized for the design of actuating systems or active structures. Direct application of SMA technology to the field of robotics is based, however, not only on the actual mechanical performance obtainable, but also on the very small dimensions and weights of practical SMA-based actuating systems.

Moreover, the possibility of making the shape of the actuator element conform to the desired geometry, renders SMA technology suitable for the design of different scales of actuators. Thus SMA could be employed both for actuating systems to be inserted inside common robots, e.g. for commanding the motion of the interphalangeal joints of robot fingers, or for the actuation of the joints of miniature structures like gnat-robots, of active microtools positioned, for example, on a catheter tip.

At present, several works on SMA actuators for robotic applications have been presented. A complete analysis for the use of SMA for robotic actuators has been performed by Hirose et al. [3] who studied a particular configuration, named ξ array, of SMA coil-springs arranged mechanically in parallel, in order to obtain a larger total force, but connected electrically in series, in order to achieve a more uniform heating of the material. The same configuration has been considered also by Ikuta et al. [4] for the development of an SMA servoactuator system with electric resistance feedback control, implemented in an active endoscope. Other important issues connected with the design of SMA actuators, such as the cooling problem of the active elements and the control problem of the whole actuating system, have been studied by Hashimoto and coworkers [5] and Kuribayashi [6], respectively.

We have addressed our work on SMA not only just to obtain prototypes of SMA robotic actuators, but rather to acquire expertise and knowledge on the material behavior in order to fully exploit its potential properties.

In this paper we summarize the results of the study we have carried out on the three main research domains we have identified as fundamental for the design of SMA actuating systems: (i) the material characterization phase; (ii) the definition and characterization of the active element geometry; and (iii) the design and optimization of the particular actuating configuration.

2. The Material Characterization Phase

The specimen we used for the tests on SMA material was a Ti–50.0 Ni (at.%) alloy wire, of 0.5 mm diameter and initial length $l = 100.0$ mm. The study of the thermomechanical behavior of the SMA material was conducted by performing three different tests: (i) isometric, (ii) isothermal and (iii) isotonic.

Before starting the above experiments, a procedure intended to determine the specimen length at high temperature (at $T > A_f$, where A_f is the austenite finish temperature) was accomplished. We refer to this length, during all the following experimental tests, as an indication of correct SME behavior.

The results of the isometric test prompted us to build up the concept of the limit curve for the SMA material. The thermomechanical behavior of the SMA material can be associated, in 3D space (stress, strain, temperature), to a functioning surface F (stress, strain, temperature) $= 0$. This surface is the locus of all the possible equilibrium points of the material and contains the limit curve. The meaning of the limit curve can be understood by considering that curve, belonging to the functioning surface in the 3D space, as a boundary of the working path, which represents the successive status positions of the alloy. If the working path passes across the limit curve, the SMA material will loose its SME and shape recovery will be incomplete. For this reason knowledge of the limit curve is essential to accurately predict the performance of the active elements of

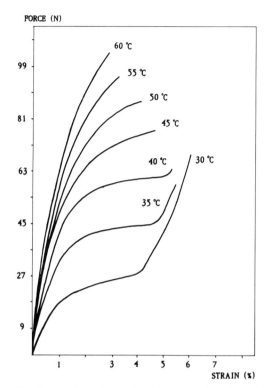

Fig. 2. Results of the isothermal tests conducted on the SMA wire.

the SMA actuating system. The usefulness of the isometric tests then obviously follows.

The three projections on the planes P1 = (strain, temperature), P2 = (strain, stress) and P3 = (stress, temperature) of the limit curve for the SMA specimen are reported in Fig. 1.

Figure 1(b) represents the course of the maximum reversion stress obtainable versus the initial deformation of the SMA sample. From the curve, two important parameters expressing the active

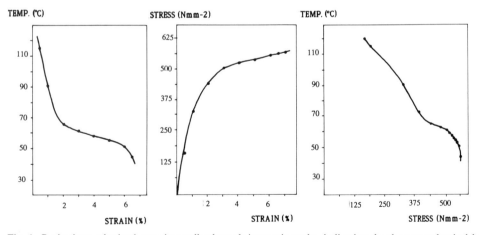

Fig. 1. Projections, obtained experimentally through isometric paths, indicating the thermomechanical behavior of the SMA wire (Ti–50.0 Ni, diameter 0.5 mm) used for the realization of the active elements of the actuating system.

element performance can be extracted: (i) the maximum deformation obtainable from the material at low temperature, and (ii) the stress relative to the apparent yield stress of the austenite at high temperature [7].

It is clear that the same limit curve could be obtained also by performing isotonic or isothermal paths to the limit conditions of the SME for the material. Results of isothermal tests are reported in Fig. 2.

3. The Active Element Design

The design phase of the active element consists in determining the appropriate geometry of the SMA element in order to obtain the desired actuation performances. The SMA wire is able to generate the largest reversion forces but, compared with an SMA spring having the same axial dimensions, the deformation is considerably smaller. It is necessary to achieve a compromise between the values of force and compliance. We have identified two particular geometries for the SMA active element: (i) a tension-loaded coil spring made by coiling an SMA wire, and (ii) a torsion-loaded coil spring made by coiling an SMA ribbon.

Once the geometry of the active element has been determined, it is necessary to make provision for the thermomechanical behavior of the active element by starting from knowledge of the material. This means that it is necessary to find a relation between the functioning surface of the material in space (stress, strain, temperature) and the surface in space (force, displacement, temperature) for the active element. In particular, the crucial problem is the theoretical determination of the limit curve for the active element.

We succeeded in solving this problem in the limit situations of thermomechanical behavior. The main difficulty was the nonlinear behavior of the material. A good validation of the theoretical results we deduced on the behavior of the active element has been obtained by experimental tests conducted with the same procedures outlined for the material. The limit curve projections obtained from isometric paths for the coil spring are represented in Fig. 3. The coil spring we used for the experimental tests has a small (1.5 mm) diameter, a length of 7.0 mm and is composed of 14 coils; it weighs 0.08 g [8].

4. General Considerations on the Design of the Actuating System

A general scheme of an SMA actuating system is represented in Fig. 4. This Figure refers to the actuation of a robotic finger joint, but the validity of the scheme is general. Two SMA active elements are arranged according to an agonistic–antagonistic configuration in order to move, by means of common tendons, a pulley situated on the joint axis. The introduction of an agonistic–antagonistic configuration, even if more complex than a bias spring-type configuration, is

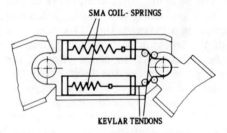

Fig. 4. Scheme of a linear SMA actuating system.

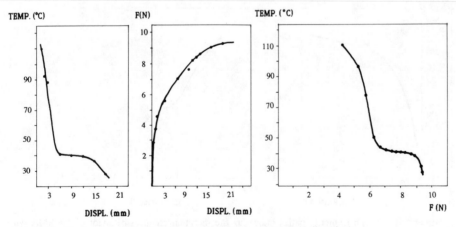

Fig. 3. The limit curve projections for the SMA active element (coil spring).

necessary in order to assure a fine control of the actuating system in both directions of motion.

The first issue to be approached in the study of an SMA actuating system is that of assuring correct SME behavior for the two active elements of the system for every working condition. The solution we found to this problem allows easy practical implementation. We have limited the working area of the active element to only a fraction of that theoretically available. Among the projections of the limit curve we have chosen the one on the plane (force, temperature) because this curve was demonstrated to be the more advantageous for our purposes. The limitation of the working area in the plane (force, temperature), as shown in Fig. 5, has been achieved by the introduction of two kinds of protection devices: (i) a mechanical and (ii) an electrical protection device in the actuating system hardware.

The mechanical protection system limits the force acting on the SMA active elements to a value F_o: this value of force corresponds to the yield stress, at high temperature, in the material of the active element. By limiting the force on the active element, the safety conditions for the SME are reached and no loss of memory occurs for every working condition. Figure 6 shows the SMA actuating system with the mechanical protection realized by means of a passive, steel spring, preloaded to F_o.

The electrical protection limits the current supplied to the active elements by the drivers. The drivers are ideal voltage-driven current generators which are calibrated at the maximum power to the value of current corresponding to the temperature T_o on the active element.

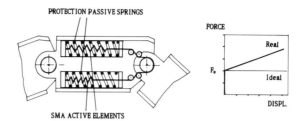

Fig. 6. Scheme of the modified actuating system incorporating the mechanical protection system; the characteristics of the passive steel spring are represented.

The determination of the value F_o, and consequently of T_o deduced from the projection of the limit curve on the plane (force, temperature), has been carried out by a graphical procedure which utilizes isotonic curves and projections of the limit curve of the active element [7].

As for the determination of the mechanical performances of the actuating system, in terms of torque and angular displacement, the problem of the optimal prestrain for the active elements has been solved.

Another important aspect to be considered is the performance of the actuating system in terms of velocity. We have tested a first version of a water cooling system capable of accelerating the heat transfer from the heated active element to the environment: a frequency cut-off of 1.25 Hz has been reached.

We include Table 1 as an example of the geometrical parameters and of the expected performances for an SMA rotatory microactuator, formed by coiling a thin SMA ribbon. The obtained configuration, composed of two counteracting SMA springs and schematically depicted in Fig. 7, will allow the actuating system to be easily incorporated in the artificial rotatory joints to be moved.

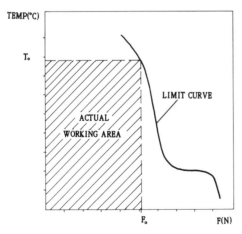

Fig. 5. The shaded domain (the region between the values of $F = F_o$ and $T = T_o$) restricts the working area of the SMA active element but allows the actuating system to operate in safety conditions.

TABLE 1. Values of the geometry and performances of a single SMA torsion spring[a]

Geometrical parameters					Performances	
R_e (mm)	R_i (mm)	N	b (mm)	L (mm)	α (deg)	M (N cm)
2.5	1	5	0.5	2.5	170	6.5
2.5	1	5	0.3	1.5	170	3.9
2.0	1	4	0.5	2.0	200	3.2
2.0	1	4	0.5	1.2	200	1.9
1.5	0.5	7	0.2	1.4	175	1.0

[a]R_e = external radius, R_i = internal radius, N = number of coils, b = thickness of the SMA ribbon, L = length of the SMA torsion spring along the axis of rotation, α = angular rotation, M = torque generated.

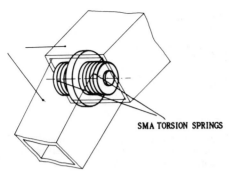

Fig. 7. Scheme of the rotatory SMA actuating system configuration.

5. Conclusions

The analysis of a complete SMA actuating system has been presented. The three phases in which we have approached the work on SMA, i.e. (i) the characterization of the thermomechanical behavior of the material, (ii) the definition of an appropriate geometry for the active elements, and (iii) the general study of all the aspects associated with the development of the actuating system, have been presented.

The interesting properties of the SMA material must be supported, in order to be fully exploited, by profound knowledge of all the above design aspects. Further work is now required for modeling the behavior of the agonistic–antagonistic configuration under all possible working conditions. Definition of the mechanical performances in terms of torque generated and angular rotation must be performed.

Acknowledgements

The authors wish to thank Fabio Vivaldi. This work has been supported, in part, by I.B.M. Italia S.p.A., and by C.N.R. (Finalized Projects on Robotics).

References

1 K. Otsuka and K. Shimizu, Pseudoelasticity and shape memory effect in alloys. *Int. Met. Rev.*, *31* (3) (1986) 93–114.
2 J. Perkins, G. R. Edwards, C. R. Such, J. M. Johnson and R. R. Allen, Thermomechanical characteristics of alloys exhibiting martensitic thermoelasticity, *Shape Memory Effect in Alloys*, Plenum, New York, 1975, pp. 273–303.
3 S. Hirose, K. Ikuta, M. Tsukamoto and K. Sato, Considerations on design of the actuator based on the shape memory effect, in E. Bautista, J. Garcia-Lomas and A. Navarro (eds.), *The Theory of Machines and Mechanisms (Proc. 7th World Congr., Seville, Spain)*, Pergamon, Oxford, 1987, pp. 1549–1556.
4 K. Ikuta, M. Tsukamoto and S. Hirose, Shape memory alloy servo actuator system with electric resistance feedback and application for active endoscope, *Proc. IEEE Robotics and Automation, Philadelphia, PA, U.S.A., 1988*, pp. 427–430.
5 M. Hashimoto, M. Takeda, H. Sagawa, I. Chiba and K. Sato, Application of shape memory alloy to robotic actuators, *J. Robotics Syst.*, *2* (1) (1985) 3–25.
6 K. Kuribayashi, A new actuator of a joint mechanism using TiNi alloy wire, *Int. J. Robotics Res.*, *4* (4), (1986) 47–58.
7 M. Bergamasco, F. Salsedo and P. Dario, A linear SMA motor as direct-drive robotic actuator, *Proc. IEEE Robotics and Automation, Scottsdale, AZ, U.S.A., 1989*.
8 M. Bergamasco, F. Salsedo and P. Dario, Shape memory alloy micromotors for direct-drive actuation of dexterous artificial hands, *Sensors and Actuators*, *17* (1989) 115–119.

Micro Actuators Using Recoil of an Ejected Mass

Toshiro HIGUCHI . Yusof HOJJAT . Masahiro WATANABE

Institute of Industrial Science University of Tokyo
7-22-1, Roppongi, Minato-ku, Tokyo 106 Japan

Abstract This paper describes a method by which the precise displacement of a micro-mechanism is possible. In this method, an object moves by the recoil of an ejected mass. Based on this concept, two different actuators are developed. In the first one, an electromagnetic impulsive force is used for ejection of the mass, and a buffer like a spring or a damper is used to return it. In the second actuator, the mass is attached to the object by inserting a piezo electric element between them. The object moves by a sudden change in the length of the piezo element. Then the piezo returns to its initial length slowly. The direction of motion depends on the direction of deformation of the piezo element. The experimental results show that the displacement of a few nano meters up to a few millimeters is possible by these actuators. The simple mechanism which is introduced, has the potential to be used for micro actuators. The fabrication of micro actuator may be achieved by means of wire cutting, etching or the technology which is used for production of semiconductor devices.

1. INTRODUCTION

Most of the actuators are devices which have two distinguished parts, and produce relative motion between these two parts. For example, these two parts are rotor and stator in an electric motor, cylinder and piston in a pneumatic or hydraulic actuator, and two ends of a piezo electric element. To use such actuators it is necessary to fix one of these two parts and use the motion of the other part. In the presented method, however, the actuator does not need any fixed part. It can be simply attached to an object and drive it. The only requirement is that the object must be held loose by a friction between the object and the guide. The main parts of this actuator are a mass and an ejector mechanism. If the ejector ejects the mass, the object will move in opposite direction by the recoil of the ejected mass.

In the ejection process, the object moves only by internal force. But when the ejected mass returns back, the friction will act as an external force. If the friction does not exist between the object and the base, the system will return back to its initial position and no displacement will be obtained.

When the ejected mass returns toward the object, excessive displacement will be obtained if a collision happens or m stops suddenly.

To eject the mass, two different methods are utilized, an electromagnetic impulsive force and a sudden deformation of a piezo electric element. In the first method a buffer such as a spring or a damper is necessary to return back the ejected mass, while in the second method the location of the ejected mass is controlled by the elongation of the piezo element.

As the construction of these actuators are very simple, this mechanism can be considered as a proper way for development of the actuators with dimensions of some millimeters.

To confirm the effectiveness of the idea, some experiments are performed by large scale actuators. This experiments show that the displacement of the order of nano meters is obtainable.

2. MICRO ACTUATOR USING THE ELECTROMAGNETIC IMPULSIVE FORCE [1,2]

2.1. Concept of the Positioning Method

In this actuator the electromagnetic impulsive force which generates between a coil and a conductive plate is utilized for ejection of the mass. The generation of electromagnetic impulsive force is shown in *Fig.1*. When an impulsive current flows in a coil which is placed closed to a conductive plate, eddy current will induce in the conductive plate, thus a repulsive force will be obtained between the current in the coil and the eddy current in the conductive plate.

The impulsive current can be generated by using a discharge circuit. Since the shape of the impact is related to the charged voltage, the capacitance of capacitor, and the inductance of the coil, the waveform of the impact can be controlled electrically.

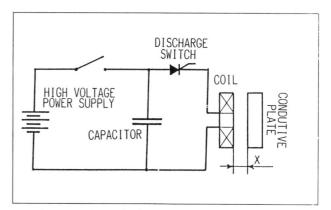

Fig.1. Generation of electromagnetic impulsive force.

To explain the mechanism of displacement we consider that a block (M) is placed on a plane surface, and a mass (m), is attached to it by inserting a spring between them (*Fig.2a*).

The friction force between M and m is supposed to be considerably less than the impulsive force, but bigger than the peak of the force in the spring. An impact between m and M will drive them in two opposite directions (*Fig.2b*). Then the spring pushes them toward each other. As the force in the spring is smaller than the friction force, only m will return and M does not move (*Fig.2c*).

When m returns toward M, as its received energy is conserved in the spring, a collision will occur between them and will cause the excessive displacement of M (*Fig.2d*).

Therefore in one period of reciprocation of m, the main body moves in two steps. By repeating this procedure, the body can move step by step without any theoretical limitation of distance.

In this method the buffer is used to return the ejected mass. If it is only a spring, the displacement happens in both ejection and collision. To prevent the mechanical collision, a damper also may be used as the buffer.

In this mechanism the acquired momenta of both objects are equal ($MV=mv$), so the received energy of smaller mass is more than the bigger one ($E_M=mE_m/M$). Thus if the ejected mass be much smaller than the main block, it is capable of pulling the main object in its direction. It has been proven that for stiff springs and relatively powerful impacts, the resulted displacement becomes in opposite direction. Because the pulled distance by the ejected mass becomes more than the displacement due to ejection and collision. Therefore the object can move in two opposite directions according to the amount of applied impact. But as the manner of displacement in opposite direction is not desirable for precise positioning, this actuator may be considered as a one directional actuator.

2.2. Theoretical Analysis

To predict the displacement of the body, it is supposed that the coefficient of friction is constant and the duration of impact can be ignored, compared to the time of movement. Under such conditions the displacement of the body due to the applied impact can be expressed as follows:

$$\Delta x = J^2/(2\mu M^2 g) \qquad (2.1.1)$$

Where, J is the impulse of the impact, Δx is the displacement of the body, μ is the coefficient of friction between the body and the surface, M is the mass of the body, and g is the acceleration of gravity. When a spring is used as the buffer, if the ejected mass does not pull the body toward itself, the displacement in the first and second steps are almost equal, therefore the total displacement in one period of reciprocation of the ejected mass, can be estimated as follows:

$$\Delta x \approx J^2/(\mu M^2 g) \qquad (2.1.2)$$

This equation shows that the displacement is not related to the ejected mass. For more accurate prediction of the displacement, the simulation of motion was performed by arrangement of the equations of motion in each step, and solving them regarding to the conjunctional conditions.

2.3. Fundamental Experiments

In order to confirm the effectiveness of the idea described in the previous sections, an experimental micro actuator has been made (*Fig.3a*). This actuator is capable of moving in two opposite directions due to the amount of the applied impact.

As the developed actuator is too small to perform the fundamental experiments, a large scale actuator has also been made, which is shown in *Fig.3b*.

The specifications of the experimental apparatus are as follows:

Mass of the slider	: 1.45 kg
Length of the slider	: 75 mm
Conductive plate	: Copper, 0.045 kg
Coil	: 96 turn, 80 μH
Friction surface	: Steel - Steel
Coefficient of friction	: About 0.19
Maximum voltage	: 500 V
Capacitance of capacitor	: 6 ~ 60 μF
Kind of discharge switch	: Thyristor
Maximum current	: 375 A
Maximum charged energy	: 7.5 J

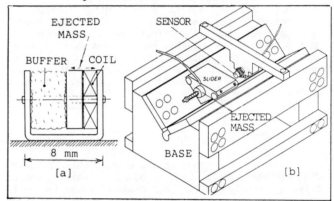

Fig.2. Mechanism of movement.

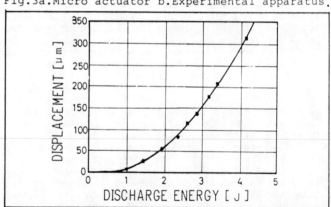

Fig.3a. Micro actuator b. Experimental apparatus.

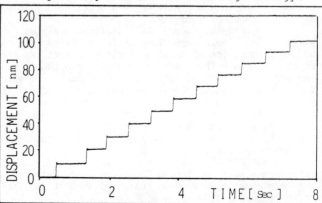

Fig.4. Displacement vs discharge energy.

Fig.5. Displacement in ten periods.

The displacement of the slider is measured by an astigmatic focus error detection sensor (range: ±1μm, resolution:2 nm) and an eddy current displacement sensor (range:1 mm, resolution:0.5 μm).

The relation between displacement and charged energy of the capacitor is shown in *Fig.4*. This result represents that the displacement is almost proportion to the square of the charged energy of the capacitor.

Fig.5 illustrates the displacement of the object in ten period of reciprocation of the ejected mass. This result shows the repeatability of the actuator and also confirms that the displacement of the order of 10 nm is obtainable.

The behavior of the movement for bigger displacements was measured by an eddy current sensor and the results are illustrated in in *Fig.6*.

Fig.6a shows the displacement of the body, when the spring is not so stiff. The movement of body for more stiff spring is shown in *Fig.6b*. In this case the ejected mass pulls the body toward itself, but the resulted displacement is positive yet. For enough stiff springs as shown in *Fig.6c*, the resulted displacement also becomes in the opposite direction.

3. MICRO ACTUATOR USING PIEZO ELECTRIC ELEMENT

3.1. Concept of Positioning Method

The second method for ejection of the mass m, is utilizing a piezo electric element. As the position, velocity, and the acceleration of m is completely controlable, it is possible to drive the body in both directions in a desirable manner, by the pattern of applied voltage.

Fig.7a is the schematic figure of the actuator. Main body (M) and the ejected mass (m) are connected by a piezo electric element. A sudden elongation of the piezo element with a force considerably more than the friction force between M and the surface, will drive M and m in two opposite directions (*Fig.7b*). This is the first step of displacement.

In the second step (*Fig.7c*), the piezo element shrinks to its initial length with a force less than the friction force between M and the surface. If it shrinks slowly, M does not move in this step. But if the piezo element shrinks with a relatively high acceleration (not so high to move M), and stops suddenly, its inertia force will push the body like a collision (*Fig.7d*). Therefore additional displacement will be obtained in the second step (*Fig.7e*).

As the variation of the length of the piezo elements is completely controlable, it is possible to arrange the mentioned procedure as is shown in *Fig.7f~j*, and drive the object in opposite direction in the same manner.

3.2. Theoretical Analysis

As the exerted force by piezo element is considerably more than the friction force between the body and the base, it is supposed that the friction force can be ignored. Under such condition, the displacement of the body in the first step can be determined by refer to the fact that the center of gravity of the system does not move in the first step. If the elongation of the piezo element is Δl, the displacement of the body becomes as follows:

$$\Delta x = \frac{m}{m+M}\Delta l \quad (3.1.1)$$

The maximum displacement in the second step happens when m returns back with the critical

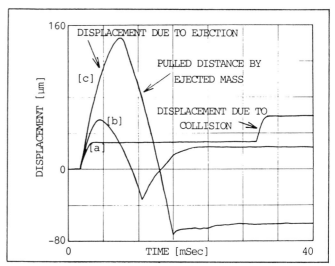

Fig.6. Behavior of movement.

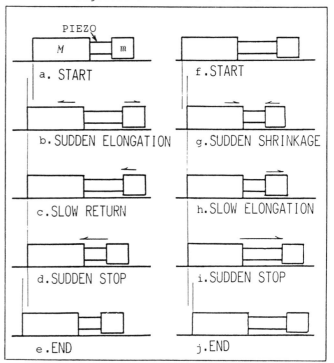

Fig.7. Mechanism of movement.

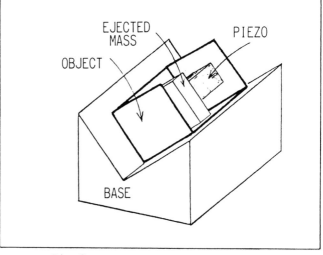

Fig.8. Experimental apparatus.

acceleration. In such condition, the force which is exerted on the m to accelerate it, must be equal to the friction force between M and the surface. Therefore, this acceleration can be determined by the following equation:

$$a_{CRITICAL} = \frac{\mu(m+M)g}{m} \quad (3.1.2)$$

Where, μ is the coefficient of friction between M and the surface. If m returns back with the critical acceleration, the displacement in the second step will be the same as the displacement in the first step. Therefore the maximum displacement in one period of reciprocation of the ejected mass can be shown as follows:

$$\Delta x = \frac{2m}{m+M}\Delta l \quad (3.1.3)$$

This equation shows that, the bigger the ejected mass is, the more displacement will be obtained.

3.3. Fundamental Experiments

The experimental apparatus which has been made to confirm the idea, is shown in Fig.8. It consists of an slider (165 gr), the ejected mass (70 gr), and the piezo element. The rated voltage of the piezo element is 150 V and its maximum elongation is 16 μm. The maximum force which is produced by this element is 350 kgf.

The relation between displacement and applied voltage is shown in Fig.9. This result shows that the displacement is directly proportion to the applied voltage.

Fig.10 illustrates the pattern of applied voltage and the resulted displacement in one period of reciprocation of the ejected mass. This result confirms the possibility of the displacement of nano meter order.

Fig.11 shows the continuous movement which is achieved by applying the voltage pattern shown in the same figure.

4. APPLICATIONS

One of the promising applications of these actuators is the development of precise X, Y, Θ stages. A symmetric stage can be achieved by using eight actuators. The concept of positioning is shown in Fig.12a. By selecting the proper pairs of the actuators, as is shown in the figure, the movements in $\pm X, \pm Y, \pm \Theta$ directions are obtained.

A more simple stage can be achieved by only four actuators which are the minimum number of required actuators, when the actuator drives the object only in one direction (Fig.12b).

In the case of piezo actuator, as it is able to move in two opposite directions, the minimum number of required actuators reduces to three. (Fig.12c).

Fig.12d shows another stage, which can be achieved through utilizing four bimorph elements, where a sudden bending is used instead of the sudden elongation.

In this research, a precise X, Y, Θ stage was developed by eight units of the electromagnetic actuators.

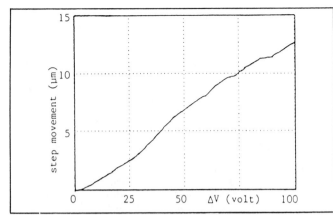
Fig.9. Displacement vs applied voltage.

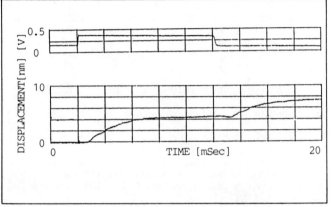
Fig.10. Displacement in one period.

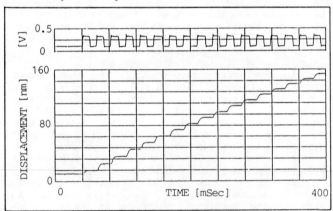
Fig.11. Continious movement.

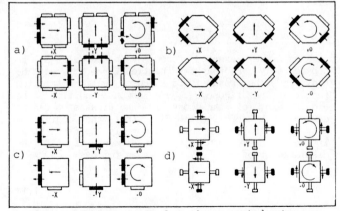
Fig.12. Mechanisms for three axial stages.

The specifications of the stage are as follows:

Size and mass of stage	: $100 \times 100 \times 30$ mm, 3kg
Guide plates	: Steel - Steel
Moving limits	: X&Y, $\pm 500 \mu m - \Theta, 4000 \mu rad$
Resolution	: X&Y, $0.5 \mu m - \Theta, 5 \mu rad$
Maximum dis. per pulse	: X&Y, $\pm 50 \mu m - \Theta, \pm 1000 \mu rad$

The developed stage is shown in Fig.13. The position of the table is detected by four eddy current displacement sensors and the results enter the personal computer. From the difference between desired position and the sensed position, computer determines the required energy and direction of movement. Then sets the voltage and applies impacts in the desired direction.

The stage is perfectly capable of adjusting the position with the accuracy of the utilized sensors.

5. MINIATURIZATION OF THE ACTUATORS

The method which was introduced and confirmed by large scale experimental apparatus, has the potential to be fabricated in micro-scales.

To miniaturize the electromagnetic actuator, an etched coil may be used together with a small buffer and a thin conductive plate. It is also possible to use wire cutting electro discharge machining and shape the end of a solid object as is shown in Fig.14, then install the etched coil inside the incision. The tin edge in this figure, plays the role of the ejected mass and the buffer at the same time. It is expected that a micro actuator with dimensions within a few millimeters is obtainable easily.

By utilizing the piezo element, further reduction of size is possible. The fabrication may be achieved by the technology which is used for production of semiconductor devices.

6. CONCLUSION

A mechanism for micro actuators which uses the recoil of an ejected mass is developed. The fundamental experiments confirm the effectiveness of the idea and show that the displacement of nano meters order is accessible.

In this method the external force exerts on the body only through the friction between the body and the base, so the body can move without any theoretical limitation of distance. As an application of the idea, a precise X,Y,Θ stage is developed. The stage is perfectly capable for automatic adjustment of the position.

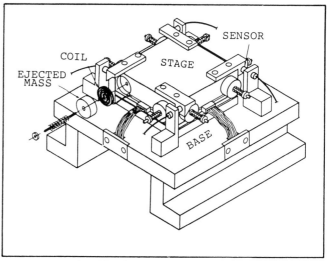

Fig.13. Three axial stage.

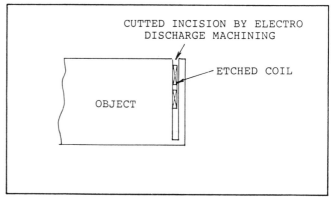

Fig.14. Miniaturization of actuator.

7. REFERENCES

[1] T.Higuchi "Application of electromagnetic impulsive force to precise tools in robotic system" Robotics Research (The second international symposium), PP.281-285, Aug 1984.

[2] T.Higuchi and Y.Hojjat, "Application of electromagnetic impulsive force to precise positioning", IFAC (10th world congress on automatic control preprints), Vol.3, PP.278-283, July 1987.

PRECISE POSITIONING MECHANISM UTILIZING RAPID DEFORMATIONS OF PIEZOELECTRIC ELEMENTS

T. Higuchi, Y. Yamagata, K. Furutani, and K. Kudoh
Institute of Industrial Science
University of Tokyo
Roppongi 7-22-1, Minato-ku, Tokyo 106, JAPAN

ABSTRACT

A new driving method suitable for micro mechanism is introduced. It utilizes friction and inertial force caused by rapid deformations of piezoelectric elements. One dimensional linear positioner using this mechanism consists of one main object put on a guiding surface, a piezo and a weight. The weight is connected to one end of the main object via the piezo. By controlling rapid extension or contraction of the piezo, it can make step-like movements of several nanometers up to ten micrometers bi-directionally against friction. Thus, repeating this step movement, it can move for a long distance. Using this mechanism, two types of joints for micro robot arm have been developed. One is a simple rotating joint with an arm of 5cm, and the other is a 3-DOF joint with 8cm arm. And minimum step movements of the two joints were smaller than $0.1\mu m$ and maximum velocities were larger than 2 mm/sec at the end of the arm. Combining two joints, a 4 degree-of-freedom micro robot arm has also been developed. Besides, other applications of this mechanism are introduced shortly, such as cell-operation, rough positioning system of STMs, and positioning device for ultrahigh vacuum.

1. INTRODUCTION

A piezoelectric element is said to be one promising actuator for micro mechanism. Ultrasonic motors and "inchworm"s are well known as mechanism using piezoelectric elements. However, most of ultrasonic motors utilizes resonance to amplify the vibration of piezos or to generate surface wave, therefore, they are not suitable for direct precise positioning and micronization. "Inchworms"[1] that have two chucks and a piezoelectric element to connect them, can make stepping motion and precise positioning, but their mechanisms, especially the chucks, seem to be difficult to micronize. Our mechanism, utilizing friction force and inertial force caused by rapid deformations of piezoelectric elements[2] (Hereafter, we call this "Impact Drive Mechanism" or simply "IDM"), can be microsized due to the simplicity of the configuration. It can make step movement of several nanometers up to a few micrometers.[3] In addition to this precise positioning ability, repeating this step action long distance movement is possible. We have already developed various positioning devices aiming for automatic assembly, micro-machining, scanning tunneling microscopy, biotechnology and other industrial applications. Other than these, a unique application is a newly developed micro robot arm using this mechanism. We have developed two types of robot arm joints and a robot arm with 4 degrees of freedom. One is a simple rotating joint with 5 cm arm, and the other is a 3 DOF joint using ball-joint. Both of them proved to have the ability to make precise positioning. And a four degrees-of-freedom micro robot arm is developed combining these two joints. In this paper, first we introduce the principle of the IDM and give some of the examples of the applications. Then, micronization of IDM is discussed about. After that, we describe about the micro robot arm.

2. IMPACT DRIVE MECHANISM

PRINCIPLE OF IDM

Fig.1 describes the mechanism of movement using one dimensional linear positioner. The main object is put on a flat surface and held by the friction force. A weight is connected to the main object via one piezoelectric element. Controlling contraction or extension of the piezo by applying a voltage waveforms to the piezo, we can move the object. Fig.1-(a) represents a control method of the deformations of the piezo for leftward movement.

I. Rapid Extension

(1): A movement cycle begins with the piezo contracted. (2): Applied a steep rising voltage, the piezo makes a rapid extension and the main object moves leftward and the weight rightward. (3): Contract the piezo slowly to return the weight.

II. Slow Contraction

(3)': While returning, the weight should be accelerated by a constant acceleration which causes inertial force less then the static friction force. Otherwise, the main object makes reverse movement. (4): By the time the piezo is contracted to the length at the beginning, a sudden stop is made. This action is just like the collision of the weight and the main object. So the whole system starts moving leftward against friction force. (5): The whole moving system runs against dynamic friction force until it loses its kinetic energy.

The cycle is completed through steps (1) to (5). Repeating this cycle, it can move for a long distance.

III. Combined drive pattern

The combined drive pattern of I and II is possible: first the piezo is contracted with a constant acceleration and just after the step (4), is extended rapidly. Using this drive pattern, the movement is almost equal to the sum of the movement of I and II. And at the same time, the moving cycle can be quicker. We mainly use this drive pattern, for this pattern can make the largest movement and is most efficient. Fig.2 shows the three voltage patterns applied to the piezo corresponding to three drive patterns.

We can make the rightward movement just in the same way as described above. It can be attained by exchanging the contractions and extensions of the piezo to each other as shown in Fig.1-(b).

The rise time of the voltage and the constant acceleration for returning weights affects on the movement. Generally one step of movement increases as the rise time

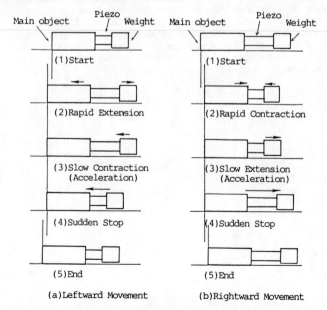

Fig.1 Principle of movement of the IDM

Reprinted from *Proceedings IEEE Micro Electro Mechanical Systems*, pp. 222-226, February 1990.

becomes shorter. This is because stronger inertial force can be generated by a faster deformation of the piezo. While, the acceleration for returning weights has an optimum value. If the acceleration is so large that inertial force caused by the acceleration exceeds static friction force, a reverse movement occurs and the movement decreases in total. And if the acceleration is too small, the kinetic energy given to the weight becomes smaller, therefore, the movement becomes small. Thus, the maximum movement can be obtained with the acceleration which cause the same inertial force as static friction.

Compared with other mechanisms using piezos, this simple mechanism has the advantages as follows[3]:(i)Controllable micro step movement can be obtained by simple configuration. We have already succeeded to get stable four nanometer micro step movements. (ii)This mechanism does not require any power to keep its position. (iii)As this mechanism utilizes inertial force, a multiple degrees-of-freedom mechanism using this method can be constructed. (iv)This mechanism consists of only weight and a piezo, so it is easy to fabricate micro-sized ones.

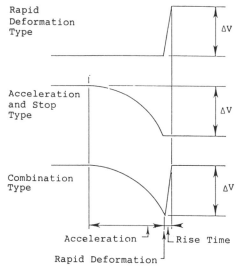

Fig.2 Voltage patterns applied to the piezo

MICRONIZATION OF IMPACT DRIVE MECHANISM

Micronization of the structure using IDM is an interesting subject. To sum up, micronization of structure to some degree is desirable to IDM, but there should be some limit. There are two reason why micronization is desirable. One reason is that the friction force becomes relatively larger compared to the weights as the size of the structure decreases. That means large friction is, to some degree, desirable in this mechanism because this mechanism uses friction force to hold the main object. The second reason is that the eigen frequency of a structure becomes higher as the size decreases. This is because IDM utilizes impulsive force, that is, if the eigen frequency of the structure were low, impulsive forces would be weakened.

For those reasons above, IDM is suitable for micronization but there should be a limit of micronization. Micronization would be limited because the friction becomes too large and unstable when microsized excessively. As the structure becomes smaller, the friction force becomes larger compared to the weight and finally friction becomes so large that it cannot be overcome by impulsive inertial force caused by piezos. And since surface finish affects on the stability of the friction, unless any special manufacturing method is developed, the surface roughness will become relatively large and the friction will be unstable. So, the smallest structure possible using IDM is estimated to be 1 mm considering the current manufacturing method.

EXAMPLES OF APPLICATIONS OF IDM

In this chapter, examples of applications of Impact Drive Mechanism are introduced.

Industrial Use: Positioning devices are most suitable applications of IDM. These positioning devices are aiming for automatic assembly, semiconductor production system, micro-machining system, and other industrial systems. A precise positioning table with three degrees of freedom using this method has already developed and tested successfully. And compact positioning stage is about to be commercialized.

Biotechnology: Advance in biology is requiring new mechanical devices. One example is a micro-manipulator for cell operation.[4] In cell operation, it is needed to insert and position a sharp capillary into a biological cell. So, precise positioning of capillary with easy operation is necessary. We have tried cell operation using IDM. It was successful to insert a capillary to desired position of an egg of a rabbit of 170 um in diameter and to inject one cell of sperm into it. Fig.3 shows the capillary positioning device and the egg of a rabbit with a capillary inserted.

Scanning Tunneling Microscopy: STM has become a powerful tool for analyzing surface of a material both in the field of science and engineering. One main problem in constructing STMs is the rough positioning system for the tunneling probe. It requires high positioning resolution of several ten nanometers, high stiffness and high stability. The IDM best satisfies these requirements and an STM has been developed successfully using this method.[5]

Ultrahigh Vacuum Equipments: Ultrahigh vacuum environment is necessary for recent scientific instruments including STMs. An XYθ positioning table for use in ultrahigh vacuum using IDM has been developed. The Impact Drive Mechanism is advantageous for ultrahigh vacuum in that the piezos can be housed inside the positioning device to reduce the outgas from piezos. The table was able to operate under pressure of 10^{-11} Torr.[6]

3.MICRO ROBOT ARM USING IDM
ONE DIMENSIONAL ROTATING JOINT

Fig.4 shows the structure of the joint. This joint consists of an arm with a shaft, a spring and a stand. Friction is generated at the contact surface of the flange of the shaft and one side of the stand. The spring pulls the shaft and press the flange to one side of the stand. The arm is aluminum and the length from the center of

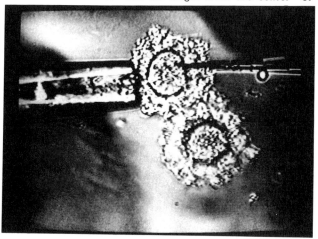

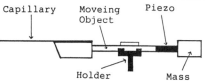

Fig.3 Capillary positioning device and the egg of a rabbit inserted capillary

the joint to the end is 50mm. A pair of piezos with weights are connected to the arm. The center axes of piezos are in right angle to the radius of the rotating center and are at the same distance (15mm) from the center. Therefore this pair of piezos and weights generate a couple of forces. The piezos are laminated type of PMN ceramics made by NEC. Its size is 2x3x10mm. When applied 150V, it makes an extension of 8 micrometer with no load and, if the extension is suppressed, generates a force of 20kgf. The weight is a 10mm cubic and its mass is 7.2g.

Fig.5 shows the three patterns of rotation of the arm and the voltage pattern applied to the piezos. The first rotation corresponds to a rapid extension, the second to a combination type. And the third rotation corresponds to a sudden stop.

Fig.6 shows the effect of the amplitude of applied voltage on the rotation. The movement increases as the amplitude of applied voltage increases. At maximum, a movement of 5 m can be obtained at the end of the arm by one step of movement. The positioning resolution at the end of the arm is considered to be lower than $0.1\mu m$. The rotating velocity of 4.8×10^{-2} rad/sec is obtained under the condition of pulse rate 1.2kHz and $\Delta V=80V$. This velocity is equivalent to 2.4mm/sec at the end of the arm.

Fig.7 shows the movement under various load of weight on the arm. Lower curve indicates the movement against the load and the movement decreases as the load weight increases. Upper curve indicates the movement for the load and it increases as the load increases. Judging from this figure, the maximum load capacity against which this joint arm can move is about 1×10^{-2}Nm. And the practical load capacity is 9×10^{-3}Nm and this is approximately equivalent to the load of 20gf at the end of the arm.

THREE DEGREES-OF-FREEDOM JOINT

Fig.8 shows the structure of 3-dim joint. This arm is supported by a ball-joint mechanism. The ball is pushed up by the spring via the ball support and friction force is generated between the ball and ball cover. Upper side of the ball support inserted a self-lubricating polyacetal liner and it reduces the friction between the ball and ball support. The arm is made of aluminum and its length from the center of the ball is 80mm. Its cross section is a 18mm regular triangle. The ball is hardened 0.35%C steel, while the ball cover is 0.45%C steel. Six piezos and six weights are attached to the arm. The piezos are the same ones that used in the one dimensional rotating arm. Their sizes are 2x3x10mm. The weight is about 8mm cubic and its mass is 4g. These six piezos and weights can generate a couple of forces around each three rotating axis, X, Y and Z.

This arm is supported by ball-joint, and has three degrees of rotating freedom. The arm can rotate for θx and θy within 20 degrees. The six piezos are driven simultaneously to generate torques for desired direction. The three voltage waveforms for rotation around X,Y and Z axes are output from computer and are distributed by analog circuits to the six piezos via power amplifiers.

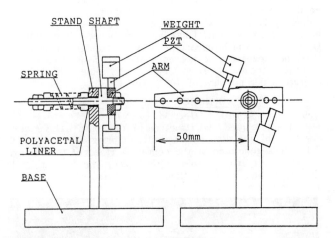

Fig.4 Structure of one dimensional rotation joint

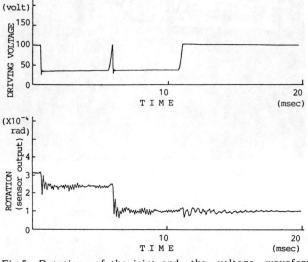

Fig.5 Rotation of the joint and the voltage waveform applied to the piezos

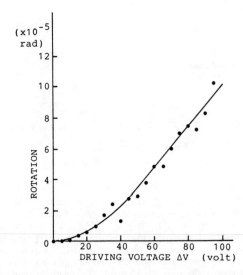

Fig.6 Effect of amplitude of voltage applied to the piezos

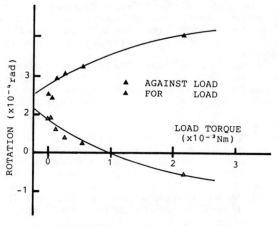

Fig.7 Effect of the load torque

Fig.9 shows the relation of the rotation and the driving voltage for the direction θx. There are interferences to the rotations for the direction Y and Z, but these are practically negligible. The rotational velocity of 6.7×10^{-2} rad/sec for θx is obtained at the pulse rate of 890Hz and $\Delta Vx=100V$. This is equivalent to the velocity of 5 mm/sec at the end of the arm.

Fig.10 shows the movement under load of weights. One end of a thin thread is connected to the end of the arm and the other end of the thread is connected to the weight through a pulley. Judging from this figure, the maximum load against which this arm can move is 1.7×10^{-2} Nm (22gf at the end of the arm), and practical load capacity is about 1.2×10^{-2} Nm (15gf at the end of the arm).

FOUR DEGREES-OF-FREEDOM MICRO ROBOT ARM

The two joints introduced above have proved that arm-like structures can be driven by Impact Drive Mechanism. There are two problems for multiple-jointed arm: (1)The driving force of IDM is not sufficient compared to the weights of the second or third arms. (2)Interferences may occur between the joints, for the joints are held only by friction force. The first problem can be solved by applying a counter weight to each arm to cancel the weight of other arms. The second problem can almost be solved under these conditions: (i)Each arm should be balanced. (ii)The static friction torque of a lower joint should be larger than that of upper joints. (iii)The piezos and weights on a arm should be arranged so that they never exert inertial force directly to the joint.

This 4 DOF robot arm has designed considering the points above. Fig.11 shows the structure of 4 degree-of-freedom micro robot arm. It has two joints, a ball-joint that have 3 degrees of freedom and a simple rotating joint. The first arm is designed so that the center of the ball-joint, second joint and the counter weight should be aligned in a line. Six piezos are attached to the first arm. They can generate torques for X,Y and T rotation. The second arm has just the same configuration as the one dimensional rotating joint introduced before, except the piezos used are 5x5x10mm size.

The interferences of the joints are tested by driving two joints separately. Fig.12 shows the result of the experiment. The rotation of the two arms are measured by gap sensors. In Fig.12-(a) only the lower joint was driven and the second joint made no rotation. While, in Fig.12-(b) only upper joint was driven and the first joint made no rotation either. Judging from these results, there were no interferences between two joints.

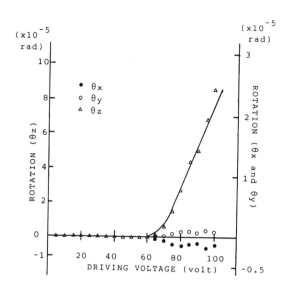

Fig.9 Rotation and the amplitude of voltage driving for θx

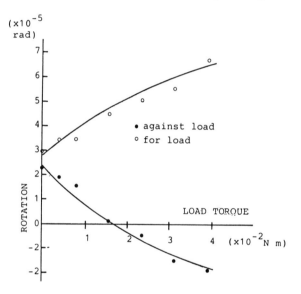

Fig.10 Effect of load torque

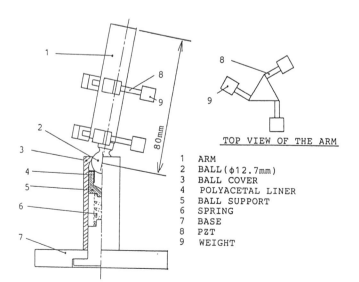

Fig.8 Structure of the three dimensional joint

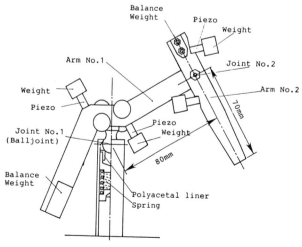

Fig.11 Structure of the four degrees-of-freedom micro robot arm

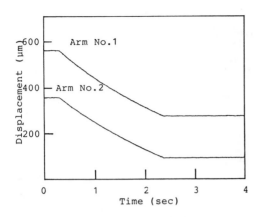

(a) Only joint No.1 is driven

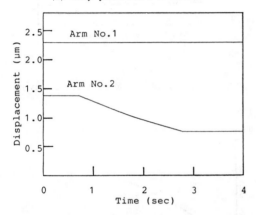

(b) Only joint No.2 is driven

Fig.12 Test of interference of the two joints

4. SUMMARY

The Impact Drive Mechanism is a unique driving method using piezoelectric elements and has applied to many industrial and scientific positioning devices. Besides, it seems to be suitable for micromechanisms. A unique micro robot arm is introduced and is proved to have the high positioning resolution and controllability with simple mechanism. And even smaller robot arm will be developed to ensure the possibility of micronization of Impact Drive Mechanism.

5. ACKNOWLEDGEMENTS

The authors would like to express thanks to NEC Co., Ltd. for giving us samples of piezos.

6. REFERENCES

[1] Catalogs of "Inchworms" are available from Burleigh Instruments, Inc; Burleigh Park, Fishers, NY 14453
[2] T.Higuchi, Y.Hojjat and M.Watanabe : Micro Actuators Using Recoils of an Ejected Mass; IEEE Micro Robot and Teleoperators Workshop Proceedings, Nov. 1987
[3] T.Higuchi, M.Watanabe and K.Kudoh : Precise Positioner Utilizing Rapid Deformations of a Piezo Electric Element; Journal of the Japan Society of Precision Engineering vol.54 No.11 Nov.1988
[4] K.SATO, K.SHIOTA, and T.GOTO: In Vitro Fertilization of Hamster Eggs by Microinjection of Rabbit Spermatozoa; J.Mamm. Ova Res., 6(2) 12-17, 1989
[5] H.Kawakatsu and T.Higuchi; A dual tunneling-unit scanning tunneling microscope; J.Vac.Sci.Technol. A Jan/Feb 1990 (to be published)
[6] Y.Yamagata, T.Higuchi, H.Saeki, and H.Ishimaru; Ultrahigh vacuum precise positioning device utilizing rapid deformations of piezoelectric elements; J.Vac.Sci.Technol.A July/Aug 1990 (to be published)

Tiny Silent Linear Cybernetic Actuator Driven by Piezoelectric Device with Electromagnetic Clamp

Koji Ikuta [1]   Satoshi Aritomi [1]   Takefumi Kabashima [2]

[1] Department of Mechanical System Engineering, Faculty of Computer Science and System Engineering, Kyushu Institute of Technology
Kawazu 680-4, Iizuka, Fukuoka 820, Japan
TEL: +81 948-29-7777   FAX: +81 948-29-7751   E-mail: ikuta@ai.kyutech.ac.jp

[2] Research Laboratory, Yaskawa Electric Mfg. Co., Ltd.
Yahatanishi-ku, Kitakyushu 806, Japan

ABSTRACT

New type of miniature linear actuators named "Cybernetic Actuator" for bio-medical applications have been developed. This actuator has four driving states such as free, increasing, decreasing and lock. Especially the state of free and lock can release excessive applied load from outside to provides safety characteristics for both human and micro mechanism. In this paper miniaturized linear Cybernetic Actuators combining piezoelectric impact drive with electromagnetic clamping are design and developed. Feasibility and basic performance of the actuators are conformed by theory and experiments. Not only four driving states as the Cybernetic Actuator but also drastic improvement of performance such as maximum speed, energy efficiency and silent drive as a miniature linear actuator are verified successfully by using "controlled friction impact drive".

1. INTRODUCTION

The final goal of our research lies on establishing new actuator systems useful for bio-medical micro/miniature robotics such as an active endoscope, miniature manipulator and artificial finger for micro surgery so on. Industrial actuators have to have large producible force, high power/speed and high accuracy. On the other hand, the needs from bio-medical field much differ from conventional industrial ones.

For example, the actuators suitable for bio-medical operation should be highly safety, medium/small range power/speed, miniature size, light weight, high cleanness and silent. Therefore it is not easy to utilize conventional industrial actuators. Authors believe that new type of miniature actuator should be developed. Recently we found that **Cybernetic Actuator** based on new concept and new principle can satisfy above requirements because its unique characteristics similar to human muscle system contributes to bring safety feature into actuator. The first prototype of practical cybernetic actuators have been reported by K.Ikuta et al in MEMS'91.[1]

In this paper, further miniaturized and improved linear cybernetic actuator employing new driving principle along with theoretical and experimental data is reported.

2. SUITABILITY OF CYBERNETIC ACTUATOR FOR BIOMEDICAL APPLICATIONS

In order to explain the suitability of cybernetic actuator for biomedical miniature/micro machine, the definition, basic concept and basic characteristics should be described in following session.

2.1 Definition and Basic Characteristics

Although the basic concept of cybernetic actuator was proposed by R.Tomovic and MacGee in 1966,[2] no successful development has been made so far.

Based on the consideration about human muscle system, they found that the actuators useful for human-like robot executing more complicated task than that of today's industrial robot should satisfy special new requirements. This type of actuators was named " Cybernetic Actuator".

The Cybernetic Actuator should have following four driving states.

State 1	:	**free**
State 2	:	**decreasing**
State 3	:	**increasing**
State 4	:	**lock (in other word "clamp")**

The state of **decreasing** means producing force/power in the opposite direction of applied load. On the contrary, the state of **increasing** means producing force/power in the same direction of applied load. The state of **lock (clamp)** means keeping position against applied load. Since the state of **free** means completely free or highly flexible against load. It is extremely important for actuator to have both states of lock and free to achieve safety feature. Because actuator can release force by slipping when applied load is excessive. It should be noted that this unique characteristics enables us to improve safety not only for human but also for the micro/miniature machine itself to be damaged by accident.

2.2. Limitation of Conventional Actuators

The driving characteristics of conventional actuators are summarized in Table 1. Unfortunately non of them is applicable. For instance, most typical industrial actuators such as geared motor cannot realize state of free due to big friction of reduction gears.

Pneumatic actuators can not achieve perfect free without high speed control by using sensory feedback. Although the ultrasonic actuator using Piezoelectric device can easily realize state of lock because of big friction, it is impossible to achieve the state of free.

The Direct-drive motor (D-D motor) can achieve the state of free in principle. However real D-D motor produces fairly large frictional torque due to suspension bearing or other rotating parts. And inertia torque is not so small. Moreover in order to keep the rotational angle under applied load, large electric current should be always applied.

Since the SMA(shape memory alloy) actuator is a one of elastic solid actuator, it is very difficult to obtain perfect state of lock and free.[3]

Therefore we have reached to the conclusion that new type of actuator based on new principle should be developed to realize cybernetic actuator.

Table 1 Driving Characteristic of Conventional Actuator

Actuator	free	decreasing	increasing	lock
motor with reduction gears	no	yes	yes	yes
D-D motor	yes?	yes	yes	yes
pneumatic	yes?	yes	yes	yes
SMA	yes?	yes	yes	yes?
ultrasonic motor	no	yes	yes	yes

2.3. First Prototype of Cybernetic Actuator

On this account, we have developed two types of cybernetic actuators in miniature size in 1991.[1] Both of the rotary type and linear type actuators were prototyped and basic performance as a cybernetic actuator was conformed successfully by several experiments. (please see the proceedings of MEMS'91).

Those prototypes were the first practical cybernetic actuator in the past. Especially linear cybernetic actuator could be miniaturized to 10 x 10 x 40 [mm] in size and 12.5 [g] in weight at that time. Maximum speed was approx.7 [mm/sec]. The maximum producible force was 200 [gf] while 100 [V]- 5[mA] was supplied to piezoelectric device for clamp. Since the configuration of actuator was very simple, it had great potential for further miniaturization.

In the following sections, further miniaturized prototype by employing new clamping method and performance will be described.

3. NEW TYPE LINEAR CYBERNETIC ACTUATOR

3.1. New Actuator Design

Fig.1(a) shows the latest model of new type linear cybernetic actuator whose dimension is 5 x 5 x 12 [mm] and weight is only 1 [g]. The experimental setup is shown in Fig.1(b). This tiny actuator is equipped with two kinds of actuator as schematically drawn in Fig.2. The miniature piezoelectric element (1 x 2 x 5 [mm]) is used to produce "impact" for moving. The miniature electromagnetic coil produces "clamping force" to control friction between actuator and guide rail made of steel.

It should be note that the method of electromagnetic clamping does not need highly precise machining nor gap adjustment between actuator unit and guide rail. This method can contribute to further miniaturization. In case of micro size, the electromagnetic clamping will be replaceable to electrostatic clamping.

By using these actuators, four driving states such as free, decreasing, increasing and lock as a "cybernetic actuator" can be realized. The capability to control friction provides safety feature for both man and the micro mechanism itself because applied force can be released by slipping when applied force is too big as mentioned before. This special feature on safety is one of the most important advantages of cybernetic actuator as well as performance improvement.

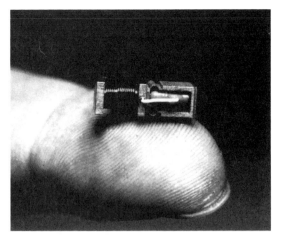

(a) Tiny Linear Cybernetic Actuator with Piezoelectric Device and Electromagnetic Clamp

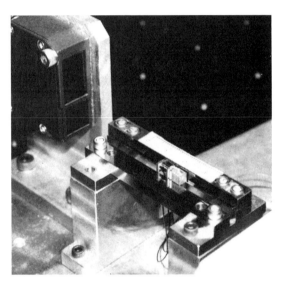

(b) Experimental Setup and Guide rail

Fig.1

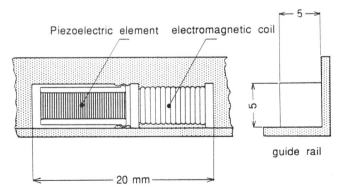

Fig.2 Design of New Linear Cybernetic Actuator

3.2. New Driving Principle

Control ability of clamping force to realize cybernetic actuator can also provide new principle of actuation and drastic improving of performance as a linear actuator.

The schematic driving sequence in one step named **"Controlled Friction Impact Drive"** in Fig.3 is summarized in six stages as follows;

stage

1. initial state
2. quick expansion of piezoelectric element can obtain small displacement under little friction
3. clamping by electromagnetic force to increase frictional force between actuator and guide rail
4. quick contraction without any slipping is possible because of large friction
5. quick release of clamping when piezoelectric element reaches to initial length
   (Inertia weight collides with the main body)
6. sliding forward by inertia until kinetic energy caused by collision is consumed by sliding friction

Although above driving sequence of our cybernetic actuator seems to be close to that of micro positioning actuator proposed by Higuchi et al [4], our linear actuator can control clamping force to avoid slip of the main body during quick contraction of piezoelectric element in stage 4.

It should be noted that sliding distance in one step (i.e. stage 6 in Fig.3) can be extended from 20 to 100 times longer than the conventional impact drive actuator. Moreover, since the frictional energy loss due to large clamping force in stage 5 and 6 of conventional impact drive can be eliminated in this actuator, energy efficiency also rises. Detail discussion on these matters will be made in the later section 5.

3.3. Comparison with Conventional Actuators

Although this configuration of actuator seems similar to the "inchworm mechanism",[5] there are a big difference between them. Since the driving principle of inchworm uses the cycle of "**static**" contraction and expansion produced by piezoelectric device, the running speed is very slow. In order to improve speed, expansion mechanism of strain is necessary. It was not easy to reduce the size and weight.

4. EXPERIMENRTAL VERIFICATION OF TINY CYBERNETIC ACTUATOR

Several experiments to verify basic feature as a cybernetic actuator and improved performance as a miniature linear actuator driven by impact with friction control.

Fig.4 shows the measured relationship between electric current into electromagnetic coil and clamping force which is equivalent to maximum producible force in case of impact drive actuator. The clamping force increases as the electric current increases. The maximum producible force reaches 70 [gf] at 1 [A]. The force/weight ratio as performance index is about 70 and very high as a miniature linear actuator.

The timing chart of the input voltage pattern into piezoelectric device to produce impact and the current into electromagnet for clamp are shown in Fig.5. By modifying these input pattern, driving sequence mentioned in previous section is obtained. Fig.6 shows the experimental data of the speed vs. driving frequency. The maximum speed at present was approximate 35 [mm/sec] under 37 [kHz] driving frequency. Since the trial to rise driving frequency up to the inaudible range (i.e."ultrasonic range") has been succeeded, this cybernetic actuator can make complete silent actuation. Even in the audible driving frequency, the electromagnetic clamping system can decrease sound level because the gap between electromagnet and guide rail is too small to generate tapping noise during clamp and release cycle.

To summarize above experimental results, newly developed linear cybernetic actuator has following characteristics.

1. safety actuation as a cybernetic actuator
2. miniature and light weight
3. simple configuration
4. adjustment free
5. silent operation
6. drastically improved maximum speed
7. drastically improved energy efficiency

By optimizing input voltage pattern into the actuators, actuator performance will be improved.

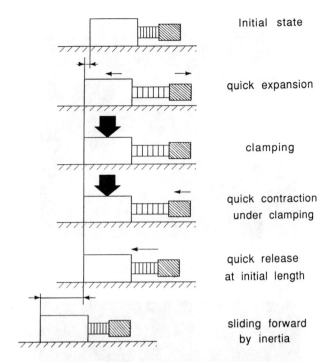

Fig.3 Driving Sequence of Linear Cybernetic Actuator

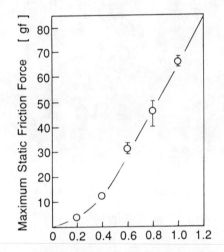

Fig.4 Measured Relationship Between Clamping Force and Electric Current into Electromagnetic Coil

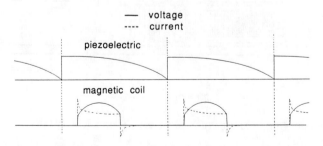

Fig.5 Timing Chart of Voltage Patterns into Piezoelectric Devices and Electromagnetic Coil

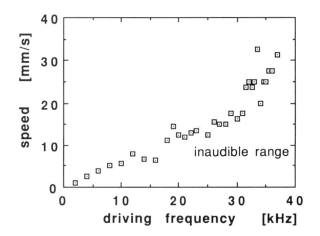

Fig.6 Frequency Dependance of Maximum Speed

5. THEORETICAL VERIFICATION OF PERFORMANCE IMPROVEMENT

The availability of proposed new impact drive named "controlled friction impact drive" is examined theoretically in comparison with conventional impact drive temporally named "constant friction impact drive". Both the maximum speed and the energy efficiency of each actuator are estimated in the following section.

5.1. Analyzing Maximum Speed

In order to estimate the maximum speed of cybernetic actuator, the total displacement which consists of two stage (stage 2 and 6 in Fig.3) in one sequence is analyzed.

5.1.1 Analysis of Total Displacement in One Step

Notation;
  M: weight of the main body of the actuator without inertia part
  m: weight of inertia part
  μ: coefficient static friction
  μ': coefficient of kinetic friction
  L: displacement of piezoelectric element
  α: acceleration of contracting piezoelectric element
  v_M: velocity of main body
  v_m: velocity of inertia weight
  v_{M+m}: velocity of total body (inertia weight and main body)
  x_1: progressive displacement in one step at stage 2 in Fig.3.
  x_2: progressive displacement in one step at stage 6 in Fig.3.
  x_{all}: total progressive displacement in one step

At stage 2

The small displacement x_1 caused by quick expansion of piezoelectric element at the stage 2 in Fig.3 can be estimated as follows;

The impulse equation at this stage is written as below,
$$M v_M = m v_m \quad (1)$$
The relationship between velocity and displacement of piezoelectric element is
$$(v_M + v_m)t = L \quad (2)$$
According to equation (1) and (2), x_1 is estimated as below.
$$x_1 = v_M t = (m/(M + m))L \quad (3)$$

At stage 5 and 6

The sliding displacement x_2 at the stage 6 caused by collision of inertia weight with the main body at the stage 5 can also be estimated as follows;
The impulse equation at stage 5 is written as below.
$$v_{M+m}(M + m) = v_m m \quad (4)$$
Based on the conservation of energy at stage 6 from stage 5, following equation is introduced.
$$(1/2)(M+m)v_{M+m}^2 = \mu'(M+m)g x_2 \quad (5)$$
On the other hand velocity v_m at stage 5 is written as below.
$$v_m = \alpha t = \sqrt{2L\alpha} \quad (6)$$
By using Eq.(4), (5) and (6), x_2 can be estimated as below.
$$x_2 = (m/(M+m))^2 L\alpha/(\mu' g) \quad (7)$$

Total displacement in one step

Hence, total displacement of the actuator in one sequence x_{all} is presented as below.
$$x_{all} = x_1 + x_2$$
$$= m/(M+m)L + (m/(M+m))^2 L\alpha/(\mu' g) \quad (8)$$

5.1.2. Comparing Total Displacement in One Step

According to Eq.8, we can compare the performance for the total displacement in one step of newly proposed "controlled friction impact drive" with conventional "constant friction impact drive".

controlled friction impact drive (newly proposed)

  Total displacement $x_{all}{}^1 = mL(1 + \mu/\mu')/(M + m)$ (9)

constant friction impact drive (conventional)

  Total displacement $x_{all}{}^2 = 2mL/(M + m)$ (10)

case 1: **tiny linear cybernetic actuator** (in this paper)
  where M=0.7[g], m=0.3[g], L=2.0[μm], μ'= 0.4
    (from the specification of latest prototype)
    $x_{all}{}^1$ = 105.6 [μm] (controlled friction)
    $x_{all}{}^2$ = 1.2 [μm] (constant friction)

case 2: **miniature linear cybernetic actuator**
  (in MEMS'91)
  where M=8.0[g], m=5.5[g], L=6.0[μm], μ'= 0.4
    $x_{all}{}^1$ = 92.9 [μm] (controlled friction)
    $x_{all}{}^2$ = 4.9 [μm] (constant friction)

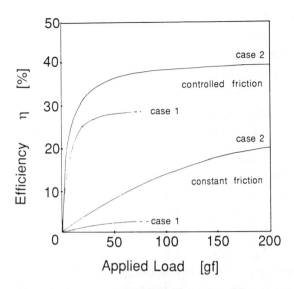

Fig.7 Comparing Energy Efficiency under Applied Load
case 1: tiny linear cybernetic actuator (in this paper)
case 2: miniature linear cybernetic actuator
(in MEMS'91)

According to above estimation, total displacement in one step which is proportional to maximum speed of actuator is improved drastically by using new impact drive.

5.2. Analyzing Energy Efficiency

In case of "controlled friction impact drive", energy loss due to friction during sliding can be decrease as mentioned before. Therefore total energy efficiency is also improved.

In this section, energy efficiency for above two kind of impact drives under certain applied load are analyzed.
The input energy into piezoelectric device W_{in} is

$$W_{in} = m\alpha L \quad (11)$$

The work toward the applied load W_{out} can be estimated as

$$W_{out} = (m^2 L\alpha/(M+m))/(\mu'(M+m)g+F) \quad (12)$$

where F: applied load

As a result, energy efficiency under applied load can be written as follows;

$$\eta = (m/(M+m))F/(\mu'(M+m)g+F) \quad (13)$$

Following Fig.7 shows the calculated results of the relationship between energy efficiency and applied load for two linear cybernetic actuators (in case 1 and 2). As similar as the result on total displacement in one step above described, efficiency of "controlled friction impact drive" is much larger than conventional impact drive.

6. CONCLUSION

New type of linear actuator so called "tiny cybernetic actuator" having a safety and silent feature suitable for biomedical tasks was developed successfully. The driving principle is based on impact drive generated by piezoelectric device with electromagnetic clamp. Since the electromagnetic clamping can control friction, basic performances such as maximum speed and energy efficiency were improved drastically too. Both theoretical and experimental result verified these unique features.

Acknowledgment
Authors would like to thank to Mr.Osamu Yamashita of NEC Corporation for presenting miniature piezoelectric elements and Mr.Ueda in Yaskawa Electric Mfg.Co.,Ltd. for his kind assistance.

Reference

[1] K.Ikuta, A.Kawahara, S.Yamazumi, " Miniature Cybernetic Actuators Using Piezoelectric Device", Proc. of International Workshop on Micro Electromechanical Systems (MEMS'91), Nara, Japan, pp.131-135,1991

[2] R.Tomovic, R.B.MacGee, "A Finite State Approach to the Synthesis of Bioengineering Control Systems", IEEE Trans. on Human Factor in Electronics,Vol.7, No.2, p65, 1966

[3] K.Ikuta, M.Tsukamoto, S.Hirose, "Shape Memory Alloy Servo Actuator System with Electric Resistance Feedback and Its Application to Active Endoscope", Proc. of IEEE International Conference on Robotics and Automation, Philadelphia, USA, pp.427-430, 1988

[4] T.Higuch et al., Proc. of IEEE International Workshop on Micro Robot and Teleoperation, 1987

[5] Patent Pending,Burleigh Instrument Inc., East Rochester, N.Y., 1986

EXPERIMENTAL MODEL AND IC-PROCESS DESIGN OF A NANOMETER LINEAR PIEZOELECTRIC STEPPER MOTOR

Jack W. Judy, Dennis L. Polla, and William P. Robbins
Department of Electrical Engineering
Institute of Technology
University of Minnesota
Minneapolis, Minnesota

ABSTRACT

A linear stepper motor capable of sub-micrometer controlled movement, or nano-actuation, has been constructed using the piezoelectric material lead-zirconate-titanate (PZT). This prototype device consists of a piezoelectric driving element measuring 25.4 mm x 12.7 mm x 1.6 mm. This device is inset in a trench to constrain motion to one dimension. An electrode on the bottom of the glider is used with an electrode on the top of the trench to implement an electrostatic clamp. A linear inertial sliding motion can be achieved by rapidly expanding or contracting the piezoelectric bar which is connected between a glider and an attached base. Glider velocities of 5.7 - 476 µm/s were measured by timing the movement of the glider over a 1.0 mm portion of the track through an optical microscope. Displacement steps of 70 - 1100 nm were calculated by dividing the measured glider velocity by the frequency of the applied voltage pulses. Displacement step size and glider velocity were controlled by the application of PZT extension voltages ranging from ±(60 - 340) V. The ability of the current prototype piezoelectric stepper motor to implement sub-micrometer motion with the large (centimeter) travel distances may have applications to electron microscopy, scanning tunneling microscopy, alignment of optical fibers, and magnetic recording. This motor has been constructed as a prototype of a version that is being integrated on a silicon wafer and a fabrication process design is given.

I. INTRODUCTION

Increasing the capability for precise manipulation of motion on a sub-micrometer scale has become evident by the progress made and the attention given to micromechanical and piezoelectric structures. The control of emission tips in electron microscopes and scanning tunnelling microscopes, the alignment of optical fibers, the control of heads in magnetic recording, and the measurement and manipulation of biological cells and air pollution particles are just some of the applications that require controlled motion on micrometer or sub-micrometer scales. These applications could potentially benefit from the control of displacements in sub-micrometer steps with a total travel range of several centimeters.

There has been much work done in the area of piezoelectric positioning devices. The characteristics of these piezoelectric positioning elements elements are reasonably well understood [1]. Scrie and Teague [2] have used piezoelectric elements and flexural pivots to accurately measure sizes of air pollution particles in the 1 - 50 µm range. Binnig et al. [3] have developed "inchworm" or "louse" piezoelectric stepping devices that use a piezoelectric element and two or more electrostatic clamps to achieve movement. These devices have been extensively used in scanning tunnelling microscopy to view materials at the atomic level [4,5,6,7,8]. Higuchi et al. [9] demonstrated linear stepping motion based on electromagnetic impulsive forces in a translator that utilizes the recoil of an ejected mass and the friction of the glider and the track. Pohl [10] has investigated piezoelectric translator devices that also use a mass recoil technique and mechanical friction to achieve motion. Higuchi et. al. [11] used piezoelectric steppers for robotics applications also for linear positioning and cell manipulation but relied solely on friction. Some limitations associated with these previous designs include, the unreliability of friction as a main operative force, and the limited range of travel.

In this paper we report the use of the piezoelectric material lead-zirconate-titanate (PZT) in the construction of a linear stepper motor that incorporates both an ejection mass as well as an electrostatic clamp. We also discuss the scalability of this device and describe a fabrication process designed to integrate a similar piezoelectric stepper motor on a silicon wafer using micromachining techniques. This stepper motor has a sub-micrometer minimum step size and has a potentially unlimited linear travel distance. In practice, the travel distance is limited by the voltage leads to the PZT element and the length of the track.

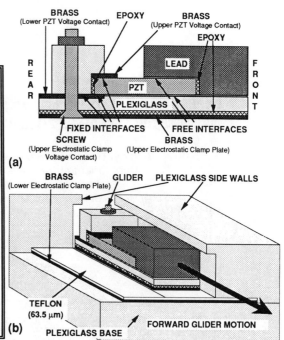

Length of PZT	25.4	mm
Width of PZT	12.7	mm
Thickness of PZT	1.6	mm
Mass of PZT	.0037	kg
Mass of Lead	.0277	kg
Mass of Glider	.0107	kg
Length of Glider	50.8	mm
Length of Track	101.6	mm
Piezoelectric Constants of PZT:		
d_{31}	-268	(m/V) x 10^{12}
d_{33}	510	(m/V) x 10^{12}
Piezoelectric Young's Moduli of PZT:		
E_{11}	6.46	(N/m^2) x 10^{-10}
E_{33}	4.78	(N/m^2) x 10^{-10}
Mechanical Q of PZT	72	-
Area of Clamp	600	mm^2
Friction Surface	Brass-Teflon	
$\mu_{s,off}$	0.36 - 0.43	-
$\mu_{s,on}$	0.62 - 0.78	-
Maximum Velocity	476.0	μm/s
Minimum Velocity	5.7	μm/s
Maximum Step Size	1.1	μm
Minimum Step Size	0.07	μm

Table 1. Electrical and physical parameters for the piezoelectric stepper motor.

Figure 1: a) Side view of glider. b) Cross section of glider and track with perspective.

II. PRINCIPLES OF OPERATION

The operation of our piezoelectric linear stepper motor is based on inertial acceleration of an attached load by a piezoelectric element and the difference between static and kinetic friction which is controlled with an electrostatic clamp. A translator device capable of moving small objects (0-30 g) has been constructed in a rugged, yet simple, scalable design.

A. Experimental Design

Our motor consists of a PZT driving element connected between a glider base and an attached lead load (Table 1, Fig 1a). The glider is seated inside a Plexiglass trench. The stepper motor is able to move smoothly forward and backward in the trench without significant binding (Fig 1b). A brass plate on the bottom of the glider is used with a brass plate and 63.5 μm Teflon® layer on top of the track to form an electrostatic clamp. The piezoelectric material is attached to the back portion of the glider allowing free expansion and contraction of the PZT element. Extension and contraction of the piezoelectric bar is governed by the well-known constitutive relations

$$D = e^S E + eS \quad (1)$$
$$T = c^E S + eE \quad (2)$$

where D is electric displacement in an electric field E, e^S is the dielectric constant with zero or constant strain, e is the piezoelectric stress constant, S is the macroscopic strain in the material, T is the externally applied stress, and c^E is the elastic stiffness in the presence of constant or zero electric field. Table I. Electrical and physical parameters for the piezoelectric stepper motor.

B. Stepper Motor Operation

The operation of our translator is based on inertial acceleration and is carried out in four simple steps. In the first step, the electrostatic clamp is activated with a large voltage (500 V max) to firmly hold the glider in place (Fig. 2a). This is followed, in step two, by an extension of the PZT element with a variable voltage to incrementally move the center of mass of the entire structure (glider and attached load) forward (to the right as shown in Fig. 2b). This causes the lead to overhang the right end of the glider base. During the third step, the electrostatic clamp is released. This reduces the friction between the glider and the track and allows the glider to move freely in response to any applied forces (Fig. 2c). The voltage across the PZT is held constant during this step. This keeps the piezoelectric material extended and the overall center of mass in the same position as at the end of the previous step. In the fourth step, the PZT element is grounded which causes the PZT element to contract towards the new center-of-mass position. This generates an inertial force that pulls the back of the glider forward by a small distance Δd (Fig. 2d) and then the cycle is repeated. Displacement can be achieved in either direction through application of voltages of the appropriate polarity.

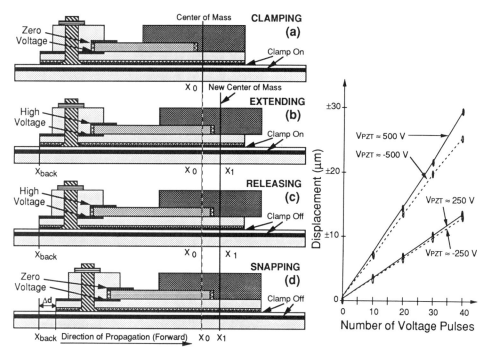

Figure 2: Four Step Movement Cycle.

Figure 3: Low Frequency Stepper Motor Movement.

C. Sliding Step Analysis

The overall motion of this device is dependent on the electrostatically controlled frictional interaction of the glider with the track. The static frictional forces between the glider and the track when the electrostatic clamp is off and on are given by

$$F_{s,off} = \mu_{s,off} [m_L + m_G] g \quad (3)$$
$$F_{s,on} = \mu_{s,on} [m_L + m_G] g \quad (4)$$
$$\mu_{s,on} > \mu_{s,off} \quad (5)$$

respectively. Where m_L is the mass of the lead, m_G is the mass of the glider, g is the acceleration due to gravity, and $\mu_{s,off}$ and $\mu_{s,on}$ are the coefficients of static friction with the electrostatic clamp off and on respectively. The driving impulse force imparted to the glider in step 4 (F_{imp}) due to the rapid expansion or contraction of the PZT element will cause the glider to move if $F_{imp} > F_{s,off}$. Unwanted forces (F_{error}) may act on the glider in step 2 and these forces will decrease the stepper motors performance. Unwanted forces in step 2 may be caused by a quick extension of the PZT element or by stepping up an incline. The unwanted forces can be neutralized if $F_{s,on} > F_{error}$ and since $\mu_{s,on}$ can be electrically controlled by the electrostatic clamp, $F_{s,on}$ can be adjusted for different applications of the stepper motor. For example, in situations where $\mu_{s,off}$ is extremely low, due to lubrication, the clamp can make movement possible by increasing the friction to $\mu_{s,on}$. Such that $F_{s,on}$ is indeed $> F_{error}$ otherwise the device would not move.

D. Electrostatic Clamp

The effectiveness of the electrostatic clamp in immobilizing the glider structure was investigated theoretically. Assuming parallel conducting plates and neglecting fringing fields, the electrostatic attractive force F_{clamp} between the glider and the track is given by

$$F_{clamp} = \frac{\varepsilon_r \varepsilon_0 A V^2}{2 d^2} \quad (6)$$

where ε_r is the relative dielectric constant of the track insulator, ε_0 is the permittivity of free space, A is the plate area, V is the applied voltage, and d is the separation between the parallel plates of the glider and the track. For a 500 V clamping voltage applied over an a 6.45 cm^2 area and with an insulator (TEFLON®) thickness d = 63.5 μm, the clamping force equals 0.37 N. This clamping force can be increased by reducing the insulator thickness to near its voltage breakdown limit.

III. RESULTS

A. Piezoelectric Material

The piezoelectric material used was a lead zirconate titanate ceramic supplied by Transducer Products Inc., Goshen, CT (their type LTZ-2M) The expansion/contraction of our piezoelectric sample, shown in Fig. 3, is approximately linear with applied voltage from 0 V to ±500 V and gives rise to a $\Delta L/L \approx 1.66 \times 10^{-7}$ 1/V. This corresponds to a $d_{31} \approx 2.66 \times 10^{-10}$ m/V, which agrees well with that specified for LTZ-2M by Transducer Products Inc., shown in Table 1.

B. Low-Frequency Movement

The piezoelectric material rapidly expands, or snaps out, with a step of +500 V, or rapidly contracts, or snaps back, with a step of -500 V. Either of these snapping motions will impart a momentum impulse to the glider which is larger than the static frictional force, and will cause the glider to move [4]. If the PZT element is then open-circuited after the application of the high voltage step, the voltage decays like that across a capacitor with a long time constant ($t_{PZT} > 1$ s). With this long t_{PZT}, the change in voltage across the piezoelectric material occurs gradually, such that the glider is unlikely to overcome the static friction between it and the track.

The propagation of the glider with four different pulsing voltage levels is shown in Fig. 3 with positive displacement due to positive V_{PZT}. Note that the glider displacement is linearly related to the number of pulses. Another important observation is that the glider moved approximately 10% less with negative voltage pulses. The reason for this asymmetry is not completely understood at this time but may be due to hysteresis in the piezoelectric element, the mechanical structure of the device, or the electrical operation of the track and glider. We recognize that the effects of piezoelectric hysteresis could be minimized by prestressing the PZT element, but we did not do this in order to keep the design of our first prototype as simple as possible. For variable stepper motor applications, a scaling factor for negative pulse operation may be needed.

C. High-Frequency Glider Movement

At high frequencies it was anticipated that the electrostatic clamp would be needed, in addition to friction, to hold the glider in place during the clamping stage of the cycle. As shown in the conceptual timing diagram in Fig. 4 any unwanted backward movement, due to a quick charge-up of the piezoelectric element, should be significantly reduced when the electrostatic clamping voltage is high. However experimental results indicated that on a horizontal surface the electrostatic clamp has little or no effect on the glider's movement. This is due

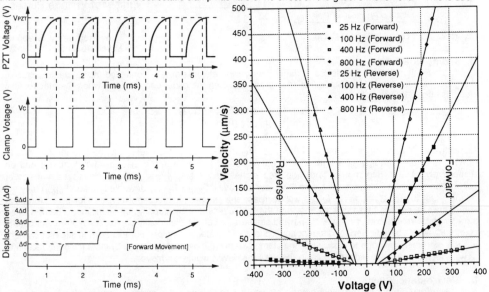

Figure 4: High Frequency Movement Timing Diagram.   Figure 5: High Frequency Movement Velocities.

to the fact that the RC charge-up delay of the piezoelectric material and the high voltage switching circuitry is gradual enough (charge-up RC time constant ≈ 1 ms) that unwanted backward movement is not present. A circuit configuration that does not have this RC charge-up delay would require a very effective electrostatic clamp to counter the movement caused by a faster charge-up. A limitation introduced by the charge-up delay is that at high frequencies, $f \approx 1/RC \approx 1000$ Hz, the delay will not allow the piezoelectric voltage to reach the maximum value. In fact as the frequency increases, the attainable piezoelectric voltage decreases.

The least-squares-fit of the data in Fig. 5 shows a linear relationship between the glider velocity and the voltage applied to the piezoelectric material for a wide range of frequencies in both the forward and reverse directions. The non-zero voltage intercept of Fig. 6 indicates that there is a minimum voltage, which varies with frequency and direction, that must be applied to the PZT element to overcome the friction between the glider and the track. In Fig. 5 the slope of best-fit 400 Hz forward movement line is nearly identical to the slope of best-fit 400 Hz backward movement line, even though the velocity of backward movement is slightly less than that of forward movement for a given voltage magnitude. This linear characteristic is valuable for controlled stepping applications.

As shown in Fig. 6, loading the stepper motor linearly reduces its velocity and the linearity between velocity and voltage remains. The effectiveness of the electrostatic clamp was evident when the track was tilted up a small degree. Without the electrostatic clamp, the glider lost almost all forward motion when the track was tilted up 5° and even began sliding backward when raised to 8°. When using the electrostatic clamp, the glider was stationary at 10° and slide backwards at 12°.

IV. DISCUSSION OF RESULTS

Consideration of the piezoelectric relationship between voltage and expansion for our device (Eqns. 1-2) predicts a linear expansion of $\Delta L \approx 4.5 \times 10^{-3}$ µm/V which corresponds to $\Delta L = 1.13$ µm at 250 V. The theoretical displacement of the glider per cycle Δd is given by equation (7).

$$\Delta d = \frac{\Delta L \, m_{(Lead)}}{m_{(Lead)} + m_{(Rest\ of\ Glider)}} \qquad (7)$$

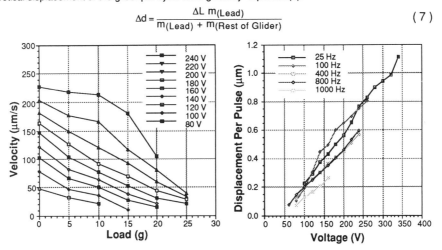

Figure 6: Effects on Velocity of Stepper Motor due to Loading.

Figure 7: Relationship between Step Size of the Stepper Motor and Voltage.

With $m_{(Lead)} = .0277$ kg and $m_{(Rest\ of\ Glider)} = .0107$ kg the resultant theoretical displacement of the glider per cycle is $\Delta d_{(Theoretical)} = 0.815$ µm/cycle at 250 V. Experimentally at frequencies of 25 Hz and 100 Hz the measured average forward displacement per cycle per volt was approximately 3.2×10^{-3} µm/cycle-V. This value corresponds to a step $\Delta d_{(Exp)} = 0.8$ µm/cycle at 250 V which is shown in Fig. 7. At frequencies of 400 Hz and 800 Hz the measured average forward displacement per cycle per volt was approximately 2.6×10^{-3} µm/cycle-V. This corresponds to a step $\Delta d_{(Exp)} = 0.6$ µm/cycle at 250 V which is also shown in Fig. 7. This 20% loss of movement is due to the fact that the charge up delay of the PZT element, approximately 1 ms, is not short enough for the PZT to reach the full applied voltage.

Although investigation of the piezoelectric extension characteristics indicated that the PZT extended and contracted equally in both directions in the static case, low frequency ($f < 1$ Hz) glider movement showed that displacement due to negative voltage pulses was approximately 10% less than that due to positive voltage pulses. This factor was also present and more pronounced in high-frequency (25 Hz $< f <$ 4000 Hz) glider movement. The difference between forward and backward velocity reaches a maximum of 70% at 25 Hz. This difference was found to decrease rapidly as the frequency increased to 400 Hz as shown in Fig. 8, but from 400 Hz to 800 Hz the difference remained relatively constant. Therefore by adding a scaling factor to the negative voltage pulses, the error in reverse movement can be reduced.

Since the square wave has two large voltage discontinuities compared to only one large discontinuity in a sawtooth waveform, a high voltage sawtooth waveform might be used to achieve more reliable movement. Even though the electrostatic clamp increased the static friction by approximately 75% and increased the tolerated maximum tilt of the track from 8° to 12°, the electrostatic clamp was not able to realize its full capability due to the extremely sharp charge-up of the square wave.

V. INTEGRATION ON A MICRO-SCALE

A large (15 mm) micro manipulator has previously been fabricated using IC compatible process [12]. We hope to scale the prototype down by 100X and fabricating on a silicon wafer by surface micromachining techniques. This should provide controlled movement in the angstrom to nanometer range. The current prototype which uses a mass recoil technique to step along will suffer in scaling since mass scales by the cube (m^3) but the electrostatic clamp will become more effective since it scales down by the square (F_{Clamp}^2). Therefore the mass ejection technique will become less effective and the clamp will become more effective relative to each other. A stepper motor device which uses two clamping plates and a piezoelectric bar to "inchworm" along may offer a better performance. The operating voltages should be scaled by V^1 so this should allow for more reasonable operating voltages. A fabrication sequence has been designed that requires 8 photolithography steps. The complete design is given in Fig. 9. A top view of the device with the parts labled is given in figure 10. This device would be able to be

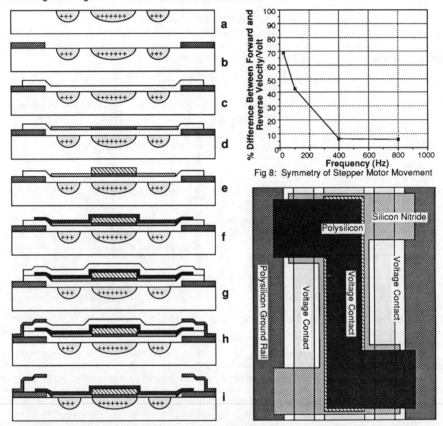

Figure 9: Fabrication Process Sequence: a. Diffused contacts, b. polysilicon ground rail, c. PSG sacrificial layer, d. polysilicon piezo contact and silicon nitride clamp dielectric, e. piezoelectric, f. polysilicon ground contact (Glider), g. PSG sacrificial layer, h. polysilicon constraining rails, i. freed structure (PSG removed).

Figure 10: Top View of Device with Parts Labeled.

Fig 8: Symmetry of Stepper Motor Movement

turned upside down without falling out of its constraining channels and should be able to step in all orientations. Three sliding contacts will not be required in this design. The three voltage contacts are buried by diffusion in the silicon wafer and a single sliding ground contact will be used. Two sliding polysilicon contacts have been proven feasible [13]. Integrating photovoltaic devices on the glider could give the system opto-to-mechanical actuation and would eliminate the requirement for sliding voltage contacts, but no work in this respect has been published to date. Thin film piezoelectric materials also impose some difficulties. Although zinc oxide (ZnO) has proven to be a very useful thin film piezoelectric material, its d_{33} strain coefficient is most likely too small to achieve a stepping motion. Currently work is being done on depositing thin films of PZT which has a much larger d_{33} coefficient.

VI. CONCLUSION

A linear piezoelectric stepper motor has been implemented that can incrementally move in average steps as small as 70 nm over linear travel distances limited only by the length of the track and currently the length of the wires. This device is based on a periodic electrostatic clamping and releasing technique applied to a sequentially activated and deactivated piezoelectric element. Glider velocities ranging from 5.7 to 476 µm/s for forward movement were measured through an optical microscope and displacement steps ranging from 0.07 to 1.1 µm for forward movement were calculated by dividing the measured glider velocity by the frequency of applied voltage pulses. Both the step size and glider velocity were varied by the application of PZT extension voltages ranging from ±(60 - 340) V. Precise movement in both forward and reverse directions can be obtained through the incorporation of a scaling factor to compensate for the slightly lower displacement per step in the reverse direction, and by choosing a linear region of a velocity/voltage line for a particular frequency. Velocities are higher at higher operating frequencies because of an increase in the number of steps per second. Even with external loading of up to 35 grams, the approximate linear relationship between displacement and applied voltage remains. The ability to move external loads was found to be limited by the effectiveness of the electrostatic clamp in creating a high static friction between the glider and track.

Application of the electrostatic clamp would be more effective with a sawtooth voltage waveform applied to the PZT element. This would minimize the generation of unwanted inertial forces during the clamping step. The piezoelectric stepper motor has possible applications to optical microscopy, control of emission tips in scanning tunneling microscopy, the precision alignment of optical fibers, and the control of micro-gap heads in magnetic recording.

ACKNOWLEDGMENT

We are grateful to Messrs. John Marchetti and Mike Leveille for their help in the construction of the stepper motor. We are grateful to Professor Hendrik Oskam for his encouragement of this work. This work was supported by the National Science Foundation (ECS-8906121 and ECS-8814651) and by the University of Minnesota Department of Electrical Engineering Honors Program.

REFERENCES

[1] H. Jaffe and D.A. Berlincourt, "Piezoelectric Transducer Materials", Proc. IEEE, vol. 53, pp. 1372, 1965.
[2] F. E. Scrie and E. C. Teague, "Piezo driven 50-µm range with sub-nanometer resolution," Rev. Sci. Instrum., vol. 49, pp. 1735-1740, 1978.
[3] G. Binnig, H. Roher, Ch. Gerber, E. Weibel, "A Piezo Drive with Course and Fine Adjustment," IBM Tech. Disc. Bull., Vol 22 (7), pp. 2897, 1979.
[4] T. Berghaus, H. Neddermeyer, S. Tosch, "A scanning tunneling microscope for the investigation of the growth of metal films on semiconductor surfaces," IBM J. Res. Develop., vol. 30, no. 5, pp. 520-524, 1980.
[5] G. Binnig, H. Roher, Ch. Gerber, E. Weibel, "Tunneling through a controlled vacuum gap," Appl. Phys. Lett., Vol 40 (2), pp. 178-180, 1982.
[6] G. Binnig, H. Roher, Ch. Gerber, E. Weibel, "Surface Studies by Scanning Tunnelling Microscopy," Phys. Rev. Lett., Vol 49, pp. 57-61, 1982.
[7] G. Binnig, H. Roher, Ch. Gerber, E. Weibel, "7x7 Reconstruction of Si(111) REesolved in Real Space," Phys. Rev. Lett., Vol 50, pp. 120-123, 1982.
[8] B.W. Corb, M. Ringger, and H.-J. Guntherodt, "An electromagnetic microscopic positioning device for the scanning tunneling microscope," J. Appl. Phys. vol. 58, pp 3947-3953, 1985.
[9] T. Higuchi, Y. Hojjat, and M. Watanabe, "Micro actuators using recoil of an ejected mass," Proc. IEEE Micro Robots and Teleoperators Workshop, Hyannis, Massachusetts, November, 1987.
[10] D. W. Pohl, "Dynamic piezoelectric translation devices," Rev. Sci. Instrum., vol. 58, pp. 54-56, 1987.
[11] T. Higuchi, Y. Yamagata, and K. Kudoh, "Precise Positioning Mechanism Utilizing Rapid Deformations of Piezoelectric Elements," IEEE MEMS Workshop, pp. 222-226, 1990.
[12] A. M. Feury, T. L. Poteat, W. S. Trimmer, "A Micromachind Manipulator for Submicron Posistioning of Optical Fibers"," Hilton Head Workshop, 1986.
[13] P. Bergstrom, T. Tamagawa, D. L. Polla, "Micro-Mechanical Memory," IEEE MEMS Workshop, vol. 58, pp. 54-56, 1990.

Zinc-Oxide Thin Films for Integrated-Sensor Applications

D.L. POLLA AND R.S. MULLER

Department of Electrical Engineering and Computer Sciences
and the Electronics Research Laboratory
University of California, Berkeley, California 94720

ABSTRACT

Integrated sensors based on the piezoelectricity and pyroelectricity in zinc-oxide thin films have been fabricated compatibly with conventional NMOS technology. A study of material and electrical properties of the zinc-oxide films, prepared by planar magnetron sputtering, has been carried out. For 1−μm thick films, the measured piezoelectric coefficient is $d_{33} = 14.4 \times 10^{-12} CN^{-1}$ and the pyroelectric coefficient is $p^{\sigma} = 1.4 \times 10^{-9} Ccm^{-2}K^{-1}$. In practical integrated sensor designs, these films have yielded voltage responses of 5.2 mV/gm (piezoelectric effect) and 150 mV/K (pyroelectric effect).

INTRODUCTION

Thin films of zinc oxide have been combined with conventional silicon planar processing techniques to form integrated sensors in which a physical or chemical variable is sensed and transduced to an on-chip amplifier circuit. Integrated sensors using zinc-oxide thin films that have been fabricated and demonstrated to date in our laboratory include:

(1) cantilever-beam accelerometer [1],
(2) SAW convolver [2],
(3) chemical vapor sensor [3],
(4) anemometer [4],
(5) infrared detector array [5],
(6) chemical reaction sensor [6],
(7) tactile sensor array [7], and
(8) infrared charge-coupled device imager [8].

The operation of all sensors was based on the piezoelectric or pyroelectric effects in the zinc-oxide films.

This paper focuses on the material properties of zinc-oxide thin films and the basic design considerations for integrated-sensor applications such as those listed above.

ZINC-OXIDE THIN-FILM DEPOSITION

Techniques which have previously been used for zinc-oxide thin-film deposition include rf and dc sputtering [9,10], chemical-vapor deposition [11], ion plating [12], planar-magnetron sputtering [13], and reactive magnetron sputtering [14]. Our work has been based on films made using planar magnetron sputtering as the deposition technique. We chose this method because of its low film damage due to electron bombardment and high deposition rate.

Zinc-oxide thin-film crystallinity, obtained under varying sputtering conditions (substrate temperature, sputter power, thickness, ambient gas mixture), has been studied by x-ray diffraction and scanning electron microscopy. Highly-oriented zinc-oxide films have been deposited on SiO_2/Si and SiO_2/Poly-Si/Si substrates. X-ray diffraction measurements indicate preferential c- axis orientation with a single diffraction peak at 33.8° in the best samples studied as shown in Fig. 1. Based on x-ray diffraction studies, we have found the best thin-film crystallinity to correspond to deposition conditions carried out at a forward sputtering power of 200 W with a 10 mTorr ambient gas mixture consisting of 50% oxygen and 50% argon. The substrate to target distance is 4 cm and the substrate temperature is maintained at 230°C during deposition.

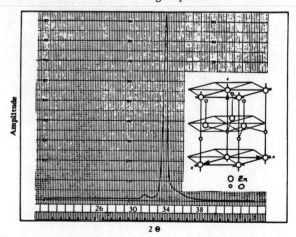

Fig. 1. Typical x-ray diffraction spectrum characterizing the preferential c- axis orientation in zinc-oxide thin films prepared by planar magnetron sputtering.

Reprinted from *Technical Digest IEEE Solid-State Sensors Workshop*, June 1986.

MATERIAL CHARACTERIZATION

Pyroelectric and piezoelectric properties have been measured by electrical techniques and correlated with the x-ray diffraction results. The measured pyroelectric coefficient at T=300 K is $p^\sigma = 1.4 \times 10^{-9}$ Ccm^{-2}K^{-1} and piezoelectric coefficient is $d_{33} = 14.4 \times 10^{-12}$ CN^{-1}, in good agreement with the range of reported values in crystalline zinc oxide [15,16]. The temperature dependence of these coefficients is shown in Figs. 2 and 3. For typical sensor applications in 1 μm-thick films at T = 300 K, these coefficients imply signal levels of 5.2 mV/gm (piezoelectric effect) and 150 mV/K (pyroelectric effect). Thin films that exhibit good piezoelectric properties were also found to exhibit good pyroelectric properties.

The ability to fabricate sensors with a near dc response is based on the fact that typical film thicknesses (1μm) are much thinner than a Debye length (120 μm). Static charge decay times (see Fig. 4) in excess of 32 days have been measured in these films when encapsulated by 0.4 μm-thick layers of SiO$_2$. Thin-film resistivities of $3 \times 10^7 \Omega$-cm and dielectric constants of $\epsilon_r = 10.3$ have been characterized in these thin-films.

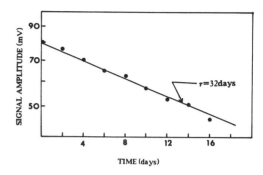

Fig. 4. Pyroelectric charge decay under conditions of constant infrared radiation. The indicated slope represents the characteristic charge retention time for zinc-oxide thin films encapsulated by silicon dioxide.

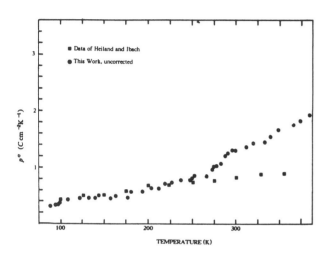

Fig. 2. Measurement of pyroelectric coefficient versus temperature in a 1.0-μm-thick zinc-oxide film.

CONCLUSIONS

Zinc-oxide thin films with very useful piezoelectric and pyroelectric properties have been prepared by planar magnetron sputtering. The material and electrical properties characterized in these films have been utilized in the design and fabrication of diverse integrated sensors compatibly fabricated using NMOS technology. The performance of these sensors demonstrates: 1) the versatility of zinc-oxide thin-film technology for integrated-sensing applications based on either the piezoelectric or pyroelectric effect, and 2) the possibility of carrying out multifunction sensing on one integrated-circuit chip.

Acknowledgements

This work was supported in part by the National Science Foundation under Grant ECS 81-20562 and in part by the State of California MICRO program.

REFERENCES

[1] P.-L. Chen, R.S. Muller, R.D. Jolly, G.L. Halac, R.M. White, A.P. Andrews, T.C. Lim, and M.E. Motamedi, "Integrated Silicon Microbeam PI-FET Accelerometer," *IEEE Electron Dev.* ED-29, 27-32 (1983).

[2] A.E. Comer and R.S. Muller, "A New ZnO on Si Convolver Structure," *IEEE Electron Dev. Lett.* EDL-3, 118-120 (1982).

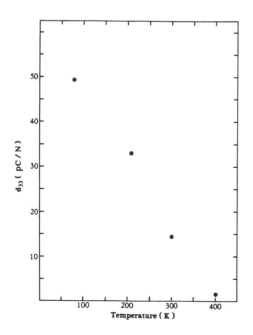

Fig. 3. Temperature dependence of the piezoelectric coefficient d_{33}.

[3] C.T. Chuang and R.M. White, "Sensors utilizing thin membrane SAW oscillators," *Proc. IEEE Ultrasonics Symposium*, Chicago, IL, 1981.

[4] D.L. Polla, R.S. Muller, and R.M. White, "Monolithic Zinc-Oxide on Silicon Pyroelectric Anemometer," *IEEE International Electron Devices Meeting*, Washington, D.C. 1983.

[5] D.L. Polla, R.S. Muller, and R.M. White, "Fully-Integrated ZnO on Silicon Infrared Detector Array," *IEEE International Electron Devices Meeting*, San Francisco, CA 1984.

[6] D.L. Polla, R.M. White, and R.S. Muller, "Integrated Chemical-Reaction Sensor," *Third International Conference on Solid-State Sensors and Actuators*, Philadelphia, PA 1985.

[7] D.L. Polla, W.T. Chang, R.S. Muller, and R.M. White, "Integrated Zinc Oxide-on-Silicon Tactile Sensor Array," *IEEE International Electron Devices Meeting*, Washington, D.C. 1985.

[8] [18] D.L. Polla, R.M. White, and R.S. Muller, "Integrated Multi-Sensor Chip," *IEEE Electron Dev. Lett.*, (to be published in 1986).

[9] H.W. Lehmann and R. Widmer, "RF Sputtering of ZnO Shear-Wave Transducers," *J. Appl. Phys. 44*, 3868-3879 (1973).

[10] G.A. Rozgonyi and W.J. Polito, "Preparation of ZnO Thin Films by Sputtering of the Compound in Oxygen and Argon," *Appl. Phys. Lett. 8*, 220-221 (1966).

[11] S.K. Tiku, C.K. Lau, and K.M. Lakin, "Chemical Vapor Deposition of ZnO Epitaxial Films on Sapphire," *Appl. Phys. Lett. 36*, 318-320 (1980).

[12] M. Matsumoto, I. Kah, and Y. Murayama, *Tech. Group Ultrasonics US 78-47, 17* IECE Japan (1979).

[13] T. Yamamoato, T. Shiosaki, and A. Kawabata, "Characterization of ZnO Piezoelectric Films Prepared by RF Planar- Magnetron Sputtering," *J. Appl. Phys. 51*, 3113-3120 (1980).

[14] B.T. Khuri-Yakub and J.G. Smits, "Reactive Magnetron Sputtering of ZnO," *J. Appl. Phys. 52*, 4772-4774 (1981).

[15] G. Heiland and H. Ibach, "Pyroelectricity in Zinc Oxide," *Solid State Commun. 4*, 353 (1966).

[16] Landolt-Borntein Tables, Volume 15, edited by O. Madelung, H. Schulz, and H. Weiss, Springer-Verlag, Berlin, 1980.

A MICROMACHINED MANIPULATOR FOR SUBMICRON POSITIONING OF OPTICAL FIBERS

A. M. Feury
Dept. of EE
Rutgers University
Piscataway, N.J. 08854

T. L. Poteat
AT&T Bell Labs
Murray Hill, N.J. 07974

W. S. Trimmer
AT&T Bell Labs
Holmdel, N.J. 07733

Abstract

Described is a micromanipulator that is fabricated from IC-compatible processes and can position in x-y coordinates to submicron accuracies. The active portion uses micromachined silicon feet and piezoelectric beams. Its "stage" is a silicon wafer with CVD insulators. Applications include an optical fiber alignment device operating in a closed loop mode with positional resolution of 500 Angstroms. Other possible applications will be discussed.

Introduction

The application of optical fiber to digital communication systems has created a need for precision optical fiber connectors. That need is currently being satisfied by components that have rigid manufacturing specifications and are therefore expensive. Switching functions generally require conversion from photonic to electronic energy due to a lack of optical switches. An active connector that can position one or more optical fibers and reposition them to an alternate (nearby) site removes the requirement for manufacturing precision and it also achieves the desired switching function. The mechanical requirements on the connector components of thin, near ideal insulating layers on very flat, smooth feet are difficult (expensive) to satisfy in conventional metal systems but are easily met in IC processing facilities. Silicon parts can be batch fabricated with high accuracy, the material is strong and well characterized, electronics can be integrated into the silicon parts, and these parts lend themselves well to small mechanical systems. We believe silicon micromachining techniques can produce unique user-driven manipulators using the versatile, ubiquitous silicon.

Device Operation and Configuration

Operation of other manipulators has been previously discussed [1]. A voltage applied between conductors that are closely spaced generates an electrostatic attraction between the conductors. This force can be on the order of pounds per square centimeter at modest voltages (~100 Volts) if the separation between the conductor plates is submicron. This spacing requirement is easily met with single crystal silicon and CVD insulators such as silicon nitride. Piezoelectric ceramics are readily available at modest cost.

We characterized an 15 mm square device by mounting an optical fiber to one foot and mounting another fiber through the substrate to form a switch. A closed-loop program allowed the fibers to be disconnected and reconnected while gathering statistics on step size and repeatability. Other configurations such as crossed-beam and L-shaped manipulators were also constructed.

Conclusion

The ability to micromachine silicon with high accuracies makes it an attractive material for manufacturing small mechanical parts. These parts do not need to be tethered to the silicon, but can be part of freely moving engines. These engines can in turn do a number of useful functions. The micromanipulator described above is useful for aligning components to high accuracies. Other possible applications for freely moving manipulators include assembly and repair of small systems. Having taken a step onto silicon, there are many intriguing possibilities.

References

[1] G. Bennig and C. Gerber, "A Piezo Drive with Course and Fine Adjustments," IBM Tech. Disc. Bull., 22(7), 2897, 1979.

Ultrasonic Micromotors: Physics and Applications

R. M. Moroney, R. M. White, and R. T. Howe

Berkeley Sensor & Actuator Center
an NSF/Industry/University Cooperative Research Center
Department of EECS and the Electronics Research Laboratory
University of California, Berkeley, CA 94720

Abstract

Recently developed *electrostatic* micromotors have received much attention. Here we present further work on an alternative, the *ultrasonic* micromotor. We have both translated and rotated small polysilicon blocks using traveling ultrasonic flexural waves in thin membranes. The 2-micron-thick blocks ranged in size from 50 µm square to 250 by 500 µm. The acoustic wavelength was 100 µm, and the few-megahertz transducer voltage was typically 5; we observed motion at voltages as low as 1.5 V. The acoustic wave power was about 1 mW. Linear motion quickly moved the blocks off of the 3 mm by 8 mm membrane region; stable rotational motion was also observed. Blocks moved linearly at about 10 mm/s and rotated at about 150 rpm, depending on the drive voltage. Motion has been observed in air, vacuum and helium ambients.

Introduction

Ultrasonic motors have been subjects of keen interest recently. Typically this type of motor has large friction between the rotor and stator. Traveling waves excited in the stator cause movement of surfaces that rub against the rotor, producing motion. This driving mechanism gives ultrasonic motors a number of interesting properties, such as high torque and the ability to have either stepped or continuous motion. Different bulk acoustic modes may be exploited for either linear [1] or rotational [2] motion. Applications for such motors include linear positioning devices [3] and autofocusing lens systems in some cameras. As a specific example, Figure 1 shows exploded views of two types of Matsushita ultrasonic motors (described by Inaba, et al. [4]). The figure highlights the design effort that has gone into the stator structure and lining materials to increase frictional coupling between rotor and stator. The Lamb-wave device we are presenting as an ultrasonic micromotor is shown roughly to scale in Figure 1c.

There are several reasons we wish to develop micro-sized ultrasonic motors. Even though other actuation schemes have been demonstrated, such as electrostatic or piezoelectric drives, we would expect different characteristics from an ultrasonic micromotor. The advantages of ultrasonic wave actuation have been described by Flynn, et al. [5]. These advantages include drive voltages considerably lower than those for electrostatic micromotors, wide choices for the rotor material, and the possibility of linear motion. Initial discussion on our Lamb-wave micromotor is presented elsewhere [6]. Here we present further work on efforts to understand the physics of the device, and we begin to investigate possible applications.

Figure 2 shows a top view and cross section of the thin membrane device we are using, and gives typical thicknesses for the membrane materials. The membrane area is about 3 by 8 mm; it is formed by a bulk micromachining process where

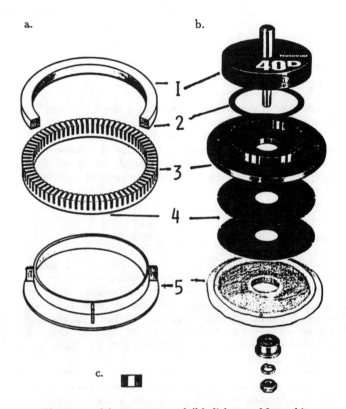

Figure 1: (a) ring type, and (b) disk type Matsushita ultrasonic motors, after Figure 15 of [4]. Typical diameter is 32 mm. (c) Lamb-wave micromotor, 3 mm by 8 mm, shown roughly to the same scale.

the back side of the silicon wafer is anisotropically etched away, leaving only the thin nitride membrane. This device was first developed for sensing applications, which are discussed along with further fabrication details by Wenzel and White [7]. The polysilicon block pictured sitting on the membrane in Figure 2 is used as the moved object. Blocks of various sizes have been made using a sacrificial layer technique used in fabricating polysilicon laterally resonant microstructures [8]. Blocks are typically 250 by 500 µm and 2 µm thick. Many have been patterned with apertures (holes 30 µm wide) or corrugations (depressions 30 µm wide and 2 µm deep) to see how they influence the block motion. These features are periodically spaced 100 µm apart, a distance equal to the typical acoustic wavelength.

Acoustic Lamb waves are excited in the membrane by interdigital transducers (IDTs) at either end. Several modes may be excited; Viktorov is a good source of information on Lamb waves, including dispersion curves for the different modes, etc. [9]. Modes are characterized as either antisymmetric or symmetric. Antisymmetric modes have primarily transverse particle motions and odd symmetry about the membrane's mid-plane, where as symmetric modes have primarily longitudinal particle motions with even symmetry about the mid-plane. The lowest order antisymmetric, or A_0, mode is most commonly used and has a typical frequency of 4 to 6 MHz. Driving a single IDT will result in linear motion, while driving both IDTs from the same signal often results in rotational motion.

There are several possible applications for an ultrasonic micromotor. Acousto-optical devices are one (see for example, the review article by Quate, et al. [10]). Repositionable elements such as mirrors or diffraction gratings might enhance or diversify currently existing optical and measuring instruments, or lead to entirely new structures. Other possible applications for ultrasonic micromotors include microrobots, friction control [11], and fluid pumping.

Theory

Before we can begin to understand the operation of our prototype ultrasonic micromotor, it is useful to get an understanding of the acoustic wave amplitude. Experimentally this may be determined using an optical diffraction technique. When an acoustic wave is present, the distorted membrane surface acts as a moving diffraction grating. The intensity of light scattered into the side lobes is proportional to the wave amplitude. Auth and Mayer [12] showed that the ratio of light intensity in the first diffraction side lobe ($\mathcal{I}_1$) to the incident light intensity ($\mathcal{I}_{inc}$) is

$$\frac{\mathcal{I}_1}{\mathcal{I}_{inc}} = J_1^2(w) \approx \left(\frac{w}{2}\right)^2, \qquad (1)$$

where J_1 is the first order Bessel function and $w = 4\pi a \cos\phi/\lambda$ is the acoustic amplitude, a, normalized using the optical wavelength λ and the angle of incidence ϕ. The approximation on the right of (1) is valid for small w and is usually accurate. Solving for the acoustic amplitude gives

$$a = \frac{\lambda}{2\pi \cos\phi} \sqrt{\frac{\mathcal{I}_1}{\mathcal{I}_{inc}}}. \qquad (2)$$

In practice, we use a photodiode to measure light intensity, which is proportional to the short-circuit photocurrent. Figure 3 shows the simple experimental setup for this measurement. We should note that the spacing of diffraction orders gives information on dimensions of the ultrasonic-wave diffraction grating.

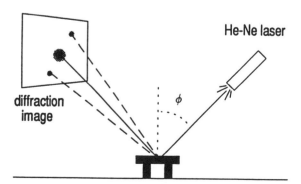

Figure 3: Schematic setup for experiment to measure the acoustic wave amplitude by light diffraction off of the Lamb-wave device.

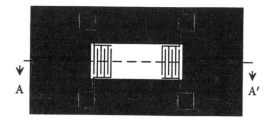

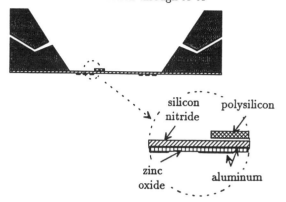

Figure 2: Lamb-wave membrane device. The enlarged cross section shows the layers of the membrane; typical thicknesses are 2 μm for the silicon nitride, 1 μm for the zinc oxide, and 0.4 μm for the aluminum layers.

To compare with the experiment, we have developed a simple theory for predicting the wave amplitude. To begin, the power (P) transmitted by a wave can be related to the displacement of the surface (ξ) by [13]

$$P = -(T_0 W)\frac{\partial \xi}{\partial x}\frac{\partial \xi}{\partial t}, \qquad (3)$$

where T_0 is the tension in the membrane and W is the width of the acoustic beam. Assuming a traveling wave displacement of $\xi = a\sin(\omega t - kx)$, it is possible to plug into (3), take a time average, and solve for the acoustic amplitude. The result is

$$a = \frac{1}{\omega}\sqrt{\frac{2P_{ave}}{\rho v W}}, \qquad (4)$$

where P_{ave} is the time averaged acoustic power, ω is the frequency, ρ is the mass density of the membrane, and v is the acoustic velocity. The time-averaged wave power may be estimated knowing the device insertion-loss and the power into a known load.

The next important step in understanding our ultrasonic micromotor is to determine how energy is transferred from the acoustic wave to the polysilicon block. We see three possibilities: 1) friction contact with the membrane; 2) wave action, like a surfer; and 3) boundary layer interaction, where the acoustic wave drags along a fluid layer which in turn drags the block. Figure 4 schematically shows these three mechanisms and qualitative expected responses for linear block velocity versus acoustic wave amplitude.

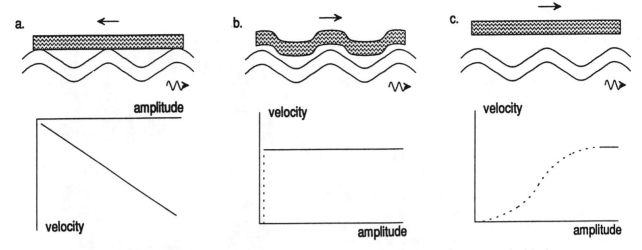

Figure 4: Sketches of mechanisms for energy transfer from the acoustic wave to the block: (a) friction contact; (b) wave action; (c) boundary layer interaction. A qualitative sketch of the velocity versus acoustic wave amplitude dependence is given for each mechanism.

The friction contact mechanism is expressed in macroscopic ultrasonic motors, as mentioned above. The velocity of a rotor is in the opposite direction of the acoustic wave propagation because of the retrograde elliptic particle motion. The rotor moves with the horizontal velocity of particles at the wave peak, or the crest velocity. This velocity is given by [4]

$$U_{crest} = -\pi \omega a \left(\frac{d}{\Lambda}\right). \quad (5)$$

Here d is the membrane thickness, and Λ is the acoustic wavelength. This linear relation between velocity and amplitude is shown in Figure 4a.

With the wave-action mechanism, a block would travel with roughly the phase velocity of the acoustic wave. This velocity is independent of wave amplitude, so we expect a constant velocity. There may be a minimum amplitude required to couple into this mechanism; this dependence is sketched in Figure 4b.

The boundary layer interaction is sketched in Figure 4c. Finding a relation between velocity and acoustic amplitude is more difficult in this case; the relation will depend on the fluid characteristics and the boundary conditions at the membrane surface. Certainly the velocity will not exceed the acoustic phase velocity. To get some understanding of the importance of viscous effects, we solve the simple problem of a transversally moving boundary at frequency ω. The decay length for such a disturbance may be shown to be [14]:

$$\delta_\nu = \sqrt{\frac{2\mu}{\rho\omega}}, \quad (6)$$

where μ is the fluid's viscosity and ρ its density. In air at a typical operating frequency of 4 MHz we find δ_ν is about 1 μm.

Results

Figure 5 shows the measured acoustic wave amplitude versus power for the A_0 mode of a Lamb-wave device. Equation (4) is used for the theoretical line and Eq. (1) is used for the experimental data. The agreement is quite close, the error is only 3% above 0.8 $\sqrt{W}$, giving us confidence in the measured amplitude of about 275 Å. This number appears small, but compared to a typical SAW wave amplitude of 15 Å it is actually quite large.

We have observed both linear and rotary motion using the flexural Lamb-wave device shown in Figure 2. As noted earlier, rotational motion is excited by driving both IDTs with the same input signal. This sets up a standing wave pattern in the membrane. We believe the rotation is due to scattered waves from the ends of the membrane that disturb the standing wave pattern and strike the block at odd angles. This conclusion is tentatively supported by the difficulty of exciting rotary motion in devices where the acoustic waves are not scattered off the back edge but simply reflected back. Figure 6 helps to visualize this concept. When the membrane has tapered ends, acoustic waves are scattered off in different directions. These scattered waves may be detected using the same optical diffraction method used to measure the acoustic amplitude. With flat ends on the membrane, the reflected waves are colinear with the incident waves and do not add any new spots to the diffraction pattern. When it is excited, rotary motion is quite stable; we rotated one sample continuously for 22 hours before turning off the rf drive.

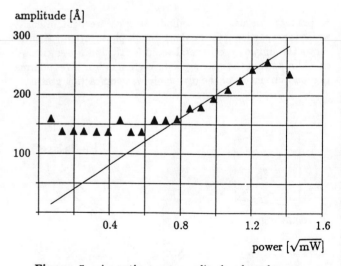

Figure 5: Acoustic wave amplitude plotted versus square root of the power in the wave. Solid triangles show data from the laser diffraction and the line is the calculated amplitude using Eq. (4). Noise from the ambient light prevented measuring acoustic amplitudes below about 150 Å.

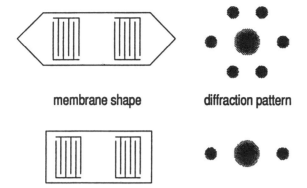

Figure 6: Qualitative comparison of optical diffraction patterns from acoustic waves traveling in two different membrane designs.

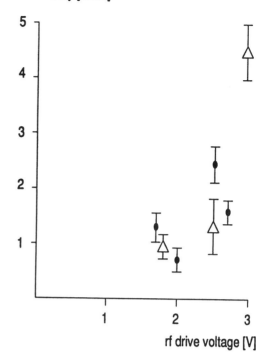

Figure 7: Linear velocity in air (preliminary data). Solid ovals indicate the block traveled to the right (driven by the left hand IDT), open triangles to the left (driven by the right hand IDT). All linear motions observed so far in air have been in the same direction as the acoustic wave. The acoustic wave amplitude is about 100 Å at an input voltage of 3 V. The driving frequency was 4.17 MHz.

Linear motion is excited by driving a single IDT. Figure 7 shows some preliminary linear velocity data for a 250 by 245 μm^2 block. This particular block was both corrugated and apertured. The acoustic wavelength was 80 μm. Note that we plotted only those data points where the block started near, but not on, the driving IDT. This is an effort to keep the initial conditions somewhat uniform; block speeds have been observed to decrease monotonically as the starting position is moved further from the driving transducer. No special cleaning procedures were done prior to taking the data in room air. The speed values are calculated from frame-by-frame analysis of a video tape of the motion. The error is due to inaccuracies in reading the block position and variation in speed between frames. The velocities all appear to have reached their final value; the block has stopped accelerating while still in view of the video camera.

Another test with linear motion is to send an rf tone burst into the IDT and observe the blocks response. Preliminary indications are that with a large driving voltage of 10 V, a 250 μm square block will require about 1,000 input cycles (a drive duration of approximately 15 ms) before it moves. With a 10,000 cycle burst (also 10 V drive amplitude), this square block moves repeatably approximately 50 μm with each burst. Occasionally the block would continue floating after the burst ended. Presumably a cushion of air became trapped beneath the block and acted as an air bearing. Studies are now underway to characterize motion in response to a tone burst.

QUALITATIVE OBSERVATIONS

The lowest antisymmetric mode (A_0 mode) is most commonly used with this device. We have also observed motion using the A_2 mode. This mode has a frequency of about 23 MHz; we confirmed its 50 μm wavelength using the diffraction measurement technique described above.

While manipulating blocks using the A_2 mode, one block became stuck to the membrane by one corner. Presumably dirt or some other defect caused this sticking. With the one corner secured, however, very interesting things happened to the block. As the acoustic drive was varied, the block tilted and turned. Shining a laser beam on the block revealed that the block's normal was not parallel to the membrane normal, in other words, the block was not resting flat on the membrane surface. This can be seen from the offset of the block diffraction from the undiffracted light spot. With an applied signal, the block angle (determined from the diffraction patterns) is stable to better than 0.01 degrees. For this particular block the angle is reproducible as the applied voltage is varied. We do not think that the A_2 mode is crucial to this effect. Figure 8 shows a diffraction pattern for a block manually placed in a similar position to the stuck block described above. Figure 8b shows the block and IDT locations to help in interpreting the diffraction spots. Preliminary tests at moving blocks with the A_3 and S_0 modes were unsuccessful.

Since blocks have been observed to diffract the incident light as well as the acoustic wave, we have used this effect to determine rotation rates of spinning blocks. The detecting photodiode is placed near, but not on, the central reflected spot. As the block spins, its diffracted light periodically illuminates the diode giving a periodic change in its voltage. The period may be determined using an oscilloscope. Observed rotation rates were near 100 rpm; more tests are underway to characterize the dependence on applied drive.

Block corrugations appear to affect the block motion. Results from the tone burst tests indicate that having the corrugations perpendicular to the acoustic wave fronts (i.e., parallel to the direction of wave propagation) may reduce the burst duration required to excite motion by as much as a factor of two.

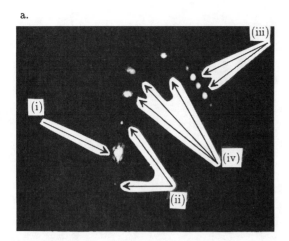 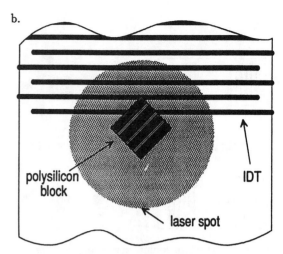

Figure 8: He-Ne laser diffraction pattern from a 250 μm square block with three corrugations spaced 100 μm apart. One corner of the block is resting on the IDT region, as shown schematically in Figure 8b. In Figure 8a, (i) points to the central undiffracted spot, and (ii) to the two diffraction orders from the interdigital transducers. The central spot also has the diffraction pattern from a square aperture; this comes from the shadowing of the incident laser light by the block. The diffraction pattern from the block itself is displaced from the central spot and pointed at by (iii). This separation gives a measure of the angle between the block normal and surface normal. The remaining spots pointed to by (iv) are not from the membrane objects, they are due to reflections from the camera housing.

DISCUSSION AND CONCLUSIONS

Our results to this point strongly indicate that a fluid boundary layer is responsible for moving the blocks in air. The data in Figure 7 reveal two very important points: first, the blocks move *in the direction* of the acoustic wave propagation vector; and second, the velocity of the block depends on the acoustic wave amplitude. The first of these points immediately rules out the friction contact mechanism for motion in air. The second point rules out the wave action mechanism. More work is needed to clarify the role of surface interaction between the block and the membrane. Tentative results with moving blocks in an SEM under vacuum (presented earlier, [6]) suggested that blocks moved against the acoustic wave in vacuum. This is presumably by the friction contact mechanism and thus supports the explanation of a fluid boundary layer giving motion in air. Charging effects in the SEM, however, make the results only suggestive, rather than conclusive.

Many possible applications exist for the ultrasonic micromotor. Optical interactions with moving blocks open the door to repositionable mirrors, diffraction gratings, and other such structures. An optical chopper could easily be constructed where a small object is used to physically block a light beam. By measuring the rotation rate of blocks using the rotating diffraction pattern, we have demonstrated this sort of concept.

Another major application for this sort of motor is a fluid pump (the present device could be used to "pump" dry powders). Lamb wave devices are able to operate in water [7]; because of the very low acoustic velocity (typically 400 m/s), the energy is not radiated into the surrounding medium. If a fluid boundary layer is indeed responsible for moving the blocks, it should be possible to fabricate an ultrasonic fluid pump. With some very simplistic approximations on the fluid flow, we predict pump rates on the order of 10 μL/s in water and 100 μL/s in air. We are beginning to fabricate devices specifically to investigate this intriguing prospect.

While investigating fundamental physics and device applications, we are making new devices to test fluid pumping and optical chopping. Also, we are incorporating electrodes to apply a normal force to the blocks. This will be interesting as an experimental parameter and possibly as a control parameter in some applications. Study of block motion in our vacuum probe station for thorough characterization of the block motions in vacuum and other controlled ambients is also planned. Surface conditions must also be investigated, as it is difficult to repeat experiments precisely. Furthermore, the ultrasonic micromotor presented here is fabricated in two separate processes and then manually placed together. Hand assembly should be avoided in most practical devices, therefore the stationary and moving parts should be fabricated together, probably by surface machining.

In conclusion, we have presented many initial observations of phenomena in a prototype ultrasonic micromotor structure. It is very important to realize that this device is not simply a scaled-down macroscopic ultrasonic motor; the physics behind its operation appear to be completely different. In a macro ultrasonic motor, friction is used to couple energy from stator to rotor, while on the micromotor a fluid boundary layer appears responsible for motion in air. Several ideas for applications have been presented, and the groundwork is laid down for future work.

ACKNOWLEDGEMENTS

This work has been supported by a National Science Foundation Graduate Fellowship and by the Berkeley Sensor & Actuator Center, an NSF/Industry/University Cooperative Research Center.

REFERENCES

[1] M. Kuribayashi, S. Ueha, and E. Mori, "Excitation conditions of flexural traveling waves for a reversible ultrasonic linear motor," *J. Acoust. Soc. Am.*, vol. 77, no. 4, pp. 1431–1435, 1985.

[2] M. Kurosawa, K. Nakamura, T. Okamoto, and S. Ueha, "An Ultrasonic Motor Using Bending Vibrations of a Short Cylinder," *Trans. on UFFC*, vol. 36, no. 5, pp. 517–521, 1989.

[3] K. Mori, T. Kumagae, and H. Hirai, "Ultrasonic Linear Motor for a High Precision X-Y Stage," 1989 IEEE Ultrasonics Symposium, October 3–6 Montreal, Quebec, Canada, paper PD-6.

[4] R. Inaba, A. Tokushima, O. Kawasaki, Y. Ise, and H. Yoneno, "Piezoelectric Ultrasonic Motor," 1987 IEEE Ultrasonics Symposium, pp. 747–756.

[5] A. M. Flynn, R. A. Brooks, and L. S. Tavrow, "Cornerstones for Creating Gnat Robots," (MIT A. I. Memo 1126, July 1989).

[6] R. M. Moroney, R. M. White, R. T. Howe, "Ultrasonic Micromotors," 1989 IEEE Ultrasonics Symposium, October 3–6 Montreal, Quebec, Canada, paper TJ-4.

[7] S. W. Wenzel, R. M. White, "A Multisensor Employing an Ultrasonic Lamb-Wave Oscillator," *Trans. Electron Devices*, vol. 35, no. 6, pp. 735–743, 1988.

[8] W. C. Tang, T. H. Nguyen, and R. T. Howe, "Laterally Driven Polysilicon Resonant Microstructures," in *Proc. IEEE Micro Electro Mechanical Systems* (Salt Lake City, UT, 1989), pp. 53–59.

[9] I. A. Viktorov, *Rayleigh and Lamb Waves*, (Plenum Press, NY, 1967).

[10] C. F. Quate, C. D. W. Wilkinson, and D. K. Winslow, "Interaction of Light and Microwave Sound," *Proc. of the IEEE*, vol. 53, no. 10, pp. 1604–1623, 1965.

[11] M. G. Lim, J. Chang, D. Shultz, R. T. Howe, and R. M. White, "Polysilicon Structures to Characterize Static Friction," to be presented at the 1990 IEEE Micro Electro Mechanical Systems Workshop, Napa Valley, CA, 1990, report #14.

[12] D. C. Auth and W. G. Mayer, "Scattering of Light Reflected from Acoustic Surface Waves in Isotropic Solids," *J. Appl. Phys.*, vol. 38, no. 13, pp. 5138–5140, 1967.

[13] W. C. Elmore, M. A. Heald, *Physics of Waves*, (Dover, 1985).

[14] M. Thompson, G. K. Dhaliwal, C. L. Arthur, and G. S. Calabrese, "The Potential of the Bulk Acoustic Wave Device as a Liquid-Phase Immunosensor," *Trans. on UFFC*, vol. UFFC-34, no. 2, pp. 127–135, 1987.

Section 8

Valves and Pumps

FLUIDS (gases and liquids) are useful in the micromechanical domain. Chemicals, biological cells, suspended particles, and numerous other constituents can be transported using fluids. And high-pressure fluids are useful to exert large forces in small regions. The problem is controlling the flow of the fluid.

Using large valves surrounding a micromechanism is usually a poor solution. The tubes between the large valves and micro device induce time delays, and slow the speed and accuracy of the micromechanism. The wiring of many tubes between the large and small can be a nightmare. Also, one of the advantages of micromechanics is the ability to design using a myriad of inexpensive micro devices. If these micro devices each have to be controlled by a large, expensive macro valve, this advantage is lost.

Micro pumps have many uses in micro systems. Critical problems in designing micro pumps are building good valves, generating the forces needed to pump the fluid, and keeping the pump free of extraneous matter that can clog the valves and chambers.

What one needs is micro valves and strong actuators that can be integrated with the micro systems. Making these micro valves is difficult using conventional macro valve designs and concepts. For example, in my kitchen sink are two valves with large, flexible washers. When the valve starts to leak, it is back to the hardware store. This scenario is unsuitable in micro systems for several reasons. Micro devices are difficult to repair. It is usually difficult to even get inside the system to find out what is wrong. Also, the amount of material in a micro washer decreases as the scale size to the third power. A washer one hundredth the size of my kitchen sink washer has one millionth the amount of material. Micro washers just do not have much mass to withstand the abrasions and cracks. If there is any mixing of gases and liquids in the micro valve, the forces generated by surface tension can be disastrous. As an example, water condensing out in a micro gas valve can fill and seal the valve orifice.

Generating the actuation force and coupling it to the valve is also difficult. The electromagnetic forces used in most macro valves can be difficult to generate, and the electrostatic forces are often too small to withstand the forces generated by the fluids. Many alternate actuation schemes are under active research.

The valve in "A Microminiature Electric-to-Fluidic Valve," boils a liquid to generate the force of closure. The washer is replaced by a thin, flexible membrane. In "The Fabrication of Integrated Mass Flow Controllers," and "Normally Closed Microvalve and Micropump Fabricated on a Silicon Wafer," a piezoelectric actuator actuates the valve seat. These papers also discuss interesting ways to couple the valve to external devices. The next paper, "A Thermopneumatic Micropump Based on Micro-Engineering Techniques," uses two valves and a thermally actuated membrane to make a micro pump. The paper "Variable-Flow Micro-Valve Structure Fabricated with Silicon Fusion Bonding," uses magnetic actuation of a micro valve to precisely control the flow of a fluid. The next paper, "A Pressure-Balanced Electrostatic-Actuated Microvalve," reduces the forces needed for actuation by partially balancing the pressure of the fluid acting on the valve against itself. The final paper, "Micromachined Silicon Microvalve," uses electrostatic forces to close a cantilevered beam against the inlet orifice.

A MICROMINIATURE ELECTRIC-TO-FLUIDIC VALVE

Mark J. Zdeblick James B. Angell

Department of Electrical Engineering
Stanford University
124 Applied Electronics Lab
Stanford, California 94305 USA

ABSTRACT

A 3mm x 3mm electro-pneumatic valve made with standard micromachining techniques is presented. The valve utilizes a sealed cavity filled with a liquid. One wall of the cavity is formed with a flexible membrane, which can press against a pneumatic nozzle. When the liquid is heated, it's pressure increases, pushing the membrane toward the nozzle, turning it off. Since one can heat the liquid to high temperatures, operation at high pressures is possible. A response time of 1ms is expected for the device under development. A valve of this size will be useful in various fields, including gas chromatography, robotics, and fuel injection. Integrating this valve with sensors and electronics will allow the development of complete pressure regulator and flow meter systems on single chips.

INTRODUCTION

Most industrial machines and many other mechanisms in our industrial society are pneumatically powered to provide the most efficient actuation. With the advent of the computer, a need has existed to provide two-way communication between electric and pneumatic systems. This link enables the computer to provide high level control of a pneumatically actuated system. For many years, the link has been provided with pressure, flow and other types of sensors alongside electro-mechano-pneumatic valves. This presentation will describe a new type of actuation mechanism, one that can be readily fabricated using planar, lithographic techniques. It will also describe, for the first time, a way of combining both actuation and sensing functions on the same piece of silicon. It should be noted that co-location of the actuator and sensor is the optimal configuration for most cybernetic systems.

PRINCIPLE OF OPERATION

While the thermodynamic properties of liquids, solids, and gasses are well established, the applicaion of these properties to novel systems continues. The thermodynamic property that is being exploited here is the positive variation of pressure with temperature for most materials. In fact, when a liquid changes state into a gas, there is an exponential relationship between vapor pressure and temperature.

$$P(T) = P_0 \exp\left(\frac{-L_0}{RT}\right)$$

Thus, if a system has a finite number of atoms contained in a roughly constant volume, the pressure of the system can be controlled by varying the temperature.

An important figure of merit for this thermodynamic relationship can be defined: $\mathcal{Z} = \frac{\delta P}{\delta E}$, or the change in pressure due to a finite input of energy. For a perfect gas with heat capacity C_v in a constant volume V, it can be shown that $\mathcal{Z} = \frac{R}{VC_{v_g}}$. Thus, miniaturization dramatically improves the efficiency of the device. For gas-liquid systems, the figure of merit is:

$$\mathcal{Z} = \frac{Mp(T_1)\left[\exp\left(\frac{L_0}{RT_1} - \frac{L_0}{RT_2}\right) - 1\right]}{gC_{v_g}\delta T + lC_{v_l}\delta T + L_0\delta l}$$

where M is the molecular weight, L_0 is the latent heat of vaporization, g and l are the initial masses in grams of the gas and liquid, respectively, and δl is the mass of liquid that changes state when the temperature rises δT. As an example, $\mathcal{Z}$ has been calculated for Methyl Chloride CH_3Cl for a volume of $1\mu l$ under various conditions (Table 1). Note that as the concentration of gas in the two-phase system increases, $\mathcal{Z}$ increases to 80000 kPa/Joule. Using this material, one can achieve a 10 psi increase in pressure by applying 200 mW for 5 msec. Similarly, one needs to dissipate the same amount of energy at the same rate to decrease the pressure 10 psi in 5 msec. By using better materials in a smaller control chamber, one can improve the efficiency and response time.

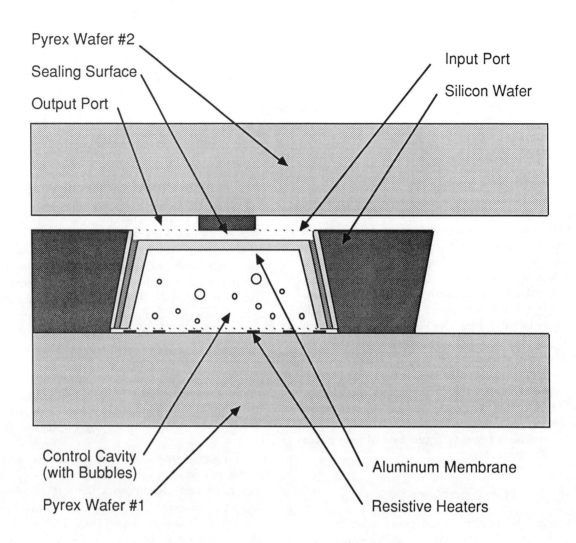

Figure 1: Cross section of Electric - to - Fluidic valve. Section is through I/O ports and control cavity.

The device being presented takes advantage of this thermodynamic relationship by heating up an encapsulated liquid, gas, or solid in order to control its pressure. One of the walls of the encapsulating vessel is thinned to the shape of a membrane. A pneumatic nozzle is positioned opposite the center of the deflecting membrane. Thus, when the encapsulated material is heated by a thin film resistor formed on one of the walls of the cavity, the pressure inside of the cavity increases, sealing the membrane against the nozzle, turning it off. As the encapsulated material cools, its pressure decreases and the membrane deflects away from the nozzle, turning the valve 'on'. The deflection is limited either by a wall of the vessel, or the contents of the vessel.

Other methods of changing the temperature of the encapsulated material include: optical heaters with fiber optic light guides; conductive, convective or radiative heating and cooling; remotely located microwave heaters; thermoelectric coolers, e.g. 'Peltier'; and 'Thompson' and other refrigeration methods.

Pressure and flow sensors on the device can monitor the output pressures and flows. Typically, silicon will act as both the sensing surface and the signal processing substrate. Together with this electrical feedback, the valve can act as either a pressure regulator or as a mass flow controller.

IMPLEMENTATION

A cross section of the device is shown in Figure 1. Wafer A is a silicon wafer bonded to two pyrex 7740 wafers. The bottom pyrex wafer seals the cavity and provides a substrate for a resistive heater. It is also transparent to light and would therefore enable optic-to-fluidic transduction. Contact to the titanium resistive heater is made through a hole in the silicon and the top pyrex plate. A 10μm thick polyimide diaphragm rests at the top of the control cavity and is separated from the silicon sealing surface above it.

% Gas	Temperature	Figure of Merit (kPa/Joule)
0	25°C	10000.
30	25°C	53000.
50	25°C	70000.
90	25°C	82000.

Table 1: Figure of Merit for Gas - Liquid System of Methyl Chloride

At the junction of the silicon and top pyrex wafers, the input and output ports are formed and meet at the sealing surface.

RESULTS

Each of the components has been separately built and tested. The idea of heating an encapsulated liquid with a miniature resistive heater in order to move a diaphragm has been demonstrated. Using a 3mm x 3mm x 30μm silicon diaphragm, deflections of 45μm have been recorded. The 'on' flow rate using Nitrogen at a supply pressure of 5 psi was found to be 60 sccm. As of the time of this writing, complete devices have been built, but they have not yet been tested.

ACKNOWLEDGMENTS

The authors would like to thank the Electronics Division of General Motors Research Laboratories for their continued financial support of this project. They also thank Dr. Phil Barth, Dr. Paul Clifford, and Bruce Fleischer for various useful conversations related to the project. A patent covering the technology presented is pending.

THE FABRICATION OF INTEGRATED MASS FLOW CONTROLLERS

Masayoshi Esashi[+], Soohae Eoh[++], Tadayuki Matsuo[+] and Segan Choi[+++]

[+]Tohoku University, Aoba Aramaki, Sendai 980, JAPAN
[++]Yeungnam Institute of Technology, TaeMyeng Dong, Taegu 634, KOREA
[+++]Yeungnam University, Tae Dong, Kyongsan 632, KOREA

ABSTRUCT

Two types of mass flow controller of which flow sensor and control valve are integrated on a silicon wafer were fabricated for fine gas control. One has a differential pressure type flow sensor and the other has a thermal flow sensor. Silicon micro valves which use small stack piezo actuator were developed for the control valve.

INTRODUCTION

The silicon micromachining technique has realized high performance of small sensors and micro mechanical devices.[1]
Not only small flow sensors[2,3] and micro valves fabricated on a silicon wafer show very fast responce, but also the integration of flow sensors and micro valves on one wafer reduces dead space. The fast response and the negligible dead space realize micro mass flow controllers for an accurate control of small gas flow.

Such micro mass flow controllers should be very useful for the very sophisticated vacuum systems used in advanced semiconductor processing.

Two types of the integrated mass flow controller are described. One uses a differential pressure sensor for the purpose of flow sensing and the other uses a thermal flow sensor. Both types have silicon micro valves which use small sized stack piezo actuator (9 X 3 X 1.4 mm^3).

DIFFERENTIAL PRESSURE TYPE MASS FLOW CONTROLLER

(a) Principle and structure

There exists the following co-relation between mass flow Q and differential pressure ΔP.

$$Q \propto \sqrt{\rho \Delta P} \propto \sqrt{\frac{P \Delta P}{T}} \quad (1)$$

,where ρ, P and T is density, pressure and temperature respectively.

sensors for these parameters are integrated with a micro valve making up mass flow controller as shown in Fig.1. The temperature sensor is a simple p-n junction diode. Piezo-resistive silicon diaphragm pressure sensors are used for the absolute pressure (P) and the differential pressure (ΔP) measurements.

Two Pyrex glass plates of 1.2 mm thickness were bonded anodically on the top and the bottom of a double side polished wafer. Small holes were drilled in thick glass plate by sand-blasting with a metal mask to make gas inlet and outlet. The diameter of the hole was controlled within 300 μm by monitoring optically during sand-blasting.

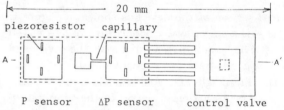

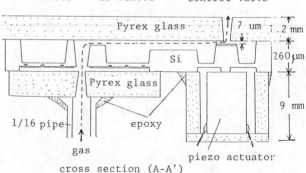

Fig.1 Differential pressure type mass flow controller.

(b) Micro valve

A silicon mesa suspended with a flexible silicon diaphragm is pressed against the glass plate by small piezo actuator when the valve is closed.

The force required to press the diaphragm is proportional to the Young's modulus of silicon and to the cube of T/L where T and L are the thickness and the width of the diaphragm respectively. This gives upper limit of T. On the other hand, there is the risk that the diaphragm will electrostatically deflected and bonded to the glass during the anodic bonding process. This determines the allowable lower limit of T. T and L adopted were about 50 μm and 700 μm respectively.

The gap between the glass plate and the silicon mesa must be smaller than the maximum stroke of the actuator (i.e. 10 μm).

The piezo actuator was first mounted in a glass tube holder which was lapped to level with the free end of the actuator. The holder was then bonded with an epoxy resin to the silicon wafer with the actuator resting at the small knob, which is located at the center of the valve to press the mesa uniformly.

An experimental result of the flow through the valve versus the voltage applied to the piezoactuator is shown in Fig.2. Maximum nitrogen flow at 1 atm was about 60 cc/min. The hysteresis of the curve in Fig.2 is due to the characteristics of the actuator itself. This hysteresis could be reduced by applying periodic open and close voltage which had variable duty ratio.

The sealing surface must be extremely smooth and flat to achieve a gas-tight seal between the silicon and the glass. KOH etchant was used to get flat silicon surface, however, some valves showed leakage. The observed leakage is believed to be caused by silicon peaks which keep the silicon and the glass (valve seat) separated.

(c) Pressure sensors and flow measurements

Pressure sensors have square diaphragms of which thickness is about 25 μm and length is about 2.5 mm. The diffusion layer for piezoresistors is buried in the silicon to prevent the instability by surface ion etc.[4] and to solve the problem of leakage current at the p-n junction feed through, which was caused by the anodic bonding.

The absolute pressure sensor which has reference cavity showed a sensitivity of about 4 ppm/mmHg. An experimental results of flow measurement by a differential pressure sensor and a capillary for pressure drop is shown in Fig.3. The parabolic curve in Fig.3 is explained by equation 1, that is, the differential pressure is proprtional to the square of flow. This small sensitivity at low flow seems to be a weak point of this type of flow sensor.

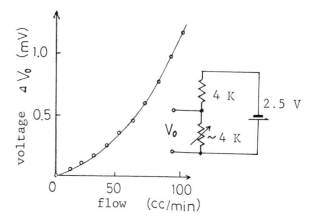

Fig.3 Flow measurement with a differential pressure sensor.

(d) Flow control characteristics

A preliminary experiments for controlling flow were made by the feedback from the differential pressure sensor to the micro valve. The source pressure versus flow characteristics of the controller at various point of flow setting is shown in Fig.4.

One suitable application of the micro valve and the flow controller could be the source gas injection into a vacuum chamber, because the valve could be fabricated directly on the Pyrex glass viewing port flange. No dead space from the valve to the chamber and very quick response would enable excellent atomic level controllability for the growth of super lattice structures for example.

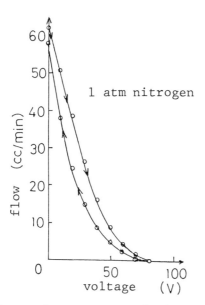

Fig.2 Flow-voltage curve of micro valve

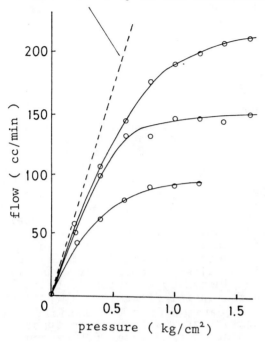

Fig.4 Source pressure-flow curves of the differential pressure type flow controller.

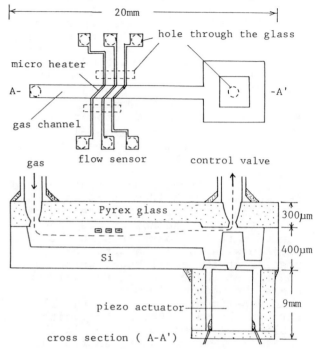

Fig.5 Integrated thermal mass flow controller

THERMAL MASS FLOW CONTROLLER

(a) Principle and structure

A micro heater, which is fabricated on a micro bridge suspended with free standing thin insulator like silicon dioxide, has extremely small thermal mass and excellent thermal isolation. This brings out low power and quick response thermal devices, and has been applied for thermal infrared sensors, thermal conductivity detectors or flow sensors.[2)3)]

Thermal mass flow sensors have been used generally for commercial mass flow controllers and their principle is as follows. The amount of heat (H) released from a heater to the fluid is expressed with the following King's equation.

$$H = (A + B \sqrt{Q})(T - T_a) \quad (2)$$

where, Q, T and Ta mean the mass flow, the heater temperature and the fluid temperature respectively, and A and B are constants. Mass flow Q is given from the measurement of H keeping the heater temperature T constant.

The structure of an integrated thermal mass flow controller is shown in Fig.5. Two controllers and a temperature sensor for ambient are on a wafer of 20 mm X 20 mm. A 400 μm thick silicon wafer, which has gas channels, flow sensors and valve diaphragms on it, was bonded anodically with a Pyrex glass of about 300 μm thickness. Gas channels and throughholes for gas inlet and outlet, for lead wire connection and for sealing were etched in the glass.

(b) Thermal mass flow sensor

Nickel micro heaters of about 4000 Å thickness were sandwiched by bottom CVD SiO$_2$ layer and top PECVD SiON layer. The heater was fabricated as a bridge on V-groove gas channel shown in Fi.6. Lead wires were connected to the nickel with a conductive epoxy. To prevent the gas leakage through the nickel pattern feed through, throughholes were etched to seal the feed through gap.

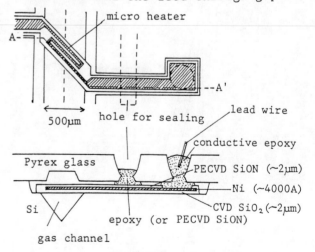

Fig.6 The structure of thermal mass flow sensor

Fig.7 shows an example of the characteristics of the flow sensor. The output voltage was roughly proportional to the square root of the mass flow and this was that predicted in equation 2.

The response of the flow sensor to a step change of source pressure is shown in Fig.8. The delay of the flow signal to the pressure waveform was less than few millisecond.

(c) Thermal mass flow controller

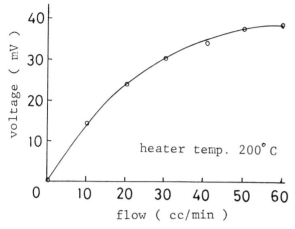

Fig.7 Characteristics of flow sensor

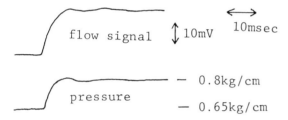

Fig.8 Transient response of flow sensor to a step change of source pressure.

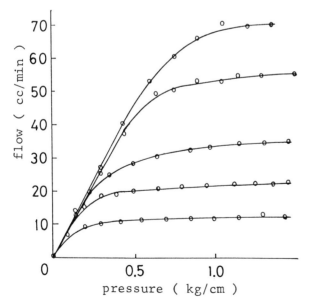

Fig.9 Source pressure-flow curves of the thermal mass flow controller.

The flow versus source pressure curve of the mass flow controller fabricated is shown in Fig.9. Fig.10 is a step response of a valve voltage and a deviation voltage to flow control signal. Though the deviation voltage shows saturation period after the step change due to a lack of valve driving capability of the circuit, this quick response proved the effectiveness of the integration of a valve and a flow sensor on a wafer.

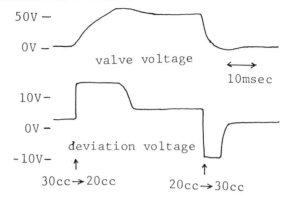

Fig.10 Transient response of valve voltage and deviation voltage to a step change of flow control signal.

CONCLUSION

Two types of integrated mass flow controller were fabricated and their preliminary experiments showed promise extremely precise control of gas flow.

Making innovations with total gas control system on a wafer would contribute for the next generation of semiconductor process, gas analysis etc..

ACKNOWLEDGEMENT

The authers wish to thank Mr. A. Nakano and Mr. E. Tobita for their assistance in the fabrication.

REFERENCES

1) S.C.Terry, J.H.Jerman & J.B.Angell, "A Gas Chromatographic Air Analyzer Fabricated on a Silicon Wafer", IEEE Trans. on Electron Devices, Vol. ED-26, 1979, pp 1880-1886.

2) K.Peterson, J.Brown & W.Renken, "High-Precision, High-Performance Mass-Flow Sensor with Integrated Linear Flow Micro-Channels", 1985 Int. Conf. on Solid-State Sensors and Actuators, 1985, pp 361-363.

3) G.B.Hocker, R.G.Johnson, R.E.Higashi & P.J.Bohrer, "A Microtransducer for Air Flow and Differential Pressure Sensing Applications", Workshop on Micromachining and Micropackaging, Ohio Nov. 7-9, 1984

4) M.Esashi, H.Komatsu & T.Matsuo, "Biomedical Pressure Sensor Using Buried Piezoresistors", Sensors and Actuators, Vol.4, 1983, pp 537-544.

NORMALLY CLOSE MICROVALVE AND MICROPUMP FABRICATED ON A SILICON WAFER

Masayoshi Esashi, Shuichi Shoji and Akira Nakano

Depertment of Electronic Engineering
Tohoku University
Aza Aoba Aramaki Sendai 980 Japan

ABSTRACT

A normally close microvalve and micropump were fabricated on a silicon wafer by micromachining techniques. Normally close microvalve has a silicon diaphragm and a small piezoelectric actuator to drive it. The controllable gas flow rate is from 0.1 ml/min to 85 ml/min at gas pressure of 0.75 kgf/cm^2. The micro pump is a diaphragm type pump which consists of two polysilicon oneway valves and a diaphragm driven by a small piezoelectric actuator. The maximum pumping flow rate and pressure are 20 μl/min and 780 mmH$_2$O/cm^2 respectively.

INTRODUCTION

Micromachining techniques have realized high performance and very small size of sensors and mechanical devices.[1] Micro flow systems can be constructed with micro flow sensors and flow control devices such as microvalve and pump. The integration of these devices enables a feedback flow control. It has advantages in high sensitivity, quick response, and negligible dead space. This enables precise control of very small flow. Another remarkable feature of this system is that it is small enough to be placed very close to the use point.

An integrated mass flow controller was fabricated on a silicon wafer for gas flow control.[2] It consists of a thermal mass flow sensor and a normally open micro valve. This type of the controller, however, needs a shutoff valve connected in series for practical use, which causes a problem of dead space at the connector. This problem could be solved by a normally close microvalve in stead of the normally open microvalve.

An integrated micro chemical analyzing system for blood gas analysis using ISFET was also developed.[3] The micro flow system of two normally open microvalves and one normally close microvalve for liquid flow control was used to calibrate the ISFET and clean it. In order to integrate conventional chemical analyzers as a flow injection analysis (FIA) system on a silicon wafer, a micro pump becomes necessary as well as microvalves.

Since flow control micro devices play important roll in the improvement of integrated micro flow systems, we developed a normally close microvalve for gas flow control and a micropump for liquid flow control. A new type of liquid flow sensor integrated with the micropump was also developed.

Glass to silicon anodic bonding has been used for micromachining. However, silicon to silicon bonding has great advantages for sophisticated micromechanical devices, because fine silicon structures can be fabricated using anisotropic etching. There are some methods to bond two silicon wafers as electrostatic bonding with intermediate sputtered Pyrex glass film and thermal bonding with sputtered low temperature glass.[4] The former has a problem of breakdown during bonding process. On the other hand, the later requires bonding temperature higher than 600 °C. A new method of bonding Si-Si anodically with low temperature glass film was developed and it was applied for the fabrication of the micro pump.

Structure, fabrication and characteristics of the normally close microvalve and the micropump are described below.

NORMALLY CLOSE MICROVALVE

1. Structure and principle

The structure of a normally close micro valve for gas flow control is illustrated in Fig. 1. The micro valve consists of a Pyrex glass and a silicon. The glass has a gas outlet throughhole. The silicon has a gas inlet throughhole, a gas channel and a movable diaphragm which has a mesa structure at the center. Two piezoelectric actuators on a glass plate are connected to the back side of the silicon mesa with a glass pipe. The gap between the silicon and the piezoactuator is about 1 μm and these components are assembled

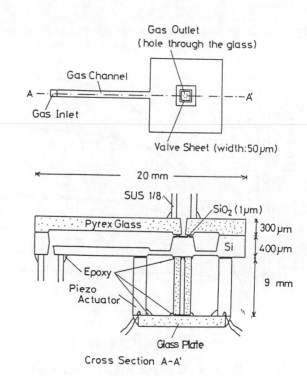

Fig. 1 Structure of the valve.
(Close mode)

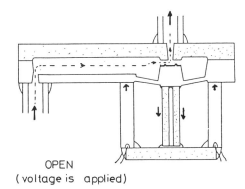

Fig. 2 Principle of the valve. (open mode)

with epoxy resin. The gas flow can be stopped with a valve sheet on the mesa as shown in Fig. 1. Applying a voltage to the actuators, the mesa is pulled down and a gap is made between the glass and the valve sheet. This open mode is schematically shown in Fig. 2.

2. Fabrication

The starting material of silicon structure was a (100) oriented 280 μm thick n-type wafer. A 300 μm thick Pyrex glass (7740) was also used. The cross sections in each fabrication steps are shown in Fig. 3.

(1) Silicon wafer (Fig. 3(a))

By wet oxidation and the following double side photolithography, silicon dioxide patterns of a gasket and a throughhole were formed on both side of the wafer. The silicon was etched to depth of 7 μm by anisotropic etching in KOH 35wt% solution. The 50 μm thick diaphragm was fabricated by etching the wafer from top side in KOH solution with Si_3N_4 mask. A gas channel was also engraved on the top side at this step. The SiO_2 on the top side except the gasket and the mesa was selectively removed by buffered HF.

(2) Glass structure (Fig. 3(b))

Thin Cr and Au film were evaporated on both side of the glass. By double side photolithography and the following HF etching, patterns for the through hole on the top side and the gas channel on the back side were engraved. Then, the throughhole were drilled using electrochemical discharge machining in KOH solution.

(3) Bonding

The silicon and the glass are anodically bonded at 390 °C, 500 V. This temperature was optimum for stress-free bonding.

(4) Actuator and tube connection

Two piezoelectric actuators were fixed to the both end of a glass plate with epoxy. A glass pipe was pasted at the center of the mesa using epoxy resin. About 1 μm thick photoresist was on the back side of the wafer except the diaphragm. After fixation of the glass plate to the glass pipe, the resist patterns were removed by immersing in acetone. A gap of about 1 μm was formed between the actuator and the glass by this process. The SUS 1/8 inch tube connectors were fixed for gas inlet and outlet with epoxy.

3. Characteristic

The gas flow characteristic of the normally close valve was measured with nitrogen gas. The relationships between the voltage applied to the actuator and the N_2 gas flow rate at gas pressure of 0.2, 0.5, and 0.75 kgf/cm^2 are shown in Fig. 4. The valve has initial leakage depending on the gas pressure. The leakage become smaller than 0.1

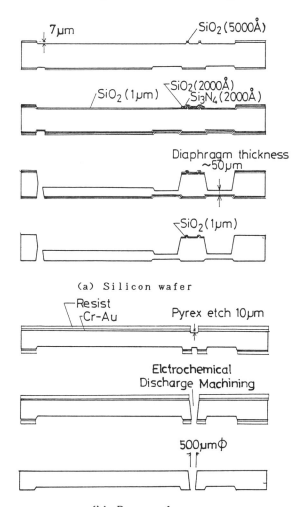

Fig. 3 Fabrication of the valve.

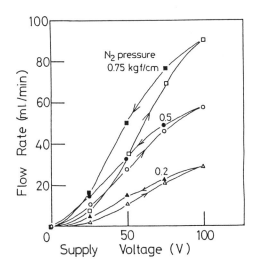

Fig. 4 Rerasionship between applied voltage and gas flow rate of the valve.

ml/min by applying a force to the mesa. The necessary force at gas pressure of 0.75 kgf/cm2 is about 0.45 kgf. The gas flow was controlled in the range of 0.1 ml/min to 85 ml/min at 0.75 kgf/cm2.

Microvalves using piezoelectric actuator have hysteresis in the flow versus supply voltage characteristic. This is originated in the characteristic of the piezoactuator.

MICROPUMP

1. Oneway valve

Two oneway valves were used in the micropump. They have poly-crystalline silicon as the material and phosphosilicate glass (PSG) as a sacrificial layer.[5] The structure of an oneway valve is illustrated in Fig.5. The valve consists of movable parts made of polysilicon and a liquid passage hole engraved in a silicon substrate. The thick circular valve lid (6 µm in thickness) is supported by four thin arms (2 µm in thickness). The arms are connected to the outside ring attached to the silicon substrate. In forward operation (pressure is applied from the back side of the substrate), the valve lid is risen by the pressure in balance with the elasticity of the support arms, and the liquid flows through the gap between the valve lid and the substrate. In the reverse operation (pressure is applied from the top side of the substrate), the valve lid is pressed against the substrate, covering the hole and the valve is closed. The valve was batch-fabricated in an IC-compatible process using polysilicon and PSG deposited by chemical vapor deposition (CVD).

An example of the relationship between the applied pressure and the liquid (D. I. water) flow rate is shown in Fig.6. The size of the valve are shown in the same figure. The characteristic is similar to that of a diode. The forward-to-reverse rectifying ratio was more than 150 at applied pressure of 1000 mmH$_2$O/cm2. The characteristics depends on the dimensions of the valve structure. The 6 µm thick valve lid realized better forward-to-revearse rectifying ratio than that having same thickness (2 µm) with the arms.

2. Structure and principle of the micropump

The micropump consists of a piezoelectric actuator, two oneway valves and a pressure chamber with a thin diaphragm moved by the piezoactuator. The structure is illustrated in Fig.7. The pump consists of two silicon wafers and a piezoactuator. A mesa suspended by a thin diaphragm and a pressure chamber are formed on one wafer. An inlet and an outlet having a oneway valve were formed on the other. The size of this pump is about 8mm x 10mm x 10mm.

By applying a voltage to the piezoactuator, the pressure of the chamber increases and the valve 2 (V2) is opened, while the valve 1 (V1) is closed. Thus the liquid in the chamber flows out through V2 to the outlet (pumping mode). When the pressure is decreased by cut off the actuator voltage, the V2 is closed while V1 is opened. New liquid flows into the chamber from the inlet (supplying mode).

The generative force of the actuator is larger than 5 kgf and is enough to deform the diaphragm. The pumping rate depends mainly on

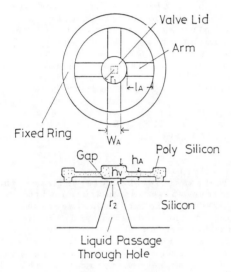

Fig. 5 Structure of the oneway valve.

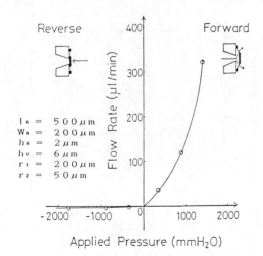

Fig. 6 Flow characteristic of the oneway valve

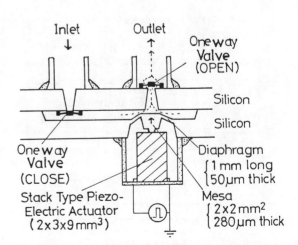

Fig. 7 Structure of the micropump

the diaphragm size and the displacement of the mesa moved by the piezoactuator. The calculated pumping volume for every one drive (voltage:90V) is about 22nl when the displacement is 5.5 μm.

3. Low temperature Si-Si bonding

The silicon to silicon bonding used is an electrostatic bonding with an intermediate sputtered low temperature glass.

The process for this bonding is as follows. (1) A low temperature glass film (Iwaki glass Co. 7570) of the thickness about 2 μm was deposited on one silicon substrate by a magnetron RF sputtering in 8 mmTorr O_2 atmosphere at 50 W RF power. (2) The silicon coated with low temperature glass was placed on another bare silicon. Two wafers were aligned by observing the patterns on both wafers under IR TV camera. (3) The assembly was heated up to a temperature of 50 °C. and negative dc voltage of 100 V was applied to the glass coated silicon wafer.

The suitable bonding condition was 2 um thick glass layer, electric field of 5×10^5 V/cm (100 V/2 μm) and bonding temperature of 50 °C.

4. Fabrication of the micro pump

The fabrication process of the micropump is briefly described below.
(1) Two one-way microvalves were fabricated on both side of a 200 μm thick double side polished silicon wafer.
(2) The pressure chamber of about 10 μm in depth and the mesa were formed on both side of the 280 μm thick silicon wafer by photolithography and following KOH anisotropic etching. Then the 50 μm thick diaphragm was formed by etching in KOH solution. The low temperature glass layer of about 2 μm thickness was deposited by sputtering on the side which pressure chamber was formed.
(3) These two silicon wafers were bonded to each other by the Si-Si bonding mentioned above.
(4) The actuator was mounted with epoxy in the glass tube, which was lapped to level with the free end of the actuator. This assembly and tube connectors for inlet and outlet were fixed with epoxy resin.

The photograph of three micropumps is fabricated on a wafer is shown in Fig. 8

Fig. 8 Photograph of the micropumps.

5. Characteristics

The characteristics of the micropump were measured by driving the piezoactuator with rectangular voltage wave. D.I. water was used as the liquid. The relationships between the pumping rate and the rectangular wave

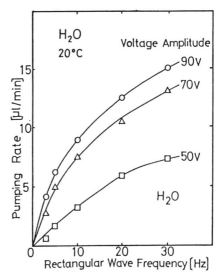

Fig. 9 Relationship between rectangular wave frequency and pumping rate without load.

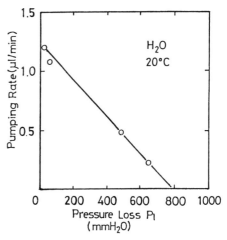

Fig. 10 Load characteristic of the pump.

frequency at voltage amplitude of 50, 70, 90 V respectively are shown in Fig. 9.

The flow rate increased in proportion with the frequency, below 10 Hz. The pumping rate for every one pulse calculated from the result was 21.6 nl/Hz. at 90 V, that shows good agreement with that of the calculated volume change of the pressure chamber (22nl/Hz). Above 10 Hz, the pumping rate was gradually saturated with increasing the frequency.

The load characteristic was measured by connecting micro flowchannel which have various flow resistance. The relationship between the pressure loss the flowchannel and the pumping rate when driven by 90V, 3Hz is shown in Fig. 10.

These results indicate that the maximum pumping rate without load is about 20 μl/min (at 90V, 20Hz) and the maximum pumping pressure is about 780 mmH_2O/cm^2 (at 90V, 3Hz).

6. Micro flow sensor

In order to monitor and control the pumping rate of the micropump, we developed a new micro liquid flow sensor which consists of a oneway valve as a orifice and a piezo resistive diaphragm pressure sensor. The structure is illustrated in Fig. 11. It can be fabricated by compatible process with the micropump. The flow rate is measure by sensing the pressure loss at the oneway valve. Since the pressure loss change ($\Delta P/\Delta Q$) of the oneway valve is larger at small flow rate than at large flow rate, this sensor has large sensitivity to small flow.

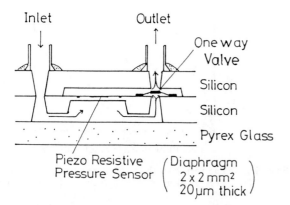

Fig. 11 Structure of the liquid flow sensor.

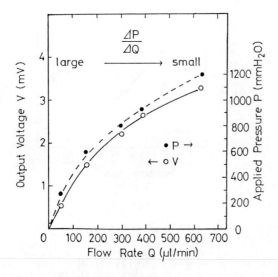

Fig. 12 Sensitivity of the flow sensor.

The relationship between the output voltage of the pressure sensor and the liquid flow rate, using D.I. water, is shown in Fig. 12. The flow rate versus the applied pressure of the oneway valve is also shown in the same figure. The sensor has sensitivity at small flow as designed.

DISCUSSION

Since the normally close microvalve has quick response, very small dead space and controllability of very small flow rate, the number of molecules through the valve by one open-close action can be limited to very small.

The microvalve can be also applied to control liquid flow as well as gas flow. With a little modification of this structure, a sample injector can be also fabricated. This will be very useful for realizing integrated chemical analyzing systems.

The combination of electrostatic bonding and low temperature glass bonding realized a low temperature silicon to silicon bonding. The bonding is attributed to the large attractive force at the silicon-glass interface, causing some deformation of the glass layer and intimate contact. The low viscosity of the low temperature glass could be useful feature for this bonding. The low bonding temperature and the small applied voltage does not damage active device during bonding process and therefore this method should be applicable to the fabrication of many integrated sensor systems.

The maximum pressure of the micropump output is limited by following mechanism. A part of the liquid in the pressure chamber flow back through the oneway valve 1 when the load resistance is compatible to the reverse resistance of the oneway valve. In order to achieve a high output pressure, it is necessary to increase the reverse flow resistance of the oneway valve. On the other hand, the saturation of the pumping rate is considered to be originated in the resistance when liquid flow into the pressure chamber. This could be improved by decreasing forward flow resistance of the oneway valve. For these reasons, the increment of the forward-to-reverse rectifying ratio of the oneway valve is important to improve the performance.

The detective flow rate of the micro flow sensor fabricated was practically limited by the sensitivity of the pressure sensor. The diaphragm thickness should be optimized to make the detective flow range comparable with the pump.

Parallel connection of two micropumps, that is dual pump, and a feedback control from the flow sensor to the actuator will make constant and rippleless flow. In chemical analyzing systems, such a stable flow dose not generate noise at the detector and then extends the detection limit.

CONCLUSIONS

The normally close microvalve fabricated is applicable to the mass flow controller for very sophisticated vacuum systems such as an MOMBE or ALE which requires precise control.

A low temperature silicon to silicon bonding technique developed is very useful for fabricating integrated sensor and actuator systems.

The micropump fabricated achieves a precise flow control of the order of 1 μl/min by changing a frequency or a voltage to drive the piezoactuator.

The microvalve and the micropump are expected for small and high performance integrated chemical analyzing system and so on.

ACKNOWLEDGMENTS

The authors would like to acknowledge Prof. T. Wakiyama and Prof. T. Ohmi of the Tohoku university for their kind encouragement.

REFERENCES

[1]. M. Esashi, et.al., "The Fabrication of a Micro Valve by Means of Micromachining", Proc. of the 6th Sensor Symposium, Tukuba, (May, 1986), pp269-272.

[2]. M. Esashi, T. Matsuo, S. Eoh and S. Choi, "The Fabrication of Integrated Mass flow Controllers", TRANSDUCERS'87, IEEE Int. Conf. on Solid state sensors and actuators, Tokyo, (June, 1987), pp830-833.

[3]. S. Shoji, M. Esashi and T. Matsuo, "Prototype Miniature Blood Gas Analyzer Fabricated on a Silicon Wafer", Sensors and Actuators, 14, (1988), pp101-107.

[4]. W. H. Ko, J. T. Suminto and G. J. Yeh, "Bonding Technique for Microsensors", Micromachining and Micropackaging of Transducers, Elsevier, Amsterdam (1985), pp41-61.

[5]. L. S. Fan, Y. C. Chong and R. S. Muller, "Integrated Movable Micromechanical Structures for Sensors and Actuators", IEEE Trans. Electron Devices, 35, (1988), pp724-730.

A Thermopneumatic Micropump Based on Micro-engineering Techniques

F. C. M. VAN DE POL, H. T. G. VAN LINTEL, M. ELWENSPOEK and J. H. J. FLUITMAN

University of Twente, Department of Electrical Engineering, P.O. Box 217, 7500 AE Enschede (The Netherlands)

Abstract

The design, working principle and realization of an electro-thermopneumatic liquid pump based on micro-engineering techniques are described. The pump, which is of the reciprocating displacement type, comprises a pump chamber, a thin silicon pump membrane and two silicon check valves to direct the flow. The dynamic pressure of an amount of gas contained in a cavity, controlled by resistive heating, actuates the pump membrane. The cavity, chambers, channels and valves are realized in silicon wafers by wet chemical etching. Experimental results are presented. Maximum yield and built-up pressure equal 34 μl/min and 0.05 atm, at a supply voltage of 6 V. Results of simulations show good agreement with the actual dynamic behaviour of the pump.

Introduction

Since 1983, research on microminiature pumps has been carried out at the University of Twente in the research group 'Transducers and Materials Science', which is part of the research unit 'Sensors & Actuators'. Starting from the principle of a peristaltic piezoelectric pump, developed in 1980 at Stanford University by Wallmark and Smits [1], several prototypes have been realized, applying micromachining of silicon and using bimorph or monomorph piezoelectric discs for the actuation [2].

A major drawback of these piezoelectrically driven pumps is the required high electrical supply voltage (about 100 V). Besides this, application of the piezoelectric discs interferes with integrated fabrication. Aiming at the realization of a pump merely applying micro-engineering techniques like thin-film technology, photolithographic techniques and silicon micromachining and considering various actuation principles, we concluded that thermopneumatic actuation was a good alternative. This actuation principle has been used by Zdelblick and Angell in a microminiature valve, using a gas/liquid system and resistive heating [3].

We realized and reported on a thermopneumatic actuator [4], comprising a cavity filled with air, a built-in heater resistor and a flexible membrane. Here, the mounting of such a thermopneumatic actuator on a pump body is discussed, and results for a complete, thermopneumatically driven, micropump are presented.

Design and Working Principle

Geometrical Design

Figure 1 shows cross sections of the pump, comprising a pump chamber and two circular, silicon check valves. The actuating part of the pump comprises a cavity filled with air, a square silicon pump membrane and a built-in aluminium meander which serves as a heater resistor. The resistor is supported by a thin silicon sheet suspended by four small silicon beams, which serve as thermal insulators. Aluminium current leads connecting the meander to the bond pads run through narrow channels, which form a restriction to gas flow. The flexible membrane can displace liquid present in the pump chamber.

Valves

Each valve consists of a circular, silicon diaphragm comprising a flexural outer ring and a

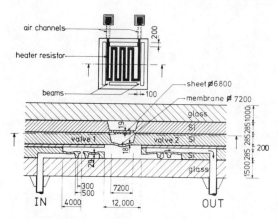

Fig. 1. Cross-sectional views of the pump (dimensions in μm).

rigid inner sealing ring. Because of the flexibility of the outer ring, a pressure acting on the valve deflects the sealing ring. A thin oxide film on the summit of the sealing ring slightly deflects the diaphragm when the valve is closed, giving the valve a small pre-tension.

Working Principle

The application of an electric voltage to the heater resistor causes a temperature rise of the air inside the cavity and a related pressure increase inducing a downward deflection of the pump membrane. As the pump membrane deflects downwards, the pressure in the pump chamber is increased. Provided this pressure exceeds the sum of the outlet pressure and the pre-tension of valve 2, this valve will open. Because of the pressure difference across the open valve 2, liquid flows through the gap underneath the sealing ring, thereby reducing the pressure in the pump chamber, while valve 1 checks the return flow to the inlet. When the pump chamber pressure equals the sum of the pre-tension of valve 2 and the outlet pressure, valve 2 closes. In the meantime, the pressure of the air inside the cavity, being higher than that outside, induces a flow of gas out of the cavity through the narrow air channels, causing this air pressure to decrease.

When the voltage is switched off, the air inside the cavity cools down, causing a pressure drop in the cavity and a related upward deflection of the pump membrane. As the pump membrane moves upwards, the pressure in the pump chamber decreases. Provided the inlet pressure exceeds the sum of the pressure in the pump chamber and the pre-tension of valve 1, this valve will open. Because of the pressure difference across the open valve 1, liquid flows through the gap underneath the sealing ring, thereby increasing the pressure in the pump chamber, while valve 2 checks the return flow from the outlet. When the pressure in the pump chamber equals the inlet pressure minus the pretension of valve 1, this valve closes. In the meantime, the pressure of the air inside the cavity, being lower than that outside, induces a flow of gas through the narrow air channels into the cavity, causing this air pressure to increase.

'Thermal' Response

'Thermal' response (warming up and cooling down of the air inside the cavity) can be described by a 'thermal' relaxation time, τ_t, which is determined by the thermal resistance of the air inside the cavity, the thermal resistance of the small suspending beams, and the total heat capacity of the cavity. Since the heat capacity of the air is negligible compared to that of the suspended thin silicon sheet supporting the resistor, the total heat capacity approximately equals the heat capacity of the sheet.

The maximum achievable temperature, T_{max}, inside the cavity is determined by the input power and the thermal resistances mentioned. For the pump described here, τ_t equals 0.1 s.

'Pneumatic' Relaxation

'Pneumatic' relaxation (increase or decrease of the air pressure in the cavity due to a flow of gas into or out of it, induced by a pressure difference over the air channels) can be described (in approximation, since the pneumatic system is non-linear) by a 'pneumatic' relaxation time, τ_p, which is determined by the volume of the cavity and the flow resistance of the narrow air channels.

The maximum achievable pressure inside the cavity, p_{max}, is determined by T_{max} and τ_p, and decreases with τ_p. However, if $\tau_p \gg \tau_t$, then T_{max} is almost independent of τ_p. For the pump described here, τ_p equals 5 s ($\gg \tau_t$).

'Hydraulic' Relaxation

'Hydraulic' relaxation (increase or decrease of the pump chamber pressure due to a flow of liquid into or out of the pump chamber, induced by pressure differences over the valves) is determined by the volume stroke of the pump membrane and the flow resistances of the valves. Since the valves show highly nonlinear behaviour, this relaxation cannot be described by a single relaxation time. Nevertheless, we can state that the hydraulic relaxation, for the pump described here, is (much) slower than the thermal response, and (much) faster than the pneumatic relaxation. The pumped volume in one pump stroke, ΔV_{ps}, is determined by p_{max}, the pump frequency, and the hydraulic relaxation. If the hydraulic relaxation is much faster than the pneumatic relaxation, ΔV_{ps} is almost independent of τ_p, and is determined by the speed of the hydraulic relaxation and the pump frequency.

The maximum liquid volume, pumped during one pump stroke, $\Delta V_{ps,max}$, equals the difference in pump chamber volumes at the end of a (hydraulically relaxed) downward stroke, and at the end of a (hydraulically relaxed) upward stroke of the pump membrane.

Simulation

Bond Graph Techniques and TUTSIM

Bond graph techniques [5] are used for physical modelling of the pump. The model comprises the thermal behaviour of the pump, the thermodynamics of the air inside the cavity, the fluid dynamics of the air and liquid channels, and the mechanics of the flexible pump membrane and valves. Simulation of the dynamic pump behaviour is carried out with TUTSIM, a simulation program developed at the University of Twente. A detailed description of the bond graph model will be reported elsewhere.

Experiments

Realization

Starting materials for the pump are three (100) 2-in silicon wafers, polished on both sides, and three Duran borosilicate glass wafers. The silicon wafers are shaped by wet chemical etching in a KOH water solution using standard photolithographic techniques for pattern definition. The glass wafers are cut out off a plate and polished (surface roughness < 0.05 μm).

The silicon wafers are attached to one another by anodic bonding, using intermediate layers of silicon oxide and sputtered borosilicate glass [6]. The silicon and glass wafers are attached by direct anodic bonding [2, 7]. The oxide on the summit of the sealing rings of the valves not only provides them with a pre-tension, but also prevents them becoming bonded to the glass: 'selective bonding' [2].

An evaporated aluminium film, patterned by wet chemical etching, is used for the heater resistor. More details on manufacturing and technology can be found in refs. 2 and 4. The dimensions of the pump are given in Fig. 1.

Priming

The pump is not self-priming. It can be filled straightforwardly by injecting water in the inlet and/or extracting air at the outlet, with the aid of a syringe. However, if this priming method is applied, air bubbles often remain inside the pump chamber or channels, obstructing the proper functioning of the valves or affecting the pump behaviour. The best priming procedure appears to be the one described in ref. 2: the pump is submerged in water in a bell jar. As the bell jar is evacuated, the air in the pump is removed. Then, as air is admitted, the pump fills with water.

Measurements

In Fig. 2, the measurement set-up is depicted. A pulsed d.c. voltage, variable in height and frequency, is applied to the bonding pads connected to the heater resistor. Pumped volume and yield are measured with the aid of a glass capillary tube and a stopwatch. Specific pressures at the inlet

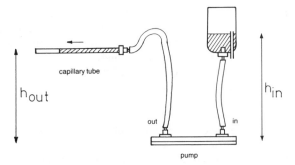

Fig. 2. Measurement set-up.

and outlet of the pump are established by adjusting h_{out} and h_{in}.

The working of the valves is characterized by measuring spontaneous flow (no voltage applied) as a function of inlet and outlet pressures.

Dust or precipitation of dissolved particles might interfere with the proper working of the valves or, in the most severe case, even cause stoppage of the channels or tubes. Therefore, the pump, the water reservoir and the capillary tube are placed in a laminar flow box. Deionized, filtered water is used. The use of degassed (in an evacuated bell jar) water helps to prevent the formation of air bubbles inside the pump.

Results and Discussion

Figure 3 shows the pump yield as a function of pump frequency, applied voltage and back pressure (outlet pressure minus inlet pressure), for a constant inlet pressure of 1 atm. Measured results appeared to be reproducible within 5%.

For a given back pressure and supply voltage, see Fig. 3(a), at first the yield increases almost linearly with pump frequency, reaches a maximum for f_m, and then decreases to zero. In the linear part, yield is determined by p_{max} (T_{max}) and pump frequency ($\Delta V_{ps} \simeq \Delta V_{ps,max}$). The frequency f_m is determined by the speed of hydraulic relaxation: at higher frequencies, there is not enough time for hydraulic relaxation ($\Delta V_{ps} < \Delta V_{ps,max}$). The frequency f_m increases with the supply voltage and decreases with the back pressure, because of the nonlinear flow resistances of the valves.

In Fig. 3(b), a number of pump curves are presented for various applied voltages and pump frequencies. The dashed curves in Fig. 3 represent the results of simulations, agreeing within some 15% with measured yields. Simulated temperature rise and pressure rise inside the cavity equal 30 °C and 0.06 atm, for a d.c. supply voltage of 6 V. Required electrical energy per pumped volume equals a few J/μl, depending on pump frequency and back pressure.

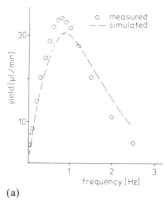

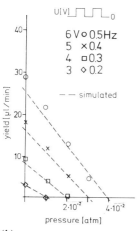

Fig. 3. (a) Pump yield as a function of pump frequency for an applied voltage of 6 V at zero back pressure (outlet minus inlet pressure). (b) Pump yield as a function of back pressure, applied voltage and pump frequency.

Figure 4 shows measured and simulated spontaneous flow as a function of inlet pressure minus outlet pressure. If outlet pressure exceeds inlet pressure minus the pre-tensions of the valves, no measurable flow occurs, indicating that the closure of the valves is very good. Calculated pre-tension of one valve equals 10^{-3} atm.

Conclusions

The design, working principle and realization of an electro-thermopneumatic micropump are described. The pump comprises a pump chamber, a thin silicon pump membrane and two silicon check valves. The pump membrane is actuated by the dynamic pressure of an amount of gas contained in a cavity, controlled by resistive heating. Results of measurements and simulations of the dynamic behaviour of the pump are presented and discussed.

The yield can be regulated by varying the pulsed d.c. supply voltage in height and/or frequency. The pump behaves as expected: simulated and measured yields, as functions of back pressure, supply voltage and pump frequency, agree within some 15%. Maximum yield and built-up pressure equal 34 µl/min and 0.05 atm, for a supply voltage of 6 V.

Important parameters are the thermal, pneumatic and hydraulic relaxation times, determining the dynamic behaviour of the pump. Temperature rise and related pressure increase inside the cavity are a few tens of °C and some hundredths of an atmosphere. The required electrical energy per pumped volume equals a few J/µl, depending on pump frequency, back pressure and supply voltage. Supply voltage and power consumption are less than 10 V and 2 W respectively.

Dust or air bubbles readily interfere with the proper functioning of the valves and strongly affect the pump behaviour. The closure of the valves is very good. The construction of the pump is simple and it can be fabricated by merely applying micro-engineering techniques like thin-film technology, photolithographic techniques and silicon micromachining.

Acknowledgements

The authors would like to thank Gert-Jan Burger, Rob Legtenberg, Dick Ekkelkamp, John Baxter and Peter Breedveld for their help and contributions.

References

1 J. G. Smits, Piezoelectric micropump with three valves working peristaltically, *Sensors and Actuators, A21–A23* (1990) 203–206.
2 H. T. G. van Lintel, F. C. M. van de Pol and S. Bouwstra, A piezoelectric micropump based on micromachining of silicon, *Sensors and Actuators, 15* (1988) 153–167.

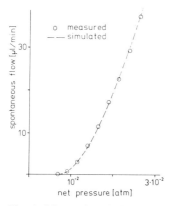

Fig. 4. Measured and simulated spontaneous flow vs. net pressure (inlet minus outlet pressure).

3 M. J. Zdeblick and J. B. Angell, A microminiature electric-to-fluidic valve, *Proc. 4th Int. Conf. Solid-State Sensors and Actuators (Transducers '87), Tokyo, Japan, June 2–5, 1987*, pp. 827–829.
4 F. C. M. van de Pol, D. G. J. Wonnink, M. Elwenspoek and J. H. J. Fluitman, A thermo-pneumatic actuation principle for a microminiature pump and other micromechanical devices, *Sensors and Actuators, 17* (1989) 139–143.
5 A. M. Bos and P. C. Breedveld, 1985 update of the bond graph bibliography, *J. Franklin Inst., 319* (1985) 269–286.
6 A. Hanneborg and P. A. Øhlckers, Anodic bonding of silicon chips using sputter-deposited Pyrex 7740 thin films, *Proc. 12th Nordic Semiconductor Meeting, Jevnaker, Norway, June 8–11, 1986,* pp. 290–293.
7 G. Wallis and D. I. Pomerantz, Field assisted glass-metal sealing. *J. Appl. Phys., 40* (1969) 3946–3949.

VARIABLE-FLOW MICRO-VALVE STRUCTURE FABRICATED WITH SILICON FUSION BONDING

Farzad Pourahmadi, Lee Christel, Kurt Petersen,
Joseph Mallon, Janusz Bryzek

NovaSensor
Fremont, CA

ABSTRACT

A new concept in silicon micro-valves has been demonstrated for a specific commercial application, current-to-pressure (I/P) converters. Valves for this application demand high gas flow rates and large deflections of the active valve element (+/- 50 microns) which are not readily attainable with silicon microstructures. However, they do not require a tightly closed condition. This micro-actuator has been fabricated using silicon fusion bonding techniques together with conventional silicon anisotropic etching. The design of the device was optimized and its operation was analysed with the aid of finite element modelling.

INTRODUCTION

Numerous types of micro-valves have been demonstrated using silicon micromechanical fabrication technology. Diaphragm-based valves can be effective for applications which require low leakage rates [1]. The maximum amount of deflection attainable with a silicon diaphragm is small, however, typically less than 20 microns. Such limited deflection magnitudes restrict the use of these valves (and other actuators) to low flow (and small motion) applications. Larger deflections can only be realized by making the diaphragms thinner, thereby decreasing the ruggedness and the operating pressures of the micro-valves.

Another difficulty with micro-valves is associated with the actuation mechanism. The most common actuation mechanisms employed in the literature are thermal [1], electrostatic [2], and piezoelectric [3]. All these mechanisms apply very large forces for small deflections, but are not effective for diaphragm or actuator deflections above about 20 microns. In other implementations, such as the fluid diode [4], the pressure of the supply fluid itself is used to actuate the valve.

The silicon micro-valve described here is designed for very large deflections (over 50 microns) and large gas flow rates. The electromagnetic actuation mechanism is also designed for such large deflections.

Micro-valves have many possible applications. The exact application of such a device is the most important consideration in the design of a practical silicon micro-valve. The device described here is specifically designed for potential use in a current-to-pressure (I/P) converter. An I/P converter accepts a dc current signal (4-20 mA) as input and produces a pneumatic pressure as output. The pneumatic pressure is used to activate and regulate a "macro"-mechanical actuator such as a process flow valve. A typical pneumatic circuit consists of a 20 psi regulated source, flowing through an orifice into an expansion chamber which mechanically operates a macro-actuator. The pressure inside the expansion chamber is controlled by adjusting the flow rate in a leakage path between the orifice and the chamber. Flow rate is regulated by a small valve, typically electro-magnetically or piezo-electrically actuated, at flow rates of about 20 cc/sec. When the valve is fully open, the pressure in the chamber is minimized (3 psi); when the valve is closed, the pressure in the chamber is maximized (15 psi). In a well-designed instrument, the output pressure over this range is linear with input current. In this application, it is unnecessary (and undesirable) to maintain a leak-tight closed condition. These small, delicate industrial instruments are pains-takingly fabricated, adjusted, and calibrated at a rate of about 200,000 per year world-wide.

FABRICATION

The fabrication sequence for the micro-valve is illustrated in Figure 1. Two silicon wafers are used in the fabrication of each micro-valve wafer. Wafer No. 1 is machined to become the valve exit port (backside) and recessed valve seat (frontside). Wafer No. 2 is bonded to wafer No. 1 and ultimately becomes the valve flexure (flapper) which closes the valve when forced against the seat by the actuator force.

The fabrication begins when wafer No. 1 is double-side polished to a thickness of about 400 um. A dielectric resistant to silicon etching is deposited on the wafer, and photolithography is used to define patterns on both sides of the

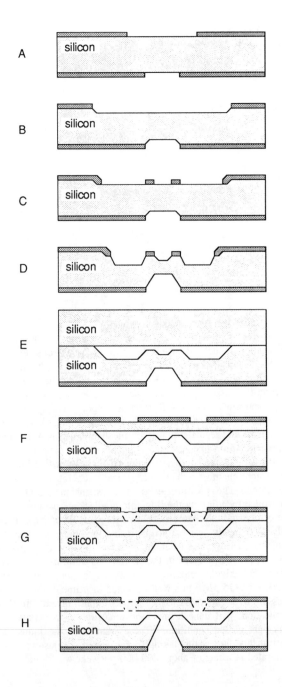

wafer. On the back, a simple square pattern is defined. This will ultimately be etched through the wafer to become the valve backside port. The area defined on the frontside comprises both the valve seat and an area surrounding the seat into which the fluid will flow from the front of the device when the valve is open.

After the first lithography step, the front of the wafer is etched in KOH or another suitable anisotropic etchant to a depth of about 50 um. This step recesses the surface that will eventually be the top of the valve seat. An additional dielectric layer is then deposited on the front of wafer No.1 and a second photolithography step is used to define the valve seat. The wafer is then etched in KOH again. During this etch step, the top of the valve seat is protected, but the area surrounding it is further etched to a depth about 75 um below the top of the valve seat. During these first two etch processes, the backside of the wafer is simultaneously etched, partially forming the valve port.

After the front of wafer No. 1 is stripped of dielectrics, wafer No. 2 is bonded to the frontside of wafer No. 1 using Silicon Fusion Bonding (SFB) [5]. The wafer sandwich is then thinned using grinding, polishing and etching until wafer No. 2 is thinned to the desired flexure thickness of about 60 um. Additional dielectric is deposited on the frontside and patterned to define the valve flexures. Additional anisotropic etching is used to etch through the thickness of wafer No. 2, forming the beams and opening the top surface to the area surrounding the valve seat. Finally, the backside of the wafer is etched until the port meets the center of the valve seat, thus completing the micro-valve fabrication. A scanning electron microscope photo of an early version of the microvalve is shown in Figure 2.

Figure 1 Fabrication sequence for the fusion bonded micro-valve. A) Deposit dilectric and pattern both sides; B) first silicon etch - 50 microns; C) redeposit dielectric on front and pattern; D) second silicon etch - 125 microns; E) strip frontside and bond second wafer to first; F) thin the top wafer, deposit dielectric, and pattern for flapper etch; G) etch through top silicon to form valve flapper; H) etch from backside to open valve port.

Figure 2 Scanning microscope photo of an early version of the micro-valve chip. The chip dimensions are 4 x 4 mm.

DESIGN AND MODELLING

The flapper part of the micro-valve is of particular design importance because its function is to travel up and down, thereby adjusting the gap, and increasing the pressure drop across the valve. The motion of the flapper, its compliance with respect to the fluid pressure and actuator force, and its maximum stress levels are of great importance to a successful design.

Extensive Finite Element Analysis was performed to optimize the geometry and the strength of the device for the specified operating conditions and expected performance. ANSYS was used for the modeling of the micro-valve. The Finite Element Model is comprised of 3288 stif45 elements with 5795 nodes.

The structure of the micro-valve consists of a central region which is supported by four folded symmetrical beams each connected to a shoulder. The lateral motion of the flapper under the force exerted by the actuator induces considerable stress in the beams, especially in the regions where the beams are connected to the central flapper and also at the shoulder connection. The stress buildup is primarily due to bending moments generated by the flapper displacement. In addition, however, a torsional torque is superimposed on the beams, due to the particular design of the flexures. The combination of these two loads results in the buildup of highly tensile stress at the corner regions [6]. The existence of sharp corners at the intersection of <111> and <100> planes (due to anisotropic etching) causes stress concentration at those locations and magnifies the level of stress which can cause breakage during loading. Therefore, a stress relieved design is deemed necessary to reach the desired performance.

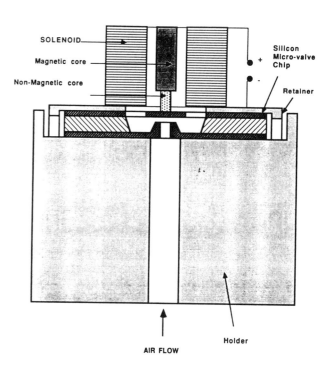

Figure 4   Schematic diagram of completed assembly for testing the performance of the micro-valve.

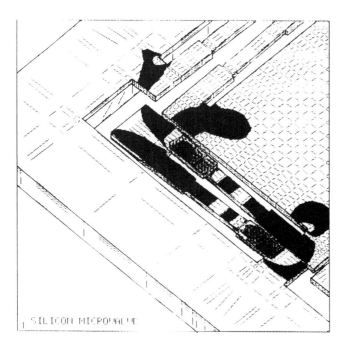

Figure 3   Stress distribution at the end of a folded cantilever beam determined by Finite Element Modelling. This beam shape was optimized for maximum vertical deflection, +/- 75 microns.

The approach here for relieving the stress at the beams' corner points has been to add multiple steps at the terminations of the beams, thus generating a gradual transition region preventing the stress from going through an abrupt change [7]. This design modification has proven successful in stress relief, resulting in an improved performance of the flexure for a given displacement. The dimensions for the step and beams have been optimized using FEM modeling.

Figure 3 shows the results of the Finite Element Model in which for 1 psi pressure applied to the top surface of the flapper element, 2928 psi maximum stress is generated. The corresponding central deflection of the flapper is calculated to be 3.4 micron/psi. This close-up shows the distribution of stress at the end of a folded beam design (different from that shown in Figure 2).

OPERATION

The operation of the micro-valve is based on flow control through pressure drop in the valve flow channel by adjusting the flapper/valve-seat gap.

The complete micro-valve assembly is shown in Figure 4. The operation and performance of this assembly was characterized over its expected operating range. High pressure air is regulated at the inlet and the flow rate is measured with a precision rotameter. The pressure drop in the micro-valve is adjusted externally by the displacement of the core of a solenoid which is proportional to the electrical power input. A pressure sensor mounted on the back of the micro-valve measures the pressure buildup which opens the valve as it acts against the actuator's exerted force. The micro-valve outlet opens up to atmospheric pressure.

Figure 5 shows typical results for flow rate versus input power to the solenoid in the low pressure region of the operating range.

CONCLUSION

The combination of a well-defined application, silicon fusion bonding fabrication technology, and detailed finite element modelling has provided a framework for implementing a new concept in silicon micro-valves. In addition, it has been shown that silicon fusion bonding is an important process, not only for sensor fabrication [5], but also for the fabrication of silicon micro-actuators. Optimization of design features with FEM increased deflection capability of the valve flapper by a least a factor of 5, over +/- 75 microns.

ACKNOWLEDGEMENTS

The authors would like to acknowledge the cooperation of the technical staff at NovaSensor for their assistance in the success of this project.

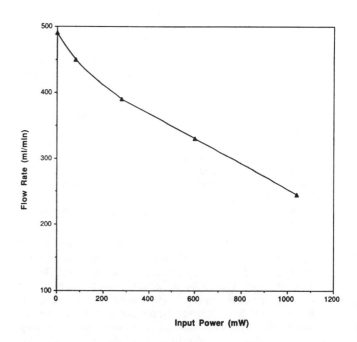

Figure 5   Typical results for flow rate versus input power to the solenoid in the low pressure region of the operating range.

REFERENCES

1] M. Zdeblick and J. Angell, "A Microminiature Electric-to-Fluidic Valve," Proceedings of The Fourth International Conference on Solid-State Sensors and Actuators, Tokyo, 1987, pg. 827.

2] T. Ohnstein, T. Fukiura, J. Ridley and U. Bonne, "Micromachined Silicon Microvalve", in the Proceedings of the IEEE Micro Electro Mechanical Systems Workshop, Napa Valley, CA, 1990, pg. 95.

3] M. Esashi, S. Eoh, T. Matsuo and S. Choi, "The Fabrication of Integrated Mass Flow Controllers," Proceedings of The Fourth International Conference on Solid-State Sensors and Actuators, Tokyo, 1987, pg. 830.

4] S. Park, W. Ko and J. Prahl, "A Constant Flow-Rate Microvalve Actuator Based on Silicon and Micromachining Technology", Proceedings of the IEEE Solid-State Sensor and Actuator Workshop, Hilton Head, SC, 1988, pg. 136.

5] K. Petersen, P. Barth, J. Poydock, J. Mallon and J. Bryzek, "Silicon Fusion Bonding for Pressure Sensors", Solid State Sensor and actuator Workshop, Hilton Head Island, S.C., pp. 144-147, (1988)

6] "Theory of Elasticity", S.P.Timoshenko, J.N. Goodier, 3rd edition, Mc-Graw Hill (1970).

7] Mark's Handbook of Mechanical Engineering, American Society of Mechanical Engineers (ASME), Mc-Graw Hill (1980).

A Pressure-Balanced Electrostatically-Actuated Microvalve

Michael A. Huff, Michael S. Mettner*, Theresa A. Lober, and Martin A. Schmidt
Microsystems Technology Laboratories
Massachusetts Institute of Technology
Cambridge, MA 02139

*Robert Bosch Company
Stuttgart, Federal Republic of Germany

Abstract

We report a new microvalve structure which is designed to enhance the limited actuation forces available in microfabricated devices by using a pressure-balancing scheme. The concept is to allow the fluid to provide a balancing force on the moving part of the device thereby reducing the force required to open the valve. Although various methods may be used to actuate the valve, we have chosen electrostatic actuation since it is readily integrated with the valve fabrication sequence. This paper will discuss the design and fabrication of this microvalve. Flow testing of the valve has not been completed and will not be presented in this paper. The process for implementing the valve concept uses multiple wafer bonding steps (three in the present prototype), and has yielded valves which have been successfully actuated in air using voltages below 350 Volts. While developing a process for the microvalve we investigated two important facts related to bonded wafers with sealed cavities. First, by bonding two silicon wafers together, one of which has a cavity etched into it and electrochemically etching back one of the wafers to a very thin layer, we have been able to measure the residual gas pressure in the cavity. Using the theory of large deflections of circular plates, we have determined that the residual pressure is approximately 0.8 atms. Second, we have found that high temperature processing of bonded silicon wafers which have a sealed cavity between the bonded layers results in a permanent set in the top layer of lightly doped silicon due to the expansion of the trapped gases within the cavity. We have determined the onset of plastic deformation to be within the temperature range of 800-850 C.

Introduction

One of the first reported microfabricated valves was designed as an injection valve for use in a integrated gas chromatograph in 1978 [1]. This valve consisted of a silicon micromachined valve seat and a nickel diaphragm which was actuated by an external solenoid. Much of the more recent work is similar to this first design except the nickel diaphragm has been replaced by a silicon micromachined diaphragm[2]. In addition to these externally actuated valve designs, there has been a number of one-way check valves[3,4], a thermally actuated flow-channel type valve capable of controlling large fluid pressures[5], and an integrated electrostatically actuated valve fabricated by surface micromachining [6].

The use of microfabrication is attractive for valves because of small dead volumes, potentially high speed, and low-cost due to batch fabrication. However, the creation of an integrated microvalve is made difficult by the inability to generate enough mechanical force to actuate the valve using conventional actuation schemes. The basic concept of the pressure-balanced valve, illustrated in Figure 1, is to have the pressure of the fluid provide a balancing force on the moving part of the valve. The fluid pressure produces an upward directed force which tends to close the valve and simultaneously a downward directed force tending to open the valve. Consequently, the total force required to actuate the valve can be designed to be only a small fraction of the total pressure force of the fluid by properly sizing the area of the top and bottom surfaces of the valve. Any number of methods may be used to actuate the valve including thermal, pneumatic, magnetic, and electrostatic. We have chosen to use electrostatic actuation in the prototype design due to the ease with which this type of actuation can be integrated into the fabrication sequence. The valve is actuated by an electrostatic force created by an applied voltage between the lower and upper conducting layers which are separated by an insulating layer. Even though it is typically only possible to electrostatically generate several psi of pressure force, the pressure-balancing design makes it possible to control fluids at much higher pressures. The limit of the

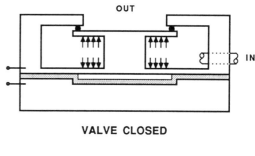

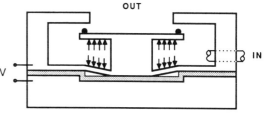

Figure 1. Valve Concept

balancing will be determined by the ability to match and control the important surface areas of the valve.

The valve is designed to be used in a "bi-stable" mode, in which the structure is either fully open or closed. In this case, what is of most interest in the operation of the valve is the voltage required to cause electrostatic "pull-in".

Design

Fluidic Behavior

Figure 2 shows the typical dimensions for a completed prototype microvalve. We can estimate the fluid flow rate through the valve by modelling the structure as an orifice. In laminar flow, the flow rate through an orifice is given by [7]:

$$(1) \quad Q = \frac{2\delta^2 D_h A_o}{\mu} (P_{IN} - P_{OUT}),$$

where δ is the discharge coefficient, D_h is the hydraulic diameter, A_o is the area of the orifice opening, μ is the viscosity of the fluid, and P_{IN} and P_{OUT} are the pressures on either side of the orifice. For a sharp-edged slit orifice $\delta = 0.157$. Substituting into equation (1) values for, the hydraulic diameter $D = 2w$ (w is the deflection of the structure into the cavity), the orifice area $A_o = 2\pi c w$, $\Delta P = 70$ bar, and assuming the fluid is silicon oil where $\mu = 100$ mPa sec, the flow rate is estimated to be 78 ml/min.

To ensure that the flow is laminar through the orifice, the Reynold's number must be calculated and compared to the transition Reynold's number R_t. If $Re < R_t$ then the flow is laminar. The

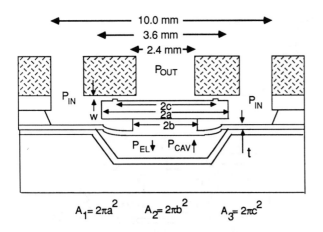

Figure 2. Valve Dimensions

Reynold's number for an orifice is defined by:

$$(2) \quad Re = \frac{\rho(Q/A_o)D_h}{\mu},$$

where ρ is the density of the fluid, $\rho = 0.8$ g/cm^3 for silicon oil. The transition Reynold's number is given by:

$$(3) \quad Re_t = \left(\frac{0.611}{\delta}\right)^2,$$

and is found to be approximately 15 for a sharp-edged slit-type orifice. Substituting values for the variables into equation (2) above it is found that $Re \ll Re_t$ and therefore the flow is laminar.

Mechanical Behavior

The mechanical behavior of the valve is examined by modelling the valve structure as shown in Figure 2. We assume that the circular plate is symmetrically loaded with respect to the axis perpendicular to the plate, the deflections are small in comparison with the thickness of the plate, and there are no residual stresses in the structure. The solution for the maximum deflection of this plate under the loading of a concentrated force, Q_T, acting on the thick center region or "plunger" and a uniform pressure force, q_T, acting on the thin annular region is given as follows[8]:

$$(4) \quad w = K_1 \frac{Q_T a^2}{Et^3} + K_2 \frac{q_T a^4}{Et^3},$$

where w is the maximum deflection of the plate, a is the outer radius of the circular plate, E is Young's modulus of silicon, t is the thickness of the plate in the thinner region, and K_1 and K_2 are constants given in tabular form for various ratios of the outer and inner diameters. This equation assumes poisson's ratio is equal to 0.3.

Referring to Figure 2, the total concentrated force, Q_T, acting on the center plunger can be written as:

$$(5) \quad Q_T = P_{EL} A_2 - P_{CAV} A_2 - P_{IN}(A_3 - A_2) + P_{OUT} A_3,$$

where P_{IN} and P_{OUT} are the pressures of the fluid inside the valve cavity and outside the valve, respectively, P_{EL} is the electrostatic force generated by application of a voltage on the structure, and P_{CAV} is the pressure inside the cavity which is dependent on the deflection of the plate into the cavity. A_1, A_2 and A_3 are the areas of the plunger cap, the plunger base, and the sealing ring, respectively, as shown in the Figure. The total uniform load on the plate can be written as:

$$(6) \quad q_T = P_{EL} + P_{IN} - P_{CAV}.$$

The electrostatic force can be written as:

$$(7) \quad P_{EL} = \frac{1}{2} \varepsilon_o \left(\frac{V}{h_{AIR} + \varepsilon_r^{-1} h_{OX}}\right)^2,$$

where ε_o is the permittivity of air, ε_r is the relative permittivity of silicon dioxide, V is the voltage applied, h_{OX} is the thickness of the oxide layer and h_{AIR} is the gap spacing between the cavity bottom and the deflecting plate. For ease of calculation, we assume that the electric pressure acting on the thin annular region is equal to the pressure acting on the plunger. This is an approximation, since in reality P depends on the electrode spacing and thus will vary over the thin annular region. The air gap spacing is related to the deflection of the plate as:

$$(8) \quad h_{AIR} = h_o - w,$$

where h_o is the initial cavity depth. The pressure inside the cavity is dependent on the deflection of the plunger into the cavity and can be written as:

$$(9) \quad P_{CAV} = \frac{P_1 V_1}{V_1 - \Delta V},$$

where P_1 is the initial pressure inside the cavity, equal to 0.8 atm, and V_1 is the initial volume of the sealed cavity. We neglect the static deflection of the structure due to the sealing ring. The mode shape assumed for the structure as it deflects into the cavity is a frustum of a right circular cone. Therefore ΔV is given by:

$$(10) \quad \Delta V = \frac{1}{3} \pi w (a^2 + b^2 + ab).$$

The variable a is the outer radius and b is the radius of the plunger. Combining the above equations and substituting in the values for the dimensions shown in Figure 2, a general equation relating the applied voltage to the deflection of the plate into the cavity is derived:

$$(11) \quad w = 8.91 \times 10^{-6} + 4.25 \times 10^{-22} \left(\frac{V}{5.41 \times 10^{-6} - w}\right)^2 - \left(\frac{5.45 \times 10^{-11}}{7.11 \times 10^{-6} - w}\right),$$

where it is assumed $t = 20$ μm, $h_o = 5$ μm, $h_{OX} = 1.6$ μm, and $P_{IN} = P_{OUT}$ = atmospheric pressure. Using equation 11, it is found that the structure has a "pull-in" voltage of 328.4 Volts.

Fabrication

The process for implementing the valve concept uses three wafer bonding steps and does not yet have the sealing ring as shown in Figure 2. The fabrication sequence (Figure 3) of the valve begins with a n-type <100> 0.5-2.0 ohm-cm 4-inch silicon wafer. The wafer is placed in a phosphorus diffusion furnace at 925 C for 1.5 hours in order to highly dope the surface. This step is done to ensure that good electrical contact can be made in order to actuate the valve. After a one hour drive-in at 950 C, the wafer is stripped of the phosphorous-doped glass and a 1000 Å thermal oxide is grown. After the masking oxide is patterned the wafer is placed in 20% KOH at 56 C for approximately 22 minutes thereby etching circular recessed electrodes 5 μm deep and 3.6 mm in diameter, Figure 2a. The masking oxide is then stripped and a thermal oxide, 1.6 μm-thick, is grown on the wafer. This thick layer of silicon dioxide acts as the dielectric isolation during electrostatic actuation and will also serve to protect the handle wafer from the subsequent silicon etching steps.

A second 4-inch wafer, <100> p-type 10-20 ohm-cm, is thermally bonded to the front side of the first wafer, Figure 2b. Prior to bonding, the two wafers are cleaned using a standard pre-oxidation clean and then hydrated by immersion into a 3:1, sulfuric acid:hydrogen peroxide solution. After a spin rinse and dry, the polished surfaces of the wafers are physically placed into intimate contact. Using an infrared inspection system, the wafers are examined for voids. Assuming the bond is void-free, the composite two-wafer structure is placed into a dry oxidation furnace for one hour at 1000 C to complete the bond. Once removed from the

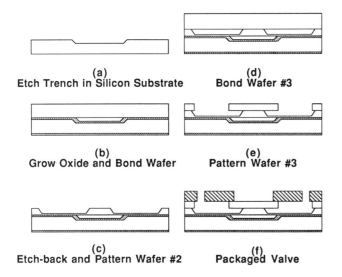

Figure 3. Valve Fabrication Sequence

We have successfully fabricated the valve without the sealing ring shown in Figure 2. Figure 5 shows a top-view SEM of a valve that has been sawed in half and Figure 6 is a SEM of the valve structure displaying the quality of the bonds between the various layers.

The complete microvalve prototype, without sealing ring, was successfully actuated in air with voltages below 350 Volts. We are currently in the process of flow testing.

furnace, the bonded wafers are inspected again using the infrared inspection system.

The bonded wafers are then placed in a 20% KOH solution at 56 C for approximately 23.5 hours and the second wafer is etched back to a resulting thickness of 75 μm. The surface of the etched back wafer is then mechanically polished to a mirror-smooth finish. The resulting thickness of the second silicon wafer is 50 μm. A masking oxide, (LTO), 5000 Å thick is deposited onto the just polished surface and subsequently patterned. The wafers is then etched using a 20 % KOH solution at 56 C for approximately 1.5 hours forming the base of the valve, Figure 2c. After the masking oxide is stripped a third 4-inch <100> silicon wafer, p-type 10-20 ohm-cm, is thermally bonded to the polished surface of the second wafer, Figure 2d. The bonding is done in exactly the same way as described above. After bonding, the composite structure is inspected using the infrared inspection system. Figure 4 is a thermal print-out of the infrared inspection system clearly indicating the plunger base bonded to the third wafer.

The third wafer is etched back in a 20% KOH solution at 56 C to a resultant thickness of 75 μm. A 5000 Å-thick oxide is deposited onto the third wafer surface using either LTO or an ACVD oxide. Subsequently, the wafer is patterned and etched in a 20% KOH solution at 56 C forming the top layer and releasing the valve, Figure 2e. The complete wafer is sawed into individual valves, which are then packaged using a capping glass plate which contains inlet and outlet ports, Figure 2f.

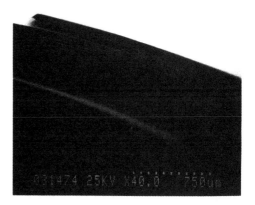

Figure 5. SEM of Valve Structure

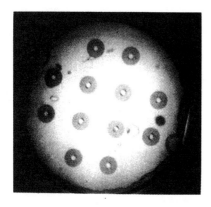

Figure 4. Infrared Image of Valve Wafer

Figure 6. Close-up SEM of Valve

Residual Pressure in Sealed Cavities

The bonding of a silicon wafer with cavities to another silicon wafer results in trapped gases within the sealed cavity. It was of interest to determine the residual pressure of this trapped gas in order to model the behavior of the valve. Starting with p-type <100> silicon wafers, we diffused phosphorous into the wafers to act as an electrochemical etch stop. These wafers where then bonded to other silicon wafers which had circular cavities of various depths ranging from 30 to 50 μm. After selectively removing the oxide from the backside of the phosphorous doped wafers, the wafers were then etched in KOH in order to thin the wafers to a resultant thickness of 50 μm. The wafers then were electrochemically etched in KOH, resulting in a smooth, 8.0 ± 0.2 μm-thick capping layer of silicon over the sealed cavity. The thickness of this layer was verified using an SEM. After etching, it is immediately noticed that the capping layer is deflected into the cavity, indicating that the residual pressure is below atmospheric. The deflections of the capping layer of silicon were measured with a calibrated microscope. Using the theory of large deflection of circular plates we can write an expression relating the maximum deflection of the circular plate to the pressure differential across the plate as [8]:

$$(12) \quad w = .662 \, a^3 \left(\frac{q \, a}{E \, t} \right)^{1/2},$$

where E is Young's modulus of silicon, a is the radius of the circular cavity, t is the thickness of the capping layer. In this equation it is assumed that the residual stress in the silicon layer is negligible.

The differential pressure across the plate is a function of the deflection of the capping layer into the cavity and is given by:

$$(13) \quad q = P_{ATM} - \frac{P_1 V_1}{V_1 - \Delta V},$$

where P_{ATM} is atmospheric pressure, P_1 and V_1 are the initial pressure and initial volume of the cavity, respectively, and ΔV is the change in cavity volume. The deflection of the capping layer is assumed to have a spherical mode shape. Calculated values of the residual pressure inside the cavities measured are given in Table I. where it is seen the average residual pressure is 0.79 atm.

Table I. Residual Pressure Inside Sealed Cavities

Wafer #	Average Depth of Cavity (μm)	Average Deflection of Capping Layer (μm)	Residual Pressure Inside Cavity (Atm)
1	53	20.1	.778
2	53	19.5	.787
3	32	12.8	.795
4	31	13.1	.782

Plastic Deformation of Lightly Doped Silicon

High temperature thermal processing of a bonded wafer with a sealed cavity results in the expansion of the trapped gas causing the capping layer to be loaded beyond the yield point. After removing the wafer from the furnace and allowing it to cool, it is observed that a residual strain is present in the capping layer, as is clearly seen in the SEM in Figure 7. The occurrence of plastic deformation of lightly doped silicon and heavily doped silicon has been reported elsewhere[9,10]. It was of interest to determine the onset temperature of plastic deformation in lightly doped silicon. Therefore, we prepared bonded wafer samples with cavities and placed the wafers in a furnace at various temperatures. Starting at 600 C the wafers were annealed in nitrogen for one hour and then

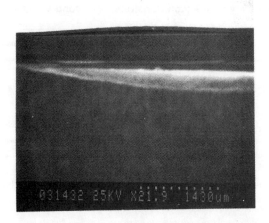

Figure 7. Plastically Deformed Silicon Over Cavity

removed. The change in deflection of the capping layer of the cavity was then observed using a calibrated microscope. The wafers were placed back in the furnace, this time with a temperature of 650 C and removed after one hour and re-measured. This sequence continued, incrementally increasing the temperature each time by 50 C until 1000 C was reached. We found the onset of plastic deformation to be between 800-850 C. It has been reported that the onset of plastic deformation is highly dependent on the amount of oxygen precipitates within the silicon and therefore the onset temperature we observed may vary considerably depending on the wafers and process[9].

As a result of this experiment, it was concluded that high temperature processing of the device wafers must be avoided whenever a sealed cavity exists inside the bonded wafers. Consequently, the masking oxide used to pattern bonded silicon layer 2 and layer 3 was a low-temperature deposited oxide. It is also noted that the bonding of layer 3 to the composite structure of layer 1 and layer 2 requires a high temperature anneal. However, in this case the pressure balancing of the valve helps to prevent the valve from being permanently deformed.

Conclusion

We report on a new type of valve microstructure using the concept of pressure-balancing which allows the control of fluids at very high pressures. The process to implement the concept of the valve uses a series of wafer bonding steps to form the structure. It has been observed that gas is trapped within a sealed cavity between two bonded wafers and it was experimentally determined that this trapped gas had a residual pressure of 0.8 atm. This would indicate that the oxygen trapped within the cavity had reacted with the exposed silicon sidewalls and that the amount of inert gases left within the cavity are in direct proportion to their content in air. By exposing a bonded wafer having a sealed cavity and a relatively thin capping layer, we observed plastic deformation of the capping layer. We determined that the onset temperature of this plastic behavior to be between 800 and 850 C. We have successfully actuated prototype microvalves using voltages below 350 Volts. We are in the process of flow testing.

Acknowledgements

We wish to thank the Robert Bosch Company, the Keithley Career Development Chair, and the Department of Justice for their support of this work.

In addition, we wish to thank Vincent M. McNeil and Suchee Wang for their assistance with the electrochemical etching.

This work was performed in the Microsystems Technology

Laboratories at MIT and we gratefully acknowledge the assistance of the MTL staff in this work.

References

[1] S.C. Terry, J.H. Jerman and J.B. Angell, "A Gas Chromatogragh Air Analyzer Fabricated on a Silicon Wafer," IEEE Transactions on Electron Devices, vol. ED-26, No. 12, p.1880, December, 1979.

[2] S. Nakagawa, S. Shoji and M. Esashi, "A Micro Chemical Analyzing System Integrated on a Silicon Wafer," Proceedings, IEEE Micro Electro Mechanical Systems, Napa Valley, CA, p. 89, February 11-14, 1990.

[3] H.T.G. Van Lintel, F.C.M. Van De Pol and S. Bouwstra, "A Piezoelectric Micropump Based On Micromachining of Silicon," Sensors and Actuators, Vol. 15, p. 153, 1988.

[4] S. Park, W.H. Ko, and J.M. Prahl, "A Constant Flow-Rate Microvalve Actuator Based On Silicon And Micromachining Technology," Technical Digest, IEEE Solid-State Sensor and Actuator Workshop, Hilton Head, S.C., p. 136, June 6-9, 1988.

[5] M.J. Zdeblick, "A Planar Process For An Electric-to-Fluidic Valve," PhD thesis, Stanford University, June, 1988.

[6] T. Ohnstein, T. Fukiura, J Ridley and U. Boone, "Micromachined Silicon Microvalve," Proceedings, IEEE Micro Electro Mechanical Systems, Napa Valley, CA, p.95, February 11-14, 1990.

[7] H.E. Merritt, Hydraulic Control Systems, New York: Wiley, p.44, 1967.

[8] S.P. Timoshenko, Theory of Plates and Shells, New York: McGraw-Hill, 1940.

[9] B. Leroy and C. Plougonven, "Warpage of Silicon Wafers," Journal of the Electrochemical Society, vol. 127, No. 4, p. 961, 1980.

[10] F. Maseeh and S.D. Senturia, "Plastic Deformation Of Highly Doped Silicon," 5th International Conference on Solid-State Sensors and Actuators & Eurosensors III, Transducers '89, Montreux, Switzerland, p. 246, June 25-30, 1989.

MICROMACHINED SILICON MICROVALVE

T. Ohnstein, T. Fukiura*, J. Ridley and U. Bonne
Sensor and System Development Center
Honeywell, Inc.
Bloomington, Minnesota

*Yamatake-Honeywell
Tokyo, Japan

ABSTRACT

An electrostatically actuated silicon microvalve which modulates a gas flow was fabricated and demonstrated. The microvalve converts an electric signal to a pneumatic signal for pressure or gas flow control. Application areas for the microvalve include pneumatic flow control for industrial, commercial and medical applications. The microvalve is integrally fabricated on a single silicon wafer using surface and bulk micromachining. The microvalve operates against pressures of up to 114 mmHg and flows of up to 150 sccm with a 30 volt signal and hold back pressures of up to 760 mmHg. The valve may be operated in dc or pulse width modulated voltage modes.

INTRODUCTION

Recently, there have been several reports of using silicon micromachining for making microvalves [1]-[4]. These microvalves for controlling a gas or fluid flow have either operated as passive devices, such as check valves [1],[2], controlled by system pressures, or active valves actuated using a piezoelectric stack mounted on the actuator assembly [3],[4]. These devices have been fabricated by the micromachining of two or more pieces of silicon or pyrex which require a final assembly to form the valve. This paper describes an electrostatically actuated microvalve that is integrally fabricated on a silicon substrate by a sequence of thin film depositions and surface and bulk micromachining techniques.

STRUCTURE AND OPERATION

Figure 1 shows a perspective view of a cross section of the microvalve. The microvalve consists of a thin film structure built up on the silicon substrate which forms the valve base plate and the valve closure plate. The closure plate is a cantilever structure that is free to move. The closure plate is released from the substrate structure by the removal of a sacrificial layer at the end of the fabrication process. The gas inlet orifice is anisotropically etched through the silicon substrate. The base plate and closure plate contain electrodes for electrostatic actuation of the valve.

The microvalve is a normally-open type valve with gas flow through the silicon substrate exhausting around the closure plate. The gas flow is modulated by applying a voltage

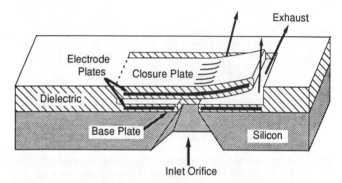

Figure 1. Perspective view of microvalve cross section (not drawn to scale).

between the electrode plates. The electrostatic force between the plates pulls the closure plate against the base plate shutting off the gas flow.

Figure 2 is a photograph of the top view of a microvalve. The device consists of a 5 x 5 array of valves as shown in Figure 1. Each element in the array has an inlet orifice through the substrate and a cantilever closure plate over the orifice. Each cantilever is attached at its left hand edge and is free on the other three sides. Each orifice is 24 x 60 microns and each

Figure 2. Photomicrograph of the top view of a microvalve. The device consists of a 5 x 5 array of cantilever closure plate/orifice valves connected in parallel. The array is truncated in the photograph.

Reprinted from *Proceedings IEEE Micro Electro Mechanical Systems*, pp. 95-98, February 1990.

cantilever closure plate is 350 x 390 microns. The overall chip size is 3600 x 3600 microns. The microvalve shown is a two terminal device with all the valves in the array connected in parallel. When actuated, all the closure plates move together to modulate the gas flow.

FABRICATION

The major fabrication steps for the microvalve are illustrated in Figure 3. Starting with Figure 3a, silicon nitride is deposited on both sides of a double-side polished (100) silicon wafer. The base plate electrode metal is deposited and patterned on the front side of the wafer. Then the electrode plate is passivated with a second silicon nitride deposition to complete formation of the valve base plate. The inlet orifice pattern is formed by plasma etching via holes through the silicon nitride layers on both the front and back sides of the wafer. The silicon substrate is anisotropically etched at the end of the process sequence to open the orifice. Next, a sacrificial layer is deposited and patterned on the front side of the wafer. The sacrificial layer can be either a metal or an oxide film that can be removed with a selective etch at the end of the device processing.

Figure 3b shows the completion of the steps necessary to form the valve closure plate structure and contacts to the electrodes. A silicon nitride layer is deposited over the sacrificial layer to

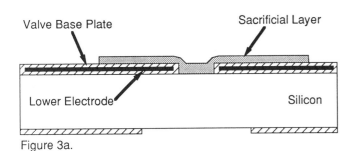

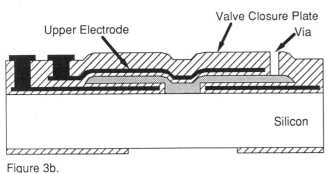

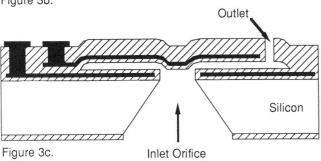

Figure 3. Cross Section views of the microvalve showing the major fabrication steps (not drawn to scale).

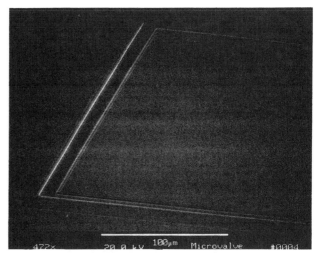

Figure 4. SEM photograph of one cantilever closure plate. The cantilever is attached at the right side, out of the photograph view, and extends to the left in the photograph.

form the bottom of the closure plate and to passivate the bottom of the closure plate electrode. The closure plate electrode metal is then deposited and patterned. The closure plate electrode is passivated by a final silicon nitride deposition. This silicon nitride layer forms the bulk of the closure plate and gives it mechanical strength. The total thickness of the closure plate is approximately 1 micron. Plasma etching is used again to cut a via hole to the sacrificial layer forming the edges of the cantilever and to the electrode plates for contact pads. The contact pad metals are then deposited and patterned.

The microvalve is then ready for the final etching steps. As illustrated in Figure 3c, the gas inlet orifice is etched through the silicon substrate using a standard KOH etch. The sacrificial layer is removed from the structure by its selective etch, etching from both sides of the wafer, to release the closure plate.

Figure 4 shows an SEM photograph of the free end of a cantilever closure plate after the final etch to release the plate. The cantilever is attached to the right side out of the field of the photograph and extends to the left. The cantilevers are very flat with no buckling or warping. The groove around the end of the cantilever was the via cut used to access and remove the sacrificial layer and which now becomes the gas flow outlet after removal of the film. Figure 5 is an SEM photograph that shows a close-up view of the hinge attachment point of the cantilever plate. In this photograph, the cantilever is attached to the substrate structure at the left side and the free-standing cantilever extends towards the right side of the photograph. The gap left by the removal of the sacrificial layer is visible under the cantilever plate.

PACKAGING AND TESTING

The microvalves were packaged for testing by mounting them with epoxy on a transistor header. The header had a machined hole aligned with the valve chip allowing gas flow through the valve. The header was sealed into a fixture that allowed connection to the test station and gas supply. The test fixture contained a filter housing so that the compressed air gas supply could be filtered to reduce the possibility of particulate contamination of the microvalve. A 0.1 micron filter was used in the testing.

Figure 6 shows a simplified diagram of the experimental test station for the microvalves. The station has the ability to control the pressure of the inlet gas supply to the microvalve. There are sensors to measure the flow through the microvalve and the pressure drop across the valve. The valve is operated with a drive circuit that can be either a dc voltage or a pulse width modulated signal. The system can be operated with the microvalve sealed with flow tubes on both the inlet and outlet sides of the valve or with the outlet vented to the room. In the latter case where the outlet is vented to the room, the valve operation can be visually observed with a microscope and camera and its performance recorded.

EXPERIMENTAL RESULTS

The open position orifice flow characteristics of the microvalve are shown in Figure 7. The measurement was taken with the valve unenergized. The flow through the valve was increased, and the pressure drop across the valve, ΔP, was measured. A best fit curve for the orifice flow data was calculated and found to give the following relationship:

$$\Delta P = 8.83e\text{-}4 * Q^2 + 0.595 * Q + 4.95$$

where ΔP is the pressure drop across the valve and Q is the mass flow through the valve.

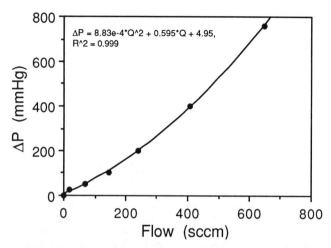

Figure 7. The microvalve orifice flow characteristic measured with the microvalve in the unenergized, "open" position.

It can be seen from this relationship that there is both a linear flow term describing laminar flow through the valve [5] and a second order flow term describing orifice flow through the valve [6]. The laminar flow term dominates the lower end of the operational range of the microvalve with the orifice flow term becoming significant above 200 sccm. Measured data showed that the valve can be operated with flows of up to 150 sccm with a relatively low pressure drop across the valve of 114 mmHg.

The operation of the valve for flow modulation was measured by applying a dc voltage to close the valve against increasing pressures and flows. Valve closure was monitored both visually and electrically by the capacitance of the valve. It was possible to close the valve against pressures of up to 110 mmHg with 30 volts applied to the valve.

Figure 8 shows a flow modulation curve for the microvalve. The gas flow was measured as the voltage applied to the valve was increased to 30 volts and then back down to 0 volts. The system supply pressure was 92 mmHg for this measurement. Operated with a dc voltage, the microvalve was able to modulate the flow between 130 sccm and 4 sccm. When the microvalve was operated with a dc voltage, it was difficult to hold in a position for flows between the maximum and minimum and the valve operated like a bistable device with an open and a closed position. In applications where proportional control of the flow is not necessary, the valve may be operated in this dc voltage mode. A large hysteresis was also observed in the flow curves. The hysteresis may be due to the coarseness of the data points. This is being investigated further to determine the extent of the actual hysteresis of the device. The microvalve may be operated with a pulse-width modulated signal, where the average flow through the valve is controlled by the driving signal frequency and pulse width variation. It has been found that proportional flow control is easier in this mode of operation than when actuated with a dc voltage.

Figure 5. SEM photograph of the hinge attachment point of the cantilever closure plate. The cantilever extends to the right from the attachment point.

Referring once again to Figure 8, the minimum flow in the closed position for this measurement was 4 sccm. It has been found that most of this leakage was due to the packaging of the microvalve. Recent improvements in sealing of the packaging interfaces have reduced the closed position leak rate to 0.1 sccm. Improvements in packaging are being implemented to further reduce the leak rate.

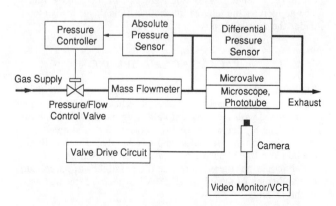

Figure 6. Diagram of the microvalve test station.

The characteristic hold-back pressure of the microvalve is a measure of the ability of the valve to remain closed against increasing pressures. The hold back pressure of the microvalve was measured by closing the valve against a relatively low pressure of 54 mmHg with 30 volts dc. The inlet supply pressure was then increased while monitoring the flow leaking through the valve. A hold back pressure of 760 mmHg (14 psi) has been demonstrated. As the pressure was increased above this point a failure of the valve to hold closed was observed. The failure mechanism is not catastrophic though. The valve fails when one or more of the cantilever closure plates fails to hold back the pressure and releases from the base plate. The closure plate is not destroyed by this and continues to be operational once the pressure is reduced.

SUMMARY

An electrostatically actuated silicon microvalve which modulates a gas flow was fabricated and demonstrated. The microvalve is integrally fabricated on a single silicon wafer using surface and bulk micromachining and requires no final assembly. The microvalve operates against pressures of up to 114 mmHg and flows of up to 150 sccm with a 30 volt signal. The microvalve will hold back pressures of up to 760 mmHg. The valve may be operated in dc or pulse width modulated voltage modes. Investigation is continuing to characterize, model and improve the microvalve performance.

REFERENCES

[1] J. Tiren, L. Tenerz and B. Hok, "A Batch-Fabricated Non-Reverse Valve With Cantilever Beam Manufactured By Micromachining Of Silicon," Senors And Actuators, Vol. 18 pp. 389-396:1989.

[2] S. Park, W. Ko and J. Prahl, "A Constant Flow-Rate Microvalve Actuator Based On Silicon And Micromachining Technology," Proc. IEEE Solid-State Sensor And Actuator Workshop, pp. 136-138:1988.

[3] M. Esashi, S. Shoji and A. Nakano, "Normally Closed Microvalve And Micropump Fabricated On A Silicon Wafer," Proc. IEEE Workshop On Micro Electro Mechanical Systems(MEMS), pp. 29-34:Feb. 1989.

[4] M. Esashi, et al, "The Fabrication Of A Micro Valve By Means Of Micromachining," Proc. 6th Japanese Sensor Symp., pp. 269-272:1986.

[5] G. Batchelor, *An Introduction To Fluid Dynamics*, Cambridge Univ. Press, pp. 179-186:1967.

[6] R. Goldstein, *Fluid Mechanics Measurements*, Hemisphere Pub. Corp., pp. 245-252:1983

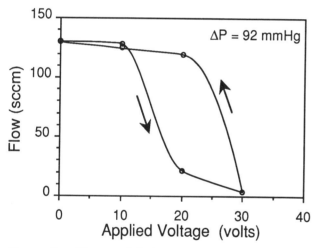

Figure 8. Flow modulation curves measured for the microvalve.

Section 9

Fluidics

As objects become smaller, their surface area to mass increases dramatically. This leaves micro parts increasingly sensitive to surface effects like fluid flow. For example, consider a bit of dirt kicked up by a little whirlwind. The pebbles rapidly fall back to the ground; the sand falls as soon as the whirlwind dies; the fine dust may remain suspended for hours.

Micro science leads us into many interesting areas; for example, the omnipresent air flowing around our micro devices. The mean free path of a molecule in air is about 1/20 of a micron. Typical clearances of micro devices are a fraction of a micron to several microns. Hence, the distance that an air molecule travels before it strikes the next object is commensurate with the micro channels. This ratio is characterized by the Knudsen number, $Kn = \lambda/L$; where λ is the mean free path and L is the characteristic length of the flow (such as the air gap in a moving micromechanical device). For $Kn < 10^{-3}$, the fluid can be considered a continuum, and our statistical mathematics adequately predicts the fluid properties. If $Kn > 10$, the gas can be considered a free molecular flow, and the motion of each molecule can be individually calculated to understand the dynamics of the gas. However, between these regions are the slip-flow region ($10^{-3} < Kn < 0.1$) and the transitional-flow region ($0.1 < Kn < 10$), where conventional theories break down. Theories are currently being developed for the slip-flow and transitional-flow cases. For the typical 1-µm gap in many micro devices, the Knudsen number, Kn, is 0.05. Our need to understand micro devices is pushing the micro sciences, and our capability to make and measure devices on the micro scale will give information and impetus to the micro scientist.

A second type of fluid is liquids. Here, viscosity is even more troubling to the micro devices trying to move in what often appears to be molasses. Also, the surface tension of liquids is enormous in the micro domain, and can be helpful, or the bane of one's design, depending on the cleverness of the designer. Much remains to be learned and exploited about liquids.

The first paper, "Microminiature Fluidic Amplifier," describes a clever amplifier that uses fluids flowing through micro channels. The next paper, "A Planar Air Levitated Electrostatic Actuator System," presents a micro air table, where small parts can be moved around without the problems of sliding friction. In the next two papers, "Liquid and Gas Transport in Small Channels" and "Squeeze-Film Damping in Solid-State Accelerometers," the properties of micro fluids are measured and analyzed. The last three papers in this section ("A Micromachined Floating-Element Shear Sensor," "A Multi-Element Monolithic Mass Flowmeter with On-Chip CMOS Readout Electronics," and "Environmentally Rugged, Wide Dynamic Range Microstructure Airflow Sensor") use micromechanical structures to measure the properties of fluid flow.

Microminiature Fluidic Amplifier

Mark J. Zdeblick   Phillip W. Barth   James B. Angell

Stanford University
Stanford, CA 94305

Abstract — **A microminiature fluidic amplifier with vertical-walled features 6 μm wide and 35 μm deep has been fabricated in silicon using dry anisotropic etching. Operation was demonstrated using nitrogen as the working fluid. DC gains greater than unity were seen over a supply pressure range of 10 to 80 psi. The amplifier is intended to be used as a circuit element in integrated fluidic circuits or, together with on-board electronics, as part of a transducer system. Dry etching techniques similar to those used here offer new opportunities for the micromachining of integrated sensors and actuators.**

Introduction

A fluidic amplifier operates on gasses or liquids to amplify differential pressures or flows. A number of commercial fluidic systems, including sensors for high temperature, pressure, air speed, angular rate change, and fuel flow, are currently being used in various aerospace applications[1]. These commercial devices, made from non-semiconductor materials, are at least ten times larger in linear dimension than the device to be discussed. The smaller size of this new device results in higher operating pressures, and should lead to faster responses and a higher degree of integration than is currently available. Moreover, the demonstrated ability to integrate electronic function with silicon-based fluidic devices would make possible new sensors and actuators.

The device was made by anisotropically etching a 35 μm deep trench in silicon and then sealing the trench with a brass manifold to form a capillary channel. This dry anisotropic etch is an extension of a recently developed dry etching technique developed by McVittie et al. [2] to etch depth 35 μm. The process yields near-vertical walls independent of crystal orientation, as shown in SEM photographs (Figures 1 and 2). The vertical etch rate of this mixed halocarbon (SF_6 and C_2ClF_5) etchant is 0.4 μm/min when used with the Drytek RIE-100, a parallel plate reactive ion etcher system. The etch depth of 35 μm is limited by the thickness of the photoresist.

The silicon wafer was also etched in KOH from the backside to provide feed-through holes for the input and output ports. Removable brass manifolds clamped on top and bottom of the silicon wafer provided ports accessible with Tygon tubing, while a 10 μm layer of Parylene on the brass acted as gasket material.

A plan view of the wafer, shown in Figure 1, helps illustrate the operation of the device. The darker areas are the floor of the trench; the labels correspond to the different input and output ports. When the supply port is at a higher pressure than the vent, $P_s > P_v$, a laminar jet emanates from the supply's nozzle. This jet traverses the amplifier region and reaches the output ports, where it is split into two flows by the splitter. The input control ports steer the jet. When either a small differential input pressure, $\Delta P_i = P_{i_r} - P_{i_l}$, or input flow, ΔF_i, is applied, the jet will be deflected toward one of the output ports, resulting in a differential output pressure & flow, ΔP_o & ΔF_o. For small variations about $\Delta P_i = 0$, the device can be modeled as $\Delta P_o = A_p * (\Delta P_i + P_{io})$, where A_p is the pressure gain of the device and P_{io} is the input offset pressure. Ideally, $P_{io} = 0$.

The device was characterized by varying the supply pressure, P_s, and resistive loads, R_l, then measuring the input and output flow rates and pressures. From these data, we were able to calculate the following: pressure gain, A_p; flow gain, A_f; power gain, A_w.

Results

Of the four amplifiers fabricated, one has been thoroughly tested to date. This amplifier was tested by varying the capillary load resistance from 0 to 14 kPa/sccm and varying the supply pressures from 62 to 236 kPa (9 to 34 psi). A_p, A_f, and A_w were determined for each P_s and R_l. A_f decreases from 5.3 to 1.2 as R_l increases from 0 to 14 kPa/sccm[1]. Operation at $P_s = 133$ kPa (19 psi) produced the highest flow gain at any given R_l. The pressure gain increases asymptotically from 0.1 to 2.7 as R_l increases. Again, operation at $P_s = 133$ kPa (19 psi) produced the highest gains. (Another amplifier produced a blocked output ($F_o = 0$) pressure gain of 4.5, operating at 133 kPa.) The power gain increases from zero to a maximum of 5.9 and then declines for higher R_l. The maximum power gain was observed at $R_l \approx 4$ kPa/sccm for most supply pressures. The A_f vs. A_p curve demonstrates some of the noise in the system. A set of smooth, parallel curves was expected; deviations are probably due to asymmetries in the load resistors.

The average flow rates in the supply, control, and output channels were also measured to determine the Reynolds number at both maximum flow and maximum power gain. The most striking observation is that the maximum output flow, F_o, is at most 31% of the supply flow, F_s, and that F_o at maximum power gain is 21% of F_s. These figures imply that most of the supply flow is going out the vents, even when the output ports vent to air. A second observation is that the Reynolds number of the supply jet is between 400 and 1300 — near the turbulent flow transition. It should be noted that since P_s is greater than twice the ambient pressure, supersonic flow at some point in the device is likely.

Conclusions and projections

This work demonstrated a 10X reduction in feature size and a 10X increase in P_s compared to 'standard' fluidic amplifiers.

---

[1] sccm is an abbreviation for standard cubic centimeters per minute

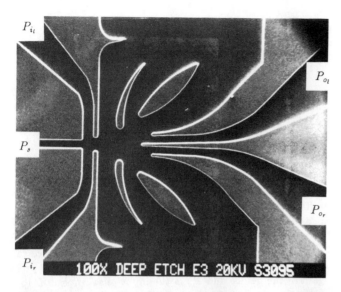

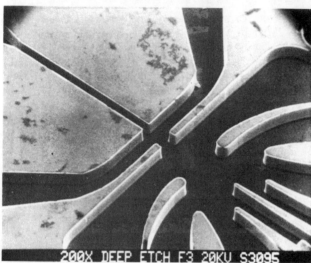

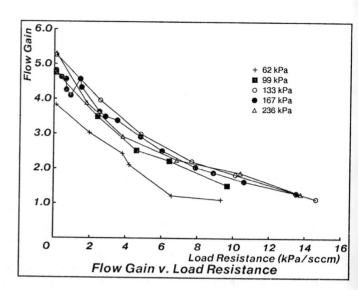

Flow Gain v. Load Resistance

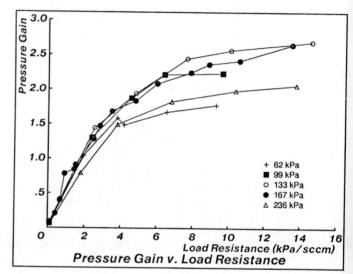

Pressure Gain v. Load Resistance

Power gain greater than one and input offset, P_{io}, less than 1% of the span of P_i were demonstrated. In addition, this work shows that dry anisotropic etching can be reliably used to produce 'vertical-walled' structures with flat bottoms. Ring oscillator structures to obtain speed data have been fabricated and are awaiting testing.

It seems likely that improvements in single stage gains are obtainable with design optimization. Integration of pressure sensors and fluidic devices on a single wafer could provide fluidic to electronic transduction with low power and high speed. Compatible fabrication processes seem feasible, but need to be developed. Future devices could take advantage of developed anodic bonding techniques [3] to reduce leakage between channels and yield 'permanent' devices. It seems feasible to pursue a level of integration for fluidic devices comparable in size and complexity to 1960's vintage electronic circuits [2].

---

[2]Mark J. Zdeblick is a Doctoral Candidate in the department of Electrical Engineering. His mailing address is Applied Electronics Laboratories Suite 16, Stanford University, Stanford CA, 94305.

Dr. Phillip W. Barth was with Stanford, he is now a Program Director at NovaSensor, 2975 Bowers Ave., Santa Clara, CA 95050.

Prof. James B. Angell is the Associate Chairman of the Department of Electrical Engineering at Stanford. His mailing address is McCullough 162, Stanford University, Stanford, CA, 94305.

Acknowledgements

The authors would like to thank Dr. James McVittie of Stanford Integrated Circuits Laboratory for his help with the anisotropic etching technique, Martin Berdahl of Jet Propulsion Laboratories (JPL) for fabricating the brass test fixtures used to characterize the fluidic amplifier and for helpful discussions about fluidics in general, and Bill Reynolds for helpful discussions of fluid flow properties. This project was initially supported by JPL grant JPL 956888 and currently by General Motors grant GM 43-5.

References

[1] *Garret Fluidic Systems* Garret Pneumatic systems division, PO box 52-17, Phoenix, AZ 85010.

[2] "Anisotropic etching of Si using SF_6 with C_2ClF_5 and other mixed halocarbons," J. McVittie and C. Gonzalez, Extended abstracts, Fall Meeting, *Electrochemical Society*, Vol 84-2, p 584.

[3] "A gas chromatography system fabricated on a silicon wafer using integrated circuit technology," Stephen Terry, Ph.D. Thesis, *Stanford University Technical Report No. 4603-1*, May 1975

A Planar Air Levitated Electrostatic Actuator System

Kristofer S. J. Pister, Ronald S. Fearing, and Roger T. Howe
Department of Electrical Engineering and Computer Sciences
Electronics Research Laboratory
University of California, Berkeley, California 94720

Abstract

A micromotor capable of moving multiple objects in three dimensions has been fabricated. The fixed surface of the motor, fabricated on a silicon wafer, contains air nozzles which levitate rigid platforms, as well as conductive plates which generate an electric field to apply forces to the platforms. Initial testing has shown the air nozzles form a stable very low friction bearing, and the motor is capable of several hundred microns of motion in two dimensions with fields generated by a 2 volt potential. Applications of this motor include a microrobotic work cell.

Background

Microfabricated sensors and actuators are abundant in laboratories and in industry. These devices present a unique opportunity for the development of robotic work cells composed of a combination of sensors and actuators on one or more platforms moving in a chip sized work space. The motion of these sensor/actuator platforms should ideally be generated by devices on the same scale as the platforms themselves. This motivates the study of micromotors capable of generating motion in two or more dimensions.

Work in the field of micro-motors has shown that frictional forces play a dominant role in the dynamics of micro-machines [1]. Recently much effort as been directed at removing friction entirely [2], or using it as an integral part of the functionality of the motor [3][4]. Unfortunately most of these motors generate either oscillatory or rotary motion which can not be used directly for translation of sensor/actuator platforms.

A relatively unexplored area involves minimizing friction by ensuring that moving parts do not come into contact. This has been done using superconducting electromagnetic levitation [5] and a combination of free fall and magnetic clamping [6]. These systems have an advantage in that they are capable of linear motion in one or more directions. Our approach is similar to these. In order to minimize friction, the moving elements of our design ride on a cushion of air supplied through nozzles in the fixed surface of the motor. These moving parts, or platforms, are then actuated by electrostatic fields generated by conductors in the surface of the motor (Figure 1).

Design

Our goal was to levitate and actuate platforms roughly 1 mm square. To generate a more uniform air cushion we chose the nozzle spacing ($100\mu m$) to be small compared to the dimension of the platform. The nozzle dimension used was $10\mu m$. Lacking the technology to fabricate $10\mu m$ nozzles through a full wafer thickness, we decided to create nozzles through membranes and supply pressure behind the membrane. Given expected supply pressures of 100-1000 pascals and the need to have a very flat surface (for the bearing to function) it became obvious that the membranes would have to either be very thick (greater than $10\mu m$) or very narrow (less than $500\ \mu m$).

Due to processing constraints the narrow membrane approach was chosen. Each bearing surface is made up of 20 parallel low stress silicon nitride membranes, each membrane is 9 mm long by $130\ \mu m$ wide and $1\ \mu m$ thick. The membranes cover air supply trenches beneath them, and air is introduced to the underside of the membrane via front surface supply holes on one side of the membranes (Figure 4).

The fabrication of the air bearings is straightforward. An oxide layer is patterned to define the future membrane locations. Silicon nitride is grown on top of the oxide and the air nozzles are patterned through the nitride, contacting the oxide layer. The oxide is etched in HF, freeing the nitride membranes. The resulting air supply channels are not sufficiently large, so the structure is then immersed in KOH to bulk etch the silicon under the membranes [7][8].

In one version of the process, after the membranes are created, aluminum is sputtered and patterned to form electrodes. Exposed electrodes can lead to problems such as shorting, and the lack of an insulating cover for the electrodes allows vertical attractive forces on the platforms to become extremely high, thereby making unwanted clamping more common. To avoid this problem, polysilicon electrodes are grown and patterned after the first nitride deposition. A second layer of nitride is then deposited to insulate the electrodes before the air nozzles are patterned and the membranes freed (Figure 2).

Air Bearing

Figure 7 shows a cross section of a 1 mm platform floating on the bearing. P_i represents the pressure above the ith nozzle, Q_i the flow between nozzles, Q_{Si} the flow through the ith nozzle. P_a is atmospheric pressure (gauge pressure 0), and P_S is the pressure supplied to the underside of the membranes (assumed constant). Assuming fully developed laminar incompressible flow (i.e. Reynolds number << 1400 and Mach number << 1), the volumetric flow rate between supply holes, Q_i, will be a linear function of the pressure drop between the holes, similarly the flow through the nozzles will be proportional to the pressure drop across the nozzle. These approximations ignore entrance and exit effects, which are typically non-linear. These relations are expressed as

$$P_i - P_{i+1} = R_i Q_i \quad (1)$$
$$P_S - P_i = R_S Q_{Si} \quad (2)$$

The flow resistances R_i will be functions of the altitude, h_i, of the platform between the ith and the $(i+1)^{th}$ nozzles. R_S is constant for a given nozzle size. For incompressible viscous flow in a finite slot the resistance is

$$R_i = \frac{12\mu L}{w h_i^3} \quad (3)$$

where L and w are the length and width of the channel, respectively, and μ is the viscosity of air ($1.78 \times 10^{-5} Kg/m \cdot s$) [9]. For a platform floating at $20\mu m$ R_I is $2.7 \times 10^{10} Pa \cdot s \cdot m^{-3}$.

The nozzle resistance can be approximated by the resistance of an inscribed circular nozzle,

$$R_S = \frac{8\mu L}{r^4} \quad (4)$$

where L is the thickness of the membrane. For a $10\mu m$ nozzle this gives a theoretical nozzle resistance of $7.3 \times 10^{10} Pa \cdot s \cdot m^{-3}$. The measured nozzle resistance at 500 pascals applied pressure was roughly $1.9 \times 10^{11} Pa \cdot s/m^3$, in reasonable agreement with the theoretical value given that entrance effects on the flow were ignored in the theoretical calculation.

Given these resistance values the pressure distribution under the platform can be found and from it the load carrying capacity of the bearing. The pressure above the ith air nozzle is

$$P_i = \alpha_i P_{i-1} + \beta_i P_{i+1} + \gamma_i P_S \quad (5)$$

where P_S is the constant supply pressure (all pressures are gauge pressures), and

$$\alpha_i = R_s R_{i+1}/d_i \quad (6)$$
$$\beta_i = R_i R_s/d_i \quad (7)$$
$$\gamma_i = R_i R_{i+1}/d_i \quad (8)$$
$$d_i = R_i R_s + R_s R_{i+1} + R_i R_{i+1} \quad (9)$$

This set of linear equations can be solved for the P_i's. The average of these pressures is the theoretical load carrying capability of the bearing. The numbers calculated from this model are consistently three or four times the experimentally determined capability. This is at least partly due to the modelling assumption that the flow is only one dimensional. The model predicts flow rates of up to $3 \times 10^{-3} cm^3 s^{-1}$ under the edge of the platform (Q_0), with peak velocities of $30 cm \cdot s^{-1}$. Flow through nozzles is predicted to be $10^{-4} cm^3 s^{-1}$ with average velocities less than $10 m \cdot s^{-1}$.

Experimental values of air flow to the entire test chip are 5 to 10 $cm^3 s^{-1}$ at 500 pascals. With roughly 7000 nozzles per chip this means that average unobstructed flow through a single nozzle at a differential pressure of 500 pascals is $10^{-3} cm^3 s^{-1}$ with a flow rate of 10 $m \cdot s^{-1}$ at the nozzle, in rough agreement with the model. This velocity corresponds to a Mach number of .03, indicating that the flow is accurately modeled as incompressible. The Reynolds number for this flow is roughly 1, indicating that the flow is laminar and viscous. Another number of interest when working at this scale is the Knudsen number, the ratio of the mean free path of air (.064 μm at STP) to a characteristic dimension of the problem. With our current design, typical numbers for bearing thicknesses and nozzle dimensions are greater than $10\mu m$, giving a Knudsen number less than .01, which indicates that the air can still be treated as a continuous fluid. Future bearings will likely have substantially smaller characteristic dimensions and will require somewhat more advanced analysis.

Electrostatics

Figure 8 shows a model of the capacitance between two electrodes with the platform above them. The assumptions of this model are that a conductive platform covers all of the left electrode, forming capacitor C_1, and only part of the right electrode, forming capacitor C_2. The total capacitance, C, between the two electrodes is C_0, the capacitance between the two electrodes with no platform present, in parallel with C_1 and C_2 in series. The total capacitance and the energy stored in the capacitor (as functions of the platform position) are

$$C(x,y) = C_0 + \frac{C_1(y)C_2(x,y)}{C_1(y)+C_2(x,y)} \quad (10)$$

$$U(x,y) = \frac{1}{2}C(x,y)V^2 \quad (11)$$

The force on the platform due to an applied potential is then

$$F_x = \frac{1}{2}V^2 \frac{\partial C(x,y)}{\partial x} \quad (12)$$
$$F_y = \frac{1}{2}V^2 \frac{\partial C(x,y)}{\partial y} \quad (13)$$

Using this model and making the approximation that C_1 and C_2 are ideal parallel plate capacitors, we can estimate the forces on the platform as a function of position. Assuming the platform is a 1 mm square conductor and the electrodes are $80\mu m$ wide and extend the length of the platform, and the applied potential is 10 volts,

$$F_x = \frac{3.2 \times 10^{-10} \epsilon_0}{y(x+8 \times 10^{-5})^2} \quad (14)$$

$$F_y = 4 \times 10^{-6} \epsilon_0 \frac{x}{y^2(x+8 \times 10^{-5})} \quad (15)$$

This system was simulated with the platform centered on the right electrode and a potential of 10 volts [10]. Cushion thickness was varied between 1 and $20\mu m$. The simulation results for forces on the platform are shown as function of float height in Figure 10. Note that the data has the expected shape: the vertical force appears to be roughly proportional to y^{-2}, and the lateral force roughly y^{-1}. The simulated vertical forces match those predicted in equation 15 to within 20%, but the predicted lateral forces are 2 to 5 times lower than their simulated counterparts. Clearly this is at least partly due to the nature of the approximations made in developing the model, but the two dimensional simulation results may not be substantially more accurate, given the three dimensional nature of the problem [11].

Simulations were also done of a dielectric platform under the same conditions. Figure 9 shows an example of the simulated equipotential lines around the platform. Simulated forces on a dielectric platform were roughly an order of magnitude lower than forces on the conductive platform.

Results

Air bearing

Our experimental setup is unfortunately rather imprecise. Currently a constriction valve and a water column barometer are used to regulate and measure pressure respectively. Despite this, we have successfully levitated a large variety of materials including $2\mu m$ thick nitride membrane fragments, a 1mm cube of silicon, aluminum foil (15 μm thick), and a scrap of binder paper.

The load carrying capability of the bearing was tested with several thicknesses of platforms from .25mm to 1mm. The platforms were stacks of 1 or more 1mm square pieces of silicon wafers, each $250\mu m$ thick. Air cushion thickness as a function of bearing pressure was determined by focusing on the top surface of the platform and recording change in focus position as a function of change in pressure. Results are shown in Figure 6. Pressures shown are recorded pressures at the gauge, not the actual pressure beneath the membranes. Membrane pressures may actually be less than one half the indicated pressure, due to flow resistance in the tube connecting the gauge with the test device.

Clearly bearing thicknesses of one or two mils are not optimal for this device, given that macro scale bearings typically have clearances in this range. Decreasing the nozzle size, or increasing the membrane thickness should improve bearing performance. Both of these would increase nozzle flow resistance, thereby decreasing the cushion thickness and increasing stability. The current bearings may yield better results if a more sophisticated pressure regulator is used.

Motor

Three types of motion have been seen as a result of applied electric fields. Surface clamping and pinned rotation have been observed in addition to the expected horizontal translation. Surface clamping occurs when the attractive force of the electric field overcomes the repulsive force of the air bearing. Pinned rotation results from dirt between the bearing surface and the platform, causing the platform to in effect become a rotor pinned at the contaminant. Clamping and rotation have been observed with both dielectric and conductive platforms. Translation has been seen only with conductive platforms.

Currently only 1 pole translational motion has been seen; i.e., we have not used the motor as a multi-pole stepper motor, because of a lack of probing equipment. Translational motion has been observed at bearing pressures between 500 and 2000 pascals. Voltages between 2 and 15 volts typically cause platform motion of between 100 and 500 microns. This motion has been observed at frequencies between .5 and 10 hz, with the largest amplitude motion occurring around 1 or 2 hz. The amplitude decrease at lower frequencies may indicate that some charging of the platform is taking place. The actual electrostatic forces on the platform have not been determined, but a rough lower bound on the lateral component is 3×10^{-9} Newtons. Based on the simulations of electrostatic force this indicates a bearing cushion thickness of approximately $10 \mu m$, which is roughly what we expect the thickness to be.

Voltages above roughly 20 volts will overcome the air bearing levitation force and clamp the platform to the surface of the bearing. Once a platform has been clamped, only a small voltage is necessary to maintain surface contact, even if bearing pressure is increased to 10 kPa. This effect is due to the decrease in surface area of the platform exposed to pressurized air (just the air nozzle area, roughly 1% of the total area of the bearing surface).

Before clamping takes place, the vertical electrostatic force on the platform attracts it to the surface. Initial experimentation has indicated that the thickness of the air cushion decreases in a roughly linear manner as the vertical force is increased. Although we have not measured the effect of the electric field on the cushion thickness, we believe that the the vertical position of the platforms will be smoothly controllable up to the point of clamping.

Long Term Goals

Given the ability to move multiple platforms in the same work space, we envision a $1 cm^2$ work surface with many different platforms on its perimeter, each platform with its own function. Perhaps the system could be used as a microassembly cell, with individual parts floated into the workspace and manipulated by two or more platforms acting together to form a gripper. Parts and platforms could be individually clamped to the surface while assembly or machining takes place. One advantage of this system is that the moving parts are easily replaced. If a platform outlives its usefulness it would simply be shuttled off the work surface and replaced by another. The idea of multiple robotic platforms cooperating in the same workspace has been succesfully engineered on the macro scale by Automatix, Inc., makers of Robotworld [12]. Robotworld has several platforms suspended from a one meter square steel sheet. The platforms ride on an air bearing, as in our system, but are actuated by linear electromagnetic stepper motors. Robotworld is typically used for pick and place operations such as IC board stuffing, but is capable of more advanced manipulation such as small parts assembly [13].

If lateral drive forces can be increased sufficiently it may be possible to have small power lines running to a platform, in which case essentially any microfabricated sensor or actuator could be placed on a mobile platform. Alternatively, it may be possible to have power available on the surface of the bearing, where platforms could be moved into position and then clamped to the surface to contact supply pads. A potential application of this idea would be a reconfigurable probe station for integrated circuit (IC) failure analysis. Platforms with microprobes or capacitive sensors could be maneuvered to their desired locations and clamped to the surface, then the system would be placed in contact with the IC to be tested.

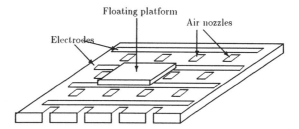

Figure 1: Illustration of a platform floating above the air nozzles and electrodes on the surface of the motor.

Conclusion

We have demonstrated a stable microfabricated air bearing capable of levitating objects between $2 \mu m$ and 1mm thick with extremely low kinetic friction, and no static friction. In addition we have shown the compatibility of this bearing with an integrated electrostatic actuation system.

Testing has been limited to verification of the fundamental principles of operation, but experimental results indicate that a two dimensional linear electrostatic stepper motor should be realizable with this process. In addition, it should be possible to control the vertical motion of the platforms by varying either the supply voltage or bearing pressure.

Electrostatic forces on the platforms are currently on the order of 1 to 10 nN with an applied potential of 2 volts. If cushion thickness can be decreased to a few microns, it should be possible to attain forces several orders of magnitude higher than this, resulting in platform accelerations on the order of 1g. Research will be geared toward understanding and improving the air bearing in an effort to realize this goal. Likely directions of investigation are increasing nozzle flow resistance and using a backside air supply to improve pressure uniformity.

Acknowledgements

Partial support of K. Pister provided by the Berkeley Sensor & Actuator Center, an NSF/Industry/University Cooperative Research Center. Many thanks to Tom Booth, Ben Costello, Leslie Field, Martin Lim, Carlos Mastrangelo, and Stuart Wenzel for their invaluable processing advice, assistance, and encouragement.

References

[1] Yu-Chong Tai, Long-Sheng Fan, and Richard S. Muller. Ic-processed micro-motors: Design, technology, and testing. In *IEEE Micro Electro Mechanical Systems, Salt Lake City, Utah*, 1989.

[2] William C. Tang, Tu-Cuong H. Nguyen, and Roger T. Howe. Laterally driven polysilicon resonant microstructures. In *IEEE Micro Electro Mechanical Systems, Salt Lake City, Utah*, 1989.

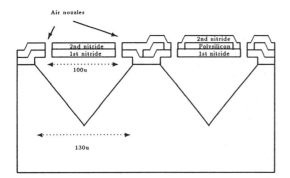

Figure 2: Cross section of the membranes and air channels showing air nozzles through the membranes and polysilicon electrodes embedded between the two layers of silicon nitride.

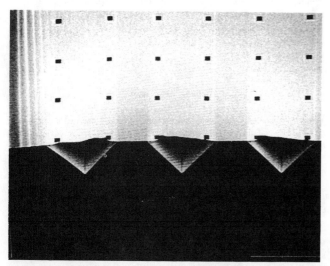

Figure 3: Scanning electron micrograph of membranes, air supply channels, and air nozzles. Faintly visible are buried polysilicon electrodes, 500μm square.

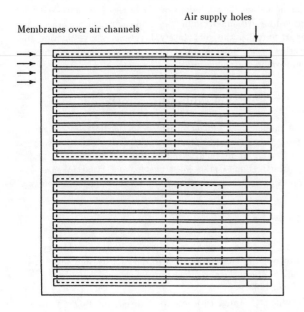

Figure 4: Basic outline of the test chip. Dashed boxes indicate areas with electrode test patterns.

Figure 5: Detail micrograph showing two platforms at rest on the motor surface. Air nozzles are 100μm center to center, platforms are 120μm and 250μm thick.

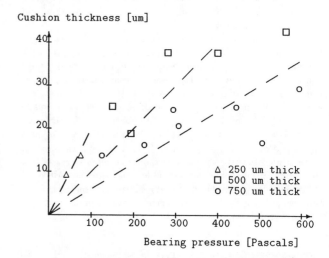

Figure 6: Bearing cushion height as a function of supply pressure. Data was taken with three platform thicknesses. All platforms were roughly 1 mm square.

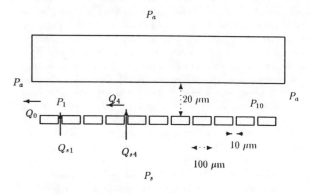

Figure 7: Cross section of platform above air nozzles showing pressures above (P_i) and flow through (Q_{si}) each nozzle, and flow under the platform (Q_i).

476

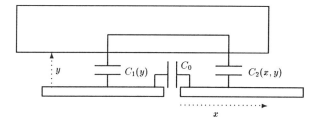

Figure 8: A model of the variable capacitance between electrodes

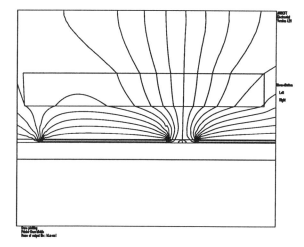

Figure 9: Simulation of device cross section showing equipotential lines. Platform is an uncharged dielectric, electrodes are at -5V and +5V above a grounded substrate.

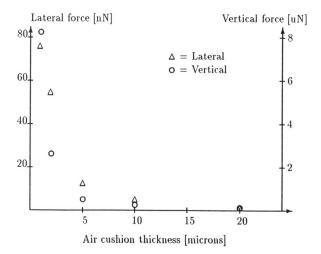

Figure 10: Simulated electrostatic forces on the platform as a function of cushion thickness.

[3] W. Trimmer and R. Jebens. An operational harmonic electrostatic motor. In *IEEE Micro Electro Mechanical Systems, Salt Lake City, Utah*, 1989.

[4] S.C. Jacobsen, R.H. Price, J.E. Wood, T.H. Rytting, and M. Rafaelof. The wobble motor: an electrostatic, planetary-armature, microactuator. In *IEEE Micro Electro Mechanical Systems, Salt Lake City, Utah*, 1989.

[5] Yongkwon Kim, Kakoto Katsurai, and Hiroyuki Fujita. A proposal for a superconducting actuator using meissner effect. In *IEEE Micro Electro Mechanical Systems, Salt Lake City, Utah*, 1989.

[6] Ron Pelrine and Ilene Busch-Vishniac. Magnetically levitated micro-machines. In *IEEE Micro Robots and Teleoperators Workshop, Hyannis, Massachusetts*, 1987.

[7] S. Sugiyama, T. Suzuki, K. Kawahata, K. Shimaoka, M. Takigawa, and I. Igesashi. Micro-diaphragm pressure sensor. In *IEEE International Electron Devices Meeting, Los Angeles, CA, Dec. 7-10*, pages 184–187, 1986.

[8] Carlos H. Mastrangelo and R. S. Muller. Vacuum-sealed silicon micromachined incandescent light source. In *IEEE International Electron Devices Meeting, Washington, D.C., Dec. 3-6*, pages 503–506, 1989.

[9] Dudley D. Fuller. *Theory and Practice of Lubrication for Engineers*. John Wiley and Sons, New York, 1956.

[10] Ansoft Corporation, 4516 Henry Street, Pittsburgh, PA 15213. *Maxwell Users Guide*, January 1989.

[11] Richard H. Price, John E. Wood, and Stephen C. Jacobsen. The modelling of electrostatic forces in small electrostatic actuators. In *Solid State Sensor and Actuator Workshop, Hilton Head, SC, June 6-9*, 1988.

[12] Automatix, Inc., 1000 Tech Park Dr., Billerica, Mass. 01821.

[13] K. Pister, R. Murray, and M. Berkemeier. Manipulating objects with robotworld, a proposal. Robotics Laboratory, Dept. of EECS, UC Berkeley, Berkeley, CA 94720, 1988. Proposal and video tape available on request.

LIQUID AND GAS TRANSPORT IN SMALL CHANNELS

J. Pfahler, J. Harley, and H. H. Bau
Department of Mechanical Engineering and Applied Mechanics

J. Zemel
Department of Electrical Engineering and Center for Sensor Technology
University of Pennsylvania
Philadelphia, Pennsylvania

ABSTRACT

An experimental investigation of fluid (liquid and gas) flow in extremely small channels is presented. Potential applications for such channels include, among other things, cooling of electronic circuits, ultra-small heat exchangers, and reactors for modification and separation of biological cells. We are carrying out a sequence of experiments to determine the friction law which governs liquid and gas flow in extremely small channels. It was found that in the relatively large flow channels, the experimental observations were in rough agreement with the predictions from the Navier-Stokes equations. However, in the smallest of the channels, deviations from the Navier-Stokes predictions are observed.

INTRODUCTION

Microfabrication techniques developed in the electronics industry enable us to manufacture flow channels ranging in depth from a few hundred angstroms to a few hundred microns. These small channels have a variety of current and potential applications such as channels for integrated cooling of electronic circuits, reactors for modification and separation of biological cells, selective membranes, and liquid and gas chromatographs. It is more than likely that new applications will be identified in the near future.

Our long term objective in this study is to obtain fundamental information about the behavior of fluids in extremely small flow channels. The questions we are trying to answer are: (i) at what length scales would the continuum assumptions break down; (ii) do the Navier-Stokes (N-S) equations adequately model the fluid flow at these very small scales or should they be modified and if so, how; (iii) do phenomena which typically are ignored in large scale channels become important in small scale channels; and (iv) is transition to turbulence affected by the small size of the channels?

In order to answer these questions, we are conducting a sequence of systematic experiments in increasingly thinner flow channels. The initial phase of our work consists of measuring the pressure drop as a function of the flow rate. The experimental results are compared with theoretical predictions based on the Navier-Stokes equations and with information available in the literature for flow in large channels. In this paper, we report some results pertaining to liquid (alcohol and silicone oil) and gas (Nitrogen) flow in shallow channels.

Since the ability to manufacture such small structures is relatively new, it is not surprising that very little has been reported on the subject in the literature. Tuckerman has pioneered the use of small channels for cooling of electronic circuits. He used channels 287-376 μm deep and 55-60 μm wide [1,2,3]. As fully developed conditions typically were not achieved in his experiments, comparison between theory and experiment is accomplished by extrapolating the data to zero Reynolds number. Usually Tuckerman's experimental data lies above and within 5% of theoretical predictions. Nakagawa et al. [4] investigated water flow in 5 μm deep and 200-800μm wide channels. Insufficient information is provided to accurately evaluate their data, but the experimental results appear to agree within ±10% with theoretical predictions, with the data being scattered on both sides of the theoretical curve. Peiyi and Little [5] investigated gas flow in channels 7.6-40.3μm deep and 136-200μm wide. They observed friction factors above those expected from classical Moody charts. They also reported transition to turbulence at Reynolds numbers as low as 350. Both the increase in the friction factor and the early transition to turbulence are attributed to the coarseness of their channels. Harley et al [6] reported experimental results for water and alcohol flow in channels of about 100μm width and depths ranging from 0.8μm to 54μm. The data for the deeper channel (54μm) was in good agreement with theoretical predictions. The data for the 1.7μm deep channel exhibited a friction coefficient consistently lower than the one predicted by theory while the data for the 0.8μm deep channel was substantially higher than predicted. The former results have been reinforced by more recent experiments which we report in this paper. The latter results are now attributed to channel plugging and have been disregarded.

On the theoretical front, in the sixties, Eringen and co-workers [7-11] advanced a theory which states that fluid flow in microchannels will deviate from that predicted by the N-S equations. We are not aware of any experimental confirmation of Eringen's theory.

EXPERIMENTAL APPARATUS AND PROCEDURE

The experimental apparatus is depicted schematically in Figure 1. The experimental apparatus consists of a flow cell or test structure and the fluid handling system. The fluid handling system is comprised of an adapter, a regulated pressurized cylinder containing the test fluid, a filter, a pressure guage and a pipette for flow rate measurement.

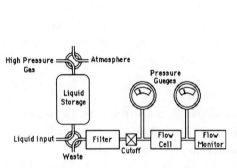

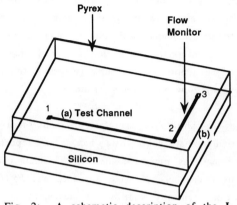

Fig. 1: Schematic layout of the fluid handling system

Fig. 2: A schematic description of the L shaped channel. The on-situ flow monitor is under development. The location of the numbers indicates where we currently measure pressure with external transducers.

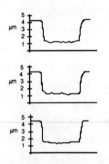

Fig 3: A trace of three cross-sections of a test channel with an approximate depth of 3μm as traced by the profilometer.

A schematic description of the channel is given in Fig. 2. The flow system consists of two channels of different widths forming a L shape etched in silicon using planar photolithographic micromachining techniques [12]. The narrower channel (denoted **a** in Fig. 2) is the test channel. The second, wider channel was added for future installation of an in-situ flow rate measurement device (currently under development). Fluid is supplied into and removed from the ends of the narrow channel through vertical bores (points 1 and 2 in Fig. 2). A third bore at the L's end (point 3 in Fig. 2) allows for pressure measurement. The upper face of the channel consists of a Pyrex glass cover which is attached to the silicon using standard anodic bonding techniques. The glass cover permits direct flow observation. This has proved to be particularly useful in identifying sources of blockage or air bubble entrapment in the thin channels.

In the experiments reported here, we use <100> silicon. The cross-section of each of the channels is approximately rectangular; and the channels' depth varies from a fraction of a micron to about 5 microns. The channel depth was obtained by using a surface profilometer (alpha-step) with an accuracy of 2%. The channels' surface is remarkably smooth. Fig. 3 shows a cross section of the channel as traced by the profilometer. The test structure is attached to an adapter through which fluid is supplied into and discharged from the test structure. Leaks are prevented through the use of O-rings. The silicon "wind tunnel" is held in place with a clamp bolted to the adapter. This arrangement simplifies replacement of the test structure.

The pressure drop is obtained by measuring the pressure at the inlet to the test section and at the L's end. Pressure losses in the fittings are estimated to be a small fraction of the total pressure drop (typically less than 1% for flow velocities of the order of 10^{-2} m/s). In the future, we are planning to integrate the pressure measurement into the silicon wafer, as the current pressure measurement technique suffers from a slow dynamic response.

In the liquid experiments, the liquid is supplied from an accumulator pressurized by a regulated nitrogen cylinder. Before entering the test structure, the liquid is passed through a 0.5 μm filter. The flow rate is monitored by the volumetric discharge as a function of time. A number of readings are taken at any given pressure to verify that the flow is reasonably steady and that there are no air bubbles in the liquid. In the gas experiments, we follow a similar procedure. Here, however, the gas is supplied directly from a regulated, pressurized cylinder.

The experimental procedure consisted of varying the inlet pressure with the regulating valve and measuring the flow rate. Two sequences of measurements were taken, one set with monotonically increasing pressure and one with monotonically decreasing pressure. This was done in order to detect hysteresis, if any.

In the experiments reported here, we used n-propanol, silicone oil and nitrogen gas. Representative properties of the different fluids are given in the table below. In our experiments, we measured the ambient temperature and we introduced appropriate corrections in the thermophysical properties to compensate for temperature variations.

Table: Thermophysical properties of the various fluids used in the experiments at 25°C.

PROPERTY	**n-propanol**	**silicone oil**	**N_2**
Molecular mass	60.1	410	28.013
Polarity	yes	no	no
Density (kg/m^3)	779	870	1.145
Viscosity (kg/m-s)	$1.95 \cdot 10^{-3}$	$1.74 \cdot 10^{-3}$	$1.78 \cdot 10^{-5}$
Surface Tension (dynes/cm)	23.4	18.7	-----

ANALYSIS OF EXPERIMENTAL DATA

The momentum equation for one-dimensional, steady, compressible fluid flow in a channel with uniform cross-section can be written as [15]:

$$\Delta p = p_1 - p_2 = \Delta p_a + \overline{f} \frac{G}{2D_H} \int_0^L u\, dx. \tag{1}$$

where p is the pressure; u is the cross-sectionally averaged velocity; G ($=\rho u$) is the mass flow rate per unit cross-sectional area; subscripts 1 and 2 denote, respectively, inlet and outlet conditions; ρ is the liquid's density;

$$D_H = \frac{4\,(\text{cross sectional area})}{\text{wetted perimeter}} \tag{2}$$

is the hydraulic diameter; $\overline{f}$ is the friction factor averaged along the channel's length L;

$$\Delta p_a = G^2 \left(\frac{1}{\rho_2} - \frac{1}{\rho_1} \right) \tag{3}$$

is the pressure loss resulting from the fluid's acceleration. In the case of incompressible flow, the velocity u is constant and $\Delta p_a = 0$.

In our experiments, we measured the flow rate and the pressure drop. The results are presented in non-dimensional form in terms of the friction factor ($\overline{f}$) and the Reynolds number,

$$\mathrm{Re} = \frac{u D_H}{\nu} = \frac{G D_H}{\mu}, \tag{4}$$

where ν and μ are, respectively, the temperature adjusted kinematic and shear viscosities. Both the mass flow rate (G) and the Reynolds number are approximately constant for any given flow.

We present most of our results by depicting the friction constant (C),

$$C = f\,\mathrm{Re} \tag{5}$$

as a function of the Reynolds number.

In our experiments, the channels were approximately rectangular with height d and width w. Typically the height d ranged from a fraction of a micron to about $5\mu\mathrm{m}$ while the width $w \sim 100\mu\mathrm{m}$. The directly measured quantities were the pressure drop Δp and the volumetric flow rate V. The friction coefficient ($\overline{f}$) was obtained from equation (1). For example, for incompressible flow:

$$C = \frac{8}{\mu}\frac{\Delta p}{LV}\frac{wd^3}{(1+\frac{d}{w})^2}.$$

(6)

We write this expression to emphasize the necessity of obtaining the channel depth (d) with great accuracy, as any error in the depth measurement triples the error in the value of C.

In all our experiments $\frac{D_H}{L} \approx 5 \cdot 10^{-4}$. Hence, the flow may be taken as being fully developed. According to the Navier-Stokes equations for fully developed, laminar, incompressible channel flow, C is a constant which depends only on the channel's cross sectional geometry. See, for example Shah and London [13], Knudsen and Katz [14] and the literature cited there. Thus, we investigate the adequacy of the Navier-Stokes equations for our particular flow conditions by comparing the predicted C with the measured one.

We were not able to find any theoretical calculations of the friction factor for compressible flow. Based on limited experiments [15], it is widely assumed in the literature that the friction factor for compressible flow is similar to the one for incompressible flow.

EXPERIMENTS WITH LIQUIDS

The results of our liquid experiments are set forth in Figs 4 and 5 where we depict C as a function of the Reynolds number. The vertical bars indicate one standard deviation of the experimental data. Fig. 4 depicts our experimental results for n-propanol. In accordance with theory, the experimental data does not show any dependence between C and the Reynolds number. The experimental value of C (~79) is, however, consistently lower than the one predicted by theory. The theoretical predictions for the the 0.5 and 3μm deep channels are, respectively, 95.4 and 92.6. Fig. 5 depicts our experimental data for the silicone oil. In the larger channel (4.65μm deep), we observe no dependence between the experimental constant (C~77) and the

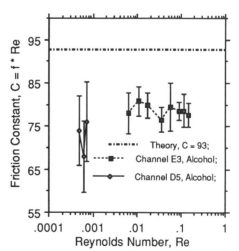

Fig. 4: C values for alcohol are depicted as a function of the Reynolds number for flow in 0.5μm·115μm·10.5mm (D5) and 3.0μm·110μm·10.5mm (E3) channels.

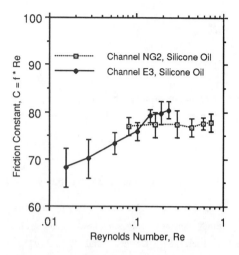
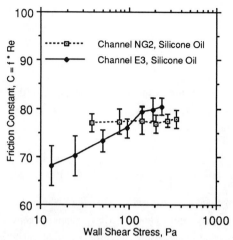

Fig 5: C values for silicone oil are depicted as a function of the Reynolds number for flow in 4.65μm·95μm·10.9mm (NG2) and 3.0μm·110μm·10.5mm (E3) channels.

Fig 6: C values are depicted as a function of the average wall shear stress for silicone oil flowing in 4.65μm·95μm·10.9mm (D5) and 3.0μm·110μm·10.5mm (E3) channels.

Reynolds number. Here again the experimental value is lower than the theoretical one (C=90). For the shallower channel (3.0μm deep), we observe dependence between C and the Reynolds number. This is a deviation from theoretical predictions.

In Fig. 6, we depict the data of Fig. 5 in a slightly different way. We show C as a function of the average wall shear stress. The purpose of this figure is to examine whether the sloping of the data in Fig. 5 may be a result of shear thickening. The figure shows that for the same shear rates, we observe sloping of the small channel data but not the larger one. Also, the silicone oil is known to be newtonian in the range of shear rates reported here and to exhibit shear thinning rather than shear thickening at higher shear rates. We do not have sufficient data yet to make definite conclusions; but thus far it seems that we may be seeing a channel size effect.

EXPERIMENTS WITH GASES

The flow of gases in microchannels presents a somewhat more complex situation than that of the flow of liquids. Due to the very high pressure gradients involved, the flow cannot be considered incompressible. In order to satisfy mass conservation, the gas must accelerate as its density decreases with decreasing pressure towards the exit of the channel. Thus the pressure difference between inlet and exit consists of two components. One component is due to the acceleration of the gas while the other is due

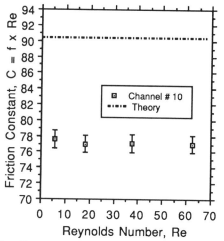
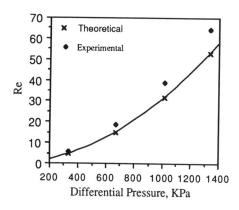

Fig. 7: The friction constant C is depicted as a function of the Reynolds number for nitrogen gas flow in a channel with cross-sectional dimensions of 4.65µm·98.73µm.

Fig. 8: The flow rate (Re) is depicted as a function of the pressure drop (Δp). The solid line represents theoretical results while the symbols represent experimental data.

to wall friction. In our experiments, we measure the total pressure loss (Δp) and we estimate Δp_a. Because of the very small size of the channel, the thermal equalization time between the gas temperature and that of the wall is very small (on the order of microseconds). Thus, to the first approximation, we assume the flow to be isothermal.

Fig. 7 depicts the experimental value of C as a function of Reynolds number. The experimental C (~77) appears to be independent of the Reynolds number but consistently lower than the theoretical prediction (C=90.4). The vertical bars in Fig. 7 correspond to one standard deviation of the measured data. In Fig. 8, we show the predicted and measured Reynolds numbers as a function of the pressure drop. The discrepancy between the prediction and the experiment is consistent with the deviation between the measured and predicted values of C (Fig. 7).

DISCUSSION

In this paper, we presented measurements of the friction factor for fully developed laminar flow in small channels. In most of the experiments, we observed that the friction law could be prescribed as $f \cdot Re = C$, with C being independent of the Reynolds number. This behavior is in line with predictions based on the Navier-Stokes equations. The experimentally obtained C was consistently lower than the theoretically predicted one. We can not state categorically, however, that the flows deviate from Navier-Stokes behavior as the experimental results depend crucially on the depth of the channel, a quantity difficult to measure at the orders of magnitude under consideration.

We also observed dependence of C on the Reynolds number of silicone oil in a 3µm deep channel. We do not have an explanation for this dependency. It could possibly be a result of non-newtonian behavior or perhaps we are observing some channel size effect.

The work presented here raises a few interesting questions. Additional research is needed to provide us with the answers.

ACKNOWLEDGEMENT

The research reported in this paper has been supported by the NSF through grant EET 88-15284.

REFERENCES

1. Tuckermann, D.B., and Pease, R.F.W., "High-Performance Heat Sinking for VLSI," *IEEE Electron Device Letters*, Vol. EDL-2, No. 5, 126-129, May 1981.

2. Tuckermann, D.B., and Pease, R.F.W., "Optimized Convective Cooling using Micromachined Structures," *J. Electrochemical Society,* Vol. 129, No. 3, C98, March 1982.

3. Tuckermann, D.B., *Heat Transfer Microstructures for Integrated Circuits*, Ph.D. thesis, Department of Electrical Engineering, Stanford University, 1984. Also, Report UCRL 53515, Lawrence Livermore National Laboratory.

4. Nakagawa S., Shoji S., and Esashi M., "A micro-chemical analyzing system integrated on silicon chip," Proceedings IEEE: *Micro Electro Mechanical Systems*, Napa Valley CA 1990 (90CH2832-4).

5. Peiyi W. and Little W.A., "Measurement of friction factors for the flow of gases in very fine channels used for microminiature Joule-Thompson refrigerators," *Cryogenics*, 273-277. 1983.

6. Harley J., Pfahler J., Bau H.H. and Zemel J., "Transport processes in micron and submicron channels," *ASME Proceedings HTD-116*, 1989 (Figliola R.S., Kaviany M., and Ebadian M.A., eds.) Also *Sensors and Actuators* A21-23, 431-440, 1990.

7. Eringen, A., and Suhubi, E., "Nonlinear Theory of Simple Micro-Elastic Solids I," *Int. J. Eng. Science*, Vol. 2, 189-203, 1964.

8. Eringen, A., " Simple Microfluids," *Int. J. Eng. Science*, 205-217, 1964.

9. Eringen, A., "Theory of Micropolar Fluids," *J. Math. and Mech.*, Vol. 16, 1-18, 1966.

10. Ariman, T., and Turk, M.A., "Microcontinuum Fluid Mechanics - A Review," *Int. J. Eng. Science*, Vol. 11, 905-930, 1973.

11. Ariman, T., Turk, M.A., and Sylvester, N.D., "Applications of Microcontinuum Fluid Mechanics," *Int. J. Eng. Science,* Vol. 12, pp. 273-293, 1973.

12. Peterson, K.E., " Silicon as a Mechanical Material," *Proceedings of the IEEE*, Vol. 70, No. 5, 420-457, May 1982.

13. Shah, R.K., and London, A.L., *Laminar Flow Forced Convection in Ducts*, Academic Press, 1978.

14. Knudsen J.G., and Katz, D.L. *Fluid Dynamics and Heat Transfer*, McGraw-Hill, 1958.

15. Shapiro A., *Comprssible Fluid Flow*, Ronald Press, 1953.

Squeeze-Film Damping in Solid-State Accelerometers

JAMES B. STARR

Solid State Electronics Center
Honeywell, Inc., Plymouth, MN

ABSTRACT

Damping of proof mass oscillations is a fundamental consideration in the design of accelerometers. The introduction of silicon micro-machining into accelerometer technology facilitates very thin fluid layers that provide sufficient damping through viscous dissipation in a gas. Thicker layers require a liquid for sufficient damping. Because the viscosity in liquids is in general more temperature dependent than in gases, the smaller micromachined devices can exhibit less dependence of damping on temperature.

In the midst of these encouraging developments, the accelerometer designer is faced with three fundamental challenges. First, one must be aware of both the nonlinear and the compressible behavior of gas films that can lead to undesirable accelerometer performance, not to mention inadequate damping. Second, one must contend with other than relatively simple geometries in predicting damping coefficients. Third, one may approach the limits of continuum flow in very thin gas layers. Fortunately, there exist in the literature solutions for squeeze-film behavior that can be consulted when designing accelerometers. In addition, finite element techniques are readily available that facilitate damping predictions for other than simple geometries. In this paper criteria are set forth that are aimed at alerting the designer to the onset of nonlinear and compressible effects in accelerometers. A finite element technique is presented that essentially solves Reynold's equation for small displacements and "squeeze" numbers by analogy with heat conduction in a solid with internal heat generation. Examples are presented that show film pressure profiles generated for complex geometries with non-uniform film thicknesses and squeeze velocities. A technique for mitigation of compressibility effects through film pressurization is suggested.

INTRODUCTION

Sensors that accurately replicate accelerations of mechanical systems require some mechanism for energy dissipation, i.e. a damper. An accelerometer is the mechanical equivalent of a spring-mass system. As such, its response is frequency dependent. In particular, without damping, the output signal will increase significantly with frequency. Indeed, at its resonant frequency, very large amplitudes may produce the electrical consequence of signal saturation or the mechanical consequence of structural failure. Silicon accelerometers are especially susceptible to these consequences by virtue of the highly elastic behavior of silicon at typical operating temperatures.

The extensive use of silicon technology in accelerometers in recent years has introduced accelerometer products that offer significantly improved performance and size for a given cost. The inherent mechanical stability of silicon along with stable force transduction technologies has resulted in batch-producibility of precision accelerometers. Piezoresistive, capacitive and beam-resonant approaches have all been shown to produce excellent accelerometer products.

The goodness of an accelerometer product does depend on considerations that are not an inherent part of either silicon material or transduction technologies. As previously mentioned, damping is a key consideration. It is needed to achieve a flat response over a wide range of frequencies. Fortunately, micromachining has been one area in which silicon sensor technology has grown significantly so as to enhance provisions for damping. It has facilitated the forming of very thin gas layers that serve as an excellent means of viscous dissipation, even when the fluid medium is a gas. This means that the accelerometer designer often does not have to resort to the use of liquid fill fluids to achieve adequate damping. The extent of damping becomes less temperature dependent because of the smaller relative change in viscosity of a gas as compared with a liquid. A simple damping configuration is illustrated in Figure 1. Damping forces result from the buildup of pressure that is needed to produce lateral motion of the viscous fluid.

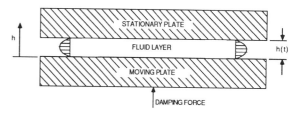

Figure 1. Squeeze-Film Damper Schematic

Full realization of the advantages of gas-film damping does require consideration of several factors affecting performances. Indeed, the designer must consider the following:

- The fluid behavior must be governed by viscous forces that are larger relative to momentum changes.
- Dimensions must be large enough to ensure the gas film exhibits continuum behavior.
- For certain conditions and geometries, damping may be adversely affected by compressibility effects.
- Displacements of the gas film may be large enough to introduce nonlinear behavior and signal distortions into the accelerometer
- Geometries of the damper may not necessarily be so simple as to utilize existing solutions for gas film pressures.

The following discussion addresses each of these factors. Results are presented in terms of concise formulations that should serve the accelerometer designer in his attempt to produce an excellent product.

REYNOLDS NUMBER CONSIDERATIONS

The behavior of squeeze films is in general governed by both viscous and inertial effects within the fluid. However, for the very small geometries encountered in silicon micromachined devices, inertial effects can be ignored. In such a case the behavior of the fluid is governed by

$$\partial[(\rho h^3/\mu)\partial P/\partial x]/\partial x + \partial[(\rho h^3/\mu)\partial P/\partial y]/\partial y = 12\partial(\rho h)/\partial t \quad (1)$$

Where P is the film pressure, ρ is density, μ is viscosity, h is film thickness, and x and y are spatial coordinates. This expression is derived from the much more complicated Navier Stokes equation under the assumption of small Reynolds numbers. According to Langlois in Reference (1), the specific condition for validity of equation (1) is

$$\omega h^2 \rho / \mu \ll 1.0 \quad (2)$$

where ω is the oscillation frequency of the damper plate. This condition is readily satisfied in typical silicon accelerometers. For example, an air-filled accelerometer with a squeeze-film thickness of 25 microns, operating at a frequency of 1.0 kHz, would have a Reynolds number defined by equation (2) of only 0.25.

CONTINUUM LIMITS

The small film geometries producible by micromachining of silicon may result in a condition where the fluid film cannot be treated as a continuum. As film thickness is reduced, eventually a condition is reached where the mean-free-path of molecules within the gas becomes significant with respect to the thickness of the film. In such a case, the effectiveness of damping is reduced due to a slip-flow condition existing at the boundaries of the film. For example, the mean-free-path for air at 25°C and 1.0 atm is about .09 microns. Slip-conditions may be encountered when the mean-free-path is about one percent of the film thickness. Thus if a film thickness of under 9 microns is used, the possibility of slip-flow should be considered. Such effects may be mitigated by using a high-viscosity gas of larger molecular weight, such as neon, or by pressurizing the gas.

SIMPLIFIED ANALYSES

The task of predicting damping coefficients for accelerometers is eased under certain conditions. First of all, thin gas films may be realistically assumed to be isothermal. This is because of the usually small conducting paths within the gas to bounding walls that have a relatively high heat capacity. For isothermal films, density is proportional to absolute pressure. Hence, all densities appearing in equation (1) can be replaced by pressure. This condition is in reality a very modest limitation on the analysis. Three additional simplifying assumptions are that pressure variations within the film are small relative to the average absolute pressure, the film thickness is uniform, and displacements of the moving surface are small relative to the film thickness. Subsequent sections of this paper will discuss conditions under which two of these assumptions are unrealistic. However, where they are realistic, there results a damping characteristic that is linear and characteristic of what is needed for precision accelerometers. The isothermal, small-pressure-variation and small-displacement assumptions applied to equation (1) result in the expression

$$\partial^2 P_*/\partial x^2 + \partial^2 P_*/\partial y^2 = 12\mu(\partial h/\partial t)/h_o^3 \quad (3)$$

where P_* is the pressure departure within the film from the ambient value, and h_o is the nominal film thickness. Equation (3) is simply a Poisson's equation that can be readily solved for simple geometries. The solution provides the pressure distribution within the film which can usually be integrated to provide an expression for the damping force. Two expressions for damping force are provided herewith for plates that move in a uniform manner transverse to the film plane. For a circular disk:

$$F_D = [3\pi\mu R^4 (\partial h/\partial t)]/(2h_o^3) \quad (4)$$

where R is the radius of the disk. For a rectangular plate of width 2W and length 2L,

$$F_D = 16f(W/L)W^3 L (\partial h/\partial t)/h_o^3 \quad (5)$$

where f(W/L) is given in Figure 2. The limiting case of W/L→0 would, of course, correspond to a very long narrow strip.

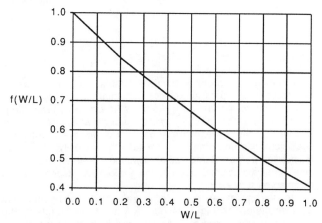

Figure 2. Effect of Aspect Ratio on Damping Produced by a Rectangular Plate

LARGE DISPLACEMENTS

The characteristics of squeeze film damping will be altered when displacements become significant with respect to the nominal film thickness. Such a phenomenon is suggested by equation (3) even though it is derived for small displacements. The term on the right-hand side includes the inverse of film thickness cubed. One should therefore expect that as amplitude increases, there will be a strong tendency for the damping coefficient to increase. The extent of this increase can be determined from equations derived by Sadd and Stiffler in Reference 2. Damping under the limitation of small density changes but with larger displacements can be calculated by simply multiplying expressions such as those in equations (4) and (5) by a displacement function defined as

$$f_D(\varepsilon) = (1 - \varepsilon)^{-3/2} \quad (6)$$

where ε is the ratio of plate displacement to nominal film thickness. A plot of this expression is shown in Figure 3. Note the very significant increase in damping with displacement; it only takes a relative displacement of 0.25 to cause a 10 percent increase in damping. It is also important to note that large displacements will result in a distorted accelerometer response to a sinusoidal input acceleration.

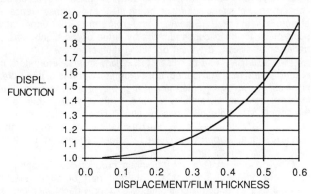

Figure 3. Effect of Displacement on Damping

COMPRESSIBILITY EFFECTS

Compressibility effects in squeeze-films have been addressed in both References (1) and (2). Use is made of a squeeze number, σ, defined by

$$\sigma = 12\mu\omega b^2/(h_o^2 P_a) \qquad (7)$$

where μ is viscosity, ω is frequency, b is a characteristic length, h_o is the nominal film thickness, and P_a is the ambient gas pressure. A power series in σ is generated so as to satisfy Equation (1) for a uniform film thickness. Langlois showed that for small values of σ the film behaved as though the fluids were incompressible. His results show that for a long narrow strip with b defined as its half-width, σ values of about 0.2 or less exhibit essentially incompressible behavior. At the other extreme, where σ is quite large, the film essentially acts as an air spring and exhibits little behavior desired of an energy dissipator. From these data one concludes that compressibility effects can be neglected by designing the accelerometer so that the squeeze number is much less than 0.2. Sadd and Stiffler in reference (2) solved the Reynolds equation also in terms of a power series in σ but with the admission of displacements that are significant relative to the nominal film thickness. Their expressions are useful for quantify effects of both the squeeze number and displacement on air-spring like behavior and on signal rectification. In general, their expressions are concluded to be useful for σ values up to about 0.3.

Air-spring behavior increases with σ because the fluid film in effect forms a restriction against free outflow from the edges of the film. This effect can be quantified in terms of the relative increase in stiffness in a simple spring-mass representation of the accelerometer. By applying expressions in Reference (2) to a spring-mass system using a spring of stiffness k, there results

$$\Delta k/k = C_k f_k(\varepsilon)(\omega/\omega_n)\zeta\sigma \qquad (8)$$

where $f_k(\varepsilon) = (1 + 3\varepsilon^2 + 3\varepsilon^4/8)/(1-\varepsilon^2)^3$,

ω is the resonant frequency and ζ is the damping relative to the critical level for the nominal spring-mass system. Values of C_k are 0.8 for a long, narrow strip and 0.33 for a circular disk. A plot of f_k is presented in Figure 4. From Equation (8) and Figure 4 we conclude that increases in stiffness due to compressibility are significant functions of both the squeeze parameter σ and relative plate displacement ε. In the limit of small displacements, assuming a critically damped accelerometer operating at resonant frequency, the relative increase in stiffness with a disk-shaped damper would be about $\sigma/3$.

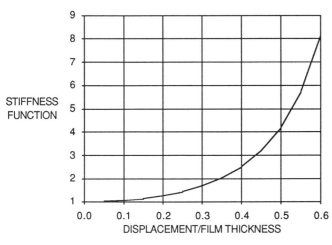

Figure 4. Fluid Stiffness Function

The inherent nonlinear characteristic of a damper may also introduce a rectifying effect into the accelerometer response. This effect exists only in connection with compressibility effects; it disappears when σ is zero. Again using expressions from Reference (2), one can derive a relationship between the time-averaged fluid force generated by the damper, designated F_R and the inertial force associated with the accelerometer's proof mass, designated F_I. The resulting expression, valid for $(\omega/\omega_n) \ll 1.0$, is

$$F_R/F_I = f_R(\varepsilon)(\omega/\omega_n)\zeta\sigma \qquad (9)$$

where for a disk

$$f_R(\varepsilon) = 5\varepsilon(4 + 3\varepsilon^2)/[48(1 - \varepsilon^2)^3]$$

and for a narrow strip

$$f_R(\varepsilon) = \varepsilon(4 + 3\varepsilon^2)/[4(1 - \varepsilon^2)^3]$$

Plots of these two functions are shown in Figure 5. Note that the rectification effect vanishes for zero relative displacement. This characteristic is very important in the design of precision accelerometers such as might be used in inertial navigational systems. Indeed, for such accelerometers provisions for limiting mass displacement would be required, e.g. through the use of a force rebalance servo system.

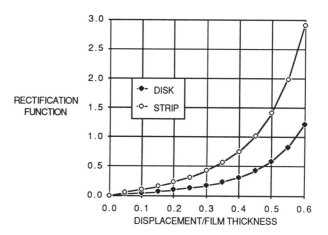

Figure 5. Rectification Function

When faced with the minimization of compressibility effects, it is important note the makeup of the squeeze number. Note that in its denominator is P_a, the ambient pressure. This suggests that one could reduce the value of σ and therewith the effects of compressibility by increasing the absolute pressure of the fill gas in the accelerometer.

COMPLEX GEOMETRIES

As previously discussed, the assumptions of incompressible behavior and uniform film thickness result in a governing equation that is in reality a Poissons Equation. As such, one may use existing finite element thermal analysis to generate solutions for film pressure. To that end, the Reynolds Equation may be written as

$$\partial^2\psi/\partial\eta^2 + \partial^2\psi/\partial\xi^2 = \lambda \qquad (10)$$

where η and ξ are spatial coordinates normalized by a reference length L, and

$$\psi = P \ast h_o^3 / [12 \, (\partial h / \partial t)_{max} \, \mu L^2]$$

The term λ facilitates the introduction of non-uniform motion of the damper as would occur with a hinged device; it is the ratio of the local plate velocity to the maximum plate velocity. Note that λ is analogous to internal heat generation per unit volume and ψ is analogous to temperature.

As an example, Figure 6 illustrates a pressure contour plot generated for a damping element. The damper is hinged at the lower edge. A hole within the device was required for routing current to a force rebalance device. With these data available, one can carry out a numerical integration of the pressure field to determine the damping force. It is also possible to determine damping action corresponding to higher order modes of oscillation that may occur. This is accomplished by evaluating damping with the plate hinged at the appropriate nodal axis.

CONCLUSIONS

In designing an accelerometer that incorporates a squeeze-film damping device, care must be taken to avoid the deleterious effects of fluid compressibility and large displacements. In general, ideal damping will result if both the squeeze number and the relative displacement are kept much less than 1.0. If those conditions are satisfied, one may then use finite element techniques to estimate damping coefficients for other than simple geometries.

REFERENCES

(1) Langlois, W. E., "Isothermal Squeeze Films", *Quarterly of Applied Mathematics,* vol.XX, no. 2, July, 1962, pp. 131-150.

(2) Sadd, M. H. and A. K. Stiffler, "Squeeze Film Dampers: Amplitude Effects at Low Squeeze Numbers", *Journal of Engineering for Industry,* Transactions of the ASME, vol. 97, Series B, Nov. 4, 1975, pp1366-1370.

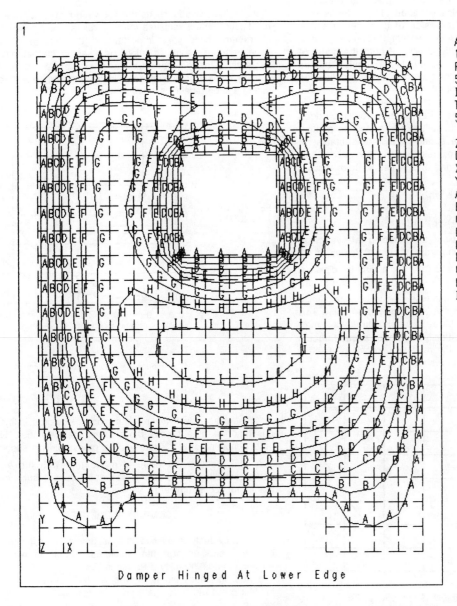

Figure 6. Contour Plots of Dimensionless Pressure (ψ) for a Hinged Damper

A MICROMACHINED FLOATING-ELEMENT SHEAR SENSOR

Martin A. Schmidt, Roger T. Howe, and Stephen D. Senturia
Microsystems Technology Laboratories
Department of Electrical Engineering and Computer Science

and
Joseph H. Haritonidis,
Turbulence Research Laboratory
Department of Aeronautics and Astronautics

Massachusetts Institute of Technology
Cambridge MA 02139 USA

ABSTRACT

This paper presents a surface-micromachined floating-element shear sensor suited for air-flow and wind-tunnel applications. The specific application intended is the characterization of both the mean and fluctuating components of wall shear stress in a turbulent boundary layer.

INTRODUCTION

A wall bounding a flowing fluid sets up a gradient of velocity near the wall. The resulting wall shear stress is one of the key parameters in bounded turbulent flows. Both the mean and fluctuating parts of the shear stress are manifestations of the structure of the flow above the wall. The mean value determines the overall drag characteristics of a particular configuration. The temporal and spatial fluctuations are of importance in sound generation, separated flows, passive or active control of turbulence and, in general, assessment of which types of flow structures are primarily responsible for momentum transport between the outer part of the boundary layer and the wall. The ultimate goal of this research is to produce a probe with a shear stress sensitivity greater than 1 Pascal, a 0-20 kHz bandwidth, and a spatial resolution of 100 μm.

A large number of researchers have studied wall shear stress in turbulent boundary layers[1-4]. Reported measurements show considerable scatter which may be attributed in part to the difficulties in measurement. There are a large variety of techniques available for measuring shear stress. Figure 1 illustrates the hot-film and floating-element shear-stress probes, which are two commonly used sensors.

The hot-film is an example of an _indirect_ measurement in which the heat convected away from the film is measured and a value of shear stress is inferred.

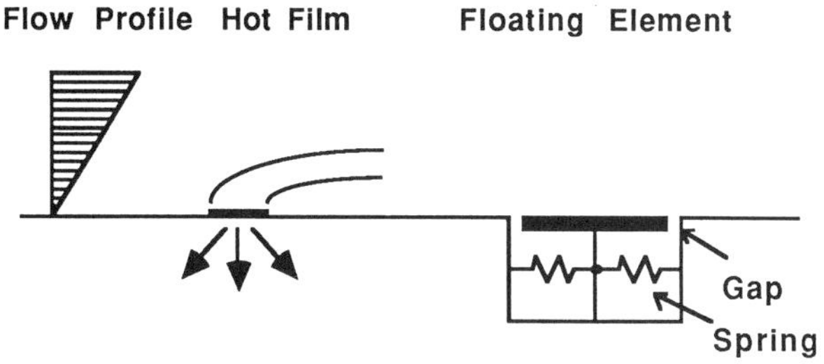

Fig.1 Two common probes used in shear stress measurements: hot film and floating element.

Calibration of this sensor requires assumptions regarding the flow profile as well as the heat transfer properties of a turbulent boundary layer. The floating element is a _direct_ measurement of the shear stress at the wall.

Microfabrication can improve the characteristics of both types of sensor. Size reduction improves the spatial resolution and allows for the use of sensor arrays. In the case of the hot-film sensor, heat loss to the substrate can be reduced using micromachined thermal isolation techniques. In the case of the floating element, micromachining will reduce gaps between the floating element and the surrounding wall, minimizing flow around the element which can substantially perturb the measurement. Also, the AC response of the floating element, which is limited by the first resonant frequency of the element, can be improved by reducing the size, and hence mass, of the element. This paper reports a microfabricated floating-element shear sensor.

MICROFABRICATED ELEMENT

Figure 2 shows the simplest version of the floating-element shear sensor. A square element suspended above the silicon surface by four tethers will deflect laterally by an amount proportional to an applied shear stress. The structure is fabricated using a surface micromaching process similar to previously

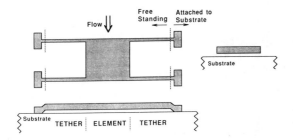

Fig.2 Illustration of microfabricated floating-element shear sensor: top view and two sections.

reported suspended polysilicon structures[5,6]. Figure 3 is a SEM photograph of a floating element, fabricated in an 8 μm thick polyimide film deposited over a 3 μm sacrificial aluminum spacer, which is completely removed in a mixture of phosphoric-acetic-nitric acids and water, releasing the suspended polyimide structure. Residual tensile stress in the polyimide[7] keeps the structure suspended above the surface. These structures have been tested in an uncalibrated laminar flow cell and have been observed to deflect several μm. We now look at the mechanical design issues and how we propose to detect deflections of this size.

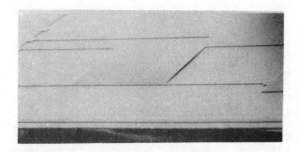

Fig.3 SEM photogragh of 1 x 1 mm polyimide floating element.

DEFLECTION ANALYSIS

The residual tensile stress in the polyimide dominates the load-deflection characteristics of the floating element. The structure is modeled as two clamped-clamped beams of length 2L with an axial force due to the residual stress in the tethers. This model assumes that the element moves rigidly under an applied force which is the product of the wall shear and the area of the element. Figure 4 shows a plot of the calculated sensitivity as a function of residual polyimide stress for a 1 mm x 1 mm x 10 μm element with 1 mm x 10 μm x 10 μm tethers. At low values of residual stress, the sensitivity is constant and

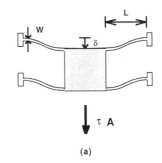

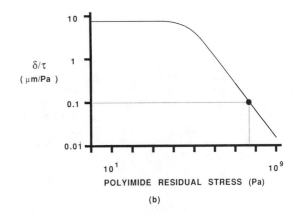

Fig.4 (a) Floating element under shear-stress loading. (b) Calculated sensitivity of sensor in Fig.3 versus polyimide residual stress.

determined only by the modulus of the material and its geometry, but at higher values, the sensitivity becomes a function of the residual stress, at which point we are in the "tensioned-wire" region. This is the region of operation given a typical polyimide residual stress of 30-40 MPa[7]. The results indicate that stresses on the order of 1 Pa are measurable given a readout scheme that resolves 0.1 μm deflections.

READOUT

Figure 5 illustrates a differential capacitance readout scheme for detecting micron-sized deflections. A thin conducting shield plate is embedded in the polyimide floating element directly above three electrodes on the chip surface. The central drive electrode is coupled to the symmetrically placed sense electrodes by the capacitance to the shield plate, C_o, and the shield-plate-to-sense electrode capacitances, C_1 and C_2. Capacitance C_o is a constant but C_1 and C_2 will vary linearly with element deflection. The sense electrodes are connected directly to the gates of a pair of matched metal-gate depletion-mode PMOS transistors to buffer the signal before going off-chip to the transresistance amplifier.

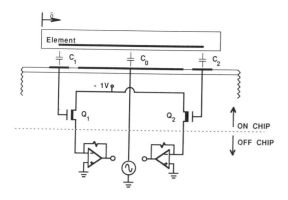

Fig.5 Schematic illustration of differential capacitor readout scheme.

The test chip layout with differential capacitor readout is 4 x 5 mm. The bonding pads have been placed sufficiently far downstream of the element so as not to perturb the flow. The element is patterned such that polyimide will be left in the field region to passivate the chip and also reduce the amount by which the element protrudes from the surface. The gaps between the element and the surrounding polyimide can easily be reduced to 10 μm or less.

FABRICATION AND TEST

The test chip is fabricated by first making the aluminum-gate depletion-mode PMOS device and passivating with CVD oxide. We then deposit and cure a 1 μm polyimide layer to improve passivation and provide a mechanical buffer layer against thermal stresses between the CVD oxide and the 3 μm evaporated aluminum spacer. Once the spacer is patterned and defined, a 1 μm coat of polyimide is deposited and partially cured at an intermediate temperature. The thin shield plate metal is vacuum evaporated to a thickness of 30 nm and patterned. The remaining polyimide is deposited to reach the desired thickness of 10-30 μm and then cured. A nonerodible aluminum etch mask is evaporated and patterned and the polyimide is etched using an O_2 plasma in a parallel-plate reactor to obtain vertical sidewalls. At the same time, the polyimide is removed over the bonding pads. The aluminum spacer is etched away to release the floating element, followed by the removal of the CVD oxide from the bonding pads by a wet chemical etch.

Control of the stress nonuniformities in a released structure such as this is critical in order to fabricate an element which is flat over the dimensions required[8]. Even more challenging is the control of stresses in a composite structure such as the element with an embedded thin conductor. We have investigated a variety of polyimides and two metals which exhibit good adhesion to polyimide (aluminum and chrome). Increasing polyimide thickness will in general reduce the curvature in a homogeneous polyimide released structure such that it is possible to fabricate a 1 x 1 mm element with a total deviation from a flat surface of less than 1 μm using polyimide thicknesses on the order of 10-30 μm. The thin metal however, can introduce substantial curvature into an already flat polyimide element. The thickness dependence of the curvature is a complex function of the relative stresses in the polyimide and metal. Aluminum evaporated in both electron-beam and filament systems exhibits very high compressive stress. This, coupled with the tensile stress in the polyimide and the asymmetric placement of the metal in the film, can generate substantial bending moments. Furthermore, this high compressive stress creates a delamination in the film from the underlying polyimide during subsequent processing.

The evaporated chrome is under tensile stress, as exhibited by tensile microcracking of the film. We have found that the magnitude of the chrome stress is sufficiently low, perhaps due to the microcracking, to make it possible to fabricate plates with less than 1 μm of curvature over 0.5 mm, and 1-2 μm over 1 mm. In spite of cracks, the plate remains a good conductor. Further study of these effects is in progress.

The fabricated structures have been tested by mechanically deflecting the plate using a probe tip and measuring the response of the sensor. Figure 6 is a plot of these measurements, with a 10V drive applied to C_0. The sensitivity, referred to the gate of the FETs, is 0.8 mV per μm of deflection. This is in good agreement with a sensitivity of 1.0 mV per μm of deflection, calculated using the circuit of Figure 5 (with parasitic capacitors to the substrate added). The resolution of this sensor will be limited by the noise in the depletion-mode FETs, which we estimate to be less than 1 μV in a 10kHz bandwidth. Given the estimated deflection sensitivity of 0.1 μm/Pa, the signal-to-noise ratio for 1 Pa will be greater than 40 dB.

PACKAGING

We have designed a first-level package for the shear sensor using silicon micromachining that allows the device to be inserted into the flow with minimal perturbation. The sensor is held in a micromachined silicon plate which

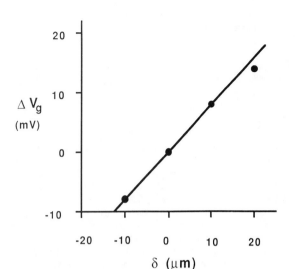

Fig.6 Measured response of test chip to mechanical loading.

includes a series of anisotropically etched pressure taps (to be used for sensor calibration), and a rectangular anisotropically etched cavity which holds the chip. The use of micromachining allows us to minimize gaps between the chip and its holder. Figure 7 shows a schematic of the package as designed for laminar flow calibration. The chip and silicon plate are flush mounted to the lucite base plate and attached with epoxy.

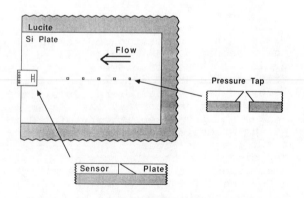

Fig.7 First-level sensor package for laminar flow calibration.

CONCLUSIONS

We have fabricated a micromachined floating-element shear-stress sensor with integrated readout. The structures have been tested, and a micromachined first-level package has been designed to calibrate the sensor in laminar flow.

ACKNOWLEDGEMENTS

The authors would like to acknowledge valuable discussions with Steven Petri and Prof. Patrick Leehey of MIT. Martin Schmidt would like to acknowledge the support of a 3M Sensor Fellowship. Samples were fabricated in the Microelectronics Laboratory of the MIT Center for Materials Science and Engineering, which is supported in part by the National Science Foundation under Contract DMR-84-18718. Valued discussion and assistance in fabrication were provided by Mehran Mehregany, Mark Allen, and Herb Neuhaus.

REFERENCES

1) K.G. Winter, "An Outline of the Techniques Available for the Measurement of Skin Friction in Turbulent Boundary Layers", Prog. Aerospace Sci., Vol. 18, 1977, pp 1-57.

2) P.H. Alfredsson, A.V. Johansson, J.H. Haritonidis, and H. Eckelmann, "On the Fluctuating Wall Shear Stress and the Velocity Field in the Viscous Sublayer", Journal of Fluid Mechanics, submitted.

3) J.M. Paros, "Application of the Force-Balance Principle to Pressure and Skin Friction Sensors", Proc. Inst. Enviromental Sci., 1970, pp 363-368.

4) J.M. Allen, "Improved Sensing Element for Skin-Friction Balance Measurements", AIAA Journal, Vol. 18, 1980, pp 1342-1345.

5) R.T. Howe, "Polycrystalline Silicon Microstructures", in C.D. Fung, P.W. Cheung, W.H. Ko, and D.G. Fleming, (eds.), Micromachining and Micropackaging of Transducers, Amsterdam, Elsevier, 1985, pp 169-187.

6) H. Guckel, D.W. Burns, J.J. Hammond, "Planar Processed, Integrated Displacement Sensors", in C.D. Fung, P.W. Cheung, W.H. Ko, and D.G. Fleming, (eds.), Micromachining and Micropackaging of Transducers, Amsterdam, Elsevier, 1985, pp 199-203.

7) M.G. Allen, M. Mehregany, R.T. Howe, and S.D. Senturia, "Microfabricated Structures for the In-Situ Measurement of Residual Stress, Young's Modulus, and Ultimate Strain of Thin Films", Appl. Phys. Lett., in press.

8) S.D. Senturia, "Microfabricated Structures for the Measurement of Mechanical Properties and Adhesion of Thin Films", this conference.

A MULTI-ELEMENT MONOLITHIC MASS FLOWMETER WITH ON-CHIP CMOS READOUT ELECTRONICS

Euisik Yoon and Kensall D. Wise

Center for Integrated Sensors and Circuits
Solid-State Electronics Laboratory
University of Michigan
Ann Arbor, MI 48109-2122

ABSTRACT

This paper reports the implementation of control circuitry for a multi-element mass flow sensor capable of simultaneously measuring five different gas-related parameters necessary for mass flow. A process has been developed to merge the micromachined sensing structures and the CMOS readout circuitry on a single chip. The total process requires thirteen masks: eight for the double-poly single-metal CMOS process and five for additional processing of the transducer elements. The on-chip circuitry includes offset-free instrumentation amplifiers, an analog multiplexer, heater drive circuits, self-test circuitry, and a band-gap temperature sensor using substrate npn bipolar transistors. The resulting multi-element chip requires only ten external pins and delivers high-level buffered output signals describing gas velocity, type, direction, pressure, and temperature. The chip dissipates $120mW$ from $\pm 5V$ supplies and measures $3.5mm$ x $5mm$ in $3\mu m$ features.

INTRODUCTION

During the last few years, one of the more important topics associated with the development of silicon sensors has been the possible integration of several different types of transducers along with appropriate interface circuitry on a common chip [1-4]. Recently, we reported a multi-element monolithic mass flow sensor capable of simultaneously measuring flow velocity, flow direction, gas type, pressure, and ambient temperature [5]. While each of these elements performs well individually, the resulting chip requires that ten separate transducers be sensed or controlled. Further optimization therefore requires on-board circuitry to boost and multiplex the transducer output signals and to reduce the number of external leads. In this paper, we report the realization of these on-chip interface electronics and discuss details of the circuit design and of the process modifications needed to allow on-chip circuitry. The resulting multi-element sensor illustrates a nearly generic structure, allowing the measurement of five independent variables with minimal cross-parameter sensitivity and with on-board signal conditioning. Our initial application for this device is in monitoring semiconductor process gases in the atmospheric pressure range, although a number of other applications are also possible. In monitoring process gases, the device, using its on-board electronics, functions as part of a VLSI sensing node, working over a standardized sensor bus with the system tool controller. Digital compensation of the device for offset, linearity, and slope errors is implemented at the controller level using coefficients uploaded from the node.

MERGED PROCESS FLOW

Figure 1 shows a cross-section and top view of the multi-element flowmeter chip. The transducing structures are identical to those of the multi-element mass flow sensor chip previously reported [5]. Flow velocity, direction, and gas type sensors are supported on dielectric windows selectively etched in the silicon wafer. The four windows are located in a common well and are supported and thermally isolated using boron-diffused support beams. These beams are formed by the boron etch-stop when the silicon is selectively removed from the back of the wafer as the final step in wafer fabrication. The windows are $0.5mm$ x $0.5mm$ x $1.5\mu m$ in size and are formed using a layered silicon dioxide/silicon nitride/silicon dioxide sandwich which is

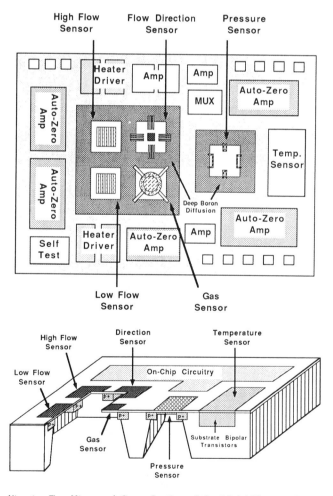

Fig. 1: Top View and Cross-Section of the Multi-Element Mass Flow Sensor with On-Chip Readout Electronics.

Reprinted from *Technical Digest IEEE Solid-State Sensor and Actuator Workshop*, pp. 161-164, June 1990.

in mild tension. The two sensors for flow velocity use a thin gold/chromium ($260nm/40nm$) film as a temperature detector and polysilicon as a heater. Flow direction is measured using a third window, which contains two orthogonal pairs of differential polysilicon-gold thermopile detectors together with a polysilicon heater. The fourth window contains a conductivity cell for the measurement of gas type [6]. A piezoresistive polysilicon pressure sensor is located in a separate well for the measurement of ambient pressure.

On-chip sensing and control feedback circuitry is fabricated on the unetched bulk portion of the chip. A standard $3\mu m$ p-well CMOS process is used with modifications for compatibility with the sensor fabrication. Five additional masks are necessary in addition to the double-poly single-metal CMOS process itself (implemented in eight masks). These additional masks define the deep boron diffusion for front-side window definition, the shallow boron diffusion for the pressure sensor window, the thin film for the gas sensor, the dielectric cuts to remove the window dielectrics over the circuit area, and the backside silicon etch wells. The deep boron diffusion used for window definition is performed simultaneously with the p-well drive-in diffusion [7]. The depth of this diffusion is important in that it determines (along with the lateral dimension) both the strength of the window support structure and the effectiveness of the beams in preventing any thermal interactions (crosstalk) between windows. At a depth of $15\mu m$, the crosstalk between windows is less than 0.4 percent.

Figure 2 shows the merged process flow, which combines the standard CMOS and the micromachined diaphragm processes. Fabrication begins by implanting the p-well areas of the active devices at a typical dose of $2.0 \times 10^{13} cm^{-2}$ with boron at $150 KeV$. A $1.5\mu m$-thick cap oxide is next deposited by LPCVD and opened to define

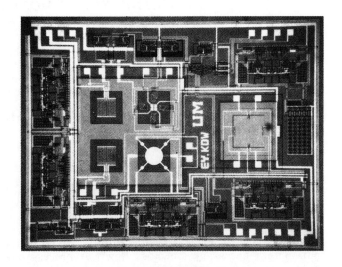

Fig. 3: Fabricated Multi-Element Flow Sensor Chip. On-Chip control circuitry is included on the unetched bulk silicon region.

the window rim areas. The deep boron diffusion is then performed at $1175°C$ for 16 hours, simultaneously driving the p-well and giving etch-stop and p-well depths of $15\mu m$ and $5\mu m$, respectively at the end of process. The pressure sensor diaphragm area is diffused next to produce a typical etch-stop thickness of $2\mu m$. This diffusion sets the sensitivity of the pressure sensor.

At this point, a pad oxide and LPCVD nitride layer are formed, the field areas are implanted, and a $0.9\mu m$ field oxide is grown, after which the window dielectrics are deposited by LPCVD. Typical windows consist of silicon dioxide, silicon nitride, silicon dioxide composite films in thicknesses of $200nm$, $300nm$, and $800nm$, respectively. The temperatures used in the formation of these films (820-$920°C$) do not significantly alter the p-well or etch-stop profiles. The window dielectrics are removed over the circuit areas, and the rest of the process flow is identical with the standard CMOS process (including all channel implants). The silicon wafer is etched from the back side in ethylenediamine pyrocatechol to form the window/diaphragm wells (and die separation) as the final step in the process. Using this etchant, an alternative to aluminum metallization is needed; gold on chromium has been used here both for circuit interconnect and for thin-film temperature sensing ($3400 ppm/°C$).

For this process, the sensor diffusions are performed first so that only the CMOS p-well diffusion needs to be adjusted as a result of the merged process. The more critical portions of the circuit process are unaffected up to final metal. Figure 3 shows a photograph of a fabricated flowmeter with on-chip readout electronics. The overall die size is $3.5mm \times 5mm$ in $3\mu m$ features.

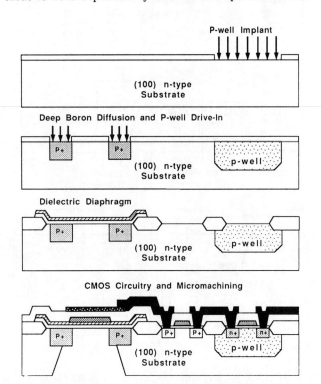

Fig. 2: Detailed Flow Chart of the Merged Process which Combines the Standard CMOS and the Micromachined Diaphragm Processes.

DESIGN OF ON-CHIP CIRCUITRY

Figure 4 shows a block diagram for the on-chip readout circuitry. The circuitry requires as inputs $\pm 5V$, GND, three multiplexer address bits (S0, S1, and S2), a self-test control signal (ϕ_{TEST}), and a two-phase nonoverlapping clock ($\phi 1$ and $\phi 2$), and delivers a multiplexed high-level output. The nominal clock frequency is $14 KHz$. Switched-capacitor offset-zeroing amplifiers are used where DC levels are important, while simple

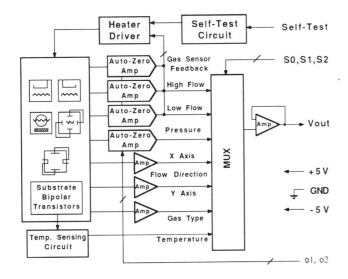

Fig. 4: Block Diagram of the On-Chip Readout Electronics.

resistively-coupled amplifiers are used elsewhere (e.g., in the thermopile direction sensor). The thermally-based sensors (for flow velocity, flow direction, and gas type) operate in constant temperature mode, with temperatures set by on-chip feedback loops and heater drivers.

Figure 5 shows the self-test circuitry implemented in one of the velocity sensors to measure changes in the electrothermal characteristics of the diaphragm window *in situ*. With the self-test signal (ϕ_{TEST}) off, the velocity sensor operates in the normal constant-temperature feedback loop for measuring flow. When ϕ_{TEST} is on, two additional circuit blocks are included in the feedback loop. These circuits form an electrothermal oscillator whose pulse duration is a function of the diaphragm time constant and the flow velocity (which continues to be monitored by the other velocity sensor during the self-testing mode). Figure 6 shows typical waveforms of the self-test outputs. The time durations t_1 and t_2 can be used to calculate the intrinsic thermal time constant (τ) of the diaphragm sensor window given by:

$$\frac{1}{\tau} = \frac{\ln\xi}{t_2} \cdot [1 - \frac{P_{fc}}{P_e} \cdot \frac{(T_l - T_f)}{(T_v - T_f)} \cdot \frac{(\xi^{(1+t_r)} - 1)}{(\xi^{t_r} - 1)}]$$

$$\text{where } \xi = \frac{T_h - T_f}{T_l - T_f}, \quad t_r = \frac{t_1}{t_2},$$

T_h and T_l are the high and low switching temperature levels of the Schmitt Trigger, respectively, T_f is the ambient flow temperature, and P_e is the power supplied to the heater during the heating time (t_1) [5, 8]. P_{fc} is the forced convection heat dissipation due to flow and is measured from the other velocity sensor which operates at the temperature T_v. This transient thermal response of the diaphragm can be used to detect the buildup of surface deposits that would lead to errors in the velocity measurements.

Substrate npn bipolar transistors are used to form a temperature sensor. Figure 7 shows the circuit schematic of the temperature sensor and the structure of the substrate transistor. The source/drain diffusion of the NMOS transistors is used as emitter, the p-well region as base, and the n-substrate as collector. The Gummel plot of the substrate bipolar transistor is shown in Fig. 8. A typical common emitter current gain of 100 has been measured. The temperature sensor output response is given by:

$$\frac{dV_{out}}{dT} = \frac{R_2}{R_1} \cdot \frac{dV_T}{dT} \cdot \ln A - \frac{dV_{BE}}{dT}$$

where V_T is the thermal voltage (kT/q), V_{BE} is the base-emitter voltage of bipolar transistor (typically, $dV_{BE}/dT \approx -2mV/°C$), and A is the ratio of emitter size between two bipolar transistors. The output response of the temperature sensor is designed to be $+4mV/°C$ using the resistor ratio, which can be set precisely.

RESULTS AND DISCUSSION

Table 1 summarizes the characteristics of the individual sensor elements including the temperature sensor realized with circuit components. Each of these transducers has been successfully implemented without any significant change in its characteristics due to the addition of the process steps necessary for the on-chip circuitry. The baseline threshold voltages are +0.8V for NMOS transistors and -0.8V for PMOS transistors, respectively. The total chip size is only 30 percent larger than the previously-reported multi-element chip [5] with

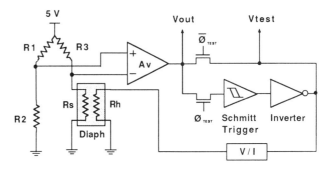

Fig. 5: Schematic of the Self-Test Circuitry Implemented in One of the Flow Velocity Sensors.

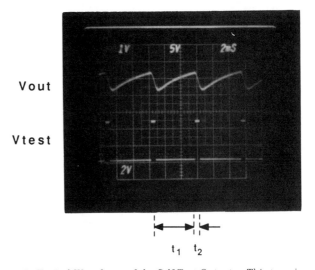

Fig. 6: Typical Waveforms of the Self-Test Outputs. This transient response can be used to detect any changes in the electrothermal characteristics of the diaphragm sensor window.

no on-chip circuitry. The number of external leads required is reduced from 24 to 10 using the analog multiplexer, considerably simplifying the bonding and packaging of the device. The on-chip circuitry also improves noise immunity by amplifying the sensed signals by a factor of 20 before presenting them to the output pads. Application of the chip to the measurement for semiconductor process gas flows is underway.

CONCLUSION

The thin diaphragm structure, responding to different pressure or to thermal events, is suitable for the measurement of a number of parameters including those involved in the determination of mass flow. This paper has reported the integration of sensing and control circuitry on the same chip with these transducers, preserving the performance of both the transducers and the circuitry and allowing functions such as amplification, multiplexing, and self-testing to be realized to improve system reliability. The resulting chip requires ten I/O pads, dissipates $120mW$ from dual $5V$ supplies and measures only $3.5mm \times 5mm$ in $3\mu m$ features.

ACKNOWLEDGMENTS

The authors would like to express their appreciation to the Semiconductor Research Corporation for their support of this research. The important work of Mr. J. Ji and Mr. S. Cho in the development of the CMOS process used as a baseline for the merged process reported here is also gratefully acknowledged.

REFERENCES

[1] D. L. Polla, R. S. Muller, and R. M. White, "Integrated Multisensor Chip," *IEEE Electron Device Lett.*, vol. EDL-7, pp. 254-256, Apr. 1986.

[2] O. Tabata, H. Inagaki, and I. Igarashi, "Monolithic Pressure-Flow Sensor," *IEEE Trans. Electron Devices*, vol. ED-34, pp. 2456-2462, Dec. 1987.

[3] D. L. Polla, H. Yoon, T. Tamagawa, and K. Voros, "Integration of Surface-Micromachined Zinc Oxide Sensors in n-well CMOS Technology," in *IEEE IEDM Tech. Digest*, pp. 495-498, December 1989.

[4] E. S. Kim, R. S. Muller, and P. R. Gray, "Integrated Microphone with CMOS Circuits on a Single Chip," in *IEEE IEDM Tech. Digest*, pp. 880-883, Dec. 1989.

[5] E. Yoon and K. D. Wise, "A Dielectrically-Supported Multi-Element Mass Flow Sensor", in *IEEE IEDM Tech. Digest*, pp. 670-673, Dec. 1988.

[6] C. L. Johnson, K. D. Wise, and J. W. Schwank, "A Thin-Film Gas Detector for Semiconductor Process Gases," in *IEEE IEDM Tech. Digest*, pp. 662-665, Dec. 1988.

[7] E. Yoon and K. D. Wise, "A Monolithic RMS-DC Converter Using Planar Diaphragm Structures," in *IEEE IEDM Tech. Digest*, pp. 491-494, Dec. 1989.

[8] G. Stemme, "A CMOS Integrated Silicon Gas-Flow Sensor with Pulse-Modulated Output," *Sensors and Actuators*, vol. 14, pp. 293-303, 1988.

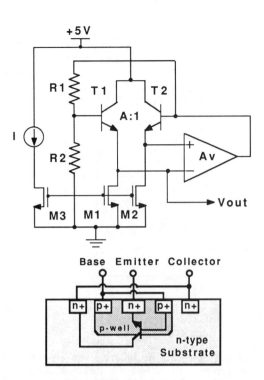

Fig. 7: Circuit Schematic of the Temperature Sensor and the Structure of the Substrate npn Bipolar Transistors Realized Using the p-well and the Source/Drain Diffused Layers in the Standard CMOS Process.

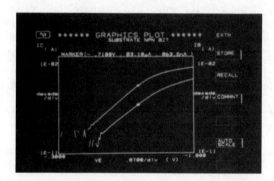

Fig. 8: Gummel Plot of the Substrate npn Bipolar Transistor.

Table 1

Sensor	Range	Resolution	Sensitivity
Flow Velocity	0 to 5 m/sec	0.5 cm/sec	34mV/V @5m/sec
Flow Direction	± 180 °	± 5 °	
Pressure	0 to 800 mmHg	± 0.5 mmHg	12ppm/mmHg
Temperature	-55 to 125 °C	0.1 °C	4 mV/V°C
Gas Type	See reference [6]		

ENVIRONMENTALLY RUGGED, WIDE DYNAMIC RANGE MICROSTRUCTURE AIRFLOW SENSOR

T. R. Ohnstein, R. G. Johnson, R. E. Higashi, D. W. Burns,
J. O. Holmen, E. A. Satren and G. M. Johnson
Sensor and System Development Center
Honeywell, Inc.
Bloomington, Minnesota

R. E. Bicking and S. D. Johnson
MICRO SWITCH
Freeport, Illinois

ABSTRACT

A silicon microstructure airflow sensor has been developed with a wide dynamic operating range and is rugged for long life operation in harsh environments. Platinum metallization is used for the airflow sensor because of its resistance to corrosion. Processing of thin film platinum has been developed to achieve a high, stable value of the platinum temperature coefficient of resistance (TCR). A first order coefficient for the platinum TCR of .003 ($\Omega/\Omega/°C$) has been achieved. The airflow sensor was designed with an exceptionally wide dynamic range, from less than 3 ft/min to over 30,000 ft/min (>10,000:1). The sensor design can be adjusted for sensitivity or range depending upon the application requirements. The airflow sensors were subjected to accelerated life testing to demonstrate the ability to maintain electrical stability and physical integrity in harsh environments. The life testing consisted of operation of the sensors with airflow in overpowered, high temperature and high humidity conditions for extended periods. The sensors were also subjected to extended periods of time during which dust particles were added to the flow to simulate a dusty environment. The sensors performed well throughout the accelerated life testing with little change in output characteristics.

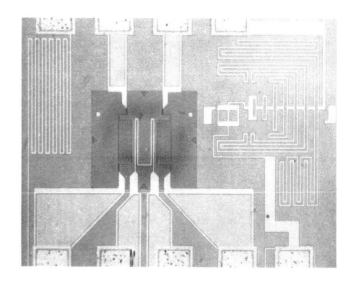

Figure 1. Photomicrograph of the airflow sensor die.

INTRODUCTION

A silicon microstructure airflow sensor has been developed for applications which require a wide dynamic response range and high volume flows. The sensor has also been designed for operation in dirty or corrosive environments where the gas flow environment cannot be filtered. This paper details the airflow sensor structure, performance and results of environmental testing.

FLOW SENSOR STRUCTURE AND OPERATION

The airflow sensor consists of a thin film structure fabricated on a silicon substrate. Figure 1 shows a photograph of the airflow sensor chip. There are three temperature sensitive resistor elements on a dielectric diaphragm near the center of the chip. The diaphragm which can be seen as a dark square area on the chip is thermally isolated from the silicon substrate. There are two other resistors on the chip that are in thermal contact with the substrate for measuring the reference ambient temperature.

The photograph in Figure 2 is a closer view of the diaphragm area and the three resistor elements that make up the active part of the airflow sensor. A cross section of the airflow sensor through the diaphragm is shown in Figure 3. The diaphragm

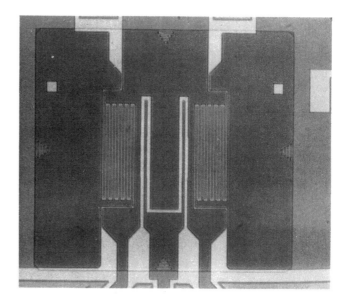

Figure 2. Photomicrograph of the airflow sensor diaphragm area showing the heater and sensor elements.

Reprinted from *Technical Digest IEEE Solid-State Sensor and Actuator Workshop*, pp. 158-160, June 1990.

has a continuous closed surface with no large steps or exposed edges which make it resistant to impact damage or buildup of particles in dusty environments. Silicon nitride is used as the diaphragm material. The resistor metallization is completely passivated by the silicon nitride. The diaphragm is thermally isolated from the silicon substrate by anisotropically etching the silicon from the backside of the wafer.

A platinum metallization is used in the airflow sensor to provide stable resistors that are resistant to corrosion in harsh environments. A thin film platinum process has been developed for this sensor which has a first order temperature coefficient of resistance(TCR) of 3.0×10^{-3} ($\Omega/\Omega/°C$) and a second order TCR of -5.0×10^{-7} ($\Omega/\Omega/°C^2$). The platinum metallization is passivated by silicon nitride and only gold is exposed at the bonding pads. However, in the event of scratches or pinholes in the nitride passivation, the platinum provides a metallization that is resistant to corrosion even if exposed.

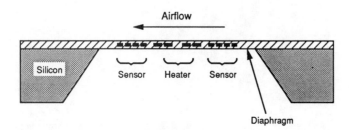

Figure 3. Cross section of the airflow sensor through the diaphragm area (not drawn to scale).

Referring again to Figures 2 and 3, the center resistor element is used as a heater to heat the diaphragm to a constant temperature above ambient during operation. The excellent thermal isolation of the diaphragm allows the operation of the heater at a temperature 160 °C above ambient with less than a 1 °C temperature rise on the sensor chip. Two sensing resistors are symmetrically positioned on either side of the heater along the axis of the airflow. Airflow is directed laterally across the chip as shown in Figure 3. The flow and direction of flow are sensed by the transfer of thermal energy primarily through the air from the upstream sensor to the downstream sensor. The airflow cools the upstream resistor and heats the downstream resistor. The flow detection circuit is a Wheatstone bridge where the upstream and downstream sensors make up two arms of the bridge. In a no-flow condition the sensors will be at the same temperature and the bridge will be balanced with no output voltage. A temperature difference between the sensors caused by an airflow unbalances the bridge circuit resulting in a voltage output. The output voltage can be calibrated to flow velocity, volume flow or mass flow. A description of the operation of the airflow sensor in greater detail can be found in previously published work (1)-(3).

PACKAGING AND TESTING

The response characteristics of the airflow sensor depend greatly upon the packaging of the sensor die. The design of the flow channel in which the sensor is mounted is critical in determining the characteristics of the flow in the channel. It will govern whether the flow that the sensor sees is laminar or turbulent in the desired flow range. For the applications of this airflow sensor it was desired that the flow is in the laminar flow regime over a large range. The flow channel was designed for laminar flow over the range from 0 - > 30,000 ft/min.

The sensor die was mounted on a ceramic thick film circuit substrate. The laminar flow channel was then mounted over the sensor. The dimensions of the flow channel were 0.1" wide, 0.025" high and 3.0" long. The response of the airflow sensor was tested by applying a calibrated flow through the laminar flow channel.

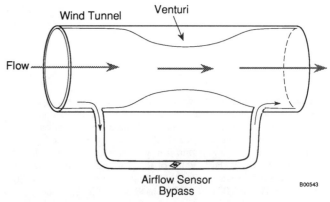

Figure 4. A section of the wind tunnel with the sensor connected as a bypass for measuring airflow.

To simulate larger volume flow applications, a wind tunnel was used for testing the airflow sensor. The airflow sensor in the laminar flow channel was mounted as a bypass element on the main duct of the wind tunnel similar to that illustrated in Figure 4. A venturi element in the main flow channel provides a pressure drop which drives the flow in the bypass. The response of the airflow sensor was calibrated to the flow in the main channel.

EXPERIMENTAL RESULTS

The flow response of the airflow sensor was measured in the laminar flow tube and is shown in Figure 5. The curve is a plot of the output voltage of the Wheatstone bridge circuit versus flow. The sensor has a wide dynamic range of response from less than 3 ft/min to over 30,000 ft/min. It has been found that the sensitivity and the range of the sensor response can be tailored to the application by varying the sensor chip design. The sensor response shown in Figure 5 is from a device that was designed for a large operating range.

An accelerated life test was performed on the airflow sensors which simulated 10 years of sensor use and is an indication of the stability of the sensor over time. For this test airflow was through the laminar flow channel directly. The heater was powered to a level 50% higher than in normal operation so that it was running at a higher than normal temperature. The ambient temperature of the airflow past the sensor was heated to 85°C with 85% RH. The test under flow conditions continued for at least 100 hours. The criteria for passage of the test was that the no-flow output or null voltage of the bridge circuit could shift by no more than 1mV and the output voltage could shift by no more than 5% of full scale at full flow from before the test to after the test. The sensors were within these specifications after 400 hours of testing.

To simulate operation of the sensor in a dusty environment over a 10 year time period, an accelerated life dust test was performed. The airflow sensor was mounted as a bypass element on the wind tunnel as described above and illustrated in Figure 4. The sensor response was calibrated to the flow in the wind tunnel and the output voltage of the Wheatstone bridge circuit was amplified to give a maximum output of 5 volts full scale. Standard Arizona "coarse" road dust was used for this test. Arizona "coarse" road dust is a standard mixture of dust particles that is used in environmental testing. The "coarse" name denotes a mixture of particles ranging in size from <5 microns to 200 microns. The test was over a 15 hour time period where 100 grams of the dust was evenly mixed and added to the airflow in the wind tunnel. The airflow was periodically cycled between 10% and 100% of full flow throughout the test. Full flow in the wind tunnel was approximately 16,000 ft/min. The flow response of the airflow

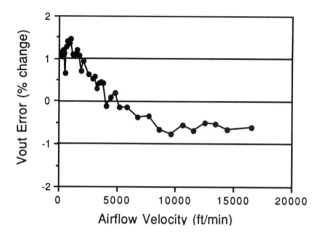

Figure 7. A plot of the output voltage shift after the dust test. The data is plotted as the percentage change in voltage from before the test to after the test.

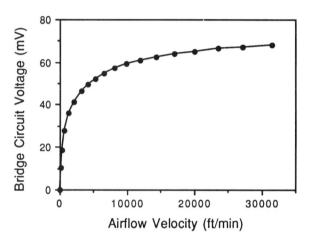

Figure 5. A plot of airflow velocity versus the bridge circuit output voltage measured in the laminar flow channel.

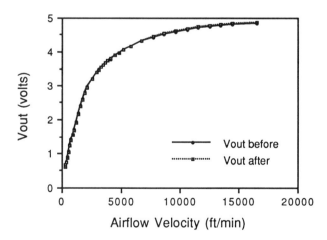

Figure 6. A plot of airflow velocity versus output voltage before and after dust testing. The data is for airflow in the wind tunnel. The output voltage of the bridge has been amplified.

sensor was measured before and after the dust test. The results are shown in Figure 6. There was very little change in the curve after the dust test. The percentage change in the output voltage is plotted in Figure 7. Over most of the range the change was less than 1%. The bypass mounting of the sensor chip gives some protection to the sensor from the dust in the main flow but it was evident that dust did pass through the sensor flow tube. There was however no significant accumulation of dust either on the sensor chip or in the flow channel.

SUMMARY

A silicon microstructure airflow sensor has been developed that has a wide dynamic range of flow response. The sensor has been shown to be environmentally rugged, passing accelerated life tests for electrical stability and physical integrity. A stable thin film platinum process has been developed for the sensor metallization that is resistant to corrosion in harsh operating environments.

REFERENCES

(1) G. Benjamin Hocker, "Solid State Sensor Research At Honeywell: Active Thin Films Plus Microstructures," Tech. Digest IEEE Solid-State Sensor Conference, Hilton Head Island, South Carolina, June 6-8, 1984, pp. 18-19.

(2) R. G. Johnson, R. E. Higashi, P. J. Bohrer and R. W. Gehman, "Design And Packaging Of A Highly Sensitive Microtransducer For Air Flow And Differential Pressure Sensing Applications," Proc. 1985 Int. Conf. Solid-State Sensors And Actuators (Transducers '85), Philadelphia, Pa., U.S.A., June 11-14, 1985, pp 358-360.

(3) R. G. Johnson and R. E. Higashi, "A Highly Sensitive Silicon Chip Microtransducer For Air Flow And Differential Pressure Sensing Applications," Sensors And Actuators, Vol. 11, 1987, pp. 63-72.

Section 10

Surface Micromachining

THE transition from our macro world to the micro world brings many surprises and changes in perspective. Micro manufacturing is one example: It is fundamentally different from our large-scale macro manufacturing plants.

In our normal macro manufacturing, human dexterity is extensively used. One picks up the part, shakes it, and feels the closeness of the mating parts. One listens to things moving, and even gets early warnings by the sense of smell. The capabilities of humans to manipulate and sense are awesome. (If you doubt this, try to design a machine that will replace a 17-year-old preparing and serving food at a fast-food restaurant.) Even our automated production equipment requires the human touch to set up, align, and fix.

In the micro world, most of these senses are denied us. The image of a microscope gives us most of our information. It is even difficult to look at micro devices from different angles. Most micro devices are as remote to our senses as objects at the bottom of the ocean, or on the moon.

How does one automate and make inexpensive the manufacture of micromechanical systems? Micro scientists and engineers have developed new techniques that rely on making whole mechanical systems in place, etching out unwanted sacrificial material, and mass assembly of thousands of devices with a single operation. This section, Surface Micromachining, and the next section, Bulk Micromachining, will explain two methods for inexpensively making and assembling micro systems. It will be interesting to see if these powerful micro assembly techniques will eventually be used by our conventional large-scale manufacturing processes.

This section of the book deals with one micro assembly technique, surface micromachining. Here, layers are deposited and patterned on a substrate. Sacrificial material is deposited where open areas or clearance tolerances are wanted; structural material is deposited wherever the final part is desired. After deposition, the sacrificial material is removed, and the desired mechanical device is left. These micromechanical devices are made without being touched by hands or assembly tooling.

In 1967, the paper "The Resonant Gate Transistor" described using sacrificial material to release the gate of a transistor. (This paper is included in Section 1, Early Papers in Micromechanics.) This work demonstrated the ability of silicon fabrication techniques to free mechanical systems from a silicon substrate. In 1983, the paper "Polycrystalline Silicon Micromechanical Beams" discussed the use of polysilicon as the structural material, and silicon dioxide as the sacrificial material. Of special importance is the ability of Hydrofluoric acid to etch silicon dioxide out of very thin, long channels. This allows large polysilicon structures to be released using thin silicon dioxide layers.

The use of structural polysilicon and sacrificial silicon dioxide to fabricate free-moving mechanical gears, springs, and sliders was demonstrated in 1988 in two papers, "Integrated Fabrication of Polysilicon Mechanisms" and "Integrated Movable Micromechanical Structures for Sensors and Actuators."

The next three papers, "Polysilicon Microbridge Fabrication Using Standard CMOS Technology," "Process Integration for Active Polysilicon Resonant Microstructures," and "Fabrication of Micromechanical Devices from Polysilicon Films with Smooth Surfaces" discuss the fabrication issues involved in using polysilicon to make mechanical parts. The final paper, "Selective Chemical Vapor Deposition of Tungsten for Microelectromechanical Structures" describes a different silicon fabrication procedure to etch away underlying material and free micromechanical components.

Some questions for future micro designers: Do we need equipment to assemble, test, and manipulate individual micro devices? Do we need to acquire a sense of touch in the micro world? Should some or all of this equipment be made out of micro devices? I believe there is still much work to be done.

Polycrystalline Silicon Micromechanical Beams

R. T. Howe and R. S. Muller

Department of Electrical Engineering and Computer Sciences, and the Electronics Research Laboratory, University of California, Berkeley, California 94720

ABSTRACT

Using the conventional MOS planar process, miniature cantilever and doubly supported mechanical beams are fabricated from polycrystalline silicon. Poly-Si micromechanical beams having thicknesses of 230 nm to 2.3 μm and separated by 550 nm to 3.5 μm from the substrate are made in a wide range of lengths and widths. Two static mechanical properties are investigated: the dependences of maximum free-standing length and beam deflection on the thickness of the beam. By annealing the poly-Si prior to beam formation, both of these properties are improved. Nonuniform internal stress in the poly-Si is apparently responsible for the beam deflection.

A novel method is described for making cantilever and doubly supported micromechanical beams from polycrystalline silicon (poly-Si). Only two masking steps are needed, as illustrated in Fig. 1. The first step, Fig. 1a, opens windows in an oxide layer grown or deposited on the silicon wafer. Next, poly-Si is deposited and plasma-etched in a second masking step, leaving the cross-section of Fig. 1b. Immersing the wafer in buffered HF removes all oxide, undercutting the poly-Si layer and creating the cantilever beam, illustrated in Fig. 1c. Doubly supported beams, Fig. 1d, are made by including a second oxide window at the opposite end of the beam.

For several reasons, this technique for making poly-Si beams is a useful addition to silicon micromechanics technology. First, poly-Si is widely used in MOS integrated circuits and a well-developed technology exists for depositing it in thin films and controlling its electrical properties. Second, the piezoresistivity of poly-Si could be utilized in integrated sensors; a monolithic pressure transducer based on a poly-Si membrane has already exploited this property (1). Finally, the simple fabrication process outlined above uses only conventional MOS planar technology which greatly simplifies the integration of micromechanical beams and MOS circuitry. This compatibility with standard processing contrasts with beam fabrication techniques based on anisotropic etchants (2, 3). MOS transistors have been fabricated adjacent to miniature cantilever beams made by anisotropic etching of the epitaxial silicon underlying a patterned, etch-resistant layer (4). However, the use of an epitaxial layer and anisotropic etchants in this earlier work represents a substantial departure from the conventional MOS process.

To assess fully the potential of this new technique, both the static and dynamic mechanical properties of poly-Si beams should be determined. This study investigates the static properties by fabricating poly-Si beams having a wide range of dimensions. In particular, the dependences of maximum free-standing beam length and cantilever beam deflection on the thickness of the beam are determined. A more complex process, including an MOS circuit for sensing the beam vibrations, is being developed to investigate the dynamic properties. Results for Young's modulus and fatigue characteristics of poly-Si micromechanical beams will be reported later.

Fabrication

We now describe poly-Si beam fabrication in more detail. In order to minimize stress concentration during nexing, the step in the poly-Si beam, shown in Fig. 1c, should be gradual. This consideration makes it desirable to have a tapered oxide-window edge. Two process runs with different oxide-layer compositions are made. Figure 2 illustrates the resulting oxide-window edges, as observed with a scanning-electron microscope. In the first run, the oxide layer is one-third wet thermal SiO_2 and two-thirds CVD SiO_2. The latter is densified at 975°C for 20 min after deposition. In the second run, a more gradual edge profile is obtained with an oxide layer consisting of 10% thermal SiO_2 and 90% phosphosilicate glass (about 8.75% phosphorus content). The thin, rapidly etching surface layer needed for a tapered oxide-window edge is created by a low-energy argon implant (5). The phosphosilicate glass is densified at 1100°C for 20 min prior to the argon implant.

The poly-Si thin film is deposited by pyrolysis of silane at a temperature of 640°C and a pressure of 600 mTorr, resulting in a deposition rate of approximately 10 nm min^{-1} and a grain size of 30-50 nm (6). No high temperature processing is done after poly-Si deposition, except for one sample hving a 3.5 μm oxide layer and a 2.0 μm poly-Si layer. In this case, the wafer is annealed at 1100°C in N_2 for 20 min prior to immersion in buffered HF.

In order to establish the dimensional limitations on poly-Si beams, a wide variety of beams have been fabricated in the two process runs. Several combinations of oxide and poly-Si thicknesses, listed in Fig. 3, were used. The first-run cantilevers are T-shaped with the cross bar at the free end (cf. Fig. 5). They range

Key words: mechanics, membrane, integrated circuits.

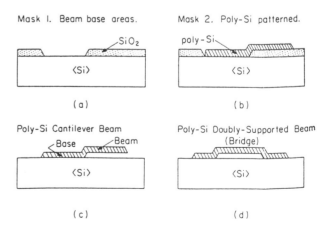

Fig. 1. Poly-Si microbeam fabrication

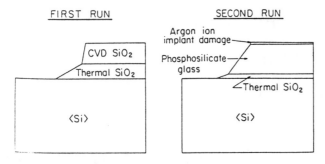

Fig. 2. Oxide-window edges for both process runs

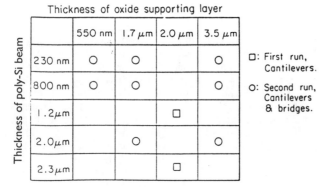

Fig. 3. Poly-Si and oxide thicknesses used to fabricate beams

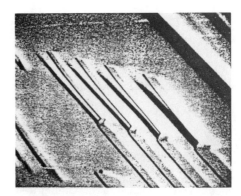

Fig. 4. Poly-Si cantilevers: 60 μm long, 800 nm thick, with a 3.5 μm beam-substrate gap.

Fig. 5. Apertured poly-Si beam: 100 μm long, 2.3 μm thick, with a 2 μm beam-substrate gap.

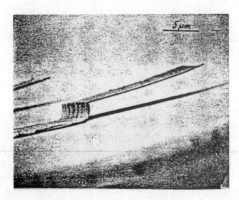

Fig. 6. Thin poly-Si cantilever: 25 μm long, 230 nm thick, with a 3.5 μm beam-substrate separation.

in length from 25 to 200 μm, with various widths. Both simple cantilevers (without cross bars) and bridges are included on second-run masks. These cantilevers are 25-400 μm long, and the bridges range from 100 to 500 μm in length. Several series of beams are included in which length or width is varied with the other dimension being held constant. All cantilevers have bases as long as the beams (cf. Fig. 1c) except for one second-run series in which the base length has been progressively shortened.

Results

All of the combinations of oxide and poly-Si thicknesses listed in Fig. 3 produce some free-standing poly-Si beams. No cracking is seen at the step in the poly-Si, even for first-run cantilevers, in which the oxide-window edge is very abrupt (cf. Fig. 2). Long bases for poly-Si cantilevers are not needed, as demonstrated by the series of cantilevers with shortened bases. For all the combinations enumerated in Fig. 3, a 5 μm long base is sufficient for a free-standing cantilever which is 45 μm long and 20 μm wide.

Scanning-electron micrographs of four representative cantilevers are shown in Fig. 4. These beams are 60 μm long, 800 nm thick, and are 3.5 μm above the silicon substrate. Widths range from 25 μm for the leftmost beam to 60 μm for the rightmost one. The bases are at the lower right, and the steps in the poly-Si are clearly visible. We will discuss later the upward deflection of these beams which can be seen in Fig. 4.

An apertured poly-Si beam is shown in Fig. 5. The T-shaped cantilever is 100 μm long, 40 μm wide, 2.3 μm thick, and has a 2 μm gap between beam and substrate. No special care in rinsing or drying the wafer after undercutting the beams is taken for the first process run, which produced this beam. An apertured cantilever is of interest for certain sensor applications (7) and the beam shown in Fig. 5 demonstrates the feasibility of this concept.

Very thin poly-Si layers can cover large oxide steps and form free-standing micromechanical beams, as can be seen in Fig. 6. The cantilever in Fig. 6 is 25 μm long and only 230 nm thick with a 3.5 μm separation between beam and substrate. Some debris from scribing is visible on top of the beam and variations in the oxide-window edge have been reproduced in the thin poly-Si layer. This beam is from a process run in which the wafers are dipped in water after etching in buffered HF and dried under an infrared lamp.

Beyond a certain length, both cantilevers and bridges are observed either to deflect downward and ultimately contact the substrate or to deflect upward severely. This limiting length, defined as the maximum free-standing length L_m, is repeatable from die to die within a wafer. It is observed that L_m depends strongly on the thickness of the beam. Width and beam-substrate gap do not have a repeatable influence on L_m in the first two process runs. However, the maximum free-standing length is probably sensitive to processing details, such as the poly-Si deposition conditions and the oxide etching procedure. The results obtained here are, nonetheless, useful as guidelines for evaluating poly-Si beam technology.

In order to determine L_m as a function of thickness for these poly-Si beams, a series of 15 μm wide cantilevers and bridges with increasing lengths were investigated. The average of the lengths of the longest repeatably free-standing beam and the next longer beam is an empirical estimate of L_m. This measurement is plotted against beam thickness for both cantilevers and bridges in Fig. 7. The curves in Fig. 7 are empirical fits to the data and are given by

$$L_m = \begin{cases} \sqrt{4500t} & \text{for cantilevers} \\ \sqrt{11,500t} & \text{for bridges} \end{cases} \quad [1]$$

where t (μm) is the beam thickness and L_m (μm) is the maximum free-standing beam length. Cantilevers or bridges located above the relevant curve in Fig. 7 cannot be used, since they will touch the substrate or be severely deflected. Note that the sample which is annealed before the beams are undercut shows substantial increases in maximum free-standing lengths for both cantilevers and bridges, as indicated by the filled symbols in Fig. 7. As mentioned above, these results are process-dependent and therefore should be considered as first-order guidelines.

Cantilever poly-Si beams with lengths less than L_m deflect upward, away from the substrate, as can be seen in Fig. 4 and 6. Scanning-electron micrographs indicate that the deflection curve is parabolic for small tip deflections (less than one-tenth of the beam length). In addition, the tip deflections of cantilevers of constant thickness are proportional to the squares of their lengths. These results suggest the presence of an internal bending-moment M_i in the poly-Si beams, since a tip deflection y given by

$$y = \frac{6 M_i L^2}{E_Y t^3} \quad [2]$$

where L is the cantilever's length, t its thickness, and E_Y its Young's modulus, is predicted for this case (8).

The physical origin for the bending moment M_i is discussed later.

It is convenient to use the quantity (y/L^2) to characterize beam deflection, since it is independent of beam length. The inset in Fig. 8 defines the tip deflection measurement, and the plot shows the variations of (y/L^2) with beam thickness for both first- and second-run cantilevers. According to Fig. 8, thicker beams deflect less at a given length, as one would expect. As before, annealing the sample affects the static mechanical properties of poly-Si beams. This is indicated by the filled circle in Fig. 8. In contrast to the unannealed sample of the same thickness no measurable deflection is observed for cantilevers shorter than L_m on the annealed wafer. These cantilever deflection results also are sensitive to processing details and should be interpreted accordingly.

Discussion

Micromechanical cantilever and doubly supported beams can be fabricated from poly-Si using a process that is compatible with MOS planar technology. This compatibility, the simple beam-fabrication process, and the piezoresistivity of poly-Si make these structures attractive for silicon micromechanics. In particular, an integrated sensor for organic vapors is being fabricated which incorporates a resonant poly-Si cantilever (7).

Two static properties of these beams have been investigated, the maximum length for a reliably free-standing beam, and the built-in deflection of cantilevers. The observed deflection is consistent with the presence of an internal bending moment which is plausible as the result of nonuniform internal stress across the poly-Si layer. Internal stress in a thin film arises from mismatches in thermal-expansion coefficients and from nucleation and growth phenomena (9). Annealing at 1100°C for 20 min before undercutting the beams with buffered HF increases the maximum free-standing length and eliminates the built-in deflection for beams shorter than L_m. This annealing procedure is sufficient to reflow the phosphosilicate glass layer underneath the poly-Si lyer and cause recrystallization of the poly-Si, both of which may help to relax the nonuniform internal stress. Diffusion of phosphorus from the glass into the poly-Si will occur during this relatively severe anneal, which may be undesirable for certain applications. Additional research is needed to determine the minimum temperature and time required for stress relaxation in the poly-Si film. However, beams of useful sizes can be made without any post-deposition treatment of the poly-Si, as is

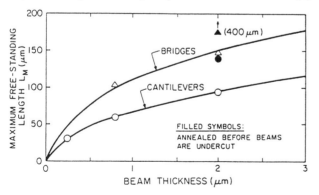

Fig. 7. Dependence of maximum free-standing length on beam thickness.

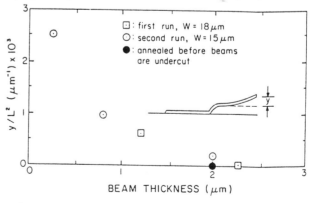

Fig. 8. Cantilever-tip deflection as a function of beam thickness

demonstrated in Fig. 4-6. The dynamic properties of poly-Si micromechanical beams are currently being investigated, and the results will be reported later.

Acknowledgments

We would like to thank Mrs. D. McDaniel for valuable laboratory assistance, and Dr. J. Shott of Stanford Integrated-Circuits Laboratory for use of a CVD oxide reactor. This work was supported by the National Science Foundation under grants ENG78-21854 and ECS-81-20562.

Manuscript submitted Aug. 16, 1982; revised manuscript received Feb. 14, 1983. This was Paper 118 presented at the Montreal, Que., Canada, Meeting of the Society, May 9-14, 1982.

Any discussion of this paper will appear in a Discussion Section to be published in the December 1983 JOURNAL. All discussions for the December 1983 Discussion Section should be submitted by Aug. 1, 1983.

REFERENCES

1. J. M. Jaffe, *Electron Lett.*, **10**, 420 (1974).
2. K. E. Petersen, *IEEE Trans. Electron Devices*, **ed-25**, 1241 (1978).
3. R. D. Jolly and R. S. Muller, *This Journal*, **127**, 2750 (1980).
4. K. E. Petersen and A. Shartel, *Tech. Digest IEDM*, 673 (1980).
5. J. C. North et al., *IEEE Trans. Electron Devices*, **ed-25**, 809 (1978).
6. C. I. Drowley, Unpublished results.
7. R. T. Howe, M. S. Report, Dept. of Elect. Eng. and Comp. Sciences, Univ. of California, Berkeley, October 1981.
8. E. P. Popov. "Mechanics of Materials," 2nd ed., p. 361, Prentice-Hall, Englewood Cliffs, NJ (1976).
9. D. S. Campbell, in "Handbook of Thin Film Technology," L. I. Maissel and R. Glang, Editors, pp. 12-22, McGraw-Hill, New York (1970).

Integrated Fabrication of Polysilicon Mechanisms

MEHRAN MEHREGANY, KAIGHAM J. GABRIEL, MEMBER, IEEE, AND WILLIAM S. N. TRIMMER, MEMBER, IEEE

Abstract—Successful implementation of simple mechanisms on silicon is a prerequisite for the design of monolithic microrobotic systems. This paper describes the integrated fabrication of planar polysilicon mechanisms incorporating lower and higher *kinematic pairs*, where the term *kinematic pair* signifies a joint. The two lower kinematic pairs (revolute and prismatic) commonly used in macrorobotic systems are precisely those two joints that are compatible with silicon microfabrication technology. The mechanisms are fabricated by surface micromachining techniques using polysilicon as the structural material and oxide as the sacrificial material. Turbines with gear and blade rotors as small as 125 μm in diameter and 4.5 μm in thickness were fabricated on 20-μm-diameter shafts. A clearance as tight as 1.2 μm was achieved between the gear and the shaft. Gear trains with two or three sequentially aligned gears were successfully meshed. A submillimeter pair of tongs with 400-μm range of motion at the jaws was fabricated. This structure incorporates a single prismatic joint and two revolute joints, demonstrating linear-to-rotary motion conversion.

I. Introduction

BULK and surface micromachining of silicon have been used extensively to fabricate a variety of micromechanical structures such as thin silicon diaphragms [1]–[3], beams [4]–[6], and other suspended structures [7]–[9] in single-crystal silicon or in films deposited on a silicon substrate. These micromechanical structures are generally limited in motion to small deformations and are physically attached to the substrate. Such elastic components may be used occasionally as flexible joints, but their overall usefulness in the design of *mechanisms* is limited. "Mechanism" as used here is a means for transmitting, controlling, or constraining relative movement and is considered as a collection of rigid bodies connected together by joints.

Mechanisms, actuators, sensors, and drive electronics are the four building blocks of any robotic system. For a robot arm to trace a specified trajectory, the actuators provide the input to move the arm mechanism. The sensors in conjunction with the control and drive electronics provide the necessary local or global feedback for correcting the in-route and final position of the end-effector. Therefore, a prerequisite to the design of any monolithic microrobotic system is the successful implementation of mechanisms on silicon. Although it is difficult to envision the variety of future applications for such microrobotic systems, the more immediate impact of these systems may be in the area of photonic component alignment.

This paper discusses the design and fabrication of conventional mechanisms scaled down to submillimeter dimensions. Initially, a brief discussion of the necessary kinematic pairs for monolithic microrobotics systems is presented. Related work is reviewed, and the integrated fabrication process for polysilicon mechanisms is discussed. Finally, fabricated mechanisms including turbines, gear-trains, and a submillimeter pair of tongs are described.

II. Kinematic Pairs

A joint, also referred to as a kinematic pair, signifies a connection between two bodies. This connection may be in the form of a point, line, or curve contact such as the straight line contact between the teeth of two meshed gears. Joints of this type are classified as higher kinematic pairs. On the other hand, the two bodies may be in contact over a large area of a surface such as a ball joint (spherical pair). These joints are classified as lower kinematic pairs and may be found in six configurations (in three dimensions) including spherical, planar, cylindrical, turning, prismatic, and screw pairs. Hunt [10] has discussed this topic in detail, which has been summarized here to establish a common terminology among researchers working in the area of microsensors, microactuators, and robotics.

Due to the planar nature of the silicon microfabrication technology, any mechanism design at this time would essentially have a planar geometry. Of the six lower kinematic pairs in space, two have appropriate planar counterparts for mechanism design on silicon. The turning and prismatic pairs in space simplify to revolute (rotary) and prismatic (linear) joints in the plane. Conventional robotics mostly relies on these two single-degree-of-freedom joints due to their single input nature. Therefore, limiting to these two forms of joints is not constraining in the design of monolithic micromechanisms. However, lower kinematic pairs do not provide all of the necessary movements as demanded by robotic systems. Gear trains (joining via higher kinematic pairs) are indispensable in many mechanism designs.

III. Related Work

We have previously reported on the fabrication of discrete silicon components in an attempt to investigate the

Manuscript received September 30, 1987; revised January 15, 1988.
M. Mehregany is with the Robotic Systems Research Department, AT&T Bell Laboratories, Holmdel, NJ 07733, and with the Massachusetts Institute of Technology, Cambridge, MA 02139.
K. J. Gabriel is with the Machine Perception Research Department, AT&T Bell Laboratories, Holmdel, NJ 07733.
W. S. N. Trimmer is with the Robotic Systems Research Department, AT&T Bell Laboratories, Holmdel, NJ 07733.
IEEE Log Number 8820665.

potential of small hybrid silicon mechanisms [11], [12]. Gears and turbine blades 40 to 50 μm in thickness and 300 to 2400 μm in diameter were fabricated. Reactive-ion etching was used to form the structures on the surface of a silicon substrate that was later dissolved to free the components. A silicon air turbine was assembled out of discrete components and operated at 400 rps. However, the final assembled systems have large tolerances due to discrete assembly requirements. In addition, handling of small discrete parts is difficult, limits component-size reduction, and reduces final yield. Finally, gears and turbine blades fabricated from single-crystal silicon are prone to cleaving along crystallographic planes.

The integrated fabrication of pin joints and bearings using surface micromachining techniques have been previously reported [12]–[17]. The following sections describe in detail the integrated fabrication of polysilicon mechanisms including: turbines incorporating appropriate flow channels and using gear or blade rotors, gear trains incorporating appropriate flow channels and having two or three meshed gears, and a pair of tongs with both revolute and prismatic joints. The integrated fabrication technique eliminates the need for discrete component assembly, and further dimensional control (component size and intercomponent clearance) is only limited by the standard integrated circuit fabrication capabilities.

IV. Fabrication

Fig. 1 illustrates the process steps for the microfabrication of a typical structure (e.g., a turbine with gear rotor):

First oxide step: Using high-pressure oxidation, a 4.0-μm-thick layer of thermal oxide is grown at 850°C. A photolithography step is used to remove the oxide where the flow channel walls are to be fabricated. A second photolithography step is performed to create circular and annular steps (2.0 to 2.5 μm deep) in the oxide (Fig. 1(a)). This provides the depressions necessary for annular bearings to be incorporated into the movable parts. The oxide etches here and in the next two steps are all performed in a reactive-ion etcher using a CHF_3 plasma.

First polysilicon step: A 4.5-μm-thick LPCVD polysilicon layer is deposited at 630°C using a silane-hydrogen mixture. The polysilicon is patterned in a reactive-ion etcher using a $Cl_2/CFCl_3/Ar$ plasma, forming the movable part and the flow channels (if needed) of the eventual structure (Fig. 1(b)). For the polysilicon etch step, a 2.0-μm-thick CVD oxide layer deposited at 700°C from tetraethylorthosilane (TEOS) and patterned in a CHF_3 plasma is used as an etch mask. Upon completion of the polysilicon etch, the remaining masking oxide is removed in a CHF_3 plasma.

Second oxide step: Additional CVD oxide (1.2 μm thick) is deposited at 700°C from TEOS, covering the entire structure. The oxide is patterned in a CHF_3 plasma opening a hole, exposing the substrate below and allowing for the attachment of the constraining member. In addition, this etch defines the flow channels in the oxide (Fig. 1(c)).

Second polysilicon step: A 3.0-μm-thick polysilicon layer is deposited under the same conditions as the first polysilicon step. This layer is patterned (using a reactive-ion etcher in a $Cl_2/CFCl_3/Ar$ plasma) to create the shaft and cap contacting the silicon substrate, and to build up the walls of the flow channels (Fig. 1(d)).

Finally, the movable part of the structure can be released by dissolving the oxide in a buffered or diluted HF solution (Fig. 1(e) shows a partially released gear sawed in half to display the cross section of the structure). When releasing the structures, the oxide about the shaft area is

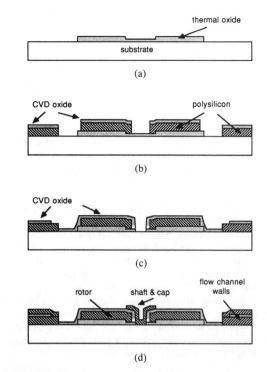

Fig. 1. Cross-sectional views of a typical structure during the fabrication process: (a) First oxide step; (b) first polysilicon step; (c) second oxide step; (d) second polysilicon step; and (e) a resulting 120-μm-radius gear which is sawed in half and partially released.

the last to dissolve. With the exception of the tongs, the length of time required in 10:1 buffered HF for the release of the structures presented in the next section is 6 to 8 h at room temperature. Shorter release times can be achieved by using straight HF-water mixtures. The release time is close to 40 min for a 1:1 HF-water solution at room temperature. The tongs require approximately 48 h in 7:1 buffered HF before they are entirely released. When the sacrificial oxide layers were fully dissolved, the structures were entirely free to move and did not require any additional manipulation to release them.

Even though stress nonuniformities through the polysilicon thickness can result in out-of-plane deformations of the structures, such deformations were not observed. This may have been due to the large thickness-to-length ratios of the structures (each polysilicon layer is at least 3 μm thick), the magnitude of stress variations in the polysilicon, or both. Therefore, it was not necessary to incorporate a step in the above fabrication process for annealing the stress in polysilicon.

Note that the clearance between the shaft and the moving part is determined by the smaller of the clearances allowed by the lithographic mask or the second oxide layer thickness. In our case, the oxide thickness was the controlling parameter giving a clearance of 1.2 μm on each side. The lithographic clearance was selected at 3.0 μm on each side of the shaft. Note that by using dry anisotropic etches throughout the patterning and definition steps the potential for undesirable lateral etching is eliminated.

V. Fabricated Mechansims

Turbines with gear or blade rotors, 125 to 240 μm in diameter, were fabricated using the above process steps. Each turbine incorporated two input ports and an output port. Fig. 2 is a SEM photograph of an entirely released turbine with a blade rotor, 125 μm in diameter. The SEM picture was taken after the turbine was spun by air. We have used a high-speed camera to measure the top speed of our turbines. However, even at 2000 frames per second (which is currently our limit), the motion of the turbine rotor cannot be resolved to allow accurate predictions of turbine speed. A lower limit of 15 000 rpm can be calculated assuming that the rotor advances by at least one eighth of a revolution (one blade) per frame. Furthermore, our current air flow control method does not allow for meaningful analytical estimation of rotor speeds based on simple fluid dynamic models.

Gear trains incorporating two or three meshed gears were fabricated with gear reduction ratios ranging from 1.4:1 to 1:1. Fig. 3 shows a partially released gear train with a 1.4:1.0:1.0 gear reduction ratios. The two flow channels on the top are connected to the two independent input ports; the two flow channels at the bottom are connected to the output port. Fig. 4 is a SEM photograph of the gear-train after the gears have been entirely released and then moved manually using a micro-probe to turn the larger gear. Fig. 5 shows a close-up of the engaged teeth of the two smaller gears. Note that the annular bearing

Fig. 2. A released polysilicon turbine with a blade rotor (10 μm per white dash).

Fig. 3. A partially released polysilicon gear-train (100 μm per white dash).

Fig. 4. The gear-train after being released and moved.

steps that are incorporated in the gears keep them approximately 2 μm above the substrate surface, reducing the surface contact area. Also note the partially etched surface near the teeth that resulted from partial failure of the first polysilicon layer etch mask. Fig. 6 is a side-view

Fig. 5. The meshed teeth of the two small gears in Fig. 4 (tooth depth is 20 μm).

Fig. 7. A partially released pair of tongs (100 μm per white dash).

Fig. 6. Side view of the cap, shaft, and the annular bearing depression of the gear on the right in Fig. 5.

Fig. 8. A released pair of tongs opened by pushing the handle forward (100 μm per white dash).

SEM photograph showing the cap, the shaft, and the annular bearing depression for the small gear on the right in Fig. 5.

Fig. 7 is a SEM photograph of a partially released pair of tongs. The jaws open when the linearly sliding handle is pushed forward, demonstrating the linear slide and the linear-to-rotary motion conversion. Fig. 8 shows an entirely relased pair of tongs with the jaws open. For this pair of tongs, the jaws open up to 400 μm wide.

The above structures are rugged and are not easily damaged by external manipulation. We have regularly used microprobes to pull and push on these structures without damaging them. However, when air is used to spin the turbines or the gear trains at high speeds, the cap portion of the shaft often breaks allowing the rotor to come loose (Fig. 9). Shaft cap fracture has been the only mode of failure observed for these structures thus far. This is most likely due to vertical motion of the rotor introduced by uncontrolled air flow in the turbine. Fig. 10 shows a close

Fig. 9. A pair of gears with damaged caps resulting from a high-speed spin (10 μm per white dash).

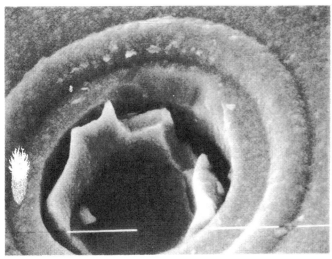

Fig. 10. A close-up view of the right shaft in Fig. 9 (10 μm per white dash).

up of the broken shaft of the right gear in Fig. 9. Note the tight clearance between the shaft and the gear (i.e., 1.2 μm on each side) and the top view of the annular bearing step near the shaft.

VI. Conclusion

Silicon microfabrication processes are used for integrated fabrication of polysilicon mechanisms capable of large rotary or linear motion in the plane of the substrate about the normal axis. The integrated fabrication process has two distinct advantages: first, it provides the potential for accurate control of the structural geometries and required clearances, avoiding undesirably large tolerances in the final structure. Second, it eliminates the need for handling individual parts, greatly reducing the processing cost and improving the final yield. Successful implementation of the lower and higher kinematic pairs incorporated in the planar mechanisms presented here is the preliminary step toward the design of monolithic microrobotic systems.

Acknowledgment

The authors wish to thank K. Orlowsky and N. Ciampa for fabrication resources and J. Walker for valuable technical assistance.

References

[1] Y. S. Lee and K. D. Wise, "A batch-fabricated silicon capacitive pressure transducer with low temperature sensitivity," *IEEE Trans. Electron Devices*, vol. ED-29, no. 1, pp. 42-48, Jan. 1982.
[2] H. Guckel, S. Larsen, M. G. Lagally, J. B. Miller, and J. D. Wiley, "Electromechanical devices utilizing thin silicon diaphragms," *Appl. Phys. Lett.*, vol. 31, no. 9, pp. 618-619, Nov. 1977.
[3] M. Mehregany, M. G. Allen, and S. D. Senturia, "The use of micromachined structures for the measurement of mechanical properties of thin films," in *Tech. Dig. IEEE Solid-State Sensors Workshop* (Hilton Head, SC, June 1986).
[4] K. E. Peterson, "Micromechanical light modulator array fabricated on silicon," *Appl. Phys. Lett.*, vol. 31, no. 8, pp. 521-523, Oct. 1977.
[5] L. M. Roylance and J. B. Angell, "A batch fabricated silicon accelerometer," *IEEE Trans. Electron Devices*, vol. ED-26, no. 12, pp. 1911-1917, Dec. 1979.
[6] H. Guckel, T. Randazzo, and D. W. Burns, "A simple technique for the determination of mechanical strain in thin films with application to polysilicon," *J. Appl. Phys.*, vol. 57, no. 5, pp. 1671-1675, Mar. 1985.
[7] M. Mehregany, R. T. Howe, and S. D. Senturia, "Novel microstructures for the *in-situ* measurement of mechanical properties of thin films," *J. Appl. Phys.* vol. 62, pp. 3579-3584, Nov. 1987.
[8] J. C. Greenwood, "Etched silicon vibrating sensor," *J. Phys. E: Sci. Instrum.*, vol. 17, pp. 650-652, 1984.
[9] M. A. Schmidt, R. T. Howe, S. D. Senturia, and J. H. Haritonidis, "A micromachined floating-element shear sensor," in *Tech. Dig. 4th Int. Conf. Solid-State Sensors and Actuators* (Tokyo, June 1987), pp. 277-282.
[10] K. H. Hunt, *Kinematic Geometry of Mechanisms*. London: Oxford, 1978, ch. 1.
[11] M. Mehregany, K. J. Gabriel, and W. S. N. Trimmer, "Micro gears and turbines etched from silicon," *Sensors and Actuators*, vol. 12, pp. 341-348, Nov./Dec. 1987.
[12] K. J. Gabriel, W. S. N. Trimmer, and M. Mehregany, "Micro gears and turbines etched from silicon," in *Tech. Dig. 4th Int. Conf. Solid-State Sensors and Actuators* (Tokyo, June 1987), pp. 853-856.
[13] K. J. Gabriel, M. Mehregany, and W. S. N. Trimmer, "Micro mechanical components," in *Proc. ASME Winter Ann. Meeting* (Boston, MA, Dec. 1987), DSC vol. 6, pp. 397-401.
[14] L. S. Fan, Y. C. Tai, and R. S. Muller, "Pin joints, gears, springs, cranks, and other novel micromechanical structures," in *Tech. Dig. 4th Int. Conf. Solid-State Sensors and Actuators* (Tokyo, June 1987), pp. 849-852.
[15] R. S. Muller, "From ICs to microstructures: Materials and technologies," in *Tech. Dig. IEEE Micro Robots and Teleoperators Workshop* (Hyannis, MA, Nov. 1987).
[16] S. F. Bart, T. A. Lober, R. T. Howe, J. H. Lang, and M. F. Schlecht, "Microfabricated electric actuators," *Sensors and Actuators*, in press.
[17] J. H. Lang, M. F. Schlect, and R. T. Howe, "Electric micromotors: Electromechanical characteristics," in *Tech. Dig. IEEE Micro Robots and Teleoperators Workshop* (Hyannis, MA, Nov. 1987).

Integrated Movable Micromechanical Structures for Sensors and Actuators

LONG-SHENG FAN, MEMBER, IEEE, YU-CHONG TAI, AND RICHARD S. MULLER, FELLOW, IEEE

Abstract—Movable pin joints, gears, springs, cranks, and slider structures with dimensions measured in micrometers have been fabricated using silicon microfabrication technology. These micromechanical structures, which have important transducer applications, are batch-fabricated in an IC-compatible process. The movable mechanical elements are built on layers that are later removed so that they are freed for translation and rotation. A new undercut-and-refill technique that makes use of the high surface mobility of silicon atoms undergoing chemical vapor deposition is used to refill undercut regions in order to form restraining flanges. Typical element sizes and masses are measured in millionths of a meter and billionths of a gram. The process provides the tiny structures in an assembled form, avoiding the nearly impossible challenge of handling such small elements individually.

I. INTRODUCTION

THE UNPRECEDENTED growth of integrated-circuit technology and computing techniques has made sophisticated data processing accurate, economical, and widely available. Today's electronic systems are capable of dealing with large numbers of physical input and output variables, but the transducers that provide interfaces between the electrical and physical world are in many cases outmoded and dependent on awkward hybrid-fabrication techniques. Many of the materials and processes used to produce integrated microcircuits, however, can be employed in new ways to produce microsensors and actuators. These structures complement the IC process and provide a means to produce new electronic systems.

Thus far, micromechanical transducer structures such as cantilevers, bridges, and diaphragms have been fabricated with IC-compatible processes for various useful applications. These structures, however, contain only bendable joints, a severe limitation on mechanical design capabilities for many applications. Microstructures with rotatable joints, sliding and translating members, and mechanical-energy storage elements would provide the basis for a more general micromechanical transducer-system design. Because such structures add important degrees of freedom to designers, we have investigated techniques to fabricate them using IC-based microfabrication processes [1]. Rotatable silicon elements, made using IC technology, have also been reported by Gabriel *et al.* [2]. The new mechanical elements use polysilicon thin-film technology combined with techniques that we describe in this paper. An important advantage of the procedures described is that they provide mechanical structures containing more than one part in a preassembled form; this avoids individually handling the very tiny structures. The initial demonstration of the technique to make these structures employs polysilicon as the structural material and phosphosilicate glass for the sacrificial layer. Other materials may, however, be used in place of these, provided that they are compatible with the overall process.

II. STRUCTURES AND PROCESSES

A. Fixed-Axle Pin Joints

A pin joint is composed of an axle around which a member (rotor) is free to rotate. Movement along the axle by the rotor is constrained by flanges. Fig. 1 shows the cross section and top view of a pin joint fixed to the substrate that has been fabricated using polycrystalline silicon. The rotor, axle, and flange are all made of polysilicon that has been deposited by a low-pressure chemical-vapor-deposition (LPCVD) process on top of a silicon substrate. The pin joint is produced using a double-polysilicon process and a phosphosilicate glass (PSG) sacrificial layer in a three-mask process as indicated in Fig. 2. In this process, openings are first made by dry etching a composite layer of polysilicon on PSG deposited by sequential LPCVD processes. Another PSG layer is deposited over the entire structure including the edges of the circular openings. Photolithography steps are then used to expose bare silicon at specific locations so that a subsequent deposition of polysilicon will anchor to the silicon substrate at these desired places. After depositing and patterning the second polysilicon layer for axles and flanges, all previously deposited PSG layers are removed in buffered hydrofluoric acid (BHF). The remaining polysilicon layers form the pin-joint structure. The rotor is free to rotate when the PSG layer is removed by BHF. An SEM photo (Fig. 3) shows a completed pin joint of this type.

B. Self-Constraining Pin Joints

A rotating-joint structure that provides several new possibilities for mechanical design can be built using only a small variation on the process described above. To differentiate joints of this type from the fixed-axle pin joints

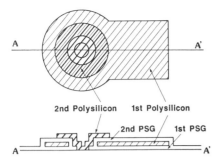

Fig. 1. Top view and cross section of a polysilicon micromechanical pin joint.

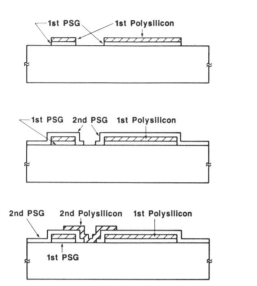

Fig. 2. Fabrication process for the anchored pin joint shown in Fig. 1.

Fig. 3. SEM photograph of anchored pin joint. The outer radius of the flange connected to the axle is 25 μm.

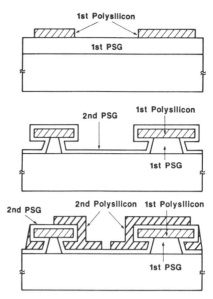

Fig. 4. Fabrication process for the self-constraining joint.

described above, we call these structures *self-constraining joints*. Self-constraining joints can, for example, allow for rotation while, at the same time, permitting translation across the silicon surface. These joints need to have a flange on the axle underneath the rotor to keep it in place. The axle can either be fixed to the substrate or else left free to translate across its surface.

Self-constraining joints are produced by a double polysilicon process with a PSG sacrificial layer. An undercut-and-refill technique is introduced to position the second-layer polysilicon both over and under the axle formed of first-layer polysilicon. Fig. 4 outlines the process for these joints, which are produced using two masks. In this process, after the PSG and polysilicon layers have been deposited by LPCVD, the polysilicon is patterned by dry etching. The next step is to use a timed etch of the first PSG layer to undercut the polysilicon. An optional mask may be used if only selected regions are to be undercut. Another PSG layer is then deposited. A second polysilicon layer that fills in the undercut regions is the patterned to produce axles and flanges. After this, all PSG layers are removed in a buffered hydrofluoric-acid solution. The remaining polysilicon layers form the self-constraining joint structure as shown in the SEM photograph of Fig. 5. The interleaved polysilicon layers, evident in Fig. 4, can be made because of the high surface mobility of silicon atoms during the LPCVD process. This permits the undercut regions to refill so that restraining flanges remain over and under the first-layer polysilicon.

For more complex structures, both pin joints and self-constrained joints can be made in the same process. Fig. 6 shows four-joint crank structures fabricated using both types of joints in a three-mask process (four, if the optional undercut mask is used). In Fig. 6, the joints at both ends of the central element are self-constraining but they are freed from the substrate. The other two joints are pinned to the substrate. Note that, except for the fixed joints, the entire structure has moved from its original position, indicated by a darkened pattern on the silicon substrate in Fig. 6. Using a surface profiler (Alpha-Step 200), we have found that in the darkened pattern there is a pit that is roughly 100 nm deep. This pit appears to be caused by enhanced etching of the silicon surface by the BHF under the polysilicon moving elements. The enhanced etching may, in turn, result from localized stress in this region. Further research to test this hypothesis is underway.

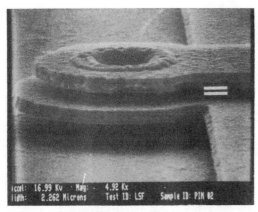

Fig. 5. SEM photograph of a self-constraining joint. The rotor is attached to a hub that turns in a collar projecting from a stationary polysilicon surface. There is a retention flange on the hub below the collar.

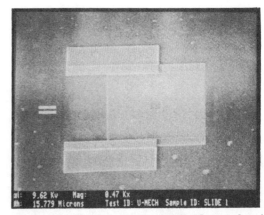

Fig. 7. Square slider with two edges restricted by flanges. One side of the slider is 100 μm and the central opening is a square of 10 μm.

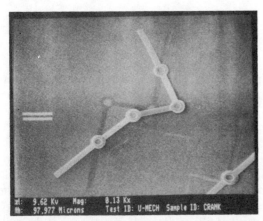

Fig. 6. SEM photograph of a four-joint crank having a central arm held by self-constraining joints that are free to translate. The original position of the crank is indicated by the darker pattern. All crank arms are 150 μm in length.

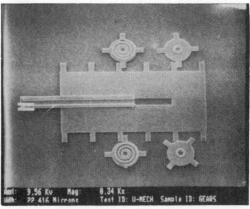

Fig. 8. Gear and slider combination. The slider is 210 by 100 μm. The toothed edges mesh with four gears, two of which have flat spiral springs attached.

C. Flanged Structures

The procedures carried out to make the two types of pin joints can be employed to produce other mechanical structures. For example, the three-mask pin-joint process can be used directly to fabricate the square slider shown in Fig. 7. The slider has a polysilicon moving element that is constrained by flanges along two of its edges so that only translational movement in one dimension is allowed. In the gear-slider combination of Fig. 8 (produced by the pin-joint process with four masks), the slider has a guide at its center to constrain movement to one dimension. Fig. 9 shows a crank-slot combination that requires five masks to produce. The slot element, formed by first-layer polysilicon, is pinned to the substrate by another element made of second-layer polysilicon so that both translational and rotational movements can take place. The joint for the crank-slot in Fig. 9 is self-constraining.

D. Design Variations

The versatility of the techniques described permits many potentially useful variations in design. As an example, Fig. 10 is a slider that can be thought of as a *dual* to the structure in Fig. 7. The flanges holding the parts together

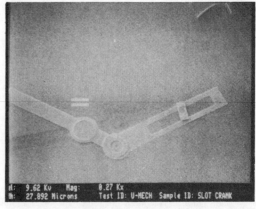

Fig. 9. Crank-slot combination with a center pin. The slot is 130 μm long and 20 μm wide. The diameters of the two joints are each 50 μm.

are on opposite members in each of these two versions of the mechanical slider. In the same sense, Fig. 11 shows a four-joint crank that is *dual* to the one in Fig. 6. All four joints in this crank are made using a self-constraining process; the two end joints are fixed and the two center joints can translate.

E. Micromechanical Bushings

The undercut-and-refill technique is useful for other applications also. When combined with pin joints, this tech-

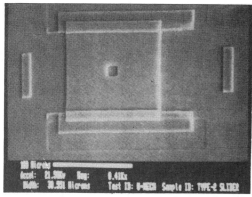

Fig. 10. Slider structure with outer edges guided by self-constraining joints. Stops limit the extent of lateral motion by the slider.

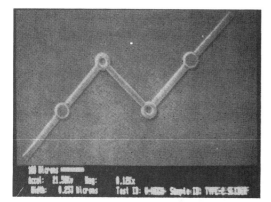

Fig. 11. Four-joint crank made with self-constraining joints.

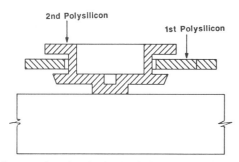

Fig. 12. Cross section of a micromechanical bushing built by the self-constraining-joint process.

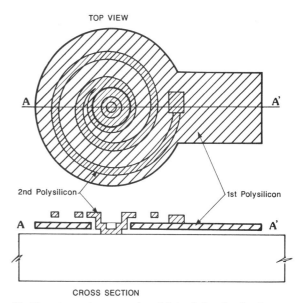

Fig. 13. Top view and cross section of flat spiral spring fixed on one end to an axle.

nique permits the fabrication of bushings that can be used, for example, to elevate a rotor away from the silicon surface. This can greatly reduce frictional forces, especially if the bushing elements are coated with or made from another material such as silicon nitride that may provide better wear properties. Fig. 12 shows a cross section for a bushing produced using the self-constraining-joint process described above and one extra mask for anchoring to the substrate (three masks in all).

F. Polysilicon Springs

Mechanical energy storage is important in many systems, and therefore it is very desirable to be able to fabricate micromechanical springs. These elements can also be produced using the process described above. The flat spiral spring attached to a pin joint, shown in Fig. 13, is made of second-layer polysilicon and connected on one end to the axle of a pin joint. The other end is attached to a movable disc made of first-layer polysilicon. The spring, produced using four masks, returns the disc to its original position after it is displaced. Figs. 14 and 15 show SEM photographs of restraining springs connecting rotors to pin-joint axles. Both springs are made of 2-μm-wide second-layer polysilicon. Shown in Fig. 15 is a beam spring that has an appreciably larger spring constant than does the flat spiral spring of Fig. 14.

G. Processing Details

In the foregoing, we have described the essential techniques for the *in situ* fabrication of assembled micro structures. The achievable dimensions for the finished elements depend on the lithography and processing steps used, but they can be roughly estimated at ten or fewer micrometers. Any of the microstructures can be fabricated separately using fewer than four masks. Six masks are needed to build all of the structures in the same run. When all are produced at one time there is an unavoidable loss in element precision because of the extra processing steps.

To illustrate the processes more completely, we describe in fuller detail the steps used to fabricate all structures on the same chip. First, a 1.5-μm-thick phosphorus-doped (8 wt.%) LPCVD silicon-dioxide layer is deposited at 450°C on a (100) silicon substrate. Photolithography and the first mask are used to open selected areas on the substrate where the first-layer polysilicon is to be anchored. Undoped LPCVD polysilicon, 1.5 μm-thick, is then deposited at 630°C and patterned with a second mask in a CCl_4 plasma. The third mask is used to define the undercut regions for self-constraining and bushing structures. Buffered HF etching creates a 2-μm undercut. Next, a 0.5-μm-thick phosphorus-doped (8 wt.%) LPCVD sil-

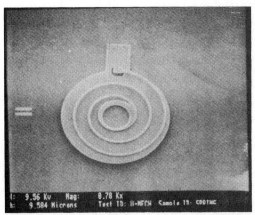

Fig. 14. SEM photograph of the spring-axle structure shown in Fig. 13. The 2.5-revolution spiral spring is made of 2-μm-wide second-layer polysilicon. Its inner end is fixed to an axle 10 μm in diameter, and its outer end is connected to a movable arm.

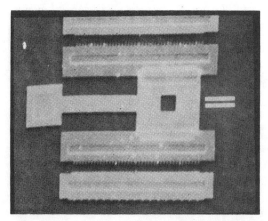

Fig. 16. Slider and bridge structure. One end of the bridge is anchored to the substrate and the other is attached to a slider. The central square opening is 20 μm on a side and the square slider is 80 μm on a side.

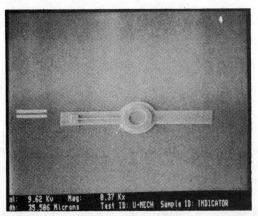

Fig. 15. SEM photograph of a beam spring attached to a central axle. The beam is 60 μm long and 2 μm wide.

icon-dioxide sacrificial layer is deposited at 450°C. The fourth and fifth masks are used to pattern the silicon-dioxide layer to anchor the spring element to the substrate and the rotor, respectively. Undoped LPCVD polysilicon, 1.0 μm thick, is then deposited at 630°C, defined and patterned using the sixth mask, and eteched in a CCl_4 plasma. Prolonged etching (and therefore thick resist films) are required to remove completely any residue of polysilicon from the regions near to topographic steps. For shorter etching times, a more isotropic plasma such as SF_6 might be used. A 1-h annealing step in nitrogen at 1000°C is used to reduce stress in the polysilicon. To release the structures from the oxide required 6 h of etching in a 5:1 buffered HF solution.

III. Study of Mechanical Properties

An important use for these structures is to carry out research on the micromechanical properties of materials. This research is especially necessary since many of the materials have thus far been applied exclusively for electronics. One means for obtaining useful data is to carry out visual inspection of high-speed magnified video-tape images that show the response of dynamically actuated elements. Analysis of these data will permit studies of frictional behavior, damping, fatique limits, and of fundamental properties, such as Young's modulus, Poisson's ratio, and the orientational dependences of these parameters.

In general, a residual stress is found in LPCVD polysilicon after deposition. Previous papers [3], [4] have described useful ways to study uniform stress distributions. The stress distribution in the direction of polysilicon film growth is, however, very likely not to be uniform and therefore to induce a bending moment across the films. The bending moment is of special concern in cases where flatness is important.

Using slider structures (Figs. 7 and 10), we expect to be able to separate the effects of uniform and bending stresses in thin-film mechanical structures. Fig. 16 shows the top view of a slider and a bridging beam. The outer edge of the flanges are defined with teeth to act as measuring scales. One end of the beam is anchored on the silicon substrate, and the other end is connected to a self-constraining slider. Since the slider allows translational movement, the compressive-stress component in the polysilicon beam can be released after freeing the whole structure, while the bending moment will be left in the beam. Detailed analytical study of such structures to determine both the compressive stress and the bending moment in deposited polysilicon is underway.

IV. Fracture-Strength Study

A flat spiral spring made of second-layer polysilicon is used to restrain a pin joint. Within its fracture limit, the spring can return the structure to its original position after it has been moved. Experiments have been done on these spring structures to estimate the lateral fracture stength of the polysilicon. A simplified mechanical analysis provides the basis for this estimate. For a 2 μm-wide 1-μm-thick spring extending 2.5 revolutions with inner radius $r_1 = 10$ μm connected to the central axle and outer radius $r_2 = 30$ μm connected to one arm, fractures occur at deflections of roughly 300°. For a spiral spring of width h and thickness t, assuming that adjacent turns do not come into contact, the energy stored is [5]

$$U = \int_0^l \frac{M^2 \, ds}{2E_Y I} \quad (1)$$

where M is the bending moment, E_Y is Young's modulus, I is the moment of inertia of the spring cross section, ds is the length of a small element of the spring, and l is the total length of the spiral. The spiral is generated using the equation $r = a\theta$ where θ is the polar-coordinate angle and a is a design constant chosen for a particular spring size. The bending moment can be shown [5] to be constant along the length of the spring. The angular deflection Φ is

$$\Phi = \frac{\partial U}{\partial M} = \int_0^l \frac{M}{E_Y I} \, ds = \frac{M}{E_Y I} \frac{r_2^2 - r_1^2}{2a}. \quad (2)$$

The moment of inertia of the rectangular cross section is $I = th^3/12$ and the maximum stress in the spring σ_{max} is $Mh/2I$. Using these equivalents in (2), we can express the maximum stress in terms of the angular deflection Φ.

$$\sigma_{max} = \frac{E_Y h a}{r_2^2 - r_1^2} \Phi. \quad (3)$$

The spiral springs were unwound using microprobes until they fractured. The bending moment in the spring loaded in this manner is constant along its length. Spring fractures occurred in all cases at deflection angles of 300 ± 30°. The fractures were observed in one or several locations and typically more then 20 μm away from the attachment points. The spiral springs (of the type shown in Fig. 13), have inner and outer radii of $r_1 = 10$ μm and $r_2 = 30$ μm, respectively. Other parameters are: thickness $t = 1$ μm, width $h = 2$ μm, and spiral constant $a = 1.27$ μm. Using the observed value of Φ_{fract} at fracture (300° or 5.24 radians) in (3), we calculate a fracture strength σ_{fract} that is 1.7 percent times Young's modulus E_Y for thin-film polysilicon. At least two simplifying assumptions underlie the conclusions made above: 1) that the spring motion is entirely in the horizontal plane (neglecting possible vertical motions that would relax stress), and 2) that the spring has sufficient turns to be treated as ideal [6]. Other studies, still in progress, indicate a slightly lower fracture strength (in the order of 1.3-1.4 percent of E_Y). For a perspective on our results, we note that the highest reported fracture strength for single-crystal silicon is 2.6 percent times E_Y [7]. We expect that values for E_Y will depend on the deposition conditions for the film and on the direction of the stress relative to the growth direction.

To estimate E_Y for our polysilicon films, we make use of Johnson's analysis [8], X-ray diffraction studies showing the distribution of crystalline orientations in our films, and published orientation-dependent elastic constants for single-crystal silicon [9]. Ignoring grain-boundary effects, we estimate that E_Y for our films is 169 GPa and the fracture strength is in the 2 to 3 GPa range. For comparison, Guckel and co-workers have published a value of E_Y for polycrystalline silicon of 22.2 Mpsi (153 GPa) [10].

Polysilicon Material Studies

An analysis to be published will detail the procedures sketched in the previous paragraph; we provide here only a few features of our studies of polycrystalline silicon to clarify the discussion. Using X-ray diffraction, we have found that the undoped LPCVD polysilicon films grown on PSG at 630°C have a preferred orientation that is generally in the (110) direction normal to the substrate. Annealing these films in nitrogen fosters grain growth but does not change their orientation. The polycrystalline-film orientation is actually described in terms of a distribution function derived from analysis of the X-ray diffraction data. This distribution function is used to calculate the effective film properties in terms of single-crystal parameters [8].

V. Conclusions

We have described a technique to build micro-scale movable mechanical pin joints, springs, gears, cranks, and sliders using a silicon microfabrication process. The ability of LPCVD polysilicon to fill undercut regions has been utilized in this research to build new structures including rotating and translating joints, bushings, and sliding elements.

The initial demonstration of the technology has employed polycrystalline silicon for the movable-joint members, but the process is not limited to using this material. The structural members might possibly be made from metals, alloys, and dielectric materials provided that these materials can be freed from their supporting substrate by selective etching of sacrificial (dissolvable) materials. Construction of these new elements gives rise to the need for research on mechanical parameters and properties for design. The process to produce these elements points up the need for further studies of sacrificial-layer etching, LPCVD growth, and remnant stresses in microfabricated systems. At the same time the realization of these structures brings new focus on the brightening prospects for producing microminiature prime movers [11].

The movable micromechanical structures can be batch-fabricated into multi-element preassembled mechanisms on a single substrate, or, if desired, they can be freed entirely from their host substrate to be assembled as separate elements. The potential uses for this new technology include the production of miniature ratchets, micro-positioning elements, mechanical logic, tuning elements, optical shutters, micro-valves, micro-pumps, and other mechanisms that have numberless applications in the macroscopic world. The method promises unheralded precision in the construction of miniature mechanical parts and systems with routine control at micrometer dimensions. Their manufacture in the world of micromechanics opens important avenues for further research and development.

ACKNOWLEDGMENT

We thank Prof. G. Johnson for valuable discussion on the characterization of polycrystalline films, and K. Voros, R. Hamilton, and the staff of the Berkeley Microfabrication Laboratory for help in experiments and fabrication.

REFERENCES

[1] L. S. Fan, Y. C. Tai, and R. S. Muller, "Pin-joints, springs, cranks, gears, and other novel micromechanical structures," in *Tech. Dig. 4th Int. Conf. Solid-State Sensors and Actuators* (Tokyo, June 1987), pp. 849-852 (U.S. patent pending).
[2] K. J. Gabriel, W. S. N. Trimmer, and M. Mehregany, "Micro gear and turbines etched from silicon," in *Tech. Dig. 4th Int. Conf. Solid-State Sensors and Actuators* (Tokyo, June 1987), pp. 853-856.
[3] R. T. Howe and R. S. Muller, "Stress in polycrystalline and amorphous silicon thin films," *J. Appl. Phys.*, vol. 54, pp. 4674-4675, Aug. 1983.
[4] H. Guckel, T. Randazzo, and D. W. Burns, "A simple technique for determination of mechanical strain in thin films with applications to polysilicon," *J. Appl. Phys.*, vol. 57, pp. 1671-1675, Mar. 1, 1985.
[5] S. Timoshenko, *Strength of Materials*, 3rd ed. Princeton, NJ: Van Nostrand, 1955.
[6] R. P. Kroon and C. C. Davenport, "Spiral springs with small number of turns," *J. Franklin Inst.*, vol. 225, p. 171, 1938.
[7] G. L. Pearson, W. T. Read, and W. L. Feldmann, "Deformation and fracture of small silicon crystals," *Acta Metallurgica*, vol. 5, pp. 181-191, Apr. 1957.
[8] G. C. Johnson, "Acoustoelastic response of polycrystalline aggregates exhibiting transverse isotropy," *J. Nondestructive Evaluation*, vol. 3, pp. 1-8, 1982.
[9] H. J. McSkimin, W. L. Bond, E. Buehler, and G. K. Teal, "Measurement of the elastic constants of silicon single crystals and their thermal coefficients," *Phys. Rev.* vol. 83, p. 1080, 1951.
[10] H. Guckel, D. W. Burns, C. R. Rutigliano, D. K. Showers, and J. Uglow, "Fine grained polysilicon and its application to planar pressure transducers," in *Tech. Dig. 4th Int. Conf. Solid-State Sensors and Actuators* (Tokyo, June 1987), pp. 277-282.
[11] R. P. Feynman, "There's plenty of room at the bottom," in *Miniaturization*, H. D. Gilbert, Ed. New York: Reinhold, 1961, pp. 282-296.

POLYSILICON MICROBRIDGE FABRICATION USING STANDARD CMOS TECHNOLOGY

M. Parameswaran, H. P. Baltes and A. M. Robinson

Dept. of Electrical Engineering and The Alberta Microelectronic Centre
The University of Alberta, Edmonton, Canada T6G 2G7

ABSTRACT

We describe a technique for fabricating polysilicon microbridges using a standard industrial CMOS process. The introduction of new layout methodologies enables us to fabricate CMOS compatible microbridge structures requiring only one additional postprocessing step. This step involves etching of the field oxide below the polysilicon layer in CMOS process which is equivalent to the sacrificial layer that is typical of any microbridge fabrication procedure.

INTRODUCTION

Within the last decade, there has been a growing interest to fabricate micromechanical structures for sensor applications using silicon micromachining technology [1]. In particular, polysilicon micromechanical structures have become popular primarily because of the favourable mechanical properties of the polysilicon material [2]. Its capability to form resilient, deformable mechanical structures has been demonstrated by Howe, through its application in vapour sensing [3]. In view of integrating the micromechanical structures with pertinent circuitry, efforts are in progress to suitably modify the standard planar integrated circuit (IC) technology [4,5]. These modifications comprise of additional steps or altering selected steps in the processing sequence, thereby customising the process for a specific application. For purposes of low cost mass production as well as high reliability, it is desirable to use CMOS processing technology for the fabrication of polysilicon based micromechanical structures. By incorporating additional lithographic steps for the etching of the sacrificial layers, polysilicon microstructures can be realised using standard CMOS process. Hence, attempts to use the standard technology with a minimum number of postprocessing steps will be most economical for commercial production.

In this paper, the technique employed in fabricating polysilicon microbridge structures using an industrial CMOS (CMOS-3DLM) process is presented. This processing service is offered by Northern Telecom Canada, to the students of Canadian universities [6]. In our attempt to fabricate the microbridge, we have introduced some unconventional layout design procedures and one additional postprocessing step.

LAYOUT DESIGN

Figure 1(A) shows the layout of a simple circuit in which a polysilicon layer is placed over two diffusions (layout terminologies are used). The cross-sectional view of the circuit shown in Figure 1(B) helps to examine the physical formation of these various layers more clearly. In this view, one can see the poly being separated by a thin gate oxide over the diffusion area and by a thick field oxide everywhere else. By removing the field oxide below the poly, we obtain the desired bridge structure.

In the standard CMOS process, a layer of oxide (usually PSG or BPSG) is used to insulate the polysilicon and the other underlying layers prior to metallization. Deposition of the passivation layer follows the metallization step. These additional layers deposited over the polysilicon should be removed in order to obtain a free-standing bridge structure. This is achieved by introducing some unconventional layout designs (see Figure 2) viz., a contact-cut and a pad-opening over the region where the microbridge is to be formed.

The dimensions of the contact-cut and pad-opening should be over-sized on the breadth side of the poly layer. These openings provide access to the region below the polysilicon for an isotropic oxide etchant to remove the field-oxide thereby forming the microbridge as illustrated in Figure 3. Since the transistors are formed by the self aligned technique, the gate material (polysilicon in this case) is also exposed to ion implantation during the source and drain definition step. Hence, the region where the polysilicon eventually becomes a microbridge is masked from ion implantation with the use of P+ and N+ mask layers [6]. This step is again an unconventionl method of layout design specially adopted for the microbridge fabrication in standard CMOS technology.

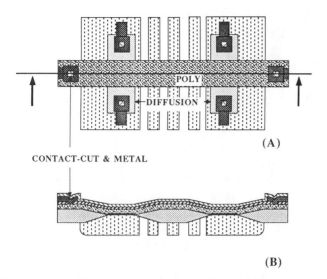

Figure 1. Layout and cross-sectional view of standard design.

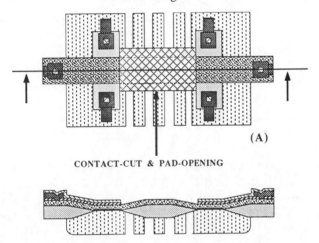

Figure 2. Layout and cross-sectional view of modified design.

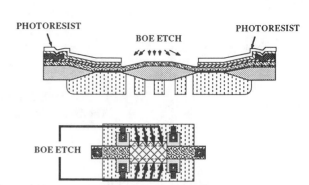

Figure 3. Postprocessing for obtaining the microbridge.

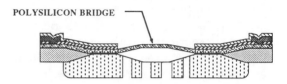

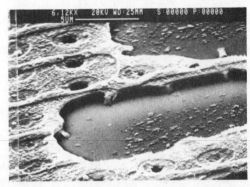

Figure 4. SEM photograph of structure obtained with etch time of 20 minutes.

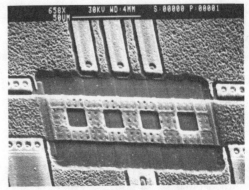

Figure 5. SEM photograph of 30 μm ×102 μm microbridge.

POSTPROCESSING

The removal of field oxide below the poly is performed by a lithographic step after the CMOS processing sequence. This includes spinning the die with a 1 μm thick photoresist, soft-baking for 45 seconds at 90°C and creating a opening directly above the microbridge (see Figure 3(A)). Isotropic etching of the oxide below the poly is performed with buffered oxide etch (BOE). Depending on the bridge geometry the etch time required is 35-45 minutes at 27 °C. To determine this time, the bridge was destroyed and inspected for oxide residues in the space underneath using a scanning electron microscope. Figure 4 shows the oxide residues of a sample etched for 20 minutes. Figure 5 and 6 show the scanning electron micrographs of two completed microbridge structures (30 μm × 102 μm and 47 μm × 180 μm).

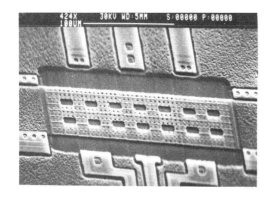

Figure 6. SEM photograph of 47 μm × 180 μm microbridge.

STRESS-RELIEF CONSIDERATIONS

Built-in stress in CVD thin films, especially in polysilicon is very common and has to be given due consideration [4,7]. The resulting bridge structure could otherwise exhibit buckling or deformation from the expected regular shape. To relieve the built-in stress, a high temperature anneal is used. The exact temperature and duration of the anneal depend on various factors such as deposition temperature, thickness, dopant concentration etc., and have to be determined expermentally [8]. The two high temperature cycles in the CMOS-3DLM process, taking place after the deposition of the polysilicon (990 °C for 45 minutes and 950 °C for 30 minutes) aid relieving the built-in stress but do not guarantee complete stress relief in polysilicon, since the typical CMOS process is not optimised for the purposes of creating micromechanical structures. Consequently, this latest attempt resulted in poor yield. Private communications with the process engineers suggest that incorporating one additional anneal step after the deposition of polysilicon will not seriously alter the CMOS process sequence, if the diffusion (well-diffusion) time, which is done prior to the deposition of polysilicon, is properly adjusted.

CONCLUSION

We have presented a technique to fabricate polysilicon microbridge structures that is highly compatible with CMOS technology and requiring only one additional postprocessing step. It is anticipated that the reliability of the microbridge with integrated electronics, fabricated using this technique will be comparable to that of a typical standard IC process. With the proper stress relief measures, low cost fabrication of microbridges appears feasible. The fabrication of annealed bridges using standard IC technology is currently in progress.

ACKNOWLEDGEMENTS

We gratefully acknowledge the services of the Canadian Microelectronic Corporation (CMC) in arranging the chip fabrication. We wish to thank G. Fitzpatrick and M. Buchbinder for the technical assistance. This work was supported by the Natural Science and Engineering Research Council of Canada (NSERC).

REFERENCES

[1] K. E. Petersen, "Silicon As A Mechanical Material", *Proc. of IEEE.*, V 70 (5), May 1982, pp. 419-457.

[2] Long-Sheng Fan, Yu-Chong Tai and R. S. Muller, "Pin Joints, Gears, Springs, Cranks and other Novel Micromechanical Structures", *Digest of Technical Papers, 4th int. Conf. on Solid State Sensors and Actuators* (Transducers'87), Japan, June 1987, pp. 849-852.

[3] R. T. Howe, "Integrated Silicon Electromechanical Vapour Sensor", Ph.D Thesis, Department of Electrical Engineering and Computer Sciences, Univeristy of California, Berkeley, CA, 1984.

[4] H. Guckel and D. W. Burns, "A Technology for Transducers", *Digest of Technical Paper, Int. Conf. on Solid State Sensors and Actuators* (Transducers'85), Japan, June 1985, pp. 90-92.

[5] H. Guckel, D. W. Burns, C. R. Rutigliano, D. K. Showers and J. Uglow, "Fine Grained Polysilicon and its Application to Planar Pressure Transducers", *Digest of Technical Paper, Int. Conf. on Solid State Sensors and Actuators* (Transducers'87), Japan, June 1987, pp. 277-282.

[6] *"Guide to the Integrated Circuit Implementation Services of the Canadian Microelectronic Corporation"*, Carruthers Hall, Queens University, Kingston, Canada, 1987.

[7] R. T. Howe and R. S. Muller, "Polycrystalline Silicon Micromechanical Beams", *J. Electrochem. Soc.*, Vol. 130, No. 6, June 1983, pp.1420-1423.

[8] S. P. Murarka and T. F. Retajczyk, Jr, "Effect of Phosphorus Doping On Stress in Silicon and Polycrystalline Silicon", *J. Appl. Phys.*, 54 (4), April 1983, pp. 2069-2072.

Process Integration for Active Polysilicon Resonant Microstructures

MICHAEL W. PUTTY and SHIH-CHIA CHANG

General Motors Research Laboratory, Warren, MI 48090 (U.S.A.)

ROGER T. HOWE

Berkeley Sensor and Actuator Center, Department of Electrical Engineering and Computer Science, University of California, Berkeley, CA 94720 (U.S.A.)

ANDREW L. ROBINSON and KENSALL D. WISE

Center for Integrated Sensors and Circuits, Solid-State Electronics Laboratory, Department of Electrical Engineering and Computer Science, University of Michigan, Ann Arbor, MI 48109 (U.S.A.)

Abstract

Microsensors based on active polysilicon resonant microstructures are attractive because of their wide dynamic range, high sensitivity and frequency shift output. In this paper, we discuss processing issues for integrating electrostatically-driven and -sensed polysilicon microstructures with on-chip nMOS devices. Surface-micromachining using sacrificial spacer layers is used to obtain released microstructures. A novel feature is the use of rapid thermal annealing (RTA) for strain relief of the ion-implanted, phosphorous-doped polysilicon. Resonance frequencies of cantilever beams indicate a lower-bound Young's modulus of about 90 GPa and an upper-bound compressive residual strain of only 0.002%, indicating that RTA is potentially useful for strain relief.

1. Introduction

Solid-state microsensors are currently being investigated for a wide variety of applications because of their potential for low cost and high performance. One promising approach uses resonant microstructures, where the resonance frequency of a mechanical element is made sensitive to a physical or chemical parameter of interest [1]. To measure the resonance frequency of a micromechanical structure, a means for producing and detecting its motion is needed. Vibration has been produced by several different methods, including thermal, piezoelectric and electrostatic drives. The resulting motion of the structure has been detected by using piezoresistors, piezoelectric thin films, capacitance variation (electrostatic detection), as well as other techniques. For conducting microstructures, including polysilicon, electrostatic drive and detection has several attractions [1]. The most important advantage of this technique is the simple mechanical structure that results. Because the microstructure itself is conducting, it can be used as one plate in the capacitor. Thus, electrostatic drive and detection does not require a complicated microstructure incorporating the additional thin films or diffusions that would be necessary for other drive and detection schemes mentioned above. This simplicity makes the mechanical properties of the microstructure much easier to control.

To vibrate polysilicon resonant microstructures, a sinusoidal electrostatic drive force is exerted on the microstructure by applying a signal consisting of an a.c. drive voltage v_d and d.c. polarization V_P to the drive electrode; the resulting microstructure motion is detected as current flowing through the device (Fig. 1). When the microstructure is driven at its first natural frequency f_1, the device current and microstructure motion reach their maximum amplitude [1]. If the same pair of electrodes is used to drive the microstructure and sense its motion, the device is referred to as a one-port resonant microstructure [2]; if two electrodes are used to provide the driving force and motion detection, then it is a two-port resonant microstructure [1]. While the devices fabricated in this work are one-part microstructures, our results are directly applicable to other configurations.

Polysilicon resonant microstructures are fabricated using a set of processes referred to as surface micromachining [3]. A key feature of surface-micromachining processes is a sacrificial spacer layer, which is used to support the microstructure layer during deposition and patterning. The sacrificial layer is then etched away, leaving a free-standing microstructure (Fig. 2).

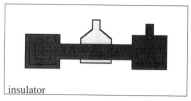

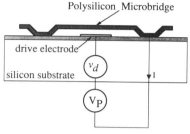

Fig. 1. Operation of resonant microstructures.

In order to integrate circuits with surface-micromachined structures, it is desirable to add the microstructures to previously fabricated circuits. A significant advantage of this approach is that the final steps in the process can be optimized for the mechanical properties of the microstructure. This partitioning generally results in a simpler overall process, because the impact of the complex circuit fabrication sequence on the micromechanical structures need not be considered. Under this scenario, it is desirable that the steps used to fabricate the microstructures have minimal impact on the electrical characteristics of the circuits. If this can be accomplished, the on-chip circuitry (except perhaps for the metallization) could be fabricated by an IC foundry.

A critical issue in the fabrication and operation of resonant microstructures is the mechanical properties of the beam material. If the material is under compression, then the resonance frequency f_1 will be decreased, with the potential for buckling in long beams. Tensile residual strain will raise f_1 and also reduce the amplitude of vibration. The control of residual strain in polysilicon by varying deposition temperature and annealing cycles has been studied extensively [4–6]. A furnace anneal in a nitrogen ambient is commonly used to relieve residual stress in phosphorus-doped polysilicon. This approach is not attractive for active microstructures with on-chip circuitry, since it will degrade the electrical characteristics.

In this paper, we address the fabrication of active polysilicon resonant microstructures, concentrating on the steps used to control the mechanical properties of the polysilicon and the interaction of these steps with on-chip circuitry. We describe a prototype fabrication sequence used to make electrostatically driven surface-micromachined polysilicon beams integrated with an nMOS depletion-mode transistor for capacitive detection of the beam vibration. We then present initial results of measurements used to estimate the plate modulus and residual strain of the polysilicon from the resonance frequencies of cantilevers and bridges.

2. Process Integration

The fabrication of polysilicon resonant microstructures with on-chip circuitry involves some important considerations that are not encountered in conventional silicon IC processing, some of which have been previously discussed by Howe [4]. In this Section, we discuss similar process integration issues, emphasizing the important interactions between the requirements of the micromechanical structures and on-chip circuits. Four areas, polysilicon strain relief, sacrificial spacer layer, active device protection and interconnect, deserve special consideration and are discussed in detail. This is followed by a description of the entire fabrication sequence.

2.1. Strain Relief of Polysilicon

An important process design consideration for active resonant microstructures is how to achieve low-strain polysilicon with good mechanical properties and still incorporate on-chip circuitry. Since the on-chip circuitry is subjected to all of the

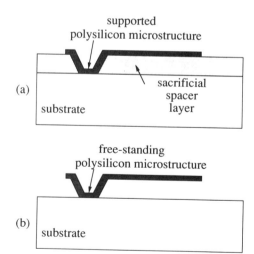

Fig. 2. Schematic representation of surface micromachining. (a) After patterning of the polysilicon microstructure; (b) after removal of the sacrificial spacer layer, leaving a free-standing polysilicon microstructure.

processing required by the polysilicon microstructures, including a thermal cycle that is generally required to relieve the high compressive stress of as-deposited polysilicon [4, 7], this strain-relief cycle and its effect on the on-chip circuitry must be carefully considered.

Many transistor and circuit parameters depend strongly on the location of dopants in the silicon substrate. For example, modern MOS processes have shallow source/drain diffusions to limit short-channel effects, and it is necessary to limit thermal processing following the diffusion implants to keep the junctions from being driven too deep. The transistor threshold voltages often depend on maintaining a threshold-control implant close to the substrate/gate oxide interface. Also, in a CMOS process latch-up immunity is influenced by careful control of the doping profile in the diffused well, as in a retrograde-well process.

Conventional furnace steps have been investigated for strain relief of polysilicon [4, 8]. However, these steps use high temperatures and large thermal budgets, and would certainly have a detrimental effect on the on-chip circuitry. To minimize this problem, we have investigated rapid thermal annealing (RTA), with its associated low thermal budget, as a strain-relief technique for polysilicon microstructures. We have found RTA, for three minutes at 1150 °C, to be effective as furnace annealing for reducing the strain in as-deposited polysilicon. The low thermal budget for this process has a minimal effect on the on-chip circuitry but is very effective for reducing the high compressive strain in the polysilicon film, allowing free-standing microstructures to be incorporated with high-quality circuitry. Experimental data supporting these conclusions will be presented below.

2.2. Sacrificial Spacer Layer

Surface micromachining relies on very high selectivity between the sacrificial spacer layer and the polysilicon used for the microstructures. Etching of the sacrificial layer from under the microstructure must take place laterally, and distances of many microns are common. Thus very long etches are required to remove the sacrificial layer completely. Some form of silicon dioxide is commonly used as the sacrificial layer, because it is etched much more rapidly in hydrofluoric acid (HF) than is polysilicon (which remains essentially untouched). Phosphosilicate glass (PSG) is often used as the sacrificial layer, because it is etched more rapidly in HF than is undoped silicon dioxide. It can also be conveniently deposited using chemical vapor deposition (CVD), and is a common material in conventional IC processing.

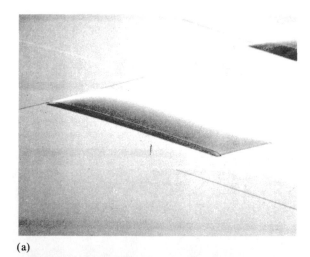

(a)

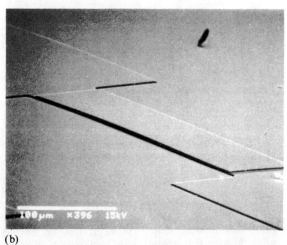

(b)

Fig. 3. Scanning electron micrographs showing effect of sacrificial spacer layer on polysilicon film. (a) PSG spacer, resulting in badly distorted microstructure; (b) composite PSG/SiO$_2$ spacer, resulting in flat microstructure.

However, when a single-layer PSG sacrificial spacer is used, we observed that the polysilicon microstructure was severely distorted after the strain-relief cycle (Fig. 3(a)). Similar effects were observed during furnace annealing at 1100 °C of polysilicon/PSG/Si$_3$N$_4$ sandwiches [9]. This distortion may have occurred because the PSG spacer flows at 1100 °C and does not adhere well to the underlying Si$_3$N$_4$; this allows the microstructure to assume a relaxed configuration before the high compressive stress has been relieved. One way to avoid this problem is to use a composite sacrificial spacer of PSG on top of an undoped CVD silicon dioxide layer. In this configuration, the PSG still flows at the anneal temperatures but adheres well to the underlying silicon dioxide layer (which does not flow), thereby holding the polysilicon layer in place during the anneal procedure. This results in a flat, undistorted microstructure (Fig. 3(b)).

Although the composite sacrificial spacer does not etch as quickly as the single-layer PSG spacer, it does etch much faster than a single-layer silicon dioxide spacer; etch times for the composite spacer are adequate for microstructure fabrication. A composite sacrificial spacer comprising 1.6 μm of silicon dioxide and 0.4 μm of PSG with 6 wt.% phosphorous is used in this process.

2.2.1. Undercut protection

In the surface micromachining processes, it is desirable to remove the sacrificial layer at the end of the process because it is difficult to perform photolithography, depositions or further processing on free-standing microstructures. However, this requires that the active on-chip circuitry be exposed to the very aggressive HF etch that is used to remove the sacrificial layer. Clearly some means of protecting the active circuitry during the undercut etch is required. In our process we employ Si_3N_4 deposited by low-pressure chemical vapor deposition (LPCVD) to protect the active devices (as in the resonant-microbridge vapor sensor process [4]); Si_3N_4 is etched relatively slowly by HF.

A problem we found in using Si_3N_4 as an etch protection layer was a degradation or complete loss of the d.c. polarization voltage, V_P, between the drive electrode and the microstructure, when the silicon nitride layer covered the drive electrode [10]. Some possible explanations for this loss of voltage include surface leakage on the silicon nitride film or charge storage at the interface between the silicon nitride and the drive electrode silicon dioxide. Whatever the cause, this problem was solved by removing the silicon nitride over the drive electrode. Because it must be located under the microstructure, it is convenient to fabricate the drive electrode from the gate polysilicon layer, which is resistant to attack by HF. As a result, the removal of the Si_3N_4 layer does not affect the passivation against HF.

2.3. Interconnect

In our process, the polysilicon microstructure layer was also used as the circuit interconnect layer (rather that the aluminum metallization used in the resonant-microbridge vapor sensor [4]) since it had sufficient conductivity for this purpose. This interconnect approach was used in the multichannel microprobe [11] and has the advantage of not needing to be protected, as aluminum would, during the undercut etch.

If more complicated circuitry requiring low-resistance interconnect were added to the resonant microstructures, there are several different options available. If the circuitry, including interconnect, is fabricated by a foundry, then the interconnect must be able to withstand the high temperature of the sacrificial spacer layer deposition and RTA strain-relief processes. A conventional plasma-CVD Si_3N_4 overglass layer can probably serve as the undercut protection layer. If bond pad openings are defined prior to microstructure formation, then the exposed material must also withstand the HF undercut etch. Tungsten meets all of these requirements, and is commonly used in the IC industry.

If, on the other hand, the interconnect layer can be deposited and defined after the microstructure polysilicon definition (but before the undercut etch), then it need not withstand the high-temperature steps. Tungsten can also be used under these circumstances. If it is necessary that the interconnect have minimal resistance, then gold is acceptable, since it is not attacked by HF.

2.4. Fabrication Sequence

The process integration issues discussed above provide the basis for a prototype fabrication sequence that builds heavily upon the resonant-microbridge vapor sensor process developed by Howe [4] and the multichannel microprobe process developed by Najafi and Wise [11]. The test vehicle was a test chip that includes several nMOS transistors to characterize the on-chip circuitry, and several active microcantilevers and active microbridges of different lengths to characterize the mechanical properties of the polysilicon.

The active resonant microstructures discussed here incorporate the one-transistor nMOS interface circuit shown in Fig. 4. The air-gap motional capacitance of the microstructure is buffered for off-chip detection by an integrated depletion-mode nMOS transistor, and the diode clamp is used to leak d.c. gate voltage of the transistor to ground.

The fabrication sequence begins by diffusing boron to form a p^+ ground plane that improves the isolation between the drive electrode and the

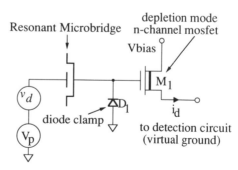

Fig. 4. On-chip interface circuit.

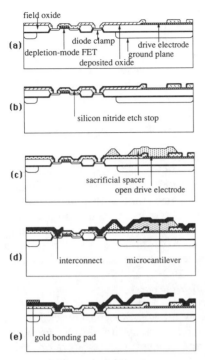

Fig. 5. Process sequence for active resonant microstructures: (a) after formation of ground plane and nMOS circuitry; (b) after patterning of the Si_3N_4 etch stop layer; (c) after patterning of the sacrificial spacer layer; (d) after patterning of the polysilicon microstructure layer; (e) after removal of the sacrificial layer.

resonant microstucture. With the p$^+$ ground plane defined, the on-chip transistor and diode clamp are fabricated by a conventional four-mask LOCOS nMOS process (Fig. 5(a)). As mentioned above, the gate polysilicon is also used as the drive electrode for the resonant microstructure; an n$^+$ diffusion was used for this purpose in the resonant-microbridge vapor sensor [4].

The next step is the deposition and patterning of the Si_3N_4 etch-stop layer used to protect the circuitry during the removal of the sacrificial layer (Fig. 5(b)). It is removed from the area over the drive electrode (as discussed above) and from the regions where the polysilicon interconnect layer makes contact to the silicon substrate at the diode in the on-chip circuit. With the active devices protected, the composite PSG/CVD SiO_2 sacrificial spacer layer itself is deposited and patterned (Fig. 5(c)).

After the sacrificial layer is defined, the polysilicon for the resonant microstructures is deposited to a thickness of 1 μm by LPCVD at a temperature of 600 °C. Fine grain polysilicon was chosen for the microstructure because of its reported highly-reproducible characteristics and superior mechanical properties [8]. The polysilicon is then heavily doped with phosphorus using ion

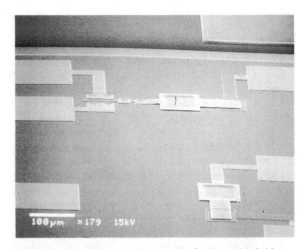

Fig. 6. Scanning electron micrograph of active microbridge.

implantation at an energy of 150 keV and a dose of 5×10^{15} cm^{-2}. Next the RTA strain-relief procedure is performed. For our process, this is carried out in an AG Associates Heatpulse 610 RTA system. The wafer was rapidly heated (approximately two seconds) in a nitrogen ambient by quartz lamps to a temperature of 1150 °C, as measured by an optical pyrometer, and held at this temperature for three minutes. After the high-temperature soak the quartz lamps were shut off and the wafer rapidly cooled to temperatures below 600 °C in approximately two seconds. After the RTA strain-relief procedure, the polysilicon microstructure is patterned (Fig. 5(d)).

Gold bonding pads are then added to the device. Gold was chosen for the bonding pads since it is compatible with the final HF undercut etch. Finally, the sacrificial layer is removed by

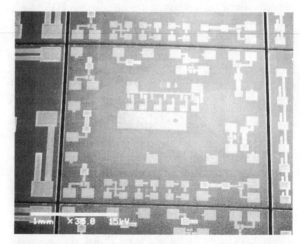

Fig. 7. Photomicrograph of test chip. Test transistors are in the middle, with active microstructures and interface circuits around the perimeter.

etching in concentrated HF and the devices are ready for packaging and testing (Fig. 5(e)). A scanning electron micrograph of an active microbridge fabricated by this process is shown in Fig. 6, and a photomicrograph of the entire test chip is shown in Fig. 7.

3. Experimental Results

In this Section we discuss some experimental results obtained from these test devices. The performance of the electrical devices (diodes and nMOS transistors) is discussed. Data on Young's modulus (E) and residual strain (ϵ) for polysilicon that has been strain relieved by RTA are then presented.

3.1. Electrical Devices

To measure the properties of the diode clamp in the on-chip interface circuit, the test chip includes a large-area diode. The measured characteristics of this device show that it is well behaved, with a clean breakdown ($BV = -39$ V). The minority carrier lifetime is 90 μs, obtained from measurements of the small-signal junction resistance [10].

Figure 8 shows the drain characteristics (I_D versus V_{DS} with V_{GS} as a parameter) for transistors of widely varying dimensions. These transistors show normal behavior. Extracted transistor parameters are summarized in Table 1, and are close to design values. The conduction parameter k' for both transistors is slightly lower than expected for a surface-channel device with a 100 nm gate oxide, but this result is consistent with the buried-channel nature of the depletion transistor [10].

3.2. Mechanical Properties of RTA Strain-relieved Polysilicon

The mechanical properties (plate modulus and residual strain) of polysilicon can be directly extracted from measurements of the resonance frequencies of microcantilevers and microbridges [12]. Our test chip included several microcantilevers and microbridges of different lengths for this purpose. The resonance frequencies of these devices were measured with a HP 4195A network analyzer and the measurement circuit shown in Fig. 9, or with a phase-locked-loop oscillator circuit that continuously vibrated the resonant microstructure at its natural frequency [10]. For these measurements the drive voltage was adjusted such that the amplitude of the microstructure motion was about ±100 nm as measured by the device capacitance. Table 2 details the dimensions and typical measured resonance frequencies of the devices on the test chip.

The plate modulus $E/(1 - v^2)$, where v is Poisson's ratio, can be found by measuring the resonance frequencies of microcantilevers of varying length. We actually measure the plate modulus rather than Young's modulus because the fabricated microcantilevers had dimensions more like plates than beams; in our devices $W/t \gg 1$, but $L/W \sim 1$ ($W \equiv$ width, $t \equiv$ thickness and $L \equiv$ length). Since microcantilevers are released structures with one free end, their resonance frequency is not affected by the average residual strain in the film. Assuming clamped boundary conditions at the fixed end, the first resonance

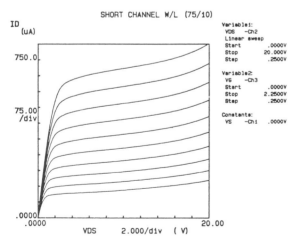

Fig. 8. Drain characteristics of test transistors.

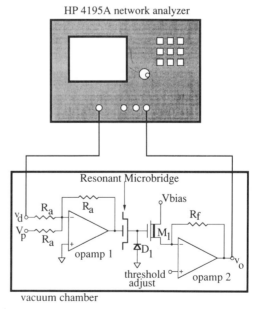

Fig. 9. Circuit for measurement of resonance frequency.

TABLE 1. Extracted transistor parameters

Parameter	Drawn transistor dimensions	
	$W = 75\ \mu m,\ L = 10\ \mu m$	$W = 10\ \mu m,\ L = 75\ \mu m$
V_T: threshold voltage	−1.64 V	−1.50 V
g_m: transconductance (in saturation, $V_{GS} = 0$ V)	140 μmho	1.5 μmho
k': conduction parameter	11.8 μA/V^2	7.5 μA/V^2
λ: channel-length modulation parameter	0.018 V^{-1}	0.002 V^{-1}

frequency f_1 of the microcantilever is given by

$$f_1 = 0.615 \frac{t}{L^2}\left(\frac{E}{\rho(1-v^2)}\right)^{1/2} \quad (1)$$

where ρ is the mass density. Unfortunately, an exact value for Young's modulus could not be obtained from the test structures due to uncertain boundary conditions at the fixed end of the microcantilever. For an ideal clamped boundary condition at the fixed end, the resonance frequency of two microcantilevers of different lengths should scale as L^{-2}. This scaling is not observed in the test devices, indicating a deviation from the perfectly clamped boundary conditions [2]. The microcantilever flares at the anchor, as shown in Fig. 6, which is one source of deviation from clamped boundary conditions. As the length of the microcantilever becomes shorter, this overlap area causes a larger reduction of the resonance frequency of the microcantilever relative to that for ideal boundary conditions. In spite of these difficulties, a lower-bounds estimate of 90 GPa can be obtained for Young's modulus from the longest microcantilever on the test chip by using the equation for clamped boundary conditions and assuming $v = 0.25$ [5]. This value for Young's modulus is 50% less than that reported by Guckel *et al.* [5] for fine-grained undoped polysilicon, but within 10% of that reported by Murarka and Retajczyk [13] for coarse-grained heavily phosphorus-doped polysilicon.

The residual strain ϵ in the polysilicon film can be found by measuring the resonance frequency of microbridge structures. Since microbridges are clamped–clamped structures, the residual strain in the microstructure film causes a shift in the resonance frequency of the microbridge in much the same way as increased tension causes a change in pitch of a guitar string. For the compressive strain that is typically found in polysilicon microstructures, the resonance frequency of microbridges is lower than the zero strain case and is given by

$$f_1 = 1.0279 \frac{t}{L^2}\left(1 - 0.147\epsilon \frac{L^2}{t^2}\right)\left(\frac{E}{\rho(1-v^2)}\right)^{1/2} \quad (2)$$

Again, the uncertain boundary conditions at the fixed ends of the microstructure make an accurate determination of residual strain in the polysilicon film problematic. Nevertheless, an upper-bounds estimate of 0.002% can be obtained by substituting the estimated plate modulus for polysilicon and using eqn. (2), which assumes clamped boundary conditions. This approach probably overestimates the residual strain, since the entire shift in resonance frequency of the microbridge is attributed to residual strain, when in actual fact a large part of this frequency shift is due to relaxed boundary conditions and a lower value of E. This value of residual strain is as low as any reported results [5, 6], indicating that RTA is a useful strain-relief technique.

4. Conclusions

Microsensors based on resonant sensing elements are a promising area of research. In this paper we have discussed integration for active polysilicon resonant microstructures, emphasizing the important considerations needed to fabricate free-standing polysilicon microstructures with previously-fabricated on-chip nMOS circuitry. In particular, RTA was employed to reduce the high intrinsic compressive strain of the as-deposited

TABLE 2. Microstructure dimensions and measured resonance frequencies

Device type	Length (μm)	Width (μm)	Thickness (μm)	Resonance frequency (Hz)
Cantilever	75	50	1.0	165 827
Cantilever	100	50	1.0	101 502
Bridge	175	50	1.0	210 487
Bridge	200	50	1.0	182 635

polysilicon and our initial measurements suggest that this technique is a viable alternative to furnace annealing. RTA produces polysilicon films with very low strain (0.002%), at a reduced thermal budget, preserving the high quality of the nMOS circuitry. A composite PSG/CVD SiO_2 sacrificial layer was also developed to permit the fabrication of flat microstructures.

Although these results are very encouraging, further work is clearly needed. In particular, the RTA strain-relief technique was not studied systematically; no effort was made to optimize the anneal times, temperatures or polysilicon deposition parameters. However, our results were quite repeatable in that the extracted mechanical properties of the polysilicon films did not vary significantly through several process cycles. A careful study of RTA on various kinds of polysilicon (i.e., course fine grained, boron/phosphorus doped) is needed before the merits of this process can be fully evaluated. Also, additional circuitry, such as a complete on-chip sustaining amplifier, could be added to the resonant microstructures in cases where this would provide lower cost or improved performance. This additional circuitry could be implemented in a higher-performance circuit technology such as CMOS. Some of the results presented here may be useful for achieving these goals.

Acknowledgement

The authors are grateful to David S. Eddy of General Motors Research Laboratories for supporting this work.

References

1 R. T. Howe, Resonant microsensors, *Proc. 4th Int. Conf. Solid-State Sensors and Actuators (Transducers '87), Tokyo, Japan, June 2–5, 1987*, pp. 843–848.
2 M. W. Putty, S.-C. Chang, R. T. Howe, A. L. Robinson and K. D. Wise, One-port active polysilicon resonant microstructures, *Proc. IEEE Micro Electro Mechanical Systems Workshop, Salt Lake City, UT, U.S.A., Feb. 1989*, pp. 60–65.
3 R. T. Howe, Surface micromachining for microsensors and microactuators, *J. Vac. Sci. Technol. B*, 6 (1988) 809–813.
4 R. T. Howe, Polycrystalline silicon microstructures, in C. D. Fung, P. W. Cheung, W. H. Ko and D. G. Fleming (eds.), *Micromachining and Micropackaging of Transducers*, Elsevier, Amsterdam, 1985, pp. 169–187.
5 H. Guckel, D. W. Burns, H. A. C. Tilmans, D. W. DeRoo and C. R. Rutigliano, Mechanical properties of fine grain polysilicon: the repeatability issue. *Tech. Digest, IEEE Solid-State Sensor and Actuator Workshop, Hilton Head Island, SC, U.S.A., June 1988*, pp. 96–99.
6 L.-S. Fan and R. S. Muller, As-deposited low-strain LPCVD polysilicon. *Tech. Digest, IEEE Solid-State Sensor and Actuator Workshop, Hilton Head Island, SC, U.S.A., June 1988*, pp. 55–58.
7 H. Guckel, T. Randazzo and D. W. Burns, A simple technique for the determination of mechanical strain in thin films with applications to polysilicon, *J. Appl. Phys.*, 57 (1985) 1671–1675.
8 H. Guckel, D. W. Burns, C. R. Rutigliano, D. K. Showers and J. Uglow, Fine grained polysilicon and its application to planar pressure transducers. *Proc. 4th Int. Conf. Solid-State Sensors and Actuators (Transducers '87), Tokyo, Japan, June 2–5, 1987*, pp. 277–282.
9 R. T Howe, Integrated silicon electromechanical vapor sensor, *Ph.D. Thesis*, University of California, Berkeley, Dec. 1984.
10 M. W. Putty, Polysilicon resonant microstructures, *Master's Thesis*, University of Michigan, Ann Arbor, MI, Aug. 1988.
11 K. Najafi, Multielectrode intracortical recording arrays with on-chip signal conditioning, *Tech. Rep. 177*, Solid-State Electronics Laboratory, Department of Electrical Engineering and Computer Science, University of Michigan, Ann Arbor, MI, May 1986.
12 D. W. DeRoo, Determination of Young's modulus of polysilicon using resonant micromechanical beams, *Master's Thesis*, University of Wisconsin at Madison, Jan. 1988.
13 S. P. Murarka and T. F. Retajczyk, Jr., Effect of phosphorus doping on stress in silicon and polycrystalline silicon, *J. Appl. Phys.*, 54 (1983) 2069–2072.

Fabrication of Micromechanical Devices from Polysilicon Films with Smooth Surfaces

H. GUCKEL, J. J. SNIEGOWSKI and T. R. CHRISTENSON

Wisconsin Center for Applied Microelectronics, Department of Electrical and Computer Engineering, University of Wisconsin, Madison, WI 53706 (U.S.A.)

S. MOHNEY and T. F. KELLY

Materials Science Center, University of Wisconsin, Madison, WI 53706 (U.S.A.)

Abstract

Micromechanical devices such as bearings require smooth surfaces. Fine-grained polysilicon can be produced with a surface roughness near 8 Å rms. The ability to anneal films of this type into tension eliminates size restrictions due to compressive buckling.

The use of these films in micromechanical devices has been restricted because hydrogen fluoride-etched structures are covered by an etch residue that leads to contact welding. Contact between opposing surfaces is induced mainly by surface tension effects. This problem may be avoided by removing the deflection mechanism. Thus, freezing of a water–methanol rinse after sacrificial etching all but eliminates surface tension. Removal of the ice mixture via sublimation at 0.15 millibar occurs readily. Free-standing structures with smooth surfaces and small gaps are next passivated by silicon nitride deposition or other techniques.

Introduction

The goals or a list of requirements for optimized construction techniques for micromechanical devices are readily established by examining the published results [1, 2] on movable joints and mechanical bearings. These devices, when made from polysilicon, are certainly not scaled versions of their conventional counterparts. They suffer from the use of large-grained polysilicon, which leads to surface roughness but avoids surface adhesion problems. They are subject to size constraints: larger than necessary film thicknesses and gap dimensions and smaller than desired lateral lengths, restrictions which are in part due to the compressive built-in strain that is typical for polysilicon. The surface adhesion problem (two freshly etched, smooth silicon surfaces will bond if brought into contact) again enters and for all practical purposes eliminates devices with small gaps. Finally, there is the machining of polysilicon films via, say, reactive ion etching. Vertical flank quality is poor for large-grained films and very often leads to shedding, which in turn causes failure in moving parts.

On the positive side, polysilicon is widely available, which allows experimentation, its strength is still increasing, and there is hope of improving those aspects that were mentioned earlier.

Smooth Surface Polysilicon

The dominant deposition technique for polysilicon is the thermal decomposition of silane, SiH_4, in a low-pressure chemical vapor reactor [3]. Grain size control is achieved in principle by temperature selection. Thus, depositions below 580 °C are reported to be amorphous, whereas depositions above 580 °C lead to increasing grain size with deposition temperature. A rapid increase in deposition rate with temperature occurs, and is to be expected.

Our work on LPCVD polysilicon agrees roughly with the above statements. We base our evaluation of deposited films on three basic measurements: (1) surface roughness with a Surfscan 200 haze measurement; (2) surface roughness increases due to attack by 48% VLSI grade hydrofluoric acid after a 12 h room-temperature immersion; (3) local strain field measurements via the buckled beam and ring technique [4]. An additional measurement, the use of a transmission electron microscope for direct measurements of grain size, is now in progress and is of course difficult and time consuming. All films are grown in a custom-built reactor, which is being improved frequently [5]. Substrate configurations are 3″

Reprinted with permission from *Sensors & Actuators*, H. Guckel, J. Sniegowski, and T. Christenson, "Fabrication of Micromechanical Devices from Polysilicon Films with Smooth Surfaces," Vol. 20, pp. 117-122, 1989. © Elsevier Sequoia.

single-crystal wafers, thermally oxidized and nitrogen annealed, or strain-free silicon nitride covered substrates.

Our experience over the last two years with this type of film has led to a steady improvement and understanding of film quality. It is perhaps good that we recently experienced processing difficulties that reflected themselves as unexplained increases in grain size and significant attack by HF during long etches. This difficulty was traced to a contaminated silane supply. A detailed analysis of the contaminant is now in progress. A new source bottle, which was prepared carefully by the gas supplier, produces very good results. The recognition that gas purity is a problem has led to the conclusion that the amorphous–polycrystalline boundary could indeed be set by a nucleation center that is part of the silane supply. This would imply that pure silane could lead to deposition temperatures above 580 °C for very fine-grained films. This idea was pursued and has lead to a shift in deposition temperature from 580 °C to 591.5 °C. The benefit, an increase of the deposition rate from 50 Å min^{-1} to 68 Å min^{-1}, is highly desirable. There is some indication that further improvements are possible.

Film quality is monitored by Surfscan measurements. Typical results are surface roughnesses of 8 Å rms with zero defects at 0.5 μm aperture for 2 μm film thickness if a 5 mm wide outer ring for 3″ material is excluded from the measurement. The built-in strain for the film is roughly 0.6% compressive. This implies that a film 2 μm thick on a 3″ wafer has a strain energy of roughly 0.1 J.

Transmission electron microscopy (TEM) on as-deposited films has recently been completed. The cross-sectional TEM samples were prepared by mechanically thinning, dimpling and ion milling until a perforation appeared. The microscopy was performed in a JEM 200CXII. Figure 1 shows the results for the as-deposited film. It indicates that the film consists of two regions: a polysilicon layer, which starts at the oxide or nitride interface, and an amorphous layer, which forms on top of the polycrystalline deposit. The polycrystalline material exhibits grains from 100 Å to 4000 Å in size. There is no preferred orientation and the silicon surface that forms when the oxide is removed is perfectly smooth. The amorphous material also has a smooth surface and is responsible for the Surfscan results mentioned earlier.

Anneal cycles for this type of film have also been studied. For very long anneals in nitrogen, 1150 °C for 3 h, the film becomes completely polycrystalline without grain size shifts. Shorter anneals partially convert the amorphous layer. This conversion involves a volume contraction. If this

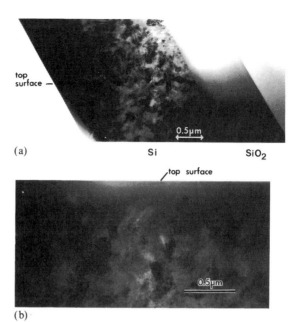

Fig. 1. TEM cross-sections showing the as-deposited polysilicon (2.1 μm) on silicon dioxide (8500 Å). (a) The amorphous polysilicon layer can be seen at the top surface of the polysilicon (left side of photo). (b) A close-up of the top polysilicon surface.

contraction is significant enough, the built-in strain field should change from compression to tension. Figure 2 contains data that support this possibility and result in the conclusion that this type of film can be converted to very significant tensile strains [6].

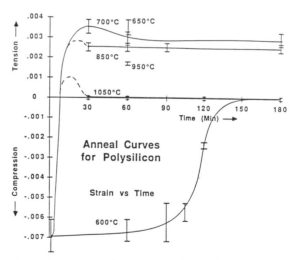

Fig. 2. Anneal curves for polysilicon deposited at 580 °C and 300 mT. Annealing was done by ramping at 15 °C min^{-1} from the deposited temperature of 580 °C to the anneal temperature. The strain field was measured using buckling criteria for micromechanical test structures that respond to tensile and compressive strain fields [6].

Mechanical Stability During Processing

The dominant application area for polysilicon films at Wisconsin involves beam structures that are fabricated by using an isoplanar silicon process. In this procedure the starting material is covered with strain-compensated silicon nitride, which is patterned with a tub mask. An oxidation of the tub area is followed by an oxide strip and reoxidation, which preserves the planarity of the wafer. Typical tub oxide thicknesses are 1 μm. The masking nitride can be removed and the entire wafer surface can be covered with fine-grained polysilicon, typically 2 μm thick. This film is next patterned by reactive ion etching in NF_3 to CF_4 to form, say, clamped–clamped beam shapes. The sacrificial oxide is removed in concentrated HF, which is followed by a deionized wafer dilution rinse. If the beam is long, say 500 μm, and slender, say 20 μm, drying after the wafer rinse normally produces a beam that is permanently attached to the tub bottom. The tendency to form this 'weld' increases as the surface roughness of the polysilicon at the tub oxide–polysilicon interface decreases. This phenomenon is, of course, not very desirable and, in a sense, is to be expected as a form of wafer-to-wafer bonding [7]. A possible explanation of the chemical nature of the bond may be based on the recent results of Gould and Irene [8]. They examine the etching of silicon dioxide in aqueous HF *in situ* by ellipsometry and contact angle measurements. Their results indicate that HF etching produces a residue that is most probably a fluorcarbon film. This film would presumably be present on the single-crystal material of the tub bottom and on the polysilicon of the beam. It produces a bond because beam deflections due to surface tension occur during rinse drying.

If this type of bonding is to be eliminated, a very necessary condition if silicon micromechanics is to take advantage of smooth surfaces, two fundamental approaches could be used. In the first technique, the residue would be eliminated by liquid chemical treatment after HF etching or after the dilution rinse. Surface tension-induced deflection would therefore only come into play after the chemically active surfaces have been passivated. This approach was pursued for some time at Wisconsin. Extensive rinses in hydrogen peroxide–ammonium hydroxide mixtures occasionally accomplish this, but not always. In a second technique, the source of the deflection is eliminated. This allows the fabrication of a dry, non-welded, chemically active structure. The residue in this procedure would be removed after the dry structure is formed. This second technique has produced reliable and repeatable results. In order to understand the rationale for the procedure, it is useful to consider a clamped–clamped beam of length $L = 500$ μm, width $b = 20$ μm, and thickness $h = 2$ μm. This beam, when loaded by the uniform pressure p, will deflect at the beam center by [9]

$$w = pbL^4/384EI \qquad (1)$$

where

$$I = bh^3/12 \qquad (2)$$

and $E = 1.65 \times 10^{12}$ dyne cm^{-2} is Young's modulus for polysilicon [10]. Evaluation for a 1 μm center deflection, which is typically the tub oxide thickness, yields to the conclusion that pb will be near 0.68 dyne cm^{-1}.

Surface tension forces are absent if all objects on the wafer surface are covered with fluid. They appear as soon as the upper surfaces of the beams are uncovered during drying. An exact analysis of the forces on the beam is very difficult because the contact angle between the fluid and the beam surface is unknown. A very pessimistic upper bound is obtained by arguing that the downward force on the beam is of the order of $2L\sigma$, where σ is the surface tension of the water. This would place the equivalent pressure below $pb = 145$ dyne cm^{-1}, or 200 times above the pressure needed to produce a 1 μm center deflection. Surely this force is important and responsible for beam to tub contact.

In order to eliminate it, the wafer must never be allowed to dry during and after HF etching. In this condition, the water-covered wafer is then frozen. This all but eliminates surface tension effects. The frozen water is removed by sublimation, which is quite rapid for water and requires vacuum pumping of the cold substrate.

Experimental work along these lines reveals some practical difficulties. The water-covered wafer can be cooled quite easily by evaporative cooling in a vacuum system that can pump readily to the low millibar vacuum range. A wafer support which is a thermal open circuit helps. If the covering fluid is deionized water, one expects supercooling and hence a rapid, nearly explosive freezing with low-temperature, hard ice. The destruction of micromechanical devices on the wafer is a normal consequence of this. This difficulty can be overcome by adding a nucleation center to the rinse water. Methanol is quite effective. It produces a mixture that freezes without supercooling and forms a soft ice at 0.15 millibar, which sublimes readily. Reheating of the wafer after sublimation is necessary to avoid condensation after removal from the vacuum system.

The dry structure must next be passivated. The test for a successful procedure is quite simple: if a test structure such as a beam is deflected on purpose until contact is established and if the structure returns to its original shape after the deflection force has been removed, the passivation scheme is a success. Experimental work along these lines has led to the conclusion that a thin LPCVD layer of silicon nitride, which is of course conformal, is always effective. Short oxidations help, but will not always work and do introduce additional strain. Heating in nitrogen reduces adhesion but does not eliminate it completely. Figures 3 and 4 are results that have been obtained recently and show structures fabricated by using strain-compensated silicon nitride as the passivation technique. Both structures used more than one sublimation cycle because the metal, 99.999% aluminum, was applied and wet etched after sacrificial oxide etching.

A second source of unwanted beam deflections involves electric fields due to built-in potentials. Opposing surfaces, the inner surface of the beam and the tub surface, form the plates of a capacitor

Fig. 4. SEM micrograph of a doubly supported tuning fork covered with 4100 Å electron-beam evaporated aluminum. The tuning fork self shadows the evaporation, producing a region under the forks without metal. The gap at the edge of the forks to the substrate is roughly 0.5 µm. Dimensions are: fork length = 400 µm, fork width = 4.0 µm, gap between forks = 2.0 µm.

Fig. 3. SEM micrograph of a doubly supported beam fabricated from fine-grained polysilicon in an isoplanar process. Note the lack of texture in the polysilicon surface. The fabrication included the freeze/sublimation technique followed by passivation of surfaces with LPCVD silicon nitride. Beam dimensions are: length = 600 µm, width = 60 µm, thickness = 2.5 µm, and air gap = 1.0 µm.

which are separated by the dielectric of the gap. The result is an attactive pressure of the form

$$p = \frac{1}{2} \varepsilon \mathcal{E}^2 \qquad (3)$$

where ε is the gap permittivity and $\mathcal{E}$ is the electric field at the appropriate surface. The importance of this pressure can again be estimated by arguing that a 500 µm long beam, as specified earlier, will deflect if pb exceeds 0.68 dyne cm^{-1}. For a uniform field in a parallel-plate configuration, this would imply that electric fields of the order of 3×10^4 V cm^{-1} and above are troublesome. Since 3 V built-in voltages are normally not available in real situations, a 1 µm gap device will be mildly affected. If, however, the gap is decreased by surface tension and if the built-in voltage persists, the field will easily be sufficient to pull the beam into contact and in fact does this via a mechanical instability. In a passivated structure, the electrostatically induced contact can always be broken by applying an external bias that opposes the built-in voltage, an observation that leads to a test for this type of adhesion.

Studies to understand the behavior of the electrostatically loaded beam have been started. They are based on the assumption that the gap supports the entire built-in potential or that

$$\varepsilon \approx -\frac{V_{FB}}{(g-w)} \qquad (4)$$

where V_{FB} is the ideal flat-band voltage, g is the non-deflected gap dimension and w is the deflection of the beam. The deflection is determined through the approximate beam deflection relationship:

$$-EI\frac{d^4w}{dx^4} = pb \qquad (5)$$

which leads to

$$\frac{d^4w}{dx^4} = -\varepsilon\frac{V_{FB}^2 b}{2(g-w)^2 EI} \qquad (6)$$

Attempts to understand this non-linear differential equation via numerical integration are now in progress. The results will be reported soon.

Vertical Flank Definition

The results reported so far are encouraging for the manufacture and processing of micromechanical devices with flat, smooth surfaces. The ability to produce vertical flanks without rough side walls is a necessity for structures such as bearings. The preferred tool for this is reactive ion etching with etch gases such as NF_3, CF_4, or SF_6 or much more complex and, unfortunately, much more corrosive, chlorine and bromine-based chemistries. In either case there is a decided advantage for fine-grained rather than coarse-grained polysilicon. The reason is simply that grain boundaries normally etch at a different rate than the crystallites.

The procedure that has been employed at Wisconsin uses a Plasma Therm 1441 reactor in the RIE mode at 150 W. The etch gas is CF_4 at 15 to 20 mTorr pressure. An etch rate of roughly $0.5\ \mu m\ h^{-1}$ is used to etch those sections of the polysilicon that are exposed. The masking material is a chromium layer 800 Å thick which has been defined via photoresist and chromium etching in CCl_4. Figure 5 shows the results, which are encouraging but need improvement. A more detailed analysis indicates that chromium is redeposited during the etch and thereby contributes to the surface roughness of the wall. For a bearing surface, this is not acceptable and must be improved.

Fig. 5. The pattern shown is in a 2 μm thick polysilicon membrane. Patterning was done with CF_4 reactive ion etching and a Cr mask. The membrane was produced by removal of the silicon substrate with EPW (ethylene diamine–pyrocatechol–water) with an etch stop of strain-compensated silicon nitride layer under the polysilicon. Silicon nitride was also used as a protection layer for the front polysilicon surface and as a backside masking layer. Subsequently, the silicon nitride was removed.

Conclusions

Fine-grained polysilicon fits well into the construction techniques needed in micromechanics. The possibility of using very smooth surfaces with small gaps has been considerably enhanced by the reported freeze–sublimation–passivation technique, which avoids the surface sticking problem that many groups have experienced. A major difficulty is still that of etching smooth vertical walls, a structure that is easier to fabricate with fine-grained rather than coarse-grained polysilicon.

Acknowledgements

One of us, H.G., is indebted to the National Science Foundation via grant EET-8815285 for research support.

S. Mohney is funded by an Office of Naval Research fellowship.

References

1 L.-S. Fan, Y.-C. Tai and R. S Muller, Pin joints, gears, springs, cranks, and other novel micromechanical structures, *Proc. 4th Int. Conf. Solid-State Sensors and Actuators (Transducers '87), Tokyo, Japan, June 2-5, 1987*, pp. 849-852.
2 M. Mehregany, K. J. Gabriel and W. S. N. Trimmer, Integrated fabrication of polysilicon mechanisms, *IEEE Trans. Electron Devices, ED-35* (1988) 719-723.
3 W. Kern and G. L. Schnable, Low-pressure chemical vapor deposition for very large-scale integration processing—a review, *IEEE Trans. Electron Devices, ED-26* (1979) 647-657.
4 H. Guckel, D. W. Burns, C. R. Rutigliano, D. K. Showers and J. Uglow, Fine grained polysilicon and its applications to planar pressure transducers, *Proc. 4th Int. Conf. Solid-State Sensors and Actuators (Transducers '87), Tokyo, Japan, June 2-5, 1987*, pp. 277-282.
5 H. Guckel, D. W. Burns, H. A. C. Tilmans, C. C. G. Visser, D. W. DeRoo, T. R. Christenson, P. J. Klomberg, J. J. Sniegowski and D. H. Jones, Processing large aspect ratio microstructures, in *Tech. Digest, Solid-State Sensor and Actuator Workshop, Hilton Head Island, SC, U.S.A., June 6-9, 1988*, pp. 51-54.
6 H. Guckel, D. W. Burns, C. C. G. Visser, H. A. C. Tilmans and D. DeRoo, Fine grained polysilicon films with built-in tensile strain. *IEEE Trans. Electron Devices, ED-35* (1988) 800-801.
7 J. B. Lasky, S. R Stiffler, F. R. White and J. R. Abernathy, Silicon-on-insulator (SOI) by bonding and etchback, *Int. Electron. Dev. Meet. (IEDM), Washington, DC, U.S.A., Dec. 1-4, 1985*, pp. 684-687.
(a) M. Shimbo, K. Furukawa, K. Fukada and K. Tanzawa, Silicon-to-silicon direct bonding method, *J. Appl. Phys., 60* (1986) 2987-2989.
(b) W. P. Maszara, G. Goetz, T. Caviglia, A. Cserhati, G. Johnson and J. B. McKitterick, Wafer bonding for SOI, *Mater. Res. Soc. Symp. Proc., 107* (1988) 489-494.
(c) R. D. Black, E. L. Hall, N. Lewis, R. S. Gilmore, S. D. Arthur and R. D. Lillquist, Silicon and silicon dioxide thermal bonding, *Mater. Res. Soc. Symp. Proc., 107* (1988) 495-500.
8 G. Gould and E. A. Irene, An *in situ* study of aqueous HF treatment of silicon by contact angle measurement and ellipsometry, *J. Electrochem. Soc., Solid-State Sci. Technol., 135* (1988) 1535-1539.
9 B. I. Sandor, *Strength of Materials*, Prentice-Hall, Englewood Cliffs, NJ, 1978, pp. 410-411.
10 H. Guckel, D. W. Burns, H. A. C. Tilmans, D. W. DeRoo and C. R. Rutigliano, Mechanical properties of fine grained polysilicon—the repeatability issue, *Tech. Digest, Solid-State Sensor and Actuator Workshop, Hilton Head Island, SC, U.S.A., June 6-9, 1988*, pp. 96-99.

Selective Chemical Vapor Deposition of Tungsten for Microelectromechanical Structures

N. C. MACDONALD, L. Y. CHEN, J. J. YAO, Z. L. ZHANG, J. A. MCMILLAN and D. C. THOMAS

School of Electrical Engineering, The National Nanofabrication Facility, Cornell University, Ithaca, NY 14853 (U.S.A.)

K. R. HASELTON

School of Applied and Engineering Physics, Cornell University, Ithaca, NY 14853 (U.S.A.)

Abstract

A selective chemical vapor deposition (CVD) tungsten process is used to fabricate three-dimensional micromechanical structures on a silicon substrate. Patterned structures are formed in silicon dioxide trenches by selective nucleation and growth of tungsten from the bottom of the trench. Examples are shown for selective growth of tungsten on single-crystal silicon, on thin films of silicon and on silicon-implanted, silicon dioxide layers. This high deposition-rate selective CVD tungsten process has been used to fabricate tungsten micromechanical structures greater than 4 μm thick. Tungsten patterns with features of 0.3 μm × 0.3 μm in 0.6 μm of tungsten are fabricated using electron beam lithography. As an example of an electromechanical structure, cantilever beams have been fabricated to make micromechanical tweezers that move in two dimensions by the application of potential differences between the beams (lateral motion), and between the beams and the silicon substrate (vertical motion). Tungsten microtweezers 200 μm in length with a cross-section of 2.7 μm × 2.5 μm closed with an applied voltage of less than 150 V.

Introduction

Recent research interest in micromechanical structures has focused on the fabrication of microactuators and micromotors. The intent of this research is to fabricate integrated, movable microstructures for precision motion and actuation. The research builds on the extensive silicon technology base developed during the past thirty years. This silicon technology has recently been extended to include the fabrication of movable microstructures using polycrystalline silicon with sacrificial layers to release (free) the microstructures [1, 2]. One disadvantage of the present polysilicon technology is that the deposition of only relatively thin layers of polysilicon is practical, and the silicon micromechanical structures are usually non-planar structures that are not easily extended to three-dimensional structures or to multilevel micromechanical devices and systems.

Only a few materials have been tested and characterized for application to silicon-based micromechanics. Thus, more materials development is needed to enhance our choice of compatible micromechanical materials.

Thin films several micrometers thick can be deposited using chemical vapor deposition (CVD) processes. Both CVD silicon dioxide and CVD tungsten processes have been developed with deposition rates greater than 0.1 μm min^{-1} [3]. Recently, selective CVD tungsten processes have been developed for integrated circuit metallization using CVD silicon and metals for seeding [3–7]. Other selective CVD tungsten processes use patterned CVD silicon dioxide trenches with implanted silicon [8] at the bottom of the trench to form patterned tungsten microstructures. The CVD tungsten selectively seeds on the material at the bottom of the trench and grows vertically; no anisotropic etching of the tungsten is required to produce high aspect-ratio fixed or released tungsten structures.

Here we describe the extension of the selective CVD tungsten process to produce micromechanical structures, including low-stress cantilever beams and other high aspect-ratio and submicron-patterned tungsten structures. We show that it is possible to produce very sharp 0.3 μm × 0.3 μm square patterns in 0.6 μm of tungsten. We also demonstrate the use of the selective CVD tungsten process by the fabrication of 'microtweezers', a research vehicle that is used to investigate the electomechanical properties of the tungsten microstructures, including mechanical fatigue [9]. The

TABLE 1. Mechanical properties of materials used in silicon integrated circuits with diamond included for reference

	Yield strength (GPa)	Knoop hardness (kg mm^{-2})	Young's modulus (E) (GPa)	Density (ρ) (10^3 kg m^{-3}; g cm^{-3})	Thermal conductivity (W m^{-1} °C^{-1})	Thermal expansion (10^{-6} °C^{-1})	Phase velocity $(E/\rho)^{1/2}$ (km s^{-1})
Diamond[a]	5.3	7000	1035	3.5	2000	1.0	17.2
Si$_3$N$_4$[a]	1.4	3486	385	3.1	19	0.8	11.1
SiO$_2$ (fibers)	8.4	820	73	2.5	1.4	0.55	5.4
Si[a]	7.0	850	190	2.3	157	2.33	9.1
W	4.0	485	410	19.3	178	4.5	4.6
Al	0.17	130	70	2.7	236	25	5.1

[a]Single crystal.

microtweezers move both laterally and vertically by the application of potential differences between the tweezers' beams, and between the tweezers' beams and the silicon substrate. Computer simulations of the electric field distribution and the mechanical bending of the tweezers' beams are used to calculate the time-resolved motion and the spatial profile of the microtweezers as a function of the applied potentials.

Materials Selection for Micromechanics

The integration of micromechanical devices with silicon integrated circuits is a key goal of microelectromechanical systems (MEMS) research. Therefore, choosing materials and processes compatible with silicon technology is one key aspect of the micromechanical materials selection process. A second criterion for materials selection is the desired electromechanical materials properties. Table 1 is a list of materials commonly used in silicon integrated circuits along with their mechanical properties [10]. Diamond is included in Table 1 only for comparison. Single-crystal silicon, polysilicon and silicon dioxide materials have been used extensively for the fabrication of microdynamic devices. Thin CVD films of silicon nitride and silicon dioxide have been used as insulators and sacrificial layers. As outlined in Table 1, all these actively used materials have a density of 2–3 g cm^{-3} and, except for CVD silicon dioxide, are presently used in layers of the order of one micrometer or less in thickness. A third and most important selection criterion for materials is that the material can maintain its shape on being released or freed. That is, the tungsten or other materials process must yield low-stress microstructures. For many materials, process compatibility with silicon processes and obtaining low-stress released microstructures are difficult criteria to meet.

Recently CVD tungsten has been considered for use as a conductor for integrated circuit interconnections. Both blanket tungsten with etching [11, 12] and selective CVD tungsten [3–7] have been investigated extensively. Because of tungsten's compatibility with silicon integrated circuit technology and unique mechanical properties relative to the actively used materials for micromechanics, we selected tungsten as a possible material for fabrication of micromechanical structures.

The selective CVD tungsten process exhibits the following advantageous properties for use in micromechanics:

(1) Tungsten is compatible with silicon integrated circuit processes.

(2) The CVD tungsten process yields a high deposition rate ($6 \mu m \, h^{-1}$).

(3) CVD tungsten produces low-stress thick ($>4 \mu m$) films.

(4) Selective CVD tungsten deposition eliminates the need to etch tungsten.

(5) CVD tungsten is a planar process that fills high aspect-ratio CVD silicon dioxide trenches.

(6) CVD tungsten surfaces replicate the smoothness of the silicon oxide trench sidewall.

(7) High-resolution submicron patterns can be made using selective CVD tungsten processes.

(8) Tungsten processes are compatible with the multiple-level tungsten interconnects and tungsten plugs used in integrated circuits.

(9) The fabrication of multilayer mechanical devices is possible using multiple layers of CVD silicon dioxide as support and sacrificial layers.

(10) Tungsten is a high-density material ($19.3 \, g \, cm^{-3}$) with a high Young's modulus.

The main disadvantage of tungsten relative to polysilicon, silicon nitride and silicon dioxide is the roughness of the top or growing tungsten surface. Examples that illustrate these properties of the selective CVD tungsten process follow.

Selective CVD Tungsten Processes and Patterns

The processing steps for the fabrication of tungsten micromechanical structures are based on selective CVD tungsten technology. Figure 1 outlines three processing sequences used to fabricate tungsten structures. The substrate material is usually a silicon wafer or other suitable flat surface, e.g., a quartz plate or CVD silicon dioxide. The first process sequence (Fig. 1(a)) produces structures seeded from and attached to a silicon substrate. First, a layer of low-pressure CVD silicon dioxide, typically $3-6 \mu m$ thick, is deposited, pattern and etched to form the channels for the selective CVD tungsten deposition. To obtain vertical resist profiles, we use contrast-enhancement material and photoresist for patterning. The channels for the tungsten pattern are formed by magnetron reactive ion etching the CVD silicon

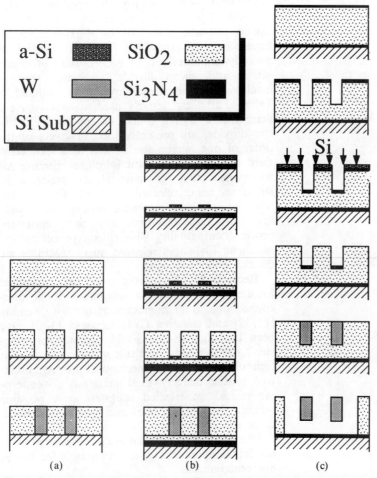

Fig. 1. Processing sequences for selective CVD tungsten, for seeding and growth of tungsten on (a) exposed substrate silicon; (b) thin-film polysilicon or metals; and (c) silicon dioxide implanted with silicon.

dioxide in a CHF$_3$ ambient. Tungsten is then selectively seeded and deposited on the exposed silicon at the bottom of the trench to fill the silicon dioxide channels. The tungsten deposition is performed in a Genus 8402 cold wall reactor with a wafer temperature of approximately 580 °C. Deposition pressure is 400 mT with gas flows of 4900 sccm H$_2$ and 180 sccm WF$_6$.

Examples of tungsten patterns seeded on single-crystal silicon are shown in Figs. 2, 3 and 4. Figure 2 shows a honeycomb structure fabricated using the process depicted in Fig. 1(a). The tungsten film thickness is 2.5 μm. Three features are evident: first, uniform seeding and growth of the tungsten on the silicon has occurred and the tungsten features replicate the silicon dioxide pattern; secondly, the sidewalls of the tungsten structures are relatively smooth; thirdly, the top tungsten surface is rough with somewhat jagged edges. A harmonic [13] or 'wobble' [14] motor pattern (Fig. 3) including the insulator to isolate the moving ring from the stator also exhibits the same surface roughness observed in Fig. 2. Note that the silicon dioxide on the stator and the tungsten sidewalls are of comparable smoothness (roughness). The tungsten replicates the sidewall topography of the silicon dioxide etch process. Selective tungsten seeded from polysilicon exhibits similar

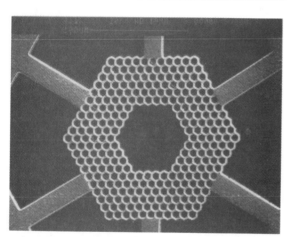

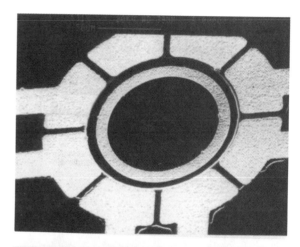

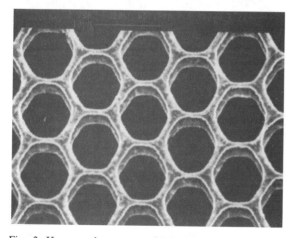

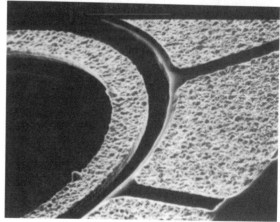

Fig. 2. Honeycomb structure fabricated using the process depicted in Fig. 1(a). Tungsten thickness is 2.5 μm. Three different magnifications are shown.

Fig. 3. Harmonic motor pattern fabricated using the process depicted in Fig. 1(a). Tungsten thickness is 3.5 μm.

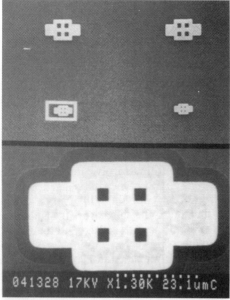

(a)

(b)

(c)

Fig. 4. Tungsten patterns fabricated using the process depicted in Fig. 1(a). Electron beam lithography is used to achieve the submicron dimensions in 0.6 μm thick tungsten.

surface features as those observed for tungsten seeded on substrate silicon.

Further indication of the excellent sidewall smoothness and pattern replication is shown in Fig. 4 for 0.6 μm thick tungsten seeded on silicon; this Figure shows high-resolution tungsten features fabricated using electron beam lithography with a trilayer resist process [15]. The exposure was performed on a JEOL JBX-5DII electron beam lithography machine with a beam energy of 50 keV and beam spot size of about 25 nm. The channels for the tungsten pattern are formed by low pressure (30 mT) reactive ion etching the CVD silicon dioxide in a CHF_3 ambient. A tungsten pattern with four 0.3 μm × 0.3 μm squares in 0.6 μm thick tungsten is shown in Fig. 4(a); these micrographs were recorded at normal beam incidence in the scanning electron microscope (SEM). The bottom frame of Fig. 4(a) shows the very sharp corners (<20 nm) on each square. However, Fig. 4(b) taken at 60° beam incidence in the SEM highlights the rough top surface of the tungsten, but also shows very smooth sidewalls. A tungsten pattern that further demonstrates the excellent sidewall smoothness of the tungsten and high-resolution pattern transfer capability of the process is shown in Fig. 4(c). Features with dimensions less than 100 nm should be practical for tungsten seeded on single-crystal silicon using the process depicted in Fig. 1(a).

The roughness of the top tungsten surface and overfilling of the trenches with tungsten (Fig. 3) for some features requires further research attention to determine the origins of these process defects.

Tungsten structures released from the substrate can be fabricated using the process depicted in Figs. 1(b) and (c). The process shown in Fig. 1(b) uses a patterned thin-film silicon layer (or metal) at the bottom of a silicon dioxide trench to seed and grow tungsten selectively [4, 5]. The remainder of the process is very similar to the growth process described in Fig. 1(a). The tungsten structure is released by partially or completely etching the CVD silicon dioxide down to the

(a)

(b)

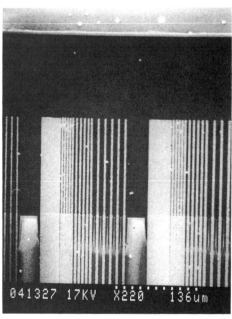

(c)

Fig. 5. Released tungsten beams using the process in Fig. 1(c). The doubly supported beams are 350 μm long by (a) 1 μm wide and (b) 2 μm wide. (c) Tungsten cantilevered beams 150 μm long by 3.5 μm thick with minimum line widths of 1.5 μm (left), 2.5 μm (middle) and 5 μm (right).

silicon or the silicon nitride etch stop/substrate isolation layer.

A novel selective CVD tungsten process uses ion-implanted silicon to nucleate and grow tungsten. This process sequence is shown in Fig. 1(c). Like the process sequence shown in Fig. 1(b), a layer of low pressure CVD silicon dioxide approximately 8 μm thick is used to form the trenches for the selective CVD tungsten deposition. A 100 nm thick silicon nitride layer is then deposited as the implantation mask. Contrast enhancement material is used with photoresist to obtain vertical resist profiles. The trenches for the tungsten beams are formed by reactive ion etching the composite dielectric structure in a CHF_3 ambient, and the silicon dioxide channels are etched to a depth of approximately 3.5 μm. After the photoresist is removed by an oxygen plasma, silicon atoms at an energy of 40 keV are implanted. A silicon dosage of 1×10^{17} cm^{-2} has been selected to initiate the deposition of tungsten on CVD silicon dioxide. The wafer is subjected to an oxygen plasma to remove any carbon contamination encountered during the implantation. To expose the peak silicon concentration in the implanted oxide channels, we use buffered HF. The silicon nitride mask is then removed in hot phosphoric acid at 155 °C.

Examples of released tungsten test structures are shown in Figs. 5 and 6. To estimate qualitatively the stress distortion, we fabricated tungsten beams that are fixed at both ends and span a 350 μm trench (Figs. 5(a) and (b)). Tungsten beams of 1.5 μm × 3.5 μm (thick) show some bowing (Fig. 5(a)). A 10× magnification of the highlighted 'subframe' shows a high magnification of the bowed tungsten lines. The 2.5 μm × 3.5 μm (thick) tungsten beams (Fig. 5(b)) show no measurable bowing for the 350 μm span. Figure 5(c) shows straight, non-distorted tungsten cantilever beams 150 μm in length, 1.5 μm, 2.5 μm and 5 μm in

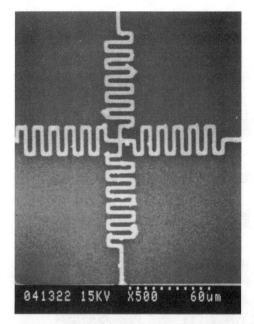

(a)

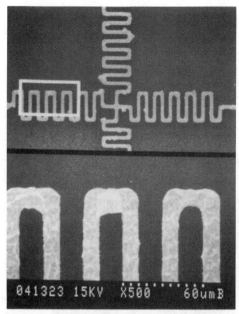

(b)

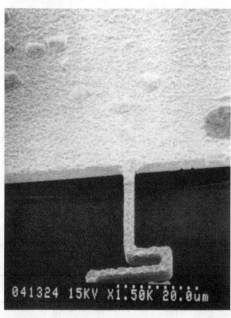

(c)

Fig. 6. Released tungsten meander structure fabricated using the process depicted in Fig. 1(c). The total length of each meander is 630 μm. The tungsten thickness is 3.5 μm.

width, and 3.5 μm thick. A released meander tungsten structure is shown in Fig. 6. The total length of each meander (Fig. 6(a)) line is 630 μm with a cross-section of 2 μm × 3.5 μm (thick). Figure 6(b) shows a high-magnification image of the serpentine features, including a nodular defect of tungsten on the top surface. A high magnification of the attached end of the meander line is shown in Fig. 6(c). Once again, these images show low-stress released tungsten patterns with smooth sidewalls. However, the top surface of the tungsten beams exhibits random nodular tungsten growth and a rough texture.

Tungsten Microtweezers

Tungsten microtweezers have been designed, fabricated and tested [9]. The nominal cross-section of the microtweezers is shown in Fig. 7. Each tungsten beam is conformally coated with a thin layer (200 nm) of silicon dioxide to prevent shorting of the tweezer beams upon closure. A computer simulation of the lateral shape of one tweezer beam after the application of a step function of voltage is shown in Fig. 8. The simulation shows that the two tweezer beams close in 20 μs.

A pair of microtweezers was fabricated using the processing sequence shown in Fig. 1(c) and the ion implantation parameters discussed in the text relative to that Figure. The three SEM micrographs in Fig. 9 show the motion of the microtweezers under test. For the tweezer operation, small deflections were first observed at 145 V, and then a sudden full deflection was observed at 150 V. This threshold behavior of the microtweezers agrees quite well with the simulated result. To reopen the microtweezers, the voltage must be set well below the threshold voltage. Figure 9(c) shows that the tweezers reopen after the voltage is reduced to 60 V. Figure 10 shows the two-dimensional free end ($L = 200$ μm) deflections of the tweezers' beams. At zero applied voltage the tweezers' beams are open (Fig. 10(a)) and the beams exhibit a uniform gray scale contrast in the

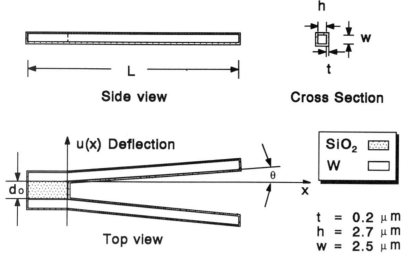

Fig. 7. The microtweezer structure: $w = 2.5$ μm, $h = 2.7$ μm, $L = 200$ μm, $d_0 = 3$ μm, $t = 0.5$ μm, and $\theta = 0.5°$. The ends of the tweezer beams make contact for $u(L) = -1$ μm.

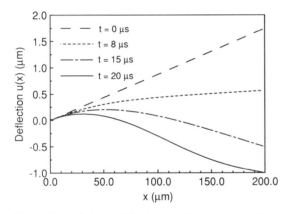

Fig. 8. Dynamic beam deflection as a function of position for one tweezer beam and for times $t = 0$ μs, 8 μs, 15 μs and 20 μs. A step function of 120 V is applied between the two tweezer beams.

SEM image; for an applied voltage between the tweezers' beams of greater than the threshold voltage (Fig. 10(b)), the tweezers close in the $x-y$ plane. The voltage contrast in the SEM image shows that the left tweezer beam (white) is biased negative with respect to the right beam (black). Figure 10 shows the downward (z) deflection of the right electrode to touch the silicon nitride isolation layer covering the silicon substrate (refer to Fig. 1); the measured vertical threshold voltage applied between the right tweezer beam and the silicon substrate for full vertical deflection was 150 V.

Summary

A selective chemical vapor deposition (CVD) tungsten process has been used to fabricate three-dimensional tungsten micromechanical structures on a silicon substrate. Examples of CVD tungsten micromechanical structures described in this paper show the versatility of the CVD tungsten process for the formation of fixed and released micromechanical structures. Stress test patterns show the released tungsten beams and meander patterns can span distances of greater than 350 μm. Since the CVD tungsten process uses CVD silicon dioxide as the patterning layer, multiple levels of micromechanical structures can be fabricated on the same chip by depositing and patterning successive layers of CVD silicon dioxide. The planar high aspect-ratio tungsten process is excellent for fabricating laterally driven microstructures.

The key process defect that requires additional research is the roughness of the top surface of the tungsten and the tendency for randomly dispersed tungsten nodules to form on the top surface.

Two insulated tungsten beams have been used to make micromechanical tweezers that move in two dimensions when potential differences are applied between tweezers' beams (lateral motion) and the tweezers' beams and the silicon substrate (vertical motion). We use the microdynamical tweezer structure as a research vehicle to investigate the micromechanical properties of CVD tungsten; to evaluate the accuracy and completeness of our electromechanical simulations and models; to study the time-resolved microdynamics of the tweezers; and to provide directions for the development of new CAD and simulation tools for microdynamics.

Submicron patterning of CVD micromechanical structures has been demonstrated using electron beam lithography. This high-resolution

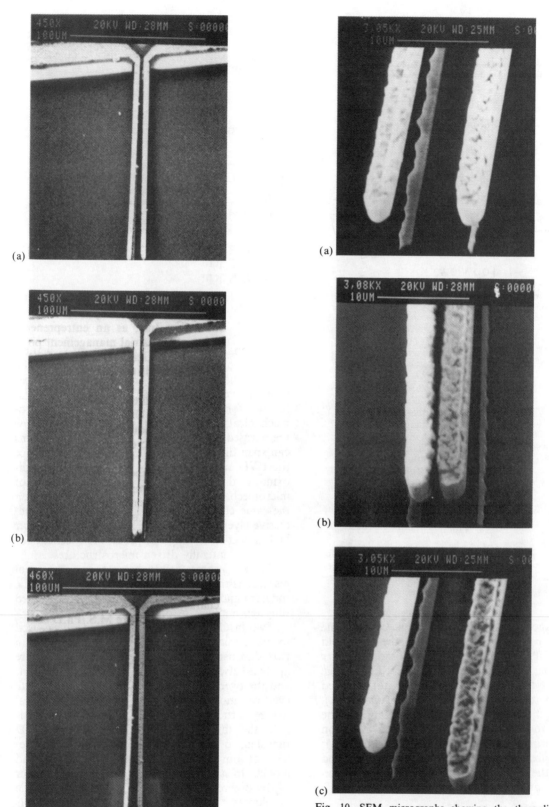

Fig. 9. SEM micrographs of tweezers for an applied voltage of (a) 0 V, tweezers opened; (b) 150 V, tweezers closed; and (c) 60 V, tweezers reopened.

Fig. 10. SEM micrographs showing the three-dimensional free-end deflection of the tweezer beams. (a) 0 V applied, no deflection; (b) 150 V applied between the two beams showing $x-y$ (in the plane direction) deflection; (c) 150 V applied between the silicon substrate and one beam (black), showing z deflection.

patterning capability provides an opportunity to fabricate low-mass nanomechanical structures in CVD tungsten.

Acknowledgements

This work was supported by the National Science Foundation under Grant Nos. ECS-8805866 and ECS-8815775. The structure fabrication was performed at the National Nanofabrication Facility (NNF), which is supported by the National Science Foundation, Cornell University, and Industrial Affiliates. The authors would like to thank the staff of NNF for their technical assistance.

References

1 S. F. Bart, T. A. Lober, R. T. Howe, J. H. Lang and M. F. Schlecht, Design considerations for microfabricated electric actuators, *Sensors and Actuators*, 14 (1988) 269.
2 R. S. Muller, From ICs to microstructures: materials and technologies, *Proc. IEEE Micro Robots and Teleoperators Workshop, Hyannis, MA, U.S.A., Nov. 1987*.
3 I. Beinglass, Selective CVD tungsten deposition—A new technology for VLSI and beyond, in R. S. Blewer (ed.), *Tungsten and Other Refractory Metals for VLSI Applications*, Materials Research Soc., Pittsburgh, PA, 1985, p. 13.
4 R. H. Wilson, R. W. Stoll and M. A. Calacone, Highly selective, high rate tungsten deposition, in R. S. Blewer (ed.), *Tungsten and Other Refractory Metals for VLSI Applications*, Materials Research Soc., Pittsburgh, PA, 1985, p. 35.
5 S. L. Ng, S. J. Rosner, S. S. Laderman, T. I. Kamins, D. R. Bradbury and J. Amano, Interaction of CVD tungsten with underlying metal layers, in E. K. Broadbent (ed.), *Tungsten and Other Refractory Metals for VLSI Applications II*, Materials Research Soc., Pittsburgh, PA, 1986, p. 93.
6 S. Kang, R. Chow, R. H. Wilson, B. Gorowitz and A. G. Williams, Application of selective CVD tungsten for low contact resistance via filling to aluminum, *J. Electron. Mater.*, 17 (1988) 213.
7 N. E. Miller, CVD tungsten technology: a brief history, in R. S. Blewer (ed.), *Tungsten and Other Refractory Metals for VLSI Applications*, Materials Research Soc., Pittsburgh, PA, 1985, p. 375.
8 D. C. Thomas and S. S. Wong, A planar multilevel tungsten interconnect technology, *IEDM Tech. Digest*, 1986, p. 811.
9 L. Y. Chen, Z. L. Zhang, J. J. Yao, D. C. Thomas and N. C. MacDonald, Selective chemical vapor deposition of tungsten for microdynamic structures, *Proc. IEEE Micro Electromechanical Systems Workshop, Salt Lake City, UT, U.S.A., Feb. 1989*, p. 82.
10 K. E. Petersen, Silicon as a mechanical material, *IEEE Proc.*, 70 (1982) 420.
11 S. Tandon and G. W. Jones, Reactive ion etching of tungsten in SF_6 and CF_4, in R. S. Blewer and C. M. McConica (eds.), *Tungsten and Other Refractory Metals for VLSI Applications IV*, Materials Research Soc., Pittsburgh, PA, 1988, p. 165.
12 E. K. Broadbent, J. M. Flanner, W. G. M. Van Den Hoek and I-W. H. Connick, High density, high reliability tungsten interconnection by filled interconnect groove (FIG) metallization, in V. A. Wells (ed.), *Tungsten and Other Refractory Metals for VLSI Applications II*, Materials Research Soc., Pittsburgh, PA, 1987, p. 191.
13 W. Trimmer and R. Jebens, An operational harmonic electrostatic motor, *Proc. IEEE Micro Electromechanical Systems Workshop, Salt Lake City, UT, U.S.A., Feb. 1989*, p. 13.
14 S. C. Jacobsen, R. H. Price, J. E. Wood, T. H. Rytting and M. Rafaelof, The wobble motor: an electrostatic, planetary-armature microactuator, *Proc. IEEE Micro Electromechanical Systems Workshop, Salt Lake City, UT, U.S.A., Feb. 1989*, p. 17.
15 T. C. Mele, A. H. Perera and J. P. Krusius, High resolution trilayer electron beam resist system employing P[MMA/MAA] and reliable reactive ion etch processes, *SPIE Proc.*, 923 (1988) 217.

Section 11

Bulk Micromachining

IN most manufacturing, the metric is defined by the tooling used to machine the part. In bulk micromachining, the metric is also contained in the part being made.

The wafers used by the electronics industry are sliced out of a single pure crystal of silicon. Under the correct conditions, chemicals etch the different crystallographic planes within the silicon at different rates. These anisotropic etches usually remove material from the faces of {1 1 1} planes much more slowly than other planes. This differential etching and ingenuity allow the fabrication of an amazingly wide range of parts.

During manufacture of a part, all of the wafer is coated with a protective layer, except for carefully designed open areas. When the wafer is placed in the anisotropic etch, the exposed silicon is etched away until a slow etching, or stop plane, is reached. Changing how the silicon wafer is cut from the silicon crystal changes how these stop planes are orientated in the silicon wafer.

Using different chemicals and conditions produces etches that are isotropic; that is, they have the same etch rate in all directions. These etchants tend to leave rounded cavities on the surface of the wafer. They also undercut the protective mask.

Changes in chemical composition in the wafer or applied voltages can also be used to determine the final shape of the etched part. For example, a boron-doped layer etches slowly, and can be undercut to form diaphragms and beams.

Clever design, corner compensation, and careful control of the rate at which different planes etch into each other can make a variety of shapes. More complicated shapes can be made by bonding different wafers and other materials, such as glass plates, together. Anodic bonding and silicon fusion bonding are the two most popular methods for bonding wafers together. These techniques produce very strong bonds. Both anodic and silicon fusion bonding are high-temperature processes.

The papers in this section show the wide variety of fabrication processes and structures available in bulk micromachining. Devices range from hemispheres encapsulating gas, to sharp probes, to pressure-sensing diaphragms, to valves. Fabrication techniques demonstrate isotropic and anisotropic etching, boron diffusion, and several bonding techniques.

Fabrication of Hemispherical Structures Using Semiconductor Technology for use in Thermonuclear Fusion Research

K. D. Wise, T. N. Jackson, N. A. Masnari, and M. G. Robinson

Electron Physics Laboratory, Department of Electrical and Computer Engineering, The University of Michigan, Ann Arbor, Michigan 48109

D. E. Solomon, G. H. Wuttke, and W. B. Rensel

KMS Fusion, Inc., Ann Arbor, Michigan 48106

(Received 17 October 1978; accepted 2 February 1979)

Initial investigations and laboratory experiments in the area of target fabrication using solid state circuit processing techniques to create special target components have indicated the feasibility of a unique advanced manufacturing concept. A question of the applicability of silicon integrated circuit technology and how it might be applied to the fabrication of inertial confinement fusion targets was posed. The combined efforts of the University of Michigan Electron Physics Laboratory, and the Division of Material Sciences of KMS Fusion, Inc. have provided some relatively quick and very encouraging answers and results. The initial efforts to demonstrate electron beam patterning and etching of micron-high letters through the walls of glass microballoons, the fabrication of free-standing thin-walled flanged hemispheres joined to form spherical structures, and a pellet support membrane have been the proof-of-principle milestones and goals.

PACS numbers: 52.50.Jm, 85.40. — e

INTRODUCTION

Laser-induced thermonuclear fusion has been produced by the implosion of glass-shell pellets 50 to 150 μm in diameter, internally pressurized with up to 100 atm of gaseous deuterium–tritium (DT) mixtures.[1] More advanced, structured targets for laser and electron beam fusion have been discussed in the literature.[2-5] Such theoretical designs suggest the directions pellet research must take in developing methods for pellet production that are economically practical as well as feasible from an engineering standpoint.

Two primary goals for inertial-confinement fusion-pellet technology are (1) developing a broad technology base capable of meeting the expected requirements for fusion reactor pellets over the long term, and (2) providing in the near term an adequate supply and variety of pellets for research. In general, a pellet will consist of (1) a container to hold the thermonuclear fuel and (2) by hyperstructure of shells, layers, and/or coatings concentric with the container, designed to tailor and enhance the response of the pellet to irradiation by the laser or other driver. The technology base must include both the means for producing the container and the microfabrication techniques for creating the hyperstructure. Hollow, spherical shells made of glasses, polymers or metals are currently used as containers.

Advances in pellet technology are increasingly directed toward improvements in fabrication techniques that will be needed to solve the problems of the next generation of experimental targets and reactor-pellet production. These problems include high-volume production and automated complex microassembly. A very promising approach is by way of silicon based technology, drawing on the advanced photolithographic and etching techniques developed for integrated-circuit production. Few other materials have been as extensively studied and characterized as silicon. This paper considers the application of silicon-based techniques for producing laser fusion pellets.

PROCEDURES

Two approaches to fabricating pellets using silicon have been considered. Both approaches lead to the production of spherical pellets by joining pairs of hemispherical shells, and both share a common first step—the formation of cavities in silicon by isotropic etching. This first step is shown schematically in Fig. 1. A silicon wafer is oxidized at 1200°C in wet oxygen to form a layer of silicon dioxide approximately 1 μm thick. Chromium and gold in that order are then evaporated

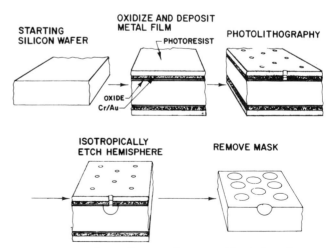

FIG. 1. Sequence of steps illustrating formation of hemispherical cavities in silicon.

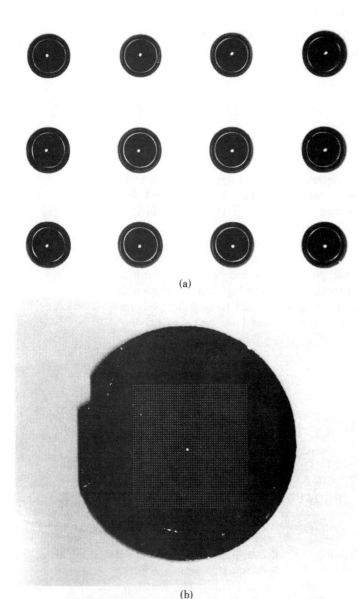

(a)

(b)

FIG. 2. Hemispherical cavities in silicon: (a) two-dimensional array of cavities etched in the central portion of a 2-in. silicon wafer; (b) magnified view.

FIG. 3. Cross-sectional view of a hemispherical cavity.

onto both sides of the oxidized wafer and circular windows are opened extending through both metal layers on the front wafer surface using a standard photolithographic procedure. The remaining metallized areas are then electroplated with gold to a thickness of several micrometers and photolithography is again used to open windows in the exposed silicon dioxide. The double photolithography is employed to minimize pinholing. The silicon wafer is next subjected to an isotropic etch consisting of 90% nitric acid and 10% hydrofluoric acid[6] under continuous agitation for the time required to form the desired hemispherical shape. The etch rate in this etchant is strongly dependent on thorough agitation and is about 10 μm per minute at room temperature.

Two factors are of primary importance in determining the quality of the hemispheres obtained using this procedure. First, the size of the opening in the silicon dioxide chromium/gold etch mask relative to the desired cavity must be chosen carefully. If the opening is too large, the cavity will have a flat bottom, appearing bathtub-like rather than hemispherical. If the etch window is too small, it will not be possible to move etchant in and out of the cavity satisfactorily. A window size equal to about 25% of the final diameter after etching has been found to result in well defined hemispherical cavities.

The second factor determining the quality of the hemispheres is the agitation method. Since this etch is rate limited by the transport of the acid to the silicon surface,[6] and since a relatively large amount of silicon must be removed through the small mask opening, the agitation should be vigorous, and should result in an etchant velocity that is uniform and normal to the wafer surface.

Proper selection of etch-window diameter plus vigorous agitation of the etchant results in the creation of well-defined hemispherical cavities (of nearly constant radius and excellent surface finish). For example, Fig. 2(a) illustrates a two-dimensional array of such cavities, approximately 250 μm in diameter, etched in the central portion of a 2-in. wafer. The uniformity of the etched cavities is apparent in Fig. 2(b), which is a magnified view of one small section of the array. Careful examination of this wafer revealed that the cavities were smoothly etched and nearly hemispherical. This excellent quality is substantiated by Fig. 3, a cross-sectional view obtained by cleaving a wafer along a line through one row of etched cavities.

In our first pellet-fabrication approach, the silicon cavities are used as molds to form pellet halves of various materials. For example, Fig. 4 shows how hemispherical shells of poly-(methyl methacrylate) (PMMA) or metal can be realized. The metal-and-oxide etch mask on the wafer is first removed and a thin layer of gold is evaporated onto the silicon cavities and then electroplated to a thickness of one to two micrometers. Next, a film of PMMA is spun on the wafer, coating the

FIG 4. Fabrication sequence for PMMA or metal hemisphere formation.

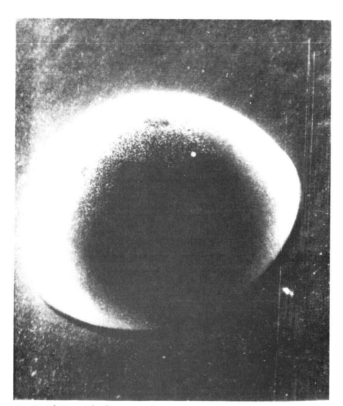

FIG. 5. Photograph of a PMMA hemisphere.

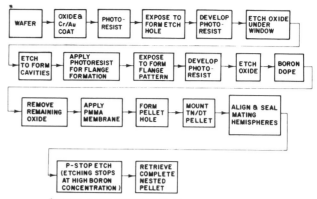

FIG. 7. Flow chart illustrating silicon pellet formation using the p-stop etch technique.

cavities. The PMMA dilution (in trichloroethylene) and spinning speed determine the PMMA film thickness. (The PMMA can be selectively removed between cavities at this stage, using a photolithographic operation to leave a flanged PMMA lining in the cavity.) Since gold has a low adhesion to silicon, the gold film can be easily lifted, taking with it the PMMA structures, which conform to the cavity shape. The gold can be etched away leaving flanged hemispheres of PMMA with wall thicknesses of one micrometer or less. Figure 5 shows a PMMA hemisphere produced by this procedure.

The molding technique just described has two primary advantages. First, since the silicon mold is used nondestructively and can be used repeatedly, great care and considerable expense can be justified to produce a high-quality structure. Second, this method can be used to produce pellets from a wide variety of materials, including metals, glasses, and plastics, which can be evaporated, sputtered, plated, spun, or otherwise deposited in thin uniform layers. Also, it should be relatively simple to produce multilayer pellets using this approach.

Our second approach to producing pellet structures from etched silicon cavities involves the use of etchants whose etch rate is sensitive to doping in the silicon. In particular, an etchant composed of ethylene diamine, pyrocatechol, and water[7] is sensitive to the amount of boron in the silicon crystal. The etch rate drops nearly to zero for boron levels exceeding about 5×10^{19} cm^{-3} (about 0.1% boron in the silicon lattice). This method is shown schematically in Fig. 6 and further

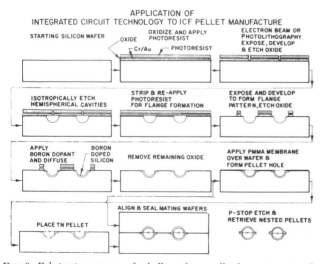

FIG. 6. Fabrication sequence for hollow silicon pellet formation using the p-stop etch technique.

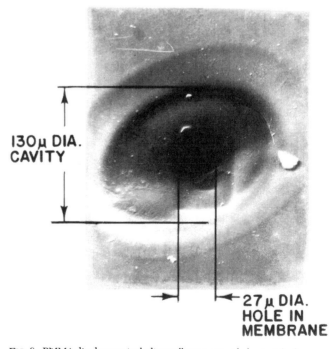

FIG. 8. PMMA diaphragm, including pellet mounting hole, stretched across a silicon hemispherical cavity. (Visible rectangular pattern represents electron beam damage of PMMA film during focusing step.)

defined by the flow chart in Fig. 7. The Cr/Au etch mask for the cavities is first removed and the silicon dioxide layer is photolithographically patterned so that silicon is exposed both in the cavity and in a narrow ring, or flange, around the cavity rim. The remaining oxide acts as a mask when a high concentration of boron is introduced into the exposed silicon via high-temperature diffusion. After this doping step the oxide is stripped and the wafer is etched in the ethylene diamine-pyrocatechol-water mixture (17 ml ethylene diamine, 3 cm pyrocatechol, 8 ml water). Since this etchant is a p-stop etchant (i.e., it stops on the diffused boron layer), a boron-doped single-crystal p-type silicon shell having the shape of the initial cavity and flange remains following the etch. The wall thickness can be precisely controlled by controlling the time and temperature of the diffusion.

Spherical pellets can be achieved by aligning the hemispherical-cavity arrays of two identically processed wafers and sealing the junction prior to the p-stop etching step. The aligned wafers may be sealed electrostatically[8] or by other appropriate sealing techniques. The joined wafers are etched in the ethylene diamine-pyrocatechol-water mixture until the n-type (undoped) silicon has been dissolved leaving only individual hollow, p-type (boron-doped) silicon pellets with flanges.

These techniques can also be adapted to allow incorporation of other structures within hollow spherical shells. For example, we have investigated the production of nested pellets in which an inner sphere is "levitated" within a larger sphere and concentric with it by supporting the smaller sphere or pellet on a diaphragm stretched across a hemispherical cavity. Such a diaphragm can be produced with PMMA by using the procedure illustrated in Fig. 8. A thin layer of gold is evaporated onto a polished wafer after which a thin layer of PMMA is spun onto the gold layer. The wafer is then thermally joined to a second wafer in which hemispherical cavities have been etched, sandwiching the PMMA/gold layer. Since evaporated gold has very poor adhesion to silicon, the polished silicon wafer can be easily detached, after which the gold layer can be etched away leaving the PMMA film stretched across the etched cavities. If desired, circular holes or other patterns can

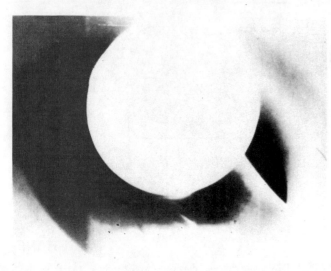

FIG. 9. Small pellet supported by PMMA diaphragm stretched across a hemispherical cavity.

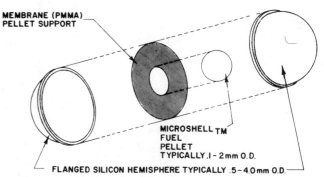

FIG. 10. Nested pellet (exploded view).

be produced in the PMMA diaphragm by exposure to a low-intensity electron beam. The pellet to be supported by the diaphragm can then be positioned at the desired location. For example, Fig. 9 illustrates a diaphragm stretched across a hemispherical cavity and supporting a small pellet. The internal fuel pellet can be "bonded" to the diaphragm or held between two diaphragms stretched across mating hemispherical cavities. The pellet halves are then joined and individual pellets separated as previously described. The final product is shown in Fig. 10. As experimental requirements evolve to include larger pellets (~1-mm o.d. and larger), the flanges described and depicted can be made negligibly small; in fact, the wall itself presents an adequately sized interface sealing surface.

CONCLUSION

The proposed fabrication techniques are capable of producing inertial confinement fusion pellets from a variety of materials in a variety of configurations. The potential for mass production can be appreciated by noting that a single 4-in. silicon wafer can yield as many as 100,000 250-μm pellets. If processing costs are assumed to be comparable with those for integrated-circuit processes, the economic advantages are apparent.

ACKNOWLEDGMENTS

This work was supported by the United States Department of Energy under Contract ES-77-C-02-4149. Work at the Electron Physics Laboratory was performed under KMS Fusion, Inc. Subcontracts PO 19872 and PO 23056.

[1] G. Charatis, J. Downward, R. Goforth, B. Guscott, T. Henderson, S. Hildum, R. Johnson, K. Moncur, T. Leonard, F. Mayer, S. Segall, L. Siebert, D. Solomon, and C. Thomas, "Plasma Physics and Controlled Nuclear Fusion Research," Proceedings of the International Conference, Tokyo, 1974 (IAEA, Vienna, 1975).
[2] G. Yonas, J.W. Poukey, K. R. Prestwich, J. R. Freeman, A. J. Toepfer, and M. J. Clauser, Nucl. Fusion **14**, 731 1974.
[3] J. Nuckolls, Laser Fusion Targets, Lawrence Livermore Laboratory, Report UCIR-607, 1972.
[4] G. H. McCall and R. L. Morse, Laser Focus **10**, 37 December (1974).
[5] R. C. Kirkpatrick, C. C. Cremer, L. C. Madsen, H. H. Rogers, and R. S. Cooper, "Structured Fusion Target Design," Nucl. Fusion **15**, 333 April (1975).
[6] D. J. Klein and D. J. D'Stefan, J. Electrochem. Soc. **109**, 37 (1962).
[7] A. Bohg, J. Electrochem. Soc. **118**, 401 (1971).
[8] A.D. Brooks, R. P. Donovan, and C. A. Hardesty, J. Electrochem. Soc. **119**, 545 (1972).

Micromachining of Silicon Mechanical Structures

G. KAMINSKY [a]

AT&T Bell Laboratories, Murray Hill, New Jersey 07974

(Received 25 January 1985; accepted 9 April 1985)

Refined and very exact processing procedures for producing geometrically highly precise silicon structures are described. The structures are micromachined utilizing wet chemical orientation and/or concentration etching techniques. The use of a specially designed and highly effective etching system is a significant refinement in the processing procedure. The structures have excellent large and small scale uniformity. Relief structures having features as small as 0.2 μm have been produced. We report on devices for three specific research applications: (1) Very high Q torsional oscillators useful for high sensitivity measurements in the study of the mechanical properties of single crystal silicon and also in the study of thin freely suspended liquid crystal films, (2) unstrained focusing x-ray mirrors, and (3) μm level mechanical "shadow masks" useful for noncontaminative, *in situ* patterning of thin films deposited in MBE and/or other deposition systems.

I. INTRODUCTION

The availability of high quality silicon crystals, precise lithographic means of patterning the surfaces of this material, and well studied means of deeply anisotropically etching silicon, allows the "micromachining" of three dimensional structures unique for their small size, precision, and/or mechanical properties. Research on the micromachining processes or on the devices exploiting such processes include etching of deep narrowing grooves in silicon,[1–5] silicon chip separation or dicing,[1,6] etch thinning of silicon,[7] preparation of very thin silicon membranes,[8–11,18] isotropic etching of silicon,[6,7,12,13] orientation dependent etching,[10,11,17,19] wet chemical etching agitation systems,[6,7,9,12,14] shadow masks,[1–5,16] x-ray masks,[8,9,14,16] substrates for electron beam and x-ray lithography,[8,18] connector and terminal blocks,[2] high precision nozzles,[8,15] and gratings.[2,3,16,20]

In this paper we describe the application and refinements of silicon micromachining to fabricate, over wafer size area, devices for three research applications: very high Q torsional oscillators useful for high sensitivity measurements in the study of the mechanical properties of (1) single crystal silicon in the temperature range 5 mk–10 K and (2) thin freely suspended liquid crystal films as they go from an isotropic liquid into a hexatic phase; unstrained, focusing x-ray mirrors; and μm level mechanical "shadow" masks useful for noncontaminative, *in situ* patterning of thin films deposited in the high vacuum environment of molecular beam epitaxy (MBE) and/or of other deposition systems. These and other structures have been made on large areas either because of the nature of the device or for efficiency in making many devices. We describe our techniques and precautions for obtaining satisfactorily uniform results in these cases of micromachining.

II. ETCHING CONDITIONS, FABRICATION TECHNIQUES, AND FIXTURING

Numerous investigators[1,6–9,12,13] have reported on the common use of mixtures of hydrofluoric, nitric, and acetic acid for rapid isotropic etching of silicon. Frequently, silicon oxide or silicon nitride layers on the surfaces of the silicon wafers are patterned by standard photolithographic techniques and are used as an etch masking material while etching silicon wafers. For micromachining, strongly anisotropic etches used in the same way provide unique opportunities for a limited range of special geometries in deep etches. Bassous[2] and Bean[3] have reviewed the most popular etchants and attainable geometries for the anisotropic etching of silicon.

Typically the lower silicon index planes (100), (110), and (111) are of interest, and etch rate ratios of several hundred to one are possible for the weakly bound (100), (110) planes relative to (111). The available geometries here and in most work are obtained by etching either (100) or (110) oriented crystals through masking windows with edges parallel to the (111) planes.

A (100) silicon wafer has four convergent self-limiting (111) etch planes that are 54.74° to the surface plane of the wafer. These (111) low etch rate planes are 90° to each other and define the angular sidewalls of the anisotropically etched cavity. The pattern of the sidewalls at the (100) silicon surface is a square or rectangle whose sides are parallel to the [110] directions of the crystal. Therefore any etch pattern on the (100) silicon wafer surface should be closely aligned parallel to the [110] reference flat. The resulting anisotropically etched structure of a (100) silicon wafer[2,15] is a well-defined pyramidal cavity. The cavity could be a self-terminating pit or a truncated pyramidal hole depending on the side dimensions of the etch pattern at the wafer surface and the wafer thickness.

A (110) silicon wafer has two parallel sets of self-limiting (111) etch planes that are 90° to the surface plane of the wafer. The two sets are not 90° to each other but intersect at angles of 109.48° and 70.52°. Therefore any etch pattern on the (110) silicon surface[2,3] must be precisely aligned parallel to these angles in order to insure the formation of a vertical-walled structure during anisotropic etching and/or to obtain deeply etched grooves or cavities with extremely flat and steep sidewalls defined by the (111) planes.

Large etch rate ratios may be obtained in both cases using alkaline-based etches or etches based on ethylenediamine or hydrazine. We have used all three varieties of etchants for various applications, including commercially available versions: PSE-100,[23] PSE-200,[24] and PSE-300.[25] Appendix A contains some descriptions and information regarding various etchants and conditions. However, two etchants were employed primarily for most of our work described here. A caustic mixture[3,4] of 350 g potassium hydroxide in 650 cm^3 of water maintained at 85 °C used usually with SiO_2 masking for both Si (100) and (110) etching were used in two of our applications. In the third application, a mixture[2,15] of 45 g pyrocathechol, 255 cm^3 of ethylene diamine, in 120 cm^3 of water (P-ED solution) was used. In addition to its orientation dependent etching characteristic, the P-ED etchant is concentration dependent such that a p^{++} silicon layer doped to 7×10^{19}/cm^3 or higher will completely inhibit[17,22] (etch-stop) all etching. These etching characteristics have been used in fabricating thin membrane windows in supporting silicon frames in earlier work and are used for producing overhanging membranes as knife edges in our shadow mask fabrication.

In our early experience, aside from problems in obtaining precision oriented silicon material and pattern alignment, the primary difficulty in making wide area patterns or collections of single devices over wide areas was in obtaining uniform etching so that prolonged deep etching can be terminated in a reasonable predictable time while also yielding etch front surfaces that are flat and smooth. We found it to be of great importance to follow special handling and cleaning procedures after patterning so as to obtain a uniform etch start of the silicon structure and it was found imperative to establish sufficient agitation between the etch solution and/or between the solution and silicon sample during etching. Unlike most handling procedures that are programmed for processing on only one side of a wafer, we implement special care and procedures to insure both sides of the wafers are kept scratch and defect free during all processing steps. A procedure using *in situ* annealing[29] designed to reduce the pinholes and processed-induced defects in the masking oxide was also implemented. The silicon starting material ranged in size 25–76 mm diam and 0.1–0.7 mm thickness. We now generally use wafers chem-mechanically polished on both sides, whose surface planes are oriented within ± 0.1° and whose reference flats are also accurate within ± 0.1°. This specification has been fulfilled by Western Electric on special request, therefore, simplifying the critical alignment procedure by eliminating several extra and undesirable processing steps required in fabricating an etched orientation feature.

The "Syton" polished wafers were initially cleaned by a conventional chemical preoxidation procedure, thermally oxidized in steam at 1150 °C for 2 h, stripped of all oxide in heated buffered oxide etchant (BOE),[30] recleaned in the preoxidation solutions, and annealed for 30 min in argon (or 99% N_2 : 1% O_2) in the same furnace and at the same temperature a subsequent thermal oxidation is carried out. This thermal oxidation was done in steam at 1150 °C for periods of time as required to grow a SiO_2 layer of thickness needed for oxide masking the particular pattern and structure being made. An oxide layer 0.5–1.4 μm was suitable for most cases. After a resist coating was applied to both sides of the wafers,

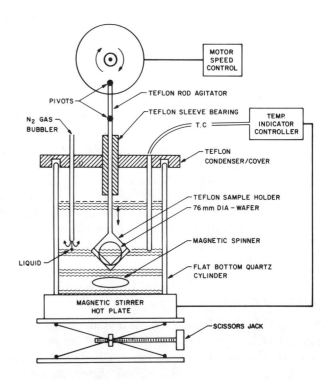

FIG. 1. Schematic diagram of system designed and constructed for anisotropic etching of silicon structures.

patterns were aligned and defined parallel to the reference flats to within 0.1° by optical lithography. Patterned openings in the SiO_2 layer were then etched to bare silicon by etching in BOE. The wafers were stripped of resist first by the stripper recommended by the manufacturer and then by A30 commercial stripping cleaning solution and/or an oxygen plasma clean to insure no resist material was left on the wafers. Again a standard preoxidation chemical cleaning procedure was employed in combination with a BOE dip. This prepares the wafers for the final step of anisotropically etching the desired structure.

First attempts to fabricate our structures resulted in only partial success and poor reproducibility mainly due to inadequate agitation that existed within the etchant solution and/or between the solution and sample during etching. Finne and Klein[14] have reported the use of a glass refluxing system to prevent composition change by loss of volatile matter and bubbling N_2 through the etch solution to prevent air oxidation and provide etch solution agitation. Derick[12] describes a temperature controlled etching procedure with mechanical motion resulting in a swirling type agitation of sample and etchant. Craft[6] describes an etching machine that uses vertical-pulse agitation of a basket containing eight 1 1/2 in. diam wafers in an isotropic etch for chip separation (dicing) with edge control to ~ 5 μm and scalloping of ± 10 μm. Cheung[9] uses a system designed to give a rotating and bubbling motion achieved by feeding pressurized O_2 and N_2 above the sample surface while the sample is rotated in a breaker at an angle. Stoller[7] describes a method that yields uniform chemical thinning of silicon over the entire wafer area. This approach involves the use of gas bubbles inpinging on the sample surface during etching to insure a continuous supply of fresh etchant and the continuous removal of reaction products. Our experience has shown that none of the above cited methods of etching were suitable for our needs. Unfavorable results were also achieved by using ultrasonic agitation, in which the vigorous and erratic formation of gas bubbles at the etchant–sample interface proved to be very ineffective.

An etching system proven to have a very effective agitation and highly efficient wet chemical etching capability was designed and constructed and is shown in Fig. 1. The system features a variable-speed vertical motion of the sample in the etch solution which is stirred and heated by a commercial magnetic-stirrer hot plate. The gas bubbler, Teflon condenser/cover, motor speed controller, and temperature indicator-controller provided a satisfactory means for maintaining and controlling the chemical composition and temperature for etches lasting several hours.

Several designs for supporting the samples during etching were employed so as to minimize the risk of breakage from such vigorous agitation during etching. Provisions for observation during etching were also included. Initially the primary support in all cases is obtained by arranging the device pattern on a standard size silicon wafer such that a substantial size of supporting border of full wafer thickness is always present with, if possible, some intermediate grid or structure as additional support. One of the simplest and most useful sample fixtures is a standard open-back 2 and 3 in. commercial wafer holder[26] designed for manual etching and modified for attachment and use in our "chugging" system. Since almost all etching is performed at elevated temperatures (as high as 118 °C), mounting waxes and cements are avoided, and instead, physical clamping of wafers and/or test chips are employed. One such design, [shown in Fig. 2(a)], provides a very secure and convenient arrangement for supporting various size and irregular shaped test chips while minimizing the risk of breakage. A flat Teflon disk 2–3 in. diam × 1/16 in. thick, with many randomly tapped holes, is used, with nylon screws, to hold the wafer and/or chips in place. The number, arrangement, and fit of the screws can thus be easily adjusted to form a tailor-made clamp for a variety of shaped samples. In situations where the back surface of a wafer or chip must be protected during etching, one may use deposited silicon nitride, deposited silicon oxide, or inert metal layers. All of these films, if pinhole free, are compatible with most MOS processing techniques and can subsequently be easily removed by selective etching. Otherwise a Teflon sample holder with a solid back and chemically inert o-ring seal [such as shown in Fig. 2(b)] may be employed. This design relies on sealing and isolating the back of the wafer from the etchant, at the outer supporting border, by clamping the silicon wafer with a Teflon retaining ring and nylon screws. An alternative method to clamping is to rely on vacuum between a double o-ring seal.

We found no significant difference in the quality of structures fabricated from zone-refined silicon as compared to Czochralski-grown silicon, or from the use of either n- or p-type silicon in the resistivity range 0.024 Ω cm–30 KΩ cm. Our experience has shown that the aforementioned handling, cleaning, and processing procedures must almost always be followed to insure achieving good high quality structures reproducibly. Good dimensional control of the

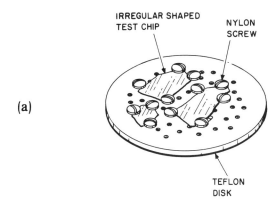

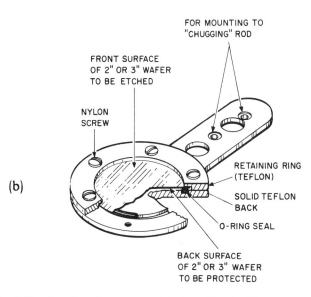

FIG. 2. Sketches of sample supports used during etching. (a) Silicon test chip etching holder, and (b) silicon wafer etching holder.

micromachined silicon structures during orientation-dependent etching is possible only when accurate crystallographic alignment of the surface pattern has been achieved. The straight edges of the pattern openings must be aligned parallel to the intersection of the (111) planes to within 0.1°.

III. ETCH RATE AND UNIFORMITY MEASUREMENTS

In our early work, several preliminary etching experiments were performed on relatively wide area etch patterns as an aid in processing and structure design. Both the KOH and the P-ED etchants were tested in addition to the isotropic etch of HNO_3 : HF (21 cm^3 : 4 cm^3) included for comparison. The degree of etch-rate linearity at the crucial etch-start period and the flatness of the etch-front surfaces were the qualities of interest and importance here because a failure in these aspect considerations ruled out the possible success in the more demanding deep etching experiments. The tests also served to verify rather quickly the results reported elsewhere in the literature and the adequacy of crystal preparation procedures. The (100) silicon samples were prepared as previously described with 1-μm-thick thermally grown SiO_2 masking and mechanical agitation except for the isotropic

TABLE I. CHARACTERISTICS OF SHALLOW ETCHED-LARGE AREA SILICON STRUCTURES

WAFER – CHIP NO.	ETCH MASK	ETCH	TEMP °C	ETCH TIME	ETCH DEPTH μm	ETCH RATE	CHARACTERISTICS AND REMARKS
I – 1	AZ-1350J	HNO$_3$-HF	25	4 sec	1.71		
I – 2	"	(21 cc HNO$_3$ ·	"	8 "	1.78		1) ETCHED RECESS CORNER IS DEEPLY UNDERCUT.
I – 3	"	4 cc HF)	"	16 "	3.68		2) LOCAL SURFACE ROUGHNESS IS ~ ± 1,000 Å.
I – 4	"		"	32 "	6.17	2300 Å/sec	3) LARGE AREA FLATNESS (SCALLOPING) IS ~ ± 2,000 Å.
II – 5	SiO$_2$		"	4 sec	1.64		
II – 6	"		"	8 "	2.19		
II – 7	"		"	16 "	3.71		
II – 8	"		"	32 "	8.39		
III – 9	"	KOH-H$_2$O	85	1 min	1.77		1) ETCHED RECESS CORNER IS SQUARE AND SHARP WITH NO UNDERCUTTING.
III – 10	"	(35%-65%)	"	2 "	3.71	2.2 MICRONS /min	2) LOCAL SURFACE ROUGHNESS IS ~ ± 200 Å.
III – 11	"		"	4 "	10.68		3) LARGE AREA FLATNESS (SCALLOPING) IS ~ ± 300 Å.
III – 12	"		"	8 "	18.62		
IV – 13	"	P.-E.D.-H$_2$O	110	1 min	1.12		1) ETCHED RECESS CORNER IS SQUARE AND SHARP WITH NO UNDERCUTTING.
IV – 14	"	(45g: 255cc: 120 cc)	"	2 "	2.11	1.1 MICRONS /min	2) LOCAL SURFACE ROUGHNESS IS ~ ± 200 Å.
IV – 15	"		"	4 "	3.72		3) LARGE AREA FLATNESS (SCALLOPING) IS ~ ± 1,500 Å.
IV – 16	"		"	8 "	9.90		

etch tests where both photoresist and oxide masking were employed and only vertical hand agitation was used for the very short periods involved.

A summary of the etching conditions and characteristics of the shallow-etched-large area structure is listed in Table I, while plots of the etching time vs etch depths are shown in Fig. 3. Note in Fig. 3(a) the isotropic (HNO$_3$–HF) etch has an etch rate of 2300 Å/s and is fairly linear up to 32 s of etching time and an etch depth of 8 μm. As shown in Fig. 3(b), the alkali etch has a linear etch rate of 2.1 μm/min and an etch depth of 19 μm after 8 min of etching. The P-ED etch rate of 1.1 μm/min is linear up to 8 mins etching time and a depth of 10 μm as shown also in Fig. 3(b). The photoresist used as an etch mask for shallow etching at room temperature compares favorably with the oxide mask. Sloan DEKTAK-II[27] profile patterns of these shallow etched surfaces were recorded to determine flatness and smoothness. Severe undercutting at the recessed corner for structures etched in the isotropic etch were observed while sharp square recessed corners and flat bottoms were achieved in structures etched with the anisotropic etch. An 18.6-μm-deep recessed structure, obtained by etching in the KOH etch for 8 min, indicated large area roughness (scalloping) at ± 300 Å with local surface irregularities of only 200 Å. This level of surface finish is comparable to that of the original (unetched) surface but also appears to be the measurement limit due to building vibration.

The results suggest in these ideal circumstances that etch-front roughness to etch-depth ratios of order of 0.002 or better can be achieved with the anisotropic etch and that uniform etching characteristics on large scales can be obtained from the start on properly prepared surfaces. These results have been confirmed in most of our practical useful structures while variations up to 10% were obtained in our most demanding applications involving deep, narrow grooves over a 76 mm diam wafer.

IV. HIGH Q TORSIONAL OSCILLATORS

Highly pure and low defect density silicon is potentially a high performance material for low loss mechanical oscillators. The oscillators made here are a two dimensional, paddle-shaped version of metal alloy torsional oscillators used as a substrate in low temperature measurements of viscous losses and in high temperature experiments in thin liquid crystal films deposited on oscillators. Sensitivity is limited by losses in the oscillator itself, arising in polycrystalline material or relatively poor crystal from the stress induced motion of grain boundary or crystal defects. Theoretically, there are no losses in a perfect crystal and the paddles made here come sufficiently close to this ideal that, relative to Be Cu oscilla-

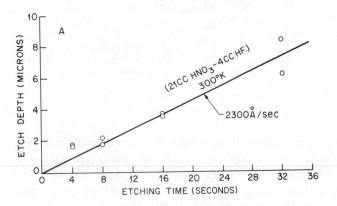
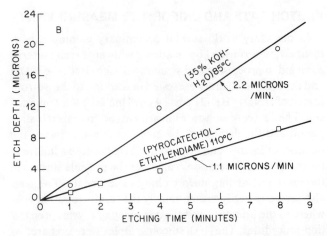

FIG. 3. Plots of etch depth as a function of etching time for shallow-etched large-area silicon structures. (a) Isotropic etch rate, and (b) anisotropic etch rates.

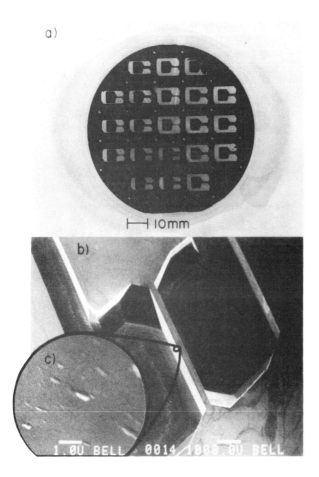

FIG. 4. Several aspects of micromachined high Q torsional oscillators. (a) Photomacrograph as seen through transmitted light of 21 etched paddles before being separated from their wafer support. (b) SEM photomicrograph of etched structure shown at a high tilt angle. (c) Highly magnified SEM photomicrograph portion of side wall illustrating the high degree of smoothness.

tors, an order of magnitude improvement in oscillator Q is obtained at room temperature and a three order of magnitude improvement is observed at 10 mk: $Q_{273} \approx 10^5$; $Q_{002} \approx 10^8$.

Figure 4 shows the etched paddles in different aspects before being separated from their wafer support. They were micromachined in 0.5-mm-thick (110) silicon oriented wafers having a resistivity range 0.024 Ω cm–30 KΩ cm. Preparation and etching of the paddles was conducted as previously described using 1.4 μm thickness of SiO_2 masking layers patterned such that the sides of the paddles were parallel to the [111] reference flat direction to better than 0.1°. The reverse side of the wafer, also polished, was protected by the same oxide plus a coating of photoresist. Excellent results were achieved by employing the chugging motion of our newly designed etching machine (Fig. 1) using the caustic mixture etchant kept in motion by the magnetic stirrer while the sample, held in a standard open-back Teflon holder, was agitated vertically at a rate of 400 strokes/min. The etching time of 3 h produced a thickness loss of 0.7 μm in the patterning SiO_2 and negligible loss (1 μm) in the (111) silicon sidewall faces for this application, thus representing a 110 : 111 etching ratio of 500 : 1. Comparison of the dimensions of the width of the structure's supporting arm at one surface relative to the opposite surface reveal a difference of 2.8 μm measured to an accuracy of $\pm$ 1.2 μm. This taper represents a slope of the sidewall which is 1.4 μm through the 500 μm thickness of the arm or somewhere between 1 to 18 mins slope. As shown in Fig. 4, the sidewalls of the structure display excellent edge acuity and surface smoothness of a size level 1 μm.

The wafer format allows production of a selection of otherwise identical oscillators with different operating frequencies as can be seen from the range of dimensions. Most importantly the integrity of the crystals is maintained through the damage-free chemical machining of the paddles as is confirmed by their high performance. The planar nature of the oscillators facilitate deposition of capacitive excitation electrodes which may be lithographically or otherwise implanted on the paddle surfaces for implementation. More than 300 torsional oscillators involving 18 separately processed and micromachined silicon wafers and of three different geometric designs have been fabricated.

V. STRAIN-FREE FOCUSING X-RAY REFLECTORS

Silicon crystals are often used as x-ray mirrors while focusing versions can be produced by bending a sufficiently thin crystal or wafer. However, in bending the crystal, the lattice is distorted. To avoid this strain at the reflecting surface, one may provide a mirror of adjacent unstrained crystal in the appropriate geometry. As suggested by Moncton, this can be achieved using the "silicon anisotropic etching technology" to form a deep linear grating in the silicon reflecting surface thus providing a structure suitable for one dimensional focusing. Bending the crystal produces strain only on the membrane side of the crystal which also serves as a uniform, strong backing for the composite lens. Therefore the final micromachined silicon grating structure should have a uniform etch depth to provide distortionless bending and to maintain the integrity of the supporting back and should also have as small grid spacing as is consistent with being able to develop the desired lens curvature resembling a continuous mirror and/or while not appreciably reducing the reflecting area.

Shown in two different perspectives in Fig. 5 is a portion of an overly deeply etched silicon lens grating after being cleaved in cross section. The illustration therefore reveals an exaggerated or "worst-case" geometry of a structure. The grating was micromachined in a (110) silicon wafer whose surface plane and reference flat are oriented to within 0.1° and has a diameter of 76 mm and thickness of 0.5 mm. Preparation and etching of the grating was conducted as previously described for the silicon (110) anisotropic etching technology using 1.4 μm thickness of a SiO_2 masking layer patterned such that the sides of the slit pattern were aligned parallel to the [111] reference flat direction to better than 0.1°. The patterned windows in the SiO_2 masking layer consisted of a linear array of 220 slits, each 34 μm wide spaced on 220 μm centers and encompassing a 64 mm diam circle. The reverse polished side of the wafer was also protected by oxide and/or a coating of photoresist as needed. The "Linear Grating" lens structure shown was micromachined by anisotropic etching as described previously, and in this case,

(a)

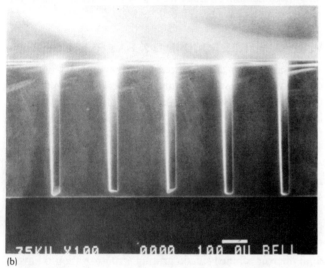

(b)

FIG. 5. Two perspectives of a precisely oriented and aligned silicon lens structure after deep etching and being cleaved in cross section.

using a vertical chugging of the sample at 350 strokes/min in the KOH : H$_2$O solution maintained at 80–85 °C for a total etching time of 3 1/2 h. During this micromachining period, a thickness loss of 1 μm in the patterning oxide and a thickness loss of 2 1/2 μm in the (111) silicon sidewall faces was experienced. The final grating structure, as shown, had a 41 μm wide opening at the top surface compared with a 36 μm width at the bottom of the groove and represents a slope of only 18 min, thus the achievement of extremely steep channel sidewalls. This grating structure has a channel depth to width ratio of 14 to 1 while the ratio of the channel depth bottom variation (roughness) to channel depth is no more than 6% over the area of the portion of the crystal shown. This structure is an illustration of the excellent uniformity and symmetry obtainable even in the extreme etching case cited here where the depth of the grooves are overly etched by ∼17% compared to the ideal depth required for the lens design.

In a similarly processed and overly etched structure and again employing a silicon wafer having a 76 mm diam and 0.5 thickness, the grating was etched to a remaining back

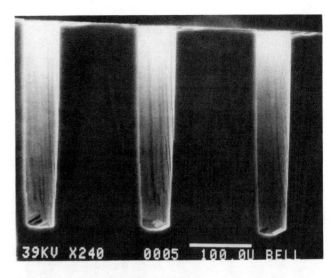

FIG. 6. SEM photomicrograph displays in cross section the lens structure obtained when the crystal orientation and/or pattern alignment is poor. Excessive sloping and scalloping is evident.

thickness of ∼20 μm and required 4 h of etching time. The final etched channel opening in this case of only 29 μm, represents again an increase of 7 μm on the oxide slit opening as noted in the aforementioned lens structure. This wafer, patterned and structured in a much smaller slit width scale, also displayed excellent overall uniformity and symmetry but appeared to fall short of that found in our earlier work of wide area experiments, presumably due to difficulty in achieving adequate etching in these narrow grooves as the channel deepens. Indications of this limitation were observed by transmitted light and also by transmission scans using a 15-μm krypton laser probe ($\lambda = 0.67$ μm) to measure the remaining silicon back thickness. Variations in this silicon membrane thickness were up to 40 μm or ∼10% of the average channel depth measure across the entire 64 mm grating area. Although these variations are tolerable for our lens application, an implied limit on the aspect ratio of grooves in similar applications is suggested.

Further evidence of the requirement for maintaining very precise crystal orientation and pattern alignment is illustrated in Fig. 6. This SEM photomicrograph displays in cross section the lens structure obtained when the crystal orientation and/or pattern alignment control is poor (>1°). Although the symmetry shown is quite good, excessive sloping (>1°) and scalloping of the groove walls occurred and in turn led to a decrease of the etching efficiency and finally a practical limit of channel depth obtainable (that is, the ability to etch a structure with deep grooves without excessive widening effects).

Gratings having submicron periods as small as 0.2 μm and channel depth to channel width ratios of 40 : 1 have been achieved and is shown in the SEM photomicrograph of Fig. 7. These grating structures also utilized the silicon (110) anisotropic etching technology employing highly precise crystal orientation and pattern alignment techniques. The method of providing etch masking in this case was to define, by eb lithography and selective metal etching, the grating pattern of 100 Å thickness Cr with a 500 Å thickness Au top layer. The final silicon grating shown was micromachined by an-

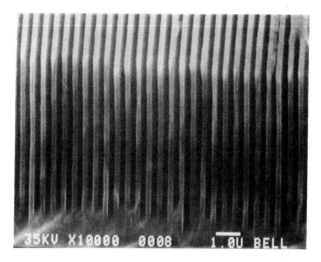

FIG. 7. SEM photomicrograph of submicron grating anisotropically etched in (110) silicon; 0.2 μm period; 8 μm deep; channel depth to width ratio is 40:1.

isotropically etching in a solution of KOH in H$_2$O (350 g : 650 cm^3) at either 24 °C for 15 min or 60 °C for 4 min and while the sample was vigorously agitated vertically at 450 strokes/min.

VI. NONCONTAMINATING MICRON SIZE SHADOW MASKS

Shadow masking is generally an obsolete method of patterning vacuum deposited thin films. However for simple geometries and modest registration requirements, such methods can offer simplification in processing relative to

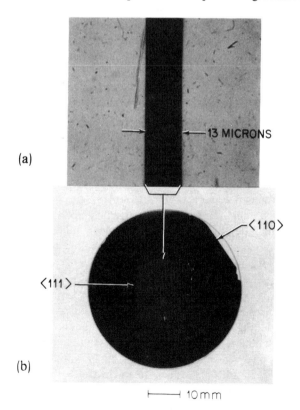

FIG. 8. Macrograph of vertical-walled (110) silicon shadow mask structure. (a) 39 slits, each 13 μm wide supported in a silicon wafer 51 mm diam and 150 μm thickness. (b) Photomicrograph of slit feature having excellent edge uniformity; edge acuity is better than 1 μm.

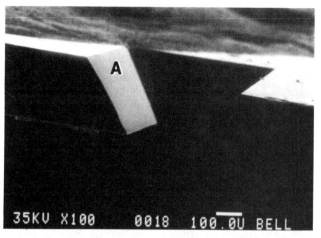

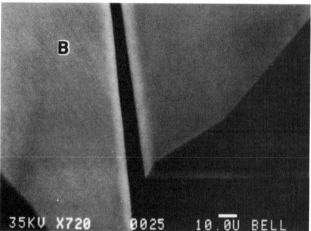

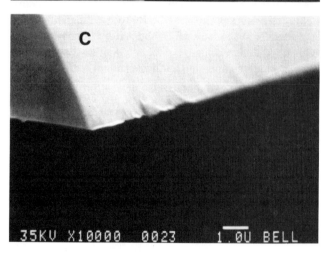

FIG. 9. SEM photomicrograph of angle-walled (100) silicon shadow mask structure. (a) Slit opening 240 μm wide anisotropically etched through a 600-μm-wide window in masking oxide of 1 μm thickness and supporting silicon frame of 267 μm thickness. (b) Angle-walled silicon shadow mask with knife-edge slit opening that varies in width 5–8 μm. (c) Knife edge of angle-walled shadow mask opening which typically has a slit feature edge acuity of much better than 1 μm.

modern lithographic techniques using wet processing intervening between subsequently deposited films. More importantly, shadow masking affords the possibility of *in situ* deposition of patterned layers without disturbance or contamination of the high quality of UHV formed material in MBE and other systems. Of course the mask structure itself, which may be heated and also be in close proximity to the substrate, must not be a source of contamination. Additionally, the mask must be structurally strong enough for *in situ* manipulation. Such a structure can be micromachined of ready available high quality silicon by employing, with several refinements, the anisotropic etching technology of orientation dependent and/or concentration dependent etching. In our work, three different shadow mask structure designs were investigated.

In the first design utilizing the silicon (110) anisotropic etching technology, we were able to fabricate precision vertical-walled mask structures with slit openings as narrow as 13 μm wide supported by a 150 μm thickness of silicon and is shown in Fig. 8. Although this design produces a slit structure that has excellent definition, uniformity and edge acuity (better than 1 μm), self-shadowing properties severely limit the use of this mask structure at the level of slit widths illustrated here.

The second design utilized the silicon (100) anisotropic etching technology to produce slit openings as narrow as 5 μm in a structure having slant-angle sidewalls. The walls intersect the masking surface at 54 3/4° and is shown in several perspectives in Fig. 9. This structure is useful as a non-contaminating shadow mask whose knife-edge openings eliminate self shadowing effects for most applications involving normal deposition angles. However, uniform slit openings much less than 25 μm wide are difficult to control without imposing severe restrictions on the specification of thickness and surface uniformity across the starting silicon wafer. Additionally, the center to center feature spacing (packing density) is limited by the *V*-groove design and such parameters as slit feature width and wafer thickness.

Our third design utilizes a P-ED etching mixture serving simultaneously as an orientation-dependent and a concen-

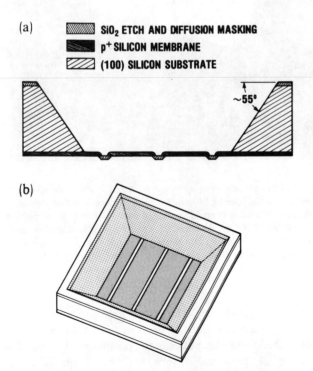

FIG. 10. (a) and (b) are sketches of the composite Si-doped Si membrane mask structure before removal of the pattern defining SiO$_2$ stripes.

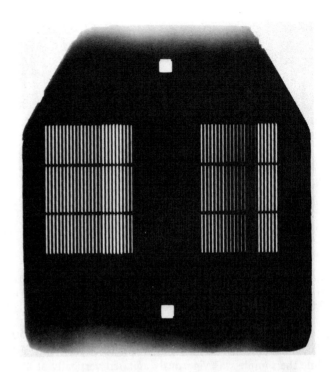

FIG. 11. Photomicrograph of two 14 × 18 mm slit arrays in an actual mask. The holes above and below the array are for convenience in fixturing. A range of slit openings (not resolved) is contained in the test mask.

tration-dependent etchant. The P-ED etchant micromachines a composite mask of micron size slits in a p^+ silicon membrane (doped $> 7 \times 10^{19}$ cm^{-3}) supported on a frame of stronger, thick silicon locally etched away from the membrane. What is desired is a structure with an effectively thin pattern defining layer but with overall structural strength. Figure 10 shows sketches of our mask near completion. Figure 11 shows a micrograph of a finished mask. The masks are formed as follows.

As a starting material we used silicon wafers chem-mechanically polished on both sides whose (100) surface planes and (110) reference flats were oriented to within ± 0.1° although this precision is not required for this application. The (110) reference flat precisely defines the intersection of (111) planes with the (100) surface as required for precision alignment for most silicon micromachining. A typical mask pattern is to consist of an array of parallel openings spanning area of order cm^2. The wafers are first oxidized to a depth 0.5–1.4 μm. A reverse tone strip pattern is photolithographically produced on one side of the wafer, i.e., only narrow stripes of SiO$_2$ remain. A boron diffusion is then carried out to produce a diffused layer in those areas not protected by SiO$_2$. It is important to note that the diffused layer extends laterally beneath the stripes for a distance comparable to the diffusion depth. In a typical process using a Boron$^+$ planar diffusion source at 1150 °C and for a time of 8 h, a concentration of the required value of $\gtrsim 7 \times 10^{19}$ cm^{-3} is obtained to a depth of 6 μm with a consequent line width loss of twice this amount. This occurs because finally it is the undoped Si stripes which becomes the shadow mask openings. After the diffusion, a window of size larger than the stripe array and encompassing it is etched in the SiO$_2$ on the backside of the wafer. Using 90-g pyrocatechol 510-cm^3 ethylene diamine in

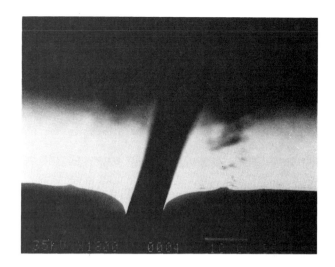

FIG. 12. Cross section of a slit opening showing the doped layer with its diffusion determined contour. The slit width opening is 4 μm. The Si membrane thickness is 6 μm.

240 cm³ of water at 110 °C, a time of 4 h is required to etch this larger cavity through the 380-μm-thick Si where it is stopped by the highly doped Si layer producing the inverted pyramidal cavity of Fig. 10. About 0.2 μm of SiO$_2$ is consumed during the etch so that a starting thickness greater than this value is desired on the wafer back. The etch resistance of the doped Si on the front side is sufficient that no additional protection is required. Figure 10(a) shows the mask after completion of this deep etch with the SiO$_2$ stripes in place. Dissolution of the stripes in buffered HF completes the mask. Removal of the stripes prior to the deep etch is also acceptable and has the advantage of presenting less stress to the thin membrane in its final state. As shown in Fig. 11 additional features may be simultaneously formed in the frame to fit into mask manipulation fixturing.

The shape and size of the mask slits is determined by the diffusion process primarily and controlling the diffusion reproducibly is the most crucial part of the fabrication process. Figure 12 shows in cross section a slit opening from a broken wafer. The shape of the membrane edge is characteristic of the diffusion profile with its shape fortuitously favorable to the present application. The slit edge acuity is considerably better than 1 μm and at line widths of a few microns, is sufficient for good line definition in the deposition process. The line spacing of 508 μm was set by other considerations and can be reduced in principle to twice the linewidth.

The aforementioned mask design and fabrication procedures have produced mechanical shadow masks for *in situ* stripe definition of thin films in MBE or other deposition systems. The masks are reasonably durable and capable of a few micron line widths in the restricted geometry considered. They have particular appeal because of their noncontaminating nature. Numerous mask structures have been made and are presently awaiting implementation.

In summary we have described the production of three types of silicon structures for research applications. Their fabrication involved exploitation of micromachining principles by earlier workers with practical modifications of their procedures to obtain excellent large and small scale uniformity. In the process, we have set some guide lines on the precision to which these aspects of uniformity can be reliably controlled as well as determining that structures can be produced at the micron level and even the 0.1 μm level given the processing control we have demonstrated.

We gratefully acknowledge G. J. Dolan who initially suggested and encouraged this work as a research project and also performed the eb lithography defining the submicron grating, suggestions from A. Y. Cho in regard to the noncontaminating mask structures, and particularly F. K. Reinhart, who suggested the need for and the potential for such masks. We also gratefully acknowledge the contributions of R. L. Batdorf and T. Weidner who performed the boron diffusion and provided much useful information regarding the diffusion process.

APPENDIX A: Some descriptions and information regarding various etchants used in Silicon Micromachining.

ETCHANTS USED IN SILICON MICROMACHING

ETCHANT	TEMP (°C)	RATE (μm/hrs) (100)	RATE (μm/hrs) (110)	RATE (μm/hrs) (111)	REFERENCE AND REMARKS
PSE-100 (HYDRAZINE-BASED)	95-120	200	NEG	NEG	(24) NO ATTACK ON Aℓ
PSE-200 (ALKALI-BASED)	75-100		500	NEG	(23) ATTACKS Aℓ; NO ATTACK ON Ag, Au
PSE-300 (ETHYLE DIAMINE-BASED)	100	25		NEG	(25) NO ATTACK ON Aℓ, Ag, Au, Cu, Ta
HNO$_3$: HF (21CC : 4CC)	25	(800 ISOTROPIC)			(*) SHALLOW-ETCHED, LARGE AREAS
HNO$_3$: HF : CH$_3$CO$_2$H (27CC : 27CC : 46CC)	25	(1500 ISOTROPIC)			(6) ON LARGE AREAS
(75CC : 8CC : 17CC)	25	(300 ISOTROPIC)			(3) ON LARGE AREAS
(71CC : 2CC : 27CC)	25	(12 ISOTROPIC)			(3,13) ON LARGE AREAS
KOH : H$_2$O : PROPANOL (200g : 800CC : 200CC)	80-100	60		<01	(2,3,16,21) FOR DEEP ETCHING; ETCH-STOPS AT p** CONC (>7x10^{19} cm^{-3})
KOH : H$_2$O (35g : 65CC)	80-85	55		NEG	(*) SHALLOW-ETCHED ON VERY LARGE AREAS
(35g : 65CC)	80-85		165	0.3	(*) DEEPLY-ETCHED ON ~250μm WIDE AREAS
(35g : 65CC)	80-85		135	0.6	(*) DEEPLY-ETCHED IN NARROW (~30μm) GROOVES
(35g : 65CC)	60		120	NEG	(*) 0.2μm GROOVES ETCHED 8μm DEEP
(44g : 56CC)	24		2	NEG	(4,5) 0.6μm GROOVES ETCHED 44μm DEEP
(35-50g : 65-50CC)	80	25	50	NEG	(3,4) ON LARGE AREAS
P : E.D. : H$_2$O (45g : 255CC : 120CC)	110-118	100-130		NEG	(*) DEEP ETCHING; ETCH-STOPS AT p** Si; ETCHES SiO$_2$ ~350Å/hrs
(45g : 255CC : 120CC)	100	65		NEG	(3) DEEP ETCHING; ETCH-STOPS AT p** Si; ETCHES SiO$_2$ ~500Å/hrs; HILLOCKS EASILY FORMED ON (111) WALLS
(40g : 250CC : 80CC)	118	50	30	1	(2,14,15) DEEP ETCHING; ETCH-STOPS AT p** Si; ETCHES SiO$_2$ ~150Å/hrs
(*) FROM OUR WORK					

a) Present address: Princeton University, Department of Electrical Engineering, Engineering Quadrangle, Princeton, NJ 08544.

[1] A. I. Stoller, RCA Rev. **31**, 271 (1970).
[2] E. Bassous, IEEE Trans. Electron Devices ED-**25**, 1178 (1978).
[3] K. E. Bean, IEEE Trans. Electron Devices ED-**25**, 1185 (1978).
[4] D. L. Kendall, Annu. Rev. Mater. Sci. **9**, 373 (1979).
[5] D. L. Kendall, Appl. Phys. Lett. **26**, 195 (1975).
[6] W. H. Craft, Circuits Manuf. **17**, 45 (1977).
[7] A. I. Stoller, R. F. Speers, and S. Opresko, RCA Rev. **31**, 265 (1970).
[8] C. J. Schmidt, P. Lenzo, and E. G. Spencer, J. Appl. Phys. **46**, 4080 (1975).
[9] N. W. Cheung, Rev. Sci. Instrum. **51**, 1212 (1980).
[10] C. L. Huang and T. Van Duzer, Appl. Phys. Lett. **25**, 753 (1974).
[11] C. L. Huang and T. Van Duzer, IEEE Trans. Electron Dev. ED-**23**, 579 (1976).
[12] L. Derick, AT&T Bell Laboratories, memoradum No. MM 55-113-51, October 3, 1955.
[13] B. Schwartz and H. Robbin, J. Electrochem. Soc. **108**, 356 (1961).
[14] R. M. Finne and D. L. Klein, J. Electrochem. Soc. **114**, 965 (1967).
[15] E. Bassous and E. F. Baran, J. Electrochem. Soc. **125**, 1321 (1978).
[16] N. Tsumita, J. Melngailis, A. M. Hawryluk, and H. I. Smith, J. Vac. Sci. Technol. **19**, 1211 (1981).
[17] A. Bohg, J. Electrochem. Soc. **2**, 401 (1971).
[18] I. Adesida, T. E. Everhart, and R. Shimizu, J. Vac. Sci. Technol. **16**, 1743 (1979).
[19] D. J. Coe, Solid State Electron. **20**, 985 (1977).
[20] Won-Tien Tsang and Shyh Wang, J. Appl. Phys. **46**, 2163 (1975).
[21] J. B. Price, in *ECS Semiconductor Silicon 1973* (Electrochemical Society, Princeton, NJ, 1973), pp. 339–353.
[22] E. D. Palik, J. W. Faust, Jr., H. F. Gray, and P. F. Green, J. Electrochem. Soc. **129**, 2051 (1982).
[23] Preferential Silicon Etchant, type PSE-200, Transene Co., Inc., Bulletin No. 207, Nov., 1982.
[24] Preferential Silicon Etchant, type PSE-100, Transene Co., Inc., Bulletin No. 207, Nov., 1982.
[25] Preferential Silicon Etchant, type PSE-300, Transene Co., Inc., Bulletin No. 207, Nov., 1982.
[26] Fluoroware, Inc., Chaska, Minnesota.
[27] Slone "DEKTAK II" surface profiler with a 12μm size probe.
[28] D. Bishop, *Elastic Constants Measurements of Liquid Crystals and Low Temperature Properties of Solids* (to be published).
[29] S. P. Muraka, H. J. Levenstein, R. B. Marcus, and R. S. Wagner, J. Appl. Phys. **48**, 4001 (1977).
[30] Buffered Oxide Etchant (7 : 1), Transene Co., Inc., Bulletin No. 208.
[31] G. Agnolet, J. D. Reppy, G. Kaminsky, and D. J. Bishop, Bull. Am Phys. Soc. **29**, 223 (1984).
[32] D. Bishop *et al.*, *Mechanical Measurements of Smectic Freely Suspended Liquid Crystal Films* (to be published).
[33] Nano Spec/AFT model SDP-2000T Optical Film Thickness Computer.

STRINGS, LOOPS AND PYRAMIDS-
BUILDING BLOCKS FOR MICROSTRUCTURES

H.H. Busta
Amoco Research Center
Naperville, IL 60566

A.D. Feinerman, J.B. Ketterson
Northwestern University
Evanston, IL 60201

and

R.D. Cueller
Johnson Controls, Inc.
Milwaukee, WI 53201

Abstract

By combining the selective deposition of tungsten on silicon with micromachining techniques, several microelectronic devices have been developed. These include ultrathin tungsten strings with cross-sections ranging from $300 \times 300 A^2$ to $600 \times 2000 A^2$, tungsten loops, and tungsten cladded pyramids.

Initial interest of the strings and loops concentrated in studying quantum mechanical effects such as localization and the Aharonov-Bohm effect at cryogenic temperatures. However, the strings can be made free-standing, thus becoming available for vibration sensing and the loops could be used as bearings. The pyramids could find applications in scanning tunneling microscopes, as surgical scalpels, and as Fowler-Nordheim field emitters in flat panel displays and in high current sources.

Introduction

Micromachining techniques of silicon and thin films are used to fabricate three dimensional devices finding applications for pressure and flow sensing, as accelerometers, alignment structures for fiber optical devices, micro relays, etc. The processing techniques consist of the conventional IC processes such as photolithography, diffusion, ion implantation, oxidation, wet and dry chemical etching, combined with anisotropical etching and the use of sacrificial layers.

In this paper, we describe three structures that use the selective nature of LPCVD tungsten and the displacement reaction

$$2WF_6 + 3Si \rightarrow 2W + 3SiF_4 \quad (1)$$

to form novel devices.

Strings

Recently, there has been a great interest in the formation of thin wires. First, fundamental physical effects, such as localization of electron states in a disordered potential and electron-electron interaction in one dimension, can be studied at cryogenic temperatures [1]. Second, submicron device structures can be formed, thus allowing the investigation of the fundamental limits to very large scale integration (VLSI) in silicon technology and also the formation of high-speed GaAs devices [2]. Third, limits to metal interconnects in VLSI can be investigated. These include RC time constant limits, electromigration effects at high-current densities and contact formation between metal-to-semiconductor and metal-to-metal.

Fourth, with additional processing steps, these wires can be isolated from their support structures, thus making them free-standing. Possible applications lie in high resonant frequency vibration sensors.

Electron beam [3] and x-ray lithographical [4] methods are used to fabricated small metal wires. These techniques are accessible to relatively few investigators due to the high cost of the necessary equipment.

A method to form rectangular shaped tungsten filaments is reported. It is based on the selective deposition of tungsten on the sidewall of undoped polysilicon. The technique utilizes conventional optical lithography. The width of the filament is determined by the processing conditions, and the thickness by the thickness of the polysilicon film.

Figure 1 illustrates the processing steps. The starting material is a 3-inch, (100), n-type or p-type silicon wafer which is oxidized in steam at 1000°C to form a 1000-A thick SiO_2 layer. On top of this layer, 700A of undoped polysilicon is deposited at 620°C in an LPCVD reactor. The wafer is then oxidized at 900°C in dry oxygen to form a 200-A SiO_2 layer. The thickness of the polysilicon is now 600A, since 100A of silicon is consumed in forming the oxide. Rectangular shaped patterns are then defined by optical lithography. The top SiO_2 layer is wet chemically etched in buffered HF, and the polysilicon layer is removed in a plasma etcher using CF_4 as the etch gas. The resulting pattern is shown in Fig. 1(a). Two contact areas are formed at the ends of the rectangular pattern and the SiO_2 is removed as shown in Fig. 1(b). The structure is now ready for the selective deposition of tungsten. This takes place in a commercially available 4 inch hot-wall LPCVP quartz reactor whose plumbing system has been modified to allow argon and WF_6 to enter the system through mass flow meters. Argon is used as a carrier gas with a flow rate of 335 cm^3/min. The tungsten deposition parameters are 400°C, 0.3 Torr total pressure of argon and WF_6, and a WF_6 flow rate of 12 cm^3/min. The deposition time is 5 min.

The resulting structure is shown in Fig. 1(c). The subfigure on the right side has the protective SiO_2 layer on top of the polysilicon removed for visual clarification. The filamentary nature of the tungsten film along the sidewall as well as the contact pad area can be seen.

Aluminum is then deposited on top of the wafer and contact pads are photoshaped. The aluminum is wet chemically etched in phosphoric acid. One more process can be employed to sever one of the two tungsten filaments. The resulting structure is shown in Fig.

1(d). To obtain free-standing wires, the supporting polysilicon layer between the wires can be etched in EPW consisting of 660ml ethylenediamine, 140g pryocatechol and 330ml H_2O. Etching takes place at 100°C. This is followed by etching the SiO_2 support layer on the bottom of the wires in buffered HF.

Wires with cross-sections ranging from 300 x 300A^2 to 600 x 2000A^2 and lengths of 67.5, 140 and 265μm have been fabricated using this method.

Loops

A structure that has generated recent interest to observe theoretically predicted quantum mechanical effects is a small metallic loop. The magnetoresistance of this structure undergoes periodic oscillations with respect to a flux quantum h/2e. Here h is Planck's constant and e is the charge of the electron. This behavior was predicted by Aharonov-Bohm [5]. It is a consequence of the phase dependence of electrons on the magnetic vector potential encircled by the loop and of the nonequilibrium nature of the transport in this loop if its diameter is smaller than the inelastic scattering length of an electron in the metal or metal alloy. In Ref. 6, loops consisting of Au and $Au_{60}Pd_{40}$ were processed using e-beam lithography. The approximate shape of such a loop is shown on the left side of Fig. 3. The contact area of the two wire elements to the loop itself is rather large. Since the phase difference along either side of the loop depends on the path the electrons can take, it is desirable to minimize the contact area to obtain the maximum amplitude of oscillations due to the constructive/destructive interference behavior. The desired loop structure is shown on the right side of Fig. 3. By improving their lithographic techniques, the authors in Ref. 6 produced loops similar in shape to the one shown on the right side of Fig. 3 [7].

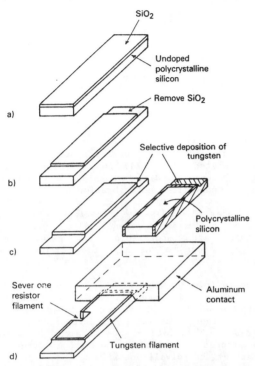

Fig. 1 Processing procedure for tungsten filament formation. (a) Definition of a rectangular SiO_2/polysilicon island on top of SiO_2. (b) Removal of SiO_2 to define contact areas to the filaments. (c) Selective deposition of tungsten. The filaments form along the sidewalls of the polysilicon. No deposition takes place on SiO_2. (d) Aluminum contact formation and possible severing of one filament.

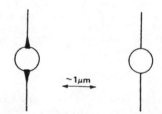

Fig. 3 Loop configuration for experimental investigation of the Aharonov-Bohm effect. The left loop depicts the geometry given in Ref. 6; the right loop is the desired configuration.

Figure 2 shows the SEM micrograph of a small segment of a 67.5μm long filament with a width of approximately 2000A and a height of 600A.

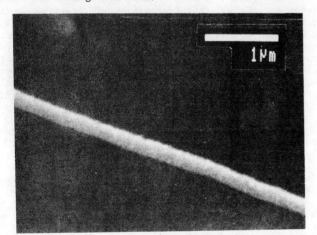

Fig. 2 SEM micrograph of a short section of a tungsten filament with a width of approximately 2000A and a height of 600A.

A different approach to form small loops of minimum contact area, without the use of e-beam lithography, is described here. The processing sequence is depicted in Figs. 4 and 5. The starting material is a 3-inch, (100), n-type or p-type silicon wafer which is oxidized in steam at 1100°C to form a 3000A-thick SiO_2 layer. On top of this layer, 3000-5000A of polysilicon (Poly I) is deposited by LPCVD at 620°C and subsequently oxidized to form a 200-300A thick SiO_2 layer. This layer is then photoshaped and etched to form a structure shown in Fig. 4(a). A second polysilicon layer (Poly II) approximately 800A thick is deposited next and oxidized at 1000°C to form a 200-300A thick oxide layer. This is shown in Fig. 4(b). A photolithographical step is then performed and the SiO_2 and Poly II are etched by HF and plasma, respectively, to form the structure shown in Fig. 4(c). The overlap of Poly II with the first SiO_2 layer is approximately 0.5-1.0μm. The loop geometry is defined next. This is shown in Fig. 5. After removal of the oxide layers the poly Si is etched in a plasma etcher.

The contact windows are now defined and the SiO_2 layer is etched. The structure is now ready for the conversion of the surface area of the exposed poly Si regions into tungsten. LPCVD deposition parameters were the same as for filament formation. Aluminum is deposited next and photoshaped. After packaging and wirebonding, the unit is ready for testing. The optical micrograph of an array of loops processed by this technique is shown in Fig. 6. The loop to the right has a diameter of 4.5μm.

By increasing the thicknesses of the two polysilicon layers and employing RIE etching techniques, the circular loops can be processed which, in principle, can form wear-resistant bearings for movable microstructures.

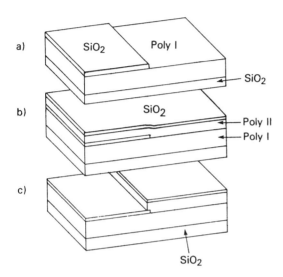

Fig. 4 Partial processing procedure for tungsten loop formation. (a) Definition of SiO_2 edge on top of polysilicon layer. (b) Deposition of a second layer of polysilicon and oxide formation. (c) Formation of a SiO_2/polysilicon edge which overlaps the SiO_2 edge in (a) by approximately 0.5-1.0μm.

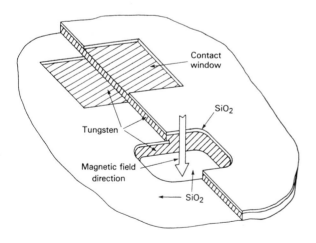

Fig. 5 Partial processing procedure for tungsten loop formation. After loop definition, the SiO_2 layers and the polysilicon are etched in HF and in a plasma etcher, respectively. Contact windows are then defined and the SiO_2 is removed in HF. This is followed by the conversion of the surface area of the exposed polysilicon regions into tungsten.

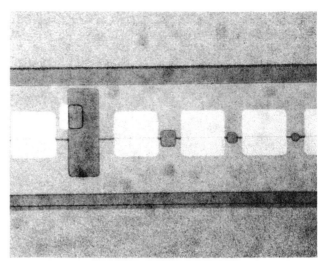

Fig. 6 Optical micrograph of an array of tungsten loops which has been processed according to the procedures outlined in Figs. 4 and 5. The loop to the right has a diameter of 4.5μm.

Pyramids

Micromachining techniques are being employed to form sharp objects with radii ranging from several to a few thousand Angstroms. Applications lie in tunneling tips to be used in the scanning tunneling microscope [8], and in high current density field emitter sources [9], and in large area displays [10]. Processing techniques include chemical thinning of thin tungsten wires for the tunneling tip and a form of shadow masking deposition for the high current density field emitter [9, 10].

Micromachined silicon pyramids are an alternative way to form structures with small tips, but problems occur when trying to deposit a uniform metal layer near the vicinity of the tip using sputtering and e-beam deposition techniques. Since in the LPCVD method of tungsten, silicon is displaced in the reaction at a rate of 2:1 [11], conformal coverage can be obtained. By combining anisotropic etching to form silicon pyramids with LPCVD of tungsten, we formed individual pyramids and arrays. In addition to the applications described above, these tungsten cladded pyramids can be mounted on appropriate support structures and be used as scalpels to perform microsurgical procedures. Figure 7 describes the processing procedure to form these pyramids.

The starting material is a 3 inch, (100), n-type or p-type silicon wafer. After initial cleaning, a 2000Å thick SiO_2 layer is grown and 5μm x 5μm squares are photoshaped (Fig. 7a) with their diagonals aligned parallel to the flat of the wafer which is oriented along the <110> direction. After BOE etching and photoresist removal, the wafer is etched in an EPW solution at 100°C for approximately 15 min. (Fig. 7b). The wafer is then inserted into a 10:1 HF solution to remove the SiO_2 layer on top of the pyramids. This step is necessary to ensure uniform tungsten coverage. It was found that after EPW etching, a residual SiO_2 layer of approximately 20Å is formed. This layer, in some cases could inhibit the silicon reduction of WF_6 [12]. The selective deposition takes place next with the same parameters as described above. Figure 7c shows the final configuration. To check for uniform tungsten coverage of the pyramids, scanning Auger microprobe analysis was performed at locations near the top and the bottom of a pyramid. This is depicted as Position 1 and Position 2 in Fig. 8. Figure 9 shows the Augur profile at Position 1 and Fig. 10 at Position 2. Sputtering times to remove the tungsten are comparable, thus indicating near conformal coverage. Figure 11 shows the SEM micrograph of an array of field emitters prior to oxide removal and tungsten cladding. Figure 12 shows the top of one pyramid after tungsten cladding and Fig. 13 shows an array of tungsten cladded silicon pyramids.

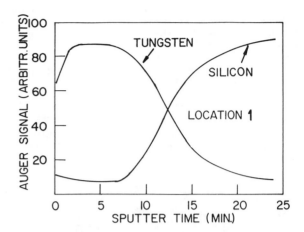

Fig. 9 Auger profile of tungsten-cladded pyramid at Position 1 in Fig. 8.

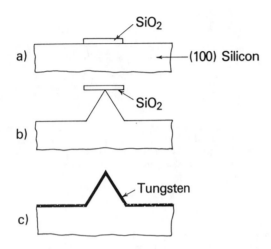

Fig. 7 Processing sequence for a tungsten-cladded pyramid. (a) shows the structure after photoshaping and etching of a 5μm x 5μm square whose diagonal is aligned parallel to the <110> direction, (b) shows the structure after EPW etching and (c) after tungsten deposition.

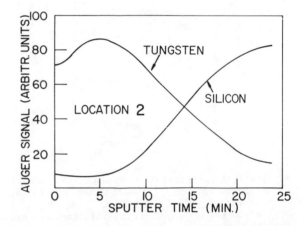

Fig. 10 Auger profile of tungsten-cladded pyramid at Position 2 in Fig. 8.

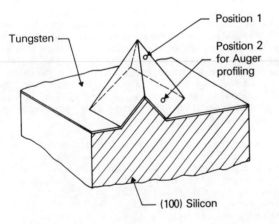

Fig. 8 Perspective view of a tungsten-cladded pyramid indicating the location where scanning Auger microprobe analysis was performed.

Fig. 11 SEM micrograph of a field emitter structure at processing step (b) in Fig. 7.

Fig. 12 SEM micrograph of a field emitter after tungsten deposition.

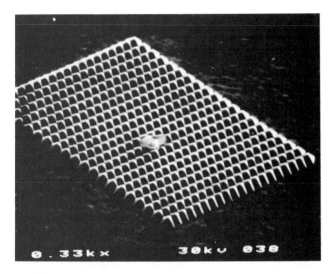

Fig. 13 SEM micrograph of an array of tungsten cladded field emitters.

Summary

By combining micromachining processing techniques with the selective deposition of tungsten by low pressure chemical vapor deposition, strings, loops and pyramids were fabricated. These devices find applications in investigating fundamental quantum mechanical effects (strings, loops), as well as for micromechanical structures such as vibration sensors and wear resistance bearings. The pyramids find applications as electron emitters and/or as scalpels.

References

1. N. Giordano, Phys. Rev., B22, 5635 (1980).

2. J.A. Krumhansl and Yoh-Han Pao, Physics Today, 32, 25 (1979): E.D. Wolf, ibid, 32, 34 (1979); A.N. Broers, ibid., 32, 38 (1979); J.L. Moll and D. Hammond, ibid., 32, 46 (1979).

3. A.N. Broers, J.M.E. Harper and W.W. Molzen, Appl. Phys. Lett., 33, 392 (1978).

4. D.C. Flanders, Appl. Phys. Lett., 36, 93 (1980).

5. Y. Aharonov and D. Bohm, Phys. Review, 115, 485 (1959).

6. C.P. Umbach, S. Washburn, R.B. Laibowitz and R.A. Webb, Phys. Review B., 30, 4048 (1984).

7. R.A. Webb, S. Washburn, C.P. Umbach and R.B. Laibowitz, Phys. Rev. Letters, 54, 2696 (1985).

8. G. Binnig and H. Rohrer, Scientific American, 50 (August 1985).

9. R. Greene, H. Gray and G. Campisi, IEDM Meeting, Washington, DC, 172 (Demcember 1985).

10. R.T. Gallagher, Electronics, 18 (June 16, 1986).

11. K.Y. Tsao and H.H. Busta, J. Electrochem. Soc., 131, 2702 (1984).

12. H.H. Busta and C.H. Tang, J. Electrochem. Soc., 133, 1195 (1986).

CORNER COMPENSATION STRUCTURES FOR (110) ORIENTED SILICON*

D. R. Ciarlo
Electronics Engineering Department
Lawrence Livermore National Laboratory
P.O. Box 5504, L-156
Livermore, California 94550

Abstract

We have studied the use of (110) oriented silicon wafers for the fabrication of microscale mechanisms such as cantilever beams and latching mechanisms. Using this orientation, cantilever beams can be fabricated such that they deflect in a parallel direction the wafer surface. It might be possible to fabricate structures such as setback accelerometers and latching mechanisms with (110) silicon cantilevers. We have designed and fabricated corner compensation structures for use with (110) oriented cantilever beams.

Introduction

Most of the reports in the literature on micromechanical silicon structures discuss the use of (100) oriented silicon wafers.[1-3] Either orientation dependent etching (ODE) or concentration dependent etching (CDE) is used to form microstructures such as grooves, cavities, thin membranes and cantilever beams. On (100) oriented wafers, the alignment flat is on a (110) plane. If a rectangular shaped hole is etched on the surface of these wafers with the edges of the rectangle aligned parallel or perpendicular to this flat, using an anisotropic etchant such as potassium hydroxide, a cavity is formed whose sidewalls are the (111) crystal planes. If the mask opening is sufficiently large to prevent a self stopping condition, this procedure can be combined with CDE to fabricate long thin cantilever beams which deflect in a direction perpendicular to the surface of the (100) oriented silicon wafer.[4] The following study explores the use of corner compensation structures which are required for the fabrication of cantilever beams from (110) oriented wafers. Such beams can be oriented so they deflect in a direction parallel to the wafer surface.

Etching (110) Oriented Silicon Wafers

With silicon wafers, the (111) crystal planes are the slowest etching planes when anisotropic etchants such as KOH or ethylenediamine pyracatechol and water (EPW) are used. The etch rate ratios of 44% KOH for the crystal directions (110):(100):(111) have been measured to be 600:300:1.[5] This high etch rate ratio makes possible the fabrication of interesting microstructures.

If the surface of a (110) oriented silicon wafer is etched through a square mask opening one gets a six sided structure. There are four (111) planes which are perpendicular to the (110) wafer surface and two (111) planes which make an angle of 35.3° to the wafer surface. If this pattern is etched long enough, a self stopping condition will be achieved as the two low angle (111) planes meet along a line. This self stopping condition is why a straight through four sided hole cannot be etched through a (110) oriented silicon wafer as might first be expected.

Although (110) oriented wafers are not as common as the other orientations because they aren't used in IC processing, they are readily available. They are desired when structures with smooth vertical walls are to be fabricated. Wafers with a (110) orientation can be purchased with an orientation flat in any desired position. A convenient position is along a (111) plane which is perpendicular to the (110) surface.

Long narrow slits in (110) oriented wafers can be fabricated if the mask is accurately oriented to the proper crystal direction to minimize mask undercutting. Silicon nitride, deposited by LPCVD is an excellent etch mask to use when silicon is anisotropically etched because of its essentially zero etch rate to KOH. A 400Å film of silicon nitride is sufficient to mask against the KOH etchant. This nitride can be easily patterned with photoresist and etched in a CF_4 plasma.

To achieve an accurate alignment to the crystal planes, we developed an alignment target.[7] The lines in this target are 3mm long, 8μ wide, and they fan out at an angle of 0.1 degrees. This target is printed near the perimeter of a two-inch wafer and then etched 100μ into the wafer surface. By evaluating the undercut in this target as a function of the alignment angle, the correct crystal direction can be determined. With two such targets etched into a two-inch wafer near the perimeter and by using a contact mask aligner with split field alignment optics, we can align to the crystal to better than 0.05 degrees. With such good alignment, it is possible to etch deep narrow structures such as shown in Fig. 1 which is an SEM of the cross section of 60μ wide grooves with a 120μ period. The grooves are 200μ deep. The atomically smooth (111) vertical sidewalls are apparent in this SEM.

Cantilever Structures Using (110) Silicon

Looking at the silicon slabs in Fig. 1, one can imagine etching all the way through the silicon wafer so that each of these slabs becomes a silicon cantilever beam, anchored at one end, which would deflect in a direction parallel to the wafer surface. Realizing that such deep vertical structures can be etched in (110) silicon we attempted to fabricate cantilever beams using this procedure. Such beams might be used in a

---

* Work performed under the auspices of the U.S. Department of Energy by the Lawrence Livermore National Laboratory under contract number W-7405-ENG-48.

Reprinted from *Proceedings IEEE Micro Robots and Teleoperators Workshop*, November 1987.

sequential setback accelerometer or in a latching structure as shown in Fig. 2. For these applications, all the straight edges on the mask are aligned so that they are parallel to the line of intersection between the (110) surface and a perpendicular (111) crystal plane. This results in vertical profiles. Inside corners are well behaved and their etched configuration can be accurately predicted. Outside corners are more difficult to deal with because at an outside corner, many crystal planes exist and some etch faster than others, leading to corner rounding. Since the devices to be built require one surface to rub against another, it is necessary to have smooth flat surfaces.

Corner Compensation for (110) Etching

Test structures were designed to study (110) corner compensation. These structures consisted of 500μ wide bars, with and without a 100μ four sided area attached to each corner. All edges are aligned to either 0°, + 70.5° or - 70.5° to the wafer flat so that they will coincide with the intersection of the (110) surface and a perpendicular (111) crystal plane. Figure 3 shows SEMs of a 500μ wide bar which has been etched into the surface of a (110) wafer. This bar does not have any corner compensation structures. The etch depth is 100μ. As can be seen the corners are rounded so that approximately 100μ of material from the exact edge of the corner is not flat.

Test structures with corner compensation are shown in Fig. 4. Here there is an extra area of silicon, 100μ on a side, which extends out from the corner in a direction parallel or at an angle to the beam depending on which surface is required to be smooth. Again all edges are parallel to the line of intersection between a perpendicular (111) crystal plane and the (110) surface. These were etched into the (110) surface to a depth of 100μ.

In addition to corner compensation, corner rounding can be minimized by etching from both surfaces so as to minimize the etch time required to achieve a cantilever beam of a given width. This requires accurate front to back alignment and double polished wafers. Fixtures have been built and alignment marks designed to make this double sided alignment possible. By using such a fixture, a (110) oriented wafer can be etched 100μ deep from both surfaces. With corner compensation, we speculate that a latching cantilever beam can be etched which is 200μ wide (equal to the wafer thickness) and 50μ thick (in a direction along the wafer surface).

Discussion

This study has explored the use of (110) oriented silicon wafers for the fabrication of cantilever beams. With proper corner compensation and double sided etching, it appears that cantilever beams which are 50μ thick and 200μ wide can be fabricated so that they deflect in a direction parallel to the wafer surface instead of perpendicular to the surface as is usually the case. With such cantilevers, it may be possible to fabricate integrating setback accelerometers and latching mechanisms since the cantilevers can be simultaneously etched so that they slide along and interlock with each other.

Acknowledgement

The author wishes to acknowledge R. K. Warner of Harry Diamond Laboratories for his helpful comments and support.

Figure 1. Deep grooves etched in (110) silicon wafers showing the atomically smooth (111) crystal planes which make up the sidewalls. These grooves are 60μ wide and 200μ deep.

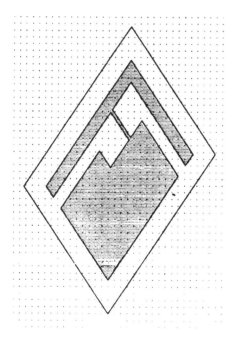

Figure 2. Possible latching mechanism which might be fabricated with (110) oriented silicon wafers. Shaded area gets etched. All edges are aligned to a vertical (111) crystal plane.

References

[1] K. E. Bean, "Anisotropic Etching of Silicon," *IEEE Transactions on Electron Devices*, Vol. ED-25, No. 10, Oct. 1978.

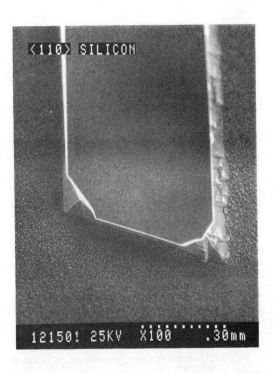

Figure 3. SEMs of test patterns with no compensation structures. Etch depth is 100. Outside corners are rounded.

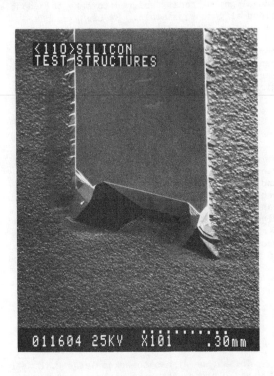
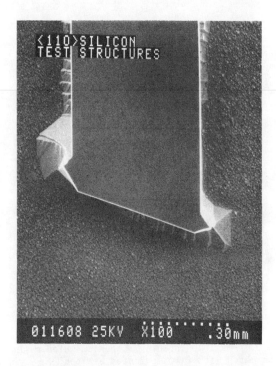

Figure 4. SEMs of test patterns with corner compensation structures. Etch depth is 100μ.

[2] E. Bassous, "Fabrication of Novel Three-Dimensional Microstructures by the Anisotropic Etching of (100) and (110) Silicon," IEEE Transactions on Electron Devices, Vol. ED-25, No. 10, Oct. 1978.

[3] K. E. Petersen, "Silicon as a Mechanical Material," Proceedings of the IEEE, Vol. 70, No. 5, May 1982.

[4] R. D. Jolly and R. S. Muller, "Miniature Cantilever Beams Fabricated by Anisotropic Etching of Silicon," J. Electochem. Soc., Vol. 127, No. 12, Dec. 1980, p. 2750.

[5] D. L. Kendall, "Vertical Etching of Silicon at Very High Aspect Ratios," in Annual Review of Materials Science, Edited by Robert A. Huggins, Vol. 9, 1979.

[6] D. R. Ciarlo and D. E. Miller, "Silicon Diffraction Gratings for Multilayer Structures," Proceedings of the SPIE, Vol. 688, Edited by N. M. Ceglio and P. Dhez, San Diego California, Aug. 19-20, 1986.

A STUDY ON COMPENSATING CORNER UNDERCUTTING IN ANISOTROPIC ETCHING OF (100) SILICON

Xian-ping Wu and Wen H. Ko[*]

Electronic Engineering Department
Fudan University
Shanghai, China

[*]Electronics Design Center
Case Western Reserve University
Cleveland, OH 44106 U.S.A.

ABSTRACT

Convex corner undercutting in anisotropic etching of (100) silicon is discussed. It is determined that the main beveling planes at undercut corners are all {212} planes, whether KOH or hydrazine or EPW solutions are used as etchant. A novel design of mask compensation patterns for corner undercutting is presented. This design fits all three types of anisotropic etchant. A simple and efficient calculating method of this design is presented to obtain satisfactory compensation.

INTRODUCTION

It is known that there is corner undercutting at convex corner in anisotropic etching of (100) orientated silicon using any known anisotropic etchant such as hydrazine-water solution[1,2] or potassium hydroxide (KOH)-water solution or EPW (Ethylenediamine-Pyrocatechol-Water)[3,4], if the edge of mask is parallel to ⟨110⟩ orientation. Some methods to restrict or compensate the corner undercutting have been presented. M.J.Declercq et al have researched effect of components ratio of hydrazine-water solution and etching temperature to corner undercutting . But it was found that the best condition for reducing corner undercutting can not be keeping with the best condition for getting the flatest and the smoothest etched surface. K.E.Bean and M.M.Abu-Zeid[3,4] added little square or superimposed rectangular stripes geometries to convex corner of mask as mask compensating patterns. But there are serious distortions at the etched convex corner using both above designs.

In this paper a detailed investigation is presented of beveling planes at undercut convex corner using three types of etchant mentioned above. Comparing previously reported studies, some contradictions about beveling plane have been exposed. To make this problem clear is important. Further more, to find a satisfactory technique to compensate the corner undercutting is an interesting undertaking.

EXPERIMENTAL TECHNIQUE

The samples are (100) orientated N-type silicon wafers with single or double side polished. The risistivity is about 4Ωcm and thickness is about 250μm. A thermoxide layer is used as etching mask which is about 0.6μm thick. The test pattern is defined using conventional photo-resist and oxide-etching procedures. The edge of mask is parallel to ⟨110⟩ orientations, except indicating in later text specially.

The etching equipment is a quartz refluxing system with a temperature controller.

The etchants and their compositions are as follows:

Hydrazine etchant
---80% hydrazine + 20% water (in volume)
KOH etchant
---100g KOH+320g water+80g normal propanol
"B etch" EPW etchant[5]
---250 c.c.Ethylenediamine + 40g.Pyrocatechol + 80c.c.Water
"F etch" EPW etchant
---250 c.c.Ethylenediamine + 80g.Pyrocatechol + 80c.c.Water + 1.5g Pyrazine

The "F etch" EPW solution is a catalyzed EPW etchant.

An Alpha step depth profiler and a mechanical micrometer are used to measure the etched depth. A reading microscope is used to measure the width of mesa and sidewall.

RESULTS AND DISCUSSIONS

I. BEVELING PLANES AT UNDERCUT CONVEX CORNER

When the mask edge is aligned to parallel to $\langle 110 \rangle$ orientation on (001) silicon wafer, in anisotropic etching the corner undercutting at convex corner of mesa starts and increases gradually. See Fig.1a. At last, a square mesa becomes an octagonal-shaped one. See Fig. 1b. The length of square mask on Fig.1 equals 1.5mm. "F etch"EPW etchant is used at temperature 115°C.

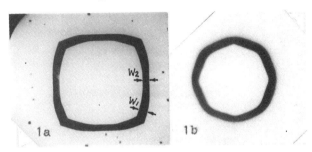

1a. etched 3 hours. 1b. etched 5 hours.
etched depth:240μm. mesa height:260μm.
Fig.1 Corner undercutting

What is the beveling faceting planes? D.B.Lee indicated that they are $\{211\}$ planes using hydrazine-water solution. K.E.Bean identified that the beveling planes are $\{331\}$ planes using KOH and EPW etchants. But M.M.Abu-Zeid reported that they are best represented by $\{212\}$ beveling planes using ethylenediamine-water solution (no pyrocatechol added).

In our study, three same masked samples are etched by hydrazine, KOH or "F etch"EPW etchants respectively. Their photomicrographs are shown in Fig.2. Although the extents of corner undercutting are different with each other, but the included angles of the bevel measured from photographs are all the same which is about 127°. So the intersecting lines should be $\langle 210 \rangle$ directions, not $\langle 310 \rangle$ directions which is intersected by $\{331\}$ and (001). Included angle by $\langle 310 \rangle$ directions is about 143°. See Fig. 3. However it is intersting to indicate that the geometric shape surrounded by $\langle 210 \rangle$ or $\langle 310 \rangle$ are alike except their positions, that are shown on Fig.4.

Since the edges of mask are parallel to $\langle 110 \rangle$ directions, not $\langle 100 \rangle$ directions, so only Fig.4a can be used to represent the corner undercutting as Fig.1. or Fig.2.

In other hand, although both of $\{211\}$ and $\{212\}$ planes intersect the (001) surface at $\langle 210 \rangle$ directions, but their included angles are different. Former is 65.91° that is larger than 54.74° included by $\{111\}$ and (001) planes. Later is 48.19° which is smaller than 54.74°. It is shown in Fig.5.

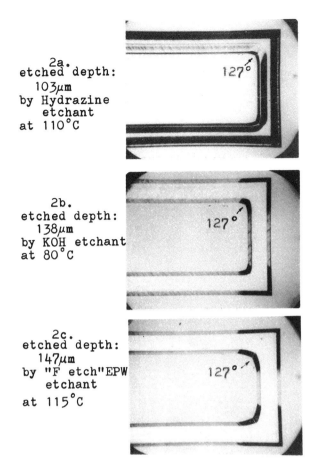

2a. etched depth: 103μm by Hydrazine etchant at 110°C

2b. etched depth: 138μm by KOH etchant at 80°C

2c. etched depth: 147μm by "F etch"EPW etchant at 115°C

Fig.2 Corner undercutting photomicrographs, by varied etchants.

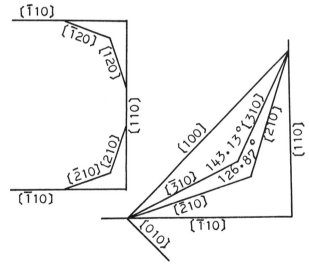

Fig.3 Analysis for included angles of the bevel

In our experiments, see Fig.1, the projecting width w_1 of beveling plane is about 215μ, when the etched depth d is 240μ. Meanwhile the projecting width w_2 of sidewall $\{111\}$ plane is about 170 μ. The included angle α_1 and α_2 can be gotten by the experimental measurement.

575

$\alpha_1 = tg^{-1}\frac{d}{W_1} = 48.1°$

$\alpha_2 = tg^{-1}\frac{d}{W_2} = 54.7°$

They are in accord with expecting theoretical values that are shown in Fig.5.

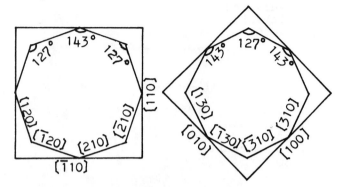

Fig.4 Geometric shapes surrounded by ⟨210⟩ or ⟨310⟩ orientations

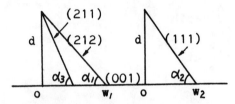

Theoretical values:
$\alpha_1 = 48.19°$, $\alpha_2 = 54.74°$, $\alpha_3 = 65.19°$

Fig.5 Included angles by (001) plane and (212) or (111) or (211) plane

II. RESULT OF TILTING MASK EXPERIMENT

Since the beveling faceting planes of mesa are {212} planes, the edges of mask can be tilted to parallel to ⟨210⟩ orientations to avoid corner undercutting.

An experiment is processed. Sample A and sample B are masked using the same mask. The mask edges of sample A are parallel to [110] and [$\bar{1}$10] orientations, while the mask edges of sample B are parallel to [210] and [$\bar{1}$20] orientations. Both the sample A and B are etched in "B etch" EPW solution for 200 minuts at temperature 115°C. The etched depth is about 190μ. See Fig.6.

It can be found that there is no corner undercutting at convex corner of sample B. But it can also be found that there is corner undercutting at concave corner of sample B and the lateral etch is serious. This case can be explained according to crystallography. On etching a concave surface as hole the limiting shape is bounded by the slowest etching planes, and on etching a convex surface as mesa the limiting shape is bounded by the fastest etching planes. The {111} plane is the

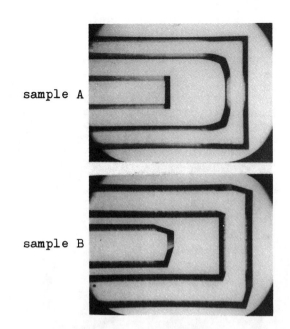

Fig.6 Photomicrographs of tilting mask experiment

slowest etching plane by all reported investigations and the {212} plane is supposed to be the fastest etching plane for (001) silicon wafer. However, using the method of tilting mask to avoid convex corner undercutting is not satisfying for most sensors designs because of its weakness mentioned above.

A NOVEL DESIGN OF MASK COMPENSATION PATTERNS

This paper presents a novel design of mask compensation pattern. Since the beveling planes {212} intersect (001) surface at ⟨210⟩ orientations, so four angles bounded by ⟨210⟩ orientations are added to the convex corner of a square mask pattern with edges parallel to ⟨110⟩ orientations, as shown in Fig.7.

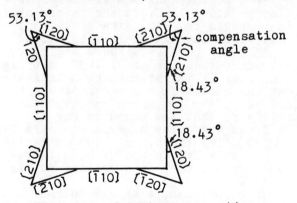

Fig.7 Design of mask compensation

In etching process these compensation angles become smaller and smaller, but there are no obvious distortion. The photomicrographs of etched samples with or without mask compensation are shown in Fig.8. In this experiment, the "F etch"EPW solution is used at 115°C for 120 minutes. The etched depth is about 192μm. The distance between the point of square corner and the top of the compensation angle is designed to equal 250√2μ. In this case, the compensation is just. If the etched depth is larger than desired value, the corner will be undercut again.

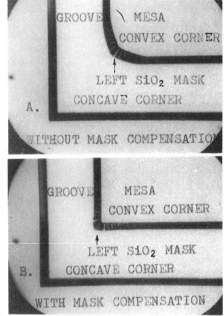

Fig.8 Photomicrographs of etched samples using "F etch"EPW

To calculate the dimension of compensation angle, a normalized factor $U = \frac{\delta}{H}$ is defined to describe the extent of corner undercutting, where δ is the width of undercutting in the direction perpendicular to the $\langle 210 \rangle$ intersection line. It means the etched quantity of plane $\{212\}$ on (001) surface. H is the etched depth. In our design, the compensation angles are bounded by $\langle 210 \rangle$ orientations. Based on symmetry of silicon crystal, the etch rate to every crystal plane of $\{212\}$ planes is the same. So the width δ_c of compensation angle should be equal to the width δ_u of corner undercutting. This design rule fits any type of anisotropic etchant. See Fig.9. The d_u and d_c is the distance between points of corner before and after etching, without or with mask compensation respectively. The P is the intersecting point of undercutting line and compensation line on $\langle 110 \rangle$ side of mask pattern. As $\delta_c = \delta_u$, following relations can be found: $d_c = \sqrt{5}\delta_c$, $d_u = \frac{\sqrt{5}}{2}\delta_u$
$d_c = 2d_u$, $\overline{OP} = \sqrt{2}d_c$

Since the extent of corner undercutting depends on such factors as type of etchant, components ratio of etchant and etching temperature etc.. The normalized factor U is usually found by experiments.

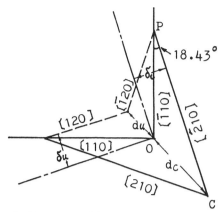

Fig.9 Compensation angle

Some data about undercutting factor U have been reported, except using conventional "B etch"EPW and catalyzed "F etch"EPW. As compared with KOH etchant, EPW has such advantages as no sodium or potassium contamination, slow etching rate to SiO_2 mask etc..[5] Further more the surface etched quality by EPW (especially, "F etch"EPW) is much better than by hydrazine or KOH. However, the corner undercutting using EPW is larger than that using KOH or hydrazine. The normalized factor U is about 0.42 for "Betch"EPW at 115°C and is about 0.85 for "F etch"EPW.

CONCLUSION

(1) The beveling planes at undercorner all $\{212\}$ planes, whether KOH or Hydrazine or EPW etchant is used.
(2) Compensation angles bounded by $\langle 210 \rangle$ orientations is very effective.
(3) When the width of compensation angle equals the width of corner undercutting the compensation is perfect, no obvious distortion appears on etched bottom.
(4) By this mask compensation, a square mesa corner with excellent etched surface can be formed using EPW solution.

ACKNOWLEDGMENT

The authors are grateful to Dr.C.C. Liu for his support of this work done partly at Electronics Design Center, CWRU., USA. They also wish to thank Prof. Min-hang Bao for helpful discussions at Fudan University, China where this work is finally completed.

REFERENCES

1. D.B.Lee, J.Appl.Phys. Vol.40,No.11, 1969, pp. 4569-4574.
2. M.J.Declercq et al., J.Electrochem. Soc. Vol.122, No.4, 1975, pp.545-552.
3. K.E.Bean. IEEE Tranc.on Electron Devices, Vol.ED-25, No.10, 1978, p.1185-
4. M.M.Abu-Zeid, J.Electrochem.Soc. Vol.131, No.9, 1984, pp.2138-2142.
5. X.P.Wu et al., Sensors and Actuators Vol.9, 1986, pp.333-343.

A New Silicon-on-Glass Process for Integrated Sensors

L. J. Spangler and K. D. Wise

Center for Integrated Sensors and Circuits
Solid-State Electronics Laboratory
University of Michigan
Ann Arbor, Michigan 48109-2122

ABSTRACT

This paper reports a process for the formation of high-performance thin single-crystal silicon films on glass substrates. The process utilizes the electrostatic bonding of a silicon wafer to glass and subsequent etching of the silicon to form films having thicknesses controlled from less than 2 μm to over 20 μm. The use of Corning 1729 glass substrates yields an excellent thermal expansion match to the silicon film and allows the use of post-bond processing temperatures for the films of as high as 800°C, allowing the formation of both MOS and bipolar device structures. Thus, integrated circuitry can be incorporated in dissolved-wafer sensing structures. A variety of related processes are also possible where some or all of the silicon device processing is performed at high temperature before bonding to the glass.

INTRODUCTION

For many sensing applications, the use of a substrate which is electrically insulating, thermally insulating, or optically transparent is desirable. In most such applications, monolithic active circuitry is also useful with the transducer array to allow signal amplification, multiplexing, or package simplification. Recently, an ultraminiature silicon-on-glass capacitive pressure sensor was reported [1] for application in a biomedical catheter. This sensor was based on a simple four-mask single-sided bulk-silicon process featuring a batch wafer-to-glass electrostatic seal and subsequent unmasked wafer dissolution. The process reported here significantly extends that work by allowing the monolithic integration of high-performance circuitry in such dissolved wafer-processes. The process produces higher-quality silicon films than do recrystallized SOI processes [2] and has the added advantage of allowing the use of relatively thick silicon films which can be pre- and post-bond processed to produce structures not possible using other techniques. This approach appears very attractive for applications such as active-matrix LC displays, large-area tactile imagers, and ultra-miniature presure/flow sensors which require active readout electronics and/or distributed signal processing.

PROCESS DESCRIPTION

The overall process for creating the SOI film is shown in Fig. 1. The process begins with a <100>-oriented silicon wafer having an epitaxial layer in which the eventual devices will be created. A silicon etch-stop is realized at the epi-substrate

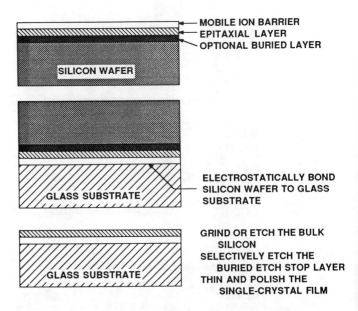

Fig. 1: Process Flow for Producing Single-Crystal Silicon Films on Glass.

junction using any of several techniques. Two approaches have been used in this research. In the first, a heavily-doped diffused boron buried layer is used between the epi film and a lightly-doped substrate. Epitaxial growth should be optimized here to minimize out-diffusion into the film and dislocations due to boron-induced stress in the buried layer. The second approach has utilized a heavily-doped (n- or p-type, $> 10^{18} cm^{-3}$) substrate with an epitaxial layer of the desired thickness and resistivity grown directly on it. Such p/p+ "CMOS" wafers are widely available for use in CMOS VLSI. An electrochemical etch-stop at an n-p epi-substrate junction or an ion-implanted etch-stop might also be used.

After formation of the epitaxial layer and etch-stop, the silicon wafer is electrostatically bonded to a glass substrate. Both Corning 1729 and 7740 substrates have been used. Corning type 1729 glass offers a better match to the thermal expansion coefficient of silicon than do other glasses [3], including 7740, and its high anneal point (853°C) allows device fabrication after bonding. Figure 2 shows the time required to achieve a bonded area of approximately one square inch for various bonding voltages and glass thicknesses using a point contact for the potential on the glass. While the required voltage is generally higher for 1729 glass than for 7740, scaling the 1729 glass thickness shown to 0.5mm should reduce the voltage by nearly a factor of two,

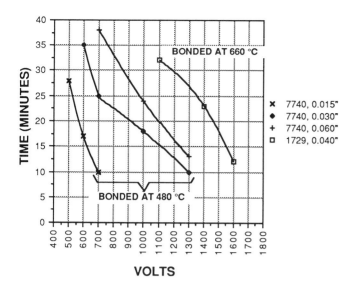

Fig. 2: Bonding Times and Voltages associated with forming a Bonded Area of One Square Inch for Various Glass Thicknesses.

placing it in a range similar to that for 7740. In alkali-containing glasses such as 7740, the mobile alkali cations drift toward the anode (silicon), causing an increased fraction of the applied potential to be dropped near the silicon-glass bonding interface and resulting in a lower bond voltage [2]. In a low-sodium glass such as 1729, the voltage distribution in the glass is thought to be more uniform, increasing the required potental but also allowing the voltage to scale more directly with thickness. While sodium thus plays an important role in determining the potential distribution in the bonding materials, it is not thought to otherwise contribute to the bond formation. Oxygen ions generated from defects in the glass, particularly E-centers, are thought to drift towards the anode, where they form an irreversible bond with the silicon [2]. Due to the high local forces generated by the electrostatic field, a bonded area coverage of over 99.8 percent on three-inch wafers has typically been observed.

As with 7740 electrostatic bonds, the 1729 silicon bond is stronger than the silicon itself, with fracture occurring in the bulk of the silicon wafer rather than at the interface. This supports the idea that oxygen from the glass acts as the bridging atom between the two bonding surfaces since the Si-O bond is about 2.5 times stronger than the Si-Si bond.

In device applications, a stress-compensated multilayer film of silicon dioxide, silicon nitride, and silicon dioxide [4] is deposited on the silicon surface via pyrolytic CVD prior to bonding to the glass. This film functions as a mobile ion barrier, preventing sodium migration into the silicon during subsequent device processing. Electrostatic bonds between dielectrically-coated silicon wafers and 1729 glass have typically been performed at 670°C, and subsequent heating to as much as 850°C has not noticeably altered the bond or its strength.

After bonding, the silicon wafer is subjected to an unmasked etch to the etch-stop. With a boron etch-stop, an EDP final etch [5] has been used, whereas with a heavily-doped substrate, an 8:3:1 acetic, nitric, hydrofluoric etch has been employed [6]. The EDP etch effectively reduces the etch rate to zero, while the 8:3:1 etch slows by a factor of at least 100 when encountering lightly-doped material, effectively yielding an etch-stop as well. In this work, these etching processes have allowed the formation of 75mm SOI wafers. The boron etch-stop has yielded slightly better silicon film thickness uniformity, while the "CMOS" substrates have yielded better epitaxial film quality. Figure 3 shows the contour plot for an SOI film formed using a 27 µm-thick epitaxial layer and the boron etch-stop process. The plot was derived from surface profilometry after preferentially removing the buried layer with a chemical etch. Over most of the wafer area, the film thickness is uniform to within five percent, and some of the variation noted is likely due to nonuniformity in the epi and buried-layer thicknesses themselves.

Following the removal of the silicon bulk, the remaining SOI film is polished and etched to remove the etch-stop layer (if present) and yield the final film. While this step contributes additional film nonuniformities (in our work, about ten percent of the etched thickness), the overall process is capable of producing final films which are uniform in thickness to within ten percent or better (e.g., ±0.5 µm on a 5 µm final film), which is adequate for most sensing applications.

X-ray diffraction measurements have shown these SOI films to be <100>-oriented and single-crystal, essentially identical to the starting substrates. Surface defect-sensitive etchants have indicated defect densities similar to those of the starting material as well. Hall measurements on 1.2-5µm-thick films produced using the boron etch-stop on 1729 glass with an ion barrier have produced n-type mobilities of 893 and p-type mobilities of 354 as compared with values of 1105 and 342 obtained in the bulk, respectively.

DEVICE STRUCTURES

A wide variety of device structures are possible with this process or variations of it. One of the principal advantages of this process over other SOI approaches is that high-temperature processing of the silicon is possible prior to bonding so that unique structures can be created using the ability to double-side process the film. Figure 4 shows an MOS structure realized using single-sided processing of the film. Both n- and p- channel MOSFETs have been realized having this structure. The n-channel devices resulted in electron surface mobilities of 640 cm^2/V-sec, the highest ever reported for an SOI transistor on glass [7]. The off-state drain leakage current was less than 0.1pA/µm. The upper limit of 800°C on post-bond processing temperature allows relatively standard gate oxidation processes to be used as well as the activation of ion-implanted dopants.

Figure 5 shows a bipolar transistor structure realized using double-sided film processing. A high-temperature diffusion was used prior to bonding to form a highly doped n^+ layer next to the ion barrier. Standard processing was then used to form the film, after which the p-type base was implanted using boron and a polysilicon emitter was formed using CVD. As with the MOS devices above, device islands were

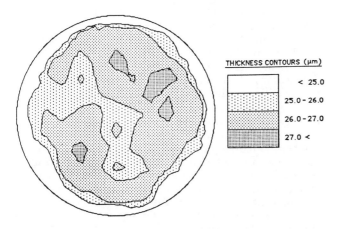

Fig. 3: Thickness Contours for an SOI Film on a Three-Inch Glass Wafer. The SOI film has a mean thickness of 26.8μm after removal of the boron etch-stop layer.

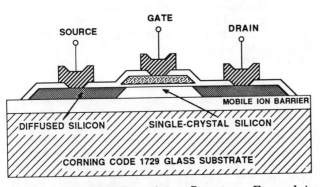

Fig. 4: An MOS Transistor Structure Formed in Single-Crystal Silicon on Glass.

formed using a selective anisotropic etch, and deposited oxides were used to passivate the device. Aluminum was used for interconnect. These devices have not yet been fully functional due to difficulties with the polysilicon emitter technology; however, the entire process has been run without other difficulty. As the island thickness increases, thicker resists are required for post-island processing, limiting feature sizes to 5 μm and above. These are adequate for most sensing applications.

Figure 6 shows the proposed process for an ultaminiature silicon pressure sensor with on-chip electronics. A "CMOS" wafer is deep diffused with boron to form a rim for the transducer, while a shallow high-temperature boron diffusion is used for the recessed diaphragm. A barrier dielectric is deposited and the wafer is sealed to the glass, after which the bulk of the wafer is dissolved to the etch-stop. Relatively-standard device processing (<800°C) is then used to form readout circuitry [1] in the remaining circuit islands. All of the individual steps in this fabrication sequence have been successfully run separately, and the overall combined process is now being implemented.

A variety of other alternative processes could also be implemented with this approach to sensor fabrication. Particularly attractive for some applications are those in which most or all of the circuit processing is done prior to glass bonding, so that the devices are on the lower surface of the silicon, next to the glass. In this case an electrostatic bond to the glass via deposited pillars or other elevated areas on the silicon is required along with a conducting shield to protect the devices from the electrostatic fields encountered during bonding. Such structures are more complex than that shown in Fig. 6 but avoid the necessity of processing the glass wafers. When an encapsulating layer of CVD silicon nitride is used over the glass, however, no cross-contamination problems have been observed, and the ability to perform double-sided processing on semiconducting SOI films of almost arbitrary thickness allows the formation of a number of structures impossible to realize using other techniques.

ACKNOWLEDGMENTS

The authors wish to thank Dr. Khalil Najafi and Ms. Qian Shi of the Solid-State Electronics Laboratory, University of Michigan, for their assistance. The encouragement, assistance, and financial support provided by the General Motors Research Laboratories is also gratefully acknowledged.

REFERENCES

1. H. L. Chau and K. D. Wise, "An Ultraminiature Solid-State Pressure Sensor for a Cardiovascular Catheter," Digest 4th Int. Conf. on Solid-State Sensors and Actuators (Transducers '87), IEE Japan, Tokyo, June 1987, pp. 344-347.

2. L. J. Spangler, "A Process Technology for Single-Crystal Silicon-on-Insulator Sensors and Circuits," Ph.D. Dissertation, The University of Michigan, April 1988.

3. F. P. Fehlner, W. H. Dumbaugh, and R. A. Miller, "Refractory Glass Substrates for Thin-Film Transistors," Proc. 6th Annual Int. Display Research Conf., Tokyo, 1986, pg. 100.

4. I. H. Choi and K. D. Wise, "A Silicon Thermopile-Based Infrared Sensing Array for Use in Automated Manufacturing," IEEE Trans. Elect. Dev., 33, pp. 72-79, January 1986.

5. A. Reisman, etal., "The Controlled Etching of Silicon in Catalyzed Ethylenediamine-Pyrocatechol-Water Solutons," J. Electrochem. Soc., 126, pg. 1406, Aug. 1979.

6. H. Muraoka, T. Ohhashi, and Y. Sumitomo, "Controlled Preferential Etching Technology," in Semiconductor Silicon 1973, Princeton, N. J.: Electrochem. Soc., pg. 327.

7. L. J. Spangler and K. D. Wise, "A Technology for High-Performance Single-Crystal Silicon-on-Insulator Transistors," IEEE Electr. Dev. Lett., 8, pp. 137-139, April 1987.

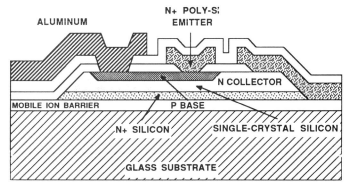

Fig. 5: A Vertical NPN Bipolar Transistor Structure on Glass. The n+ buried collector is formed using a pre-bond diffusion from the front of the silicon wafer.

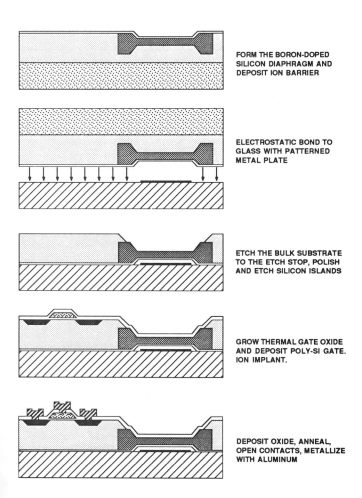

Fig. 6: A Proposed Process for an Active Ultraminiature Capacitive Pressure Sensor on Glass.

MECHANISMS OF ANODIC BONDING OF SILICON TO PYREX® GLASS

Kevin B. Albaugh
Paul E. Cade
IBM General Technology Division
Essex Junction, VT 05452

Don H. Rasmussen
Department of Chemical Engineering
Clarkson University
Potsdam, NY 13676

The mechanisms of formation of anodic bonds between glasses and metals are examined. Prior workers have suggested electrochemical, electrostatic, and thermal mechanisms for bond formation, but the dominant mechanism has not been clearly defined. The process is found to be an electrochemical analog to thermal glass-to-metal seals, where the metal surface is oxidized into the glass due to the development of large electric fields across the anodic depletion layer.

Introduction

Anodic bonding is commonly used for joining glass to silicon for micromechanical applications (1). The utility of anodic bonding arises from the low process temperature. Since the glass and silicon remain rigid during the process, it is possible to bond glass to a silicon surface preserving grooves in either the glass or silicon, which allows formation of devices such as pressure transducers. The physical processes occurring during bonding are of importance in determining the surface conditions required and the process conditions (temperature, voltage, and time) required to form a permanent, high quality bond. The current vs time transient at constant voltage contains a significant amount of information regarding the process mechanisms, which are predominantly electrochemical.

Experiment

Bonding was accomplished between discs of silicon (57mm diameter wafers) and Pyrex 7740 glass (50 mm diameter, 3.18 mm thick) in an apparatus previously described (2). The temperature range was 250° to 330°C, and the voltage ranged from 500 to 1000 V. The bonding was performed at constant applied voltage after the temperature had stabilized. Constant current could also be used, but was avoided since dielectric breakdown occurs at long times. The apparatus is shown in Fig. 1.

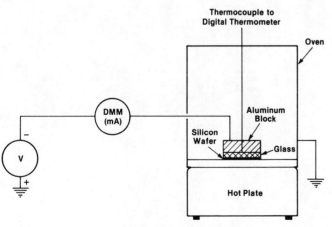

Figure 1. Apparatus

PYREX is a registered trademark of the Corning Glass Works, Inc.

Since no external pressure was applied, the bonding surfaces are joined by electrostatic attraction (3). In this case, bonding occurs by a two-step process: first, the contact area increases due to electrostatic attraction, and second, the permanent bonding forces develop. The extent of bonding is determined visually.

The primary measurements are the current vs time characteristic, the sample temperature measured at the center of the cathode, and the extent of bonding (area bonded/total area).

Current Characteristic

The variation of current with time gives a significant amount of information about the process. The current decays rapidly at the start of the cycle, due to the initial charging of the depletion layer. A model for the transient current response has been developed (4), which treats the depletion layer as a capacitor and the bulk glass as a series resistor. For the short-time regime, the leakage across the depletion layer is neglected. The model current characteristic is shown in Fig. 2, which approximates well the actual response shown in Fig. 3.

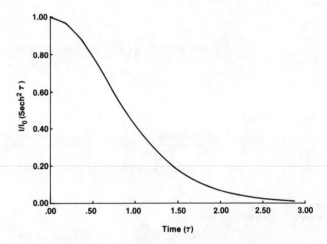

Figure 2. Current response—theoretical

The actual response has an exponential tail which is predicted by the long-time model of Carlson (5-8). This tail is due to leakage within the depletion layer.

The area under the initial charging peak gives the charge removed from the depletion layer. The curvature is determined by the series resistance and the charge concentration in the depletion layer. The charge concentration observed experimentally is very large, and indicates that most of the ionizable material in the depletion layer is removed. Since effectively no current is observed when the applied voltage is removed, the depletion layer is charge neutral, which indicates that any oxygen counterions present in the depletion layer are delivered to the anode.

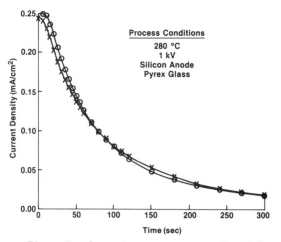

Figure 3. Current response—experimental

The experience with thermal sealing indicates that seals of glass to metal form when the metal surface is oxidized into the glass with proper control over the metal oxide to provide an adherent metal oxide scale. In the case of silicon to glass bonding, one merely needs to oxidize the silicon surface into the glass network, and one would not expect a large amount of oxygen to be delivered to the anode to do so. In the present experiments, bonding occurs after the delivery of 2 mC/cm2 to the anode, which suggests that <20 nm of oxide needs to be grown on the silicon to form a good bond to glass (as long as the surfaces are optically flat).

From the model for the current characteristic, the kinetic parameters for bonding are those that govern the rate of charge transfer: the voltage, temperature, and ion content of the glass. The process window is, as the delivery of oxygen to the anode will occur (albeit at widely varying rates) for a variety of temperatures and applied potentials. High temperature and high bias leads to rapid bonding.

Practical Implications

Surface preparation

The total amount of oxidation is very small for bonds to silicon. In some material systems, such as aluminum/glass or iron/glass, the metal ions are mobile and are transported into the glass, but for silicon/glass a mobile silicon ion does not exist. This means that surface preparation must be very good, since no oxidation of high spots into the glass can occur (9,10). The surfaces remain rigid during the bonding process and so little toleration of point asperities (particles) is possible as well.

A variety of surface cleans were evaluated during the study. Most cleans work quite well, such as acetone rinsing, sulfuric/nitric acid cleans, and chromic etch (11). Merely washing with sodium lauryl sulphate then rinsing is inadequate, since the soap leaves a residue which consumes oxygen at the anode, scavenging it prior to oxidation of silicon. The surface should be free of organic residues and as free of particulates as possible, but does not appear to be particularly sensitive to the presence of other residues (e.g., chromium ion).

Applied Pressure

The two conditions for bonding described above are that the surfaces to be bonded must be in intimate contact, and that after intimate contact sufficient oxidation occurs at the anode to provide permanent bridging bonds. If the surfaces are not in intimate contact at the start of the process, some of the charge transfer is "wasted": any anodic oxidation does not lead to bridging between the anode and the glass. Use of externally applied pressure avoids this difficulty, which is somewhat peculiar to the silicon anode since the total amount of charge transfer available is lower than for systems with mobile metal cations (e.g., aluminum). Even for aluminum, though, the process benefits from relatively large amounts of compression, as large as 70 MPa (12).

Conclusions

The formation of anodic bonds is satisfactorily explained by oxidation of the metal surface into the glass. Sufficient charge transfer occurs to allow the oxidation, and the observation of charge neutrality in the depletion layer indicates that oxygen ions are delivered to the anode in the case of silicon anodes. It is also possible in some cases that metal ions are delivered from the anode to the depletion layer, but this will result in formation of metal oxide as the metal ions combine with oxygen ions remaining in the depletion layer. Electrostatic attraction is important when the parts are not initially brought in intimate contact by an applied pressure. Joule heating of the depletion layer is not required for bonding to occur, and is inconsistent with minimal filling of surface topography on silicon anodes.

REFERENCES

1. Petersen, K. E., Proc. IEEE, V70(5) 1982 p420

2. Albaugh, K. B., Mat. Lett., 4 (11-12) 1986 p.465

3. Wallis, G., J. Amer. Ceram. Soc., V53(10) 1970 p563

4. Albaugh, K. B., submitted to J. Electrochem. Soc.

5. Carlson, D. E., et. al, J. Amer. Ceram. Soc., V57(7) 1974 p291

6. Carlson, D. E., J. Amer. Ceram. Soc., V57(11) 1974 p461

7. Carlson, D. E., et. al, J. Amer. Ceram. Soc., V57(7) 1974 p295

8. Carlson, D. E., K. W. Hang, and G. F. Stockdale, J. Amer. Ceram. Soc., V55(7) 1972 p337

9. Borom, M. P., J. Amer. Ceram. Soc., V56(5) 1973 p254

10. Brownlow, J. M., IBM Technical Report RC 7101, May 3, 1978

11. Skoog, D. A., and D. G. West, Fundamentals of Analytical Chemistry, 2nd Ed., Holt, Rinehart, and Winston, NY 1969

12. Arata, Y., A. Ohmori, S. Sano, and I. Okamoto, Trans. JWRI, 13(1) 1984 p35

Silicon Fusion Bonding for Pressure Sensors

Kurt Petersen, Phillip Barth, John Poydock,
Joe Brown, Joseph Mallon Jr., Janusz Bryzek

NovaSensor
1055 Mission Court
Fremont, CA 94539
(415) 490-9100

ABSTRACT

Two novel processes for fabricating silicon piezoresistive pressure sensors are presented in this paper. The chips described here are used to demonstrate an important new silicon/silicon bonding technique, Silicon Fusion Bonding (SFB). Using this technique, single crystal silicon wafers can be reliably bonded with near-perfect interfaces without the use of intermediate layers. Pressure transducers fabricated with SFB exhibit greatly improved performance over devices made with conventional processes. SFB is also applicable to many other micromechanical structures.

INTRODUCTION

Micromachining technology is often severely constrained 1) because of the structural limitations imposed by the traditional backside etching process and 2) because few practical methods have yet been demonstrated for true silicon/silicon direct bonding. Devices fabricated using techniques such as thermomigration of aluminum, eutectic bonding, anodic bonding to thin films of pyrex, intermediate glass frits and "glues" suffer from thermal expansion mismatches, fatigue and creep of the bonding layer, complex and difficult assembly methods, unreliable bonds and/or expensive processes. In addition, none of these processes provide the performance and versatility required for advanced micromechanical applications. For example, the (backside) cavity sidewalls slope outward from the diaphragm in anisotropically etched pressure sensor chips, forcing the overall chip area to be substantially larger than the active sensing diaphragm area. Another problem area is that narrow, hermetic gaps between two single crystal wafers are difficult to fabricate with precision because of the intermediate layers required. Finally, the most exciting class of new micromechanical structures, which have recently been developed, are actually thin film movable structures that are incompatible with current silicon/silicon bonding techniques.

Workers have addressed these problem by fabricating beams [1], diaphragms [2], and (recently) more complex structures such as mechanical springs and levers [3] from polycrystalline silicon using sacrificial-spacer etching. The silicon/silicon bonding technique described here has been previously discussed in several papers [4, 5, 6, 7]. SFB (Silicon Fusion Bonding) is a very powerful process for creating a wide range of **all single crystal** mechanical structures, that can replace polysilicon in many of the currently used sacrificial-etching methods. In this paper, several pressure sensor designs will be presented that illustrate the practical implications of this technology. Another paper at this meeting describes an acceleration sensor realized with the same processing techniques [8].

FABRICATION OF LOW PRESSURE SENSORS

One important application of silicon/silicon bonded sensors is ultra-miniature catheter-tip transducer chips for in vivo pressure measurements. A typical chip currently in use for this product is shown as Design 1 in Figure 1. This device has dimensions of 2400 µm by 1200 µm by 175 µm thick. It has a 650 µm square active diaphragm which is 8.0 µm thick. Diaphragm thickness is controlled to within 0.5 µm by an etch-stop process. The silicon wafer is anodically bonded to a glass constraint wafer which is only 125 µm thick. Much of the processing of this product requires lithography, etching, bonding, and general handling of these extremely thin wafers. A piezoresistive half-bridge is located on two edges of the diaphragm for a pressure sensitivity of 18 µV/V/mmHg. Linearity of these chips is better than about 1.2 % of the full scale output. These chips have been in production at NovaSensor for about a year and a half.

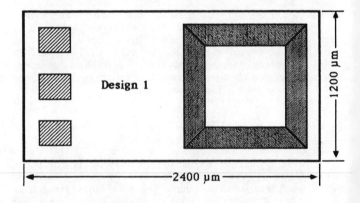

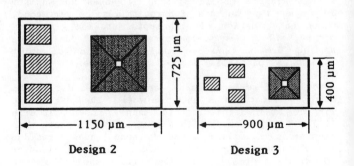

Figure 1 The three designs shown schematically here represent three generations of cathetar-tip sensors fabricated at NovaSensor. Design 1 uses a "conventional" silicon pressure sensor process sequence. Designs 2 and 3 use the SFB process.

A new generation of silicon fusion bonded (SFB) chips makes it possible to fabricate much smaller chips with equivalent or better overall performance. Schematic drawings of these chips are shown in Figure 1 and the fabrication procedure is outlined in Figure 2. The bottom, constraint substrate is first anisotropically etched with a square hole which has the desired dimensions of the diaphragm. In the most recent generation of chips, the bottom wafer has a (standard) thickness of 525 µm and the diaphragm is 250 µm square, so the anisotropic etch forms a pyramidal hole with a depth of about 175 µm. At the same time the pattern for these holes is defined, an alignment pattern is produced on the backside of the wafer in a double-sided aligner. Next, the etched constraint wafer is SFB bonded to a top wafer consisting of a p-type substrate with an n-type epi layer. The thickness of the epi layer corresponds to the required thickness of the final diaphragm for the sensor.

The bulk of the second wafer is removed by a controlled-etch process, leaving a bonded-on single crystal layer of silicon which forms the sensor diaphragm. Next, resistors are ion-implanted, contact vias are etched, and metal is deposited and etched. These patterns are aligned to the buried cavity using a double-sided mask aligner, referenced to the marks previously patterned on the backside of the wafer at the same time as the cavity. All these operations have a high yield because they are performed on wafers which are standard thickness. In the final step, the constraint wafer is ground and polished back to the desired thickness of the device; about 140 µm. The bottom "tip" of the anisotropically etched cavity is truncated during this polishing operation, thereby exposing the backside of the diaphragm for a gauge pressure measurement.

In an optional configuration, the initial cavity need not be etched to completion. Figures 3 and 4 show SEM's of a device with a sealed reference cavity only 30 µm deep. This design can be directly compared to the polysilicon sacrificial-layer pressure sensor techniques currently under investigation [2].

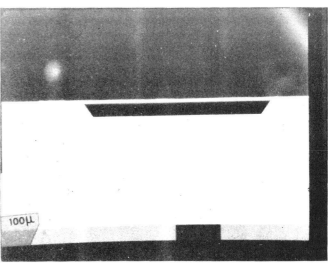

Figure 3 SEM cross-section of a test structure consisting of a 30 µm deep cavity anisotropically etched in the constraint wafer, and a 6 µm thick, 400 µm wide diaphragm suspended over the reference cavity. The fabrication procedure follows the general sequence outlined in Figure 1.

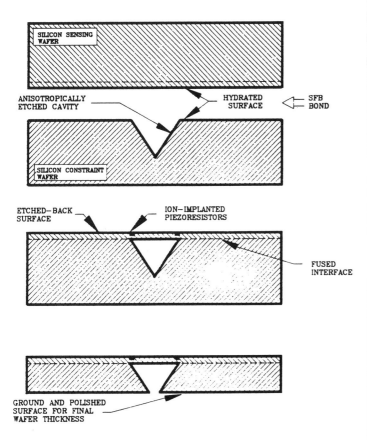

Figure 2 Fabrication process of SFB bonded low pressure sensor suitable for ultra-miniature cathetar-tip applications.

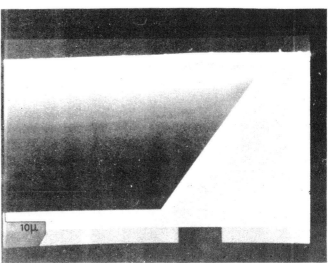

Figure 4 Magnified SEM cross-section of the bond region, reference cavity, and 6 µm diaphragm in Figure 2.

Despite the fact that all dimensions of this chip are about half those of the conventional chip described above, the pressure sensitivity of the SFB chip is identical to the larger device and its linearity is actually improved to 0.5% of full scale. The implementation of SFB technology increases wafer yield because the wafers are not as thin (fragile) during most wafer processing steps. In addition, of course, the small size of these chips is advantageous since each wafer contains almost 16,000 chips.

A comparison of conventional and SFB technology (as applied to ultra-miniature catheter-tip transducers) is shown in Figure 5. For the same diaphragm dimensions and the same overall thickness of the chip, the SFB device is almost 50% smaller. In this special application, chip size is critical. SFB fabrication techniques make it possible to realize extremely small chip dimensions, which may permit this catheter-tip pressure sensor to be used with less risk to the patient in many situations.

FABRICATION OF HIGH PRESSURE SENSORS

High pressure absolute sensors have been constructed in a modified SFB process to create much thicker diaphragms and to provide a larger mass of silicon in the constraint wafer, thereby optimizing materials compatibility and minimizing thermal mismatch problems.

First, the oxidized constraint wafer is aligned and exposed on both sides. The bottom surface has a "marker" pattern which will be used as an alignment reference later in the process in a manner similar to the low pressure sensor. The top surface has a round cavity pattern which will correspond to the diameter of the sensing diaphragm. After the oxide is wet etched, the silicon is plasma etched in CF_4 or anisotropically etched in KOH or EDP on both sides of the wafer simultaneously to a depth of about 10 µm.

Next, all photoresist and oxide layers are stripped from the constraint wafer and the top surface of this wafer (the surface with the round depression) is bonded to another n-type wafer at 1100°C. This operation is illustrated in the top part of Figure 6. After bonding, the top wafer is mechanically ground and polished back to a thickness corresponding to the desired pressure range. For a 900 um diaphragm with a thickness of 200 µm, this chip will result in a pressure sensor with a full scale output of 130 mV at 4000 psi. Linearity is exceptional at about 0.2% of full scale.

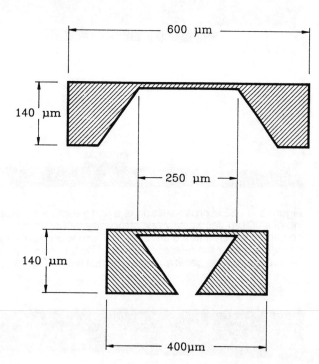

Figure 5 Comparison of miniature, low pressure silicon sensors fabricated using a "conventional" process, and the SFB process described here. For the same diaphragm dimensions and design groundrules, the SFB process results in a chip which is 50% smaller than the conventional process.

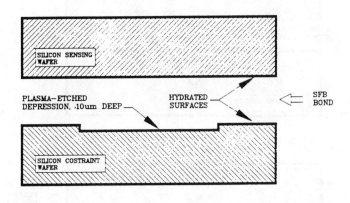

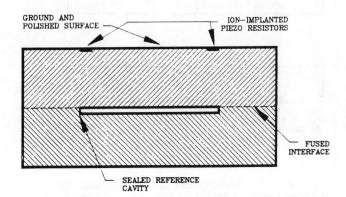

Figure 6 Fabrication process for SFB bonded high pressure sensors suitable for pressure ranges from 1,000 psi to 10,000 psi.

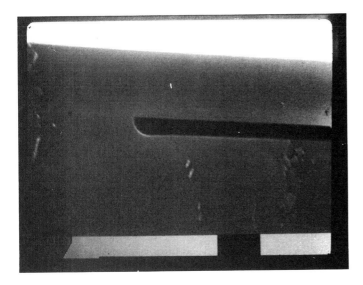

Figure 7 SEM cross-section of a silicon/ silicon bonded reference cavity of the type used in the high pressure sensor. The narrow cavity is only 2 μm deep in this figure. After plasma-etching the shallow cavity, the top wafer is SFB bonded as outlined in Figure 6.

Finally, an insulator is deposited, piezoresistors are implanted and annealed, contact holes are opened, and metal is deposited and etched. This completes the process for the pressure sensor. No back-side etching and no anodic bonding are required. In addition, 100% of the bulk of the chip is single crystal silicon. A monolithic single crystal silicon sensor chip provides important performance advantages in temperature coefficient of offset. The SFB chip, a total silicon thickness of over 600 μm, exhibits a very low temperature coefficient of bridge offset of about 0.3%/100°C of full scale output. A similar chip made with conventional backside etching and anodic glass/silicon bonding for the constraint, exhibits a typical TC of zero higher by a factor of 3.

CONCLUSION

The successful development of these two products shows that SFB bonding can be applied on a commercial basis. Not only are the processes employed here simpler than those currently used to build conventional pressure transducers for the same applications, but the yields are higher, the costs are lower, and the chip performance is improved compared to conventional technologies.

Beyond the two devices shown here, the potential of SFB bonding in other micromechanical structures is enormous. This process eliminates most of the disadvantages of previous silicon/silicon bonding methods. It is hermetic; it does not require any intermediate bonding layers; it accurately preserves any pattern previously etched in either or both bonded wafers; it has a high yield strength, as much as double that of anodic bonds; it can be used to create vacuum reference cavities; it can be used in place of sacrificial layer technology in many applications; it eliminates thermal expansion and Young's modulus mismatches in bonded wafer structures. The interface itself is a true single-crystal/single-crystal boundary. We have demonstrated, for example, that pn junctions with low leakage and sharp break-down voltages can be formed by bonding a wafer with a p-type diffusion to an n-type wafer.

We have only begun to exploit this extremely powerful new technology. During the next few years, silicon fusion bonding will revolutionize the field of silicon microstructures and will have a vast impact on high performance silicon microsensors.

ACKNOWLEDGEMENTS

The authors would like to acknowledge the important contributions of Ted Vermeulen, Rose Scimeca, Van Nguyen, and Terry Cookson in the development of SFB for silicon sensor applications.

REFERENCES

1] R.T. Howe and R.S. Muller, **Integrated Resonant-Microbridge Vapor Sensor**, International Electron Devices Meeting, December 1984, pg. 381.

2] H. Guckel, D.W. Burns, C.R. Rutigliano, D.K. Showers, and, J. Uglow, **Fine Grained Polysilicon and Its Application to Planar Pressure Transducers**, International Conference on Solid-State Sensors and Actuators, June 1987, pg. 277.

3] L.S. Fan, Y.C. Tai, R.S. Muller, **Pin Joints, Gears, Springs, Cranks, and Other Novel Micromechanical Structures**, International Conference on Solid-State Sensors and Actuators, June 1987, pg. 843.

4] J. Lasky, S. Stiffler, F. White, and J. Abernathey, **Silicon-on-Insulator (SOI) by Bonding and Etch-back**, International Electron Devices Meeting, December 1985, pg. 684.

5] H. Ohashi, K. Furukawa, M. Atsuta, A. Nakagawa, and K. Imamura, **Study of Si-Wafer Directly Bonded Interface Effect on Power Device Characteristics**, International Electron Devices Meeting, December 1987, pg. 678.

6] L. Tenerz, and B. Hok, **Silicon Microcavities Fabricated with a New Technique**, Electronics Letters, 22, pg. 615, (1986).

7] J. Ohura, T. Tsukakoshi, K. Fukuda, M. Shimbo, and H. Ohashi, **A Dielectrically Isolated Photodiode Array by Silicon-Wafer Direct Bonding**, IEEE Electron Device Letters, EDL-8, pg. 454, (1987).

8] P.W. Barth, F. Pourahmadi, R. Mayer, J. Poydock, K. Petersen, **A Monolithic Silicon Accelerometer with Integral Air Damping and Over Range Protection**, Solid-State Sensors Workshop, Hilton Head, S.C., 1988.

Low-temperature Silicon-to-silicon Anodic Bonding with Intermediate Low Melting Point Glass

MASAYOSHI ESASHI, AKIRA NAKANO, SHUICHI SHOJI and HIROYUKI HEBIGUCHI

Department of Electronic Engineering, Tohoku University, Aza Aoba, Aramaki, Aoba ku, Sendai 980 (Japan)

Abstract

Room-temperature silicon-to-silicon bonding has been performed. It is an electrostatic bonding using sputtered low melting point glass as an intermediate layer. Wafers can be bonded at room temperature with an applied voltage of about 50 V. This technique is useful for the fabrication of intelligent sensors and microelectromechanical systems.

Introduction

Microsensors and microelectromechanical devices have been fabricated by micromachining [1]. An etching technique, especially anisotropic etching of silicon, realizes a fine structure. Bonding substrate to substrate is also a key technology in micromachining. Glass-to-silicon anodic bonding has generally been used for making pressure sensors or other devices, [2, 3]. Silicon-to-silicon bonding realizes more sophisticated microdevices, because fine structures can be made on both substrates.

Electrostatic bonding with an intermediate sputtered Pyrex glass film [5] and thermal bonding with sputtered low melting point glass have been reported as silicon-to-silicon bonding methods. These methods, however, have the following disadvantages. The former has a problem of breakdown during the bonding process. The latter requires a bonding temperature higher than 600 °C. Silicon-to-silicon direct bonding, with or without SiO_2 in between, is also possible [4], but it requires temperatures higher than 1000 °C. A low-temperature and low-voltage bonding method will give much more flexibility in fabrication processes.

We have investigated a new silicon-to-silicon bonding method, which involves electrostatic bonding with an intermediate thin layer of a sputtered low melting point glass, and can be performed at room temperature.

Anodic Bonding

In anodic bonding, the electrostatic force is generated by the applied voltage at the gap between the silicon and the glass layer. As the d.c. voltage is applied, a polarized region in the glass extends from the glass–silicon interface because mobile ions are moved by the electric field. The charge and the voltage across the silicon–glass–silicon structure are shown in Fig. 1. The voltage across the polarized region, V_p, is given from Poisson's equation by

$$V_p = \frac{\rho X_p^2}{2\varepsilon' \varepsilon_0} \quad (1)$$

where ρ is the charge density in the glass and $\varepsilon' \varepsilon_0$ is the permittivity of the glass. The charge per unit area of the polarized region, σ_s, is given by

$$\sigma_s = \rho X_p \quad (2)$$

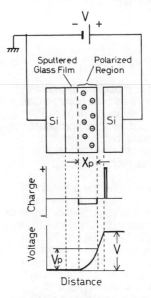

Fig. 1. Silicon-to-silicon anodic bonding using sputtered low melting point glass.

The thickness of the polarized region, X_p, is

$$X_p^2 = \frac{2\varepsilon'\varepsilon_0 V_p}{\rho} \qquad (3)$$

The electrostatic force P between the glass and the silicon is given by

$$P = \frac{1}{2}\varepsilon_0 E^2 = \frac{1}{2}\frac{\sigma_s^2}{\varepsilon_0} \qquad (4)$$

where E is the electric field in the gap.

Combining eqns. (2), (3) and (4), the electrostatic force P is given as

$$P = V_p \rho \varepsilon' \qquad (5)$$

In silicon-to-silicon bonding with an intermediate glass film, V_p cannot be large enough, because it is limited by breakdown in the glass film, and the electrostatic force P is restricted by the V_p value. Equation (5) means that a large breakdown voltage, large charge density and large permittivity are required for the intermediate glass film.

Intermediate Layer

A soft glass can be deformed to fill the roughness at the surface, so it is considered to be suitable as the intermediate layer for the low-temperature bonding. That is why we chose a low melting point glass (Iwaki Glass Co., #7570) that is soft at low temperature. The properties of Iwaki #7570 glass are compared with those of Pyrex glass (#7740) in Table 1. The low softening point of #7570 ensures its softness, and the large volume resistivity contributes to a large breakdown voltage. In addition, the large relative permittivity is also an advantage.

Low melting point glass was deposited on a silicon wafer by magnetron r.f. sputtering. Thus sputtering was carried out at 30% oxygen concentration in argon ambient. The total pressure was 6 mTorr. Film thicknesses ranging from 0.5 μm to 4 μm were used for this study. The surface roughness of the deposited film was within 200 Å peak to peak.

Softening point (4.5 × 10⁷ poise)   10¹⁷ Ω cm   10¹⁵ Ω cm

TABLE 1. Properties of #7570 glass and #7740 glass

Property	#7570	#7740
Strain point (4×10^{14} poise)	342 °C	510 °C
Softening point (4.5×10^7 poise)	440 °C	821 °C
Volume resistivity	10^{17} Ω cm	10^{15} Ω cm
Relative permittivity	15	4.6

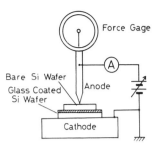

Fig. 2. Schematic of bonding apparatus.

Bonding Process

The bonding apparatus is illustrated in Fig. 2. A silicon wafer was placed on top of the other silicon wafer with a low melting point glass film. The wafer can be aligned by using an IR TV camera. Some pressure was applied to obtain enough contact at the surfaces. A negative d.c. voltage was applied to the silicon wafer with a glass film on it for 10 min, and the current was monitored.

Experimental Results

An example of the current during the bonding process is shown in Fig. 3. When voltage was applied, a transient current was observed. It decreased, but a small current remained.

Bonding took place for all thicknesses of the intermediate layer from 0.5 μm to 4 μm. The minimum bonding voltage was from 30 to 60 V.

The effect of the pressure and the voltage on the bonding with glass 2 μm thick is shown in Fig. 4. The relation between the thickness of the glass film and the minimum bonding voltage at a pressure of 160 kPa is shown in Fig. 5.

The bonding strength is measured by a tensile test as shown in Fig. 6. It depends strongly on the surface roughness and particles, but was larger than 1.5 MPa.

In addition, this new method could bond the glass-coated silicon to an aluminum film or an

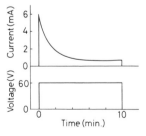

Fig. 3. Applied voltage and current during the bonding.

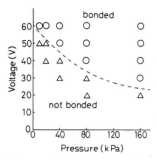

Fig. 4. Effect of applied voltage and pressure on the bonding (glass thickness is 2 μm).

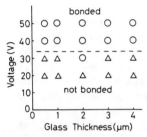

Fig. 5. Effect of applied voltage and glass thickness on the bonding (applied pressure is 160 kPa).

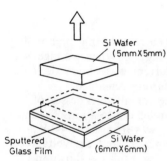

Fig. 6. Tensile test.

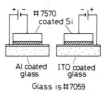

Fig. 7. Examples of low-temperature bonding: Si to Al-coated glass and Si to ITO-coated glass.

Fig. 8. The bonding interface observed through ITO-coated glass.

indium–tin oxide (ITO) film on a glass substrate (Fig. 7). The glass substrate was Corning non-alkali glass #7059, which has a different thermal expansion coefficient (4.6 ppm/°C) from silicon (2.5 ppm/°C). The bonded interface can be observed through the ITO-coated glass (Fig. 8). In Fig. 8, the dark part was bonded, but the interference fringe part was not.

The results obtained are as follows:

(1) The larger the pressure applied, the lower the bonding voltage.

(2) Bonding is performed at 60 V independently of the pressure (Fig. 4).

(3) No dependency of the glass thickness on the minimum bonding voltage was observed (Fig. 5).

(4) The silicon wafer with the glass film could be bonded to a bare silicon wafer, aluminum-coated glass and ITO-coated glass.

Discussion

The low viscosity of the #7570 glass at low temperature promotes the deformation of the glass, so that it expands in the intimate contact area. Bond formation is attributed to the large electrostatic force at the interface, and to the field-induced reaction of silicon with the oxygen ions from the glass. The large relative permittivity of #7570 contributes to the electrostatic force.

Other mechanisms, such as the local heating at the interface caused by the current, have been considered and should be studied.

Since the yield of the bonding is influenced strongly by the surface flatness and cleanness, particle-free surfaces are important for bonding large wafers.

Conclusions

This bonding method is not only safe for active devices but is also free of thermal stress during the bonding process. Hence, this method enables materials with different thermal expansion coefficients to be bonded and many applications will be expected.

Acknowledgements

The authors thank Professor T. Wakiyama and Professor T. Ohmi for their encouragement.

References

1 C. D. Fung, P. W. Cheung, W. H. Ko and D. G. Fleming (eds.), *Micromachining and Micropackaging of Transducers*, Elsevier, Amsterdam, 1985, pp. 41–61.
2 T. A. Knecht, Bonding techniques for solid state pressure sensors, *Tech Digest, 4th Int. Conf. Solid-State Sensors and Actuators (Transducers '87), Tokyo, Japan, June 2–5, 1987*, pp. 95–98.
3 T. R. Anthony, Anodic bonding of imperfect surfaces, *J. Appl. Phys.*, 54 (1983) 2419–2428.
4 M. Shimbo, K. Fukukawa, K. Fukuda and K. Tanzawa, Silicon-to-silicon direct bonding method, *J. Appl. Phys.*, 60 (1986) 2987–2989.
5 A. A. Brooks and R. P. Donovan, Low-temperature electrostatic silicon-to-silicon seals using sputtered borosilicate glass, *J. Electrochem. Soc.*, 119 (1972) 545–546.

Fusing Silicon Wafers with Low Melting Temperature Glass

LESLIE A. FIELD and RICHARD S. MULLER

Berkeley Sensor & Actuator Center*, Department of Electrical Engineering and Computer Sciences, and the Electronics Research Laboratory, University of California, Berkeley, CA 94720 (U.S.A.)

Abstract

This paper describes research on a low-temperature technique for fusing two silicon wafers together. The work has focused on integrating the bonding process with MOS IC processing. A seal that is apparently hermetic is achieved at 450 °C between wafers which have gone through the individual processing steps needed to fabricate MOS capacitors and electrical leads through the sealed package. In order to use this process to bond at low temperature, the bonding surfaces must be smooth, clean, and phosphorus free.

1. Introduction

Silicon–silicon bonding, the process of fusing two silicon wafers together, is of growing importance for microstructure fabrication [1], silicon-on-insulator processes [2] and wafer-scale packaging [3].

Bonding of silicon wafers with oxide by purely thermal means was first reported in 1985 by a group at IBM as a step for producing silicon-on-insulator (SOI) [4]. A variation on this technique using low melting temperature glasses appeared promising to us for the packaging and fabrication of sensors. A brief report of the basic technique for low-temperature oxide bonding with boron-doped glass has been presented by our group at an earlier conference [5], while more complete processing details are available elsewhere [6].

2. Experimental

To achieve bonding at low temperature, we first grow approximately 1 μm of undoped thermal oxide and deposit approximately 0.1 μm of boron-containing glass on a silicon wafer. The two wafers to be bonded are placed with their flats aligned and a 120 g quartz weight is put on top of them. Experiments are usually performed in our laboratory's annealing and wet-oxidation Tylan furnaces, although we have demonstrated that bonding can also be successfully conducted on a hot plate. All experiments include a 30 min step at the test temperature.

The boron-glass layer has been demonstrated to bond strongly at 450 °C to clean silicon or to thermal silicon dioxide; the layer is only required to coat one of the two wafers. A summary of the bonding procedure is given in Table 1.

Our boron-oxide deposition uses 100 mm solid-source boron wafers (type GS-139, manufactured by Owens-Illinois, Inc.) at atmospheric pressure in a Tylan furnace. The silicon wafers are at a surface-to-surface distance of 2.5 mm from the solid-source doping wafers during deposition. When new solid-source wafers are installed in the furnace, they are pretreated at high temperature in a wet ambient for at least 60 h at 1075 °C with gas flows of 2.5 slpm N_2, 2.5 sccm O_2 and 5 sccm of 10% H_2 in N_2 and then idled at 600 °C for approximately 16 h before being used.

Boron-glass-coated silicon wafers must be held at high temperature or they must be used for bonding immediately after the boron glass deposition step if good low-temperature bonding is to be achieved. When a boron-glass-coated wafer is allowed to stand in room air for any appreciable time, a crystalline material forms on the surface, which acts to prevent bonding. We observe this material on the wafer surface within hours of removal of the wafer from the furnace; we believe it to be boric acid (its crystalline appearance and our thermodynamic calculations are consistent with this assumption). The material probably begins to form as soon as the wafers are removed

TABLE 1. Boron-glass bonding: typical procedure

1.0	Deposit boron glass on bonding wafer: 1 h at 1075 °C 2.5 slpm N_2, 2.5 sccm O_2, 5 sccm H_2/N_2
2.0	Bond wafers
2.1	Clean wafer which has no deposited boron glass
2.2	Place the two wafers face-to-face on quartz holder
2.3	Bonding: 30 min at 450 °C, O_2 or H_2/N_2

---

*An NSF/Industry/University Cooperative Research Center.

from the furnace. Therefore, the wafers are placed in the bonding furnace as soon as they can be loaded into the required wafer holder. This takes approximately five minutes.

The effectiveness of bonding is determined by scribing the top wafer of a bonded pair and breaking. If the break occurs through both wafers, and one wafer cannot be peeled apart from the other, the bond is considered successful. Typically, many scribe and break tests are performed on a bonded-wafer pair, resulting in the pair being broken into many small pieces. The bonding percentages quoted in this report identify the percentage of the wafer area that remains bonded after the scribe-and-break test.

The seals appear to be hermetic. We checked for large leaks by pressurizing a transparent lucite vessel containing fragments of bonded wafers which contained cavities made of etched trenches. After pressurizing to 60 PSIG for 1.5 h, we rotated the vessel so that the wafer fragments were submerged in a low-viscosity oil, and then the vessel was depressurized. If there were large leaks, we would have observed bubbles emerging from the bonded interface through the oil. No evidence of leaks was seen. We also checked for small leaks by testing He-filled sealed trenches with a He leak detector, which could detect a leak as low as $5 \times 10^{-4}\%$ of the trench volume per minute. Again, no evidence of a leak was detected.

No degradation of seal quality is evident when bonded wafer fragments are immersed in water at room temperature for a few weeks or when they are immersed in piranha (H_2SO_4/H_2O_2) or NH_3OH overnight. The bonded oxide–oxide interface of a wafer pair survived immersion in a hot (105 °C) EDP (ethylene diamine, pyrocatechol, water and pyrazine) etch overnight.

3. Results

The major requirements for wafer bonding at low temperature (450 °C) using the boron-glass process are:

(1) the surface must be phosphorus free;
(2) the surface must be free of particulate matter;
(3) a smooth surface of phosphorus-free silicon or silicon-dioxide must be in contact with the boron-glass bonding wafer.

3.1. Bonding to SiO_2

With a 1 μm thermal oxide layer underlying the boron glass on one wafer and a thermal oxide layer on the other, a 30 min treatment at 450 °C reproducibly gives strong bonding over most of the area of the silicon-wafer pair. It seems to make no difference in bonding to SiO_2 whether the bonding is carried out in wet or dry O_2, forming gas (H_2/N_2) or vacuum. We have bonded to wafers covered with very thin (~30 nm) or rather thick (2 μm) thermally grown SiO_2. The bonding can be conducted quite effectively at temperatures higher than 450 °C. From one test, bonding at 430 °C also seems to give acceptable results. However, bonding with the boron glass at 420 °C yielded very little bonded area in two separate tests.

When an oxide contains phosphorus, the temperature required for bonding is increased. According to phase diagrams for the $SiO_2/B_2O_3/P_2O_5$ ternary glass system, small amounts of P_2O_5 in the presence of B_2O_3 will lead to the formation of a BPO_4 phase. This phase has a melting point that is several hundred degrees centigrade higher than that of B_2O_3. The P_2O_5 phase itself has a melting point even higher than that of the BPO_4 phase [7, 8].

We have been unable to bond wafers coated with undoped LPCVD oxide (LTO) at 450 °C by the methods described. We have also been unsuccessful in bonding to a PECVD oxide, and have found very limited success bonding to boron-doped spin-on-glass (SOG) at 450 °C. Bonding took place on only a small area of the interface between two wafers in two out of seven tests using three different spin-on-glasses with varying application, bake and cure methods. The best results were obtained with Allied's B50 SOG, with a mild bake at 120 °C and either no cure or a cure at 450 °C.

3.2. Bonding to Silicon

A bare silicon wafer can be bonded to a boron-glass-coated wafer. It seems likely that the thin layer of native oxide that is known to grow rapidly under ambient conditions plays a role in the bonding.

In a furnace dedicated to phosphorus-free silicon deposition, smooth amorphous silicon can be deposited to which bonding is good. However, if the furnace has been contaminated with phos-phorus from running wafers with exposed phosphorus-doped oxide, the amorphous silicon subsequently deposited will not bond at 450 °C.

3.3. Bonding to Wafers that have Undergone Prior Processing

Wet-chemical etching of LPCVD Si_3N_4, phosphorus-doped polysilicon, d.c. magnetron sputtered aluminum/2% silicon, LPCVD-deposited phosphorus-doped oxide, or PECVD-deposited oxide films, followed by a piranha clean and HF dip, returns the wafers to the condition required for good bonding. When the LPCVD Si_3N_4 or phosphorus-doped polysilicon films are plasma etched and cleaned in piranha and HF dipped, the bonding is poor. This may be due to surface roughening or contaminants introduced by the plasma-etching process. When plasma polysilicon etching is followed by wet-chemical silicon etching and wafer cleaning, and additional surface SiO_2 is wet etched, the bonding ability is recovered. With unprotected aluminum on the non-bonding areas of a wafer, piranha clean cannot be used and some debris remains on the surface (presumably silicon from the 2% Si/Al film), making bonding unsuccessful.

A typical photolithographic process is outlined in Table 2. The steps involving HMDS, photoresist, developer and acetone are compatible with subsequent low-temperature bonding. However, bonding is adversely affected if the photoresist has been hard baked.

Bonding is found to be excellent after LOCOS oxidation if the process shown in Table 3 is followed. Hard bake and plasma descum of the photoresist are avoided. A smaller aperture opening on the wafer stepper is used during this step to ensure that the recessed inter-die areas will extend slightly into the die area, avoiding any spikes of material at the die edges. These spikes result from patterning misalignments and prevent bonding. In the nitride plasma-etch step both the front and back of the wafers are etched so that both sides of the wafers are oxidized to equalize stress and keep the wafers flat. The field oxidation following the plasma etch leaves the wafer with its required smooth surface for bonding.

Wafers that have undergone ion implantation with arsenic, nitride deposition, photoresist hard baking and plasma etching have been successfully bonded by the process. Imperfections left by photoresist or plasma processing on the oxide surface are probably removed by the subsequent oxidation. The process flow is summarized in Table 4. Wafers implanted with phosphorus rather than arsenic have not bonded successfully.

Successful bonding has been carried out on wafers after windows for contacts were plasma etched and all phosphorus-doped silicon removed. These wafers had 1 μm of thermal oxide on them initially. During the contact-window etch, this thermal oxide that will be used for the bonding is underneath the polysilicon, and so is not roughened or contaminated by the etching. The procedure used in this experiment is shown in Table 5.

TABLE 2. Photoresist process

1.0	Pattern
1.1	Clean wafers
1.2	Dehydrate wafers, 120 °C for 10 min
1.3	HMDS: 1.5 min
1.4	Spin photoresist; includes 1 min bake at 120 °C
1.5	Expose: GCA 6200 10 × Wafer Stepper
1.6	Develop
1.7	Descum: O_2 plasma, 300 mTorr, 50 W, 1 min
1.8	Hard bake: 20 min 120 °C in air
2.0	Etch or implant as process flow requires
3.0	Remove photoresist and clean wafers

TABLE 3. Modified LOCOS process

1.0	Grow 100 nm pad oxide at 950 °C
2.0	Deposit 500 nm LPCVD nitride
3.0	Pattern for capacitors, <100% aperture opening
4.0	Nitride plasma etch in SF_6/He at 50 W
5.0	Remove photoresist
6.0	Grow 2 μm field oxide at 1100 °C
7.0	Nitride removal with hot phosphoric acid etch
8.0	Pattern for circuitry recess
9.0	Oxide etch: 6 min 1/5 BHF
10.0	Remove photoresist and clean wafers

Result: 95% bonding at 450 °C in O_2

TABLE 4. Ion implantation for leads

1.0	Grow 40 nm pad oxide at 950 °C
2.0	Deposit 150 nm LPCVD nitride
3.0	Pattern for implantation. Hard bake and plasma descum
4.0	Nitride plasma etch
5.0	Leads implant with As, 55 keV, 5.0×10^{15}/cm^2
6.0	Oxidation (40 nm) and activation at 950 °C
7.0	Pattern for capacitors. Hard bake and plasma descum used
8.0	Nitride plasma etch
9.0	Remove photoresist and clean wafers
10.0	Grow 1 μm field oxide at 1000 °C
11.0	Pattern for recess. Hard bake and plasma descum
12.0	Oxide wet etch for circuitry recess: 1/5 BHF 4.5 min
13.0	Remove photoresist

Improved processing: no hard bake or plasma descum. Repeat steps 10.0 (grow 300 nm more oxide)–13.0.

14.0	Nitride removal with hot phosphoric acid etch (155 °C)
15.0	Pattern for leads etchdown
16.0	Oxide wet etch for leads etchdown in 1/5 BHF
17.0	Remove photoresist and clean wafers

TABLE 5. Contact-hole patterning

1.0	Phosphorus-doped LPCVD polysilicon deposition at 650 °C
2.0	Pattern for contact holes
3.0	Polysilicon plasma etch: 300 W, 280 mTorr, He/CCl_4/O_2
4.0	Oxide plasma etch
4.1	Fast etch: 850 W, 2.8 Torr, He/CHF_3/CF_4
4.2	Slow etch: 700 W, 3.0 Torr, He/CHF_3/CF_4
5.0	Remove photoresist and clean wafers
6.0	Phosphorus-doped LPCVD polysilicon deposition at 650 °C
7.0	Wet etch all polysilicon
8.0	Clean wafers

4. Discussion

Boron-glass bonding seems to occur via a simple melting of the glass once its melting temperature (~450 °C) has been reached. The fact that the bonding does not work at low temperatures for wafers with significant phosphorus content provides a strong argument in support of melting being the phenomenon responsible for boron-glass bonding. The lowest melting point in the $B_2O_3/P_2O_5/SiO_2$ system is hundreds of degress centigrade higher than that for B_2O_3/SiO_2. Above 450 °C, a liquid phase exists for B_2O_3/SiO_2 mixtures. The temperature must exceed 900 °C in order for a liquid phase to exist for $B_2O_3/P_2O_5/SiO_2$, as, with even a small amount of P_2O_5 in the system, a high-melting BPO_4 phase is formed. If it is only necessary to contaminate the top monolayer of B_2O_3/SiO_2 sufficiently to form BPO_4, (where the actual adhesion between the wafers is made to happen), the amount of phosphorus that could contaminate the system is small.

5. Conclusions

The most severe limitation of the low-temperature boron–glass bonding process is the constraint that the processing be phosphorus free. If phosphorus is in the system, bonding can still be made to occur, but the temperature required is much higher (~900 °C).

The boron glass must be deposited on the sealing wafer immediately prior to bonding. This means that any structures on the capping wafer must be rugged enough to be subjected to the boron-doping process.

Plasma processing and hard baking of photoresist cause some disturbance on the SiO_2 surface, which makes bonding difficult to achieve. If photoresist hard baking or plasma processing (except for plasma etching of contact holes) are used with no subsequent oxidation step, satisfactory bonding may not be possible.

Arsenic rather than phosphorus should be used for diffused leads. Amorphous silicon can be used, but it must have a smooth surface and can only be doped with boron or arsenic.

If phosphorus-containing polysilicon is present, it must be capped with nitride to protect against phosphorus contamination of the bonding.

Acknowledgements

The authors appreciate the ready help and advice of Katalin Voros, Bob Hamilton and the UC Berkeley Microlab staff. The advice and encouragement given by Professors Roger T. Howe and Richard M. White is also appreciated.

References

1 K. Petersen, P. Barth, J. Poydock, J. Brown, J. Mallon, Jr. and J. Bryzek, Silicon fusion bonding for pressure sensors, *Solid State Sensor and Actuator Workshop, Hilton Head Island, SC, U.S.A., June 6–9, 1988*, pp. 144–147.

2 R. Black, S. Arthur, R. Gilmore, N. Lewis, E. Hall and R. Lillquist, Silicon and silicon dioxide thermal bonding for silicon-on-insulator applications, *J. Appl. Phys., 63* (1988) 2773–2777.

3 S. Shoji and M. Esashi, Micro-pump for integrated chemical analyzing systems, *Tech. Digest, 7th Sensor Symp., Tokyo, Japan, May 1988*, pp. 217–220.

4 J. Lasky, S. Stiffler, F. White and J. Abernathey, Silicon-on-insulator by bonding and etch-back, *Int. Electron Devices Meet., Washington, DC, U.S.A., Dec. 1985*, pp. 684–687.

5 L. Field and R. Muller, Low-temperature silicon–silicon bonding with oxides, *Ext. Abstr., 171st Meet. Electrochem. Soc., Philadelphia, PA, U.S.A., May 1987*, Vol. 87-1, pp. 333–334.

6 L. Field, Low-temperature silicon–silicon bonding with oxides, *Master's Rep.*, Department of Electrical Engineering and Computer Sciences, University of California at Berkeley, Berkeley, CA, U.S.A., Dec. 1988.

7 E. Levin, C. Robbins and H. McMurdie, *Phase Diagrams for Ceramists*, American Ceramic Society, Columbus, OH, 1964, pp. 142, 262, 263.

8 E. Levin, C. Robbins and H. McMurdie, *Phase Diagrams for Ceramists 1969 Supplement*, American Ceramic Society, Columbus, OH, 1969, p. 98.

Silicon Fusion Bonding for Fabrication of Sensors, Actuators and Microstructures*

PHILIP W. BARTH

Vice President/Chief Scientist, NovaSensor, 1055 Mission Court, Fremont, CA 94539 (U.S.A.)

Abstract

Silicon fusion bonding (SFB) is the joining together of two silicon wafers without the use of intermediate adhesives. The technology has been used to fabricate silicon-on-insulator (SOI) substrates and silicon power devices, and also has wide applications in the fabrication of silicon sensors, actuators and other microstructures. This paper reviews the development and current status of SFB. A history of the technology from the early 1960s to the present is presented. Process techniques necessary to incorporate SFB successfully into silicon micromachining processes are discussed, and examples of successful SFB structures are presented. Comparisons to competing techniques are made, and the potential for future development of SFB structures is discussed.

Introduction

Two pieces of single-crystal silicon, with or without surface layers of thermally grown SiO_2, can be fused together at high temperature to form a bi-crystal structure. This simple fact, rediscovered several times in the past three decades, has only recently been wholeheartedly adopted for fabricating useful silicon structures, and forms the basis for exciting new areas of investigation in silicon microstructures. This paper discusses the work accomplished to date in this silicon fusion bonding technology and outlines some of the broad possibilities for future work.

The term silicon fusion bonding (SFB) has been introduced to avoid confusion with other silicon-to-silicon bonding techniques, which use polymer glues, melting glass layers, solid electrolyte glass layers or melting alloys. Unlike those techniques, SFB is completely compatible with subsequent high-temperature process operations such as oxidation and diffusion, and it introduces little or no thermal stress because of the well-matched thermal expansion of the bonded layers. In addition, precisely machined cavities can be enclosed by SFB, dopant diffusion can occur through SFB interfaces and the interface region can be used as an etch-stop layer. In consequence, SFB can be used to create useful, stable, high-quality silicon microstructures.

The technique has been investigated for creating dielectrically isolated substrates for silicon-on-insulator technology, and for fabricating abrupt p–n junctions between lightly-doped silicon regions for power devices. In the author's view the economics of the technique are not favorable for very-large-scale integration (VLSI) as compared to competing techniques, but the economic considerations are favorable for both power devices and silicon microstructures.

The Sections below give a brief history of the technology, discuss the basic process techniques and present examples of sensors fabricated with SFB. The ultimate lower limits of pressure sensor size are discussed and are shown to be limited by bonding pad size. Comparisons are made to competing surface micromachining techniques, and arguments are made concerning the relative utility of SFB versus other techniques for silicon-on-insulator, power devices and microstructures.

A Short History

The first application of SFB known to the author occurred at NEC in the early 1960s [1], when it was used to bond discrete transistor chips together. If anecdotal evidence can be believed, the phenomenon of silicon wafers bonding together was rediscovered and ignored as a nuisance many times by many workers over several decades. In the 1970s IBM workers began looking seriously at the phenomenon [2], and in 1985 they published a seminal article on the technique for SOI at the International Electron Devices Meeting (IEDM) [3]. Almost simultaneously, Toshiba began publishing information on direct wafer bonding for power devices [4–7].

NovaSensor was founded in October 1985 and made an early commitment to develop SFB as a commercially viable technology for silicon micro-

---

*Invited paper.

machining. The credit for the development of SFB at NovaSensor must go in large measure to Kurt Petersen, who recognized the importance of the technology at the 1985 IEDM. Taking post-1985 publications as evidence, other organizations that took an early interest in the technology included the University of Uppsala [8] and Allied Signal [9].

NovaSensor began to publish the results of its SFB work at the 1988 Solid-State Sensors and Actuators Workshop [10, 11]. These publications stimulated interest in the technology for silicon micromachining, and new work in the field has spread to several research centers. Publications on SFB were also presented at the Transducers '89 Conference at Montreux, Switzerland [12, 13].

Process Basics

Bonding two silicon wafers together requires surface cleanliness, surface flatness, surface hydration and high temperature. The bonding chemistry is not yet completely understood, but its broad outlines are evident. The following is a compendium of the current conventional wisdom regarding the process [3–5, 7–9, 11], with very rough definition of the temperature ranges at which different processes occur.

Initially, both bonding surfaces must be polished, then treated to make them hydrophilic, for example, by boiling in nitric acid. This treatment leaves a high density of OH groups attached to surface silicon atoms. When the two bonding surfaces are placed in contact at room temperature, an immediate weak bond forms due to hydrogen bonding (similar in concept to the hydrogen bonds that hold together complimentary strands of DNA).

The bonded wafers are then heat treated. Both oxidizing and non-oxidizing ambients can be used. As the temperature rises, chemical reactions begin. The first, occurring at temperatures below approximately 300 °C, is the conversion of OH groups to water, loosely describable as:

$$Si-O-H + H-O-Si \rightarrow Si-O-Si + H_2O$$

(where the additional satisfied bonds of the silicon atoms are implicitly considered to be unaffected).

During the bonding process voids tend to form in some areas of the wafer and then to disappear. The generated water vapor pressure may be responsible for creating these voids. As the temperature increases further (approximately 300–800 °C) the water dissociates and additional oxygen becomes free to bond to silicon, while hydrogen diffuses through the silicon lattice. Gas pressure in any voids drops, and these voids tend to be pressed together by external atmospheric pressure.

(The situation is similar to the 'reactive sealing' phenomenon occurring in the fabrication of some poly-Si pressure sensors [14].)

At yet higher temperatures (800–1400 °C), oxygen tends to diffuse into the crystal lattice. It should be noted that a typical oxygen concentration in Czochralski-grown silicon can be 10^{18} cm^{-3}, so that the addition of a small amount of oxygen from the gap between wafers has little effect on wafer electrical properties. Finally, silicon atoms fill in any microscopic surface-roughness voids by thermal diffusion.

To the author's knowledge, no direct measurement of the bond tensile strength has been conducted. Burst pressures for a 6 mm diameter cavity in a chip 10 mm square by 1 mm thick bonded to another chip have been measured to be as high as 200 kg force cm^{-2} (approximately 20 MPa) [5]. Surface energy has been measured as high as 2300 erg cm^{-2} [9]. A promising observation conducted at NovaSensor is that burst pressures for silicon diaphragms fabricated using SFB are approximately equal to those for diaphragms of the same dimensions fabricated by traditional orientation-dependent etching, for pressure application from either side of the diaphragm. This observation indicates that the bond strength approaches the yield strength of single-crystal silicon, typically on the order of 1 GPa.

It should be noted that an alternative form of fusion bonding avoids the need for the hydration step and substitutes the use of an electrostatic attractive force applied at high temperature [15]. In this approach, a silicon dioxide layer is required on both wafer surfaces and the electrical fixturing required must operate at the bonding temperature. Strong bonding begins to occur around 1100 °C. The relative complexity of this process makes it economically less attractive than the hydrated-surface approach discussed above.

Surface cleanliness is perhaps the most important requirement for SFB. It is instructive to consider SFB in relation to the well-known silicon-to-glass anodic bonding technique [16, 17], in which the bonded-on glass is pulled into contact by electrostatic forces. The electrostatic anodic bonding force is both large and comparatively long range, and can pull the glass and silicon together even in the presence of a dust particle, leaving a small 'tent' area with the dust particle as the center pole.

In SFB the forces are both smaller and shorter range, so that the effect of dust particles is much more deleterious. Particles may prevent bonding altogether, or may cause large 'tent' areas several mm in diameter. These problems can be tolerated in small to medium scale integration, such as that used for sensor fabrication processes and power devices, but may be unacceptable for VLSI processes.

Several sensors have been developed incorporating SFB in their fabrication processes. These include cantilever accelerometers [10] with over-range protection, high-temperature pressure sensors [13], high over-range pressure sensors [12] and ultra-miniature pressure sensors [11]. These examples are only the first of many expected to emerge from different research centers.

Silicon Fusion Bonding for Pressure Sensors

The procedures required for fabricating several types of differential and absolute pressure sensors using SFB have been outlined in other publications [11–13]. Briefly, a typical fabrication sequence for an absolute pressure sensor runs as follows:

(a) Cavities that define diaphragm areas are batch-fabricated on one silicon wafer.

(b) A second silicon wafer is fusion bonded over these cavities.

(c) A controlled etching technique is used to remove the bulk of this second wafer, leaving only thin diaphragm areas over the cavities.

(d) Conventional microcircuit techniques are used to form piezoresistors, interconnects and bonding pads.

(e) The composite wafer is diced into individual chips.

(f) Each chip is tested and packaged.

One of the great economic benefits of SFB is the ability to produce small pressure sensor chips. This ability is due to the reduction of both etched cavity area beneath the diaphragm and the die-attach area required to support an applied pressure.

Traditional pressure sensor fabrication techniques require diaphragm formation by orientation-dependent etching from the backside of a silicon wafer. The result is a set of sloping sidewalls, reaching outward from the diaphragm edge at an angle of 54.7° from the horizontal. For a square diaphragm of edge length L the minimum etched cavity edge length is $L + t\sqrt{2}$, where t is the wafer thickness. In contrast, the cavity beneath a fusion-bonded diaphragm can have sidewalls that slope inward. For an absolute pressure sensor the cavity beneath the diaphragm can be shallow, and the required cavity area is approximately the same as the diaphragm area. For a differential pressure sensor fabricated by SFB, the cavity can be etched from the frontside of a thin wafer, or from both sides of a thicker wafer, and in most cases the cavity size on the backside of the wafer is smaller than the diaphragm size.

A differential pressure sensor chip must have some die-attachment area around the edge of the etched cavity, and for application of pressure to the backside of the diaphragm this die-attachment area must be large enough so that the adhesive used can support the applied pressure in tension. For a pressure P applied to a backside cavity of area A_c, and for an adhesive with a tensile strength S, the die-attachment area A_d must be

$$A_d > A_c P / S \qquad (1)$$

Because A_c is smaller for an SFB-fabricated sensor, A_d can also be smaller, and overall chip size can in most cases be determined by diaphragm size and bonding pad size with reduced concern for die-attachment considerations.

A full-bridge silicon pressure sensor only 1 mm × 1 mm × 400 μm has been fabricated at NovaSensor using SFB (Fig. 1). This chip approaches the ultimate lower limits in size for a discrete silicon pressure sensor. Considering these ultimate lower limits is instructive for future development efforts, and leads to the conclusion that the size and spacing of bonding pads almost completely determine die size.

A useful pressure sensor must meet established performance specifications. A typical set of industrial sensor requirements includes the following parameters:

Full-scale output voltage	100 mV at 5V supply
Bridge resistance	5 kΩ
Output linearity	0.1%
Temperature coefficient of sensitivity	−20%/100 °C
Temperature coefficient of resistance	> +22%/100 °C

Assuming that a sheet resistance of 1000 Ω/square is chosen for the piezoresistive bridge, each sensing resistor will be five squares long. Assuming a minimum feature size of 2 μm, resistors can then be 10 μm long. Such resistors are small enough to be placed on a 100 μm square diaphragm and retain good sensitivity (without loss due to spread of the resistor area to less-stressed regions away from the diaphragm edge). There are problems of thermal dissipation for such small resistors (with power densities of 1250 W cm^{-2} for a 5 V bridge voltage), but these problems will be ignored for the present analysis.

Given the existence of a 100 μm square sensing diaphragm with a 5 kΩ piezoresistive bridge, the choice of bonding pads becomes the major determinant in overall chip size. For example, use of thermosonic wire bonding places a working lower limit on bonding pad size on the order of 100 μm on an edge. If bonding pads 100 μm square spaced on 175 μm centers are chosen, and if these pads are spaced a minimum of 50 μm from the

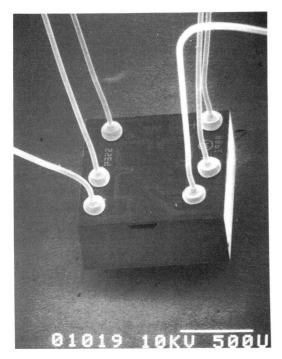

Fig. 1. Silicon pressure sensor fabricated by SFB. Die area is 1 mm × 1 mm. Note that the size of wire-bond balls is approximately equal to half the center-to-center minimum bond spacing.

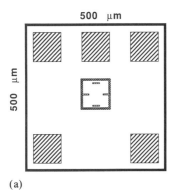

Fig. 2. Minimum-area pressure sensors if bond pads 100 μm square on 200 μm centers are used. (a) Square chip geometry used to preserve symmetry. 100 μm pads on 200 μm centers; symmetry maintained 23 500 chips per 4″ wafer. (b) Symmetry abandoned, rectangular chip geometry used. 100 μm pads on 200 μm centers; symmetry abandoned 32 500 chips per 4″ wafer.

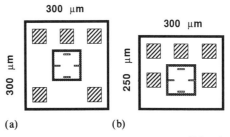

Fig. 3. Minimum-area pressure sensors if bond pads 50 μm square on 100 μm centers are used. (a) Square chip geometry used to preserve symmetry. 50 μm pads on 100 μm centers; symmetry maintained 58 500 chips per 4″ wafer. (b) Symmetry abandoned, rectangular chip geometry used. 50 μm pads on 100 μm centers; symmetry abandoned 68 500 chips per 4″ wafer.

diaphragm edge, the square chip geometry of Fig. 2(a) results, and the overall chip size is 500 μm square. If a decision is made that the chip does not have to be square, the size can be reduced further as in Fig. 2(b) to 350 μm × 500 μm.

If a decision is made to go to TAB bonding techniques for lead attachment, the bonding pad size might be reduced to 50 μm square on 100 μm centers. In this case the resulting chip size would be 300 μm square for the assumption of a square chip or 300 μm × 250 μm for a rectangular chip, as shown in Fig. 3.

Silicon Fusion Bonding for Accelerometers

Silicon accelerometer performance benefits greatly from the availability of SFB technology. A cantilever accelerometer chip only 3.3 mm × 3.3 mm × 1.5 mm has been developed at NovaSensor (Figs. 4–6); an earlier incarnation of this chip was reported previously [10]. The use of SFB for this chip permitted the development of a unique, bi-directional over-range-protection mechanism consisting of interdigitated, overlapping mechanical stops, schematically illustrated in Fig. 7. The actual implementation of these stops is visible in Figs. 4 and 5 as a set of three rounded extensions at either end of the seismic mass.

Traditionally, cantilever accelorometers have been difficult to fabricate without unacceptable yield loss during fabrication due to breakage of the cantilever beams. As a result, several manufacturers have avoided the cantilever beam approach and have introduced silicon accelerometers employed silicon suspending members on two or more opposite edges of a silicon seismic mass. These approaches have the virtue of greater ruggedness during fabrication, but are not optimized for both high sensitivity and high over-range capability. The accelerometer chip pictured

Fig. 4. Plan photomicrograph of a silicon cantilever accelerometer chip surrounded by rice grains. Chip size 3.3 mm square. Over-range protection features are present at top and bottom of photomicrograph.

Fig. 5. Scanning electron micrograph of a silicon cantilever accelerometer chip. A poppy seed is placed on top of the seismic mass.

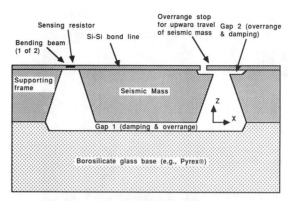

Fig. 6. Schematic cross-section of a cantilever, showing one section of the over-range protection mechanism at the left side of the upper chip surface.

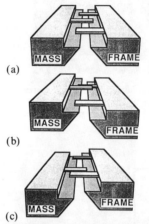

Fig. 7. Schematic perspective view of the over-range protection mechanism. (a) Seismic mass and suspending frame in a neutral position. (b) Mass rises, tabs on frame stop motion. (c) Mass falls, tab on mass stops motion. Note that the actual tabs are thin enough to be flexible and so do not shatter at contact.

in Figs. 4–6 has a high sensitivity of $10 \text{ mV}/g^{-1}$ at a supply voltage of 5 V, and also withstands over-range loads of several hundred g in any axis. This sensor possesses a unique combination of high sensitivity, high over-range capability, small die size and low fabrication cost per die, which permits its use in applications such as automobile active suspension systems in an efficient cost-effective manner.

Silicon Fusion Bonding and Surface Micromachining Technologies

A major research thrust in micromachining technology, especially in the emerging field of microdynamics, has been the development of

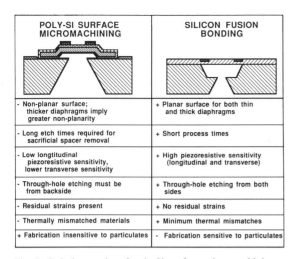

Fig. 8. Relative merits of poly-Si surface micromachining vs. SFB for pressure sensors.

polycrystalline silicon structures fabricated by surface micromachining (SMM) techniques, in which structures are defined in a shallow area on the silicon surface using deposited layers, photolithography and sacrificial etching techniques [14, 18, 19]. These techniques, like SFB, can result in pressure sensor structures whose size is determined by bonding pad size rather than diaphragm size. However, SFB has a number of advantages in this application. Foremost is the inherent process simplicity of an all single-crystal structure, with none of the variability introduced by the many process parameters inherent in polycrystalline silicon technology. In addition, SFB pressure sensors can be built for either low or high pressure ranges for a given diaphragm size by providing diaphragm thicknesses ranging from a few micrometers to hundreds of micrometers, whereas SMM pressure sensors can provide diaphragm thicknesses no greater than a few micrometers. A related characteristic is that SFB structures are more planar than SMM structures, which translates to less scratching of photomasks during contact photolithography. Figure 8 lists additional relative merits of SFB versus poly-Si SMM technology for this application.

Silicon-on-Insulator Techniques for VLSI

Technological development for silicon sensors, actuators and microstructures has traditionally been piggy-backed on technology developed for the circuit world. This circumstance has radically simplified the development efforts required for non-circuit silicon devices. If the VLSI world invests efforts in SFB, the benefits will flow to micromechanics research as well.

In addition, technology developed for power devices has relied heavily on the research investment in general-purpose circuitry, and the use of SFB for power devices would also benefit from investment in SFB by the VLSI world.

Both silicon microstructures and silicon power devices began as small-scale integration (SSI) devices, with one or a few devices per chip. More recently, 'smart' sensors and 'smart' power devices have been developed, carrying both fields into medium-scale integration (MSI). The economics of SFB as it presently exists are favorable for both fields, because yield losses (bond failures induced by surface dirt) on the order of 5–10% can be tolerated. However, in VLSI a yield loss of 1% at any step can prove disastrous. Thus the cleanliness requirements for SFB in VLSI fabrication greatly exceed those in other fields, and may be even more severe than the typical VLSI cleanliness requirements for lithography.

SFB shows promise in providing thin silicon-on-insulator layers for VLSI circuits in a bond-and-etchback silicon-on-insulator (BESOI) technology [3, 9]. In order to succeed in this application, SFB must be combined with a commercially viable etch-back technique, high yields must be maintained for large production volumes and costs must be low as compared to competing techniques. Unfortunately for the sensor and microstructure industry, zone melting recrystallization (ZMR) and separation by implanted oxygen (SIMOX) presently appear to have a lead over SFB for MOS VLSI applications, and SFB suffers from a relative lack of funding by VLSI researchers. In addition, in the author's opinion there are technological and economic reasons for SFB's lag as compared to other techniques. As a result, it appears necessary for the sensor industry and research community to take the lead in developing SFB technology.

Zone melting recrystallization [20] has been developed in several research centers. In this technique a layer of polycrystalline silicon is deposited on an insulating oxide that covers a single-crystal silicon wafer. The polycrystalline layer is 'seeded' by contact with the single-crystal wafer at one or more regions etched through the oxide layer, and is then recrystallized by localized scanned heating techniques so that a single-crystal layer of thin silicon results. The silicon quality is fairly good. Layer thickness and uniformity are determined by chemical vapor deposition techniques, and can be excellent. In addition, the quality of the Si/SiO_2 interface at the lower boundary of the thin layer is good. The expenses involved include the original silicon wafer, oxidation of the wafer, chemical vapor deposition, additional oxide deposition and thermal scanning for recrystallization.

Separation by implanted oxygen [21] uses a high-energy, high-dose oxygen implant into a silicon wafer to result in an oxygen-rich zone, which can be heat treated to create a buried SiO_2 region beneath the single-crystal surface of the wafer. The single-crystal layer thickness can subsequently by built up using conventional epitaxy if desired [22]. As with ZMR, layer thickness and uniformity appear very good, but the properties of the lower Si/SiO_2 interface and the silicon near this interface are not as good because of oxygen straying outside the buried oxide layer. The expenses for this technique include the starting silicon cost, one or more expensive oxygen implant and annealing steps and any chemical vapor deposition step. The technique appears to be somewhat more expensive than the ZMR technique and to produce silicon of equivalent or worse quality.

Silcion fusion bonding can produce layers of excellent crystal quality, but the available etch-back techniques limit either the crystal quality, the layer thickness and uniformity, or both. If, for example, a heavy p^+ doping layer is used as an etch stop (and then removed from the remaining silicon layer) as in [9], the resulting lightly-doped layer contains 10^2 to 10^3 threading dislocations per cm^2 (which may still be higher quality than either ZMR or SIMOX layers). Electrochemical etch-stop techniques can produce higher-quality silicon but are difficult to use in the presence of an insulating oxide layer between the supporting wafer and the device layer, and in any case lateral voltage drops encountered in using electrochemical etch stops would require an undesirably thick silicon layer, or one that is doped to a high level, in order to obtain a uniform surface. It should be noted, however, that thicker silicon layers may be acceptable for bipolar SOI as opposed to MOS SOI, and that SFB may have an advantage in this area.

The expenses involved in SFB for BESOI as in [9] include two starting silicon wafers, heavy boron doping, epitaxy, an oxidation step, lapping and etching of one wafer, a second etch step to remove the p^+ etch-stop layer and a third step (etch or oxidation) to remove a stain region left by the second etch. In addition, the technique requires the use of two etchants (ethylenediamine–pyrocatechol–water followed by hydrofluoric–nitric–acetic acids), which present handling and disposal difficulties. The overall complexity and expense of fabricating SOI wafers by this technique appear greater than in the ZMR process (assuming automated production techniques are used for both processes). The author has had extensive personal experience with similar dielectric isolation technology of somewhat greater complexity [23], and it is his considered opinion that process complexity will limit the serious pursuit of SFB for VLSI technology. He would enjoy being proven wrong in this opinion.

Summary

Silicon fusion bonding presents major new possibilities in the design of silicon micromachined structures when combined with other available processing techniques. SFB has already been used in novel accelerometers, high-temperature pressure sensors, ultraminiature pressure sensors and high over-range pressure sensors.

SFB does not appear to be the technique of choice for VLSI SOI technology, but it is highly viable for use in silicon microstructures, and it is incumbent on the micromachining community to pursue further development of the technology. With the development of 'smart' power devices occurring in parallel with the development of 'smart' sensors, it is to be hoped that evolution of SFB for both microstructures and power devices will continue and will provide cross-fertilization between the two fields.

Acknowledgements

This paper would not have been possible without the help of the people who have participated in SFB technology development at NovaSensor including Joe Brown, Janusz Bryzek, Sean Cahill, Lee Christel, Joe Mallon, Kurt Petersen and Rose Scimeca.

References

1 T. Nakamura, Semiconductor device, *U.S. Patent 3 288 656* (Nov. 29. 1966).
2 G. E. Brock, D. DeWitt, W. A. Pliskin and J. Riseman, Fusion of silicon wafers, *IBM Tech. Discl. Bull. 19* (1977) 3405–3406.
3 J. B. Lasky, S. R. Stiffler, F. R. White and J. R. Abernathy, Silicon-on-insulator (SOI) by bonding and etch-back, *IEDM Tech. Digest Int. Electron Devices Meet., Washington, DC, U.S.A., Dec. 1–4, 1985*, pp. 684–687.
4 Simple wafer fusion builds better power chips, *Electronics*, (Dec. 23) (1985) 20.
5 M. Shimbo, K. Furukuwa and K. Fukuda, A newly developed silicon to silicon direct adhesion method, *Abstract No. 232, Ext. Abstr., Electrochemical Soc., 169th Meet., Spring 1986, Boston, MA, U.S.A.*, Vol. 86-1, pp. 337–338.
6 M. Shimbo, K. Furukuwa, K. Fukuda and K. Tanazawa, Silicon-to-silicon direct bonding method, *J. Appl. Phys., 60* (1986) 2987.
7 H. Ohasi, K. Furukawa, M. Atsuta, A. Nakagawa and K. Imamura, Study of Si-wafer directly bonded interface effect on power device characteristics, *IEDM Tech. Digest, Int. Electron Devices Meet., Washington, DC, U.S.A., Dec. 6–9 1987*, pp. 678–681.

8 L. Tenerz and B. Hök, Silicon cavity structures fabricated with a new technique, *Electron. Lett.*, 22 (1986) 615–616.
9 W. P. Maszara, G. Goetz, A. Caviglia and J. B. McKitterick, Bonding of silicon wafers for silicon-on-insulator, *J. Appl. Phys.*, 64 (1988) 4943–4950.
10 W. Barth, F. Pourahmadi, R. Mayer, J. Poydock and K. Petersen, A monolithic silicon accelerometer with integral air damping and overrange protection, *Tech. Digest, IEEE Solid-State Sensor and Actuator Workshop, Hilton Head Island, SC, U.S.A., June 6–9, 1988*, pp. 35–38.
11 K. Petersen, P. Barth, J. Poydock, J. Brown, J. Mallon Jr., and J. Bryzek, Silicon fusion bonding for pressure sensors, *Tech. Digest, IEEE Solid-State Sensor and Actuator Workshop, Hilton Head Island, SC, U.S.A., June 6–9, 1988*, pp. 144–147.
12 L. Christel, K. Petersen, P. Barth, J. Mallon and J. Bryzek, Single-crystal silicon pressure sensors with 500X overpressure protection, *Sensors and Actuators*, A21–A23 (1990) 84–88.
13 K. Petersen, J. Brown, P. Barth, J. Mallon and J. Bryzek, Ultra-stable high temperature pressure sensors using silicon fusion bonding, *Sensors and Actuators*, A21–A23 (1990) 96–101.
14 H. Guckel and D. W. Burns, A technology for integrated sensors, *Tech Digest 3rd Int. Conf. Solid-State Sensors and Actuators, Transducers '85, Philadelphia, PA, U.S.A., June 11–14, 1985*, pp. 90–92.
15 R. C. Frye, J. E. Griffith and Y. H. Young, A field-assisted bonding process for silicon dielectric isolation, *J. Electrochem. Soc.*, 133 (1986) 1673.
16 D. I. Pomerantz, Anodic bonding, *U.S. Patent 3 397 278* (Aug. 13, 1968).
17 S. Johnsson, K. Gustafsson and J.-Å. Schweitz, Strength evaluation of field assisted bond joints and joint systems in glass–silicon microstructures, to be published.
18 R. S. Hijab and R. Muller, Micromechanical thin-film cavity structures for low pressure and acoustic transducer applications, *Tech. Digest 3rd Int. Conf. Solid-State Sensors and Actuators (Transducers '85), Philadelphia, PA, U.S.A., June 11–14, 1985*, pp. 178–181.
19 H. Guckel and D. W. Burns, Laser-recrystallized piezoresistive micro-diaphragm sensor, *Tech. Digest 3rd Int. Conf. Solid-State Sensors and Actuators (Transducers '85), Philadelphia, PA, U.S.A., June 11–14, 1985*, pp. 182–185.
20 P. M. Zavracky, D. P. Vu, L. Allen, W. Henderson, H. Guckel, J. J. Sniegowski, T. P. Ford and J. C. C. Fan, Large diameter SOI wafers by zone-melting recrystallization, *Mater. Res. Soc. Symp. Proc.*, 107 (1988) 213–219.
21 R. L. Bates, SOI epitaxial technology perspective: ISE or SIMOX? *Microelectron. Manuf. Testing*, (Mar.) (1989) 11.
22 Hon Wai Lam, Epitaxial growth on SIMOX wafers, *Abstr. No. 80, Ext. Abstr. Electrochem. Soc. Spring Meeting, Boston, MA. U.S.A., May 4–9, 1986*, Vol. 86-1.
23 P. W. Barth, Dielectric isolation technology for bipolar and MOS integrated circuits, *Ph.D. Dissertation*, Stanford University Electrical Engineering Department, Stanford, California, Mar. 1980, 219 pp.

SCALING AND DIELECTRIC STRESS COMPENSATION OF ULTRASENSITIVE BORON-DOPED SILICON MICROSTRUCTURES

S. T. Cho, K. Najafi, and K. D. Wise

Center for Integrated Sensors and Circuits
Solid-State Electronics Laboratory
Department of Electrical Engineering and Computer Science
University of Michigan
Ann Arbor, Michigan 48109-2122

ABSTRACT

The scaling of boron-doped silicon membranes based on diaphragm dimensions and stress compensation has been characterized. Devices with varying edge length and plate thickness were fabricated and tested for sensitivity. The stress for p++ silicon, LPCVD silicon dioxide and LPCVD silicon nitride has been measured using an electrostatic technique which utilizes silicon microbridges. Silicon membranes with varying thicknesses of oxide and nitride were characterized for sensitivity. The results confirm a previously reported analytical scaling theory for these structures. Based on this theory, scaled experimental devices have been found to show sensitivities within three percent of the calculated design targets.

INTRODUCTION

Silicon micromachined membranes are an integral part of sensing structures for a wide range of devices, including catheter-based pressure sensors [1], accelerometers [2], and flowmeters [3]. Many of the applications for these devices demand both low cost and high sensitivity. Boron-doped silicon membrane structures are simple to fabricate, high yield, and compatible with batch processing. However, for ultrasensitive applications, the tensile stress due to the boron doping reduces the sensitivity considerably.

Figure 1 shows the cross-section of a capacitive pressure sensing structure. In order to improve sensitivity, the edge dimensions can be increased or the plate thickness can be reduced. However, size restrictions or fracture limits can make either of these solutions impractical. The use of dielectrics to compensate the diaphragm stress is an attractive alternative for improving device performance. Although a scaling theory which takes into account intrinsic stress has been reported [4], it has not been verified experimentally. This paper reports experimental results on the scaling of boron-doped silicon membranes. The stress for p++ silicon, LPCVD silicon dioxide, and LPCVD silicon nitride has been determined using an electrostatic technique involving silicon microbridges [5]. Sensitivity as a function of edge dimension, plate thickness and stress was examined. A design example is also presented to verify the theory.

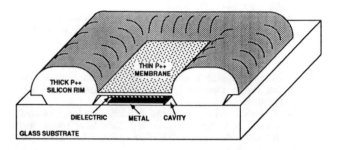

Fig. 1: Cross-Section of a Capacitive Pressure-Sensing Structure.

FABRICATION

Membrane structures and microbridges were fabricated using a versatile, single-sided dissolved wafer process. The fabrication sequence involves both silicon and glass processing and is shown in Fig. 2. Processing starts with a p-type (100) silicon wafer of moderate doping ($>1\Omega\text{-}cm$). A recess is etched into the silicon using KOH (Fig. 2a). This recess defines the cavity of the pressure sensor. Two high temperature (1175°C) boron diffusions follow. The first is a deep diffusion (12-15μm) which defines the rim of the membrane (Fig. 2b). The second is a shallow diffusion (2.1-5.7μm) which determines the thickness of the diaphragm (Fig. 2c). By varying the KOH etch depth and the shallow boron diffusion time, a wide variety of operating ranges and sensitivities for the capacitive pressure sensor can be obtained. LPCVD oxide or nitride films are then deposited on the membrane to compensate its stress and to provide protection against the metal plate and diaphragm shorting together. The silicon dioxide was deposited at 920±10°C, with a pressure of 450 *mtorr*, and a nitrous oxide: dichlorosilane gas ratio of 2:1. The nitride was deposited at 820±10°C, with a pressure of 250 *mtorr* and an ammonia:dichlorosilane gas ratio of 4:1.

The glass processing involves depositing and patterning a multi-metal system (10nm Ti/ 20nm Pt/ 250nm Au) on a #7740 Corning glass wafer. The metal forms the bottom plate of the capacitor and the output leads from the transducer. The silicon and glass are then electrostatically bonded together [6]. The bonding takes place at 400°C with a potential of 1000V applied between the glass and silicon. In order to make electrical contact to the silicon, a metal lead on the glass is allowed to overlap the silicon rim over a small area. The gold is removed from the metal area here to reduce the step height between the glass and silicon. During the electrostatic bonding process, the silicon and glass are drawn tightly together, ensuring a low-resistance Pt/Ti contact to the silicon (40Ω for a 40μm x 20μm area). The final step (Fig. 2d) in the process is a selective etch in ethylenediamine-pyrocatechol-water (EDP). EDP dissolves the silicon and stops on the heavily-doped (p+) diffused layers. The overall fabrication sequence requires only single-sided processing with four non-critical masking steps on silicon. Precise dimensional control is obtainable and the process is high yield as well as compatible with batch processing.

MATERIALS CHARACTERIZATION

In order to understand the effect of stress on membrane performance, it is critical to be able to measure the material properties of the diaphragms. An electrostatic technique involving boron-doped silicon microbridges was used to do the characterization. In Fig. 3, the test structure is shown. When a voltage is applied across the free standing bridge, an electrostatic normal force causes the beam to deflect. If this electrostatic force exceeds the intrinsic stress of the beam, then the bridge pulls in and collapses. The p++ silicon and the metal plate on the glass form a capacitor; by monitoring the change in capacitance for applied voltage, the capacitance should increase sharply at pull-in.

The pull-in voltage is given by [5]:

$$V_{PI}^2 = \frac{8}{27} d_o^3 \frac{kP}{\varepsilon_o A} \frac{1}{[\frac{kl}{4} - tanh(\frac{kl}{4})]} \quad (1)$$

$$k = \sqrt{P/EI} \quad (2)$$

where d_o is the zero voltage gap spacing, l is the length of the beam, A is the area, E is Young's modulus, P is the axial force created by the intrinsic stress and I is the moment of inertial, given by $I = bh^3/12$ with b being the width and h being the thickness. If the capacitive gap and the area of the capacitor are kept constant, then the pull-in voltage is dependent only on the beam length. By fabricating two identical bridges, except for beam length, E and P can be calculated. The two pull-in voltages are related by:

$$\frac{V_{PI}^2(1)}{V_{PI}^2(2)} = \frac{d_o^3(1)}{d_o^3(2)} \frac{\frac{l_2}{l_1} - tanh(\frac{l_2}{l_1}x_1)}{x_1 - tanh(x_1)}; \text{ where } x_1 = \frac{kl_1}{4}. \quad (3)$$

Variables x_1 and k are first calculated; from this, P can be obtained from equation 1. The intrinsic stress is simply given by $\sigma = P/bh$. If P is known, then the Young's modulus can be calculated using equation 2.

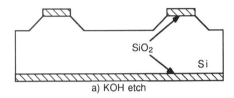

a) KOH etch

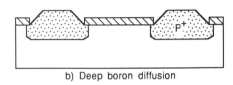

b) Deep boron diffusion

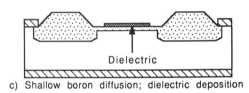

c) Shallow boron diffusion; dielectric deposition

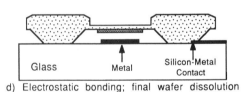

d) Electrostatic bonding; final wafer dissolution

Fig. 2: Dissolved-Wafer Process Flow.

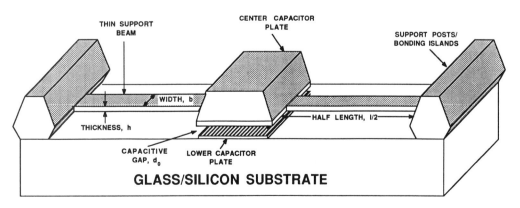

Fig. 3: Test Structure for the Measurement of Mechanical Properties of Thin Films.

For composite structures, the stress and Young's modulus can be derived. The stress relationship of a beam containing two different materials is given by [7]:

$$\sigma_c t_c = \sigma_1 t_1 + \sigma_2 t_2 \qquad (4)$$

where σ_c is the total composite stress, t_c is the total thickness, $\sigma_{1,2}$ is the stress for each layer and $t_{1,2}$ is the corresponding thickness of each material. For a composite beam structure, the pull-in voltage technique determines σ_c. If σ_1 and each layer thickness is known in addition to σ_c, then the stress of the remaining layer (σ_2) is known.

In order to determine the stress of p++ silicon, LPCVD silicon dioxide and LPCVD silicon nitride, several structures were fabricated. A bridge composed of just boron-doped silicon was first tested and the stress was determined. Composite structures of boron-doped silicon/silicon dioxide and boron-doped silicon/silicon nitride were also fabricated to determine the dielectric stresses. Since all the layer thicknesses and the stress of the boron-doped silicon is known, the pull-in voltage measurement determines the composite stress and equation 4 can be used to derive the dielectric stresses.

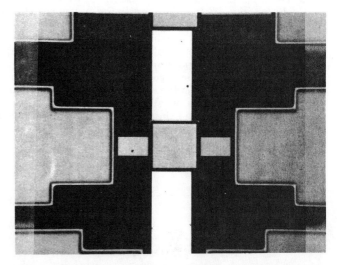

Fig. 4: Top View of a Fabricated Test Bridge for the Measurement of Thin-Film Mechanical Properties.

Figure 4 shows a fabricated beam. The bridge width is 100 μm; the two lengths used for measurements are 330μm and 730μm. The center capacitor plate is 250 μm x 250 μm; the beam thickness is 2.5μm. The stress of the p++ silicon was found to be 2.4×10^7 Pa, while the stress for the oxide and nitride layers is given in Fig. 5 as a function of dielectric thickness.

Young's modulus from composite structures can be derived in a way similar to the stress calculation. The relationship of Young's modulus to its component layers for a composite beam is given as [8]:

$$E_c I = \frac{w t_b^3}{12(t_a E_a + t_b E_b)} t_a E_a E_b \, K_1, \text{ where}$$

$$K_1 = 4 + 6\frac{t_a}{t_b} + 4(\frac{t_a}{t_b})^2 + \frac{E_a}{E_b}(\frac{t_a}{t_b})^3 + \frac{E_b}{E_a}(\frac{t_b}{t_a}) \qquad (5)$$

where E_c is the composite Young's modulus, t_a, t_b, E_a, E_b are the corresponding thicknesses and Young's moduli, respectively. By solving for E_b given that E_a and E_c are constants, equation 5 becomes a quadratic and E_b can be solved for.

There are a few important points that need to be made concerning this technique. In order to accurately measure the response of the beam, the gap spacing must be less than 0.4 times the thickness of the beam. If the gap spacing is larger than this number, the beam does not truly bend when it pulls in; it stretches, which distorts the measurement. Since several of the terms are squared or cubed, deriving consistent results for Young's modulus has also been found to be difficult. It is possible that the Young's modulus values can vary as much as an order of magnitude. The data presented here was taken over several sets of beams, including oxide/nitride composites; the most consistent numbers were averaged together and are presented in Fig. 5.

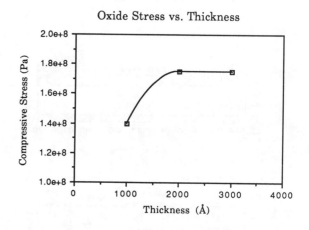

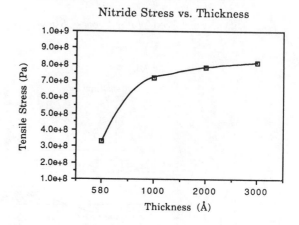

Fig. 5: Dielectric Stress as a Function of Dielectric Thickness for LPCVD Silicon Dioxide and Silicon Nitride Films.

SCALING OF BORON-DOPED MEMBRANES

In the same way that it is critical to develop electrical models for silicon, it is important for sensor technology to be able to derive models which describe mechanical behavior. Scaling theory for the pressure sensitivity of square, stress-free membranes has already been reported [9] and is given by:

$$S_{bending} = \frac{1}{C_o} \frac{\Delta C}{\Delta P} = 0.0746 \frac{1-v^2}{E} \left(\frac{a^4}{h^3 d} \right) \quad (6)$$

where v is Poisson's ratio, $2a$ is the membrane length, h is the membrane thickness, d is the plate separation and C_O is the zero pressure capacitance. In general, diaphragms that are either very thick (>10 μm) or very small (< 1mm in length) can be considered to be stress free. However, for many emerging applications, the intrinsic stress cannot be ignored. Using a rigidly clamped circular membrane of radius a, the sensitivity derived taking into consideration the intrinsic stress is given by [4]:

$$S_{tension} = 192 \left[\frac{1}{k^4} + \frac{1}{8k^2} - \frac{1}{2k^3} \frac{I_0(k)}{I_1(k)} \right], \text{ where}$$

$$k^2 = \frac{12(1-v^2)|\sigma_i|a^2}{Eh^2} \quad (7)$$

I_n is the modified Bessel function of the first kind of order n; σ_i is the intrinsic stress and $S'_{tension}$ is a normalized sensitivity (the sensitivity with intrinsic stress divided by the stress-free sensitivity). A dimensionless intrinsic stress term (σ_i') based on k can be defined as

$$\sigma_i' = \frac{12(1-v^2)|\sigma_i|a^2}{Eh^2}. \quad (8)$$

If the gap spacing is kept constant, then the sensitivity is a function of the thickness (h), the edge length (a) and the internal stress (σ_i). Square capacitive pressure sensors were fabricated and tested to examine how each of these factors scales the sensitivity. Two sets of devices were fabricated: one set without dielectrics and one set with dielectrics. The set without dielectrics had two edge lengths (1mm, 2mm) and four different thicknesses (2.1, 3.1, 4.0, 5.7μm). The set with dielectrics had one length (2mm) and one thickness (2.6μm). Three sets of dielectrics were deposited: oxide (1234, 2037, and 3200Å), nitride (580, 1000Å) and one composite oxide/nitride (1000Å/1000Å). The fabricated test chip is shown in Fig. 6.

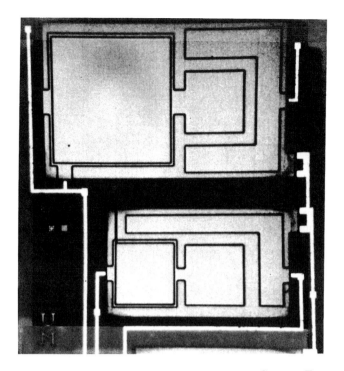

Fig. 6: Top View of Capacitive Pressure-Sensor Test Structures used for Studies of Membrane Sensitivity and Stress Compensation.

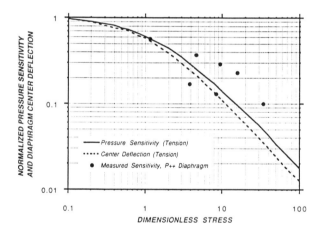

Fig. 7: Theoretical and Measured Pressure Sensitivity of Boron-Doped Silicon Diaphragms as a Function of Dimensionless Stress.

Figure 7 shows the normalized sensitivity versus dimensionless stress for the case of no dielectrics. The data points track the theoretical curve, but the points are scattered. The greatest potential sources of error are: i) assumed values for Young's modulus and Poisson's ratio (E=1.7 x 10^{11} Pa, v=0.066) ii) uncertainty in the membrane thickness (this term is cubed in equation 6), and iii) the theory is derived for circular membranes.

For the dielectric case (Fig. 8), the composite stress was calculated by using stress values presented in Fig. 5 in equation 4. In this case, the values track the curve very tightly. For $\sigma_i' > 10$, the sensitivity can be approximated as

$$S_{tension} = \frac{a^2}{8\sigma_i h d}. \qquad (9)$$

Table 1 lists the theoretical sensitivity based on equation 9 and the actual measured sensitivity. The errors are typically less than 20%; the greatest difference in measured and theoretical values occurs for 3200Å of oxide. However, the stress reduction at this point is so high that an oxide thickness variation of 200Å can create a 40% variation in the sensitivity. In general, a stress reduction of a factor of 6 is attainable before membrane buckling occurs.

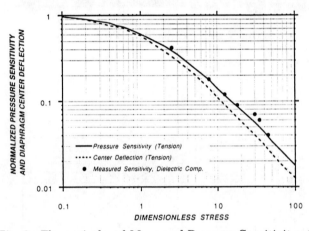

Fig. 8: Theoretical and Measured Pressure Sensitivity of Boron-Doped Silicon Diaphragms with Deposited Dielectric Compensation as a Function of Dimensionless Stress.

DESIGN EXAMPLE

In order to verify the usefulness of the models presented, an actual design problem was solved. A capacitive pressure sensor for a pressure-based microflowmeter required a resolution $(\Delta C / \Delta P)$ of $1 fF/mtorr$. The design target parameters are listed in Table 2.

An arbitrary gap spacing of 3.5μm was chosen. The first design criteria was mechanical strength; the membrane should be able to withstand overpressures of 760 $Torr$ (1 ATM). Using a conservative approach, a fracture pressure of 1140 $Torr$ was chosen (50% overprotection). The fracture pressure (p_f) is given by [10]

$$p_f = 0.347 \frac{\sigma_f^2 h}{E d} \qquad (10)$$

where σ_f is the fracture stress of silicon (3×10^9 Pa). Solving for h, this gives a membrane thickness of about 2.8μm.

The remaining parameters were designed according to equation 9. The intrinsic stress for a sensitivity of about 1.1 was appproximately 7.7×10^6 Pa. Using equation 4 and assuming an oxide stress of -1.75×10^8 Pa (compressive) and 2.4×10^7 Pa for silicon, an oxide thickness of 2500Å is required.

The actual measured parameters are also listed in Table 2. In general, membrane thickness and gap spacing can be controlled to within 0.1-0.2 μm and the dielectric thickness to <50Å. The membrane was subjected to overpressures in excess of 720 $Torr$; the membrane did not fracture and exhibited a hysteresis of less than 0.3% full scale. The measured sensitivity is within 3% of the target, which illustrates the design capability possible with accurate models and process control.

Dielectric Thickness	S_{measured}	S_{theoretical}
None	73	71
1234Å SiO$_2$	100	96
2037Å SiO$_2$	149	175
3200Å SiO$_2$	410	683
580Å Si$_3$N$_4$	74.9	69.4
1000Å Si$_3$N$_4$	29.9	35.2
1000Å/1000Å SiO$_2$/Si$_3$N$_4$	50.7	43.1

Table 1: Measured Sensitivity of Thin p++ Silicon Membranes Coated with Various Deposited Dielectrics.

Table 2: Design Parameters and Measured Performance for an Ultrasensitive Capacitive Pressure Sensor.

	Target	Measured
Membrane Thickness (μm)	2.8	2.9
Gap Spacing (μm)	3.5	3.62
Dielectric Thickness (Å)	2500	2530
Initial Capacitance (pF)	5.589	5.413
Sensitivity (ppm/mtorr)	203	209
Initial Slope (fF/mtorr)	1.13	1.13
Overpressure (torr)	1102	>720

CONCLUSIONS

Experimental results for the scaling of boron-doped membrane structures has been reported. Scaling as function of membrane length, thickness and stress has been characterized and previously reported scaling theory has been confirmed. An electrostatic technique was used to measure material parameters and provided accurate results. The scaling theory was applied to an actual design problem, and the resulting device performance was within 3% of the specification.

ACKNOWLEDGEMENTS

The authors would like to thank Mr. Clark Lowman for many useful discussions. The authors would also like to thank Ms. Terry Hull and Mr. Jeff Lucas for their assistance in the fabrication. This work was supported by the Semiconductor Research Corporation under Contract #88-MC-085.

REFERENCES

[1] H.L. Chau and K.D. Wise, "An Ultraminiature Solid-State Pressure Sensor for a Cardiovascular Catheter," *IEEE Trans. Electron Devices*, pp. 2355-2362, December 1988.

[2] J.T. Suminto, G.-J. Yeh, T.M. Spear, and W. Ko, "Silicon Diaphragm Capacitive Sensor for Pressure, Flow, Acceleration and Attitude Measurements", *Transducers '87*, pp. 336-339, July, 1987.

[3] S.T. Cho, K. Najafi, C.L. Lowman, and K.D. Wise, "An Ultrasensitive Silicon Pressure-Based Flowmeter," *IEDM Tech. Digest*, pp. 499-502, December, 1989.

[4] H.L. Chau and K.D. Wise, "Scaling Limits in Batch-Fabricated Silicon Pressure Sensors," *IEEE Trans. Electron Devices*, Vol. ED-34, No.4, pp. 850-858, April 1987.

[5] K. Najafi and K. Suzuki, "A Novel Technique and Structure for the Measurement of Intrinsic Stress and Young's Modulus of Thin Films," *Proc. IEEE Workshop on Microelectromechanical Systems*, Feb. 1989, pp. 96-97.

[6] L. Spangler and K.D. Wise, "A New Silicon-on-Glass Process for Integrated Sensors", *IEEE Sensor and Actuator Workshop*, Hilton Head, pp. 140-142, June 1988.

[7] O. Tabata, K. Kawahata, S. Sugiyama, and I. Igarashi, "Mechanical Property Measurements of Thin Films Using Load-Deflection of Composite Rectangular Membrane", *Proc. IEEE Workshop on Microelectromechanical Systems*, Feb. 1989, pp. 152-156.

[8] R.J. Roark and W.C. Young, <u>Formulas for Stress and Strain</u>, New York: McGraw-Hill, 1965.

[9] S.K. Clark and K.D. Wise, "Pressure Sensitivity in Anistropically Etched Thin-Diaphragm Pressure Sensors," *IEEE Trans. Electron Devices*, Vol. ED-26, pp. 1887-1896, Dec. 1979.

[10] K. Chun, "A High Performance Si Tactile Imager Based on a Capacitive Cell," Ph.D. Dissertation, The University of Michigan, May, 1986.

Field Oxide Microbridges, Cantilever Beams, Coils and Suspended Membranes in SACMOS Technology

D. MOSER, M. PARAMESWARAN* and H. BALTES**

Physical Electronics Laboratory, Institute of Quantum Electronics, Swiss Federal Institute of Technology (ETH) Zurich, CH-8093 Zurich (Switzerland)

Abstract

Standard industrial CMOS integrated circuit fabrication technology combined with unconventional layout design and a single post-processing anisotropic silicon etching step can be used to produce field oxide based micromechanical structures for sensor applications. In this note, we present the extension of our previous technique [*Sensors and Materials*, 1 (1988) 115–122; *Sensors and Actuators*, 19 (1989) 289–307] to further microstructures and its transfer to the self-aligned contact CMOS (SACMOS) process of FASELEC Corporation. We also report numerical modelling of cantilever beam oscillations.

1. Introduction

Present day CMOS technology yields highly reliable integrated circuits (IC). The power of this technology can be extended further to produce not only integrated sensors such as magnetic field sensors, but even microstructures that are potentially useful for mechanical and chemical sensors. Features inherent in the usual CMOS fabrication process sequence in combination with a single post-processing etching step allow either sandwiched oxide [1] or polysilicon [2] microstructures to be batch manufactured inexpensively. This is brought about by an unconventional layout design methodology [1–4] that allows for the technological requirements of both the CMOS IC processing sequence and the post-processing etching step (anisotropic silicon etching [5] in the case of oxide structures or isotropic sacrificial oxide etching [6] in the case of polysilicon structures). The anisotropic or isotropic final etching step selectively removes material in order to form the desired free-standing microstructure. Our approach is a contribution to current attempts to integrate micromechanical structures and signal-conditioning circuitry on the same die.

Our previous oxide beams and bridges [1, 3, 4] were based on the 3 µm CMOS process of Northern Telecom Electronics, Ltd. In this paper, we report the first successful transfer of our method to a different process and manufacturer, namely the 3 µm SACMOS process offered to Swiss universities by FASELEC Corporation, Zurich and present further structures, such as coils and suspended membranes. SACMOS stands for self-aligned contact CMOS, a technology with minimal area used for the metal contact lines. Portability of our microstructure design and fabrication method may be expected in principle. But the mechanical quality of the microstructures depending on the detailed processing procedures and parameters involved is hard to predict, and the portability of our approach has to be demonstrated experimentally. Besides design and fabrication, modelling and testing are essential parts of device development. We report predictions of stress patterns and resonant modes of cantilever beams obtained by using the general-purpose finite-element code ANSYS.

2. Design Methodology

In the case of sandwiched oxide structures, the layout design method [1, 3, 4] can be summarized as follows. The structures are fabricated in two steps. The first step is designing and implementing a suitable layout by the industrial CMOS process. The second step is the anisotropic etching step forming a pit on the silicon substrate. To this end, a portion of the substrate has to be exposed to the anisotropic etchant. Removal of the pertinent field and CVD oxide layers during the CMOS processing itself is achieved by a unique layout design, i.e., a device well, contact cut and pad opening all one above the other defined in the area where the

---

*Present address: Department of Electrical Engineering, University of Alberta, Edmonton, Alberta, T6G 2G7, Canada.
**Author to whom correspondence should be addressed.

silicon substrate has to be exposed. N^+ and n-guard exclusion masks surround the entire area in order to maintain a minimum dopant concentration in the portion of the substrate to be exposed to anisotropic etching.

The typical layout for the sandwiched oxide microbridge is shown in Fig. 1(a), where the silicon substrate is exposed to the ambient through two triangular-shaped areas formed by the design of a corresponding pad opening over contact cut over active area. The cross-section shown in Fig. 1(b) is expected to result after the standard CMOS process with the CVD oxide layers removed in the active area, but with doping prevented by the n^+ and n-guard exclusion masks. The oblique direction of the bridge with respect to the vertical edge parallel to the (110) direction on the (100) wafer is essential for the anisotropic etchant to undercut [5] and form the microbridge during the post-processing step. The resulting bridge has a sandwiched structure consisting of CVD oxide layers on top of the field oxide. The layout for the corresponding cantilever beam structure is shown in Fig. 2. The convex cantilever edge surrounded by exposed silicon substrate is undercut [5, 7] by the anisotropic etchant. Free-standing sandwiched layers consisting of CVD on polysilicon (or aluminium metal) on field oxide can be designed in a similar fashion.

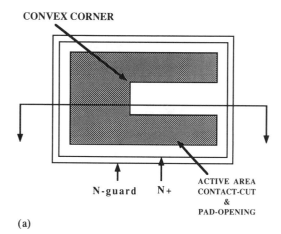

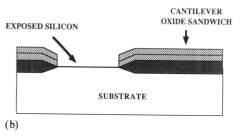

Fig. 2. (a) Layout design for fabrication of oxide cantilever with CMOS process. (b) Cross section of structure obtained after CMOS process, before anisotropic etching postprocessing.

3. Post-processing and Results

The anisotropic etching was performed with KOH or EDP [5, 7]. As the oxide layers are not significantly attacked by the post-processing etchant, no additional masking step is required. EDP is the preferable etchant for this post-processing step, since its etch rate of silicon dioxide is much lower than that of KOH. We have also used hydrazine for that purpose. The etchant attacks silicon in those regions where the silicon substrate is exposed by virtue of the special design and the standard CMOS process. The oxide microstructures form upon completion of the etching process. Typical etch temperatures and times with 30% KOH solution were 90 °C and 60 min. Any circuits formed by the CMOS process on the same die are protected during the post-process etching because they are covered by the CVD oxide film. Only the metallized contact pads would require special protection.

Figure 3 shows an SEM photograph of a 10 μm by 200 μm sandwiched oxide microbridge similar to our previous structure [1], but based on the SACMOS process of FASELEC. Figure 4 shows a bridge with a metal layer sandwiched between the field oxide and the CVD films. Figures 5, 6, and 7 show various cantilever beams.

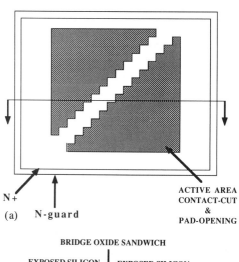

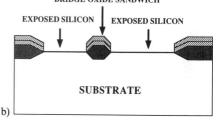

Fig. 1. (a) Layout design for fabrication of oxide microbridge with CMOS process. (b) Cross section of structure obtained after CMOS process, before anisotropic etching postprocessing.

Fig. 3. SEM photo of oxide microbridge based on SACMOS process.

Fig. 6. SEM photo of oxide paddle based on SACMOS process.

Fig. 4. SEM photo of oxide-based bridge with metal layer sandwiched between field oxide and CVD oxide.

Fig. 7. SEM photo of oxide-based cantilever with metal contact line sandwiched between field oxide and CVD oxide.

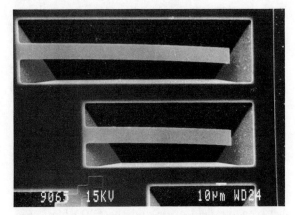

Fig. 5. SEM photo of oxide cantilevers based on SACMOS process.

Fig. 8. SEM photo of oxide coil based on SACMOS process.

Fig. 9. Suspended sandwiched oxide membrane with four diagonal supports.

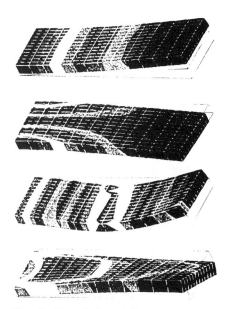

Fig. 10. Numerical modelling of cantilever resonant modes with stress pattern.

The beam in Fig. 7 has a metal connection sandwiched between the field oxide and CVD films, similar to our previous beam [1].

Using similar layout design and post-processing etching, a sandwiched oxide coil or spiral (see Fig. 8) and a sandwiched oxide membrane suspended above a cavity by four diagonal supports (Fig. 9) have been fabricated. The undercut at the convex corners is crucial in the case of the coil, and the post-processing etching proceeds downwards until stopped. A resistive element, to be used as a heater, can be sandwiched between the oxide layers in the case of the suspended membrane. The resulting structure could function as a gas-flow sensor similar to the one reported by Honeywell researchers [8].

4. Modelling

The general-purpose finite-element code AN-SYS can be used to predict stress patterns and vibration modes of free-standing microstructures [9]. We have applied the ANSYS code for modelling our oxide cantilever beams. Figure 10 shows the wave forms of the fundamental and three higher-order resonant modes and their stress patterns that we obtained for a 50 μm by 200 μm oxide beam, under the assumption of a uniform oxide layer of 3 μm thickness. The fundamental frequency is predicted to be 68 kHz. Laser interferometric observations of the corresponding vibrations are in progress.

Acknowledgements

It is our pleasure to thank Mr S. Linder for performing the post-processing etching, Mr E. Bolz for assisting with setting up the silicon etching equipment and Mr S. Rudin for performing the ANSYS simulations.

References

1 M. Parameswaran and H. P. Baltes, Oxide microstructure fabrication: a novel approach, *Sensors Mater.*, *10* (1988) 115–122.
2 M. Parameswaran, H. P. Baltes and A. M. Robinson, Polysilicon microbridge fabrication using standard CMOS technology, *Tech. Digest, IEEE Solid-State Sensors and Actuators Workshop, Hilton Head Island, SC, U.S.A. June 6–9, 1988*, pp. 148–150.
3 M. Parameswaran and H. P. Baltes, Fabrication of micromechanical structures with standard CMOS technology, *AMA Seminar Mikromechanik, Heidelberg, F.R.G., Mar. 14–15, 1989*, pp. 237–252.
4 M. Parameswaran, H. P. Baltes, Lj. Ristic, A. C. Dhaded and A. M. Robinson, A new approach for the fabrication of micromechanical structures, *Sensors and Actuators*, *19* (1989) 289–307.
5 K. E. Petersen, Silicon as a mechanical material, *Proc. IEEE*, *70* (1982) 420–457.
6 R. T. Howe and R. S. Müller, Polycrystalline silicon micromechanical beams, *J. Electrochem. Soc.*, *130* (1983) 1420–1423.
7 A. Heuberger (ed.), *Mikromechanik*, Springer, Berlin, 1989, Chs. 3.2. and 4.1.
8 R. G. Johnson and R. E. Higashi, A microbridge air-flow sensor, *Scientific Honeyweller (Sensor issue)*, *8* (1987) 23–28.
9 P. W. Barth, F. Pourahmadi, R. Mayer, J. Poydock and K. Peterson, A monolithic silicon accelerometer with integral air damping and overrange protection, *Tech. Digest, IEEE Solid-State Sensor and Actuator Workshop, Hilton Head Island, SC, U.S.A., June 6–9, 1988*, pp. 35–38.

Micromachining of Quartz and its Application to an Acceleration Sensor*

J. S. DANEL, F. MICHEL and G. DELAPIERRE

LETI, a Division of Commissariat à l'Energie Atomique, CENG, 85 X, 38041 Grenoble Cédex (France)

Abstract

This paper is an attempt to describe the chemical etching of quartz crystals and its interest. First, we shall give some general considerations about the quartz crystal, its crystallographic and other properties. We shall make a brief com- parison with silicon, and show the interest of quartz when the design of a sensor is required. A non-exhaustive survey on the concept of micromachining, a method developed to manufacture so-called three-dimensional components, will be made.

In the case of quartz, we shall give typical data on the composition of etching products, the behaviour with temperature and etching diagrams.

As a conclusion, we shall describe the practical realization of a micromechanical sensor, a performant closed-loop accelerometer. We shall show how some knowledge about the etching of quartz can lead to the realization of an industrial product, using standard microelectronics technologies.

1. Introduction

Nowadays more and more people have become interested in the techniques described as micromachining. Initially these techniques were developed to manufacture three-dimensional components by people working for the semiconductor industry. This is why most of the work reported in this field has been carried out on silicon [1–3]. But in fact the concept of micromachining is valid for many materials.

The object of this paper is to give some notions about another material, monocrystalline quartz. First of all we shall give some general considerations about the quartz crystal. Then a comparison with silicon will be made, showing the possible interest of quartz. We shall explain what ideas lie behind the concept of micromachining and display a few results obtained for quartz. Finally an industrial application, a quartz accelerometer, developed in our laboratories, will be presented.

*Invited paper.

2. Quartz Crystal [4, 5]

Quartz belongs to the trigonal trapezohedral class (32) of the rhombohedral subsystem. The lattice type is hexagonal. This class is characterized by one axis of three-fold symmetry and three polar axes of two-fold symmetry perpendicular thereto and separated by angles of 120°. There is no centre of symmetry and no plane of symmetry. The axes of reference (X, Y, Z) may be chosen such that X is one of the axes of two-fold symmetry and Z is the axis of three-fold symmetry.

The chemical composition is SiO_2. The Si atoms are in four-coordination with oxygen and constitute the (SiO_4) tetrahedron, which is the basic unit of the structure. Each oxygen is shared with two Si atoms. At a temperature of 573 °C, quartz undergoes a modification of structure: α-quartz becomes β-quartz with a hexagonal structure. Various properties of quartz are listed in Table 1. They are useful for the conception and dimensioning of sensors.

A property of quartz that has found important applications is piezoelectricity [6]. This effect is used to control the frequency of electronic circuits. Time control is a crucial problem in modern electronics. This is why quartz has been studied in depth although most of this work was done to achieve better resonators. From these few data, it can be seen that, due to the rather poor crystallographic symmetry of quartz, many of its properties are anisotropic: its behaviour is not the same in the Z-direction and along the plane perpendicular to it. Silicon, on the contrary, belongs to a crystallographic class with high symmetry.

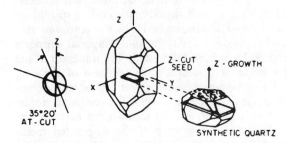

Fig. 1. A view of a natural quartz crystal. X is the electrical axis and Y the mechanical axis.

TABLE 1. Some properties of quartz

Physical property	Value $\|Z$	Value $\perp Z$	Remarks
Thermal conductivity (cal/cm s °C)	29×10^{-3}	16×10^{-3}	decreases with T
Dielectric constant	4.6	4.5	decreases with T
Thermal expansion coefficient (°C)	7.1×10^{-6}	13.2×10^{-6}	increases with T
Electrical resistivity (Ω cm)	$0.1 \times 10^{+15}$ (ionic)	$20 \times 10^{+15}$ (electronic)	decreases with T
Young's modulus (N m^{-2})	$9.7 \times 10^{+10}$	$7.6 \times 10^{+10}$	
Density	2.65		

3. A Brief Comparison between Quartz and Silicon

Some properties of the two materials are given in Table 2. We have indicated those for which quartz can be interesting and explain the success of silicon.

3.1 Mechanical Properties
On this point the two materials do not seem to exhibit many differences.

3.2. Electrical Properties
Quartz is an insulator. As a result if conducting tracks are deposited onto its surface, there is no leakage current between the different conductors. This may be useful in some applications.

Quartz is piezoelectric. This effect is used for one sensor family in which the basic component is a quartz resonator. The external data to be measured are considered as a perturbation that modifies the frequency of the system.

Silicon is a semiconductor and is mainly used for integrated circuits. This is one advantage of silicon over other materials when designing a sensor: the sensor itself and the associated electronic circuit can be realized on the same chip.

3.3. Why Silicon for a Sensor?
Most of the work on micromachined sensors has been done with silicon. There are several reasons for this: the main one is the one-chip concept; this material has been deeply studied, and its physical properties are well known, as well as the technologies enabling it to be used; silicon is easily available, the monocrystalline substrates are large (over five inches in diameter), rather cheap and of good quality.

This is why, when conceiving a sensor, one has to look carefully if silicon cannot be used. However, the desire to obtain on the same chip a micromachined sensor and microelectronic circuit may lead to a technology so complex that the production yield is near zero. When the one-chip concept is not really necessary, it can be interesting to look for other materials. We think that quartz could be one of these materials.

Before giving some results on the micromachining of quartz, we shall explain what micromachining is.

4. The Concept of Micromachining [7]

This notion includes all the processes used to manufacture mechanical components on a micrometric scale. It has originated with the evolution of integrated circuits. A basic principle is batch-fabrication, which means that a lot of similar components can be fabricated at the same time on a single wafer.

The micromachining of a wafer is done by etching away some parts of this wafer. We shall restrict ourselves to chemical etching, although other means are available, dry-etching for example.

The etching speed of a single-crystal wafer depends on the crystalline orientation of this wafer [8]. The behaviour is related to the atomic density of the etched surface, the slowest speed corresponding generally to the densest planes. To have a thorough knowledge of the dependence of etching speed on orientation, wafers with different surface orientations may be used. Once these data

TABLE 2. Mechanical properties of quartz and silicon

Property	Quartz	Silicon
Density	2.65	2.32
Young's modulus (10^{+10} N m^{-2})	(001):10 $\perp$(001):8	(100):13 (110):17 ($1\bar{1}1$):19
Bending mechanical strength (10^{+7} N m^{-2})	9	7–20

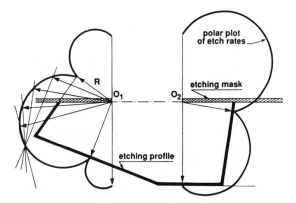

Fig. 2. Two-dimensional Wulff diagram.

are known, it is possible to foresee the etching profile that is obtained when masking a wafer with a given design. The method has been described by Wulff [9–11], and is illustrated in Fig. 2.

Points O_1 and O_2 are the limits of the etching mask. The origin of the etching diagram (i.e., the variation of etching rate with direction) is placed at O_1 and O_2, the left side of the diagram being taken for O_1 and the right side for O_2. The perpendiculars to the rate vectors, R, are drawn. The etching profile is in fact the envelope curve of these perpendiculars. It is the set of points on these lines that can be reached from O_1 and O_2 without crossing any other line. This method may be used for a three-dimensional case when the etching mask has a complex shape.

The main problem is to obtain the three-dimensional etching diagram, which requires a lot of data and consequently of experiments. However, the diagram can be obtained by observing etching profiles on a few substrates. The observation is made on grooves or beams. Planar facets appear on the side-walls, corresponding to minimum etch-rate directions.

The complete etch-rate diagram is then estimated by joining the ends of the etch-rate vectors by arcs of a circle. These arcs are determined using minimum criteria [10, 12].

5. Micromachining of Quartz

5.1. Etching Procedure

The etchants used for quartz are warm solutions of hydrofluoric acid, HF, and ammonium fluoride, NH_4F [13].

The masking material generally used is a chromium–gold thin film obtained by vacuum evaporation or sputtering. The thin-film etching patterns are made by photolithographic techniques.

Experiments have shown that the effect of the etching solution on a mechanically lapped surface differs from the effect observed on a chemically polished surface [14, 15]. The quality of surface finish prior to etching is thus important for micromachining. Great care must be given to the preparation of this surface before attempting to perform any etching on it. This is especially true if the micromachined structures must have surfaces of good quality, such as those required for membranes.

5.2. Properties due to Quartz Symmetry

Etching diagrams have the quartz symmetry: Z is a three-fold axis and X a two-fold axis. This reduces the number of data that need to be obtained to know the complete diagrams. The observation can be limited to an angular sector of 120° around Z and to an angular sector of 180° around X. The study of the etching profiles of beams on a Z-cut wafer can be reduced to an angular sector of 60°.

Sectors 1, 3, 5 in Fig. 3. are equivalent in a rotation of 120° around Z; so are sectors 2, 4, 6, and the profiles in sector 4 are in fact the same as in sector 1 (with a left–right inversion). The profiles on the minus-Z face are deduced by a rotation of 180° around X.

5.3. Variation with Etchant Composition and Temperature

It is interesting to know the variation of the etch rate when the amounts of NH_4F and HF are varied in the etching solution, as the profiles can vary greatly with this parameter (see Fig. 4).

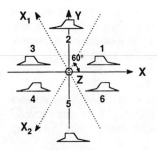

Fig. 3. Etching profiles on a Z-cut wafer.

	Y - section	X - section
HF	x $\uparrow Z$ 28°	32° $\uparrow Z$ 0° $\rightarrow Y$
$NH_4 HF_2$	65° 90° 28° 28°	32° 0°

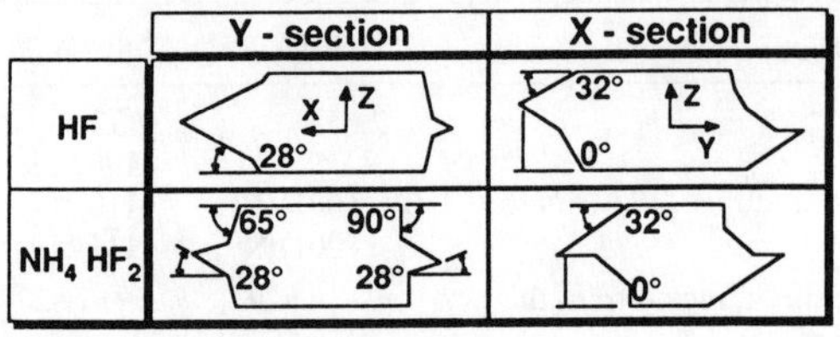

Fig. 4. Examples of profiles on a Z-cut wafer.

TABLE 3. Variations of etching rate ($\Theta = 25\ °C$)

Bath composition (mole/l)	Etch rate (μm h^{-1})				Ratios		
	R_x	R_y	R_z	R_{AT}	R_z/R_x	R_z/R_y	R_z/R_{AT}
10.9 HF	0.02	<0.005	9.6	1.65	240	1000	2.9
7.2 HF + 4 NH$_4$F	0.025	0.005	2.55	1.35	102	500	1.9
5.4 NH$_4$NH$_2$	0.015	0.015	1.1	0.48	73	73	2.3
5.4 NH$_4$HF$_2$ + 1.8 NH$_4$F	0.015	0.015	0.75	0.3	50	100	2.5

We indicate in Table 3 published results [13] as, to our knowledge, no work has yet been done on varying systematically the wafer orientation and the etchant concentration.

The etching rate is always much faster along the Z-axis. In fact, for most applications where a mechanical structure must be obtained, Z-cut wafers are used.

Despite the lack of more thorough experiments, it seems that both the etching and the anisotropy increase with the proportion of HF.

The expected variation with temperature is described by the Arrhenius equation [17, 18]:

$$R = A \exp[-E/kT]$$

where E is the apparent activation energy of the reaction, A includes concentration terms, k is the Boltzmann constant and T is the absolute temperature. The etch rate has an exponential variation with temperature. A few scientists think that the activation energy E should depend on the orientation [19]. Some results are given in Table 4.

However, most publications [13–15] consider that the activation energy depends only on the chemical reaction and is then constant with the cut. It is the same as the activation energy for the dissolution of SiO$_2$ [20] (see Table 5).

For a concentration (NH$_4$F)/(HF) around one, the activation energy is about 40 kJ/mole, which gives a multiplication by 1.6 in etching rate for every ten degrees for temperatures around 50 °C.

TABLE 4. Observed variation of activation energy (etchant 40% HF)

Direction of etching	E (kJ/mole)
+x	21
y	15
z	48

TABLE 5. Activation energy for silica

Solution composition (F)		[NH$_4$F]/[HF]	$\Delta E_{app.}$ (kcal/mole)
NH$_4$F	HF		
0	10.0	0	6.7
2.00	10.0	0.2	8.1
1.00	2.50	0.4	8.2
8.00	10.0	0.8	8.8
8.00	2.50	3.2	10.3
10.5	3.53	3.0 (7:1 buffered HF)	11.0

5.4. Etching Diagrams

To our knowledge, the only three-dimensional diagrams published were obtained by Ueda et al. [16, 21] (see Fig. 5). The etching rates on quartz wafers with 21 different angles of cut were measured and the final diagrams were interpolated. The etchant was ammonium bifluoride, NH$_4$HF$_2$,

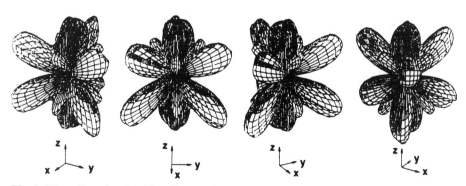

Fig. 5. Three-dimensional etching diagrams for quartz.

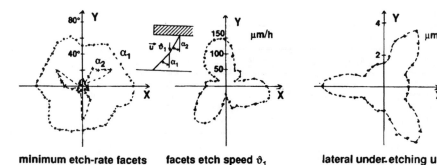

minimum etch-rate facets α_1, α_2 facets etch speed ϑ_1 lateral under-etching u

Fig. 6. Simplified two-dimensional diagrams.

TABLE 6. Mechanical strength of micromachined beams

Mechancial strength	Tensile	Torsion	Bending
Macroscopic value (10^7 N m^{-2})	13		9
Micromachined (theoretical limit)	60 (110)	130 (160)	170 (230)

and the temperature 82 °C. By applying the Wulff method to such diagrams, the cross-sectional shapes of devices fabricated on wafers with any orientation may be predicted.

It would surely be interesting to have the diagrams for every temperature an every concentration. This would permit the best conditions for manufacturing a given structure to be chosen.

For the design of a sensor, we needed micromachined beams with steep sides. Experiments were conducted on easily available quartz wafers with Z-orientation. The profiles of beams etched by HF/NH$_4$F (proportion 3/2) at a temperature of 82 °C were observed. The result given in Fig. 6 represent the orientation of the minimum etch-rate facets and their etching rates. These two-dimensional diagrams are sufficient to predict the final shape resulting from an etching mask deposited on the wafer.

A noteworthy fact about the beams obtained by chemical micromachining is that they offer a very good surface quality. As a result, they display a mechanical strength higher than that obtained by classical means [16] (see Table 6).

6. Practical Realization: an Acceleration Sensor [7, 22]

Our laboratory became interested in the micromachining of quartz because of our involvement in the conception of a batch-fabricated acclerometer. A study was carried out on quartz and silicon in order to obtain thin flexible foils, and finally quartz was thought better for two main reasons: (a) it is an insulator, so there is no leak resistance between the conductor tracks; (b) the micromachining technology appears less complicated than for silicon and is more relevant to obtaining beams with steep edges (great anisotropy between the Z-direction and all directions perpendicular thereto).

From the experimental data, the steepest edge is obtained for a beam along the X-axis. As only one side is steep, it is necessary to use a double-face etching, taking advantage of the fact that X is a two-fold axis.

The accelerometer in itself is a rather complex structure (see Fig. 7). Two foils parallel to the X-axis are flexible in the x direction parallel to the wafer plane. They support a seismic mass m which gives its sensitivity to the system. The mass displacement due to an acceleration A_x is measured by a capacitive detector. The surface of the mass is covered by a resistive coil, which allows the acceleration force mA_x to be conteracted by a Laplace force IB_z when placed in a permanent magnetic field B_z and driving an electric current I.

Typical dimensions of this design are: wafer thickness, 125 µm; beam width, 5 to 8 µm; beam length, 2 mm; mass dimensions, 2 × 2 mm; detector gap, 50 µm; coil material, gold; thickness, up to 3.5 µm; track width, 8 µm; track interspace, 7 µm. The coil is obtained by electroplating. The remetallization of the electrodes is done by evaporation, a mechanical mask protecting the parts of the wafer where no metallization is required.

The structure has the following features:

(a) Good mechanical properties obtained by micromachining of a monocrystal.

(b) An excellent directivity, for the width of the beam (5 µm) is small compared to its thickness (125 µm).

(c) The conception is monolithic, with the mechanical and detection functions on the same substrate.

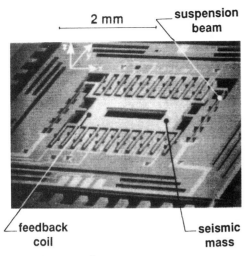

Fig. 7. Quartz accelerometer.

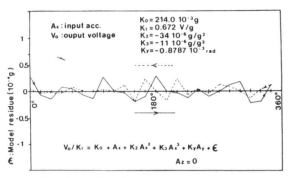

Fig. 8. Experimental results.

(d) The components of any acceleration perpendicular to z can be obtained by micromachining two similar structures orientated along two different X-axes.

The sensor output is expressed by the following equation:

$$V_o/K_1 = K_o + A_x + K_2 A_{x2} + K_3 A_{x3} + K_y A_y + K_z A_z + \varepsilon$$

where K_1 = scale factor and K_o = offset, A_x, A_y, A_z = acceleration components.

Figure 8 represents the value of the residuum ε given by experiments in the gravitation field ($A_x = \pm 1$ G).

7. Conclusions

A few data on the micromachining of quartz have enabled us to manufacture a reliable sensor. Quartz seems quite a good material for the realization of mechanical microstructures. Very complex shapes combining beams, membranes, grooves, etc., can be achieved.

We now intend to study these possibilities more thoroughly, particularly by looking at the variations of etching diagrams when parameters such as temperature and etchant concentration are varied.

References

1 J. B. Angell, S. C. Terry and P. W. Barth, Silicon micromechanical devices, *Sci. Am.*, 248 (Apr.) (1983) 36–47.
2 K. E. Petersen, Dynamic micromechanics on silicon: techniques and devices, *IEEE Trans. Electron Devices*, ED-25 (1978) 1241–1250.
3 K. E. Petersen, Silicon as mechanical material, *Proc. IEEE*, 70 (1982) 420–457.
4 C. Frondel, *Dana's system of Mineralogy, Silica Minerals*, Vol. 3, Wiley, New York, 7th edn, 1962, pp. 9–170.
5 P. Pascal, *Nouveau Traité de Chimie Minérale*, Vol. 8, *Silicium*, Masson, Paris, 1965, pp. 38–74.
6 V. Bottom, *Introduction to Quartz Crystal Unit Design*, Van Nostrand Reinhold, New York, 1982, pp. 50–62.
7 G. Delapierre, Micro-machining: a survey of the most commonly used processes, *Sensors and Actuators*, 17 (1989) 123–128.
8 R. B. Heimann, Principals of chemical etching, the art and science of etching crystals, in *Crystals*, Vol. 8, Springer, Berlin, pp. 214–224.
9 G. Wulff, Zur Frage der Geschwindigkeit des Wachsthums und der Auflösung der Krystallflächen, *Z. Krist*, 34 (1901) 449–530.
10 C. Herring, Some theorems on the free energies of crystal surfaces, *Phys. Rev.*, 82 (1951) 87–93.
11 R. J. Jacodine, Use of modified free energy theorems to predict equilibrium growing and etching shapes, *J. Appl. Phys.*, 33 (1962) 2643–2647.
12 D. W. Shaw, Morphology analysis in localized crystal growth and dissolution, *J. Cryst. Growth*, 47 (1979) 509–517.
13 J. K. Vondelig, Fluoride-based etchants for quartz, *J. Mater. Sci.*, 18 (1983) 304–314.
14 J. R. Vig, J. W. Lebus and R. L. Filler, Chemically polished quartz, *Proc. 31st Ann. Symp. Frequency Control, Monmouth, NJ, U.S.A. 1977*, Electronic Industries Association. Washington, DC, 1977, p. 131.

15 C. R. Tellier, Some results on chemical etching of AT-cut quartz wafers in ammonium bifluoride solutions, *J. Mater. Sci.*, 17 (1982) 1348–1354.
16 T. Ueda, F. Kohsaka and D. Yamazaki, *Proc. 3rd Int. Conf. Solid-State Sensors and Actuators (Transducers '85), Philadelphia, PA, U.S.A., June 11–14, 1985*, pp. 113–116.
17 P. J. Holmes, *The Electrochemistry of Semi-conductors*. Academic Press, New York, 1962, Chs. 6 and 8.
18 H. Eyring and E. M. Eyring, *Modern Chemical Kinetics*, Reinhold, New York, 1965, pp. 5–9.
19 E. I. Lazorina and V. V. Soroka, Etching of quartz and some features of the surface layer, *Sov. Phys. Crystallogr.*, 18 (1974) 651–653.
20 J. S. Judge, A study of the dissolution of SiO_2 in acidic fluoride solutions, *J. Electrochem. Soc.*, 118 (1971) 1772–1775.
21 T. Ueda, F. Kohsaka, T. Iino and D. Yamazaki, Theory to predict etching shapes in quartz and application to design devices, *Trans. Soc. Instrum. Contr. Eng.*, 23 (1987) 1–6.
22 G. Delapierre, J. S. Danel, F. Michel, J. L. Bost, A. Boura and O. Aujay, A quartz micro-machined closed loop accelerometer, *Proc. Eurosensors 1987, Cambridge, U.K., Sept. 22–24, 1987*, pp. 223–224.

Section 12

LIGA

THE LIGA process (in German, <u>L</u>ithographie, <u>G</u>alvanoformung, <u>A</u>bformung) exposes Poly-methyl methacrylite (PMMA) plastic with synchrotron radiation through a mask. The exposed PMMA is then washed away, leaving vertical wall structures with spectacular accuracy. Structures a third of a millimeter high and many millimeters on a side are accurate to a few tenths of a micron. Metal is then plated onto the structure, replacing the PMMA that was washed away. This metal piece can become the final part, or can be used as an injection mold for parts made out of a variety of plastics.

FABRICATION OF MICROSTRUCTURES USING THE LIGA PROCESS

W. Ehrfeld, P. Bley, F. Götz, P. Hagmann, A. Maner, J. Mohr,
H. O. Moser, D. Münchmeyer, W. Schelb, D. Schmidt, E. W. Becker

Kernforschungszentrum Karlsruhe GmbH,
Institut für Kernverfahrenstechnik, Postfach 3640,
D-7500 Karlsruhe, Federal Republic of Germany

Abstract. For fabricating microstructures with extreme structural heights a new microfabrication technique has been developed by the Karlsruhe Nuclear Research Center. The so-called LIGA method is based on a combination of deep-etch X-ray lithography, electroforming and molding processes (in German: Lithographie, Galvanoformung, Abformung). The minimum lateral dimensions of the microstructures which can be fabricated from metallic and polymeric materials without any restrictions in the cross-sectional shape, are in the micrometer range and the structural heights are some hundred micrometers. A large number of innovative microstructure products may be produced in the field of sensors, compound materials, filtration, fluid dynamics, microoptics, synthetic fibers, communication techniques and microelectronics. Various concepts of such products, e. g. microfiltration modules, spinneret plates, electrical microconnectors, optical multiplexers, micronozzles, and various sensors, have been worked out.

1. Introduction

Advanced processes for fabricating large-scale integrated circuits more and more include special fabrication steps which allow to generate microstructures with large structural heights. The aspect ratio which is defined as the structural height or depth to the minimum lateral dimension may easily reach a value of 10 when using dry etching methods like reactive ion etching or reactive ion beam etching. Under special conditions, an aspect ratio of nearly 10 may also be obtained in optical microlithography. Much higher values of some 100 have been demonstrated for X-ray lithography [1,2] which correspond to the maximum aspect ratios attainable by wet-chemical anisotropic etching of monocrystalline silicon [3].

The increasing variety of semiconductor-based methods of microfabrication which are suitable for high aspect ratio and, hence, semi three-dimensional structures, has resulted in a growing number of innovative microstructure products appearing in various disciplines apart from semiconductor technology. In the frame-work of this development, also new process schemes have been worked out and optimized which combine VLSI fabrication methods with those of other technological fields.

This paper deals with the so-called LIGA process [1], its application potential and, in particular, its specific advantages in the fabrication of microsensors. The process is based on a combination of deep-etch X-ray lithography, electroforming and micromolding processes and the abbreviation LIGA is simply derived from these major process steps (in German: Lithographie, Galvanoformung, Abformung). It has been demonstrated by a great number of laboratory-scale experiments that metallic and polymeric microstructures with minimum lateral dimensions in the micrometer range and structural heights of several hundred micrometers can be fabricated by the LIGA process. The lateral tolerances are in the submicrometer range over the total height of the microstructures and, in contrast to anisotropic etching of monocrystalline materials, the cross-sectional shape of the high aspect ratio structures can be designed without any restrictions. The micromolding process step which utilizes subminiaturized mold inserts generated by deep-etch X-ray lithography and electroforming may be directly regarded as a new branch in the well-known mass production of polymeric structures by injection molding. Consequently, a promising basis exists for favorably low costs in series fabrication.

The possible range of applications of the LIGA method covers various technological disciplines like micromechanics, integrated optics, sensor technology, microelectronics, and chemical, medical and bio-engineering. Interesting examples for potential future microstructure products are electrical and optical microconnectors, optical multiplexers, highly porous microfiltration systems, devices for plasmapheresis, spinneret plates, slide bearings and, in particular, various microsensors for measuring vibration, position, radiation and composition.

The development work on the LIGA process has been initiated by the Karlsruhe Nuclear Research Center (KfK) for the production of micron-sized nozzles for the separation of the uranium isotopes [4]. Under contract with KfK, the Siemens AG and the Fraunhofer-Gesellschaft have made substantial contributions in demonstrating the feasibility of the LIGA method by extensive experimental work. The current development work on process technology goes ahead as a cooperation between KfK and the industrial companies Steag AG, Essen, and Degussa AG, Frankfurt. Further firms participate in co-operative projects with KfK, Steag and Degussa for developing a variety of microstructure products.

2. Principle of the Fabrication Process

The principle of the LIGA method is evident from the process sequence shown in Figure 1. A polymeric material (resist) which changes its dissolution rate in a liquid solvent (developer) under high-energy irradiation, is exposed through an X-ray mask to highly intensive parallel X-rays. The mask consists of a thin membrane of low atomic number material which is largely transparent to X-rays and an absorbing pattern of high atomic number material applied onto the membrane [5]. The radiation source is an electron synchrotron or an electron storage ring which, at present, is the only possibility to generate the highly collimated photon flux in the spectral range required for precise deep-etch X-ray lithography in thick resist layers. If one considers a thickness between 10 and 1000 μm for the resist layers to be structured, the optimum critical wavelength of the synchrotron radiation source ranges from some 0.1 to 1 nm for typical resist materials [6]. Usually, cross-linked poly-methylmethacrylate (PMMA) with suitable adhesion promoting agents is used as an X-ray resist which is polymerized as a thick layer directly on an electrically conductive substrate [1,2,7].

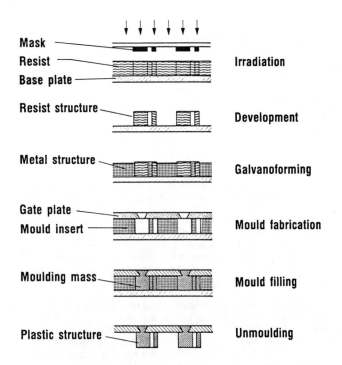

Fig. 1: Schematic representation of a process sequence of the LIGA method for mass fabrication of metallic microstructures: A primary resist template is generated by synchrotron radiation lithography. In the second step, a complementary metallic structure is obtained by electroforming. This structure is used as a tool for a multiple reproduction of polymer templates by reaction injection molding. Finally, metallic microstructures are produced using these secondary templates in a secondary electroforming process. It stands to reason that, depending on the specific aspects attributed to the various microstructure products, the process sequence may end with an earlier process step or may also be repeated partially or completely.

The irradiation of PMMA results in a chain scission of the polymer. Accordingly, the irradiated regions can be dissolved in a proper solvent or solvent mixture (developer) and a high aspect ratio relief structure of PMMA is obtained. Swelling of unexposed material and resulting defects or dimensional distortions during the developing process can be avoided by using multi-component developers whose solubility parameters are near the periphery of the solubility range of PMMA [2].

In the next step, the PMMA structure is used as a template in an electroforming process where metal is deposited onto the electrically conductive substrate in the spaces between the resist structures. In this way, a complementary metallic structure is obtained which can be either the final microstructure product or can be used as a microtool (mold insert) for multiple reproduction by means of a molding process.

The micromolding process has been optimized using methacrylate based casting resins with a special internal mold release agent [8]. In the process sequence shown in Fig. 1 the mold material is introduced into the mold cavities through the holes of a gate plate. This plate which has a formlocking connection with the polymeric microstructures after hardening of the resin serves as an electrode in a second electroforming process for generating secondary metallic microstructures. It has been demonstrated by many experiments that, in spite of an aspect ratio of about 100 and minimum lateral dimensions of only some micrometers, a yield of approximately 100 % can be obtained in the micromolding process. In addition, optimization studies on the electroforming process have given a promising basis for mass production of metallic microstructures [9].

Figure 2 demonstrates the performance and the technological potential of the LIGA method by some scanning electron micrographs of separation nozzle structures. They were taken from the primary (resist) template, the primary metallic structure, the secondary (molded) template, and the secondary metallic structure. It is evident that the secondary structures are perfect copies of the primary structures. Consequently, mass production of polymeric and metallic microstructures should be feasible without a continuous utilization of a synchrotron radiation source which is only necessary for fabricating mold inserts.

The precision in generating high aspect ratio resist structures by deep-etch X-ray lithography is illustrated by Figure 3. It shows a scanning electron micrograph of a cross-shaped PMMA test structure (Fig. 3a) where the width of a bar has been determined by means of a critical dimension measuring system. It is evident from Fig. 3b that the width varies only by ± 0.1 µm over the total height of 400 µm. This variation results from a combined influence of various mechanisms like Fresnel diffraction, the range of high-energy photoelectrons generated by X-rays, the finite divergence of synchrotron radiation and the time evolution of the profiles during the developing process [6].

3. Potential Fields of Application and Concepts for Microstructure Products

In contrast to microelectronics where a broad experience in a giant market exists and leading products like microprocessors and memory chips are definitely established, other disciplines are predominantly in a very early stage in using microfabrication techniques. Although silicon-based pressure sensors, thin film temperature sensors, magnetic bubble memories and several other products are fabricated in a large number of pieces and even though microstructures in general have a high potential for reducing fabrication costs, industry is still very cautious in the acceptance of microstructure technology beyond microelectronics.

Consequently, the LIGA process which is still in a laboratory stage cannot be introduced with a clearly defined leading product. It is rather necessary to demonstrate its technological potential by a variety of microstructure products which may form the basis for future development. Potential fields of application and concepts for such microstructure products will be described in the following.

a)

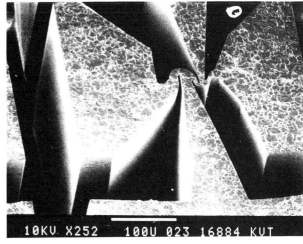

c)

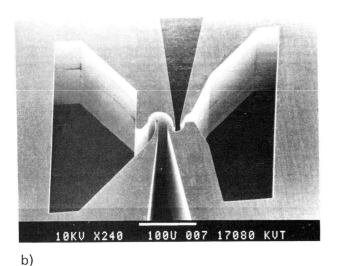

b)

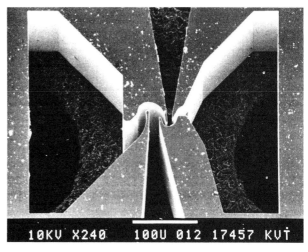

d)

Fig. 2: Scanning electron micrographs of separation nozzle structures produced at different steps of the process sequence of the LIGA method:
a) Resist structure produced by synchrotron radiation lithography (primary template).
b) Metallic structure produced by electrodeposition of nickel into the free spaces between the walls of the resist structure (mold insert).
c) Polymer structure produced by reaction injection molding using the mold insert produced by lithography and electroforming (secondary template).
d) Metallic structure produced by electrode position of nickel into the free spaces between the walls of the polymer structure (final product).

The length of the white bars at the lower edges of the micrographs represent a lateral dimension of 100 µm. The structure height is 300 µm, the minimum channel width of the curved separation nozzle is 3 µm. The stripes on the surface of the resist structure (a) result form preceding machining of the surface. The scars in the surface of the polymer structure (c) reproduce the surface of the substrate which forms the bottom of the mold insert.

3.1 Micro- and Ultrafiltration

Today's microfiltration membranes are preferably fabricated from polymers by phase inversion technique or by sintering of metallic or ceramic materials. They are characterized by a relatively broad distribution in respect to the size and position of the pores and are normally used as depth filters. Surface filters are produced e. g. by stretching of polymeric membranes or by nuclear track technique where the tracks of high energy particles traversing a foil are enlarged by etching. However, all these fabrication procedures include statistical processes which result either in a broad pore size distribution or, in the case of isoporous nuclear track membranes, in a limited porosity.

In contrast to the standard methods for membrane fabrication, the LIGA process is capable of producing microporous structures of extremely high porosity and absolutely uniform pore size [10]. This is illustrated by the scanning electron micrographs of honeycomb structures shown in Figure 4. The microstructure of the individual pores can be designed without any restrictions provided that the minimum lateral dimensions are greater than 0.5 µm. Because of the high aspect ratio a high mechanical stability is achievable. The membranes can be prepared from a wide variety of materials, e. g. of cross-linked polymers, thermoplastic materials, metals, metallic alloys and, probably, also from special glasses and ceramic materials. Accordingly, such membranes should successfully be applicable at elevated temperatures and in aggressive environments.

Many of todays reverse osmosis or gas separation membranes and almost all pervaporation membranes are manufactured as composite structures having a thin selective layer on top of a microporous support. Such supporting structures must have a certain thermal, chemical and mechanical stability, a narrow pore size distribution and, to ensure a low hydrodynamic resistance, a high porosity. All these properties may be ideally obtained in microporous structures fabricated by the LIGA process.

Besides plate shaped filtration structures, threedimensional microfiltration elements can be fabricated by the LIGA process. This is illustrated by the schematic representation of a filtration module shown in Figure 5. Microbars generated by deep-etch lithography, electroforming or molding processes are arranged in a Venetian blind pattern, where the bar arrays are forming the filtrating structure. The filtrate and the concentrate are withdrawn in a countercurrent flow. Since the concentrate is flowing parallel to the filtrating surface, plugging is supposed to be reduced considerably.

Such microfiltration elements might be particularly useful for a careful separation of living cells from biological fluids. In order to avoid damage to the cells by shear stresses and a deposition of damaged cells in dead end volumes, the feed stream might be splitted stepwise into partial streams like the branching of blood vessels and fed into the filtrating regions. The resulting partial streams of the concentrate might be collected correspondingly by means of a branched collecting unit, where both units could be fabricated as integral parts of the microfiltration module.

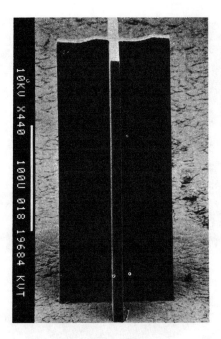

a)

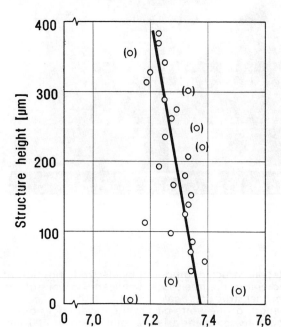

b)

Fig. 3: Demonstration of the precision achievable by deep-etch X-ray lithography in the lateral dimensions of high aspect ratio structures.
a) Scanning electron micrograph of a cross-shaped test structure.
b) Bar width b of the test structure determined by a critical dimension measuring system. The variation is within ±0.1 µm over the total structure height of 400 µm.

a)

b)

c)

Fig. 4: Honeycomb structures fabricated by the LIGA process.
a) Honeycomb structure fabricated from PMMA by deep-etch X-ray lithography. The structural height amounts to 350 μm, the size of the openings is 80 μm and the wall thickness is 4 μm.
b) Honeycomb structure fabricated from nickel by deep-etch X-ray lithography and electroforming. The structural height amounts to 300 μm, the size of the openings is 80 μm and the wall thickness is 7 μm.
c) Honeycomb and prism structures fabricated by micromolding.

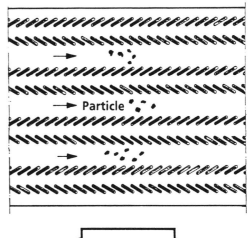

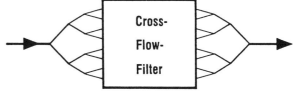

Fig. 5: Schematic representation of a counter flow microfiltration structure and a corresponding module with integrated feed and discharge units.

3.2 Compound and Carrier Structures

Compound materials which are fabricated by sintering of several components or by infiltration of a molten phase into a porous sintered body are generally characterized by an irregular microscopic distribution of the various materials and, hence, a random variation of the local material properties. Accordingly, the specific properties of the different materials cannot be utilized with the same efficiency as in regular microstructures which have been designed and optimized according to the specific application.

In contrast to such fabrication processes, the LIGA method allows to produce completely regular layers of compound materials which should have distinct advantages compared to irregularly structured layers. For example, an electrical contact plate which should have a high mechanical stability as well as a high electrical conductivity might consist of a honeycomb microstructure of a mechanically stable material filled with a second material with high electrical conductivity. By analogy, a slide bearing layer might be fabricated by filling a regularly-shaped supporting structure with a solid or pasty lubricant which allows to define exactly the shape and size of lubricating and supporting regions.

An optimum carrier structure for a catalyst may be simply characterized by a large ratio of the contact surface to the total mass of carrier material. In addition, the pressure losses of the chemically reacting fluid flowing through the carrier structure should be as low as possible and, moreover, the consumption of the expensive catalytically active material should be minimized. Again, regular microstructures are technically superior to irregular structures, however, a commercial success is only achievable by a low-cost microfabrication process.

3.3 Synthetic Fiber Production

In the production of synthetic fibers, a molten polymer or a polymer solution is pressed through capillaries (nozzles) in a spinneret plate and hardened downstream of the nozzles by cooling, by evaporation of the solvent, or by means of a precipitating agent. For many reasons, capillaries with non-circular cross-sections are required. The minimum characteristic dimensions are of the order of 50 μm and the length of the capillaries is usually considerably larger than the characteristic dimensions. Such "profiled" capillaries are normally produced by spark erosion which process establishes a practical lower limit of 20 to 50 μm for the minimum characteristic dimensions.

In fabricating spinneret plates with profiled capillaries the LIGA process has a number of technical advantages compared with the spark erosion process [11]. In particular, the minimum characteristic dimensions can be reduced by an order of magnitude and a high capillary length with an excellent surface finish can be obtained without difficulties. In contrast to a successive fabrication of the capillaries in spark erosion, using the LIGA process a large number of capillaries can be generated in parallel. A simple commercial comparison is not possible since both processes have completely different cost structures with regard to the number of spinneret plates to be produced, the number and the shape of the capillaries, the plate material, etc.

In a laboratory scale some test specimens of spinneret plates have been fabricated. The scanning electron micrographs shown in Fig. 6 represent an array of resist templates of octolobal capillaries made by X-ray lithography, a partial view of the corresponding spinneret plate made by electroforming, and an enlarged view of a single capillary. Templates have also been produced by means of micromolding.

In addition to the fabrication of spinneret plates, the LIGA process might be favorably applied in a number of micromechanical parts required in synthetic fiber production. Typical examples are guide structures, special modules for polymer filtration and devices for fiber texturation.

3.4 Electrical and Optical Microconnectors

The continuous increase of logic functions in microprocessor chips and the decrease in critical dimensions results in a corresponding demand for high density electrical connections with subminiaturized dimensions. A similar trend will probably be observable in the near future in optical communication technology, where the interconnection of optical waveguides, e. g. in the case of multiple monomode fiber connectors, requires micromechanical fabrication methods with submicron tolerances.

A typical example for a possible fabrication of a manifold microplug is shown in Fig. 7. In the first step, templates of the plugs are fabricated by deep-etch X-ray lithography or by micromolding using a radiation sensitive mold material. In the second step, the templates are filled with metal by electrodeposition and, in the third step, the insulating material of the templates is partially removed by irradiation and subsequent development so that the coupling region is uncovered.

Besides a detachable microconnection, flexible permanent microconnections can be fabricated by the LIGA process. Technical advantages may be achievable in compensating differences in thermal expansion coefficients of the chip and the substrate or in generating structures with a more efficient dissipation of heat.

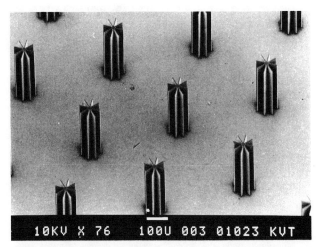
a)

b)

c)

Fig. 6: Fabrication of spinneret plates with octolobal capillaries illustrated by scanning electron micrographs.
a) Resist templates of the capillaries.
b) Nickel plate with octolobal capillaries.
c) Single capillary with a slit width of 5 μm.

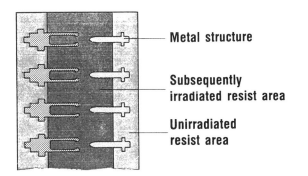

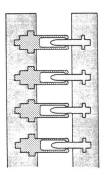

Fig. 7: Schematic representation of a process sequence for fabricating a manifold microplug. The fan-out of the plugs has been omitted for reasons of simplicity.

3.5 Microoptics

In the field of microoptics, the LIGA process could be used for fabricating waveguides, gratings, small prisms and small cylindrical lenses, zone plates, polarizers and spectral filters for the infrared range, spatial filters, modulators and many other optical devices and structures. More or less arbitrarily, one may discriminate between two basic configurations. One is represented by a perforated, self-supporting membrane of plate structure which is put into the path of rays and where the optical effect is determined by the special pattern of perforation. The other configuration may be characterized by the fact that optical structures are generated on a stable substrate where the path of rays is parallel to the surface of the substrate and partially determined by a waveguide structure. In the following, two special examples of these basic configurations will be described.

Figure 8 shows a scanning electron micrograph of a self-supporting Fresnel zone plate made by electroforming with nickel using a template generated by deep-etch X-ray lithography. In a similar way, self-supporting gratings or infrared polarizers might be fabricated.

Figure 9 illustrates the principle of a wavelength division multiplexer/demultiplexer for optical communication systems which utilizes the Rowland configuration [12]. A curved reflective grating (echelette grating) effects a spectral resolution of the incoming light and a selective focusing to the outputs.

Fig. 8: Scanning electron micrograph of a self-supporting Fresnel zone plate. The smallest characteristic dimensions are 5 µm, the structural height is 80 µm. The zone plate is still connected with the electroforming electrode.

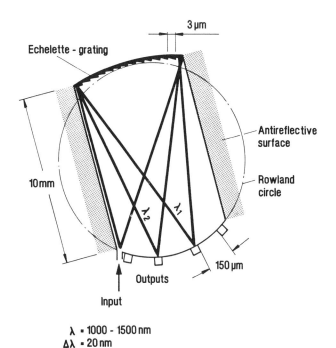

Fig. 9: Principle of a LIGA configuration for wavelength division multiplexing in optical communication systems.

Two basic designs of such LIGA devices are shown in Figure 10. The first consists of a slab waveguide of PMMA where the grating pattern and the input and output structures are simultaneously fabricated by deep-etch X-ray lithography. The grating is covered with a reflecting metallic layer and the waveguide is cemented between cover plates with a refractive index lower than that of PMMA. The other device consists of a corresponding metallic structure which has a higher mechanical stability but also a higher attenuation than the slab waveguide. Both configurations are presently optimized with respect to the geometrical parameters, the fabrication procedure and the material to be applied.

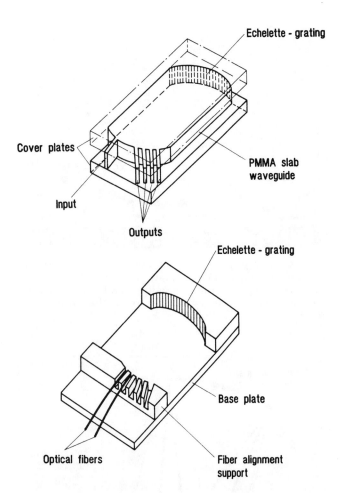

Fig. 10: Design of PMMA slab waveguide and metallic multiplexer/demultiplexer devices.

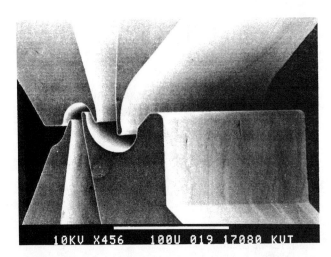

Fig. 11: Scanning electron micrograph of a double deflecting separation nozzle electroformed from nickel. The minimum slit width of the curved nozzle is 3 µm, the slit length is about 300 µm.

3.6 Flow Devices

In fluid and gas dynamics a number of devices and configurations inherently must have dimensions in the micrometer range or can be improved with respect to efficiency and performance by miniaturization. The outstanding example for the LIGA process is the separation nozzle which initiated the development of this microfabrication method [4]. In the separation nozzle, the optimum operating pressure is inversely proportional to the characteristic dimensions of the flow. Since a higher gas pressure results in a corresponding reduction of the compressor size and the piping system for a given throughput of a separation nozzle plant, the reduction of the dimensions of the nozzle has a direct economic advantage. The precision in fabricating such nozzles is demonstrated by Fig. 11 which shows a scanning electron micrograph of a double deflecting nozzle.

Furthermore, the LIGA method could be applied for the fabrication of ink-jet nozzle arrays, pneumatic sensors and pneumatic logic devices. In all cases, complex micron-sized flow schemes might be realized and a comparatively high degree of integration would be achievable.

3.7 Sensors

The development of modern sensors aims at miniaturization, more complex structures and integration with electronic signal conditioning. Furthermore, the expenditure for adjustment, trimming and replacement should be minimized and, accordingly, the manufacturing tolerances should be as small as possible. In this respect, the LIGA method has a number of advantages which will be illustrated by a description of several sensor devices in the following.

3.7.1 Vibration and Acceleration Sensors

For measuring vibration or acceleration a so-called seismic mass is required which is connected to spring suspension and displaced by the accelerating forces. The displacement is determined in such devices by several physical mechanisms, e. g. by piezo-electric, piezo-resistive, capacitive or inductive means. Since the signal level of miniaturized devices is generally relatively low, the electronic amplification of the small signal should be performed as near as possible to the place of signal generation.

A LIGA configuration for measuring vibration or acceleration is shown schematically in Figure 12. By means of deep-etch X-ray lithography a template is generated which allows to manufacture a spring plate and a rigid stationary electrode directly upon a microelectronic circuit or, correspondingly, a multitude of such arrangements upon a completely processed silicon wafer. The change in capacity and the corresponding voltage change, respectively, may change directly the electric potential of the gate electrode of a MOSFET circuit.

Compared to anisotropic etching of monocrystalline silicon which is a well-known and proven technique for fabricating similar integrated sensors for vibration measurement, the LIGA technique could have several advantages. In particular, the fabrication of the electronic circuits can be fully completed prior to the fabrication of the sensor structure. Therefore, no mixing between micromechanical and microelectronic process steps occurs and a separate optimization is possible for the different fabrication procedures. Since the sensor element is arranged vertically upon the electronic circuit, the space requirement is minimized. This should be a further advantage compared to silicon micromechanics where the spring plates are usually fabricated in a plane parallel to the surface of the wafer. Nevertheless, the LIGA method allows also to produce cantilever structures which are arranged parallel to the wafer surface.

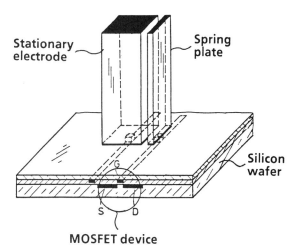

Fig. 12: Schematic representation of a LIGA sensor for measuring vibration or acceleration. An elastic plate and a stationary, rigid electrode are arranged directly on an electronic circuit which measures a change in capacity when the elastic plate is bended by accelerating forces.

3.7.2 Position, Displacement and Magnetic Field Sensors

For measuring position, displacement, small distances or changes of a magnetic field, sensor devices using inductive circuits are used in a large number of configurations. The LIGA method allows to generate small helical coils or complex coil arrays from materials with high electrical conductivity as well as microstructures with high aspect ratio from ferroelectric materials. Because of the large structural height achievable by the LIGA process, a low electrical resistance can be realized for small coils, which is an obvious advantage compared to corresponding configurations generated by thin film techniques whereas, compared to thick film techniques, much smaller dimensions and tolerances can be obtained.

Such microstructures can be fabricated on various insulating substrate materials provided with a thin plating base which is removed after the electroforming process by selective etching. Similar to the sensor element shown in Fig. 12, the coils can also be connected directly to a microelectronic circuit. By means of the magnetic force generated by a microcoil a mechanical force acting upon a micromechanical element can be compensated. Consequently, the capacitive change of the acceleration sensor shown in Fig. 12 could be registered by means of a reset control circuit. Thus, hysteresis errors would be minimized and the dynamic range would be extended.

3.7.3 Optical sensors

The optical multiplexer/demultiplexer shown in Figs. 9 and 10 is a miniaturized spectrometric system which allows to measure simultaneously a spectral distribution [12]. Although the spectral resolution of such a device is relatively low, it can be used for various applications, e. g. in photometric analysis of mixtures, chromatic testing, optical film thickness measurement, and as a spectral detector in liquid chromatography.

Furthermore, microoptic elements as discussed in Section 3.5 and fiber optic devices should be mentioned when regarding the LIGA method in connection with optical sensors.

3.7.4 Radiation sensors

The microchannel plate is an imaging electron multiplier which is capable of extremely high gain. It usually consists of a glass plate with a thickness of 0.5 to 1 mm which is perforated by closely packed channels with a diameter of some 10 μm. The inside surface of each channel has a high but finite resistivity and secondary emission properties such that the surface can function as the dynodes and voltage divider of an electron multiplier. In each side of the plate all channels are connected electrically in parallel through a metallization layer. A further type of a microchannel plate consists of a stack of perforated metallic foils which are insulated against each other and kept at different electrical potentials.

The LIGA method has been proposed for fabricating both types of microchannel plates [13]. Microchannel plates from glass can be made by a molding process using a metallic mold insert with a matched coefficient of thermal expansion. In addition, ceramic structures can be generated by oxidation of a microchannel structure made from aluminium by electroforming. Laminated microchannel plates can be produced by depositing alternately layers of electrically conducting and insulating materials into a LIGA template consisting of closely adjacent columns. The major advantage of producing microchannel plates by the LIGA method can be derived from the fact that, compared to other fabrication methods, the tolerances of the dimensions and positions of the channels can be greatly reduced. As a result, single channels or groups of channels can be matched directly to other discrete microstructures at the input or output of a microchannel plate.

A further configuration of an electron multiplier is shown in Fig. 13. Metallic microstructures forming dynode arrays are produced by the LIGA method on an insulating substrate and the dynodes are connected to a voltage divider. In this case, the dynodes can simply be arranged in a ring-pattern enclosing a measuring volume. If a photocathode is placed in front of the first dynodes of the electron multiplier array, such a device can be used for determining simultaneously the azimuthal distribution of scattered light. There are many further applications for miniaturized electron multiplier arrays as well as for LIGA microchannel plates which range from particle detection in high energy physics to high sensitivity electronic cameras and night vision equipment.

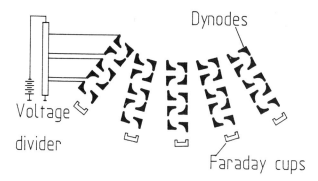

Fig. 13: Schematic representation of a miniaturized electron multiplier array which can be manufactured by the LIGA method.

4. Conclusions

The feasibility of the LIGA method which is based on a combination of deep-etch X-ray lithography, electroforming and molding processes has been successfully demonstrated by laboratory experiments. This new microfabrication method is expected to be superior to other processes if, in mass production of microstructures, specific requirements are imposed on the spatial resolution, the aspect ratio, the structural height, the parallelism of the structure walls and, in particular, if an unrestricted design of the cross-sectional shape and an optimum selection of the material required for each microstructure product has to be ensured.

The large variety of LIGA microstructure products which have been discussed in this paper may be regarded as a good basis for a broad commercial utilization of the LIGA process in the future. In particular, microsensors should be regarded as a promising field of application but also other areas which utilize the mass fabrication potential of the electroforming and molding techniques, e. g. microfiltration, compound material layers, microelectronic interconnection and packaging as well as synthetic fiber production, should be interesting for several industries.

In order to get ready for an industrial implementation of the LIGA method, extensive development work on process equipment is performed. A prototype of a cleanroom electroforming machine has been produced which allows a completely automatic fabrication of metallic microstructures [9]. In the field of micromolding, an injection molding machine for thermoplastic materials, which largely corresponds to the machines used for the fabrication of compact audio disks, has been installed and will be qualified for production purposes. Furthermore, a prototype of a reaction injection molding machine has been designed.

The experiments on deep-etch X-ray lithography have been carried out to date at the 2.5 GeV electron synchrotron of the University of Bonn. However, for a broad industrial utilization of the LIGA method a dedicated synchrotron radiation source must be available. Therefore, the design of a compact synchrotron radiation source has been worked out at the Karlsruhe Nuclear Research Center [14]. This design is based on an 1.5 GeV electron storage ring with low energy injection which is equipped with four superconducting magnets. The space required for this machine is 8 by 8 m without concrete shielding and, as a consequence of the compactness, favorably low costs should be attainable.

In conclusion, it can be stated that the development efforts for industrializing the LIGA method are steadily increasing. The authors are convinced that a successful commercialization should be possible within the next few years.

References

[1] E.W. Becker, W. Ehrfeld, P. Hagmann, A. Maner, D. Münchmeyer: Fabrication of microstructures with high aspect ratios and great structural heights by synchrotron radiation lithography, galvanoforming, and plastic moulding (LIGA process). Microelectronic Engineering 4 (1986) 35-56.

[2] W. Ehrfeld, H.J. Baving, D. Beets, P. Bley, F. Götz, J. Mohr, D. Münchmeyer, W. Schelb: Progress in Deep-Etch Synchrotron Radiation Lithography. Proc. 31st. Int. Symp. on Electron, Ion and Photon Beams, Woodland Hills, USA, 1987. J. Vac. Sci. Techn., to be published.

[3] K.E. Peterson: Silicon as a Mechanical Material. Proc. IEEE 70 (1982), 420-457.

[4] E.W. Becker, H. Betz, W. Ehrfeld, W. Glashauser, A. Heuberger, H.J. Michel, D. Münchmeyer, S. Pongratz, R.v. Siemens: Production of Separation Nozzle Systems for Uranium Enrichment by a Combination of X-Ray Lithography and Galvanoplastics. Naturwissenschaften 69 (1982) 520-523.

[5] W. Ehrfeld, W. Glashauser, D. Münchmeyer, W. Schelb: Mask Making for Synchrotron Radiation Lithography. Proc. Int. Conf. on Microlithography, Interlaken, Schweiz, 23.-25. 9. 1986. Microcircuit Engineering 5 (1986) pp. 463-470.

[6] E.W. Becker, W. Ehrfeld, D. Münchmeyer: Untersuchungen zur Abbildungsgenauigkeit der Röntgentiefenlithografie mit Synchrotronstrahlung bei der Herstellung technischer Trenndüsenelemente. KfK-Bericht 3732, Kernforschungszentrum Karlsruhe, 1984.

[7] J. Mohr, W. Ehrfeld, D. Münchmeyer, A. Stutz: Resist Technology for Deep-Etch Synchrotron Radiation Lithography. European Symp. on Polymeric Materials, Lyon, France, Sept. 14-18, 1987 (to be published).

[8] P. Hagmann, W. Ehrfeld, H. Vollmer: Fabrication of Microstructures with Extreme Structural Heights by Reaction Injection Molding. European Symp. on Polymeric Materials, Lyon, France, Sept. 14-18, 1987 "Die Makromolekulare Chemie/Macromolecular Symposia Series 1988" (to be published).

H. Vollmer, W. Ehrfeld, P. Hagmann: Untersuchungen zur Herstellung von galvanisierbaren Mikrostrukturen mit extremer Strukturhöhe durch Abformung mit Kunststoff im Vakuum-Reaktionsgießverfahren. KfK-Bericht 4267, Kernforschungszentrum Karlsruhe, 1987.

[9] A. Maner, S. Harsch, W. Ehrfeld: Mass Production of Microstructures with Extreme Aspect Ratios by Electroforming. Int. Techn. Conf. (SUR/FIN'87), American Electroplaters and Surface Finishers Soc., July 13-16, 1987, Chicago; AESF annual techn. conf. proc. 74th, Chicago (1987) K-4.

W. Becht, W. Ehrfeld, A. Maner, D. Schmidt: Galvanoformung metallischer Mikrostrukturen mit großer Strukturhöhe. Berichtsband über das 8. Ulmer Gespräch am 24.-25.4.1986 in Neu-Ulm (Donau), Eugen G. Leuze Verlag, Saulgau, S. 56-65 (1986). Galvanotechnik 77 (1986) S. 2695-2703.

[10] W. Ehrfeld, R. Einhaus, D. Münchmeyer, H. Strathmann: Microfabrication of Membranes with Extreme Porosity and Uniform Pore Size. Proc. 5th Int. Symp. on Synthetic Membranes in Science and Industry, Tübingen, 2.-5.9.86, in press.

[11] E.W. Becker, W. Ehrfeld, P. Hagmann, A. Maner, J. Mohr, D. Münchmeyer: Konzepte für die Herstellung von Spinndüsenplatten durch Röntgenlithografie mit Synchrotronstrahlung, Galvanoformung und Kunststoffabformung (LIGA-Verfahren). KfK-Bericht 3961, Kernforschungszentrum Karlsruhe, 1985.

[12] D. Münchmeyer, W. Ehrfeld: Accuracy Limits and Potential Applications of the LIGA Technique in Integrated Optics. Proc. 4th Int. Symp. on Optical and Optoelectronic Applied Science and Engineering, The Hague, Netherlands, 1987. Micromachining of Elements with Optical and Other Submicron Dimensional Surface Specifications, SPIE, Vol. 803, in press.

[13] E.W. Becker, F.S. Becker, W. Ehrfeld: Konzepte für die Herstellung von Vielkanal-Bildverstärkerplatten durch Röntgentiefenlithografie und Mikrogalvanoplastik. KfK-Bericht 3750, Kernforschungszentrum Karlsruhe, 1984.

[14] D. Einfeld, O.F. Hagena, P.R.W. Henkes, R. Klingelhöfer, B. Krevet, H.O. Moser, G. Saxon, G. Stange: Entwurf einer Synchrotronstrahlungsquelle mit supraleitenden Ablenkmagneten für die Mikrofertigung nach dem LIGA-Verfahren. KfK-Bericht 3976, Kernforschungszentrum Karlsruhe, 1985.

DEEP X-RAY AND UV LITHOGRAPHIES FOR MICROMECHANICS

H. Guckel, T.R. Christenson, K.J. Skrobis, D.D. Denton, B. Choi,
E.G. Lovell, J.W. Lee, S.S. Bajikar and T.W. Chapman

College of Engineering
University of Wisconsin-Madison
1415 Johnson Drive
Madison, Wisconsin 53706

ABSTRACT

Micromechanical devices are three-dimensional structures. They therefore require processing sequences which differ from standard IC-processing which is essentially two-dimensional. The use of very thick photoresist layers is but one method to achieve this. However, when thick photoresist processes are combined with electroplated metal technology, major advances in micromechanics can occur because the available material base expands greatly.

The photoresist for our work is poly(methyl methacrylate) or PMMA. It is sensitive to x-ray photons and deep ultraviolet radiation. It can be used for spin coating or can be applied by casting and in situ polymerization. This implies that there are at least four possible versions of the deep lithography process. All of them hinge on the availability of a developer with very high selectivity and low stress formation during liquid processing.

The work which is reported here deals first of all with a bi-level photoresist system which uses optical processing only. This process, DUVL, results in PMMA definitions with excellent resolution for photoresist thicknesses of up to 10 μm. It can be extended to 30 μm or so by a simple process change. The second approach uses thicker, in our case larger than 40 μm, photoresist layers. These films involve cross-linked PMMA and are exposed by synchrotron radiation. Both films are developed in an excellent wet developing system which was formulated by a German research group. The results show good pattern definition.

The exposed and developed substrates are electroplated with either gold or nickel films. Gold is used to produce x-ray masks on polysilicon mask blanks. The patterns are optically defined and produce good results above 0.6 μm. Synchrotron-exposed samples are plated with nickel. They produce LIGA-like structures with good results.

INTRODUCTION

X-ray lithography has two major application areas: submicron VLSI and micromechanics. The XRL technology difference for these two fields arises primarily from the dramatically different photoresist thicknesses which are used. Thus, XRL for VLSI employs spun-on layers in the one micrometer range. XRL for micromechanics, on the other hand, becomes interesting at photoresist thicknesses of more than ten micrometers and may employ thicknesses as large as several hundred micrometers. The term: deep x-ray lithography or DXRL is therefore richly deserved.

The photoresist thickness issue causes a second major difference for the two application areas. VLSI x-ray lithography can be implemented with rather modest x-ray flux densities and is therefore practical with traditional point and plasma sources. DXRL requires the high brightness which is currently only available from a synchrotron source. This is somewhat unfortunate because synchrotron access is required for process research in this potentially important field. This difficulty is somewhat reduced by W. Ehrfeld's work in a process which he calls LIGA [1]. This procedure, which is the earliest example of DXRL, utilizes synchrotron exposure only once to produce a master pattern. It then copies the master by injection molding and therefore avoids further synchrotron use for DXRL device production. A somewhat different approach for modest photoresist thicknesses has been used at Wisconsin. It is based on the fact that the most important photoresist for DXRL is poly(methyl methacrylate) or PMMA. This resist can be exposed with a deep UV source at 2300 Å where it has an absorption length of roughly 4 μm. Thus, if deep lithography or DL is defined to start at 10 μm or so, many aspects of DXRL can be studied and useful devices can be derived from ultraviolet deep lithography or DUVL which of course will not involve synchrotron access. The implications are then as follows: DUVL is accessible to many research groups. It is also a tool which leads to much more rapid process development for DXRL.

Micromechanics based on DXRL involves four major areas. The source, as stated earlier, is a synchrotron with a suitable beam line and fixturing for the mask and substrate. The photoresist process is the second area. Since very large photoresist thicknesses are involved, photoresist application via spinning may not be feasible. Furthermore, since developing of thick layers may involve long immersion times, the selectivity of the developer for exposed and unexposed photoresist films must be nearly infinite. Swelling and distortions must be avoided if mechanical damage and geometric distortions are not allowed. The third topic involves a suitable mask for DXRL. The mask blank should not absorb any photons. This requires a low atomic number membrane with a thickness in the micrometer range as a compromise solution. The absorber is formed by patterned high atomic number material, i.e., gold or tungsten. The desired contrast ratio sets the absorber thickness for a given mask blank. For very thick photoresist and therefore long exposures, absorber thicknesses of several micrometers are required. This normally implies that the absorber must be electroplated. Bath compatibility with the photoresist system, built-in strain and deposit uniformity are some but not all of the difficulties which must be addressed.

The previous discussion indicates that process development for DXRL involves a "Catch-22" condition: a mask is needed for DXRL. Its construction requires DXRL. This dilemma must therefore be resolved if a viable research program in DXRL is to proceed.

MASKS FOR DEEP X-RAY LITHOGRAPHY

Mask blanks for x-ray lithography as stated earlier are fully supported membranes of low atomic number materials. In our case a suitable mask blank for VLSI x-ray lithography had already been developed by using silicon nitride and tensile polysilicon films before work on DXRL was initiated [2,3,4]. Because DXRL exposures are long and radiation damage was of concern, a DXRL mask blank made from one micrometer thick tensile polysilicon was selected as the blank of choice. A diaphragm size of 2.5 cm $\times$ 0.8 cm acknowledges the fact that synchrotron beam scanning via x-ray mirrors which is used in VLSI x-ray lithography would not be used in order to reduce exposure time. Blanks of this type are now being produced quite routinely in a high yield process.

The absorber for a DXRL mask was chosen to be electroplated gold. The argument for this is found in the fact that absorber thicknesses for the polysilicon mask blank become quite large for good contrast ratios if long exposure times are anticipated. The plating occurs into photoresist recesses with high edge acuity on a properly prepared mask blank. The mask blank preparation involves a suitable plating base which is currently produced by sputtering 50 Å of chromium followed by 150 Å of nickel.

The photoresist for mask fabrication has to be at least 7.5 μm thick to accommodate absorber thicknesses of up to 4 μm. The requirement on edge acuity is very high because any non-vertical flank on the plated absorber is in effect a contrast change and therefore causes geometric drift between the mask and the work piece. At this photoresist thickness the acuity issue can be satisfied by using PMMA as the photoresist of choice. This material has the capability to produce essentially vertical flanks as demonstrated by the LIGA process. However, in the present situation, x-ray exposure is not feasible because this would require a mask which does not exist. Either E-beam or optical patterning is needed. Since one of us, H.G., had worked on a high resolution bi-level optical approach, this procedure was adopted and led to the concept of an optically defined x-ray mask.

The actual process, the DUVL process, involves first of all a substrate with a plating base. The substrate material is normally silicon. However, there are many other possibilities because the maximum process temperature is only 180°C. The PMMA photoresist layer is formed from 496,000 M.W. PMMA in a 9% by weight chlorobenzene solvent. This material can be bought from KTI. The optical characteristics of the material can be improved by adding a dye which in the present case is Kodak Laser Grade Coumarin 6. The mixture is 50mg of dye per 25cc of PMMA. The material is spun at 2000 RPM for 45 seconds. It is then annealed very carefully in a precision bake-out oven by ramping to 180°C at 1°C/minute, holding at 180°C for 1 hour and ramping to room temperature at -1°C/minute. This results in a PMMA thickness of 1.6 μm. The process is repeated until the desired photoresist thickness is obtained. The prepared substrate is next covered with a compatible optical photoresist. This material must have the following properties: it must be a good optical receiver for mercury arc exposure, it must be opaque at deep UV wavelengths and it must not cause interface problems. The commercial KTI 809 system satisfies all of these requirements. The material is spun at 5000 RPM for 30 seconds. It is then pre-baked at 90°C for 15 minutes. Exposure in hard contact on a Canon 501 follows. Developing in a mixture of 1:1 KTI 809 developer and DI-water produces easy pattern transfer at 1 μm line width and optimized pattern transfer at 0.6 μm. Blanket exposure at 2300 Å at 15 mw/cm² for 31 minutes in an HTG system at 8 μm PMMA thickness completes the photoresist processing phase.

The next processing step is that of developing the PMMA. It was stated earlier that this field owes a lot to Dr. W. Ehrfeld. His progress in turn is based very much on a nearly perfect developing system [5]. The developer is

    60 vol%   2-(2-butoxyethoxy)ethanol
    20 vol%   tetrahydro-1, 4-oxazin
                    (morpholine)
     5 vol%   2-aminoethanol
    15 vol%   water

typically used at 35°C in a recirculating spray developing system. The selectivity of the developer is nearly perfect and stress build-up due to the developing process is extremely low. The importance of this system cannot be overstated: without this type of performance, there would be no DXRL.

The developing cycle is followed by a 20 minute rinse at 35°C. The rinse consists of 80% 2-(2-butoxyethoxy)ethanol and 20% water. A second rinse for 5 minutes in DI-water completes the process. Figure 1 illustrates typical results.

Plating follows a mild oxygen plasma de-scum cycle. Typical non-optimized plating conditions for gold are 50 mA/cm² at 60°C with a plating rate of 0.3 μm/min. The plating composition is a commercially available solution which is manufactured by Technic Inc.

Photoresist removal is accomplished by blanket UV exposure and developing. Figure 2 illustrates the mask result.

The processing sequence which has been discussed here is stable and can be extended to thicker PMMA layers. Thus,

Fig. 1 Partially exposed PMMA test pattern. The pattern has been developed throughout the exposed PMMA. The pattern height is 8 μm.

Fig. 2 Polysilicon x-ray mask. The figure contains an optical micrograph of a backlit x-ray mask. The light areas are x-ray transmissive; the dark areas are covered by 3.5 μm of plated gold. The polysilicon thickness is 1 μm.

spun-on films of 30 μm thickness have already been produced. They can be exposed with multiple deep UV exposures and can be processed with several developing cycles if a slightly different optical transfer system is used. This is necessary because the PMMA developer removes the KTI 809. The cycles are needed because the absorption length of the PMMA at 2300 Å is roughly 4 μm and therefore exposure to 8 μm is a practical limit at UV for a single cycle. It is therefore possible to claim that DUVL for micromechanics is readily implemented in-house for most research groups.

THICK PHOTORESIST TECHNOLOGY FOR X-RAY EXPOSURE

It is possible to argue that DXRL can become significant with spun-on films only. However, there are many advantages for cast films. Perhaps the two most important issues are planarization and flexibility in polymer composition and, in particular, the fact that cross-linked films can be produced. The difficulty in this area is simply the fact that this type of processing has not been studied in detail and that therefore a learning curve is required. The learning cycle has been shortened considerably by the publications of the German LIGA group [6,7,8].

The procedures which have been developed here are based on a casting syrup which is made from pre-polymerized PMMA with a number-averaged molecular weight of 100,000. It is dissolved in MMA and a cross-linking agent, ethylene glycol dimethacrylate (EGDA) is added. Two initiators are used. Thus, solution I contains dimethyl aniline or DMA and solution II contains benzoyl peroxide or BPO. The two solutions are mixed at room temperature which causes free radical formation via a reaction between the two initiators. Polymerization starts and proceeds at a rapid pace. Since the two solutions have to be well mixed and degassed, special equipment was designed and constructed to do this.

Typically, 1 to 2 ml of the syrup are applied to a casting fixture which supports the wafer. The function of this device is that of forcing the syrup into a uniform film. This implies flat plane contact to the upper surface of the film which is accomplished by using a sheet of Mylar as a release film. Mechanical spacers and film shrinkage set the final film thickness. The cast films are next annealed at 110°C. Precision positive ramping at 20°C/hour and negative ramping at 5°C/hour together with a 1 hour hold at 110°C complete the process.

X-RAY EXPOSURE

PMMA exposure via synchrotron radiation is not a simple issue. The reason for this is, in part, the spectrum of the source: it is broad band as demonstrated in Figure 3. The second problem is found in the absorption length of PMMA: it is very wavelength-sensitive as indicated by Figure 4. Direct exposure with this type of source and absorber will evidently be nonuniform in the direction normal to the substrate plane. This nonuniformity typically involves overexposure of the PMMA surface and underexposure at the substrate/PMMA interface. One can estimate proper exposing conditions by using data from reference [9]. The threshold of exposure for the developing system which has been mentioned here is 1.6 kJ/cm^3. Reported resist damage occurs at a dose of 20 kJ/cm^3. These numbers are based on the Bonn synchrotron which operates at higher energy and much lower beam current. ALADDIN, the synchrotron at Madison, is a different machine and test exposures indicate that the upper exposure limit may be quite a bit higher. There are also some indications that exposure levels on a higher current machine are rate dependent. In any case, since both an upper and lower exposure limit exist, one has no choice but to change the spectral output of the machine. This can be done quite easily by inserting a filter between the incoming beam and the mask-substrate combination. Figure 5 shows the result. Figure 6 shows the corrected spectrum after a 50 μm Be filter and the 1 μm silicon mask blank.

Initial exposures with the polysilicon mask were accomplished by using the beam line ES-O which was constructed by the Wisconsin Center for X-Ray Lithography

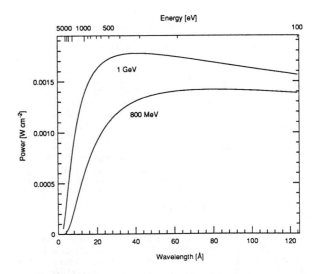

Fig. 3 Spectrum of ALADDIN at 800 MeV and 1 GeV, with a beam current of 100 mA and horizontal acceptance of 4.858 mrad.

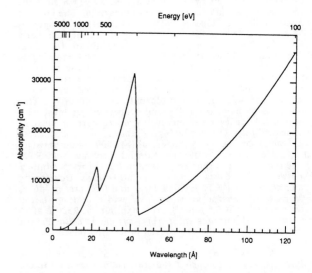

Fig. 4 Absorptivity vs. wavelength of poly(methyl methacrylate).

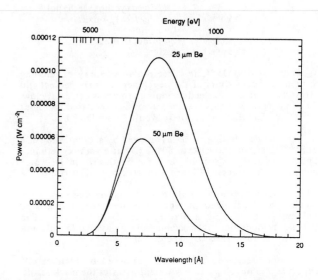

Fig. 5 Spectrum of ALADDIN modified by beryllium filters, at 800 MeV, with a beam current of 100 mA and horizontal acceptance of 4.858 mrads.

group. Maximum PMMA thickness was 40 μm and exposures were done by mechanically scanning the wafer-mask combination at 2.7 scans/min. with a 2 cm scan amplitude. Mask to wafer separation was 50 μm. The wafer-mask assembly was exposed to 20 torr of helium for cooling purposes. Typical exposure levels were 50 J/cm^2 with exposure times always less than 1 hour.

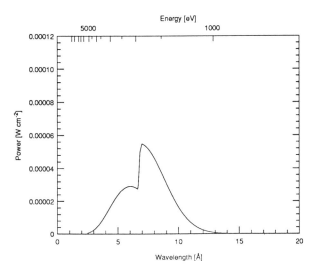

Fig. 6 Spectrum of ALADDIN after a 50 μm beryllium and a 1μm polysilicon mask at 800 MeV, with a beam current of 100 mA and horizontal acceptance of 4.858 mrads.

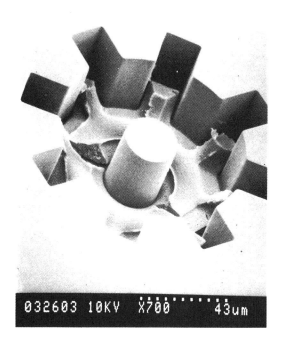

Fig. 8 X-ray exposed cast PMMA test pattern. The PMMA thickness in Fig. 8 is 40 μm. The x-ray exposure was too short for complete developing.

Fig. 7 X-ray exposed PMMA test pattern. This pattern was produced from an 18 μm thick spin-coated PMMA film and x-ray exposure. It was then fully developed.

Fig. 9 Nickel plated LIGA-like test structure. Fig. 9 illustrates plating resolution. The structures are nickel columns of 5 μm diameter and 18 μm height.

DXRL RESULTS

The x-ray exposed wafers were developed as indicated earlier. Figure 7 indicates the resist performance. Particularly noteworthy are the smoothness of the unexposed sections and the edge acuity. The absence of crazing is recognized. Figure 8 involves a cast PMMA sample for which the exposure was too short. It demonstrates the selectivity of the developer and of course the edge acuity for a cross linked film.

Some of the samples were electroplated. For this purpose a nickel sulfamate plating system was constructed. The bath is operated at 50°C at a pH of 4.3 and uses a plating current density of 50 mA/cm^2. It resulted in the structures which are shown in Figure 9.

CONCLUSIONS

The progress which has been reported here indicates that DUVL is a viable tool for micromechanics which is accessible without synchrotron access. This process can be used with many metals which can be plated and therefore extends the micromechanics material base considerably. This is good even though it is an admission that silicon does have limitations which one of us, H.G., finds somewhat disturbing.

The Wisconsin x-ray lithography effort has passed the stage of relying on LIGA information. It is also past the feasibility stage. It will be applied to new and interesting structures in the near future.

ACKNOWLEDGEMENTS

We are indebted to Dr. W. Ehrfeld and Dr. U. Ehrfeld for extensive conversations about the LIGA process and their generous help which allowed us to proceed at a rapid pace. The cooperation of the Wisconsin Center for X-Ray Lithography, Professor F. Cerrina, Director, is gratefully acknowledged. The help which Mr. T. Martin of Silicon Sensors extended to us in gold plating is appreciated. The encouragement of G. Hazelrigg of NSF contributed to the research progress. One of us, H.G., is indebted to his friends for constant support with many difficult problems.

This work was supported by the National Science Foundation under Grant EET-8815285.

REFERENCES

(1) W. Ehrfeld, F. Götze, D. Münchmeyer, W. Schelb and D. Schmidt, "LIGA Process: Sensor Construction Techniques via X-Ray Lithography," Technical Digest, IEEE Solid-State Sensor and Actuator Workshop, pp. 1-4, 1988.

Reference (1) was an invited presentation at the 1988 Sensor and Actuator Workshop. The LIGA literature is quite extensive and cannot be fully cited here due to space limitations. The interested reader should note that a full list of LIGA references is available from the authors. Other key LIGA references are cited below.

(2) C.C.G. Visser, J.E. Uglow, D.W. Burns, G. Wells, R. Redaelli, F. Cerrina and H. Guckel, "A New Silicon Nitride Mask Technology for Synchrotron Radiation X-Ray Lithography: First Results," Microelectronic Engineering, Vol. 6, pp. 299-304, 1987.

(3) H. Guckel, D.W. Burns, T.R. Christenson and H.A.C. Tilmans, "Polysilicon X-Ray Masks," Microelectronic Engineering, Vol. 9, pp. 159-161, 1989.

(4) H. Guckel and D. Burns, "Polysilicon Thin Film Process," U.S. Patent No. 4,897,360, Jan. 30, 1990.

(5) V. Ghia and W. Glashauser, "Verfahren für die Spannungsfreie Entwicklung von bestrahlten Polymethylmethacrylat-Schichten," Offenlegungsschrift DE 3039110, Siemens AG, Munich.

(6) J. Mohr, "Analyse der Defektursachen und der Genauigkeit der Strukturübertragung bei der Röntgentiefenlithographie mit Synchrotronstrahlung," Diplomarbeit, Universität Karlsruhe, F.R.G., 1986.

(7) A. Stutz, "Untersuchungen zum Entwicklungsverhalten eines Röntgenresists aus vernetztem Polymethylmethacrylat" Diplomarbeit, Universität Karlsruhe, F.R.G., 1987.

(8) S. Harsch, "Untersuchungen zur Herstellung von Mikrostrukturen großer Strukturhöhe durch Galvanoformung in Nickelsulfamatelektrolyten," Diplomarbeit, Universität Karlsruhe, F.R.G., 1987.

(9) D. Münchmeyer, "Untersuchungen zur Abbildungsgenauigkeit der Röntgentiefenlithografie mit Synchrotronstrahlung bei der Herstellung technischer Trenndüsenelemente, "Diplomarbeit, Universität Karlsruhe, F.R.G., 1984.

Section 13

Computer-Aided Design

COMPUTER-AIDED design (CAD) tools can save substantial time and money in developing a silicon fabrication process. To fabricate a particular design in the clean room requires many serial fabrication steps. When the device is complete, one cannot go back and tweak a process, or add a hole somewhere. One must usually start all over again. Appropriate CAD tools allow the designer to ask what if, to modify, and to optimize the design without having to spend weeks or months fabricating the device.

The difficulty is that CAD tools for micromechanical devices are hard to make. Simultaneously, one must deal with complex and interrelated fabrication processes, three-dimensional structures, the dynamics of moving parts, and the generation of forces that depend on all of the above. Changing one parameter affects the others. This coupling of many different parameters ensures that designers of micromechanical CAD systems have many interesting years ahead. The payoff for the micromechanics community of these efforts will be large.

OYSTER, a 3D Structural Simulator for Micro Electromechanical Design

George M. Koppelman

Manufacturing Research
IBM T.J.Watson Research Center
Yorktown Heights, NY 10598

ABSTRACT

As micro elctromechanical systems become more complex, designers will find it useful to derive models of the geometry of their device structures directly from the process description and planar mask patterns. OYSTER simulates the geometric effects of sequential IC process stages, including patterning of photoresists with planar masks, in order to produce 3D polyhedral representations of all material structures in a design cell after each process stage. It has been developed for IC simulation but is applicable to micro electromechanical systems manufactured using similar processes. In addition, because it uses basic functions of a solid modeling program developed for mechanical systems, it provides access to features like dimensioning and tolerancing and kinematics.

The polyhedral models may be used with various analytic procedures as sources of geometric data for finite element calculations, or they may be subjected to interference calculations, or inspected to detect structural anomalies. As used in IC simulation, OYSTER provides the ability to introduce worst case or stochastic manufacturing variations in mask alignment, etch depth, or deposition thicknesses. Related specifically to micro mechanical design, we demonstrate examples of variations in mask alignment and calculate resultant variations in center of gravity and moments of inertia for simple micro mechanical objects.

INTRODUCTION

Modern semiconductor devices are becoming more three dimensional as vertical aspects of the structures become more pronounced and the mask feature dimensions decrease. The importance of 3D structure is reflected both in the device's operating electrical characteristics and in manufacturability considerations. We have developed a method of simulating the geometric results of semiconductor processes using solid modeling techniques generally associated with mechanical CAD tasks. This allows us to create three dimensional solid polyhedral models of each material component of a device after each stage in the processing sequence. Our system, called OYSTER [5], is based on GDP, a robust polyhedral solid modeling engine that was originally developed for interactive modeling of macro mechanical structures [7, 8].

OYSTER has created a new interface for users familiar with the domain of process technology and does not require, or even permit, direct geometric interaction. OYSTER presents its users with a set of parameterized Process Description Language (PDL) statements that are invoked, in sequence, to simulate a process run. During execution of an OYSTER step called by a PDL statement, underlying geometric algorithms analyse the current multi-layer device topology to determine which portions of designated materials are to be effected by the type of process simulation that has been invoked, and then calculate the resulting geometric shapes in modified existing layers or added new layers. The designer's mask shapes are used to to expose photoresist layers as part of the lithography step sequence which then affect the geometry of subsequent etch or lift-off processes.

Micro electromechanical devices, formed using essentially the same processes as semiconductor devices, of course naturally exhibit 3D structures. Customary mechanical design practices used in the macro mechanical world no longer apply to the micro design and fabrication domain [4]. In the micro domain, geometric design data, defined in the macro domain by drawing views, is separated into explicit planar mask patterns and implicit vertical dimensions and other structural relationships that must be derived from the description of the manufacturing processing sequence. A tool such as OYSTER can help designers to evaluate both electrical and mechanical device properties that derive from the process sequence.

We shall illustrate the range of OYSTER techniques using a bipolar device example, then attempt to show how these techniques, together with some experimental GDP features, might be applied to micro mechanical applications.

OYSTER GEOMETRIC MODELING OF SEMICONDUCTOR PROCESSES

OYSTER takes only two types of data input on which it bases its complete simulation. These are the device mask artwork and the sequence of process steps as a list of statements following the OYSTER command syntax. These two inputs determine the complete manufacturing sequence. However, it takes the full set of intermediate shape interrelationships formed by OYSTER during the complete simulation sequence to determine the geometry of the resulting device model.

Reprinted from *Proceedings IEEE Micro Electro Mechanical Systems*, pp. 88-93, February 1989.

We simulate only the geometric effects of the physical processes and in a given step require the designer to give an explicit extent parameter defining the end result of the process. In a later version we plan to accept process parameter data such as duration of etch and etch rate by material and calculate final extents using empirical tables. While OYSTER is not a process simulator such as SUPREM [2], it can use 2D profiles resulting from such simulators as input templates from which to extrude 3D shapes. A robust set of operators for performing Boolean operations on polyhedral shapes has been augmented with new operators for special profile shaping and global growing and shrinking of polyhedral solids [3]. These global offsetting operators also allow us to simulate approximate shape rounding at edges and vertices with a pre-specified number of facets per edge.

Generally, the costs in space and time of simulating a step go up with the level of shape detailing requested. For many applications, OYSTER can be effectively used with square masks and square edges and corners on shapes. In other cases, the user can request rounded mask corners, up to two edge facets on all edges, and apply special edge profiles shapes for certain etch and deposition results. There are features for selecting sub regions of a device when more detailed edge effects are needed. Solid shapes that are used to derive geometric input for electrical device calculations may require other constraints.

We are able to run the simulation with mask overlay misalignment and variations in layer thicknesses. The cumulative effect of these variations, seen during simulation of subsequent process steps, can expose structural and electrical yield dependencies. A range of graphical and model library facilities are also provided to facilitate interactive user sessions.

OYSTER SEMICONDUCTOR EXAMPLE

We have based our simulation example on a standard text exposition of a bipolar device to illustrate the execution of an OYSTER PDL sequence [1]. The mask set for the example device is shown in Fig 1a. The complete PDL statement sequence has 48 steps. The first eight statements with annotations are given below. These PDL statements specify nominal extents.

Grow thermal oxide:
GOX OX1 SUBSTR 250an 50

OX1 is the name of new oxide material. SUBSTR is the name of the wafer material. 250an is extent (thickness) of oxide in angstroms. 50 is the percentage of oxide grown into the host surface. The oxide (OX1) will be grown on those surfaces of the host material (SUBSTR) that are exposed to the atmosphere.

Apply photoresist:
APR PR1 1um

PR1 is the name of the new photoresist material. 1um is extent (thickness) of photoresist in microns. The photoresist (PR1) will be applied with a planar top surface conforming to the underlying device topology that exists. The level of the top surface will be 1um above the highest underlying feature.

Expose with biased mask:
EXPB PR1 LEVEL1 DF

PR1 is the name of the photoresist material to be exposed. LEVEL1 is the mask name. DF is a keyword abbreviation specifying dark field exposure. A biased version of the nominal mask (LEVEL1) with faceted corners will be used to expose the photoresist (PR1).

Develop photoresist:
DEV PR1

PR1 is the name of the photoresist material to be exposed. The exposed portion of the photoresist (PR1) will be removed and the resulting photoresist material will consist only of the unexposed portion.

Implant dopant:
IMP N+BURIED SUBSTR 1um 1 THRU OX1

N+BURIED is the name of the implanted region. SUBSTR is the host material to receive the implant. 1um is the extent (depth) of the implanted region. 1 is a type code specifying the implant species. THRU is a keyword indicating that the following material names (in this case just OX1) serve as screening layers. Only those surface regions of the host material (SUBSTR) that are exposed to the atmosphere (in this case, the SUBSTR surface that is exposed through the photoresist pattern) and ignoring any screening layer(s), will be transformed into an implant region (N+BURIED) of type (1).

Reactive ion etch:
RIE (OX1 N+BURIED) 1250an

OX1 and N+BURIED are the combined materials to be RIE'd. 1250an is the extent (depth) of the RIE in angstroms. The named materials (OX1) and (N+BURIED) will be etched vertically to a depth of 1250 angstroms where their combined upper surface is exposed to the atmosphere. Any overhanging materials that shadow the exposed surfaces will restrict the effective etch region(s) accordingly.

Strip:
STRIP PR1

PR1 is the name of the material to be stripped. The entire material (PR1) will be removed from the device.

Fig 1b. shows the device model after the first eight PDL statements have been run. Sectioning and quadranting, as shown in the figure, can be performed interactively after any of the steps. Fig 1c. shows the complex topology of the completed device model.

In addition to the PDL statements described above, there are OYSTER PDL statements for various forms of material *deposition*, isotropic *etching*, and *lift-off* An actual user session would combine such PDL sequences with interactive commands that define faceting, profiling, model storage and retrieval, a variety of sectioning routines and hardcopy output. The PDL sequences themselves can be entered interactively or executed from a previously stored listing that may specify an entire process technology.

Another run of the complete PDL sequence was made using randomly generated extents. The same initial PDL sequence is shown below with variational syntax.

GOX OX1 SUBSTR (250an +/- 25an) 50
APR PR1 (1um +/- 1000an)
EXPB PR1 LEVEL1 DF GAUSSIAN .5um
DEV PR1
IMP N+BURIED SUBSTR (1um +/- 1000an) 1
    THRU OX1
RIE (OX1 N+BURIED) (1250an +/- 125an)
STRIP PR1

Fig. 1d shows the bipolar model that results from the variational run. During the run, the randomly selected extents are printed on the run log, along with the original random seed value.

OYSTER GEOMETRIC MODELING OF MICRO-MECHANICAL PROCESSES

The ability of OYSTER to transform a planar mask set together with its PDL description of a process step sequence into solid 3D representations of micro mechanical objects is a natural replacement for the standard drawing and blueprint visualizations used in the macro mechanical domain. The interactive simulation feature would allow rapid prototyping of solid part models which, when stored in the OYSTER model data base, then provide part sources for experimental recombinations and assembly not envisaged for IC circuits.

Modeling requirements and benefits for the micro mechanical domain differ from the semiconductor domain in several other ways as well. Current logic and memory device fabrication involves many more levels of lithography and subsequent material deposition and erosion processes which can result in critical topological dependencies. In addition, for circuit devices, feature dimensions continue to push manufacturing tolerances, forcing estimates of statistical variations in the manufacturing process to be taken into consideration during the design phase. This effect is not yet apparent in the micro mechanical world although in our example we show that a modeling tool such as OYSTER can track a shaft-gear lithography misalignment in terms of changes in measured mechanical parameters. For logic and memory device fabrication, the basic unit of yield is the chip. A single defect can cause a dense VLSI chip to be rejected whereas individual micro mechanical parts can be rejected while adjacent parts may well be good. Packing density of part templates on a wafer ought to be such that average yields would allow a large portion of a given run to be discarded if necessary.

There are also differences in mechanical design for the micro and macro domains that will have a bearing on the future design of CAD tools. In the micro domain, all parts produced in a given wafer run would tend to be similar in relative dimension and position of features. Many or most of the parts needed for a given product could be produced at once and therefore be relatively uniform in the sense given above. Non-uniformity would show up as anomalies which could simply be discarded before assembly if there were good detection. A well integrated design system that includes a solid modeling capability would allow a designer to make full use of features of silicon processing that provide an advantage in the micro domain over the macro domain. For a given device requiring assembly of parts, the production design could be maximally arranged so that parts and features that require mutual alignment could be created in common lithography steps, thereby assuring tolerance except for random chip anomalies or wafer edge effects. Automatic tolerancing approaches being developed for mechanical CAD systems would be applicable in sorting through the choices presented by a multi-part assembly.

OYSTER MICRO-MECHANICAL EXAMPLE

We have created a simulation example based on recent reports of experimental micro mechanical parts [6]. It is meant as a conceptual exercise to demonstrate OYSTER and GDP features and is not intended to portray an actual process proposal. Fig 2a. shows a gear and shaft that have been created using the mask set shown and a possible process sequence given below:

DPOS SHAFT 1 30um
APR PR1 2um
EXP PR1 LEVEL1 LF
DEV PR1
RIE SHAFT 30um RIEF2
STRIP PR1
GOX SACOX (SUBSTR SHAFT) 1.0um 50
DPOS GEAR 1 20um
APR PR2 2um
EXP PR2 LEVEL2 LF
DEV PR2
RIE GEAR
STRIP PR2
STRIP SACOX

The RIEF2 parameter designates a 2D undercut profile developed during the RIE of the shaft that serves as a flange. The designer enters the profile points as empirically determined or obtains them from a 2D process simulator. The profile is then extruded into a 3D shape to fit the RIE'd vertical surfaces. In this conceptual process, a piece of the gear material is formed on top of the shaft column. Fig 2b. presents a close-up quadranted view before removal of the SACOX. The removal of the SACOX that frees the gear from the substrate would also cause the piece of the gear material on top of the shaft to float off and probably lodge in one of the gear teeth. We have found that OYSTER frequently detects such unexpected remnants of deposition and etch processes. Fig 2c. shows a cross section of the gear and flanged shaft after removal of SACOX. A gravity function in the modeler could cause the gear to rest on the new substrate surface.

Once the gear material is created as a solid model, it can be treated as a mechanical part. The volume of the gear, its center of gravity, and X,Y, and Z moments of inertia can be calculated by GDP directly from the polyhedral solid. GDP also has a dimensioning feature that calculates and displays dimensions between selected points. The dimension notations themselves then form part of the model data and are redisplayed at the correct positions if the model is scaled or rotated. Fig 2d. shows the dimensioned gear in micron units.

By running the process example with Gaussian misalignment of the mask exposure steps for both the shaft and gear mask levels, we create a slightly off center gear. The values for volume, center of gravity, and moments are given in Table I for the nominal and worst case misalignments of 1.0um and 2.0um.

SUMMARY

We have developed a structural simulator for generalized IC process steps and have shown that it can be used to model micro-electromechanical devices that are manufactured using similar technologies. A combination of features from the semiconductor and macro mechanical domains can be used by designers to develop new techniques suited to the micro domain.

ACKNOWLEDGEMENTS

I would like to thank Ralph Hollis of Manufacturing Research for suggesting that OYSTER might be of interest to the micro mechanical design community and for bringing me up to date on the literature. I also wish to thank Linda Wilson of the GTD division, IBM, for her assistance in developing the bipolar OYSTER example and William Fitzgerald of Manufacturing Research for his help in applying the dimensioning to the gear model. And I must especially acknowledge the key contributions of the OYSTER development group in Manufacturing Research that are reflected in every aspect of the current system.

REFERENCES

[1] S. M. Sze, editor. *VLSI Technology* McGraw-Hill Book Company, 1983.

[2] Dimitri A. Antoniadis and Robert W. Dutton, "Models for Computer Simulation of Complete IC Fabrication Process," *IEEE Transactions on Electron Devices*, vol. ED-26, no. 4, April 1979.

[3] R. C. Evans, G. Koppelman, and V. T. Rajan, "Shaping geometric objects by cumulative translational sweeps," *IBM Journal of Research and Development*, vol. 31, no. 3, May 1987.

[4] K. Gabriel, J. Jarvis, and W. Trimmer, editors. Small Machines, Large Opportunities: A Report on the Emerging Field of Microdynamics, NSF Workshop on Microelectomechanical Systems Research, 1988.

[5] George M. Koppelman and Michael A. Wesley, "OYSTER: A Study of Integrated Circuits as Three-Dimensional Structures," *IBM Journal of Research and Development*, vol. 27, no. 2, March 1983.

[6] Richard S. Muller, "From ICs to Microstructures: Materials and Technologies," *Proceedings of the IEEE Micro Robots and Teleoperators Workshop*, November 1987.

[7] M. A. Wesley, T. Lozano-Perez, L. I. Lieberman, M. A. Lavin, and D. D. Grossman, "A Geometric Modeling System for Automated Mechanical Assembly," *IBM Journal of Research and Development*, vol. 24, no. 1, January 1980.

[8] R. N. Wolfe, M. A. Wesley, J. C. Kyle, Jr., F. Gracer, and W. J. Fitzgerald, "Solid Modeling for Production Design," *IBM Journal of Research and Development*, vol. 31, no. 3, May 1987.

Table I.

Alignment	Volume	C of Mass (X,Y)	M of Inertia (Z) about C of Mass
Nominal	7.85E+5	0.0 -0.0023	5.64392E+9
Shaft: X -1.0um Y -1.0um Gear: X 1.0um Y 1.0um	same	0.0108 0.0108	5.64369E+9
Shaft: X -2.0um Y -2.0um Gear: X 2.0um Y 2.0um	same	0.0216 0.0216	5.64294E+9

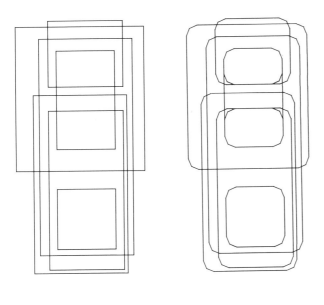

Fig. 1a Mask set for bipolar IC example. The faceted mask set on the right was used in this example.

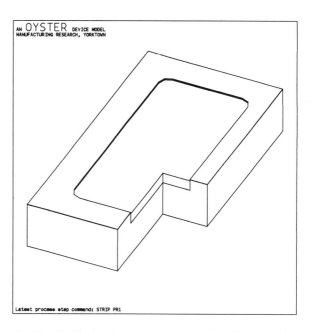

Fig. 1b The bipolar device model after the first 8 process steps in quadranted view.

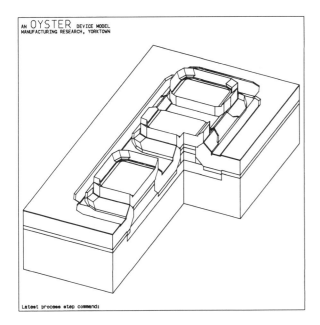

Fig. 1c The complete OYSTER model of the bipolar device after 48 steps, shown in quadranted view. Nominal mask alignment and process extents have been used.

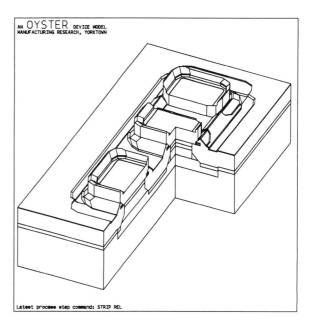

Fig. 1d The complete OYSTER model of the bipolar device after 48 steps, shown in quadranted view. Random variations in mask alignment and process extents have been used.

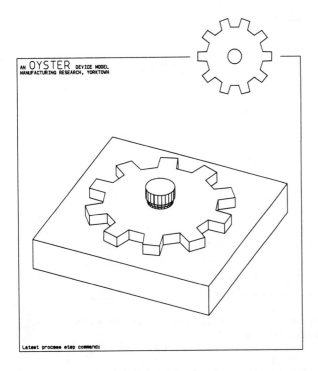

Fig. 2a A gear and shaft created using the process sequence given in the example. The masks are shown above the model.

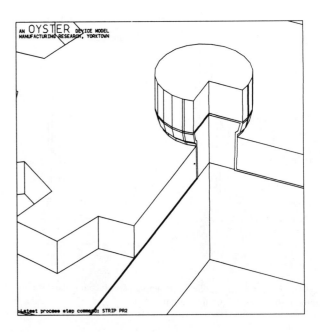

Fig. 2b A quadranted close-up view of Fig. 2a. The SACOX layer can be seen together with the undercut shaft and the gear. Note the portion of the gear material resting on the shaft column.

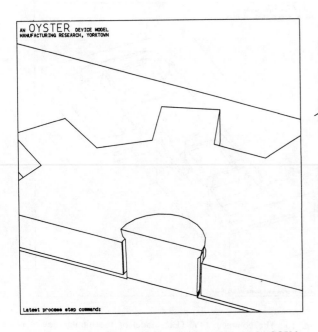

Fig. 2c A sectioned view of gear and shaft after the SACOX has been etched away. The gear is shown in its original position rather than on the surface of the substrate.

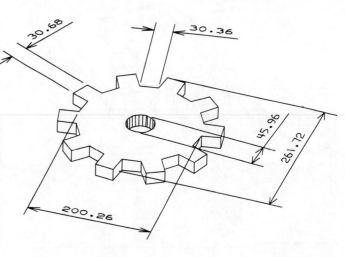

Fig. 2d The gear part shown with dimensions in microns.

A CAD ARCHITECTURE FOR MICROELECTROMECHANICAL SYSTEMS

Fariborz Maseeh[1], Robert M. Harris[2], Stephen D. Senturia[2]

[1]Department of Civil Engineering
[2]Department of Electrical Engineering and Computer Science
Microsystems Technology Laboratories
Massachusetts Institute of Technology
Cambridge, MA 02139

ABSTRACT

A CAD architecture for microelectromechanical systems is presented in which conventional mask layout and process simulation tools are linked to three-dimensional mechanical CAD and finite-element tools for analysis and simulation. The architecture is exercised by an elementary example on the stress-induced curvature of an oxidized silicon wafer. An architecture for an object-oriented material property simulator is also presented in which material properties and their process dependence are stored and are accessed based on the specific process conditions.

I. INTRODUCTION

With the development of increasingly sophisticated microelectromechanical devices, including microsensors, pumps, valves, and micromotors, and with the increasing performance demands being placed on these devices, notably in the precision and accuracy of microsensors, there is a critical need for CAD tools which will permit rational design of these devices. There are two fundamental problems that confront the designer [1,2]: (A) the need to construct a three-dimensional solid model from a description of the mask set and process sequence to be used in fabrication of a micromechanical device; and (B) the need to be able to predict the material properties of each of the constituent components in a device, including possible process dependences of these properties. At the present time, there is no CAD system, either mechanical or microelectronic, which successfully addresses these problems in a coherent way. Koppelman [3] has developed a program called OYSTER which permits construction of a 3-D polyhedral-based solid model from a mask set and primitive process description, but as yet, there is no provision for linking to FEM tools or to standard layout and process modeling tools, and no database for prediction of material properties from the process sequence.

This paper presents an architecture for a microelectromechanical CAD system which addresses these problems, and reports on the first implementation of this architecture. There are two critical functions not presently available in commercial packages: (1) a solid modeling tool (the "Structure Simulator"), which takes mask layout data and a process description and builds a 3-D solid model in a format compatible with the mechanical CAD system; and (2) a "Material Property Simulator", which takes process sequence information from the Structure Simulator, extracts the material property values from a database, and merges the material property information with the 3-D solid model for subsequent mechanical analysis.

The CAD architecture is explained in Section II. In Section III, we illustrate the successful analysis of an example, the thermal-stress-induced curvature of an oxidized silicon wafer, using elementary versions of the Structure Simulator and Material Property Simulator. This example exercises the architecture is some detail. One conclusion of this first attempt is that simple table-lookup for material properties is impossibly cumbersome in the general case due to the process dependence of material properties. Therefore, an object-oriented database approach has been developed for the Material Property Simulator, which is presented in Section IV.

II. CAD ARCHITECTURE

Fig.1 shows the architecture of our CAD system. It consists of three sections outlined by dashed blocks, the Microelectronic CAD section, the Mechanical CAD section, and the Material Property Simulator. The interactions among sections and their various constituents are shown by arrows, the directions of which specify the flow of information. The "User Interfaces" denote the various user's direct access to specific units in each section.

We have implemented this architecture in a Sun-4 host, drawing on existing codes wherever possible. The primary interface for mechanical modeling is through PATRAN [4], a mechanical CAD package which provides for interactive construction of 3-D solid models, graphical display, and interfacing with FEM packages (we are using ABAQUS [5]). The 3-D geometry resides in the PATRAN Neutral File with additional model information in a separate file. Initially, we have used the material-property format of the Neutral File as a first version of the Material Property Simulator while the more elaborate object-oriented version is being developed.

All of the commercially available codes in Fig. 1 are installed and operating. The Structure Simulator has been implemented at an elementary level, and interfaced with the Mechanical CAD. The first entries of material property data (for silicon, silicon dioxide [6], and silicon nitride [7]) have been made in a PATRAN readable file. A simplified example has been successfully tested using this architecture. We now describe the CAD architecture in detail, and then present the example.

A. MICROELECTRONIC CAD

In the Microelectronic CAD section, mask layout is created in CIF [8] format using KIC [9], and the process sequence is created in the MIT-developed Process Flow Representation (PFR) [10] using a standard file editor. SUPREM-III [11] and SAMPLE [12,13] are installed to provide depth and cross-sectional modeling capabilities.

Reprinted from *Proceedings IEEE Micro Electro Mechanical Systems*, pp. 44-49, February 1990.

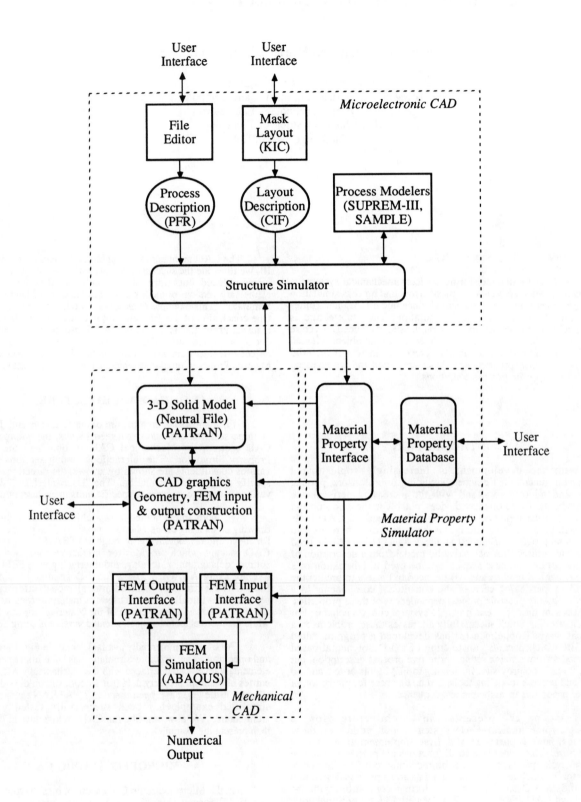

Fig. 1 CAD architecture for microelectromechanical design. Arrows denote the flow of information.

The critical block in this section is the Structure Simulator, which must merge the mask layout and process information to construct a three-dimensional solid model. Two kinds of information must be tracked, the <u>geometry</u> of the structure (position, shape, and connectivity of each component), and the <u>material type</u> and associated process conditions used to create each component of the structure. The geometry is passed directly to the Mechanical CAD section by creating a PATRAN Neutral File. The material type and associated process conditions for each component are passed to the interface portion of the Material Property Simulator through what we call the Process History File.

Physically, a process step causes a change in the wafer (e.g. diffusion, deposition, etching). The final structure is the result of concatenating a sequence of such changes. This suggests that the Structure Simulator should perform its function in a process-step by process-step basis, making appropriate modifications to the solid model for each process step. The operation of the Structure Simulator is summarized in Table I, and is illustrated in Fig. 2. The current process step is read from the PFR process sequence, and is interpreted as a set of tasks that must be done to create construction operators. When appropriate, the PFR information is passed to the process modelers (SUPREM-III, SAMPLE), and the simulation results from the process modelers are combined with the mask data to create the appropriate construction operators. These are then used to modify the solid model from the previous step, both the geometry portion in the Neutral File, and the material-type information in the Process History File.

The construction operators must be implemented to ensure that the resulting solid model is physically valid (i.e., describing a reasonable approximation to the actual structure). Without careful attention to the robustness of construction algorithms, invalid solid models could result (e.g., an unphysical topology such as a Klein bottle, or two objects occupying the same place). Implementation of robust construction operators is considerably simplified if the operators are required to be a combination of a small set of primitive construction operators. For microelectromechanical design, the following primitive operators constitute a useful minimal set: film deposition and growth, film etching, masking, impurity introduction and diffusion, and wafer joining. For initial implementation, we have selected a restricted subset: conformal deposition, and masked etching. These two primitives provide significant geometric flexibility, and permit the simulation of many interesting microelectromechanical systems.

Table I. STRUCTURE SIMULATOR OPERATIONS

1. • Interpret process step
2. • Determine primitive construction operators
   > Consult layout information and process modelers
3. • Use the primitive operators to modify solid model
4. • Output results
   > Geometric information to Neutral File
   > Material information to Process History File
5. • Repeat steps 1-4 for next process step

B. MATERIAL PROPERTY SIMULATOR

The Material Property Simulator reads the process sequence for each component of the solid model from the Process History File and generates a set of material property data. The material properties are passed to the Mechanical CAD section, either into the PATRAN Neutral File, the PATRAN interactive (graphics) section which creates loads and boundary conditions as it builds the FEM input file, or the FEM input file itself (see Fig. 1). The FEM input file path allows for the use of FEM simulator independent of the PATRAN graphics software; furthermore, it provides for a way to introduce <u>intrinsic stress</u> into the mechanical model. (Note: because of the detailed organization of PATRAN, intrinsic stress must be treated differently than <u>thermal mismatch stress</u>; the latter can be created by thermal loads and dissimilar coefficients of thermal expansion, as in the example to follow.)

Initially, we have elected to use PATRAN-readable file formats to enter material property data (manually) for each type of material, which might become a constituent layer of the device, using dimensions and units compatible with the geometric model in the Neutral File. This zeroth-order approach is used in our example. However, manual construction of data sets for every permutation of process sequences for every material is impractical. An object oriented Material Property Simulator is designed to organize the process dependence of properties in an effective way, and is explained in detail in Section IV.

For simplicity, only thermal stresses are considered in this zeroth-order approach; intrinsic stress is neglected. The object oriented Material Property Simulator, however, will account for both intrinsic stress and thermal stress.

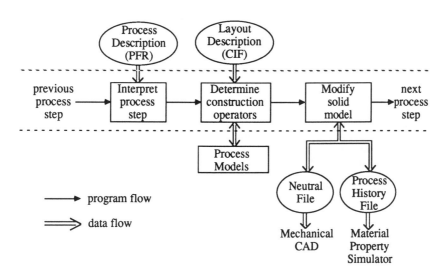

Fig. 2 Single process step operation of Structure Simulator.

C. MECHANICAL CAD

In the Mechanical CAD section, the geometric information from the Microelectronic CAD section is discretized into finite elements, material properties from the Material Property Simulator are associated with their proper geometries, and the appropriate loads and boundary conditions are specified. The solid model geometry and material properties are read and manipulated in PATRAN graphics, and a complete input model suitable for finite element analysis is generated and optimized interactively (see Fig. 1). The FEM model is then translated into an ABAQUS input file for FEM analysis. (Intrinsic stress would also be entered at this point). The results of FEM analysis are then translated back into a form readable by PATRAN graphics for display. PATRAN can then be used to examine the simulated mechanical behavior of the device.

The particular FEM simulator used (ABAQUS in this case) can be replaced by any other for which the PATRAN interfaces exist. Such a replacement would require modification of the details of the Material Property Simulator-to-FEM interface, but no major changes in architecture.

III. EXAMPLE

We next present an elementary example, the wafer curvature technique for thin film stress measurement [14], which exercises our CAD architecture in some detail. This technique, which is based on measuring the pure bending deformation of a supporting substrate due to the residual stress in a deposited film, has been abundantly used in measuring thin film stress. The method works equally well for tensile and compressive films and requires simple setup for measurements. As long as the stiffness of the film is negligible compared with that of the substrate and the out-of-plane deformation of the composite system is smaller than the half of the substrate thickness, this technique can be effectively utilized to measure stress in thin films. Film stress can be related to out-of-plane bending of the substrate via the Stoney equation [14]:

$$\sigma_f = (\frac{4}{3})(\frac{E_s}{1 - v_s})(\frac{t_s^2}{t_f})(\frac{z}{L^2})$$

where σ_f, E_s, v_s are film stress, substrate Young's modulus, and substrate Poisson's ratio; t_s and t_f are substrate and film thickness; and z and L are the out-of-plane deflection and length over which the deflection is measured.

The specific process considered in this example is: A 4-inch silicon wafer is thermally oxidized at 1030°C in wet ambient for four hours. The oxide is optionally removed from the back side using a wet etch, which results in wafer bending due to the stress in the front side oxide. This bending can be compared with the Stoney equation.

A menu-driven interface to the Structure Simulator allows the user to create a solid model in PATRAN neutral file format, based on the oxidation process conditions and wafer geometry. (There is no masking in this example.) The user specifies wafer thickness, wafer diameter, oxidation temperature, ambient, and time. The interface calls SUPREM-III, reports the resulting oxide thickness, and incorporates this thickness into the solid-model geometry. Due to the axisymmetric nature of our problem, the resulting model is two dimensional. The model geometry is displayed by the PATRAN graphics, where it is then discretized into finite elements, and appropriate boundary conditions are specified. Material properties are associated with each layer, in this case, by reading from a preconstructed data file readable by PATRAN graphics containing information on silicon, silicon dioxide, and silicon nitride. A thermal cooling load of -1000°C was applied to the entire model to create oxide residual stress due to thermal coefficient mismatch with the silicon substrate. The generated model is optimized and translated to an ABAQUS input file through the PATRAN-to-ABAQUS translator. The resulting input file is then linked to ABAQUS for FEM analysis. The analysis results are translated through the ABAQUS-to-PATRAN translator to files readable in PATRAN graphics, and the final results of the analysis are examined.

In the case of the wafer with double sided oxide, the oxide film is under compressive stress, but there is no out-of-plane deformation due to symmetric loading. When oxide is removed from the backside, the wafer bends to regain equilibrium. Solid models for both double-sided oxide and single-sided oxide were generated in the Structure Simulator as described above, and analyzed under the same thermal load. The wafer bending from the FEM simulation, for the latter case, is in close agreement with the Stoney equation as shown in Fig. 3.

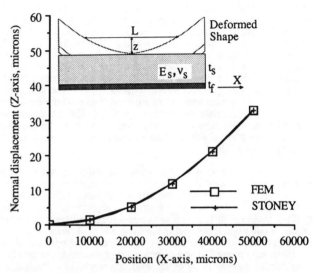

Fig.3 Out-of-plane displacement of an oxidized silicon wafer due to -1000°C thermal load.

IV. OBJECT ORIENTED MATERIAL PROPERTY SIMULATOR

Many of the properties of microelectronic materials, such as stress and density, are process dependent. This necessitates collection and extraction of properties of various materials in an organized way. An object-oriented database environment enables the definition and organization of data sets by their process dependence. This advantage is utilized in the development of the current Material Property Simulator.

Fig. 4 illustrates the Material Property Simulator architecture. It consists of a material property interface unit, and a material property database unit which holds tables of materials properties vs. process conditions. The interface unit consists of an interpolation routine and different interfaces to the mechanical and microelectronic sections, as shown. Process sequence information is read from the Process History File generated by the Structure Simulator. The process information includes the associativity to the corresponding components of the geometry and also a specification of dimensions and units. This information is handed to the interpolation routine, which accesses the material property database to retrieve the required data tables (described below), interpolates (using a multidimensional cubic spline) the resulting information to the values appropriate to the precise process conditions, and converts the dimensions and units to be compatible with the Structure Simulator geometry scale and the applied loads. The neutral file and graphics interface relays the material property information to the PATRAN Neutral File or the graphics unit. The intrinsic stress induced by process is interfaced directly with the FEM simulator input file. This same link can be used to transfer all of the material properties directly to the FEM simulator input file, if it is desired to use the FEM analysis independent of PATRAN graphics.

In the material property database section, the object oriented database environment GESTALT has been selected for storing and retrieving the material properties [15]. GESTALT has several advantages. First, it shields application programs from the many details of the underlying database (currently INGRES, a commercial relational database system [16,17]). It also provides for an environment in which application programs are written in different higher level languages (currently C and LISP). For example, the LISP interface could be used for adding rule-based applications for "intelligent" manipulation of data. GESTALT is flexible enough so that the underlying database can be replaced/upgraded without affecting existing application programs. Furthermore, it is being developed as part of the integrated circuit computer-integrated-manufacturing (IC-CIM) effort at MIT which provides for a natural link to the manufacturing environment. GESTALT can be accessed directly through a user interface for direct input and output of the stored data; it can also be linked to different simulators through the material property interface.

Fig. 5 illustrates the schema of the material property management through the GESTALT interface. The FILM is represented by a name (film_name), its fabrication process (film_process), and the applicable property names (material_property_names). The FILM PROCESS is organized by a name (process_name) and a list of the parameters (process_parameters) used in fabrication. The PROCESS captures the dependency of the material property data with process parameters stored in the relational database INGRES. A material property (material_property_value) versus process parameter (parameter_value) relation can be uniquely retrieved from INGRES by a process_name (with corresponding process_parameter_name) and a film_name (with the corresponding material_property_name). The retrieved values are in the form of a table (or function) for each property vs parameter, and a unique value of the material property can be selected by interpolating in this table for the given process parameter.

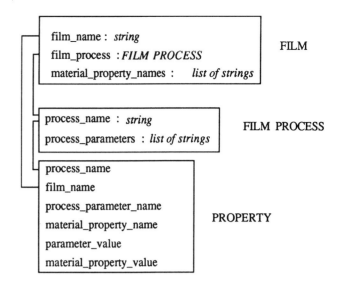

Fig. 5 Architecture of object oriented database organization.

At present, a first version of the Material Property Simulator has been written in C, and initial data for the process dependence of some of the properties of silicon dioxide have been entered into the database. Retrieval and interpolation has been demonstrated, but the interfaces to the rest of the system have not yet been implemented.

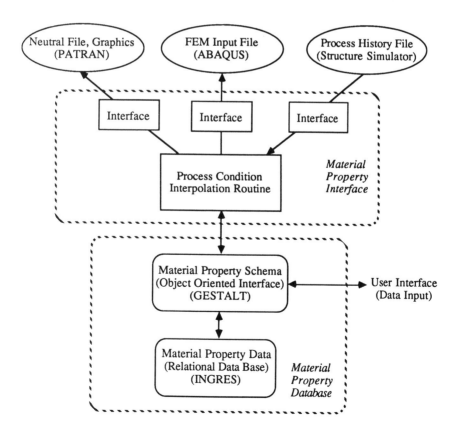

Fig. 4 Material Property Simulator architecture.

V. CONCLUSION

We have presented a CAD architecture for microelectromechanical systems, and have demonstrated the viability of that architecture in an elementary example. An object-oriented approach to the material property database problem is described.

It is anticipated that this architecture, which links the process dependence of material properties with a solid modeling capability, will find broad applications in conventional microelectronic device design, in device packaging, and related fields.

ACKNOWLEDGEMENTS

The authors gratefully acknowledge the helpful discussions and technical assistance of Mr. Michael Heytens, author of GESTALT, in the development of the material property database, Duane Boning for providing the some of the components of the Structure Simulator interface and for his assistance in adapting them to our use, and Prof. Donald Troxel for his many valuable suggestions. We thank Mr. Sean M. Gelston and Mr. Miles Arnone for their help in the literature search of material properties. The microelectromechanical system and material property simulator design is supported in part by DARPA under contract MDA972-88-K-0008, and the Department of Justice under contract J-FBI-88-067. The research is carried out in the Microsystems Technology Laboratories of the Massachusetts Institute of Technology.

REFERENCES

[1] S. D. Senturia, "Microfabricated structures for the measurement of mechanical properties and adhesion of thin films", Transducers '87, Tokyo, 1987, pp. 11-16.

[2] S. D. Senturia, "Can we design microrobotic devices without knowing the mechanical properties of materials?", IEEE MicroRobots and Teleoperators Workshop, Hyannis, 1987.

[3] G. M. Koppelman, "OYSTER : a 3D structural simulator for microelectromechanical design", MEMS '89, Salt Lake City, 1989, pp.88-93.

[4] PDA Engineering, Costa Mesa, CA.

[5] Hibbitt, Karlsson, and Sorensen, Inc., Providence, RI.

[6] F. Maseeh, S. M. Gelston, S. D. Senturia, "Mechanical properties of microelectronics thin films : Silicon Dioxide", MIT VLSI Memo No. 89-575, Oct. 89.

[7] F. Maseeh, M. Arnone, S. D. Senturia, "Mechanical properties of microelectronics thin films: Silicon Nitride", MIT VLSI Memo No. 89-576, Oct. 89.

[8] L. Conway and C. Mead, Introduction to VLSI Systems, Sec. 4.5, Addison-Wesley : Reading, MA, 1980.

[9] G. C. Billingsley, "Program reference for KIC", Memo No. UCB/ERL M83/62, U. C. Berkeley, Oct. 1983.

[10] M. B. McIlrath and D. S. Boning, "Integrated process design and manufacture", presented at SRC workshop on Integrated Factory Management for Integrated Circuits, Nov. 16-17, 1989, College Station, TX.

[11] C. P. Ho, J. D. Plummer, S. E. Hansen, and R. W. Dutton, "VLSI process modeling SUPREM III", IEEE Tran. Elect. Dev., vol. ED-30, no.11, pp.1438-1452, Nov. 1983.

[12] W. G. Oldham, S. N. Nandgaonkar, A. R. Neureurather, and M.M. O'Toole, "A general simulator for VLSI lithography and etching processes: Part I - application to projection lithography", IEEE Trans. Electron Dev., vol. ED-26, no.4, pp.717-722, Apr. 1979.

[13] W. G. Oldham, A. R. Neureurather, C. Sung, J. L. Reynolds, S. N. Nandgaonkar, and M.M. O'Toole, "A general simulator for VLSI lithography and etching processes: Part II - application to deposition and etching", IEEE Trans. Electron Dev., vol. ED-27, no.8, pp.1455-1559, Aug. 1980.

[14] G. G. Stoney, "The tension of metallic films deposited by electrolysis", Proc. R. Soc. London, Ser. A., vol. 82, pp. 172, 1909.

[15] M. L. Heytens and R. S. Nikhil, "GESTALT : An expressive database programming system", ACM SIGMOD Record, vol.18, no.1, pp.54-67, Mar. 1989.

[16] M. Stonebraker, G. Wong, P. Kreps, and G. Held, "The design and implementation of INGRES", ACM Transaction on Database Systems vol.1, no.3, pp.189-222 1976.

[17] INGRES Corp., Alameda, CA.

CAEMEMS: AN INTEGRATED COMPUTER-AIDED ENGINEERING WORKBENCH FOR MICRO-ELECTRO-MECHANICAL SYSTEMS

Selden Crary and Yafan Zhang

Center for Integrated Sensors and Circuits
Electrical Engineering and Computer Science Department
The University of Michigan, Ann Arbor, Michigan 48109-2122

ABSTRACT

We are constructing a computer-aided engineering system for micro-electro-mechanical systems. The system will take advantage of existing high-level integrated mechanical-engineering software tools and add application-specific software for the system to function as a MEMS-specific designers' workbench environment. The system, which we have called CAEMEMS, will include database structures (MEMS Databases); a process modeler that can model MEMS-relevant process sequences (MEMS CAD); and a device modeler that can perform finite-element simulations, link individual devices together through appropriate matrix operations, and perform simulations of entire mechanical systems (MEMS Sim).

SUMMARY

A block diagram of the CAEMEMS system is shown in Figure 1. The MEMS database will incorporate existing knowledge of the isotropic and anisotropic etch rates of crystalline and polycrystalline silicon, silicon dioxide, silicon nitride, quartz, and other materials, such as metallic thin films. This database will contain etch-rate information as a function of material, doping, orientation, etchant, and temperature, and it will be updated as new information becomes available. Mask information and process steps will be the inputs that MEMS CAD will process, using appropriate diffusion models and etching simulators, such as the Wulff-Jaccodine construction, its derivatives, or plasma-etching models under development in our Center. The output will be an accurate three-dimensional representation of the micro-electro-mechanical device, including its detailed composition. This representation will be available to the user for design verification and visualization as a projection on the screen of the workstation, and the user will be able to manipulate the representation by scaling, rotation, and planar-cut operations.

MEMS Sim will have the capability of capturing the three-dimensional geometries of devices in the form of finite-element meshes, to which the user will be able to apply appropriate boundary conditions and forcing functions, to be used for finite-element and systems simulations of the devices. Finite-element simulations that will be implemented include displacement, stress, and thermal analyses, both static and dynamic, as well as modal analyses and fluid flow. Systems simulations

CAEMEMS

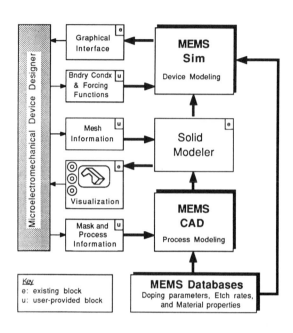

Figure 1. The CAEMEMS block diagram includes high-level, pre-existing components, as well as MEMS-specific software.

will include frequency and transient responses to excitations, forces of constraint, component load, energy distributions. Outputs from the simulations will be available as color graphics.

Because much of CAEMEMS is being built out of, or linked to, existing blocks of software; such as a solid modeler, an automatic mesh generator, systems for computing finite-element solutions to partial differential equations, an automatic mesh refiner, and a mechanical-engineering system simulator; our research is focusing on the computer-aided design and computer-aided engineering tasks unique to MEMS. At present, we are developing a simulator, along with a friendly user interface, for the broad class of diaphragm pressure sensors, including bossed and corrugated structures, see Figures 2 and 3.

When complete, CAEMEMS will be a valuable tool for the micro-electro-mechanical–device community. In addition, it will form the basis for the next stage of automated design for this field, which we envision to be the creation of an integrated set of compilers and simulators of micro-mechanical parts. These parts will

Figure 2. Visualization of microstructures for design verification and interpretation is an integral part of the developing CAEMEMS concept. Here, one of the authors (YZ) is seen viewing an image of a corrugated, bossed diaphragm structure designed in an early prototype CAEMEMS environment.

be accessed from a library of structures and linked together at the design stage, in conjunction with appropriate circuitry. In this way, designers will be able to design and simulate complete micro-electro-mechanical systems.

In our oral presentation we discuss general issues regarding merged CAD/CAE tools for MEMS simulation, which, in addition to the above, include easy accessibility for novice users and availability and transportability of code. We also show results of our initial results in constructing the CAEMEMS system.

ACKNOWLEDGMENTS

Support for this work was provided by the National Science Foundation under grant #ECS-8915215. Additional funding was provided by the Sensors and Signal Processing Laboratory of Honeywell, Inc., General Motors Research Laboratories, and Monolithic Sensors, Inc. We thank Khalil Najafi and Kensall D. Wise of our Center for valuable discussions. Also contributing to our work were Joseph M. Andrusiak, Omar S. Juma, and Ann A. Gordon, all of the University of Michigan.

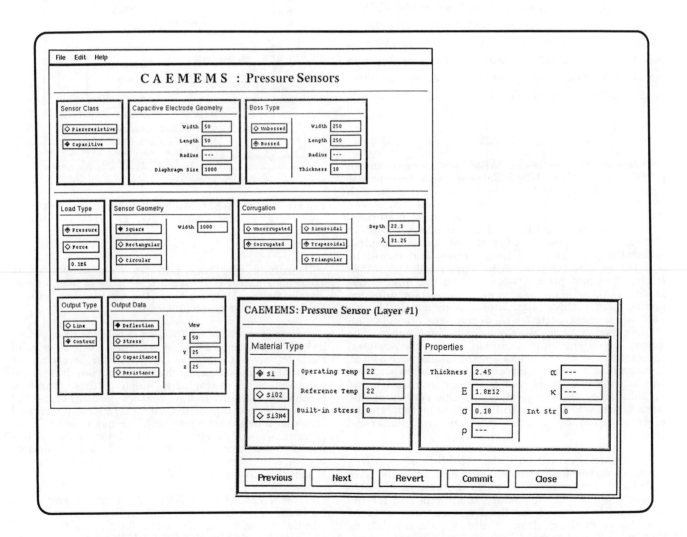

Figure 3. A self-explanatory user interface, built from developing software tools, is an integral feature of CAEMEMS.

CAD for Silicon Anisotropic Etching

R.A. Buser, N.F. de Rooij
Institute of Microtechnology, University of Neuchâtel,
Rue A.L. Breguet 2, CH-2000 NEUCHATEL, Switzerland

Abstract. A computer program is proposed, which simulates silicon single crystal etching in KOH. Starting from a two-dimensional mask the program finds the relevant etching planes and delivers a projected three-dimensional output of the etched structure with the etchtime (or etchdepth) as parameter.

Simulation programs have become an essential tool for micromachining, too. They consist typically of FEM- and design-programs as used in classical engineering[1]. Few effort has been done however to simulate the etching process itself. In order to be able to predict easily the three-dimensional shape of etched silicon structures from a two-dimensional mask, we have developed a simulation program in the language pascal, which has proved its usefulness in many simulations of micromachined silicon structures[2,3,4.]. Recently a simulation program for a special situation has been announced by Sato[5.] Our program runs for etching of (100) oriented silicon in KOH, but etch simulation programs for other etchants and other crystals might be constructed using the same principle. As long as the order of the etch rates of the different planes are not changed it works for any temperature supposed the etch rates are known. Such a program represents a considerable knowledge on anisotropic etching.

The program runs on a HP UNIX system. For the graphical subroutines we used Starbase, but the core of the program is written in standard Pascal. The system was built up with Computer HP 330, Harddisk (150 Mbyte), Cartridge tape, Graphic tablet, plotter (A3) and printer.

The program is file oriented. So each task ends with a data file which can be easily stored and read by the same or other programs (for example input/output auxiliaries). Thus, a mask file is created, then a model file (records which contain all additional planes) and finally a structure file (real), where the etched structure is represented by polygons, which can be plotted in a colour according to the miller plane family they belong. The input can be done directly or by using a simple CIF file produced by VTI (mask layout program from VLSI Technology).

The leading idea of the modelling stems from the observation that in spite of an infinite number of planes in a crystal, in etched structures only a few are relevant: (111),(100) and (311), each with its proper etch rate. As can be seen from the discussion above, a mask design makes only sense with rectangular lines(100) planes and lines at 45° (111) planes. In addition the polygons have to be closed. From the sense of rotation of digitizing, the program decides, wether the inside of the polygon will be exposed to etch (open masking layer) or the outside. The program identifies the lines of the mask as crystal directions. Because only one specific crystal plane can be revealed by etching, this assignment is unequivocal (see Fig.1.). Special problems occur at the convex corners, where the fast etching planes are revealed. The program classifies thus different types of corners and adds the necessary planes. Once the crystal planes to be etched are identified, these planes are moved along their normal vector according to their specific etch rate until either

i) etching is over

ii) it is "eaten" by its neighbor planes (faster etching planes, convex shape)

iii) it disappeared when bounded by slower etching planes (concave shape)

Intersection of the planes in the new position reveals the etched structure (see Fig.2.).When the etching gets deeper, some planes will disappear, which is taken into account by the program, too (see Fig.3.). In a KOH etching solution there is one situation to be considered with care, namely the case where two (vertical) 100) planes move towards each other and disappear. This phenomenon can be used as a corner compensation as reported earlier[6]. The program checks for this situation and produces a warning, when the corner is overetched (see Fig.3.).

---

[1] see e.g. OYSTER, by G. M. Koppelman, IEEE MEMS Workshop 1989, Salt Lake City.
[2] Buser, R.A. Silicon anisotropic etch simulation program, IMT internal report No 253 IC 01/89, 1989.
[3] R.A. Buser, and N.F. de Rooij, Silicon Microetching and FOS: Properties, Compatibility and Synthesis of Transducers, Proccedings of the Sixth European Fibre optic Communications and Local Area NEtworks Exposition, held June 29-July 1, 1988, pp.223, RAI Amsterdam, NL.
[4] R.A. Buser; Theoretical and Experimental Investigations on Silicon Single Crystal Resonant Structures, Ph.D. Thesis University of Neuchâtel, IMT, May 1989.
[5] Kazuo Sato, Akira Koide, and Shinji Tanaka; Measurement of anisotropic etching rate of single crystal silicon to the comple orientation. JIEE Technical Meeting Micromachining and Micromechatronics, 1989, Japan., p 19.
[6] R.A. Buser, B. Stauffer, N.F. de Rooij, Realization of a mesa array in (001) oriented silicon wafers for tactile sensing applications, Extended Abstracts of the Fall Meeting of the Electrochemical Society, San Diego, Cal. October 19-24, 1986.

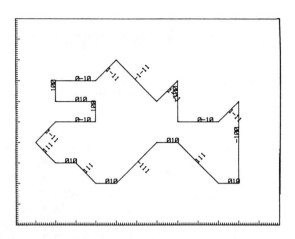

Fig.1. shows a layout with various possible combinations of adjacent planes. They are labeled by the corresponding Miller indices. (100μm/ unit)

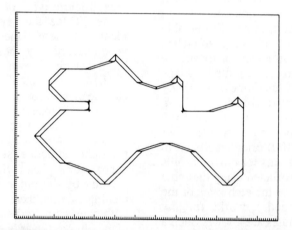

Fig.2. shows the result of the simulation program after 280μm deep etching.

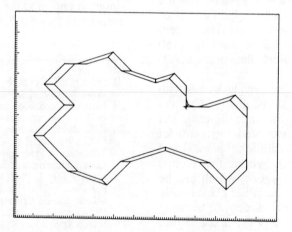

Fig.3. shows the result of the simulation program after 500μm deep etching, where a convex corner formed by 111 planes is visible.

Section 14

Metrology

BASIC to all sciences is the ability to measure and characterize physical phenomena. The emerging science of micromechanics is presently learning how to develop the standards and measurement techniques needed by researchers and engineers.

In the first paper, "Can We Design Microrobotic Devices Without Knowing the Mechanical Properties of Materials," Stephen Senturia critically examines the metrology needs of micromechanics. The next five papers ("The Use of Micromachined Structures for the Measurement of Mechanical Properties and Adhesion of Thin Films," "Mechanical Property Measurements of Thin Films Using Load-Deflection of Composite Rectangular Membrane," "Fracture Toughness Characterization of Brittle Thin Films," "Spiral Microstructures for the Measurement of Average Strain Gradients in Thin Films," and "Polysilicon Microstructures to Characterize Static Friction") describe how to use micromechanical tools to measure the properties of micro components. The paper "Study on the Dynamic Force/Acceleration Measurements" describes a high-accuracy experiment using macro equipment to accurately characterize micro devices. The final paper, "Anomalous Emmissivity from Periodic Micro-Machined Silicon Surfaces," describes how periodic micro structures can modify macro phenomena, such as the emissivity from a surface.

CAN WE DESIGN MICROBOTIC DEVICES WITHOUT KNOWING
THE MECHANICAL PROPERTIES OF MATERIALS?

Stephen D. Senturia
Microsystems Technology Laboratories
Department of Electrical Engineering and Computer Science
Massachusetts Institute of Technology
Cambridge, MA, 02139, USA

ABSTRACT

The tools of microfabrication are now being applied to the design and fabrication of small components which can be used in sensors and actuators, and, eventually, as parts of microbotic systems. The successful design of micromechanical parts depends on the accuracy with which the behavior of the structures can be predicted, fabricated, and reproduced. This, in turn, depends on three critical factors: control of the mechanical properties of the materials used for fabrication, dimensional control of the structure during fabrication, and structural analysis of the primary part and its support or encapsulation. Tools for handling the latter two issues are relatively well developed. However, for many of the materials used in microfabricated structures, basic data on mechanical properties and their control through process variables is lacking. This paper reviews briefly some of the methods that have been used to develop such data, and reports in more detail on some new approaches to this problem that are particularly suited to the study of thin films having residual tensile stress. The use of these methods is illustrated with measurements on polyimide films spin-cast onto oxidized silicon substrates.

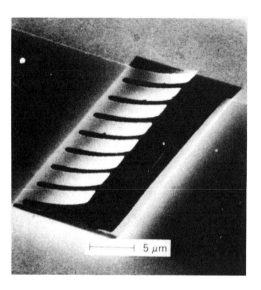

Fig. 1   Oxide cantilevers showing curvature due to nonuniform residual stress (courtesy of Kurt Petersen, Novasensor).

INTRODUCTION[1]

The question posed in the title of this paper has an obvious answer. The thought of attempting a serious _design_ without reasonable knowledge of key properties of constituent materials is clearly absurd. Yet, as practitioners of the microfabrication art, we are regularly placed in this situation by our ignorance of the behavior of the materials with which we routinely work. And it is not because we are lazy. Determination, documentation, and control of the mechanical behavior of microelectronic materials remains a significant probem area, and the problem is aggravated by the other demands on designers in this field. Not only is the designer required to know all about integrated-circuit and microfabrication technology, but he or she must also be knowledgeable about measurement science and mechanics, and about the specific type of transducer principle being used in the design. Furthermore, because measurement devices and actuators usually cannot be critically tested until they have been packaged, the designer must also understand the constraints involved in device encapsulation.

Microsensors and microactuators place new and stringent requirements on the _mechanical properties_ of the component parts.   Figure 1 illlustrates the problem of trying to construct a flat cantilever beam[2]. Lateral stresses, if uniform, would relax when the cantilever is etched free from its support; however, if the stresses are nonuniform (or not symmetrically balanced) through the thickness of the beam, the free-standing beam will curl, as shown in the Figure. Thus, not only must the designer know the normal mechanical properties of the materials, such as Young's modulus and Poisson's ratio, but the designer must be able to predict residual stresses. This turns out to be a difficult problem, which affects the entire design process for microbotic devices.

THE ANALOGY TO INTEGRATED-CIRCUIT CAD

To paraphrase Polya[3], when confronting a new problem, one should attempt to discover an analogy that transforms the problem into one for which you already know the answer. In this case, the powerful methods developed for integrated circuit design provide useful guidance. Indeed, CAD tools now exist for every step of the design process, from specifying the basic process with SUPREM, to understanding electrical device performance (many programs exist), to modeling circuit performance (including parasitics) with circuit simulators like SPICE. Furthermore, design checkers have been developed to verify that the final mask set actually results in the required circuit, and automatic logic checking and timing estimates can be made, all before the circuit is built. As a result, while the design cost for a new circuit may be significant, the final device can be expected to perform according to specifications.

It is instructive to ask how such a support structure for design came about. The process modeling program SUPREM required collection of an enormous database of experimental information on oxidation and diffusion rates under a wide variety of circumstances, and the incorporation of that information into a digital environment suitable for solving the diffusion equation with moving boundaries. Similarly, programs for device and circuit modelling required, first, the creation of a mathematical environment for modelling, and, second, the encoding of appropriate data -- representing the cumulative experimentation of many workers over many years -- into modules of code that represented the physical behavior of the device or circuit subsection.

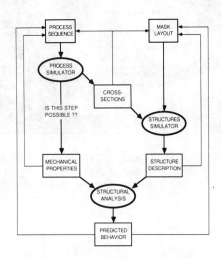

Fig. 2   Schematic   CAD   sequence   for microsensors.

Now let us examine the case for microsensor design. Figure 2 schematically illustrates the design process and the CAD components one would like to have. A design consists of a _process sequence_ and a _mask layout_. Three simulation steps are required to go from the design to predicted behavior. The PROCESS SIMULATOR, for a given process sequence, gives the residual stress, elastic modulus, Poisson's ratio, and tensile strength of each constituent material, as well as the usual device cross-section produced by SUPREM. The STRUCTURE SIMULATOR combines the cross-section output from the PROCESS SIMULATOR and mask data and fabrication tolerances (possibly expressed in terms of lithographic design rules) to produce a three-dimensional geometric description of the fabricated structure. Finally, the STRUCTURAL ANALYSIS program combines the structure description with the mechanical property data and performs a prediction of the mechanical behavior of the complete structure, including effects of the support or encapsulation, and incorporating the basic transduction principle (such as piezoresistance) into predictions of the electrical behavior of the device under various conditions.

Is this fanciful? Not necessarily. Most of these tools are already developed to some degree. Specifically, the paths leading to the structure description could be developed by combining SUPREM-like cross-sections with mask-analysis programs. Further, there are many finite element codes that can perform structural analysis, provided the model used for the structure is appropriate (in terms of selection of elements, and the handling of geometric non-linearity and large deflections). While these finite element codes are not without problems, and must be used with great care and with continuous check against experiment, there is, nevertheless, a vast set of resources for the designer to draw on.

What is missing is the ability to predict the basic mechanical property data from the process sequence!

Many investigators have developed methods for measuring various mechanical properties of microelectronic materials, and several comprehensive reviews are available [4,5]. To cite a few key works, modulus and/or residual stress values have been reported for thermal oxides and CVD dielectrics [2,6-8], polysilicon [9,10], polyimides [1,11,12,16-19], and metals [13]. The techniques used include wafer curvature [14], cantilever studies [7,9], membrane load-deflection studies [12,15], and buckling of clamped structures due to residual stress [9,10]. However, even with the data that has been obtained from these works and from the many additional references too numerous to cite here, we do not have the ability to do reliable prediction of mechanical properties of microelectronic materials in the actual state in which they are deposited, or as they are modified by subsequent processing (which is so well handled by SUPREM for oxidation and diffusion).

Further, the available data are incomplete. It is true that for a material under very specific deposition conditions (possibly specific to one apparatus) one may know the state of stress and may also know the modulus. However, the tensile strength and Poisson's ratio are less well known. (Consider, for example, the number of reliability problems that are related to brittle fracture of dielectrics.) Finally, except possibly for well-characterized individual pieces of equipment, we do not know how to predict how these properties will change if we modify the process. Until we have such insight, and can routinely draw on it during the design process, each new design will be burdened by the need to develop the relevant mechanical property data as part of the design process. This is a serious problem.

SUSPENDED MEMBRANES AND RELEASED STRUCTURES

Recent work in our laboratory has resulted in some new techniques for the study of thin-film mechanical properties [12,16-19]. One motivation for this work has been the hope that as our knowledge improves, prediction of such properties for a given process sequence may become routinely possible. The techniques are particularly well suited to in-situ measurements on films with residual tensile stress, and, hence, are very appropriate for polymer layers, such as polyimides, used as interlevel dielectrics and passivants. The remainder of this paper briefly surveys this new work.

Figure 3 shows the process sequence used for the structures. Suspended membranes (see Fig. 4) are prepared by first creating a silicon diaphragm by conventional anisotropic etching, then applying the coating, and, finally, removing the supporting silicon

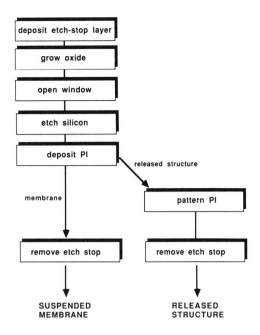

Fig. 3 Process sequences for suspended membranes and released structures.

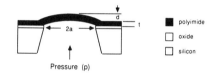

Fig. 4 Schematic cross section of the suspended membrane.

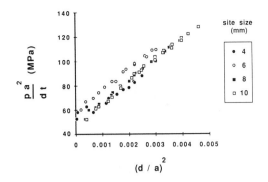

Fig. 5 Load-deflection behavior of a polyimide membrane, illustrated schematically at the top. Data for different diaphragm sizes illustrates scaling according to equation (1).

from the back with an SF_6 plasma. Suspended square polyimide membranes of thickness between 1 and 15 μm and between 1 and 25 mm on a side have been fabricated using this sequence. By pressurizing one side of the membrane, as shown in Fig. 4, and measuring the deflection, one can extract both the residual stress and Young's modulus of the membrane. The load-deflection curve for these structures is of the form

$$pa^2/dt = C_1 + C_2(d/a)^2 \quad (1)$$

where p is the pressure, d is the center deflection, a is the square size, t is the membrane thickness, C_1 is a constant that is proportional to the residual stress, and C_2 is a constant that is proportional to Young's modulus.

The data in Figure 5, which were taken from a single coated wafer containing four membrane sites of different sizes, obey the scaling implied by Equation (1)[16]. The intercept yields the stress and the slope yields the modulus. For this particular sample, the stress is 17 MPa, and the modulus is 6 GPa, resulting in a residual stress-to-modulus ratio of 0.003. This stress-to-modulus ratio, also referred to as the _residual strain_, is what must be compared to the _ultimate strain_ (see below) when evaluating the potential reliability problems associated with cracking of films. We have examined a variety of polyimides, and have observed residual strains in the range from 0.0002 to 0.012, depending on the details of the polyimide chemistry. These results are being reported separately.

A _released structure_[12,18] is made by the same process as the suspended membrane (see Fig. 3) except that the polymer film is patterned into an asymmetrical structure before removing the thin silicon support. An example is shown in Fig. 6. Once released, the wide suspended strip (width W_1) pulls on the thinner necks (total width W_2), resulting in a deflection δ from its original mask position toward the right to its final position after release. The residual tensile stress in the film is the driving force for the deformation. By varying the geometry, it is possible to create structures in which the strain in the thinner sections is very small, to others in which the _ultimate strain_ of the film is exceed- ed. For structures in which the strain is small enough to be modeled with linear elastic behavior, the deflection d can be related to the stress-to-modulus ratio (σ_0/E) as follows:

$$\sigma_0/E = \delta(W_1/L_1 + W_2/L_2)/(W_1 - W_2) \quad (2)$$

where the geometry terms are defined in Fig. 6.

Figure 7 shows a photograph of two such released structures, one with thicker necks, the other with necks so thin that the film fractured when released. Based on the residual tensile strain of the film and the geometry of the structures that failed, it was determined that the ultimate strain of these particular films was 4.5%.

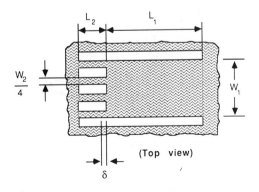

Fig. 6 Schematic top-view of axial beam released structure.

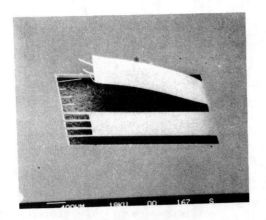

Fig. 7   Two released structures, one of which has exceeded ultimate strain of film, resulting in fracture of the necks.

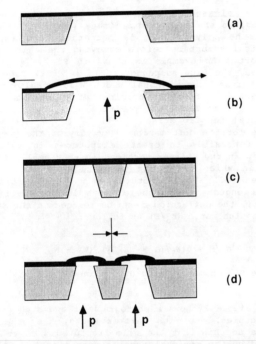

Fig. 8   Illustrating the adhesion test sites. (a) suspended membrane; (b) outward peel; (c) island structure; (d) inward peel.

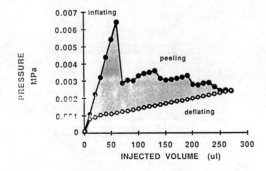

Fig. 9   Peel adhesion data for a sample of the type in Fig. 8(b). The shaded area is the work done to create the peeled surface.

ADHESION MEASUREMENTS

Adhesion of various films to one another is as important as the mechanical properties of the individual films in overall device performance and reliability. Figure 8 shows schematically how two different suspended structures are used for adhesion measurements [16,17,19]. Fig. 8(a) shows a suspended membrane, and Fig. 8(b) shows the same membrane after it has been peeled from its substrate by an applied load. Figure 9 illustrates the P-V cycle for such an experiment, in which the original membrane is inflated, then peels, then is deflated. The P-V work in creating the new surface is given by the shaded portion of Fig. 9. It is equal to the average work of adhesion for the film-substrate interface times the area peeled during the test.

Figure 9 (c) and (d) illustrate a more versatile structure in which an island is left in the center of the suspended membrane. Peel is induced toward the center of the island. This structure is better suited for studying systems with very good adhesion [17,19]. The fracture mechanics of such structures, however, must be analyzed with care in order to relate the critical pressure at which peel occurs to the strength of the adhesive bond. In preliminary measurements however, reasonably good agreement has been obtained between the P-V analysis and the analysis of the critical debond pressure.

Our work on adhesion is still in a very preliminary phase. However, microfabrication does show promise as a method of studying adhesion in-situ with a variety of very thin films.

CONCLUSION

The ability to predict mechanical properties is essential to microactuator design, yet no CAD tool presently exists that permits such prediction from a given process sequence. Further, it is not yet clear that such a CAD tool could be created without further advances in the basic science of thin film mechanics and morphology. The creation of the CAD tool and its supporting data require quantitative, reproducible experimental methods. Many good methods exist, and we have elaborated on some new methods specifically directed toward materials with residual tensile stress. It is critical, however, that the process dependence of these mechanical properties be studied, documented, and understood in order to provide a basis for intelligent design of the next generations of microrobotic devices.

ACKNOWLEDGEMENT

The author is indebted to Mark G. Allen, Mehran Mehregany, Martin Schmidt, Fariborz Maseeh, and Prof. Roger Howe for many useful discussions of mechanical property issues, and to Prof. Dimitri Antoniadis for discussions of the development of SUPREM and other CAD tools. Thanks also to Dr. Kurt Petersen for providing the photograph in Figure 1.

REFERENCES

1) The present paper is based in large part on an invited paper presented at the 4th International Conference on Solid State Sensors and Actuators, Tokyo, June, 1987; S. D. Senturia, "Microfabricated structures for the measurement of mechanical properties and adhesion of thin films, *Proc. Transducers '87*, IEE of Japan, pp. 11-16, June 1987.

2) K. E. Petersen, "Silicon as a mechanical material", *Proc. IEEE*, vol. 70, pp. 420-457, 1982.

3) G. Polya, *How to Solve It*, Princeton, Princeton University Press, 1973.

4) R. W. Hoffman, "The mechanical properties of thin condensed films", in G. Haas and R. E. Thun (eds.), *Physics of Thin Films*, vol. 3, pp. 211-273, 1966.

5) D. S. Campbell, "Mechnanical properties of thin films", in L. I. Maissel and R. Glang (eds.), *Handbook of Thin Films*, New York, McGraw Hill, 1970, Chapter 12.

6) E. P. EerNisse, "Stress in thermal SiO_2 during growth", *Appl. Phys. Lett.*, vol. 35, pp. 8-10, 1979.

7) K. E. Petersen and C. R. Guarnieri, "Young's modulus measurements of thin films using micromechanics", *J. Appl. Phys.*, vol. 50, p. 6761, 1979.

8) A. K. Sinha, H. J. Levinson, and T. E. Smith, "Thermal stresses and cracking resistance of dielectric films (SiN, Si_3N_4, and SiO_2) on Si substrates", *J. Appl. Phys.*, vol. 49, pp. 2423-2416, 1978.

9) R. T. Howe and R. S. Muller, "Stress in polycrystalline and amorphous silicon thin films", *J. Appl. Phys.*, vol 54, pp. 4674-4675, 1983.

10) H. Guckel, T. Randazzo, and D. W. Burns, "A simple technique for the determination of mechanical strain in thin films with applications to polysilicon", *J. Appl. Phys.*, vol. 57, pp. 1671-1675, 1985.

11) P. Geldermans, C. Goldsmith, and F. Bedetti, "Measurement of stresses generated during curing and in cured polyimide films", in K. Mittal (ed.), *Polyimides*, New York, Plenum Press, 1984, pp. 695-711.

12) M. G. Allen, M. Mehregany, R. T. Howe, and S. D. Senturia, "Microfabricated structures for the *in-situ* measurement of residual stress, Young's modulus, and ultimate strain of thin films", *Appl. Phys. Lett.*, vol. 51, pp. 241-243, 27 July 1987.

13) R. W. Hoffman, "Internal stresses in thin films", in B. Schwartz and N. Schwartz (eds.), *Measurement Techniques for Thin Films*, New York, The Electrochemical Society, 1967, pp. 312-333.

14) R. Glang, R. A. Holmwood, and R. L. Rosen-feld, "Determination of stress in films on single crystalline silicon substrates", *Rev. Sci. Instr.*, vol. 36, pp. 7-10, 1965.

15) E. I. Bromley, J. N. Randall, D. C. Flanders, and R. W. Mountain, "A technique for the determination of stress in thin films", *J. Vac. Sci. Technol.*, vol. B1, pp. 1364-1366, 1983.

16) M. G. Allen and S. D. Senturia, "Microfabricated test structures for adhesion measurement", Tech. Digest, Adhesion Society Meeting, Williamsburg, VA, Feb., 1987.

17) M. G. Allen and S. D. Senturia, "Microfabricated structures for the measurement of adhesion and mechanical properties of thin films", *Proc. ACS Division of Polymeric Materials: Science and Engineering*, vol. 56, pp. 735-739 April 1987.

18) M. Mehragany, R. T. Howe, and S. D. Senturia, "Novel microstructures for the *in-situ* measurement of mechanical properties of thin films, *J. Appl. Phys.*, in press.

19) M. G. Allen and S. D. Senturia, "Analysis of critical debonding pressures of stressed thin films in the blister test, *Journal of Adhesion*, in press.

THE USE OF MICROMACHINED STRUCTURES FOR THE MEASUREMENT OF MECHANICAL PROPERTIES AND ADHESION OF THIN FILMS

Mehran Mehregany, Mark G. Allen, and Stephen D. Senturia

Microsystems Technology Laboratories
Massachusetts Institute of Technology
Cambridge, Massachusetts 02139

This paper reports the application of silicon micromachining to the measurement of mechanical properties of thin films such as intrinsic stress, Young's modulus, and adhesion. The measurement is based on the deflection and subsequent peeling of suspended membrane sections of the film. The original goal of the work was to make a quantitatively reproducible adhesion test by applying micromachining techniques to the blister peel test described by Hinkley [1]. Our initial measurements demonstrated the importance of residual stress in the films, which resulted in an expanded emphasis on the basic mechanical properties of the membrane as a prelude to accurate adhesion measurements. We will briefly discuss the process for micromachining suspended membranes, the theory leading to the determination of mechanical properties of the films, our results, and the present status of the adhesion work.

Square suspended membrane sites are fabricated using standard micromachining processes [2]. Figure 1 gives a schematic of the fabrication process. First, a p+ etch stop layer is formed by boron deposition from high temperature solid sources [3]. The deposition is at 1175 °C for 120 minutes in an environment of 90% nitrogen and 10% oxygen. A thermal oxide is grown at 990 °C for a total time of 75 minutes (15 minutes dry O_2, 45 minutes steam, 15 minutes dry O_2), giving a resulting oxide thickness of 3200 Å. Using standard photolithographic techniques, the test site pattern is defined on the back oxide while the front oxide is protected with photoresist. A 50/50 hydrazine/water solution [4] is used under reflux at 115 °C to form the silicon diaphragms, which are approximately 4.7 microns thick at this boron doping level. The polyimide (a BTDA-ODA/MPDA polymer obtained from Dupont) is spin cast on the wafer in multiple coats. After each coat, a prebake is done in air at 135 °C for 14 minutes. After the final coat, the film is cured in nitrogen at 400 °C for 45 minutes. Finally, the diaphragm supporting the film is etched away in a SF_6 plasma to form the free-standing polyimide membranes. Films ranging in thickness from 5 to 11 microns have been produced in this manner.

Theoretical analyses of the load-deflection behavior of elastic membranes have been done by many authors [5], [6], [7]. In this work, we have used the energy method described by Timoshenko [5]. Preliminary load-deflection results indicated that the intrinsic tensile stress in the film (due to shrink-

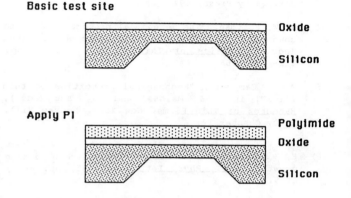
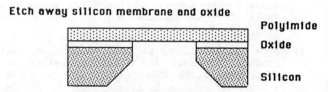

Figure 1. Process schematic

age during cure and/or thermal expansion coefficient mismatch between film and substrate) is not negligible, so the energy method was modified to include the contribution of this stress. In the energy method, which is necessarily approximate, functional forms for the displacements of the deflected surface are assumed. These functions contain several undetermined constants. The constants are found by minimizing the total system energy, leading to an expression for the deflected surface as a function of pressure and intrinsic stress.

Figure 2 shows the coordinate system for the deflected membrane. The origin is located in the plane of the oxide surface at the center of the square membrane. The functional forms we chose for the deflections are the lowest order components of the Fourier series expansion of the true solution:

$$u = c \sin(\frac{\pi x}{a}) \cos(\frac{\pi y}{2a})$$
$$v = c \sin(\frac{\pi y}{a}) \cos(\frac{\pi x}{2a})$$
$$w = w_0 \cos(\frac{\pi x}{2a}) \cos(\frac{\pi y}{2a})$$

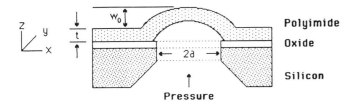

Figure 2. Definitions of blister parameters

where u, v, and w are the deflections along the x, y, and z axes respectively (see Fig. 2); w_0 and c are the two arbitrary constants to be determined; and a is the halflength of the square. These assumed functional forms satisfy the zero-strain boundary conditions at the edges of the membrane and lead to nearly hemispheric deflection near the membrane center. Performing the energy minimization leads to the following expressions for w_0 and c:

$$c = \frac{51\pi^2}{315\pi^2+320} \frac{w_0^2}{a}$$

$$E\left[\frac{w_0}{t}\right]^3 + 1.6649 \left[\frac{a}{t}\right]^2 \frac{N_0 w_0}{t} = 0.5469 \frac{pa^4}{t^4}$$

where N_0 is the intrinsic stress, E is Young's modulus, p is the differential pressure and t is the film thickness. The expression for w_0, the deflection at the center of the film, is of particular interest since this is the quantity we measure. The expression predicts a linear dependence of deflection on pressure at low pressure due to intrinsic film stress, and a cubic dependence at higher pressure. What constitutes low and high pressure is determined by the magnitude of the intrinsic stress. The linear term depends only on residual stress and the cubic term depends only on elastic film constants (i.e., Young's modulus). In this model, therefore, the residual stress can be determined by the initial slope of the load-deflection curve, while Young's modulus can be determined by the curvature of this same curve. However, it must be recognized that more exact solutions for the membrane deflection may not separate so cleanly into two independent terms.

Load-deflection measurements of the suspended polyimide films are made by mounting the wafer in a specially designed chuck which seals the trapezoidal cavity in the wafer and permits the application of differential pressure by injecting volumes of air into that cavity with a microliter syringe. The cavity pressure is measured with a silicon pressure transducer which is built into the chuck. The deflection of the film at the center of the test site (x=y=0 in Fig. 2) is measured using an optical microscope. Typical data are shown in Figure 3 (circles), to which we have fit the model previously described (solid line). As can be seen, the data appear mostly linear with slight curvature. The almost linear behavior demonstrates the dominance of the intrinsic stress component of the deflection over the elastic component. Thus, we expect to be able to extract values for intrinsic film stress with greater precision than values for Young's modulus. This is indeed what is found.

Table I gives a summary of the data we have collected on various sized test sites. In this first study, each wafer contained only one test site at its center. Thus, each entry in the table represents a

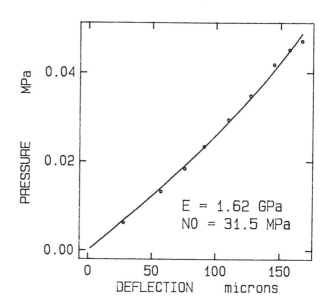

Figure 3. Model fitted to typical load-deflection data

different wafer. As can be seen, except for the 3x3 wafers, values for the intrinsic film stress agree rather well for various geometries. However, the values of Young's modulus obtained, although of the correct order of magnitude, vary substantially. We attribute this to poor resolution of the curvature measurement in such a residual stress dominated regime. This has led us to seek alternative approaches to determine Young's modulus. Note that the ratio of intrinsic stress to Young's modulus, which can be thought of as an 'intrinsic strain', is large, of the order of 1%. In general, the elastic component of the load-deflection behavior will be small until the elastic strain reaches 1%. This is dif-

SIZE (2a)	Thickness (microns)	E (GPa)	N_0(MPa)
2	10.6	0.71	29.7
2	10.5	-	31.4
3	10.6	6.9	43.8
3	6.3	8.6	49.4
3	10.3	3.4	45.3
4	10	1.6	31.5
6	10	-	33.2
6	10	1.5	28.7

Table I. Intrinsic stress and modulus data

ficult to achieve in a membrane deflection. In other work which is being reported separately [8], we have developed an independent method which exploits this "intrinsic strain" in micromachined structures to determine the ratio of Young's modulus to intrinsic stress. With that ratio and with the value for the intrinsic stress determined by the method described here, a value for Young's modulus of these films can be determined with better precision than with the present measurements.

As mentioned above, the suspended membranes are also suited for a measurement of the adhesion of a polymer film to a silicon wafer. By increasing the differential pressure on the test site, the film will peel off the substrate, forming a blister. Such a test has been reported by Hinkley [1], using test sites fabricated by a non-lithographic process. However, the test performed by Hinkley used a constant pressure source to peel the film. Since the critical pressure for peel (the pressure at which peel initiates) decreases with increasing radius, the constant pressure peel is inherently unstable. Once initiated, the blister either will peel to the edge of the wafer, or burst if the ultimate tensile stress of the film is exceeded. Our test chuck permits the use of controlled-volume peel. Since we are increasing pressure by injection of known volumes, the volume of the blister formed is constrained and controllable, both at peel initiation and incrementally thereafter. We have demonstrated that this method can be used to follow the peeling of polyimide on silicon. Quantitative measurements of adhesion using this technique are now under way.

ACKNOWLEDGEMENT

This work was supported in part by the Office of Naval Research and Dupont, and by 3M Corporation through fellowship of one of the authors (MM). Sample fabrication was carried out in the MIT Microsystems Technology Laboratories, and in the Microelectronics Central Facility of the Center for Materials Science and Engineering, which is supported in part by the National Science Foundation under contract DMR-81-19295. The authors acknowledge valued technical discussions with Martin Schmidt, and assistance in process development from Paul Maciel, Technical Director of the Microsystems Technology Laboratories.

REFERENCES

[1] J.A. Hinkley, "Adhesion of Polymer Films to SiO_2" Journal of Adhesion, 16, 115 (1983)

[2] K. Peterson, "Silicon as a Mechanical Material" Proceedings of the IEEE, 70, 420 (1982)

[3] Boron+ solid sources, model GS-245 Owens-Illinois, Inc.

[4] Silicon Anisotropic Etchant, PSE-100, Transene Company, Inc.

[5] S. Timoshenko, Theory of Plates and Shells McGraw-Hill (1940) Ch. 8,9

[6] S. Way, "Uniformly Loaded, Clamped, Rectangular Plates With Large Deflection", Proceedings of the Fifth International Conference for Applied Mechanics, Cambridge, Ma. 123 (1938)

[7] J.G. Williams, Fracture Mechanics of Polymers Holsted Press, 1984, Ch. 2

[8] M. Mehregany, R. Howe, S. Senturia, "Novel Microstructures for the Study of Residual Stress in Polyimide Films", Paper submitted to the Electronic Materials Conference, Amherst, Ma. June, 1986

MECHANICAL PROPERTY MEASUREMENTS OF THIN FILMS USING LOAD-DEFLECTION OF COMPOSITE RECTANGULAR MEMBRANE

Osamu Tabata, Ken Kawahata, Susumu Sugiyama, and Isemi Igarashi

Toyota Central Research and Development Laboratories Inc.
Nagakute-cho, Aichi-gun, Aichi-ken, 480-11, Japan

ABSTRACT

Internal stress and Young's modulus of thin films are determined by measuring the deflection versus pressure of the rectangular membranes of materials. In order to reduce the measurement error for the Young's modulus due to unknown Poisson's ratio, 2 mm x 8 mm rectangular membrane is adopted. Measurements are mede by using a computerized measurement system. Low pressure chemical vapor deposited (LPCVD) silicon nitride films are characterized and found to have an internal stress of 1.0 GPa and a Young's modulus of 290 GPa, showing that the rectangular membrane load-deflection technique could be utilized to measure the internal stress and Young's modulus of films deposited onto the LPCVD silicon nitride membranes. By using this composite membrane technique, an LPCVD polysilicon film and a plasma-CVD silicon nitride film are characterized. The Internal stress and Young's modulus were found to be -0.18 GPa and 160 GPa for the LPCVD polysilicon film, 0.11 GPa and 210 GPa for the plasma-CVD silicon nitride film, respectively.

INTRODUCTION

Thin films are considered to play an important role as construction materials for Micro Electro-Mechanical Systems (MEMS). Accordingly, there has been increasing interest in the mechanical properties of thin films. However, it is impossible to predict them from a process sequence at present, since they are strongly affected by the deposition conditions and the possible subsequent processing steps such as ion implantation and annealing. In order to achieve desired functions of MEMS, it is important to know at the design stage the mechanical properties of thin films and to control them by processing steps. Hence, many data concerning the mechanical properties of thin films under various process conditions should be measured.

There are several techniques for measuring the mechanical properties of thin films, such as substrate curvature technique for internal stress measurement and acoustic wave technique for Young's modulus measurement. Especially, the membrane load-deflection measurement is the most attractive technique, since by using this technique both internal stress and Young's modulus, which are the most important properties for the MEMS, can be measured simultaneously. However, this technique has several disadvantages;(1) Poisson's ratio should be known to determine the Young's modulus, (2) It is not applicable to compressively stressed films due to membrane buckling and films which are difficult to be made into membranes.

In order to overcome these disadvantages, a new measurement technique based on load-deflection of composite rectangular membrane has been developed. In this technique, a film to be measured is deposited on a well-characterized rectangular substrate membrane, making a composite membrane. The composite membrane technique can be applied to various films, such as compressively stressed films and films difficult to be made into membranes. Although the composite membrane technique was originally proposed by Bromley[1], the applicabilty of the composite membrane technique to the measurement of Young's modulus and compressive stress has not been confirmed yet. We have shown that these properties of the film can be calculated from the properties of the composite membrane and the substrate membrane. Furthermore, by using the rectangular membrane, Young's modulus can be determined more accurately without knowing the Poisson's ratio of the films compared with the technique using a circular[1] or a square[2] membrane.

This paper describes: (1) Measurement principle of load-deflection of the composite rectangular membrane, (2) Sample preparation process and the configuration of the measuring apparatus, (3) Measured results for low pressure chemical vapor deposition (LPCVD) silicon nitride films, LPCVD polysilicon films, and plasma-CVD silicon nitride films.

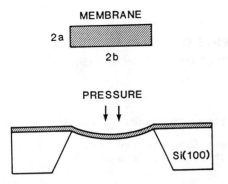

Fig.1 Schematic view of a composite rectangular membrane sample.

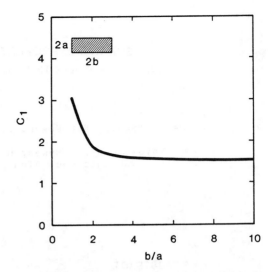

Fig.2 Dependence of the constant C_1 on membrane shape.

MEASUREMENT PRINCIPLE

Load-Deflection Analysis of Rectangular Membrane

An analysis of the load-deflection of the rectangular membrane, schematically shown in Fig.1, with sides 2a and 2b (a=<b) has been made by using an energy minimization approach considering the contribution of internal tensile stress[2,3]. The total strain energy of the rectangular membrane is obtained by adding the strain energy of deformation to the elastic strain energy due to internal tensile stress. The strain energy due to bending can be neglected in the case of very thin films with deflection many times larger than their thickness. The Fourier expansion of the true solution with undetermined two constants are chosen for the functional form of the displacement of a point in the membrane. Then the work input into the membrane is calculated to obtain the total potential energy of the membrane. Subsequently, the total potential energy is minimized with respect to the two constants in the displacement equations, making use of the principle of virtual displacements, resulting in a relationship between the load and the deflection. The obtained load-deflection relationship of the rectangular membrane is

$$p = C_1 \sigma t h/a^2 + C_2 E t h^3/a^4 \quad (1)$$

$$C_1 = \pi^4(1+n^2)/64 \quad (2)$$

$$C_2 = \{\pi^6/32(1-\nu^2)\}\{(9+2n^2+9n^4)/256 - (4+n+n^2+4n^3-3n\nu(1+n))^2/2 / (81\pi^2(1+n^2)+128n+\nu(128-9\pi^2(1+n^2)))\} \quad (3)$$

where p is the applied pressure, E the Young's modulus, σ the internal stress, t the membrane thickness, and h the membrane deflection. C_1 and C_2 are the constant values determined by the membrane shape b/a=1/n and Poisson's ratio ν. The internal stress and Young's modulus appear in the linear term of h and in the cubic term of h, respectively. Hence, independent

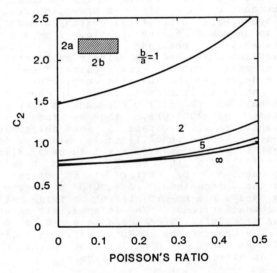

Fig.3 Dependence of the constant C_2 on Poisson's ratio and membrane shape.

determination of both quantities is readily accomplished by measuring the deflection versus pressure, and fitting the data to the Eq.(1).

C_1 and C_2 are calculated to 3.04 and 1.83, respectively, for square membrane (b/a=1) with assumed Poisson's ratio of 0.25. These values coincide with the values described by Allen[2]. Figure 2 shows the dependence of C_1 on b/a. As b/a increases, C_1 decreases to approach the constant value of 1.52, which is half of the C_1 for square membrane. Figure 3 shows the dependence of C_2 on Poisson's ratio for various values of b/a. It is clear that C_2 increases with the increase of Poisson's ratio, and the dependence on Poisson's ratio weakens with the increase in b/a. This means that calculated Young's modulus is less affected by Poisson's ratio for larger b/a.

Table 1 Deposition conditions.

	TEMP. (°C)	PRESSURE (Torr)	GAS (sccm)
LPCVD Si_3N_4	790	0.6	$SiH_2Cl_2:NH_3$ 20:170
LPCVD poly-Si	630	0.4	$SiH_4:N_2$ 20:110
plasma-CVD Si_3N_4	300	0.5	$SiH_4:N_2$ 15:535

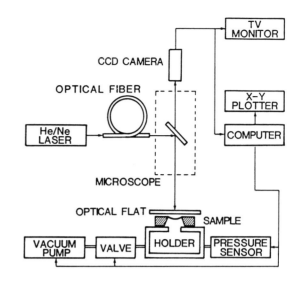

Fig.4 Block diagram of a computerized measurement system.

Composite Membrane Technique

The load-deflection technique can be applied only to films in tensile stress which are easily made into membranes. We have extended this technique to measurements of various films by using composite membranes. The composite membrane is formed by depositing an object film to be measured onto a well-characterized LPCVD silicon nitride membrane. Since the LPCVD silicon nitride membrane is in tensile stress, the total internal stress of the composite membrane can be maintained at tensile stress. The total stress at the center of the pressurized rectangular membrane in the direction parallel to the 2a side is given by

$$(\sigma_c + \varepsilon E_c/C_{3c})(t_s + t_o)$$
$$= (\sigma_s + \varepsilon E_s/C_{3c})t_s + (\sigma_o + \varepsilon E_o/C_{3o})t_o \quad (4)$$

$$C_{3x} = 1 - \nu_x(n+\nu_x)/(1+n\nu_x) \quad (x=c,s,o) \quad (5)$$

where subscripts c, s, and o denote the composite membrane, the substrate membrane, and the object membrane of measurement, respectively, ε is strain of the membrane induced by pressurizing and C_3 is the constant value. C_3 equals $1-\nu$ for square membrane (n=1) and $1-\nu^2$ for infinitely long membrane (n=0). Since Eq.(4) is valid for any pressurized conditions, internal stress of the object film can be calculated by Eq.(6).

$$\sigma_o = ((t_s+t_o)\sigma_c - t_s\sigma_s)/t_o \quad (6)$$

By assuming the same value of Poisson's ratio, Young's modulus can be calculated by Eq.(7).

$$E_o = ((t_s+t_o)E_c - t_s E_s)/t_o \quad (7)$$

The calculated error due to the above assumption for Poisson's ratio can be reduced by using the rectangular membrane.

MEASURING TECHNIQUE

Sample Preparation

The characterized materials used are LPCVD silicon nitride, LPCVD polysilicon, and plasma-CVD silicon nitride. Silicon nitride and polysilicon are the dominant film materials for MEMS, such as micro diaphragm pressure sensors[4]. Deposition conditions are summarized in Table.1. In the first step of the sample fabrication process, LPCVD silicon nitride is deposited on a 3 inch (100) oriented silicon wafer with thickness of 360 μm. Then rectangular window patterns are opened in the silicon nitride on the backside of the wafer by conventional photolithography and reactive ion etching (RIE) using a mixture of CF_2Br_2 and O_2 gas. By using this wafer as substrate, the object film to be measured is deposited on it. Then the exposed silicon of the backside wafer is etched away using an anisotropic etching solution containing potassium hydroxide (KOH). The frontside of the wafer is protected with wax during the etching process. The completed membrane size was 2 mm x 8 mm. The deposition of the object film might be preceded by backside etching of the silicon wafer.

Measurement Apparatus

A computerized measuring apparatus has been made. The diagram of the apparatus is shown in Fig.4. The sample is mounted on a holder which is connected to a vacuum pump. The holder is placed on a microscope stage and an optical glass is placed on it. Subatmospheric pressure is applied to the holder, and resultant deflection of the membrane is observed interferometrically as a function of the holder pressure. The holder pressure is measured with an accuracy of 0.1 mmHg by using silicon piezoresistive pressure sensor. A He/Ne laser is used as a light source and connected to the microscope through a single mode fiber. The well known Newton interference fringes are detected by a CCD camera and displayed on a TV monitor. The membrane anisotropy may be simultaneously evaluated by their departure from a symmetrical shape. Since the membrane deflection at the center of the membrane is

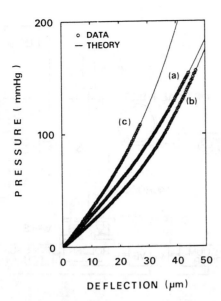

Fig.5 Experimental load-deflection characteristics for substrate rectangular membrane (a) and composite rectangular membranes (b), (c). :(a) LPCVD silicon nitride 2000 Å, (b) LPCVD polusilicon 2000 Å on LPCVD silicon nitride 2000 Å, (c) plasma-CVD silicon nitride 5000 Å on LPCVD silicon nitride 2000 Å.

Table 2 Calculated internal stress and Young's modulus.

	INTERNAL STRESS(GPa)	YOUNG'S MODULUS(GPa)
LPCVD Si_3N_4	1.0	290
LPCVD poly-Si	-0.18	160
plasma-CVD Si_3N_4	0.11	210

measured by counting the interference fringes, the high measurement accuracy can be realized with the measurement resolution of 3164 Å.

The measurements are made as follows. First, subatmospheric pressure is applied to the holder, then the holder pressure is gradually increased from subatomospheric pressure to atmospheric pressure. The pressure increase reduces the number of fringes of interference pattern. As every instance of a fringe is vanished, the holder pressure is measured by the pressure sensor and the measured data is stored in a computer. The detection of fringe vanishment is done by image processing. The measurement is completed when the holder pressure equals atmospheric pressure. Both data sampling and processing are accomplished through the computer automatically. The measured data and the calculated results are monitored on a display and also fed to an x-y plotter.

MEASUREMENTS RESULTS AND DISCUSSION

The composite rectangular membrane technique has been verified with different thickness of LPCVD silicon nitride substrate membrane. The membranes of 2 mm x 8 mm were used, and the numerical constants C_1 and C_2 were found through calculation to be 1.617 and 0.829, assuming the Poisson's ratio to be 0.25. The substrate membrane thickness of 1000, 2000, and 3000 Å were used. The LPCVD polysilicon with thickness of 2000 Å and plasma-CVD silicon nitride with thickness of 5000 Å were deposited on the substrate membrane before backside etching. Since the LPCVD polysilicon had compressive stress, the thickness of polysilicon was selected to maintain the stress of the composite membrane at tensile stress. Figure 5 shows load-deflection data obtained from three types of rectangular membranes. The solid lines through the points represent the least-squares fit of Eq.(1). It is seen that the measured data for composite rectangular membrane could be well characterized by Eq.(1). Calculated internal stress and Young's modulus are summarized in Table 2. As for the LPCVD silicon nitride membrane, there was no apparent variation of the measured internal stress and Young's modulus in the thickness range of 1000 to 3000 Å. The standard deviation of the data obtained from repeated measurements of the sample was 0.01 GPa for internal stress and 5 GPa for Young's modulus. The distribution of the calculated values in the 3 inch silicon wafer was 0.02 GPa for internal stress and 10 GPa for Young's modulus on standard deviation. From these results, the lower measurement range of the composite membrane technique utilizing LPCVD silicon nitride as substrate membrane is considered to be not less than 0.02 GPa for internal stress and not less than 10 GPa for Young's modulus. It is expected that the lower the internal stress in substrate membrane, the more sensitive the measurement of the internal stress and Young's modulus in the deposited film. As we have already demonstrated, the ion implantation technique seems to be attractive for lowering the tensile stress and Young's modulus of the LPCVD silicon nitride film[5].

Figure 6 and Figure 7 show the calculated internal stress and Young's modulus for LPCVD polysilicon and plasma-CVD silicon nitride, respectively, as a function of the thickness of the substrate membrane. Measured results were compared with those obtained from square membranes of 2 mm x 2 mm. The solid lines in Fig.6 and Fig.7 show the dependencies of internal stress and Young's modulus on thickness of substrate membrane predicted from Eq.(6) and Eq.(7), respectively, by using values summarized in Table 2. As for the internal stress, the rectangular and square membranes gave the same calculated values. Furthermore, the calculated and predicted values for internal stress showed good agreement. As regards the Young's modulus, the calculated values for

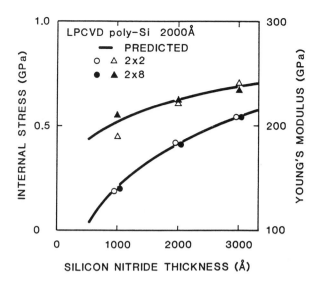

Fig.6 Calculated internal stress and Young's modulus for composite membrane composed of LPCVD polysilicon and CVD silicon nitride.

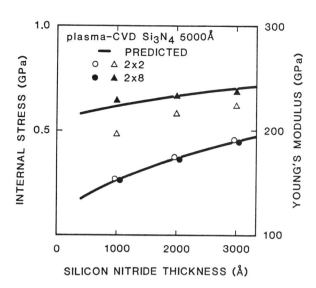

Fig.7 Calculated internal stress and Young's modulus for composite membrane composed of plasma-CVD silicon nitride and CVD silicon nitride.

the rectangular membranes were different from the calculated values for the square membranes. However, for the rectangular membranes, the calculated Young's modulus values and predicted values showed approximately good agreement. The further investigation of Young's modulus for composite membrane are being made presently.

CONCLUSION

Internal stress and Young's modulus of thin films were determined by measuring the deflection versus pressure of rectangular membranes of the materials. In order to reduce the measurement error for the Young's modulus due to unknown Poisson's ratio, 2 mm x 8 mm rectangular membrane was adopted. An energy method was used to obtain the load-deflection relationship of a rectangular membrane. Measurements were made by using a computerized measurement system. Low pressure chemical vapor deposited (LPCVD) silicon nitride films were characterized and found to have an internal stress of 1.0 GPa and a Young's modulus of 290 GPa. It is shown that the rectangular membrane load-deflection technique could be extended to measure the internal stress and Young's modulus of films of interest deposited onto LPCVD silicon nitride membranes. This composite membrane technique can be applied to various films, such as compressively stressed film and the film which is difficult to be made into membrane. An LPCVD polysilicon film and a plasma-CVD silicon nitride film were deposited to make the composite membrane. The internal stress of -0.18 GPa and a Young's modulus of 160 GPa for the LPCVD polysilicon film, internal stress of 0.11 GPa and Young's modulus of 210 GPa for the plasma-CVD silicon nitride film were obtained.

Further studies should be continued to make clear the process dependence of the mechanical properties of the thin films. The measuring apparatus described in this paper will be useful as a characterization tool for this purpose.

REFERENCES

(1) E.I.Bromley,J.N.Randall,D.C.Flanders, and R.W.Mountain," A Technique for the Determination of Stress in Thin Films," J.Vac.Sci.Technol.B, vol.1, No.4, pp1364-1366,1983.
(2) M.G.Allen, M.Mehregany, R.T.Howe, and S.D.Senturia,"Microfabricated Structures for the in situ Measurement of Residual Stress, Young's Modulus, and Ultimate Strain of Thin Films," Appl. Phys.Lett., vol.51, No.4, pp241-243, 1987.
(3) S.Timoshenko and S.Woinowsky-Krieger, "Theory of Plates and Shells," McGraw-Hill, New York, 1959.
(4) S.Sugiyama, T.Suzuki, K.Kawahata, K.Shimaoka, M.Takigawa, and I.Igarashi, "Micro Diaphragm Pressure Sensor," Technical Digest, IEEE IEDM, pp184-187, 1986.
(5) O.Tabata, K.Kawahata, S.Sugiyama, H.Inagaki, and I.Igarashi, "Internal Stress and Young's Modulus Measurements of Thin Films Using Micromachining Technology," Technical Digest of the 7th Sensor Symposium, pp173-176, 1988.

Fracture Toughness Characterization of Brittle Thin Films

L. S. FAN, R. T. HOWE and R. S. MULLER

University of California at Berkeley, Department of Electrical Engineering and Computer Sciences and the Electronics Research Laboratory, Berkeley Sensor & Actuator Center*, Berkeley, CA 94720 (U.S.A.)

Abstract

This paper describes the design and fabrication of a series of test structures for measuring the fracture toughness of brittle thin films, with application to low-pressure chemical-vapor-deposited (LPCVD) Si_3N_4 and low-stress, silicon-rich LPCVD Si_xN_y. Crack-initiating features are patterned in clamped-clamped microbridges, which are loaded in tension after release from the substrate due to the tensile residual stress in the nitride films. Initial measurements yield an upper bound $K_{IC} < 14$ MPa m$^{1/2}$ for Si_3N_4 and $K_{IC} = 1.8 \pm 0.3$ MPa m$^{1/2}$ for low-stress nitride, using typical values for the residual stress.

Introduction

Thin-film mechanical properties are essential for designing micromechanical sensors or actuators [1]. A variety of *in situ* techniques has been developed for measuring tensile or compressive residual stress [2]. Ultimate strengths of thin films have been studied by the snapping of spiral springs [3], bridges loaded in compression [4], and bridges loaded in tension [5]. This paper describes a simple technique for extracting the fracture toughness of brittle thin films using an array of surface-micromachined structures.

Inorganic microstructural thin films, such as polycrystalline silicon or silicon nitride, are typically brittle materials. The fracture mechanisms of these films differ from those of ductile materials. Here the plastic deformation region is small enough that linear elasticity mechanics applies throughout the structure [6]. In brittle materials, cracks propagate when the release of strain energy G_I is larger than the increase in surface energy of the crack. Stress intensity factors are conveniently used to evaluate fracture in brittle materials. It can be shown that stress is proportional to the inverse of the square root of the distance from the crack tip [6], and this proportionality factor, which has units of Pa m$^{1/2}$, is used to define the stress intensity factor K_I. Fracture occurs when K_I is larger than a critical value called the fracture toughness K_{IC}, which relates to strain energy release rate as $K_{IC} = EG_I$, where E is the appropriate modulus for either plane strain or plane stress problems [6].

Single-mask Fracture Test Structures

The visual inspection of an array of easily fabricated test structures is attractive for in-line monitoring of mechanical properties [2]. Microbridges fabricated from a film with high tensile strain, such as LPCVD Si_3N_4, are appropriate for designing stress-destructing fracture test structures. A series of microbridges having stress-intensifying features which generate a range of values of K_I will enable upper and lower bounds to be placed on the fracture toughness K_{IC}, by observing which structures fracture after being released from the substrate. Figure 1 shows the top view and cross section of a microbridge with such a feature. For compressive films such as coarse-grain polysilicon, the ring/crossbar structure developed by Guckel *et al.* [2] might possibly be used to generate a tensile axial load on the crossbar, which would incorporate stress-concentrating features.

Three types of test structures are used: microbridges with a notch on one edge [7], a C-shaped tension specimen [8] (Fig. 2), and the tapered-double-cantilever-beam specimen [9] (Fig. 3). Since

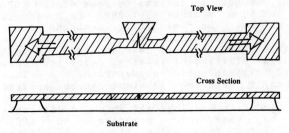

Fig. 1. Top view and cross section of a fracture specimen, showing axial load generated by residual tensile stress in the microstructural film.

---

*An NSF/Industry/University Cooperative Research Center.

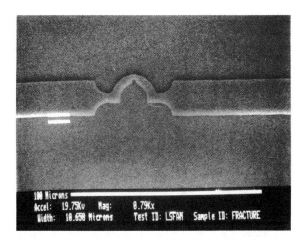

Fig. 2. SEM of a 15 μm wide bridge with a C-shaped stress concentrator in the center.

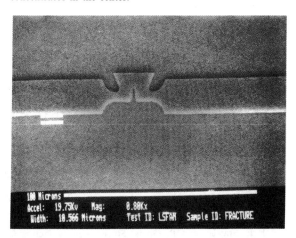

Fig. 3. SEM of a 15 μm wide bridge with a tapered-double-cantilever-beam structure in the center.

LPCVD Si_3N_4 or low-stress, silicon-rich LPCVD Si_xN_y [10] is deposited as a structural layer on a 2 μm thick sacrificial layer of phosphosilicate glass (PSG). The microstructures are defined by photolithography and plasma etching, with concentrated hydrofluoric acid being used to etch the PSG sacrificial layer. Those structures which fracture are observed to do so immediately upon release from the underlying PSG layer.

For the 200 nm thick Si_3N_4 film, all of the microbridges with stress concentrators fractured at their midpoints at the moment of release, which was recorded on videotape. From these observations, the critical geometry parameter is smaller than 14 μm$^{1/2}$ (or K_{IC} is smaller than 14 MPa m$^{1/2}$ assuming a typical residual stress of 1 GPa for LPCVD Si_3N_4.) On the other hand, for the silicon-rich, low-stress Si_xN_y film, only tapered-double-cantilever-beam structures with the highest stress concentrations fractured, as shown in Fig. 4(a)–(c). The resulting critical geometry parameter is 60 ± 10 μm$^{1/2}$ (or K_{IC} around 1.8 ± 0.3 MPa m$^{1/2}$ assuming a stress level of 30 MPa.) It is worth noting that Fig. 4(b) shows one partially fractured structure. The crack tip stopped propagating because the residual stress of the film had been partially released and the stress intensity factor on the crack tip dropped below the critical value. A careful calculation could potentially yield an accurate value of fracture toughness. A numerical analysis using boundary-element techniques is currently underway to establish a more accurate value.

this methodology involves only variations in geometric dimensions, only geometric parameters or dimensionless (e.g. strain instead of stress) film parameters can be obtained. However, these parameters will be adequate for predicting the critical geometrical dimensions for microstructures made of the same films. An array of these structures is designed which is characterized by a critical geometry parameter (defined as the ratio of fracture toughness to residual stress) ranging from 14 to 290 μm$^{1/2}$. Corresponding K_I values are from 4.2 to 87 MPa m$^{1/2}$, assuming a residual stress of 300 MPa. For comparison, a typical value of fracture toughness for steel is around 60 MPa m$^{1/2}$ and a typical value for glass is around 0.6 MPa m$^{1/2}$.

Fabrication and Test Results

Fracture specimen arrays are made with the sacrificial-layer technique. Either conventional

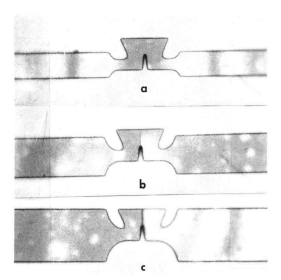

Fig. 4. Optical micrographs of released fracture test structures: (a) an unfractured specimen, (b) a partially fractured specimen, (c) a fractured specimen.

Conclusions

A simple one-mask technique for characterizing fracture design parameters has been applied to tensile-stressed LPCVD silicon nitride films. The design parameter obtained is the critical geometry parameter, which can be converted into fracture toughness by multiplying by the residual stress. The latter should be measured using other test structures on the same sample [1, 2]. However, the critical geometry parameters are useful for predicting the fracture behavior. One limitation of the technique is the possibility that the fracture toughness of the film is affected by the micromachining etchant. In Fig. 4, the surface of the low-stress nitride is discolored, perhaps indicating attack by hydrofluoric acid during micromachining.

In this technique, lithography and etching processes limit the radii of curvature of the crack-initiating features to the order of 1 μm. If the feature is too rounded, then failure of the test structure may occur due to yielding of the material rather than due to crack propagation. For all the specimens studied, fracture occurs from the tip of the notch, indicating the crack propagation is the failure mechanism. However, further experimental observations of the failed specimens and numerical simulation of the stress field in the area of the feature are needed to support this interpretation. The fracture toughness calculations further assume that the initial residual stress causes a tensile, axial load on the microbridge upon its release from the substrate. By means of boundary-element analysis, a more accurate model is being studied which accounts for both body forces and surface tractions on the released structure. Initial indications are that the effect of surface tractions can be significant.

Acknowledgements

We thank Professors G. C. Johnson and A. P. Pisano of the UC Berkeley Department of Mechanical Engineering for valuable discussions and C. H. Mastrangelo for help in processing.

References

1 S. D. Senturia, Microfabricated structures for the measurement of mechanical properties and adhesion of thin films, *4th Int. Conf. Solid-State Sensors and Actuators (Transducers '87), Tokyo, Japan, June 2–5, 1987*, pp. 11–26.
2 H. Guckel, D. W. Burns, C. C. G. Visser, H. A. C. Tilmans and D. DeRoo, Fine-grained polysilicon films with built-in tensile strain, *IEEE Trans. Electron Devices, ED-35* (1988) 800–801
3 L. S. Fan, Y. C. Tai and R. S. Muller, Integrated movable micromechanical structures for sensors and actuators, *IEEE Trans. Electron Devices, ED-35* (1988) 724–730.
4 Y. C. Tai and R. S. Muller, Fracture strain of LPCVD polysilicon, *IEEE Solid-State Sensor and Actuator Workshop, Hilton Head Island, SC, U.S.A., June 6–9, 1988*, pp. 88–91.
5 M. Mehregany, R. T. Howe and S. D. Senturia, Novel microstructures for the *in situ* measurement of mechanical properties of thin films, *J. Appl. Phys., 62* (1987) 3579–3584.
6 M. F. Kanninen and C. H. Popelar, *Advanced Fracture Mechanics*, Oxford University Press, New York, 1985.
7 W. F. Brown, Jr. and J. E. Srawley, Plane strain crack toughness testing of high strength metallic materials, *Am. Soc. Test. Mater. Spec. Tech. Publ., 410* (1966) 12.
8 J. A. Kapp, J. C. Newman, Jr. and J. H. Underwood, A wide range stress intensity factor expression for the C-shaped specimen, *J. Test Eval.*, (1980) 314–317.
9 J. E. Srawley and B. Gross, Stress intensity factors for crackline-loaded edge-crack specimens, *NASA Tech. Note, TN D-3820*, 1967.
10 M. Sekimoto, H. Yoshihara and T. Ohkubo, Silicon nitride single-layer X-ray mask, *J. Vac. Sci. Technol., 21* (1982) 1017–1021.

SPIRAL MICROSTRUCTURES FOR THE MEASUREMENT OF AVERAGE STRAIN GRADIENTS IN THIN FILMS

Long-Sheng Fan*, Richard S. Muller, Weijie Yun, Roger T. Howe, and Jiahua Huang

Berkeley Sensor & Actuator Center
Department of Electrical Engineering and Computer Sciences
and the Electronics Research Laboratory
University of California, Berkeley, California 94720

ABSTRACT

Spiral microstructures which are fabricated from thin-film materials having a residual strain that varies through the thickness of the film are observed to wind up and to curl upon release from the substrate. The angle of rotation, the change in the lateral size (projected onto the plane of the substrate), and the out-of-plane height of the spiral free end can be measured easily from optical micrographs. In this paper, we develop a theory for this behavior and compare its predictions with experiment.

INTRODUCTION

Surface-micromachined spiral springs sometimes deform upon release from the substrate. Spirals made from LPCVD polycrystalline silicon [1] and sputtered-tungsten [2] films have been observed both to wind up and to curl after etching of underlying sacrificial oxide layers. Understanding the source for these phenomena is important since it could lead to use of the spiral as a diagnostic structure for film strain and possibly to the application of strained spiral structures in micromechanical systems.

This paper develops a theory which identifies the gradient of in-plane residual strain (through the thickness of the film) as the cause of the deformation of spiral springs. The angular rotation (wind-up) and lateral contraction of Archimedean spirals are predicted as a function of the average strain gradient in the film. The lateral contraction and rotation can be easily and accurately determined by means of an optical micrograph, making the Archimedean spiral attractive for measurement of average strain gradient. Since the deformation of spirals with a given strain gradient can now be predicted, they may be used as micromechanical components or for the *in situ* assembly of nonplanar microstructures.

ANALYSIS

The variation of internal stress through the thickness of a film can lead to an internal bending moment. To analyze this, consider a structural member having width w, thickness t and residual stress $\sigma(y)$. In general, there will be an internal bending moment M_1 (the subscript 1 will soon be related to a principal axis for a specific structure) which is given by

$$M_1 = \int_0^t \sigma(y) w (y - t/2) \, dy \qquad (1)$$

where $y = 0$ is at the bottom surface of the film. From Eq. (1), the moment is zero unless $\sigma(y)$ varies with y. We assume a linear variation of stress through the film thickness.

$$\sigma(y) = \bar{\sigma} + \Gamma E (y - t/2) \qquad (2)$$

where $\bar{\sigma}$ is the mean stress, E is the elastic modulus, and Γ is the strain gradient through the thickness. Inserting Eq. (2) into Eq. (1) and integrating, we calculate

$$M_1 = \frac{\Gamma E w t^3}{12} \qquad (3)$$

Our analysis will consider the effect of this uniformly varying stress on Archimedean spiral structures. The equation of such a spiral in polar coordinates is

$$r = \alpha \theta \qquad (4)$$

where r represents the radius to the centerline of the spiral cross section and α is the parameter denoting the radius increment per angle θ.

At the outset, we can establish that a uniform strain field ($\Gamma = 0$) does not generate any rotation of the free end of the spiral. To do this, consider an unconstrained spiral suddenly subjected to a uniform temperature increase; this creates a uniform strain field and the spiral expands causing the total length and radius of the spiral to increase. Since the spiral length is linearly related to radial size, there will not be any rotation of its free end. To show this mathematically, consider an incremental length ΔL along the spiral where ΔL is related to the angles θ_1 and θ_2 at its end points,

$$\Delta L = \int_{\theta_1}^{\theta_2} r \, d\theta \qquad (5)$$

If we multiply both sides of Eq. (4) by $(1+\varepsilon)$, where ε is the uniform strain, Eq. (4) is valid without changing the range of integration. Thus, there is no rotation. The spiral is just scaled up or down dependent upon the sign of ε. Since ε is very much smaller than unity, this scaling is, in fact, very slight.

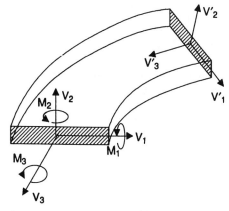

Figure 1. Local coordinate system.

---

* This author is currently with IBM Research Division, Almaden Research Center, 650 Harry Road, San Jose, CA 95120-6099

Reprinted from *Proceedings IEEE Micro Electro Mechanical Systems*, pp. 177-181, February 1990.

Not only uniform, but also vertical variation of residual strain can easily occur as a result the processing cycle of deposited thin-film material. The strain variation as expressed in Eq. (2) is considered here. To formulate the three-dimensional analysis that can occur owing to this strain field, the local coordinate system shown in Fig. 1 is used. As shown in the figure, at any cross section of the spiral, the two principal axes of the cross section v_1, v_2 and the unit normal vector v_3 form an orthonormal bases set. Along the spiral, the principal axes rotate. Consider the effect on a vector $\mathbf{F}$ of this axial rotation. If the rotation is around the basis vector v_3 for an angle $\Delta\phi_3$, the new vector $\mathbf{F}'$ can be expressed as $R_3\mathbf{F}$, where R_3 is the appropriate rotation operator [4]. In matrix form,

$$R_3 = \begin{bmatrix} \cos(\Delta\phi_3) & \sin(\Delta\phi_3) & 0 \\ -\sin(\Delta\phi_3) & \cos(\Delta\phi_3) & 0 \\ 0 & 0 & 1 \end{bmatrix} \quad (6)$$

If the rotational angle $\Delta\phi_3$ is infinitesimal ($d\phi_3$), the matrix R_3 becomes

$$R_3 = \begin{bmatrix} 1 & d\phi_3 & 0 \\ -d\phi_3 & 1 & 0 \\ 0 & 0 & 1 \end{bmatrix} \quad (7)$$

Thus, the infinitesimal variation of vector $\mathbf{F}$ is

$$d\mathbf{F} = \mathbf{F}' - \mathbf{F} = \begin{bmatrix} 0 & d\phi_3 & 0 \\ -d\phi_3 & 0 & 0 \\ 0 & 0 & 0 \end{bmatrix} \mathbf{F} \quad (8)$$

For more general cases where there are infinitesimal rotations $d\phi_1$, $d\phi_2$, and $d\phi_3$ around axes v_1, v_2 and v_3, respectively, Eq. (8) becomes

$$d\mathbf{F} = (R_1 R_2 R_3 - I)\mathbf{F} = \begin{bmatrix} 0 & d\phi_3 & -d\phi_2 \\ -d\phi_3 & 0 & d\phi_1 \\ d\phi_2 & -d\phi_1 & 0 \end{bmatrix} \mathbf{F} \quad (9)$$

where I is the unit matrix. Since the infinitesimal operators R_1, R_2, and R_3 commute with each other, the sequence of rotations in Eq. (9) is inconsequential. In classical mechanics, the incremental rotational angles around the basis vectors v_1, v_2, and v_3 per incremental arc length (ds) are known as "curvatures", and are denoted by κ_1, κ_2, and κ_3 [5]. Mathematically,

$$\kappa_i = \frac{d\phi_i}{ds} \quad (10)$$

where $i = 1, 2, 3$, and ds is the differential arc length along the spiral centerline.

From Eqs (9) and (10), we have

$$\frac{d\mathbf{F}}{ds} = \begin{bmatrix} 0 & \kappa_3 & -\kappa_2 \\ -\kappa_3 & 0 & \kappa_1 \\ \kappa_2 & -\kappa_1 & 0 \end{bmatrix} \mathbf{F} \quad (11)$$

To determine how the individual basis vectors vary, we substitute v_1, v_2, and v_3 for $\mathbf{F}$ to obtain

$$\frac{dv_1}{ds} + \kappa_3 v_2 - \kappa_2 v_3 = 0 \quad (12a)$$

$$\frac{dv_2}{ds} + \kappa_1 v_3 - \kappa_3 v_1 = 0 \quad (12b)$$

$$\frac{dv_3}{ds} + \kappa_2 v_1 - \kappa_1 v_2 = 0 \quad (12c)$$

The solution of Eqs. (12) will permit us to infer the deformation in the shape of a geometrical structure from knowledge of the changes in the curvature values before and after it is released from a sacrificial supporting layer. We denote the curvatures of a structure before it is released from the substrate by κ_{10}, κ_{20} and κ_{30}.

If we consider now the specific case of an Archimedean spiral formed in a plane and released from a sacrificial layer, we can make considerable simplifications in Eqs. (12). First, the curvatures can be directly related to the moment in a structural cross section by [6]

$$M_i = (\kappa_i - \kappa_{i0}) E_i I_i \quad (13)$$

where E_i is the elastic modulus and I_i is the moment of inertia. Comparing Eq. (3) with Eq. (13), we establish that the strain gradient $\Gamma = \kappa_1$ since $I_1 = wt^3/12$ and $k_{10} = 0$ (original structure is planar). The original planarity of the structure also assures that $\kappa_{30} = 0$. Since the residual strain does not vary in the plane of the spiral surface, $\kappa_3 = \kappa_{30} = 0$, and $\kappa_2 = \kappa_{20}$ which is a function of θ. Using these values in Eqs. (12) we have

$$\frac{dv_1}{ds} - \kappa_{201} v_3 = 0 \quad (14a)$$

$$\frac{dv_2}{ds} + \Gamma v_3 = 0 \quad (14b)$$

$$\frac{dv_3}{ds} + \kappa_{20} v_1 - \Gamma v_2 = 0 \quad (14c)$$

The experimental spirals are fixed at either the center or outermost end. Under these boundary conditions, we solve Eqs. (14) by numerical methods (fourth-order Runge-Kutta method, specifically) and use computer-generated graphics to provide 3-dimensional perspective views of the deformed spirals. Figures 2(a) and (b) show two such computer-generated spirals with plus and minus 2mm^{-1} strain gradients, respectively. The deformed spirals are mirror images of each other. A family of these deformed spiral images showing a range of degrees of curling can be generated by varying Γ, the strain gradient in the films. These deformed spirals are characterized by contractions in spiral diameter, rotations of their inner ends, and curling either up or down.

Figure 2(a). Computer simulated spiral with strain gradient of 2.0mm^{-1}.

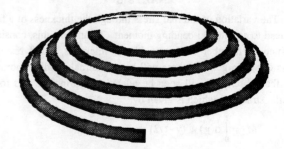

Figure 2(b). Computer simulated spiral with strain gradient of -2.0mm^{-1}.

The diameter contraction (in μm) is defined as the width difference between the widest caliber of the original spiral and the corresponding quantity of the released spiral projected onto the substrate. The end-point rotation (in degrees) is defined by the angle between this projected end point and the original position of the end point before the spiral is released from the sacrificial layer. The height (in μm) is defined as the elevation of the highest end above the substrate. Figures 3, 4, and 5 show the theoretical results for end-point rotation, diameter contraction, and height for spirals with $\alpha = 10\mu m$ and $3\pi < \theta < 11\pi$. The results are calculated for strain gradients Γ varying in magnitude from 0 to 3mm^{-1} and they are applicable for either sign of Γ. The rotation, shown in Fig. 3, increases monotonically with Γ and becomes greater than 180° when it is 2.2 mm^{-1} as shown in Fig. 3. The spiral-diameter contraction behaves almost linearly for Γ between 0.5 mm^{-1} and 3 mm^{-1}. The height dependence is linear for small Γ, but gradually saturates when Γ becomes large, as shown in Fig. 5. As shown in Fig. 2(a) and (b), the sign of the height is determined by the sign of Γ.

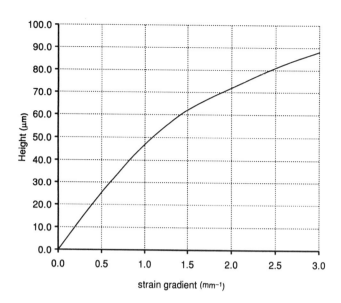

Figure 5. The theoretical result of the end-point height for a spiral with the same geometry as in Fig. 3.

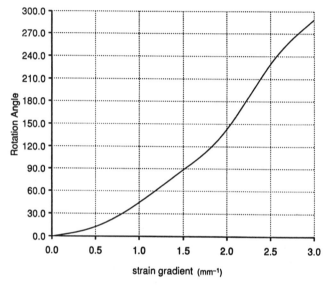

Figure 3. The theoretical result of the end-point rotation for a spiral with $\alpha = 10\mu m$ and $3\pi < \theta < 11\pi$.

Figure 6(a). Nomarski image of a spiral with negative strain gradient before releasing. ($\alpha = 10\mu m, 3\pi < \theta < 11\pi$).

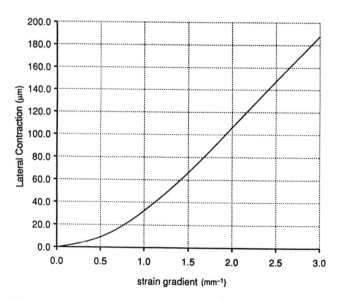

Figure 4. The theoretical result of the lateral contraction for a spiral with the same geometry as in Fig. 3.

Figure 6(b). Nomarski image of a spiral in Figure 6(a) after releasing.

EXPERIMENTAL OBSERVATIONS

Archimedean spirals were included in many test depositions of polycrystalline silicon. The spirals were 4-turn structures anchored to the silicon substrate at either their centers or their outside ends. They were formed from 1.0-to-2.0 μm-thick unannealed LPCVD polycrystalline silicon deposited at 600°C. Some films were found to be under compressive stress and others had an average tensile stress. The sacrificial layers, consisting of 100 nm-thick thermally oxidized silicon dioxide, were removed in concentrated HF, and the etching process and release of the spiral was recorded on videotape.

Figures 6(a) and (b) are Nomarski photographs of a spiral before and after its sacrificial supporting layer has been etched. The spiral, which is anchored at its outer end, has rotated 160°. From Fig. 3, we infer that this film has an average strain gradient of 2.2 mm^{-1}. The contraction of this spiral sample can be measured on Fig. 6 to be 110 μm, which is consistent with the theoretical value shown on Fig. 4. Figure 7 is a SEM photograph of the spiral pictured in Fig. 6 showing that it has sprung up from the substrate, indicating that Γ is negative. (If spirals with negative strain gradients are anchored at their center points, they will be pinned to the substrate when the sacrificial layer is removed.) The similarity between the SEM photograph in Fig. 7 and the computer simulation in Fig. 2(b) for a spiral with a strain gradient of -2.0 mm^{-1} is striking. As a final check, we can measure the height of the spiral by using optical focus/defocus techniques. Although tedious, the accuracy of this method can be of the order of ±1 μm. Using this method, we measure the height of the inner end of the spiral in Fig. 7 to be 76 μm. The measurement agrees well with the theoretical prediction of 73 μm from Fig. 5.

Figure 8 shows a spiral supported at its inner edge that has been made from a polysilicon film deposited at a different deposition condition. This film has a positive strain gradient, and the spiral looks like a cone resting on the substrate. It can be compared qualitatively with the computer-generated form shown in Fig. 2(a).

CONCLUSIONS

We have discussed the effects of a uniform strain gradient on spiral microstructures and derived theory to show that they deform into helical springs with accompanying contractions and rotations. Numerical results are given for four-turn spirals with $\alpha = 10$ μm. The predicted results are shown to correspond well with the behavior of an experimental sample. Research now underway will provide data on an array of samples to be described at the conference.

By applying the theory presented here, the average strain gradients in thin films can be measured easily from optical micrographs of spirals before and after their release from the substrate. The inferred gradient value can be checked by observing three different variables, the height of the free end of the spiral, its rotation, and the contraction of the spiral diameter. Although the strain gradient can also be inferred by observing the behavior of cantilever beams, more information is available from study of the spirals. With spirals, it is also possible to observe both positive and negative strain gradients by making some structures attached at the spiral centers and others attached at their ends. This will assure that at least one of the structures will extend from the surface, a situation that cannot be guaranteed for cantilevers.

The analysis presented here also suggests that strain gradients in deposited films may possibly be exploited to extend structures made using surface micromechanics into greater dimensions away from the substrate.

REFERENCES

1. J. Huang, R. T. Howe, and S. D. Senturia, "Mechanical Technology CAD," *Research in Microsystems Technology*, Massachusetts Institute of Technology, May 1988, p 11,77.

2. W. Yun, "CMOS Metallization for Integration with Micromachining Processes," *M.S. Report*, University of California at Berkeley, May 1989, p 37-38.

3. R. T. Howe and R. S. Muller, "Polycrystalline silicon micromechanical beams," *J. Electrochem. Soc.*, **130**, 1420-1423, (1983).

4. Herbert Goldstein, Classical Mechanics, p. 107-109, Addison Wesley, 1957.

5. James M. Gere and Stephen P. Timoshenko, Mechanics of Materials, 2nd Edition, p 205-207, Brooks/Cole Engineering Division, Monterey, CA, 1976.

6. *ibid.* p 214.

Figure 7. A SEM micrograph of the same spiral as in Fig. 6(b).

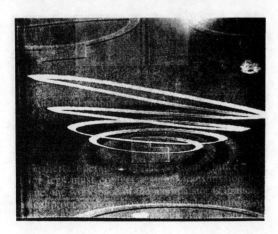

Figure 8. A SEM micrograph of a spiral with positive strain gradient.

POLYSILICON MICROSTRUCTURES TO CHARACTERIZE STATIC FRICTION

M. G. Lim, J. C. Chang, D. P. Schultz, R. T. Howe, and R. M. White

University of California at Berkeley
Department of Electrical Engineering and Computer Sciences
and the Electronics Research Laboratory
Berkeley Sensor & Actuator Center
Berkeley, California 94720

Abstract

A polysilicon microstructure to characterize static friction has been developed. A normal force is applied to a displaced suspended structure by an underlying electrode. The tangential force to measure the frictional force is produced by a restoring force from the displaced suspension. The spring constant is found empirically by resonating the structure. The μ for course-grained polysilicon-polysilicon interfaces was found to be 4.9 ± 1.0. Silicon nitride-polysilicon surfaces exhibited less friction with a μ of 2.5 ± 0.5. Friction was found to be independent of apparent area. Tests were also conducted on a piezoelectric crystal resulting in a reduction of friction and improved reliability of data extraction.

I. INTRODUCTION

Friction has been the subject of scientific inquiry for centuries [1]. Amontons reported in 1699 that friction was independent of contact area and proportional to the load between the contacting surfaces. Coulomb verified these observations in 1781 and clearly distinguished between static and dynamic friction. He proposed two physical explanations for friction: molecular adhesion between the surfaces and interlocking asperities on the two surfaces. The first he rejected because of the observed independence of friction coefficients on contact area. The latter hypothesis was accepted until the 1940s, when it was recognized that the real area of contact was much less than supposed from macroscopic measurements. A molecular adhesion model could then be constructed which was consistent with the observation that friction was independent of the apparent contact area between the surfaces. Modern friction models require extensive information of the properties of the contacting materials, such as elastic moduli, yield strength, shear strength, and penetration hardness [2]. In addition, such surface properties as chemical reactivity and surface energy are useful for improving the friction model.

Recent developments in micro electromechanical systems (MEMS) have focused attention on the critical need to characterize the frictional properties of thin films [3,4,5]. Fundamental research on the mechanical properties, process, and characterization of polysilicon has become essential in order to fabricate more reliable structures [6,7]. Adhesion of microstructures to each other or to the substrate sometimes occurs during or after fabrication [8]. This phenomenon reduces the yield of free-standing structures and may also complicate the interpretation of data from measurements of static friction. Measurements of dynamic friction [9,10] and static friction [11] have been reported recently. For application to MEMS design and fabrication, a friction-test microstructure is desirable for *in situ* measurements of friction locally on films which have undergone a specified sequence of fabrication steps. Such a test structure would make it possible to measure basic properties which would then be used to model friction in microbearings fabricated on the same substrate.

This paper describes the design, fabrication process, and initial experimental results for a test microstructure for measuring static friction. In order to measure static friction, (i) a normal load N must be applied to the contacting surface and (ii) the critical tangential force F must be measured at which the contacting surfaces slip. The ratio of the two forces defines the static friction coefficient: $\mu = F/N$. The approach taken is to use electrostatic attraction between a suspended microstructure and an underlying electrode patterned on the substrate to generate the normal load N. Corrugations in the microstructure (dimples) define the apparent contact area. No voltage difference appears across the dimple and the contacting surface, both of which are electrically grounded or surrounded by grounded conductors for the case of insulating surfaces. The critical tangential force is determined indirectly by means of a previously characterized spring suspension. In practice, the structure is displaced a specified distance Δx with a microprobe or an electrostatic linear comb actuator [12], after which it is brought into contact with the substrate by means of a clamp-down voltage V_c applied to the underlying electrode. The tangential force on the contacting areas is $F = k \Delta x$, where k is the spring constant of the suspension. The latter is found in a separate experiment by resonating the same microstructure with the comb actuator [13]. By ramping down V_c, we find the threshold voltage V_t at which the contacting surfaces slip and the structure returns to its original position. For this voltage on the underlying electrode, we determine the normal force N which corresponds to the tangential force F and we can calculate the coefficient of static friction.

In this paper, we describe the detailed design and fabrication of this friction-test microstructure. The processing sequence is compatible with surface-micromachined polysilicon mechanisms and micromotors. The apparent contact area was varied to investigate its effect on friction measurements in the micro domain. Experimental measurements of μ are reported for contact (under ambient conditions) between the bottom surface of the heavily phosphorus-doped, coarse-grained polysilicon microstructure and underlying films of both polysilicon and Si_3N_4. Measurements are made with the sample mounted on a bulk-mode piezoelectric crystal in order to observe the effect of dithering on static friction. We evaluate the sources of error in the initial measurements and consider the development of an improved friction-test microstructure.

II. FRICTION MICROSTRUCTURE DESIGN & PROCESS

Design

Figure 1 is a layout of a typical friction microstructure and Figure 2 is an SEM of such a structure. The shuttle is suspended over an underlying electrode by a folded beam suspension [12]. It is electrically connected to the ground plane via the beams and suspension anchors. Four dimples are fabricated into the shuttle to prevent the shuttle from contacting the underlying electrode. Their sizes are varied to change the apparent contact area. A bias voltage, V_c, applied to the underlying electrode provides the normal force to pull down the shuttle so that the dimples contact the substrate. An external application of a normal force on the contact area is required because of the negligible gravitational forces in the micro-domain. The substrate beneath the dimples may be either polysilicon or coated with a film of silicon nitride to study the effects of different interfaces. The truss suspension adopted is used to suspend the shuttle over the substrate and to provide a restoring force to overcome friction. The suspension allows only uniaxial motion in the plane. Displacement of the shuttle is achieved by applying a bias voltage, V_{DC}, to the electrostatic comb drive. The structure is grounded via the ground plane pad and all bias voltages are referenced to this ground.

On contact, the frictional force is $F_f = \mu N$ where N is the applied normal force and μ is the coefficient of static friction. The net normal electrostatic force can be estimated by a parallel-plate capacitor analysis,

$$N = -\frac{\partial U}{\partial z} = \frac{1}{2} \frac{\varepsilon_o w l (V_t^2 - V_{cz}^2)}{z^2} \quad (1)$$

where ε_o is the permittivity of air, w is the width of the electrode, l is the length, V_t is the threshold voltage, and z is the distance between the electrode and the shuttle which is defined by the depth of the dimples. V_{cz} is the required voltage to bring the shuttle into contact with substrate accounting for the restoring force, $F_z = k_z \Delta z$, in the normal direction. An electrostatic finite element package improves our estimation of N by incorporating fringing fields [14] in the capacitance analysis. Correlation factors, α_z, are found by comparing the simulation result with the parallel plate model for various gap distances. The spring constant of the truss suspension is found from the resonant frequency, using Rayleigh's method:

$$k = 4\pi^2 f^2 (M_o + .37 M_k) \quad (2)$$

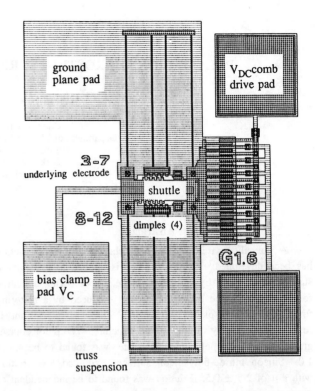

Figure 1. Layout of a typical test structure. The shuttle is suspended above the underlying electrode by a folded beam spring structure. The shuttle is laterally displaced by V_{DC} and is normally displaced and clamped by V_c.

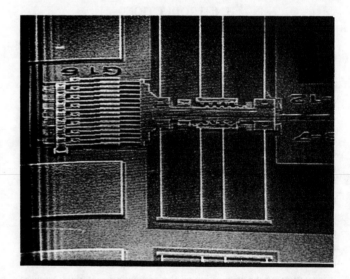

Figure 2. SEM of a structure suspended by 200 μm long beams.

where f is the resonant frequency, M_o is the shuttle mass, and M_k is the suspension mass. We compare the empirically found k_x with the k_x predicted by simple beam theory to deduce what dimensional changes have occurred through processing. The masses are calculated from the product of their volumes with the density of polysilicon, $\rho = 2300 \ kg/m^3$. The static friction coefficient μ is determined by equating the friction force F_f at the threshold voltage, V_t, to the restoring force from the spring suspension, $F = k\Delta x$. Solving for μ yields

$$\mu = \frac{k\Delta x}{N} \quad (3)$$

$$= \frac{1}{\alpha_z} \frac{8\pi^2 f^2 (M_o + .37 M_k) z^2}{\varepsilon w l (V_t^2 - V_{cz}^2)} \Delta x \quad (4)$$

$$= \frac{0.2 f^2}{(V_t^2 - V_{cz}^2)} \Delta x \quad (5)$$

for typical dimensions of M_o=0.03 μg, M_k=0.017 μg, w=10 μm, l=120 μm, z=0.95 μm, and α_z=1.5. We estimate that the V_t component in (5) comprises 90% of the uncertainty in our μ values due to N evaluation and V_t detection uncertainties.

Initial results from the resonant frequency measurements imply that the suspension's beam widths are less than the mask-defined dimensions. In fact, we have found using an electrical line width test structure that the 2 μm beams have been reduced to 1.6 μm due to the photolithography and etching processes. Compensating for the line width reduction, we estimate $k_z = 0.04$ μN/μm using a strain energy analysis. A $V_{cz} = 1.6$ V is required to overcome k_z and displace the shuttle down 1 μm. Substituting for V_{cz} in (5) yields,

$$\mu = \frac{0.2 f^2}{(V_t^2 - 2.5)} \Delta x \quad (6)$$

The derivation of (6) involves numerous approximations and serves to identify sources of error for experimental improvement.

Process

The structures are fabricated with the five-mask process illustrated in Fig. 3. A significant advantage of this technology is that all the critical features are defined with one mask, eliminating registration errors. The process begins with a 1500 Å-thick low-pressure chemical-vapor-deposition (LPCVD) nitride deposited on top of a layer of 5000 Å-thick thermal SiO_2 (Fig. 3(a)). The next step involves the deposition and definition of the first polysilicon layer. A layer of 4000 Å-thick, *in situ* phosphorus-doped polysilicon is deposited by LPCVD at 650 °C patterned with the first mask (Fig. 3(b)). The layer serves as the underlying electrode and as the ground plane to all the structures. A 800 Å-thick LPCVD Si_3N_4 is the third step of the process (Fig. 3(c)). This layer of silicon nitride is used as contact pads to examine friction between polysilicon and nitride. An oxide mask for the nitride allows isotropic etching in hot phosphoric acid. This eliminates residual stringers of Si_3N_4 associated with an anisotropic RIE process.

A 2um-thick LPCVD sacrificial phosphosilicate glass (PSG) layer is deposited and patterned to define dimple areas. In Fig. 3(d) a hemispherically shape dimple in the PSG is achieved by a time wet etch in HF. An alternative dimple process is a complete RIE etch and then a redeposition of another PSG layer (not shown). The depth of the dimple is then better known since the control of film deposition is greater than the control of wet etching. Anchor openings are then defined into the sacrificial layer with RIE (Fig 3(e)). The 2 um-thick polysilicon structural layer is then deposited by LPCVD (undoped) at 610 °C. The structural layer is doped by depositing another layer of 4000 Å-thick

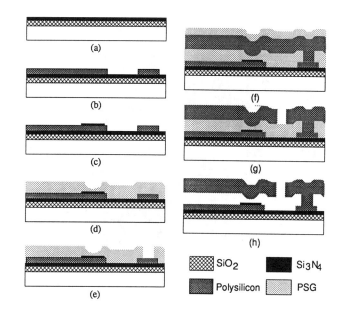

Figure 3. Process sequence of friction test structure (a) Deposition of LPCVD nitride on top of a layer of thermal SiO_2. (b) Deposition and definition of *in situ* phosphorus-doped LPCVD polysilicon. (c) Nitride contact pad definition. (d) Bushing formation in PSG after HF etch. (e) RIE to open anchors in PSG. (f) Deposition of undoped polysilicon and second PSG. (g) Definition of poly 2 (microstructure) after PSG removal. (h) Cross section of released structure after sacrificial PSG etch.

PSG (Fig. 3(f)). and then annealed at 1050 °C for one hour. The annealing process is designed to dope the polysilicon symmetrically by diffusion from the top and bottom layers of PSG. This high temperature annealing produces a rough friction surface. After removing the top PSG layer, the final masking step (Fig. 3(g)) defines our structures which are anisotropically patterned in a CCL_4 plasma. Figure 3(h) illustrates the final cross section after the wafer is immersed in 5:1 dilute HF to etch the sacrificial PSG. The wafer is rinsed repeatedly with DI water for at least 30 minutes and then dried under an IR lamp.

III. EXPERIMENTAL METHODS

Figures 4a and 4b illustrate the experimental procedure for measuring static friction. First, the shuttle is displaced either mechanically by using a microprobe or electrostatically by applying V_{DC} at the comb drive. A bias, V_c, is then applied to the underlying electrode to clamp the structure's dimples down onto the substrate. The minimum clamping voltage, $V_{c,min}$, is found by trial and error. The probe or V_{DC} is then removed. The clamping voltage is then ramped down from V_c using an HP4140B until the structure springs back. Figure 5a illustrates the timing of the voltages V_{DC} and V_c. At t_0, V_{DC} is applied. V_c is applied at t_1, pulling the structure down until it contacts the substrate. Then at t_2 V_{DC} is removed and the shuttle is held in place by V_c. At t_3 the clamping voltage is ramped down until t_4, when the structure snaps back. The value of V_c at t_4 is defined as the threshold voltage V_t. Figure 5b represents displacement of the shuttle during the period from t_0 to t_4.

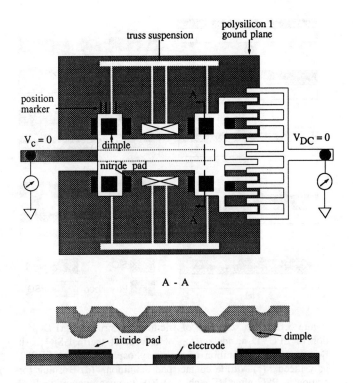

Figure 4a. Initial state - shuttle position prior to displacement and clamping (not to scale)

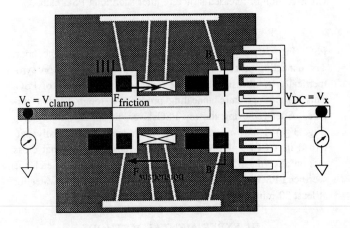

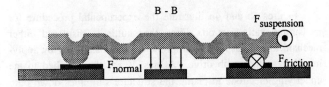

Figure 4b. Shuttle after displacement and clamping. V_{DC} laterally displaces shuttle and V_c pulls it down. (not to scale)

Careful attention must be made during the retraction of the probe after V_c is applied. Frequently, the shuttle slips back during the retraction due to the structure adhering to the probe. At this point the friction is greater than the original contact point and is no longer a reliable location for a test. In some cases the shuttle would not spring back even after the bias was completely removed. In fact, tests which were continued after slipping had

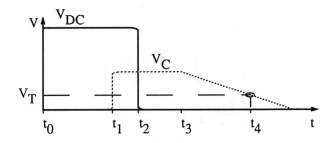

Figure 5a. Timing of displacement and clamping voltages. V_t is the threshold voltage used to determine the applied normal force.

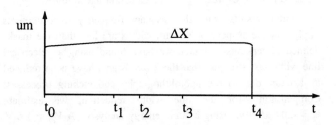

Figure 5b. Shuttle displacement resulting from application of biases in Fig 5a.

occurred results in a larger friction coefficient. This problem is not observed when the shuttle is displaced electrostatically using V_{DC} at the comb drive. We were able to reproduce deflections by maintaining the proper bias, which enabled repetitive testing of our devices.

The effects of dithering were investigated by mounting our sample on a 3-inch diameter, 0.5-inch thick piezoelectric material. We dithered the sample by exciting the piezoelectric material at its resonant frequency of 160 kHz with 5 V ac. Two identical friction tests were conducted, one involving dithering the other test without dithering.

IV. RESULTS

Table I summarizes initial data collected from tests conducted in ambient conditions at room temperature. Measurements were taken from a single wafer. Structures A-D have polysilicon-nitride interfaces whereas structures E-K have poly-poly contacting surfaces. The threshold voltage, V_t, is averaged over many trials. The resonant frequency, f_r, is determined visually through a microscope by observing maximum amplitude. The displacement, Δx, is achieved by aligning the shuttle to a position marker defined by the same mask which defines the shuttle. The coefficient of friction, μ, is then calculated using Eq. (6).

We observe in Table I that the friction forces between poly and nitride surfaces are less than for poly-poly interfaces. This is consistent with measurements made by [15] and with the surface energy theory where similar materials have greater adhesion properties than do dissimilar materials [2]. The dependency of μ on the apparent contact area, A_{app}, is negligible as indicated in Table I.

Table I Extraction of Parameters to Calculate Friction Coefficients

Structure	A_{ap} (μm)	f_r (kHz)	Δx (μm)	V_t (V)	μ (μN/μm)
poly-nitride contact					
A	113	6.2	3.0	3.6	2.2
B	314	6.0	5.0	4.3	2.3
B	314	6.0	7.0	4.9	2.3
B	314	6.0	10.0	5.1	3.0
C	8	4.8	3.0	2.7	2.9
C	8	4.4	7.0	4.3	1.7
D	314	4.7	7.0	3.6	2.9
poly-poly contact					
E	256	5.1	3.0	2.3	5.6
F	4	6.1	10.0	4.4	4.4
G	50	5.8	10.0	3.9	5.3
H	8	5.1	10.0	3.2	6.7
I	4	6.7	3.0	2.8	5.0
I	4	6.5	7.0	4.2	3.9
J	50	6.7	6.0	4.2	3.6

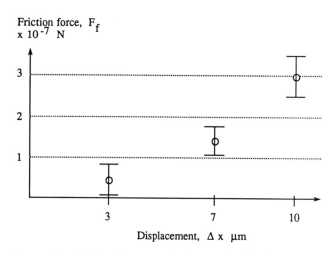

Figure 7. Friction force required to maintain displacement. Values are calculated from V_t in Fig.6. and spring constant, k_x. $F_{f_r} = 0.34, 1.4, 3.0 \times 10^{-7} N$

Figure 6 depicts V_t data for $\Delta x = 3, 7,$ and 10 μm. The friction forces required to sustain these displacements are graphed in Fig. 7 by reducing the data in Fig 6. The relationship confirms our expectation that a larger Δx requires a larger frictional force to balance the greater restoring force associated with Δx.

By dithering the wafer on a bulk piezoelectric crystal we observe in Figure 8 that the excitation reduced the amount of friction and The bold line represents the dithered measurements. improve the reproducibility of our tests. Random surface asperities that present adhesion or mechanical retarding forces may be less prevalent due to excitation of the surface. The standard deviations from our measurements indicate that data from tests performed on the piezoelectric crystal had 30% less scatter.

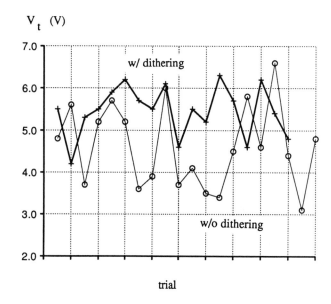

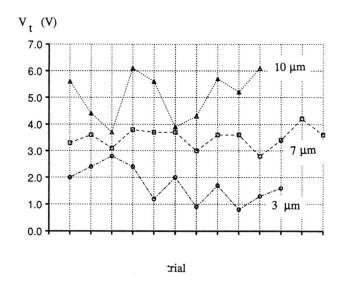

Figure 6. Graph of recorded V_t for a number of trials. Each line represents a different displacement, $\Delta x = 10, 7,$ and 3 μm from top to bottom.

Figure 8. Graph illustrating the effects of dithering. The thin line represents an undithered test procedure whereas the bold line reflects tests conducted on a bulk piezoelectric crystal. Dithering reduces friction and improves testing reproducibility.

V. DISCUSSION

The measured values for μ in Table I are higher than expected. They may reflect other retarding forces in our system that are not directly associated with the basic friction law. Clamping bias dependency leads us to believe that there are surface effects that may play a role. A large part of the surface effects are caused by contaminants introduced by the ambient. The contamination may include oxide buildup, dust particles, water vapor, and other unknowns in the environment. Testing in a vacuum chamber will provide better control of the testing conditions. System errors related to measurement uncertainties would undoubtably yield higher friction values. Accurate measurements of V_t depend on the detection of release. Visually it is difficult to observe the initial slippage which would constitute the release point. Currently we rely on the substantial spring return to verify when the static friction cannot sustain the displacement. A smaller recorded V_t value leads to a erroneously high μ. We believe it is this error that contributes to the high μ values. High spatial and temporal resolution is needed to detect small motions associated with slipping.

The method in which we displace and then pull down the shuttle is another source of systematic uncertainty. The V_{DC} used to displace the shuttle also induces a lifting effect on it [12]. Application of V_c will tilt the shuttle because only one end is lifted. As the clamp down voltage is applied the edge without the comb fingers will contact the substrate first. The whole shuttle will eventually be clamped down but only after one end was driven in at an angle into the substrate. The unequal load distribution on the four contact pads will result in a higher A_{real}. The forces at this initial contact surface are both normal and shear resulting in a greater real contact area [2]. The frictional force will increase with A_{real} resulting in higher μ values. This effect may be minimized by designing a shuttle with only one dimple at its center.

Another source of error is the uncertainty of the net normal force, N. Both the applied normal force, F_z, and the normal spring force, F_{k_z}, are not well known. A larger underlying electrode coupled with smaller z displacements would make the calculation of μ less sensitive to N estimation error

One way to reduce the F_{k_z} would be to fabricate the suspension with thinner beams or increase the dimple heights. Both would the tendency for the structure to stick to the substrate immediately after fabrication. An alternative method to keep the testing motion planar is to minimize Δz. Higher contact pads located away from the dimples would lie beneath the dimples only after the shuttle was displaced. This reduces Δz and preserves the vertical clearance between the suspended structure and the substrate to prevent post-release sticking. Improved comb drives have been designed to reduced the lifting effect. A differentially balanced drive [16] with alternating fingers biased with opposite polarity has been fabricated and tested. The new comb drive exerted no observable lifting forces, however, a reduction in lateral force was observed.

The minimum clamping voltage, $V_{c_{min}}$, necessary to pull down and hold the shuttle against the restoring force is appreciably greater than the eventual V_t. Ideally $V_{c_{min}} = V_t$. The disparity in voltage may indicate that the friction mechanism is dependent on "intimate contact". A certain force might be required to attain this "intimate contact" at which the adhesion occurs between the mating surfaces. This phenomena motivates the investigation of the dependency of V_t on the application of the V_c bias.

We investigated whether the clamping force has any effect on the amount of friction. One motivation of this particular test is to determine whether the materials in contact creep. [2] found that for creeping materials, the value of A_{real} will increase with time of application of the load. We have observed that as we increased the clamping voltage, V_c, the friction force increases as evidenced by the reduction of V_t necessary to sustain the displacement. The friction between the two surfaces has increased due to the larger clamping force. One explanation is that we are digging deeper into a surface whether it be the poly substrate, native oxide, or an unknown contaminant layer. Similarly, the load application time of the clamping voltage increases friction. We held the clamping bias for various times before decrementing our voltages and found that as the hold time increases the friction increases. This behavior is indicative of creeping which might lead us to believe that there exists some pliable contaminant layer that exhibits creep like behavior.

VI. CONCLUSIONS

We have designed and fabricated a polysilicon microstructure for *in-situ* measurements of static friction on thin films. The viability of this testing device has been demonstrated and problem areas identified. Furthermore, we have outlined an experimental method to routinely extract data.

In order to measure friction a normal force was applied to induce friction. The restoring force of a displaced spring was used to measure the frictional force. We measured static coefficient values for poly-poly and poly-nitride surfaces and have found the former to have a greater friction coefficient. The magnitude and duration of V_c has been qualitatively observed to increase friction. Apparent contact area does not appear to have any influence on friction. Dithering has been observed to reduce friction and to improve reproducibility of our measurements.

We have identified possible sources of uncertainty in the experimental measurements. Suggestions for improvements have been made. The problem of motion detection must be addressed in order to extract accurate data. Moreover, testing in a controlled ambient, such as a vacuum, will lend insight on surface reactions and vastly improve reproducibility of measurements.

Acknowledgements

The authors are grateful to Evan Stateler, Tom Booth, and the rest of the UC Berkeley Microfabrication Facility for their assistance in fabricating the devices. We also like to thank the many students of the Berkeley Sensor and Actuator Center who have contributed in various ways.

REFERENCES

[1] F.P. Bowden and D. Tabor, *Friction and Lubrication*, Methuen & Co., 1967.

[2] E. Rabinowicz, *Friction and Wear of Materials*, John Wiley & Sons, 1965.

[3] Y.C. Tai, L.S. Fan and R.S. Muller, "IC-processed micromotors", *Proceedings: IEEE Conference on Micro Electro Mechanical Systems*, Salt Lake City, UT, February, 1989.

[4] L.S. Tavrov, S.F. Bart, J.H. Lang, and M.F. Schlecht, "A LOCOS process for an electrostatically fabricated motor," *Technical Digest, Fifth International Conference on Solid-State Sensors and Actuators (Transducers '89)*, Montreux, Switzerland, June 1989.

[5] S.C. Jacobson, R.H. Price, J.E. Wood, T.H. Rytting, and M. Rafaelof, "The Wobble Motor: design, fabrication, and testing of an eccentric-motion electrostatic microactuator," *Proc. IEEE Robotics and Automation Conference,* Scottsdale, Arizona, May, 1989.

[6] S.D. Senturia, "Microfabricated structures for the measurement of mechanical properties and adhesion of thin films, *Technical Digest, Fourth International Conference on Solid-State Sensors and Actuators (Transducers '87)*, Tokyo, Japan, June 1987, pp. 11-16.

[7] H. Guckel, J.J. Sniegowski, and T.R. Christenson, "Advances is Processing Techniques for Silicon Micromechanical Devices with Smooth Surfaces," *Proc. IEEE Workshop on Micro Electro Mechanical Systems,* Salt Lake City, Utah, February 1989.

[8] H. Guckel, J.J. Sniegowski, and T.R. Christenson, "The Application of Fine Grained, Tensile Polysilicon to Mechanically Resonant Transducers'" *Transducers 89,* Montreux, Switzerland, June, 1989.

[9] K.J. Gabriel, F. Behi, and R. Mahadevan, "*In-Situ* Measurement of Friction and Wear in Integrated Polysilcon Microstructures", *Transducers' 89, The 5th Int'l. Conference on Solid-State Sensors and Actuators,* Montreux, Switzerland, June, 1989.

[10] Y.C. Tai and R.S. Muller, "Frictional Study of IC-Processed Micromotors", *Sensors and Actuators,* to be published.

[11] Y.C. Tai, "IC-Processed Polysilicon Micromechanics: Technology, Material, and Devices," *PhD Thesis,* U.C. Berkeley, 1989.

[12] W.C. Tang, T.H. Nguyen, M.W. Judy, and R.T. Howe, "Electrostatic-comb drive of lateral polysilicon resonators", *Transducers 89,* Montreux, Switzerland, June, 1989.

[13] W.C. Tang, T.H. Nguyen, and R.T. Howe, "Laterally Driven Polysilicon Resonant Microstructures", *Proc. IEEE Micro Electro Mechanical Systems Workshop,* Salt Lake City, UT, Feb. 1989.

[14] Ansoft Corporation, Pittsburgh, PA 15123 *Maxwell Users Guide,* January, 1989

[15] L.S. Fan, Y.C. Tai, and R.S. Muller, "IC-Processed Electrostatic Micro-motors", *IEEE IEDM Technical Digest 1988,* San Francisco, CA, December, 1988, pp. 666-669.

[16] W.C. Tang, University of California at Berkeley, personal communication.

Study on the Dynamic Force/Acceleration Measurements

AKIRA UMEDA and KAZUNAGA UEDA

National Research Laboratory of Metrology, Agency of Industrial Science and Technology, Ministry of International Trade and Industry, 1-4, 1-Chome, Umezono, Tsukuba, Ibaraki 305 (Japan)

Abstract

This paper presents a new method of estimating the dynamic characteristics of a shock accelerometer using elastic wave propagation in a metal bar. The output from the accelerometer attached to a Davis bar is compared with the theoretical acceleration calculated using the relationship derived from the one-dimensional wave equation of motion in elastic media. The experiment was carried out using the 2 m long stainless steel bar, a solenoid valve air gun, a transient recorder, and a personal computer for signal processing. Details are given of the experimental equipment, the comparison of the accelerometer output with the differentiation of the strain gauge signal, the comparison of the integration of the acceleration signal with the strain gauge signal, and the response curve of the accelerometer calculated using the fast Fourier transform.

Introduction

Recently, interest has been focused on the accuracy of acceleration measurements, including shock. This is because accurate and reliable measurement is required in many fields of engineering. A typical example is the shock measurement of dummy heads in the collision test during the development of new automobiles.

At present, the acceleration standard is only up to 100 m/s². The acceleration in this range is calibrated using a laser interferometer and shaker [1]. However, the acceleration measurement carried out in a practical situation is far out of the range of this standard and the users must rely on the data offered by manufacturers [2, 3]. In the traditional measurement, the filter frequency of the amplifier is adjusted to eliminate the signal fluctuation coming from the resonance characteristics in the accelerometer, but this results in a phase shift. To overcome these difficulties, the establishment of reliable calibration methods and acceleration standards are necessary.

Theory

Figure 1 shows the principle of this method. When a projectile strikes a bar, the elastic wave propagates along the bar as a compression wave. The motion of the elastic wave, according to the theory of linear elasticity, is expressed by the following equation

$$\frac{\partial^2 U}{\partial t^2} = C^2 \frac{\partial^2 U}{\partial x^2} \quad (1)$$

where C is the velocity of the longitudinal wave in the bar and U is the displacement of the particle parallel to the axis of the bar. The wave reflects at the other end of the bar and propagates in the opposite direction as a tensile wave. At the instant of reflection, the wave produces the acceleration (Fig. 1(a)):

$$a(t) = 2C\dot{\varepsilon}(t) \quad (2)$$

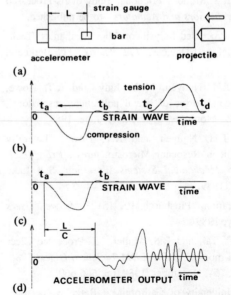

Fig. 1. The principle of the method of estimating the dynamic characteristics of an accelerometer: (a) method of generating an elastic wave; (b) strain wave observed with the strain gauge; (c) strain wave as an input signal to an accelerometer; and (d) output from an accelerometer.

Reprinted with permission from *Transducers '89, Proceedings of the 5th International Conference on Solid-State Sensors and Actuators and Eurosensors III*, A. Umeda and K. Ueda, "Study on the Dynamic Force/Acceleration Measurements," Vol. 2, pp. 285-288, June 1990. © Elsevier Sequoia.

where $a(t)$ and $\dot{\varepsilon}(t)$ are the acceleration of the end of the bar and the strain rate at the end, respectively. Equation (3) can be obtained by integration of eqn. (2) [4]

$$\varepsilon(t) = \frac{1}{2C} \int a(t)\, dt \qquad (3)$$

However, it is not possible to measure the strain at the end of the bar with a strain gauge. The gauge is attached to the bar at distance L from the end. As a result, the observed signal on the strain gauge appears as indicated in Fig. 1(b). The strain pulse in the interval $(t_a - t_b)$ is a compression wave, working as the stimulation to the accelerometer. On the other hand, the strain wave observed in the interval $(t_c - t_d)$ is a reflected wave and does not stimulate the accelerometer. The dynamics of the accelerometer is considered to be the relationship between the wave as an input signal, indicated in Fig. 1(c), and the output signal, indicated in Fig. 1(d) [5]. As a result, the acceleration generated at the end of the bar by the strain wave is expressed as follows:

$$a(t) = 2C\dot{\varepsilon}_m\left(t - \frac{L}{C}\right) \qquad (4)$$

where ε_m is the measured strain at distance L from the end of the bar.

The signal in Fig. 1(d) is an example of an output signal from the accelerometer that is screwed to the end of the bar. The delay time is L/C. The ringing observed in the signal comes from the resonance characteristics in the accelerometer. Therefore, the transfer function, $[g_a(t)]$, of the accelerometer can be expressed as follows:

$$F[g_a(t)] = \frac{F[a_m(t)]}{4\pi C f j F\left[\varepsilon_m\left(t - \frac{L}{C}\right)\right]} \qquad (5)$$

where F, f, a_m and j are the Fourier transform operator, frequency, output from the accelerometer and the imaginary unit, respectively.

Experimental

Figure 2 shows the block diagram of the experimental equipment. The bar is made of stainless steel (SUS304), 2 m in length and 30 mm in diameter. The projectile with a tapered tip is made of pure aluminum, 14.6 mm in diameter and 100 mm in length. The angle of the linearly tapered tip is 150°. The geometry of the launching tube is 1.5 m in length and 14.8 mm in inner diameter. The projectiles are driven by air pressure, ranging from atmospheric pressure up to

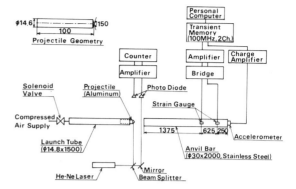

Fig. 2. Block diagram of the experimental equipment.

0.7 MPa, and collide with one end of the bar. The signals from the accelerometer and the strain gauge amplifier are digitized and analyzed by a personal computer.

In order to find the best way of supporting the bar, three methods were tested: in the first method, ball bearings were placed on the V-shaped channel to support the bar; in the second method, a coil of a rubber belt was formed on the bar and clamped in the tapered bore of the support; and in the third method, O-rings and simple supports were used. The decay of the elastic wave during the multi-reflection process was observed for 20 ms after the projectile struck the end of the bar: in the first method, the final amplitude was 60% of the initial value; and in the second and third methods, the final amplitude was 10% and 40%, respectively. Thus, the ball bearings were selected to support the bar.

In order to check the performance of the strain gauge, we compared the observed strain wave with the theoretical strain wave form in Skalak's theory where the lateral inertia of the bar is considered [6]. Figure 3 shows the comparison of the experiment with the theory. The observed strain wave is slightly smaller than the theoretical wave. The phase delay can also be seen. The gauge length of the strain gauge was 1 mm and the gauge factor was 2.18. The resistivity was

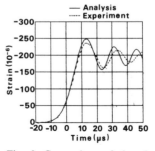

Fig. 3. Comparison of the observed strain wave with the theoretical wave.

120 Ω and a two-gauge method was used. The bandwidth of the strain gauge amplifier was 200 kHz.

It is not possible to estimate the wave velocity by measuring the time interval between the two consecutive wave heights, because dispersion occurs during the propagation. However, according to Skalak's theory, a theoretically calculated wave intersects the ideal stepwise strain wave at the point where the value is constant to the wave height. The estimated velocity of the elastic wave in the bar was $(5.01 \pm 0.05) \times 10^3$ m/s.

The accelerometers used in the experiment were of the piezoelectric type and screwed to the end of the bar with a stud bolt supplied by the manufacturers. The speed of the projectile was measured with the counter triggered by the signal from a photodiode that senses the flight of the projectile. The maximum sampling rate of the transient recorder was 100 MHz and storage capacity was 2 kbytes.

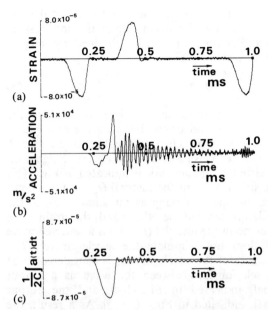

Fig. 4. Integration of the acceleration signal: (a) strain wave observed; (b) output from the accelerometer; and (c) integration of the acceleration signal.

Results

Figure 4 shows the strain gauge signal, the accelerometer output and the integration of the acceleration. The numerical integration was carried out using the trapezoidal formula. The velocity of the projectile was 14.6 m/s and the sampling speed was 0.5 μs. It was apparent that the reflected strain wave did not influence the accelerometer output signal. The wave form of the integrated acceleration signal agreed well with the strain wave form except for the time delay. The peak value of the integrated acceleration was slightly larger than that of the strain wave.

Table 1 shows the comparison of the strain multiplied by $2C$ and the acceleration integration. In every case, the phase of the integrated acceleration was slightly behind the strain signal. It was found that the charge amplifiers did not seriously influence the peak of the acceleration signal. Figure 5 shows the differentiated strain signal time delayed numerically by L/C, the acceleration signal, and their overlap. The differentiation was carried out numerically using the 7-point least squares smoothing method [7], after the 200 kHz rectilinear window processing of the fast Fourier

TABLE 1. Comparison of the strain wave peak value with the peak of the integration of acceleration. Three kinds of accelerometers and two kinds of amplifiers are used

Accelerometer	Charge amplifier	Strain gauge output $2C\varepsilon$		Accelerometer output $\int a(t)\,dt$	
		Peak velocity (m/s)	Time (μs)	Peak velocity (m/s)	Time (μs)
Kistler 8005	Kistler 5007	0.79	118.4	0.81	121.6
	B & K 2635	0.72	112.9	0.73	113.1
Endevco 2271A	Kistler 5007	0.71	108.4	0.77	117.9
	B & K 2635	0.73	107.9	0.77	114.4
B & K 4393	Kistler 5007	0.72	109.6	0.77	113.8
	B & K 2635	0.74	112.4	0.82	118.2

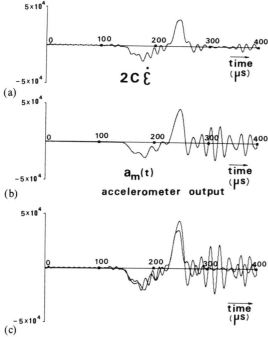

Fig. 5. Differentiation of a strain wave: (a) differentiated strain wave form (200 kHz low-pass filter and 7-point least squares smoothing differentiation); (b) output from the accelerometer (B & K 4393); and (c) overlap drawing of the above signals.

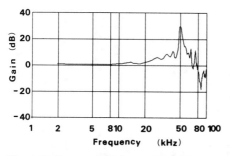

Fig. 6. Gain response curve calculated for the accelerometer (B & K 4393). The strain wave signal is filtered at frequency 200 kHz and cut off for the fast Fourier transform. Acceleration signal is also cut off for fast Fourier transform, but not filtered.

transform to the strain wave [8]. Ringing was observed in the differentiated strain wave. The rising phase and the falling phase of the acceleration agreed well with the differentiated strain signal. However, two phases of the peaks in the acceleration signal were different from those of the differentiated strain signal. The ringing observed in the acceleration signal was not derived from the differentiation of the strain signal. This clearly explains the resonance characteristics existing in the accelerometer.

Figure 6 illustrates the response curve calculated using eqn. (5). The resonance frequency of the accelerometer was 53 kHz according to the data. The diagram shows a resonance frequency of 50 kHz. At frequencies below 10 kHz the gain was 1–2 dB larger than 0 dB. This was explained by the fact that the gain of the strain gauge transfer function was not high enough, resulting in the underestimation of the acceleration calculated from eqn. (5). It is theoretically possible to calculate the phase relationship from eqn. (5).

Conclusions

1. A new method of evaluating the dynamic characteristics of an accelerometer was proposed.
2. The calculated results can be interpreted to show that the theory holds fundamentally. Prospects for the reliable and quantitative characterization of a shock accelerometer are obtained.
3. In this experiment, it was assumed that the strain gauge can measure the dynamic strain accurately. However, as Fig. 3 shows, there is ambiguity in the dynamic characteristics of strain gauges. Therefore, a new non-destructive method of evaluating the dynamic characteristics of a strain gauge and a theoretical compensation for the wave dispersion are required for a future shock acceleration standard.

References

1 Methods for the calibration of vibration and shock pick-ups, *ISO 5347-0*, International Standards Organization, Geneva, 1987, Parts 0–19.
2 D. C. Robinson, Requirements for the calibration of mechanical shock transducers, *N.B.S. Tech. Note 1233*, 1987.
3 A. Umeda, Present state of dynamic force/acceleration measurements in Japanese industries, *Proc. 34th Ann. Technical Meet. Institute of Environmental Sciences*, King of Prussia, PA, U.S.A., May 3–5, 1988, p. 237.
4 G. W. Brown and G. A. Drago, Shock calibration of accelerometers using Hopkinson's bar, *ME #77-3*, College of Engineering, University of California, Berkeley, 1977.
5 R. D. Sill, Shock calibration of accelerometers at amplitudes to 100 000 G using compression waves, *Endevco TP283*.
6 R. Skalak, Longitudinal impact of a semi-infinite circular elastic bar, *J. Appl. Mech., Trans. ASME*, 24 (1957) 59–64.
7 A. Savitzky and M. J. E. Golay, Smoothing and differentiation of data by simplified least squares procedures, *Anal. Chem.*, 36 (1964) 1627–1639.
8 R. Bracewell, *The Fourier Transform and its Applications*, McGraw-Hill, New York, 1965.

Anomalous Emissivity from Periodic Micromachined Silicon Surfaces

PETER J. HESKETH*, BENJAMIN GEBHART**, AND JAY N. ZEMEL*

Center for Sensor Technology
and
*Department of Electrical Engineering
**Department of Mechanical Engineering and Applied Mechanics
University of Pennsylvania
Philadelphia PA 19104-6390

Abstract—Thermal processes involving small micromachined surface elements have begun to attract an increasing amount of attention. While much of the attention has been directed toward the cooling of integrated circuits, there remains a fundamental question about the nature of the interactions when the characteristic scale of the physical process becomes comparable to the dimensional scale of the local geometry. One of the simpler processes to investigate is the radiant heat transfer from a solid whose surface has been micromachined to form a regular periodic structure, for example, a deep grating or a two-dimensional (2-D) array of pits or hillocks. Such surfaces will be referred to as microconfigured surfaces. Microconfigured surfaces have potential applications as adsorbers of gases, absorbers of radiation, and microheat exchangers. The results reported here are, to the best of our knowledge, not only the first experimental study of the thermal emissivity of deep gratings, but is also a subject for which essentially no direct theoretical study has been performed. A deep grating is defined as one where the ratio of the depth of the grating, h, to repeat distance or characteristic geometric scale, S, is equal to or greater than 0.5. The emissivity measurements were conducted in the range $0.14 < Z = \lambda/S < 0.68$ where λ is the measured electromagnetic radiation wavelength. The resulting data demonstrated that surface geometry plays a highly significant role in the emissivity.

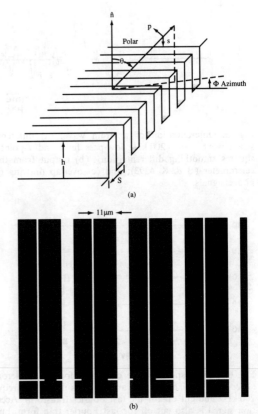

Fig. 1. a) Schematic layout of the deep grating showing the repeat distance, S; amplitude of the grating, h; and the angular measurement coordinates. b) Electron micrograph of the microgrooves, (A = 2), at a magnification of 1250× and a 10 µm wide marker.

THE experiments employed (110) oriented silicon wafers on which a 22µm "square wave" grating is etched to various depths using standard photolithography and a 40%/wt. solution of KOH at 52°C. The width of the silicon ridge is 11µm as shown schematically in figure 1a. A scanning electron micrograph of a 22µm deep grating is shown in figure 1b. The peak amplitude to half width ratio, A, for this structure is A = 2. Another grating with A = 4 has also been investigated. After etching, the microconfigured wafers are then spin-coated with a phosphorus based dopant which is heated to 1250°C for 10 hours. The phosphorus surface concentration is in the $0.6 - 1.1 \times 10^{20}/\text{cm}^3$ regime which yields an effective skin depth for $\lambda > 5\mu\text{m}$ of 2µm or less. The back surface of the silicon wafer was also heavily doped to act as the heater element. A blackbody was constructed and its temperature was stable to better than 10m°C during the course of the measurements. The temperature of the blackbody was used as a reference against which the temperature of the silicon sample was stabilized with a feedback control loop. Temperature fluctuations were typically less than 0.1°C during the course of measurement.

The initial studies were on silicon chips containing the microconfigured surface that had been sliced from the wafer. These chips were mounted in a specially designed low thermal loss holder. Calculations indicated that the thermal losses other than radiation are about 40% of the total energy input. In the temperatures range used, 300°C < T < 400°C, significant difference in the energy input required to reach operating temperatures were observed for smooth and microconfigured surfaced samples. The emissivity measurement consisted of the blackbody source,

Reprinted from *Technical Digest IEEE Solid-State Sensors Workshop*, June 1986.

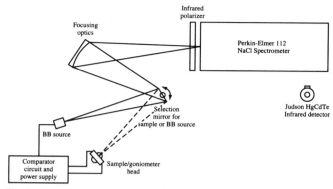

Fig. 2. Schematic layout of the emissivity measurement system.

The basic measurements are shown in figures 3 and 4. In figure 3a, the wavelength dependence of the emissivity for both s and p polarization is shown for $\Phi = 0°$ and $\theta = 0°$. Note the

the sample, and a reflection system that permitted viewing of the silicon surface, both smooth and microconfigured, down to azimuthal angles of 75°C, as shown in figure 2. The polar angle drive was controlled by a computer-driven stepping motor thereby allowing the polar angle dependence of the emissivity to be measured with good precision. Measurements were carried out for azimuthal directions parallel ($\Phi = 0°$) and perpendicular ($\Phi = 90°$) to the grooves. The polarization measurements employed a Perkin-Elmer infrared polarizer at the slits of a Perkin-Elmer 112 NaCl prism spectrometer. A Judson HgCdTe infrared detector was used to measure the intensity of the radiation. The data was logged on a computer-controlled system and that also controlled the stepping motor.

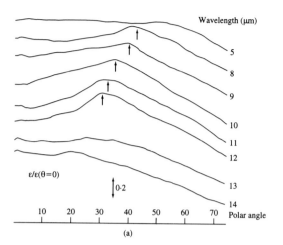

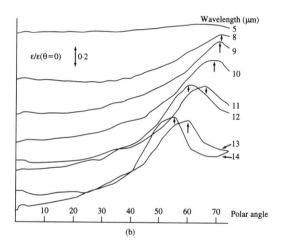

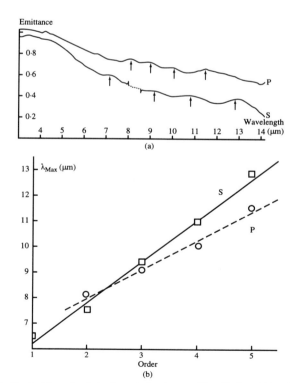

Fig. 3. a) Normal polarized spectral emittance, ($\theta = 0°$, $\Phi = 90°$), of the microconfigured surface shown in Figure 1(b). (b) Variation of the wavelength maxima versus mode number for p and s polarizations.

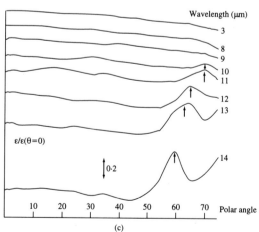

Fig. 4. Polar angular dependence of the emittance as a function of wavelength normalized to $\theta = 0°$. Sample temperature is 400°C.
a) s polarization, $\Phi = 90°$
b) p polarization, $\Phi = 0°$
c) s polarization, $\Phi = 0°$
d) p polarization, $\Phi = 90°$
e) Polar angle dependence of the peak emittance on wavelength for data in Figures 4a, 4b, and 4c.

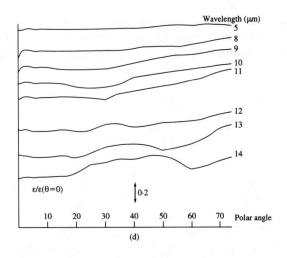

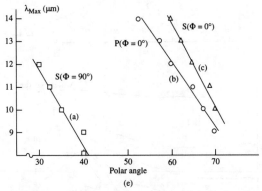

Fig. 4. (continued)

oscillatory and complementary character of the emissivity. A plot of the peak λ positions versus an integer yields a linear rather than a parabolic relation as seen in figure 3b. The polar angular dependence of the emissivity is quite extraordinary. In figures 4a ($\Phi = 90°$, s polarized radiation), 4b, ($\Phi = 0°$, p polarized radiation), and 4c, ($\Phi = 0°$, s polarized radiation), the polar angle position of the maxima in the emissivity decrease with increasing wavelength, as shown in figure 4e. However, in figure 4d, ($\Phi = 90°$, p polarized radiation) the emissivity does not show any pronounced peaks. Essentially, the same behavior was observed on a sample with the same S value but $A = 4$. The only plausible explanation available at present for both the mode and angular dependence of these measurements is that the grating couples a surface polariton to a radiative mode of the sample.

Acknowledgment

Two of the authors (P.H. and J.N.Z) wish to acknowledge the partial support of the EEMS Program of the National Science Foundation, D. Silversmith, Program Manager, under Grant ECS-84-12241. The authors with to thank Professor H. Baltes for useful discussions and one of us (J.N.Z) wishes to thank Professor A. Maradudin for a helpful conversation.

Author Index

A
Albaugh, K. B., 582
Allegretto, W., 359
Allen, M. G., 257, 664
Angell, J. B., 38, 437, 471
Aritomi, S., 411

B
Bajikar, S. S., 634
Baltes, H. P., 521, 610
Bart, S. F., 74, 161, 294
Barth, P. W., 38, 471, 584, 596
Bau, H. H., 478
Bean, K. E., 50
Becker, E. W., 623
Benecke, W., 355
Bergamasco, M., 396
Bicking, R. E., 499
Bley, P., 623
Bonne, U., 464
Bornhauser, H. P., 390
Brennen, R. A., 246
Brown, J., 584
Bryzek, J., 455, 584
Burns, D. W., 499
Busch-Vishniac, I., 309
Buser, R. A., 655
Busta, H. H., 565

C
Cade, P. E., 582
Chang, J. C., 679
Chang, S. C., 524
Chapman, T. W., 634
Chau, K., 359
Chen, L. Y., 538
Cho, D., 300
Choi, B., 634
Choi, S. T., 440, 604
Chou, A. T., 246
Christel, L., 455
Christenson, T. R., 532, 634
Ciarlo, D. R., 570
Crary, S., 653
Cueller, R. D., 565

D
Daniel, J. S., 614
Dario, P., 396

Davis, J. R., Jr., 21
Delapierre, G., 614
Denton, D. D., 634
de Rooij, N. F., 655

E
Egawa, S., 251
Ehrfeld, W., 623
Elwenspoek, M., 450
Eoh, S., 440
Esashi, M., 440, 444, 588

F
Fan, L-S., 147, 151, 514, 672, 675
Fearing, R. S., 473
Feinerman, A. D., 565
Feury, A. M., 427
Feynman, R., 3, 10
Field, L. A., 592
Fluitman, J. H. J., 450
Fujimasa, I., 368
Fujita, H., 209, 276, 314, 339
Fukiura, T., 464
Fukuda, T., 261
Furutani, K., 406

G
Gabriel, K. J., 117, 213, 272, 372, 509
Gebhart, B., 690
Götz, F., 623
Grace, K. W., 231
Guckel, H., 532, 634

H
Hagmann, P., 623
Haritonidis, J. H., 491
Harley, J., 478
Harris, R. M., 647
Haselton, K. R., 538
Hatazawa, Y., 339
Hebigushi, H., 588
Hesketh, P. J., 690
Heuberger, A., 355
Higashi, R. E., 499
Higuchi, T., 251, 401, 406
Hochuli, T., 390
Hojjat, V., 401
Holmen, J. O., 499
Howe, R. T., 157, 187, 194, 286, 428, 473, 491, 505, 524, 672, 675, 679

Author Index

Huang, J., 675
Huff, M. A., 459

I
Igarashi, I., 411

J
Jackson, T. N., 551
Jacobsen, S. C., 231, 267, 331
Jarvis, J., 117
Jebens, R., 327
Jerman, H., 363
Johnson, G. M., 499
Johnson, R. G., 499
Johnson, S. D., 499
Judy, J. W., 417
Judy, M. W., 194

K
Kabashima, T., 411
Kaminsky, G., 555
Katsurai, M., 314
Kawahata, K., 667
Kelly, T. F., 532
Ketterson, J. B., 565
Kim, C-J., 203
Kim, Y-K., 314
Ko, W. H., 574
Koppelman, G. M., 641
Kudoh, K., 406
Kudoh, T., 339
Kumar, S., 300
Kuribayashi, K., 379, 385

L
Lang, J. H., 74, 161, 168, 286, 294, 344
Lee, J. W., 634
Lim, M. G., 198, 203, 679
Lintel, H. T. G., 450
Lober, T. A., 157, 459
Lovell, E. G., 634

M
MacDonald, N. C., 538
Mahadevan, R., 272
Mallon, J., Jr., 455, 584
Maner, A., 623
Maseeh, F., 647
Masnari, N. A., 551
Matsuo, T., 440
McMillan, J. A., 538
Mehregany, M., 74, 161, 168, 344, 509, 664
Mettner, M. S., 459
Michel, F., 614
Mohney, S., 532
Mohr, J., 623

Moroney, R. M., 428
Moser, D., 610
Moser, H. O., 623
Muller, R. S., 147, 151, 203, 424, 505, 514, 592, 672, 675
Münchmeyer, D., 623

N
Nagarkar, P., 344
Najafi, K., 604
Nakajima, N., 368
Nakano, A., 588
Nathanson, H. C., 21
Newell, W. E., 21
Nguyen, T-C. H., 187, 194
Nikolich, A. D., 257

O
Ogawa, K., 368
Ohnstein, T. R., 464, 499
Omodaka, A., 209, 276, 339

P
Parameswaran, M., 359, 521, 610
Paravicini, R., 390
Pelrine, R., 309, 320
Petersen, K. E., 58, 455, 584
Pfahler, J., 478
Pisano, A. P., 203, 246
Pister, K. S. J., 473
Polla, D. L., 417, 424
Poteat, T. L., 427
Pourahmadi, F., 455
Poydock, J., 584
Price, R. H., 267
Putty, M. W., 524

R
Rafaelof, M., 331
Rasmussen, D. H., 582
Rensel, W. B., 551
Ridley, J., 464
Riethmüller, W., 355
Robbins, W. P., 417
Robinson, A. L., 524
Robinson, A. M., 359, 521
Robinson, M. G., 551
Rytling, T. H., 331

S
Sakata, M., 339
Salsedo, F., 396
Satren, A., 499
Scheidl, M., 257
Schelb, W., 623
Schlecht, M. F., 161, 286

Schmidt, D., 623
Schmidt, M. A., 459, 491
Schnakenberg, U., 355
Schultz, D. P., 679
Schwarz, G., 390
Senturia, S. D., 74, 161, 168, 344, 491, 647, 659, 664
Shoji, S., 444, 588
Skrobis, K. J., 634
Smith, R. L., 257
Sniegowski, J. H., 532
Solomon, D. E., 551
Spangler, L. J., 578
Starr, J. B., 487
Sugiyama, S., 667

T
Tabata, O., 667
Tai, Y-C., 147, 151, 514
Tanaka, T., 261
Tang, W. C., 187, 194, 198
Tavrow, L. S., 74, 161
Terry, S. C., 38
Thomas, D. C., 538
Trimmer, W. S. N., 96, 117, 213, 272, 327, 372, 427, 509

U
Ueda, K., 686
Umeda, A., 686

V
Van de Pol, F. C. M., 450

W
Walker, J. A., 372
Watanabe, M., 401
White, R. M., 428, 679
Wickstrom, R. A., 21
Wise, K. D., 495, 524, 551, 578, 604
Wood, J. E., 267, 331
Wu, X-P., 571
Wuttke, G. H., 551

Y
Yamagata, Y., 406
Yao, J. J., 538
Yoon, E., 495
Yoshitake, M., 385
Yun, W., 675

Z
Zdeblick, M. J., 437, 471
Zemel, J. N., 690
Zhang, Y., 653
Zhang, Z. L., 538

Subject Index

A
Acceleration, 104
Accelerometers, 39, 48, 79, 87, 487, 599, 618, 630, 686
Accuracy, 97
Acoustic, wave, 428
Actuators, **96,** 131
   boiling, 437
   comb drive, **187, 194, 198, 203**
   electrostatic, 14, 22, 84, 103, 115, 131, **147, 151, 157, 161, 187, 194, 198, 203, 207, 209, 213, 231, 237, 246, 251, 257, 261, 267, 272, 276, 286, 294, 300, 320, 327, 331, 339, 344,** 460, 465, 474, 524, 535, 679
   electrostatic comb drive, **187, 194, 198, 203,** 680
   harmonic, 325, **327, 331, 339, 344**
   impulse or impact, 353, **401, 406, 411, 417**
   magnetic, 102, 110, 267, 287, 307, **309, 314, 320,** 457
   piezoelectric, 353, **424, 427–28**
   shape memory alloys (SMA), **237,** 353, **373, 379, 385, 390, 396**
   side drive, **147, 151, 157, 161, 168, 174**
   thermal, 79, **353, 355, 359, 363, 368, 372, 379, 385, 390, 396, 437, 450**
   ultrasonic, **428**
Adhesion, 664
Advantages of micromechanics, **97,** 331
Aharonov-Bohm effect, 566
Air bearing, 322, 348, 473
Amplifier, 471
Anneal, 190, 505, 523–24, 532
Ant picture, 99–100
Applications, 11, 13, 58, 100, 117, 120, **122,** 272, 407, 624
Architecture, 134
Assembly, 100, 128
Atoms, 8

B
Beam, 22, 43, 81, 84, 188, 231, 246, 355, 360, 505, 509, 514, 521, 524, 532, 538, 544, 570, 610
Bearings, 209, 309, 314, 320, 347, 473, 509, 514, 637
Bibliography, 92, 108, 140, 292
Bimetallic effect, 355, 363
Biological, 103
Biomedical, 411
Bonding, 460, 551, 555, 578, 582, 584, 588, 592, 596
   anodic (electrostatic bonding), 47, 67, 444, 447, 578, 582, 588
     low temperature, 447, 588
   Mallory, 67
   silicon fusion, 584, 592, 596
Boron, *see* doped silicon
Breakage, 60, 81, 394, 513, 518, 539, 672
Brownian motion, 17
Bulk micromachining, **551, 555, 565, 570, 578–79, 582, 584, 588, 592, 596, 604, 610, 614**

C
Cantilever beam, 22, 43, 81, 84, 188, 231, 246, 355, 360, 505, 509, 514, 521, 524, 532, 538, 544, 570, 610
Capacitance, 164, 300
Capillary, 45, 47
Cavities, sealed, 462
Charge, 269, 300
Charge-coupled device (CCD), 76
Chemical, 5, 8
Chemical-vapor deposition (CVD), 505, 509, 514, 521, 524, 532, 538
Chromatography, 38, 45, 72
Cleaning, 556
CMOS, 359, 495, 521, 579, 610
Comb drive, **187, 194, 198, 203,** 680
Computer
   micro, 5, 17
   quantum, 18
   reversible, 18
Computer aided design (CAD), 639, **641, 647, 653, 655**
Connections, 70, 134, 527, 565, 628
Coolers, 73, 368
Corner compensation, 457, 570, 574
Couette flow, 191, 195
Crab leg suspension (folded cantilever), 185, 188
Cranks, 514
Current density, 102, **110,** 312
Cybernetic actuator, **401, 406, 411, 417**

D
Damping, 15, 80, 290, 487
Deflection calculations, 364, 668, 675
Density, 539
Design tools, 641, 647, 653, 655
Diamagnetic, 309, 315, **320**
Diaphragms, 43, 69, 75, 84, 363, 428, 460, 532, 604, 610, 660, 664, 672
Dielectric, 269
Dielectric constant, 330
Disk memory, 61, 90
Displacement current, 114
Dissolved wafer process, 578, 604
Doped silicon, 84, 356, 527, 604

E
Earnshaw's Theorem, 309
Edges, 68, 567
Efficiency, 277

Subject Index

Ejected mass, **401, 406, 411, 417**
Electrets, 231, 268, 270
Electro discharge machining (EDS), 335
Electrodes, curved, 240, 327
Electromagnetic fields, 309, 327
Electromagnets, 309, 312, 401
Electromigration, 565
Electron microscope, 5
Electron multiplier, 631
Electroplating, 623, 634
Electrostatic breakdown, 288
Electrostatic fields, 131, 198, 210, 213, 232, 241, 267, 282, 298, 300, 340, 419, 535, 588
Emissivity, 690
Energy, 328
Energy loss, 17, 278
Epitaxial, 66, 75
Etching, 61, 128, 551, 555–65, 570, 578–79, 582, 584, 588, 592, 596, 604, 610, 614, 655
  anisotropic, 40, **50,** 61, 68, 356, 360, 456, 460, 555, 565, 570, 574, 610, 655
  dry, 471
  electrochemical, 64
  ethylenediamine-pyrocatechol-water (EPW), 574
  hydrazine-water, 574
  isotropic, 50, 62, 551
  laser, 65
  photochemical, 66
  potassium hydroxide (KOH)-water, 556, 570, 574, 655
  quartz, 614
  wet, 50
Etching rate, 557–58, 563, 570
Etching spheres, 552

F
Fatigue, 60, 81, 394, 513, 518, 539, 672
Feedback, 233
Ferroelectrics, 270
Fiber optic, 70, 231, **237, 427;** *See also,* optics
Field emitters, 565
Filling cavities with liquid, 452
Film actuator, 251
Filters, 34, 627
Flow, liquids, 483
Flow gas, 364–65, 459, 473, **478,** 482, 491, 495, 630
Fluidics, **471, 473, 478, 487, 491, 495, 499**
Flux quantum, 566
Force, 102, 110, 198, 214, 232, 247, 251, 273, 277, 309, 320, 381, 392, 399, 401, 406, 679; *See also,* actuators
Fracture, 60, 81, 394, 513, 518, 539, 672
Frequency response, 161, 361, 372
Friction, 17, 61, 126, 145, 147, **168,** 309, 314, 320, 327, 333, 349, 401, 406, 419, 430, 473, 679, 683

G
Gas seals, 551

Gasket, 435, 471
Gear ratio, 327, 333, 345
Gears, 509, 514, 637
Glass, 437, 440, 444, 578, 582, 584, 589, 604
Gratings, 559, 629, 690
Gripper, 203
Grooves, 63, 68, 72, 560

H
Hands, tiny, 7, 8, 11
Hardness, 539
Harmonic motor, 61, 90, 166, 168, 325, **327, 331, 339, 344**
Heat transfer, 690
Holes, 68, 445
Hydraulic, 103
Hydrofluoric acid (HF), (BHF), (LHF), (VHF), 157, 190, 348, 505, 509, 514, 521, 524, 532

I
Image charge, 252, 269
Impulse, 402
Impulse actuator, 353, **401, 406, 411, 417**
Inchworm, 401, 406, 422, 427
Induced current, 253
Inductance, 310
Induction motor, 251, 294
Inertial forces, 104
Infinitesimal machine, 6
Infrastructure, 136
Injection molding, 623
Ink jet, 43–44, 69, 91
Insulators, 270
Integration of electronics and micro devices, 495
Interconnections, 70, 134, 527, 565, 628

K
Knudsen number, 469

L
Lagrangian, 115
Lamb wave, 428
Levitation, 199, **309, 314, 320,** 348
Linear actuator, 187, 194
Lithographie Galvanoformung, Abformung (LIGA), **623, 634**
Lithography, 13, 38, 75, 127
Logic gates, 17
Lorentz force, 314–15
Low-pressure chemical-vapor deposition (LPCVD), *see* chemical-vapor deposition (CVD)
Lubrication, 6

M
Magnet, permanent, 114, 321
Magnetic, 309, 314, 320
Magnetic field, 110, 316, 321
Magnetized, uniform sphere, 321

Manufacturing, 503
Mass, 104
Mass flow controllers, 440
Material usage, 97, 101
Maxwell's equations, 110, 310
McLellan electric micro motor, 9, 11
Mean free path of air molecules, 290
Measurements (metrology), 133, 478, 555, 607, **659, 664, 667, 672, 675, 679, 686, 690**
Meissner effect, 314
Membranes, 43, 69, 75, 84, 363, 428, 460, 532, 604, 610, 660, 664, 672
Memory, 3, 10
   disk, 61, 90
Metrology, 133, 478, 555, 607, **659, 664, 667, 672, 675, 679, 686, 690**
Micro pellets, 551
Micro robot, 98, 262, 310, 379, 385, 396, 407, 509
Micro satellite, 101
Micro science, 133
Micro surgery, 15, 101
Microchannel plates, 631
Microdynamical systems, 119
Mirror, 80, 82; *See also,* optics
Mobile microrobots, 15
Modeling, 300, 335
Molding, 623, 634
Moment of inertia, 290
Motor dynamics, 152, 174, 177, 204, 233
Muscle (actin and myosin filaments), 331

N
Nozzles, 38

O
Optics, 70, 80, 82, 429

P
Packaging, 91
Parallel plate capacitor, 116, 214
Parylene, 60
Paschen's law, 288
Perturbation method for solutions, 302
Phosphosilicate glass (PSG), 189, 515
Photolithography, 13, 38, 75, 127
Piezoelectric, 401, 403, 406, 411, 417, 424, 427–28, 445, 614
Piezoresistive, 49, 78–79, 441, 584
Pin joint, 505, 514
Plastic deformation, 462
Plating, 623, 634
Pneumatic, 103, 437
Points, 68, 567
Poisson's ratio, 529, 607, 667
Polarization of dielectrics, 269
Polyimide, 257, 493

Poly-methyl methacrylate (PMMA), 552, 623, 634
Polysilicon, 151, 157, 161, 163, 189, 347–48, 356, 360, 509, 514, 521, 524, 532, 679
Polysilicon grain, 532
Porous silicon, 65
Potassium hydroxide (KOH), 556, 570
Power, 104, 219, 291, 310
Precision, 7, 16, 97, 309
Pressure, 459
   differential sensor, 459
   vapor, 437
Pull-in voltage, 25, 84, 90, 605; *See also,* actuators, electrostatic
Pump, 435, **444,** 446, **450**
Pyramids, 567
Pyrex glass, 437, 440, 444, 578, 582, 584, 589, 604
Pyroelectric, 425

Q
Q, resonance, 21, 188, 197, 558
Quantum mechanical effects, 565
Quartz, 614–15

R
Radiation, 690
Range of mechanical systems, 99
Recoil, 353, **401, 406, 411, 417**
Reliability, 135
Rent's law, 14
Research areas, 125
Research needs, 124
Residual stress and strain, 530, 664, 667, 675
Resistance, 291, 296, 391, 499
Resonance, 21, 22, 24, 29, 80, 85, 90, 188, 195, 200, 524, 681
Resonant Gate Transistor, **21,** 82
Response time, 97, 102, 437
Reynold's number, 290, 459, 487
Rotational rate, maximum, 290

S
Sacrificial techniques, 1, 21, 28, 83, 129–30, 151, 157, 169, 189, 347, 465, 505, 509, 514, 521, 524, 532, 660, 673
Scaling, **96,** 101, 109, 307, 310
Scaling matrices, 104
Scanning tunneling microscopes, STM 565
Sensor accelerometer, 39, 48, 79, 87, 487, 599, 618, 630, 686
Sensor
   capacitive, 49, 492
   gas, 47
   gas flow, 442, 491, 495, 499
   pressure, 38, 48, 77, 440, 584, 598, 608
   stress and strain, 604, 662
   temperature, 76
Shadow masks, 561
Shape Memory Alloy (SMA), **237, 372, 379, 385, 390, 396**
   reversible or two way, 385
Shielding, electrostatic, 151, 154, 168, 176, 198, 347

Side drive electrostatic motors, **147, 151, 157, 161, 344, 349**
Silicon, 38, **58**, 462, 538
Silicon compounds, 86
Silicon crystal, 40, 51, 555, 575
Silicon-crystal silicon (SCS), 76
Silicon microfabrication, 11–12, 38, 125
Silicon on glass, 578
Silicon properties, 59
Simulations, 641, 647, 653, 655
Single-point diamond machining, 238
SiO_2, 82
Sliders, 505, 514
SOI wafers, 579, 601
Space, 101
Speed 97, 104
Spring, 401, 514, 517, 544, 610, 679
Squeeze film, 323, 487
Stability, 145, 232, 289
Stirling engines, **368**
Stress, residual, 530, 664, 667, 675
Stroboscopic, 174
Sublimation, release of polysilicon, 534
Superconductor, 314
Surface micromachining, **509, 514, 521, 524, 532, 538**
Surface tension, 103, 310, 435, 534
Surface texture, 158, 532, 542
Surgery, 7
Switch, 88
Synchrotron radiation, 623, 634

T
Thermal conduction, 47, 103, 111, 310, 539
Thermal coolers, 73
Thermal expansion, 393, 539
Thermal mass flow sensors, 442
Thermomigration, 67
Thermonuclear fusion, 551
Thermopneumatic actuator, 450
Three-dimensional fabrication, 14
Three-dimensional motion, 261, 408
Time, cycle, 97, 102, 437

Top drive, 161
Top motor, 339; *See also,* harmonic motor
Torque, 147, 169–70, 174, 189, 218, 275, 290, 299, 309, 327, 329, 337, 340, 344, 346, 349
Transistor, 21, 26, 36
Transit time, 97, 104
Traveling waves, 428
Tribology, friction, 17, 61, 126, 147, **168,** 309, 314, 320, 327, 333, 349, 401, 406, 419, 430, 473, 679, 683
Tungsten, 538, 565
Tuning, 21
Turbine, 377, 511
Tweezers, 203, 220, 377, 538, 544

U
Unit cube scaling, 104

V
Valves, 38, 46, **363**, 435, **437, 440, 444, 450, 455, 459, 464**
Van Sant's art, 11
Viscosity, 15, 80, 290, 487

W
Waveguide, 72; *See also,* optics
Wear, 168, 171
Wires, thin, 565
Wobble motor, *see* harmonic motor

X
X-ray, 623, 634
X-ray lithography, 623, 634
X Y stage, 404, 427, 473

Y
Yield strength, 390
Young's modulus, 86, 191, 196, 200, 391, 519, 524, 534, 539, 606, 615, 664, 667

Z
Zinc oxide, 424

About the Editor

WILLIAM Trimmer received a B.A. in 1966 from Occidental College (Los Angeles, CA) in physics, and a Ph.D. in 1972 in experimental relativity from Wesleyan University (Middletown, CT), where he was graduate student body president. He has taught at Montclair State College (Montclair, NJ) and The College of Wooster (Wooster, OH), where he was department chairman. He also has worked in Singer's Corporate Laboratory developing displays for sewing machines, and at Johnson & Johnson developing acoustical imaging systems for early detection of breast cancer.

In 1982, Dr. Trimmer joined AT&T Bell Laboratories. One of his programs developed a robotic system to test the small and fragile chips used to detect light in fiber optic systems. This robot was used for years on the factory floor to test chips. In 1984, he started a program on micro robots that evolved into a research effort in micromechanics and MEMS. His first paper in micromechanics was presented at the 1986 IEEE Hilton Head Conference.

In 1987, he co-chaired the first workshop on micromechanics called "Micro Robots and Teleoperators". This was the initial workshop in the international series of MEMS workshops. In 1989, he started, and was editor of, the Micromechanics section of the journal *Sensors and Actuators.* Next, he organized the first joint IEEE and ASME publication, the *Journal of Microelectromechanical Systems,* and was editor for the first six years of the journal.

In 1990, he started his own company, Belle Mead Research. BMR develops products involving micromechanics, helps facilitate new micromechanics business ventures, and has several science and engineering research programs.

Dr. Trimmer has published many papers and given numerous talks on micromechanics. He lives with his wife Ann and sons Scott and Mark, ages 16 and 13. He likes reading, gardening, walking and, in general, enjoys learning and doing new things.

Editor's Notes on the Second Printing

Nature abhors a vacuum. So also do authors abhor blank pages.

The first printing of this book went to press while I was in Texas. When I saw the final version, there were three blank pages. Since that time, those three pages have bothered me every time I looked through the book. At last, the second printing gives the opportunity to put ink on those white spaces.

One more change to the second printing. The IEEE Electron Devices Society was an early supporter of this book. The original proposal for this book was approved by the EDS, and during the two years it took to prepare the book, they gave their encouragement and support. Unfortunately, the first edition did not list them as a sponsor. I apologize for the oversight, and thank the IEEE Electron Devices Society and the IEEE Industrial Electronics Society for their support and encouragement.

This book gives a historical perspective on Micromechanics and MEMS, and makes available the insights of the founders of this field. These last three pages are devoted to my current thoughts on this field of the small and on what directions we are taking. This material is adapted from a talk given at the Tenth Annual International Workshop on Micro Electro Mechanical Systems in Nagoya, Japan, [Ref 1] and from an Editorial in the *Journal of Microelectromechanical Systems*, December 1997 [Ref 2].

Introduction

How a field progresses is determined by people's perceptions of the appropriate and possible. How our field prospers is determined by the areas we pursue and what we believe is possible.

Micromechanics is an extremely broad field, a field that will touch most aspects of our grandchildren's lives. This field encompasses all of the current technologies—only it is concerned with a smaller dimensional scale. Micromechanics also promises applications in all disciplines. Richard Feynman well conveyed the excitement of our new discipline:

> I imagine experimental physicists must often look with envy at men like Kamerlingh Onnes, who discovered a field like low temperature, which seems to be bottomless, and in which one can go down and down. Such a man is then a leader and has some temporary monopoly in a scientific adventure. Percy Bridgman, in designing a way to obtain high pressures, opened up another new field and was able to move into it and lead us all along. The development of ever higher vacuum was a continuing development of the same kind.
>
> I would like to describe a field, in which little has been done, but in which an enormous amount can be done in principle. This field is not quite the same as the others in that it will not tell us much of fundamental physics (in the sense of, "what are the strange particles?") but it is more like solid-state physics in the sense that it might tell us much of great interest about the strange phenomena that occur in complex situations. Furthermore, a point that is most important is that it would have an enormous number of technical applications.
>
> What I want to talk about is the problem of manipulating and controlling things on a small scale. [Ref 3]

Below are discussed the earlier perceptions that hindered the development of small mechanisms, our current field and needs, and a look at our future.

The Genesis

Perhaps things normally start small and grow. Man's habitats have grown from houses to buildings to skyscrapers. Our ability to travel has increased from a few miles on foot to horses to trains, and now we can encircle the world in a few days. Individually, we work to make large accomplishments in hopes of enormous success. We are enthralled with the big and significant and substantial.

The insignificant, insubstantial, and minuscule is usually beneath our concern.

And yet.

A dozen years ago, I was trying to persuade a machinist to build a very small structure. He listened patiently for awhile and then said, "Why do you want something small, a toy? I can make you something that is big and good." In his mind, most people's minds, small things were cheap and no more than a toy. When H. A. Rowland (1848–1901, professor of physics at the Johns Hopkins University, Baltimore) went to make very small and accurate grooves for diffraction gratings, he used large machines and buried them in even larger vaults for thermal stability. Ten years ago an eminent colleague at Bell Laboratories looked me in the eye and said, "Your micro things will never amount to anything. Large objects will always do a better job at a lower cost." This was very strongly the feeling at that time.

Even Feynman responded with good-natured jesting to critics of small machines. In his famous talk, "There's Plenty of Room at the Bottom," given at the American Physical Society meeting in 1959 he says, "What would be the utility of such machines? Who knows? Of course, a small automobile would only be useful for the mites to drive around in, and I suppose our Christian interests don't go that far." [Ref 4] And in his 1983 talk, "Infinitesimal Machinery," at the Jet Propulsion Laboratory he says, "I also talked in the 1960 lecture about small machinery and was able to suggest no particular use for the small machines. You will see there has been no progress in that respect." [Ref 5]

Originally, the ingenious and intelligent systems were mechanical, things such as clocks that chimed and displayed dancing figures on the hour. Electronics is now doing a superb job of providing this intelligence. Complex calculations and decisions have now become inexpensive. Presently, it is the mechanical devices required to interface electronics to the world that are expensive. Fortunately, the new micromechanical devices integrate well with electronics: one providing the intelligence and one providing the hands.

Electronics has led much of the recent development of micromechanical devices by providing many of the tools and techniques, making the rapid advances possible. This partnership is to great advantage.

Surprisingly, mechanical systems can now be smaller and less expensive than electrical systems. (If you have trouble with your electronics, don't worry, we will just throw more mechanics at it.)

There is an increasing breadth of microfabrication techniques enriching our capabilities. Examples include LIGA, EDM, precision machining, plating, and molding. To ignore the wide range of fabrication techniques available is to limit oneself.

Yet, how did things insignificant in size gain a purpose?

Perhaps Johann Gutenberg gave an indication of the usefulness of small mechanical devices. Gutenberg means good mountain, and indeed, in 1456 he set in motion a mountain of small mechanical devices (individual movable type) for the good of mankind. One interesting aspect of his work is interchangeability—a few standardized units are made that can be combined to meet most needs. This concept may be useful for our micro devices.

Ideas, excellent ideas, often seem to gain a life of their own. Like grass growing through the pavement, they seem to search for fertile minds and the correct opportunities. Often key ideas are invented in several different places. For example, Pi Sheng of China made movable type of Chinese characters from clay in 1040. And Korea molded metal type in sand in 1361. Gutenberg was just one of several who expressed the idea of movable type. If you have a good idea, I encourage you to develop it now; there is a high probability others share your idea.

Until recently, minute mechanical systems have developed at a stately pace. For years the watch makers' art has represented the limits of our micro excursion. And the practitioners of the watch industry have succeeded admirably. For example, the motor in a wrist watch has high efficiency, runs for years (even after being dropped), and costs less than a cup of coffee. Yet, when I was talking with a gentleman who had designed many of the watches we wear, he said, "I have spent my life trying to make smaller mechanisms, and when you show me something really smaller, I do not know what to do with it." This was a common response to motors the diameter of a human hair.

THE PRESENT

The rapid race to more clever micro machines has just begun.

The earlier disdain for the small and insignificant is gone. Now there is a growing excitement about the micro.

Gone are my fears that the micro field would grow on "isn't that neat" and then die when no purpose was found. Enough people now recognize the importance of micro science and engineering and product development to ensure the field.

Things insignificant in size do have a grand purpose.

Yet, how should we proceed?

First, we should share our excitement in the micro exploration.

Second, we should recognize the enormous breadth of our field. There is something special about this field of small mechanical systems.

Most advances represent a specific technology. The Scanning Tunneling Microscope, for example, gives us the ability to detect and perhaps manipulate atoms. High-temperature superconductors hold the promise of efficient power transmission and novel electronic circuits. The diesel engine gives us a source of mechanical power. Each of these is an important advance of a single thing.

The field we are contemplating here today is vast beyond our normal concerns. It is the science and engineering and development and commercialization of a whole new realm of human enterprise.

I defy you to think of a large-scale, macro discipline in science or engineering that does not have a small-scale, micro equivalent. Your challenge, should you decide to accept it, is the imaging of the macro into the micro.

Third, this field is by nature composed of disparate disciplines, and most problems require expertise from many fields. Collaborations have enabled much of the excellent work in our field.

Fourth, the science and metrology of small devices should be developed. These are logical areas for university research and government funding. I strongly encourage government and foundation support for a broad range of scientific and metrology research on micro and nano devices. Understanding the basic principles and properties will pay far higher dividends than "one more widget."

Akira Umeda of the National Research Laboratory of Metrology in Japan aptly stated the need for uniform and well-defined ways to measure micro phenomena, "If there is no measurement standard—It is speculation."

In our large-scale, macro world, we use our senses of touch and smell and taste and hearing and vision to help solve problems. We feel the play in the gears, run our fingers over the surface, shake the device, smell the motor burning, and turn the object for a better view. In the small-scale, micro world, most of our measurements are made visually through a microscope.

Fifth, good science depends on carefully written and well-considered articles. This requires careful work, knowledge of the previous literature, and honesty. Many articles in the literature are hasty rewrites of the author's previous work. Even more horrid are articles claiming as new, results that have been in the literature for years. As long as tenure and reputation depend on paper poundage, the literature will be clogged. Equally important is the openness of editors and reviewers to new or controversial ideas. One purpose of the scientific literature is to enable debate and controversy. Science grows by the unfettered competition of ideas, not people.

Sixth, we are fortunate to be in such a dynamic and growing field. It is important for members of the field to give public service back to the community. By this, I do not mean trying to get one's name on one more committee. I mean choosing something that will substantially benefit the community and quietly doing it.

And seventh, this field has been blessed with good friendships, helpfulness, and comradeship. We can continue to make this a congenial field.

You, gentle readers, are the pioneers. How you interact with your colleagues, what areas you pursue, your understanding of how to develop a new science, and the clarity of your vision will define how this field explodes.

THE FUTURE

What will the field of micro devices be like in a thousand years?

The first thing to do after such a question is to recover from the shock of being asked.

Yet, it is useful to ponder the potential.

To stir the debate, here are several predictions.

One, micromechanical devices will become omnipresent. They will fill the niches of our lives. Can you find fundamental limitations that will keep micro devices from becoming inexpensive and readily available? If not, why not their proliferation?

Clayton Teague of the National Institute of Standards and Technology in the United States gave an interesting presentation [Ref 6] on Feynman's tiny hands. Feynman proposed small hands manufacturing smaller hands, which in turn manufacture even smaller hands. This ever smaller procession of tiny hands can then be used to manufacture large numbers of useful micro devices. In this talk, Clayton Teague also discussed John von Neumann's conjecture on the self-replication of complex systems. [Ref 7] At what point can our micro devices start to self-replicate? A self-replicating micro system needs careful consideration.

Two, unless there is a need for something to be large, it will be small. There are many reasons for this. Material costs are less for small systems. The systems use less space, and small systems can perform functions more rapidly. Because of the small size and low cost, multiple systems can be used for one function, increasing the robustness and reliability. Because of their size, these microsystems can be dispersed; instead of the large devices used now, many micro devices can be used to give a finer grained sensing and manipulation of our world. Very few things need to be large. But hopefully in the year 3000, dinner will still be large.

Three, the worlds of the micro (millimeter to micron) and the nano (micron to Angstrom) and genetic engineering will evolve into closely interrelated fields. Already micro devices are helping to image and handle molecules, and nano technology is making small tubes and balls that hold the promise of mechanical structures.

Conclusion

As a child, I read about the great scientists, men and women who have structured our understanding of the universe by their discoveries. I wished I could have been with these great scientists and shared in their adventure.

We are fortunate. We are sharing in a great adventure.

Though our work from day to day may seem insignificant in size, together our work is grand in purpose.

I wish you well.

William Trimmer

References

[1] William Trimmer, "Grand in Purpose, Insignificant in Size," given at the Tenth Annual International Workshop on Micro Electro Mechanical Systems, Nagoya, Japan, January 26–30, 1997; please see page 9 of the proceedings.

[2] William Trimmer, "Editorial," *Journal of Microelectromechanical Systems,* Volume 6, Number 4, pp. 290–293, December 1997.

[3] Richard Feynman, "There's Plenty of Room at the Bottom," Caltech's *Engineering & Science* magazine, February 1960. (Reprinted in *Micromechanics and MEMS: Classic and Seminal Papers to 1990,* p. 3.)

[4] Richard Feynman, "There's Plenty of Room at the Bottom," Caltech's *Engineering & Science* magazine, February 1960. (Reprinted in *Micromechanics and MEMS: Classic and Seminal Papers to 1990,* p. 7.)

[5] Richard Feynman, "Infinitesimal Machinery," *Journal of Microelectromechanical Systems,* Vol. 2, No. 1, March 1993. (Reprinted in *Micromechanics and MEMS: Classic and Seminal Papers to 1990,* p. 11.)

[6] E. Clayton Teague, "1/N Feynman Machines as a Path to Ultraminiaturization?," SPIE Conference on Microelectronic Manufacturing, Austin, Texas, October 23–24, 1995.

[7] John von Neumann, ed. and Arthur W. Burks, "Theory of Self-Reproducing Automata," University of Illinois Press, Urbana, IL, 1966.